ADSORPTION

SURFACTANT SCIENCE SERIES

ADDITIONAL VOLUMES IN PREPARATION

ADSORPTION
Theory, Modeling, and Analysis

edited by
József Tóth
University of Miskolc
Miskolc-Egyetemváros, Hungary

CRC Press
Taylor & Francis Group
Boca Raton London New York

CRC Press is an imprint of the
Taylor & Francis Group, an **informa** business

First published 2002 by Marcel Dekker, Inc.

Published 2019 by CRC Press
Taylor & Francis Group
6000 Broken Sound Parkway NW, Suite 300
Boca Raton, FL 33487-2742

© 2002 by Taylor & Francis Group, LLC
CRC Press is an imprint of Taylor & Francis Group, an Informa business

First issued in paperback 2019

No claim to original U.S. Government works

ISBN 13: 978-0-367-44706-9 (pbk)
ISBN 13: 978-0-8247-0747-7 (hbk)

Visit the Taylor & Francis Web site at
http://www.taylorandfrancis.com

and the CRC Press Web site at
http://www.crcpress.com

Preface

This book presents some apparent divergences, that is, its content branches off in many directions. This fact is reflected in the titles of the chapters and the methods applied in discussing the problems of physical adsorption. It is not accidental. I aimed to prove that the problems of physical adsorption, in spite of the ramified research fields, have similar or identical roots. These statements mean that this book is 1) diverse, but still unified and 2) classical, but still modern. The book contains monographs at a scientific level and some chapters include parts that can be used by Ph.D.-level students or by researchers working in industry. Here are some examples. According to the classical theories of adsorption (dynamic equilibrium or statistical mechanics), the isotherm equations (Langmuir, Volmer, Fowler–Guggenheim, deBoer, Hobson, Dubinin, etc.) and the corresponding thermodynamic functions of adsorption (entropy, enthalpy, free energy) include, in any form, the expression $1 - \Theta$, where Θ is the coverage and, therefore, $0 < \Theta < 1$. This means that if the expression $1 - \Theta$ appears as denominator in any of the above-mentioned relationships, then in the limiting case

$$\lim_{\Theta=1} (1 - \Theta) = 0$$

these functions tend to infinity. Perhaps the oldest thermodynamical inconsistency appears in Polányi's equation, which expresses the adsorption potential with the following relationship:

$$P_a = RT \ln \left(\frac{p_0}{p} \right)$$

where p_0 is the saturation pressure. It is clear that

$$\lim_{p \to 0} P_a = +\infty$$

The mathematical and thermodynamical consequences of these facts are the following:

1. The monolayer adsorption can be completed only when the equilibrium pressure is infinitely great.
2. The change in thermodynamic functions are also infinitely great when the monolayer capacity is completed.
3. The adsorption potential tends to infinity when p tends to zero.

iii

All consequences are physically and thermodynamically nonsense; however, in spite of this fact, the functions and isotherms having these contradictions can be applied excellently in practice. This statement is explicitly proven in Chapters 2, 3, 4, 6, 13, and 15, in which the authors apply Langmuir's and/or Polányi's equation to explain and describe the experimentally measured data. The reason for this is very simple: Because the measured data are far from the limiting cases ($\Theta \to 1$ or $p \to 0$), the deviations caused by the unreal values of thermodynamic functions are not observable. This problem is worth mentioning because in all chapters of this book—explicitly or implicitly—the question of thermodynamic consistency or inconsistency emerges, and the first chapter tries to answer this question. However, independent of this problem, every chapter includes many new approaches to the topics discussed.

The chapters can be divided into two parts: Chapters 1–9 deal mostly with gas–solid adsorption and Chapters 10–15 deal with liquid–solid adsorption. Chapter 2 discusses the gas–solid adsorption on heterogeneous surfaces and provides an excellent and up-to-date overview of the recent literature, giving new results and aspects for a better and deeper understanding of the problem in question. The same statements are valid for Chapters 4–7. In Chapters 8 and 9, the problems of adsorption kinetics, using quite different methods, are discussed; however, these methods are successful from both a theoretical and a practical point of view. The liquid–solid adsorption discussed in Chapters 10–15 can be regarded as developments and/or continuations of Everett's and Shay's work done in the 1960s and 1970s.

In summary, I hope that this book gives a cross section of the recent theoretical and practical results achieved in gas–solid and liquid–solid adsorption, and it can be proved that the methods of discussion (modeling, analysis) have the same root. The interpretations can be traced back to thermodynamically exact and consistent considerations.

József Tóth

Contents

Contributors

Zbigniew Adamczyk Institute of Catalysis and Surface Chemistry, Polish Academy of Sciences, Cracow, Poland

Kamal Al-Malah Department of Chemical Engineering, Jordan University of Science and Technology, Irbid, Jordan

Ferenc Berger Department of Colloid Chemistry, University of Szeged, Szeged, Hungary

Małgorzata Borówko Department for the Modelling of Physico-Chemical Processes, Maria Curie-Skłodowska University, Lublin, Poland

Gianfranco Cerofolini Discrete and Standard Group, STMicroelectronics, Catania, Italy

Francisco Cuadros Departmento de Fisica, Universidad de Extremadura, Badajoz, Spain

Imre Dékány Department of Colloid Chemistry, University of Szeged, Szeged, Hungary

Vladimir Nikolajevich Kislenko Department of General Chemistry, Lviv State Polytechnic University, Lviv, Ukraine

Seung-Mok Lee Department of Environmental Engineering, Kwandong University, Yangyang, Korea

Johannes Lützenkirchen Institut für Nukleare Entsorgung, Forschungszentrum Karlsruhe, Karlsruhe, Germany

Hasan Abdellatif Hasan Mousa Department of Chemical Engineering, Jordan University of Science and Technology, Irbid, Jordan

Angel Mulero Departmento de Fisica, Universidad de Extremadura, Badajoz, Spain

Salil U. Rege*   Department of Chemical Engineering, University of Michigan, Ann Arbor, Michigan

Alexander A. Shapiro   Department of Chemical Engineering, Technical University of Denmark, Lyngby, Denmark

Erling H. Stenby   Department of Chemical Engineering, Technical University of Denmark, Lyngby, Denmark

Etelka Tombácz   Department of Colloid Chemistry, University of Szeged, Szeged, Hungary

József Tóth   Research Institute of Applied Chemistry, University of Miskolc, Miskolc-Egyetemváros, Hungary

Ralph T. Yang   Department of Chemical Engineering, University of Michigan, Ann Arbor, Michigan

Li Zhou   Chemical Engineering Research Center, Tianjin University, Tianjin, China

---

Current affiliation: Praxair, Inc., Tonawanda, New York.

ADSORPTION

1

Uniform and Thermodynamically Consistent Interpretation of Adsorption Isotherms

JÓZSEF TÓTH Research Institute of Applied Chemistry, University of Miskolc, Miskolc-Egyetemváros, Hungary

I. FUNDAMENTAL THERMODYNAMICS OF PHYSICAL ADSORPTION

A. The Main Goal of Thermodynamical Treatment

It is well known that in the literature there are more than 100 isotherm equations derived based on various physical, mathematical, and experimental considerations. These variances are justified by the fact that the different types of adsorption, solid/gas (S/G), solid/liquid (S/L), and liquid/gas (L/G), have, apparently, various properties and, therefore, these different phenomena should be discussed and explained with different physical pictures and mathematical treatments. For example, the gas/solid adsorption on heterogeneous surfaces have been discussed with different surface topographies such are arbitrary, patchwise, and random ones. These models are very useful and important for the calculation of the energy distribution functions (Gaussian, multi-Gaussian, quasi-Gaussian, exponential) and so we are able to characterize the solid adsorbents. Evidently, for these calculations, one must apply different isotherm equations based on various theoretical and mathematical treatments. However, as far as we know, nobody had taken into account that all of these *different* isotherm equations have a *common* thermodynamical base which makes possible a *common* mathematical treatment of physical adsorption. Thus, the main aim of the following parts of this chapter is to prove these common features of adsorption isotherms.

B. Derivation of the Gibbs Equation for Adsorption on the Free Surface of Liquids. Adsorption Isotherms

Let us suppose that a solute in a solution has surface tension γ (J/m^2). The value of γ changes as a consequence of adsorption of the solute on the surface. According to the Gibbs' theory, the volume, in which the adsorption takes place and geometrically is parallel to the surface, is considered as a separated phase in which the composition differs from that of the bulk phase. This separated phase is often called the Gibbs surface or Gibbs phase in the literature. The thickness (τ) of the Gibbs phase, in most cases, is an immeasurable value, therefore, it is advantageous to apply such thermodynamical considerations in which the numerical value of τ is not required. In the Gibbs phase, n_1^s are the moles of solute and n_2^s are those of the solution, the

1

free surface is A_s (m^2), the chemical potentials are μ_1^s and μ_2^s (J/mol), and the surface tension is γ (J/m^2). In this case, the free enthalpy of the Gibbs phase, $G^s(J)$, can be defined as

$$G^s = \gamma A_s + \mu_1^s n_1^s + \mu_2^s n_2^s \tag{1}$$

Let us differentiate Eq. (1) so that we have

$$dG^s = \gamma\, dA_s + A_s\, d\gamma + \mu_1^s\, dn_2^s + n_1^s\, d\mu_1^s + \mu_2^s\, dn_2^s + n_2^s\, d\mu_2^s \tag{2}$$

However, from the general definition of the enthalpy, it follows that

$$dG^s = s^s\, dT + v^s\, dP + \mu_1^s\, dn_1^s + \mu_2^s\, dn_2^s \tag{3}$$

where s^s is the entropy of the Gibbs phase (J/K) and v^s is its volume (m^3). Lete us compare Eqs. (2) and (3) so we get for constant values of A_s, T, and P;

$$A_s\, d\gamma + n_1^s\, d\mu_1^s n_2^s\, d\mu_2^s = 0 \tag{4}$$

The same relationship can be applied to the bulk phase with the evident difference that here

$$A_s\, d\gamma = 0$$

that is,

$$n_1\, d\mu_1 + n_2\, d\mu_2 = 0 \tag{5}$$

where the symbols without superscript s refer to the bulk phase.

     For the sake of elimination, let us multiply $d\mu_2$ from Eq. (4) by n_2^s/n_2 and take into account that

$$d\mu_1 = \frac{n_2}{n_1}\, d\mu_2 \tag{6}$$

and at thermodynamical equilibrium,

$$d\mu_1 = d\mu_1^s \quad \text{and} \quad d\mu_2 = d\mu_2^s \tag{7}$$

so we have from Eq. (4),

$$A_s\, d\gamma + \left(n_1^s - n_2^s \frac{n_1}{n_2} \right) d\mu_1^s = 0 \tag{8}$$

The second term in the parentheses is the total surface *excess amount* of the solute material in the Gibbs phase in comparison to the bulk phase. In particular, in the Gibbs phase, n_1^s mol solute is present with n_2^s mol solution, whereas in the bulk phase $n_2^s(n_1/n_2)$ mol solute is present with n_2 mole solution. The difference between the two amounts is the *total surface excess* amount, n_1^σ. So, according to the IUPAC symbols [1]

$$n_1^\sigma = n_1^s - n_2^s \frac{n_1}{n_2} \tag{9}$$

Dividing Eq. (8) by A_s, we have the Gibbs equation, also expressed by IUPAC symbols,

$$-\left(\frac{\partial \gamma}{\partial \mu_1} \right)_T = \Gamma_1^\sigma \tag{10}$$

where the surface excess concentration (mol/m^2) is

$$\Gamma_1^\sigma = \frac{n_1^\sigma}{A_s} \tag{11}$$

If the chemical potential of the solute material is expressed by its activity, that is,

$$d\mu_1 = RT \, d \, \ln a_1 \tag{12}$$

then the Gibbs equation (10) can be written in the practice-applicable form

$$\Gamma_1^\sigma = -\frac{a_1}{RT}\left(\frac{\partial \gamma}{\partial a_1}\right)_T \tag{13}$$

In Eq. (13), the function γ versus a_1 is a measurable relationship because the activities of most solutes are known or calculable values, therefore, the differential functions, $(\partial \gamma / \partial a_1)_T$, are also calculable relationships. So, we can introduce a *measurable* function $\psi_F(a_1)$, defined as

$$\psi_F(a_1) = a_1 \left(\frac{\partial \gamma}{\partial a_1}\right)_T \tag{14}$$

Thus, the substitution of Eq. (14) into Eq. (13) yields

$$\Gamma_1^\sigma = \frac{1}{RT}\psi_F(a_1) \tag{15}$$

The function $\psi_F(a_1)$ has another and clear thermodynamical interpretation if it is written in the form

$$\psi_F(a_1) = RT\Gamma_1^\sigma \tag{16}$$

Equation (16) is very similar to the three-dimensional gas law, namely, in this relationship, instead of the gas fugacity and the gas concentration (mol/m^3), the function $\psi_F(a_1)$(J/m^2) the surface concentration (mol/m^2), respectively, are present. It means that Eq. (16) can be regarded as a two-dimensional gas law.

Equation (15) can also be considered as a general form of adsorption (excess) isotherms applicable for liquid free surfaces. For example, let us suppose that the differential function of the measured relationship γ versus a_1 can be expressed in the following explicit form:

$$-\frac{d\gamma}{da_1} = \frac{\alpha}{\beta + a_1} \tag{17}$$

where α and β are constants. So, taking Eq. (17) into account and substituting Eq. (14) into Eq. (15), we have

$$\Gamma_1^\sigma = \frac{1}{RT}\frac{\alpha a_1}{\beta + a_1} \tag{18}$$

Equation (18) is the well-known Langmuir isotherm, applicable and measurable for liquid free surfaces. It is evident that any measured and calculated explicit form of the function $\psi_F(a_1)$—according to Eq. (15)—yields the corresponding explicit excess isotherm equation.

C. Derivation of the Gibbs Equation for Adsorption on Liquid/Solid Interfaces. Adsorption Isotherms

The derivation of the Gibbs equation for S/L interfaces is identical to that for free surfaces of liquids if the following changes are taken into account:

1. Instead of the measurable interface tension (γ), the free energy of the surface, A^s (J/m^2), is introduced and applied because, evidently, γ cannot be measured on S/L interfaces. From the thermodynamical point of view, there is no difference between A^s and γ.

2. In several cases, the surfaces A_s (m²) of solids cannot be exactly defined or measured. This statement is especially valid for microporous solids. According to the IUPAC recommendation [1], in this case the monolayer equivalent area $(A_{s,e})$ determined by the Brunauer–Emmett–Teller (BET) method (see Section VI) must be applied. $A_{s,e}$ would result if the amount of adsorbate required to fill the micropores were spread in a close-packed monolayer of molecules.

Taking these two statements into account, instead of Eq. (8) the following relationship is valid for S/L adsorption when the liquid is a binary mixture:

$$a_s m \, dA^s + \left(n_1^s - n_2^s \frac{n_1}{n_2} \right) d\mu_1^s = 0 \tag{19}$$

where a_s is the specific surface area of the adsorbent (m²/g) (in most cases determined by the BET method), m (g) is the mass of that absorber and A^s is the free energy of the surface. Here, it is also valid that

$$n_1^\sigma = \left(n_1^s - n_2^s \frac{n_1}{n_2} \right) \tag{20}$$

Dividing Eq. (19) by $a_s m = A_s$ and applying again the relationship $d\mu_1 = RT \, d \ln a_1$, we obtain

$$-\frac{a_1}{RT} \left(\frac{\partial A^s}{\partial a_1} \right)_T = \frac{n_1^\sigma}{a_s m} = \Gamma_1^\sigma \tag{21}$$

If the function $\psi(a_1)$, similar to Eq. (14), is introduced, then we have

$$\psi_{S,L}(a_1) = -a_1 \left(\frac{\partial A^s}{\partial a_1} \right)_T \tag{22}$$

That is,

$$\Gamma_1^\sigma = \frac{\psi_{S,L}(a_1)}{RT} \tag{23}$$

or

$$\psi_{S,L}(a_1) = RT\Gamma_1^\sigma \tag{24}$$

From Eq. (21), it follows that

$$\Gamma_1^\sigma = \frac{n_1^\sigma}{a_s m} = \frac{n_1 \sigma}{A_s} \tag{25}$$

Equation (25) defines the surface excess concentration, Γ_1^σ, where the surface of the solid adsorbent, in most cases, is determined by the BET method.

In L/S adsorption, Eq. (23) or (24) cannot be applied directly for the calculation of the excess adsorption isotherm because the function A^s versus a_1, as opposed to the function γ versus a_1, is not a measurable function. Therefore, another method is required to measure the excess surface concentration; however, this measured value must be compared with the value of Γ_1^σ present in the Gibbs equation (24).

The basic idea of this method is the following. Let the composition of a binary liquid mixture be defined by the mole fraction of component 1; that is,

$$x_{1,0} = \frac{n_{1,0}}{n_{1,0} + n_{2,0}} = \frac{n_{1,0}}{n_0}, \tag{26}$$

where $n_{1,0}$ and $n_{2,0}$ are the moles of the two components *before* contacting with the solid adsorbent and n_0 is the sum of the moles.

When the adsorbent equilibrium is completed, the composition of the bulk phase can again be defined by the mole fraction of component 1:

$$x_1 = \frac{n_1}{n_1 + n_2} = \frac{n_{1,0} - n_1^s}{n_{1,0} - n_1^s + n_{2,0} - n_2^s} = \frac{n_{1,0} - n_1^s}{n_0 - \left(n_1^s + n_2^s\right)} \tag{27}$$

where n_1^s and n_2^s are the moles adsorbed into the Gibbs phase (i.e., these amounts disappeared from the bulk phase). From Eqs. (26) and (27), we obtain

$$n_0(x_{1,0} - x_1) = n_1^s(1 - x_1) - n_2^s x_1 \tag{28}$$

The left-hand side of Eq. (28) includes measurable parameters only and is defined by the relationship

$$n_1^{n(\sigma)} = n_0(x_{1,0} - x_1) \tag{29}$$

where $n_1^{n(\sigma)}$ is the so-called *reduced excess amount*, because $n_1^{n(\sigma)}$ is the excess of the amount of component 1 in a reference system containing the same total amount, n_0, of liquid and in which a constant mole fraction, x_1, is equal to that in the bulk liquid in the real system. Equations (28) and (29) were derived for first time by Bartell and Ostwald and de Izaguirre [2, 3]. The importance of Eq. (29) is in the fact that it permits the measurement of the $n_1^{n(\sigma)}$ versus x_1 excess isotherms directly. However, the exact thermodynamical interpretation of S/L adsorption requires that the measured value of $n_1^{n(\sigma)}$ in Eq. (29) be compared with the surface excess concentration, Γ_1^σ, present in Gibbs equation (24). In order to this comparison, let us introduce in Eqs. (28) and (29) the reduced surface excess concentration, (i.e., let us divide those relationships by A_s). Thus, we obtain

$$\Gamma_1^{n(\sigma)} = A_s^{-1}\{n_1^s(1 - x_1) - n_2^s x_1\} \tag{30}$$

where

$$\Gamma_1^{n(\sigma)} = A_s^{-1} n_1(\sigma) = A_s^{-1}\{n_0(x_{1,0} - x_1)\} \tag{31}$$

It has been proven by Eqs. (25) and (20) that

$$\Gamma_1^\sigma = \frac{n_1^\sigma}{A_s} = A_s^{-1}\left\{n_1^s - n_2^s \frac{n_1}{n_2}\right\} \tag{32}$$

Let us write Eq. (32) in the form:

$$\Gamma_1^\sigma = A_s^{-1}\left\{n_1^s - n_2^s \frac{x_1}{1 - x_1}\right\} \tag{33}$$

From Eqs. (30) and (33), we obtain the relationship between the reduced surface excess contraction, $\Gamma_1^{n(\sigma)}$, and the one present in the Gibbs equation (21):

$$\Gamma_1^\sigma = \frac{\Gamma_1^{n(\sigma)}}{1 - x_1} \tag{34}$$

Taking Eqs. (21) and (34) into account, we obtain the following Gibbs relationship:

$$\Delta A_1^s = -\frac{RT}{A_s} \int_0^{a_1(\max)} \frac{\Gamma_1^{n(\sigma)}}{(1 - x_1)} \frac{da_1}{a_1} \tag{35}$$

Equation (35) provides the possibility for calculating the change in free energy of the surface, ΔA_1^s, if the activities of component 1 are known. In dilute solutions, $a_1 \approx x_1$; therefore, in this case, the calculation of ΔA_1^s by Eq. (35) is very simple.

The most complicated problem is to calculate or determine the composite (absolute) isotherms n_1^s versus x_1 and n_2^s versus x_2 because, in most cases, we do not have any information about the thickness of the Gibbs phase. If it is supposed that this phase is limited to a monolayer, then it is possible to calculate the composite isotherms.

We can set out from the relationship

$$n_1^s \phi_1 + n_2^s \phi_2 = A_s \tag{36}$$

where ϕ_1 and ϕ_2 are the areas effectively occupied by 1 mol of components 1 and 2 in the monolayer Gibbs phase (m$_2$/mol). From Eqs. (36), (28), and (29), we obtain the composite isotherms

$$n_1^s = \frac{A_s x_1 + \phi_2 n_1^{n(\sigma)}}{\phi_1 x_1 + \phi_2 (1 - x_1)} \tag{37}$$

and

$$n_2^s = \frac{A_s (1 - x_1) - \phi_1 n_1^{n(\sigma)}}{\phi_1 x_1 + \phi_2 (1 - x_1)} \tag{38}$$

Equations (37) and (38) can be applied when—in addition to the monolayer thickness—the following conditions are also fulfilled: (1) The differences between ϕ_1 and ϕ_2 are not greater than $\pm 30\%$, (2) the solution does not contain electrolytes, and (3) lateral and vertical interaction do not take place between the components. In Fig. 1 can be seen the five types of isotherm, $n_1^{n(\sigma)}$ versus x_1, classified for the first time by Schay and Nagy [4]. In Fig. 2 are shown the corresponding composite isotherms calculated by Eqs. (37) and (38).

It should be emphasized that the fundamental thermodynamics of S/L adsorption is exactly defined by (35) and are also the exact measurements of the reduced excess isotherms based on Eq. (29). However, the thickness of the Gibbs phase (the number of adsorbed layers), the changes in the adsorbent structure during the adsorption processes, and interactions of composite molecules in the bulk and Gibbs phases are problems open for further investigation. More of them are successfully discussed in Chapter 10.

D. Derivation of the Gibbs Equation for Adsorption on Gas/Solid Interfaces

This derivation essentially differs from that applied for the free and S/L interfaces, because, in most cases, the bulk phase is a pure gas (or vapor) (i.e., we have a one-component bulk and Gibbs phase; therefore, the excess adsorbed amount cannot be defined as it has been taken in the two-component systems). This is why we are forced to apply the fundamental thermodynamical relationships in more detail than we have applied it earlier at the free and S/L interfaces.

The first law of thermodynamics applied to a normal three-dimensional one-component system is the following:

$$dU = T\, dS - P\, dV + \mu\, dn \tag{39}$$

where U is the internal energy (J), S is the entropy (J/K), V is the volume (m^3), μ is the chemical potential (J/mol), P is the pressure (J/m^3), and n is the amount of the component (mol).

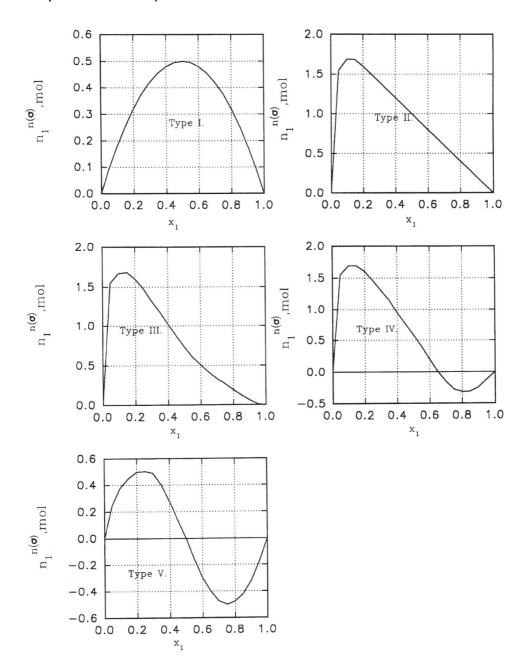

FIG. 1 The five types of excess isotherm $n_1^{n(\sigma)}$ versus x_1 classified by Schay and Nagy [4].

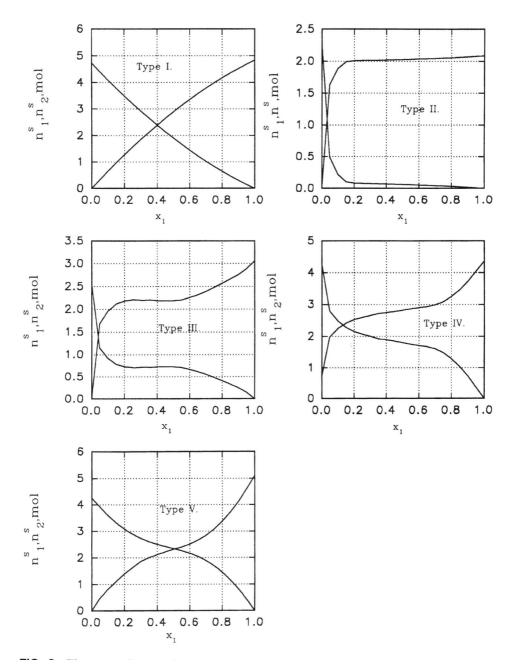

FIG. 2 The composite monolayer isotherms corresponding to the five types of excess isotherm and calculated by Eqs. (37) and (38).

Let us apply Eq. (39) to the Gibbs phase; thus, it is required to complete Eq. (39) with the work (J) needed ot make an interface; that is,

$$dU^s = T\,dS^s - P\,dV^s + \mu^s\,dn^s - A^s\,dA_s \tag{40}$$

where the superscript s refers to the Gibbs (sorbed) phase (i.e., U^s is the inside energy of the interface, S^s is the entropy, and A^s is the free energy of the interface [Gibbs phase]). Let us express the total differential of Eq. (40):

$$dU^s = T\,dS^s + S^s\,dT - P\,dV^s - V^s\,dP + \mu^s\,dn^s + n^s\,d\mu^s - A^s\,dA_s - A_s\,dA^s \tag{41}$$

Equations (40) and (41) must be equal, so we obtain

$$n^s\,d\mu^s = -S^s\,dT + V^s\,dP + A_s\,dA^s \tag{42}$$

Dividing both sides of Eq. (42) by n^s, we have the chemical potential of the Gibbs phase:

$$d\mu^s = -s^s\,dT + v^s\,dP + \frac{A_s}{n^s}\,dA^s \tag{43}$$

where s^s and v^s are the *molar* entropy and volume, respectively, of the Gibbs phase. The chemical potential of the bulk phase (one-component three-dimensional phase) is equal to Eq. (43), excepted for the work required to make an interface. Thus, we obtain

$$d\mu^g = -s^g\,dT + v^g\,dP \tag{44}$$

where the superscript g refers to the bulk (gas) phase. The condition of the thermodynamical equilibrium is

$$d\mu^g = d\mu^s \tag{45}$$

Taking Eqs. (43)–(45) into account, we have

$$A_s\left(\frac{\partial A^s}{\partial P}\right)_T = n^s(v^g - v^s) \tag{46}$$

Equation (46) is the Gibbs equation valid for S/G interfaces. As it can be seen, the thickness [i.e., the molar volume of the Gibbs phase (v^s)] is an important parameter function here.

On the right-hand side of Eq. (46), $v^g n^s$ is the volume (m³) of n^s in the bulk (gas) phase and $n^s v^s$ is the volume of n^s in the Gibbs phase. It means that the difference

$$n^s(v^g - v^s) = V^\sigma \tag{47}$$

is the *surface excess volume* of adsorptive (expressed in m³), which, according to the IUPAC symbols, is called V^σ; that is, the exact form of Gibbs equation (46) is

$$V^\sigma = A_s\left(\frac{\partial A^s}{\partial P}\right)_T \tag{48}$$

Let us express Eq. (48) as the *surface excess amount* (in mol), n^σ; it is necessary to divide Eq. (48) by the molar volume of the adsorptive, that is,

$$n^\sigma = \frac{V^\sigma}{v^g} \tag{49}$$

or, taking Eq. (47) into account,

$$n^\sigma = n^s\left(1 - \frac{v^s}{v^g}\right) \tag{50}$$

Thus, Eq. (48) can be written in the modified form

$$n^\sigma = \left(\frac{A_s}{v^g}\right)\left(\frac{\partial A^s}{\partial P}\right)_T \tag{51}$$

Let us integrate Eq. (51) between the limits P and P_m, where P_m is the equilibrium pressure when the total monolayer capacity is completed. Thus, from Eq. (51) we obtain

$$A^s(P) = \frac{1}{A_s}\int_P^{P_m} n^\sigma v^g \, dP \tag{52}$$

Suppose that the absorptive in the gas phase behaves like an ideal gas; we can then write

$$A^s(P) = \frac{RT}{A_s}\int_P^{P_m} \frac{n^\sigma}{P} \, dP \tag{53}$$

If the condition

$$v^g \gg v^s \tag{54}$$

is fulfilled, then taking Eq. (50) into account, we obtain

$$A^s(P) = \frac{RT}{A_s}\int_P^{P_m} \frac{n^s}{P} \, dP \tag{55}$$

In spite of the simplifications leading to Eq. (55), this relationship is the well-known and widely used form of the Gibbs equation.

It may occur that the absorptive in the gas phase does not behave as an ideal gas. In this case, instead of pressures, the fugacities should be applied or the appropriate state equation

$$v^g = f(P) \tag{56}$$

must be substituted in Eq. (55), that is,

$$A^s(P) = \frac{1}{A^s}\int_P^{P_m} n^s f(P) \, dP \tag{57}$$

Evidently, Eq. (57) is valid only if condition (54) is fulfilled. In the opposite case, the equation

$$A^s(P) = \frac{1}{A^s}\int_P^{P_m} n^\sigma f(P) \, dP \tag{58}$$

must be taken into account.

E. The Differential Adsorptive Potential

The Gibbs equations derived for free, S/L, and S/G interfaces provide a uniform picture of physical adsorption; however, they cannot give information on the structure of energy [i.e., we do not know how many and what kind of physical parameters or quantities influence the energy (heat) processes connected with the adsorption]. As it is well known these heat processes can be exactly measured in a thermostat of approximately infinite capacity. This thermostat contains the adsorbate and the adsorptive, both in a state of equilibrium. We take only the isotherm processes into account [i.e., those in which the heat released during the adsorption process is absorbed by the thermostat at constant temperature ($dT = 0$) or, by converse processes (desorption), the heat is transferred from the thermostat to the adsorbate, also at constant temperature]. Under these conditions, let dn^s-mol adsorptive be adsorbed by the adsorbent and, during this process, an amount of heat δQ (J) be absorbed by the thermostat at constant T. Thus, the general definition of the differential heat of absorption is

$$\left(\frac{\partial Q}{\partial n^s}\right)_{X,Y,Z} = q^{\text{diff}} \tag{59}$$

where X, Y, and Z are physical parameters which must be kept constant for obtaining the exactly defined values of q^{diff}. Let us consider the parameters $X = T$, $Y = v^s$ and v^g, and $Z = A_s$; we can now discuss the problems of the adsorption mechanism as in Ref. 5.

The molecules in the gas phase have two types of energy: potential and kinetic. During the adsorption process, these energies change and these changes appear in the differential heat of adsorption. The potential energy of a molecule of adsorptive can be characterized by a comparison: A ball standing on a table has potential energy related to the state of a ball rolling on the Earth's surface. This potential energy is determined by the character and nature of the adsorbent surface and by those of the molecule of the adsorptive.

The kinetic energies of a molecule to be adsorbed are independent of its potential energy and can be defined as follows. Let us denote the rotational energy of 1 mol adsorptive as U^g_{rot} and U^s_r is that in the adsorbed (Gibbs) phase. So, the change in the rotational energy is

$$\Delta U_r = U^g_r - U^s_r \tag{60}$$

Similarly, the change in the translational energy is

$$\Delta U_t = U^g_t - U^s_t \tag{61}$$

The internal vibrational energy of molecules is not influenced by the adsorption; however, to maintain the adsorbed molecules in a vibrational movement requires energy defined as

$$\Delta U^s_v = U^s_v - U^s_{v,0} \tag{62}$$

where U^s_v is the vibrational energy of 1 mol adsorbed molecules and $U^s_{v,0}$ is the vibrational energy of those at 0 K. If the above-mentioned potential energy is denoted by U_0, then we obtain

$$q^{\text{diff}}_h = U_0 + \Delta U_r + \Delta U_t - \Delta U^s_v + U^s_l \tag{63}$$

where the subscript h refers to homogeneous surface and U^s_l is the energy which can be attributed to the lateral interactions between molecules adsorbed. Equation (63) can be written in a shortened form if the two changes in kinetic energies are added:

$$\Delta U_k = \Delta U_r + \Delta U_t \tag{64}$$

that is, Eq. (63) can be written

$$q^{\text{diff}}_h = U_0 + \Delta U_k - \Delta U^s_v + U^s_l \tag{65}$$

The energy connected with the lateral interactions, U^s_l, depends on the coverage (i.e., the greater the coverage or equilibrium pressure, the larger is U^s_l. This is why the differential heat of adsorption, in spite of the homogeneity of the surface, changes as a function of coverage (of equilibrium pressure). However, in most cases, the adsorbents are heterogeneous ones; therefore, it is very important to apply Eq. (65) for these adsorbents too. For this reason, let us consider the heterogeneous surface as a sum of N homogeneous patches having different adsorptive potential, U_{0i} (patchwise model). According to the known principles of probability theory, one can write

$$W_i = \Delta \delta_i \tau_i (1 - \Theta_i) \tag{66}$$

where W_i is the probability of finding a molecule adsorbed on the ith patch, $\Delta \delta_i$ is the extent of the patch (expressed as a fraction of the whole surface), τ_i is the relative time of residence of the molecule on the ith patch, and Θ_i is the coverage of the same patch. In this sense, it can be defined an average or *differential adsorptive potential*, formulated as follows:

$$U_0^{\text{diff}} = \frac{\sum_i^N W_i U_{0,i}}{\sum_i^N W_i} \tag{67}$$

Similar considerations yield

$$\Delta U_v^{s,\text{diff}} = \frac{\sum_i^N W_i \Delta U_{v,i}^s}{\sum_i^N W_i} \tag{68}$$

Because the kinetic energies and U_l^s do not change from patch to patch (i.e., they are independent of $U_{0,i}$), we can write

$$q^{\text{diff}} = U_0^{\text{diff}} + \Delta U_k - \Delta U_v^s, \text{diff} + U_l^s \tag{69}$$

If the heterogeneity of the surface is not too small, then it can be estimated that

$$U_0^{\text{diff}} + U_s^l \gg \Delta U_k - \Delta U_v^{s,\text{diff}} \tag{70}$$

From relationship (70), it follows that the differential potential is approximately equal to the difference between the differential heat of adsorption and the energy of lateral interactions; that is,

$$U_0^{\text{diff}} = q^{\text{diff}} - U_l^s \tag{71}$$

As will be demonstrated in the next section, the thermodynamic parameter functions, A^s and U_0^{diff} are the bases of a uniform interpretation of S/G adsorption. However, before this interpretation, a great and old problem of S/G adsorption should be discussed and solved.

II. THERMODYNAMIC INCONSISTENCIES OF G/S ISOTHERM EQUATIONS

A. The Basic Phenomenon of Inconsistency

In Section I.D., it has been proven that the exact Gibbs equation (48) contains the *surface excess volume*, V^σ, defined by the relationship

$$V^\sigma = n^s(v^g - v^s) \tag{72}$$

where n^s is the *measured* adsorbed amount (mol) and v^g and v^s are the *molar* volume (m³/mol) of the measured adsorbed amount in the gas and in the adsorbed phase, respectively. Equation (72) means that n^s should be equal to the equation

$$n^s = \left(\frac{V^\sigma}{v^g - v^s} \right) \tag{73}$$

Let us calculate the function $n^s(P)$ for methane (i.e., for the methane isotherms by concrete model calculations). From the literature [6], we obtain the following data. The critical pressure, P_c, is 4.631 MPa and the critical temperature, T_c, is 190.7 K. Thus, the reduced pressure (π) and reduced temperature (ϑ) are $\pi = P/P_c$ and $\vartheta = 1.56$ if the calculation is made for isotherms at 298.15 K (25°C). Also from the literature [6], at $\vartheta = 1.56$ in the range of $8 \leq \pi \leq 30$ (i.e., 37 MPa $\leq P \leq$ 139 MPa), the compressibility factor Z varies approximately as a linear function:

$$Z(\pi) = 0.0682\pi + 0.356 \tag{74}$$

Taking into account that

$$v^g = Z(\pi)\frac{RT}{P} \tag{75}$$

we can calculate the molar volume of the gas phase in the pressure range $8 \leq \pi \leq 30$. The functions $v^g(P)$ can be seen in Fig. 3.

Together with the function $v^g(P)$, is plotted the surface excess volume function $V^\sigma(P)$ is calculated on the real supposition that in this range of pressure, $V^\sigma(P)$ decreases (see Fig. 4). In the left-hand side of Fig. 3, two linear functions $V^\sigma(P)$ are plotted:

$$V^\sigma(P) = -0.1 \times 10^{-6}P + 45 \tag{76}$$

(Fig. 3, top) and

$$V^\sigma(P) = -0.2 \times 10^{-6}P + 45 \tag{77}$$

(Fig. 3, bottom), where P is expressed in MPa. Equations (76) and (77) mean that a smaller and greater decreasing of $V^\sigma(P)$ have been taken into account. In the right-hand side of Fig. 3, the functions $n^s(P)$ can be seen. These functions have been calculated using Eq. (73), assuming different values of v^s (30 cm^3/mol and 20 cm^3/mol). Evidently, in the whole domain of

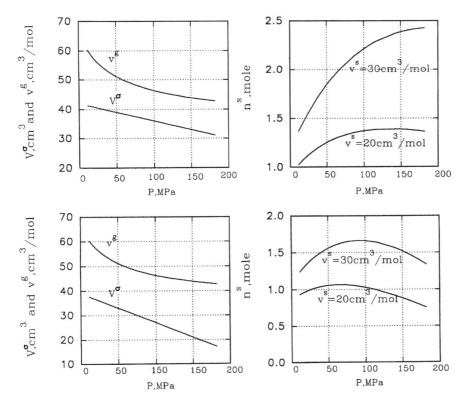

FIG. 3 Model calculations prove that in a high equilibrium pressure range, the gas/solid adsorption isotherms have maximum values.

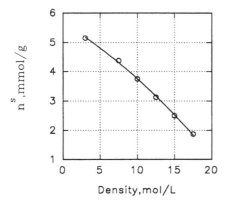

FIG. 4 Direct measurement proves that in a high pressure range, the adsorption isotherm of methane measured on GAC activated carbon at 298 K decreases approximately linearly. (From Ref. 8.)

pressure, $v^g > v^s$ is valid. The functions $n^s(P)$ (i.e., the form of isotherms) demonstrate where and why the measured adsorbed amount has the maximum value. The reality of this model calculation has also been proven experimentally by many authors published in the literature [7]. The last of those is shown in Fig. 4 [8].

As a summary of these considerations, it can be stated that according to the Gibbs thermodynamics, a plateau of isotherms in the range of high pressures, especially when P tends to infinity ($P \to \infty$), cannot exist.

B. Inconsistent G/S Isotherm Equations

In spite of the proven statements mentioned in Section II.A, there are many well-known and widely used isotherm equations which contradict the Gibbs thermodynamics (i.e., these equations are thermodynamically inconsistent). The oldest of these is the Langmuir (L) equation [9], having the following form:

$$\Theta = \frac{P}{1/K_L + P} \tag{78}$$

or

$$P = \frac{1}{K_L} \frac{\Theta}{1 - \Theta} \tag{79}$$

where

$$\Theta = \frac{n}{n_m^s} \tag{80}$$

and

$$K_L = k_B^{-1} \exp\left(\frac{U_0}{RT}\right) \tag{81}$$

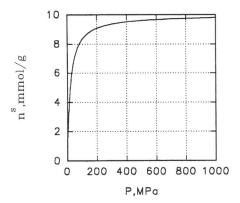

FIG. 5 The Langmuir equation (78) is thermodynamically inconsistent because it has a plateau as the great equilibrium pressure goes to infinity: $n_m^s = 10.0$ mmol/g, $K_L = 0.05$ MPa^{-1}, $P_m \to \infty$.

In Eqs. (80) and (81), n_m^s is the total monolayer capacity, U_0 is the constant adsorptive potential, and k_B is defined by de Boer and Hobson [10]:

$$k_B = 2.346(MT)^{1/2} \times 10^5 \tag{82}$$

where M is the molecular mass of the adsorbate and T is the temperature in Kelvin. The numerical values in Eq. (82) are correct if P is expressed in kilopacals.

The inconsistent character of Eq. (78) or Eq. (79) appears in their limiting values. In particular,

$$\lim_{P \to \infty} \Theta = 1 \tag{83}$$

or

$$\lim_{P \to \Theta} P = \infty \tag{84}$$

These limiting values mean that the total monolayer capacity is only completed if P tends to infinity (i.e., P decreases without limits while the isotherm has a plateau, as is shown in Fig. 5).

In Section II.A, it has been proven that according to the Gibbs thermodynamics, a plateau in the range of great pressure cannot exist; therefore, the Langmuir equation is thermodynamically inconsistent. This statement is valid for all known and used isotherm equations having limiting values (83) or (84). The most important of those are discussed in Section III and it is demonstrated there how this inconsistency can be eliminated in the framework of a uniform interpretation of G/S adsorption.

III. THE UNIFORM AND THERMODYNAMICALLY CONSISTENT TWO-STEP INTERPRETATION OF G/S ISOTHERM EQUATIONS APPLIED FOR HOMOGENEOUS SURFACES

The elimination of the thermodynamical inconsistency of the isotherm equations can be done in two steps. the first step is a thermodynamical consideration and the second one is a mathematical treatment. Both can be made *independently* of one another; however, a connection exists between them and this connection is the main base of the uniform and consistent interpretation of G/S isotherm equations.

A. The First Step: The Limited Form and Application of the Gibbs Equation

Equation (55) is the limited form of the Gibbs equation because it includes the suppositions $v^g \gg v^s$ and the applicability of the ideal-gas law.

Let us introduce in Eq. (55) the coverage defined by Eq. (80); we now obtain

$$A^s(P) = A^s_{id} \int_P^{P_m} \frac{\Theta}{P} \, dP \tag{85}$$

where

$$A^s_{id} = \frac{RT}{\varphi_m} \tag{86}$$

In Eq. (86),

$$\varphi_m = \frac{A_s}{n^s_m} \tag{87}$$

that is, φ_m is equal to the surface covered by 1 mol of adsorptive at $\Theta = 1$. It is easy to see that Eq. (86) is the free energy of the surface when the total monolayer is completed ($n^s = n^s_m$) and this monolayer behaves as an ideal two-dimensional gas. Therefore, A^s_{id} can be applied as a reference value; that is,

$$A^s_r(P) = \frac{A^s(P)}{A^s_{id}} \tag{88}$$

So, from Eq. (85), we obtain

$$A^s_r(P) = \int_P^{P_m} \frac{\Theta}{P} \, dP. \tag{89}$$

Equation (89) defines the change of the *relative free energy of the surface*, $A^s_r(P)$, in the pressure domain $P–P_m$. Equation (89) is thermodynamically correct if, in the pressure domain $P–P_m$, the ideal-gas law is applicable and the supposition $v^g \gg v^s$ is valid. The applicability of Eq. (89) may be extended if instead of pressures, the fugacities are applied (i.e., the limits of integration are f and f_m, corresponding to pressures P and P_m, respectively). This extension of Eq. (89) is supported by the fact that the supposition $v^g \gg v^s$ in most cases is still valid when instead of the ideal-gas state equation the relationship (56) should be applied.

B. The Second Step: The Mathematical Treatment and the Connection Between the First and Second Steps

Let us introduce a differential expression having the form

$$\psi(P) = \frac{n^s}{P} \left(\frac{dn^s}{dP} \right)^{-1} \tag{90}$$

It is important to emphasize that the numerical values of the function $\psi(P)$ can be calculated from the *measured* isotherm (viz. dn^s/dP is the differential function of the isotherm). It is also evident that this differential relationship can be calculated as a function of n^s; that is,

$$\psi(n^s) = \frac{n^s}{P} \left(\frac{dn^s}{dP} \right)^{-1} \tag{91}$$

The values of functions (90) and (91) also do not change when the adsorbed amounts are expressed in coverages, Θ:

$$\psi(\Theta) = \frac{\Theta}{P}\left(\frac{d\Theta}{dP}\right)^{-1} \tag{92}$$

Let us write Eq. (92) in this form:

$$\frac{dP}{P} = \frac{\psi(\Theta)}{\Theta}\,d\Theta \tag{93}$$

From Eq. (93), we obtain

$$\int_{P}^{P_m} \frac{dP}{P} = \int_{\Theta}^{1} \frac{\psi(\Theta)}{\Theta}\,d\Theta \tag{94}$$

or

$$P = P_m\,\exp\left\{-\int_{\Theta}^{l} \frac{\psi(\Theta)}{\Theta}\,d\Theta\right\} \tag{95}$$

Similarly, integration of Eq. (90) yields

$$\int_{n^s}^{n_m^s} \frac{dn^s}{n^s} = \int_{P}^{P_m} \frac{dP}{\psi(P)P} \tag{96}$$

or

$$\frac{n^s}{n_m^s} = \Theta = \exp\left\{-\int_{P}^{P_m} \frac{dP}{\psi(P)P}\right\} \tag{97}$$

If the integration is performed between limits $P-P_0$ and $\Theta-\Theta_0$, where P_0 is the saturation pressure and Θ_0 is the corresponding coverage, then we have

$$\frac{P}{P_0} = P_r = \exp\left\{-\int_{\Theta}^{\Theta_0} \frac{\psi(\Theta)}{\Theta}\,d\Theta\right\} \tag{98}$$

and

$$\frac{n^s}{n_0^s} = \Theta_0 = \exp\left\{-\int_{P}^{P_0} \frac{dP}{\psi(P)P}\right\} \tag{99}$$

Equations (95) and (97)–(99) are implicit integral isotherm equations with general validity because functions $\psi(\Theta)$ or $\psi(P)$ from any measured isotherms can be calculated. These relationships are only the results of a pure mathematical treatment. However, it is easy to prove the connection between the implicit integral isotherms and the limited Gibbs equation (89). In particular, let us substitute Eq. (92) into Eq. (89); then, we have

$$A_r^s(\Theta) = \int_0^{\Theta} \psi(\Theta)\,d\Theta \tag{100}$$

Equation (100) permits a simple numerical or analytical calculation of the relative change in free energy of the surface when the coverage changes in domains $0-\Theta$. If Θ is expressed in n^s, then

$$A_r^s(n^s) = \frac{RT}{A_s} \int_{n^s}^{n_m^s} \psi(n^s)\,dn^s \tag{101}$$

From Eq. (100) follows the exact thermodynamical meaning of the function $\psi(\theta)$:

$$\psi(\Theta) = \left(\frac{\partial A_r^s}{\partial \Theta}\right)_T \tag{102}$$

It is important to emphasize again that

$$\psi(\Theta) = \psi(P) = \psi(n^s) \tag{103}$$

when P, n^s, or Θ are conjugated pairs of the measured isotherms. It means that the functions in Eq. (103) are thermodynamically equivalent. It is evident that the applicability of Eq. (100) and the validity of Eq. (102) are equal to those of the Gibbs equation (89). Equations (95) and (97)–(99) permit a consistent and uniform interpretation of G/S isotherms. The thermodynamical consistency is assured by the integration to a *definite* upper limit which can guarantee that the isotherm equations do not have limiting values equal to limits (83) or (86) and it is also guaranteed that all conditions leading to Eq. (55) are fulfilled. The uniformity assured that (1) all equations have the same implicit mathematical form, (2) in all equations, the functions ψ having directly or indirectly the thermodynamical meaning defined by Eq. (102) and (3) the functions ψ can be calculated from every measured isotherm so that can always be selected for these isotherms the mathematically and thermodynamically correct equation. (For example see Fig. 18 in Section III.H).

C. The Uniform and Consistent Interpretation of the Modified Langmuir Equation, General Considerations

Let us apply the implicit integral relationships (96) for derivation of the Langmuir equation. For this reason, it must be demonstrated that the function $\psi(P)$ belonging to the Langmuir equation is

$$\psi_L(P) = \frac{n^s}{P}\left(\frac{dP}{dn^s}\right) = K_L P + 1 \tag{104}$$

Equation (104) can both be mathematically and experimentally proven; namely, if an isotherm *measured* on a homogeneous surface is fitted with a polinome and it is differentiated analytically, then the function $\psi_L(P)$ is equal to Eq. (104) (i.e., the slope of the straight line is equal to K_L and the zero point is equal to 1). Let us insert Eq. (104) into Eq. (97); we then obtain

$$\ln\left(\frac{n^s}{n_m^s}\right) = \int_{P_m}^{P} \frac{dP}{(K_L P + 1)P} \tag{105}$$

The integration on the right-hand side of Eq. (105) can be performed analytically:

$$\int_{P_m}^{P} \frac{dP}{(K_L P + 1)P} = \ln\left(\frac{K_L P}{K_L P + 1}\right) - \ln\left(\frac{K_L P_m}{K_L P_m + 1}\right) \tag{106}$$

Introducing the integration constant, χ_L,

$$\chi_L = 1 + \frac{1}{K_L P_m} \tag{107}$$

we obtain from Eqs. (106) and (107) the modified Langmuir (mL) equation

$$n^s = \frac{n_m^s \chi_L P}{1/K_{mL} + P} \tag{108}$$

In Eq. (109), the condition that at $P = P_m n^s$ is equal to the total monolayer capacity is fulfilled [i.e., the limiting values (83) and (84) are eliminated. The original Langmuir equation (78) does not contain the constant χ_L. *Mathematically*, this fact means that the integration in Eq. (97) is perfomed between the limits P and infinity [i.e., the total monolayer capacity is completed at an infinitely great equilibrium pressure ($\chi_L = 1$)].

Before demonstrating other properties of the mL equation, it is necessary to prove the validity of Eq. (103); that is,

$$\psi(n^s) = \psi(P) = \psi(n\Theta)$$

It is easy to calculate the following function $\psi_{m,L}(\Theta)$ belonging to the modified Langmuir equation:

$$\psi_{mL}(\Theta) = \frac{\chi_L}{\chi_L - \Theta} \tag{109}$$

Taking Eqs. (107) and (109) into account, we have

$$\frac{\chi_L}{\chi_L - \Theta} = K_{mL}P + 1 \tag{110}$$

It is important to remark that Eq. (104) belongs to the *original* Langmuir equation; however, from Eq. (110), we obtain the *modified* Langmuir equation (108).

This result, demonstrated with the example of the mL equation, is of general validity and can be drafted as it follows:

1. If it is required to transform an inconsistent isotherm equation into a consistent one, then Eq. (97) or (99) should be applied, where $\psi(P)$ belongs to the *inconsistent* equation.
2. The function $\psi(\Theta)$ of the *inconsistent* equation cannot be applied for this transformation [i.e., the integration of Eq. (95) or (98) with the inconsistent functions $\psi(\theta)$ does not lead to consistent isotherms equations]. However, the inconsistent functions $\psi(\theta)$ are applicable to prove the inconsistency of the thermodynamical functions [see Eq. (112)].
3. The reason for statement (1) is the fact that the function $\psi(\Theta)$ has a concrete thermodynamical meaning defined by Eq. (102). Therefore, all thermodynamical consistencies or inconsistencies are *directly* reflected by the function $\psi(\Theta)$.
4. From statements (2) and (3), it follows that in Eqs. (95), (98), and (100), only the consistent form of the function $\psi(\Theta)$ can be applied.

How these consistent forms of $\psi(\Theta)$ can be calculated or determined are discussed in the following subsections. However, before this discussion, it is required to demonstrate other inconsistencies of the original Langmuir equation. The change in relative free energy of the surface is defined by Eq. (100). To calculate this change, the explicit form of the function $\psi(\Theta)$ is required. This function, belonging to the original Langmuir equation has the following form:

$$\psi_L(\Theta) = \frac{1}{1 - \Theta} \tag{111}$$

Let us substitute Eq. (111) into Eq. (100) and perform the integration; we thus obtain

$$A_r^s(\Theta = 1) = \int_0^1 \frac{d\Theta}{1 - \Theta} = \left[\ln\left(\frac{1}{1 - \Theta}\right)\right]_0^1 = \infty \tag{112}$$

Equation (112) reflects a thermodynamic inconsistency because the change in free energy of the surface never can be infinite. However, if we substitute the function $\psi_{mL}(\theta)$ [Eq. (109) into Eq. (100)], we have

$$\Delta A_r^s(\Theta = 1) \int_0^1 \frac{\chi_L}{\chi_L - \Theta} \, d\Theta = \chi_L \left[\ln\left(\frac{1}{\chi_L - \Theta}\right) \right]_0^1 = \chi_L \, \ln\left(\frac{\chi_L}{\chi_L - 1}\right) \tag{113}$$

Because $\chi_L > 1$, the change in relative free energy of the surface always has a finite value. It is evident that if the integration in Eq. (100) is performed between the limits zero and a finite value of Θ, then we have

$$A_s^r(\Theta) = \chi_L \left\{ \ln\left(\frac{\chi_L}{\chi_L - \Theta}\right) \right\} \tag{114}$$

Summarizing all considerations relating to the thermodynamic consistency of the mL equation, a statement of general validity can be made: The consistent form of the mL equation (and others) can be derived because Eq. (97) requires integration with a *finite* value of the upper limit. If this upper limit, P_m, is not so great that instead of n^s, the surface excess volume, V^σ, or surface excess amount, n^σ, ought to apply, then, according to Eq. (102), the thermodynamical interpretation of the function $\psi(\Theta)$ is correct. Therefore, the isotherm equations derived from Eq. (97) or from Eq. (98) are also thermodynamically consistent. From this statement, it follows that the inconsistencies of the well-known monolayer isotherm equations are such that the original Langmuir equation and all those discussed in following sections are connected with the fact that these relationships were not derived from consistent differential equations requiring integration. Thus, these relationships include the limiting value

$$\lim_{P \to \infty} \Theta = 1$$

which is thermodynamically inconsistent. It is also proven that all inconsistencies are reflected by the function $\psi(\Theta)$. The discussions and relationships proving the consistency of the mL equation, inconsistencies of the original Langmuir equation, and Eqs. (95)–(100) providing the derivation of consistent isotherm relationship permit a general method for interpretation of any isotherm equation.

The calculation method of this interpretation is demonstrated in detail with the example of the mL equation; however, this method can be (and is) applied for every isotherm equation discussed in this chapter. The results of these calculations are shown in Fig. 6 and details of those are in particular the following. For Fig. 6 (top, left), the applied mL equations are

$$P = \frac{1}{K_{mL}} \frac{\Theta}{\chi_L - \Theta} \tag{115}$$

where χ_L and K_{mL} may be varied; $\chi_L = 1$ (solid line) is the original Langmuir isotherm. This figure represents the *measured* adsorption isotherms, assuming that the total monolayer capacity (n_m^s) is known (see BET method and others). For Fig. 6 (top right), the functions $\psi_{mL}(P)$ are linear,

$$\psi_{mL}(P) = K_{mL}P + 1 \tag{116}$$

and it can be calculated directly because the mL equation can also be expressed in the form $\Theta = f(P)$. However, most of the isotherm equations cannot emplicitly be expressed in this form,

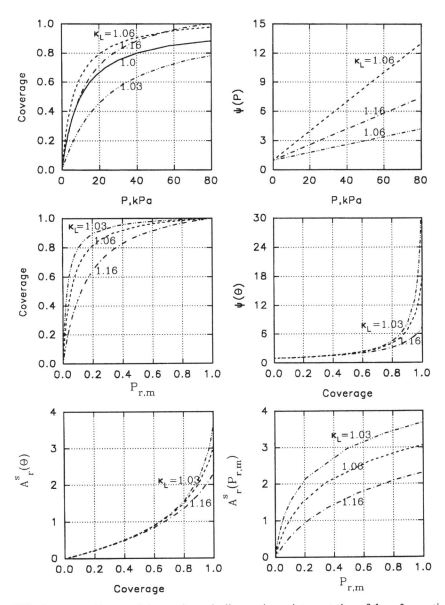

FIG. 6 The uniform and thermodynamically consistent interpretation of the mL equation. The values of the parameters are as follows: $K_L = 0.1$ kPa^{-1}, $\chi_L = 1.0$, $P_m \to \infty$ (solid line, original Langmuir equation); $K_{mL} = 0.04$ kPa^{-1}, $\chi_L = 1.03$, $P_m = 833.3$ kPa ($- \cdots - \cdots -$); $K_{mL} = 0.15$ kPa^{-1}, $\chi_L = 1.06$, $P_m = 111.11$ kPa ($---$); $K_{mL} = 0.08$ kPa^{-1}, $\chi_L = 1.16$, $P_m = 78.13$ kPa ($- \cdot - \cdot -$).

as only the term $P = \vartheta(\Theta)$ exists. In these cases, the calculation of the function $\psi(P)$ is the following. First, the function $\psi(\Theta)$ is calculated, which, in this case, has the form

$$\psi_{mL}(\Theta) = \frac{\chi_L}{\chi_L - \Theta} \tag{117}$$

Because

$$\psi_{mL}(\Theta) = \psi_{mL}(P) \tag{118}$$

to every value of P calculated by Eq. (115) can be attributed a value of $\psi_{mL}(\Theta)$. Thus, we obtain the function $\psi_{mL}(P)$ numerically. It is evident that if the isotherm investigated can be expressed both in terms of $\Theta = f(P)$ and $P = \varphi(\Theta)$, then the two methods lead to the same relationship, $\psi(P)$.

The practical importance of the function $\psi(P)$ is discussed in Section III.A. For Fig. 6 (middle, left), this interpretation of the *measured* isotherms is thermodynamically exact, because if reflects the fact that the integration in Eq. (97) has been performed to the finite upper limit P_m and, therefore, the equilibrium pressure should be expressed in a relative pressure defined as

$$P_{r,m} = \frac{P}{P_m} \tag{119}$$

The value of P_m can be calculated form the integration constant χ_L defined by Eq. (107). Thus, we obtain

$$P_m = [K_L(\chi_L - 1)]^{-1} \tag{120}$$

Here, it can also be seen that at $\chi_L = 1$, $P_m \to \infty$. This limiting value is thermodynamically inconsistent.

For Fig. 6 (middle, right), the function $\psi(\Theta)$ is very important form two standpoints. First, the analytical or numerical integration of this function permits the calculation of the relative free energy of the surface [see Eq. (100)]:

$$A_r^s(\Theta) = \int_0^\Theta \psi(\Theta) \, d\Theta \tag{121}$$

Second, the function $\psi(\Theta)$ is required to calculate the function $\psi(P)$ numerically if the isotherm equation cannot be expressed in terms of $\Theta = f(P)$. In Fig. 6 (bottom, left), the functions $A_r^s(\Theta)$ calculated analytically or numerically by Eq. (121) are represented. In Fig. 6 (bottom, right) the functions $A_r^s(P_{r,m})$ calculated similar to the function $\psi(P)$ are shown. In particular, to every value of $P_{r,m} = P/P_m$ calculated by Eqs. (115) and (120) are attributed the values of $A_r^s(\Theta)$, so we obtain the functions $A_r^s(P_{r,m})$. These two types of function in the bottom of Fig. 6 characterize thermodynamically the adsorption process and thus seem to complete the uniform interpretation of the mL and other isotherm equations. The thermodynamic consistency is best reflected by the functions $A_r^s(\Theta)$ and $A_r^s(P_{r,m})$ because both functions have finite values at $\Theta = 1$ or at $P_{r,m} = 1$.

D. The Uniform and Consistent Interpretation of the Modified Fowler–Guggenheim Equation

Fowler and Guggenheim [11] derived an isotherm equation which takes the lateral interaction of the adsorbed molecules into account. It has the following explicit form:

$$P = \frac{1}{K_F} \frac{\Theta}{1 - \Theta} \exp(-B_F\Theta) \tag{122}$$

where

$$B_F = \frac{C\omega}{RT} \tag{123}$$

In Eq. (123), ω is defined as the interaction energy per pair of molecules of nearest neighbors, and C, is a constant. Thus, the orignal Langmuir equations is transformed into Eq. (122). Equation (122) contains all thermodynamic inconsistencies mentioned in connection with the original Langmuir equation because the limiting values

$$\lim_{P \to \omega} \Theta = 1 \quad \text{and} \quad \lim_{\theta \to 1} P = \infty$$

are also valid for Eq. (122).

To obtain a consistent form of Eq. (122), let us calculate its function $\psi(\theta)$:

$$\psi_F(\Theta) = \frac{1}{1 - \Theta} - B_F \Theta \tag{124}$$

It has been proven that the consistent Langmuir equation has the function $\psi_{mL}(\Theta)$ defined by Eq. (109); therefore, the consistent (modified) Fowler–Guggenheim (mFG) equation should have the following function:

$$\psi_{mF}(\Theta) = \frac{\chi_F}{\chi_F - \Theta} - B_F \Theta \tag{125}$$

Let us substitute Eq. (125) into Eq. (95). After integration, we have

$$P = P_m (\chi_F - 1) \, \exp(B_F) \frac{\Theta}{\chi_F - \Theta} \, \exp(-B_F \Theta) \tag{126}$$

where the constant of integrations, I_F, is

$$I_F = P_m (\chi_F - 1) \, \exp(B_F) \tag{127}$$

Let us compare Eqs. (126) and (127) with Eq. (122); we have

$$P = \frac{1}{K_{mF}} \frac{\Theta}{\chi_F - \Theta} \, \exp(-B_F \Theta) \tag{128}$$

where

$$K_{mF} = I_F^{-1} = [P_m (\chi_F - 1) \, \exp(B_F)]^{-1} \tag{129}$$

or

$$P_m = I_F [(\chi_F - 1) \, \exp(B_F)]^{-1} \tag{130}$$

Therefore, the consistent form of the FG equation is relationship (128). For example, the limiting value of Eq. (128) is

$$\lim_{\Theta \to 1} P = P_m \tag{131}$$

It is expected that the original and modified FG equations, which take explicitly the lateral interactions between molecules adsorbed into account, can describe different types of isotherms. In Fig. 7 can be seen the four types of isotherm which the modified FG equation can describe and explain. In the following, the limits and values of parameters (χ_F, B_F) determining the types of isotherms are interpreted.

From the physical meaning of the mFG relationships, it follows that they have to reflect the two-dimensional condensation too, similar to the three-dimensional van der Waals equation. In this case, the isotherm equation,

$$P = \varphi(\Theta)$$

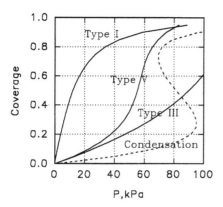

FIG. 7 The modified Fowler–Guggenheim equation can describe four types of isotherm (Types I, III, and V, and condensation).

should have local maximum and local minimum values. These values of θ exist when the condition

$$\frac{d\varphi(\Theta)}{d\Theta} = 0 \tag{132}$$

is met. Let us differentiate Eq. (128) and calculate the values of B_F for which Eq. (132) is fulfilled; we thus obtain

$$B_F = \frac{\chi_F}{\Theta(\chi_F - \Theta)} \tag{133}$$

Equation (133) determines all values of B_F at which two-dimensional condensation takes place. The coverages, Θ, present in Eq. (133) are the places of minima and maxima mentioned earlier (see also the S-shape condensation isotherm in Fig. 7). The functions $B_F(\Theta)$ are shown in Fig. 8 by solid lines.

In Fig. 8, it can be seen also that functions $B_F(\Theta)$ have absolute minimum values. After differentiation of Eq. (133), we have the values of coverage where these minima occur:

$$\Theta_{\min} = 0.5\chi_F \tag{134}$$

How these places of minima, $\Theta_{\min}$, increase according to Eq. (134) are shown in Fig. 8; however; by inserting Eq. (134) into Eq. (133), the decreasing character of $B_{F,\min}$ can be calculated explicitly:

$$B_{F,\min} = \frac{4}{\chi_F} \tag{135}$$

In Fig. 8, the values of $B_{F,\min}$ are represented by horizontal dotted lines.

For the determination of other types of isotherm corresponding to Eq. (128), it is essential to calculate the function $B_F(\Theta)$ which fulfils the condition

$$\psi_{mF}(\Theta) = 1 \tag{136}$$

According to Eq. (125), condition (136) is met by the following values of B_F and Θ:

$$B_F = \frac{1}{\chi_F - \Theta} \tag{137}$$

In Fig. 8, functions (137) are shown with dash-dot-dot lines.

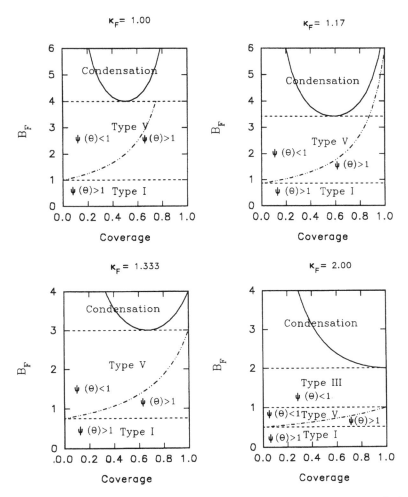

FIG. 8 Limiting values of parameters B_F and χ_F determining the types of FG and mFG isotherm.

Mathematically, the condition $\psi(\Theta) = 1$ means that all values of Θ present in Eq. (137) can be drawn from the origin proportional lines. One of these situations is represented in Fig. 9. The regions of coverages where $\psi(\Theta) > 1$ and $\psi(\Theta) < 1$ and the point where the proportional line drawn from the origin is a tangent can be seen. Evidently, $\psi(\Theta) = 1$ is also valid when the initial domain of an isotherm is a proportional line (i.e., the isotherm begins with a Henry section).

The above analysis is also represented in Fig. 8. The first figure ($\chi_F = 1$) relates to the original FG equation, which can describe Types I and V and condensation isotherms. However, for Type V isotherms, the value of B_F tends to infinity when Θ tends to 1. In this fact is also reflected the thermodynamical inconsistency of the original FG equation. In the top (right) of Fig. 8, $\chi_F = 1.17$, the place of minimum, Θ_{min}, according to Eq. (134) has been increased. So, from the analysis and figures above, it follows that the Type I isotherm is described when $0 < B_F < 1/\chi_F$, Type V isotherms can occur when $1/\chi_F < B_F < 4/\chi_F$, and two-dimensional condensation takes place when $B_F > 4/\chi_F$. A very interesting limiting case is shown at the

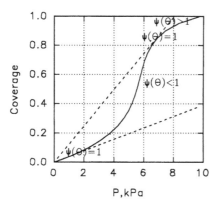

FIG. 9 The values of function $\psi(\Theta)$ corresponding to the Type V isotherm.

bottom (left) of Fig. 8 when $\chi_F = 1.333$: In this case, the minimum value of $B_{F,\min}$; that is, Eq. (134) at $\Theta = 1$ is equal to the value of B_F corresponding to Eq. (137):

$$\frac{4}{\chi_F} = \frac{1}{\chi_F - 1} \tag{138}$$

Solving Eq. (138) for χ_F, we obtain

$$\chi_F = 1.333 \tag{139}$$

Equation (139) means that if the values of χ_F are greater than 1.333, then the modified FG equation can also describe isotherms of Type III. This situation is represented at the bottom (right) of Fig. 8 when $\chi_F = 2.00$. Thus, the extended applicability of the modified FG equation is the following. If χ_F is greater than 1.333, then the types of isotherm are determined by the following limiting values of B_F:

Type I when $0 < B_F < 1/\chi_F$
Type V when $1/\chi_F < B_F < 1/(\chi_F - 1)$
Type III when $1/(\chi_F - 1) < B_F < 4/\chi_F$
Type condensation when $B_F > 4/\chi_F$

Thus, it is proven why the four types of isotherm shown in Fig. 7 can be described by the modified FG equation.

The analysis made above is limited to mathematical considerations. In Figs. 10–12 the thermodynamically consistent and uniform interpretation of isotherms Types I, III, and V, respectively, are shown. In these figures, quite equal to Fig. 6, are represented the functions $\Theta(P)$, $\psi(P)$, $\Theta(P_{r,m})$, $\psi(\Theta)$, $A_r^s(\Theta)$, and $A_r^s(P_{r,m})$. The calculations of these functions have been made as follows:

Top (left): The applied mFG equations are

$$P = \frac{1}{K_{\mathrm{mF}}} \frac{\Theta}{\chi_F - \Theta} \exp(-B_F \Theta) \tag{140}$$

The types of isotherms in Figs. 10–12 have been determined by the corresponding values of χ_F and B_F (see Fig. 8).

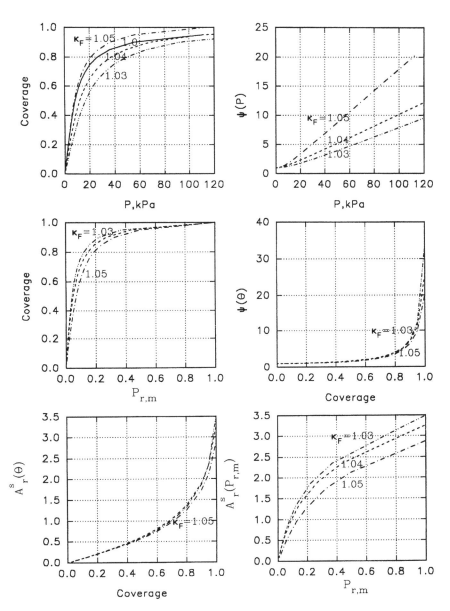

FIG. 10 The uniform and thermodynamically consistent interpretation of the mFG isotherms of Type I. The values of the parameters are as follows: $K_F = 0.08$ kPa^{-1}, $\chi_F = 1.0$, $B_F = 0.8$, $P_m \rightarrow \infty$ (solid line, original FG equation); $K_{mF} = 0.04$ kPa^{-1}, $\chi_F = 1.03$, $B_F = 0.7$, $P_m = 413.8$ kPa ($-\cdot\cdot-\cdot\cdot-$); $K_{mF} = 0.06$ kPa^{-1}, $\chi_F = 1.04$, $B_F = 0.5$, $P_m = 252.7$ kPa ($---$); $K_{mF} = 0.08$ kPa^{-1}, $\chi_F = 1.05$, $B_F = 0.8$, $P_m = 112.3$ kPa ($-\cdot-\cdot-$).

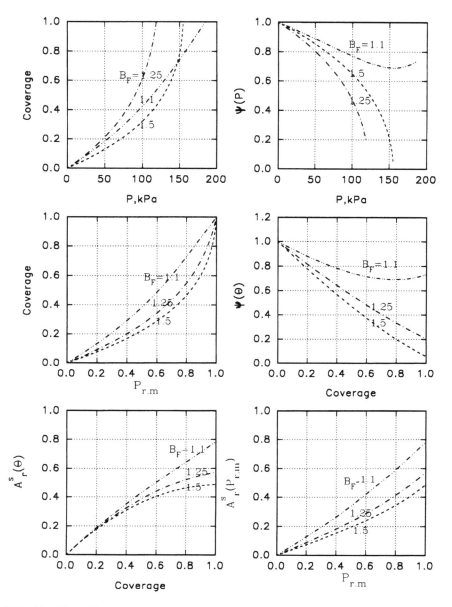

FIG. 11 The uniform and thermodynamically consistent interpretation of the mFG isotherms of Type III. The values of the parameters are as follows: $K_{mF} = 15 \times 10^{-4}$ kPa, $\chi_F = 2.2$, $B_F = 1.1$, $P_m = 184.9$ kPa $(-\cdot\cdot-\cdot\cdot-)$; $K_{mF} = 8 \times 10^{-4}$ kPa, $\chi_F = 2.8$, $B_F = 1.50$, $P_m = 155.0$ kPa $(---)$; $K_{mF} = 11 \times 10^{-4}$ kPa, $\chi_F = 3.2$, $B_F = 1.25$, $P_m = 118.4$ kPa $(-\cdot-\cdot-)$.

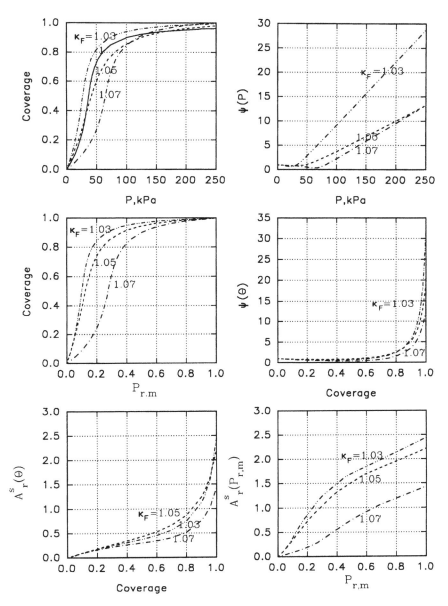

FIG. 12 The uniform and thermodynamically consistent interpretation of the mFG isotherms of Type V. The values of parameters are as follows: $K_F = 6 \times 10^{-3}$ kPa^{-1}, $\chi_F = 1.0$, $B_F = 3.0$, $P_m \rightarrow \infty$ (solid line, original FG equation); $K_{mF} = 0.01$ kPa^{-1}, $\chi_F = 1.03$, $B_F = 2.5$, $P_m = 273.6$ kPa $(-\cdot\cdot-\cdot\cdot-)$; $K_{mF} = 8 \times 10^{-3}$ kPa^{-1}, $\chi_F = 1.05$, $B_F = 2.0$, $P_m = 338.3$ kPa $(---)$; $K_{mF} = 3 \times 10^{-3}$ kPa^{-1}, $\chi_F = 1.07$, $B_F = 3.0$, $P_m = 237.1$ kPa $(-\cdot-\cdot-)$.

Top (right): The functions $\psi(P)$ by the relationships

$$\psi_{mF}(\Theta) = \frac{\chi_F}{\chi_F - \Theta} - B_F\Theta \tag{141}$$

and

$$\psi_{mF}(\Theta) = \psi_{mF}(P) \tag{142}$$

Middle (left): The thermodynamically consistent and uniform interpretation of the measured isotherms in form of $\Theta(P_{r,m})$; that is

$$P_{r,m} = \frac{P}{P_m} \tag{}$$

and

$$P_m = K_{mF}^{-1}[(\chi_F - 1)\ \exp(B_F)]^{-1} \tag{143}$$

have been calculated.

Middle right: The functions $\psi(\Theta)$ are bases for calculation of functions $A_r^s(\Theta)$ and $A_r^s(P_{r,m})$. Bottom: The functions $A_r^s(\Theta)$ and $A_r^s(P_{r,m})$ are calculated numerically by the integral equation

$$A_r^s(\Theta) = \int_0^\Theta \psi(\Theta)\ d\Theta \tag{144}$$

These functions (i.e., the relative changes in free energy of the surface shown in Figs. 10–12) characterize thermodynamically the adsorption processes. In particular, in Fig. 11, (isotherms of Type III)

$$A_r^s(\Theta) < 1 \tag{145}$$

is valid in the whole domain of coverage; that is,

$$A_r^s(\Theta) < A_{id}^s \tag{146}$$

This means that the change in free energy of the surface is always less than that would have been caused by a two-dimensional ideal monolayer completed on a homogeneous surface. However, for isotherms of Type I and V (in Figs. 10 and 12, respectively), there always exists a definite coverage where

$$A_r^s(\Theta) = A_{id}^s \tag{147}$$

The values of Θ and $P_{r,m}$ when Eq. (147) is valid are excellent and very simple parameters characterizing the adsorption system investigated.

For practical applications of the mFG equation and for calculations of constants $K_{m,F}$, χ_F, B_F, and P_m, a three-parameter fitting procedure is recommended. In particular, Eq. (126) can be written in the form

$$P = I_F \frac{\Theta}{\chi_F - \Theta}\ \exp(-B_F\Theta) \tag{148}$$

Equation (148) can be fitted to the measured points (Θ, P) with parameters I_F, χ_F, and B_F. The average percentile deviation ($\Delta\%$) has been calculated using the following relationship:

$$\Delta\% = \frac{1}{N} \sum_{i=1}^{N} \left| \frac{(P - P_c)}{P} \times 100 \right| \tag{149}$$

where P is the measured equilibrium pressure, P_c is calculated equilibrium pressures, and N is the number of the measured points (Θ, P).

We have the constants I_F, χ_F, and B_F as results of the fitting procedure. According to Eq. (129), we obtain

$$K_{\mathrm{mF}} = (I_F)^{-1} \tag{150}$$

Finally, the pressure when the total monolayer capacity is completed, P_m yields Eq. (130) or (127),

$$P_m = I_F[(\chi_F - 1)\ \exp(B_F)]^{-1} \tag{151}$$

E. The Uniform and Consistent Interpretation of the Modified Volmer Equation

Volmer [12] was the first scientist to take the mobility of the adsorbed molecules into account. His considerations were based on the dynamic equilibrium between the gas and adsorbed phase and obtained the following relationship:

$$P = \frac{1}{K_V} \frac{\Theta}{1 - \Theta}\ \exp\left(\frac{\Theta}{1 - \Theta}\right) \tag{152}$$

where the exponential term reflects the mobility of molecules in the adsorbed layer. The function $\psi(\Theta)$ of the Volmer equation has the form

$$\psi_V(\Theta) = \left(\frac{1}{1 - \Theta}\right)^2 \tag{153}$$

Equation (153) means that the mobility of the monolayer, in comparison to the immobile Langmuir monolayer, is expressed by the relationship

$$\psi_V(\Theta) = [\psi_L(\Theta)]^2 \tag{154}$$

Unfortunately, Eq. (152) is thermodynamically inconsistent, because in it, the limiting values

$$\lim_{P \to \infty} \Theta = 1 \quad \text{and} \quad \lim_{\Theta \to 1} P = \infty$$

are on contradiction with the Gibbs equation (51) and it is also unacceptable that according to Eq. (100),

$$\lim_{\Theta \to 1} A_r^s = \infty \tag{155}$$

namely,

$$A_r^s(\Theta = 1) = 1 \int_{\Theta}^{1} \frac{d\Theta}{(1 - \Theta)^2} = \infty \tag{156}$$

These are the reasons why a consistent form of the Volmer equation should be derived. This derivation is quite similar to Eqs. (109) and (125); that is, instead of Eq. (153), a modified Volmer (mV) equation can be written:

$$\psi_{\mathrm{mV}}(\Theta) = \left(\frac{\chi_V}{\chi_V - \Theta}\right)^2 \tag{157}$$

Let us substitute Eq. (157) into Eq. (195); after integration, we obtain

$$P = P_m(\chi_V - 1) \, \exp\left(-\frac{1}{\chi_V - 1}\right) \frac{\Theta}{\chi_V - \Theta} \, \exp\left(\frac{\Theta}{\chi_V - \Theta}\right) \tag{158}$$

where the constant of integrations, I_v, has the form

$$I_V = P_m(\chi_V - 1) \, \exp\left(-\frac{1}{\chi_V - 1}\right) \tag{159}$$

Thus we obtain the modified Volmer equation in a simple form:

$$P = \frac{1}{K_{mV}} \frac{\Theta}{\chi_V - \Theta} \, \exp\left(\frac{\Theta}{\chi_V - \Theta}\right) \tag{160}$$

where

$$K_{mV} = (I_V)^{-1} = \left[P_m(\chi_V - 1) \, \exp\left(-\frac{1}{\chi_V - 1}\right)\right]^{-1} \tag{161}$$

or

$$P_m = I_V\left[(\chi_V - 1)\exp\left(-\frac{1}{\chi_V - 1}\right)\right]^{-1} \tag{162}$$

For practical applications of Eq. (158) and for calculations of constants K_{mV}, χ_v, P_m, and n_m^s, a three-parameter fitting procedure is proposed. In particular, Eq. (158) can be written as

$$P = I_V \frac{\Theta}{\chi_V - \Theta} \, \exp\left(\frac{\Theta}{\chi_V - \Theta}\right) \tag{163}$$

or

$$P = I_V \frac{n^s}{\chi_V n_m^s - n^s} \, \exp\left(\frac{n^s}{\chi_V n_m^s - n^s}\right) \tag{164}$$

The parameters to be fitted are I_V, χ_V, and n_m^s. The constants K_V and P_m can be calculated similar to Eqs. (150) and (151),

$$K_{mV} = (I_V)^{-1} \tag{165}$$

and equal to Eq. (162),

$$P_m = I_V\left[(\chi_V - 1) \, \exp\left(-\frac{1}{\chi_V - 1}\right)\right]^{-1} \tag{166}$$

It is very important to remark that the original and the modified Volmer equation can only describe isotherms of Type I. It is obvious, because, from Eq. (157), it follows that

$$\chi_V > \chi_V - \Theta \tag{167}$$

that is, in the whole domain of coverage, it is valid that

$$\psi_{mV}(\Theta) > 1 \tag{168}$$

This means that the mobility of adsorbed molecules cannot cause a change in the type of isotherm. This change may only happen if the interactions between the adsorbed molecules are taken into account, as it is done by the mFG equation.

The mathematically uniform and thermodynamically consistent interpretation of the mV equation is represented in Fig. 13. The functions $\Theta(P)$, $\psi(P)$, $\Theta(P_{r,m})$, $\psi(\Theta)$, $A_r^s(\Theta)$, and $A_r^s(P_{r,m})$ are calculated the same way as the calculations were performed for the mL and mFG equations shown in Figs. 6 and 10. The most important difference between the interpretation, of mFG and mV equations is the functions $\psi(P)$ (top right in Figs. 10 and 13). These functions, reflect best the differences between interactions of the adsorbed molecules and the mobility of them.

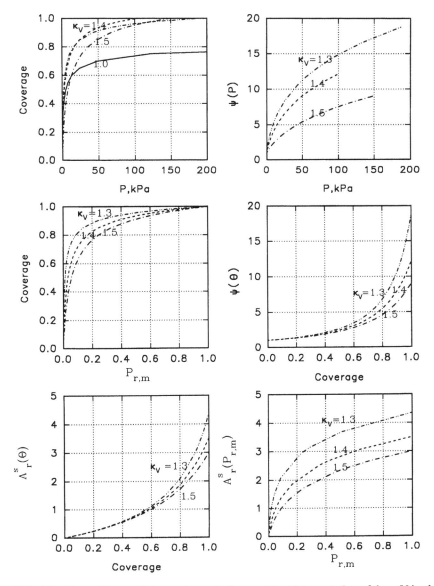

FIG. 13   The uniform and thermodynamically consistent interpretation of the mV isotherm of Type I. The parameters are as follows: $K_V = 0.50$ kPa^{-1}, $\chi_V = 1.0$, $P_m \to \infty$ (solid line, original Volmer equation); $K_{mV} = 0.50$ kPa^{-1}, $\chi_V = 1.3$, $P_m = 186.9$ kPa ($-\cdot\cdot-\cdot\cdot$); $K_{mV} = 0.30$ kPa^{-1}, $\chi_V = 1.4$, $P_m = 101.5$ kPa ($---$); $K_{mV} = 0.10$ kPa^{-1}; $\chi_V = 1.5$, $P_m = 147.8$ kPa ($-\cdot-\cdot-\cdot$).

F. The Uniform and Consistent Interpretation of the Modified de Boer–Hobson Equation

Both the mobility and the interactions are taken into account by the de Boer–Hobson (BH) equation [10], having the following explicit form:

$$P = \frac{1}{K_B - \Theta} \exp\left(\frac{\Theta}{1 - \Theta} - B_B\Theta\right) \tag{169}$$

The corresponding function $\psi_B(\Theta)$ is

$$\psi_B(\Theta) = \left(\frac{1}{1 - \Theta}\right)^2 - B_B\Theta \tag{170}$$

Equation (170) is also thermodynamical inconsistent because the limiting value is $\Theta = 1$ if $P \to \infty$, with a plateau on the isotherms. To derive the consistent form, the modified form of the function $\psi_B(\Theta)$ is also required:

$$\psi_{mB}(\Theta) = \left(\frac{\chi_B}{\chi_B - \Theta}\right)^2 - B_B\Theta \tag{171}$$

Substitution of Eq. (171) into Eq. (95) and integration yields

$$P = P_m(\chi_B - 1) \exp\left(B_B - \frac{1}{\chi_B - 1}\right) \frac{\Theta}{\chi_B - \Theta} \exp\left(\frac{\Theta}{\chi_B - \Theta} - B_B\Theta\right) \tag{172}$$

where the constant of integration has the form

$$I_B = P_m(\chi_B - 1) \exp\left(B_B - \frac{1}{\chi_B - 1}\right) \tag{173}$$

Thus, we obtain the modified BH (mBH) equation in a simple form:

$$P = \frac{1}{K_{mB}} \frac{\Theta}{\chi_B - \Theta} \exp\left(\frac{\Theta}{\chi_B - \Theta} - B_B\Theta\right) \tag{174}$$

where

$$K_{mB} = (I_B)^{-1} = \left[P_m(\chi_B - 1) \exp\left(B_B - \frac{1}{\chi_B - 1}\right)\right]^{-1} \tag{175}$$

or

$$P_m = I_B\left[(\chi_B - 1) \exp\left(B_B - \frac{1}{\chi_B - 1}\right)\right]^{-1} \tag{176}$$

Because Eq. (174) takes the interactions between the adsorbed molecules into account, it is again expected that the mBH equation can describe different types of isotherms and reflect the two-dimensional condensation also. The limits and values of B_B and χ_B determining the applicability of Eq. (174) to different types of isotherm can be calculated quite similar to the limits and values shown in Fig. 8. However, the numerical values of relationships corresponding to the mBH equation differ from those plotted in Fig. 8.

The starting point of our considerations is again the calculation of the values of B_B which meet the condition

$$\frac{d\varphi(\Theta)}{d\Theta} = 0$$

Thus, we obtain

$$B_B = \left(\frac{\chi_B}{\chi_B - \Theta}\right)^2 \Theta^{-1} \tag{177}$$

Equation (177) corresponds to Eq. (133).

The absolute minimum points, corresponding to Eq. (134) are

$$\Theta_{\min} = 0.3333\chi_B \tag{178}$$

The minima of B_B, corresponding to Eq. (135) are

$$B_{B,\min} = \frac{6.75}{\chi_B} \tag{179}$$

The values of B_B which meet the condition $\psi_B(\Theta) = 1$ are

$$B_B = \frac{2\chi_B - \Theta}{(\chi_B - \Theta)^2} \tag{180}$$

Equation (180) corresponds to Eq. (137).

According to Eqs. (177)–(180), the limits and values of B_F mentioned are the following. The mBH equation describes isotherms of Type I when $0 < B_B < 2\chi_B$. Isotherms of Type V can occur when $2/\chi_B < B < 6.75/\chi_B$, and two-dimensional condensation takes place when $B_B \geq 6.75/\chi_B$.

Similar to Eq. (138), it may occur that

$$\frac{6.75}{\chi_B} = \frac{2\chi_B - 1}{(\chi_B - 1)^2} \tag{181}$$

Solving the second-power equation (181), we obtain

$$\chi_B = 1.873 \tag{182}$$

This means that for all values of χ_B which are greater than 1.873, isotherms of Type III also are described by the mBH equation. So, the limits and values of B_B and χ_B are modified if $\chi_B > 1.873$: isotherms of Type I when $0 < B_B < 2/\chi_B$; isotherms of Type V when $2/\chi_B < B_B < (2\chi_B - 1)/(\chi_B - 1)^2$; and those of Type III when $(2\chi_B - 1)/(\chi_B - 1)^2 < B_B < 6.75/\chi_B$. These limits are interpreted in Fig. 14.

For practical applications of Eq. (172) and for calculations of constants K_{mB}, χ_B, B_B, and P_m, a three-parameter fitting procedure is again recommended. In particular, Eq (172) has the following form:

$$P = I_B \frac{\Theta}{\chi_B - \Theta} \exp\left(\frac{\Theta}{\chi_B - \Theta} - B_B\Theta\right) \tag{183}$$

Equation (183) can be fitted to the measured points (Θ, P), with parameters I_B, χ_B, and B_B. Therefore, as for Eqs. (150), (151), (165), and (166), we obtain

$$K_{\mathrm{mB}} = (I_B)^{-1} \tag{184}$$

and

$$P_m = I_B \left[(\chi_B - 1) \exp\left(B_B - \frac{1}{\chi_B - 1}\right)\right]^{-1} \tag{185}$$

If the value of n_m^s is not known (i.e., Θ cannot be calculated), then a four-parameter fitting procedure may be tried. However, the calculation of n_m^s is discussed in detail in Section VI.

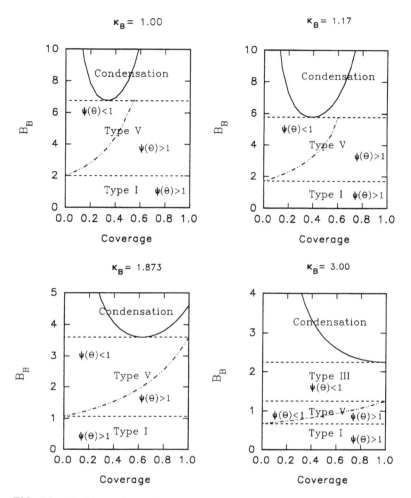

FIG. 14 Limiting values of the parameters B_B and χ_B determining the types of BH and mBH isotherms.

The mathematically uniform and thermodynamically consistent interpretation of the mBH equation are shown in Figs. 15–17. The function $\Theta(P)$, $\Theta(P_{r,m})$, $\psi(\Theta)$, $A_r^s(\Theta)$, and $A_r^s(P_{r,m})$ in these figures are similar to the corresponding relationships in Figs. 10–12. The essential and immediately observable difference can be seen in the function $\psi(P)$ (see Figs. 10 and 15, top right). The importance of this difference is discussed in Section III.H.

G. Physical Interpretation of Constants K_x Present in the Modified Isotherm Equations Applied to Homogeneous Surfaces

Equations (120), (129), (161), and (175) mathematically define the constants K_L, K_{mF}, K_{mV}, and K_{mB} present in the corresponding isotherm equations. In addition to this interpretation, a physical one is possible. According to de Boer–Hobson's theory [10], the Henry constant (H) of an isotherm measured on homogeneous surface is

$$H = k_B^{-1} \exp\left(\frac{U_0}{RT}\right) \tag{186}$$

where k_B is defined by Eq. (83).

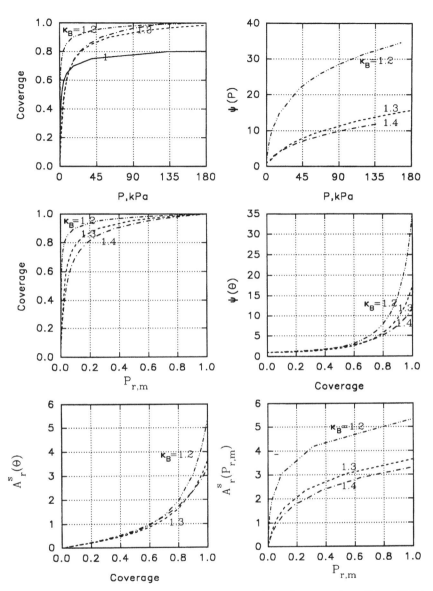

FIG. 15 The uniform and thermodynamically consistent interpretation of the mBH isotherms of Type I. The values of parameters are as follows: $K_B = 0.50$ kPa^{-1}, $\chi_B = 1.00$, $B_B = 1.50$, $P_m \to \infty$ (solid line, original BH equation); $K_{mB} = 1.1$ kPa^{-1}, $\chi_B = 1.20$, $B_B = 1.40$, $P_m = 166.4$ kPa $(-\cdot\cdot-\cdot\cdot-)$; $K_{mB} = 0.1$ kPa^{-1}, $\chi_B = 1.30$, $B_B = 1.40$, $P_m = 230.4$ kPa $(---)$; $K_{mB} = 0.15$ kPa^{-1}, $\chi_B = 1.4$, $B_B = 0.40$, $P_m = 136$ kPa $(-\cdot-\cdot-)$.

The original Langmuir equation has the form

$$P = \frac{1}{K_L}\frac{\Theta}{1 - \Theta} \tag{187}$$

The Henry constant (H) is defined by the limiting value

$$\lim_{P \to 0} \frac{\Theta}{P} = H \tag{188}$$

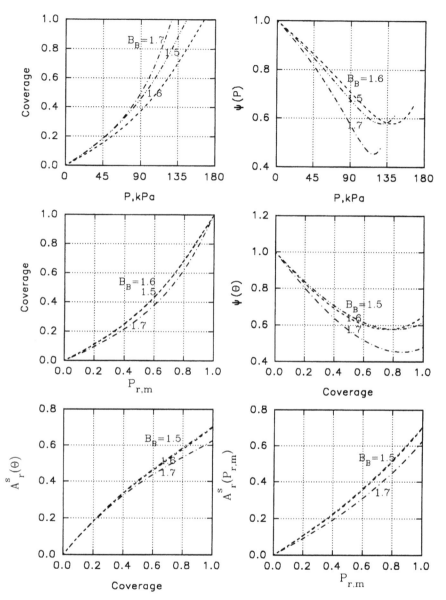

FIG. 16 The uniform and thermodynamically consistent interpretation of the mBH isotherms of Type III. The values of parameters are as follows: $K_{mB} = 0.75 \times 10^{-3}$ kPa^{-1}, $\chi_B = 3.1$, $B_B = 1.70$, $P_m = 186.7$ kPa $(- \cdot - \cdot -)$; $K_{mB} = 10^{-3}$ kPa^{-1}, $\chi_B = 3.0$, $B_B = 1.60$, $P_m = 166.4$ kPa $(---)$; $K_{mB} = 1.1 \times 10^{-3}$ kPa^{-1}, $\chi_B = 3.2$, $B_B = 1.50$, $P_m = 145.3$ kPa $(- \cdots - \cdots -)$.

It means that in the original Langmuir equation,

$$K_L = H \qquad (189)$$

(i.e., K_L represents the Henry constant). At the modified Langmuir equation,

$$P = \frac{1}{K_{mL}} \frac{\Theta}{\chi_L - \Theta} \qquad (190)$$

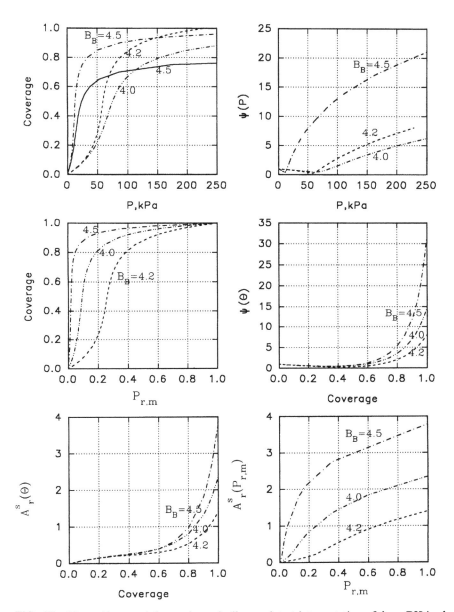

FIG. 17   The uniform and thermodynamically consistent interpretation of the mBH isotherms of Type V. The values of parameters are as follows: $K_B = 0.012$ kPa^{-1}, $\chi_B = 4.50$, $P_m \to \infty$ (solid line, original BH equation); $K_{mB} = 0.012$ kPa^{-1}, $\chi_B = 1.2$, $B_B = 4.50$, $P_m = 687.0$ kPa $(- \cdot - \cdot -)$; $K_{mB} = 2 \times 10^{-3}$ kPa^{-1}, $\chi_B = 1.4$, $B_B = 4.20$, $P_m = 228.4$ kPa $(---)$; $K_{mB} = 2.1 \times 10^{-3}$ kPa^{-1}, $\chi_B = 1.3$, $B_B = 4.00$, $P_m = 815.0$ kPa $(- \cdot \cdot - \cdot \cdot -)$.

the limiting value is

$$\mathrm{H} = \lim_{P \to 0} \frac{\Theta}{P} = \mathrm{K_{mL}} \chi_L = k_B^{-1} \exp\left(\frac{U_0}{RT}\right) \tag{191}$$

that is,

$$K_{\mathrm{mL}} = (k_B \chi_L)^{-1} \exp\left(\frac{U_0}{RT}\right) \tag{192}$$

Taking Eqs. (140), (160), and (174) into account, we also obtain the physical interpretation of constants, K_{mF}, K_{mV}, and K_{mB}, respectively:

$$K_{\mathrm{mF}} = (k_B \chi_F)^{-1} \exp\left(\frac{U_0}{RT}\right) \tag{193}$$

$$K_{\mathrm{mV}} = (k_B \chi_V)^{-1} \exp\left(\frac{U_0}{RT}\right) \tag{194}$$

$$K_{\mathrm{mB}} = (k_B \chi_B)^{-1} \exp\left(\frac{U_0}{RT}\right) \tag{195}$$

It is easy to see and explain that these constants, representing the slopes of isotherms in very low equilibrium pressures, differ in value from the parameter χ only if the absorptive potential, U_0, and k_B are identical.

H.   Properties of the Function $\psi(P)$ Corresponding to the Modified Langmuir, FG, Volmer, and BH Isotherm Equations

In previous subsections, the thermodynamic properties, especially the change in relative free energy of the surface, have been discussed. However, the calculation of the functions $A_r^s(\Theta)$ and $A_r^s(P_{r,m})$ require integration according to Eq. (100); therefore, it needs time. This time is necessary when we are interested in the thermodynamic properties of the adsorbed phase. Nevertheless, after measuring an isotherm, the following question should be answered immediately: Which isotherm equation can be applied to describe and explain the measured data? This problem may be solved by calculation the function $\psi(P)$ defined by relationship (90):

$$\psi(P) = \frac{n^s}{P} \left(\frac{dn^s}{dP}\right)^{-1}$$

This calculation can be made without knowledge of the specific area of the absorbent and it needs only the differentiation of an explicit function fitted to the measured points. These function $\psi(P)$ corresponding to the modified Langmuir, FG, Volmer, and BH equations are shown in Fig. 18.

   In the top (left) of Fig. 18 are plotted the functions $\psi(P)$ corresponding to isotherms Type I, which can be seen in the top (right) of Fig. 18. These isotherms of Type I are very similar; therefore, with a simple fitting procedure the thermodynamically correct isotherm equation cannot be selected. However, the function $\psi(P)$, especially in lower domain of pressure (see bottom in Fig. 18), are very different and characteristic for the corresponding isotherm equation. So, after calculation of the function $\psi(P)$, the correct isotherm equation can be selected and applied for the isotherm measured on a homogeneous surface.

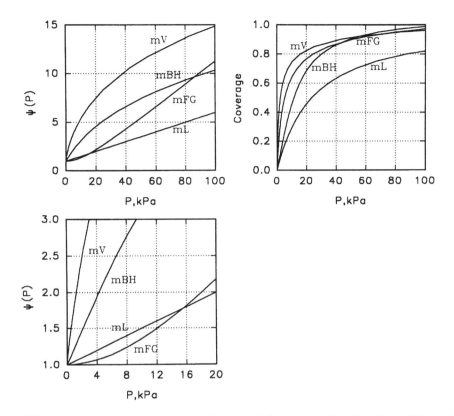

FIG. 18   Comparison of isotherms of Type I and the corresponding functions $\psi(P)$ calculated by the mL, mV, mFG and mBH equations. The values of parameters are as follows: mL: $K_{mL} = 0.04$ kPa^{-1}, $\chi_L = 1.03$, $P_m = 833.3$ kPa;   mV: $K_{mV} = 0.50$ kPa^{-1}, $\chi_V = 1.30$, $P_m = 187.0$ kPa;   mFG: $K_{mF} = 0.08$ kPa^{-1}, $\chi_F = 1.05$, $P_m = 112.3$ kPa; mBH: $K_{mB} = 0.15$ kPa^{-1}, $\chi_B = 1.40$, $P_m = 136.1$ kPa.

IV.   THE UNIFORM AND THERMODYNAMICALLY CONSISTENT INTERPRETATION OF G/S ISOTHERM EQUATIONS APPLIED FOR HETEROGENEOUS SURFACES

A.   The Tóth Equation

An old problem of adsorption theories is how the very complex effects of energetic heterogeneity of solids can be taken into account. One of these attempts have been made by Tóth [13]. The starting point of his theory is the following observation. A heterogeneous surface uptakes more adsorptive, and at the same, *relative* equilibrium pressure, than a homogeneous surface with specific surface area equal to that of the heterogeneous adsorbent. Consequently, in this case, there is no difference in the monolayer capacities (n_m^s) of the homogeneous and heterogeneous surfaces. This requirement can be taken by one parameter, t, applied to the coverage as a power into account; that is,

$$\Theta^t > \Theta \quad \text{if} \quad 0 < t < 1 \tag{196}$$

It is also an experimental observation that if the lateral interactions between the adsorbed molecules are greater than the adsorptive potential at the same coverage, then

$$\Theta^t < \Theta \quad \text{for} \quad t > 1 \tag{197}$$

First, let us consider relationship (196). According to the modified Langmuir equation, its function $\psi(\Theta)$ can be defined by Eq. (129):

$$\psi_{mL}(\Theta) = \frac{\chi_L}{\chi_L - \Theta}$$

Taking requirement (196) into account, we obtain

$$\psi_T(\Theta) = \frac{\chi_T}{\chi_T - \Theta^t} \quad 0 < t < 1 \tag{198}$$

Substitution of Eq. (198) into Eq. (95) and integration yields

$$P = P_m (\chi_T - 1)^{1/t} \frac{\Theta}{(\chi_T - \Theta^t)^{1/t}} \tag{199}$$

where the constant of integrations is

$$I_T = P_m (\chi_T - 1)^{1/t} \tag{200}$$

The simple form of Eq. (199) is

$$P = \left(\frac{1}{K_T}\right)^{1/t} \frac{\Theta}{(\chi_T - \Theta^t)^{1/t}} \tag{201}$$

where

$$K_T = (I_T)^{-t} \tag{202}$$

and

$$P_m = I_T [(\chi_T - 1)^{1/t}]^{-1} \tag{203}$$

It is evident that at $\Theta = 1$ and $P = P_m$.

Equation (199) is thermodynamically consistent, because the change in relative free energy of the surface, when the total monolayer capacity is completed, is a *finite* value; namely, according to Eq. (100),

$$A_r^s(\Theta = 1) = \int_0^1 \frac{\chi_T}{\chi_T - \Theta^t} \, d\Theta = \text{finite value} \tag{204}$$

The constant K_T can also be expressed by physical parameters. In particular, it is known that for homogeneous surfaces, Eqs. (186) and (188) are valid; that is

$$H = \lim_{P \to 0} \frac{\Theta}{P} = k_B^{-1} \exp\left(\frac{U_0}{RT}\right) \tag{205}$$

where U_0 is the constant adsorptive potential. However, Eq. (199) relates to heterogeneous surfaces; therefore, instead of U_0, the corresponding value of the differential adsorptive potential, U_0^{diff}, should be present in Eq. (205). In the domain of the very low equilibrium pressures and coverages (i.e., when P and Θ tend to zero), it is also valid that

$$\lim_{\Theta \to 0} U_0^{\text{diff}}(\Theta) = \text{const} \tag{206}$$

In this sense, for heterogeneous surfaces, it is also correct that

$$H = \lim_{P \to 0} \frac{\Theta}{P} = k_B^{-1} \exp[U_0^{\text{diff}}(\Theta = 0)] \tag{207}$$

For Eq. (201), it is valid that

$$H = \lim_{P \to 0} \frac{\Theta}{P} = (K_T \chi_T)^{1/t} \tag{208}$$

So, comparing Eq. (208) with Eq. (207), we obtain

$$(K_T \chi_T)^{1/t} = k_B^{-1} \exp\left(\frac{U_0^{\text{diff}}(\Theta = 0)}{RT}\right) \tag{209}$$

that is,

$$K_T = \frac{k_B^{-t}}{\chi_T} \exp\left(\frac{t U_0^{\text{diff}}(\Theta = 0)}{RT}\right) \tag{210}$$

For practical applications of Eq. (201), a three-parameter fitting procedure is proposed. The parameters to be fitted are K_T, χ_T, and t. The value of P_m is calculated from Eq. (203). For the calculation of the total monolayer capacity, a four-parameter iteration may also be possible because Eq. (201) may also be written in this form:

$$P = \frac{1}{(K_T)^{1/t}} \frac{n^s}{[\chi_T (n_m^s)^t - (n^s)^t]^{1/t}} \tag{211}$$

However, for calculating n_m^s, a general method is discussed in Section VI.

The mathematically uniform and thermodynamically consistent interpretation of Eq. (199) is shown in Fig. 19, where the functions $\Theta(P)$, $\psi(P)$, $\Theta(P_{r,m})$, $\psi(\Theta)$, $A_r^s(\Theta)$, and $A_r^s(P_{r,m})$ are plotted. The calculations of these functions have been made as follows:

Top (left):

$$P = \left(\frac{1}{K_T}\right)^{1/t} \frac{\Theta}{(\chi_T - \Theta^t)^{1/t}} \tag{212}$$

or

$$\Theta = \frac{(\chi_T)^{1/t} P}{[(1/K_T) + P^t]^{1/t}} \tag{213}$$

Top (right):

$$\psi_T(P) = K_T P^t + 1 \tag{214}$$

or the values of the function

$$\psi_T(\Theta) = \frac{\chi_T}{\chi_T - \Theta^t} \tag{215}$$

are conjugated with values of P caculated from Eq. (212).
Middle (left): From Eq. (200), it follows that

$$P_m = [K_T(\chi_T - 1)]^{-1/t} \tag{216}$$

so

$$\frac{P}{P_m} = P_{r,m} \tag{217}$$

can be calculated.
Middle (right): Equation (215) has been applied.

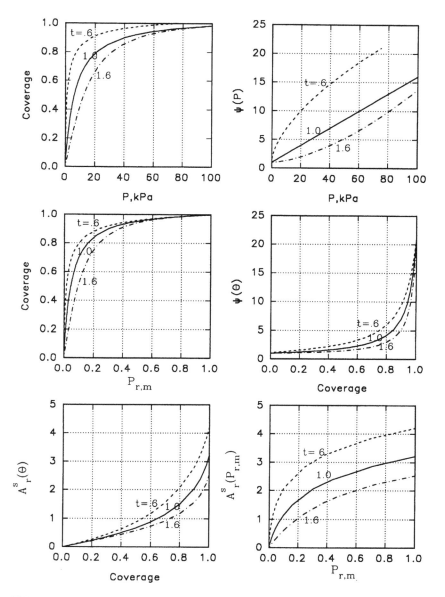

FIG. 19 The uniform and thermodynamically consistent interpretation of the Tóth isotherms of Type I. The values of parameters are as follows: $K_L = 0.15$ kPa^{-1}, $\chi_L = 1.05$, $t = 1.0$, $P_m = 133.3$ kPa (solid line, mL equation); $K_T = 1.5$ kPa^{-1}, $\chi_T = 1.05$, $t = 0.6$, $P_m = 75.0$ kPa ($---$); $K_T = 8 \times 10^{-3}$ kPa^{-1}, $\chi_T = 1.05$, $t = 1.6$, $P_m = 133.0$ kPa ($- \cdot - \cdot -$).

Bottom (left):

$$A_r^s(\Theta) = \int_\Theta^1 \frac{\chi_T}{\chi_T - \Theta^t} \, d\Theta \tag{218}$$

The integration has been performed numerically.
Bottom (right): The values of Eqs. (218), (212), and (217) have been conjugated.

In Figure 19, the Tóth equations are represented with different parameters t (0.6, 1.0, 1.6). It is evident that $t = 1$ represents the modified Langmuir equation. The values $t < 1$ relate to heterogeneous surfaces where the adsorbent–absorptive interactions are greater than those between the molecules adsorbed. The values $t > 1$ relate to the reverse situation. From these suppositions, it follows that the mL equation cannot be applied to homogeneous surfaces only, but it can describe isotherms measured on heterogeneous surfaces on the condition that the two interactions adsorbent-adsorptive interactions and those between adsorbed molecules are approximately equal. This suppositions implicit include the fact that the parameter t in the Tóth equation expresses not only the heterogeneity of the surface but also reflects the interactions and mobility (immobility) of the molecules adsorbed. This statement is proven in Section IV.F.

It is also remarkable that the isotherms of Type I with different parameters t are very similar relationships; however, the corresponding functions $\psi(P)$ in the top (right) of Fig. 19 are very different and characteristic. Thus, the selection of the appropriate parameter t can be made with the help of function (214).

B. The Modified Fowler–Guggenheim Equation Applied to Heterogeneous Surfaces (FT Equation)

The modified, therefore, the consistent, mFG equation is applicable to homogeneous surfaces. As has been demonstrated, its explicit form is

$$P = \frac{1}{K_{\mathrm{mF}}} \frac{\Theta}{\chi_F - \Theta} \exp(-B_F \Theta) \tag{219}$$

The corresponding function $\psi_{\mathrm{mF}}(\Theta)$ has the form

$$\psi_{\mathrm{mF}}(\Theta) = \frac{\chi_F}{\chi_F - \Theta} - B_F \Theta. \tag{220}$$

According to Tóths idea Eq. (220) expresses the heterogeneity of the surface if the parameter as a power is introduced; that is,

$$\psi_{\mathrm{FT}}(\Theta) = \frac{\chi_F}{\chi_F - \Theta^t} - B_F \Theta^t \quad t > 0 \tag{221}$$

Substituting Eq. (221) into Eq. (95) and integrating, we obtain

$$P = P_m(\chi_F - 1)^{1/t} \exp\left(\frac{B_F}{t}\right) \frac{\Theta}{(\chi_F - \Theta^t)^{1/t}} \exp\left(-\frac{B_F \Theta^t}{t}\right) \tag{222}$$

where the constant of integration has the form

$$I_{\mathrm{FT}} = P_m(\chi_F - 1)^{1/t} \exp\left(\frac{B_F}{t}\right) \tag{223}$$

Thus, we have the modified FG equation applicable for heterogeneous surfaces:

$$P = \left(\frac{1}{K_{\mathrm{FT}}}\right)^{1/t} \frac{\Theta}{(\chi_F - \Theta^t)^{1/t}} \exp\left(-\frac{B_F \Theta^t}{t}\right) \tag{224}$$

where

$$K_{\mathrm{FT}} = (I_{\mathrm{FT}})^{-t} = [P_m^t(\chi_F - 1) \exp(B_F)]^{-1} \tag{225}$$

or

$$P_m = I_{\text{FT}}\left[(\chi_F - 1)^{1/t} \, \exp\left(\frac{B_F}{t}\right)\right]^{-1} \tag{226}$$

From Eq. (224), as it is also valid that the Henry constant can be expressed

$$\lim_{P \to 0} \frac{\Theta}{P} = (K_{\text{FT}}\chi_F)^{1/t} = H \tag{227}$$

Taking Eq. (206) into account, we have

$$K_{\text{FT}} = \frac{k_B^{-t}}{\chi_F} \exp\left(\frac{tU_0^{\text{diff}}(\Theta = 0)}{RT}\right) \tag{228}$$

that is, according to Eq. (209), $K_{\text{FT}} = K_T$.

It is evident that Eq. (224) is a thermodynamically consistent equation because

$$\lim_{\Theta=1} P = P_m \tag{229}$$

The types of isotherms described by Eq. (224) are equal to those described by the FG and mFG equations. The principle of calculations of the limiting values of B_F and χ_F at which the different types of isotherm can occur are also identical to those applied by the mFG equation in Section III.D. This is the reason why only the results are summarized here. In this sense, the value of B_F where Eq. (132) is met is

$$B_F = \frac{\chi_F}{\Theta^t(\chi_F - \Theta^t)} \tag{230}$$

The values of coverages where minima of Eq. (230) occur are

$$\Theta_{\min} = (0.5\chi_F)^{1/t} \tag{231}$$

The corresponding minima of values of B_F are unchanged [see Eq. (135)]:

$$B_{F,\min} = \frac{4}{\chi_F} \tag{232}$$

The values of B_F where condition $\psi(\Theta) = 1$ is fulfilled are

$$B_{\text{F}} = \frac{1}{\chi_F - \Theta^t} \tag{233}$$

Taking Eqs. (230)–(233) into account, it can be verified that the limiting values of B_F and χ_F determining the types of isotherms do not change. These values are shown in Fig. 8. Taking these limiting values into account the different types of isotherm can be uniformly interpreted, as it has been done in Figs. 10–12. In Fig. 20, the functions $\Theta(P)$, $z\psi(P)$, $\Theta(P_{r,m})$, $\Psi(\Theta)$, $A_r^s(\Theta)$, and $A_r^s(P_{r,m})$ corresponding to isotherms of Type I are shown only because these functions occur more frequently in practice than other types.

The calculations relating to Fig. 20 are summarized as follows:

Top (left):

$$P = \left(\frac{1}{K_{\text{FT}}}\right)^{1/t} \frac{\Theta}{(\chi_F - \Theta^t)^{1/t}} \, \exp\left(-\frac{B_F\Theta^t}{t}\right) \tag{234}$$

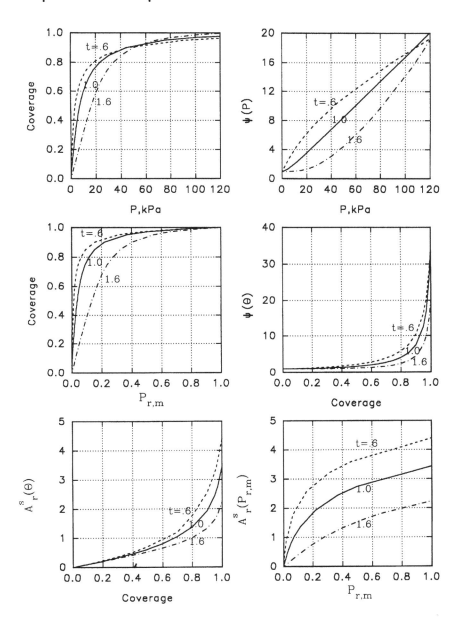

FIG. 20 The uniform and thermodynamically consistent interpretation of the FT isotherms of Type I. The values of parameters are as follows: $K_{mF} = 0.10$ kPa^{-1}, $\chi_F = 1.03$, $B_F = 0.5$, $t = 1.0$, $P_m = 291.5$ kPa (solid line, mF equation); $K_{FT} = 0.45$ kPa^{-1}, $\chi_F = 1.03$, $B_F = 0.90$, $t = 0.6$, $P_m = 202.2$ kPa ($---$); $K_{FT} = 4 \times 10^{-3}$ kPa^{-1}, $\chi_F = 1.05$, $B_F = 0.80$, $t = 1.6$, $P_m = 124.4$ kPa ($-\cdot-\cdot-$).

Top (right):

$$\psi_{FT}(\Theta) = \frac{\chi_F}{\chi_F - \Theta^t} - B\Theta^t \qquad (235)$$

The corresponding pairs (ψ_{FT}, P) of Eqs. (234) and (235) are plotted.

Middle (left):

$$P_{r,m} = \frac{\text{Eq. (224)}}{\text{Eq. (226)}}.$$ (236)

Middle (right): Equation (235) has been applied.
Bottom (left):

$$A_r^s(\Theta) = \int_\Theta^l \psi_{\text{FT}}(\Theta) \, d\Theta$$ (237)

The integration has been performed numerically.
Bottom (right): The conjugated pairs of $A_r^s(\Theta)$ and Eq. (236) are plotted.

For practical applications of Eq. (4.29), a four-parameter fitting process is proposed, where the parameters to be iterated are I_{FT}, χ_T, t, and B_F. I_{FT} is defined by Eq. (223), P_m is calculable from Eq. (226), and K_{FT} is determined by Eq. (225). The calculation of the total monolayer capacity, n_m^s, is discussed in detail in Section VI.

C. The Modified Volmer Equation Applied to Heterogeneous Surfaces (VT Equation)

The thermodynamically consistent (modified) mV equation and its function $\psi_{\text{mV}}(\Theta)$ have been defined by Eqs. (160) and (157), respectively. Introducing the parameter t proposed by Tóth, we obtain

$$\psi_{\text{VT}}(\Theta) = \left(\frac{\chi_V}{\chi_V - \Theta^t}\right)^2, \quad t > 0$$ (238)

Substitution of Eq. (238) into Eq. (157) and integration yields

$$P = P_m(\chi_V - 1)^{1/t} \, \exp\left\{-\frac{1}{t(\chi_V - 1)}\right\} \frac{\Theta}{(\chi_V - \Theta^t)^{1/t}} \, \exp\left\{\frac{\Theta^t}{t(\chi_V - \Theta^t)}\right\}$$ (239)

where the constant of integration is

$$I_{\text{VT}} = P_m(\chi_V - 1)^{1/t} \, \exp\left\{-\frac{1}{t(\chi_V - 1)}\right\}$$ (240)

Thus, we obtain the mV equation applicable for heterogeneous surfaces:

$$P = \left(\frac{1}{K_{\text{VT}}}\right)^{1/t} \frac{\Theta}{(\chi_V - \Theta^t)^{1/t}} \, \exp\left\{\frac{\Theta^t}{t(\chi_V - \Theta^t)}\right\}, \quad t > 0$$ (241)

where

$$K_{\text{VT}} = (I_{\text{VT}})^{-t} = \left[P_m(\chi_V - 1) \, \exp\left\{-\frac{1}{\chi_V - 1}\right\}\right]^{-t}$$ (242)

or

$$P_m = I_{\text{VT}}(\chi_V - 1)^{1/t} \, \exp\left\{-\frac{1}{t(\chi_V - 1)}\right\}$$ (243)

Taking Eq. (206) into account, we again obtain that

$$\lim_{P \to 0} \frac{\Theta}{P} = (K_{VT}\chi_V)^{1/t} = H \tag{244}$$

that is, according to Eq. (209),

$$K_{VT} = \frac{k_B^{-t}}{\chi_V} \exp\left\{\frac{tU_0^{diff}(\Theta = 0)}{RT}\right\} \tag{245}$$

It is evident again that Eq. (241) is a thermodynamically consistent relationship because

$$\lim_{\Theta=1} P = P_m \tag{246}$$

For practical applications of Eq. (241), a three-parameter fitting process is proposed, where the parameters to be iterated are I_{VT}, χ_V, and t. I_{VT} is defined by Eq. (240), P_m and K_{VT} can be calculated from Eqs. (243) and (242), respectively. Also a four-parameter fitting procedure may be tried because substituting n_m^s in Eq. (239), we obtain

$$P = I_{VT} \frac{n^s}{[\chi_V (n_m^s)^t - (n^s)^t]^{1/t}} \exp\left\{\frac{(n^s)^t}{t[\chi_V(n_m^s)^t - (n_s)^t]}\right\} \tag{247}$$

The mathematically uniform and thermodynamically consistent interpretation of functions $\Theta(P)$, $\psi(P)$, $\Theta(P_{r,m})$, $\psi(\Theta)$, $A_r^s(\Theta)$, and $A_r^s(P_{r,m})$ are shown in Fig. 21. The calculations of these functions are summarized as follows: top (left), Eq. (241); top (right), the conjugated pairs of Eqs. (241) and (238); middle (left),

$$P_{r,m} = \frac{\text{Eq. (241)}}{\text{Eq. (243)}} \tag{248}$$

middle (right), Eq. (2.38); bottom (left),

$$A_r^s(\Theta) = \int_\Theta^1 \left(\frac{\chi_V}{\chi_V - \Theta^t}\right)^2 d\Theta \tag{249}$$

bottom (right), the conjugated pairs of Eqs. (249) and (248) have been applied. The integration of Eq. (249) has been performed numerically.

D. The Modified de Boer–Hobson Equation Applied to Heterogeneous Surfaces (BT Equation)

The thermodynamically consistent mBH equation and its function $\psi_{mB}(\Theta)$ have been defined by Eqs. (174) and (171), respectively. If it is introduced the parameter t proposed by Tóth, then we have

$$\psi_{BT}(\Theta) = \left(\frac{\chi_B}{\chi_B - \Theta^t}\right)^2 - B_B\Theta^t \tag{250}$$

Substituting Eq. (250) into Eq. (95) and performing the integrations, we obtain

$$P = P_m(\chi_B - 1)^{1/t} \exp\left\{-\frac{1}{t(\chi_B - 1)} + \frac{B_B}{t}\right\} \frac{\Theta}{(\chi_B - \Theta^t)^{1/t}} \exp\left\{\frac{\Theta^t}{t(\chi_B - \Theta^t)} - \frac{B_B\Theta^t}{t}\right\} \tag{251}$$

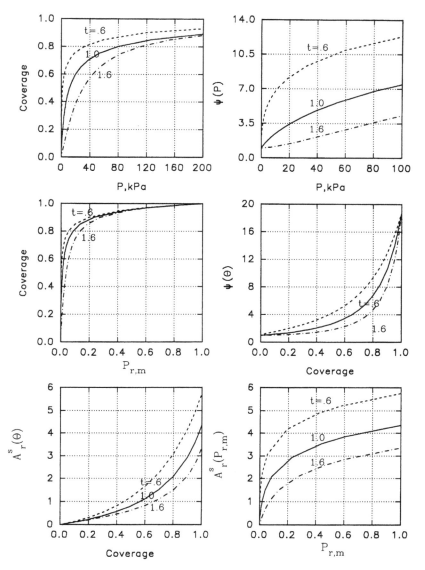

FIG. 21 The uniform and thermodynamically consistent interpretation of the VT isotherms of Type I. The values of parameters are as follows: $K_V = 0.1\,\text{kPa}^{-1}$, $\chi_V = 1.3$, $t = 1.0$, $P_m = 934.9\,\text{kPa}$ (solid line, mV equation); $K_{VT} = 2.0\,\text{kPa}^{-t}$, $\chi_V = 1.3$, $t = 0.6$, $P_m = 6.06\,\text{kPa}$ (---); $K_{VT} = 2 \times 10^{-3}\,\text{kPa}^{-t}$, $\chi_V = 1.3$, $t = 1.6$, $P_m = 828.8\,\text{kPa}$ ($- \cdot - \cdot -$).

where the constant of integration is

$$I_{BT} = P_m(\chi_B - 1)^{1/t} \exp\left\{-\frac{1}{t(\chi_B - 1)} + \frac{B_B}{t}\right\} \tag{252}$$

Thus, we have the mBH equation applicable for heterogeneous surfaces:

$$P = \left(\frac{1}{K_{BT}}\right)^{1/t} \frac{\Theta}{(\chi_B - \Theta^t)^{1/t}} \exp\left\{\frac{\Theta^t}{t(\chi_B - \Theta^t)} - \frac{B_B\Theta^t}{t}\right\} \tag{253}$$

where

$$K_{BT} = (I_{BT})^{-t} \tag{254}$$

or

$$P_m = I_{BT}\left[(\chi_B - 1)^{1/t}\,\exp\left\{-\frac{1}{t(\chi_B - 1)} + \frac{B_B}{t}\right\}\right]^{-1} \tag{255}$$

Taking Eq. (206) into account, we again obtain that

$$\lim_{P \to 0}\frac{\Theta}{P} = (K_{BT}\chi_B)^{1/t} = H \tag{256}$$

that is, according to Eq. (209),

$$K_{BT} = \frac{k_B^{-t}}{\chi_B}\,\exp\left\{\frac{tU_0^{\mathrm{diff}}(\Theta = 0)}{RT}\right\} \tag{257}$$

The limiting value of Eq. (251),

$$\lim_{\Theta=1} P = P_m \tag{258}$$

proves again that the BT equation is thermodynamically consistent.

For practical applications of Eq. (253), a four-parameter fitting process is recommended, where the parameters to be iterated are I_{BT}, χ_B, t, and B_B. I_{BT} is defined by Eq. (252); P_m and K_{BT} can be calculated from Eqs. (255) and (254), respectively.

The customary uniform and consistent interpretation of functions $\Theta(P)$, $\Psi(P)$, $\Theta(P_{r,m})$, $\psi(\theta)$, $A_r^s(\Theta)$, and $A_r^s(P_{r,m})$ are shown in Fig. 22. The calculation of these functions are summarized as follows: Top (left), Eq. (253); top (right), the conjugated pairs of Eqs. (250) and (253); middle (left),

$$P_{r,m} = \frac{\text{Eq. (253)}}{\text{Eq. (255)}} \tag{259}$$

middle (right), Eq. (250); bottom (left),

$$A_r^s(\Theta) = \int_\Theta^1\left[\left(\frac{\chi_B}{\chi_B - \Theta^t}\right)^2 - B_t\Theta^t\right]d\Theta \tag{260}$$

bottom (right), the corresponding pairs of Eqs. (260) and (259) have been applied. The integration of Eq. (260) has been performed numerically.

E. The Mathematically Generalized Form of mL, mFG, mV, mBH, Tóth, FT, VT, and BT Equations

The mL, mFG, mV, and mBH equations, Eqs. (115), (128), (160), and (174), respectively, are the modified thermodynamically consistent relationships applicable to *homogeneous* surfaces. Also, the consistent relationships for Tóth, FT, VT, and BT equations [Eqs. (201), (224), (241), and (253), respectively] are applicable to *heterogeneous* surfaces. Those eight equations have been discussed in detail and interpreted in separated subsections of this section. This separated interpretation has proven that the eight relationships mentioned can be combined in a *single* equation.

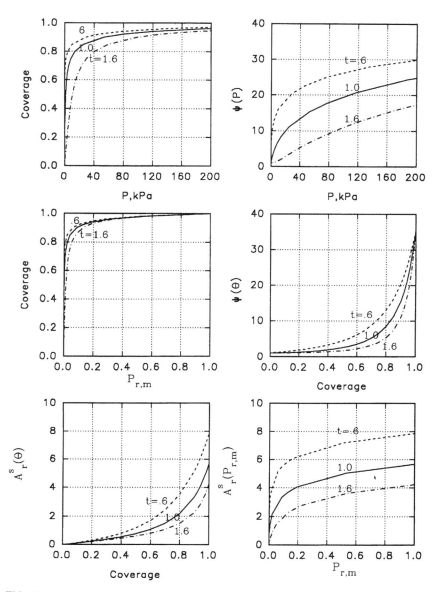

FIG. 22 The uniform and thermodynamically consistent interpretation of the BT isotherms of Type I. The values of parameters are as follows: $K_{mB} = 0.6$ kPa^{-1}, $\chi_B = 1.2$, $B_B = 0.7$, $t = 1.0$, $P_m = 614.2$ kPa (solid line, mBH equation); $K_{BT} = 9.0$ kPa^{-t}, $\chi_B = 1.2$, $B_B = 0.7$, $t = 0.6$, $P_m = 486.4$ kPa $(---)$; $K_{BT} = 0.01$ kPa^{-t}, $\chi_B = 1.2$, $B_B = 0.6$, $t = 1.6$, $P_m = 760.6$ kPa $(-\cdot-\cdot-)$.

This equation has the form

$$P = I \frac{\Theta}{(\chi_x - \Theta^t)^{1/t}} \exp\left(\frac{\alpha \Theta^t}{t(\chi_x - \Theta^t)} - \frac{B_x \Theta^t}{t}\right) \tag{261}$$

where

$$I = P_m(\chi_x - 1)^{1/t} \exp\left(\frac{-\alpha}{t(\chi_x - 1)} + \frac{B_x}{t}\right) \tag{262}$$

In Eqs. (261) and (262), a new parameter, α, is introduced which may have only two values; namely, if the adsorbed layer is mobile, then $\alpha = 1$; in other cases, $\alpha = 0$. Let us see how Eq. (261) transforms into the eight equations:

1. If $t = 1$, $B_x = 0$, and $\alpha = 0$ (homogeneous surface, no interactions, immobile layer), then we have the mL equation:

$$P = P_m(\chi_L - 1)\frac{\Theta}{\chi_L - \Theta}$$

See Eqs. (115) and (120).

2. If $t = 1$, $B_x = B_F$, and $\alpha = 0$ (homogeneous surface, interactions, immobile layer), then we have the mFG equation:

$$P = P_m(\chi_F - 1)\ \exp(B_F)\frac{\Theta}{\chi_F - \Theta}\ \exp(-B_F\Theta)$$

See Eq. (126).

3. If $t = 1$, $B_x = 0$, and $\alpha = 1$ (homogeneous surface, no interactions, mobile layer), then we have the mV equation:

$$P = P_m(\chi_V - 1)\ \exp\left(-\frac{1}{\chi_V - 1}\right)\frac{\Theta}{\chi_V - \Theta}\ \exp\left(\frac{\Theta}{\chi_V - \Theta}\right)$$

See Eq. (158).

4. If $t = 1$, $B_x = B_B$, and $\alpha = 1$ (homogeneous surface, interactions, mobile layer), then we have the mBH equation:

$$P = P_m(\chi_B - 1)\ \exp\left(B_B - \frac{1}{\chi_B - 1}\right)\frac{\Theta}{\chi_B - \Theta}\ \exp\left(\frac{\Theta}{\chi_B - \Theta} - B_B\Theta\right)$$

See Eq. (172).

5. If $0 < t < 1$, $B_x = 0$, and $\alpha = 0$ (heterogeneous surface, no interactions, immobile layer), then we have the Tóth equation:

$$P = P_m(\chi_T - 1)^{1/t}\frac{\Theta}{(\chi_T - \Theta^t)^{1/t}}$$

See Eq. (199).

6. If $t > 0$, $B_x = B_{FT}$, and $\alpha = 0$ (heterogeneous surface, interactions, immobile layer), then we have the FT equation:

$$P = P_m(\chi_F - 1)^{1/t}\ \exp\left(\frac{B_F}{t}\right)\frac{\Theta}{(\chi_F - \Theta^t)^{1/t}}\ \exp\left(\frac{-B_F\Theta^t}{t}\right)$$

See Eq. (2.22).

7. If $t > 0$, $B_x = 0$, and $\alpha = 1$ (heterogeneous surface, no interactions, mobile layer), then we have the VT equation:

$$P = P_m(\chi_V - 1)^{1/t}\ \exp\left\{-\frac{1}{t(\chi_V - 1)}\right\}\frac{\Theta}{(\chi_V - \Theta^t)^{1/t}}\ \exp\left\{\frac{\Theta^t}{t(\chi_V - \Theta^t)}\right\}$$

See Eq. (239).

8. If $t > 1$, $B_x = B_B$, and $\alpha = 1$ (heterogeneous surface, interactions, mobile layer), then we have the BT equation:

$$P = P_m(\chi_B - 1)^{1/t} \exp\left\{-\frac{1}{t(\chi_B - 1)} + \frac{B_B}{t}\right\} \frac{\Theta}{(\chi_B - \Theta^t)^{1/t}} \exp\left\{\frac{\Theta^t}{t(\chi_B - \Theta^t)} - \frac{B_B\Theta^t}{t}\right\}$$

See Eq. (251).

F. Applicability of the Tóth Equation to all Isotherms of Type I Measured on Heterogeneous Surfaces

Let us compare the functions $\psi(P)$ corresponding to Tóth, FT, VT, and BT equations shown in the top right of Figs. 19–22, respectively. Thus, it is observable that if the values of parameter t are between 0 and 1, that is,

$$0 < t < 0$$

then all functions $\psi(P)$ can be described by simple power functions similar to that of the function $\psi_T(P)$ [see Eq. (214)]:

$$\psi_T(P) = \psi_{FT}(P) = \psi_{VT}(P) = \psi_{BT}(P) = K_x P^{t_x} + 1 \tag{263}$$

It is evident that the values of K_x and t_x are not equal in the four relationships; this fact is reflected by the subscript x. For example,

$$\psi_{FT}(P) = K_{FT}P^{t_{FT}} + 1 \tag{264}$$

or

$$\psi_T(P) = K_T P^t + 1 \tag{265}$$

and so forth. If

$$1 > t_x \tag{266}$$

then it is valid that

$$\psi_T(P) = \psi_{FT}(P) = \psi_{VT}(P) = K_x P^{t_x} + 1 \tag{267}$$

and

$$\psi_T(P) = \psi_{BT}(P) = K_x P^{t_x} + 1 \tag{268}$$

Equations (263), (267), and (268) mean that the Tóth equation with appropriate parameters K_x and t_x can express all properties of an adsorbent–adsorptive system such as the heterogeneity mobility, immobility, and lateral and vertical interactions experimentally reflected by isotherms of Type I. This statement is proven and shown in Figs. 23–25. The functions shown in these three figures are calculated as follows: At the top (left), the functions $\psi(P)$ are plotted; the symbol $\triangledown$ refers to the function $\psi_T(P)$ corresponding to the Tóth equation (214) and the symbol $\bigcirc$ refers to the function $\psi_x(P)$. The top (right) functions $\psi(P)$ are shown in the domain of very low equilibrium pressures. At the bottom (left) are plotted the isotherms $\Theta(P)$ in question (symbol: $\bigcirc$) and the Tóth equation (symbol: $\triangledown$) fitted to this isotherm $\Theta(P)$. At the bottom (right) are shown these two isotherms in the domain of very low equilibrium pressures. The deviations between the two isotherms in each figure have been calculated using Eq. (149). The values for $\Delta\%$ are not greater than the errors of the isotherm measurements ($\pm 3\%$). The corresponding (however, the different) values of K_x, t_x, χ_x, and U_0^{diff} ($\Theta = 0$) calculated using Eqs. (82), (228), (245), 257), and (210) are shown in Table 1.

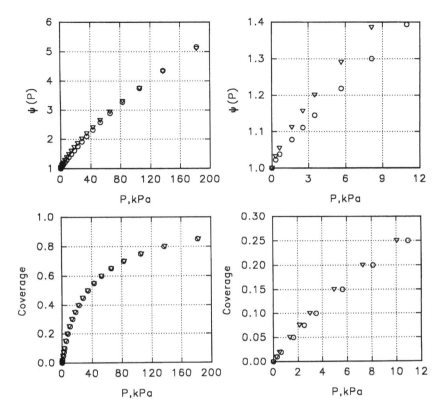

FIG. 23 The Tóth isotherms of Type I can be applied instead of FT isotherms of Type I. The values of parameters of the FT isotherm are as follows: $K_{FT} = 0.12$ kPa^{-t}, $\chi_F = 1.1$, $B_F = 0.6$, $t = 0.6$, $P_m = 584.9$ kPa (○); The values of the parameters of the Tóth equation are as follows: $K_T = 0.0897$ kPa^{-t}; $\chi_T = 1.104$, $t = 0.7334$, $P_m = 585.7$ kPa (▽).

In these equations, the following parameters and values have been applied:

$$M = 16,$$
$$T = 273 \text{ K} \tag{269}$$

Therefore, according to Eq. (82),

$$k_B = 2.346(16 \times 273)^{1/2} \times 10^5 = 155.049 \times 10^5 \text{ kPa} \tag{270}$$

and

$$RT = 2.27 \text{ kJ/mol} \tag{271}$$

So,

$$U_0^{\text{diff}}(\Theta = 0) = \frac{RT}{t} \ln(K_x \chi_x k_B) \tag{272}$$

From the values shown in Table 1, the following conclusions of general validity can be drawn. The FT equation (localized, with interactions) can be substituted with a Tóth equation (localized, no interactions) applied to a surface with less heterogeneity ($t_T < t_{FT}$) and with a lower value of $U_0^{\text{diff}}(\Theta = 0)$ than those corresponding to the FT relationship. It means that the energies of

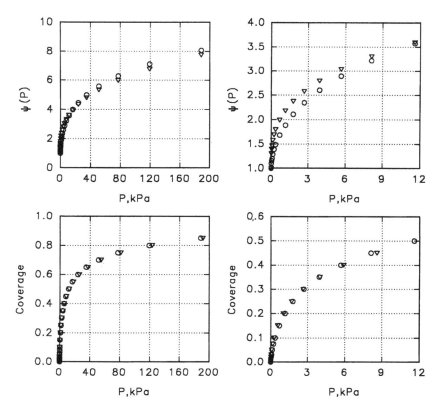

FIG. 24 The Tóth isotherms of Type I can be applied instead of VT isotherms of Type I. The values of parameters of the VT isotherms are as follows: $K_{\mathrm{VT}} = 0.50$ kPa^{-t}, $\chi_V = 1.4$, $t = 0.6$, $P_m = 943.0$ kPa ($\bigcirc$); The values of parameters of thte Tóth equation are as follows: $K_T = 1.1255$ kPa^{-t}, $\chi_T = 1.0854$, $t = 0.342$, $P_m = 946.6$ kPa ($\triangledown$).

lateral interactions can be replaced with a localized adsorption interaction on a less hetero-geneous surface with a less differential adsorptive potential. The same statement is valid for the Tóth and BT equations ($t_T < t_{\mathrm{BT}}$) with the only difference being that the Tóth equation can be applied instead of an adsorption reflecting a mobile layer with interactions. Entirely opposite is the situation in the case of Tóth and VT equations. The mobility without interactions (VT equation) can only be replaced with a localized adsorption on much larger heterogeneous surface and differential adsorptive potential (Tóth equation) than those corresponding to the VT relationship. It can be explained with the following supposition: The large adsorbent–adsorptive (vertical) interaction energies may replace the energies of mobility.

G. Inapplicability of the Tóth Equation to Isotherms of Type I Measured on Heterogeneous Surfaces

The isotherm equations applicable to *heterogeneous* surfaces interpreted in this chapter have been derived by the following method: First, the known equations were demonstrated (the thermodynamically inconsistent equations applicable to homogeneous surfaces), then the

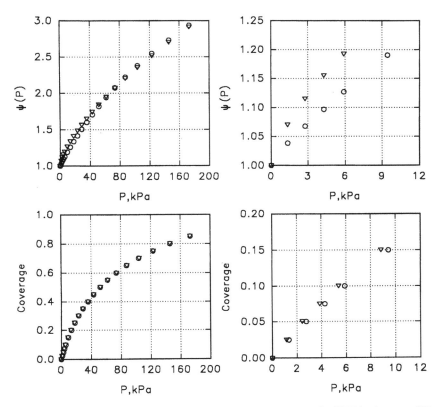

FIG. 25 The Tóth isotherms of Type I can be applied instead of BT isotherms of Type I. The values of parameters of the BT isotherm are as follows: $K_{BT} = 0.05$ kPa^{-t}, $\chi_B = 1.9$, $B_B = 0.8$, $t = 0.6$, $P_m = 295.0$ kPa ($\bigcirc$); The values of parameters of the Tóth isotherm are as follows: $K_T = 0.0575$ kPa^{-t}, $\chi_T = 1.364$, $t = 0.680$, $P_m = 295.7$ kPa ($\triangledown$).

inconsistencies were eliminated by application of Eqs. (95) or (97), and, finally, the *heterogeneity* of the surface was taken into account by the parameter t proposed by Tóth [13]. However, there are some known isotherm equations which were derived (or applied) especially for heterogeneous surfaces, therefore, the introducing of parameter t would be unnecessary. In Section IV.G these equations are discussed.

TABLE 1 Parameters of FT, VT, BT, and Tóth Equations when the Tóth Can Be Applied Instead of FT, VT, and BT Equations

Name of equation	K_x (kPa$^{-t_x}$)	t_x	χ_x	$U_0^{diff}(\Theta = 0)$ (kJ/mol)
FT	0.120	0.600	1.100	54.97
Tóth	0.090	0.734	1.104	44.06
VT	0.500	0.600	1.400	61.29
Tóth	1.125	0.342	1.085	111.2
BT	0.0500	0.600	1.900	53.73
Tóth	0.0575	0.680	1.364	46.77

1. The Freundlich Isotherm

More than 100 years ago, the first isotherm equation was defined by Freundlich, who empirically found a simple relation between the adsorbed amount and the equilibrium pressure:

$$n^s = k_{FR} P^{1/n}, \quad n > 0 \tag{273}$$

where k_{FR} and n are empirical constants. Let us suppose that the total monolayer capacity, n_m^s, is a known value, so from Eq. (273), we obtain

$$\Theta = C_{FR} P^{1/n} \tag{274}$$

or

$$P = C_{FR}^{-n} \Theta^n \tag{275}$$

where

$$C_{FR} = \frac{k_{FR}}{n_m^s}$$

Let us calculate the functions $\psi(\Theta)$ and $\psi(P)$; we get

$$\psi_{FR}(\Theta) = \psi_{FR}(P) = n \tag{276}$$

that is, the power n, according to Eqs. (102) and (103), has a concrete thermodynamical meaning. Thus, the application of Eq. (97) yields

$$\Theta = (P_{r,m})^{1/n} \tag{277}$$

or

$$P_{r,m} = \Theta^n \tag{278}$$

where, evidently,

$$P_{r,m} = \frac{P}{P_m}$$

From the slope of linear form of Eq. (277), n can be calculated:

$$\ln(n^s) = \frac{1}{n} \ln P + \ln\left\{ \left(\frac{1}{P_m}\right)^{1/n} n_m^s \right\} \tag{279}$$

It means that P_m and n_m^s can be calculated by a two-parameter fitting procedure. It is also possible that Eq. (279) is neglected and a three-parameter (n_m^s, n, P_m) fitting is applied. Equation (277) is thermodynamically consistent because

$$\lim_{\Theta=1} P = P_m \tag{280}$$

The mathematically uniform and thermodynamically consistent interpretation of the Freundlich equation is shown in Fig. 26, where the functions $\Theta(P)$, $\psi(P)$, $\Theta(P_{r,m})$, $\psi(\Theta)$, $A_r^s(\Theta)$, and $A_r^s(P_{r,m})$ have been calculated with the following relationships. Top (left), Eq. (275) has been applied with different values of n; top (right), Eq. (276); middle (left), Eq. (277); middle (right); Eq. (276); bottom (left), according to Eq. (100), the function $A_r^s(\Theta)$ should be calculated using

$$A_r^s(\Theta) = \int_0^\Theta \psi(\Theta)\, d\Theta = n \int_0^\Theta d\Theta = n\Theta \tag{281}$$

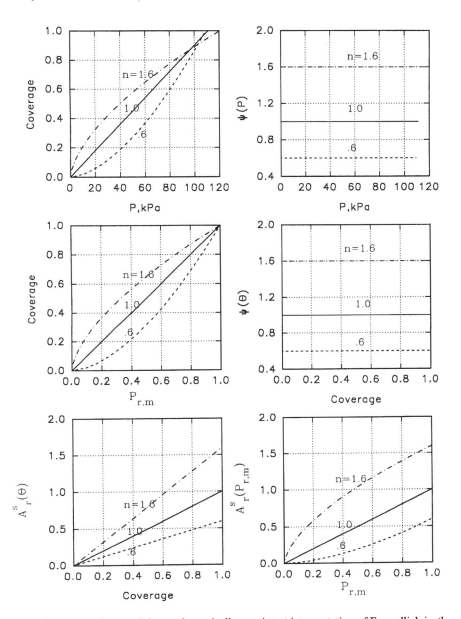

FIG. 26 The uniform and thermodynamically consistent interpretation of Freundlich isotherms of Types I and III. The values of parameters are as follows: $C_{FR} = 4 \times 10^{-4}$ kPa^{-n}, $n = 0.6$, $P_m = 109.3$ kPa (---); $C_{FR} = 5 \times 10^{-2}$ kPa^{-n}, $n = 1.6$, $P_m = 120.7$ kPa $(-\cdot-\cdot-)$; $C_{FR} = 9 \times 10^{-3}$ kPa^{-1}, $n = 1.0$, $P_m = 111.11$ kPa (solid line, Henry equation).

Bottom (right), the values of Eq. (281) and $P_{r,m}$ were conjugated. It is clear that $n = 1$ represents the Henry (H) equation; that is,

$$\Theta = P_{r,m} \tag{282}$$

2. The Generalized Freundlich Isotherm

The Generalized Freundlich (GF) equation supposes that the heterogeneity and other properties of the surface and the adsorbed layer can take into account similarly to Tóth's idea: The coverage present in the Langmuir equation should be raised to a power n. Thus, we have

$$\Theta^n = \frac{P}{(1/K_{GF}) + P}, \quad n > 0 \tag{283}$$

or, expressed for P,

$$P = \frac{1}{K_{GF}} \frac{\Theta^n}{1 - \Theta^n}. \tag{284}$$

The functions $\psi(P)$ and $\psi(\Theta)$ corresponding to Eq. (283) are

$$\psi_{GF}(P) = (K_{GF}P + 1)n \tag{285}$$

and

$$\psi_{GF}(\Theta) = \frac{n}{1 - \Theta^n} \tag{286}$$

Equations (283) and (284) are thermodynamically inconsistent because

$$\lim_{\Theta = 1} P = \infty \tag{287}$$

To obtain a consistent form, let us substitute Eq. (285) into Eq. (97). Performing the integrations, we have

$$\Theta^n = \frac{\chi_{GF} P}{1/K_{GF} + P} \tag{288}$$

or

$$P = \frac{1}{K_{GF}} \frac{\Theta^n}{\chi_{GF} - \Theta^n} \tag{289}$$

χ_{GF} present in these equations is the integration constant with the following explicit form:

$$\chi_{GF} = 1 + \frac{1}{K_{GF} P_m} \tag{290}$$

that is,

$$P_m = [K_{GF}(\chi_{GF} - 1)]^{-1} \tag{291}$$

Equations (288) and (289) are consistent relationships because

$$\lim_{\Theta - 1} P = P_m \tag{292}$$

The functions $\psi(P)$ and $\psi(\Theta)$ corresponding to Eq. (288) are the following:

$$\psi_{GF}(P) = (K_{GF}P + 1)n \tag{293}$$

and

$$\psi_{GF}(\Theta) = \frac{n\chi_{GF}}{\chi_{GF} - \Theta^n} \tag{294}$$

For practical use of Eq. (288) or (289), a three-parameter (χ_{GF}, K_{GF}, and n) fitting procedure is proposed if the total monolayer capacity, n_m^s, or the specific surface area of the adsorbent are known values. In the opposite case, the method discussed in Section VI is recommended. The customary uniform and thermodynamically consistent interpretations of the GF equation are shown in Fig. 27. The calculations of functions $\Theta(P)$, $\Psi(P)$, $\Theta(P_{r,m})$, $\psi(\Theta)$, $A_r^s(\Theta)$, and $A_r^s(P_{r,m})$ are summarized as follows: top (left): isotherms of Type I, III, and V have been calculated from Eq. (284) with different parameters K_{GF}, χ_{GT}, and n; top (right): Eq. (285); middle (left): Eqs. (289) and (291); middle (right): Eq. (286); bottom (left): Eq. (121) with numerical integration; bottom (right): the conjugated pairs of Eq. (121) and P_{rm}, have been applied.

In Fig. 27, can be seen the main differences between the GF and Tóth equations. In particular, the functions $\psi_{GF}(P)$ are linear relationships having different intersection points at $P = 0$. Because most of the isotherms of Type I measured on heterogeneous surfaces have functions $\psi(P)$ concave from below, the applicability of GF equations are rather limited. However, the GF equation—in limited domains of equilibrium pressure—can describe isotherms of Types I, III, and V. Thus, it is also demonstrated that a power on the converage (Θ^n) can reflect not only the heterogeneity of the surface but the different interactions between the molecules existing in the adsorbed layer.

3. The Sips Isotherm

Sips took the heterogeneity of the adsorbents and the interactions in the adsorbed layer into account by raising to a power the equilibrium pressure only. Thus, his equation has the following form:

$$P^n = \frac{1}{K_S}\frac{\Theta}{1 - \Theta} \tag{295}$$

or

$$\Theta = \frac{P^n}{(1/K_S) + P^n}, \quad n > 0 \tag{296}$$

The functions $\psi(P)$ and $\psi(\Theta)$ corresponding to Eqs. (295) and (296) are

$$\psi_S(P) = (K_S P^n + 1)\frac{1}{n} \tag{297}$$

and

$$\psi_S(\Theta) = \frac{1}{n(1 - \Theta)} \tag{298}$$

Equation (295) is thermodynamically inconsistent because

$$\lim_{\Theta=1} P = \infty \tag{299}$$

For the sake of having a consistent form, let us substitute Eq. (291) into Eq. (97). Performing the integrations, we obtain

$$P^n = \frac{1}{K_S}\frac{\Theta}{\chi_S - \Theta} \tag{300}$$

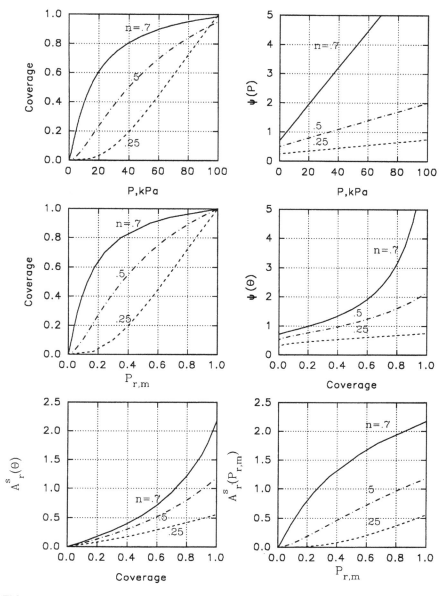

FIG. 27 The uniform and thermodynamically consistent interpretation of the GF isotherms of Types I, III, and V. The values of parameters are as follows: Type I: $K_{GF} = 0.09$ kPa^{-1}, $\chi_{GF} = 1.10$, $n = 0.70$, $P_m = 111.11$ kPa (—); Type III: $K_{GF} = 0.02$ kPa^{-1}, $\chi_{GF} = 1.50$, $n = 0.25$, $P_m = 100.0$ kPa($- \cdot - \cdot -$); Type V: $K_{GF} = 0.03$ kPa^{-1}, $\chi_{GF} = 1.30$, $n = 0.50$, $P_m = 111.11$ kPa (---).

or

$$\Theta = \frac{\chi_S P^n}{(1/K_S) + P^n} \tag{301}$$

where χ_S is the integration constant with the following explicit form:

$$\chi_S = 1 + \frac{1}{K_S P_m^n} \tag{302}$$

From Eq. (302) it follows that

$$P_m = [K_S(\chi_X - 1)]^{-1/n} \tag{303}$$

The function $\psi_S(\Theta)$ corresponding to the consistent equation (301) is

$$\psi_S(\Theta) = \frac{\chi_S}{n(\chi_S - \Theta)} \tag{304}$$

The function $\psi(P)$ is equal to Eq. (297).

For the practical use of Eq. (295) or (296), a three-parameter (χ_S, K_S, and n) fitting procedure is recommended. This recommendation has a reality only if the function $\psi(P)$ calculated from the measured isotherm is one of the relationships shown in the top (right) of Fig. 28. In this figure are represented the functions $\Theta(P)$, $\psi(P)$, $\Theta(P_{r,m})$, $\psi(\Theta)$, $A_r^s(\Theta)$, and $A_r^s(P_{r,m})$ corresponding to the modified (consistent) Sips equation. The calculations of these functions are briefly summarized as follows: top (left): isotherms of Types I, III, and V have been calculated from Eq. (300) with different values of n, K_S and χ_S; top (right): Eq. (297); middle (left): Eqs. (300) and (303); middle (right): Eq. (304); bottom (left): Eq. (121) with numerical integration; bottom (right): the conjugated pairs of Eq. (121) and $P_{r,m}$ have been applied. The main differences between the Tóth and Sips equations are that by the Sips equation, the limiting vlaues

$$\lim_{P \to 0} \psi_S(P) \neq 1 \tag{305}$$

is valid; that is, the Sips equation cannot mathematically describe the Henry section of an isotherm; however, it can describe isotherms of Types III and V.

4. The Dubinin–Radushkevic Isotherm

Dubinin and his co-workers [14,15] have experimentally observed that for isotherms measured on microporous adsorbents (especially on activated carbons), the following relationship can be applied:

$$n^s = n_m^s \exp\{-B_D \varepsilon^2\} \tag{306}$$

where B_D is a constant and ε is a function of temperature and equilibrium pressure. It was a great recognition that if ε is regarded equal to the adsorptive potential defined by Polanyi and having the form

$$\varepsilon = RT \, \ln\left(\frac{P_0}{P}\right) \tag{307}$$

then the function $n^s(\varepsilon)$, calculated for an adsorbate on microporous adsorbent, is independent of the temperature [i.e., the function $n^s(\varepsilon)$ has a characteristic feature]. From this fact, it follows that substituting Eq. (307) into Eq. (306), we obtain an isotherm equation applicable to microporous adsorbents:

$$n^s = n_m^s \, \exp\left\{-B_D(RT)^2\left[\ln\left(\frac{1}{P_r}\right)\right]^2\right\} \tag{308}$$

Dubinin has proven that the constant B_D reflects the lateral interaction energies existing between the adsorbed molecules. Because in Eqs. (307) and (308), P_0 is the saturation pressure (i.e., the temperature is below the critical temperature of the absorptive), n_m^s may lose its physical meaning. Thus, it is more convenient to define a hypothetical or equivalent monolayer capacity (also denoted by n_m^s) which would result if the amount of adsorbate required to fill the pores were spread in a close-packed monolayer of molecules. To avoid this "hypothetical" monolayer

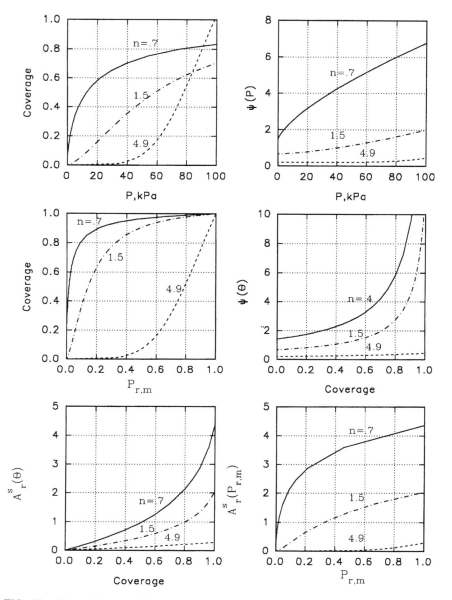

FIG. 28 The uniform and thermodynamically consistent interpretation of Sips isotherms of Types I, III, and V. The values of parameters are as follows: Type I: $K_s = 0.15$ kPa^{-n}, $\chi_S = 1.06$, $n = 0.70$, $P_m = 836.6$ kPa (——); Type III: $K_S = 0.02$ kPa^{-n}, $\chi_S = 1.85$, $n = 4.9$, $P_m = 98.6$ kPa (– – –); Type V: $K_S = 0.2 \times 10^{-2}$ kPa^{-n}, $\chi_S = 1.06$, $n = 1.5$, $P_m = 411.0$ kPa (– · – · –).

capacity, it is theoretically possible to calculate the adsorptive potential ε valid also for the real specific surface area (i.e., valid for the real monolayer capacity), n_m^s and the corresponding equilibrium pressure P_m. In this case,

$$\varepsilon = RT \ \ln\left(\frac{P_m}{P}\right)$$

(309)

Substituting Eq. (309) into Eq. (306), we obtain

$$n^s = n_m^s \, \exp\left\{-B_D(RT)^2\left[\ln\left(\frac{P_m}{P}\right)\right]^2\right\}$$ (310)

Equation (310) met the requirement that

$$\lim_{P \to P_m} n^s = n_m^s$$ (311)

however, in spite of this fact, the Dubinin–Radushkevic (DR) equation is thermodynamically inconsistent. The basic reason for this inconsistency is the definition of the adsorptive potential, ε, because

$$\lim_{P \to 0} \varepsilon = +\infty$$ (312)

This limiting value is a thermodynamically irrational, which also will be proven by the function $A_r^s(\Theta)$ (see Eq. (319). Originally, Dubinin supposed that Eq. (308) or (310) is applicable only to isotherms of Type I measured on microporous adsorbents. However, it can be proven that these relationships are valid only for isotherms of Type V. In order to prove this statement, let us calculate the function $\psi(P)$ according to Eq. (90); thus, we obtain

$$\psi_D(P_r) = \left\{2B_D(RT)^2 \, \ln\left(\frac{P_0}{P}\right)\right\}^{-1}$$ (313)

and according to Eq. (92),

$$\psi_D(\Theta) = [2(B_D)^{1/2}RT]^{-1}\left[\ln\left(\frac{n_m^s}{n^s}\right)\right]^{-1/2}$$ (314)

From Eq. (313), it follows that the DR equation can only describe isotherms of Type V because with decreasing values of P_r, the condition

$$\psi_D(P_r) < 1$$ (315)

will surely be satisfied. Condition (315) can be obtained from Eq. (313) for the values of B_D:

$$B_{D,\psi=1} = \left[2(RT)^2 \, \ln\left(\frac{P_0}{P_{\psi=1}}\right)\right]^{-1}$$ (316)

where $P_{\psi=1}$ is the equilibrium pressure by which $\psi_D(P_r) = 1$ is realized and $B_{D,\psi=1}$ is the corresponding value of B_D. From Eq. (316), it also follows that very small values of $B_{D,\psi=1}$ (i.e., by very small interactions energies), we can measure only the convex part of the isotherm of Type V. However, the isotherms in the domain of relative pressures less than 10^{-4} is often not measured. Assuming a room temperature of 298 K and a domain of relative pressures less than 10^{-4} from Eq. (316), we have

$$B_{D,\psi=1} = 8.8 \times 10^{-3} \, \text{mol}^2/\text{kJ}^2$$ (317)

Pressures corresponding to this small value of B_D, the convex part of isotherms passes unnoticed and we measure only the concave part of isotherms of Type V (i.e., we measure isotherms of Type I). On microporous active carbons, the lateral interactions are negligible to the adsorptive potential; therefore, the values of B_D are small and the measured isotherms are of Type I. These statements are proven in Fig. 29, where the usual uniform interpretation of the DR equation can be seen. The calculations of functions (P), $\psi(P)$, $\Theta(P_{r,m})$, $\psi(\Theta)$, $A_r^s(\Theta)$, and $A_r^s(P_{r,m})$ have been performed with the following relationships: Top (left): Eq. (310) in the form

$$\Theta = \exp\left\{-B_D(RT)^2 \ln\left(\frac{P_m}{P}\right)\right\} \tag{318}$$

where

$$\Theta = \frac{n^s}{n_m^s}$$

has been applied; top (right): Eq. (313) with $P_0 = P_m$; middle (left): Eq. (318) with $P_{r,m} = P_m/P$. Middle (right): Eq. (314); bottom (left): Eq. (121) with numerical integration of Eq. (314); bottom (right): the conjugated pairs of $A_r^s(\Theta)$ and $P_{r,m}$ have been applied. In Fig. 29 can be seen that at values of $B_D = 0.3$, the isotherm $\Theta(P)$ still has a small convex domain $[\psi_D(P) < 1]$. From the function $A_r^s(\Theta)$ the thermodynamical inconsistency of DR equation can be stated; namely, by all values of B_D, the limiting value

$$\lim_{\Theta \to 1} A_r^s(\Theta) = \infty \tag{319}$$

is valid. Because of this fact, the numerical integration of the function $\psi_D(\Theta)$ has been performed with upper limits $\Theta = 0.99$.

In spite of the above-mentioned inconsistencies, the DR equation has great importance in the practice of adsorption measurements. In particular, this relationship is the only one to which belongs a function $\psi(P)$ having a point of inflexion and after this inflexion is an increasing domain with a concave (exponential) form [see Fig. 29, top (right)]. This means that the DR equation describes the *plateau* of isotherms of Type V (I) excellently because function $\psi(P)$ increases rapidly whereas n^s tends to be a limiting value. This plateau is due to *multilayer* adsorption on the small external area of microporous adsorbents, if such an area exists at all. For practical applications, the linear form of Eq. (308) is recommended:

$$\ln(n^s) = -B_D(RT)^2\left[\ln\left(\frac{1}{P_r}\right)\right]^2 + \ln(n_m^s) \tag{320}$$

If P_m is applied in Eq. (318), then a three-parameter (n_m^s, P_m, B_D) fitting procedure is proposed.

H. Practical Application of the Function $\psi(P)$ Corresponding to All Isotherms of Type I Measured on Heterogeneous Surfaces

In most cases, isotherms of Type I can be measured on heterogeneous and porous adsorbents. However, the form of the isotherms are very similar, therefore, it is very difficult to select the appropriate isotherm equation which both mathematically and thermodynamically describe and explain the measured isotherm.

For example, an arbitrary selected equation may be fitted to the measured data with satisfying accuracy; however, this equation may by a thermodynamically inconsistent relationship. At the same time, there may exist another equation which both mathematically and thermodynamically is applicable for the same measured isotherm. To avoid this problem, it is

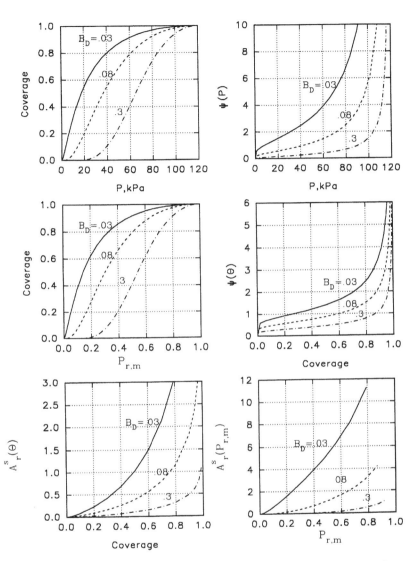

FIG. 29 The uniform interpretation of DR isotherms of Type V. The values of parameters are as follows: $B_D = 0.03$, $P_m = 120$ kPa, $RT = 2.48$ J/mmol (——); $B_D = 0.08$, $P_m = 120$ kPa, $RT = 2.48$ J/mmol (———); $B_D = 0.30$, $P_m = 120$ kPa, $RT = 2.48$ J/mmol (— · — · —).

proposed to first calculate the function $\psi(P)$ corresponding to the measured isotherm and then, based on the function, to select the appropriate isotherm equation. This method is advantageous because the function $\psi(P)$—as opposed to the isotherms of Type I—is very different and characteristic, so the selection of the corresponding and appropriate isotherm equation is very simple. In Fig. 30 are represented and summarized the functions $\psi(P)$ corresponding to all isotherms of Type I which may occur on porous heterogeneous surfaces. This statement should be completed with the fact proven in Section IV.F, that the Tóth equation can be applied instead of the FT, VT, and BT relationships.

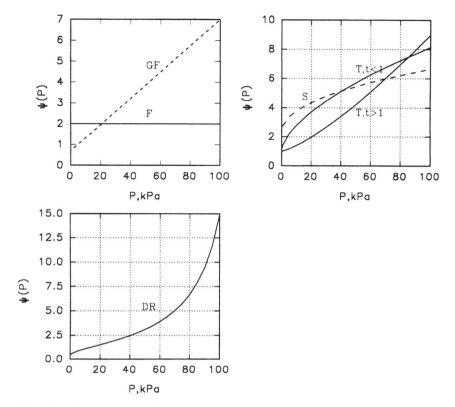

FIG. 30 The functions $\psi(P)$ corresponding to Freundlich, Sips, Tóth, and DR isotherms of Type I determined on heterogeneous surfaces.

V. MULTILAYER ADSORPTION ON G/S INTERFACES

A. The Brunauer–Emmett–Teller Equation and Its Thermodynamical Properties

The derivation of the Brunauer–Emmett–Teller (BET) equation can be found in most text books, so here the starting point of our discussions is its final and well-known form:

$$n^s = \frac{n_m^s c P_r}{[1 + (c - 1)P_r][1 - P_r]} \tag{321}$$

where

$$c = \exp\left\{\frac{U_0 - \lambda}{RT}\right\} \tag{322}$$

In Eq. (322), U_0 is the adsorptive potential of the homogeneous surface and λ is the heat of condensation of the adsorptive. From this statement can be seen the simple suppositions of the BET theory; in particular, (1) the surface is homogeneous, (2) the multilayer adsorptive potential theoretical is equal to the forces of condensation, and (3) at limiting value $P_r \to 1$, the condensation takes place (i.e., the number of layers is infinite). Equation (321) can also be written in the form

$$n^s = \frac{n_m^s c P_r}{1 + (c-1)P_r} + \frac{n_1^s P_r}{1 - P_r},$$ (323)

where n_1^s is the amount of adsorptive adsorbed only in the *first* layer [i.e., it is the first term in the right-hand side of Eq. (323)]:

$$n_1^s = \frac{n_m^s c P_r}{1 + (c-1)P_r}$$ (324)

It is easy to see that Eq. (324) is a Langmuir equation with the following limiting value:

$$\lim_{P_r \to 1} n_1^s = n_m^s$$ (325)

According to Eq. (325), the total monolayer capacity is only completed when $P_r = 1$ (i.e., when the condensation takes place). It also means that $P_0 = P_m$.

It is evident that the second term on the right-hand side of Eq. (323) represents the multilayer (second and other layers) adsorption; therefore, the BET relationship can be regarded as the sum of a Langmuir monolayer adsorption and of a pure multiplayer adsorption. In Fig. 31 are shown both the monolayer and the multilayer components of a BET relationship with a value of $c = 50$. The usual uniform interpretation of the BET equation can be seen in Fig. 32.

The function $\psi(P_r)$, according to Eq. (326), has been calculated with the following relationship:

$$\psi_{\text{BET}}(P_r) = \frac{(1-c)P_r^2 + (c-2)P_r + 1}{(c-1)P_r^2 + 1}$$ (326)

The function $\psi(\Theta)$ with conjugated pairs of Θ and Eq. (326), the function $A_r^s(\Theta)$ with the numerical integration of function $\psi(\Theta)$, and the function $A_r^s(P_r)$ with conjugated pairs of $A_r^s(\Theta)$ and P_r have been calculated. The most important properties of the functions $\psi(P_r)$ and $\psi(\Theta)$ are that they have a maximum value where the dominantly monolayer adsorption is finished and the

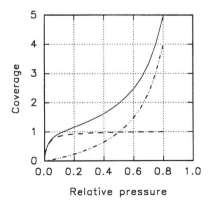

FIG. 31 The BET isotherm and its monolayer and multilayer component isotherms. (—); (− − −); (− ·· − ·· −); $c = 50$.

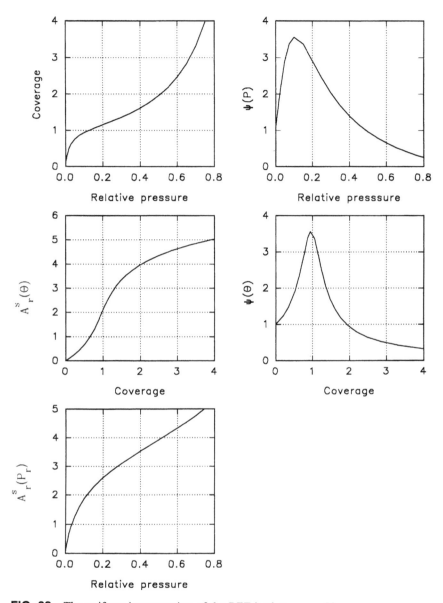

FIG. 32 The uniform interpretation of the BET isotherm; $c = 50$.

dominantly multilayer adsorption begins. (See also in Fig. 31 the monolayer and multilayer component isotherms.) These maximum values are also reflected by the function $A_r^s(\Theta)$ having a point of inflexion where the dominantly monolayer adsorption is finished. This phenomenon is thermodynamically evident. The change in relative free energy of the surface in the domain of the monolayer adsorption is relatively much greater then in the multilayer region. In Section VI.A, it will be proven that these maximum values of the function $\psi(P_r)$ or $\psi(\Theta)$ can be applied for distinction of the monolayer and multilayer domains of the isotherms.

B. Modifications of the BET Equation

The modifications of the BET theory have the goal of eliminating the simplifications on which the equation is based. Let us look at the simplifications and elimination of those in order of their importance.

First, one of the significant simplifications is that in the *whole* multilayer region, the adsorptive potential is equal to the heat of condensation of the adsorptive. In order to modify this simplification, let us change the physical interpretation (322) of constant c:

$$\bar{c} = \exp\left\{\frac{U_0 - (\lambda \pm \lambda')}{RT}\right\} = \exp\left\{\frac{U_0 - \lambda}{RT}\right\}\exp\left\{\frac{\pm\lambda'}{RT}\right\} \tag{327}$$

where the value of λ' is due to the assumption that the adsorptive potential in the multilayer is less than or greater than the heat of condensation. We introduce a new parameter k defined as

$$k = \exp\left\{\frac{\pm\lambda'}{RT}\right\} \tag{328}$$

Introducing the parameter k leads to a modified form of the BET equation [13]:

$$n^s = \frac{n_m^s c P_r}{[1 + (c - k)P_r][1 - kP_r]} \tag{329}$$

or again separating Eq. (329) into the monolayer and multilayer components, we have

$$n^s = \frac{n_m^s c P_r}{1 + (c - k)P_r} + \frac{n_1^s P_r}{(1/k) - P_r} \tag{330}$$

With a three-parameter (n_m^s, c, k) fitting procedure, Eq. (329) can be applied in practice.

The second modification takes into account the possibility that to a definite relative equilibrium pressure, $P_{r,e}$, only (not dominantly!) monolayer adsorption takes place. In this case, the modified BET equation has the following form [13]:

$$n^s = \frac{n_m^s c P_r}{[1 + kP_{r,e} + (c - k)P_r][1 - k(P_r - P_{r,e})]}, \quad P_r > P_{r,e} \tag{331}$$

Equation (331) can also be separated into monolayer and multilayer components:

$$n^s = \frac{n_m^s c P_r}{1 + kP_{r,e} + (c - k)P_r} + \frac{n_1^s(P_r - P_{r,e})}{1 - k(P_r - P_{r,e})}, \quad P_r > P_{r,e} \tag{332}$$

The function $\psi(P_r)$ of the BET equation modified with parameters k and $P_{r,e}$ has the following form [13]:

$$\psi_{\text{BET},m}(P_r) = \frac{(k - c)kP_r^2 + [(1 - kP_{r,e})(c - k - 1)]P_r + (1 + kP_{r,e})^2}{(c - k)kP_r^2 + (1 + kP_{r,e})^2}, \quad P_{r,e} > P_r \tag{333}$$

It is evident that for $k = 1$ and $P_{r,e} = 0$, Eq. (333) transforms into Eq. (326).

The practical application of Eq. (331) also requires a three-parameter (n_m^s, c, k) fitting procedure because $P_{r,e}$ (the relative equilibrium pressure where the multilayer adsorption begins) corresponds to the maximum value of the function $\psi(P_r)$. Both from the mathematical form and from the physical meaning of Eq. (331), it follows that this relationship can only be applied

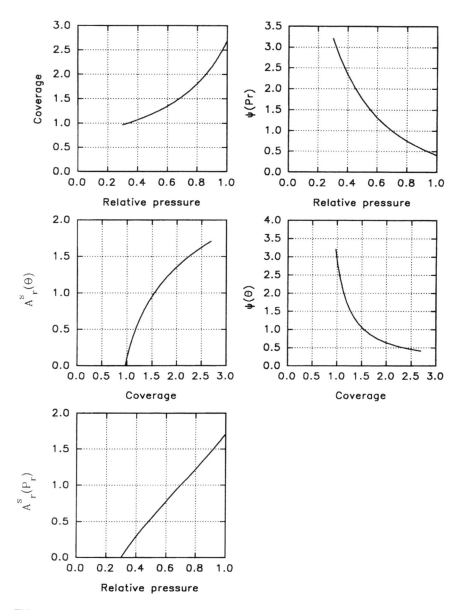

FIG. 33 The uniform interpretation of the multilayer domain of the modified BET isotherm ($P_r > P_{r,e}$ and $k \neq 1$). The values of parameters are $c = 100$, $P_{r,e} = 0.3$, $k = 0.90$.

to isotherms having a pressure domain in which the function $\psi(P_r)$ decreases. The uniform interpretation of Eq. (331), shown in Fig. 33, relates to values of functions where $P_r > P_{r,e}$. The calculations of functions $\Theta(P_r)$, $\psi(P_r)$, $\psi(\Theta)$, $A_r^s(\Theta)$, and $A_r^s(P_r)$ have been made as usual: top (left): Eq. (331); top (right): Eq. (333); middle (right): the conjugated pairs of Eqs. (331) and (333); middle (left): numerical integration of the function $\psi(\Theta)$; bottom: the conjugated pairs of $A_r^s(\Theta)$ and P_r have been applied. From the above-mentioned facts also, it follows that the change in relative free energy of the surface is calculated only for the range of coverages greater than 1, so at total monolayer coverage, $A_r^s(\Theta = 1)$ is equal to zero.

C. The Harkins–Jura Equation and Its Thermodynamical Properties

Harkins and Jura (HJ) derived a new isotherm equation about 60 years ago [16]. This equation is also very important because it was certified by ASTM for the calculation of the thickness (of number of layers) of the adsorbed phase. The authors assumed that the adsorbed layer may be regarded as a two-dimensional liquid corresponding to a definite two-dimensional state of equation. According to this idea, the following isotherm equation has been derived:

$$\ln(P_r) = B - \frac{A}{\Theta^2} \tag{334}$$

or

$$P_r = \exp\left\{ B - \frac{A}{\Theta^2} \right\} \tag{335}$$

where B and A are constants. Expressing Eq. (334) for the coverage, Θ, we have

$$\Theta = \left[\frac{A}{B - \ln(P_r)} \right]^{1/2} \tag{336}$$

Let Θ_0 be the coverage when $P_r \to 1$; so, the limiting value of Eq. (336) is

$$\lim_{P_r \to 1} \Theta = \left(\frac{A}{B} \right)^{1/2} = \Theta_0 \tag{337}$$

that is,

$$\Theta_0 = \left(\frac{A}{B} \right)^2 \quad \text{or} \quad B = \frac{A}{\Theta_0^2} \tag{338}$$

Substituting Eq. (338) into Eqs. (334)–(336), we obtain

$$\ln P_r = A\left\{ \frac{1}{\Theta_0^2} - \frac{1}{\Theta^2} \right\} \tag{339}$$

$$P_r = \exp\left\{ A\left(\frac{1}{\Theta_0^2} - \frac{1}{\Theta^2} \right) \right\} \tag{340}$$

and

$$\Theta = \left\{ \frac{A}{(A/\Theta_0^2) - \ln(P_r)} \right\}^{1/2} \tag{341}$$

For practical applications of Eqs. (340) and (341), a two-parameter (A and Θ_0) fitting procedure is recommended. The value of Θ_0 represents the number of layers at $P_r \to 1$ (i.e., knowing the volume of adsorbate molecules, the thickness of the adsorbed phase can be calculated). If the specific surface area is an unknown value, then the fitting procedure should be extended to n_m^s [i.e., a three-parameter (A, Θ_0, and n_m^s) fitting is required].

Because Eq. (337) represents a definite limiting value, Θ_0, the HJ equation in this respect is thermodynamically consistent. The limits and conditions of the applicability of the HJ equation can again be determined with the help of the function $\psi(\Theta)$ or $\psi(P_r)$.

According to Eq. (92), from Eq. (341), we get

$$\psi_{\mathrm{HJ}}(\Theta) = \frac{A}{\Theta^2} \tag{342}$$

and from Eq. (3.40), we obtain

$$\psi_{\mathrm{HJ}}(P_r) = 2\left\{\frac{A}{\Theta_0^2} - \ln(P_r)\right\} \tag{343}$$

The change in relative free energy of the surface can be explicitly calculated:

$$A_r^s(\Theta) = \int_\Theta^{\Theta_0} \psi_{\mathrm{HJ}}(\Theta)\, d\Theta = 2A\left\{\frac{1}{\Theta} - \frac{1}{\Theta_0}\right\} \tag{344}$$

Therefore, so all relationships are available for the uniform interpretation of the HJ equation shown in Fig. 34.

From the practical point of view, the function $\psi(P_r)$ shown on the top (right) of Fig. 34 is very important. In particular, this function decreases in the whole domain of coverage (i.e., the HJ equation can describe the multilayer adsorption *only*). This fact is reflected by the limiting value of the function $\psi_{\mathrm{HJ}}(P_r)$:

$$\lim_{P_r \to 0} \psi_{\mathrm{HJ}}(P_r) = \infty \tag{345}$$

This limiting value is also valid for the function $\psi(\Theta)$ shown on the bottom (left) of Fig. 34. From Eq. (342), it follows that at the same value of A, the functions $\psi(\Theta)$ differ only from the

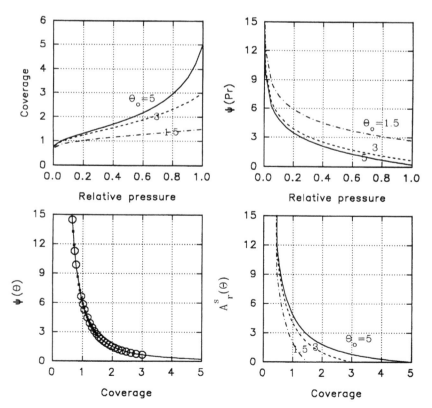

FIG. 34 The uniform interpretation of the HJ isotherms. The values of the parameters are $A = 3$; $\Theta_0 = 5$ (—), $\Theta_0 = 3$ (- - -), $\Theta_0 = 1.5$ (- · - · -); bottom left: $\Theta_0 = 5$ (—), $\Theta_0 = 3$ (O), $\Theta_0 = 1.5$ (●).

values of Θ_0. This fact is reflected on the bottom (left) part of Fig. 34. The function $A_r^s(\Theta)$ on the bottom (right) of Fig. 34 also includes this limiting value:

$$\lim_{\Theta \to 0} A_r^s(\Theta) = \infty \tag{346}$$

In this respect the HJ equation is thermodynamically inconsistent; however, this inconsistency cannot be eliminated by the method discussed in this chapter.

These properties of the functions $\psi(P_r)$ and $\psi(\Theta)$ are physically and thermodynamically uncertain because the first molecule adsorbed on the surface cannot be formed in a multilayer phase. In spite of this contradiction, the HJ equation in the domain of the relative pressure $0.05 < P_r < 0.7$ can be applied well for the calculation of the value Θ_0 (Section VII).

D. The Cloud Model of Multilayer Adsorption

The source of this model is the BET theory; in detail, mathematical forms follow. If $k = 1$ and $P_{r,e} = 0$, then—according to Eq. (323)—the original BET equation can be written as

$$n^s = \frac{n_1^s}{1 - P_r} \tag{347}$$

where

$$n_1^s = \frac{n_m^s c P_r}{1 + (c - 1)P_r} \tag{348}$$

If $k \neq 1$ and $P_{r,e} = 0$, then

$$n^s = \frac{n_1^s}{1 - k P_r} \tag{349}$$

where

$$n_1^s = \frac{n_m^s c P_r}{1 + (c - k)P_r} \tag{350}$$

Finally, if $k \neq 1$ and $P_{r,e} \neq 0$, then

$$n^s = \frac{n_1^s}{1 - k(P_r - P_{r,e})} \qquad P_r > P_{r,e} \tag{351}$$

where

$$n_1^s = \frac{n_m^s c P_r}{1 + k P_{r,e} + (c - k)P_r} \tag{352}$$

Let us compare Eqs. (347), (349), and (351); these relationships can be written in a general form:

$$n^s(P_r) = \frac{n_1^s(P_r)}{1 - k_n f(P_r)} \tag{353}$$

where $n^s(P_r)$ is the total multilayer isotherms, $n_1^s(P_r)$ is the monolayer component isotherm and the function $k_n f(P_r)$ expresses the multilayer character of adsorption.

Equations (348), (350), and (352) indicate that, according to the BET theory, the monolayer component isotherm is always a Langmuir equation.

Connected with Eq. (353), the following two problems have emerged:

1. Can a model be constructed which is independent of the BET theory but its mathematical expression is identical to Eq. (353)?
2. In this new model, instead of the Langmuir equation, can other monolayer equations, $n_1^s(P_r)$, be applied, especially those which are also valid for heterogeneous surfaces?

These two problems may be solved by the following model of multilayer adsorption. Let us write the dynamic equilibrium for the multilayer adsorption onto the first adsorbed layer:

$$k_1 n^s P_r = k_2 \ \exp\left(-\frac{Q_m}{RT}\right) n_t^s \tag{354}$$

In Eq. (354), k_1 and k_2 are kinetic constants of the rate of adsorption and desorption, respectively, Q_m is the average heat of adsorption taking place on the first (already adsorbed) layer, n^s is the total (monolayer plus multilayer) adsorbed amount, and n_t^s is the amount of gas adsorbed onto the first layer only. From the definitions of n_t^s, n^s and n_1^s it follows that

$$n_t^s = n^s - n_1^s \tag{355}$$

The physical meaning of Eq. (354) is that the rate of the *pure* multilayer adsorption (adsorption onto the first adsorbed layer) is proportional to the total amount of adsorbed gas, n^s, whereas the desorption rate is proportional to the amount of gas adsorbed onto the first layer, n_t^s. This physical picture may be real for the following conditions:

1. In the adsorbed phase, *all molecules* (i.e., n^s) remain in the uncovered state for a definite short time; however, this time is enough to attract and bind additional molecules from the gas phase.
2. This uncovered state is assured by the movements of molecules in the second and subsequent layers. These moments take place both in the lateral and vertical directions and the velocities of those are the faster, the greatest the distance from the adsorbent surface.
3. Condition 1 can only be met if the number of layers is limited (maximum of three or four).
4. The movements of the molecules in the multilayer is similar to a "cloud swirling" in air. This is the source of the name of the model.

Let us introduce the following definition:

$$k_n = \left(\frac{k_1}{k_2}\right) \ \exp\left(\frac{Q_m}{RT}\right) \tag{356}$$

Taking Eqs. (355) and (356) into account, we have, from Eq. (354),

$$n^s P_r = k_n^{-1}(n^s - n_1^s) \tag{357}$$

Equation (357) may be written in the general form

$$n_s(P_r) = \frac{n_1^s(P_r)}{1 - k_n P_r} \tag{358}$$

which is totally identical to Eq. (353) because the monolayer isotherm, $n_1^s(P_r)$ is quite independent of the total isotherm, $n^s(P_r)$. Thus, our most important condition relating to the model is met: The monolayer isotherm can be chosen corresponding to the properties (especially to the heterogeneity) of the adsorbent.

The constant k_n, in addition to Eq. (356), has another important and measurable physical interpretation. In particular, suppose that the condensation does not take place when $P_r \to 1$; then the limiting value of Eq. (358) is the following:

$$\lim_{P_r \to 1} n^s(P_r) = \frac{n_m^s}{1 - k_n} \tag{359}$$

that is,

$$k_n = \frac{n^s(P_r \to 1) - n_m^s}{n^s(P_r \to 1)}, \quad P_r \to 1 \tag{360}$$

or

$$k_n = \frac{n^s(P_r \to 1)/n_m^s - 1}{n^s(P_r \to 1)/n_m^s}, \quad P_r \to 1 \tag{361}$$

Because n_m^s represents the amount of adsorptive adsorbed in the total monolayer,

$$\frac{n^s(P_r \to 1)}{n_m^s} = N \tag{362}$$

where N is number of layers when $P_r \to 1$. Thus, from Eq. (361) we obtain

$$k_n \to \frac{N - 1}{N} \tag{363}$$

This means that the number of layers can be calculated from Eq. (363):

$$N = \frac{1}{1 - k_n}, \quad 1 \leq k_n \tag{364}$$

From Eq. (364), it follows that k_n is always less than 1. If $k_n = 1$, (i.e., $N \to \infty$), then, according to the BET theory, condensation takes place. Comparing Eq. (364) with Θ_0 defined by the HJ theory [see Eq. (337)], we obtain

$$N = \Theta_0 \tag{365}$$

It may also occur that the multilayer adsorption begins only at a definite relative pressure, $P_{r,e}$, as has been assumed in the modifications of the BET theory. In this case, the basic dynamic equilibrium equation of the cloud (C) model, Eq. (354), should be modified as

$$k_1 n^s(P_r - P_{r,e}) = k_2 \exp\left(-\frac{Q_m}{RT}\right) n_t^s, \quad P_r > P_{r,e} \tag{366}$$

Finally, instead of Eq. (358), we obtain

$$n^s(P_r) = \frac{n_1^s(P_r)}{1 - k_n(P_r - P_{r,e})}, \quad P_r > P_{r,e} \tag{367}$$

In this case, evidently, it is valid that

$$N = \frac{1}{1 - k_n(1 - P_{r,e})} \tag{368}$$

Equation (365) or (368) may be an empirical control of the cloud model if a measured multilayer isotherm or its multilayer domain can be described both with the HJ equation and with one of the isotherm equations belonging to the C model. These C-model equations may be *all* relationships in which the monolayer composite isotherm is one of those discussed earlier in this chapter

(Tóth, FT, VT, BT, GF, Sips, DR equations). For example, if the Tóth equation is the appropriate monolayer composite isotherm, then the corresponding C-model equation—according to Eq. (367)—is the following:

$$n^s = \frac{n_m^s (\chi_T)^{1/t} P_r}{(1/K_t + P_r^t)^{1/t} [1 - k_n (P_r - P_{r,e})]}, \quad P_r > P_{r,e} \tag{369}$$

If k and $P_{r,e}$ are zero, then, evidently, only monolayer adsorption takes place in the whole domain of P_r.

The same is the situation if $P_{r,e} \neq 0$ (i.e., in domain $0 \leq P_r \leq P_{r,e}$) is valid that $k_n = 0$ ($N = 1$); so, in this region, again the monolayer component isotherm can only be applied. The distinction between the monolayer and multilayer sections of an isotherm will be discussed in detail in Section VI.A. For this reason, it is necessary to calculate the function $\psi(P_r)$ corresponding to all C-model isotherm equations. The implicit form of those is Eq. (367). To this relationship should be applied the general and well-known form of the function $\psi(P_r)$:

$$\psi(P_r) = \frac{n^s}{P_r} \left(\frac{dn^s}{dP_r} \right)^{-1} \tag{370}$$

Applying Eq. (370) to Eq. (367), we obtain the general and implicit form of the function $\psi(P_r)$ corresponding to all C-model (CM) isotherm equations:

$$\psi_{CM}(P_r) = \left\{ \frac{1}{\psi_1(P_r)} + \frac{k_n P_r}{[1 - k_n (P_r - P_{r,e})]} \right\}^{-1}, \quad P_r > P_{r,e} \tag{371}$$

In Eq. (371), $\psi_1(P_r)$ corresponds to the monolayer component isotherm, $n_1^s(P_r)$, present in Eq. (367). Therefore, for example, the function $\psi(P_r)$ corresponding to Eq. (369) has the following explicit form [see Eq. (214) too]:

$$\psi_{CM}(P_r) = \left\{ \frac{1}{K_T P_r^t + 1} + \frac{k_n P_r}{[1 - k_n (P_r - P_{r,e})]} \right\}^{-1}, \quad P_r > P_{r,e} \tag{372}$$

Thus, all relationships are available to calculate the usual and uniform interpretation of an isotherm belonging to the C model. This uniform interpretation of Eq. (369) are shown in Fig. 35 with the assumption that $P_{r,e} = 0$ [i.e., it is possible to compare the properties of the HJ equation (336) with those of Eq. (369)]. Therefore, comparing Fig. 34 with Fig. 35, the following important differences can be stated. Top (left): Opposite to HJ equation, by Eq. (369), the monolayer component isotherm can be calculated; however, the whole function $\Theta(P)$ is very similar to the HJ isotherm. The greatest differences can be seen in the functions $\psi(P_r)$, $\Psi(\Theta)$, and $A_r^s(\Theta)$. All of these functions corresponding to Eq. (369) have definite values when $P_r \to 0$ or $\Theta \to 0$. So, in this respect, the cloud model represents thermodynamically consistent relationships. The P_r and Θ domains of applicability of the two isotherm equations can also be determined. In particular, according to Fig. 36, the HJ equation is applicable only in the domain of coverage $1.0 \leq \Theta \leq 3$, because in this region, the function $\psi_{HJ}(\Theta)$ has a typically decreasing character. This statement is also valid for the function $\psi_{HJ}(P_r)$.

The comparison of the two isotherms in different domains of P_r is shown in Fig. 36.

On the left-hand side of Fig. 36, Eq. (369) is represented by the solid line. The circles represents the HJ equation (336) fitted to the values of coverage calculated by Eq. (369). The deviations are much greater than the errors of measurements; however, the value of Θ_0 calculated by the HJ equation is equal to 3.14, which satisfactorily coincides with $N = 3.03$, corresponding to Eq. (369). The right-hand side of Fig. 36 shows Eq. (369) and the HJ equation

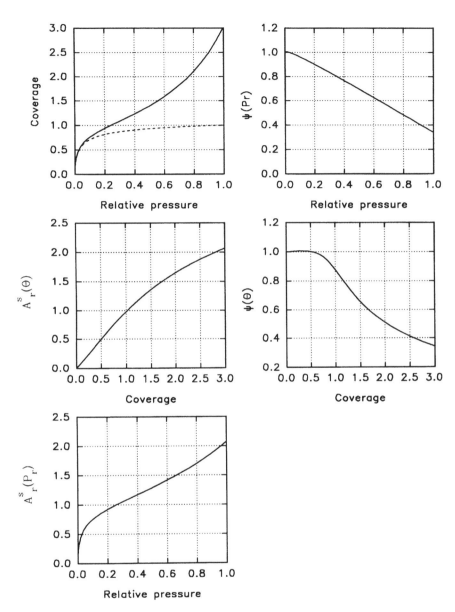

FIG. 35 The uniform and thermodynamically consistent interpretation of the C model with the Tóth equation [Eq. (369)]. The values of parameters are $K_T = 10.0$ kPa^{-t}, $\chi_T = 1.10$, $t = 0.45$, $k = 0.67$, and $P_{r,e} = 0$.

(336) in the domain $0.2 \leq P^r \leq 1$, where the two isotherms coincide, and the number of layers are approximately identical:

$$\Theta_0 = 3.08 \quad \text{and} \quad N = 3.03 \tag{373}$$

These calculations have proved the following important statements:

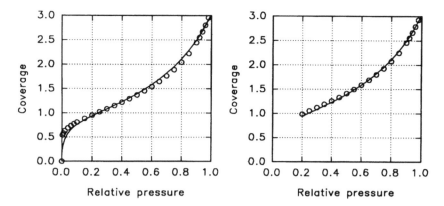

FIG. 36 The comparison of a HJ isotherm with a C model isotherm. Left: The values of parameters of the C model with the Tóth equation: $K_T = 10.0$ kPa^{-t}, $\chi_T = 1.1$, $t = 0.45$, $k_n = 0.67$, $P_{r,e} = 0$ (—). The values of parameters of the HJ equation are $A = 1.6131$, $B = 0.1642$, $(A/B)^{1/2} = \Theta_0 = 3.14$. Right: $A = 1.7484$, $B = 0.1837$, $(A/B)^{1/2} = \Theta_0 = 3.08$.

1. The cloud model includes isotherm equations which are also applicable to the calculation of the thicknesses of adsorbed layers.
2. The C-model isotherms are thermodynamically consistent.
3. The applicability of one of the C-model equations can be determined by the function $\psi_{CM}(P_r)$ calculated using Eq. (371).
4. If the measured values of Θ are conjugated with those of the functions $\psi_{CM}(P_r)$, we obtain the functions $\psi_{CM}(\Theta)$, which also characterize the appropriate C-model equation.
5. In the case of multilayer adsorption, the function $\psi(P_r)$ or $\Psi(\Theta)$ is again the most important choice of the thermodynamically and mathematically consistent and appropriate isotherm equation.

VI. CALCULATION OF THE SPECIFIC SURFACE AREA OF ADSORBENTS

Gas adsorption measurements are widely used for determining the surface area of a variety of different solid materials, such as industrial adsorbents, catalysts, pigments, ceramics, and building materials. The measurements of adsorption at the G/S interface also forms an essential part of many fundamental and applied investigations in the behavior of solid surfaces and of thermodynamical properties of the adsorption process. This latter topic was discussed in detail and investigated in the previous sections.

The calculation of the specific surface area from isotherms is based on the relationship

$$a^s = N_a a_m n_m^s \tag{374}$$

where n_m^s is the total monolayer capacity (expressed in mmol/g), a_m is the molecular cross-sectional area occupied by the adsorbate molecule in the complete monolayer (nm^2), and N_a is Avogadro's number. Thus, a^s, expressed in square meters per gram, is

$$a^s = 602.3 a_m n_m^s. \tag{375}$$

For calculating the value of a_m, it is assumed that the monolayer of molecules is a close-packed array and the molecules have approximately spherical form. Therefore,

$$a_m = 0.02\sqrt{3}\left(\frac{M\rho}{4\sqrt{2}Na}\right)^{2/3} \tag{376}$$

where M is the molecular mass and ρ is the specific volume of the liquid adsorptive at the temperature of the isotherm (expressed in g and cm^3/g, respectively).

In most cases, the monolayer capacity can be calculated from isotherms of Types I and II. In recent years [1], a change has been taking place in the interpretation of the Type I isotherms for microporous adsorbents (especially for activated carbons). According to this new interpretation, the initial (steep) part of the isotherms of Type I measured on microporous adsorbents represents micropore filling, and the low slope of the plateau is due to multilayer adsorption on the small external area. In Section VI.B, it will be shown how this multilayer part of the isotherms can be determined and taken into account by the calculation of the specific surface area. Because the micropore filling does not represent exactly the surface coverage, it is suggested [1] that the term "monolayer equivalent area" should be applied to microporous solids. This equivalent area is equal to that which would result if the amount of adsorbate required to fill the micropores were spread in a close-packed monolayer of molecules.

A. Calculation of the Specific Surface Area from Isotherms of Type II Without and With the Monolayer Domain

As is well known, the BET gas adsorption method has become the most widely used standard method for the determination of the specific surface area of solids; however, the BET theory has some simplifications which should be taken into account when this procedure is applied. First, the BET theory assumes that the monolayer and multilayer adsorptions begin *together* (i.e., the measured isotherm does not have a pure monolayer domain). The second assumption is that, because of the condensation, the following limiting value is valid:

$$\lim_{P_r \to 1} n^s = \infty \tag{377}$$

The limiting value (377) may cause the differences between the measured and calculated data in the domain of $P_r > 0.5$, and in the domain of low relative pressures, the pure monolayer adsorption and the heterogeneity of the absorbents may cause deviations between the measured and calculated data of isotherms. These are the reasons why the original BET equation can only be applied in the domain of relative pressures $0.05 < P_r < 0.3$. During the last 60 years, many thousands of measurements have proven that from nitrogen isotherms measured at 77 K and in the domain of relative pressures $0.05 < P_r < 0.3$, with the help of the original BET equation, real surface areas can be calculated. The following cross-sectional area of nitrogen is generally accepted:

$$a_m(N_2, 77) = 0.162 \text{ nm}^2 \tag{378}$$

The different modifications of the original BET equation discussed in Section V.A.1. extended its applicability for the relative pressure domain $P_{r,e} < P_r < 0.9$.

The general use of nitrogen as the adsorptive at 77 K can be explained with two facts:

1.   Liquid nitrogen is relatively inexpensive, clear, and is easy to obtain.
2.   A very important property of nitrogen isotherms measured at 77 K is that these isotherms very rarely have a pure monolayer domain ($P_{r,e} \approx 0$), and if this were so, the value of $P_{r,e}$ from the function $\psi(P_r)$ can be determined easily.

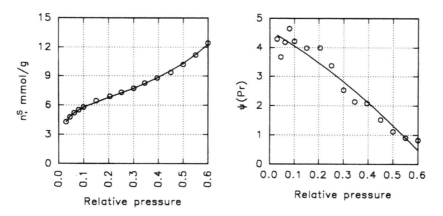

FIG. 37   Left: Nitrogen isotherm measured on silica gel at 77 K [17] (○); the modified BET equation (—).
Right: The calculated data of function $\psi(P_r)$ (○); the fitted function $\psi(P_r)$ (—).

These statements are supported below by concrete data of measured isotherms. A typical
nitrogen isotherm *without* a pure monolayer domain measured at 77 K and its function $\psi(P_r)$ are
shown in Fig. 37.

On the left-hand side of Fig. 37, the circles are the measured points on silica gel [17]. On
the right-hand side, the calculated data of function $\psi(P_r)$ are plotted; the solid line is a second-
power function fitted to the calculated data of the function $\psi(P_r)$. It is easy to see that the
function $\psi(P_r)$ does not have an increasing domain [i.e., the isotherm does not have a pure
monolayer domain $(P_{r,e} \approx 0)$]. Thus, the *original* BET equation in the domain of
$0.05 < P_r < 0.3$ can be applied well. In this case, the values of constants are the following:
$n_m^s = 5.814$ mmol/g, $c = 110.4$, $P_{r,e} = 0$,   and $k = 1$.   The calculated surface area is
$a^s = 567$ m^2/g. However, if a modified BET equation ($P_{r,e} = 0$, $k \neq 1$) is fitted to the measured
points, then the following values of constants are obtained: $n_m^s = 5.866$ mmol/g, $c = 81.36$,
$k = 0.879$, and $a^s = 572$ m^{2}g. As can be seen, the value of a^s did not change; however, the
whole isotherm ($0 < P_r \leq 0.6$) can be described well by the modified BET equation (solid line
on left-hands side of Fig. 37).

In Fig. 38 is plotted the argon isotherm of Type II *with* the monolayer domain measured on
rutile at 85 K [18]. At the top (left) of Fig. 38 is shown the whole isotherm measured in the
relative pressure domain of $0.0006 < P_r \leq 0.8$. At the top (right) of the figure is plotted the
enlarged part of the whole isotherm in domain of $0.0006 < P_r < 0.04$. The bottom (left)
represents the calculated function $\psi(P_r)$ of the whole isotherm and the bottom (right) shows its
enlarged part in the domain of $0.0006 < P_r < 0.04$. It can be seen excellently that in the domain
of $0.0006 < P_r < 0.04$, the isotherm has a pure monolayer section: The isotherm is Type I and
the corresponding function $\psi(P_r)$ increases. However, in the domain of $0.04 < P_r < 0.8$, the
isotherm is a multilayer one: The funcion $\psi(P_r)$ decreases and the isotherm increases in the form
of a function concave from below. Therefore, with help of the function $\psi(P_r)$, the monolayer and
multilayer sections of the isotherm can be distinguished $(P_{r,e} \approx 0.03)$. This distinction
determines the domains of relative pressures in which the appropriate isotherm equation
should be applied to calculate the specific surface area.

Only the multilayer domain of argon isotherm measured on rutile at 85 K is plotted in Fig.
39. All circles indicate the measured data. At top (left), the solid line represents the original BET
equation ($P_{r,e} = 0$, $k = 1$) applied in the domain of $0.03 < P_r < 0.4$. The corresponding

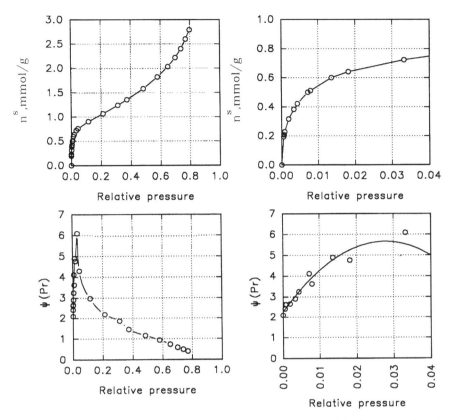

FIG. 38 The argon isotherm measured on rutile at 85 K [18] and its function $\psi(P_r)$. Distinction of the monolayer and multilayer parts of isotherm are based on the function $\psi(P_r)$.

constants are $c = 118.87$ and $n_m^s = 0.865$ mmol/g. In all parts of the figure, the dotted line represent the monolayer component isotherms. The solid line at the top (right) represents the modified BET equation ($P_{r,e} = 0.03$; however, k=1); we have obtained constants $c = 112.78$, $n_m^s = 0.904$ mmol/g, and $P_{r,e} = 0.03$. The bottom figure proves that the whole multilayer isotherm can only be described with the double-modified BET equation ($P_{r,e} = 0.03$ and $k < 1$). Thus, we have obtained the following parameters: $c = 79.26$, $n_m^s = 0.970$ mmol/g, $k = 0.851$, and $a^s = 84.0$ m^2/g. (The cross-sectional area of argon at 85 K is equal to 0.142 nm^2.) The physical reality of the separation of the monolayer and multilayer domains proves the fact that from the monolayer domain of the argon isotherm, approximately the same specific surface area can be calculated as has been made from the whole multilayer domain with the help of the double-modified BET equation. In Fig. 40 is shown the monolayer domain ($0.006 < P_r < 0.04$) of the argon isotherm, where the solid line represents the Tóth equation applied to this monolayer region of P_r.

The parameters of the Tóth equation are $\chi_{T,0} = 1.102$, $t = 0.283$, $K_{T,0} = 9.855$, and $n_m^s \approx n_0^s = 1.192$ mmol/g. The calculated specific surface area is 102 m^2/g. Drain and Morrison [18] have applied the classic and original BET methods to determine the special surface area; they obtained $a^s(N_2, 77 \text{ K}) = 95$ m^2/g. The agreements among the three different methods (85, 102, and 95 m^2/g) are satisfying. This statement is of general validity in spite of fact that the

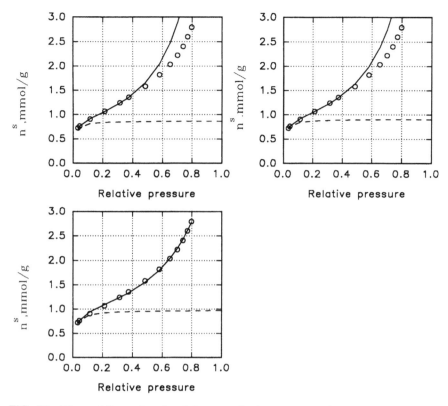

FIG. 39 The multilayer domain of the argon isotherm measured on rutile at 85 K: measured data: (○); calculated data (—).

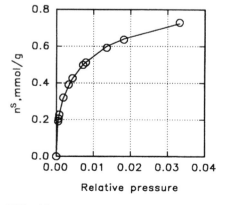

FIG. 40 The monolayer domain of argon isotherm measured on rutile at 85 K: measured data (○); calculated data (—).

Tóth equation should be derived from Eq. (210) when it is applied to a monolayer domain of an isotherm measured *below* the critical temperature. In this case,

$$n^s = \frac{n_0^s (\chi_{T,0})^{1/t} P_r}{[(1/K_{T,0}) + P_r^t]^{1/t}} \tag{379}$$

where

$$\chi_0 = 1 + 1/K_{T,0} \tag{380}$$

and

$$n_\infty^s = n_0^s (\chi_0)^{1/t} \tag{381}$$

n_0^s is the adsorbed amount when $P_r \to 1$.

This problem, connected with the general determination of BET surface area, is discussed in detail in Section VI.C.

B. Calculation of the Specific Surface Area from Isotherms of Type I with a Multilayer Domain and Measured Below the Critical Temperature

In several cases, it may occur that an isotherm is apparently Type I, however the function $\psi(P_r)$ proves that this isotherm has a definite multilayer domain [i.e., the function $\psi(P_r)$ has a maximum value]. Such an isotherm and its function $\psi(P_r)$ are shown in Fig. 41. At the top (left)

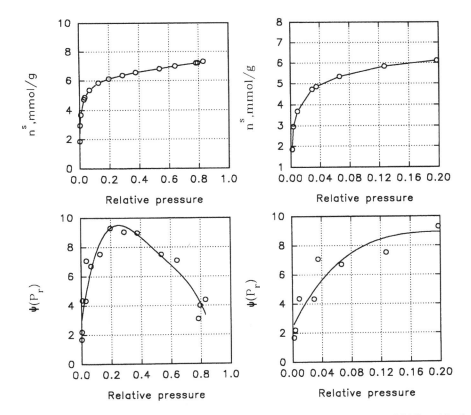

FIG. 41 The ethane isotherm measured on BPL activated carbon at $-60.5°C$ [19] and its function $\psi(P_r)$. Distinction of the monolayers and multilayers parts of the isotherm are based on the function $\psi(P_r)$.

is shown the isotherm apparently of Type I; however, at the bottom (left), the function $\psi(P_r)$ indicates that the function $\psi(P_r)$ has a maximum value. The top (right) and bottom (right) show the enlarged parts of the isotherm $n^s(P_r)$ and the function $\psi(P_r)$ in the domain of $0 < P_r < 0.20$. Thus, the distinction of the monolayer and multilayer domains makes it possible to apply the appropriate isotherm equations to the corresponding domains of the isotherm.

In Fig. 42 are represented an incorrect (top, left) and a correct (top, right) application of the original and modified BET equation. At the top (left), the original BET equation ($P_{r,e} = 0$, $k = 1$) is applied to the usual domain of $0.03 < P_r < 0.2$. The parameters are $c = 380.0$, $n_m^s = 5.083$, and the calculated specific surface area is $719 \ m^2/g$. The dotted line represents the monolayer component isotherm and the solid one represents the calculated isotherm. At the top (right), the double-modified BET equation is applied to the domain of $P_{r,e} < P_r < 0.8$ (i.e., to the whole multilayer domain of P_r). The parameters of the modified BET equation are $c = 57.27$, $n_m^s = 6.651 \ mmol/g$, $P_{r,e} = 0.03$, and $k = 0.166$. The calculated specific surface area is $941 \ m^2/g$. In Fig. 42, it can be seen that the modified BET equation describes well the multilayer domain of the isotherm. The circles represent the measured data and the solid lines represent the calculated ones.

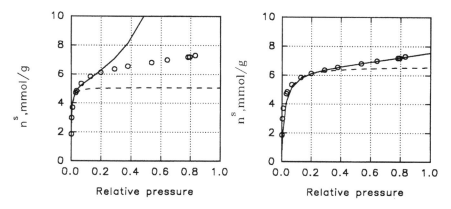

FIG. 42 Application of the original and the double-modified ($P_{r,e} \neq 0$, $k \neq 1$) BET equations to the ethane isotherm measured on the BPL activated carbon at $-60.5°C$ [19]: measured data (○); calculated data (—); monolayer component isotherm (---).

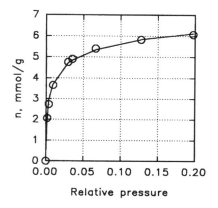

FIG. 43 The monolayer domain of the ethane isotherm measured on BPL activated carbon at $-60.5°C$: measured data (○); calculated by the Tóth equation (—).

Finally, in Fig. 43 is represented the monolayer domain of the ethane isotherm to which the Tóth equation (379) has been applied. The parameters are $\chi_{T,0} = 1.044$, $t = 0.461$, $K_{T,0} = 22.79$, and $n_0^s \approx n_m^s = 6.704$ mmol/g. The calculated specific surface is 949 m^2/g. By all calculations of the specific surface area, the cross-sectional area of ethane at $-0.60.5°$C (213 K) calculated from Eq. (376) is $a_m(C_2H_6, 213$ K$) = 0.235$ nm^2. Comparing the values of a^s calculated correctly from the multilayer and monolayer domains of the ethane isotherm (941 and 949 m^2/g, respectively) can be stated that the distinction of the two domains with help of the function $\psi(P_r)$ is both a mathematically and physically correct method. To this statement can be added that the specific surface area given by the producer of the BPL activated carbon is equal to 988 m^2/g.

C. Calculation of the Specific Surface Area from Isotherms of Type I Without a Multilayer Plateau and Measured Below the Critical Temperature

As was proven earlier in this chapter, if an isotherm of Type I has both a monolayer and a multilayer domain, then there are two possibilities for calculating the specific surface area. The modified BET equation to the multilayer domain and one of the monolayer equations to the monolayer domain can be applied. The appropriate monolayer equation and the value of $P_{r,e}$ (the relative equilibrium pressure where the multilayer adsorption begins) can be calculated from the function $\psi(P_r)$. In most cases, the Tóth equation is the simplest monolayer relationship because it can be applied instead of the FT, VT, and BT equations (see Section IV.F). However, if the isotherm of Type I does not have a multilayer plateau, then the approximation

$$n_m^s \approx n_0^s \tag{382}$$

may not be valid. This is the reason why it is necessary to derive the Tóth equation from Eq. (97) too (i.e., why the integration between limits $P–P_m$ and $n^s–n_m^s$ is necessary). Thus, performing the integration according to Eq. (97), we obtain

$$n^s = \frac{n_m^s (\chi_T)^{1/t} P_{r,m}}{[(1/K_{T,m}) + P_{r,m}^t]^{1/t}} \tag{383}$$

where

$$P_{r,m} = \frac{P}{P_m} \tag{384}$$

$$\chi_T = 1 + \frac{1}{K_{T,m}} \tag{385}$$

and

$$K_{T,m} = K_T P_m^t. \tag{386}$$

If the integration is performed between the limits $P–\infty$ and $n^s–n_\infty^s$, then we have

$$n^s = \frac{n_\infty^s P}{[(1/K_T) + P^t]^{1/t}} \tag{387}$$

where the limiting value

$$\lim_{P \to \infty} n^s = n_\infty^s \tag{388}$$

is mathematically correct; however, it is thermodynamically inconsistent.

Finally, if the integration is performed between the limits $P-P_0$ and $n^s-n_0^s$, we have Eq. (379):

$$n^s = \frac{n_0^s (\chi_{T,0})^{1/t} P_r}{[(1/K_{T,0}) + P_r^t]^{1/t}} \tag{389}$$

where

$$P_r = \frac{P}{P_0} \tag{390}$$

$$\chi_{T,0} = 1 + \frac{1}{K_{T,0}} \tag{391}$$

and

$$K_{T,0} = K_T P_0^t \tag{392}$$

This means that the Tóth equation in three forms, Eqs. (383), (387), and (389), can be applied depending on the aim of the application and on the temperature of the isotherm. If the temperature is below the critical one $(T_c)(T < T_c)$, then all three forms of the Tóth equation can be applied to *one* isotherm because the values of the parameters n_∞^s and t are independent of the dimension of the equilibrium pressure and the parameters, $K_{T,m}$ and $K_{T,0}$, as defined by Eqs. (386) and (392). In these circumstances, the following

$$n_\infty^s = n_0^s (\chi_{T,0})^{1/t} \tag{393}$$

and

$$n_\infty^s = n_m^s (\chi_T)^{1/t} \tag{394}$$

that is,

$$(\chi_{T,0})^{1/t} = \frac{n_\infty^s}{n_0^s} \tag{395}$$

and

$$(\chi_T)^{1/t} = \frac{n_\infty^s}{n_m^s} \tag{396}$$

It is evident that Eq. (383) and, therefore, Eqs. (394) and (396) can only be applied if n_m^s and P_m are known values. However, in most case, these values are unknown because we do not know the BET surface area (N_2, 77 K) of the adsorbent. Thus, our aim is to prove that it is possible to calculate the values of a^s(N_2, 77 K) form the isotherms of Type I without a multilayer plateau; thus, the separate determination of the nitrogen isotherm at 77 K can be omitted. This omission does not mean that the BET(N_2, 77 K) method is no longer necessary. To the contrary, the following proposed calculation of a^s is based on the BET(N_2, 77 K) method, but the proposed method refers to the possibility that a^s(N_2, 77K) can be substituted by $a^s(X, T)$, where X is the absorptive to be investigated and T, $(T < T_c)$, is the temperature of the isotherm of the adsorptive X.

From Eqs. (393) and (394), it follows that

$$(\chi_T)^{1/t} = \frac{n_0^s}{n_m^s} (\chi_{T,0})^{1/t} \tag{397}$$

In Eq. (397), the values of n_0^2 and $(\chi_{T,0})^{1/t}$ from Eqs. (387) and (391)–(393) can be calculated for any adsorptives X measured at temperature T $(T < T_c)$. This fact may be expressed by the following of Eq. (397):

$$[\chi_T(X, T)]^{1/t} = \frac{n_0^s(X, T)}{n_m^s(X, T)}[\chi_{T,0}(X, T)]^{1/t} \tag{398}$$

Taking Eq. (375) into account, the value of $n_m^s(X, T)$ in Eq. (398) can be calculated:

$$n_m^s(X, T) = \frac{a^s(N_2, 77\ K)}{a_m(X, T)(602.3)} \tag{399}$$

Equation (399) includes the following suppositions:

1. The molecules of adsorptives X are approximately spherical, so Eq. (376) can be applied to calculate the value of $a_m(X, T)$.
2. The specific surface area determined by the BET method, $a^s(N_2, 77\ K)$, is a known value.

Taking Eq. (393) into account, Eq. (398) can also be written in the following:

$$[\chi_T(X, T)]^{1/t} = \frac{n_\infty^s(X, T)}{n_m^s(X, T)} \tag{400}$$

Because $n_\infty^s(X, T)$ from Eq. (387) and $n_m^s(X, T)$ from Eq. (399) can be calculated, the following questions may be asked. Does a function $[\chi_T(X, T)]^{1/t}$ versus $[\chi_0(X, T)]^{1/t}$ of *general validity* exist from which, in the domain of the validity of suppositions (1) and (2), the BET specific surface area, $a^s(N_2\ 77\ K)$, from any isotherms (X, T) can be calculated? In particular,

$$a^s(X, T) = n_m^s(X, T)a_m(X, T)(602.3) \tag{401}$$

Comparing Eq. (401) with Eq. (399), we have that

$$a^s(X, T) = a^s(N_2, 77\ K) \tag{402}$$

Equations (401) and (402) mean that if the function $[\chi_T(X, T)]^{1/t}$ versus $[\chi_{T,0}(X, T)]^{1/t}$, calculated with help (with determination) of the value $a^s(N_2, 77\ K)$, is of *general validity*, then from any isotherms of Type I, the BET surface $a^s(N_2, 77\ K)$ can be determined without measuring the nitrogen isotherm at 77 K.

The first step to determine the function $[\chi_T(X, T)]^{1/t}$ versus $[\chi_{T,0}(X, T)]^{1/t}$ later, this function is briefly denominated with $\varphi(\chi)$] is the calculation of values $a_m(X, T)$. In Fig. 44 are shown the values of a_m of ethane, ethylene, propane, propylene, and carbon dioxide as a function of temperature (expressed in °C). These values have been calculated from Eq. (376); however, for a better applicability of those to all data, polynomials were fitted with the following results:
For ethane,

$$a_m(C_2H_6\ \text{and}\ C_2H_4, T) = -2.55 \times 10^{-7}T^3 + 9.4 \times 10^{-4}T^2 + 0.275 \tag{403}$$

For carbon dioxide,

$$a_m(CO_2, T) = 3.93 \times 10^{-6}T^2 + 5.4 \times 10^{-4}T + 0.205 \tag{404}$$

For propane,

$$a_m(C_3H_8, T) = 2.38 \times 10^{-6}T^2 + 6.1 \times 10^{-4}T + 0.288 \tag{405}$$

For propylene,

$$a_m(C_3H_6, T) = -1.78 \times 10^{-7}T^2 + 7.3 \times 10^{-4}T + 0.272 \tag{406}$$

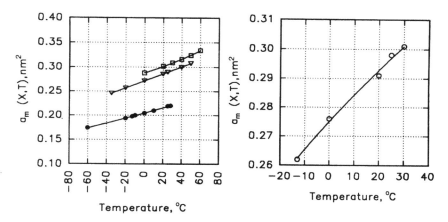

FIG. 44 The cross-sectional area (a_m) of different gases as functions of the temperature (in °C): carbon dioxide (○); ethane and ethylene (●); propane (▽); propylene (□).

The regression constants, R are in domain of $1 \leq R \leq 0.997$. The next step in calculating the function $\varphi(\chi)$ was the fitting of the Tóth equation (387) to isotherms of ethane, ethylene, carbon dioxide, propane, and propylene measured on microporous adsorbents (activated carbons) at different temperatures below the critical one. These measured data are collected in Valenzuela and Myers' handbook [19]. The adsorbents are activated carbons BPL, Nuxit, Columbia L, BPL-P, and Fiber Carbon KF-1500. All isotherms are Type I without a multilayer plateau. The temperature domains are those shown in Fig. 44. The specific surface areas determined by the BET method, $a^s(N_2, 77\,K)$ and the saturation pressures P_0 are also collected in Ref. 19. The parameters n_∞^s, K_T, and t are also known from the fitting procedure of the Tóth equation (387).

The constraints of the fittings procedures are

$$1 < [\chi_0(X, T)]^{1/t} < 2.5 \tag{407}$$

and

$$\sigma = \frac{1}{N} \sum_{i=1}^{N} \left| \frac{n^s - n_c^s}{n^s} \right| \times 100 \leq 2 \tag{408}$$

where σ is the average absolute error (%) of the fitting procedure, N is the number of the measured points, n^s is the measured adsorbed amount, and n_c^s is measured adsorbed amount calculated by the Tóth equation (387).

Thus, all data and parameters were available to calculate the pairs of $[\chi_T(X, T)]^{1/t}$ versus $[\chi_{T,0}(X, T)]^{1/t}$ from Eqs. (391), (399), and (400). These pairs are shown in Fig. 45 and they are marked with circles. To these circles was fitted a polinome (solid line in Fig. 45), which is the function $\varphi(\chi)$ with the following explicit form:

$$\varphi(\chi) = -0.537x^3 + 3.1421x^2 - 3.6754x + 2.0877 \tag{409}$$

where, evidently

$$x = [\chi_{T,0}(X, T)]^{1/t} \tag{410}$$

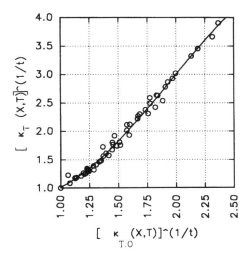

FIG. 45 Calculated data of $[\chi_T(X, T)]^{1/t}$ and $[\chi_{T,0}(X, T)]^{1/t}$ for microporous adsorbents ($\bigcirc$); the fitted polinome (—).

and

$$\varphi(\chi) = [\chi_T(X, T)]_c^{1/t} \tag{411}$$

where the subscript c refers to the fact that $[\chi_T(X, T)]_c^{1/t}$ is calculated from Eq. (409). Because the function $\varphi(\chi)$ fits to the data with regression factor 0.996, the general validity and applicability of Eq. (409) to microporous adsorbents may be regarded as a proven fact. Thus it may be written that $[\chi_T(X, T)]_c^{1/t} \approx [\chi_T(X, T)]^{1/t}$. This general validity and applicability means that the function $\varphi(\chi)$ is independent of the quality of adsorptives, temperatures, and the values of $a^s(N_2, 77\,K)$. It depends on the structure of the adsorbents only. In particular, the function $\varphi(\chi)$ essentially differs from that belonging to isotherms measured on nonmicroporous adsorbents. In Fig. 46, the circles represent the pairs of values $[\chi_T(X, T)]^{1/t}$ and $[\chi_{T,0}(X, T)]^{1/t}$ calculated from isotherms of carbon dioxide, propane, propylene, ethylene, argon, oxygen, and sulfur dioxide measured on nonmicroporous adsorbents (silica gels, graphitized carbon blacks, rutile) at the domain of temperatures $-188\,°C < T < 40\,°C$.

The measured data, specific surface areas, and saturation pressures are also collected in Ref. 19. The calculations of $[\chi_T(X, T)]^{1/t}$ and $[\chi_{T,0}(X, T)]^{1/t}$ have been made quite similar to those applied to isotherms measured on microporous adsorbents. The function $\varphi(\chi)$ fitted to the circles in Fig. 46 has the following explicit form:

$$\varphi(\chi) = 0.0158x^2 + 0.8044x + 0.2338 \tag{412}$$

where $x = [\chi_{T,0}(X, T)]^{1/t}$.

The main difference between the functions $\varphi(\chi)$ corresponding to the microporous and nonmicroporous adsorbents is reflected by Eq. (390). In particular;

$$n_0^s(X, T) \approx n_m^s(X, T) \tag{413}$$

then by Eq. (390),

$$[\chi_T(X, T)]^{1/t} \approx [\chi_{T,0}(X, T)]^{1/t} \tag{414}$$

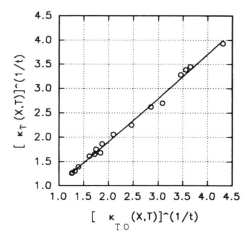

FIG. 46   Calculated data of $[\chi_T(X, T)]^{1/t}$ and $[\chi_{T,0}(X, T)]^{1/t}$ for nonmicroporous adsorbents ($\bigcirc$); the fitted polinome (—).

is valid [i.e., the function $\varphi(\chi)$ is a proportional, linear function]. As in Fig. 45, it can be seen the function $\varphi(\chi)$ differs from linearity because

$$n_0^s(X, T) > n_m^s(X, T) \tag{415}$$

and this difference, defined by Eq. (415) increases as the value of $[\chi_{T,0}(X, T)]^{1/t}$ increases. It means that the microporous structure and the micropore filling are reflected by relation (415). To the contrary, in Fig. 46 can be seen that the function $\varphi(\chi)$ is approximately a proportional and linear one; therefore,

$$n_0^s(X, T) \approx n_m^s(X, T) \tag{416}$$

(i.e., the total monolayer capacity is completed close to the saturation pressure).

Finally, below is briefly summarized, in steps, how the unknown specific surface area, $a^s(N_2, 77\,K)$, from any Type I isotherms without multilayer plateau measured on microporous or nonmicroporous adsorbents at temperatures below the critical one can be calculated.

Step 1.   Fitting the Tóth equation to the measured data under constraints (406) and (407); determination of parameters n_∞^s, K_T, and t.

Step 2.   Calculation of the value of $[\chi_{T,0}(X, T)]^{1/t}$ from Eqs. (391) and (392).

Step 3.   Taking into account that in both Eqs. (409) and Eq. (412),

$$x = [\chi_{T,0}(X, T)]^{1/t}$$

the value of $\varphi(\chi)$, i.e., $[\chi_T(X, T)]^{1/t}$ can be calculated.

Step 4.   Taking Eq. (400) into account, we obtain

$$n_m^s(X, T) = \frac{n_\infty^s(X, T)}{[\chi_T(X, T)]^{1/t}}$$

Step 5.   Applying Eqs. (401) and (402).

Summarizing this method, it should again be underlined that this calculation of $a^s(N_2, 77\,K)$ is only a substitution of the original BET method. The occasional but suitable omission of the nitrogen isotherm at 77 K does not mean that the well-known and widely used BET method is no longer necessary. To the contrary, to widen the applicability of functions $\varphi(\chi)$, more and more calculations of $a^s(N_2, 77\,K)$ are required. This requirement is also supported by the fact proven by the investigations made until now that the differences between the values of $a^s(N_2, 77\,K)$ calculated by the method proposed above and measured by the original BET method are not greater than the deviations (errors) of the original method (maximum $\pm 10\%$).

VII. PROBLEMS

This section of the chapter has an aim which differs from that of books dealing with original and new research works. In particular, some possibilities of mistakes which may occur when somebody applies the results and methods discussed in this chapter to his research works are interpreted. The readers' interest may be aroused about the fields of physical adsorption existing which are not dealt with in this chapter; however, the methods and calculation discussed earlier can be extended to these fields too. Some of these problems are now discussed.

A. Consequences of the Calculations Made with Data of Isotherms of Type II Measured in a Narrower Domain of Equilibrium Relative Pressure

In the top (left) of Fig. 47 is shown only a domain $0 \le P_r \le 0.35$ of an oxygen isotherm measured on rutile at 85 K [18]. The circles represent the measured data. At the top (right), the function $\psi(P_r)$ can be seen. Here, the solid line represents a polinome fitted to the calculated value of function $\psi(P_r)$ (circles). From the function $\psi(P_r)$ and from the measured data $P_r = 0.0405$ (i.e., to this relative pressure dominantly monolayer adsorption takes place); therefore, the Tóth equation can be applied to this part of the isotherm. Because the saturation pressure of oxygen at 85 K is a known value and is equal to 57.33 kPa, the pressures can be expressed as relative ones. This means that Eq. (389) can be directly fitted to the measured data in the domain of $0 \le P_r \le 0.0405$. So, we obtain that

$$n_\infty^s = n_0^s (\chi_{T,0})^{1/t} = 1.2313 \tag{417}$$

and

$$\begin{aligned} t &= 0.3805, \\ K_{T,0} &= 17.9164 \end{aligned} \tag{418}$$

From Eq. (391), we have

$$(\chi_{T,0})^{1/t} = 1.1534 \tag{419}$$

Therefore, from Eq. (417), we obtain the total monolayer capacity of adsorbent:

$$n_m^s \approx n_0^s \approx \frac{n_\infty^s}{(\chi_{T,0})^{1/t}} = 1.0675 \text{ mmol/g} \tag{420}$$

At the bottom (left) of Fig. 47 can be seen the six measured points (circles) of the monolayer domain. The dotted line represents the isotherm calculated by the Tóth equation with parameters (417) and (418). The agreements between the measured and calculated data are less than or equal to the errors of experimental measurements ($\pm 2\%$). To the multilayer domain of the oxygen

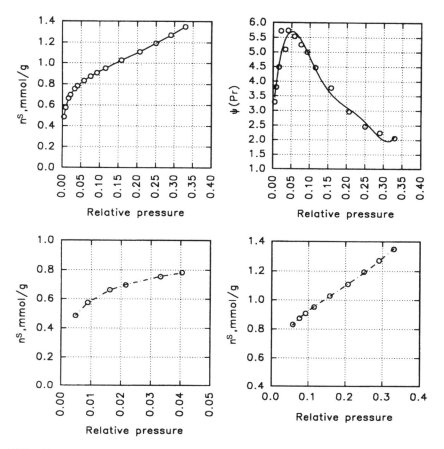

FIG. 47 Top: Oxygen isotherm measured on rutile at 85 K [18] and its function $\psi(P_r)$. Bottom: The monolayer and the multilayer domain of the isotherm.

isotherm, $0.0405 < P_r < 0.35$, the double-modified BET equation (531) was applied. The fitting procedure yielded the following parameters:

$$n_m^s = 0.9403 \text{ mmol/g} \tag{421}$$

$$c = 108.140 \tag{422}$$

$$k = 1.1005 \tag{423}$$

The value of $P_{r,e}$ (0.0405) was known from the function $\psi(P_r)$. At the bottom (right) of Fig. 47 are shown the measured data (circles) in the multilayer domain of the oxygen isotherm. The dotted line indicates the isotherm calculated by Eq. (331) with parameters (421)–(423). The agreements between the measured and calculated data are equal to those obtained for the monolayer domain. The cross-sectional area occupied by an oxygen molecule is

$$a_m(O_2, 85 \text{ K}) = 0.139 \text{ nm}^2 \tag{424}$$

Thus, we can calculate the specific surface area of rutile both from the monolayer and multilayer domains of the oxygen isotherm. From the Tóth equation, we have

$$a^s(O_2, 85 \text{ K}) = 602.3 \times 0.139 \times 1.0675 = 89 \text{ m}^2/\text{g} \tag{425}$$

and from the double-modified BET equation, we get

$$a^s(O_2, 85\ K) = 602.3 \times 0.139 \times 0.9403 = 79\ m^2/g \tag{426}$$

The original BET method yields [19]

$$a^s(N_2, 77\ K) = 91\ m^2/g \tag{427}$$

Because the value of a^s is small, the agreements between the different values of a^s are satisfying.

In spite of this fact, the parameter k present in Eq. (333) cannot be explained by the cloud model. According to Eq. (364), the number of layers (N) are

$$N = \frac{1}{1-k} \tag{428}$$

that is, if $k > 1$, then the negative value of N is an unexplainable result. However, according to the modified BET equation (331), k means that the adsorptive potential in the second and other layers are greater than the heat of condensation [see Eq. (329)]:

$$k = \exp\left(\frac{\lambda'}{RT}\right) \tag{429}$$

where, evidently, λ' is a possible value. Thus, the problem can exactly be defined. Is the double-modified BET equation or the cloud model false?

To answer this question, let us apply the HJ equation (341) to the multilayer domain of the oxygen isotherm (i.e., to the domain of $0.0405 < P_r < 0.35$). The following form of the HJ equation has been applied:

$$P_r = \exp\left\{A\left(\frac{1}{\Theta_0^2} - \frac{1}{\Theta^2}\right)\right\} \tag{430}$$

where

$$\Theta = \frac{n^s}{n_m^s} \tag{431}$$

The results of the three-parameter (n_m^s, Θ_0, and A) fitting procedure are

$$n_m^s = 0.9814\ mmol/g \tag{432}$$

$$\Theta_0 = 110181.7 \tag{433}$$

and

$$A = 2.0523 \tag{434}$$

In Fig. 48 are plotted (with circles) the measured data of the multilayer domain; the dotted line represents the isotherm calculated by the HJ equation (430) with parameters (432)–(434).

The agreement is not greater than the errors of measurements, and the calculated specific surfaces area is, again, a real value:

$$a^s(O_2, 85\ K) = 602.3 \times 0.139 \times 0.9814 = 82\ m^2/g \tag{435}$$

However, the value of Θ_0 is unexplainable, similar to the value k discussed for the cloud model. Based on these results, the problem to be solved can be formulated as follows: The values of the *monolayer* capacity (i.e., those of the specific surface areas) are real values and are approximately independent of the model and equations applied to *monolayer* and *multilayer* domains of the oxygen isotherm (89, 79, 91, and 82 m²/g). In spite of this fact, the parameters relating only to the *multilayer* domain (k, Θ_0) are divergent and unexplainable. The conclusion

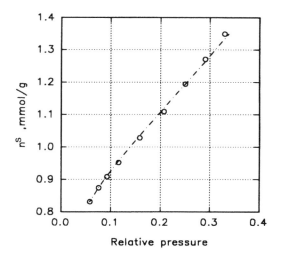

FIG. 48 Application of the HJ equation for the oxygen isotherm measured on rutile at 85 K. Measured data (○); calculated data (– – –). (From Ref. 18.)

that can be drawn from these facts may be the following: The *multilayer* domain of the oxygen isotherm, $0.0405 < P_r < 0.35$, is too narrow to obtain physically explainable and relevant parameters. This supposition is proven with functions plotted in Fig. 49. At the top (left) of Fig. 49, the whole measured oxygen isotherm in the domain of $0 < P_r \leq 0.9$ is shown, and at the top (right), its function $\psi(P_r)$ is plotted. It is evident that the monolayer domain, $0 < P_r < 0.0405$, is equal to that shown in Fig. 47. However, the multilayer domain is much wider: $0.0405 < P_r < 0.9$. The application of the double-modified BET equation to this wider multilayer domain yields the following parameters:

$$n_m^s = 1.2333 \text{ mmol/g} \tag{436}$$
$$c = 25.1233 \tag{437}$$

and

$$k = 0.6651 \tag{438}$$

The application of HJ equation (433) to this multilayer domain yields the parameters

$$n_m^s = 1.17520 \text{ mmol/g} \tag{439}$$
$$\Theta_0 = 3.4 \tag{440}$$
$$A = 1.6474 \tag{441}$$

At the bottom (left and right) of Fig. 49 are shown the measured (circles) and the calculated isotherms (dotted lines) obtained by applying the double-modified BET equation and the HJ equation (433) to the multilayer domain. Agreements are equal to those shown in Fig. 48. The most interesting result is that the parameters corresponding to the multilayer domain are real, relevant, and approximately equivalent ones. So, from Eqs. (438) and (428)—according to the cloud model—we have the number of layers:

$$N = \frac{1}{1 - 0.6653} = 3.0 \tag{442}$$

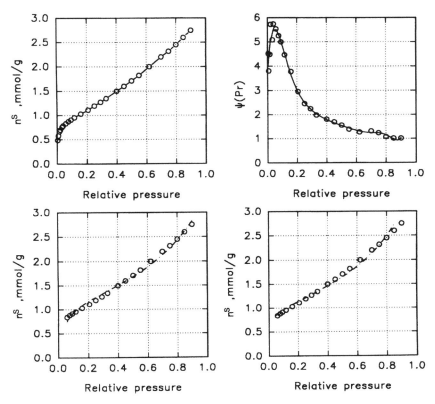

FIG. 49 Top: Oxygen isotherm measured on rutile at 85 K [18] and its function $\psi(P_r)$; bottom (left): measured data ($\bigcirc$); data calculated by the modified BET equation ($---$); Bottom (right): measured data ($\bigcirc$); data calculated by the HJ equation ($---$).

This value is approximately equal to Eq. (440) (i.e., the cloud model and the HJ model led to the same result). This statement is also valid for the total monlayer capacity (i.e., for the values of the specific surface area). In particular, according to Eq. (436),

$$a^s(O_2, 85\ K) = 602.3 \times 0.139 \times 1.2333 = 103\ m^2/g \tag{443}$$

and according to Eq. (439),

$$a^s(O_2, 85\ K) = 602.3 \times 0.139 \times 1.17520 = 98\ m^2/g \tag{444}$$

So the values of specific surface area of rutile calculated from oxygen isotherm measured at 85 K and by different models and equations are the following:

1. Original BETE method (N_2, 77 K):   91 m^2/g
2. Tóth equation for the monolayer domain: 89 m^2/g
3. Double-modified BET equation for the narrow multilayer domain: 79 m^2/g
4. HJ equation for the narrow multilayer domain: 82 m^2/g
5. Double-modified BET equation for the wide multilayer domain: 103 m^2/g
6. HJ equation for the wide multilayer domain: 98 m^2/g

Such large deviations are used with the application of the original BET method when the value of a^s is less than 100 m^2/g.

In summary, it can be stated that the calculation of the monolayer capacity (the specific surface area) is independent of the domain of the *multilayer* adsorption; however, the parameters relating to the multilayer domain, especially the number (thickness) of layers, depend on the width of the P_r interval corresponding to the multilayer adsorption. For this reason, it is proposed to apply both the double-modified BET equation and the HJ relationship to the multilayer section. The two equations can control each other.

B. Consequences of the Calculations Made with Data of Isotherms of Type I Measured in a Narrower Domain of Relative (or Absolute) Equilibrium Pressure

Let us investigate the isotherm of Type I of carbon dioxide measured on Nuxit-AL activated carbon at 20°C (273.15 K) [19] and in a wide domain of equilibrium pressure, between 10 and 600 kPa. [This isotherm and its function $\psi(P)$ are shown at the top of Fig. 50.)

The function $\psi(P)$ proves that the Tóth equation should be applied to this isotherm; however, the following problems also should be solved: Does the accuracy of the application of Tóth equation depend on the domains of equilibrium pressure to which this equation is applied? This question can be formulated another way: Do the physical and thermodynamical properties of the Tóth equation depend on the maximum equilibrium pressures (or maximum coverage) to which the isotherm has been applied? To answer these questions, the carbon dioxide isotherm mentioned earlier has been divided into four parts (see Fig. 50): the domains of equilibrium pressures 0–111 kPa, 0–250 kPa, 0–380 kPa, and 0–600 kPa. To these domains of pressures, the Tóth equation of the form

$$n^s = \frac{n_\infty^s P}{[(1/K_T) + P^t]^{1/t}} \tag{445}$$

was applied [i.e., in these four domains, the three-parameters (n_∞^s, K_T, t) fitting procedures were performed]. By these calculations, the parameters n_∞^s were input values, and from these values, the parameters K_T and t were calculated. With the change of n_∞^s, the parameters K_T and t also changed; that is, for *one* domain of the carbon dioxide isotherm, more Tóth equations with different parameters n_∞^s, K_T, and t were obtained. It is evident that these different Tóth equations described the isotherm in every domain of pressures mentioned earlier with different accuracies. The deviations between the calculated (n_c^s) and measured (n^s) values of the adsorbed amount were calculated using the relationship

$$\text{Deviation } \% = \frac{1}{N} \sum_{i=1}^{N} \left| \frac{n_c^s - n^s}{n^s} \times 100 \right| \tag{446}$$

where N is the number of the measured data.

The parameters of the Tóth equations applied to the four parts of isotherms is summarized in Table 2, and the deviations (446) as a function of n_∞^s corresponding to the four part of isotherms are shown in Fig. 51. From Table 2 and Fig. 51, the following conclusions can be drawn:

1. The wider the measured pressure intervals of the isotherm, the greater the accuracy (the less the deviation) of the Tóth equation.
2. The functions shown in Fig. 51 prove that in all intervals, there exists a value of n_∞^s (and the corresponding values of parameters t and K_T) which assures a minimum deviation % between the calculated and measured values of n^s.

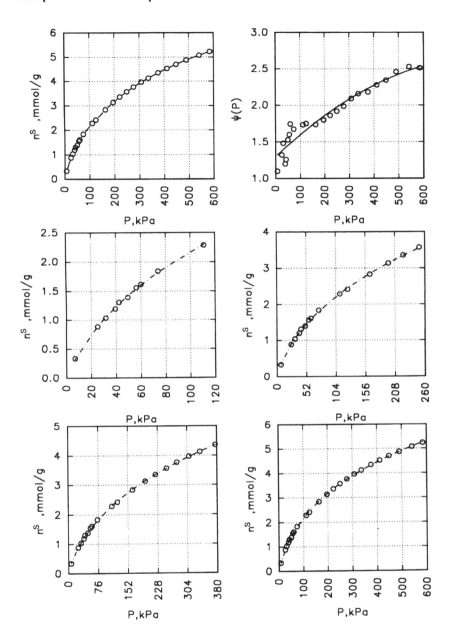

FIG. 50 Top: Carbon dioxide isotherm measured on Nuxit AL activated carbon at 20°C and its function $\psi(P_r)$; middle and bottom: The four parts of the isotherms. Measured data ($\bigcirc$); calculated data by the Tóth equation ($---$).

TABLE 2 Parameters of the Tóth Equation Applied to Different Pressure Intervals of the Carbon Dioxide Isotherm

Pressure intervals (kPa)	No. of measured data	n_∞^s (mmol/g)	t	K_T (kPa^{-t})	Deviation % [Eq. (446)]
0–111	10	10.0	0.5359	0.0669	1.594
		12.5	0.4785	0.0843	1.477
		15.0	0.4403	0.0982	1.393
		17.5	0.4105	0.1107	1.338
		18.5	0.4032	0.1140	1.403
		20.0	0.3910	0.1197	1.464
		25.0	0.3598	0.1356	1.624
		30.0	0.3378	0.1481	1.727
0–250	15	12.5	0.4962	0.0739	1.474
		15.0	0.4489	0.0918	1.274
		17.5	0.4154	0.1070	1.175
		18.7	0.4026	0.1135	1.134
		20.0	0.3903	0.1201	1.184
		22.5	0.3706	0.1315	1.283
		25.0	0.3546	0.1415	1.365
0–380	19	15.0	0.4631	0.0822	1.415
		20.0	0.3977	0.1130	0.996
		21.25	0.3862	0.1194	0.955
		21.85	0.3812	0.1224	0.939
		22.5	0.3760	0.1255	0.963
		23.7	0.3672	0.1310	1.000
		25.0	0.3586	0.1366	1.039
		26.5	0.3496	0.1427	1.078
0–600	24	18.0	0.4216	0.0998	0.948
		19.0	0.4089	0.1066	0.882
		20.0	0.3976	0.1131	0.841
		20.5	0.3924	0.1162	0.832
		21.0	0.3874	0.1192	0.825
		21.5	0.3827	0.1222	0.842
		22.0	0.3782	0.1251	0.871

3. The lower these minimum values of deviations, the wider the measured pressure intervals are.
4. The wider the measured pressure intervals, the greater the values of n_∞^s corresponding to the minimum deviations are.
5. In spite of the facts mentioned in points 1–4, every deviation shown in Table 2 is less than the errors of measuring of an adsorption isotherm ($\pm 2\%$). This means that every Tóth equation with all of the parameters tabulated in Table 2 can describe the carbon dioxide isotherm with good accuracy.

Statement 5 is proven in Fig. 50, where the four parts of carbon dioxide isotherm are plotted in middle and bottom parts and the dotted lines represent the calculated isotherms with an average deviation of 1.6%.

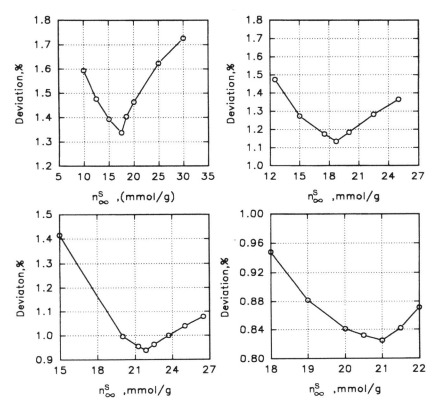

FIG. 51 Deviations between the measured and calculated data as a function of n_∞^s. Calculations for the four parts of the carbon dioxide isotherm have been made by the Tóth equations.

The parameters in Table 3 prove that *all* of the Tóth equations tabulated in Table 2 have physical reality based on the fact that all of these equations describe the isotherm with satisfactory accuracy. The parameters in Table 3 have been calculated with the following relationships [see Eqs. (391), (392), (395), and (409)]:

$$(\chi_{T,0})^{1/t} = \left(1 = \frac{1}{K_T p_0^t}\right) \tag{447}$$

$$n_0^s = \frac{n_\infty^s}{(\chi_{T,0})^{1/t}} \tag{448}$$

$$(n_{T,m}^s)^{1/t} = -0.537x^3 + 3.1421x^2 - 3.6754x + 2.08774 \tag{449}$$

where

$$x = (\chi_{T,0})^{1/t}$$
$$n_m^s = \frac{n_\infty^s}{(\chi_{T,m})^{1/t}} \tag{450}$$

TABLE 3 Parameters of the Tóth Equations Required to Calculate the Specific Surface Area

Pressure intervals (kPa)	n_∞^s (mmol/g)	$(\chi_{T,0})^{1/t}$	n_0^s (mmol/g)	$(\chi_{T,m})^{1/t}$	n_m^s (mmol/g)	a^s (m²/g)
0–111	10.0	1.2869	7.7708	1.4170	7.0574	918
	12.5	1.4352	8.7098	1.6973	7.3646	958
	15.0	1.5870	9.4519	1.0220	7.4183	965
	17.5	1.7524*	9.9866*	2.4061*	7.2733*	946*
	18.5	1.8014	10.2698	2.5240	7.3297	954
	20.0	1.8933	10.5635	2.7478	7.2787	947
	25.0	2.1985	11.3715	3.4881	7.1673	932
	30.0	2.5016	11.9926	4.1498	7.2294	941
0–250	12.5	1.4072	8.8830	1.6413	7.6170	991
	15.0	1.5686	9.5626	1.9811	7.5716	985
	17.5	1.7327	10.1001	2.3591	7.4181	965
	18.7*	1.8120*	10.3201*	2.5496*	7.3344*	954*
	20.0	1.8979	10.5381	2.7589	7.2492	943
	22.5	2.0634	10.9044	3.1641	7.1110	925
	25.0	2.2288	11.2168	3.5590	7.0244	914
0–380	15.0	1.5394	9.7441	1.9168	7.8255	1018
	20.0	1.8726	10.6801	2.6972	7.4151	965
	21.25	1.9567	10.8601	2.9032	7.3196	952
	21.85*	1.9971*	10.9410*	3.0022*	7.2781*	947*
	22.5	2.0407	11.0258	3.1087	7.2377	942
	23.7	2.1216	11.1710	3.3049	7.1712	933
	25.0	2.2090	11.3175	3.5127	7.1170	926
	26.5	2.3101	11.2549	3.7451	6.9425	903
0–600	18.0	1.7328	10.3877	2.3595	7.6288	992
	19.0	1.8027	10.5399	2.5271	7.5186	978
	20.0	1.8728	10.6791	2.6977	7.4138	965
	20.5	1.9078	10.7552	2.78332	7.3653	958
	21.0*	1.94303*	10.8078*	2.8696*	7.3180*	952*
	21.5	1.9782	10.8685	2.9559	7.2737	946
	22.0	2.0134	10.9269	3.0420	7.2320	941

Note: The asterisk indicates data calculated from the Tóth equations.

and

$$a^s = 602.3 \times 0.216 \times n_m^s \tag{451}$$

The saturation pressure of carbon dioxide at 20°C is

$$P_0 = 5734.21 \text{kPa} \tag{452}$$

The data in Table 3 prove the following:

1. The calculated specific surface areas differ from the BET(N_2, 77 K) values (900 m²/g) less than +10%, which is a possible error of the original BET method.
2. The parameters and data marked with an asterisk are calculated from the Tóth equations corresponding to the minimum values of deviation %. These most important data are as follows:
   $n_0^s = 9.9866, 10.3201, 10.941,$ and $10.8078 \text{ mmolg}^{-1}$
   $a^s = 946, 954, 947,$ and $952 \text{ m}^2/\text{g}$, respectively

The values of n_0^s correspond to the total empty volume of the adsorbent (pore-filling, micropores and mesopores). It can be seen that these important values approximately are independent of the domains of pressure in which the isotherm has been measured.

REFERENCES

1. KSW Sing, et al. Pure Appl Chem *57*:603, 1985.
2. FE Bartell. J Am Chem Soc *44*:1866, 1922.
3. W Ostwald, R deIzaguirre. Kolloid Zeitschr *30*:279, 1922.
4. G Schay, L Nagy. Acta Chim Hung *39*:365, 1963.
5. S Ross, IP Olivier. *On Physical Adsorption*. New York: Interscience Publishers, 1964, pp 123–137.
6. *International Critical Tables*. (EW Washburn, ed.). New York: McGraw-Hill, 1928, Vol III, p 14.
7. PG Menon. Chem Rev *68*:277, 1968.
8. G. Aranovich, M Donohue. J Colloid Interf Sci *194*:392, 1997.
9. I Langmuir. J Am Chem Soc *38*:2267, 1916.
10. *The Thermodynamical Character of Adsorption*. (JH deBoer, ed.). Oxford: Clarendon Press, 1953, p 28.
11. *Statistical Thermodynamics*. (RH Fowler, EA Guggenheim, eds.). Cambridge: Cambridge University Press, 1949, pp 431–450.
12. MZ Volmer. Z Phys Chem *115*:23, 1925.
13. J Tóth. Adv Colloid Interf Sci *55*:1, 1995.
14. MM Dubinin, Zsurn Prine Chim *14*:906, 1941.
15. MM Dubinin. Dokl Akad Nauk SSSR. Ser Chim *55*:331, 1947.
16. WD Harkins, G. Jura. J Am Chem Soc *66*:1362, 1944.
17. S Brunauer, PH Emmett, E Teller. J Am Chem Soc *60*:309, 1938.
18. LT Drain, JA Morrison. Trans Faraday Soc *48*:840, 1952.
19. *Adsorption Equilibrium Data Handbook*. (DP Valenzuela, AL Myers, eds.). Englewood Cliffs, NJ: Prentice-Hall, 1984, pp 1–205.

2
Adsorption on Heterogeneous Surfaces

MAŁGORZATA BORÓWKO   Maria Curie-Skłodowska University, Lublin, Poland

I. INTRODUCTION

During the second half of this century, the study of adsorption on solid surfaces has steadily gained considerable interest. Adsorption at the solid–fluid interface plays a significant role in various disciplines of the natural science and underlies a number of technological processes.

The applications of adsorption are widespread; among the many fields of practical importance based on this process, one can mention heterogeneous catalysis, flotation, material science, microelectronics, ecology, separation of mixtures, purification of air and water, electrochemistry, chromatography, and so forth.

Understanding the mechanism of adsorption is timely and important from a fundamental scientific perspective. Adsorption is defined as a change in concentration of a given substance at the interface with respect to its concentration in the bulk part of the system. Such a perturbation in the local concentration is the most characteristic feature of nonuniform fluids. Adsorption is one of the fascinating phenomena connected with the behavior of fluids in a force field extorted by the solid surface. This process has a great influence on the structure of thin films and it affects phase transitions and critical phenomena near the surface. Briefly, adsorption dictates the thermodynamical properties of nonuniform fluids.

Within this context, it is not surprising that adsorption has been intensively investigated. The progress in this field has gained a strong impetus due to the introduction of numerous experimental techniques, such as scanning tunneling microscopy (STM), low-energy electron diffraction (LEED), x-ray diffraction, Raman spectroscopy, small-angle x-ray spectroscopy (SAXS), nuclear magnetic resonance (NMR), temperature-programmed desorption (TPD), and many others. These methods provide detailed information about physicochemical properties of solid surfaces and adsorbates.

Adsorption on solid surfaces is certainly a challenging scientific field for the development of theories, including recent advances in statistical physics. In spite of the fact that modern statistical mechanics gives a general foundation for description of many-particle systems, the development of the theory of interfacial phenomena is difficult because of their great variety and complexity. The majority of problems in surface science are not exactly solvable. Therefore, additional assumptions are indispensable in obtaining analytical expressions describing adsorption equilibrium. The key to theoretical studies of adsorption is a precise formulation of the model. However, adsorption depends on many parameters connected with the nature of the adsorbate and adsorbent. In this situation, the choice of basic elements of the model system is extremely important. It is now becoming widely recognized that all real surfaces are inhomogeneous to a greater or lesser extent, so the adsorbent heterogeneity should be included in advanced research. This assumption introduces considerable difficulties in the theoretical description of adsorption. Nevertheless, many original articles and reviews devoted to theory of adsorption on heterogeneous surfaces and its application to interpret experimental data have been published [1–12].

In this chapter, the key findings of such studies are briefly summarized and an outlook of promising trends in the treated fields is provided. In Section II, an introduction to the theory of adsorption on solids and a summary of basic models of heterogeneous surfaces will be given. The next sections are devoted to different theoretical methods used in this field.

The history of the search for a theory of adsorption on heterogeneous solid surfaces goes back more than 80 years. Consequently, we have accumulated a great deal of information on this phenomenon. On the other hand, some aspects are still being explored. However, there is no full theoretical synthesis of these studies yet. The first attempts to formulate a general theory of adsorption on solid have involved the concept of "the integral equation of adsorption isotherm." This idea has inspired many investigations and new directions of research. Notable among these is the characterization of adsorbents by means of the adsorption energy distribution functions. One should point out a universal character of the integral equation approach. It can be used to describe various systems, namely single gases, gas mixture, and multicomponent solutions. From a practical point of view, it is important that this treatment leads to simple, analytical expressions. One can easily apply them in any laboratory and industry. This is a still growing field of research. An overview of some of the activities will be given in Section III. In Section IV, selected applications of lattice models to study of the adsorption of gases and solutions on heterogeneous surfaces will be analyzed. Section V provides a review of results obtained by using the density functional theory. In particular, its applications to capillary condensation in chemically heterogeneous pores will be considered. Section VI reviews Monte Carlo simulations of adsorption on heterogeneous surfaces. Our final focus in this chapter is on phase transitions in surface films. This problem has been a subject of intensive investigations recently. The results will be discussed in Section VII. Section VIII concludes with a summary and outlook of open problems.

II. FUNDAMENTALS

We consider the system consisting of a fluid in contact with an unperturbed surface of a solid. In a first step the adsorbent may be treated as an ideally smooth plane, being a source of force field that depends only on distance from the surface. As it has been mentioned already, in a real situation, the surface is never free of various geometrical dislocations, defects, or chemical impurities and the surface heterogeneity is a considerable property of the system. There are many sources of surface heterogeneity (e.g., the polycrystalline character of solid, growth steps, crystal edges and corners, vacancies, existence of various atoms or functional groups at a surface, irregularities in a crystalline structure of a surface, chemical contaminants, etc.). In the case of microporous solids, the main source of surface heterogeneity is the complex geometrical structure, containing pores of different sizes and shapes. For all of these reasons, a potential field generated by a real solid at a given distance from an adsorbent depends on the position with regard to the surface. Adsorption properties of such an adsorbent are different at various points, of the surface. In this situation, we use the term "absolute heterogeneity" of the adsorbent. The surface can be viewed as composed of a certain number of different adsorption sites (active centers). Each kind of adsorption site is characterized by its own value of energy of adsorbate–adsorbent interactions. It is clear that these interactions depend on the chemical nature of adsorbate molecules. Therefore, from adsorption data measured for a definite system, we can obtain information concerning only the "relative heterogeneity" of the adsorbent. It provides information about active centers, which can be detected by molecules of a given adsorbate [5].

Generally, the interaction energy of an isolated, spherical molecule with the adsorbent is a function of Cartesian coordinates $v(x, y, z)$. The relative heterogeneity of solids may be characterized by the adsorption potential energy $v(x, y, z)$. The surface is taken to be the (x, y) plane, and the z axis is normal to the surface. The form $v(x, y, z)$ has been used as a classification criterion of surfaces by many authors who considered various types of adsorbent [5,6]. In this review, for simplicity, we distinguish only two classes of solid surface: homogeneous and heterogeneous. The ground for this classification is the adsorption energy corresponding to the minimum of $v(z)$: $\varepsilon = -v_{\min}(z)$. When adsorption energies of all of adsorption sites are identical, the adsorbent is homogeneous, otherwise it is heterogeneous. In the case of the adsorption of mixtures, the set of adsorption energies of all components is usually used to define the adsorbent heterogeneity [5]. However, for liquid–solid adsorption the heterogeneity is often characterized by the differences of adsorption energies of components with respect to the adsorption energy of a selected one [8].

One of the most important features of the heterogeneous absorbent is its topography (i.e., the way in which different adsorption sites are distributed over the surface). Two extreme models are usually considered. The first model has been proposed by Langmuir, who assumed that the surface is composed of isotropic domains consisting of the same adsorption sites (patchwise model). According to this model, adsorption processes occurring on different patches are totally uncorrelated. Adsorbate–adsorbate interactions between molecules adsorbed on different patches are usually neglected. It is a shortcoming of the patchwise model. The other concept of surface topography assumes that sites of different energies are completely randomly distributed over the surface. This is the so-called random model of a heterogeneous surface. The latter seems to be very realistic for numerous adsorbents. The random and patchwise topographies may be treated as limiting cases of a more general model which assumes partial correlation between probabilities of finding sites of given energies at a certain distance [13–15]. Readers seeking more detailed information on the surface models are referred to Refs. 5–8.

A considerable element of the model is the assumption connected with the possibility of the kinetic motion of adsorbed molecules. When the motion of molecules in the z direction is restricted but molecules are able to move freely in the (x, y) plane, the process is classified as mobile adsorption. However, if the lateral translation is also hindered, the process is classified as localized adsorption. The motion of admolecules is controlled by the energetic topography of the surface, molecular interactions, and thermal energies. The adsorbed molecule is considered as localized on a surface when it is held at the bottom of a potential well with a depth that is much greater than its thermal energy. Except for extreme cases, adsorption is neither fully localized nor fully mobile and can be termed partially mobile [8]. Because temperature strongly affects the behavior of the system, adsorption may be localized at low temperatures and become mobile at high temperatures.

In theoretical studies, two different concepts of the adsorption system are considered. The first assumes that the adsorbed film forms an individual thermodynamic phase, being in thermal equilibrium with the bulk uniform part of the system. This model has been very effectively used to describe various adsorption systems [5,6]. It allows one to derive the relatively simple equations for adsorption equilibrium by utilizing the quality of the chemical potentials of a given component in both phases. Numerous models of the surface (adsorbed) phase are considered; it may be assumed to be a monolayer or multilayer and either localized, mobile, or partially mobile, molecular interaction can be taken into account or neglected, and so on. However, the thermodynamical correctness of the concept of surface phase is controversial.

The other model of the adsorption system, the so-called three-dimensional model, seems to be more realistic and promising. In this model, we do not assume the existence of a distinct surface phase but consider the problem of a fluid in an external force field. From a theoretical point of view a solution of this task requires only knowledge of the adsorbate–adsorbate and adsorbate–adsorbent interactions. However, in practice, we encounter difficulties connected with the mathematical complexity of the derived equations. The majority of the statistical–mechanical theories of nonuniform fluids have been formulated for adsorption on homogeneous surfaces. Nevertheless, the recent results obtained for heterogeneous solids are really interesting and valuable [16].

A new opportunity, which creates good prospects for avoiding many problems connected with the theoretical description of adsorption on heterogeneous surfaces, has appeared as a result of the introduction of computer simulation methods [17,18]. Over the last three decades, computer simulations have grown into a third fundamental discipline of research in addition to experiment and theory. The study of adsorption on heterogeneous solid surfaces has especially benefited from the molecular simulation method, first of all, because of the complexity of interactions of adsorbate molecules with differently distributed active centers that are not easily described by the methods of statistical mechanics. Computer simulation can, in principle at least, provide an exact solution of the assumed model.

In all of the above-mentioned theoretical approaches, the main ingredients of the model are potentials describing the molecular interaction in the whole system. The adsorbate–adsorbate interactions can be described by using numerous equations. The most popular are the following potentials:

- The hard-sphere potential

$$U = \begin{cases} \infty, & r \leq \sigma \\ 0, & r > \sigma \end{cases} \tag{1}$$

- the square-well potential

$$U = \begin{cases} \infty, & r \leq \sigma \\ -\varepsilon, & \sigma < r < \lambda\sigma \\ 0, & r > \lambda\sigma \end{cases} \tag{2}$$

- the Lennard–Jones potential

$$U = u^{LJ}(r) = 4\varepsilon^{LJ}\left[\left(\frac{\sigma}{r}\right)^{12} - \left(\frac{\sigma}{r}\right)^{6}\right] \tag{3}$$

where ε^{LJ} is the characteristic energy, σ is the molecule diameter, r denotes the distance between molecules, and ε and λ are measure of strength and range of the interaction, respectively.

It is quite obvious that the behavior of fluids near the surface is sensitive to the model used for the fluid–solid interactions. In a general case, one should take into account the interactions of a given adsorbate molecule with all surface atoms. It is common to assume pairwise additivity of the intermolecular potentials so that the total potential of interaction of a fluid particle with the surface is obtained by summing up the interactions with all atoms from the solid:

$$v(\mathbf{r}) = \sum_i v_i(|\mathbf{r} - \mathbf{r}_i|) \tag{4}$$

where $\mathbf{r} = (x, y, z)$ denotes the position of the adsorbate molecule and $\mathbf{r}_i$ is the position of the ith atom.

For a crystalline surface the potential is a periodic function in the surface plane. The periodicity of the potential $v(\mathbf{r})$ has prompted Steele [19] to represent it in the form of the Fourier series

$$v(x, y, z) = v_0(z) + \sum_{\mathbf{q} \neq 0} v_{\mathbf{q}}(z) \exp(-i\mathbf{q}\boldsymbol{\tau}) \tag{5}$$

where i is the imaginary factor, $\boldsymbol{\tau} = (x, y)$ is the two-dimensional vector specifying the location of the admolecule in the plane parallel to the surface, $v_0(z)$ is the interaction potential averaged over the entire surface, and the sum runs over the nonzero two-dimensional reciprocal lattice vectors $\mathbf{q}$

$$\mathbf{q} = n_1 \mathbf{b}_1 + n_2 \mathbf{b}_2 \tag{6}$$

where $\mathbf{b}_1$ and $\mathbf{b}_2$ are the basic reciprocal lattice vectors and n_1 and n_2 are integers. Steele has assumed [19] that pair potentials are given by the Lennard–Jones (12,6) function [Eq. (3)] with parameters σ_{fs} and ε_{fs} determined using the standard mixing rules [20]

$$\sigma_{fs} = \tfrac{1}{2}(\sigma_{ff} + \sigma_{ss}) \quad \text{and} \quad \varepsilon_{fs} = \sqrt{\varepsilon_{ff}\varepsilon_{ss}} \tag{7}$$

where σ_{ff} (σ_{ss}) and ε_{ff} (ε_{ss}) are the Lennard–Jones potential parameters for the fluid (solid). Using the above assumptions, Steele has derived analytic expressions for the potential $v(\mathbf{r})$, which can be applied to surfaces of different symmetry [19].

Some general features of crystalline surfaces can be pointed out. In this case, the adsorption potential depends on both the plane vector $\boldsymbol{\tau}$ and on the distance from the surface. The force field generated by the crystalline surface has a periodic character. One can observe distinct minima of different depths separated by barriers of different heights. The zone of atomic size located on the minima constitute the "adsorption site." These sites are separated from one another by a "saddle point," so that an activation energy is required for surface migration.

Sometimes, it is possible to treat the surface as structureless. Then, the fluid–solid potential is obtained by replacing the sum over fluid–solid interactions by a sum of integrals over solid atoms in a given plane. In this way, Steele has derived the potential for graphitic carbons [19]:

$$v(z) = \bar{\alpha}\varepsilon_{fs}\left[\frac{2}{5}\left(\frac{\sigma_{fs}}{z}\right)^{10} - \left(\frac{\sigma_{fs}}{z}\right)^{4} - \frac{\sigma_{fs}^4}{3\Delta(z + 0.61\Delta)^3}\right] \tag{8}$$

where z is the distance from the surface and $\bar{\alpha}$, and Δ are parameters. If, in addition to integrating over each graphite plane, one integrates over all of the graphite planes in the z direction, the (9,3) potential is obtained:

$$v(z) = \varepsilon_{fs}\alpha'\left[\frac{2}{15}\left(\frac{\sigma_{fs}}{z}\right)^{9} - \left(\frac{\sigma_{fs}}{z}\right)^{3}\right] \tag{9}$$

where α' denotes a constant.

In general, the (9,3) potential is a poorer approximation to the real solid–fluid interactions than Eq. (7) and underestimates the depth of the potential well. Apart from this fact, it is one of most frequently used potentials describing interactions with the solid surface.

III. THE INTEGRAL EQUATION APPROACH

A. General Formulation

In this section, theoretical treatments based on the concept of the surface phase are discussed. The fundamentals of the formalism are presented for the adsorption of pure gases. Its adaptation to mixed-gas adsorption, adsorption from solution of nonelectrolytes, as well as applications to chromatographic systems are presented in successive subsections.

The basic relationship used in the theory of adsorption on heterogeneous solid surfaces is the so-called integral equation of adsorption isotherm, which can be written as [21,22].

$$\Theta_t(p) = \int_{\varepsilon_{\min}}^{\varepsilon_{\max}} \Theta(p, \varepsilon)\chi(\varepsilon)\, d\varepsilon \tag{10}$$

In light of Eq. (10), the overall adsorption isotherm (p), being a function of bulk gas pressure, is considered to be a weighted average of adsorption taking place on the surface elements characterized by a given value of the adsorption energy ε. The "local adsorption isotherm" is described by the function $\Theta(p, \varepsilon)$, and $\chi(\varepsilon)$ is the probability distribution function for finding the adsorption energy ε at the surface. The function $\chi(\varepsilon)$ is customarily called the adsorption energy distribution function. In the majority of applications, the minimal and maximal values of the adsorption energy are assumed to be equal to $\varepsilon_{\min} = 0$ and $\varepsilon_{\max} = \infty$. The adsorption energy distribution function characterizes only the global heterogeneity of the surface and does not give any information on its topography.

The basic assumption underlying Eq. (10) is that adsorption processes occurring on different adsorption sites are completely uncorrelated. It is fully satisfied only in the case of a lack of molecular interactions in the film or for the patchwise heterogeneous surface consisting of macroscopically large, uniform areas when the boundary effects connected with interactions between molecules adsorbed on different patches can be neglected. For strongly heterogeneous surfaces, the effects of heterogeneity dominate over the influence of molecular interactions. Many industrial applications involve adsorption on various natural and synthetic materials exhibiting very high heterogeneity. For this reason, numerous studies have been focused on this class of adsorbents. In this case, the integral equation leads to really interesting results of great practical importance.

On the other hand, the integral equation approach has some limitations. It is not the best formalism for describing the effects of weak surface heterogeneity on such processes as two-dimensional condensation and other phase transitions in adsorbed films. In such cases, the results following from the integral equation should be treated very cautiously.

The integral equation in question has been widely used over the last 50 years. Its numerous applications are summarized in several already available reviews and popular books [1–7]. Here, only the main directions of research that use the integral equation approach are discussed.

From the integral equation one can obtain the following [9]:

The overall adsorption isotherm $\Theta_t(p)$ for the distribution function $\chi(\varepsilon)$ and the local adsorption isotherm $\Theta(p, \varepsilon)$ assumed a priori (the direct problem)

The energy distribution function $\chi(\varepsilon)$ for the assumed local adsorption isotherm $\Theta(p, \varepsilon)$ and experimentally determined $\Theta_t(p)$ (the inverse problem)

The local adsorption isotherm $\Theta(p, \varepsilon)$ for the experimentally measured $\Theta_t(p)$ and a given distribution function $\chi(\varepsilon)$ (the problem of the unknown kernel).

The latter problem is usually ignored. The local adsorption isotherms can be represented by one of the many existing equations derived for adsorption on homogeneous surfaces and there are good reasons for choosing a correct adsorption model.

The direct and inverse problems are frequently encountered in practice. A solution of the integral equation allows one to find the relation between the shape of the overall adsorption isotherm and the adsorption energy distribution function.

When the model involves molecular interactions in the adsorbed film or adsorption mobility, the topography of surface should be strictly defined.

B. Adsorption of Gases

1. Adsorption Isotherms for Homogeneous Surfaces

The first theoretical treatment of adsorption on solid surfaces was the famous theory of Langmuir, which still occupies a central position in surface science [23]. His well-known equation describes localized, monolayer adsorption on homogeneous adsorbent when attractive interactions between adsorbed molecules are neglected:

$$\Theta = \frac{Kp}{1 + Kp} \tag{11}$$

where $K(T)$ is the Langmuir constant given by [2]

$$K = \alpha \exp\left(\frac{\varepsilon}{k_B T}\right) \tag{12}$$

In the above, $\alpha(T)$ is a constant connected with the ratio of partition functions of adsorbate molecules in the surface and gas phases and k_B denotes the Boltzmann constant.

The Langmuir approach was a starting point for developing the more realistic formalism in the framework of the lattice gas theories based on the Ising model [24]. It seems intuitively obvious that the lattice gas model is well suited for representing localized adsorption. The adsorbed phase is considered a two-dimensional lattice gas. The most popular isotherm involving molecular interaction effects is the Fowler–Guggenheim equation [25]

$$Kp = \frac{\Theta}{1 - \Theta} \exp\left(\frac{-\bar{z}u\Theta}{k_B T}\right) \tag{13}$$

where u is the interaction energy between admolecules, $\bar{z}$ is the number of nearest neighbors in the lattice. The Fowler–Guggenheim equation has been derived using the Bragg–Williams [26] approximation. In this treatment, the configurational degeneracy and the average nearest-neighbor interaction energy are both handled on the basis of a random distribution of molecules among lattice sites. A much more accuracy quasichemical approximation [27,28] has also been applied to obtain the adsorption isotherm equation [25]. For many years, the exact solution of two-dimensional Ising problem has been known and it was used in adsorption research. The comparison of the exact adsorption isotherms with those obtained in terms of various approximations has been presented by Steele [29]. Although the Bragg–Williams mean field theory is rather crude, the Fowler–Guggenheim equation is often used as the local adsorption isotherm because, in most cases, the effects of surface heterogeneity are much stronger than errors due to this approximation [6]. The main advantage of the Fowler–Guggenheim theory is the fact that it predicts two-dimensional condensation. The critical conditions are $\Theta_c = 0.5$ and $u_c^* = \bar{z}u/k_B T_c = 4$, where T_c is the critical temperature.

Over 40 years ago, Kiselev [30] presented an interesting concept of the associating adsorbate." He assumed that all interactions in the monolayer might be described as a series of reversible quasichemical reactions between admolecules and adsorption sites and between adsorbate molecules in the monolayer. These interactions were characterized by means of suitable reaction constants. This theory was extended by Berezin and Kiselev [31]; their final isotherm involves dispersive interactions according to the Fowler–Guggenheim model and specific interactions which cause formation different associates in the surface phase.

Many isotherm equations for submonolayer localized adsorption on homogeneous surfaces may be rewritten as follows [32]:

$$Kp = g_L(\Theta)g_I(\Theta) \qquad (14)$$

where $g_L(\Theta)$ denotes the Langmuir term and $g_I(\Theta)$ is a function connected with lateral interactions in the adsorbed layer. The form of Eq. (14) is very convenient for theoretical analysis of the role of surface topography in adsorption on heterogeneous solids. A detailed discussion of other equations for monolayer localized adsorption on homogeneous surfaces can be found in the text book of Jaroniec and Madey [5] and references therein.

The other class of isotherms follows from the theory of mobile or partially mobile adsorption [8,33]. One of the most frequently used equations of this type is the Hill–de Boer isotherm [34,35]. Moreover, an important attribute of gas adsorption is the formation of multilayer surface films. One particularly popular and successful equation describing multilayer adsorption is the Brunauer–Emmet–Teller isotherm [36]. Studies of many authors have been focused on an extension of various modifications of the BET equation to heterogeneous surfaces [5,6]. This problem will be discussed in Section III.B.2.c.

The choice of the model of "local adsorption" is always the important and quite critical step in theoretical considerations. As has probably become obvious already, adsorption on heterogeneous surfaces cannot be investigated within truly realistic models. Theorists have usually resorted to the simple models, which capture the features of the systems believed to be essential for the analyzed problem.

One important line of study has been the use of experimental adsorption data to extract information about the energy distribution function. Extensive theoretical and numerical investigations were performed to answer the question of how the chosen local isotherm affects the evaluated energy distribution [5,37–39]. It follows from these studies that the functions calculated for localized and mobile adsorption are analogous [32]. On the other hand, the results obtained by Jaroniec and Brauer [37] suggest that lateral molecular interactions and the multilayer nature of the surface phase may play a more significant role. However, in practice, the choice of local isotherm has been usually treated quite casually; in the majority studies, it is assumed that the local isotherm has a limited influence on the shape of the adsorption energy distribution function.

2. Adsorption Isotherms for Heterogeneous Surfaces

(a) Ideal Monolayer Adsorption.   Intensive investigations of the adsorption on real solids has led to the development of a large number of empirical isotherm equations. Their main advantage has been a good representation of experimental data, which they often provide. It is an important achievement of theoretical studies to find a link between empirical isotherms and heterogeneity of the surfaces. It has been shown that these equations correspond to distributions of adsorption energy, which are characteristic for the majority of the systems used in practice.

Numerous methods have been proposed to solve the integral equation with the Langmuir local adsorption isotherm. A short summary of these studies will be presented in Section III.B.3. Here, some practically important results are discussed.

We can distinguish two main classes of the overall adsorption isotherms:

- Equations reducible to the Langmuir isotherm
- Isotherms generated by the exponential isotherm equation

The overall adsorption isotherms of the first type become a form of the Langmuir equation for suitable values of the heterogeneity parameters. The isotherms of such a type can be expressed as [40–42]

$$\Theta_t = \left(\frac{(K^*p)^m}{1 + (K^*p)^m} \right)^{q/m} \tag{15}$$

where K^* is the constant defined by means of the "characteristic adsorption energy" determining the position of the energy distribution function on the energy axis, whereas q and m are the heterogeneity parameters characterizing the width and asymmetry, respectively, of the distribution $\chi(\varepsilon)$. Jaroniec and Marczewski [43] have also found an analytical form of the energy distribution function corresponding to the adsorption isotherm (15).

Numerous well-known empirical isotherms can be viewed as special cases of Eq. (15):

The Langmuir relation for $q = m = 1$
The Langmuir–Freundlich (LF) isotherm for $q = m \neq 1$ [40]
The generalized Freundlich equation (GF) for $m = 1$ and $q \neq 1$ [22]
The Tóth isotherm (I) for $q = 1$ and $m \neq 1$ [44–46].

These isotherms and corresponding energy distribution functions are presented in Fig. 1.

The general function $\chi(\varepsilon)$ connected with the overall adsorption isotherm (15) describes a continuous energy distribution with one peak [5,43]: a decreasing exponential distribution ($m = 1$), a symmetrical quasi-Gaussian distribution ($m = q$), and an asymmetrical distribution ($m \neq q$). When $q < m$ the asymmetrical distributions are widened in the direction of high

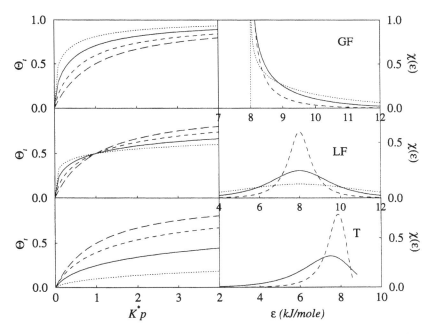

FIG. 1 Adsorption isotherms generated by Eq. (15) (the left panels) and corresponding energy distribution functions (the right panels): the generalized Freundlich isotherm (the upper panels), Langmuir–Freundlich isotherm (the middle panel), and the Tóth isotherm (the lower panel). The parameters are $k_B T = 0.6485$ kJ/mol, $\varepsilon_m(\varepsilon_a) = 8$ kJ/mol, and the heterogeneity parameter m (or q): 0.3 (dotted line), 0.5 (solid line) and 0.75 (short dashed line), respectively. The long dashed line refers to the Langmuir curve.

adsorption energies; however, for $q > m$, they have a broadening at the left-hand side (i.e., for low adsorption energies).

The results presented in Fig. 1 give clear evidence of the significant impact of the adsorbent heterogeneity on the overall adsorption isotherms. However, the influence may be of a various nature. The generalized Freundlich equation corresponds to the decreasing exponential distribution. In this case the overall adsorption isotherms lie over the Langmuir curve; for a fixed value of pressure, the surface coverage $\Theta_t(p)$ increases with the decreasing value of the heterogeneity parameter q. However, an opposite behavior is observed for the Tóth isotherm relating to the energy distribution with a broadening in the direction of low energies. For such a distribution, the number of sites with adsorption energies less than the most probable energy value is considerably greater in comparison with the number of strongly adsorbing sites. As a consequence, Tóth isotherms always lie below the Langmuir isotherm. The mediate results are obtained for Langmuir–Freundlich equation connected with a symmetrical Gaussian-type distribution. There is an identical number of high- and low-energy sites. The adsorption isotherms plotted for different values of the heterogeneity parameter exhibit small changes in shape and they intersect in one point. In the region of low pressures, they behave similarly to the GF isotherms, but for high values of pressure, this behavior is analogous to the Tóth relation. The surface heterogeneity considerably affects adsorption heats; a valuable analysis of these effects can be found in Refs. 5 and 6.

The exponential isotherm has the following form [47]:

$$\Theta_t = \exp\left\{-\sum_{j=1}^{r^*} B_j\left[k_B T \ln\left(\frac{p_a}{p}\right)\right]^j\right\} \tag{16}$$

where B_j ($j = 1, 2, \ldots, r^*$) are temperature-independent coefficients and p_a is the parameter connected with minimum adsorption energy. The exponential isotherm was proposed by Knowles and Moffat [48] and was derived from statistical mechanics by Jaroniec [47], who also evaluated the energy distribution corresponding to the isotherm by using the Sips procedure [21,22]. Because this function contains a large number of adjustable parameters ($B_1, B_2, \ldots$), it is sufficiently flexible to represent distributions of various shapes—from skewed ones, to trough quasi-Gaussian functions, to distributions having more extremes [5]. The overall isotherm (16) has been used frequently to interpret experimental data [47–55].

For special sets of the adsorption parameters, the exponential isotherm (16) reduces to well-known and popular equations:

The classical Freundlich (F) isotherm when $B_1 > 0$ and $B_j = 0$ for $j \geq 2$ [56–65]

The Dubinin–Radushkevich (DR) isotherm when $B_2 > 0$ and $B_j = 0$ for $j \neq 2$ [38,46, 66–78]

The Dubinin–Astakhov (DA) isotherm when $B_j > 0$ and $B_i = 0$ for $i \neq j$ [79,80]

The Freundlich–Dubinin–Radushkevich (FDR) isotherm for $B_1, B_2 > 0$ and $b_i = 0$ for $i > 2$ [81]

The examples of these isotherms are shown in Fig. 2.

The classical Freundlich equation has been proposed by Cerofolini [60]. This isotherm corresponds to a decreasing-exponential adsorption energy distribution function [5]. The various forms of the Freundlich equation were applied to approximate experimental isotherms [56–65,69].

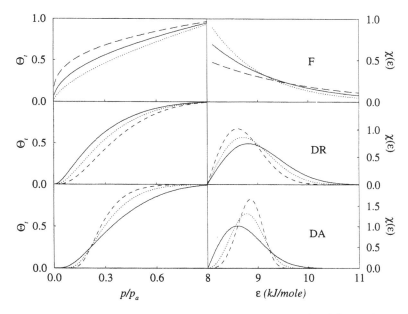

FIG. 2 Adsorption isotherms generated by Eq. (16) (the left panels) and corresponding energy distribution functions (the right panels): the Freundlich isotherm (the upper panels), the Dubinin–Radushkevich (the middle panel), and the Dubinin–Raushkevich isotherm (the lower panel). The parameters are $k_B T = 0.6485$ kJ/mol, $\varepsilon_a = 8$ kJ/mol, $B_j^0 = B_j (k_B T)^j$ (mol/kJ)^j [$j = 1$ (F) and $j = 2$ (DR)]. $B_j = 0.5$ (long dashed line), 0.75 (solid line), 1 (dotted line), and 1.4 (short dashed line) (mol/kJ)^j. For the DA isotherm, $B_j = 1$ (mol/kJ)^j, $j = 4$ (dashed line), 3 (dotted line), 2 (solid line).

The Dubinin–Radushkevich equation plays a special role in surface science:

$$\Theta_t = \exp\left\{-B_2\left[k_B T \ln\left(\frac{p_a}{p}\right)\right]^2\right\} \quad \text{for } p \le p_a \tag{17}$$

The isotherm analogous to Eq. (17) was presented by Dubinin and Radushkevich as the empirical isotherm for adsorption of vapors on porous and microporous solids [69]. During the last 50 years, the DR isotherm has been used as a standard equation for characterizing microporous adsorbents. On the other hand, numerous studies clearly proved that the equation in question can also describe gas adsorption on mesoporous, macroporous, and nonporous solids [70–78] (see Table 2.5 of Ref. 6). One of the sources for the success of DR isotherm in analysis of experimental results is probably a special form of the energy distribution function connected with this equation. It seems to be typical for many real adsorption systems. Sokołowski [79] has obtained the analytical form of this energy adsorption distribution. Its simplified version was evaluated by Hobson and Armstrong [72] from potential theory and by Cerofolini by means of the condensation–approximation method [66–68,78]. The DR isotherm is related to quasi-Gaussian distribution with a broadening at high adsorption energies. The history of the DR equation was analyzed in the review by Cerofolini [78].

A promising equation for the adsorption isotherm has been proposed by Dubinin and Astakhov [80,81]:

$$\Theta_t = \exp\left\{-B_j\left[k_B T \ln\left(\frac{p_a}{p}\right)\right]^j\right\} \quad \text{for } p \leq p_a \tag{18}$$

where B_j and j are parameters connected with the shape of the adsorption energy distribution function. When $j = 1$, Eq. (18) becomes the Freundlich isotherm; for $j = 2$ it becomes the DR isotherm. The general form of the energy distribution corresponding to the AD isotherm can be found in Ref. 5. The special form of this distribution has been obtained by Stoeckli [76] and Kadlec [82] by using the condensation–approximation method. An influence of the hetero- geneity parameter j on the energy distribution function is presented in Fig. 2. The DA equation has the same number of adjustable parameters as the Marczewski–Jaroniec equation. The correlations between these isotherms and the parameters were discussed in Ref. 83.

Because numerous solid–gas systems can be described by the Freundlich or Dubinin– Radushkevich equation, Cerofolini introduced the general expression, which is a linear combination of these isotherms (FDR) [84]. The theoretical foundation for the Freundlich and DR isotherms was a subject of intensive studies [60,64,67,68,78,85]. The results were summarized in several reviews [6,7,78].

As follows from Figs. 1 and 2, many simple equations give adsorption isotherms of similar shape. In this situation, the determination of a correct energy distribution function from adsorption data is uncertain. More detailed information on heterogeneity of the surface can be deduced from an analysis of calorimetric measurements. It seems to be quite obvious that the surface heterogeneity considerably affects adsorption heats; valuable analyses of these effects can be found in Refs. 5 and 6. In spite of all the limitations and doubts, the above-discussed equations allow one to show a relation between the adsorbent heterogeneity and the magnitude of adsorption. The important trend in theoretical investigations has been connected with an extension of these isotherms to adsorption with lateral interactions and multilayer adsorption. The essence of the idea is to rewrite the complex isotherm describing local adsorption in quasi- Langmuir form and to use the known solutions of the integral equation [e.g., Eq. (15) or (16)]. Examples of such a treatment are presented in successive subsections.

(b) Molecular Interactions in the Monolayers Formed on Surfaces of Various Topographies. As it has been already mentioned, in the case of lack of molecular interactions in the surface phase the solution of Eq. (10) is independent of the geometrical distribution of different adsorption sites. However, the surface topography is a property of a great importance when lateral interactions are taken into account. Let us consider the overall isotherm equations for localized adsorption using as the local adsorption isotherm [the general equation (14)]. In the theory of physical adsorption of gases on solids, two models of the heterogeneous surfaces are used most frequently: the first corresponding to patchwise distribution of active centers and the other assuming the random topography of an adsorbent.

In the first model, the surface consists of absolutely independent homogeneous domains. The interactions between molecules adsorbed on different patches are excluded. In this situation, the function g_I in the local isotherm (14) describes molecular interactions on a definite surface patch. In the other words, this function depends on the local surface coverage: $g_I = g_I(\Theta)$. The exact, analytical solution of the integral equation (10) is impossible for adsorption with lateral interactions. It may be solved numerically by means of a two-step procedure: (1) for each value of the adsorption energy, the local surface coverage is evaluated by solving the equation of the Eq. (14) type; (2) The integration in Eq. (10) is performed for the assumed energy distribution function.

Other relations are used for the random distribution of adsorption sites on the surface [86]. In this case, the term which characterizes the contribution of lateral interactions in the adsorption process depends on the overall surface coverage [i.e., $g_{I,t}(\Theta_t)$]. The local adsorption may be rewritten as

$$Kp = g_L(\Theta)g_{I,t}(\Theta_t) \tag{19}$$

or, in the quasi-Langmuir form,

$$\Theta = \frac{Kp^*}{1 + Kp^*} \tag{20}$$

where

$$p^* = \frac{p}{g_{I,t}(\Theta_t)} \tag{21}$$

It means that the analytical solutions of Eq. (10) for the local isotherm given by Eq. (20) are analogous to those obtained for the Langmuir local isotherm. Thus, the overall adsorption isotherms generated by Eq. (20) may be obtained from the isotherm equations discussed in the previous subsection by replacing the pressure p by the variable p^*. This trick makes calculations considerably simpler than for the patchwise surfaces. Obviously, the above procedure is correct when the mean-field type approximation for molecular interactions is involved.

The molecular interactions in the surface films formed on heterogeneous adsorbents have been taken into account in the studies of numerous scientists [5]. The Fowler–Guggenheim local adsorption isotherm has been used for the patchwise and random topographies for different energy distributions. The systematic analysis of the results can be found in Refs. 6–8.

Here, we will discuss only one problem, which has been intensively investigated; it refers to the role of heterogeneity in the critical behavior of monolayer films. It has been evidently proved that surface heterogeneity affects the values of critical parameters. However, the nature of this influence is different for various types of energy distribution function. The surface heterogeneity changes the shape of the phase diagrams considerably. As an illustration, we briefly discuss two examples. In the case of a quasi-Gaussian energy distribution, which corresponds to the Langmuir-Freundlich isotherm, $\Theta_{c,t} = 0.5$ is independent of the heterogeneity parameter m, whereas $u_c^* = \bar{z}u/k_B T_c = 4/m$ [87]. An increase in the surface heterogeneity causes an increase in the critical value of u_c^*. The phase transition is observed at a higher u_c^* than for the homogeneous surface ($u_c^* = 4$). Different values of critical parameters are obtained for the decreasing-exponential distribution (GF isotherm). For such adsorbents, both critical parameters depend on the heterogeneity parameter, and $\Theta_{c,t} > 0.5$ and $u_c^* > 4$ [88].

Apart from the Fowler–Guggenheim local adsorption isotherm, the Berezin–Kiselev equation has been extended to adsorption on heterogeneous surfaces by Jaroniec and Borówko [89]. Their results have an instructive character; the method allow us to investigate the influence of various geometrical distributions of active sites on adsorption in the framework of simple and clearly constructed model. They have considered localized monolayer adsorption on the surface consisting of two types of adsorption site. The lateral interactions caused the formation of double associates. The total adsorption was the sum of the surface coverage on adsorption sites of both kinds, which may be calculated from

$$\Theta_i = K_i p(s_i - \Theta_i) + 2L_{ii}K_i^2 p^2 (s_i - \Theta_i)^2 + L_{12}K_1 K_2 p^2 (s_i - \Theta_i)(s_2 - \Theta_2), \quad i = 1, 2 \tag{22}$$

where s_i is the fraction of adsorption sites of the ith type, K_i is the Langmuir constant, and L_{ij} denotes the association constant for the reaction between molecules located on sites i and j. The values of the association constants L_{ij} for various geometrical distributions of active sites were different. Three topographical models of heterogeneous surface were studied:

1. The random model for which all constants L_{11}, L_{22}, and L_{12} were greater than zero
2. The patchwise model when L_{11}, $L_{22} > 0$ but $L_{12} = 0$ because interactions between molecules adsorbed on various patches were neglected
3. The regular (chessboard) distribution of adsorption sites ($L_{11} = L_{22} = 0$ but $L_{12} > 0$).

In Fig. 3, the total adsorption isotherms calculated for various topographies are presented. One can conclude that even for the surface consisting of the same number of different adsorption sites ($s_1 = s_2 = 0.5$), their distribution on the surface is an important feature of the system and affects the adsorption equilibrium considerably. A sequence of adsorption isotherms calculated for various topographies depends on the pressure region.

One of the most frequently used local adsorption isotherm is the Hill–de Boer equation [34,35]. It should be pointed out that for mobile adsorption, even when lateral interactions are neglected, the additive assumptions about surface topography are necessary [6–8].

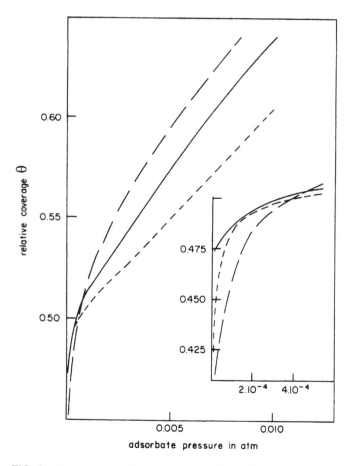

FIG. 3 Comparison of the adsorption isotherms for the patchwise (short-dashed line), random (solid line), and regular crystalline (long-dashed line) models of the surface. Parameters: $K_1 = 114,798$ atm^{-1}; $K_2 = 16.745$ atm^{-1}; $L_1 = L_2 = L_{12} = 4.424$, $s_1 = s_2 = 0.5$. (From Ref. 89.)

(c) Multilayer Adsorption. In spite of the well-known limitations of the BET model [36], it is still very popular and frequently used to determine the surface area of adsorbents. For this reason, numerous authors have published works on the extension of the BET equation to heterogeneous surfaces. According to the BET model, the status of molecules in the second and the higher surface layers is the same. As a consequence, in the majority of the generalizations of the BET theory to adsorption on heterogeneous solids, the influence of surface heterogeneity on the second and the higher adlayers is neglected. Usually, the lateral interactions are also neglected; therefore, in the methods presented here, the geometrical distribution of adsorption sites may be arbitrary.

In the first articles connected with the considered extensions of the BET isotherm, discrete energy distributions were used [90,91]. The later studies, based on the integral equation (10), have a more universal character [6]. For multilayer adsorption, this equation can be easily solved in two cases [67]: (1) when the local adsorption isotherm is a product of the monolayer coverage and a function describing the formation of higher adsorbed layers, namely

$$\Theta^M = g_1(h)\Theta(h, \varepsilon) \tag{23}$$

where $h = p/p_s$ is the relative pressure and p_s denotes the saturation vapor pressure and (2) when Θ^M is a sum of the monolayer coverage and the term describing multilayer effects in adsorption,

$$\Theta^M = g_2(h) + \Theta(h, \varepsilon) \tag{24}$$

After simple algebra from the integral equation (10) and the isotherm (22) or (24), we obtain

$$\Theta_t^M = g_1(h) \int_\Delta \Theta(h, \varepsilon)\chi(\varepsilon)\, d\varepsilon = g_1(h)\Theta_t(h) \tag{25}$$

or

$$\Theta_t^M = g_2(h) \int_\Delta \Theta(h, \varepsilon)\chi(\varepsilon)\, d\varepsilon = g_2(h) + \Theta_t(h) \tag{26}$$

where $\Theta_t^M(h)$ denotes an arbitrary overall isotherm for monolayer adsorption. This presents an elegant way of generalizing any monolayer adsorption on a heterogeneous surface to its multilayer counterpart. The functions $g_1(h)$ and $g_2(h)$, for various local multilayer isotherms, are collected in Ref. 6.

Let us focus our attention on the most general form of the BET equation, which may be rewritten in the form of Eq. (23) as the product of the function describing the monolayer adsorption on a homogeneous surface

$$\Theta(h, \varepsilon) = \frac{CH(h)}{1 + CH(h)} \tag{27}$$

with

$$H(h) = \frac{ah[1 - (ah)^l]}{1 - ah}, \quad a \in (0, 1) \tag{28}$$

and the function connected with the multilayer formation

$$g_1(h) = \frac{d \ln H(h)}{d \ln (ah)} \tag{29}$$

where l is the number of adsorbed layers and a is a constant from the interval $(0,1)$ characterizing the attractive force field of the adsorbent. From eqs. (23) and (27)–(29), one may obtain: (1) the customary BET equation ($a = 1$, $l = \infty$) [36], (2) the BET isotherm with a finite number of adlayers ($a = 1$, $1 < l < \infty$) [36,92], and (3) the modified BET equation [93] ($l = \infty$).

The overall isotherms for the local BET-type adsorption are represented by Eq. (25). These isotherms may be obtained from equations generated by Langmuir local behavior [Eq. (15) or (16)] in the following way: In the monolayer isotherms $\Theta_t(p)$, one replaces the pressure p by the functions $H(h)$, and the parameters K^* and p_a by $C^* = K^* p_s$ and $p'_a = p_a/p_s$, respectively, and then multiplies the redefined isotherm by the function $g_1(h)$ [5].

The BET model has been generalized by using of various monolayer isotherms for heterogeneous surfaces: Langmuir–Freundlich [94], Tóth [95], generalized Freundlich (GF) [94], Dubinin–Radushkevich [67,95,96], and others [5]. These equations have been applied to the interpretation of experimental data [5,6]. The above-discussed procedure has also extended to the adsorption with lateral interactions on randomly heterogeneous surfaces also [5].

Apart from analytical solutions of the integral equation for multilayer adsorption, several numerical results have been discussed [54,97–101].

Ending this section, we should mention the alternative theories of multilayer adsorption on heterogeneous surfaces, which involve the influence of adsorbent heterogeneity on the state of admolecules in the higher adlayers [102–104].

3. Methods of Evaluating the Adsorption Energy Distribution

(a) Exact Methods.   Over the last 50 years, efforts of numerous scientists have been focused on attempts to extract information about the energy distribution function from empirical adsorption data. In the literature, one can find various articles connected with this problem and many computer programs, which may be used to characterize the solid surfaces. Because of space limitation, the present discussion will be restricted to the most popular methods of evaluating the energy distribution. This subsection serves only as a guide for readers seeking more detailed information in the cited references.

The methods available for the evaluation of energy distribution functions can be divided into two main classes. In the procedures of the first type, a general form of the distribution function is assumed and the parameters are calculated from the experimental data. For other classes, no a priori assumption is made about the shape of the energy distribution.

Numerous exact and approximate methods have been proposed to solve the integral equation (10) with Langmuir local adsorption isotherm. Then, an application of a given analytical equation to represent the overall adsorption isotherm determines automatically the form of energy distribution function, the parameters of which are obtained by fitting the isotherm to experimental data [6].

One of the most effective methods of evaluation of the energy distribution function $\chi(\varepsilon)$ relating to the overall adsorption isotherm assumed a priori was proposed by Sips [21,22]. He proved that the integral equation (10) with the Langmuir local isotherm [Eq. (11)] could be rewritten as the Stieltjes transform [105]:

$$f_2(t) = \frac{\int_0^\infty f_1(s)\, ds}{s + t} \tag{30}$$

If the function $f_2(t)$ is known one can find the function $f_1(s)$ from the relationship

$$f_1(s) = (2\pi i)^{-1} [f_2(se^{-\pi i}) - f_2(se^{\pi i})] \tag{31}$$

where i is the imaginary factor. The original procedure of Sips [21] was widely discussed; the controversy was connected with the integration region used in Eq. (10) [5]. The method was later modified [22,79,106,107] and it was the basis for numerous studies. A detailed discussion of various versions of this procedure can be found in Ref. 5.

Two different, exact methods have been also proposed to solve the integral equation (10) with respect to the energy distribution function, namely the Landman–Montrol technique [108] and the Jagiello method [9,109–112].

(b) Approximate Methods. Sometimes, an approximate method should be used to solve the integral equation (10). One of the most popular procedures is based on the so-called "condensation approximation" [66–68,78,113–117]. The spirit of the method is the replacement of the true local isotherm by a simple step function

$$\Theta^C(p) = \begin{cases} 0 & \text{for } 0 \leq p < p^C(\varepsilon) \\ 1 & \text{for } p^C \leq p \end{cases} \tag{32}$$

where $p^C(\varepsilon)$ denotes the gas pressure at which all parts of the surface characterized by a given value of adsorption energy are instantaneously covered by adsorbate. This "condensation pressure" may be evaluated for any continuous local isotherms using the method of Harris [116,117] or the Cerofolini formula [66,67,117] $\Theta(p^C, \varepsilon) = 0.5$. After introducing the condensation local isotherm to the integral equation (10), we obtain

$$\Theta_t(\varepsilon') = \int_{\varepsilon'}^{\infty} \chi(\varepsilon) \, d\varepsilon \tag{33}$$

where ε' is the lowest value of adsorption energy for condensation to occur at pressure p. This value may be obtained by inverting the function $p^C(\varepsilon)$. The useful collection of the functions $\varepsilon'(p^C)$ for different adsorption models can be found in Refs. 5 and 6.

Differentiation of Eq. (33) with respect to ε' gives the energy distribution function

$$\chi^C(\varepsilon') = \frac{\partial \Theta_t(\varepsilon')}{\partial(\varepsilon')} \tag{34}$$

The above relationship may be used either when an analytical form of the overall adsorption isotherm is known or for experimental data that are a set of pairs (p, Θ_t). Frequently, the condensation approximation method has been applied to determine the energy distribution function [5,6,9]. We can point out two main advantages of the method: its simplicity and its universality. The numerical computation of the energy distribution function is very easy. Moreover, this method may be used for different local isotherms. The shape of the distribution function is independent of the form of the local isotherm, which changes only the position of the distribution on the energy axis. However, the most important limitation is that the adsorption energy defined in the condensation–approximation method is exact at $T = 0$, so the procedure may be used for the analysis of low-temperature adsorption isotherms only, because the accuracy of the evaluation energy distribution function becomes satisfactory in this region.

The various efforts to improve the effectiveness of the condensation–approximation method were made [5,6,9]. A more exact solution of the integral equation gives the asymptotically correct approximation" method, developed by Hobson [118] for mobile adsorption and later refined by Cerofolini [66] for localized adsorption. In this treatment, the local isotherm is assumed to be a combination of a linear and a condensation isotherm. Hsu et al. [119] and Rudziński et al. [109,110,120] adapted the Sommerfeld expansion method [121] to the solution of the integral equation in question. Although numerous modifications of the condensation–approximation method are known, all improvements to this method make it more complicated and introduce additional numerical problems, but they do not change its approximate nature. Therefore, many authors prefer the original condensation method rather than its modifications [5].

(c) Numerical Methods.   For numerous distribution functions, the integral equation (10) does not give an analytical expression for representing the overall adsorption isotherm. Then, various numerical methods are used. The parameters that lead to the best agreement of the calculated and experimental isotherms are determined by means of different optimization procedures. One of the most known methods is that presented by Ross and Olivier [1]. They proposed a numerical algorithm to calculate the Gaussian distribution function from the experimental adsorption isotherm. The method is based on the assumption that the solid surface has a patchwise topography and adsorption is mobile so that adsorption on the individual patches may be described by the Hill–de Boer equation. Ross and Olivier [1] applied their method to relatively homogeneous adsorbents, such as heat-treated carbon black, boron nitride, and rutile, which can be characterized by a single-peak Gaussian distribution function. However, a different distribution may be also used to study the heterogeneity effects. Here, we mention only several most interesting examples. House and Jaycock [122] assumed that the surface might be represented by a sum of two Gaussian-type distributions, Hory and Prausnitz [123] used a log-normal function, Kindl et al. [124,125] utilized the Maxwell–Boltzmann distribution, whereas Van Dongen [98] applied an exponential higher-degree polymonial as the energy distribution function.

From a practical point of view, the methods which do not introduce assumptions connected with the shape of energy distribution function are even more interesting than those discussed in the previous paragraph. The advanced numerical procedures used to derive the form of the energy distribution function from the measured adsorption isotherms are of great importance for modern investigations of solid surfaces. We presented a brief discussion of the most popular algorithms: HILDA, ALINDA, CAEDMON, CESAR, EDCAIS, and EM.

The HILDA method developed by House and Jaycock [100] may be considered a modified, numerical version of the iterative procedure proposed by Adamson and Ling [126]. An excellent short presentations of the method can be found in the review by House [127] or in the monograph by Rudziński and Everett [6]. This procedure can be outlined as follows: The form of local isotherm is assumed and the distribution function is evaluated by using the iterative routine; for each iterative step appropriate adjustments in distribution are made to bring the calculated and experimental isotherms into the best possible coincidence; the condensation–approximation is used to determine the first approximation of the distribution. The Adamson–Ling method was widely applied to evaluate the energy distribution function from the measured adsorption isotherm [97,122,128–135].

The improvement of the HILDA algorithm in comparison to the above method is connected with the following features: evaluation of the monolayer capacity by systematic normalization of $\chi(\varepsilon)$, application of an odd-ordered quadratic smoothing routine for the adsorption isotherm and distribution function, and evaluation of $\chi(\varepsilon)$ in a finite region of adsorption energies corresponding to the measured region of the adsorption isotherm [127].

The HILDA procedure offers a choice among four local isotherms: Langmuir, Fowler–Guggenheim, Hill–de Boer, and the virial isotherm. This method was employed to determine the distribution function by numerous authors for the study of a variety of solids [39,54,55, 100,127,136–141].

An interesting development of HILDA is due to Koopal and Vos [142], who proposed a modified algorithm called ALINDA (Adamson–Ling Distribution Analysis). The procedures in question differ only with respect to subtle, numerical details [5]. The smoothing routine used in ALINDA is a little better than that implemented in the HILDA program.

Apart from the original method mentioned above, Morrison and co-workers [143,144] formulated a new iterative technique called CAEDMON (Computed Adsorption Energy Distribution in the Monolayer) for the evaluation of the energy distribution from adsorption data without any a priori assumption about the shape of this function. In this case, the local adsorption is calculated numerically from the two-dimensional virial equation. The problem is to find a discrete distribution function that gives the best agreement between the experimental data and calculated isotherms. In this order, the optimization procedure devised for the solution of non-negative constrained least-squares problems is used [145]. The CAEDMON algorithm was applied to evaluate $\chi(\varepsilon)$ for several adsorption systems [137,140,146,147]. Wesson et al. [147] used this procedure to estimate the specific surface area of adsorbents.

A similar character has CESAR (Computed Adsorption Energies, SVD analysis) algorithm prepared by Koopal and Vos [142,148]. This procedure may be used for the same local isotherm as in the HILDA program. The procedure has been carefully tested by using the model adsorption isotherm generated for a bi-Gaussian distribution. It has been shown that the CESAR algorithm can reproduce the original distribution with excellent accuracy [148].

The EDCAIS (Energy Distribution Computation from Isotherms utilizing Spline function) algorithm was proposed by the Brauer and co-workers [149,150]. They used the BET and Fowler–Guggenheim equations in the Langmuirian form as the local isotherm. The energy distribution function is obtained by using of the inverse Stieltjes transform [105]. The essential point in the method is the application of various kinds of spline function for approximating the experimental monolayer coverage. The procedure takes into account the errors in adsorption measurements. The algorithm was tested for the model adsorption isotherms generated for different types of energy distribution function [149]. The authors proved that simple energy distributions were excellently reproduced, whereas for functions with three and five Gaussian peaks, the agreement between the original and calculated distribution was less satisfactory but still quite good.

Stanley and Guichon [151] have recently proposed the expectation–maximization (EM) method for numerical estimation of adsorption energy distributions. This method does not require prior knowledge of the distribution function or any analytical equation for the total isotherm. Moreover, it requires no smoothing of the adsorption isotherm data and coverages with high stability toward the maximum-likelihood estimate.

In the solution of Eq. (10), the distribution function $\chi(\varepsilon)$ is evaluated directly from the data at M^* grid points in energy space. Equation (10) is rewritten as

$$\Theta_t(p_j) = \sum_{\varepsilon_{\min}}^{\varepsilon_{\max}} \Theta(p_j, \varepsilon_i)\chi(\varepsilon_i)\Delta\varepsilon_i \tag{35}$$

where $\Delta\varepsilon_i$ is the grid spacing around ε_i. The distribution function is updated interactively with the correction step

$$\chi^{k+1}(\varepsilon_i) = \chi^k(\varepsilon_i) \sum_{p_{\min}}^{p_{\max}} \Theta(p_j, \varepsilon_i)\Delta\varepsilon_i \frac{\Theta_{t,\exp}(p_j)}{\Theta_{t,\mathrm{cal}}(p_j)} \tag{36}$$

where k is the iteration number, $\Theta_{t,\exp}$ is the experimental data, and $\Theta_{t,\mathrm{cal}}$ is the data estimated at iteration k. The correction vector $\Theta_{t,\exp}/\Theta_{t,\mathrm{cal}}$ is reconvoluted with the model, $\Theta_{t,\mathrm{cal}}$, before the previous estimate is updated. The correction is normalized by dividing the sum over pressure points in Eq. (36) by $\sum_p \Theta(p_j, \varepsilon_i)$.

The apparent monolayer capacity is obtained at covergence by integrating the final distribution function. As the first approximation, the "total ignorance" guess is used:

$$\frac{\chi(\varepsilon_i) = \Theta_{t,\exp}(p_N)}{M^*} \tag{37}$$

where the total amplitude of adsorption observed is divided evenly among all energy points taken into account. This estimate guards against any possible experimental artifacts and introduces minimum bias into the calculated adsorption energy distribution function.

The EM algorithm is warranted to converge to the global optimum at every interaction step for Poisson- and Gaussian-distributed data, and negative values are impossible because positive multiplicative correction is always applied. Oscillatory behavior of the solution is not a problem as long as the initial estimate is plausible. Stanley and Guichon [151] have compared the results obtained by using the HILDA algorithm with those evaluated according to the EM method. They have evidently proved that the latter is superior in terms of robustness, accuracy, and information theory.

The expectation–maximization method has been recently applied to characterize peat soils. Figure 3 presents the examples of adsorption energy distribution functions together with the plots of nitrogen adsorption isotherms on various samples of soils [152]. The original soil samples were first thermally dried at different temperatures and then outgassed in a vacuum at the temperature of the previous thermal treatment. In such a way, the adsorptive properties of the soils were changed. This is clearly visible in Fig. 4. The energy distribution functions evaluated for samples modified at different temperatures have different shapes. The theoretical approach used to analyze these experimental data involved both fractal scaling and energetic heterogeneity [152]. The EM is an elegant, fast, and effective method which can be used to investigate complex, natural materials.

A large number of studies have dealt with the applications of numerical methods for the determination of energy distribution functions for selective adsorption systems. These studies have focused on the following topics:

The comparison of the energy distribution functions obtained by means of different numerical algorithms [5,47,49,97,98,127,139,142,144,153–162] and techniques

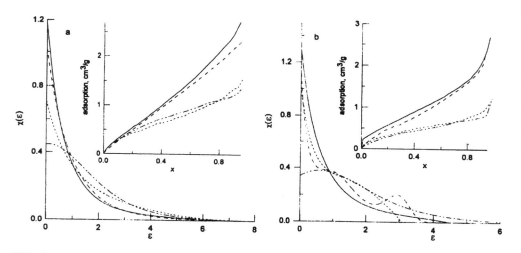

FIG. 4 Energy distributions $\chi(\varepsilon)$ and adsorption isotherms (the inset). Solid, long-dashed, short-dashed, and long-short dashed lines are for nonmodified and modified soil samples at 323, 373, and 423 K, respectively. $x = p/p_s$. The physical properties of the muck soil: sample a (b): P density, 0.25 (0 34) g/cm^3; total porosity, 84.6 (81.4) vol%; pH in H_2O, 5.48 (5.44); pH in 1 N KCl, 5.18 (4.97); P_2O_5: 60.0 (34.0) mg/g; K_2O: 19.3 (15.7) mg/g; Mg, 60.0 (21.0) mg/g; NH$_4$, 1.07 (2.05) mg/g N in NO$_3$, 37.76 (18.52) mg/g. (From Ref. 152.)

such as calorimetric measurements [110,153,163] and chromatographic experiments [5,164]

The physicochemical interpretation of the energy distribution functions evaluated from adsorption isotherms; the structural aspects such as the question of which functional groups on the surface are connected with the particular peaks on the energy distribution [39,140,162]

The sensitivity of the energy distribution function to random errors in the isotherm [5,39,162]

The monitoring of changes in energy distribution functions when the surface is subjected to a special treatment (e.g. heating, aging, annealing, or chemical modification) [39,54,55,136,137,140,142–144,150,165].

Numerous adsorption systems investigated by means of advanced numerical algorithms in order to evaluate the energy distribution function are collected in Ref. 5. The wide, critical discussions of various attempts to characterize adsorptive properties of solid surfaces are presented in Refs. 5–7, 37, and 127.

C. Adsorption of Gas Mixtures

1. General Integral Equation for Mixed-Gas Adsorption

The adsorption of gas mixtures on solid surfaces underlies a number of the processes of utilitarian significance. In spite of the great progress in the improvement of new techniques for adsorption measurements, the experiments for gas mixtures are still more time-consuming than for single gases. In this situation, predicting mixed-gas equilibria by using information extracted from single-gas isotherms is the fundamental task for theoreticians dealing with the problem in question. In the studies focused on practical applications, relatively simple solutions are always preferred. On the other hand, it is quite apparent that mixed-gas adsorption is a very complex phenomenon and its theoretical description should involve many molecular parameters characterizing the system. For all of these reasons, our knowledge of mixed-gas adsorption mechanism is still fragmentary. However, several reviews dealing with mixed-gas adsorption are available in the literature [5,8,9,38,166–168]. Here, we will focus our attention on the integral approach that has been frequently used in the hitherto theoretical studies of mixed-gas adsorption.

In the early stage of theories of the adsorption process, the majority of authors assumed the energetic homogeneity of adsorbent surface [5]. One of the first important achievements in the theory of mixed-gas adsorption has been the extension of the Langmuir isotherm [169]. Let the symbol $\Theta_{i(\mathbf{n})}$ denote the partial relative monolayer adsorption of the ith component from an n-component gas mixture on a homogeneous surface. The subscript $\mathbf{n} = (1, 2, \ldots, n)$ refers to the n-component gas mixture. According to the Langmuir model, the individual adsorption isotherm has the following form:

$$\Theta_{i(\mathbf{n})} = \frac{K_i p_i}{1 + \sum_{j=1}^{n} K_j p_j} \tag{38}$$

where p_i is the partial pressure of the ith component and K_i denotes its Langmuir constant defined for the single gas. The total adsorption isotherm may be expressed in the quasi-Langmuirian form [170]

$$\Theta_{(\mathbf{n})} = \sum_{i=1}^{n} \Theta_{i(\mathbf{n})} = \frac{\mathbf{K} \cdot \mathbf{p}}{1 + \mathbf{K} \cdot \mathbf{p}} \tag{39}$$

where $\mathbf{p} = (p_1, p_2, \ldots, p_n)$ and $\mathbf{K} = (K_1, K_2, \ldots, K_n)$ are the vectors of partial pressures and Langmuir constants, respectively; the symbol $\bullet$ denotes a scalar product.

Some authors have extended the above adsorption model by including lateral, attractive interactions between admolecules, the multilayer nature of the surface film, and adsorption mobility (see Refs. 5,9,167,168 and references therein).

Nowadays, the heterogeneity of the solid surface is almost commonly taken into account in the articles devoted to mixed-gas adsorption. The majority of these studies are based on the ideas and concepts which will be discussed next [5,8,9,167].

In the case of adsorption from an n-component mixture, each type of adsorption site should be characterized by the vector of adsorption energies of all components $\boldsymbol{\epsilon} = (\varepsilon_1, \varepsilon_2, \ldots, \varepsilon_n)$. Therefore, Jaroniec [171] has proposed the n-dimensional energy distribution function $\chi_{(\mathbf{n})}(\boldsymbol{\epsilon})$ to characterize the surface heterogeneity in the considered system:

$$\Theta_{(\mathbf{n})t} = \int_{\Delta_\mathbf{n}} \Theta_{(\mathbf{n})}(\mathbf{p}, \boldsymbol{\epsilon}) \chi_{(\mathbf{n})}(\boldsymbol{\epsilon}) \, d\boldsymbol{\epsilon} \tag{40}$$

where $\chi_{(\mathbf{n})}(\boldsymbol{\epsilon})$ satisfies the normalization condition

$$\int_{\Delta_\mathbf{n}} \chi_{(\mathbf{n})}(\boldsymbol{\epsilon}) \, d\boldsymbol{\epsilon} = 1 \tag{41}$$

In the above, $\Delta_\mathbf{n}$ is the n-dimensional integration region.

Analytical solutions of Eq. (40) are unknown; however, Jaroniec et al. [172,173] presented numerical solutions of the general integral equation (40) for a binary gas mixture and for Gaussian and log-normal distributions.

2. Simplified Integral Equation for Mixed-Gas Adsorption

The integral equation (10) can be transformed to more convenient mathematical forms; a general strategy is to reduce the n-dimensional integral to a one-dimensional integral by using various physically realistic simplifications. Most often, it is done by introducing the special assumptions about correlation between energies of different components. Two particularly interesting cases should be discussed: (1) linear correlation and (2) a lack of correlation.

The assumption of a linear correlation between adsorption energies allows one to find the analytical adsorption isotherms from gas mixtures on heterogeneous surfaces. Generally, when we can express the adsorption energies of all components as functions of the adsorption energy of a reference component (e.g., the first component),

$$\varepsilon_i = \varepsilon_i(\varepsilon_1) \quad \text{for } i = 1, 2, \ldots, n \tag{42}$$

the general integral equation (10) may be transformed to the following single integral [174–176]:

$$\Theta_{(\mathbf{n})t} = \int_{\Delta_1} \tilde{\Theta}_{(\mathbf{n})}(p, \varepsilon_1) \chi_1(\varepsilon_1) \, d\varepsilon_1 \tag{43}$$

where

$$\chi_1(\varepsilon_1) = \int_{\Delta_2} \cdots \int_{\Delta_n} \chi_{(\mathbf{n})}(\boldsymbol{\epsilon}) \, d\varepsilon_2 \cdots d\varepsilon_n \tag{44}$$

and

$$\tilde{\Theta}_{(\mathbf{n})} = \Theta_{(\mathbf{n})}(p, \varepsilon_1, \varepsilon_2(\varepsilon_1), \ldots, \varepsilon_n(\varepsilon_1))$$

Because the simplified equation (43) is considerably simpler than the general integral equation (10), it has been used to obtain the analytical adsorption isotherms for mixed-gas adsorption [174–185]. In pioneering works of Roginsky and Todes [178] and Glueckauf [179], the linear relation between adsorption energies was assumed. However, the most intensive studies were performed for the case when the difference between the adsorption energies ε_1 and ε_i is constant and independent of the type of adsorption site; that is

$$\varepsilon_i = \varepsilon_1 + \Delta_{i1} \tag{45}$$

where Δ_{i1} is a constant ($i = 1, 2, \ldots, n - 1$)

Obviously, Eq. (45) is a special case of the general dependence (42). From a physical point of view, assumption (45) seem to be quite realistic because it means that all components analogously interact with the surface and their adsorption energies vary in the same direction—they are high for the strong adsorption sites and become low for the weak active centers. As a consequence, the energy distributions for single-gas systems have the same shapes but are shifted on the energy axis. In other words, the heterogeneity parameters of these distributions are equal.

The simplified integral (43) for the local isotherm (39) may be expressed in the following form [174]:

$$\Theta_{(n)t} = \int_{\Delta_1} \frac{K_1 y}{1 + K_1 y} \chi_1(\varepsilon_1) \, d\varepsilon_1 \tag{46}$$

where

$$y = p_1 + \sum_{i=2}^{n} K_{i1} p_i \tag{47}$$

and

$$K_{i1} = \frac{K_i}{K_1} = (\alpha_{i1}) \exp\left(\frac{\Delta_{i1}}{k_B T}\right) \tag{48}$$

for $\alpha_{i1} = \alpha_i/\alpha_1$ ($i = 2, 3, \ldots, n - 1$). When the condition (45) is fulfilled, the variable y does not depend on the type of adsorption type; it is a function of the temperature and the partial pressures only. Thus, the simplified adsorption isotherm (46) has, formally, the same form as the original integral equation for the single gas [Eq. (10)]. It means that mixed-gas adsorption isotherms may be obtained from the single-gas equations discussed in Section III.B.2.a. One can repeat, somewhat modified, the scheme of evaluation used for adsorption of single gases with lateral interactions. The total adsorption for mixed-gas adsorption on heterogeneous solids may be obtained from expressions derived for gas adsorption by replacement of the pressure p by the variable y and the constant K by the constant K_1.

When the total adsorption isotherm is known, the partial adsorption isotherms can be calculated from the following relation:

$$\Theta_{(n)t} = \left(\frac{K_{i1} p_i}{y}\right) \Theta_{(n)t} \quad \text{for} \quad i = 1, 2, \ldots, n \tag{49}$$

The generalization of Marczewski–Jaroniec equation (15) for mixed-gas adsorption leads to the following expression:

$$\Theta_{(n)t} = \left(\frac{(K_1^* y)^m}{1 + (K_1^* p)^m}\right)^{q/m} \tag{50}$$

For certain values of the heterogeneity parameters (m, q), the Langmuir–Freundlich, general Freundlich (GF), and Tóth equations for gas mixtures have been derived and verified experimentally by numerous authors [175,176,182,186,187].

In the same way, the Dubinin–Astakhov isotherm has been extended to the adsorption of gas mixtures [176]:

$$\Theta_t = \exp\left\{-B_j\left[k_B T \ln\left(\frac{p_{a_1}}{y}\right)^j\right]\right\} \quad \text{for } y \le p_{a1} \tag{51}$$

where B_j is the heterogeneity parameter and p_{a1} is a constant defined for the first component. From Eq. (51), we can derive the adsorption isotherms of the Freundlich and the Dubinin–Radushkevich types [176].

When adsorption energies of components are completely independent (i.e., in the case when the correlation coefficient is equal to zero), the n-dimensional energy distribution is a product of n one-dimensional distributions $\chi_i(\varepsilon_i)$ characterizing the surface with respect to a given single gas:

$$\chi_{(\mathbf{n})}(\boldsymbol{\epsilon}) = \prod_{i=1}^{n} \chi_i(\varepsilon_i) \tag{52}$$

However, Eq. (10) for the local adsorption isotherm of the Langmuir type and the physically realistic distributions $\chi_i(\varepsilon_i)$ is still difficult to solve. It is worth mentioning that valuable results have been obtained for the Jovanovich-type local isotherm [171]. The assumption that the adsorption energies are not correlated at all has been used also to derive the Langmuir–Freundlich-type isotherm for the overall adsorption [174]:

$$\Theta_{i(\mathbf{n})} = \frac{(K_i^* p_i)^{m_i}}{1 + \sum_{j=1}^{n}(K_j^* p_j)^{m_j}} \tag{53}$$

where $m_i(T)$ is the heterogeneity parameter for single-gas adsorption.

The alternative approach to the adsorption of gas mixtures has been proposed by Jaroniec [188], who utilized the earlier kinetic considerations of Crickmore and Wojciechowski [189]. As the result, he proposed the following linear relation:

$$\ln\left(\frac{\Theta_{i(\mathbf{n})}}{\Theta_{j(\mathbf{n})}}\right) = m \ln\left(\frac{K_i^*}{K_j^*}\right) + m \ln\left(\frac{p_i}{p_j}\right) \tag{54}$$

which may be simply applied to evaluate the heterogeneity parameter and the adsorption constants [5,87,188,190]. The latter equation may be obtained from the isotherm (53) by setting $m = m_i$ for $i = 1, 2, \ldots, n$.

Figure 5 presents the linear dependences [Eq. (54)] for binary mixtures of hydrocarbons on Nuxit-AL charcoal [191]. The experimental data have been measured by Szepesy and Illes [192]. The solid lines in Fig. 5 correspond to the mixtures with ethylene, and the dashed lines refer to the mixtures with ethane. It is clearly visible that Eq. (54) approximated the experimental points excellently. In the case of the mixtures in question, the heterogeneity parameter m is equal to 0.9 for alkanes and 1 for alkenes. Further studies have shown that Eq. (54) gives a satisfactory representation of adsorption from hydrocarbons mixtures on polystyrene, silica gel, and various activated carbons [5,190,191].

The above-derived equations have been extended by the inclusion of lateral interactions in the mixed monolayers [193] formed on surfaces with the random distribution of adsorption sites

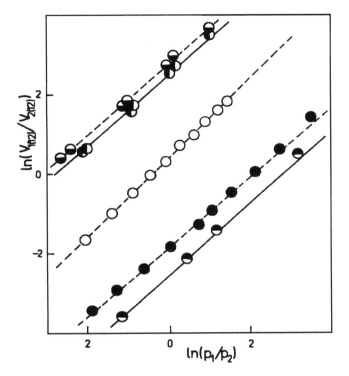

FIG. 5   The linear dependences [Eq. (54)] for the adsorption of binary mixtures of hydrocarbons on active carbon (Nuxit-AL) at 293 K: ethylene–methane (right semiclosed circles), ethylene–propylene (top semiclosed circles), ethane–methane (bottom semiclosed circles), ethane–ethylene (open circles), and ethane–propane (closed circles). The symbol $i(12)$ denotes the partial adsorbed amount of the ith component. [Reprinted from M. Jaroniec, Adsorption from multicomponent gas mixtures on solid surfaces. Thin Solid Films, *71*:273–304 (1980), with permission from Elsevier Science.]

[5,8,9,167]. The essence of the procedure is to replace in the adsorption isotherms (50) and (51), the partial pressures p_i by the variables $p_i^* = p_i\gamma_i$, where γ_i are the functions describing average molecular interaction effects in the surface films. Analogously, the kinetic equation (53) has been generalized [87]. The effect of multilayer adsorption in mixed-gas adsorption on heterogeneous surfaces has been taken into account (see Refs. 5 and 9 and references therein).

Experimental studies of adsorption of n-component gas mixture require the evaluation of a large number of parameters (strictly speaking, $2n + 1$ parameters), n partial pressures, the surface-phase composition (i.e., n relative coverages), and temperature. The measurements are time-consuming and an analysis of these data is a difficult task. Therefore, we try to limit the number of variables by introducing additional constrains to the studied systems. For this reason, the adsorption isotherms are frequently measured only for selected experimental points which correspond to one or more fixed variables. Under such conditions, it is possible to derive much less complicated expressions for adsorption isotherms. Numerous studies of mixed-gas adsorption have been performed for the following parameters held constant [5]: the total pressure [184], the partial pressure of a selected component [194], the mole fraction of components in the gaseous phase [176], the adsorbed amount of a selected component [50], and the mole fractions of components in the surface phase [195].

D. Adsorption from Solution of Nonelectrolytes

1. Introductory Remarks

Adsorption from solutions plays an important role in a number of applications in technology and poses many challenging scientific problems. There are a great variety of liquid mixtures with extremely different properties. In this section, the discussion will be limited to adsorption from ideal solutions because our main purpose is to analyze the heterogeneity effects in adsorption from liquid mixtures, which are particularly visible in such systems. Moreover, we want to show analogies between theories of adsorption from gaseous and liquid phases on heterogeneous solids. On the other hand, it should be pointed out that molecular interactions in liquid mixture have a fundamental significance for the adsorption equilibrium. In real adsorption systems, we observe an interesting interplay between interactions in the bulk and interfacial regions. In the more subtle approaches, the solution nonideality should be simultaneously considered. Adsorption from solution has been a subject of numerous articles, so the interested reader is advised to consult already available reviews [4,5,8,196].

In this chapter, we can answer a general question of how the adsorbent heterogeneity affects properties of fluid near the surface. Many studies showed conclusive evidence that the behaviors of gases and liquids are considerably different. The presence of a solid surface causes a change in the density of fluid at the solid–fluid interface in comparison to the bulk fluid density. This change may be significant for gases, which, in a wide pressure region, do not form a complete layer on the solid, and still unoccupied adsorption sites can attach molecules from the gaseous phase. Such a process, sometimes called "pure adsorption," depends mainly on the adsorption energies of gases. For a gas–solid interface, the considered change in fluid density is the most important feature of the adsorption systems. However, liquids cover completely wettable surfaces, moreover, the compressibility of liquids is small. Therefore, the change in fluid density at the interfacial region is slight. The solutions consist of at least two components, which have different adsorption properties. Because there are no vacancies in the surface film, molecules of the component with the higher adsorption energy displace molecules of the component with the lower adsorption energy. This process is stronger for the large difference of adsorption energies of components. In this case, we can say about "competitive adsorption," which is characteristic for liquid–solid interface. For sufficiently high pressures, analogous effects may appear in mixed-gas adsorption.

The primary experimental quantity characterizing the adsorption equilibrium is the excess adsorption, which may be expressed as

$$N_i^e = N_s(x_i^s - x_i^l) \tag{55}$$

where x_i^ρ denotes the mole fraction of the ith component in the ρth phase ($\rho = l, s$), the subscript l refers to the liquid phase and s is connected with the surface film; N_s is the total number of molecules of all components in the surface phase (a monolayer capacity). When all components have the same molecular size, the total number of moles in the surface phase is constant, independent of the solution composition. It is easy to show that the sum of excesses N_i^e calculated for all components is equal to zero.

Obviously, the excess adsorption is measured for both liquid–solid and gas–solid interfaces. However, in the experiments carried out for the gaseous phase, the concentration of the adsorbate in the bulk phase is usually negligible in comparison with that in the surface phase, so the excess adsorption is assumed to be equal to the absolute adsorbed amount.

E. Adsorption from Binary Solutions

The majority of articles devoted to physical adsorption at the liquid–solid interface concerns adsorption from binary solutions. The Langmuir-type model for this process has been discussed by Everett in terms of statistical thermodynamics [197]. He has proposed the well-known equation of the adsorption isotherm:

$$x_1^s = \frac{K_{12} x_{12}^l}{1 + K_{12} x_{12}^l} \tag{56}$$

where $x_{12}^l = x_1^l / x_2^l$ and K_{12} is the ratio of the Langmuir constants

$$K_{12} = a_{12} \exp\left(\frac{\varepsilon_{12}}{k_B T}\right) \tag{57}$$

where $\varepsilon_{12} = \varepsilon_1 - \varepsilon_2$ is the difference of adsorption energies and $\alpha_{12} = \alpha_1 / \alpha_2$. It is clearly visible from Eq. (57) that adsorption from binary solutions on the homogeneous surfaces depends solely on the difference in adsorption energies of components.

Let us consider adsorption from an ideal binary mixture on the heterogeneous surface. According to the treatment developed for mixed-gas adsorption, each active site is characterized by adsorption energies of both components (ε_1 and ε_2) and the global heterogeneity is described by the two-dimensional distribution function $\chi_{(1,2)}(\varepsilon_1, \varepsilon_2)$. The total mole fraction of the first component can be expressed by the integral equation [198,199]

$$x_{1,t}^s = \int_{\Delta_1} \int_{\Delta_2} x_1^s(x_1^l, \varepsilon_1, \varepsilon_2) \chi_{(1,2)}(\varepsilon_1, \varepsilon_2) \, d\varepsilon_1 \, d\varepsilon_2 \tag{58}$$

where x_1^s is the local adsorption isotherm and Δ_i denotes the integration region ($i = 1, 2$). When the Everett equation (56) is applied as the local adsorption isotherm, the latter integral may be rewritten as

$$x_{1,t}^s = \int_{\Delta_{12}'} \frac{K_{12} x_{12}^l}{1 + K_{12} x_{12}^l} \tilde{\chi}(\varepsilon_{12}) \, d\varepsilon_{12} \tag{59}$$

where the function $\tilde{\chi}(\varepsilon_{12})$ is defined as

$$\tilde{\chi}(\varepsilon_{12}) = \int_{\Delta_2} \chi_{(1,2)}(\varepsilon_{12} + \varepsilon_1, \varepsilon_2) \, d\varepsilon_2 \tag{60}$$

In the above approach, each adsorption site is characterized by the difference in adsorption energies, and the global heterogeneity is described by means of the one-dimensional energy distribution $\tilde{\chi}(\varepsilon_{12})$. The important role of the difference in adsorption energies follows from a competitive mechanism of the process. The physical sense of the notion "homogeneous surfaces" required some comments for the liquid–solid interface. In light of our definition (Section II) for a homogeneous surface, the adsorption energy of a given component is identical for all adsorption sites. Let us consider the difference in adsorption energies (ε_{12}). This difference is constant for the whole surface when (1) the adsorption energies of both components are the same for all active centers or (2) the distribution functions characterizing adsorption of single components have the same shape and are shifted along the energy axis only. In the latter case, the adsorbent is heterogeneous with respect to single components 1 and 2. However, for adsorption from their binary mixture (1,2), the energy distribution is the Dirac function, so the adsorbent surface is recognized as homogeneous. In spite of an existence of various adsorption sites, an analysis of the experimental data does not show any heterogeneity effects. The measured adsorption isotherms may be approximated with satisfactory accuracy by the equations derived for homogeneous adsorbents. Similar results are obtained when the difference in adsorption energies depends slightly on the type of adsorption sites. Thus, at the liquid–solid interface, the adsorbent heterogeneity is masked to a certain degree. These conclusions refer to solutions consisting of compounds, which interact with the solid surface in a similar way. However, for liquid mixtures containing completely different substances, the influence of adsorbent heterogeneity on the adsorption equilibrium cannot be neglected.

The integral equation (59) is quite analogous to the origin expression (10) derived for single-gas adsorption. Thus, one can adapt an already known procedure and generalize the equations proposed for gas adsorption to adsorption from binary solutions. In this order, in Eqs. (15) and (16), the pressure p should be replaced by the ratio x_{12}^l and the constant K by the constant K_{12}, that characterizes the position on the energy distribution on the energy axis. For a binary solution, the Marczewski–Jaroniec (MJ) isotherm (15) becomes

$$x_{1,t}^s = \frac{(K_{12}^* x_{12}^l)^m}{[1 + (K_{12}^* x_{12}^l)^m]^{q/m}} \tag{61}$$

The special forms of the MJ isotherm (61) have been frequently used to study adsorption from solution: generalized Freundlich equation [199], Langmuir–Freundlich isotherm [170,200–205], and the Tóth equation [186].

The Dubinin–Astakhov isotherm (18) has also been extended to adsorption from solution [200,206]; moreover, the equations obtained for the values $j = 1$ and $j = 2$ have been applied to interpret experimental data [207–209].

The above-discussed isotherms have been used in systematic investigations of influence the adsorbent heterogeneity on the excess adsorption from ideal solutions [209–213]. This effect is made beautifully clear in the model calculations carried out on the basis of the Marczewski–Jaroniec equation (61) for $K_{12}^* = 1$ (i.e., for $\varepsilon_{12}^* = 0$) [210]. In the case of the homogeneous surface, the excess adsorption is equal to zero in the whole concentration region, but for the heterogeneous surface, the excess adsorption differs from zero even if $K_{12}^* = 1$. Figure 6 shows that the type of energy distribution dictated the course for the excess adsorption isotherm. The contrary behavior of the excess isotherm corresponding to general Freundlich and Tóth equations for the same value of the heterogeneity parameter is caused by the different shapes of energy distribution functions connected with these equations. A very important conclusion is the statement that the adsorbent heterogeneity may be a reason of adsorption azeotropy ($N_1^e = 0$). The first attempt to explain the azeotropic character of the excess isotherms for ideal solutions by means of a concept of the surface heterogeneity was made by Suchowitzky [214]. This problem has been carefully analyzed by Dabrowski and Jaroniec [200].

The theory presented here was the basis for the investigations dealing with numerous problems connected with adsorption from binary solutions: the determination of energy distribution function from the excess adsorption isotherms [154,215], the role of adsorbent heterogeneity and molecular interactions in the adsorption process at liquid–solid interface [202,203,216–219], the influence of the difference in molecular sizes of components on the adsorption equilibrium [212,220,221], the multilayer effects in adsorption from solutions [222–225], the relations between heats of immersion and excess adsorption isotherms [216,219,225], and many other interesting subjects.

1. Adsorption from Multicomponent Solutions

Adsorption from multicomponent solutions is a basis of a number of industrial processes, such as separation, purification, recovery of chemical compounds, and so forth. Numerous practical applications of adsorption from multicomponent solutions require the formulation of theoretical foundations for interpretation of the experimental data. However, the difficulties mentioned

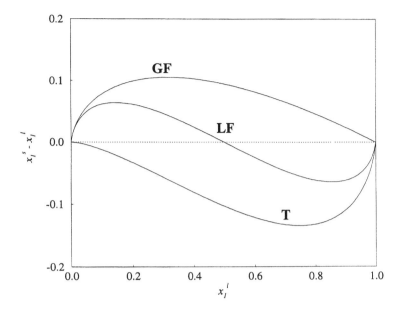

FIG. 6 Excess adsorption isotherms calculated according to Eq. (61): the generalized Freundlich isotherm (GF), Langmuir–Freundlich isotherm (LF), and the Tóth isotherm (T). Parameters: $K_{12}^* = 1$, $q = 0.5$, $m = 1$ (GF), $q = m = 0.5$ (LF), and $m = 0.5$, $q = 1$ (T).

already for mixed-gas adsorption become more pronounced in the case of adsorption from multicomponent liquid mixtures. Therefore, relatively few theoretical papers concerning this problem have been published [5,226,227]. In this subsection, we will focus on the description of the heterogeneity effects in these adsorption systems.

The competitive adsorption from the n-component ideal liquid mixture on an energetically homogeneous solid surface may be represented by a series of the following quasi-chemical reactions:

$$(i)^l + (n)^s \rightleftarrows (i)^s + (n)^l, \quad i = 1, 2, \dots, n - 1 \tag{62}$$

where $(i)^\rho$ denotes a single molecule of the ith component in the ρth phase. Using the mass law for each reaction, one can obtain the adsorption isotherm [228]

$$x_{i(\mathbf{n})}^s = \frac{K_{in} x_{in}^l}{1 + \sum_{j=1}^{n-1} K_{jn} x_{jn}^l} \tag{63}$$

where $K_{in} = K_i / K_n = \alpha_{in} \exp(\varepsilon_{in}/k_B T)$.

In the case of a heterogeneous surface, each adsorption is characterized by a set of adsorption energy differences, $\boldsymbol{\epsilon}_n = (\varepsilon_{1n}, \varepsilon_{2n}, \dots, \varepsilon_{n-1,n})$. The $(n-1)$-dimensional energy distribution $\tilde{\chi}(\boldsymbol{\epsilon}_n)$ fully describes the heterogeneity in adsorption from an n-component solution. The general integral equation has the following form [228]:

$$x_{i(\mathbf{n}),t}^s = \int_{\Delta'_{(\mathbf{n})}} x_1^s(\mathbf{x}_n^l, \boldsymbol{\epsilon}_n) \tilde{\chi}_{(\mathbf{n})}(\boldsymbol{\epsilon}_n) \, d\boldsymbol{\epsilon}_n \tag{64}$$

where $x_1^s(\mathbf{x}_n^l, \boldsymbol{\epsilon}_n)$ is the local adsorption isotherm [e.g., Eq. (63)] and $\mathbf{x}_n^l = (x_{1n}^l, x_{2n}^l, \dots, x_{n-1,n}^l)$.

Let us consider a liquid mixture in which all of the components show a similar type of interaction with the adsorbent surface except that a behavior of the nth component is completely different. In this case, condition (45) is fulfilled for $i = 1, 2, \ldots, n - 1$. For such an adsorbent, the "total adsorption" (adsorption of all components except the reference one) is given by [187]

$$x_{(n),t}^s = \sum_{i=1}^{n-1} x_{i(n),t}^s = \int_{\Delta'_{(n)}} \frac{K_{1n}y^*}{(1 + K_{1n}y^*)} \tilde{\chi}_1(\varepsilon_{1n}) \, d\varepsilon_{1n} \tag{65}$$

where

$$y^* = x_{1n}^l + \sum_{i=2}^{n-1} K_{i1}x_{in}^l \tag{66}$$

The individual isotherm can be calculated from the relation

$$x_{i(n),t}^s = \left(\frac{K_{in}x_{in}^l}{y^*}\right)x_{(n),t}^s \tag{67}$$

Because the above equations are analogous to that proposed for the mixed-gas adsorption, one can apply the identical procedure for generalization of the well-known isotherm discussed in the previous subsections. In this way, the Marczewski–Jaroniec equation (50) and the Dubinin–Astakhov equation (51) isotherms have been extended to adsorption from a multicomponent solution. Similarly, Eq. (54) has been adapted for such systems [229]. These relations were the basis for the interpretation of experimental data. As an interesting example, we present the results obtained by means of the Dubinin–Radushkevich-type equation for adsorption from ternary solutions. In this case, measurements were performed for a constant ratio of mole fractions of two components in the mobile phase. The parameters of the isotherm (the heterogeneity parameter, the parameter connected with minimal value of energy and monolayer capacity) are obtained by the best-fit procedure described in Ref. 230. As shown in Fig. 7, the agreement of the theoretical predictions and experimental point is really very good. Numerous studies have confirmed the usefulness of the discussed theoretical treatment for adsorption from multicomponent solutions in the whole concentration region, as well as for adsorption from dilute solutions [5,41].

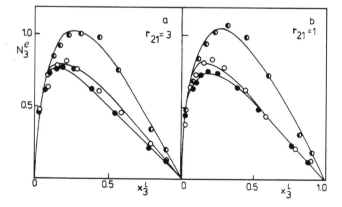

FIG. 7 Comparison of the experimental values for the excess adsorption of toluene (open circles), benzene (left semiclosed circles), and p-xylene (filled circles) from the mixed solvent n-heptane–cyclohexane with the theoretical adsorption isotherm. The ratio $r_{21} = x_2^l/x_1^l$ is equal to 3(a) and 1(b).

2. Liquid–Solid Adsorption Chromatography

Liquid adsorption chromatography is one of the most popular analytical methods. From a practical point of view, it is important to optimize the chromatographic process to guarantee a good separation of the sample components in a sufficiently short time. Such conditions may be warranted by using a mixed mobile phase in the liquid–solid chromatography LSC process because it allows us to modify the retention parameters over a wide range. The elution properties of a mixed solvent depend on molecular interactions in the liquid and adsorption mechanism in the system. The behavior of the chromatographed substance is well characterized by its capacity ratio

$$k'_c = q'\left(\frac{x_c^{s(\infty)}}{x_c^l}\right) \tag{68}$$

where q' is a parameter characteristic for a given adsorbent and $x_c^{s(\infty)}$ is the mole fraction of the chromatographed substance in the stationary (surface) phase for its infinitely low concentration in the mobile phase. It is quite clear from the definition of the capacity ratio that the theory of adsorption from multicomponent solutions is a natural basis for the development of theoretical treatment of liquid adsorption chromatography.

For a homogeneous surface, one can obtain from Eqs. (63) and (68) the expression for the capacity ratio in the n-component mixed mobile phase [226,227]:

$$k'_c = \left(\sum_{i=1}^{n}\frac{x_i^l}{k'_{(i)c}}\right)^{-1} \tag{69}$$

where $k'_{(i)c}$ is the capacity ratio of the substance c and a pure solvent i.

When the adsorbent is heterogeneous, the capacity ratio may be obtained from the simplified integral equation (64) [231]:

$$k'_{c,t} = \left(\frac{k'_{(1)c}}{x_n^l y^*}\right)x_{(n),t}^s \tag{70}$$

The latter has been derived for the case when the sample and all solvents, except solvent n, interact in a similar way with the surface. The particularly interesting results have been obtained for an ideal binary mobile phase from the Tóth-type adsorption isotherm [231]:

$$k'_{c,t} = \left[\left(\frac{x_1^l}{k'_{(1)c}}\right)^m + \left(\frac{x_2^l}{k'_{(2)c}}\right)^m\right]^{-1/m} \tag{71}$$

Figure 8 presents the results of model studies performed according to Eq. (71). It follows from these calculations that the heterogeneity of the solid surface may be the reason for the appearance of a deep minimum on the curve $k'_{c,t}$ versus x_1^l. This shape is characteristic for numerous chromatographic systems. An increase of the adsorbent heterogeneity causes a rapid decrease in the capacity ratio of the chromatographed substance in the region of middle concentrations. Equation (71) has been successfully used to analyze the experimental data [232]. Numerous further studies confirmed a significant role of the adsorbent heterogeneity in the chromatographic separation [227,232,233].

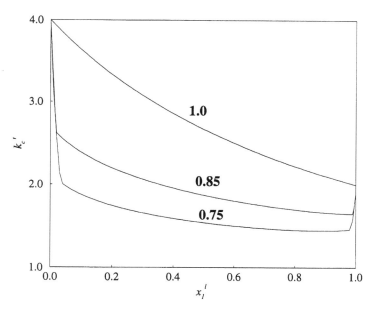

FIG. 8 The capacity ratios k_c' as a function of mole fraction of more polar eluent (1) calculated from Eq. (71) for $m = 1$ (the Snyder relation), 0.85, and 0.7.

IV. LATTICE MODELS

The lattice models have been used frequently in theoretical studies of adsorption. The wide popularity of this approach results from its flexibility and simplicity. It may be applied for systems of various dimensionalities, for many lattice geometries, for different models of adsorbate–adsorbate molecular interactions, and many spatially varying external fields. Most studies have focused on a two-dimensional or three-dimensional cubic lattice, with only isotropic nearest-neighbor couplings. The isomorphism between the Ising model and the classical lattice gas or a coarse model of binary mixture is well known and very helpful for theoretical analysis. The lattice models can be also applied to describe the systems involving polymers.

    Let us consider the monolayer adsorbed film formed on a heterogeneous surface in equilibrium with a gas of the chemical potential (μ). In the simplest version of the lattice model, every site i is characterized by the occupation variable, n_i, which equals 1 when the site is occupied by the gas atom and equals 0 when it is empty. A simple square lattice is assumed. In this case, the Hamiltonian for the adsorbed monolayer can be written as [234]

$$\mathscr{H} = -\frac{1}{2}u \sum_{\langle i,j \rangle} n_i n_j - \sum_i \varepsilon_i n_i - \sum_i n_i \tag{72}$$

where ε_i is the adsorption energy for the ith site, u denotes the energy of the molecular interactions between a pair of nearest neighbors. The summation in the first term runs over all pairs of nearest neighbors. By assuming different forms of the distribution of adsorption energy over the lattice sites, we can model various heterogeneity effects. When the energy distribution function is discrete, one can write the following normalization condition:

$$\sum_{k=1}^{M'} s_k = 1 \tag{73}$$

where s_k represents the fraction of sites with energy ε_k and M' is the number of types of adsorption sites. It follows from Eq. (10) that the total surface coverage is given by

$$\Theta_t = \sum_{k=1}^{M'} \Theta_k s_k \tag{74}$$

where Θ_k is the degree of surface coverage calculated for the kth type. The route of calculation Θ_k depends on the assumed topography of the surface.

In the case of the patchwise adsorbent, the mean field approximation leads to the set of equations (Fowler–Guggenheim type) representing adsorption equilibrium on the individual patches

$$\mu = -\varepsilon_i + k_B T \ln\left(\frac{\Theta_i}{1 - \Theta_i}\right) - 2u\Theta_i \quad \text{for } i = 1, 2, \ldots, M' \tag{75}$$

When random topography is assumed, we also have a set of equations describing adsorption on each type of site, but in a slightly different form:

$$\mu = -\varepsilon_i + k_B T \ln\left(\frac{\Theta_i}{1 - \Theta_i}\right) - 2u\Theta_t \quad \text{for } i = 1, 2, \ldots, M' \tag{76}$$

Using Eqs. (75) [or Eq. (76)] and (74), we can easily obtain the adsorption isotherm assumed for the adsorption energy distribution. In the framework of the mean field approximation expressions for any thermodynamic quantity (e.g. internal energy, heat capacity) can be readily derived [234]. Adsorption on randomly heterogeneous surfaces has been studied in terms of the above-described approach. It has been demonstrated that this mean-field-type theory was valid only at very high temperatures. Below the critical two-dimensional temperature, the predictions of theory seriously underestimate the heterogeneity effects on phase transitions in adsorbed monolayers [12,234].

The extension of the lattice gas formalism from monolayer adsorption to three-dimensional systems is straightforward. In this case, the adsorbate–adsorbent interaction energy depends not only a kind of site but also on distance from the surface. The Hamiltonian for such a system is calculated by replacing the constant value ε_i by appropriate functions in Eq. (72). In consequence, either for the patchwise model or for randomly distributed sites, for successive layers we obtain the equations analogous to those previously discussed. A numerical solution of such a set of equations is not a difficult problem, but to achieve the stable results, a rather large number of layers in the system should be assumed.

De Oliviera and Griffits [235] have studied multilayer adsorption on a homogeneous surface. They have obtained stepwise adsorption, which proved that the surface films grow in a layer-by-layer mode in the series of the successive first-order phase transitions. A general classification of possible scenarios for the film growth has been presented by Pandit et al. [236] in the framework of a mean field theory for the lattice gas model.

The heterogeneity effects in the formation of multilayer adsorbed films have not been very intensively studied in terms of the lattice models. In general, one can expect a certain rounding of the sharp steps occurring on adsorption isotherms. We will discuss the problem of possible phase transitions in the surface films in Section VII.

Here, we mention only the precursory results obtained by Nicholson and Silvester [103] for adsorption on the surface with random and patchwise surface topographies. This work led to the conclusion that the "smooth sigmoid isotherms are not necessarily associated with surface heterogeneity nor are stepped isotherms indicative of homogeneous surfaces." Nicolson and

Silvester emphasized the influence of adsorbate–adsorbate interaction on the shape of adsorption isotherm [103]. Moreover, the important role of surface topography has been shown.

In the above-discussed theories, the adsorbate molecules can occupy only one active site. However, in real systems, the adsorbate molecules, quite frequently, consist of a number of single, monomeric segments. Even the simple gases such as oxygen, nitrogen, and carbon monoxide are composed of more than one atom. The lattice gas model has been recently used to describe monolayer adsorption of polyatomic molecules on heterogeneous surfaces. The works of Nitta et al. [237] and Marczewski et al. [238] were the beginning of a systematic study of multisite occupancy adsorption on heterogeneous solid surfaces [6]. Although the original theory [237] has been formulated for adsorbates consisting of various segments and surfaces built of many different sites, the simplest case of homogeneous dimers adsorbed on the surface composed of two kinds of site has been usually studied. The key problem in statistical mechanical calculation of such a kind is to evaluate the combinatory factor of the partition function. This task is difficult for bulk systems and becomes much more complicated in the case of adsorption on heterogeneous surfaces. This problem cannot be solved exactly for adsorption of polyatomic molecules. Therefore, various approximations should be applied. Nitta and co-workers [237,239] have used two versions of the quasichemical approximation [240] to represent the combinatory factor for distributing dimer molecules on heterogeneous lattices. In the first version of the theory, each active site was treated independently, whereas in the second case, each bond was treated independently. As a result, they have derived two analytical expressions for the adsorption isotherm of dimers. The theoretical predictions have been compared with a Monte Carlo simulation. The first version of the theory was found to overestimate the occupancy of more energetically stable sites, whereas the other underestimated them. In this approach, adsorbate–adsorbate molecular interactions are neglected. It is an important shortcoming of the theory. The Nitta model has been widely discussed by Rudziński and Everett and the interested reader can find all the details in their book [6]. Ramirez-Pastor et al. [241] have compared the Nittas theoretical isotherm with a Monte Carlo simulation and experimental data. They have stated that the model gave correct predictions for weakly heterogeneous adsorbent, but qualitative differences with simulations were appeared when the adsorbent is strongly heterogeneous. Similarly, although a very good fitting of the experimental data has been achieved, the adjusted size of molecular was sometimes unrealistic. Apart from all of these drawbacks, one should note that Nitta's model was the first theoretical approach to multisite occupancy adsorption and gave a strong impetus to further studies [6].

Ramirez-Pastor et al. [242,243] have proposed a modified form of the adsorption isotherm for linear, rigid homoatomic and noninteracting molecules on heterogeneous surfaces. They have introduced the rigorous expression for the Helmholtz free energy of the adlayer formed on a homogeneous one-dimensional lattice into the multisite occupancy model of Nitta et al. [237]. This, a little striking, idea leads to the theoretical isotherms, which quite well described simulation points and fitted experimental data [242,243]. The deviations from the simulation results are, obviously, greater at low temperature. The adsorption of dimers and trimers on a lattice with two kinds of adsorption site has been also studied by means of a Monte Carlo simulation performed for different distributions of patches on the surface. These investigations have clearly shown that the adsorption depends on the relation between the size of the molecule and the patch, as well as on the topographical distribution of the patches on the surface. The adsorption of polyatomic particles on heterogeneous surfaces still requires further studies.

For many years, the lattice model has been treated as a "natural" tool for performing studies of liquid mixtures [240]. This approach also has been used to investigate the surface heterogeneity effects in adsorption from binary, regular solutions on adsorbents of various topographies [244]. The three-dimensional model of the system has been assumed. The sets of equations analogous to Eqs. (75) and (76) have been derived by means of the standard statistical mechanical method. The theoretical predictions are compared with Monte Carlo simulations. It has been evidently proved that the sizes of homogeneous patches affect the total adsorption isotherm.

Similar calculations have been performed for the dimer–monomer solutions [245]. The "classical" theory of adsorption of polymers has been extended to adsorption on heterogeneous surfaces of the patchwise topography. The adsorbed dimers could be parallel as well as normal to the surface. The three-dimensional model, thought somewhat more difficult from a mathematical point of view, is much more realistic than the "parallel model" used in the previously discussed articles. The theory has been compared with Monte Carlo simulations.

The results are quite promising. The volume fractions of dimers in the successive layers over particular patches can be obtained from the following equations:

$$
-\frac{1}{2}\ln(f_r^{(i)}) - \frac{1}{2}\sum_{n=-1}^{1}\left[c_n \Phi_{1,r}^{(i-n)}\left(\sum_{k=-1}^{1}\Phi_{1,r}^{(i-n+k)}\right)^{-1}\right]
$$
$$
- \omega \sum_{k=-1}^{1} c_k(\Phi_{1,r}^{(i+k)} + \Phi_{2,r}^{(i+k)}) - \hat{\varepsilon}_{12,r}\delta(1-i) = g, \quad i = 1, 2, \ldots l, r = 1, 2, \ldots, M'
$$

$$(77)$$

where $\phi_{j,r}^{(i)}$ denotes the volume fraction of the jth component (1 refers to dimer and 2 corresponds to monomer) in the ith layer over the rth patch,

$$
f_r^{(i)} = \sum_{k=-1}^{1} c_k \Phi_{1,r}^{(i+k)}
$$

$$(78)$$

and

$$
\hat{\varepsilon}_{12,r} = \varepsilon_{1,r} - \varepsilon_{2,r} - (u_{11} - u_{22})z^*
$$

$$(79)$$

in the case of cubic lattice $c_0 = 2/3, c_1 = c_{-1} = 1/6, z^* = \bar{z}/2k_BT$, $\varepsilon_{j,r}$ is the adsorption energy on the rth patch u_{ij} is the energy of interaction between segments i and j, and δ is the Kronecker delta function.

In Eq. (77), g is the function defined for the bulk system:

$$
g = \ln\left(\frac{\Phi_1}{\Phi_2}\right) - \frac{1}{2}(\ln \Phi_1 + 1) + \omega(\Phi_2 - \Phi_1)
$$

$$(80)$$

where Φ_j is the volume fraction of the jth component in the mixture,

$$
\Phi_j = \lim_{i \to \infty} (\Phi_j^{(i)})
$$

$$(81)$$

and ω is the well-known Flory–Huggins parameter given by

$$
\omega = -\frac{\bar{z}(u_{12} - \frac{1}{2}(u_{11} + u_{22}))}{k_BT}
$$

$$(82)$$

From the set of Eq. (77), we can obtain the "local concentrations" for the particular homogeneous patches. The volume fraction for the whole layer is defined as the average

$$
\Phi_j^{(i)} = \langle \Phi_{j,r}^{(i)} \rangle = \sum_{r=1}^{M'} \Phi_{j,r}^{(i)} s_r, \quad j = 1, 2
$$

$$(83)$$

where s_r is the fraction of the sites of the rth type.

The surface excess of dimers is defined as

$$\Gamma = \sum_{i=1}^{l}(\Phi_1^{(i)} - \Phi_1) \tag{84}$$

In Fig. 9, the excess adsorption isotherms obtained for homogeneous surfaces are presented. The agreement between the theoretical results and Monte Carlo simulations is surprisingly good even for high values of the parameter ω. Figure 10 shows theoretical and simulation results for heterogeneous surfaces. The simulations have been performed for chessboard surfaces of different sizes of patches. One can conclude from Fig. 10 that the theory may be used only when homogeneous patches are sufficiently large. The Monte Carlo results obtained for a crystalline surface ($s^* = 1$) are considerably different from theoretical predictions.

V. DENSITY FUNCTIONAL THEORY

A. Fundamentals

Among different theories, the density functional theory has been shown to be one of the most promising and powerful theoretical tools for performing studies of the structure and thermodynamical properties of surface films formed on the heterogeneous adsorbents [246–257].

In the density functional (DF) theory, the equilibrium density profile is obtained by minimizing a free-energy functional [258]. The usual DF approach starts from the definition of the grand potential as a functional of the number density of fluid:

$$\Omega = F + \int \rho(\mathbf{r})(v(\mathbf{r}) - \mu)\, d\mathbf{r} \tag{85}$$

where F is the intrinsic Helmholtz free energy of the fluid (i .e., the Helmholtz energy in the absence of the external field), $v(\mathbf{r})$ is the potential field generated by the adsorbent, and μ is the configurational chemical potential. The functional F is split into a sum of an ideal part and an excess part:

$$F = F^{\mathrm{id}} + F^{\mathrm{ex}} \tag{86}$$

where

$$F^{\mathrm{id}} = k_B T \int \rho(\mathbf{r})[\ln \rho(\mathbf{r}) - 1]\, d\mathbf{r} \tag{87}$$

The excess free energy is divided into two parts, representing the free energy due to repulsive (F^{rep}) and attractive (F^{att}) forces between the molecules:

$$F^{\mathrm{ex}} = F^{\mathrm{rep}} + F^{\mathrm{att}} \tag{88}$$

For Lennard–Jones (LJ) and similar fluids, the first step toward the development of appropriate expressions is the decomposition of the fluid–fluid potential into repulsive and attractive terms. The Weeks–Chandler–Anderson (WCA) division is usually used [259], according to which the attractive part of the Lennard-Jones potential is given by

$$u^{\mathrm{att}}(r) = u^{\mathrm{WCA}}(r) = u^{\mathrm{LJ}}(r) + \varepsilon^{\mathrm{LJ}} \tag{89}$$

The repulsive interactions, $u^{\mathrm{rep}}(r) = u^{\mathrm{LJ}}(r) - u^{\mathrm{WCA}}(r)$, are usually modeled by using a hard-sphere potential with an effective hard-sphere diameter d_{hs}. There are several possible routes for obtaining the d_{hs} (e.g., the Barker–Henderson method [16]). However, in the case of nonuniform

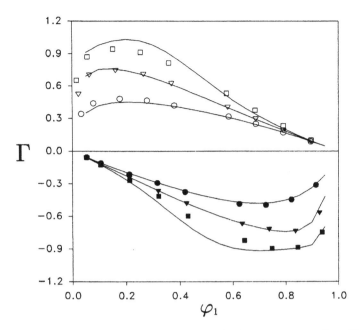

FIG. 9 Excess adsorption isotherms for the homogeneous surface. The points are the result of a Monte Carlo simulation and the lines are obtained from the theory. $\varepsilon^* = \varepsilon/u_{22}$, $u_{ij}^* = u_{ij}/u_{22}$, and $T^* = k_B T/u_{22}$. Parameters: $\varepsilon_1^* = 6$, $\varepsilon_2^* = 4$ (open symbols) and $\varepsilon_1^* = 1$, $\varepsilon_2^* = 3$ (filled symbols); ω: -0.6 (circles), 0.6 (triangles), and 1.2 (squares). (From Ref. 245.)

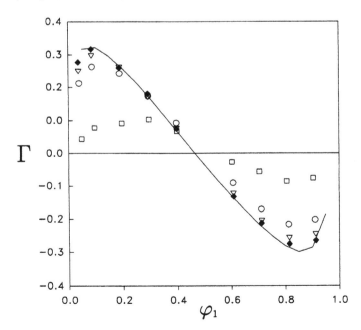

FIG. 10 Comparison of theoretical excess adsorption isotherms for the patchwise surface composed from two kinds of active center: one type adsorbing dimers ($\varepsilon_{1,1}^* = 6$, $\varepsilon_{2,1}^* = 4$), the other strongly interacting with monomers ($\varepsilon_{1,2}^* = 1$, $\varepsilon_{2,2}^* = 3$). Parameters: $u_{12}^* = 0.9$ ($\omega = 0.6$); size of patches s^*: 1 (squares), 2 (circles), 5 (triangles) and 12 (filled diamonds). (From Ref. 245.)

fluids, this parameter is simply set to be equal to the Lennard–Jones diameter σ. However, the attractive forces are treated in a mean field approximation

$$F^{\text{att}} = \tfrac{1}{2} \int d\mathbf{r} \, d\mathbf{r}' \rho(\mathbf{r})\rho(\mathbf{r}')u^{\text{att}}(|\mathbf{r} - \mathbf{r}'|) \tag{90}$$

where u^{att} is an attractive part of the fluid–fluid potential.

The excess free energy of the hard-sphere fluid may be evaluated by means of methods following from different versions of the density functional theory. Numerous approximations can be used:

Local density approximation
Tarazona's weighted-density approximation
The Kierlik and Rosinbergs' density approximation

In the case of the local density approximation, the excess free energy of hard-sphere fluid is assumed to be dependent only on the local density, so that

$$F^{\text{rep}} = F_{\text{hs}}^{\text{ex}}(\rho(\mathbf{r})) = \int d\mathbf{r} \, \rho(\mathbf{r}) f_{\text{hs}}^{\text{ex}}(\rho(\mathbf{r})) \tag{91}$$

where $f_{\text{hs}}^{\text{ex}}$ is the excess free energy per molecule of the bulk hard-sphere fluid which can be evaluated from some equation of state. The Carnahan–Starling equation is most often used [260]; hence,

$$\frac{f_{\text{hs}}^{\text{ex}}(\rho)}{k_B T} = \frac{\eta(4 - 3\eta)}{(1 - \eta)^2} \tag{92}$$

where $\eta = \pi d_{\text{hs}}^3 \rho / 6$.

The main shortcoming of this theory is that it does not account for even short-range correlations in the fluid, so it is not able to mimic the structures of thin surface films.

In the density functional theory of Tarazona, the excess hard-sphere free energy is calculated in a nonlocal manner by employing the concept of smoothed density [261–263]:

$$F^{\text{rep}} = F_{\text{hs}}^{\text{ex}}(\rho(\mathbf{r})) = \int d\mathbf{r} \, f_{\text{hs}}^{\text{ex}}(\tilde{\rho}(\mathbf{r})) \tag{93}$$

$f_{\text{hs}}^{\text{ex}}$ is given by Eq. (92) for a density equal to $\tilde{\rho}$.

The smoothed density $\tilde{\rho}(\mathbf{r})$ is obtained by averaging the local density with a weight function chosen so that the theory gives a good description of the direct correlation functions of the bulk fluids:

$$\tilde{\rho}(\mathbf{r}) = \int d\mathbf{r}' \, \rho(\mathbf{r}')W(|\mathbf{r} - \mathbf{r}'|, \tilde{\rho}(\mathbf{r})) \tag{94}$$

According to Tarazona et al. [261–263], the weight function $\tilde{\rho}(r)$ is given as a power series

$$(r, \rho) = W_0(r) + W_1(r)\rho + W_2(r)\rho^2 \tag{95}$$

and the coefficients W_0, W_1, and W_2 of this expression can be found in Ref. 263. This version of the density functional theory is very popular; among the numerous applications, one can find investigations of surface heterogeneity effects in interfacial phenomena [251,252,255].

The Kierlik–Rosinberg [264] theory is formally similar to the treatment of Tarazona. They use four weight functions, which are chosen such that this theory exactly reproduced the Percus–Yevick result for the direct correlation function in the uniform hard-sphere fluid. However, these functions are independent of the density. In consequence, the calculation of the weighted density is considerably simpler. Moreover, this version of the density theory can be easily extended to adsorbed mixtures [265].

The equilibrium density profile minimizes the grand canonical potential Ω; thus, the local density can be evaluated from the condition

$$\frac{\partial \Omega(\rho(\mathbf{r}))}{\partial \rho(\mathbf{r})} = 0 \tag{96}$$

The adsorption isotherm is defined as

$$\Gamma = \int d\mathbf{r}(\rho(\mathbf{r}) - \rho_b) \tag{97}$$

where ρ_b is the bulk density.

The above-discussed equations can be used for arbitrary spatially varying adsorbing potentials. Moreover, one can also employ a continuous density functional representation of the fluid as its version for the lattice gas model. For heterogeneous surfaces, a considerable limitation may be numerical difficulties connected with solving these equations in particular cases.

At this point, we will briefly discuss differences between the density functional theory and "classical" integral equation approach. The differences seem to be rather extensive. In the theories discussed in Section III, the fundamental function characterizing surface heterogeneity is the adsorption energy distribution. The surface topography is usually considered only in the presence of attractive intermolecular interactions. The geometrical distribution is taken into account by introducing some assumptions about correlations of pairs, triples, and so forth of adsorbing sites [6]. However, according to the present model, the complete information about the surface contains the external potential $v(\mathbf{r}) = v(x, y, z)$. It means that the surface topography is directly introduced into the theoretical considerations. The topography effects may be fully described in terms of the consistent framework. The adsorption isotherm is evaluated by solving the local density equation (96). The form of this equation indicates that the effect of surface topography may be really important. One can note that, even for hard spheres, the evaluation of the local density at a given point on the surface requires the knowledge of local densities at all points located within the distance characteristic for the weight function used. In the case of attractive molecular interactions, the integration limits in the above equations depend also on a range of intermolecular forces; thus, the role of the topography increases. The application of density functional theories may be helpful in explaining the microscopic structure of adsorbed films and in the description of surface phase transitions.

B. Selected Applications

Investigations of interfacial phenomena based on the DF theory have generally focused on ideally flat and energetically homogeneous surfaces. However, during last few years, numerous interesting applications of the method to describe a fluid behavior in an external field extorted by a heterogeneous solid have been reported [246–257].

In pioneer studies, Łajtar and Sokołowski [246] have applied the density functional theory of Tarazona to investigate monolayer adsorption on energetically heterogeneous surfaces. The random and partially correlated intermediate models of the surface topography have been studied. Unfortunately, it appeared to be impossible to generate, in a computer memory, the patchwise surface having the required total heterogeneity characterized, in this case, by the Gaussian distribution. It has been proven that the shape of the external potential $v(x, y)$, also termed "surface topography," considerably affects adsorption; the role of adsorbent topography may be even more important than the role of global surface heterogeneity characterized by the adsorption energy distribution function. The density functional theory has been also used to describe adsorption in energetically heterogeneous slitlike pores with walls characterized by the Gaussian energy distribution and random topography of the surface. The results have been compared with Monte Carlo simulations [247].

During last years, several interesting articles devoted to adsorption in heterogeneous slitlike pores have been published [248–252]. These studies have focused on the phase behavior of confined fluids. This series begins with the work Chmiel et al. [248], who have studied adsorption in energetically and geometrically nonuniform slitlike pores. They have considered the Lennard–Jones fluid. In the case of energetically heterogeneous surfaces, the slit was built of the walls consisting of two kinds of adsorption site grouped into stripes. The weakly and strongly adsorbing strips were arranged by turns. For geometrically heterogeneous surfaces, the walls were built of atoms placed on a square lattice and the pore had a step of a certain height at the center. The calculations have been directed toward the study of the influence of both types of surface heterogeneity on the capillary condensation. It has been shown that the pore nonuniformity may qualitatively change phase diagram of confined fluids. In particular, geometrical heterogeneity may lead to a splitting of the hysteresis loop into two separate parts, corresponding to condensation in different parts of the pore. This phenomenon occurs only for sufficiently long necks.

The next studies have extended our knowledge of the peculiar phase behavior of fluid confined in chemically structured slit pores. Roecken and Tarazona [249] have investigated capillaries with periodically corrugated walls along one of the transverse directions. They evaluated phase diagrams for capillary condensation characterized by two separated first-order transitions from the confined "gas" over "liquid bridges" to liquid. The two branches of the phase diagram may join together at the triple point. They have found the new stepwise mechanism for capillary condensation that was due to periodic modulation of the fluid–wall potential. The confined "liquid" condenses first where the attraction is strongest; the "liquid drops" are distributed periodically over the walls. Then, the confined "gas" may evolve in two ways along an isotherm or isochore. For weak heterogeneity (i.e., if the attraction potentials of sites are not too different), wetting layers of liquid may form along the walls and, finally, the pore fills up with liquid for the chemical potential lower than the chemical potential of bulk saturated vapor. We observe the usual capillary condensation. However, when strongly attractive sites alternate with weakly (or even repulsive) attractive ones, the liquid drops at the two opposing walls may link to form "liquid bridges" separated by "gas gaps." The mechanism of capillary condensation and the stability of the liquid-bridges phase depend on interplay between thermodynamic conditions and pore structure, particularly the ratio of the corrugation period and the pore width. These conclusions have confirmed the previous results obtained for the Ising lattice gas model solved in the mean field approximation [250]. The phase transitions in pores, the walls of which consist of alternating strips of weakly and strongly adsorbing substrates, were studied in Monte Carlo simulations [266–269]. The valuable discussion of the capillary condensation in chemically structured slits can be found in the review by Schoen [270].

Here, we will describe in more detail the interesting example of an application of density functional theory to the very complicated system, namely the four-bonding Lennard–Jones associating fluid confined in slitlike pore [255]. The fluid is confined in the slitlike pore of width H^*. The total external field is the sum of contributions from two pore walls, located at $z = 0$ and $z = H^*$:

$$v(x, y, z) = v_1(x, y, z) + v_2(x, y, H^* - z) \qquad (98)$$

Each of the wall–fluid potentials has the following form:

$$v_i(x, y, z) = v^{hom}(z) + A_i(x)v^{het}(z) \qquad (99)$$

where v^{hom} is the homogeneous part of the potential and the v^{het} describes changes in the adsorbing field along the x axis. The functions $v_{\text{hom}}(z)$ and $v^{\text{het}}(z)$ have the form of the Lennard–Jones (9-3) potential [Eq. (9)] rewritten as

$$v^{\alpha}(z) = \varepsilon_s^{\alpha}\left[\left(\frac{z_0}{z}\right)^9 - \left(\frac{z_0}{z}\right)^3\right] \tag{100}$$

where $\alpha = \text{hom}$ or het. The function $A_i(z)$ is given by

$$A_i(x) = \cos\left(\frac{2\pi}{\lambda} + \varphi_i\right) \tag{101}$$

where i denotes the wall index ($i = 1, 2$), λ characterizes the periodicity range, and φ_i is the phase shift of the slit walls.

Huerta et al. [255] have studied a model associating fluid with four bonding sites $\bar{M} = 4$, $A, A' = S_1, S_2, S_3, S_4$. Such a structure of adsorbate molecules permits the formation of a network of bonds in the bulk phase. The bonding effects have an essential influence on the thermodynamic properties of the fluid. However, it has been assumed that only $S_1 S_3$, $S_2, S_3, S_1 S_4$, and $S_2 S_4$ bonding was allowed. The association energy was assumed to be the same for all possible bonds.

A route of calculations of the free-energy functional for the associating fluid has been described in Ref. 271 and can be found in the original article [255]. Therefore, only main stages of calculations will be shown. In the case of associating fluids, the fluid–fluid interaction potential is supplemented by the associative part. The "chemical term" of the potential usually has the form of a variously located deep mound. As a consequence, the new term, F^{as}, should be added to the free-energy functional [Eq. (88)]. For the considered model of association, F^{as} has the following form:

$$F^{\text{as}} = \int d\mathbf{r}\, f_{\text{as}}^{\text{ex}}(\tilde{\rho}(\mathbf{r})) \tag{102}$$

where $f_{\text{as}}^{\text{ex}}$ is the excess free energy per particle arising from associative interactions; the average density $\tilde{\rho}$ is given by the Tarazona recipe [261–263]. Application of the Wertheim thermodynamic perturbation theory [272] to nonuniform systems leads to the following expression [273]:

$$\frac{f_{\text{as}}^{\text{ex}}(\tilde{\rho})}{k_B T} \sum_{A=1,\bar{M}}\left(\ln \chi_A(\tilde{\rho}) - \frac{\chi_A(\tilde{\rho})}{2} + 0.5\right) \tag{103}$$

where χ_A is the fraction of molecules "1" not bonded at a site A. The function χ_A follows from the solution of the equation

$$\chi_A(\tilde{\rho}) = [1 + 2\chi_A(\tilde{\rho})\tilde{\rho}g^{\text{hs}}(\sigma, \tilde{\rho})K_{AA'}f_{AA'}^{-1}] \tag{104}$$

where $f_{AA'} = \exp[-\varepsilon^{\text{as}}/k_B T] - 1$ is the bonding volume, $K_{AA'} = 0.25(1 - \cos\Theta_c)^2\sigma^2(r_c - \sigma)$, ε^{as} is the association energy, g^{hs} denotes the contact value of the pair distribution function of a hard-sphere fluid at density $\tilde{\rho}$, and Θ_c and r_c are parameters of the associative potential. The equilibrium density profile is obtained from the appropriate form of Eq. (96).

The aim of the calculations performed by Huerta et al. [255] was to elucidate the effects of surface heterogeneity, association, and confinement on the phase behavior of the fluid. They have performed calculations for various values of the parameters characterizing the system. As an illustration, we discuss the phase diagrams obtained for different model pores. In studies of such a kind, the energies are measured in units of the Lennard–Jones energy ε^{LJ} and the common definition of the reduced temperature is introduced:

$$T^* = \frac{k_B T}{\varepsilon^{LJ}} \tag{105}$$

The average density in the pore is given by

$$\langle \rho \rangle = \frac{\int d\mathbf{r}\, \rho(x, z)}{\int d\mathbf{r}} \tag{106}$$

Figure 11 presents the phase diagrams obtained for a series of model systems. Each part shows the effect of changes of a certain parameter on the phase transitions. The bulk gas–liquid coexistence is treated as a reference, so the temperature is scaled with respect to the bulk critical temperature (T_c^*). In particular, Fig. 11a presents the coexistence curves evaluated for fluid confined in the various pores of width $H^* = 5$ and for the bulk system. For homogeneous walls ($A_i = 0$), the phase diagrams consist of two branches. The first describes the layering transitions and the other corresponds to capillary condensation. Both branches seemingly join together at the triple point. The critical temperature of capillary condensation is, obviously, lower than the critical temperature of the bulk fluid. However, the critical temperature of the layering transition is almost equal to the critical temperature for capillary condensation. This behavior differs from common trends observed for confined nonassociating fluids, where the first layering temperature is usually lower than the capillary condensation temperature. In the case of heterogeneous walls, one can observe substantial changes of the fluid behavior due to modulation of the external field potential. The phase diagrams for heterogeneous systems consist of three branches. This reflects the possibility of the stepwise filling of the pore. In this case, the formation of an intermediate, "bridge" phase precedes the "true" capillary condensation, which leads to the adsorbate condensation through the entire pore. The intermediate branch of the phase diagram corresponds to the formation of a "bridge" phase that is built up over the most attractive part of the pore walls. Figure 11b shows the phase diagrams for a slightly narrower pore than those presented in Fig. 11a. In the case of homogeneous pore, even such a small change in pore width causes a considerable increase in the triple-point temperature. However, for heterogeneous systems, we observed that in a narrower pore, the entire branches of the phase diagrams, corresponding to the layering transition and to the formation of the bridge phase, are contained "inside" the layering branch for the homogeneous pore. Figure 11c illustrates the influence of the periodic term of the fluid–wall energy on the phase behavior for the pore with $H^* = 5$. For a lower value of ε_s^{het}, the triple-point temperature separating the bridge phase and capillary condensation is observed. In all of the above-discussed examples, variations of the periodic potential due to both walls has been "in phase" or "commensurate" ($\varphi_i = 0$). However, Fig. 11d presents the effect of the phase shift of the potential from a wall with respect to each ($\varphi_i = \pi/2$). The shift in the periodic part of the potential causes substantial changes in the phase diagram: namely it leads to disappearance of the bridge phase. There must exist a "critical" shift for the fluid–wall potentials, smaller than $\pi/2$, causing the disappearance of the bridge phase.

One can find a more detailed discussion of the phase behavior of associating fluids in heterogeneous pores in Refs. 251, 252, and 255. These studies have been focused on various aspects of the phenomena, which turn out to be excessively rich and complex.

The density theory has been also applied to study the effects of nonplanarity of the substrate on fluid structure. The well-defined geometrical structures such as wedges have been considered [254,256,270].

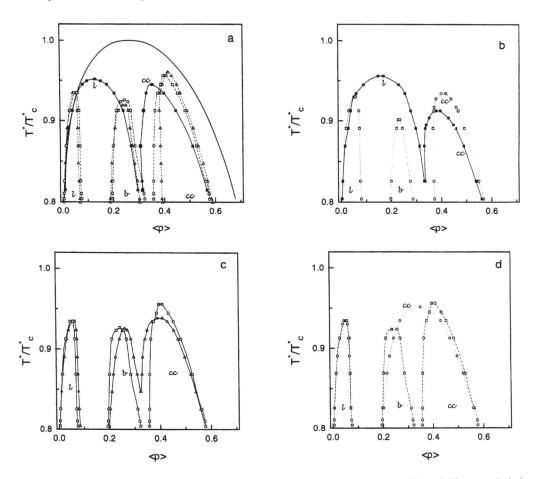

FIG. 11 Phase diagrams of the bulk associating fluid and the same fluid confined in a slitlike pore. Labels l, b, and cc refer to layering, bridging, and capillary condensation transitions, respectively. (a) Comparison of the phase diagrams for: the bulk fluid (solid line), fluid in a homogeneous pore (filled squares), and heterogeneous pores for $\varepsilon^{het} = 12$, $\varphi = 0$, $\lambda = 12$ (dashed lines with open triangles), and $\lambda = 9$ (dashed lines with open squares). The pore width $H^* = 5$. (b) Phase diagrams for the fluid in the pores with: homogeneous walls (solid line with filled squares) and heterogeneous walls for $\varepsilon^{het} = 12$, $\varphi = 0$, and $\lambda = 9$ (dashed line with open squares). The pore width is $H^* = 4.5$. (c) Phase diagrams for the heterogeneous pore of the width $H^* = 5$, $\varepsilon^{het} = 12$ (line connecting empty squares) and for $\varepsilon^{het} = 9$ (line connecting empty triangles). In both cases, the periodicity is the same, $\lambda = 9$, and $\varphi = 0$. (d) Phase diagrams for commensurate, $\varphi = 0$ (line connecting open squares), and incommensurate, $\varphi = \pi/2$ (line connecting open circles), systems. The pore width is $H^* = 5$ and the heterogeneity parameters are $\varepsilon^{het} = 12$ and $\lambda = 9$. (From Ref. 255.)

The above-discussed examples prove clearly that the heterogeneity of the walls plays a key role in phase behavior of the confined fluid. On the other hand, the above discussion shows that the density functional theory can be effectively used to investigate the subtle interfacial phenomena, which take place in very complicated systems.

VI. COMPUTER SIMULATIONS

A. Introductory Remarks

Nowadays, computer simulations are treated as the third fundamental discipline of interface research in addition to the two classical ones, namely theory and experiment. Based directly on a microscopic model of the system, computer simulations can, in principle at least, provide an exact solution of any physicochemical problem. By far the most common methods of studying adsorption systems by simulations are the Monte Carlo (MC) technique and the molecular dynamics (MD) method. In this chapter, a description of simulation methods will be omitted because several textbooks and review articles on the subject are available [274–277]. The present discussion will be restricted to elementary aspects of simulation methods. In the deterministic MD method, the molecular trajectories are computed by solving Newton's equations, and a time-correlated sequence of configurations is generated. The main advantage of this technique is that it permits the study of time-dependent processes. In MC simulation, a stochastic element is an essential part of the method; the trajectories are generated by random walk in configuration space. Structural and thermodynamic properties are accessible by both methods.

The main ingredients of a molecular simulation are a model of the system and a set of rules to change its state. The model consists of assumptions concerning a simulation geometry (e.g, a cubic box or spherical cell), potentials describing interactions between particles, and an external force field. In a classical MC simulation, a system of particles is assigned a set of arbitrarily chosen initial coordinates. Then, a sequence of configurations of the particles is generated by the successive random displacement. However, not all possible configurations are accepted. The acceptance/rejection rules should satisfy directly formulated conditions. The appropriate algorithm proposed by Metropolis et al. [278] is still more commonly used. According to this scheme, the transition from the old (o) to new (n) configuration (o/n) is accepted with probability

$$P(o/n) = \min\left\{1, \frac{T(n/o)f(n)}{T(o/n)f(o)}\right\} \tag{107}$$

where $T(i/j)$ denotes the possibility of proposing a transition from state i to state j, whereas $f(i)$ is the probability of being at state i [279]. This probability depends on the change of energy connected with a proposed transition (ΔU):

$$\frac{f(n)}{f(o)} \sim \exp\left(\frac{-\Delta U}{k_B T}\right) \tag{108}$$

The main problem in simulation is to construct functions $T(n/o)$ that result in efficient and correct sampling of the phase space. Then, the ensemble averages of needed physical quantities are obtained as weighted averages over the resulting set of configurations. For a sufficiently long chain of generated configurations, the system is equilibrated and the wanted averages may be calculated.

Simulation of adsorption has been performed in various ensembles: canonical, grand canonical, isobaric–isothermal, and Gibbs ensemble. The choice of the ensemble depends on the nature of the investigated system and the aim of the simulations. In the case of adsorption on heterogeneous surfaces, usually the grand canonical Monte Carlo simulation method (GCMC) has been used.

During computer pseudoexperiments, one can estimate various quantities characterizing interface systems. It is possible to calculate adsorption isotherms, isobars, and isosteres, to evaluate heats of adsorption and heat capacity, isothermal compressibility of the surface films, and spreading pressure, to determine the local density profiles, various structure factors, orientational distributions, and so on. Computer simulations are ideally suited to the study of structural properties of surface films. The results of simulation may be compared with experimental data as well as with predictions of theories. As a result, our understanding of adsorption mechanism becomes more deep and wide ranging.

In the following subsection, the key findings of computer simulations directed toward surface heterogeneity effects in adsorption are summarized. The analyzed examples are obtained by means of Monte Carlo method.

B. Modeling of the Adsorbent Surface

The adsorbent surface is a source of an external force field which strongly affects the behavior of adsorbate molecules. In contrast to the experiments, the molecular structure of the adsorbent surface is completely known and assumed a priori in computer simulations. However, computer generation of physically realistic surfaces is not simple and requires answers to many important questions. There are three possible approaches to this problem: (1) attempt to build the relatively primitive model which is able to reflect basic features of the system; (2) try to construct a model that is as similar as possible to the real material carefully studied by means of modern experimental techniques; (3) in the simulation procedure, attempt to mimic the manufacturing process used to produce the real adsorbent [280].

In numerous studies, the models of heterogeneous surfaces based on the idealized regular lattice model have been considered. Each lattice site on the surface is characterized by its adsorption energy. The procedure to generate random sequences of energies according to a given probability distribution $\chi(\varepsilon)$ can be found in the popular books on simulation methods [274]. The key problem is to assign a given value of energy to a chosen site. As it has been already mentioned, various active sites may be differently distributed on the surface. All models of surface topography can be readily used in computer simulation studies.

Implementation of the patchwise model is really simple if one recalls that adsorption on each patch occurs independently of adsorption on other patches and that boundary effects are neglected in this model. The aim of Monte Carlo simulation is often to evaluate an adsorption isotherm for the heterogeneous surface with a given energy distribution $\chi(\varepsilon)$ and then compare the result with experimental data. When the patchwise model is assumed, one can calculate only one adsorption isotherm on homogeneous surface. Then, we can evaluate the overall adsorption isotherm [Eq. (10)] using the adsorption isotherm for a homogeneous surface weighted with the chosen energy distribution function $\chi(\varepsilon)$ [12].

However, the application of computer simulation to the study of adsorption on randomly heterogeneous surfaces is much more difficult. We should point out that, in principle, the creation of a truly random surface characterized by the a priori assumed energy distribution is impossible. We always consider finite systems, which may show some remnants of correlations between different values of energy. The special averaging method that allows us to overcome this limitation has been applied [281,282]. The main problems connected with the generation of random surfaces are caused by the small sizes of systems we can use. Each such small system has a somewhat different realization of the spatial distribution of various active sites. The self-averaging of extensive thermodynamical quantities is possible at the thermodynamic limit of a truly macroscopic system. In this situation, it is necessary to generate many (usually of the order of $10^2 - 10^3$) replicas of the system, characterized by the same $\chi(\varepsilon)$ but different spatial distribution of adsorption energies, and repeat simulation for all of them. The final results are obtained by averaging over all samples [283–285]. It is quite clear that investigations of systems with random disorder are more time-consuming than calculations for the patchwise model.

The additional problems appear when partially correlated distribution of adsorption energies is considered [13–15,286–288]. According to this concept, for a given surface, there is a certain characteristic length r_{cor} which defines the size of domains characterized by the same value of adsorption energy. The correlation between different energies of adsorption on different sites, ε_i and ε_j, can be described by a pair density distribution function $\chi'(\varepsilon_i, \varepsilon_j)$:

$$\chi'(\varepsilon_i, \varepsilon_j) = \alpha'(r_{cor})\chi(\varepsilon_i)\chi(\varepsilon_j), \quad i \neq j \tag{109}$$

and

$$\chi'(\varepsilon_i, \varepsilon_j) = \chi(\varepsilon_i) - \sum_{j \neq i} \alpha'(r_{cor})\chi(\varepsilon_i)\chi(\varepsilon_j) \tag{110}$$

where $\alpha'(r_{cor})$ is a factor related to the correlation of different energy values. For the random model, $\alpha'(r_{cor}) = 1$; in the opposite extreme case of a patchwise surface, $\alpha'(r_{cor}) = 0$. The method for assigning different values of energy to lattice sites with the assumed form of the single-site distribution function has been described in Ref. 14 with all particulars. Here, we want to show only one step of the procedure. When the energy to be already assigned to the chosen site, the sites within the circle of the characteristic length r_{cor} around this site are considered, and the predominant energy within the circle is assigned to all of the sites encompassed by the circle. After all of the sites obtained the values of energy, the relaxation procedure is applied using Eqs. (109) and (110).

The above-presented types of heterogeneous surface are only extremely idealized models of adsorbents. The surface layers of real solids have very complicated molecular structures. The observed heterogeneities may have originated not only from the chemical impurities built into the crystalline surface but also from various geometric defects, or they may be connected with the intrinstic nature of amorphous materials. Earlier reviews of adsorption addressed various aspects of surface modeling [6,12]. Here, our emphasis is on several characteristic examples. Benegas et al. [289] have modeled a random lattice by perturbing the position of constituting atoms in an initially perfect square lattice. The positions of the atoms in a disordered lattice were obtained by using stochastic vectors taken randomly from the Gaussian distribution of the specified parameters. The adsorption sites were located at the centers of each ring of solid atoms. However, the values of energy at a given adsorption site have not been exactly those corresponding to the disordered lattice used in simulation but assigned randomly from the Gaussian distribution of predetermined mean and dispersion. This means that Benegas et al. [289] have specified the energy distribution function independently of the geometry of their random lattice. It seems to be a certain shortcoming of this treatment. Because of the randomness of the lattice, the interactions between admolecules also are randomly distributed, so this model can he mapped onto random Ising model with, also random, an external field [290–293]. One can easily construct a more realistic model of the surface by assuming that adsorption energy is given by the minimum of the summed interactions between the incoming gas molecule and all atoms constituting a given cell.

The above-mentioned example clearly shows that the energetic heterogeneity can be considered as having originated from various geometrical irregularities present in the system. Numerous works confirmed this statement [12]. The simplest model involving geometry-induced surface heterogeneity is the stepped surface consisting of macroscopically long terraces, characterized by a finite width [294–301]. The energetic heterogeneity arises from the fact that the gas–solid interactions are different at both terrace edges and in the terrace interior [294,295]. The results obtained for the stepped surfaces are very interesting, especially in the context of general studies of phase transitions in systems with restricted geometry.

For the majority of real existing solid surfaces, the picture of an amorphous surface phase seems to be more suitable than that of a crystalline lattice with defects. The simple model of amorphous solids has been proposed by Bernal [302], who constructed the adsorbent by dense, random packing of hard spheres. The surface of such a model may be visualized as a heap of ball bearings on a plate. This idea has been used frequently in the simulation of adsorption on amorphous solids [303–308]. An interesting example of purely *ab initio* results for argon adsorbed on the oxide surface can be found in the article by Bakaev [303], who has calculated the energy distribution assuming the Bernal model and compared his functions with Drain and Morrison's adsorption energy distribution function for argon adsorbed on rutile [309]. The results were quite promising. Further studies have given more precise methods for the modeling of oxides surfaces. A good example is the method used by Bakaev and Steele [305]. The heterogeneous surface has been built by a random packing of hard balls by means of the algorithm proposed by Bonissent and Mutafschiev [310]. The balls formed the simulation cell of about five thickness. Then, the surface was divided into a small rectangular mesh, and each element of the surface was treated as a single adsorption site. The Monte Carlo simulations have been performed for argon adsorption in the region of submonolayer coverages using the grand canonical ensemble method. In low-coverage parts of the adsorption isotherms, linear dependences between adsorption and the logarithm of the gas pressure have been obtained. Such a behavior corresponds to the Temkin equation of adsorption isotherm [311]. The evaluated distributions of gas–solid interactions have shown interesting properties. The important conclusion was that an increase of the surface coverage affects the obtained energy distribution function; for concreteness, it causes a gradual widening with a shift toward lower-energy values. This results from the fact that for higher coverages, adsorption on weak active sites becomes more probable, whereas it does not influence the contributions due to strongly attracting sites.

A valuable modification of the original Bernel model of amorphous surfaces has been proposed by Cascarini de Torre and Bottani [312]. The essence of the improvement was the introduction of the possibility of the deletion of groups of atoms from the surface according to some rules. Depending on the choice of those rules, the surfaces of different structures can be obtained.

The very interesting so-called "rumpled graphite basal plane model" for the surface of nongraphized carbons has been proposed by Bakaev [313]. The experimental studies have shown that the nongraphized carbons, in spite of an evident presence of domains corresponding to the graphite basal plane, are characterized by broad energy distribution functions. These two facts cannot be consistently interpreted in terms of the popular "paracrystalline model" [314]. Electron microscopy studies suggest that the interior of the sample exhibits an irregular structure and the surface is much more ordered [315]. In order to describe both of these features, Bakaev has assumed that the surface was not a rigid, flat basal graphite plane, but it exhibited distortions; figuratively speaking, it was similar to the wrinkled sheet of paper.

All of the above-presented models of amorphous surfaces are of great importance in theoretical studies directed toward the development of methods for the evaluation of the energy distribution function.

When the model surface has been already generated, one can perform various "computer adsorption experiments." In the next subsection, we will discuss selected examples of such investigations."

C. Monte Carlo Simulation of Adsorption

Monte Carlo simulation of adsorption on heterogeneous solid surfaces can be used to solve the following problems:

Investigations of the properties of the modeled surfaces
Comparison of the theoretical predictions with computer simulations performed for a
    strictly the same system models
Determination of structural properties of surface layers
Investigations of phase transitions in the surface films

One of first computer simulations of adsorption on heterogeneous solid surfaces has been performed by O'Brien and Myers [316]. They simulated the system consisting of nine kinds of adsorption site placed in the square lattice at random. That randomness was accomplished by interchanging the site energies of two randomly chosen sites at each step, biased according to the Boltzmann factor of the resulting change. The energies and their proportions were chosen in such a way that the assumed energy distribution was the Gaussian-like function. The obtained adsorption isotherms have been treated as "experimental data" and used to evaluate the energy distribution. This function was back-calculated by using the CAEDMON algorithm. Their further investigations have clearly shown that the choice of local adsorption isotherm affects the form of the obtained adsorption energy distribution.

Adsorption on specially generated heterogeneous surface has been systematically studied by Benegas et al. [289]. They reported adsorption isotherms calculated for different values of the parameters characterizing the system. They have proved that a change in geometrical surface heterogeneity can cause a considerable change of the shape of the adsorption isotherm. Similar simulations of adsorption have been performed for the Bakaev model surfaces [303,304].

Monte Carlo simulations can be applied to the testing of the theory because they give the exact solutions for a given model, whereas theories involve various, sometimes crude and rather artificial, assumptions. For instance, it has been already well-known that the mean-field-type theories can predict the behavior of the complex systems at high temperatures quite well, but considerable differences between theoretical predictions and experimental data can be observed in the region of low temperatures. In such a situation, a computer pseudoexperiment can give more precise results than the theory. Moreover, using the computer simulations, one can estimate the parameters for which theory may be effectively applied.

The computer simulations of many authors confirmed the decisive role of the surface heterogeneity in the adsorption process [12,103,234,237,241–245,247,266–270,289,303,304, 317–321].

Here, we discuss only one example of Monte Carlo simulations carried out for adsorption from solutions of monomers and dimers on the heterogeneous solids. For the solid surface, the model with patchwise regular heterogeneity was used. The striplike and chessboard surfaces have been considered. The simulations have been performed using the standard Metropolis sampling method in the canonical ensemble. The simulation box was assumed to be a cubic fragment of three-dimensional cubic lattice. The surface is located at the bottom of the box and a reflecting wall has been placed at the top of the system. The standard boundary conditions were applied in both directions parallel to the solid surface. All lattice sites were occupied by monomer or dimer segments. The possible changes of the system state in a single move consisted of (1) the displacement of a monomer by one segment of a dimer to an adjacent site or (2) the displacement of two monomers by the whole dimer molecule to the pair of neighboring sites. This scheme corresponds to 14 types of possible motion of a dimer molecule: 6 translations and 8 rotations. Each dimer molecule can be translated parallel the to the x, y or z axis. During the rotation, one segment of a dimer remains at rest while the other is rotated by the angle equal to $\pi/2$. There are four possible different rotations for each segment.

The calculations have been performed for different parameters characterizing molecular interactions in the solutions and different sizes of homogeneous domains. The surface has been assumed to be composed of two evidently different kinds of adsorption site: one of them preferentially adsorbing dimers and the other strongly attracting monomers. It has been found that the adsorption isotherms obtained for a square crystalline surface (chessboard topography with $s^* = 1$) differ considerably from the isotherms obtained for surfaces consisting of greater homogeneous patches (see Fig. 10).

The main findings have been connected orientational effects in the surface films formed on heterogeneous adsorbents of various distributions of adsorption sites. In the case of the surface composed of different active centers placed alternately, the perpendicular of orientation of dimers is predominant in the whole concentration region. For all of the remaining systems, however, the parallel orientation is more probable.

Monte Carlo simulation gives us insight into microscopic structure of the system. Let us consider the density in the first adsorbed layer as a function of the coordinate x. In the case of a striplike surface, the average density for all possible values of y at fixed x has been calculated. For the chessboard topography, we consider the profiles of local densities in two places: (1) in the interior of the homogeneous patch and (2) on the boundary of the homogeneous domains. As an illustration, the examples of the density profiles are presented in density in Figs. 12–14. One can conclude from these figures that the local orientation of the dimers is directly connected with the surface topography. The local orientational structure of the mixed film is reflection of the surface structure. Generally, the changes in orientational ordering of dimers in the vicinity of the boundary between the two different surface domains are important only for relatively low patches. Let us compare the local densities obtained for the striplike and chessboard surfaces with $s^* = 3$. In these cases, two types of site on each adsorbing patch may be distinguished; the sites lying at the boundaries between different patches and the sites in the patch interiors The densities of dimers parallel to the x axis ($\varphi_x^{(1)}$) and y axis ($\varphi_y^{(1)}$) appeared to be different for adsorption on these sites. In the case of a striplike surface, the y orientation of dimers (i.e., orientation parallel to surface stripes) is preferred in the bordering positions, whereas the probabilities of both x and y orientations are similar for dimers located in the patch interior, although $\varphi_x^{(1)} > \varphi_y^{(1)}$ in all cases. Analogous results have been obtained for a chessboard surface

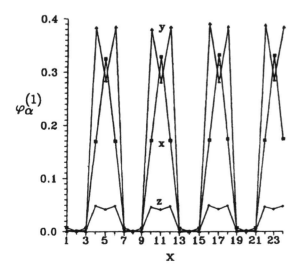

FIG. 12 The local density profile for differently oriented dimers, along the x axis, for the striplike surface with the strip width $s^* = 3$, $\omega = 0$, and the total dimer concentration $\Phi_1 = 0.1$. (Reprinted with permission from Langmuir *13*:1073–1078, March, 1997 © American Chemical Society.)

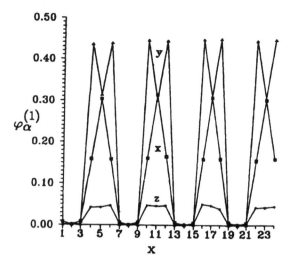

FIG. 13 The local density profile for differently oriented dimers, along the x axis and passing through the center of the patch, for the chessboard surface with $s^* = 3$, $\omega = 0$, and the total dimer concentration $\Phi_1 = 0.1$. (Reprinted with permission from: Langmuir *13*:1073–1078, March, 1997 © American Chemical Society.)

when the profiles in the middle of the patch are studied. In the case of a chessboard topography, the profiles evaluated along the boundaries show that the x-type orientation is preferred for the central adsorption site. Similar simulations were performed for the trimers adsorbed from the solutions in the monomers [318]. It has been found that the adsorbate molecule is sensitive to differences in surface topography when the patch sizes are not very much larger than the size of the molecule.

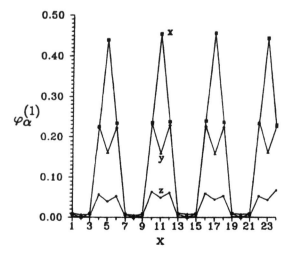

FIG. 14 The local density profile for differently oriented dimers, along the x axis and passing along the patch boundary for the chessboard surface with $s^* = 3$, $\omega = 0$, and total dimer concentration $\Phi_1 = 0.1$. (Reprinted with permission from: Langmuir *13*:1073–1078, March, 1997 © American Chemical Society.)

Notable among the applications of the Monte Carlo method are investigations of the surface phase transitions. Section VII is devoted to this subject.

VII.   PHASE TRANSITIONS

A.   Monolayer Films

The rapid progress in modern experimental techniques has shown a great variety of phase transitions in surface films. One of the major features of each physical system is its dimensionality. Although our surrounding space is three-dimensional, a monolayer, adsorbed film may be considered as being a two-dimensional one. The experiments have demonstrated that the phase diagrams evaluated for such adsorbed films are similar, to some extent, to those obtained for three-dimensional bulk matter. Monolayer films exhibit the existence of two-dimensional counterparts of gas, liquid, and solid phases [322,323]. Further studies have led to conclusion that phase behavior of adsorbed films might be even much richer. Now, it is evident that many of the observed phenomena do not have simple bulk counterparts (e.g., the commensurate–incommensurate transition or different surface order–disorder transitions) [324–328]. Much effort has been devoted to the explanation of the fundamental problems of phase transitions on uniform surfaces. These aspects have been less intensively studied for films formed on heterogeneous adsorbents [234,319,329–334]. The critical analysis of these investigations has been recently presented in the review articles [12,333]. Therefore, only a short summary will be given here.

Let us consider a monolayer film formed on the macroscopically large surface. The chemical potential of the film on the uniform surface can be written in the following form [334]:

$$\mu_a = (\varepsilon, \Phi, T) = -\varepsilon + \mu_a^*(\Phi, T) \tag{111}$$

where $\mu_a^*(\Phi, T)$ is just the two-dimensional equation of state, with Φ being the spreading pressure. In adsorption equilibrium, we have

$$\mu_g(p, T) = \mu_a(\varepsilon, \Phi, T) \tag{112}$$

where μ_g is the chemical potential of the bulk gas.

Now, assume that the adsorbed layer is not a homogeneous phase but exhibits two-phase coexistence. When two phases are separated by the first-order transition, the spreading pressure and chemical potentials of both phases should be the same (Φ_{eq} and $\mu_{a,eq}$). It is clear that the chemical potential of the bulk gas is the same as the chemical potential of the adsorbed film ($\mu_g = \mu_{a,eq}$). It means that the bulk pressure is fixed, as long as the system remains within the two-phase coexistence region. The values Φ_{eq} and μ_{eq} are also constant along the coexistence line and is independent of the number of adsorbed molecules. Thus, for the first-order phase transition, a discontinuity (vertical step) at the adsorption isotherm is observed. Moreover, the constancy of the equilibrium pressure that causes the isosteric heat of adsorption is also constant in the two-phase region. A characteristic feature of the first-order phase transition is also an appearance of a finite discontinuity in the heat capacity [329].

In the case of adsorption on a heterogeneous surface, the problem in question becomes much more difficult. For such an adsorbent, the interface between the dilute and dense phases changes its position as adsorption increases and encounters surface regions characterized by different values of adsorption energy ε. Therefore, the chemical potential of the film is no longer constant along the coexistence line but changes with the surface coverage. At the same time, the spreading pressure must still be unchanged within the two-phase region. Then, if the adsorption energy is changed and Φ_{eq} remains constant, Eqs. (111) and (112) can be satisfied only for a different value of the chemical potential of the bulk gas, thereby for a different value of gas pressure. Thus, the adsorption isotherm no longer exhibits a discontinuity but only a finite slope which is a function of the surface coverage (see Fig. 15). One can state that the heterogeneity causes smoothing of the adsorption isotherm. The experiments performed by Coulomb clearly confirmed this conclusion [335]. It is easy to show that for a heterogeneous surface, the isosteric heat of adsorption should change with the surface coverage within the two-phase region. The heterogeneity effects are also clearly visible in the case of heat capacity. The discontinuity disappears, because the boundary line is very sensitive to the local density variations, so that is not only the function of the average film density but also depends on variations in local densities of both phases due to changes in the surface field.

The above-discussed picture of the first-order phase transitions is realistic only for sufficiently weak heterogeneity effects. For highly heterogeneous surfaces, an existence of completely different adsorption sites causes considerable changes in lateral molecular interactions. In consequence, severe variations can appear in the local density of the film. It may lead to qualitative changes in the system's behavior. In particular, there will be nothing like a two-phase coexistence, as the adsorbate nucleation may be driven by the presence of strongly attracting

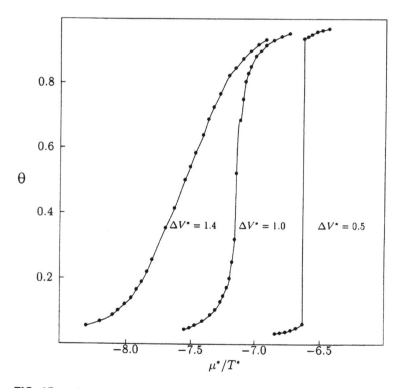

FIG. 15 Adsorption isotherms for the surface with square patches of size 3×3 at $T^* = 0.49$ for several values of the difference in adsorption energies on both kinds of active centers ΔV^*: 1.4, 1.0, and 0.6. The reduced values of energies and chemical potential are used: $\mu^* = \mu/u$ and $\Delta V^* = \Delta V/u$. [Reprinted from: A. Patrykiejew, Monte Carlo studies of adsorption IV. Phase transitions in localized monolayers on patchwise heterogeneous surfaces. Thin Solid Films *223*:39–50 (1993), with permission from Elsevier Science.]

active centers, not by mutual interactions that are responsible for homogeneous nucleation. In such a situation, adsorption isotherms become similar to the Langmuir-type relation [289].

However, the division into weakly and strongly heterogeneous surfaces is indistinct. How does one define "measure of heterogeneity"? Does the change between the weak and strong heterogeneity regimes occur gradually? Numerous important questions connected with heterogeneity effects on surface phase transitions require still answers [12].

Apart from purely energetic heterogeneity, some consequences of various geometrical irregularities should be discussed. Let us proceed with the analysis of the simplest geometrical effects following from the finite size of crystallites.

According to the finite-size scaling theory of the first-order phase transitions, the finite size of the crystallite causes a rounding of singularities in the behavior of isothermal compressibility and heat capacity. The boundary of the two-phase region formed on finite crystallites has length proportional to its one-dimensional size (L). This introduces an additional term to the chemical potential of the surface film [336]:

$$\Delta\mu_a = 2\gamma_l[L(\Theta_2 - \Theta_1)]^{-1} \tag{113}$$

where γ_l is the interfacial line tension, whereas Θ_1 and Θ_2 are densities of the dilute and dense phase, respectively. It shifts the location of the transition point in comparison with a macroscopically large surface. When the system exhibits a certain distribution of crystallite sizes, the transition point is met at different values of the bulk gas pressure for each kind of crystallite, and analogously to the case of energetic heterogeneity, the adsorption isotherm should be a continuous curve.

At a critical point, the phase transition under consideration is the second-order transition. For the uniform system, in the critical region the correlation length of statistical fluctuations, the isothermal compressibility and the heat capacity diverge to the infinity, according to the well-known power laws [337]. Moreover, the order parameter, which, for the gas–liquid transition, is defined as the difference between the densities of both coexisting phases, approaches zero at the critical temperature.

In the case of purely geometrical heterogeneity, an influence of the finite size of domains on critical behavior of the system can be deduced from the finite scaling theory [338]. In particular, for the finite system, the critical temperature is changed in comparison with that for the macroscopic one and the shift depends on the linear size of the system. The usual critical divergences of the heat capacity and compressibility are replaced by finite peaks. Moreover, in the finite system, the development of long-wavelength fluctuations is suppressed by the system size limitation, so the correlation length of statistical fluctuations can be, at most, of the same order as the size of the system. A more complex behavior can be observed on real surfaces, which consist of crystallites of different sizes.

An interesting example of geometrically induced heterogeneity is found in stepped surfaces [294–301]. Merikoski et al. [300] have used the transfer-matrix method to study adsorption on the stepped surfaces characterized by substrate steps of infinite length and small widths. Their results have given clear evidence that even a very weak potential acting along the step edges changes the behavior of surface films considerably, and they show properties characteristic for one-dimensional systems. For instance, the adsorption isotherms do not exhibit discontinuities even at very low temperatures, thus the adsorbate condensation does not occur via the first-order transition, as in the two-dimensional system, but is a continuous process.

The effects of the finite size of crystallites and the external force field have been studied with the help of Monte Carlo simulation by Patrykiejew [339]. The surface has been modeled as a collection of finite, two-dimensional homogeneous patches. A certain distribution of the patch sizes has been assumed. The patchwise model of surface has been implemented (i.e., the patches were independent one of another). The behavior of an adsorbate on a single patch has been described by the familiar two-dimensional lattice gas model Hamiltonian, with the term resulting from the presence of an external field. In the case of a single patch, the size dependence of the system follows directly from the scaling theory. The critical-point temperature is independent of an external potential. However, the phase diagrams evaluated for systems with different boundary fields are different and pronounced of asymmetry of phase boundary. The magnitude of the external field affects also the chemical potential at which the condensation occurs. When the surface consist of patches of different sizes and different magnitudes of the boundary field, the properties of the adsorption isotherm depend on the form of patch size distribution function as well as on the magnitudes of the boundary fields. It has been demonstrated that the first-order adsorbate condensation is not observed in these systems.

The Monte Carlo method has also been used to study phase transitions on energetically heterogeneous surfaces consisting of two types of active centers [319]. Sites of the same adsorption energy are grouped into finite patches and arranged in a regular, checkerboard way on the surface. The Hamiltonian has the following form:

$$\mathscr{H} = -\frac{1}{2} u \sum_{i,j} n_i n_j - \sum_{k=1}^{2} \varepsilon_k \sum n_i - \mu \sum_i n_i \tag{114}$$

where n_i denotes the number of occupied sites of the ith type, ε_i is the adsorption energy, and u is energy of interactions between gas molecules. The main parameters of the models are the size of the homogeneous patches M and the difference between the adsorption energies, $\Delta V = \varepsilon_1 - \varepsilon_2$.

From the ground-state analysis of the model, it follows that for a given size of homogeneous clusters M, there exists a certain limiting value of ΔV_0:

$$\Delta V_0 \sim \frac{1}{M} \tag{115}$$

which delimits the regions exhibiting qualitatively different behavior. For ΔV less than ΔV_0, the condensation in the surface film is qualitatively the same as in a purely homogeneous system (i.e., the first-order transition occurs at sufficiently low temperatures and only the location of the critical point depends on ΔV and M). However, when ΔV exceeds the value of ΔV_0, the condensation is a continuous process. In Fig. 15, we show adsorption isotherms for different energy heterogeneities; these examples clearly demonstrate that there is a qualitative change of the system's behavior when the difference of adsorption energies becomes greater than ΔV_0. From an analysis of the nucleated cluster size distribution functions, it follows that the difference between those two regimes is due to the changes in the mechanism of the adsorbate nucleation. For small differences in the adsorption energies, the cluster size distribution function is similar to that for a homogeneous system whereas for large values of ΔV, the first stages of condensation are connected with the preferential adsorption on strongly attracting sites. Then, the nucleation spreads gradually on the whole surface. Figure 16 presents examples of phase diagrams for systems with different values of $\Delta V^* = \Delta V/u$ and different sizes of homogeneous patches. One can conclude, from Eq. (115) that in the case of very large homogeneous patches, even very small heterogeneities should completely destroy any phase transition associated with the adsorbate condensation [12].

The adsorbed film behavior is slightly different when two types of adsorption site are randomly distributed on the surface [234]. When contribution of sites of one type is sufficiently low, then the qualitative behavior of the film resembles that observed for adsorption on a homogeneous surface. The surface heterogeneity effects lead only to the lowering of the critical point of the adsorbate condensation and to the rounding of the low- (high-) density part of the adsorption isotherms when impurity sites have higher (lower) adsorption energy than the

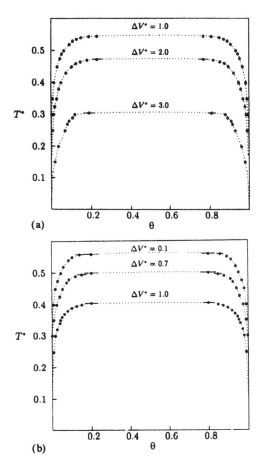

FIG. 16 Phase diagrams for systems with different surface structures [(a) 1×1 (b) 3×3] and for different values of $\Delta V^* = \Delta V / u$. [Reprinted from: A. Patrykiejew, Monte Carlo studies of adsorption IV. Phase transitions in localized monolayers on patchwise heterogeneous surfaces. Thin Solid Films *223*:39–50 (1993), with permission from Elsevier Science.]

remaining part of the surface. The critical temperature decreases linearly with the concentration of impurities (see Fig. 17).

Figure 18 shows the difference in the system behavior due to surface topography for the dual adsorbent when both types of adsorption site have equal contributions. In the case of patchwise topography, the adsorption isotherm exhibits well-pronounced steps corresponding to the subsequent occupation of patches of different sites. These steps can vanish for random topography.

The above examples clearly proved that the surface heterogeneity can change the character of the gas–liquid transitions in adsorbed monolayers and mechanism of the adsorbate condensation. In surface layers, one can also observe the commensurate–incommensurate-type transitions. Unfortunately, our knowledge of the behavior of incommensurate phases on nonuniform surfaces is still very limited. Only theoretical articles relevant to this problem are available [12].

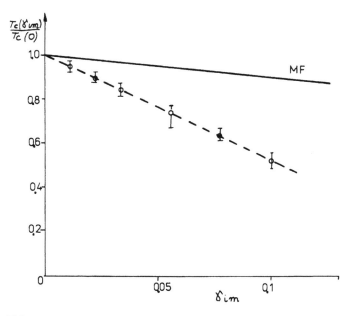

FIG. 17 Changes of the monolayer film critical temperature with the concentration of impurities (γ_{im}) obtained from the mean field theory (solid line) and from Monte Carlo simulations (circles). [Reprinted from: A. Patrykiejew, Monte Carlo studies of adsorption III. Localized monolayers on randomly heterogeneous surfaces. Thin Solid Films *208*:189–196 (1992), with permission from Elsevier Science.]

B. Multilayer Adsorption

It is quite apparent that the weak surface heterogeneity has the strongest influence on the adsorbate molecules located in the first layer because they "feel" even small local variations of the adsorbent potential. In general, the fluid–solid potential decays with the distance from the surface z, as z^{-3}, whereas the energy of interaction between adsorbate molecules decreases with their mutual distance r, as r^{-6} [see Eqs. (8) and (9)]. The influence of any local heterogeneities, such as the impurity atom built into the crystal lattice or slight geometrical defects, causes the perturbation to the surface potential that also decay with the distance from the surface, as z^{-6}. Thus, a direct influence of such heterogeneities on adsorbate molecules located, in the second and further layers is really very small, or even negligible. However, the adsorbent heterogeneity leads to considerable nonuniformities in the local fluid density in the first layer and these effects can be transmitted, via molecular interactions, to subsequent layers. As a consequence, the influence of substrate heterogeneity on higher adlayers can be quite important.

As it has been mentioned already, the systematic analysis of possible phase transitions was done by Pandit et al. [236] in terms of the lattice gas model. The Hamiltonian has a form analogous to Eq. (114). In the case of a homogeneous surface, the potential $v(l)$ depends only on the distance from the surface (i.e. on the number of layer l) and can be given by

$$v(l) = \frac{\varepsilon_0}{l^3} \tag{116}$$

where ε_0 is the energy of adsorption for adsorbate particles located in the first layer adjacent to the surface.

The mechanism of multilayer film formation depends on the relation between adsorbate–adsorbate and adsorbate–solid interactions. The controlling parameter for the surface phase

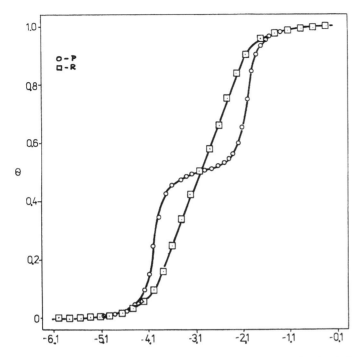

FIG. 18 Adsorption isotherms obtained from Monte Carlo simulations for the surface consisting of the same amount of two types of active center ($s_1 = s_2 = 0.5$) and $T^* = 0.5$, $\varepsilon_1/k_B = 1$, and $\varepsilon_2/k_B = 3$ for random (R) and patchwise topography (P). [Reprinted from: A. Patrykiejew, Monte Carlo studies of adsorption III. Localized monolayers on randomly heterogeneous surfaces. Thin Solid Films *208*:189–196 (1992), with permission from Elsevier Science.]

transitions is the ratio ε_0/u. When this ratio is sufficiently high, the film grows in a layer-by-layer mode. This means that at low temperatures, we observe a series of layering transitions which terminate at the corresponding critical points $T_c(l)$. According to the mean field theory,

$$\lim_{l \to \infty} T_c(l) = T_c^{3D} \tag{117}$$

where T_c^{3D} is the critical temperature of the bulk gas. However, the exact result of Binder and Landau [340] is

$$\lim_{l \to \infty} T_c(l) = T_R \tag{118}$$

where T_R is the roughening temperature, which is much lower than T_c [340]. In this case, the adsorbate wets the solid surface at any temperature down to $T = 0$.

When the ratio ε_0/u becomes lower and satisfies the inequality

$$\frac{\varepsilon_0}{u} < \left[\sum_{j=1}^{\infty} l^{-3} \right]^{-1} \tag{119}$$

the adsorbate does not wet the surface at $T = 0$, but it may still exhibit complete wetting at temperatures above the wetting temperature T_W. Thus, whereas for $T < T_W$, adsorption increases up to a finite value at the bulk condensation point, for $T > T_W$ one observes again a divergence of adsorption isotherm as the chemical potential approaches its bulk coexistence value. This divergence is termed "complete wetting." The change in the film behavior at the wetting temperature corresponds to the so-called wetting transition. According to the value of the ratio ε_0/u and the range of interaction in the system [236], the wetting transition can be the first-order transition as well as the continuous one (critical transition). When the wetting transition is first order, it is frequently preceded by the prewetting transition between thin and thick films. Then, the adsorption isotherm exhibits one jump, where the film thickness increases from a small value to a slightly larger value. As the temperature increases, the magnitude of the jump smoothly vanishes and the prewetting transition terminates at the corresponding critical point. The very instructive discussion of surface phase transitions can be found in the review by Patrykiejew et al. [333].

In the above discussion, the great variety of possible phase transitions is clearly shown. However, the aim of this chapter is to analyze the surface heterogeneity effects in various processes, which take place on the solids. Thus, we focus our attention on the changes in the behavior of multilayer films caused by nonuniformity of the surface.

The first article connected with the surface heterogeneity effects on the shape of the interface between a thick adsorbed layer of liquid and bulk gas had been published 25 years ago [341]. Then, the various aspects of the thick film's formation on heterogeneous solid surfaces have been subject of intensive theoretical [342–345] and computer simulation [320,321,346] studies. Here, we present some illustrative examples.

Let us begin with the heterogeneous adsorbent characterized by the adsorption energy distribution function

$$\varepsilon_{0,i} = \varepsilon_0 + \delta\varepsilon_{0,j} \tag{120}$$

where $\delta\varepsilon_{0,j}$ is a random variable of zero mean. In this model the surface heterogeneity is treated as a small perturbation in the system. The mean field approach to the model has been presented by Forgacs et al. [347]. They have considered the Gaussian distribution of $\delta\varepsilon_{0,j}$ and studied the effect of the heterogeneity on the order of the wetting transition. They have concluded that the prewetting transition critical point depends on the width of the Gaussian distribution. An increase in the distribution width leads to a decrease in the critical temperature of the prewetting point. Moreover, they found that the wetting transition, which is first order for a homogeneous solid, can be continuous in the presence randomly distributed heterogeneities. This is possible for sufficiently strong heterogeneity only. In terms of the mean field theory, the infinitesimal heterogeneity does not change the order of the wetting transition.

An influence of surface heterogeneity on layering transitions has been studied by means of Monte Carlo simulations [321]. Three different models of the surface have been considered, namely the surface composed from only two types of adsorption site (strong and weak) and the surface characterized by the discrete Gaussian distribution of adsorption energies and by the discrete uniform energy distribution. The adsorbate–solid potential was assumed to be given by Eq. (116), with different values of the parameter ε_0 for each adsorption site. In the case of the surface consisting of two kinds of active center, and when both types are strong enough to lead to the sequence of layering transitions, the main effect of surface heterogeneity was found to be a gradual decrease in the critical temperature for each layering transition. Examples of the adsorption isotherms for such systems are shown in Fig. 19. The effects of impurities become more important in systems characterized by weaker adsorbate–adsorbent interactions. It has been found that the first layering transition can split into two layering transitions occurring separately in the first and second layers. The situation changes in the case of a truly heterogeneous surface, consisting of many types of adsorption site. For the Gaussian and uniform distributions, the smearing of the phase transition region in the first and higher layers was observed. In this case, the adsorption isotherms do not have any vertical jump and remain continuous functions even at low temperatures.

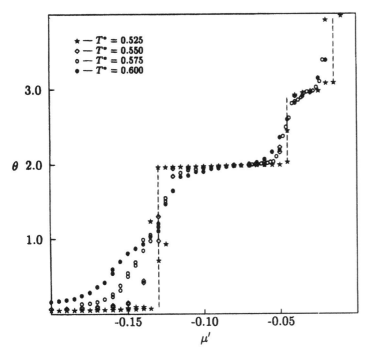

FIG. 19 Adsorption isotherms for systems characterized by $\varepsilon_{0,1}^* = 1$ and $\varepsilon_{0,2}^* = 2$ ($\varepsilon_0^* = \varepsilon_0/u$) at different temperatures (shown in the figure) and the contribution of strongly adsorbing sites equal to 0.1. (Reprinted with permission from: Langmuir *12*:159–169, January, 1996 © American Chemical Society.)

The film growth on a special type of dual surface has been also studied by Monte Carlo simulation in the lattice system. The model assumed that a narrow strip of weakly adsorbing sites surrounded by strongly attracting sites was on the surface. These calculations have evidently shown that the surface heterogeneity plays a very important role in the first stage of film growth. Two regimes of this growth have been observed. For small differences in adsorption energies on both types of site and when the weakly adsorbing strip is narrower than a certain limiting value of its width, the film spreads over the entire surface and shows only small lateral changes. On the other hand, in the case of large differences between adsorption energies and for a sufficiently wide weakly adsorbing strip, a drop of liquidlike film forms only over the strongly adsorbing domains, whereas the weakly adsorbing parts of the surface remain almost empty. The transition between these regimes is quite sharp. Similar results have been reported in the work of Schoen et al. [267,268,270], connected with adsorption in the pores with chemically heterogeneous walls.

Generally, the surface heterogeneity can change the character of surface phase transitions, shift the location of transition points, and change critical temperatures. It can completely change a molecular mechanism of the transitions. Indeed, the importance of the surface heterogeneity for phase behavior of the adsorption systems can hardly be overestimated.

VIII. CONCLUDING REMARKS

This chapter gives the review of developments in the theory of adsorption on heterogeneous solid surfaces. Theoretical developments presented in this study are focused on the problem of energetic heterogeneity of adsorbents. We have attempted to present the wide spectrum of topics which have been investigated in this field and the methods which have been used to explore them. In these studies, one can distinguish two branches: the investigations directed to practical, industrial applications and the basic research performed in order to discover new phenomena and formulate a refined theory of interfacial systems. Each of the proposed theoretical treatments is devised to address a different type of problem.

Our main focus in the first part of the chapter was on the application of the integral equation approach to describe many vastly different interfacial systems involving heterogeneous solid surfaces. Contrary to popular opinion, this treatment is not restricted to simple systems but can deal with quite complicated situations in a systematic way. This theory has been shown to yield valuable results for the adsorption of single gases, for gas and liquid mixtures, and for the chromatographic process. Most important is that the approach provides the analytical expressions for adsorption isotherms and thermodynamic functions of the surface phase. One can easily use derived relations in various technological applications. Moreover, it has now been apparent that the determination of the adsorption energy distribution is one of the most important methods for the characterization of adsorbents. In conclusion, the integral equation approach is an effective method for the situation when global heterogeneity effects should be taken into account but a simple theoretical interpretation is preferred. This means that from a technological perspective, this type of research is still very interesting.

Unfortunately, this approach does not give a deeper insight into a structure of surface films at the molecular level. The theory involves a concept of a certain averaging effects connected with heterogeneity of solid surfaces. Moreover, molecular interactions are usually described in terms of a mean field approximation. As a consequence, the integral equation approach cannot elucidate many experimental findings. In particular, various phase transitions in adsorbed layers, such as the order–disorder transition, cannot be explained in the framework of this theory.

The other class of theoretical approaches subsumes the treatments, which involve the three-dimensional model of adsorbate–adsorbate and assumes potentials of adsorbate–adsorbate and adsorbate–adsorbent interactions solely. Then, equilibrium density profiles are calculated by minimizing the free energy of the system. We have discussed two theories of such a type: the lattice model and the density functional theory.

The lattice models have been frequently used to introduce investigations of complicated phenomena. They allow us to catch the essential features of the systems. The method is relatively simple and effective and gives interesting information on the structural properties of films formed on energetically heterogeneous solid surfaces.

The density functional theory appears to be a powerful tool for studying adsorption on heterogeneous surfaces. In particular, the valuable results have been obtained for adsorption in pores with chemically structured walls. The interesting, new phenomena, such as bridging, have been discovered in this way. The phase diagrams characterizing various phase transitions in surface layers can be determined quite quickly from the functional density theory. In this context, we emphasize the economy of the computational efforts required for the application of the functional density methods. For this reason, the density theory can be under certain conditions, competitive with computer simulations. However, many applications of the density functional theory are based on rather crude, oversimplified assumptions, so the conclusions following from the calculations should be treated very cautiously.

Because of the lack of a general theory of adsorption, computer simulations have greatly contributed to the understanding of adsorption on heterogeneous solids. Monte Carlo simulations have already shown their ability to perform with the success in this area of research. In general, computer simulations are ideally suited to answer fundamental questions from a theoretical perspective. Based on a microscopic model of the system, computer simulations provide an exact solution of the model. The only necessary input for the Monte Carlo algorithm are the potentials representing all interactions in the system. We try to mimic the real solid surface and feed into the computer. Then, the adsorption process is simulated as in a real laboratory. However, in contrast to the standard experiments, simulations are carried out on a well-defined surface. Therefore, the behavior of the system can be elucidated in great detail. In many cases, the insight gained by computer simulations is unattainable by real measurements. This methodology can, in particular, clarify the role of surface topography in the adsorption on heterogeneous surfaces. In the case of adsorption on the well-defined nonuniform surfaces, one can observe a subtle interplay, defined by the competition between forces acting between the adsorbate molecules and interactions with different parts of the surface. As a consequence, the rich scenario of possible adsorption behaviors can be realized. In this respect, it is worth mentioning that Monte Carlo simulations can contribute much to the studies of phase transitions in surface films.

Nowadays, the phase transitions are probably the most intensively studied aspects of adsorption on heterogeneous surfaces. The recent developments presented in the previous section are focused on the role of surface heterogeneity in these phenomena. Although many interesting conclusions have been formulated, much room remains to explore, even for simple surface models, either in the framework of theoretical treatments or by computer simulation. In this case, computer simulations can be viewed as a promising tool for investigation because analytical approaches are limited to crude approximations and laboratory experiments are really difficult.

This short overview can only give a general idea of what has been done in the investigations of adsorption on heterogeneous surfaces. Many interesting problems had to be omitted. The important and fascinating area of adsorption in random porous media is missing [348]. Furthermore, we focus on energetic heterogeneity, and different aspects, such as the geometrical heterogeneity, are treated in a cursory way.

Here, we would like to call attention to one especially interesting topic. Many adsorbents can be considered as the fractal surfaces From a mathematical point of view, a fractal set exhibits the property that the "whole" can be represented as the collection of several parts, each one obtainable from the "whole" by a contracting similitude. The identification of an irregular surface with a fractal set has strict constraints on the surfaces' characteristics; namely it implies the recurrence of the same irregularity details when the surface is magnified successively. Evidence for the fractal behavior of a few real surfaces has been provided and an influence of their fractal nature on adsorption process has been shown [349,350]. The so-called fractal dimension should be treated as one of most important parameters characterizing such surfaces.

Ending this chapter, we should mention that further work is needed to answer all of the questions connected with adsorption on heterogeneous surfaces. Probably, a combined application of theoretical methods and simulation would provide faster progress in the studies of surface films formed on heterogeneous surfaces. At present, the modern microcontact printing techniques permit one to generate the real surfaces directly corresponding to the idealized models used in theoretical treatments. In this situation, we will find a link among theory, simulation, and experiment. A comparison of simulation results and laboratory observations should gain a deeper insight into the molecular mechanism of the studied phenomena.

REFERENCES

1. S Ross, and JP Olivier. On Physical Adsorption. New York: Interscience, 1964.
2. A Clark. The Theory of Adsorption and Catalysis. New York: Academic Press, 1970.

3. JG Dash. Films on Solid Surfaces. New York: Academic Press, 1975.
4. J Ościk. Adsorption. Warsaw/Chichester: Polish Science Publishers/Horwood, 1982.
5. M Jaroniec, R. Madey. Physical Adsorption on Heterogeneous Solids. Amsterdam: Elsevier, 1988.
6. W Rudziński, DH Everett. Adsorption of Gases on Heterogeneous Surfaces. New York: Academic Press, 1992.
7. W Rudziński, G Zgrablich, eds. Equilibria Dynamics of Gas Adsorption on Heterogeneous Solid Surfaces. Studies in Surface Science and Catalysis Vol. 104. Amsterdam: Elsevier, 1997.
8. M Jaroniec, A Patrykiejew, M Borówko. In: DA Cadenhead, JF Danielli, eds. Progress in Surface and Membrane Science, Vol. 14. New York: Academic Press, 1981, pp 1–68.
9. GF Cerofolini, W Rudziński. In: W Rudziński, G Zgrablich, eds. Equilibria Dynamics of Gas Adsorption on Heterogeneous Solid Surfaces. Studies in Surface Science and Catalysis Vol. 104. Amsterdam: Elsevier, 1997, pp 1–103.
10. M Jaroniec, J Choma. In: W Rudziński, G Zgrablich, eds. Equilibria Dynamics of Gas Adsorption on Heterogeneous Solid Surfaces. Studies in Surface Science and Catalysis Vol. 104. Amsterdam: Elsevier, 1997, pp 715–744.
11. A Dąbrowski. Adv Colloid Interf Sci 93:135, 2001.
12. A Patrykiejew, M Borówko. In: M Borówko, ed. Computational Methods in Surface and Colloid Science. Surfactant Science Series Vol. 89. New York: Marcel Dekker, 2000, pp 245–292.
13. P Ripa, G Zgrablich. J Phys Chem 79:2118, 1975.
14. AJ Ramirez-Pastor, MS Nazzaro, JL Riccardo, G Zgrablich. Surface Sci. 341:249, 1995.
15. JL Riccardo, MA Chade, VD Pereya, G Zgrablich. Langmuir 8: 1518, 1995.
16. D Henderson, ed. Fundamentals of Inhomogeneous Fluids. New York: Marcel Dekker, 1992.
17. D Nicholson, ND Parsonage. Computer Simulation and the Statistical Thermodynamics of Adsorption. New York: Academic Press, 1983.
18. M Borówko, ed. Computational Methods in Surface and Colloid Science. Surfactant Science Series Vol. 89. New York: Marcel Dekker, 2000.
19. WA Steele. Surface Sci. 36:317, 1973.
20. JO Hirschfelder, CF Curtiss, RB Bird. Molecular Theory of Gases and Liquids. New York: Wiley, 1954.
21. R Sips. J Chem Phys 16:490, 1949.
22. R Sips. J Chem Phys 118:1024, 1950.
23. I Langmuir. J Am Chem Soc 40:1361, 1918.
24. K Huang. Statistical Mechanics. New York: Wiley, 1963.
25. RH Fowler, EA Guggeineim. Statistical Thermodynamics. London: Cambridge University Press, 1949.
26. WL Bragg, EJ Williams. Proc Roy Soc London A 145:699 1934.
27. HA Bethe. Proc Roy Soc London A 150:552, 1935.
28. R Peierls. Proc Cambridge Phil Soc 32:471, 1936.
29. WA Steele. The Interaction of Gases with Solid Surfaces. Oxford: Pergamon, 1974.
30. AV Kiselev. Koll Zh 20:338, 1958.
31. GI Berezin, AV Kiselev. J Colloid Interf Sci 38:227, 1972.
32. M Jaroniec, S Sokołowski, GF Cerofolini. Thin Solid Films 31:321, 1976.
33. A Patrykiejew, M Jaroniec. Surface Sci 77:365, 1978.
34. TL Hill, J Chem Phys 14:441, 1946.
35. JH de Boer. The Dynamical Character of Adsorption. Oxford: Oxford University Press, 1953.
36. S Brunauer, PH Emmett, E Teller. J Am Chem Soc 60:309, 1938.
37. M Jaroniec, P Brauer. Surface Sci Rep 6:65, 1986.
38. A Patrykiejew, M Jaroniec. Adv Colloid Interf Sci. 20:273, 1984.
39. P Brauer, WA House, M Jaroniec. Thin Solid Films 97:369, 1982.
40. AW Marczewski, M Jaroniec. Monatsch Chem 114:711, 1983.
41. M Jaroniec. In: AL Myers, G Belford, eds. Fundamentals of Adsorption. New York: American Institute of Engineers, 1984.
42. J Tóth. In: AL Myers, G Belford, eds. Fundamentals of Adsorption. New York: American Institute of Engineering, 1984.

43. M Jaroniec, AW Marczewski. Monatsch Chem *115*:997, 1987.
44. J Tóth. Acta Chim Hung *32*:31, 1962.
45. J Tóth. Acta Chim Hung *69*:311, 1971.
46. J. Tóth, W Rudziński, A Waksmundzki, M Jaraoniec, S Sokołowski. Acta Chim Hung *82*:11, 1974.
47. M Jaroniec. Surface Sci *50*:553, 1975.
48. AJ Knowles, JB Moffat. J Colloid Interf Sci *41*:116, 1972.
49. M Jaroniec, W Rudziński, S Sokołowski, R Smarzewski, Colloid Polym Sci *253*:164, 1975.
50. M Jaroniec. J Colloid Interf Sci *52*:41, 1975.
51. M Jaroniec, JA Jaroniec. Carbon *15*:107, 1977.
52. S Ozawa, S Kusumi, Y Ogino. J Colloid Interf Sci *56*:83, 1976.
53. Y Wakasugi, S Ozawa, Y Ogino. J Colloid Interf Sci *79*:399, 1976.
54. WA House, M Jaroniec, P Brauer, P Fink. Thin Solid Films *85*:87, 1981.
55. WA House, G Born, P Brauer, S Frankle, KH Henneberg, P Hoffer, M Jaroniec. J Colloid Interf Sci *99*:493, 1984.
56. H Freundlich. Trans Faraday Soc *28*:195, 1932.
57. FC Tompkins, DM Young. Trans Faraday Soc *47*:88, 1951.
58. GD Halsey. Adv Catal *4*:259, 1952.
59. JI Joubert, I Zwiebel. Adv Chem *101*:209, 1971.
60. GF Cerofolini. Vuoto *8*:178, 1975.
61. MP Rosynek. J Phys Chem *79*:1280, 1975.
62. P Esser, W Gopel. Surface Sci *97*:309, 1980.
63. JMD Tascon, LG Tejuca. J Chem Soc Faraday I *77*:591, 1981.
64. GF Cerofolini. Z Phys Chem *262*:289, 1981.
65. GF Cerofolini. J Colloid Interf Sci *86*:204, 1982.
66. GF Cerofolini. Surface Sci *24*:391, 1971.
67. GF Cerofolini. J Low Temp Phys *6*:473, 1972.
68. GF Cerofolini. Thin Solid Film *23*:129, 1974.
69. MM Dubinin, LV Radushkevich. Dokl Akad Nauk SSSR Ser Khim *55*:331, 1947.
70. JP Hobson. J Chem Phys *34*:1850, 1961.
71. JP Hobson, J Phys Chem *73*:2720, 1969.
72. JP Hobson, RA Armstrong. J Phys Chem *67*:2000, 1963.
73. R Haul, BA Gottwald. Surface Sci *4*:334, 1966.
74. L Rosai, TA Giorgi. J Colloid Inter Sci *51*:217, 1975.
75. NI Ionescu. Surface Sci *62*:294, 1976.
76. HF Stoeckli. Carbon *19*:325, 1981.
77. D Morel, HF Stoeckli, W Rudziński. Surface Sci *114*:85, 1982.
78. GF Cerofolini. In: DH Everett, ed. Specialist Periodical Reports. Colloid Science Vol. 4. London: 1982, pp 59–83.
79. S Sokołowski. Vuoto *8*:45, 1975.
80. MM Dubinin, VA Astakhov. Izv Akad Nauk SSSR Ser Khim *71*:5, 1971.
81. MM Dubinin. Prog. Surface Membr Sci *9*:1, 1975.
82. O Kadlec. In: MM Dubinin, VV Serpinsky, eds. Adsorption and Porosity. Moscow: Nauka, 1976, p 279.
83. M Jaroniec, AW Marczewski. J Colloid Interf Sci *101*:280, 1984.
84. GF Cereofolini. J Low Temp Phys *23*:687, 1976.
85. GF Cerofolini. Surface Sci *61*:678, 1976.
86. M Jaroniec, A Patrykiejew. Phys Lett *67A*:309, 1978.
87. MM Dubinin, TS Jakubov, M Jaroniec, VV Serpinsky. Polish J Chem *54*:1721, 1980.
88. JK Garbacz, M Jaroniec, A Deryło. Thin Solid Films *75*:307, 1981.
89. M Jaroniec, M Borówko. J Colloid Interf Sci *63*:362, 1978.
90. WG McMillan. J Chem Phys *15*:390, 1947.
91. AC Zettlemoyer, WC Walker. J Phys Chem *52*:58, 1948.
92. TL Hill. J Chem Phys *14*:263, 1946.
93. S Brunauer, J Skalny, EE Bodor. J Colloid Interf Sci *30*:546, 1969.

94. JM Honig. J Phys Chem *57*:349, 1953.
95. W Rudziński, S Sokołowski, M Jaroniec, J Tóth. Z Phys Chem *255*:273, 1975.
96. CC Hsu, W Rudziński, BW Wojciechowski. Phys Lett *54A*:763, 1976.
97. LM Dormant, AW Adamson. J Colloid Iinterf Sci *38*:285, 1972.
98. RH van Dongen. Surface Sci *39*:341, 1973.
99. CC Hsu, W. Rudziński, BW Wojciechowski. J Chem Soc Faraday I *72*:453, 1976.
100. WA House, MJ Jaycock. Colloid Polym Sci *256*:52, 1978.
101. A Patrykiejew, M Jaroniec, R Smarzewski. Monatsch Chem *110*:601, 1979.
102. GD Halsey. J Am Chem Soc *73*:2693, 1951.
103. D Nicholson, RG Silvester. J Colloid Interf Sci *62*:447, 1977.
104. M Jaroniec, M Borówko, JK Garbacz. Thin Solid Films *54*:L1, 1978.
105. EC Titchmarch. Introduction of the Theory of Fourier Transforms. London: Oxford University Press (Clarendon), 1959.
106. DN Misra. Surface Sci *18*:367, 1969.
107. DN Misra. J Chem Phys *52*:5499, 1970.
108. U Landman, EW Montroll. J Chem Phys *64*:1762, 1976.
109. W Rudziński, J Jagiełło. J Low Temp Sci *45*:1, 1981.
110. W Rudziński, J Jagiełło, Y Grillet. J Colloid Interf Sci *87*:478, 1982.
111. J Jagiełło, G Ligner, E. Papirer. J. Colloid Iinterf Sci *137*:128, 1989.
112. J Jagiełło, JA Schwarz. J Colloid Interf Sci *146*:415, 1991.
113. SZ Roginsky. Dokl Akad Nauk SSSR *45*:61, 1944.
114. LB Harris. Surface Sci *10*:126, 1968.
115. LB Harris. Surface Sci *13*:377, 1969.
116. LB Harris. Surface Sci *15*:182, 1969.
117. GF Cerofolini. Surface Sci *47*:469, 1975.
118. JP Hobson. Can J Phys *43*:1941, 1965.
119. CC Hsu, BW Wojciechowski, W Rudziński, J Narkiewicz. J Colloid Interf Sci *67*:292, 1978.
120. W Rudziński, J Jagiełło. Vacuum *32*:577, 1982.
121. A Sommerfeld, H Bethe. Handbuch der Physik, Bd. 24. Berlin: Springer-Verlag 1943.
122. WA House, MJ Jaycock. J Colloid Interf Sci *47*:50, 1974.
123. SE Hory, JM Prausnitz. Surface Sci *6*:377, 1967.
124. B Kindl. RA Pachovsky, BA Spencer, BW Wojciechowski. J Chem Soc Faraday I *69*:1162, 1973.
125. B Kindl, BW Wojciechowski. J Chem Soc Faraday I *69*:1926, 1973.
126. AW Adamson, I Ling. Adv. Chem Ser *33*:53, 1961.
127. WA House. In: DH Everett, ed. Specialist Periodical Reports. Colloid Science Vol. 4. London: The Chemical Society, 1982, chap 1.
128. AW Adamson, I Ling, L. Dormant, M Orem. J Colloid Interf Sci *21*:445, 1966.
129. PY Hsieh. J Phys Chem *68*:1068, 1965.
130. J Wahlen. J Phys Chem *71*:1557, 1967.
131. JB Sorrel, R Rowan. Anal Chem *42*:1712, 1970.
132. DJ Jackson, JCR Waldax. J Colloid Interf Sci *37*:462, 1971.
133. MJ Jaycock, BW Davis. J Colloid Interf Sci *47*:499, 1974.
134. A Waksmundzki, S Sokołowski, M Jaroniec, J Rayss. Vuoto *8*:113, 1975.
135. M Rozwadowski, J Siedlewski, R Wojsz. Z Phys Chem *262*:341, 1981.
136. WA House, JM Jaycock. J Colloid Interf Sci *59*:252, 1977.
137. WA House, JM Jaycock. J Chem Soc Faraday I *73*:942, 1977.
138. WA House, J. Colloid Interf Sci *67*:166, 1978.
139. WA House, J Chem Soc Faraday I *74*:1045, 1978.
140. WA House, M Jaroniec, P Brauer, P Fink. Thin Solid Films *87*:323, 1982.
141. EW Sidebottom, WA House, JM Jaycock. J Chem Soc Faraday I *72*:2709, 1976.
142. LK Koopal, K Vos. Colloid Surfaces *14*:87, 1985.
143. S Ross, ID Morrison. Surface Sci. *52*:103, 1975.
144. RS Sacher, ID Morrison. J Colloid Interf Sci *70*:153, 1979.

145. CL Lawson, RJ Hanson. Solving Least Squares Problems. Englewood Cliff, *NJ*: Prentice-Hall, 1974.
146. DC Hinaman, GD Halsey. J Phys Chem *81*:739, 1977.
147. SP Wesson, JJ Vajo, S Ross. J Colloid Interf Sci *94*:552, 1983.
148. K Vos, LK Koopal. J Colloid Interf. Sci *105*:183, 1985.
149. P Brauer, M Fassler, M Jaroniec. Thin Solid Films *123*:245, 1985.
150. P Brauer, M Fassler, M Jaroniec. Chem Phys Lett *125*:241, 1986.
151. BJ Stanley, G Guichon. J Phys Chem *97*:8098, 1993.
152. Z. Sokołowska, M Hajnos, M Borówko, S Sokołowski. J. Colloid Interf Sci. *219*:1, 1999.
153. LE Drain, JA Morrison. Trans Faraday Soc *49*:654, 1953.
154. J Papenhuijzen, JK Koopal. In: RH Ottewill, CH Rochester, AL Smith, eds. Adsorption from Solutions London: Academic Press, 1983, p 211.
155. W Rudziński, M Jaroniec. Surface Sci *42*:552, 1974.
156. BK Oh, SK Kim. J Chem Phys *67*:3416, 1977.
157. S Sokołowski, M Jaroniec, GF Cerofolini. Surface Sci *47*:429, 1975.
158. M Jaroniec, W Rudziński. Colloid Polym *253*:683, 1975.
159. VY Davydov, AV Kiselev, LT Zhuravler. Trans Faraday Soc *60*:2254, 1964.
160. R Leboda, S Sokołowski. J Colloid Interf Sci *61*:365, 1977.
161. W Rudziński, A Waksmundzki, R Leboda, Z Suprynowicz. J Chromatogr *92*:25, 1974.
162. P Brauer, M Jaroniec. J Colloid Interf Sci *108*:50, 1985.
163. K Tsutsumi, Y Mitani, H Takahashi. Colloid Polym Sci *263*:383, 1985.
164. Z Suprynowicz, M Jaroniec, J Gawdzik. Chromatographia *9*:161, 1976.
165. MH Poley, WD Schaeffer, WR Smith, J Phys Chem *57*:469, 1953.
166. AL Myers. In: AI Liapis, ed. Fundamentals of Adsorption. New York: Engineering Foundation, 1987.
167. W Rudziński, K Nieszporek, H Moon, HK Rhee. Heterogen Chem Rev *1*:275, 1994.
168. Yu K Tovbin. In: W Rudziński, G Zgrablich, eds. Equilibria Dynamics of Gas Adsorption on Heterogeneous Solid Surfaces. Studies in Surface Science and Catalysis Vol. 104. Amsterdam: Elsevier, 1997, pp 105–152.
169. EC Markham, AF Benton. J Am Chem Soc *53*:497, 1931.
170. M Jaroniec. Adv Colloid Interf Sci *18*: 149, 1983.
171. M Jaroniec. J Colloid Interf Sci *53*:422, 1975.
172. M Jaroniec. Phys Lett *56A*:53, 1976.
173. M Jaroniec, M Borówko. Surface Sci *66*:652, 1977.
174. M Jaroniec. Colloid Polym Sci *255*:176, 1977.
175. M Jaroniec, J Narkiewicz, M Borówko, W Rudziński. Polish J Chem *52*:197, 1978.
176. M Jaroniec, J Narkiewicz, W Rudziński. J Colloid Interf Sci *65*:9, 1978.
177. BW Wojciechowski, CC Hsu, W Rudziński. Ca J Chem Eng *63*:789, 1985.
178. S Roginsky, O Todes. Acta Physicochim SSSR *20*:696, 1945.
179. E Glueckauf. Trans Faraday Soc *47*:96, 1951.
180. MI Tiemkin. Kinet Kataliz *17*:1416, 1975.
181. S Snagovsky. Kinet Kataliz *17*:1436, 1975.
182. M Jaroniec, J Tóth. Colloid Polym Sci *254*:643, 1976.
183. M Jaroniec. Colloid Polym Sci *255*:32, 1977.
184. M Jaroniec. J Colloid Interf Sci *59*:230, 1977.
185. M Jaroniec, J Tóth. Colloid Polym Sci *256*:690, 1978.
186. A Patrykiejew, M Jaroniec. A Dąbrowski, J Tóth. Croatica Chem Acta *53*:9, 1980.
187. M Borówko, M Jaroniec, W Rudziński. Z Phys Chem *260*:1027, 1979.
188. M Jaroniec. J Res Inst Catal. Hokkaido Univ *28*:31, 1980.
189. PJ Crickmore, BW Wojciechowski. J Chem Soc Faraday I *73*:1216, 1977.
190. M Jaroniec, R Madey, D Rothstein. Chem Eng Sci *42*:2135, 1987.
191. M Jaroniec. Thin Solid Films *71*:273, 1980.
192. L Szepesy, V Illes. Acta Chim Hung *35*:245, 1963.
193. M Jaroniec, A Patrykiejew, M Borówko. Surface Sci *78*:L501, 1978.
194. M Jaroniec. Vuoto *9*:57, 1976.

195. M Jaroniec, J Piotrowska, M Bulow. Thin Solid Films *106*:219, 1983.
196. GD Parfitt, CH Rochester, eds. Adsorption from Solutions. New York: Academic Press, 1983.
197. DH Everett. Trans Faraday Soc *60*:1803, 1964.
198. A Dąbrowski, M Jaroniec. Mater Chem Phys *12*:339, 1985.
199. M Jaroniec, A Patrykiejew. J Chem Soc Faraday I *76*:2486, 1980.
200. A Dąbrowski, M Jaroniec. Adv Colloid Inter Sci *27*:211, 1987.
201. A Dąbrowski, M Jaroniec. J Colloid Interf Sci *73*:475, 1980.
202. A Dąbrowski, M Jaroniec. J Colloid Interf Sci *77*:571, 1980.
203. A Dąbrowski, Chem Scripta *25*:182, 1985.
204. A Dąbrowski, Z Phys Chem *267*:494, 1986.
205. A Deryło-Marczewska, M Jaroniec, J Ościk, AW Marczewski, R Kusak, Chem Eng Sci *42*:2143, 1987.
206. A Dąbrowski, M Jaroniec, J Ościk. Surface and Colloid Science Vol. 14. New York: Plenum Press, 1987, pp 83–213.
207. A Dąbrowski, M Jaroniec. Acta Chim Hung *99*:255, 1979.
208. A Dąbrowski, M Jaroniec. Acta Chim Hung *104*:183, 1979.
209. A Dąbrowski, J Ościk, W Rudziński, M Jaroniec. J Colloid Interf Sci *69*:287, 1979.
210. M Jaroniec, AW Marczewski, WD Einickie, H Herden R Schollner. Monatsch Chem *114*:857, 1983.
211. M Jaroniec, AW Marczewski. Monatsch Chem *115*:541, 1984.
212. S Sircar. J Chem Soc Faraday I *79*:2085, 1983.
213. S Sircar. Surface Sci *148*:478, 1984.
214. A Suchowitzky. Acta Physicochem SSSR *8*:531, 1983.
215. J Ościk, A Dąbrowski, M Jaroniec, W Rudziński. J Colloid Interf Sci *65*:403, 1976.
216. W Rudziński, S Partyka. J Chem Soc Faraday I *78*:2361, 1982.
217. M Borówko, J Goworek, M Jaroniec. Monatsch Chem *113*:669, 1982.
218. M Borówko, M Jaroniec. J Chem Soc Faraday I *79*:363, 1983.
219. W Rudziński, J Zając, CC Hsu, J Colloid Interf Sci *103*:528, 1985.
220. A Dąbrowski, Monatsh Chem *114*:875, 1983.
221. A Dąbrowski, Monatsch Chem *117*:139, 1986.
222. A Dąbrowski, M Jaroniec, J Tóth. Acta Chim Hung *111*:311, 1982.
223. J Tóth. Acta Chim Hung *63*:67, 179, 1974.
224. W Rudziński, J Narkiewicz-Michałek, K Plilorz, S Partyka. J Chem Soc Faraday I *81*:999, 1985.
225. W Rudziński, J Jając. E Wolfram, I Paszli. Colloids Surface *22*:317, 1987.
226. M Borówko, M Jaroniec. Adv Colloid Interf Sci *19*:137, 1983.
227. M Jaroniec, DE Matire, M Borówko. Adv Colloid Interf Sci *22*:177, 1985.
228. M Borówko, M Jaroniec, W Rudziński. Monatsch Chem *112*:59, 1980.
229. M Jaroniec, J Ościk, A Deryło. Acta Chim. hung. *106*:257, 1981.
230. M Borówko, M Jaroniec, J. Ościk, R Kusak. J Colloid Interf Sci *69*:311, 1979.
231. M. Borówko, M Jaroniec. Chromatographia *12*:672, 1979.
232. M Jaroniec, JK Różyło, W Gołkiewicz. J Chromatogr *178*:27, 1979.
233. M Jaroniec, B Ościk-Mendyk. J Chem Soc Faraday Trans I *77*:1277, 1981.
234. A Patrykiejew. Thin Solid Films *208*:189, 1992.
235. MJ de Oliviera, B Griffits. Surface Sci *71*:687, 1978.
236. R Pandit, M Schick, M Wortis. Phys Rev B *26*:5112, 1982.
237. T Nitta, M Kuro-oka, T Katayama. J Chem Eng Japan *17*:45, 1984.
238. AW Marczewski, A Deryło-Marczewska, M Jaroniec. J Colloid Interf Sci *109*:310, 1986.
239. T Nitta, H Kiriyama, T Shigeta. Langmuir *13*:903, 1997.
240. EA Guggenheim. Mixtures. Oxford: Clarendon Press, 1952.
241. AJ Ramirez-Pastor, MS Nazzaro, JL Riccardo, G Zgrablich. Surface Sci *341*:249, 1995.
242. AJ Ramirez-Pastor, VD Pereyra, JL Ricardo. Langmuir *15*:5707, 1999.
243. AJ Ramirez-Pastor, JL Ricardo, VD Pereyra. Langmuir *16*:682, 2000.
244. M Borówko, W. Rżysko. J Chem Soc Faraday Trans *91*:3195, 1995.
245. M Borówko, W. Rżysko. J Colloid Interf Sci *128*:268, 1996; 185 (Suppl): 557, 1997.

246. L Łajtar, S Sokołowski. J Chem Soc Faraday Trans *88*:2545, 1992.
247. G Chmiel, L. Łajtar, S Sokołowski, A Patrykiejew. J Chem Soc Faraday Trans *90*:1153, 1994.
248. G. Chmiel, K Karykowski, A Patrykiejew, W Rżysko, S Sokołowski. Mol Phys *81*:691, 1994.
249. P Roecken, P Tarazona. J Chem Phys *105*:2034, 1996.
250. P Roecken. A Somoza, P Tarazona, G Findenegg. J Chem Phys *105*:2034, 1996.
251. B Millan Malo, L Salazar, O Pizio, S Sokołowski. J Phys Condens Matter *12*:8785, 2000.
252. B Millan Malo, O Pizio, A Patrykiejew, S Sokołowski. J Phys Condens Matter *13*:1, 2001.
253. G. Chmiel, A Patrykiejew, W Rżysko, S Sokołowski. Mol Phys *83*:19, 1994.
254. DH Henderson, S. Sokołowski, D Wasan. Phys Res E *57*:5539, 1998.
255. A Huerta, O Pizio, P Bryk, S Sokołowski. Mol Phys *98*:1859, 2000.
256. S Deitrich. In: C Caccano et al, eds. New Approaches to Problems in Liquid State Theory. Kluwer Academic Publishers, 1999, pp 197–244.
257. W Koch, S Dietrich, M Mapiórkowsi. Phys Rev E *51*:3300, 1995.
258. R Evans. D Henderson, ed. Fundamentals of Inhomogeneous Fluids. New York: Marcel Dekker, 1992.
259. JD Weeks, D Chandler, HC Andersen. J Chem Phys *54*:5237, 1971.
260. NF Carnahan, KE Starling. J Chem Phys *51*:635, 1969.
261. P Tarazona. Phys Rev A *31*:2672, 1985.
262. P Tarazona. Phys Rev A *32*:3148, 1985.
263. P Tarazona, UMB Marconi, R Evans. Mol Phys *60*:573, 1987.
264. E Kierlik, M Rosinberg. Phys Rev A *42*:3382, 1990.
265. E Kierlik, M Rosinberg. Phys Rev A *44*:5025, 1991.
266. M Schoen, DJ Diestler. Phys Rev E *56*:4427, 1997.
267. M Schoen, DJ Diestler. Chem Phys Lett *270*:339, 1997.
268. M Schoen, DJ Diestler. J Chem Phys *109*:5596, 1998.
269. H Bock, M Schoen. Phys Rev E *59*:4122, 1999.
270. M Schoen. In: M Borówko, ed. Computational Methods in Surface and Colloid Science. Surfactant Science Series Vol 89. New York: Marcel Dekker, 2000, pp 1–75.
271. M Borówko, S. Sokołowski, O. Pizio. In: M Borówko, ed. Computational Methods in Surface and Colloid Science. Surfactant Science Series Vol. 89. New York: Marcel Dekker, 2000, pp 167–244.
272. MS Wertheim. J Chem Phys *87*:7323, 1987.
273. CJ Segura, WG Chapman, KP Shukla. Mol Phys *90*:759, 1997.
274. MP Allen, DJ Tidesley. Computer Simulation of Liquids. New York: Oxford University Press, 1987.
275. B Smit, D Frenkel. Understanding Molecular Simulations. From Algorithms to Applications. San Diego: Academic Press, 1996.
276. K Binder, ed. The Monte Carlo Methods in Statistical Physics. 2nd ed. Berlin: Springer-Verlag, 1986.
277. K Binder, ed. The Monte Carlo Methods in Condensed Matter Physics. Topics in Applied Physics Vol. 71. Berlin: Springer-Verlag, 1992.
278. N Metropolis, AW Rosenbluth, MN Resenbluth, AH Teller. J Chem Phys *21*:1087, 1953.
279. J de Pablo, Q Yan, F Escopedo. Rev Phys Chem *50*:377, 1999.
280. LD Gelb, KE Gubbins, R Radhakrishman, M Śliwińska-Bartkowiak. Rep Prog Phys *62*:1573, 1999.
281. A Patrykiejew, K Binder. Surface Sci *73*:414, 1992.
282. H Rieger, AP Young. J Phys *26*:5279, 1993.
283. A Milchev, K Binder, DW Heermann. Z Phys B *63*:521, 1986.
284. A Patrykiejew. Polish J Chem *68*:1405, 1994.
285. J Cortés, E Valencia, P Araya. J Chem Phys *100*:7672, 1994.
286. V Mayagoitia, F Rojas, VD Pereyra, G Zgrablich. Surface Sci *221*:394, 1989.
287. V Mayagoitia, F Rojas, JL Riccardo, VD Pereyra, G Zgrablich. Phys Rev B *41*:7150, 1990.
288. JL Riccardo, VD Pereyra, G Zgrablich, F Rojas, V Mayagoitia, I Kornhauser. Langmuir *9*:2730, 1993.
289. EI Benegas. VD Pereyra, G Zgrablich. Surface Sci *187*:L647, 1987.
290. SE Edwards, PW Anderson. J Phys *F 5*:965, 1975.
291. T Schneifder, E Pytte. Phys Rev B *15*:1519, 1977.
292. I Morgernstern, K Binder, RM Hornreich. Phys Rev B *23*:287, 1981.

293. W Kinzel. Phys Rev B *27*:5819, 1983.
294. K Christmann, G Ertl. Surface Sci *60*:365, 1976.
295. R Miranda. S Daiser, K Wandelt, G Ertl. Surface Sci *131*:61, 1983.
296. R Miranda, AV Albano, S Daiser, K Wandelt, G Ertl. J Chem Phys *80*:2931, 1984.
297. EV Albano, K Binder, DW Heermann, W Paul. Z Phys B *77*:445, 1989.
298. EV Albano, K Binder, DW Heermann, W Paul. Surface Sci *223*:151, 1989.
299. EV Albano, K Binder, DW Heermann, W Paul. J Chem Phys *91*:3700, 1989.
300. J Merikoski, J Timonen, K Kaski. Phys Rev B *50*:7925, 1994.
301. J Merikoski, SC Ying. Surface Sci *381*:L623, 1997.
302. JD Bernal. Proc Roy Soc London A *284*:299, 1964.
303. VA Bakaev. Surface Sci *198*:571 (1988).
304. VA Bakaev, AV Voit. Izv Akad Nauk SSSR Ser Khim 2007, 1990.
305. VA Bakaev, WA Steele. Langmuir *8*:148, 1992.
306. VA Bakaev, WA Steele. Langmuir *8*:1379, 1992.
307. VA Bakaev, WA Steele. J Chem Phys *98*:9922, 1993.
308. VA Bakaev, WA Steele. In: A Dąbrowski, VA Tertych, eds. Adsorption on New and Modified Inorganic Sorbents. Studies in Surface Science and Catalysis Volume 99. Amsterdam: Elsevier, 1996, pp 335–355.
309. LE Drain, JA Morrison. Trans Faraday Soc *48*:316, 1952.
310. A Bonissent, B Mutaftschiev. Phil Mag *35*:65, 1977.
311. M Temkin. Zh Fiz Khim *4*:573, 1933.
312. LE Cascarini de Torre, JE Botani. Langmuir *13*:3499, 1997.
313. VA Bakaev. J Chem Phys *102*:1398, 1995.
314. S Ergun. Carbon *6*:141, 1968.
315. WM Hess, Ch R Herd. In: JB Donnet, RC Bansal, MJ Wang, eds. Carbon Black. New York: Marcel Dekker, 2nd ed. 1993.
316. JA O'Brien, AL Myers, J Chem Soc Faraday Trans I *81*:355, 1985.
317. M Borówko, A Patrykiejew, W Rżysko, S Sokołowski. Langmuir *13*:1073, 1997.
318. M Borówko, W Rżysko. Ber Bunsen Phys Chem *101*:84, 1997.
319. A Patrykiejew. Thin Solid Films *223*:39, 1993.
320. G Chmiel, A Patrykiejew, W Rżysko, S Sokołowski. Phys Rev B *48*:14,454, 1993.
321. W Gac, A Patrykiejew, S Sokołowski. Surface Sci *318*:413, 1994.
322. A Thomy, X Duval. J Chim Phys (Paris) *67*:286, 1970.
323. A Thomy, X Duval. J Chim Phys (Paris) *67*:1101, 1970.
324. M Grunze, PH Kleban, WN Unertl, FS Rhys. Phys Rev Lett *51*:582, 1983.
325. M Schick. Prog Surface Sci *11*:254, 1985.
326. A Patrykiejew, S Sokołowski, T Zientarski, K Binder. Surface Sci *421*:308, 1999.
327. RJ Behm, K Christmann, G Ertl. Surface Sci *99*:320, 1980.
328. R Imbihl, RJ Behm, K Christmann, G Ertl, T Matushima. Surface Sci *117*:257, 1982.
329. JG Dash. Films on Solid Surfaces. New York: Academic Press, 1975.
330. OP Mahajan, PL Walker Jr. J Colloid Interf Sci *31*:79, 1969.
331. DJ Callaway, M Schick. Phys Rev B *23*:3494, 1981.
332. NM Svrakič. J Phys A *18*:L891, 1985.
333. A Patrykiejew, S Sokołowski, K Binder. Surface Sci Rep *37*:207, 2000.
334. JG Dash, RD Puff. Phys Rev B *24*:295, 1981.
335. JP Coulomb. In: H Taub, G Torzo, HJ Lauter, SC Fain Jr, eds. Phase Transitions in Surface Films. New York: Plenum Press, 1991, Vol. 2, pp 113–134.
336. L Landau, EM Lifshitz, Statistical Physics. 3rd ed. New York: Pergamon Press, 1980, Part I.
337. HE Stanley. Introduction to Critical Phenomena and Phase Transitions. London: Pergamon Press, 1976.
338. V Privman. In: V Privman, ed. Finite Size Scaling and Numerical Simulation of Statistical Systems. Singapore: World Scientific, 1990, pp 1–98.
339. A Patrykiejew. Langmuir *9*:2562, 1993.

340. K Binder, DP Landau. Surface Sci *108*:503, 1981.
341. MW Cole, E Vittoratos. J Low Temp Phys *22*:223, 1976.
342. MO Robbins, D Adelman, JF Joanny. Phys Rev A *43*:4344, 1991.
343. JL Harden, D Adelman. Langmuir *8*:2547, 1992.
344. M Napiórkowski, W Koch, S Dietrich. Phys Rev A *45*:5760, 1992.
345. A Korociński, M Napiórkowski. Mol Phys *84*:171, 1995.
346. M Borówko, R Zagórski, S Sokołowski. J Colloid Interf Sci *225*:147, 2000.
347. G Forgacs, H Orland, M Schick. Phys Rev B *32*:4683, 1985.
348. P Pizio. In: M Borówko, ed. Computational Methods in Surface and Colloid Science. Surfactant Science Series Vol. 89. New York: Marcel Dekker, 2000, pp 293–345.
349. P Pfeifer, D Anvir. J Chem Phys *79*:3558, 1983.
350. D Anvir, M Jaroniec. Langmuir *5*:1431, 1983.

3

Models for the Pore-Size Distribution of Microporous Materials from a Single Adsorption Isotherm

SALIL U. REGE* and RALPH T. YANG   University of Michigan, Ann Arbor, Michigan

I. INTRODUCTION

With the recent interest in the development of new adsorbents and catalysts for gas separation as well as chemical reaction applications, the problem of accurately estimating the pore-size distribution (PSD) of microporous materials has achieved a certain imperative significance. This is particularly so because of the strong dependence of adsorption and transport characteristics on the structure and pore size of a material. By common convention, microporous materials are those with pore sizes of width less than 2.0 nm. A large number of such materials are in use today for a wide variety of applications. These include activated carbons, molecular sieve carbons in the form of pellets as well as cloth fibers, silica-based materials such as MCM-41, activated alumina, aluminophosphates, pillared clays, polymeric resins, and zeolites. The wide variety of microporous materials occurring in chemical processes is equally matched by the number of experimental techniques for PSD estimation which have been developed over the years. Some examples are mercury porosimetry [1], BET analysis [2], nuclear magnetic resonance (NMR) spin-lattice relaxation [3], x-ray diffraction (XRD), and calorimetry. However, there are certain restrictions on the type of material and the minimum pore size that can be analyzed by these techniques. By far, the most general experimental method today concerns the measurement of an adsorption isotherm of a fluid (usually nitrogen or argon at its normal boiling point) on the sorbent material.

One of the earliest and most straightforward adsorptive methods for PSD determination was molecular probing [4,5]. In this method, the saturated adsorbed amounts of probe molecules with a broad range of dimensions was experimentally measured on the material for which the PSD was desired. However, the requirement of a large number of isotherms to be measured using different sorbates was found to be tedious and cumbersome to perform. There is also an uncertainty about the results due to the pore networking effects; for example, a large pore within the sorbent may be accessible only through smaller pores which may exclude the molecular probe and thus is likely not to be detected. The problem of pore networking somewhat continues to plague even present-day PSD estimation techniques, as will be discussed later.

Current affiliation: Praxair, Inc., Tonawanda, New York.

The other attractive option was to determine the pore size by measuring a single adsorption isotherm on the material. In order to utilize the measured adsorption data and extract the PSD of the material accurately, certain physical models are required. One of the earliest models used for this purpose was the Kelvin equation [6], which has its bearings in classical thermodynamics:

$$\ln\left(\frac{P_c}{P_0}\right) = -\frac{2\gamma}{\rho RTL} \tag{1}$$

Here P_c is the condensation pressure for a slit-shaped pore of width L, P_0 is the saturation pressure of the adsorbate fluid, ρ is the fluid density, γ is the surface tension of the sorbate, and R and T are the ideal gas constant and absolute temperature, respectively. However, the Kelvin equation seemed to fail as pore size approached very low values because it does not take the thickness of the monolayer formed within the pore into account. Several models for the estimation of a statistical film thickness have been proposed (cf. Refs. 7–9) with the view of improving the PSD prediction. A commonly used method based on the modified Kelvin equation is the technique by Barrett et al. (the BJH method) [10,11]. A detailed description of such methods is available in the reviews by Gregg and Sing [1] and Webb and Orr [9].

The Kelvin-equation-based methods were found to perform reasonably well for macroporus and some mesoporous materials. However, it was found that the classical approach does not hold true for micropores, in which case the intermolecular attractive forces between the sorbate and sorbent molecules predominate over bulk fluid forces such as surface tension. The potential energy fields of neighboring sorbent surfaces are known to overlap when the pores are only a few molecular dimensions wide. This results in a substantial increase in the interaction energy of an adsorbed molecule [12], which is not accounted for by simple classical thermodynamic models such as the Kelvin equation.

Other models have been proposed in the recent past which have attempted to model the micropore adsorption on a more molecular level. In general, the suggested models fall in three categories: (1) those employing Dubinin-type equations or the theory of volume-filling models [5,13–15], (2) those relying on statistical mechanics methods for local isotherm determination [16–35], and (3) pore-filling models following the Horvath–Kawazoe (HK) methodology [4,36–58]. Of these, the first two categories have the same starting basis, namely the generalized adsorption isotherm which couples a local adsorption isotherm for a pore group with a mathematical PSD function and equates the integrated product to the total amount adsorbed. The difference in methods of types 1 and 2 solely lies in the procedures for obtaining the local isotherm. Type 3 methods (i.e., the HK methods) follow a slightly different approach. By assuming micropore filling and equating the free-energy change upon adsorption to the average interaction energy of the adsorbing molecules, the HK models translate the "step" in the isotherm data into a PSD.

In light of the large number of such models proposed to date, it is important to periodically review the suggested improvements, as well as their underlying assumptions and limitations. The present study reviews the logical concepts behind the above types of PSD model. A rigorous analysis of each method would be beyond the scope of this work and only a general overview is given. Adequate references have been provided if additional details are required. An exhaustive literature search of the commonly used microporous materials and the techniques used for their PSD estimation has also been provided. It is hoped that such a critical review will help place the range of applicability of each model in proper perspective and assist in the judicious selection of a model for PSD for a given application.

II. MODELS BASED ON THEORY OF VOLUME FILLING

A. Dubinin–Stoeckli Method

An empirical model for the adsorption of gases and vapors on microporous carbon based on the theory of volume filling (TVF) was put forth by Dubinin and Radushkevich [59]. This theory was generalized to fit a wider variety of materials by employing a more flexible exponent to the adsorption energy term and the Dubinin–Astakhov (DA) model was proposed [60]:

$$v_a = v_0 \left[-\left(\frac{A}{\beta E_0} \right)^n \right] \tag{2}$$

where v_a is the specific volume of the gas adsorbed at a relative pressure of P/P_0, v_0 is the total specific volume of the pores in the material, β is the similarity coefficient (0.33 for N_2), and E_0 is the characteristic adsorption energy for a reference substance (conventionally benzene). n is an exponent which is related to the heterogeneity of the micropore-size distribution. The adsorption potential A is defined as

$$A = -RT \ln \left(\frac{P}{P_0} \right) \tag{3}$$

where R is the gas constant and T is the absolute temperature.

The DA equation forms the basis for the Dubinin–Stoeckli (DS) method for PSD determination. As in other TVF methods, it is assumed that microporosity is composed of different pore groups and the local adsorption in each can be described by the DA equation. The exponent n is chosen to be 3, which gives sufficient flexibility to fit the isotherm equation. It must be noted, however, that DS equations derived by using $n = 2$ also frequently occur in literature [22,50].

An empirical discovery was made by Dubinin and Stoeckli [61] concerning a correlation between E_0 and the width L of ideal slit-shaped micropores:

$$E_0 = \frac{K}{L} \tag{4}$$

where K is a proportionality constant, approximately equal to $130\,kJ\,\text{Å}/\text{mol}$ for most activated carbons. The value of the mean DA characteristic energy E_0 is determined by applying the DA equation [Eq. (2)] to the measured isotherm data. The Stoeckli method suggests that instead of using different values of K for each pore group, a mean value K_0 be used. An empirical correlation has been proposed for the calculation of this mean value from E_0 as follows:

$$K_0 = 10.8 + \frac{123.1}{E_0 - 11.4} \tag{5}$$

With these assumptions, the local isotherm thus assumes the form

$$n(P, L) = \frac{v_a}{v_0} = \exp \left[-\left(\frac{AL}{\beta K_0} \right)^3 \right] \tag{6}$$

For slit-shaped pores, the PSD can be estimated by the generalized adsorption isotherm (GAI):

$$N(P) = \int_{L_{\min}}^{L_{\max}} n(P, L) f(L)\, dL \tag{7}$$

where $N(P)$ is the experimentally measured adsorbed amount at pressure P, $n(P, L)$ is the local isotherm given by Eq. (6), and $f(L)$ is the pore-size distribution function. In the Stoeckli method, the γ distribution is assumed for the mean pore widths of the different pore groups:

$$f(L) = \frac{3v_0 a^m L^{3m-1} \exp(-aL^3)}{\Gamma(m)} \tag{8}$$

where v_0 is the total micropore volume and the constants a and m are related to the mean and dispersion of the distribution, respectively. By inserting Eqs. (6) and (8) in the GAI [Eq. (7)], we obtain what has come to be known as the Dubinin–Stoeckli (DS) equation:

$$\theta = \left(\frac{a}{a + (A/\beta K_0)^3} \right)^m \tag{9}$$

The DS method thus proceeds by first fitting the measured isotherm data to the DA equation using $n = 3$. The characteristic energy E_0 is determined from the isotherm fit, and the value of K_0 is calculated using Eq. (5). Next, Eq. (9) is used to fit the isotherm data and the values of the PSD function parameters a and m are obtained. These specify the position and the spread of the required PSD of the material.

The DS method has been successfully employed for the PSD determination of micro-porous carbons [14,22,23,50,62,63]. An example is shown in Fig. 1, which shows the PSD of an activated carbon fiber (KF-1500) calculated using the DS model with CO_2 adsorption data (253–323 K) and using molecular simulations of CH_4 isotherms calculated at 308 K.

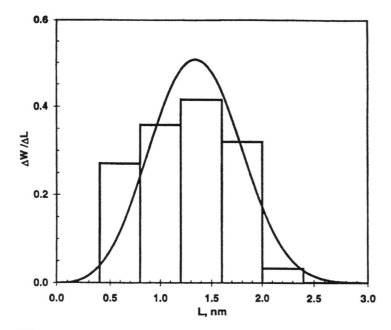

FIG. 1 Comparison of the PSD of active carbon fiber KF-1500 obtained by applying the DS equation (line curve) to CO_2 adsorption data (253, 273, 298, 323 K), and by using molecular simulations (histogram) of CH_4 (308 K). (From Ref. 62.)

B. Jaroniec–Choma Method

Based on a similar strategy as the DS method discussed earlier, Jaroniec et al. [15,63,64], have proposed another model for the determination of PSD which has come to be known as the Jaroniec–Choma (JC) method. The main difference between the JC and DS methods is that whereas the DS method requires the assumption that $E_0 = K/L$ [Eq. (4)], the JC method is flexible in the sense that any arbitrary relationship of L and E_0 is permitted. This fact is particularly important, as it has been shown by various experimental studies that the relation between L and E_0 [Eq. (4)] is only a rough approximation and more accurate correlations exist [15].

The JC method defines a new variable z defined as the inverse of characteristic energy:

$$z = \frac{1}{E_0} \tag{10}$$

Instead of directly assuming a function for the PSD of the material, a γ-type distribution function is assumed for $F(z)$ instead:

$$F(z) = \frac{n\rho^\upsilon}{\Gamma(\upsilon/n)} z^{\upsilon-1} \exp[-(\rho z)^n] \tag{11}$$

Here, the parameters ρ and υ define the shape and location of the γ distribution, respectively, and n is the exponent of the adsorption energy term in the DA equation as defined in Eq. (2). When Eq. (11) is integrated with the DA model [Eq. (2)] as the local isotherm and z varying from 0 to ∞, the following overall adsorption is obtained:

$$a_{mi} = a_{mi}^0 \left[1 + \left(\frac{A}{\beta\rho} \right)^n \right]^{-\upsilon/n} \tag{12}$$

Equation (12) is known as the JC isotherm. The parameters a_{mi} and a_{mi}^0, correspond to the amount adsorbed in the micropores and the micropore adsorption capacity at saturation, respectively. The meaning of the adsorption potential A and similarity coefficient β is as previously explained for the DA isotherm.

The JC method thus proceeds by initially extracting the micropore adsorption terms a_{mi} and a_{mi}^0 from the total adsorbed amount using a method such as the α_s-plot method [5,15,64]. The micropore adsorption isotherm data is then fit to the JC model [Eq. (12)] and the γ distribution parameters ρ and υ are extracted. Typically, the DA exponent of $n = 3$ is chosen. The other DA parameters β and E_0 are easily obtained by fitting the DA isotherm to the measured adsorption data. Thus, the distribution of $F(z)$ is realized, which can be used to obtain the PSD function $f(L)$. Next, a relationship between L and z (i.e., E_0) is assumed from literature. As mentioned earlier, there is no restriction on the type of relationship used in the JC method and the correlation suiting the system best is chosen. One example is the relation $L = Kz$. The value of K can be determined using the method described by Baksh et al. [5]. The evaluation of the PSD function $f(L)$ is then fairly simple:

$$f(L) = \left(\frac{dL}{dz} \right)^{-1} F(z) \tag{13}$$

An example of the PSD calculated for five pillared clays using the JC method is shown in Fig. 2. The elements (Zr, Al, Cr, Fe, and Ti) in the figure denote the corresponding oxides used to pillar the clays. A good agreement was observed between these predictions and molecular probe data [5].

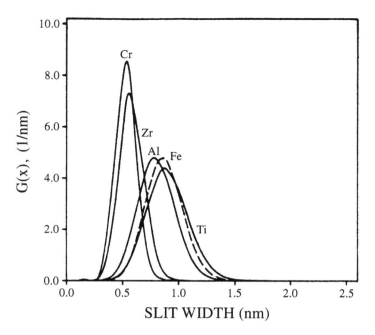

FIG. 2 PSD of five pillared clays containing oxide pillars of Cr, Zr, Al, Fe, and Ti using the JC model. (From Ref. 5.)

C. Limitations of the TVF Methods

Although the analysis of the TVF models such as the DS and JC methods is fairly simple and can be easily implemented, they are known to be susceptible to some problems. It is obvious that the γ-distribution functions [Eqs. (8) and (11)] used in their analysis have a single maximum and are hence not suited for the description of materials with a multimodal PSD. Moreover, the use of a Dubinin type of model such as the DA equation for the description of local adsorption for a pore group is also a cause for concern [47,50]. Modern simulation methods such as density functional theory and grand-canonical Monte Carlo (to be described in detail later) have shown that the filling of a sorbent micropore is stepwise and proceeds either as a single sharp filling step or two steps: an initial monolayer formation on the pore walls followed by the condensation of the sorbate in the inner pores. However, Jaroniec et al. [47] have pointed out that the DA equation predicts a gradual increase in the amount adsorbed over several orders of magnitude of relative pressure, which results in a broad PSD peak. Thus, the assumption of considerable surface or geometrical heterogeneity is inherent to the DA equation; hence, it is an unphysical model for describing the local isotherm of a micropore. The possibility of using a DA model for describing the local isotherm with a more homogeneous energy distribution by using a higher exponent ($n > 3$) is worth exploring. Nevertheless, the JC and DS methods should still apply for highly heterogeneous systems.

III. MODELS BASED ON STATISTICAL MECHANICS

For materials with an assumed single-pore geometry and an energetically homogeneous surface, the measured experimental adsorption isotherm on a microporous solid can be considered to be an aggregate of the isotherms for its individual pores. In other words, the experimental isotherm is the integral of the single-pore isotherm multiplied by the PSD. As explained previously, for slit-shaped pores, the amount adsorbed can thus be written in form of a generalized adsorption isotherm as follows:

$$N(P, T) = \int_{w_{min}}^{w_{max}} f(w)\rho(P, T, w)\, dw \tag{14}$$

where $N(P)$ is the number of moles adsorbed at a pressure P, w_{min}, and w_{max} are the limits of the pore widths, $f(w)$ is a PSD function, and $\rho(P, T, w)$ is the molar density of the adsorbate (usually N_2 at 77.3 K) at pressure P and temperature T in a pore width w [16].

The method for obtaining the PSD by such methods proceeds by first determining $\rho(P, w)$ using some tools based on statistical mechanics. A number of mathematical operations may be used to solve the integral equation given in Eq. (14) to calculate the PSD function $f(w)$ from experimentally measured isotherm data $N(P, T)$. In principle, the determination of $f(w)$ from the Fredholm-type integral equation (14) is a well-known ill-posed problem because an infinite number of data points would be required to find a unique solution. However, it has been shown that for a sufficiently large number of data (about 30–50 sample points), the function $f(w)$ is sufficiently constrained, provided the chosen functional form is flexible enough [18]. Typically, the mathematical form of $f(w)$ is either the γ distribution of log-normal distribution with multimodes, depending on the nature of the expected PSD. The forms of these functions are given as follows:

γ Distribution

$$f(w) = \sum_{i=1}^{m} \frac{\alpha_i(\gamma_i w)^{\beta_i}}{\Gamma(\beta_i)w} \exp(-\gamma_i w) \tag{15}$$

Log-normal distribution

$$f(w) = \sum_{i=1}^{m} \frac{\alpha_i}{\gamma_i w(2\pi)^{1/2}} \exp\left(\frac{-[\ln(w) - \beta_i]^2}{2\gamma_i^2}\right) \tag{16}$$

where m is the number of modes of the distribution (typically three at most) and α_i, β_i, and γ_i are adjustable parameters that define the amplitude, mean, and variance, respectively, of mode i.

Seaton et al. [16] used a multilinear least-squares fitting of the parameters of the assumed PSD function so as to match measured isotherm data. A similar method was employed by Lastoskie et al. [18] in their analysis using the nonlocal density functional theory (NL-DFT). Later, an important contribution toward the numerical deconvolution of the distribution result was made by Olivier et al. [35]. They developed a program based on the regularization method [65], in which no restrictions were imposed on the form of PSD. Moreover, this method was found to be numerically robust. Also, a simpler optimization technique has recently been suggested by Nguyen and Do [66].

It can be seen from Eq. (14) that the problem basically boils down to the accurate and efficient determination of $\rho(P, w)$. In the last decade, two main methods based on statistical mechanics have emerged. These are the density functional theory (DFT), and molecular simulation methods such as grand-canonical Monte Carlo (GCMC) simulations. Both of these methods are steadily gaining popularity for PSD estimation and are discussed briefly in the following subsections.

A. DFT Method

The mean field density functional theory (DFT) approach was primarily developed by Evans and co-workers [67,68] for studying the interactions of fluids in pores at the molecular level. Recently, DFT methods have been developed specifically with the objective of the estimation of PSD of carbon-based as well as other types of microporous materials. This technique was first proposed by Seaton et al. [16], who used the local-DFT approximation. Later, the theory was modified by Lastoskie et al. [17,18] to incorporate the smoothed or nonlocal DFT approach. The rigorous statistical mechanics basis behind the DFT model has been recently reviewed by Gubbins [34]. Some salient features of the theory are discussed later in this subsection. The DFT method initially proceeds by estimating the properties of a fluid directly from intermolecular forces such as that between sorbate–sorbent and sorbate–sorbate molecules. The interactions are divided into a short-ranged repulsive part and a long-ranged attractive part, which are both determined separately.

The DFT was primarily developed to study the structural features of carbon-based sorbents, whose pores typically have a slit-shaped geometry. The individual pore is represented by two semi-infinite parallel graphitic slabs which are considered to be uncorrugated, rigid, and chemically homogeneous. It is assumed that the pore width (i.e., the distance between the surface sorbent atoms) is w, and the distance between the graphitic layers of each slab are separated by a uniform spacing of Δ. The fluid–fluid interaction potential ϕ_{ff} can be determined by the Lennard–Jones (LJ) 6–12 potential:

$$\phi_{ff}(r) = 4\varepsilon_{ff}\left[\left(\frac{\sigma_{ff}}{r}\right)^{12} - \left(\frac{\sigma_{ff}}{r}\right)^{6}\right] \tag{17}$$

where r is the separation distance, ε_{ff} is the potential well depth for sorbate–sorbate interaction, and σ_{ff} is the sorbate molecular diameter. The Steele 10-4-3 potential provides the sorbate–sorbent interaction ϕ_{sf} with a single graphitic slab:

$$\phi_{sf}(r) = 2\pi\varepsilon_{sf}\rho_s\sigma_{sf}^2\Delta\left[\frac{2}{5}\left(\frac{\sigma_{sf}}{z}\right)^{10} - \left(\frac{\sigma_{sf}}{z}\right)^{4} - \frac{\sigma_{sf}^4}{3\Delta(z + 0.61\Delta)^3}\right] \tag{18}$$

where z is the distance from the graphite surface, σ_s is the solid density, and ε_{sf} and σ_{sf} are the sorbate–sorbent potential well depth and effective diameter, respectively. For a slit pore, the fluid molecule will interact with both sides of the pore wall; hence, the total potential V_{ext} is given by

$$V_{ext}(z) = \phi_{sf}(z) + \phi_{sf}(w - z) \tag{19}$$

The interaction parameters have to be fitted from separate experimental observations. For example, ε_{ff} and σ_{ff} are chosen such that the saturation pressure and liquid density of bulk adsorbate (N_2) calculated by mean field theory are equal to experimental values at the normal boiling point. The sorbate–sorbent interaction parameters (ε_{sf} and σ_{sf}) are obtained by fitting experimental adsorption measurements (or t curves) on nonporous carbons whose surface has chemical characteristics similar to those of the material whose PSD is desired.

The DFT proceeds by considering an individual pore exposed to bulk sorbate fluid at its boiling point. The local fluid density $\rho(r)$ of the adsorbate confined in the pore at a given chemical potential μ and temperature T is determined by a minimization of the grand thermodynamic canonical ensemble:

$$\Omega[\rho(r)] = F[\rho(r)] - \int dr\, \rho(r)[\mu - V_{ext}(r)] \tag{20}$$

where $F[\rho(r)]$ is the intrinsic Helmholtz free-energy functional, ρ is the local fluid density at position r, and the integration is over the pore volume V. The Helmholtz free energy is divided in a perturbative fashion to obtain the contribution from a reference system of hard spheres and the contribution from attractive interactions. The attractive portion of the fluid–fluid potential, represented by the Weeks–Chandler–Andersen formulation of the LJ potential, is treated by invoking the mean field approximation in which the correlations due to attractive forces are neglected. This hard-sphere part of the potential is further split into an ideal gas component and an excess component. The excess Helmholtz free energy of the molecule in a hard sphere fluid is calculated by the Carnahan–Starling equation of state using a fluid density smoothed by the use of a weighting function.

Various types of weighting function were proposed which has resulted into different versions of the DFT model. In the *local density approximation* (L-DFT model), a delta-function weighting function is chosen which implies that the hard-sphere excess free energy is evaluated using the local density profile of the fluid. Although this approximation provides useful results for fluid–fluid interfaces, it is unrealistic for solid–fluid interfaces because of the very large density fluctuations near the wall [34].

When the *smoothed* or *nonlocal density approximation* (or NL-DFT model) is used, the weighting function is chosen so that the hard-sphere direct pair-correlation function is well described for the uniform fluid over a wide range of densities. One example of such a weighting function is the model proposed by Tarazona [69], which uses the Percus–Yevick theory for approximating the correlation function over a wide range of density. In this case, the weighting function is expanded as a power series of the smoothed density. The use of a smoothed density in NL-DFT provides an oscillating density profile expected of a fluid adjacent to a solid surface, the existence of which is corroborated by molecular simulation results [17,18].

To solve for the equilibrium density profile, the grand potential functional [Eq. (20)] is minimized with respect to the local density:

$$\left.\frac{\partial\Omega[\rho(r)]}{\partial\rho(r)}\right|_{\rho=\rho_{eq}} = 0 \tag{21}$$

The density profile usually varies in one direction only. The grand ensemble is minimized with respect to $\rho(z)$ over a range of chemical potentials μ for selected values of H. The bulk relative pressure P/P_0 is related to μ through a bulk equation of state. The mean density ρ for a pore of width w is then obtained as follows:

$$\rho(w) = \frac{1}{w}\int_0^w \rho(z)\,dz \tag{22}$$

By performing DFT calculations over a wide range of pore sizes, the function $\rho(P, w)$ is obtained. Thus, by a combination of this function with an assumed PSD function as described for Eq. (14), the PSD of a sorbent can be determined. The method has also been modified to include pore curvature effects, although the analysis is much more involved than for the slit-pore case (cf. Ref. 24). For example, for cylindrical surfaces, the solid–fluid interactions are modeled with the LJ potential as follows:

$$V_{ext}(r, R_p) = \pi^2 \rho_s \varepsilon_{sf} \sigma_{sf}^2 \left[\frac{63}{32}\left(\frac{r}{\sigma_{sf}}\left[2 - \frac{r}{R_p}\right]\right)^{-10} F\left[-\frac{9}{2}, -\frac{9}{2}, 1, \left(1 - \frac{r}{R_p}\right)^2\right] \right.$$
$$\left. - 3\left(\frac{r}{\sigma_{sf}}\left[2 - \frac{r}{R_p}\right]\right)^{-4} F\left[-\frac{3}{2}, -\frac{3}{2}, 1, \left(1 - \frac{r}{R_p}\right)^2\right] \right] \tag{23}$$

where $F[\alpha, \beta, \gamma, \chi]$ is the hypergeometric series and ρ_s is the density of sorbent atoms in the pore wall. For an infinitely large pore radius, this potential reduces to the 10-4 potential.

The PSD predicted by using the L-DFT model for an activated carbon (BP71) is shown in Fig. 3. A comparison of the PSD given by the L-DFT and NL-DFT models is shown in Fig. 4. The local theory is seen to overestimate the PSD peak relative to the nonlocal model. Finally, the PSD of an A1/MCM-41 catalyst evaluated by using a cylindrical-pore version of the NL-DFT model is shown in Fig. 5.

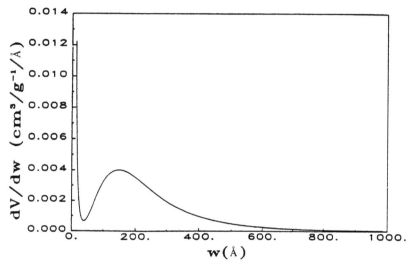

FIG. 3 PSD of an activated carbon (BP71) obtained by using the L-DFT model. (From Ref. 16.)

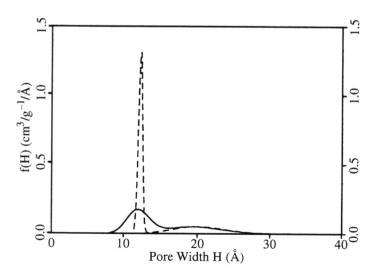

FIG. 4 Comparison of the PSD prediction of an activated carbon (AC610) using the L-DFT (dashed line) and NL-DFT (solid line) models. (From Ref. 18.)

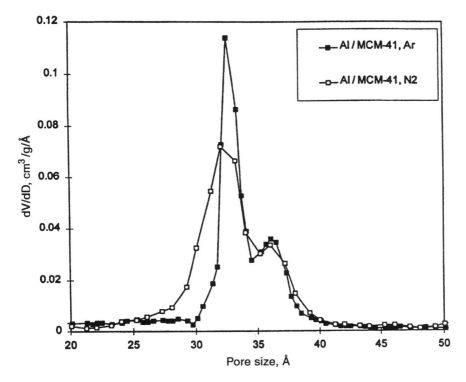

FIG. 5 PSD of an Al/MCM-41 catalyst calculated from the NL-DFT model using Ar and N_2 desorption isotherms. (From Ref. 24.)

B. Limitations of DFT and Suggestions for Improvement

The DFT method provides a practical alternative to more accurate but computation-intensive methods such as molecular simulations. However, certain drawbacks of the method have to be taken into account. Lastoskie et al. [18] have summarized some of these limitations. One practical problem, which is more of an experimental error rather than a flaw in the actual method, is the nonreliability of isotherm data in the very low-pressure range. The lowest pressure of the measured isotherm dictates the lower bound of the PSD calculation beyond which it needs to be truncated. Frequently, the measured isotherms are improperly equilibrated due to diffusional limitations at low temperatures, which can result in the underestimation of the PSD in the micropore range [18].

In order to correctly predict the adsorption properties, it is necessary to obtain an accurate estimate of the sorbate–sorbent interaction parameters. This is carried out by fitting the experimentally measured isotherm on a nonporous sample with nearly the same surface characteristics as the material whose PSD is sought. This poses a difficulty in the case of materials for which such an accurate match may not be available (e.g., zeolites, etc.). In general, there is a dearth of DFT interaction parameters for a wide range of sorbent materials because most of the studies in literature have considered only a few types of material, such as microporous carbons and silica-based materials (e.g., MCM-41) with limited sorbates (mainly N_2 and Ar). Also, there is a need to introduce electrostatic forces in the analysis because interactions of a quadrupolar molecule like N_2 is likely to have strong interactions with any existing polar groups on a sorbent surface. One common example of such interactions is N_2

adsorbed in a cation-exchanged zeolite cage. Monte Carlo studies have shown that including the quadrupolar interaction in the model resulted in a wider PSD for a microporous carbon membrane than without the coulombic interaction [70]. Involving such interactions will, undoubtedly, further complicate the DFT calculations.

Further suggestions for the improvement of the model have been made by Olivier [20]. One of the observations related to PSD predicted by DFT models for activated carbons was that they consistently predicted a relatively low population of pores near 6 Å width. Also, a near absence of pores in the range of two or three molecular diameters pore width seems to be reported, which does not seem to change even when surface heterogeneity is included in the model. The double minima thus seen in the DFT-predicted PSD is probably a model-induced artifact caused by the strong packing effects displayed by the rigid parallel walls. Such strong packing effects may be reduced by considering alternative pore geometries involving nonparallel walls. Olivier [20] also suggested that the assumption of adsorbent inertness be relaxed for micropores below 2.0 nm, allowing for a dilation or swelling of the pore structure due to the packing effects.

The pore connectivity aspect has been neglected in the DFT analysis. It is possible that the adsorption in an individual pore is affected by the adsorption in an adjacent or a networked pore, which can complicate the adsorption integral. Better insight into the connectivity phenomenon has been provided by Seaton and co-workers using Monte Carlo simulations [32,33] and efforts to develop pore-junction models are on.

The problem of the chemical heterogeneity of the sorbent surface remains to be addressed. A weighted combination of the isotherms calculated by DFT for a given pore size and a distribution of adsorptive potentials to this end has been suggested [20]. Other strategies to include surface heterogeneity are also worth exploring. These include allowing for edge effects and closed sides in the pore, as well as assigning patchwise heterogeneity to the pore walls [31].

The DFT method is such a complex in nature compared to other simple methods such as the Horvath–Kawazoe model, despite the fact that its PSD predictions for mesoporous sorbents are more accurate. Elaborate computer codes, available as software packages commercially, are needed for determination of PSD. The extension of the model to cylindrical [24] or spherical pores would further add to the complexity of the model.

C. Molecular Simulation Techniques

The DFT approach was the first statistical mechanics approach for calculation of adsorption and determination of PSD in regular-shaped pores. However, this method simplifies the interactions between sorbate molecules so as to economize the computation required for approximating the adsorption in a pore. The DFT becomes increasingly difficult to use when the pores have exotic nonregular structures or when the sorbate molecules are complex in nature. For example, in the case of polar molecules such as water and for polymeric molecules, it is easier and more accurate to use a numerical solution to basic statistical mechanic equations. With the easy availability of fast computers, alternative methods based on molecular simulations have recently become increasingly prevalent [30–32,34]. Provided that correct input parameters are selected, the molecular simulations are capable of providing essentially exact results and can be extended to predict adsorption in more complex pore shapes. The greater accuracy of the predictions is, of course, at the cost of considerably more calculation time than methods such as DFT. It needs to be mentioned, however, that the determination of adsorption isotherms and calculation of the PSD are two independent functions. Once a set of molecular simulations have been conducted for a given adsorbate in a set of model pores of a given type of sorbent material, the PSD can be determined relatively quickly for any microporous material with nearly similar structural characteristics. The availability of commercial software packages for performing molecular simulations for a wide variety of applications has made such methods increasingly attractive as the reference methods to test the predictions of other proposed adsorption theories [17,49,62].

The molecular simulation methods are broadly divided into two categories: molecular dynamics (MD) and Monte Carlo (MC) methods. In the MD method, the position and velocity of individual sorbate molecules is followed over short time intervals (of the order of 10^{-14} s) by solving Newton's equations of motion. After a suitable equilibration period, the molecular motion is averaged using statistical mechanics to obtain the properties of the system. The advantage of the method is that dynamic properties such as diffusivities are also calculated in addition to equilibrium values. Moreover a "real" picture of the mobile molecules can be visualized as compared to the artificial picture afforded by MC methods. However, the computational cost of performing MD calculations for even a few hundred molecules is prohibitive. In contrast, the MC methods are much more economical as far as computational cost is concerned, in addition to being more flexible. Hence, MC methods are more popular because they yield nearly similar results to MD methods. Detailed reviews of the molecular simulation methods are available in literature [9,15,21,30–32,70,71].

The Monte Carlo (MC) method for predicting isotherms in a range of pore sizes at different pressures at a given temperature consists of randomly moving and rotating molecules in a small section of a model pore. The equations for describing the sorbate–sorbate and sorbate–sorbent interactions are usually the same as those described for the DFT method (LJ potential and Steele's 10-4-3 potential respectively).

The probability of a molecular arrangement is given by the Boltzmann distribution law when the temperature and volume of the system as well as the number of molecules therein are constant. Three types of trial are used during the simulations: attempts to translate/reorient the molecule, attempts to delete particles, and attempts to create a molecule in the simulation cell. The decision regarding the acceptance of each trial or its rejection to return to the old configuration is based on the probability:

$$P(\text{move}) = \min\left[\exp\left(-\frac{\Delta U}{kT}\right),\ 1\right] \tag{24}$$

where $\Delta U = U_{\text{new}} - U_{\text{old}}$ is the difference in the potential energies of the new of old configurations, k is Boltzmann's constant, and T is the absolute temperature. After a large number of such moves, the statistical average number of molecules within a small section of the model pore is determined at the specified temperature and pressure. The algorithms for deciding the fate of a move as well as the sampling of the final molecular configuration are well established [21].

The result of the molecular simulation is the number of molecules adsorbed in a section of a model pore which subsequently needs to be converted into an "excess" adsorption isotherm. By excess isotherm, it is meant that the difference between calculated adsorbate density and the bulk fluid-phase adsorbate density is taken into account. This is done to make the results compatible with the experimental measurements, which, in reality, measure the excess adsorption isotherms on a sorbent. The effective volume of the simulation cell is determined taking the protrusion of the carbon atoms into the cell into account. This effective volume can be used to convert the excess adsorption into the corrected adsorbate density to be used for determination of PSD as explained previously using Eq. (14). The effect of the choice of pore model used in the PSD predictions using the GCMC technique has been discussed by Davies and Seaton [21].

The overall PSD of a Carbosieve-G sorbent predicted from MC-predicted adsorption isotherms of CH_4, CF_4, and SF_6 at three temperatures is shown in Fig. 6. The use of multiple adsorbate molecules of different sizes is said to provide a more complete PSD picture compared

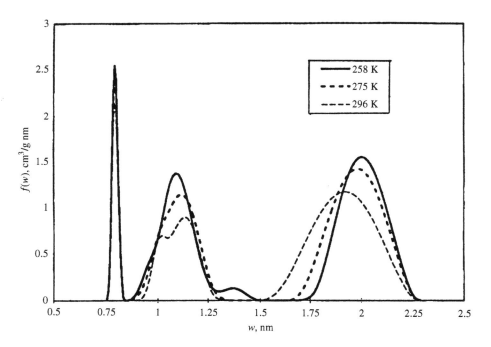

FIG. 6 PSD of Carbosive-G sorbent obtained by a combination of PSDs obtained by using the GCMC simulations of CH_4, CF_4, and SF_6 sorbates at three temperatures. (From Ref. 32.)

to the use of a single sorbate, in addition to giving some information of pore connectivity [32]. A comparison of the Stoeckli and the GCMC predicted PSDs for two activated charcoal cloths with different degrees of burn-off are shown in Fig. 7. It can be seen that the results from the two methods are in close agreement; however, the GCMC has the capability of detecting multi-modality in the PSD, which the Stoeckli method, by definition, cannot predict. Another

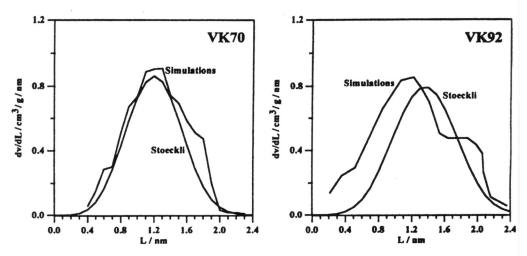

FIG. 7 Comparison of the PSD of two types of activated charcoal cloths (VK70 and VK92) calculated using GCMC simulations and DS method ($n = 3$) with N_2 (77 K) as adsorbate. (From Ref. 23.)

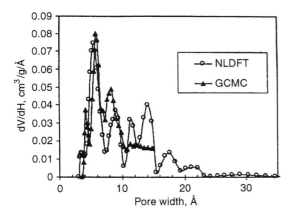

FIG. 8 PSD of a wide-pore carbon B using NL-DFT and GCMC models with a high-pressure CO_2 (273 K) adsorption isotherm. (From Ref. 25.)

comparison is shown in Fig. 8 in which the NL-DFT and GCMC PSD predictions for a wide-pore carbon are juxtaposed. In this case, a high-pressure CO_2 adsorption isotherm at 273 K was used. Again, the predictions agree well, especially for pore sizes <1.0 nm, which, incidentally, is the limit of sensitivity of CO_2 at subatmospheric pressure.

IV. MODELS BASED ON THE HORVATH–KAWAZOE APPROACH

A. HK Methods

A theoretical model was developed by Horvath and Kawazoe [4] which is based on the progressive pore filling of microporous materials with an increase in adsorbate pressure. The Horvath–Kawazoe (HK) model, as it has come to be known, is conceptually simple and the calculation of the pore sizes is facile compared to the statistical-mechanics-based approaches. The model relates the free energy of adsorption to the interaction energy of the adsorbate molecule in the pore using the Lennard–Jones (LJ) type of interactions. Furthermore, it is assumed that only pores with dimensions lower than a certain unique value will be filled for a given relative pressure of the sorbate. This is generally true for sorbents in the micropore range (less than 2.0 nm); hence, the model should be applicable for this size range. The HK models have also been shown to predict the adsorption isotherm of another sorbate or the same sorbate but at a different temperature with reasonable accuracy [58,72].

Over the last couple of decades, different types of HK model have been developed, depending on pore geometry. The original HK model discussed slit-shaped pores [4], whereas models for cylindrical pores [36] and spherical pores [37,40] have also been proposed. A brief description of the general HK concept is given next.

The molar integral change of free energy at a given temperature is given by

$$\Delta G^{ads} = \Delta H^{ads} - T\Delta S^{ads} \tag{25}$$

The molar enthalpy change on adsorption can be shown to be given by

$$\Delta H^{ads} = -q^{diff} - RT + \left(\frac{T\beta}{\theta}\right)\left(\frac{\partial \Pi}{\partial T}\right) \tag{26}$$

Assuming that the change in the entropy on adsorption is negligible compared to the other terms in the equation (notably q^{diff}), and noting that the free-energy change on adsorption (ΔG^{ads}) and q^{diff} can be respectively written as

$$\Delta G^{\text{ads}} = RT \ln\left(\frac{P}{P_0}\right) \tag{27}$$

and

$$-q^{\text{diff}} = U_0 + P_a - \Delta H^{\text{vap}} \tag{28}$$

the following relation is obtained:

$$RT \ln\left(\frac{P}{P_0}\right) + \left[RT - \left(\frac{T\beta}{\theta}\right)\left(\frac{\partial \Pi}{\partial T}\right)\right] = U_0 + P_a \tag{29}$$

Assuming the adsorbed phase to be a two-dimensional ideal gas and assuming the adsorption occurs in the Henry's law region (linear isotherm), the following equation of state can be substituted in Eq. (29):

$$\Pi = \frac{kT}{\beta}\theta \tag{30}$$

The resulting equation is the basic framework for the different HK models:

$$RT \ln\left(\frac{P}{P_0}\right) = U_0 + P_a \tag{31}$$

where U_0 and P_a denote the sorbate–sorbent and sorbate–sorbate interaction energies, respectively. Thus, the right-hand side of Eq. (31) is a function of the pore geometry and dimension, which is related to the relative pressure of the adsorbate. The calculation of the PSD is now relatively simple. By using different values of a pore dimension (pore width in the case of a slit pore, pore diameter in the case of pores with curvature), the threshold sorption relative pressure (P/P_0) at which the pore filling will occur (or "filling" pressure) can be obtained over the expected pore-size range. From the adsorption measurements using a suitable sorbate such as N_2 (77 K) or Ar (87 K), the fractional adsorbed amount ($\theta = w/w_\infty$) is obtained as a function P/P_0. From a combination of the above two functional relationships, the adsorbed amount w/w_∞ can be plotted as a one-to-one function of pore dimension L, thus giving the cumulative PSD. A differential PSD can be further obtained by calculating the derivative $d(w/w_\infty)/dL$ as a function of L. Another approach based on the adsorption potential distribution (APD), also yields a similar result [73]. The calculation of the interaction energies on the right-hand side of Eq. (31) for the different pore geometries is explained in the following subsections.

1. HK Slit-Pore Model

Different types of interaction exist for sorbate molecules in a micropore. For example, van der Vaals type of interaction or dispersive force, inductive interactions, and electrostatic interactions (e.g., ion–dipole or ion–quadrupole) can exist depending on the nature of the adsorbent surface as well as the physical properties of the adsorbate molecule. For the sake of simplicity, only the van der Vaals type of interaction (computed by LJ equations) have been considered in the HK model.

The basis for obtaining the interaction energy of a sorbate molecule in a pore is the Lennard–Jones 6-12 potential:

$$\varepsilon_{12}(z) = 4\varepsilon_{12}^{*}\left[\left(\frac{\sigma}{z}\right)^{12} - \left(\frac{\sigma}{z}\right)^{6}\right] \tag{32}$$

For the specific case of a slit-shaped pore, Halsey and co-workers gave the interaction energy of one adsorbate molecule with a single infinite-layer plane of adsorbent molecules as follows [74]:

$$\varepsilon(z) = \frac{N_S A_S}{2\sigma^4}\left[-\left(\frac{\sigma}{z}\right)^{4} + \left(\frac{\sigma}{z}\right)^{10}\right] \tag{33}$$

Everett and Powl [12] extended the above result to two parallel infinite lattice planes whose nuclei are spaced at a distance L apart:

$$\varepsilon(z) = \frac{N_S A_S}{2\sigma^4}\left[-\left(\frac{\sigma}{z}\right)^{4} + \left(\frac{\sigma}{z}\right)^{10} - \left(\frac{\sigma}{L-z}\right)^{4} + \left(\frac{\sigma}{L-z}\right)^{10}\right] \tag{34}$$

where the internuclear distance at zero-interaction energy $\sigma = (2/5)^{1/6}d_0$, d_0 being the average of the adsorbate and adsorbent molecule diameters [i.e., $(d_0 + d_A)/2$], N_S is the number of sorbent molecules per unit area, and z is the internuclear distance between the adsorbate and adsorbent molecules. The dispersion constants A_S and A_A are calculated by the Kirkwood–Müller formulas as follows:

$$A_S = \frac{6mc^2\alpha_S\alpha_A}{\alpha_S/\chi_S + \alpha_A/\chi_A} \tag{35}$$

$$A_A = \tfrac{3}{2}mc^2\alpha_A\chi_A \tag{36}$$

Further, Horváth and Kawazoe [4] proposed that the potential is increased by the interaction of adsorbate molecules within the pore. They included this additional interaction by adding the adsorbate dispersion term ($N_A A_A$) term in the numerator of the depth of potential energy minimum ($N_S A_S/2\sigma^4$) in Eq. (34) as follows:

$$\varepsilon(z) = \frac{N_S A_S + N_A A_A}{2\sigma^4}\left[-\left(\frac{\sigma}{z}\right)^{4} + \left(\frac{\sigma}{z}\right)^{10} - \left(\frac{\sigma}{L-z}\right)^{4} + \left(\frac{\sigma}{L-z}\right)^{10}\right] \tag{37}$$

However, it must be noted that this approximated, although excellent for small pores measuring less than twice the adsorbate diameter, fails for larger-sized pores [57].

The next step in the derivation was to obtain the average interaction energy by volumetrically averaging the above interaction energy over the entire free space in the slit pore:

$$\bar{\varepsilon}(z) = \frac{\int_{d_0}^{L-d_0} \varepsilon(z)\, dz}{\int_{d_0}^{L-d_0} dz} \tag{38}$$

For a slit-shaped pore, the average interaction energy was calculated as:

$$\bar{\varepsilon}(z) = \frac{N_S A_S + N_A A_A}{\sigma^4(L - 2d_0)}\left[\frac{\sigma^4}{3(L-d_0)^3} - \frac{\sigma^{10}}{9(L-d_0)^9} - \frac{\sigma^4}{3d_0^3} + \frac{\sigma^{10}}{9d_0^9}\right] \tag{39}$$

It must be noted here that a better estimate of the average interaction energy can be obtained by incorporating a weighting function based on the Boltzmann law of energy distribution [45,57]. However, this aspect was ignored in the original derivation.

Finally, the average potential energy so calculated was related to the free-energy change upon adsorption, $RT \ln(P/P_0)$.

Noting that, in molar units, $U_0 + P_a = N_{Av} \cdot \bar{\varepsilon}$, the resulting slit-pore HK model is obtained as follows:

$$RT \ln\left(\frac{P}{P_0}\right) = N_{Av} \frac{N_S A_S + N_A A_A}{\sigma^4(L - 2d_0)} \left[\frac{\sigma^4}{3(L - d_0)^3} - \frac{\sigma^{10}}{9(L - d_0)^9} - \frac{\sigma^4}{3d_0^3} + \frac{\sigma^{10}}{9d_0^9}\right] \quad (40)$$

2. HK Cylindrical-Pore Model

The interaction energy of a single gas molecule in a cylindrical pore composed of a single layer of sorbent molecules was derived by Everett and Powl [12] to be as follows:

$$\varepsilon(r) = \frac{5}{2}\pi\varepsilon^* \left[\frac{21}{32}\left(\frac{d_0}{L}\right)^{10} \sum_{k=0}^{\infty} \alpha_k \left(\frac{r}{L}\right)^{2k} - \left(\frac{d_0}{L}\right)^4 \sum_{k=0}^{\infty} \beta_k \left(\frac{r}{L}\right)^{2k}\right] \quad (41)$$

In the above equation, the constants α_k and β_k are given by

$$\alpha_k = \left(\frac{-4.5 - k}{k}\right)^2 \alpha_{k-1} \quad (42)$$

$$\beta_k = \left(\frac{-1.5 - k}{k}\right)^2 \beta_{k-1} \quad (43)$$

where $\alpha_0 = \beta_0 = 1$. Here, r is the distance of the nucleus of the gas molecule from the central axis of the cylinder and ε^* is the potential energy minimum given by

$$\varepsilon^* = \frac{3}{10} \frac{N_S A_S}{d_0^4} \quad (44)$$

Using this result, Saito and Foley [36] developed a pore-size distribution model using the HK methodology. In their work as well, the adsorbate–adsorbate interaction was incorporated in the model by appending the $N_A A_A$ term to the denominator of the potential energy minimum ε^* as follows:

$$\varepsilon^* = \frac{3}{10} \frac{N_S A_S + N_A A_A}{d_0^4} \quad (45)$$

The resulting radial energy potential profile was then averaged over the cylindrical cross-sectional area to give the following original HK equation for cylindrical pores:

$$RT \ln\left(\frac{P}{P_0}\right) = \frac{3}{4}\pi N_{Av} \frac{N_S A_S + N_A A_A}{d_0^4} \sum_{k=0}^{\infty} \left[\frac{1}{k+1}\left(1 - \frac{d_0}{L}\right)^{2k}\left(\frac{21}{32}\alpha_k\left(\frac{d_0}{L}\right)^{10} - \beta_k\left(\frac{d_0}{L}\right)^4\right)\right] \quad (46)$$

3. HK Spherical-Pore Model

The interaction between an adsorbate molecule and a single layer of sorbent molecules located along the wall of a spherical cavity has been studied in detail in literature [37,45,75,76]. If L is the radius of the cavity, r is the radial distance of a gas molecule from the center of the cavity, and N is the number of molecules on the enveloping cavity surface, the energy potential $\varepsilon(r)$ was given as

$$\varepsilon(r) = 2N\varepsilon_{12}^* \left[-\left(\frac{d_0}{L}\right)^6 \frac{1}{4(r/L)} \left(\frac{1}{(1-r/L)^4} - \frac{1}{(1+r/L)^4}\right) \right.$$
$$\left. + \left(\frac{d_0}{L}\right)^{12} \frac{1}{10(r/L)} \left(\frac{1}{(1-r/L)^{10}} - \frac{1}{(1+r/L)^{10}}\right) \right] \tag{47}$$

Using the same approach for incorporating the adsorbate–adsorbate interaction as for the earlier cases, the effective potential of an adsorbed molecule in a spherical pore resulted in the following form:

$$\varepsilon(r) = 2(N_1\varepsilon_{12}^* + N_2\varepsilon_{22}^*) \left[-\left(\frac{d_0}{L}\right)^6 \frac{1}{4(r/L)} \left(\frac{1}{(1-r/L)^4} - \frac{1}{(1+r/L)^4}\right) \right.$$
$$\left. + \left(\frac{d_0}{L}\right)^{12} \frac{1}{10(r/L)} \left(\frac{1}{(1-r/L)^{10}} - \frac{1}{(1+r/L)^{10}}\right) \right] \tag{48}$$

where N_1 and N_2 are the total number of sorbent and sorbate molecules in the respective layers and were given by

$$N_1 = 4\pi L^2 N_s \tag{49}$$
$$N_2 = 4\pi (L - d_0)^2 N_A \tag{50}$$

The minimum energies were given as

$$\varepsilon_{12}^* = \frac{A_s}{4d_0^6} \tag{51}$$

$$\varepsilon_{22} = \frac{A_A}{4d_A^6} \tag{52}$$

By integrating over the volume of the spherical cavity, Cheng and Yang [40] arrived at the following HK equation for spherical pores:

$$RT \ln\left(\frac{P}{P_0}\right) = N_{Av} \frac{6(N_1\varepsilon_{12}^* + N_2\varepsilon_{22}^*)L^3}{(L - d_0)^3} \left[-\left(\frac{d_0}{L}\right)^6 \left(\frac{1}{12}T_1 - \frac{1}{8}T_2\right) \right.$$
$$\left. + \left(\frac{d_0}{L}\right)^{12} \left(\frac{1}{90}T_3 - \frac{1}{80}T_4\right) \right] \tag{53}$$

where the dimensionless terms T_1, T_2, T_3, and T_4 are given by

$$T_i = \frac{1}{[1 + (-1)^n(L - d_0)/L]^n} - \frac{1}{[1 - (-1)^n(L - d_0)/L]^n} \tag{54}$$

where the value of n for T_1, T_2, T_3, and T_4 are 3, 2, 9 and 8, respectively [9].

B.   Sensitivity of PSD Predictions to Physical Parameters

A number of physical parameters pertaining to both the sorbate as well as sorbent are required as inputs for the HK models described. These include the size of the sorbate and sorbent molecules as well as the surface density of the molecules. This information is available in literature [77–80]. The depth of the potential energy minimum is determined by the Kirwood–Müller formalisms, which, in turn, are dictated by the polarizability (χ) of the molecules of the substrate and the adsorbing fluid as well as their magnetic susceptibilities (a). Of these, the values of polarizability and magnetic susceptibility of molecules are reported in literature in a fairly uniform manner, although a few discrepancies do exist [40]. However, the largest variation in physical parameters used in the models is observed for the size of the sorbate molecule as well as the surface density of sorbent molecules. There is a large difference in the reported values of these parameters in literature, primarily because of the difficulty in their estimation as well as the tendency to treat them as "fitting parameters" to obtain the correct pore-size distribution as desired. For example, the value of the diameter of N_2 molecule reported in literature varies from 0.30 nm [4] to 0.372 nm [71]. Similarly the values of the diameter of Ar molecule used in PSD predictions were found to vary from 0.289 nm [41] to 0.382 nm [57,71,81]. The sorbent surface density of oxide ions in the case of zeolites also show a similar variation [40,57]. In light of the discrepancies in the reported values of these parameters, it is important to examine the sensitivity of the predicted PSD with respect to the variation in input parameters to the HK model.

Such a parametric sensitivity study was conducted by Saito and Foley [36]. Considering the cylindrical HK model, a ±30% variation in each of the above input physical parameters was introduced and the effect on the calculated pore size was studied. The results of their study are shown in Table 1. It can be seen from Table 1 that the diameter of the sorbent molecule (in this case, oxide ion) has a considerable effect in predicted pore size, probably because of the higher powers to which it has been raised in the model equations. By the same reasoning, the predicted PSD should be equally sensitive to the diameter of the sorbate molecule. However, this aspect was not studied by Saito and Foley [36].

The density of the sorbate (Ar) and the sorbent (oxide ion) surface densities, as well as the magnetic susceptibility of the oxide ion, also seems to have a moderate influence. The predicted pore size was the least sensitive to polarizability of the oxide ion. These results emphasize the need for standardizing physical parameters before discriminating between different models for their merit for predicting the PSD of microporous materials.

TABLE 1 Variation in the Peak of the PSD of Zeolite Y Predicted by Using the Saito–Foley Cylindrical-Pore HK Model due to a 30% Variation in Physical Parameters

Physical parameter	Value of parameter	Calculated pore size (nm) Variation in parameters		
		−30%	0	+30%
Argon				
Density (molecules/cm^2)	8.52×10^{14}	0.68	0.74	0.78
Oxide ion				
Diameter (nm)	0.276	0.99	0.74	0.55
Polarizability (cm^3)	2.5×10^{-24}	0.71	0.74	0.75
Magnetic susceptibility (cm^3)	1.3×10^{-29}	0.65	0.74	0.81
Density (molecules/cm^2)	1.31×10^{15}	0.63	0.74	0.83

Source: Ref. 36.

C. Limitations of HK Models and Suggested Improvements

The HK models seem to provide useful results for prediction of PSD with relative ease of computation. However, by virtue of the simplifying assumptions made in order to keep the calculations to a minimum, these models have a limited range of applicability and are found to generate artifacts which undermine the fidelity of the results. Considerable research has been devoted to studying the HK method for various applications and suggestions for their improvement continue to be made to date.

One of the most often cited criticisms of this method is the assumption of discontinuous filling of the micropores and the complete filling of a pore at a specific pressure characteristic of its size [9,45,47,50]. This assumption is commonly referred to as the "condensation approximation" in literature. However, other theoretical models for the prediction of adsorption such as Monte Carlo simulations or NL-DFT theory show that this picture is valid only for very small micropores ($L < 2.0$ nm) For larger-sized pores, the filling process is shown to be stepwise and proceeds by the formation of an initial monolayer on the pore walls and the subsequent condensation of the sorbate in the inner part of the pore [47]. This aspect of the model often results in the calculated PSD to be more polydisperse than the true distribution and is quite difficult to correct. One of the remedies suggested by Kaminsky et al. [45] could be to consider the calculated PSD to be a convolution of the true PSD. If a "smearing" function characteristic of the method can be determined, the true distribution can be mathematically deconvoluted from the HK-predicted PSD.

It is a common observation that the predicted PSD seems to reflect the true PSD only for micropores sized lower than 1.3 nm. For larger-sized pores, the HK methods seem to greatly underestimate the pore size of a material [46,47,49]. The reason for this drawback is attributed to an inaccurate relation between the adsorption potential and pore size in the HK models [46,57]. In particular, there appears to be a lack of emphasis on the adsorbate–adsorbate interaction energy in the original HK models, although it is accounted for to some extent [49]. The assumption of the adsorbed fluid to behave as a two-dimensional ideal gas also is also questionable [9]. Furthermore, the averaging of the interaction energy of the molecule throughout the pore width is conducted without regard to the distribution of energy in the pore [9]. The averaging of the local potential $\varepsilon(r)$ leading to $\bar{\varepsilon}$ may be improved by accounting for adsorbate particles favoring regions of strong attractive potentials [45]. In other words, a weighting function related to the Boltzmann law of energy distribution should be incorporated while averaging the potential. Improved HK models have been proposed by Rege and Yang [57] which attempt to correct some of these inaccuracies in the determination of the interaction energy of an adsorbed molecule in a filled micropore and will be discussed separately in a subsequent section.

Studies conducted with the objective of calculating HK-predicted PSDs by employing isotherms generated by computer simulations of sorbents with known mathematical PSD functions provide further interesting insights [46,47,50]. For monodisperse pores larger than 0.9 nm, two peaks were exhibited by the HK model. This fact is obvious from Fig. 9, which shows the HK PSD predictions for two model pores of size 1.09 nm and 1.48 nm. The first peak, resulting from the monolayer formation on the pore walls, is not an indication of microporosity and, therefore, is an artifact of the method [46]. The second peak arises from the condensation of the adsorbate between already formed sorbate layers and, in principle, can be used for PSD determination. Thus, results obtained by employing the HK method for mesoporous or nonporous materials, although not recommended, should be cautiously analyzed for the presence of artificial peaks in the microporous region.

Other drawbacks of the method include the high sensitivity of predicted PSD with respect to values of interaction parameters and pore geometry [36], which has been discussed in detail in

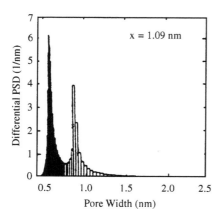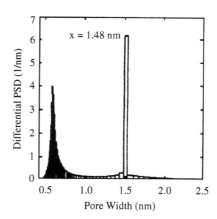

FIG. 9 PSD obtained by applying the HK model to N_2 (77 K) adsorption isotherms calculated by DFT simulations for model slit-shaped graphite pores of width 1.09 nm and 1.48 nm. (From Ref. 46.)

a previous section. Also, Cheng and Yang [40] have noted that the assumption of linearity of the adsorption isotherm (or Henry's law) made in the derivation of the model is invalid for the entire range of adsorption data and therefore needs correction. There is also some ambiguity in deciding the proper value of the saturation pressure of the bulk fluid P^{sat} when applying the method [45]. Finally, the effects of surface heterogeneity, pore networking, and, most importantly, the absence of any electrostatic forces in determining the interaction potential in the HK models remain to be addressed.

D.   Corrections to HK Methods

1.   Cheng–Yang Correction for Isotherm Nonlinearity

One of the assumptions made by Horvath and Kawazoe [35] in their original derivation was that the adsorbate behaved as a two-dimensional ideal gas. This implies that the isotherm obeys Henry's law and is therefore linear in nature. The equation of state shown in Eq. (30) was thus substituted in Eq. (29), which caused the term $(T\beta/\theta)(d\Pi/dT) = RT$ to cancel out with the other RT term in the expression, resulting in Eq. (31). However, the isotherms for the typical sorbates used in HK analysis, such as N_2 at 77 K and Ar at 87 K, clearly show a type I adsorption behavior. An example of such an isotherm is shown in Fig. 10. As is obvious from the figure, the assumption of linearity is only valid for the steeply rising portion of the isotherm, whereas the concave portion of the steep rise may also provide useful information. For this reason, Cheng and Yang [40] proposed the use of a Langmuir-type equation of state in place of Eq. (30) because it is known to represent type I isotherms in the best manner:

$$\Pi = \frac{kT}{\beta} \ln\left(\frac{1}{1-\theta}\right) \tag{55}$$

When this value of Π is substituted in Eq. (29), the following corrected relation results:

$$RT \ln\left(\frac{P}{P_0}\right) + \left[RT - \frac{RT}{\theta}\ln\left(\frac{1}{1-\theta}\right)\right] = U_0 + P_a \tag{56}$$

By applying this correction to the different pore geometry models, improved PSDs can be obtained. The so-called Cheng–Yang correction factor is negligible at low relative pressures but becomes considerable at greater pressures, when the deviation from nonlinearity becomes significant. A consequence of applying the correction for isotherm curvature is that the predicted PSD is sharpened and the peak of the distribution curve is shifted toward a smaller size compared to the original HK model. A few examples of PSD improvements brought about by the use of the Cheng–Yang modification to the HK equation are shown in Figs. 11–13 for slit-, cylindrical-, and spherical-shaped pores, respectively.

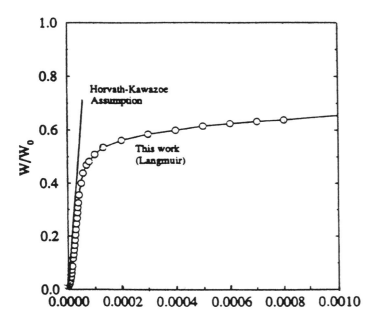

FIG. 10 Region of the adsorption isotherm of Ar (87 K) on ZSM-5 zeolite which can be used for PSD determination using the original HK model and the Cheng–Yang-corrected HK model. (From Ref. 40.)

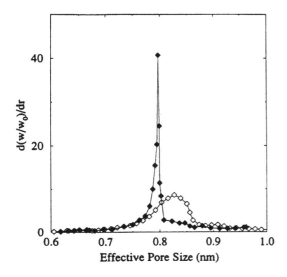

FIG. 11 PSD of faujasite zeolite calculated using the slit-pore version of the HK model. The open symbols denote the original HK model and the filled symbols show the Cheng–Yang-corrected HK model. (From Ref. 40.)

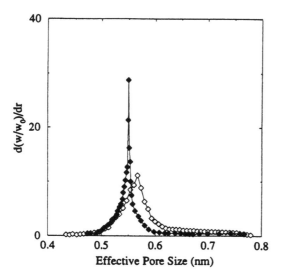

FIG. 12 PSD of ZSM-5 zeolite calculated using the cylindrical-pore version of the HK model. The open symbols denote the Saito–Foley HK model and the filled symbols show the Cheng–Yang-corrected HK model. (From Ref. 40.)

2. Rege–Yang-Corrected Models for an Improved Interaction Potential in Micropores

The original HK models considered a simplified view of the interaction of an adsorbate molecule in a pore. In general, the strategy adopted considered the sorbate–sorbent interaction of a single fluid molecule in a pore of a given geometry and then the sorbate–sorbate interaction was

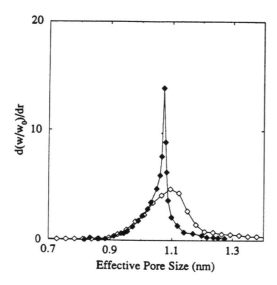

FIG. 13 PSD of 5A zeolite calculated using the spherical-pore version of the HK model. The open symbols denote the original HK model and the filled symbols show the Cheng–Yang-corrected HK model. (From Ref. 40.)

accounted for by appending the adsorbate dispersion term (i.e., $N_A A_A$, for slit and cylindrical pores $N_2 \varepsilon_{22}^*$ for spherical pores) to the sorbent dispersion term [35,36,40]. This approach provides a reasonable approximation when the pore width is not greater than two adsorbate molecule diameters. However, it fails for larger pore widths, which results in an underestimation of the true PSD peaks. Also, for pores of the size of a few molecular diameters in width filled with sorbate molecules, a more discrete picture of the molecules is required rather than a mean field approach with a continuum of fluid throughout the pore. Such a picture would require a major revision in the methods for calculating the mean interaction energy of a sorbate molecule in a pore.

Rege and Yang [57] proposed new models for determining the average potential of a molecule in a sorbate-filled micropore. For all three pore geometries discussed earlier, the general strategy used is the same. First, the number of sorbate molecule layers within the filled micropore is estimated. Each molecule is assumed to rest preferentially at a position at which its energy potential within the pore would be minimum, in accordance with the Boltzmann law of energy distribution. Furthermore, a gas molecule is assumed to interact most effectively only with its laterally immediate molecular layer. This fact is in agreement with experimental observations which show that the energy of adsorption of a monolayer of adsorbate molecules on a clean sorbent surface is the highest and that it is a nearly constant lower value on subsequently adsorbed layers of gas molecules. Any interaction with molecules not in the immediate proximity of the molecule or with those lying above or below the molecule in the same layer is assumed to be negligible. The average interaction energy is then calculated by averaging the energy potentials of the individual gas molecule layers. For cylindrical and spherical models, the average was calculated by weighing the interaction energy by an approximate molecular population of each layer. Then, the usual HK protocol is invoked to calculate the PSD from the adsorption isotherm. The complete derivation of the corrected HK models is available in literature [57] and is not reproduced here. Only the proposed algorithms for calculation of the PSD for the three pore geometries are highlighted in Tables 2–4. The improved interaction energy plots were seen to approach the correct nonzero limits at large pore size, unlike the potential relations used in the original HK models. Also, the corrected HK models showed significant improvement in the predicted PSDs compared to the original equations. Moreover, the pore-size range for which the model is applicable was reported to be substantially increased compared to the mesoporous range.

Figures 14–16 demonstrate the usefulness of the corrected HK models for slit-shaped, cylindrical pores, and spherical pores, respectively. The PSD calculated using the original HK models with and without Cheng–Yang (CY) correction are also shown for comparison. It can be seen that the corrected HK models provide an improved PSD estimate compared to the original models.

V. SURVEY OF PSD TECHNIQUES USED IN LITERATURE

An exhaustive review of PSD techniques published in literature is given in Table 5. Some interesting observations can be inferred from the table. It appears that microporous carbons are the most well-studied materials suiting a wide variety of PSD methods. This is probably due to the wide variety of carbon forms in use today and the importance of these materials in separation and purification. The relative inertness of graphitic surfaces toward quadrupolar sorbate molecules such as N_2 also makes their PSD estimation by adsorptive methods particularly attractive. The adsorbates which are most commonly employed seem to be N_2 (77 K) and Ar

TABLE 2 Algorithm (Pseudo-code) for Obtaining Effective Pore Size of Slit-Shaped Pores Using the Rege–Yang-Corrected HK Model

(1) Guess slit width L corresponding to adsorbate relative pressure P/P_0.
(2) Calculate number of adsorbate molecular layers M:

$$M = \frac{L - d_s}{d_A}$$

(3) Calculate ε_1, ε_2, and ε_3:

$$\varepsilon_1(z) = \frac{N_s A_s}{2\sigma_s^4}\left[-\left(\frac{\sigma_s}{d_0}\right)^4 + \left(\frac{\sigma_s}{d_0}\right)^{10} - \left(\frac{\sigma_s}{L-d_0}\right)^4 + \left(\frac{\sigma_s}{L-d_0}\right)^{10}\right]$$

$$\varepsilon_2(z) = \frac{N_s A_s}{2\sigma_s^4}\left[-\left(\frac{\sigma_s}{d_0}\right)^4 + \left(\frac{\sigma_s}{d_0}\right)^{10}\right] + \frac{N_A A_A}{2\sigma_A^4}\left[-\left(\frac{\sigma_A}{d_A}\right)^4 + \left(\frac{\sigma_A}{d_A}\right)^{10}\right]$$

$$\varepsilon_3(z) = 2\frac{N_A A_A}{2\sigma_A^4}\left[-\left(\frac{\sigma_A}{d_A}\right)^4 + \left(\frac{\sigma_A}{d_A}\right)^{10}\right]$$

(4) If $M < 2$, then

$$\bar{\varepsilon} = \varepsilon_1$$

Else ($M \geq 2$)

$$\bar{\varepsilon} = \frac{2\varepsilon_2 + (M-2)\varepsilon_3}{M}$$

(5) If $[RT\ln(P/P_0) = N_{Av}\bar{\varepsilon}]$, then
    Effective pore width $= (L - d_s)$
    Else
    Guess new L and iterate from (1) again.

Source: Ref. 57.

(87 K). Their small molecular diameter and ease of availability explain their popularity. Other sorbates such as CO_2 [25,42,62,86] and organic vapors such as C_6H_6 [15,52,53,84] and CH_3Cl [43,44] have also been proposed. The use of organic vapors such as CH_3Cl is said to lower the requirements for high vacuum, and equilibrium is achieved faster because the adsorption temperature is higher (273–295 K) [44]. For GCMC simulations, a wide variety of sorbates were seen to be used ranging from N_2 [11,25,31,50] and CO_2 [25,62] to more exotic species such as CH_4, CF_4, and SF_6 [30,32].

The most popular method for microporous sorbents with pores less than 2.0 nm is evidently the HK theory. For higher pore sizes, the NL-DFT and GCMC models seem to be preferred. Also, the HK methods are observed to be used for a wide variety of sorbents from carbons to zeolites and even porous glasses. In contrast, the DFT methods seem to be restricted to mainly carbons and silica-based materials such as MCM-41. Lack of interaction parameters for other types of materials in literature is probably a reason for their limited use.

These statements are valid in spite of the fact that in Eq. (3), the limiting value of the adsorption potential A tends to negative infinity as the equilibrium pressure tends to zero. Although this is thermodynamically inconsistent, the relation $RT\ln(P/P_0)$ does hold true to a certain limit of low pressure. There is a lower bound on the interaction energy provided by a

TABLE 3 Algorithm (Pseudo-code) for Obtaining Effective Pore Size of Cylindrical Pores Using the Rege–Yang-Corrected HK Model

(1) Guess pore *radius L* corresponding to adsorbate relative pressure P/P_0.

(2) Calculate number of concentric adsorbate molecular layers M:

$$M = \text{int}\left[\frac{[(2L - d_S/d_A)] - 1}{2}\right] + 1$$

(3) Calculate ε_1:

$$\varepsilon_1 = \frac{3}{4}\pi\frac{N_s A_s}{d_0^4}\left[\frac{21}{32}a_1^{10}\sum_{k=0}^{\infty}\alpha_k(b_1^{2k}) - a_1^4\sum_{k=0}^{\infty}\beta_k(b_1^{2k})\right]$$

$$a_1 = \frac{d_0}{L}; \qquad b_1 = \frac{L - d_0}{L}$$

(4) Calculate ε_i for $i = 2$ to M:

$$\varepsilon_i = \frac{3}{4}\pi\frac{N_A A_A}{d_A^4}\left[\frac{21}{32}a_i^{10}\sum_{k=0}^{\infty}\alpha_k(b_i^{2k}) - a_i^4\sum_{k=0}^{\infty}\beta_k(b_i^{2k})\right]$$

$$a_i = \frac{d_A}{L - d_0 - (i - 2)d_A}; \qquad b_i = \frac{L - d_0 - (i - 1)d_A}{L - d_0 - (i - 2)d_A}$$

(5) Calculate the population of the layers: N_i, for $i = 1$ to M:

If $(d_A \leq 2[L - d_0 - (i - 1)d_A]$, then

$$N_i = \frac{\pi}{\sin^{-1}\{d_A/2[L - d_0 - (i - 1)d_A]\}}$$

Else $(d_A > 2[L - d_0 - (i - 1)d_A])$

$$N_i = 1$$

(6) Calculate $\bar{\varepsilon}$:

$$\bar{\varepsilon} = \frac{\sum_{i=1}^{M} N_i \varepsilon_i}{\sum_{i=1}^{M} N_i}$$

(7) If $[RT\ln(P/P_0) = N_{Av}\bar{\varepsilon}]$, then

Effective pore diameter $= (2L - d_s)$

Else

Guess new L and iterate from (1) again.

Source: Ref. 57.

sorbent surface for an adsorbing molecule. Hence, practically, there should be a lower bound for the pressure at which there is any detectable adsorption on a sorbent surface.

The corrected Horvath–Kawazoe models discussed in this work provide a simple picture for adsorbate interactions in a micropore and are reasonably fast and accurate methods provided the desired pore size is within the microporous range (<2.0 nm). Some distinct advantages of these methods is the facile extension to different pore geometries compared to other methods and the ease of availability of required physical parameters from literature. However, these methods are also known to be very sensitive to certain input parameters and introduce artifacts in the PSD results for large-pore materials.

TABLE 4 Algorithm (Pseudo-code) for Obtaining Effective Pore Size of Spherical Pores Using the Rege–Yang-Corrected HK Model

(1) Guess pore *radius* L corresponding to adsorbate relative pressure P/P_0.

(2) Calculate number of concentric adsorbate molecular layers M:

$$M = \text{int}\left[\frac{[(2L - d_s)/d_A] - 1}{2}\right] + 1$$

(3) Calculate ε_1:

$$\varepsilon_1 = 2N_0\varepsilon_{12}^*\left[-\frac{c_1^6}{4f_1}\left(\frac{1}{(1-f_1)^4} - \frac{1}{(1+f_1)^4}\right) + \frac{c_1^{12}}{10f_1}\left(\frac{1}{(1-f_1)^{10}} - \frac{1}{(1+f_1)^{10}}\right)\right]$$

$$c_1 = \frac{d_0}{L}; \quad f_1 = \frac{L - d_0}{L}$$

$$N_0 = 4\pi L^2 N_s$$

(4) Calculate N_i for $i = 1$ to M:

$$N_i = 4\pi[L - d_0 - (i-1)d_A]^2 N_A$$

(5) Calculate ε_i for $i = 2$ to M:

$$\varepsilon_i = 2N_{i-1}\varepsilon_{22}^*\left[\frac{-c_i^6}{4f_i}\left(\frac{1}{(1-f_i)^4} - \frac{1}{(1+f_i)^4}\right) + \frac{c_i^{12}}{10f_i}\left(\frac{1}{(1-f_i)^{10}} - \frac{1}{(1+f_i)^{10}}\right)\right]$$

$$f_i = \frac{d_A}{L - d_0 - (i-2)d_A}; \quad f_i = \frac{L - d_0 - (i-1)d_A}{L - d_0 - (i-2)d_A}$$

(6) Calculate $\bar{\varepsilon}$:

$$\bar{\varepsilon} = \frac{\sum_{i=1}^{M} N_i\varepsilon_i}{\sum_{i=1}^{M} N_i}$$

(7) If $[RT\ln(P/P_0) = N_{Av}\bar{\varepsilon}]$, then

    Effective pore diameter $= (2L - d_S)$

Else

    Guess new L and iterate from (1) again.

Source: Ref. 57.

In conclusion, it needs to be stressed that there is still room for perfecting currently available PSD models. For example, the electrostatic forces of interaction can be significant in the case of an ionic sorbent (e.g., zeolite) and a highly quadrupolar sorbate (e.g., N_2), but they are not accounted for in the models. The accuracy of the physical parameters used in a model is an important issue also, especially because the predicted PSD is known to be highly sensitive to some of the input parameters. The accuracy of the experimentally measured adsorption isotherm data, especially at low relative pressures, is also equally significant. A valid PSD determination is only possible when the measured data are reliable and accurate physical parameters are chosen as input.

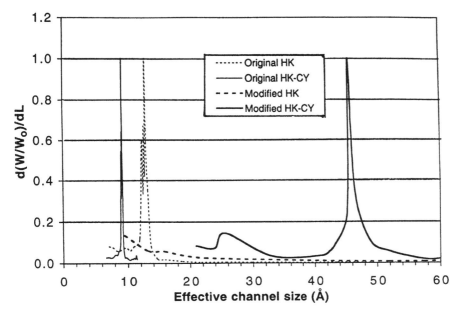

FIG. 14 PSD of MCM-41 calculated using the original and Rege–Yang-corrected HK models for cylindrical pores with and without Cheng–Yang (CY) correction. (From Ref. 57.)

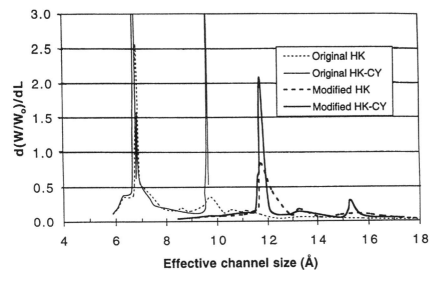

FIG. 15 PSD of VPI-5 calculated using the original and Rege–Yang-corrected HK models for cylindrical pores with and without Cheng–Yang (CY) correction. (From Ref. 57.)

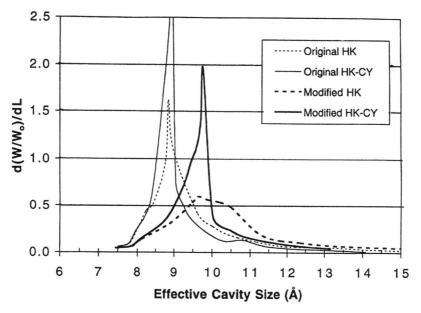

FIG. 16 PSD of 5A zeolite calculated using the original and Rege–Yang-corrected HK models for spherical pores with and without Cheng–Yang (CY) correction. (From Ref. 57.)

VI. NOTATION

a_i, b_i	Parameters in corrected cylindrical pore model for the ith layer of molecules
a_{mi}	Micropore adsorbed amount
A	Adsorption potential
A_s, A_A	Dispersion constants given by Kirkwood–Müller formalism
c	Speed of light
c_i, f_i	Parameters in corrected spherical pore model for the ith layer of molecules
d	Diameter of molecule
d_0	Mean of the diameters of adsorbate and adsorbent molecules
E_0	Characteristic adsorption energy in the DA equation
ΔG^{ads}	Free energy change on adsorption
ΔH^{ads}	Enthalpy change on adsorption
i	Index for sorbate molecular layer numbered starting from outermost layer
k	Summation counter for cylindrical pore energy profile equation, or Boltzmann constant
K, K_0	Empirical proportionality constant
L	Internuclear distance between parallel layers of slit-shaped pores, likewise defined as the radius (not diameter) for cylindrical and spherical pore
m	Mass of an electron
M	Number of layers of adsorbate molecules accommodated in pore
n	Local adsorbed amount for a pore group
N	Total adsorbed amount
N_i	Density of molecules based on surface area or number of molecules in a layer
N_1	Number of sorbent molecules on cavity surface
N_2	Number of molecules in adsorbate layer

TABLE 5 Literature Survey of Different Methods Used for Obtaining the PSD of Adsorbents

Sorbent	Sorbate	Pore-size range (nm)	Method(s) used	Ref.
Microporous carbon AC610, AX21, CXV	N_2 (77 K)	1–3	NL-DFT, γ-distribution and log-normal distribution (trimodal)	18
Microporous carbon BP 71, AC610, CXV	N_2 (77 K)	0–60	DFT (MFT), γ-distribution and log-normal distribution (bimodal)	16
Activated carbons AX21, CXV, AC610, D52	N_2 (77 K)	1.0–6.0	Modified Kelvin equation	7
Activated carbon: BPL	N_2 (77 K), H_2O (298–398 K)	0.6–2.0	MP, DA, HK, DFT (MFT), water desorption	22
Activated carbon: BPL	Ar (87 K)	0–40	NL-DFT	20
Activated carbon: BPL	CH_4 (308 K)	0.5–3.0	GCMC	30
Carbosieve-G	CH_4, CF_4, SF_6 (258–296 K)	0.75–2.25	GCMC	32
Carbosieve-G	Ar (87 K)	0–40	NL-DFT	20
Activated carbon cloths	N_2 (77 K)	0–2.4	HK, DR-Stoeckli, DS, GCMC	23
Activated charcoal cloth	N_2 (77 K)	0–3.0	DS	14
Activated carbon fibers, PAN based	N_2 (77 K)	0.4–1.4 0–100	HK (slit) NL-DFT	29, 54
Activated carbon fiber: CM, KF-1500	CH_4 (273–323 K), CO_2 (253–323 K)	0.5–3.0	GCMC, DS	62
MSC: Takeda HGM 366, HGS 638	N_2 (77 K)	0.4–1.4	HK (slit)	4, 51
MSC (HGS-638)	N_2 (77 K)	0.4–6.0	Corrected HK (slit)	57
MSC: Takeda 5A, Carbosieve-G, Ambersorb-1500	CH_3Cl (273 K)	0.4–1.2	HK (cylindrical pore)	44
MSC: T3A, P15, P10	N_2 (77 K) Ar (77 K), CO_2 (273 K)	0.3–4.0	NLDFT, GCMC	25
Activated carbon: Takeda HGS-538, Bayer K154	CO_2 (273 K)	0.35–1.4	HK (slit, cylindrical)	42
Carbon, polymer-derived: Ambersorb-600, Ambersorb-563	CH_3Cl (273 K)	0.4–1.2	HK (cylindrical pore)	43
Activated carbon (A type)	N_2 (77 K) Ar (77.5 K), C_6H_6 (293 K)	0.2–1.5	HK (slit), DA, DS	53, 54

TABLE 5   (*continued*)

Sorbent	Sorbate	Pore-size range (nm)	Method(s) used	Ref.
Carbon black, graphitized (Sterling FT-G, Black Pearl), active carbons (ROW, Darco)	N_2 (77 K)	0.5–2.0	NL-DFT, HK, DA	46, 47, 82, 83
Activated carbon	C_6H_6 (293 K)	0.2–1.0	JC	15, 84
Microporous carbon (PRC-16, phenolic resin-based)	N_2 (76.1 K), CH_4 (313–343 K)	0.5–4.0	NL-DFT	85
Carbon mixed oxide nanocomposite: C/H₃–NDS, C/Na₂–NDS	N_2 (77 K) CO_2 (298 K), CH_4 (250–290 K), SF_6 (250–290 K)	0.2–1.2	*t* Plot	86
Activated carbon: NORIT R-2	N_2 (77.5 K)	0.3–2.0	DFT, HK, (slit, cylindrical, spherical)	87
Activated carbon: CWNS, RKD4, AG, BH	N_2 (77 K)	APD[a]	α_s Method, JC	64
Activated carbon: Merck	N_2 (77 K)	0.4–1.5	HK (slit)	48
Activated carbons (modified by burn-off)	N_2 (77 K)	0.3–3.0	DS, HK (slit), L-DFT, GCMC	50
MCM-41	N_2 (77 K)	2.8–4.6	GCMC	31
MCM-41	N_2 (77 K) Ar (87 K)	1.0–8.0	NL-DFT	24, 26, 28
MCM-41	N_2 (77)	0.8–2.0	Kelvin equation, H-NMR, HREM	3
MCM-41, silicas	N_2 (77 K)	0.2–2.0	BJH	88
MCM-41	Ar (87 K)	1.0–5.0	HK (slit)	55
MCM-41	N_2 (77 K)	0.4–6.0	Corrected HK (cylindrical)	57
M41S-based catalyst supports	N_2 (77 K) Ar (87 K)	2.0–6.0	NL-DFT	27
Pillared clay	N_2 (77 K)	0.3–1.5	α_s Method, JC	5
Pillared clays	N_2 (77 K)	0.5–3.0	HK (slit)	38
Al-pillared Montmorillonite	N_2 (77 K),	0.4–1.5	HK (slit)	48
Zeolites (ZSM-5, CaA, NaX), AlPO₄-5/11, VPI-5	N_2 (77 K), Ar (87 K)	0.4–1.7	HK (slit, cylindrical, spherical pore)	40
Zeolites (CaA, ZSM-5, NaY)	N_2 (77 K) Ar (87 K)	0.5–1.4	HK (slit)	41
Zeolites (CaA, NaY, ZSM-5, NaX), VPI-5, AlPO₄-5/11	Ar (87 K)	0.6–0.8	HK (cylindrical)	36
Zeolite (NaX)	N_2 (77 K)	0–2.0	HK (slit)	56

TABLE 5 (*continued*)

Sorbent	Sorbate	Pore-size range (nm)	Method(s) used	Ref.
Zeolite (faujasite)	N_2 (77 K) Ar (87 K)	0.5–1.5	HK (slit)	89
Zeolite (Y), amorphous SiO_2	Ar (87 K)	1.0–5.0	HK (slit)	55
Zeolite (5A)	C_4H_{10} (393 K)	0–1.2	HK (spherical)	72
Zeolite (NaX, ZSM-5, CaA) $AlPO_4$-5/11, VPI-5	N_2 (77 K), Ar (87 K)	0.4–2.0	Corrected HK (slit, cylindrical, spherical)	57
Zeolite (NaX, clinoptilolite)	H_2O (295 K)	0.4–2.0	HK (cylindrical, spherical)	90
Porous glass (Vycor)	N_2 (77 K)	0–7.0	GCMC, BJH	11
Porous glass	N_2 (77 K) Ar (87 K)	0.5–1.5	HK (slit)	89

[a] APD (adsorption potential distribution) was calculated instead of PSD.

N_{Av}	Avogadro's number
P	Pressure of adsorbate in the gas phase
P_a	Sorbate–sorbate interaction energy
P_0	Saturation vapor pressure of adsorbate
q^{diff}	Differential heat of adsorption
r	Radial distance of an adsorbate molecule from the central axis for cylindrical pores and from the center in case of spherical pores
R	Gas constant
R_p	Radius of cylindrical pore
ΔS^{ads}	Entropy change on adsorption
T	Absolute temperature
ΔU	Change in potential energy after a random trial
U_0	Sorbate–sorbent interaction energy
V_{ext}	Total potential in the pore
v_a	Specific volume of adsorbed gas
v_o	Total specific micropore volume of sorbent
w	Pore width
z	Internuclear distance between an adsorbate molecule and a surface molecule

Greek Letters

α	Polarizability
β	Similarity coefficient in DA equation
γ	Surface tension of sorbate
Δ	Distance between two consecutive graphitic layers
ε_{12}	Intermolecular interaction energy potential
ε_{12}^*	Depth of potential energy minimum for adsorbate–adsorbent interaction
ε_{22}^*	Depth of potential energy minimum for adsorbate–adsorbate interaction

$\bar{\varepsilon}$	Average interaction energy
ϕ_{ff}	Fluid–fluid interaction potential
ϕ_{sf}	Solid–fluid interaction potential
Π	Spreading pressure
ρ	Sorbate fluid density
σ	Internuclear distance between two molecules at zero interaction energy
θ	Degree of void filling in the sorbent
Ω	Grand thermodynamic canonical ensemble
χ	Magnetic susceptibility (cm^3)

Subscripts

A	Corresponding to adsorbate
S	Corresponding to the adsorbent

REFERENCES

1. SJ Gregg, KSW Sing. Adsorption, Surface Area, and Porosity. New York: Academic Press, 1982.
2. S Brunauer, LS Deming, WS Deming, E Teller. J Am Chem Soc 62:1723, 1940.
3. R Schmidt, EW Hansen, M Stöcker, D Akporiaye, OH Ellestad. J Am Chem Soc 117:4049, 1995.
4. G Horváth, K Kawazoe, J Chem Eng Japan 16:470, 1983.
5. MSA Baksh, ES Kikkinides, RT Yang. Ind Eng Chem Res 31:2181, 1992.
6. S Brunauer, RSh Mikhail, EE Bodor. J Colloid Sci 24:451, 1967.
7. C Nguyen, DD Do. Langmuir 15:3608, 1999.
8. KSW Sing. Colloids Surfaces 38:113, 1989.
9. PA Webb, C Orr. Analytical Methods in Fine Particle Technology. Norcross, GA: Micromeritics Inc., 1997, Chap. 3.
10. EP Barrett, LG Joyner, PP Halenda. J Am Chem Soc 73:373, 1951.
11. LD Gelb, KE Gubbins. Langmuir 15:305, 1999.
12. DH Everett, JC Powl. J Chem Soc Faraday Trans 72:619, 1976.
13. HFJ Stoeckli. J Colloid Interf Sci 59:184, 1977.
14. PJM Carrott, MMLR Carrott. Carbon 37:647, 1999.
15. M Jaroniec, J Choma, X Lu. Chem Eng Sci 46:3299, 1991.
16. NA Seaton, JPRB Walton, N Quirke. Carbon 27:853, 1989.
17. C Lastoskie, KE Gubbins, N Quirke. Langmuir 9:2693, 1993.
18. C Lastoskie, KE Gubbins, N Quirke. J Phys Chem 97:4786, 1993.
19. JP Olivier. J Porous Mater 2:9, 1995.
20. JP Olivier. Carbon 36:1469, 1998.
21. GM Davies, NA Seaton. Carbon 36:1473, 1998.
22. BP Russel, MD LeVan. Carbon 32:845, 1994.
23. PJM Carrott, MMLR Carrott, TJ Mays. In: F Meunier, ed. Proceedings of the Sixth International Conference on the Fundamentals of Adsorption. Paris: Elsevier, 1998, p 677.
24. PI Ravikovitch, GL Haller, AV Neimark. Adv Colloid Interf Sci 76–77:203, 1998.
25. PI Ravikovitch, A Vishnyakov, R Russo, AV Neimark. Langmuir 16:2311, 2000.
26. PI Ravikovitch, D Wei, GL Haller, AV Neimark. J Phys Chem B 101:3671, 1997.
27. PI Ravikovitch, GL Haller, AV Neimark. In: F Meunier, ed. Proceedings of the Sixth International Conference on the Fundamentals of Adsorption. Paris: Elsevier, 1998, p 545.
28. PI Ravikovitch, SCO Domhnaill, AV Neimark, F Schüth, KK Unger. Langmuir 11:4765, 1995.
29. Z Ryu, J Zheng, M Wang, B Zhang. Carbon 37:1257, 1999.
30. GM Davies, NA Seaton, VS Vassiliadis. Langmuir 15:8235, 1999.
31. MW Maddox, JP Olivier, KE Gubbins. Langmuir 13:1737, 1997.
32. MVL Ramón, J Jagiello, TJ Bandosz, NA Seaton. Langmuir 13:4435, 1997.

33. KL Murray, NA Seaton, MA Day. Langmuir *14*:4953, 1998.
34. KE Gubbins. In: J Fraissard, CW Conner, eds. Physical Adsorption: Experiment, Theory and Applications. NATO ASI Series C, Vol. 491. Boston: Kluwer Academic Publishers, 1997, p 65.
35. JP Olivier, WB Conklin, MV Szombathely. In: J Rouquerol, F Rodriguez-Reinoso, KSW Sing, KK Unger, eds. Characterization of Porous Solids III. Studies in Surface Science and Catalysis, Vol. 87. Amsterdam: Elsevier, 1994.
36. A Saito, HC Foley, AIChE J *37*:429, 1991.
37. MSA Baksh, RT Yang. AIChE J *37*:923, 1991.
38. MSA Baksh, RT Yang. AIChE J *38*:1357, 1992.
39. RT Yang, MSA Baksh. AIChE J *37*:679, 1991.
40. LS Cheng, RT Yang. Chem Eng Sci *49*:2599, 1994.
41. AF Venero, JN Chiou. MRS Symp Proc *111*:235, 1988.
42. G Horváth, V Halász-Laky, I Dékány, F Berger. In: F Meunier, ed. Proceedings of the Sixth International Conference on the Fundamentals of Adsorption. Paris: Elsevier, 1998, p 611.
43. MS Kane, JH Bushong, HC Foley, WH Brendley. Ind Eng Chem Res *37*:2416, 1998.
44. RK Mariwala, HC Foley. Ind Eng Chem Res *33*:2314, 1994.
45. RD Kaminsky, E Maglara, WC Conner. Langmuir *10*:1556, 1994.
46. M Kruk, M Jaroniec, J Choma. Adsorption *3*:209, 1997.
47. M Kruk, M Jaroniec, J Choma. Carbon *36*:1447, 1998.
48. A Gill, P Grange. Langmuir *13*:4483, 1997.
49. D Valladares, G Zgrablich. Adsorp Sci Technol *15*:16, 1997.
50. DL Valladares, FR Reinoso, G Zgrablich. Carbon *36*:1491, 1998.
51. G Horvath. Colloids Surfaces A *141*:295, 1998.
52. A Swiatkowski, BJ Trznadel, S Zietek. Adsorp Sci Technol *14*:59, 1996.
53. BJ Trznadel, S Zietek, A Swiatkowski. Adsorp Sci Technol *17*:11, 1999.
54. Z Ryu, J Zheng, M Wang. Carbon *36*:427, 1998.
55. JS Beck, JC Vartuli, WJ Roth, ME Leonowicz, CT Kresge, KD Schmitt, CT-W Chu, DH Olson, EW Sheppard, SB McCullen, JB Higgins, JL Schlenker. J Am Chem Soc *114*:10,835, 1992.
56. J Seifert, G Emig. Chem Eng Technol *59*:475, 1987.
57. SU Rege, RT Yang. AIChE J *46*:734, 2000.
58. LS Cheng, RT Yang. Adsorption *1*:187, 1995.
59. MM Dubinin, LV Radushkevich. Dokl Akad Nauk SSSR *55*:327, 1947.
60. MM Dubinin, VA Astakhov. Adv Chem Soc *102*:69, 1971.
61. MM Dubinin, HF Stoeckli. J Colloid Interf Sci *75*:34, 1980.
62. F Stoeckli, A Guillot, DH Cleary, AM Slasli. Carbon *38*:929, 2000.
63. M Jaroniec, R Madey. Physical Adsorption on Heterogeneous Solid. Amsterdam: Elsevier, 1988.
64. J Choma, M Jaroniec. Mater Chem Phys *20*:179, 1988.
65. MV Szombathely, M Brauer, M Jaroniec. J Comput Chem *13*:17, 1992.
66. C Nguyen, DD Do. Langmuir *16*:1319, 2000.
67. R Evans, P Tarazona. Phys Rev Lett *52*:557, 1984.
68. R Evans. Adv Phys *28*:143, 1979.
69. P Tarazona. Phys Rev A *31*:2672, 1985.
70. S Samios, AK Stubos, NK Kanellopoulos, R Cracknell, G Papadopoulos, D Nicholson. In: B McEnaney, TJ Mays, J Rouquerol, F Rodriguez-Reinoso, KSW Sing, KK Unger, eds. Characterization of Porous Solids IV. Cambridge: The Royal Society of Chemistry, 1977.
71. DM Razmus, CK Hall. AIChE J *37*:769, 1991.
72. JP Padin, RT Yang, C Munson. Ind Eng Chem Res *38*:3614, 1999.
73. M Jaroniec, KP Gadkaree, J Choma. Colloids Surfaces A *118*:203, 1996.
74. JR Sams Jr, G Constabaris, GD Halsey. J Phys Chem *64*:1689, 1960.
75. PL Walker Jr., ed. Chemistry and Physics of Carbon. New York: Marcel Dekker, 1966, Vol. 2, p 362.
76. LJ Soto, PW Fisher, AJ Glessner, AL Myers. J Chem Soc Faraday Trans *77*:157, 1981.
77. ME Davis, C Montes, PE Hathaway, JP Arhancet, DL Hasha, JE Graces. J Am Chem Soc *111*:3919, 1989.

78. ME Davis, C Saldarriaga, C Montes, J Garces, C Crowder. Zeolites *8*:362, 1988.
79. A. Dyer. An Introduction to Zeolite Molecular Sieve. New York: Wiley, 1988.
80. WM Meier, DH Olson. Atlas of Zeolite Structure Types. London: Butterworth, 1964.
81. DW Breck. Zeolite Molecular Sieves: Structure, Chemistry, and Use. New York: Wiley, 1974.
82. M Jaroniec, KP Gadkaree, J Choma. Colloids Surface A *118*:203, 1996.
83. M Kruk, M Jaroniec, KP Gadkaree. Langmuir *15*:1442, 1999.
84. M Jaroniec, J Choma. Mater Chem Phys *18*:103, 1987.
85. N Quirke, SRR Tennison. Carbon *34*:1281, 1996.
86. K Putyera, TJ Bandosz, J Jagiello, JA Schwarz. Carbon *34*:1559, 1996.
87. J Choma, M Jaroniec. Adsorp Sci Technol *15*:571, 1997.
88. A Sayari, Y Yong, M Kruk, M Jaroniec. J Phys Chem B *103*:3651, 1999.
89. WS Borghard, EW Sheppard, HJ Schoennagel. Rev Sci Instrum *62*:2801, 1991.
90. SU Rege, RT Yang, MA Buzanowski. Chem Eng Sci *55*:4789, 2000.

4

Adsorption Isotherms for the Supercritical Region

LI ZHOU   Tianjin University, Tianjin, China

I. INTRODUCTION

When I received the letter from Professor József Tóth inviting me to prepare a chapter of the book in the area of "adsorption isotherms for the supercritical region," I hesitated to accept. As mentioned by several well-known researchers [1–4], there are not many articles that present original experimental data or discussions relating to the mechanism of the physical adsorption of fluids at above-critical temperatures. As consequence, there may be as many points of view regarding the nature and theory of supercritical adsorption as there are researchers. To construct the text of the chapter was indeed a difficult job because several serious questions have to be answered without a sophisticated textbook on which to rely. The context will inevitably reflect mainly the opinions and biases of the author. Nevertheless, the invitation was attractive and despite initial reservations, I concluded that such an exercise was indeed useful, especially if it prompts others to think about these issues and perhaps to challenge the opinions expressed here.

Physical adsorption of gases at above-critical temperatures and elevated pressures is a field of growing importance in both science and engineering [5,6]. It is the physicochemical basis of many important engineering processes and potential industrial applications. For example, separation or purification of light hydrocarbons and several other gases [7,8], storage of fuel gases in microporous solids [9–12], adsorption from supercritical gases in extraction processes, and chromatography [13–15]. Additionally, knowledge of the gas/solid interface phenomenon at high pressures is fundamental to heterogeneous catalysis. However, the limited number of reliable high-pressure adsorption data has resulted in delayed progress of the theoretical study. Important observations of high-pressure adsorption reported before 1930 was covered in studies by McBain and Britton [16]; all of the important articles on this subject published between 1930 and 1966 have been reviewed by Menon [17]. During the last 20 years, a growing interest in high-pressure adsorption research has been observed, mainly under the impetus of the quest for clean alternative fuels. Although many technological processes based on high-pressure adsorption have been operating industrially, the underlined theoretical basis is very weak. To realize the state of supercritical adsorption theory, we need just mention the fact that any thermodynamic calculations are impossible for a multicomponent separation task by adsorption process until today because we could not yet experimentally determine the total mass in the adsorbed phase, on which thermodynamics is based. Considerable progress has been made in both adsorption measurement techniques [18,19] and molecular simulation of adsorption on computers [3,10,20,21], rendering new insights into the nature of high-pressure adsorption. The intent of the chapter is not to make a comprehensive review to be a succeeding work of Menon [17].

Instead, the focus will be on how do we collect the adsorption isotherms better and on how can we ascertain information through manipulation of the experimental isotherms for the supercritical region. Because the theory of multicomponent adsorption equilibrium relies on the development of the adsorption theory of single gases, only the latter has been considered here.

II. BASIC CONCEPTION AND TERMINOLOGY

It is not uncommon to meet different opinions relating to the fundamentals of adsorption at above-critical temperatures, even the symbols used are sometimes confusing. Therefore, it seems useful to clarify some concepts and terminology used in the chapter before making any interpretation of isotherms.

A. Surface *Excess* and *Absolute* Adsorption

Adsorption itself is, according to Gibbs formalism, an *excess* quantity. The situation at the gas/solid interface is schematically represented in Fig. 1. As consequence of the adsorption potential, the average number of molecules in an element volume near the surface is larger than in an element volume of equal size in the bulk gas. A density profile is thus established from the solid surface to the ambient gas phase. Both molecular simulation of adsorption and experimental proof reveal that this density profile vanishes over quite a short range. Therefore, the layer on the solid surface, where adsorbate molecules are concentrated, is referred to as the "adsorption space" or "adsorbed phase." It thus assumes a definite thickness (τ) for the adsorbed phase.

Based on the above schematic representation, the amount adsorbed, n^{σ}, can be determined using either one of the following: First, if the density profile $\rho(z)$ can be determined or assumed, then

$$n^{\sigma} = \int_{\text{all adsorption space}} [\rho(z) - \rho_b] \, dV \tag{1}$$

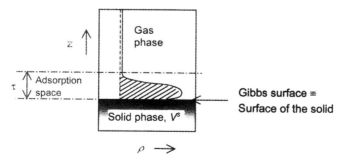

FIG. 1 Schematic representation of a gas–solid adsorption system. The solid-line curve indicates the density profile (ρ) as a function of distance (z) normal to the surface in the real system; the broken line is in the case without adsorption, and the dash-dotted line is the boundary between phases. The shadowed area marks the excess amount of adsorbed substance. (Courtesy of D. H. Everett, Butterworth, London.)

where ρ_b is the density of bulk phase. The integral clearly indicates that n^σ is a *density excess* amount of the adsorbed substance, which is usually referred to as "surface excess adsorption" or simply "excess adsorption."

Although the density profile can be determined by molecular simulation [3], it can hardly be measured experimentally. Therefore, another way to calculate n^σ is often cited in literature. Suppose the volume of the adsorption space (or the adsorbed phase) in Fig. 1 is V_a, then the excess adsorption must be

$$n^\sigma = n^s - \rho_b V_a = V_a(\rho_a - \rho_b) \qquad (2)$$

where n^s is the total mass confined in the adsorbed phase, which is usually referred to as *absolute* adsorption. That the density of the adsorbed phase ρ_a is uniform is the implicit assumption of Eq. (2). Although a density profile $\rho(z)$ was logically assumed for the adsorption space, it does not have much significance for supercritical adsorption because it is dominated up to rather high densities by the strength of the gas–solid interaction [22]. Therefore, the strength of gas–gas interaction is too weak to keep more than one layer of adsorbate molecules together. This monolayer mechanism was proven by the adsorption of methane on activated carbon [23]. The adsorption isotherms were successfully explained by a model based on the monolayer assumption. The intermolecular distance in the adsorbed phase was evaluated from the model. Such intermolecular distance in the gas phase was also determined from the gas-phase density. Both intermolecular distances decrease linearly with temperature and intersect at a point, which gives the value of $\sigma_{ff} = 0.34\,\mathrm{nm}$ [21], the distance between molecules when the Leonard–Jones potential reaches minimum, as shown in Fig. 2.

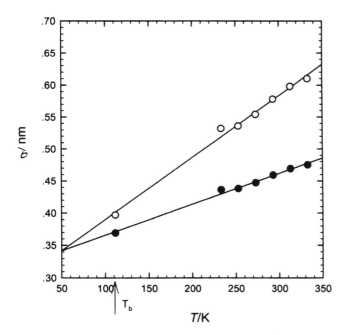

FIG. 2 A comparison of intermolecular distance: $\bigcirc$ in the bulk gas phase; $\bullet$ in the adsorbed phase. (From Ref. 23.)

B. Supercritical Adsorption

Supercritical adsorption is the adsorption at above-critical temperatures, but it is sometimes referred to as the adsorption of *supercritical fluids*. However, there are different tacit understandings of supercritical fluids. For example, "a fluid is said to be 'supercritical' when its temperature and pressure exceed the temperature and pressure at the critical point" [24]. In the supercritical extraction studies, however, "supercritical fluid" is especially applied for a narrow temperature region of $(1–1.2)T_c$ or T_c to $T_c + 10$ K, which is sometimes called the *supercritical region* [6]. According to the adsorption behavior, the adsorption of gases on solids can be classified into three typical temperature ranges relative to the critical temperature:

1. Subcritical region ($T < T_c$)
2. Near-critical region ($T_c < T < T_c + 10$)
3. The region $T > T_c + 10$

Isotherms in the first region will show the feature of subcritical adsorption and that in the third region will definitely show the feature of supercritical adsorption. Isotherms in the second region will show the feature of mechanism transition. The transition will take a more or less continuous way if the isotherms in both sides of the critical temperature belong to the same type, as the adsorption on microporous activated carbon. However, discontinuous transition could be observed on isotherms in the second region if there is a transformation of isotherm types, as observed on mesoporous silica gel [25]. The summation of the last two regions (i.e., for $T > T_c$) is currently named the "supercritical region." The decisive factor in such a classification of adsorption is merely temperature, but irrespective of pressure. This is because a fluid cannot undergo a transition to a liquid phase at above-critical temperatures, regardless of the pressure applied. This fundamental law of physics determines the different adsorption mechanisms for the subcritical and supercritical regions. For the subcritical region, the highest equilibrium pressure of adsorption is the saturation pressure, p_s, of adsorbate, beyond which condensation happens. It is, therefore, reasonable to assume that the adsorbate in the adsorbed phase is largely in the liquid state, based on which different adsorption and thermodynamic theories as well as their applications were developed. However, condensation cannot happen, no matter how great the pressure being applied for the supercritical region. It seems that no upper limit applied for pressure and, hence, "high-pressure adsorption" is synonymous with "supercritical adsorption." The assumption of liquidlike adsorbate cannot be true, and the saturated liquid can no longer be the standard state of adsorbate in the evaluation of thermodynamic quantities of the adsorbed phase at supercritical temperatures. As consequence, the available adsorption theories developed for the subcritical region are insufficient when addressing the supercritical region. A shift of the critical temperature of adsorbate is possible in small pores due to the strong interaction between adsorbate molecules and substrate. However, the shift of T_c must be quite limited; yet, there is an immense region of $T > T_c$.

III. ACQUISITION OF SUPERCRITICAL ADSORPTION ISOTHERMS

An adsorption isotherm depicts the relation between the quantity adsorbed (as defined in Section II.A) and the bulk phase pressure (or density) at equilibrium for a constant temperature. It is a dataset of specified adsorption equilibrium. Such equilibrium data are required for optimal design of processes relying on adsorption and are considered fundamental information for theoretical studies. Adsorption isotherms can be obtained either experimentally or by molecular simulation on computers. However, it should be noted that only the *excess* adsorption can be experimentally measured. The purpose of this section is not to describe the details of adsorption measurement using different techniques, but to briefly introduce their features and conceivable impact on isotherms.

A. Measurement of Gas/Solid Adsorption Equilibria

1. Manometric/Volumetric Method

This method was used in the early days of adsorption sciences by Langmuir, Dubinin, and others. It basically comprises a gas expansion process from a storage vessel (the reference cell) to an adsorption chamber including adsorbent (the adsorption cell) through a controlling valve C, as schematically shown in Fig. 3. The reference cell with volume V_{ref} is kept at a constant temperature T_{ref}, which is usually as close to room temperature as possible. The value of V_{ref} includes the volume of the tube between the reference cell and valve C. The adsorption cell, where adsorbent might be confined, is kept at the specified equilibrium temperature T_{ad}. The volume of the connecting tube between the adsorption cell and valve C is divided into two parts: one part with volume V_t exposures in room and, therefore, of the same temperature as the reference cell. Another part is buried in an atmosphere of temperature T_{ad} and, hence, its volume is added to the volume of adsorption cell, V_{ad}, which is determined by helium at T_{ref}.

The amount adsorbed can be calculated from the pressure readings before and after opening valve C based on the p–V–T relationship of real gases. A dry and degassed adsorbent sample of known weight was enclosed in the adsorption cell. An amount of gas is let into V_{ref} to maintain a pressure p_1. The moles of the gas confined in V_{ref} are calculated as

$$n_1 = \frac{p_1 V_{ref}}{z_{f1} R T_{ref}} \tag{3}$$

The pressure drops to p_2 after opening valve C. The amount of gas maintained in V_{ref}, V_t, and V_{ad} are respectively

$$n_2 = \frac{p_2 V_{ref}}{z_{f2} R T_{ref}} \tag{4}$$

$$n_3 = \frac{p_2 V_t}{z_{f2} R T_{ref}} \tag{5}$$

$$n_4 = \frac{p_2 V_{ad}}{z_{d2} R T_{ad}} \tag{6}$$

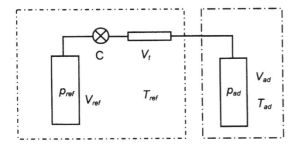

FIG. 3 Schematic structure of a volumetric setup.

The amount adsorbed or the excess adsorption n^σ is then obtained:

$$n^\sigma = n_1 + n_3' + n_4' - n_2 - n_3 - n_4 \tag{7}$$

where n_3' and n_4' are the moles of the gas remaining in V_t and V_{ad}, respectively, before opening valve C. All of the compressibility factor values are calculated by a proper equation of state, which can usually generate appropriate z values for temperatures not close to the critical zone. The main advantages of this volumetric method are simplicity in procedure, commercial availability of highly automated instruments, and the large ranges of pressure and temperature in which this method can be realized. The disadvantage of volumetric measurements is the considerable amount of adsorbent sample needed to overcome adsorption effects on the walls of the vessels. However, this may be a positive aspect if the sample is adequate. A larger amount of sample results in considerable adsorption and usually provides a larger void space in the adsorption cell, rendering the effect of uncertainty in "dead space" to a minimum. In some cases, the accuracy and reproducibility of volumetric measurements are not always satisfying. The precision of pressure readings is a decisive factor. Better than 0.1% is, hopefully, guaranteed for a reliable measurement. Pressure transmitters of precision 0.05% even 0.01% are commercially available presently. The constancy of temperature is of importance also. The amplitude of fluctuation in ± 0.1 K is usually allowable, although ± 0.05 K is also possible today. However, to keep the temperature constant within ± 0.1 K is very difficult, if not impossible, in the near-critical region. A density gradient may exist in the bulk gas phase, even outside the adsorption cell if the surrounding substance has similar thermodynamic properties as the adsorptive. As consequence, it takes a very long period to reach a stable reading. However, even ± 0.5 K of temperature fluctuation in such cases might cause a relative error of about $\pm 2\%$ for a single measurement [25]. Other factors affecting the precision and reliability of adsorption measurements will be discussed later.

2. Gravimetric Method

In this method, the weight change of the adsorbent sample in the gravity field due to adsorption from the gas phase is recorded. Various types of highly sensitive microbalance have been developed for this purpose; today, the instruments equipped with a magnetic suspension are considered favorable [26,27]. A continuous-flow gravimetric technique coupled with wavelet rectification allows for higher precision, especially in the near-critical region, as recently reported [28]. Major advantages of the gravimetric method include sensitivity and accuracy of up to 10^{-7}, the possibility of checking the state of activation of an adsorbent sample, and the approach to equilibrium of an adsorption process—even in gas mixtures. Adsorbent samples of much smaller mass, usually of the order of milligrams, are enough for gravimetric measurement. This is very favorable for studying the adsorption property of new materials that is developing. As is often the case, a dual character is noted—namely, the less the sample, the larger the effect of relevant factors. For example, the preparation of the sample before measurement is especially important when small quantities are employed; the residue moisture or other impurities might cause considerable error in the results of measurement on carbon nanotubes.

Consideration must also be given to buoyancy correction in gravimetric measurement and a counterpart is used for this purpose. The solid sample is placed in a sample holder on one arm of the microbalance while the counterpart is loaded on the other arm. Care must be taken to keep the volume of the sample and the counterpart as close as possible to reduce the buoyancy effect. The system is vacuumed and the balance is zeroed before starting experiments. Buoyancy is measured by introducing helium and pressurizing up to the highest pressure of the experiment. It is assumed that helium does not adsorb and any weight change (ΔW) is due to buoyancy.

Knowing the density of helium (ρ_{He}), one can determine the difference in volume (ΔV) between the sample and the counterpart:

$$\Delta V = \frac{\Delta W}{\rho_{He}(pT)} \tag{8}$$

The measured weight can then be corrected for the buoyancy effect at a specified temperature and pressure:

$$W = W_{exp} - \Delta V \rho_b(p, T) \tag{9}$$

where W_{exp} is the weight reading before correction.

3. Other Measurement Techniques

The maximum pressure to which the gravimetric technique has been applied is about 15 MPa. The volumetric method is very seldom applied above 100 MPa [29]. Bose and co-workers [30] developed a precision dielectric method for the determination of gas–solid adsorption. This method is particularly suitable for adsorption measurements up to 200 MPa; actually, adsorption data up to 650 MPa were reported [31]. The great advantage of the dielectric method is that it is self-sufficient up to the highest pressures and does not depend on the availability of compressibility factor values, as in volumetric or gravimetric measurements [29]. Other new, yet less widespread, adsorption measurement techniques include oscillometry [32,33], calorimetry [34,35], and electromagnetic measurements [36,37].

B. Principal Factors Affecting Adsorption Measurements

To what extent helium is adsorbed has been of major concern in adsorption studies for both volumetric and gravimetric methods. Until recently, the experimental error was often attributed to the finite adsorption of helium at high pressures, and different remedial methods were suggested [38–40]. The effect of helium adsorption on the gravimetric technique is clearly shown in Eq. (8). The volume difference, ΔV, will be overestimated if the adsorption of helium is not negligible. Its effect on the volumetric technique can be explained in terms of Fig. 1. The volume of the "solid phase" of adsorbent, V^s, is experimentally determined by helium. This volume is sometimes called "dead space" or "helium volume" of the adsorption cell, which is, indeed, the volume of adsorbent inaccessible to the helium molecules. However, this value is usually taken for the volume of adsorbent inaccessible to the adsorbate molecules. The difference in molecular dynamic size and shape between helium and adsorbate is logically a source of error. The irregular solid surface and/or the complex structure of micropores inevitably render uncertainty in the determination of V^s. As a consequence of helium adsorption, the dead volume is underestimated.

Nevertheless, the trustworthiness of experimentally measured isotherms should never vacillate for the effect of helium adsorption. It was pointed out that the helium volume can be correctly measured at 400°C [3]. In fact, not only helium but also the strongly adsorbed substances like water cannot be physically adsorbed at such a high temperature. However, the thermal expansion of adsorbent from room temperature to 400°C may result in a larger effect on "dead space" than helium adsorption at room temperature. Initial unreasonable adsorption measurements may not be caused by helium adsorption, but rather by the improper pretreatment of adsorbent sample or adsorptive gases. Monte Carlo simulation [3] also proves that "for the heterogeneous interface, the values of the excess surface adsorption for helium are identical to zero in the limit of the statistical uncertainties." The effect of uncertainty in helium volume on the results of volumetric measurements can be ameliorated if the gas-phase volume is much larger than the helium volume. The volume value used in the adsorption calculation is the volume of the adsorption cell, V_{ad} [refer to Eq. (6)], which is the volume of the void space in the adsorption cell. If the total volume of the cell is V, then $V_{ad} = V - V^s$. The relative error in V_{ad} caused by the uncertainty in V^s is usually less than the error caused by other factors for at least an order of magnitude if V_{ad} is sufficiently large.

Measurement error for both the volumetric and gravimetric methods may also come from the values of the compressibility factor of real gases at high pressures and/or low temperatures, especially in the near-critical region [41]. Any errors in z values affect the calculated value of n^σ directly, as shown by Eqs. (3)–(7) for the volumetric method. This effect functions through ρ_b for the gravimetric method, as shown in Eq. (9). Generally, compressibility factors can be reliably determined from the published p–V–T data [42–44] of adsorptive gases or calculated from a proper equation of state for most (p, T) conditions. However, available original data of the gas state are not abundant and the precision as well as the reliability of them become less dependable as the critical zone is approached, especially within the range of T_c to $T_c + 10\,\mathrm{K}$ [6,41]. The volumetric apparatus used for measuring adsorption can be also used to measure the compressibility factor if reliable z values are not available [25]. It is concluded that reasonably reliable values of the compressibility factor can be yielded by the Lee–Kesler equation [45] for the near-critical region, at least for nitrogen and methane [41]. A slight change in the z value may cause considerable difference in isotherms at conditions $\rho_b/\rho_c > 0.3$ (ρ_c is the critical density), which may be the cause of inconsistent reports for the same adsorption system. Unfortunately, how the values of the compressibility factor were evaluated in adsorption measurement has hardly been mentioned, even in very carefully designed experiments with high-precision instruments.

C. Generating Isotherms by Molecular Simulation of Adsorption

Monte Carlo and molecular dynamic approaches became a useful tool for theoretical calculations aiming at the a priori prediction of adsorption equilibria and diffusivities in small pores of various simple geometries (slits, cylinders, and spherical cavities) due to the dramatic increase in the speed and power of computers in the 1980s and 1990s [3,10,20,21,46–48]. The interactions between adsorbate molecules is represented by the Lenard–Jones potential:

$$\phi_{ff}(r) = 4\varepsilon_{ff}\left[\left(\frac{\sigma_{ff}}{r}\right)^{12} - \left(\frac{\sigma_{ff}}{r}\right)^6\right] \tag{10}$$

where r is the interparticle distance, σ_{ff} is the point at which the potential is zero, and ε_{ff} is the well depth. The adsorbent–adsorbate interactions are represented by the 10-4-3 potential [49] for graphitelike adsorbents:

$$\phi_{sf}(z) = A\left[\frac{2}{5}\left(\frac{\sigma_{sf}}{z}\right)^{10} - \left(\frac{\sigma_{sf}}{z}\right)^4 - \left(\frac{\sigma_{sf}^4}{3\Delta(0.61\Delta + z)^3}\right)\right] \tag{11}$$

where $A = 2\pi\rho_s\varepsilon_{sf}\sigma_{sf}^2\Delta$, Δ is the separation between graphite lattice planes, ρ_s is the solid density, and ε_{sf} and σ_{sf} are the cross-parameters for adsorbent–adsorbate interaction. For a slit pore of given width H, the external potential $\phi_{ext}(Z)$ experienced by a fluid molecule at z is calculated as the superposition of ϕ_{sf} for the two walls:

$$\phi_{ext}(z) = \phi_{sf}(z) + \phi_{sf}(H - z) \tag{12}$$

Molecular parameters used in the above formulation are usually taken from Ref. 49. The grand-canonical Monte Carlo method of Adams [50] is commonly used in simulation.

Such studies have provided many useful insights, for example, into the density profile of the adsorbed phase [3] and into the storage capacity of natural gas in activated carbon [10,46]. The simulation reports of Wang and Johnson [51,52] attracted the attention of many researchers in the field of hydrogen storage by carbon nanotubes/nanofibers. The conclusion of their studies is, indeed, a serious challenge to those purporting nanotubes or nanofibers as promising hydrogen carriers. However, the inadequacy of our understanding of the intermolecular forces and the approximations that have to be made for adsorbent geometry and for the local field around the adsorbate molecules remain a significant problem.

IV.   EXPERIMENTAL ISOTHERMS AND DATA ANALYSIS

A.   Experimental Datasets of the Adsorption Equilibria for the Supercritical Region

High-pressure adsorption, as mentioned previously, is an area of growing research interest. Adsorption measurements have been carried out for different types of adsorbents and adsorbates under different conditions during the past 30 years after Menon [17]. The following works with experimental data in relatively large ranges of conditions are well-known contributions to the study of high-pressure adsorption; following which other researchers advanced new experiments or even new opinions. (I must ask for a pardon from those authors whose work was not, unfortunately, accessible to me and, thus not here mentioned.) Menon [17] measured the adsorption of N_2 and Ar on activated carbon for the range $-76°C$ to $20°C$ and pressures up to 40 MPa; the adsorption of N_2 and CO on alumina for $-76°C$ to $100°C$ and pressures up to 300 MPa. Ozawa and co-workers reported on the adsorption of several gases on molecular sieves: N_2 and Ar on MSC-5A for the range $25–75°C$ and pressures up to 20 MPa, and CH_4 and CO_2 pressures to 10 MPa [53]; adsorptions of N_2, Ar, CH_4, C_2H_6, C_2H_4, CO, CO_2, and N_2O on MS-5A zeolite, N_2, Ar on MS-13X zeolite, most for the range $0–75°C$, some are at $-78°C$ and $-196°C$, the highest pressure is 20 MPa [4]. Findenegg and co-workers measured the adsorption of ethylene on graphitized carbon black (Graphon) for the range $-10°C$ to $50°C$ (T_c included) and pressures up to 12 MPa [6], the adsorption of Ar and CH_4 on Graphon for $-20°C$ to $50°C$ and pressures up to 15 MPa [54], and the adsorption of Kr on Graphon for $253–373$ K [$(1.2–1.8)T_c$] and pressures up to 15 MPa [22]. Bose and co-workers reported the adsorption equilibrium data for the highest pressure range: CH_4 on activated carbon at $0–180$ MPa [29] and that of Ar, Kr, Ne, N_2, and CH_4 on activated carbon at $0–650$ MPa [31]. Agarwal and Schwarz [55] measured adsorption equilibria on activated carbon for both sides of the critical temperature: propane for $T < T_c$, methane and nitrogen for $T > T_c$, and ethane, ethylene, and carbon dioxide for a temperature range including T_c. The highest pressure was 6 MPa.

To observe the isotherms' behavior over a wide temperature range, the author designed a cryostat, whose temperature can be held constant to within ±0.1 to ±0.5 K at any specified value in the range 77 K to room temperature. Details of the cryostat were previously presented [56]. The lower the temperature, the less the temperature constancy could be maintained, especially for the near-critical region, where long-term fluctuation of gas-phase density bears a considerable effect on temperature control. The author and co-workers used a volumetric apparatus coupled with the cryostat to measure the adsorption equilibria of hydrogen [56] and methane [57] on activated carbon AX-21. The temperature range of the hydrogen adsorption was 77–298 K with an increment of 20 K and pressures up to 7 MPa; that of the methane adsorption was 233–333 K with a 20-K increment and pressures up to 10 MPa. The temperature range of the methane isotherms was recently extended down to 158 K [58]. To observe the possible change of

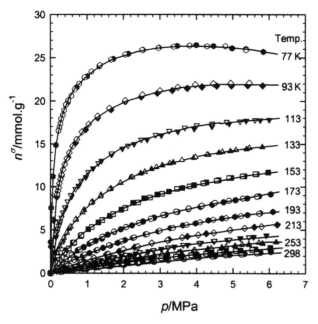

FIG. 4 Adsorption/desorption isotherms of H_2 on activated carbon AX-21. Solid symbols: adsorption; open symbols: desorption. (From Ref. 56.)

adsorption mechanism via isotherms, the adsorption of nitrogen on microporous activated carbon [59] and on a mesoporous silica gel [60] for both the subcritical and the supercritical region was measured. Adsorption isotherms of methane on the silica gel were also measured for 158–298 K and pressures to 10 MPa [41]. These isotherms are presented respectively in Figs. 4–8 and will be referred to in subsequent sections.

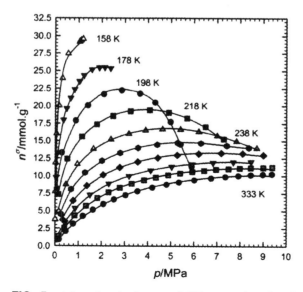

FIG. 5 Adsorption isotherms of CH_4 on activated carbon. Data points: experimental; curves: model predicted. (From Ref. 58.)

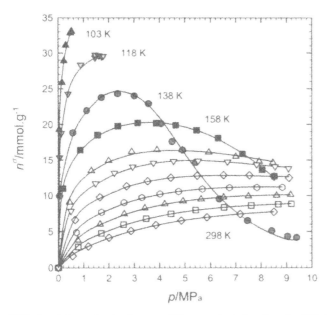

FIG. 6 Adsorption isotherms of N_2 on activated carbon for 103–298 K. Data points: experimental; curves: model predicted. (From Ref. 59.)

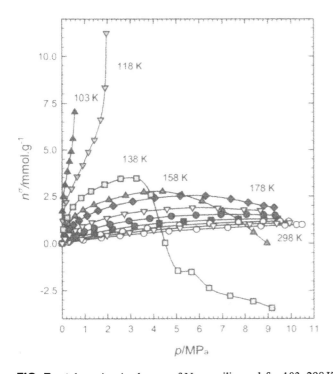

FIG. 7 Adsorption isotherms of N_2 on silica gel for 103–298 K. (From Ref. 41.)

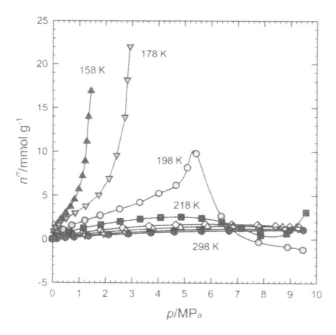

FIG. 8 Adsorption isotherms of CH_4 on silica gel for 158–298 K. (From Ref. 41.)

Measurements of the adsorption isotherms for merely the near-critical region are intimately related to studies on supercritical extraction or adsorbent regeneration. Findenegg [6] reported peculiar isotherm behavior near the critical region measured on graphitized carbon black. The maximum of the isotherms is sharply peaked, and a rapid decrease from the maximum was observed within a narrow temperature range of T_c to $T_c + 10$ K. This phenomenon is sometimes regarded as supercritical depletion and has attracted more research interests both experimentally [15,28,61–67] and by means of the density functional theory or Monte Carlo simulation [68,69]. Precision of compressibility factor values is essential for the near-critical region; some special equations of state were proposed in Ref. 70.

B. Determination of Gas-Phase State

An adsorption isotherm is usually presented as a plot of the excess adsorption n^σ versus gas-phase pressure p, which is directly recorded during measurements and, hence, is the most convenient to apply. At higher pressures, the gas-phase density is a better basis than pressure to use when comparing different adsorption isotherms, because the pressure for almost the same density rapidly increases with increasing temperature, as was shown numerically by Menon [17]. In some literature, instead of pressure, fugacity is used to identify the gas-phase state, but it does not mean that only fugacity is the appropriate variable to represent the gas-phase state for an isotherm. It depends on what is targeted. If the intention is to apply a law that is normally used under ideal-gas condition or to make a comparison with ideal-gas behavior, fugacity, as the "corrected pressure," has to be used. Otherwise, it is completely legitimate to apply pressure when representing the gas-phase state of an isotherm.

Beside density (ρ_b) and fugacity (f), the compressibility factor (z) of the gas phase is also a quantity that must be evaluated in adsorption calculations for both volumetric and gravimetric measurements. These three quantities are related to each other. The compressibility factor is usually evaluated by an equation of state; the density and fugacity is evaluated afterward. However, the gas-phase density can be directly determined if the p–V–T data are available; either z or f can then be evaluated. This is especially the case for near-critical temperatures, when the equation of state becomes less dependable.

1. Evaluation of Compressibility Factor from Published or Measured p–V–T Data

For most experimental conditions, the p–V–T data of adsorptive gases are available [42,44]. Updated p–V–T data are published in *Journal of Physical and Chemical Reference Data* from time to time. However, all of the published data are also calculated by an equation of state and arranged in tables. To obtain the data for a specified experimental condition, one has to establish a functional relationship between molar volume V_m and T for a specified p first, and between V_m and p for a specified T afterward. Although the readers are warned that any interpolation is unreliable for the near-critical region [44], an interpolated entry yielded by a locally smooth curve does make sense. For every molar volume that is calculated from the above-established correlation under the experimental condition, there is a definite value for the compressibility factor z as calculated by

$$z = \frac{pV_m}{cRT} \tag{13}$$

where c is a unit conversion factor.

In situ measurement of the compressibility factor is useful especially in the near-critical region. It is emphasized that, even in the IUPAC *Tables of the Fluid State*, the listed data are not as precise as expected near the critical point for some gases. The in situ measurement of the compressibility factor may serve as vacant runs of adsorption for a volumetric setup. No adsorbent was loaded in the adsorption cell, and z_{d2} in Eq. (6) is unknown for the vacant runs. Because no adsorption is assumed, the moles of gas confined in the space of V_{ad} is calculated as

$$n_4 = n_1 + n_3' + n_4' - n_2 - n_3 \tag{14}$$

Because T_{ref} is far from the critical temperature, both z_{f2} and z_{f1} can be reliably calculated by a proper state equation. The compressibility factor at p_2 and T_{ad} was thus calculated as

$$z_{d2} = \frac{p_2 V_{ad}}{n_4 R T_{ad}} \tag{15}$$

Such in situ measurements were carried out for nitrogen and methane for the near-critical regions [25], which were proven helpful for a better choice of reliable values of the compressibility factor.

2. Evaluation of the Compressibility Factor by an Equation of State

Most commonly applied equations of state [e.g., the third virial equation, the Soave–Redlich–Kwong (SRK) equation, the Bennedict–Webb–Rubin (BWR) equation, and the Lee–Kesler equation (LK)] can generate very reliable z values for temperatures 10 more degrees away from the critical point [41,71]. However, one has to select the appropriate equation of state for the specified gas [45]. The calculation procedure of the SRK and BWR equation for hydrogen is illustrated in Ref. 71, whereas that of the LK equation is illustrated in Ref. 45. After comparing z values of different sources, including IUPAC *Chemical Data* and those measured in situ, it was concluded that the LK equation generates the most reasonable values of the compressibility factor z for the near-critical region, at least for nitrogen and methane. According to the Lee–Kesler model, the compressibility factor, z, is calculated as

$$z = z^{(0)} + \frac{\omega}{\omega^{(r)}} (z^{(r)} - z^{(0)}) \tag{16}$$

where the superscript (0) denotes "simple fluid" $(\omega = 0)$, (r) denotes the reference fluid (octane), and ω and $\omega^{(r)}$ are the eccentric factors of the real fluid of interest and the reference fluid, respectively ($\omega^{(r)}$ was taken as 0.3978). $z^{(0)}$ and $z^{(r)}$ are calculated by the revised BWR equation expressed in the reduced form

$$\frac{p_r V_r}{T_r} = 1 + \frac{B}{V_r} + \frac{C}{V_r^2} + \frac{D}{V_r^5} + \frac{c_4}{T_r^3 V_r^2} \left(\beta + \frac{\gamma}{V_r^2} \right) \exp\left(-\frac{\gamma}{V_r^2} \right) \tag{17}$$

where

$$B = b_1 - \frac{b_2}{T_r} - \frac{b_3}{T_r^2} - \frac{b_4}{T_r^3} \tag{18}$$

$$C = c_1 - \frac{c_2}{T_r} + \frac{c_3}{T_r^3} \tag{19}$$

$$D = d_1 + \frac{d_2}{T_r} \tag{20}$$

$$V_r = \frac{p_c V}{R T_c} \tag{21}$$

Parameter values for the simple and the reference fluids (octane) are listed in Table 1.

There is only one variable in Eq. (17), V_r. A convergence strategy (e.g., the secant method), can be used to search for the solution of V_r^0 for simple fluid first, and of V_r^r for the reference fluid afterward. The values of $z^{(0)}$ and $z^{(r)}$ can then be determined:

$$z^{(0)} = \frac{p_r V_r^0}{T_r} \tag{22}$$

$$z^{(r)} = \frac{p_r V_r^{(r)}}{T_r} \tag{23}$$

TABLE 1   Parameter Values of the L-K Equation

Parameter	Simple fluid ($\omega = 0$)	Reference fluid (octane)
b_1	0.1181193	0.2026579
b_2	0.265728	0.331511
b_3	0.154790	0.027655
b_4	0.030323	0.203488
c_1	0.0236744	0.0313385
c_2	0.0186984	0.0503618
c_3	0.0	0.016901
c_4	0.042724	0.041577
$d_1 \times 10^4$	0.155488	0.48736
$d_2 \times 10^4$	0.623689	0.0740336
β	0.65392	1.226
γ	0.060167	0.03754

Source: Ref. 45.

The effect of the precision of z values on the determination of the amount adsorbed is significant because an adsorption isotherm is measured by an accumulative manner in the course of pressure increments. In other words, we can only measure the increment of the amount adsorbed for a given increment of pressure. The equilibrium amount adsorbed n^σ at a pressure p is the sum of all the adsorption increments up to pressure p. Therefore, the effect of the precision of z values is not local, but accumulative, for an isotherm. In addition, a stable state is very difficult to recognize in the vicinity of the critical point: Some slow periodic variation of pressure readings adds more uncertainty to the measurement. This might be the cause of the different results reported in the literature.

3. Determination of the Fugacity Coefficient

Fugacity of real gases should replace pressure in the application of laws or theoretical formulations that are defined for the ideal gas. For example, fugacity has to be used in evaluating adsorption potential at high pressures and/or low temperatures. The fugacity of a pure gas is a physical quantity that complies with the following relation:

$$\mu(g, T, p) = \mu^0(g, T) + RT \ln\left(\frac{f}{p^0}\right) \tag{24}$$

where g denotes gas, $\mu^0(g, T)$ is the standard chemical potential at temperature T and the reference pressure p^0, which is taken for 10^5 Pa, and f is the fugacity. However, the fugacity coefficient, defined as $\phi \equiv f/p$ is independent of the reference pressure and is more convenient to use. It was derived from the definition of fugacity that

$$\ln \phi = \frac{1}{RT} \int_0^p \left(V_m - \frac{RT}{p}\right) dp \tag{25}$$

because

$$z = \frac{pV_m}{RT}$$

$$\ln \phi = \int_0^p \left(\frac{pV_m}{pRT} - \frac{1}{p}\right) dp = \int_0^p \frac{(z-1)}{p} dp \tag{26}$$

Because the compressibility factor can be correlated with pressure for a constant temperature as

$$z = 1 + \sum_{i=1} c_i p^i \tag{27}$$

we can obtain, upon substitution of Eq. (27) into Eq. (26), the result of the integration as a polynomial function of pressure:

$$\ln \phi = \sum_{i=1} \frac{c_i p^i}{i} \tag{28}$$

Equation (28) shows that so long as the compressibility factor can be evaluated, so can the fugacity coefficient. The p–f transformation can be easily completed by

$$f = \phi p \tag{29}$$

Now, we can have the adsorption isotherms presented with respect to either pressure, fugacity, or the gas-phase density.

C. Determination of Henry Constants

Excess adsorption is usually proportional to pressure for the low-pressure region. The proportionality coefficient is called the Henry constant. It is a fundamental piece of data for an adsorption system, from which the limiting heat of adsorption can be determined. The Henry constant can also serve as a scaling factor to transform isotherms for different temperatures into a single generalized one. Furthermore, any consistent thermodynamic interpretation of high-pressure adsorption data must rely on the initial values of adsorption (i.e., on the Henry region [1]). However, the experimental excess adsorption data for this region usually bear larger uncertainty because the adsorption instruments commercially available presently do not guarantee precise measurement at low pressures. An extrapolation method was proposed to determine Henry constants, which is based on all data measured, rendering larger reliability of the Henry constant values. The adsorption isotherm can be expressed in virial form [72]:

$$\frac{Kp}{n^s} = \exp\left(2A_1 n^s + \frac{3}{2}A_2(n^s)^2 + \cdots\right) \tag{30}$$

where $A_1, A_2, \ldots$ are the virial coefficients. Equation (30) comes from an assumption that the adsorbate state in the adsorbed phase obeys a general equation of state of the virial form. Because the state of the adsorbed phase is determined by total molecules in the phase, absolute adsorption, n^s, appears in Eq. (30), according to which a plot of $\ln(p/n^s)$ versus n^s should be linear as n^s approaches zero. What is experimentally measured is not n^s, but the excess adsorption n^σ; however, $n^\sigma \to n^s$ as n^s approaches zero. Therefore, the plot of $\ln(p/n^\sigma)$ versus n^σ should also be linear as n^σ approaches zero. It follows for the linear region that

$$\ln\left(\frac{p}{n^\sigma}\right) = 2A_1 n^\sigma - \ln K \tag{31}$$

When n^σ approaches zero, we have

$$n^\sigma = Kp \tag{32}$$

Therefore, K is the Henry constant. If the intercept of the plot is ξ, then

$$K = \exp(-\xi) \tag{33}$$

As an example, the plot of $\ln(p/n^\sigma)$ versus n^σ is shown in Fig. 9 for the adsorption isotherms of hydrogen on activated carbon AX-21 (Fig. 4). The intercepts of the plot and the Henry constants calculated are listed in Table 2.

D. Determination of Isosteric Heat of Adsorption

The enthalpy change on adsorption is commonly referred to as the isosteric heat of adsorption. It was derived from the condition of phase equilibrium (i.e., the equality of chemical potentials between the adsorbed phase and the ambient gas phase):

$$\mu_a = \mu_g \tag{34}$$

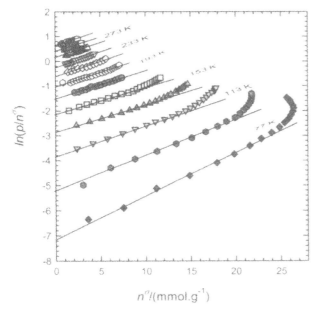

FIG. 9 The virial plot of H_2 adsorption data on activated carbon AX-21.

The chemical potential of the ambient gas phase was defined previously by Eq. (24). Differentiating at a constant adsorbed phase concentration and applying the Gibbs–Helmholtz relation $\partial(\mu/T)/\partial T = -\bar{H}/T^2$, we obtain

$$-\frac{\bar{H}_a}{T^2} = -\frac{\bar{H}_g^0}{T^2} + R\left(\frac{\partial \ln f}{\partial T}\right)_{n^s} \tag{35}$$

$$\left(\frac{\partial \ln f}{\partial T}\right)_{n^s} = \frac{\bar{H}_g^0 - \bar{H}_a}{RT^2} \approx \frac{H_g - \bar{H}_a}{RT^2} = -\frac{\Delta H}{RT^2} \tag{36}$$

TABLE 2 Evaluated Henry Constants for the H_2 Adsorption on Activated Carbon AX-21

T (K)	$1/T$	Intercepts	K	$\ln K$
298.15	3.354×10^{-3}	0.6050	0.5461	-0.6050
273.15	3.661×10^{-3}	0.3402	0.7116	-0.3402
253.15	3.950×10^{-3}	0.1057	0.8997	-0.1057
233.15	4.289×10^{-3}	-0.2481	1.282	0.2481
213.15	4.692×10^{-3}	-0.6148	1.849	0.6148
193.15	5.177×10^{-3}	-1.042	2.835	1.042
173.15	5.775×10^{-3}	-1.533	4.632	1.533
153.15	6.530×10^{-3}	-2.122	8.348	2.122
133.15	7.510×10^{-3}	-2.859	17.44	2.859
113.15	8.838×10^{-3}	-3.824	45.79	3.824
93.15	1.074×10^{-2}	-5.166	175.2	5.166
77.15	1.296×10^{-2}	-7.026	1126	7.026

The molar enthalpy of the ambient gas phase H_g is supposed to equal the partial molar enthalpy of gas phase at the reference state. If the difference in the heat capacity of the sorbate in two phases can be neglected, ΔH is independent of temperature and Eq. (36) may be integrated directly to yield

$$\ln f = \text{const.} - \frac{\Delta H}{RT} \tag{37}$$

The plot of $\ln f$ versus $1/T$ should be linear according to Eq. (37). Under the experimental condition of Ref. 56, fugacity, f, was simply replaced by pressure, p. The plot of $\ln p$ versus $1/T$ is linear, as shown in Fig. 10, for some constant amount adsorbed. The isosteric heats of adsorption for different amounts adsorbed can be evaluated from the slope of these linear plots and are presented in Fig. 11. The isosteric heat of adsorption will remain constant during the process of surface covering if the surface is energetically uniform; however, it will vary (usually decrease) with the increasing amount adsorbed for an energetically heterogeneous surface. This is the situation observed at the activated carbon AX-21. It drops from about 6.6 kJ/mol at $n^\sigma = 0.5$ mmol/g to about 6.0 kJ/mol at $n^\sigma = 8.0$ mmol/g. If a pertinent model of isotherms is available for the range of experimental condition, the isosteric heat of adsorption can be analytically obtained according to Eq. (36) [56,72].

The isosteric heat of adsorption is called the limiting heat of adsorption when the amount adsorbed approaches zero. It is usually taken as the representative thermal effect of the adsorption under studies. The limiting heat of adsorption can be evaluated from Henry constants if they are available for several temperatures, because the Henry constant varies with temperature following the van't Hoff equation [72]:

$$\frac{d \ln K}{dT} = \frac{\Delta H_0}{RT^2} \tag{38}$$

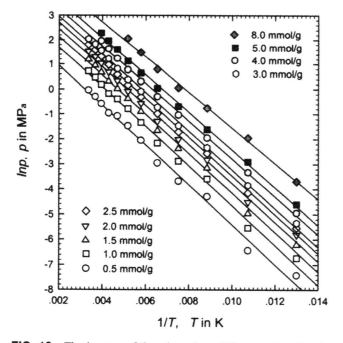

FIG. 10   The isosters of the adsorption of H_2 on activated carbon AX-21. (From Ref. 56.)

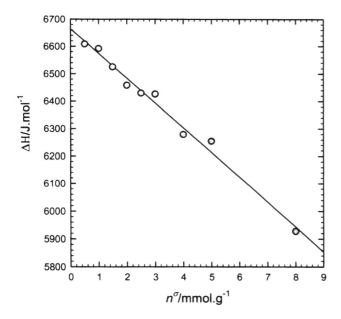

FIG. 11 Variation of the isosteric heat of adsorption with the amount adsorbed. (From Ref. 56.)

Therefore, the plot of $\ln K$ versus $1/T$ should be linear, and the limiting heat of adsorption can be evaluated from the slope of the plot. The van't Hoff plot of the hydrogen adsorption data is shown in Fig. 12. A perfect linear plot with slope of 783.7 J/mol is obtained, which yields the limiting heat of adsorption:

$$-\Delta H_0 = \frac{(783.7)(8.314)}{1000} = 6.5 \text{ kJ/mol}$$

This value is quite close to the upper value of $-\Delta H$ at $n^\sigma = 0.5 \text{ mmol/g}$ shown in Fig. 11.

E. Construction of the Characteristic or Generalized Isotherm

Correlation of the experimental isotherms by a generalized one is encouraged by the possibility that if the generalized isotherm for a given adsorbate is known, the adsorption values at any temperature and pressure can be predicted. Such a representative generalized isotherm is the Dubinin–Polanyi "characteristic curve." According to the potential theory, the plot of adsorbed volume versus adsorption potential is temperature independent and, hence, characteristic. The adsorbed volume is determined by [55]

$$V_a = n^s V_m \tag{39}$$

where V_a is the adsorbed phase volume (cm^3/g of adsorbent), n^s is the absolute amount adsorbed (mmol/g of adsorbent), and V_m is the adsorbed-phase molar volume (cm^3/mmol).

The molar work done for moving the sorbate molecules from the ambient gas phase to the standard state of the adsorbed phase is defined as adsorption potential, A, by the Polanyi theory. The standard state of adsorbate in the adsorbed phase is a saturated liquid for the adsorption at

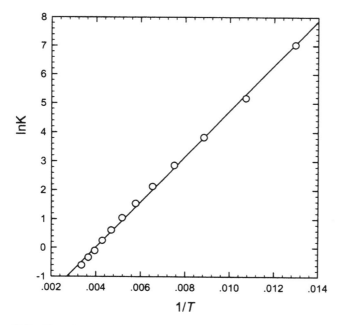

FIG. 12   The van't Hoff plot of H_2 adsorption data on activated carbon AX-21.

subcritical temperatures. The adsorption potential may be calculated directly from the ratio of the equilibrium pressure/fugacity and the saturated vapor pressure/fugacity:

$$A = -RT \ln\left(\frac{f}{f_s}\right) \quad \text{or} \quad A = -RT \ln\left(\frac{p}{p_s}\right) \tag{40}$$

Successful examples of the characteristic plot were presented by Dubinin and reproduced in Ruthven's book [73]; however, problems arise when applied to supercritical adsorption. Firstly, evaluation of adsorption potential is questionable because a saturated liquid cannot stand for the standard state of adsorbate at above-critical temperatures, and the adsorption potential cannot be properly calculated. Second, Eq. (39) cannot be evaluated because the absolute amount adsorbed, n^s, and the adsorbed phase molar volume, V_m, are also questionable for supercritical adsorption. Although empirical assumptions, as summarized by Agarwal and Schwarz [55], have been made to tackle the problem, these assumptions lack support either experimentally or theoretically.

The above-mentioned problems are not encountered in constructing the so-called generalized equilibrium isotherm [73] even for the supercritical region. Taking the Henry constant as a scaling factor, one can present the equilibrium data as plots of $\ln(n^\sigma)$ versus $\ln(Kp)$. As a consequence, isotherms for different temperatures reduce to a single generalized one. It is proven that such a generalized isotherm can be constructed for the adsorption on activated carbon, silica gel, and zeolite [57,74–76] so long as the amount adsorbed has not become very high. This means less sorbate–sorbate interaction, which corresponds to a low surface concentration of adsorbate, is decisive for constructing a successful generalized isotherm. If, however, the sorbate–sorbate interaction becomes considerable compared to the sorbate–sorbent interaction, the points representing such states will stray from the single smooth curve, as shown in Fig. 13 for the adsorption isotherms of methane on activated carbon. The smooth curve represents the condition of low surface concentration, which will be used for the formulation of *absolute*

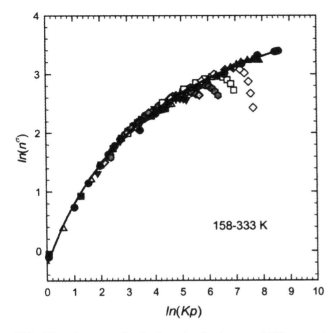

FIG. 13 The generalized adsorption isotherms of CH_4 on activated carbon.

adsorption [75] and may provide a database for thermodynamic analysis of supercritical adsorption [1]. Therefore, a model of the smooth generalized isotherm is presented:

$$\ln n^\sigma = \lambda \frac{b[\ln(Kp) - c]}{1 + [\ln(Kp) - c]} \tag{41}$$

where λ, b, and c are parameters of the model, the values of which are evaluated by nonlinear regression: $\lambda = 5.2755$, $b = 0.2149$, $c = 0.1928$, and K is the Henry constant, which is a function of temperature:

$$K = \exp\left(-4.1879 + \frac{1994.4}{T}\right) \tag{42}$$

(correlation coefficient $r^2 = 0.998$).

The model-predicted adsorption is represented by the smooth curve in Fig. 13. Based on Eqs. (41)–(42), one can predict the excess adsorption in the region tested, but only for the conditions that correspond to low surface concentration. Such conditions extend from subcritical (158 K) to supercritical temperatures (up to 333 K) and pressures up to 10 MPa.

F. Determination of the Absolute Adsorption

The significance of the determination of absolute adsorption was pointed out initially by Menon:

> The absolute adsorption is independent of the value assumed for the density of the adsorbed phase. Hence absolute adsorption is not a mere hypothetical concept; it seems to be a fair representation of *the actual situation* in the surface layer of the adsorbent. [17]

Salem and co-workers [1] emphasized the necessity that a consistent thermodynamic interpretation of high-pressure excess adsorption data relies on the Henry's region. In fact, it must rely on the quantity of absolute adsorption that contains all adsorbate molecules of the adsorbed phase, which determine the thermodynamic behavior of the phase. However, how can the absolute amount in the adsorbed phase be determined is *an essential problem* for high-pressure adsorption of supercritical gases [77]. Although we can experimentally measure excess adsorption, we have no way of measuring absolute adsorption. To estimate the total (i.e., the absolute) adsorption from the excess adsorption data is *a major challenge* in applying most modeling procedures and further thermodynamic analysis in high-pressure adsorption [28]. According to Eq. (2), either the density or volume of the adsorbed phase has to be known for the determination of n^s. Typical assumptions in the efforts of evaluating absolute adsorption include the following:

1. Assuming the density of the adsorbed phase is equal to the density of the saturated liquid [39]
2. Regarding the adsorbed phase as closely packed molecules at a specified diameter, usually corresponding to the van der Waals volume [17,78,79]
3. Evaluating the adsorbed phase density based on the linear section of the isotherm n^σ versus ρ_b plot [17]
4. Assuming the volume of the adsorbed phase is equal to that of pores [1,40]

Recently, an iteration procedure based on the Langmuir equation was proposed to evaluate the quantity of absolute adsorption [77]. The author proposed a method to establish a model for the absolute adsorption based on the experimental excess adsorption data [75,80,81]. It is seen from Eq. (2) that $n^s = n^\sigma$ if either $\rho_b \ll \rho_a$, as is the case of vapor adsorption, or V_a is not much different from zero, as is the case of supercritical adsorption under conditions of low pressures and/or high temperatures. Therefore, we can use the experimental data in the initial part of the isotherms to formulate a model for the absolute adsorption. However, we would not utilize merely a few points, instead, we would use as many data points as possible to ensure that the model established is reliable, as is the case of evaluating the Henry constants. A strategy called IST (Isotherm Space Transformation) [75,81] is thus proposed. It consists of two steps: First, the transformation of isotherms for different temperatures in the $n^\sigma - p$ space into a single generalized isotherm in the $\ln(n^\sigma) - \ln(Kp)$ space. For example, multiple isotherms of methane shown in Fig. 5 have been transformed to a single isotherm shown in Fig. 13. The smooth curve in the figure representing the adsorption data under conditions of low surface concentration is to undergo the second transformation. Equation (41), as the analytical form of the smooth curve, is expanded into the Taylor series to reach a linear representation of the adsorption data, which must be in a region with an even smaller surface concentration compared to the generalized isotherm. Therefore, the resulting linear isotherms serve to formulate the model of absolute adsorption. Taking logarithms on both sides of Eq. (41) yields

$$\ln(\ln n^\sigma) = \ln \lambda + \ln\left(\frac{b[\ln(Kp) - c]}{1 + b[\ln(Kp) - c]}\right) \tag{43}$$

By setting $y \equiv \ln(\ln n^\sigma)$, $x \equiv 1/b[\ln(Kp) - c]$, $\eta = \ln \lambda$, Eq. (43) is transformed to

$$y = \eta - \ln(1 + x) \tag{44}$$

which can be expanded to Taylor series and becomes linear for $|x| < 1$, overlooking the terms of higher order:

$$y = \eta - x \tag{45}$$

where $c = \ln \phi$, and then $x = 1/b \ln(Kp/\phi)$. Both parameters ϕ and b are to be determined and K is a constant for a given temperature. The independent variable in the new space is, thus, set to be $1/\ln p$. To avoid evaluating logarithms of a negative number, smaller units are preferred for pressure, p, and adsorption, n^σ. Pressure is better expressed in kilopascals, and adsorption, n^σ (in mmol/g), values are usually enlarged by 10 or 100 times. As expected from the above deduction, the isotherms in the $\ln(\ln n^\sigma) - 1/\ln p$ space will be linear *for the region of dilute surface concentration*, as proven by the adsorption data of hydrogen [56] and methane [57] as well as nitrogen [59]. The yielded linear isotherms of the second step transformation for the methane adsorption data are shown in Fig. 14. As argued earlier, $n^s \cong n^\sigma$ is valid for the region of dilute surface concentration; therefore, the linear isotherms must represent absolute adsorption, which can be modeled by

$$\ln \ln(\delta n^s) = \alpha + \frac{\beta}{\ln p} \tag{46}$$

where δ is the times of enlargement. The explicit expression for the absolute adsorption is thus obtained:

$$n^s = \frac{1}{\delta} \exp\left[\exp\left(\alpha + \frac{\beta}{\ln p}\right)\right] \tag{46.1}$$

Parameters α and β have been found to be functions of temperature. For example, for Fig. 14,

$$\alpha = 2.1980 - 5.4403 \times 10^{-3} T + 2.0891 \times 10^{-5} T^2 \tag{46.2}$$

$$\beta = -2.2020 + 0.036343 T - 1.7803 \times 10^{-4} T^2 \tag{46.3}$$

and $\delta = 10$ for this set of isotherms.

The isotherms of absolute adsorption, generated by Eqs. (46.1)–(46.3), as shown in Fig. 15, are always a monotonously increasing function of pressure. The excess adsorption isotherms are shown by dotted lines for comparison. It was observed that the difference between the two sets of isotherms becomes considerable only after about 2 MPa, and the difference increases with

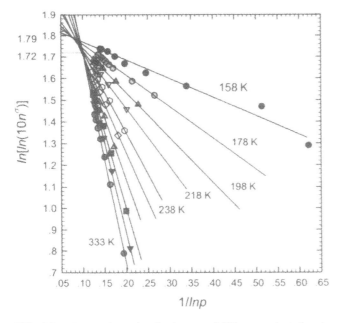

FIG. 14 Linear adsorption isotherms of CH_4 on activated carbon.

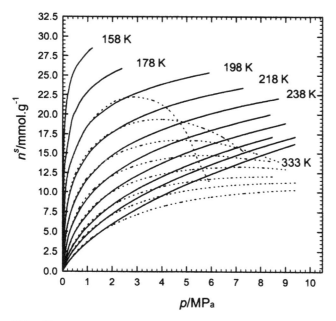

FIG. 15 Absolute adsorption isotherms of CH_4 on activated carbon.

increasing pressure and decreasing temperature. It is noted that there is almost no difference between them at 178 K and 158 K.

V. ISOTHERM MODELING

A. Introduction

The need for a mathematical model of adsorption isotherms arises primarily from understanding the mechanism of the phenomena. Starting from the consideration of the dynamic behavior of adsorptive molecules in the potential field above the adsorbent surface, Polanyi [82] advanced the adsorption potential theory, which was further developed by many succeeding scientists, especially by Dubinin and Astakhov [83]. Based on the dynamic equilibrium between adsorption and desorption, Langmuir [84] and Brunauer–Emmett–Teller [85] isotherm equations were derived. Following the ideas advanced, several modifications were made, and more isotherm equations were proposed [86,87]. The advantage of "pure" theoretical isotherm equations lies in their capability of clearly drawing a picture of what is happening in between the solid surface and the gas phase. Some properties of the adsorption system can be calculated and regarded as "reliable" based on such isotherm equations. However, the microscopic world cannot, until today, be precisely described completely by theoretical laws or parameters; therefore, empirical assumptions or parameters were introduced into the theoretical isotherms, which improved the model pertinence to the experimental isotherms. An isotherm equation with empirical assumptions, however, cannot clearly describe the microscopic process of adsorption, at least for some conditions.

An accurate description of adsorption at high pressures is of particular relevance to thermodynamic theories of multicomponent adsorption, as well as to the engineering of an adsorption process. Special consideration is needed for modeling isotherms for the supercritical adsorption because the isotherm behavior is fundamentally different from the standard (IUPAC) classification schemes [88]. All types of isotherm included in the IUPAC classification are increasing functions of pressure, but the supercritical isotherm might have a maximum, after which the isotherm becomes a decreasing function of pressure. The reason for this radical difference lies in the fact that supercritical adsorption is dominated up to rather high gas densities by the strength of the gas–solid interaction, whereas the subcritical adsorption depends mainly on the strength of the gas–gas interaction. The difference between the supercritical and the subcritical adsorption can formally be explained by the Gibbs definition of adsorption, as shown in Eq. (2). For the subcritical adsorption, there is an upper limit of pressure, the saturated pressure p_s. The density of the adsorbed phase is almost the same as that of a liquid; therefore, the density of the gas phase must be much smaller than that of the adsorbed phase for almost the whole range of pressure. Consequently, $n^\sigma = n^s$ and is always an increasing function of pressure. However, the gas-phase density always increases with pressure, but the adsorbed-phase density cannot be as large as the liquid for the supercritical region; therefore, a maximum must appear on the isotherm if the pressure reaches a high enough value. The isotherm will touch the abscissa when $\rho_b = \rho_a$, and it enters the negative area when $\rho_b > \rho_a$. The effect of the adsorbed phase volume is also important. The difference between n^σ and n^s is not important if the volume V_a has not become considerable. It is apparent now that to establish an appropriate model for supercritical adsorption isotherms, the effect of the adsorbed phase, reflected through its density or volume, must be accounted for properly [59]. The Gibbs definition formula of adsorption should thus be the framework used in modeling. There are two ways to define the excess adsorption, as presented in Eqs. (1) and (2); there are two ways to model the supercritical isotherms as well. One way is based on Eq. (1), and a density profile function has to be assumed. Another way is based on Eq. (2), and either density or volume of the adsorbed phase has to be taken into account. Peculiar isotherm behavior is observed on some adsorbents for the near-critical region (i.e., T_c to $T_c + 10\,\mathrm{K}$), for which a special modeling strategy might be needed.

B. State of the Art of Isotherm Modeling

Owing to the essential difference in isotherms as well as in the nature of adsorption between supercritical and subcritical regions, modeling supercritical isotherms has been an art for the past decades. One can find different strategies in literature, each claiming success, at least for the case presented. The initial effort in modeling the supercritical isotherms was to extend the Polanyi–Dubinin potential theory to the supercritical region [83], and empirical expressions for the saturated pressure, p_s, were recommended since then [53,89]. Typical argument for the justification of the strategy is the so-called quasiliquid [90] or overheated liquid [53], into which the supercritical gas adsorbed in micropores having a strong molecular field would transform. This method was used to interpret supercritical isotherms if maxima were not shown [91]. Although the parameters in the Dubinin–Astakhov equation have a definite physical meaning [92], problems arise in using the equation with the empirical correlation of p_s. Admitting the difference in adsorption mechanisms for both sides of the critical temperature, it was suggested to apply the volume-filling model (the D-A equation) and monolayer model (the Langmuir equation) separately for the subcritical and the supercritical isotherms [93]. However, this strategy was also questionable, especially for those isotherms with a maximum. The Dubinin–Astakhov equation was modified as a summation of a geometric series, which is capable of representing isotherms with maxima [53]. Partial success in modeling isotherms with maxima was similarly achieved by virial [73,94,95] or modified virial isotherm equations [31]. Although the virial-type equations are convenient for the evaluation of isosteric heats at zero coverage, they do not characterize adsorbents in terms of structural or energetic parameters.

Another class of supercritical isotherm models is obtained through evaluating the density profile in the adsorbed phase by grand-canonical Monte Carlo simulation [3,40,69] or by nonlocal [21] or simplified local density functional theory [61]. Such models can describe the general outline of the experimental isotherms for a range of conditions, but they can hardly fit them for all cases. Limitations in the knowledge of interactions at the molecular or atomic level and the assumptions made in modeling render this kind of model only a partial success. Similarly, a lattice model was proposed to represent the adsorption system and Ono–Kondo equations were used to relate the density in each layer; success was also achieved for describing isotherms with a maximum [2]. However, the physical meanings of the parameters in the model may not always be reasonable [96]. In fact, the Ono–Kondo equation assumes a density profile or multilayer molecules in the adsorbed phase, to which the supercritical adsorbed phase does not conform because of the nature of the monomolecular layer of adsorbate. Due to the complexity of the interactions between gaseous molecules and the solid surface, continued study for pertinent models are still actively pursued [97,98]. Presented herewith is the modeling strategy that relies on the Gibbs definition and it takes the effect of the adsorbed phase into account.

C. A New Isotherm Equation Suitable for Supercritical Adsorption

All supercritical isotherms reported until now share common features. This is basically because the dominated factor for the adsorption has shifted from a gas–gas interaction to a gas–solid interaction on crossing over the critical zone [22]. The qualitative variation of the excess surface adsorption of simple gases with pressure is rather general and is, to large extent, independent of the substrate. Therefore, it is possible to have an isotherm equation appropriate for the supercritical isotherms.

Physical adsorption of gases onto porous adsorbents at elevated pressures is of practical interest for both storage and separation/purification of the light alkanes and other gases. Porous solids with rather complex surface properties have been the major adsorbents of supercritical adsorption. Pores of different sizes will exert different interaction energies on adsorbate molecules from the point of view of Polanyi's adsorption potential; therefore, the distribution of the energetic heterogeneity of surface is largely determined by the pore-size distribution (PSD) of adsorbent. However, the two distributions are not totally the same. For example, a symmetric PSD yielded a nonsymmetric distribution of the adsorption energy [99], which is usually defined by the negative value of the potential minimum of interaction between the adsorbed molecules and the solid. On the other hand, the adsorption energy distribution is related to the experimentally measured adsorption isotherm. It follows then that a pertinent isotherm equation should be able to be derived from the adsorption energy distribution, which, in turn, is controlled by pore-size distribution. The author proposed a new isotherm equation recently based on such considerations [100].

Porous solids such as activated carbon typically have irregular pores, making it practically impossible to account for all the details of the pore structure. The half-width, r, of a pore is usually used to identify the pore size in literature. It is known from the Lenard–Jones potential function that the process of surface coverage must begin with the smallest pores accessible to the adsorbate molecules and end up with the largest ones (i.e., the flat surface). Consequently, there should be a critical pore size, r_c, for a given equilibrium condition (T, p) under which all pores with $r \leq r_c$ have been covered, but those with $r > r_c$ are still vacant. Therefore, the distribution of adsorption energy, as a function of pore size r, determines the fractional surface coverage under a given equilibrium condition. The distribution of adsorption energy with respect to r could not be symmetric because of the compounded effects of surface topology with the adsorption potentials, as shown in Fig. 16. The differential distribution function is more likely to

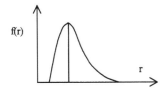

FIG. 16 Distribution of adsorption energy relative to pore size.

be steeper on the side of smaller pore size and skewed on the other side to larger pores after the peak. Such distribution is well described by the Weibull function [101]. This function, simple in its mathematical expression, has found many practical applications:

$$f(r) = \frac{q}{\alpha}(r - r_a)^{q-1} \exp\left(-\frac{(r - r_a)^q}{\alpha}\right) \quad (r \geq r_a)$$

$$f(r) = 0 \quad (r < r_a) \tag{47}$$

where r_a is the radius of an adsorbate molecule. As mentioned earlier, there is a definite critical pore size, r_c, for a given condition (T, p) of equilibrium. The fractional surface coverage under the condition could be calculated simply by

$$\theta = \frac{\int_{r_a}^{r_c} f(r)\, dr}{\int_{r_a}^{\infty} f(r)\, dr} = \int_{r_a}^{r_c} f(r)\, dr \tag{48}$$

Substituting Eq. (47) into Eq. (48) yields

$$\theta = -\int_{r_a}^{r_c} d\exp\left(-\frac{(r - r_a)^q}{\alpha}\right) = 1 - \exp\left(-\frac{(r_c - r_a)^q}{\alpha}\right) \tag{49}$$

If the temperature is constant, then r_c is determined only by pressure for a given adsorbent–adsorbate system. A simple assumption is $r_c - r_a \propto p$; Eq. (49) is thus transformed to

$$\theta = 1 - \exp(-bp^q) \tag{50}$$

where b is a parameter relating to adsorption energy change and q relates to the topology of the adsorbate in the adsorbed phase, the value of which decreases from unity for a uniform surface, down to zero for the most heterogeneous surface. If the amount adsorbed at θ is n, then n becomes n_0, the saturated capacity, at $\theta = 1$. Therefore,

$$n = n_0[1 - \exp(-bp^q)] \tag{51}$$

Equation (51) is physically reasonable because $n = n_0$ at high pressures and

$$n \approx n_0[1 - (1 - bp^q)] = K''p^q \tag{52}$$

when p approaches zero. It does not reduce to Henry's law but seems more reasonable because $n \propto p$ is valid for homogeneous surfaces, however, $n \propto p^q$ $(0 < q < 1)$ is rational for heterogeneous surfaces. Equation (52) appears to be the same as the Freundlich equation [87], but only for the region of low pressures. Equation (51) looks similar to the Jovanovich equation [102], but it is not exactly the same. As summarized by Misra [103], the Jovanovich equation, the Langmuir equation, and Henry's law can be generated from the same differential equation:

$$\frac{d\theta}{dp} = c(1 - \theta)^k \tag{53}$$

where c and k are constants and $k \geq 1$ to satisfy all the requirements of physical compatibility. The integral of Eq. (53) yields the following:

The Jovanovich isotherm:

$$\theta = 1 - \exp(-cp) \quad \text{for } k = 1 \tag{54}$$

The Langmuir isotherm:

$$\theta = \frac{cp}{1 + cp} \quad \text{for } k = 2 \tag{55}$$

Henry's law for $k = 0$.

Equations (54) and (55) reduce to Henry's law when the pressure approaches zero. However, both of the isotherms do not contain a parameter that accounts for the effect of heterogeneity of the adsorbent surface and, therefore, are not sufficient to represent the experimental isotherms obtained on porous solids. A parameter to account for the surface heterogeneity is also included in the modified Langmuir equation, such as the Langmuir–Freundlich equation:

$$n = n_0 \left(\frac{bp^q}{1 + bp^q} \right) \tag{56}$$

However, parameter values evaluated did not look as reasonable as that of Eq. (51), as will be shown by an example. The modeling procedure is given in the subsequent sections.

D. Illustration of Modeling Supercritical Isotherms

Equation (51) introduced in the previous section is used to model the adsorption isotherms of CH_4 on activated carbon for 158–333 K shown in Fig. 5. Although the critical temperature ($T_c = 190.6$ K) is covered in this range, the isotherms on each side of T_c fall into the same class because the activated carbon used as the adsorbent is highly microporous. Therefore, they can be modeled by the same isotherm equation.

1. Determination of the Absolute Adsorption

The first step of modeling is to formulate an expression for *absolute* adsorption. As elucidated in Section IV.F, the experimental excess adsorption data are represented in the coordinates of $\ln[\ln(n^\sigma)]$ versus $1/\ln(p)$, and a set of linear isotherms are obtained as shown in Fig. 14. To avoid numerical singularity, n^σ values (in mmol/g) are enlarged 10-fold and p is in kilopascals. As shown in Fig. 14, the experimental points for the condition of low surface concentration are on the linear plots. Because the relationship is linear, the formulation for absolute adsorption contains only two parameters [refer to Eq. (46)], which are evaluated by nonlinear regression analysis and shown by Eqs. (46.2) and (46.3). These equations together with Eq. (46.1) can yield the corresponding data of absolute adsorption for any experimental conditions.

2. Correlation of Gas-Phase Density

The density of methane in the gas phase can be determined by an equation of state of real gases. It was proven [41] that most state equations generate almost the same results as those measured for temperatures about 10 K above the critical. The third-order virial equation is used here to generate the density of gaseous methane under the experimental condition. For convenience of regression analysis, methane density was correlated with pressure by a polynomial function for each temperature:

$$\rho_b = \sum_{i=1} c_i p^i \tag{57}$$

The coefficient values of Eq. (57) are listed in Table 3.

3. Determination of the Adsorbed-Phase Volume, V_a

After obtaining the data for n^σ, n^s, and ρ_b one can calculate the volume of the adsorbed phase according to Eq. (2):

$$V_a = \frac{n^s - n^\sigma}{\rho_b} \tag{58}$$

The calculated volumes of the adsorbed phase, shown in Fig. 17, reveal the increase of V_a with an increasing pressure and a minor expansion with temperature. The divergence of the calculated V_a in the low-pressure region can be attributed to the larger uncertainty in the difference $(n^s - n^\sigma)$ because the experimental error is relatively large in this region. For the same reason, the values of V_a evaluated for 158 and 178 K do not make sense because $n^S \cong n^\sigma$ for the subcritical region. However, the changes of V_a with pressure can be well modeled by a polynomial function for each temperature:

$$V_a = \sum_{j=0} c_j p^j \tag{59}$$

The coefficient values are listed in Table 4.

4. Modeling the Excess Adsorption Isotherms

The excess adsorption isotherms are modeled under the framework of Eq. (2). The new isotherm, Eq. (51), is used to describe the absolute adsorption. The gas-phase density ρ_b as well as the adsorbed-phase volume V_a needed in modeling have already been expressed in Eqs. (57) and (59). Thus, we have a model for the excess adsorption isotherms:

$$n = n_0^s[1 - \exp(-bp^q)] - \left(\sum_{i=1} c_i p^i \right) \left(\sum_{j=0} c_j p^j \right) \tag{60}$$

TABLE 3 Coefficients in the CH_4–Density (mmol/cm^3) Correlation (p in MPa)

T (K)	c_1	c_2	c_3	c_4	c_5
333.15	0.36434	3.2275×10^{-3}			
313.15	0.38711	4.9239×10^{-3}			
298.15	0.40351	7.1491×10^{-3}			
278.15	0.42956	0.010983	1.9005×10^{-5}		
258.15	0.46717	0.011583	9.4601×10^{-4}	3.3939×10^{-5}	
238.15	0.51199	0.011890	2.6320×10^{-3}	0.0115×10^{-5}	
218.15	0.54333	0.037697	2.8137×10^{-3}	8.6821×10^{-4}	
198.15	0.43087	0.31486	-0.11408	1.4297×10^{-2}	5.5691×10^{-4}
178.15	0.66079	0.091646	-0.022828	0.012669	
158.15	0.76483	0.073280	0.046424		

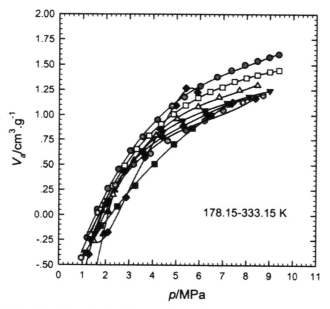

FIG. 17 The adsorbed-phase volume of methane on activated carbon. (From Ref. 58.)

There are three parameters in Eq. (60): n_0^s, b, and q. Their values are evaluated in fitting the model to the experimental excess adsorption data by nonlinear regression analysis. Each parameter is a function of temperature, as shown in Figs. 18–20. The amount adsorbed as predicted by the model is shown by the curves in Fig. 5. The model fits the experimental isotherms very well, even for the one near the critical temperature. The fitness of the model is measured by the total disagreement defined as the following:

$$d(\%) = \frac{1}{N_T} \sum_{j=1}^{N_T} \left(\frac{1}{N_m} \sum_{i=1}^{N_m} \frac{100 \times \text{abs}(n_{\text{cal}}^{\sigma} - n_{\text{exp}}^{\sigma})}{n_{\text{exp}}^{\sigma}} \right) \tag{61}$$

where n_{cal}^{σ} is the amount adsorbed that is calculated by the model, n_{exp}^{σ} is the experimentally measured amount adsorbed, N_m is the number of data at a temperature, and N_T is the number of

TABLE 4 Coefficients in the V_a (cm³/g) Correlation (p in MPa)

T (K)	c_0	c_1	c_2	c_3	c_4	c_5
333.15	-0.97921	0.74032	-0.079751	3.2295×10^{-3}		
313.15	-1.0522	0.75025	-0.083163	3.3836×10^{-3}		
298.15	-1.3899	0.92257	-0.11730	5.4298×10^{-3}		
278.15	-1.3794	1.0446	-0.18215	0.015513	-5.0848×10^{-4}	
258.15	-2.0291	1.7876	-0.48019	0.067144	-4.4231×10^{-3}	1.0249×10^{-4}
238.15	-1.4999	1.5031	-0.46992	0.080532	-6.7836×10^{-3}	2.2111×10^{-4}
218.15	-0.82740	0.43104	-0.027966	7.4067×10^{-4}		
198.15	3.7789	-6.8392	4.1785	-1.1600	0.15734	8.4023×10^{-3}
178.15	2.3657	-6.2603	3.8742	-0.68053		
158.15	—	—	—	—		

Source: Ref. 100.

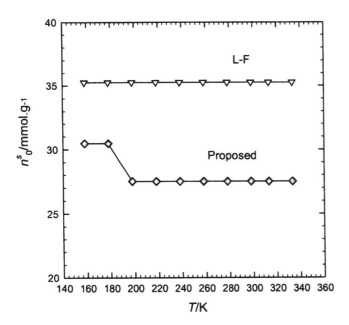

FIG. 18   Variation of parameter n_0^s with temperature. (From Ref. 100.)

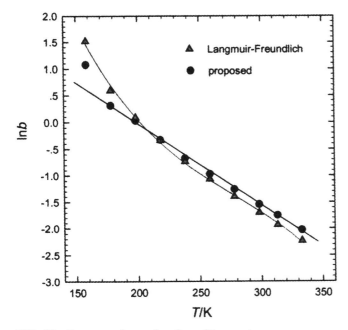

FIG. 19   Parameter b as a function of temperature.

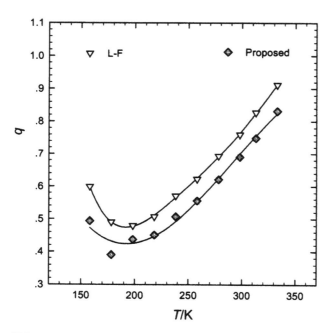

FIG. 20   Variation of parameter q with temperature. (From Ref. 100.)

isotherms. The average disagreement of the model with experimental isotherms is in the range 1–2% for the whole range tested. The adsorption isotherms of nitrogen on activated carbon shown by dots in Fig. 6 were also modeled in the same way [59], and, again, pertinent fitting was observed as shown by curves in the figure. However, if the isotherms in the subcritical region show features of other type (e.g., type II, as that shown in Figs. 7 and 8), the new isotherm equation can apply only to the supercritical region. The new isotherm equation seems physically reasonable, considering the parameter values. As shown in Fig. 18, n_0^s stays constant for the supercritical region, but it increases to a higher value for subcritical temperatures. The variation of parameter b with temperature is interesting: $\ln b$ decreases linearly with temperature T, not with $1/T$, unfortunately; the isosteric heat of adsorption could be evaluated from its slope otherwise. Parameter q shows a minimum near the critical temperature. It will approach unity at high temperatures when the surface heterogeneity is no longer important for adsorption and, hence, the surface can be regarded as "uniform." It is interesting that q takes values only in the range of 0–1 and 1 symbolizes a "uniform surface"; it is concluded that q is an appropriate index of surface heterogeneity. As a comparison with the new isotherm equation, the Langmuir–Freundlich isotherm [Eq. (56)] is also used in the modeling framework [Eq. (60)] to describe the absolute adsorption. This set of isotherms is also well modeled. The corresponding parameters of the L-F equation are also presented in Figs. 18–20. It is interesting that three corresponding parameters have similar trends of variation with respect to temperature; however, the parameter n_0^s is always constant (35.25 mmol/g) and is much higher than that for the new isotherm in the supercritical region (27.5 mmol/g). It will subsequently be shown that the latter is conformed to the other extreme values of adsorption and, hence, is more physically reasonable. Whereas all points, but the one at 158 K, of $\ln b$ are on a straight line for the new isotherm equation in Fig. 19, most points for the Langmuir–Freundlich equation are not. The exponent parameter q of both isotherm equations has almost the same trend of variation with temperature, as shown in Fig. 20.

E. Isotherm Behavior for the Near-Critical Region

Findenegg [6] reported that the maximum of the isotherms is sharply peaked near T_c, and the maximum decreases markedly within a narrow temperature range T_c to $T_c + 10$ K. This phenomenon was considered useful for adsorbent regeneration using supercritical fluids [63–66]. The peculiar isotherms behavior stimulate more experimental [15,28,61–63] and theoretical [61,69] studies on the adsorption and desorption for the near-critical region. Instead, the sharply peaked maximum reported by Findenegg, a broad maximum (a plateau), and a continuous drop near the critical point was observed, and the observations reported separately are consistent [28,61,104]. The different isotherm behavior does not mean that something wrong in observations, but that it can be traced to the different kinds of adsorbent used in various experiments. The adsorbent used by Findenegg is graphitized carbon black, but that used by the other researchers is activated carbon. The author measured the adsorption isotherms of N_2 and CH_4 on microporous activated carbon and mesoporous silica gel, respectively [41,58], and the experimental condition covered the critical region. A change of isotherm types was not observed on passing over the critical temperature on activated carbon. However, it was observed for mesoporous silica gel that type II isotherms in the subcritical region were switched to type I or to typical supercritical isotherms in the supercritical region. The peaked maximum similar to those of Ref. 6 was observed on the isotherm of methane on silica gel at 198 K ($T_c = 190.6$ K). It is, therefore, concluded that the adsorbent property, especially pore sizes, exerts considerable effect on the isotherm behavior in the near-critical region.

If the isotherms in the subcritical region, like those on microporous activated carbon, belong to type I, then the isotherms for both the subcritical and the supercritical region can be normally modeled by a unique model, as shown in the previous section. However, if a shift of isotherm types happened on crossing over the critical temperature, such as the adsorption on mesoporous silica gel and carbon black, separate models are needed for different temperature regions. Theoretical models yielded from the density functional theory quite closely described the experimental data for the near-critical region [68,69]; however, it does not seem very successful for the entire supercritical region due to the monolayer adsorption mechanism. A "crossover theory" for the surface excess adsorption of pure fluids on the solid surface was proposed by Kiselev and co-workers [105]. A good agreement between the theoretical predictions and experimental data for the CO_2–silica system is observed, but only a moderate success was achieved for the SF_6–graphite system.

VI. CONCLUDING REMARKS

Promoted by the development of industrial technology in the field of adsorption, studies on supercritical adsorption have made great progress during the past 20 years. However, a theoretical system for this class of adsorption has not been established so far. Before closing this chapter, the author would like to offer some points for discussion rather than make a broad general conclusion.

A. Is High-Pressure Adsorption Endless in the Pressure Range?

At least two problems have to be solved before a consistent system of theories for supercritical adsorption becomes sophisticated: first, how to set up a thermodynamically standard state for the supercritical adsorbed phase, so that the adsorption potential for supercritical adsorption can be evaluated? Second, how to determine the total amount in the adsorbed phase (i.e., the so-called absolute adsorption) based on experimentally measured equilibrium data. Determination of the absolute adsorption is needed for establishing thermodynamic theory because as a reflection of statistical behavior of molecules, thermodynamic rules must rely on the total, not part of, material confined in the system studied. A method is described in Section IV.F for the determination of absolute adsorption based on experimental equilibrium data, yet it is only one step toward the final solution.

The author raises here, again, the problem that has been noted, yet not been definitely answered: Is high-pressure adsorption indeed endless? Pressures were striven for the high-pressure direction in previous experiments on high-pressure adsorption studies. Adsorption at pressures as high as 650 MPa was reported [31]. Corrections for structure changes of adsorbents under high pressures were taken into account [17]. However, high-pressure adsorption is, in the author's opinion, not endless with respect to pressure or to the adsorbed phase density, and the "end" may serve to be the standard state of the supercritical adsorbed phase.

The first evidence of the existence of an end is the so-called limiting state of the supercritical adsorbate defined by the intersection of linear isotherms [80]. Such linear adsorption isotherms of CH_4 on activated carbon are shown for the region of low surface concentration in Fig. 14. All isotherms in the supercritical region, except the one at 198 K, intersect each other at one point. This point defines a target of absolute amount in the adsorbed phase per gram of adsorbent that can possibly be reached following any course of pressure increment at any temperature. Therefore, it was defined previously as limiting adsorption, n_{lim} [80]. The exact term should be *limiting absolute adsorption, n^s_{lim},* because these linear isotherms represent absolute adsorption. Its value can be estimated from the ordinate of the intersection point, as read from Fig. 14:

$$n^s_{lim} = \tfrac{1}{10} \exp[\exp(1.72)] = 26.6 \text{ mmol/g}$$

The isotherms for 198 K (near-critical temperature) and 178 K and 158 K (subcritical temperatures) intersect at another point, yielding a higher value of limiting adsorption, which is reasonable because of different mechanisms of adsorption.

The second evidence for the existence of an end can be found in the model parameter n^s_0, the saturated absolute amount, which equals 27.5 mmol/g, as shown in Fig. 18 for the adsorption system. Additional proof is presented in Table 5.

The third evidence for the existence of an end is observed on the $n^\sigma \sim \rho_b$ isotherms for the near-critical region. Because a cusplike maximum appears when the gas-phase density closely approaches the critical, the isotherms in the near-critical region can best reflect the specialty of supercritical adsorption. Two such isotherms are shown in Fig. 21. A linear section is shown at each isotherm as is seen in the figure. Both V_a and n^s are constant for the linear section according to the Gibbs definition of adsorption as expressed by Eq. (2). The extension of the linear section of the methane isotherm on activated carbon (shown by solid triangles) yields an intercept of 26.0 mmol/g, which is the value of the constant n^s. This value agrees reasonably with the limiting adsorption 26.6 mmol/g and parameter n^s_0 (27.5 mmol/g). The agreement between the

TABLE 5 Limiting Values of Two Adsorption Systems

Adsorption system	Constant n^s (mmol/g)	n_{lim} (mmol/g)	n^s_0 (mmol/g)
CH_4 on A.C.	26.0	26.6	27.5
N_2 on A.C.	28.3	28.2	29

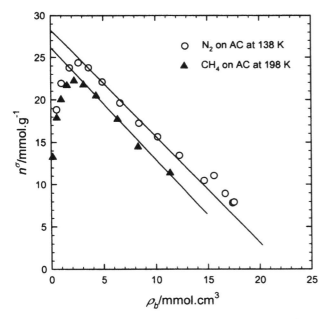

FIG. 21   Isotherms at near critical-temperatures against the gas-phase density.

limiting adsorption determined on the linear isotherm plot and the constant n^s determined on the $n^\sigma - \rho_b$ isotherm is proven by another set of isotherms of nitrogen, as shown in Table 5. Indeed, this agreement is not by coincidence.

The linear section of isotherms as those shown in Fig. 21 defines the boundary of supercritical adsorption. As argued earlier, both absolute adsorption and the volume of the adsorbed phase are constant for the scope of linearity; therefore, the density of the adsorbed phase must also be constant, The constant ρ_a equals the gas-phase density ρ_b when the linear section touches the abscissa, and it becomes less than ρ_b upon the further increase in pressure. Because the gas phase is at the top of the adsorbed phase, a thermodynamically unstable system has thus been established. The lower end of the linear section marks the boundary of adsorption, beyond which any recorded "isotherm" cannot be explained by adsorption, but rather by compression due to the density difference.

B.   What to Do Next for Supercritical Adsorption Study?

It is clear from the previous discussion that there seems to be an end in the high-pressure direction for supercritical adsorption. However, adsorbed-phase density is the decisive factor for the existence of this end. The state of adsorbate at the "end" provides the standard state of the supercritical adsorbed phase just like the saturated liquid, which is the end state of adsorbate in the subcritical adsorption. Therefore, the "end state" has to be precisely defined. It is a serious challenge to define the "end state" precisely. The three ways to determine the "end state" summarized in the previous section cannot be considered thermodynamically rigorous. It has been acknowledged that the location of the cusplike maximum is not far from the critical density of gas, and a linear section, from which the "end state" can also be evaluated, follows the maximum. Therefore, the adsorbed-phase density at the "end state" must be intimately related to the critical density. Case studies [57,59,106] show that the density of the adsorbed phase ranges from slightly lower than the critical density to an area approaching the density of a normal liquid, depending on the equilibrium pressure and temperature. The quantitative relationship with the critical density is, however, still uncertain. To establish a definite relationship for the adsorbed phase density at the end state, abundant and reliable experimental data are still required. Attention has to be paid to collecting data in the region of low surface concentration; that is, experimental points should be densely located in low-pressure region, so that the precision, with which the absolute adsorption is determined, can be improved. Next, improvement in the reliability of the compressibility factor for the near-critical region should receive more attention because of its importance in the determination of the "end state" density. When the gas-phase density approaches the critical, a minor change in the compressibility factor may cause a considerable change in the isotherm and, hence, in the determination of the "end state" [41]. However, experimental p–V–T data for the near-critical region are not as abundant and definite as for the other conditions. Theoretical analysis of the interaction strength between sorbate–sorbate and sorbate–sorbent molecules/atoms in an element space under different conditions would certainly be a complement to the experimental determination of the "end state" of the supercritical adsorbate.

VII. NOTATION

A	Adsorption potential
$A_1, A_2, \ldots$	Virial coefficients
b	A parameter in isotherm equation related to adsorption energy change
$d\%$	Total disagreement of model
$f(r)$	Differential distribution of surface energy with respect to pore size
H	Slit-pore width
$\bar{H}$	Partial molar enthalpy
ΔH	Isosteric heat of adsorption
ΔH_0	Limiting heat of adsorption
K	Henry constant
N_m	Number of data at a temperature
N_T	Number of isotherms
n	Moles of gas
n^σ	Excess adsorption
n^s	Total mass confined in the adsorbed phase, which is usually called *absolute* adsorption
n_0	Saturated capacity
p	Pressure
p_s	Saturation pressure of adsorbate
q	Exponent parameter of the isotherm equation
R	Gas constant
r	Interparticle distance; half-width of a pore
r_a	Radius of an adsorbate molecule
T	Temperature
V	Volume
V^s	Solid phase volume
W	Weight of sample
W_{exp}	Weight readings before correction in gravimetric measurement
z	Distance normal to the surface; the compressibility factor

z_d	Compressibility factor at T_{ad}
z_f	Compressibility factor at T_{ref}
Δ	Separation between graphite lattice planes
$\phi_{ext}(z)$	External potential experienced by a fluid molecule at z
ϕ_{ff}	Interaction between adsorbate molecules
ϕ_{sf}	Adsorbent–adsorbate interactions
θ	Fractional surface coverage
ρ	Density
σ_{ff}	Point at which the potential is minimum
$\sigma_{sf}, \varepsilon_{sf}$	Cross-parameters for adsorbent–adsorbate interaction
τ	Thickness of the adsorbed phase
ε_{ff}	Potential well depth
ω and $\omega^{(r)}$	The eccentric factors of the real fluid of interest and that of the reference fluid
ζ	Intercept of the virial plot

Subscripts

a	Adsorbed phase
ad	Adsorption cell
b	Bulk phase
c	Critical
exp	Experimental or readings before correction
He	Helium
m	Molar
r	Reduced
ref	Reference cell

Superscripts

s	Solid; absolute
t	Tube
0	Saturated

REFERENCES

1. MMK Salem, P Braeuer, Mv Szombathely, M Heuchel, P Harting, K Quitzsch, M Jaroniec. Langmuir *14*:3376, 1998.
2. GL Aranovich, MD Donohue. J Colloid Interf Sci *180*:537, 1996.
3. J Vermesse, D Levesque. J Chem Phys *101(10)*:9063, 1994.
4. Y Wakasugi, S Ozawa, Y Ogino. J Colloid Interf Sci *79(2)*:399, 1981.
5. JU Keller. Physica A *166*:180, 1990.
6. GH Findenegg. Fundamentals of Adsorption. New York: United Engineering Trustees, Inc., 1984, pp 277–218.
7. RT Yang. Gas Separation by Adsorption Processes. London: Butterworths, 1987.
8. DM Ruthven, F Shamasuzzaman, KS Knaebel. Pressure Swing Adsorption. New York: VCH 1994.
9. A Golovoy, EJ Blais. In: Alternate Fuels for Special Ignition Engine. SAE Conference Proceeding (C). Warrendale, PA: SAE, 1983, p 47.
10. KR Matranga, AL Myers, ED Glandt. Chem Eng Sci *47(7)*:1569, 1992.
11. O Talu. AIChE Meeting, Miami Beach, FL, 1993, p 409.

12. ND Parkyn, DF Quinn. In: JW Patrick, ed. Porosity in Carbons. London: Edward Arnold, 1995, 292.
13. LG Randall, LM Bowman Jr., eds. Separ Sci Technol *17(1)*, 1982, special issue.
14. WR Ladner, ed. Fluid Phase Equilibria, *10*(2 + 3), 1983, special issue.
15. Y-K Ryu, K-L Kim, C-H Lee. IEC Res *39*:2510, 2000.
16. JW McBain, GT Britton. J Am Chem Soc *52*:2198, 1930.
17. PG Menon. Chem Rev *68(3)*:277, 1968.
18. F Rouquerol, J Rouquerol, KS Sing. Adsorption by Powders and Porous Solids, Methodology and Applications. London: Academic Press, 1999.
19. PA Webb, C Orr. Analytical Method in Fine Particle Technology. Norcross, GA: Micromerities Instrument Corp., USA, 1997.
20. Z Tan, KE Gubbins. J Phys Chem *94*:6061, 1990.
21. SY Jiang, JA Zollweg, KE Gubbins. J Phys Chem *98*:5709, 1994.
22. S Blümel, F Köster, GH Findenegg. Chem Soc Faraday Trans 2 *78*:1753, 1982.
23. L Zhou, Y Zhou, M Li, P Chen, Y Wang. Langmuir *16*:5955, 2000.
24. PE Savage, S Gopalan, TI Mizan, CJ Martino, EC Brock. AIChE J *41(7)*:1723, 1995.
25. B Yang. Master's thesis, Tianjin University, Tianjin, China, 2000.
26. JU Keller, F Dreisbach, H Rave, R Staudt, M Tomalla. Adsorption *5*:199, 1999.
27. HW Lösch. Global Chem Process Eng Ind 104, February 1999.
28. R Humayun, DL Tomasko. AIChE J *46(10)*:2065, 2000.
29. D Vidal, P Malbrunot, L Guenggant, J Vermesse, TK Bose, R Chahine. Rev Sci Instrum *61(4)*:1314, 1990.
30. TK Bose, R Chahine, L Marchildon, JM St Arnaud. Rev Sci Instrum *58*:2279, 1987.
31. P Malbrunot, D Vidal, J Vermesse, R Chahine, TK Bose. Langmuir *8*:577, 1992.
32. JU Keller. Adsorption *1*:283, 1995.
33. H Rave, R Staudt, JU Keller. J Thermal Anal Calorim *55*:601, 1999.
34. S Sircar, MB Rao. In: JA Schwarz, CI Contescu, eds. Surface of Nanoparticles and Porous Materials. New York: Marcel Dekker, 1999, pp 501–528.
35. WF Hermminger, HK Cammenga. Methoden der Thermischen Analyse. Berlin: Springer-Verlag, 1989.
36. R Staudt, H Rave, JU Keller. Adsorption *5*:159, 1999.
37. R Staudt, S Kramer, R Dreisbach, JU Keller. In: F Meunier, ed. Proceeding of Fundamentals of Adsorption Conference. Paris: Elsevier, 1998, pp 153–158.
38. S Sircar. Ind Eng Chem Res *38*:3670, 1999.
39. R Staudt, G Saller, M Tomalla, JU Keller. Ber Bunsenges Phys Chem *97(1)*:98, 1993.
40. AV Neimark, PI Ravikovitch. In: F Meunier, ed. Proceedings of Fundamentals of Absorption. Paris: Elsevier, 1998, pp 159–164.
41. Y Zhou, Y Sun, L Zhou. An experimental study on the adsorption behavior of gases on crossing the critical temperature. The 7th International Conference on Fundamentals of Adsorption, Nagasaki, 2001.
42. NB Vargaftik. Handbook of Physical Properties of Liquid and Gases. 2nd ed. Washington, DC: Hemisphere, 1975.
43. J Phys Chem Ref Data.
44. IUPAC, Chemical Data Series No. 17. International Thermodynamic Tables of the Fluid State. New York: Pergamon, 1976.
45. TM Guo. Multicomponent Vapor–Liquid Equilibrium and Distillation. Beijing: Chemical Industry Press, 1983.
46. XS Chen, B McEnaney, TJ Mays, J Alcaniz-Monge, D Cazorla-Amoros, A Linares-Solano. Carbon *35(9)*:1251, 1997.
47. T Nitta, M Nozawa, Y Hishikawa. J Chem Eng Japan *26(3)*:266, 1993.
48. J Zhou, W Wang. Langmuir *16*:8063, 2000.
49. WA Steel. The Interaction of Gases with Solid Surfaces. Oxford: Pergamon, 1974.
50. DJ Adams. Mol Phys *28*:1241, 1974; *29*:307, 1975; *32*:647, 1979.
51. Q Wang, JK Johnson. J Chem Phys *110(1)*:577, 1999.

52. Q Wang, JK Johnson. J Phys Chem B *103*:4809, 1999.
53. S Ozawa, S Kusumi, Y Ogino. J Colloid Interf Sci *56(1)*:83, 1976.
54. J Specovius, GH Findenegg. Ber Bunsenges Phys Chem *82*:174, 1978.
55. RK Agarwal, JA Schwarz. Carbon *26(6)*:873, 1988.
56. Y Zhou, L Zhou. Sci China (Series B) *39(6)*:598, 1996.
57. Y Zhou, L Zhou. Sci China (Series B) *43(2)*:143, 2000.
58. Y Zhou, Sh Bai, L Zhou, B Yang. Chin J Chem *9(10)*:943, 2001.
59. L Zhou, Y Zhou, Sh Bai, Ch Lü, B Yang. J Colloid Interf Sci *239*:33, 2001.
60. L Zhou, Y Zhou, Sh Bai, B. Yang. Adsorption, in press.
61. JH Chen, DSH Wong, CS Tan, R Subramanian, CT Lira, M Orth. Ind Eng Chem Res *36*:2808, 1997.
62. JR Strubinger, JF Parcher. Anal Chem *61*:951, 1989.
63. M Modell, RJ Robey, VJ Krukonis, RP DeFilippi, D Oestreich. AIChE Meeting, Boston, 1979.
64. M Thommes, GH Findenegg, M Schoen. Langmuir *11*:2137, 1995.
65. RP DeFilippi, VJ Krukonis, RJ Robey, M Modell. Supercritical Fluid Regeneration of Activated Carbon for Adsorption of Pesticides. Washington, DC: EPA, 1980.
66. KM Dooley, CP Kao, RP Grambrell, CF Knopf. Ind Eng Chem Res *26*:2056, 1987.
67. CS Tan, DC Liou. Ind Eng Chem Res *27*:988, 1988; Separ Sci Technol *24*:111, 1989; Ind Eng Chem Res *28*:1222, 1989; Ind Eng Chem Res *29*:1412, 1990.
68. A Maciolek, R Evans, NB Wilding. Phys Rev E *60(6)*:7105, 1999.
69. R Subramanian, H Pyada, CT Lira. Ind Eng Chem Res *34*:3830, 1995.
70. J Milewska-Duda, J Duda, G Jodlowski, M Wojcik. Langmuir *16*:6601, 2000.
71. L Zhou, Y Zhou. Int J Hydrogen Energy *26(6)*: 597, 2001.
72. SA Al-Muhtaseb, JA Ritter. Ind Eng Chem Res *37*:684, 1998.
73. DM Ruthven. Principles of Adsorption and Adsorption Processes. New York: Wiley, 1984. Chap 3–4.
74. L Zhou, Y Zhou. Ind Eng Chem Res *35(11)*:4166, 1996.
75. L Zhou, Y Zhou. Chin J Chem Eng *9(1)*:110, 2001.
76. Y Zhou, L Zhou, Sh Bai, B Yang. Adsorp Sci Technol, in press.
77. K Murata, K Kaneko. Chem Phy Lett *321*:342, 2000.
78. AE DeGance. Fluid Phase Equilibria *78*:99, 1992.
79. M Kobayashi, E Ishikawa, Y Toda. Carbon *29(4)*:677, 1991.
80. L Zhou, Y Zhou. Chem Eng Sci *53(14)*:2531, 1998.
81. L Zhou, P Chen, M Li, Y Sun, Y Zhou. Proceedings, The 2nd Pacific Basin Conference on Adsorption Science and Technology, Brisbane, 2000, pp 717–721.
82. M Polanyi. Verh Deut Phys Ges *16*:1012, 1914; *18*:55, 1916.
83. MM Dubinin, VA Astakhov. Trans Izv Akad Nauk SSSR Ser Khim No.1: pp 5–11, 1971.
84. I Langmuir. J Am Chem Soc *40*:1361, 1918.
85. S Brunauer, PH Emmett, E Teller. J Am Chem Soc *60*:309, 1938.
86. J Toth. J Acta Chim Acad Sci Hung *32*:39, 1962; *69*:329, 1971.
87. AW Adamson. Physical Chemistry of Surfaces. New York: Wiley, 1976.
88. IUPAC Commission on Colloid and Surface Chemistry Including Catalysis. Pure Appl Chem *57*:603, 1985.
89. KAG Amankwah, JA Schwarz. Carbon *33(9)*:1313, 1995.
90. K Kaneko, K Murata. Adsorption *3*:197, 1997.
91. M Aoshima, K Fukasawa, K Kaneko. J Colloid Interf Sci *222*:179, 2000.
92. VKh Dobruskin. Langmuir *14*:3840, 1998.
93. HK Shethna, SK Bhatia. Langmuir *10*:870, 1994.
94. S Ross, JP Olivier. On Physical Adsorption. New York: Interscience, 1964.
95. L Czepiriskł, J Jagiello. Chem Eng Sci *44*:797, 1989.
96. P Bénard, R Chahine. Langmuir *13*:808, 1997.
97. DD Do, K Wang. Langmuir *14*:7271, 1998.
98. CRC Jensen, NA Seaton. Langmuir *12*:2866, 1996.
99. J Jagiello, JA Schwarz. J Colloid Interf Sci *154(1)*:225, 1992.
100. L Zhou, J-Sh Zhang, Y Zhou. Langmuir, in press.

101. W Weibull. Fatigue Testing and the Analysis of Results. New York: Macmillan, 1961.
102. DS Jovanovich. Kolloid Z Z Polym *235*:1203, 1969.
103. DN Misra. J Colloid Interf Sci *77(2)*:543, 1980.
104. S Ozawa, S Kusumi, Y Ogino. International Conference on High Pressure. Tokyo: Physical–Chemical Society of Japan, 1975.
105. SB Kiselev, JF Ely, M Yu Belyakov. J Chem Phys *112(7)*:3370, 2000.
106. Y Zhou, L Zhou. Separ Sci Technol *33(12)*:1787, 1998.

5

Irreversible Adsorption of Particles

ZBIGNIEW ADAMCZYK Institute of Catalysis and Surface Chemistry, Polish Academy of Sciences, Cracow, Poland

I. INTRODUCTION

Adsorption can be defined as a process leading to a significant increase in particle concentration occurring in a thin layer adjacent to an interface, usually liquid/solid or liquid/air. The driving force of adsorption are the short-range interactions between particles and interfaces, often called colloid or surface interactions. According to this definition, a physical accumulation of particles caused, for example, by gravity (sedimentation) is not classified as adsorption. Irreversible adsorption, treated primarily in this chapter, is often referred to as *deposition* or *adhesion*, appearing when a chemical-type contact between a particle and an interface is created. According to the colloid nomenclature, particle adsorption can be treated as a limiting form of *heterocoagulation* of particles differing widely in size.

Effective adsorption of colloid and bioparticles is important for many practical processes such as water and wastewater filtration, papermaking, xerography, self-assembling of colloid particles at patterned surfaces, protein and cell separation (affinity chromatography), immobilization of enzymes, immunological assays, and so forth. A fascinating new application of particle deposition is the "colloid bar coding" technique which enables the encoding of libraries of a million of compounds by using a few fluorescent dyes.

In other processes, such as membrane filtration, biofouling of membranes and artificial organs, flotation (slime coating formation), and production of microelectronic or optical devices, particle adsorption is highly undesirable.

Apart from practical aspects, a quantitative analysis of particle adsorption also can furnish interesting information on specific interactions under dynamic conditions between particles and interfaces, which is a crucial issue for colloid science, biophysics and medicine, soil chemistry, and so forth. Furthermore, by measuring particle adsorption in model systems (e.g., monodisperse colloid suspension), important clues can be gained concerning mechanisms and kinetics of molecular adsorption difficult for direct experimental studies. In this way, various aspects of statistical–mechanical theories can be tested (e.g., the fluctuation theory) and links between irreversible (colloid) and reversible (molecular) systems can be established. Similarly, colloid particles can be used for elucidation adsorption mechanisms of globular proteins [1,2] and to calibrate the indirect methods of measuring protein surface coverage like reflectometry [3,4] or streaming potential [3,5].

It should be pointed out, however, that despite similarities, colloid particle adsorption proceeds via more complicated paths than molecular adsorption. This is because particle transfer from the bulk to the interface is affected by a variety of interactions differing widely in magnitude and the characteristic length scale.

For separations exceeding particle dimension, the dominating transport mechanism is forced convection applied either in a controllable way, as in the impinging-jet cells [6,7] or by stirring the particle suspension [8]. This transport mechanism is especially effective for particle size above a micrometer. Particle transfer over macroscopic distances can also be provoked by external forces like gravity, leading to sedimentation effects, and electrostatic force, leading to electrophoresis. The latter mechanism is especially effective, in comparisons with sedimentation, for submicrometer-sized particles because the rate of electrophoretic motion is independent of particle size.

For separations comparable with particle dimensions, the most significant transport mechanism becomes diffusion, the rate of which increases with decreased size of particles and increased temperature. The diffusion transport mechanism was the dominating one in various experimental studies on protein and colloid adsorption [2,8–10]. However, the disadvantage of the diffusion-controlled transport is its inherently unsteady character, leading to considerable decrease in adsorption rate with time.

At separations smaller than their dimensions, adsorbing particles are subject to specific force fields generated by the interface which vary abruptly with the distance. For this range of separations, the dominant role is played by the electrostatic interactions resulting from the presence of fixed or induced charges on particles. The range of electrostatic interaction varies widely with electrolyte concentration and may become considerably larger than particle dimensions in nonpolar media or in a gas. Another interaction appearing universally at short separations is the London–van der Waals dispersion interaction, the results of attraction of induced dipoles integrated over a macroscopic volume. In contrast to electrostatic forces, the range of dispersion interactions always remains much smaller than particle dimension. These electrostatic and dispersion interactions are often referred to as the DLVO (Derjaguin–Landau, Vervey–Overbeek) forces [11,12].

At extremely small separations, after coming into physical contact with the interface, colloid particles and proteins are subject to surface deformation and reconformation processes, ion exchange, or specific chemical-type interactions, like hydrogen-bound formation or sintering. All of these time-dependent phenomena lead to particle immobilization (localized adsorption), which is responsible for the partial or total irreversibility of colloid and protein adsorption.

Additional complications in particle adsorption mechanisms appear due to the presence of particles accumulated in the vicinity of the interface, which also disturbs locally the hydrodynamic and electrostatic force fields [1,7,12,14] and interact with adsorbing particles. This leads to surface-blocking effects (which should more appropriately be called volume exclusion effects), decreasing particle adsorption rate for longer times and making the adsorption pathway nonlinear with respect to the bulk suspension concentration. The blocking effects, depending on the particle transport mechanism in the case of irreversible adsorption, become especially involved for nonspherical (anisotropic) particles [2,12,15]. An exact theoretical analysis of these many-body effects is prohibitive. However, there exist approximate models, discussed later in this chapter, which can be used for estimating particle adsorption kinetics under various transport modes.

Due to these complicated transport mechanisms and often occurring irreversibility, particle adsorption is a highly path-dependent process characterized by a wide spectrum of time scale ranging from seconds for concentrate colloid suspensions to days for dilute suspensions of larger particles adsorbing under diffusion-controlled transport [9,10,16]. Therefore, in contrast to molecular systems, the kinetic aspect of particle adsorption are of primary interest and will be extensively discussed in this chapter.

In order to perform a quantitative analysis of particle desorption kinetics, an unequivocal measure of adsorption extent is necessary. Defining a variable for quantifying particle adsorption seems simpler than for molecular systems because the thickness of the layer where a considerable concentration increase takes place is much smaller than particle dimensions [1,2]. Moreover, the location of the interface can be better defined because the density gradient of the solvent usually has a negligible extension in comparison with particle dimensions. One can, therefore, accurately measure the position of adsorbed particles relative to the interface and define the extent of adsorption as

$$N = \frac{N_p}{\Delta S} = \frac{1}{\Delta S} \int_{\Delta v_a} n \, dv \tag{1}$$

where N is the two-dimensional (surface) concentration of particles, N_p is the number of particles adsorbed over the area ΔS, Δv_a is the adsorption volume, and n is the number concentration of particles. If n depends solely on the coordinate h perpendicular to the interface, Eq. (1) can be simplified to

$$N = \int_{\delta_a} n(h) \, dh \tag{2}$$

where δ_a is the thickness of the adsorbed layer.

Equations (1) and (2) are generally valid for symmetric particles when the adsorption volume has a well-defined boundary, which is the case for colloid adsorption problems. Ambiguities may arise, however, for asymmetric particles due to the difficulty in specifying the particle's center position [1].

The proper unit of N is m^{-2} in the SI system and cm^{-2} in the CGS system. However, both units are rather inconvenient for colloid and proteins (especially the SI units) as discussed in Ref. 1. Therefore, instead of N, one uses the dimensionless surface concentration Θ, often referred to as coverage, surface coverage, or fractional coverage defined as

$$\Theta = S_g N \tag{3}$$

where S_g is the characteristic cross section of a particle. In the case of spheres, S_g equals πa^2, where a is the particle radius. The definition, Eq. (3), is unequivocal in contrast to the often used empirical definition of coverage given by

$$\Theta = \frac{N}{N_s} \tag{4}$$

where N_s is the so-called saturation or maximum coverage determined in experiments by extrapolating the data to an infinite time of adsorption. As discussed in Ref. 1, this definition is not unique and may lead to severe misinterpretation of experimental results. This is so because an apparent saturation of surfaces often occurs due to kinetic limitations or an uncontrolled cleaning procedure of the substrate (interface).

It should be mentioned that the surface concentration N or the saturation concentration N_s can only be measured by direct experimental techniques, like the microscope observation methods [1,2,6,7,9,10,12–14,16], which are difficult to apply for small colloid particles and protein. In these cases, one uses indirect, usually optical, techniques [3–5,8], giving the averaged value of the adsorbed layer thickness or the averaged mass of the adsorbed monolayer. The natural unit used in these studies to express the extent of particle adsorption is the mass of the adsorbate per unit area, which is also experimentally accessible. One uses the entire variety of units, most often mg/m², μg/cm², ng/cm², and so forth, which may sometimes lead to confusion, especially if the bulk concentration of protein is expressed in other units. The fractional coverage in this case is defined as

$$\Theta = \frac{m}{m_s} \tag{5}$$

where m_s is the mass of the monolayer (adsorbate) and m is the mass under saturation, usually found by extrapolation.

Our work is organized as follows: In the first part, we discuss electrostatic interactions between two planar double layers derived form the classical Poisson–Boltzmann (PB) equation. Then, the approximate models for calculating interactions of curved interfaces (e.g., spheres) are exposed, with emphasis on the linear superposition and extended Derjaguin summation methods. Next, the specific energy profiles originating from the superposition of electrostatic and dispersion contributions are discussed, together with the influence of surface roughness and charge heterogeneity effects. Further, the hydrodynamic driving force due to macroscopic flow fields are presented as well as the local hydrodynamic interactions particle/interface.

The next part is devoted to linear adsorption regimes appearing for low surface coverage of adsorbed particles. The phenomenological continuity equation is formulated, incorporating the convective transport in the bulk and the specific force-dominated transport at the surface. Approximate analytical models aimed at decoupling these transfer steps are described. Analytical and exact numerical results which express the initial adsorption kinetics in terms of the flow intensity and configuration, particle size, range of electrostatic interactions, and so forth are discussed.

The theoretical approaches aimed at a description of nonlinear adsorption regimes are discussed next—in particular the classical, random sequential adsorption (RSA) model. The role of polydispersity, particle shape, orientation, and electrostatic interactions is elucidated. Then, the generalized RSA model is presented, which reflects the coupling of the surface transport with the bulk transfer step governed by external force or diffusion.

The theoretical results are confronted with experimental data obtained under well-defined transport conditions like diffusion, forced convection (impinging-jet cells), and sedimentation using various experimental techniques of detecting particle monolayers, AFM, reflectometry, and electron and optical microscopy. The significance of the DLVO theory for interpreting the kinetic data obtained for the linear adsorption regime will be pointed out. Except for adsorption kinetics, the structural aspects of particle adsorption measurements will be discussed, including the pair-correlation function, maximum coverage, and fluctuations in particle density.

II. INTERACTIONS-DRIVING FORCES

A. Electrostatic Interactions

The electrostatic interactions arise due to the presence of fixed or induced (as a result of an external field) charges at particle surfaces. In polar liquid media, the main charging mechanism are (1) irreversible or preferential adsorption of ions, ionic surfactants, and charged polymers (polyelectrolytes) and (2) dissociation of ionogenic groups chemically bound to the surface. In unpolar (gaseous) media, the charge can be produced by the electrostriction or ionization phenomena. As is often the case, the net charge is located at the particle interface, forming a thin layer of negligible dimension in comparison with particle size. As a result, it can be treated as surface charge denoted by σ^0. It should be remembered, however, that the surface-charge concept is always an approximation of a real situation in which the charge layer has a finite extension, due, for example, to the surface roughness.

In order to fulfill the overall electroneutrality condition, the surface charge should be compensated for by an opposite charge distributed throughout the continuous phase. The characteristic length of the surface-charge compensation (often referred to as the screening length) depends obviously on the concentration of ions in a phase. For unpolar media (gas), the bulk ion concentration remains usually smaller than 10^{-10} mol/dm^3 (for the sake of brevity this concentration will be referred to as M), so the screening length attains macroscopic values comparable with a centimetre. On the other hand, for polar media (water), ion concentrations usually exceed 10^{-7} mol/dm^3 and the screening length becomes comparable or smaller than particle dimensions. In this case, an electrical double layer is formed which affects most of the dynamic phenomena occurring in colloid systems as well as their stability. Despite a continuous effort aimed at a quantitative description of the double layer for the colloid system [13,17,18], no complete theory emerged yet. At present, there exist two main paths of thinking about the double layer: (1) the statistical–thermodynamic approach and (2) the phenomenological approach based on the local thermodynamic balance, neglecting all ion–ion correlations, finite-size effects, and so forth. Although the latter approach is less strict, it is applicable for a broad range of situations of practical interest. In contrast, the more general statistical–thermodynamic approach produces rather specific results which cannot be generally applied for the evaluation of particle interactions.

Therefore, in our analysis, we adopt the phenomenological approach based on the Gouy–Chapmann concepts [19,20] which exploits the Poisson–Boltzmann (PB) equation for describing electric field distribution in electrified media.

1. The Poisson–Boltzmann Equation

The PB equation can be derived from the fundamental electrostatic equation formulated by Maxwell, which reads

$$\nabla \cdot \boldsymbol{D}_e = \nabla \cdot (\varepsilon_0 \varepsilon \boldsymbol{E}) = \varepsilon_0 (\nabla \varepsilon \cdot \boldsymbol{E} + \varepsilon \nabla \cdot \boldsymbol{E}) = \rho_e \tag{6}$$

where $\boldsymbol{D}_e = \varepsilon_0 \varepsilon \boldsymbol{E}$ is the dielectric displacement vector, $\boldsymbol{E} = -\nabla \psi$ is the electric field, ψ is the electrostatic potential, ε is the relative dielectric permittivity of the disperging phase (called often dielectric constant), ε_0 is the dielectric permittivity of the vacuum (equal to 8.85×10^{-12} F/m in the SI system applied in our work), and ρ_e is the charge density.

In the usual case of negligible dielectric saturation (when ε does not depend on $\boldsymbol{E}$), Eq. (6) simplifies to

$$\nabla \cdot \boldsymbol{E} = -\nabla^2 \psi = \frac{\rho}{\varepsilon_0 \varepsilon} \tag{7}$$

If charge accumulation takes place within a layer of an infinitesimal thickness (surface charge), Eq. (2) can be integrated to the important limiting form

$$\delta \boldsymbol{D}_e \cdot \hat{\boldsymbol{n}} = \varepsilon_0 (\varepsilon_2 E_2 - \varepsilon_1 E_1) = \sigma^0 \tag{8}$$

where $\delta \boldsymbol{D}_e$ is the difference of the electric displacement between the two adjacent phases (e.g., particle/electrolyte solution), E_1 and E_2 are the normal components of the field vector at the interface, and ε_1 and ε_2 are the relative dielectric permittivities in the two phases involved. Equation (8), being the limiting form of Gauss' law, represents the general electrostatic boundary condition.

In the general case, however, the Maxwell equation, Eq. (7) can only be solved if a constitutive dependence connecting the local charge density ρ_e with local electrostatic potential ψ is specified. This is usually achieved by assuming a local thermodynamic equilibrium and postulating an ideal behavior of ion solutions when the electrochemical potential of a ion $\tilde{\mu}_i$ is given by the expression

$$\tilde{\mu}_i = \tilde{\mu}_i^0 + kT \ln n_i + z_i e \psi \tag{9}$$

where, $\tilde{\mu}_i^0$ is the reference potential, k is the Boltzmann constant, T is the absolute temperature, n_i is the local ion concentration, z_i is the ion valency, and e is the elementary charge.

By inverting Eq. (9), one obtains the Boltzmann distribution

$$n_i = n_i^b e^{-z_i e \psi / kT} \tag{10}$$

where n_i^b is the ion concentration in the bulk (reference concentration).

Equation (10) can be combined with Eq. (7) by noting that $\rho_e = -e \sum_{i=1}^{N_i} z_i n_i^b$, where N_i is the total number of ions. In this way, one arrives at the PB equation, expressed in the form

$$\nabla^2 \psi = -\frac{e}{\varepsilon_0 \varepsilon} \sum_{i=1}^{N_i} z_i n_i^b e^{-z_i e \psi / kT} \tag{11}$$

As one can note, Eq. (11) is strongly nonlinear due to the presence of the exponential terms, which makes its analytical solution in the general case impractical. However, important limiting forms of the PB equation can be formulated susceptible for analytical handling.

For a symmetric electrolyte composed of two types of ions, when $z_1 = -z_2 = z$, Eq. (11) can be transformed to the useful form

$$\nabla^2 \bar{\psi} = \kappa^2 \sinh \bar{\psi} \tag{12}$$

where $\bar{\psi} = z \psi e / kT$ is the reduced electric potential and

$$\kappa^{-1} = Le = \left(\frac{\varepsilon_0 \varepsilon kT}{2e^2 I} \right)^{1/2} \tag{13}$$

is the Debye screening length, a parameter of primary interest for any particle interaction problem, $I = z^2 n^b$ is the ionic strength of the electrolyte solution, and n^b is the bulk concentration of the electrolyte.

Another frequently used form of the PB equation can be derived by applying the linearization procedure, which is justified if the maximum term $z_1 e \psi / kT < 1$. By expanding the exponential terms and exploiting the electroneutrality condition, one obtains the linear PB equation

$$\nabla^2 \psi = \kappa^2 \psi \tag{14}$$

where κ is defined by Eq. (13) and the ionic strength is now given by

$$I = \frac{1}{2} \sum_{i=1}^{N_i} z_i^2 n_i^b \tag{15}$$

As discussed in Ref. 13, the significance of Eq. (14) is increased by the fact that it is also applicable for nonlinear systems (high surface charge of particles) at distances larger than Le, where the potential decreases to low values because of the electrostatic screening. This observation was the basis of the powerful linear superposition approach (LSA) discussed later.

It is interesting to note that the screening length Le in aqueous media at room temperature varies between approximately $0.4 \, nm$ (for $0.1 \, M$ Na_3PO_4 solution) and $30 \, nm$ (for a $10^{-6} \, M$

solution of KCl). On the other hand, for nonpolar media when I becomes much lower, Eq. (14) as well as the nonlinear PB equation reduce to the Laplace equation, that is,

$$\nabla^2 \psi = 0 \tag{16}$$

This limiting form can be used for calculating electrostatic interactions between particles and electrodes in the case of adsorption from gases or unpolar solvents.

The boundary conditions for the nonlinear PB equation as well as for its limiting form [Eq. (10)] and the Laplace equation (Eq. (11)] are the same as previously expressed by Eq. (8). However, in the limiting case when particle dimensions are much larger than the screening length, the field inside particle can be neglected and Eq. (8) simplifies to

$$E_\perp = \frac{\sigma^0}{\varepsilon_0 \varepsilon} \tag{17}$$

where $E_\perp$ is the perpendicular component of the field vector (evaluated locally) and σ^0 is the local value of surface charge on particle. Equation (17) is often referred as the constant charge (c.c.) boundary condition.

For many situations occurring in practice, the surface charge may change upon particle approach as a result of ion adsorption/desorption phenomena. In this case, it is more convenient to formulate the boundary condition as [13]

$$\psi = \psi^0 \tag{18}$$

where ψ^0 is the equilibrium surface potential of a particle. Equation (18) is referred to as the constant potential (c.p.) boundary condition.

Solution of the PB equation with these boundary condition enables one to evaluate the spatial distribution of electric potential, field, and ion concentration. Knowing these quantities, one can evaluate the force and energy of interactions between colloid particles or between particles and interfaces. This can be achieved via the thermodynamic equilibrium condition (Gibbs–Duhem relationship) which can be formulated as [13]

$$\nabla \cdot (\Delta \boldsymbol{P} - \boldsymbol{T}) = 0 \tag{19}$$

where $\Delta \boldsymbol{P}$ is the osmotic pressure tensor and $\boldsymbol{T}$ is the Maxwell stress tensor given by the equations

$$\Delta \boldsymbol{P} = \left(kT \sum_{i=1}^{N_i} n_i \right) \boldsymbol{I}$$

$$\tag{20}$$

$$\boldsymbol{T} = \frac{\varepsilon_0 \varepsilon}{2} (2\boldsymbol{EE} - E^2 \boldsymbol{I})$$

where $\boldsymbol{I}$ is the unit tensor.

The force can be evaluated by integrating Eq. (20) over an arbitrary surface S enclosing the particle and making use of the Green theorem. In this way, one obtains the expression [21]

$$\boldsymbol{F} = \int\int_s \left[\left(\Delta \boldsymbol{P} + \frac{\varepsilon_0 \varepsilon}{2} E^2 \right) \hat{\boldsymbol{n}} - \varepsilon_0 \varepsilon (\boldsymbol{E} \cdot \hat{\boldsymbol{n}}) \boldsymbol{E} \right] dS \tag{21}$$

where $\boldsymbol{F}$ is the force vector. Obviously, Eq. (21) can be used in conjunction with the Laplace equation when the ion concentration remains negligible and $\Delta \boldsymbol{P} = 0$.

In the limiting case of a flat geometry (two infinite planar interfaces interacting across electrolyte solution), Eq. (21) reduces to the simple form describing the uniform force per unit area:

$$F = \Delta P(x) - \frac{\varepsilon_0 \varepsilon}{2} \left(\frac{d\psi}{dx} \right)^2 = \Delta \Pi \tag{22}$$

where x is an arbitrary position between plates, F is the force per unit area of plates which can be treated as uniform pressure, and $\Delta \Pi$ is often called the disjoining pressure [17].

The interaction energy ϕ can be most directly obtained by integrating Eq. (21) along a path starting from infinity [13]. This procedure, although generally valid, may become rather cumbersome. Therefore, in practice, one calculates the interaction energy of particles as the reversible work of the forming double layer by exploiting the Gibbs–Duhem relationship. The appropriate expression for the constant charge case (no exchange of surface charge upon particle approach) has the form [22,23].

$$\phi = \iint_{\Sigma s} dS \int_0^{\sigma^0} \psi^0 \, d\sigma' \tag{23}$$

where $\sum s$ means the summation over all the particle surfaces involved and ψ^0 is the surface potential of the particle evaluated as a function of its surface charge σ^0 (this relationship is known from the solution of the PB equation).

In the case when the surface potential remains constant upon particle approach (thermodynamic equilibrium is maintained), the expression for the interaction energy becomes [24]

$$\phi = - \iint_{\Sigma s} dS \int_0^{\psi_0} \sigma^0 \, d\psi' \tag{24}$$

It should be mentioned that due to nonlinearity of the PB equation, no analytical solutions were found in the case of multidimensional problems (e.g., two spherical particles in space being of primary practical interest). However, there exists a variety of useful limiting analytical solution (discussed next).

2. Analytical Results

There exists only a few analytical solutions of the PB which have a considerable significance for testing the accuracy of numerical solutions. One of the solutions was derived for a single plate (more precisely, half-space) immersed in an symmetric electrolyte. The electric potential distribution in this case is given by the formula derived by Vervey and Overbeek [24]:

$$\psi(x) = \frac{2kT}{ze} \ln \left(\frac{1 + \beta_0 e^{-x/Le}}{1 - \beta_0 e^{-x/Le}} \right) \tag{25}$$

where $\beta_0 = \tanh(ze\psi^0/4kT)$, x is the distance from the plate, and ψ^0 is the surface potential of the plate connected with the plate charge σ^0 by the Gouy–Chapman formula:

$$\psi^0 = \frac{2kT}{ze} \ln \left(\frac{\bar{\sigma}^0 + \sqrt{\bar{\sigma}^{0^2} + 4}}{2} \right) \tag{26}$$

where $\bar{\sigma}^0 = \sigma^0 \sqrt{(1/2\varepsilon_0 \varepsilon kTn^b)}$ is the dimensionless surface charge.

Equations (25) and (26) describe the potential distribution in the diffuse double layer by neglecting all specific adsorptions of ions in the Stern layer adjacent to the surface [24,25]. They represent practically the only exact solution of the nonlinear PB equation known in literature.

At larger separations, when $x/Le \gg 1$, Eq. (25) assumes the asymptotic form

$$\psi(x) = \frac{4kT}{ze} \beta_0 e^{-x/Le} = \bar{Y}^0 e^{-x/Le} \tag{27}$$

where

$$\bar{Y}^0 = \frac{4}{z} \tanh\left(\frac{ze\psi^0}{4kT}\right) \tag{28}$$

can be treated as the effective surface potential of the plate. Equation (28), indicating that the electric potential decreases exponentially at larger separations, has a considerable significance for predicting the electrokinetic potential of surfaces (defined as the potential in the slip plane).

In the case of the double layer around a spherical particle, an exact solution can only be derived for the linear PB equation. The potential distribution has the form

$$\psi = \psi^0 \frac{a}{r} e^{-(r-a)/Le} \tag{29}$$

where $\psi^0 = [a/(a + Le)]\sigma^0$, r is the distance from the sphere center, and a is the sphere radius. Equation (29) has a large significance because it also represents the asymptotic form of any solution to the nonlinear PB equation at distances much larger than the screening length [26]. Hence, at such distances, the electric potential distribution is given by

$$\psi = \frac{kT}{e} \bar{Y}^0 \frac{a}{r} e^{-(r-a)/Le} \tag{30}$$

where the effective potential $\bar{Y}^0$ is described in the case of the sphere by the expression [27]

$$\bar{Y}^0 = 4 \tanh\left(\frac{e\psi^0}{4kT}\right) \left(2\left\{1 + \left[1 - \frac{2a/Le + 1}{(a/Le + 1)^2} \tanh\left(\frac{e\psi^0}{4kT}\right)\right]^{1/2}\right\}^{-1}\right) \tag{31}$$

It is interesting to note that the potential as well as the field distribution around the spherical particle is symmetric, which means, according to Eq. (21) that there is no net force acting on the particle. The force only appears when the symmetry is broken (e.g., for a two-particle system). In this case, Eq. (30) can be exploited for calculating in an approximate way the interactions via the linear superposition method (LSA) discussed later.

However, exact analytical formulas can be derived for two plates bearing different surface charges σ_1^0 and σ_2^0 and immersed in an electrolyte solution of arbitrary composition. The solution of the linear PB equation with the constant charge (c.c.) boundary condition [Eq. (17)] gives the following expression for the electric potential distribution in the gap between the plates:

$$\bar{\psi} = \frac{\bar{\sigma}_2^0 + \bar{\sigma}_1^0 \cosh \bar{h}}{\sinh \bar{h}} \cosh \bar{x} - \bar{\sigma}_1^0 \sinh \bar{x} \tag{32a}$$

where $\bar{\psi} = \psi e/kT$ is the dimensionless potential, $\bar{x} = x/Le$ is the dimensionless distance from one of the plates, $\bar{h} = h/Le$, h is the gap width between the plates, and $\bar{\sigma}_1^0 = \bar{\psi}_1^0$ and $\bar{\sigma}_2^0 = \bar{\psi}_2^0$ are the reduced charges and surface potentials of the plates.

In the case when the constant potential (c.p.) boundary condition, given by Eq. (18) is valid, the potential distribution between the plates assumes the form

$$\bar{\psi} = \bar{\psi}_1^0 \cosh \bar{x} + \frac{\bar{\psi}_2^0 - \bar{\psi}_1^0 \cosh \bar{h}}{\sinh \bar{h}} \sinh \bar{x} \tag{32b}$$

Using the above expressions for the potential distribution, one can calculate from Eq. (22) the uniform pressure between plates (force per unit surface) from the equation

$$\Delta\Pi = kTI\left(\pm\left[(\bar{\psi}_1^0)^2 + (\bar{\psi}_2^0)^2\right]\cos\text{ch}^2\bar{h} + 2\bar{\psi}_1^0\bar{\psi}_2^0\frac{\cosh\bar{h}}{\sinh^2\bar{h}}\right) \tag{33}$$

where the upper sign denotes the c.c. boundary condition and the lower sign denotes the c.p. boundary conditions.

The interaction energy per unit area is accordingly given by

$$\Phi_p = Le\int_\infty^h \Delta\Pi\, d\bar{h} = kTLeI\left(\mp(1 - \coth\bar{h})\left[(\bar{\psi}_1^0)^2 + (\bar{\psi}_2^0)^2\right] + \frac{2\bar{\psi}_1^0\bar{\psi}_2^0}{\sinh\bar{h}}\right) \tag{34}$$

Equations (33) and (34) were first derived for the c.p. case by Hogg et al. [28] and will be referred to as the HHF model. Wiese and Healy [29] and Usui [30] considered the c.c. model, whereas Kar et al. [31] derived an analogous formula for the interaction energy in the case of the "mixed" case (i.e., c.p. on at plate and c.c. at the other).

It is interesting to note that the limiting forms of Eq. (34) for short separations (i.e., for $\bar{h} \to 0$) are

$$\Phi_p = kTLeI\left[\frac{(\bar{\psi}_1^0 + \bar{\psi}_2^0)^2}{\bar{h}} - (\bar{\psi}_1^0)^2 - (\bar{\psi}_2^0)^2\right] \quad \text{c.c. model}$$

$$\Phi_p = -kTLeI\left[\frac{(\bar{\psi}_1^0 - \bar{\psi}_2^0)^2}{\bar{h}} - (\bar{\psi}_1^0)^2 - (\bar{\psi}_2^0)^2\right] \quad \text{c.p. model} \tag{35}$$

It can be easily deduced that the interaction energy for the c.c. model diverges to plus infinity (repulsion) for short separations, whereas the c.p. model predicts diametrically different behavior [i.e., the interaction energy tends to minus infinity (attraction) for the same combination of surface potentials as for the c.c. case].

The difference between both models appearing at short separations seems highly unphysical. It is caused by the violation of the low-potential assumption. Indeed, in order to observe the c.c. boundary conditions, the surface potential of the plates should tend to infinity when they closely approach each other, even if these potentials were very low at large separations. As a consequence, $\bar{\psi} \gg 1$ for $\bar{h} \to 0$ and the linear PB equation is not valid. Hence, Eqs. (33) and (34) are incoherent for the c.c. model and should not be used for short separations.

The deficiency of the linear c.c. model was also demonstrated in Refs. 32 and 33 by analyzing the asymptotic behavior of the nonlinear PB equation in the limit of small plate separation. It was shown that the force and interaction energy of plates can be approximated in the c.c. model by the expressions

$$\Delta\Pi = kTI \frac{2|\bar\sigma_1^0 + \bar\sigma_2^0|}{z\bar h}$$

$$\Phi_p = kTI \left(\frac{2|\bar\sigma_1^0 + \bar\sigma_2^0|}{z}\right) \ln \bar h \tag{36}$$

As can be noted, the interaction energy remains positive at short separations and tends to infinity at a much slower rate (logarithmically) in comparison to the linear model. It can be easily estimated that for $\bar h \ll 0.01$, the differences between the linear and nonlinear models increase to an order of magnitude. It is interesting to mention, however, that in the case of the c.p. boundary conditions, the asymptotic expression for the interaction energy at short separations remains the same for the linear and nonlinear models provided that $\bar\sigma_1^0 \neq \sigma_2^0$. On the other hand, for larger separations, the expressions for the interaction energy reduce to the same asymptotic form for the two boundary conditions

$$\Phi_p = 2kTLeI\bar\psi_1^0\bar\psi_2^0 e^{-h/Le} \tag{37}$$

Thus, the interaction energy between plates decreases exponentially at large separations, similar to that of the electric potential between the plates.

It is also worthwhile noting that for equal plate potentials, the expressions for the force and interaction energy [Eqs. (33) and (34)] become

$$\Delta\Pi = 4kTI\bar\psi_0^2 \frac{e^{-h/Le}}{(1 \mp e^{-h/Le})^2}$$

$$\Phi_p = 4kTILe\bar\psi_0^2 \frac{e^{-h/Le}}{(1 \mp e^{-h/Le})} \tag{38}$$

where the upper sign denotes the c.c. model. Equation (38) was originally derived in Ref 34.

All of the discussed results are valid for metallic plates or plates of infinite thickness when the inside electric potential remains constant. The influence of the finite plate thickness on their interactions was studied in detail by Oshima [35] both under linear and nonlinear regimes. It was shown that for situations of practical interest (aqueous solutions), the correction stemming from finite plate thickness remains negligible.

3. Approximate Models (Derjaguin, LSA, ESA)

As mentioned, the analytical and numerical solutions for the plates can be exploited for constructing approximations for the spherical and anisotropic particle interactions. The anisotropic particle systems are of increasing interest considering that the shape of most bioparticles (e.g., bacteria, viruses, proteins) deviates significantly from the spherical shape [1–5]. Other examples of highly anisotropic particles are the red blood cells, blood platelets, pigments, and synthetic inorganic colloids (gold, silver iodide, silver bromide, barium sulphate, etc.) [36].

An exact determination of interaction energy for spherical and anisotropic particle systems and arbitrary electrolyte composition seems prohibitive due to the nonlinearity of the governing PB equation and the lack of appropriate orthogonal coordinate systems, except for the case of the two-sphere configuration. However, particle and protein adsorption occurs in rather concentrated electrolytes when the electrostatic interactions become short ranged in comparison with particle dimensions. This enables one to apply, for calculating particle interactions, the approximate

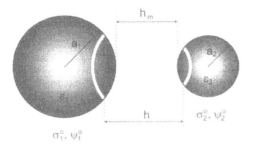

FIG. 1 Schematic representation of the Derjaguin method of calculating electrostatic interactions of dissimilar spheres.

methods such as the Derjaguin summation approach exploiting the results for plates discussed above and the LSA approach.

According to the original Derjaguin method [34], the interactions of spheres were calculated as a sum (integral) of corresponding interactions of infinitesimal surface elements (rings) having a planar geometry. The summation was carried out in the region close to the minimum separation distance h_m (see Fig. 1) by assuming a fast decay of interactions when proceeding further from this region. Thus, the Derjaguin method is only valid if the radii of the spheres a_1 and a_2 are both much larger than the double layer thickness (i.e., when the inequalities a_1/Le and $a_2/Le \gg 1$ hold). Under these assumptions, the force F and interaction energy ϕ of two unequal spheres are given by the relationships

$$F = 2\pi LeG_D\left(\int_{\bar{h}_m}^{\infty} \Delta\Pi(\bar{h}) \, d\bar{h}\right)\hat{n} = 2\pi G_D\Phi_p(h_m)\hat{n}$$

$$\phi = 2\pi LeG_D \int_{\bar{h}_m}^{\infty} \Phi_p(\bar{h}) \, d\bar{h}$$

(39)

where $\bar{h}_m = h_m/Le$ is the dimensionless minimum distance between the spheres, $G_D = a_1a_2/(a_1 + a_2)$ is the geometrical Derjaguin factor equal to 0.5a for two equal spheres and a for the plane/sphere configuration. Note that the force is acting along the vector connecting particle centers

By using Eq. (39) combined with Eqs. (33) and (34), one can derive the explicit expressions for sphere interaction energy between two dissimilar spheres (see Fig. 1) in the form

$$\phi = \pi\varepsilon_0\varepsilon\left(\frac{kT}{e}\right)^2 G_d\left[\mp\left[(\bar{\psi}_1^0)^2 + (\bar{\psi}_2^0)^2\right] \ln(1 - e^{-2h_m/Le}) + 2\bar{\psi}_1^0\bar{\psi}_2^0 \ln\left(\frac{1 + e^{-h_m/Le}}{1 - e^{-h_m/Le}}\right)\right]$$

(40)

where the upper sign denotes the c.c. boundary condition. Note that in contrast to Eq. (34), the interaction energy for the spheres does not depend explicitly on the ionic strength I.

Equation (40) was first derived by Hogg et al. [28] for the c.c. model and Wiese and Healy [29] and Usui [30] for the c.p. model.

It is interesting to note that in the limit $h_n \rightarrow 0$ Eq. (40) becomes

$$\phi = \mp\pi\varepsilon_0\varepsilon\left(\frac{kT}{e}\right)^2 G_D(\bar{\psi}_1^0 \pm \bar{\psi}_2^0)^2 \ln h_m/Le$$

(41)

where the upper sign denotes the c.c. model.

For equal sphere potentials and the c.p. model, Eq. (40) simplifies to the form derived originally by Derjaguin:

$$\phi = \pi\varepsilon_0\varepsilon a\left(\frac{kT}{e}\right)^2 (\bar{\psi}^0)^2 \ \ln(1 + e^{-h_m/Le}) \tag{42}$$

Equations (40)–(42) were commonly used in the literature for determining stability criteria of colloid suspension [28] and for describing the plane–particle interactions in particle deposition problems [37].

The Derjaguin method was generalized by White [38] and Adamczyk et al. [32] to convex bodies of arbitrary shape. The first step of these calculations was determining the minimum separation distance h_m between the two particles involved. Then, the four principal radii of curvature $(R_1', R_1'', R_2', R_2'')$ at the minimum separation region are evaluated. Finally, the interaction energy is calculated as the surface integral of the plate–plate interactions. This leads to the following expression for the generalized Derjaguin factor [38]:

$$G_D = \left(\frac{R_1'R_1''R_2'R_2''}{(R_1' + R_2')(R_1'' + R_2'') + (R_1'' - R_1')(R_2'' - R_2') \ \sin^2 \varphi}\right)^{1/2} \tag{43}$$

where φ is the angle between the x axes in the local Cartesian coordinate system.

In the case of particle–plane or two coplanar particle configurations, one has $\varphi = 0$ and Eq. (43) simplifies to the form derived by Adamczyk et al. [32]. Despite its apparent simplicity, it is very inconvenient to apply Eq. (43) for three-dimensional situations because of mathematical difficulties in finding the points of the minimum separation of the two bodies involved as a function of their mutual orientation and, consequently, to determine h_m. Even for such simple particle shapes as spheroids, one has to solve a high-order nonlinear trigonometric equations, which can only be done in an efficient way by iterative methods [15].

However, analytical results can be derived for limiting orientations of prolate spheroids as shown in Table 1. It is interesting to observe that the ratio between the Derjaguin factors (and hence of the interaction energy) for the parallel and perpendicular orientations of prolate spheroid against a planar boundary equals $1/A^2$ (where $A = b/a$ is the shorter-to-longer axis

TABLE 1 The Derjaguin $\bar{G}_D = G_D/a$ and the ESA $\bar{G}_e^0 = G_e^0/a$ Geometrical Factors for Limiting Prolate Spheroid Orientations

ratio). This means that the electrostatic attraction will be much higher for the parallel orientation (at the same separation distance h_m), so the particles will tend to adsorb parallel.

In the case of electrostatic repulsion (adsorption against an electrostatic barrier), the particles will preferably adsorb under the perpendicular orientation. The same is true for the oblate spheroid adsorption [15].

It is interesting to note that in the case of spheroid–plane interactions, the Derjaguin factor can be evaluated analytically as a function of the orientation angle α. For prolate spheroids, one has

$$G_D = a \frac{A^2}{A^2 \cos^2 \alpha + \sin^2 \alpha} \tag{44a}$$

whereas for the oblate spheroids, the solution is

$$G_D = a \frac{A}{A^2 \cos^2 \alpha + \sin^2 \alpha} \tag{44b}$$

It should be mentioned that the Derjaguin method and, in consequence, all of the data shown in Table 1 are not limited to electrostatic interactions but also describe other interactions where the range remains smaller than the particle dimension (e.g., dispersion interactions is discussed next). However, a serious limitation of the method is that it breaks down at larger particle–particle or particle–wall separations. This leads to an overestimation of the interactions and to a wrong asymptotic dependence of ϕ on the distance h_m.

The disadvantage of the Derjaguin method can be avoided by using the LSA method introduced originally by Bell et al. [26]. The main postulate of this method is that the solution of the PB equation for the two-particle system can be constructed as a linear superposition of the solutions for isolated particles in an electrolyte of infinite extension. This is justified because due to electrostatic screening, the electrostatic potential at separations larger than Le drops to very small values and its distribution can be described by the linear version of the PB equation. As a consequence, the solution of the PB equation in this region for two-particle configuration can be obtained by postulating the additivity of potentials and fields stemming from the isolated particles, that is,

$$\psi = \psi_1 + \psi_2$$
$$\boldsymbol{E} = \boldsymbol{E}_1 + \boldsymbol{E}_2 \tag{45}$$

where ψ_1 and ψ_2 are the solutions of the PB equation derived for isolated particles and $\boldsymbol{E}_1 = -\nabla\psi_1$, and $\boldsymbol{E}_2 = -\nabla\psi_2$ are the electric field vectors due to the particles.

The LSA method can, in principle, be applied for arbitrary particle shape, provided the solution of the PB for isolated particle exists. At present, however, such solutions are known for spheres in a simple 1–1 electrolyte only when the potential distribution is governed by Eqs (30) and (31). Using these solutions, with the field calculated as the gradient of the electric potential, one can derive, via Eq. (21), the following analytical expression for the interaction force between two dissimilar particles [26]:

$$\boldsymbol{F} = \phi_0 a_1 \frac{1 + r/Le}{r^2} e^{-h_m/Le} \frac{\boldsymbol{r}}{r} \tag{46}$$

where

$$\phi_0 = 4\pi\varepsilon_0\varepsilon a_2 \left(\frac{kT}{e}\right)^2 \bar{Y}_1^0 \bar{Y}_2^0,$$

$r = a_1 + r_2 + h_m$ is the distance between particle centers, and $r = r_1 - r_2$ is the vector connecting particle centers. Equation (46) indicates that the force vector is acting along the direction of the relative position vector r (i.e., parallel to the line connecting to the sphere centers).

The interaction energy can easily be calculated by integration of the force, which results in the expression [26]

$$\phi = \phi_0 \frac{a_1}{r} e^{-h_m/Le} \tag{47}$$

This equation has a simple two-parameter form, analogous to the Yukawa potential used widely in statistical mechanics [39]. It remains valid for arbitrary surface potentials and the double layer thickness provided that $h_m/Le \gg 1$. An additional advantage of this formula is that, unlike the HHF expression [Eq. (40)], it never diverges to infinity in the limit $h_m \rightarrow 0$ but is approaching the constant value $\phi_0 a_1/(a_1 + a_2)$, which can be treated as the energy at contact. For colloid particles, the value of ϕ_0 usually varies between 10 and 100 kT. Due to the simple mathematical shape, Eq. (47) is extensively used in numerical simulations of colloid particle adsorption problems.

It seems that the use of the LSA method is equivalent to acceptance of the energy additivity principle i.e., the interactions in the multiparticle systems can be calculated as the sum of contributions stemming from particle pairs (including the limiting case of particle–wall interactions). It should be mentioned, however, that the LSA method and the additivity rule is expected to break down for values of $a/Le < 1$, especially for the particle–wall configuration when the electric field from the interface is penetrating through adsorbed particle. Moreover, due to the large field prevailing in the gap between the particle and the interface, the charge migration effects are likely to appear. Due to mathematical problems in finding appropriate coordinate systems, these true many-body problems have yet not been treated in an exact way.

Another limitation of the LSA method is that it can only be used in the original form for spherical particles. Due to the increasing importance of nonspherical particle interactions, an approximate method to solve this problem has been proposed in Ref. 15. The essence of this approach, being in principle a mutation of the LSA, consists in replacing the interactions of convex bodies by analogous interactions of spheres having appropriately defined radii of curvature. As postulated in Ref. 15 these radii should be calculated as the geometrical means of the principal radii of curvature evaluated at the point of minimum separation between the bodies, that is,

$$R_1 = \frac{2R_1'R_1''}{R_1' + R_1''}$$
$$\tag{48}$$
$$R_2 = \frac{2R_2'R_2''}{R_2' + R_2''}$$

The advantage of this method, referred to as the equivalent sphere approach (ESA), consists in the fact that the known numerical and analytical results concerning sphere interactions can directly be transferred to nonspherical particles. Thus, the LSA results, [Eq. (47)] can be generalized for spheroidal particles to the form

$$\phi = \phi_0 \frac{R_1 R_2}{a(R_1 + R_2 + h_m)} e^{-h_m/Le} = \phi_0 \frac{\bar{G}_e^0}{1 + \bar{G}_e h_m/a} e^{-h_m/Le} \tag{49}$$

where

$$\phi_0 = 4\pi\varepsilon_0\varepsilon a \left(\frac{kT}{e}\right)^2 \bar{Y}_1^0 \bar{Y}_2^0$$

a is the longer semiaxis of the spheroid,

$$\bar{G}_e^0 = \frac{R_1 R_2}{a(R_1 + R_2)} = \frac{2R_1' R_2'' R_2' R_2''}{a[R_1' R_1''(R_2' + R_2'') + R_2' R_2''(R_1' + R_1'')]}$$

$$\bar{G}_e = \frac{a}{(R_1 + R_2)} = \frac{a(R_1' + R_1'')(R_2' + R_2'')}{2[R_1' R_1''(R_2' + R_2'') + R_2' R_2''(R_1' + R_1'')]}$$

(50)

are the two geometrical correction factors.

Although Eq. (49) possesses the simple Yukawa-type form, its application in the general case of spheroid interaction in space is not straightforward because of the necessity of a numerical evaluation of the geometrical functions $\bar{G}_e^0$ and $\bar{G}_e$ [12,15]. However, analogous to the Derjaguin model, these functions can be evaluated analytically for some limiting orientations collected in Table 1.

It is interesting to note that for the spheroid–plane interactions, due to the fact that $\bar{G}_e = 0$, the energy is described by the equation analogous to the Derjaguin formula [Eq. (42)] (at large separations), that is,

$$\phi = \phi_0 \bar{G}_e^0 e^{-h_m/Le}$$

(51)

where the geometrical factor $\bar{G}_e^0$ can be evaluated analytically for prolate spheroids in terms of the inclination angle α as [15]

$$\bar{G}_e^0 = 2A \frac{(G_D/a)^{3/2}}{G_D/a + A^2}$$

(52)

and G_D is the Derjaguin factor given by Eq. (44a).

The dependence of G_e^0/a and G_D/a on α for prolate spheroids is plotted in Fig. 2. As can be seen, for α approaching $90°$ (perpendicular orientation of prolate spheroids), both the Derjaguin and the ESA give similar results because $\bar{G}_e^0$ and G_D/a tend to the same limiting value $(b/a)^2$. Significant deviations occur, however, in the limit of $\alpha \to 0°$ (parallel orientation) when the Derjaguin model predicts $G_D = a$ and the ESA predicts $G_e^0 = 2b/[1 + (b/a)^2]^{-1}$. As discussed in Ref. 13, this discrepancy, increasing for small values of the b/a parameter, suggests that both models give rather inaccurate results for very elongated particles, when $A < 0.2$.

As mentioned earlier, in the case of arbitrary orientation of spheroids, one has to use numerical methods for evaluating the minimum separation distance and calculating the radii of curvature [15]. The use of efficient iterative schemes makes this task quite simple, so tedious simulations for spheroids becomes feasible [15]. Even with this complication, the use of the ESA seems considerably more efficient than any attempt at solving the PB equation for the anisotropic particle case.

4. Comparison of Exact and Approximate Results for Spheres

The estimation of the range of validity of the approximate approaches discussed is rather difficult due to the inherent difficulty of finding exact analytical solutions of the PB equation for the sphere–sphere and sphere–particle geometries. There exist approximate analytical results only, obtained by the perturbation method for two equal spheres [24,40,41]. The resulting analytical expressions are too cumbersome for direct use, however. McCartney and Levine [42] developed the approximate surface dipole integration method, which was extended by Bell et al. [26] and Sader et al. [43] to a dissimilar sphere system. The disadvantage of these analytical solutions valid for low surface potentials is that the geometrical and electrostatic factors stemming from surface potentials are coupled in a nonlinear way.

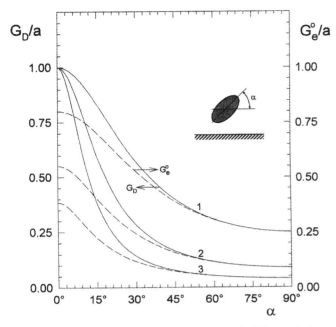

FIG. 2 Comparison of the Derjaguin geometrical factor G_D/a and the ESA factor G_e^0/a (represented by the dashed lines) for prolate spheroid–plane interactions: curve 1, $A = 0.5$, curve 2, $A = 0.3$, curve 3, $A = 0.2$.

However, many numerical solutions of the PB have become available now which can be exploited for estimating the validity of the approximate models. In these calculations, pioneered by Hoskin and Levine [21,44], one uses the finite-difference method and the PB equation is formulated in the bispherical coordinate system. The advantage of this orthogonal coordinate system is that the boundary conditions at the sphere surfaces can be accurately expressed. This coordinate system (with more mesh points) was subsequently used by Carnie et al. [45], who performed calculations of the interaction force for two spherical particles in a 1–1 electrolyte. The authors proved that the electrostatic fields distribution within the particles exerted a negligible effect on interaction force characterized by $\varepsilon < 5$ (e.g., polystyrene latex particles).

The energy of interaction for a dissimilar sphere system (including the important subcase of sphere interaction with a plane) was determined numerically in Ref. 46 using, also, the bispherical coordinate with the fine grid, enabling a large accuracy of calculations. A comparison of these numerical results with various approximate expressions is shown in Figs. 3 and 4. The normalized interaction energy $\bar{\phi} = \phi/(2\pi\varepsilon_0\varepsilon_a(\psi_2^0)^2)$ is plotted as a function of $\bar{h} = h_m/Le$ both for the two-particle system (upper part of these figures) and the particle–interface system (in this case, the energy of interactions was normalized as $\phi/2\pi\varepsilon_0\varepsilon a|\psi_1^0\psi_2^0|$). The results shown in Fig. 4 for $a/Le = 5$ correspond to colloid particles in the micrometer-size range, whereas those presented in Fig. 5 ($a/Le = 1$) reflect the interaction of nanoparticles. The exact numerical data obtained in Ref. 46 are compared with the LSA model [given by Eq. (47)], the linear HHF model [Eq. (40)], and for both the c.c. and c.p. boundary conditions. As can be seen, for $a/Le = 5$, the particle–particle energy interaction profile is well reflected by the LSA mode, whereas the HHF model shows a definite tendency to overestimate the interactions for the c.c. boundary conditions. Also, the particle–interface energy profiles are fairly well reflected by the LSA model with a slightly smaller accuracy for the c.p. boundary conditions. Similar conclusions can be drawn from a comparison of the data shown in Fig. 4 collected for

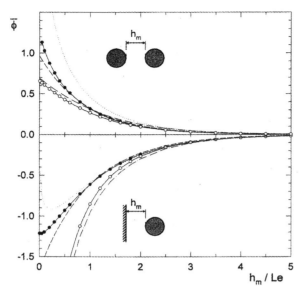

FIG. 3 The reduced interaction energy of two particles $\bar\phi = \phi/2\pi\varepsilon_0\varepsilon a(\psi_2^0)^2$ (upper part) and of particle interface $\bar\phi = \phi/2\pi\varepsilon_0\varepsilon a|\psi_1^0\psi_2^0|$ (lower part)   calculated   from   various   models   for $\bar\psi_1^0 e/kT = 3$, $\bar\psi_2^0 e/kT = -1.5$,   and $a/Le = 5$: $(- \bullet - \bullet - \bullet -)$ exact   numerical   calculations   for   the   c.c.   model, $(-\bigcirc-\bigcirc-\bigcirc-)$ exact numerical calculations for the c.p. model, $(\cdots)$ the linear HHF model, at c.c., $(- \cdot - \cdot - \cdot -)$ the linear HHF model for c.p., $(----)$ the LSA model.

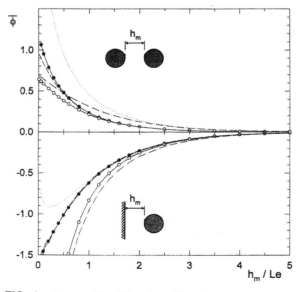

FIG. 4 Same as Fig. 3 but for $a/Le = 1$.

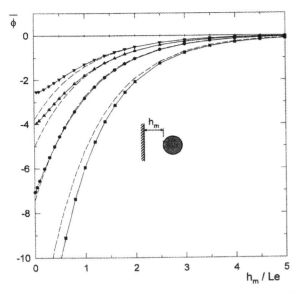

FIG. 5 The reduced particle–interface interaction energy $\bar{\phi}/2\pi\varepsilon_0\varepsilon a|\psi_1^0\psi_2^0|$ calculated for $\psi_1^0 e/kT = 3$ and $a/Le = 1$: the solid symbols denote the exact numerical solutions for the c.c. model calculated for $\psi_2^0 e/kT = -3$ (squares), $\psi_2^0 e/kT = -1.5$ (circles), $\psi_2^0 e/kT = -1$ (triangles), $\psi_2^0 e/kT = -0.75$ (reversed triangles); the dashed lines denote the results calculated from Eq. (47) according to the LSA model.

$a/Le = 1$, although much higher deviations of the HHF model from the exact data are predicted at all distances for the c.c. model, with the LSA model again performing very well.

An interesting feature of the exact numerical results shown in Figs. 3 and 4 is that the c.c. and c.p. models give very similar interaction energy values for the particle–particle case (identical surface potentials), except that for very short distances, $h_m/Le < 0.25$. Moreover, in the c.c. model, the exact energy value remains finite in the limit $h_m/Le \to 0$, which contrasts with the linear HHF model predicting a logarithmically diverging interaction energy in the limit $h_m \to 0$ [cf. Eq. (41)]. In view of these results, the long-lasting controversy in accepting the c.c. or c.p. model seems rather immaterial.

It should be mentioned, however, that the accuracy of the LSA approximation is strongly influenced by the surface potential asymmetry in the case of particle–wall interactions. This is illustrated in Fig. 5, in which one can see that the LSA reflects well the exact results if the absolute values of the surface potentials of particle and the interface do not differ too much. For the potential asymmetry exceeding $2:1$, the LSA overestimates the attraction energy for distances $a/Le < 1$.

It may be concluded from the data presented in Figs 3–5 and similar results for the interaction force discussed in Ref. 45 that significant differences between the LSA and the exact results occur at distances smaller than a/Le only where the interaction energy assumes very large absolute values, either positive (similar surface potentials) or negative (opposite surface potentials). In both cases, a relatively large uncertainty in ϕ can be tolerated. Moreover, the surface deformation, roughness, and charge heterogeneity effects are expected to play a decisive role.

One should also mention that the above results are applicable for polar media when the screening length remains comparable to particle dimensions (i.e., when $a/Le > 0.1$). For nonpolar liquids or gases the interactions are described by different models, as discussed next.

5. Interactions in Nonpolar Media

Assuming a negligible ion concentration, one can describe the force between two charged particles (e.g., liquid droplets in a gas) from the Coulomb law:

$$F = \frac{q_1 q_2}{4\pi\varepsilon_0\varepsilon r^2}\left(\frac{r}{r}\right) = 4\pi\varepsilon_0\varepsilon a_1 a_2 \frac{\psi_1^0\psi_2^0}{r^2}\left(\frac{r}{r}\right) \tag{53}$$

where q_1 and q_2 are the net charges on particles and ψ_1^0 and ψ_2^0 are the potentials of the particles. As can be noted, the force is acting along the line connecting particle centers. Integration of Eq. (53) gives the expression

$$\phi = 4\pi\varepsilon_0\varepsilon a_1 a_2 \frac{\psi_1^0\psi_2^0}{r} \tag{54}$$

for the interaction energy. It is interesting to observe that this expression becomes identical with Eq. (47) derived on the basis of the LSA, in the limit of $Le \to 0$ and small surface potentials (when $\psi_2 e/kT < 1$).

In the case when the size of one of the particles tends to infinity $(a_1/a_2 \gg 1)$, which corresponds to interactions of colloid particles with a charged macroscopic sphere (electrode), the force is given by the Lorenz equation

$$F = q_2 E = 4\pi\varepsilon_0\varepsilon a_2 \psi_2^0 E \tag{55}$$

where

$$E = \left(\frac{q_1}{4\pi\varepsilon_0\varepsilon a_1^2}\right)\frac{r}{r}$$

is the field due to the large sphere.

Obviously, Eq. (55) also describes the situation when the colloid particle is placed in a uniform electric field $|E| = (\psi_1 - \psi_2)/h$ (e.g., between two planar electrodes having the potentials ψ_1 and ψ_2 and separated by the distance h). Note that according to Eq. (55), the force does not depend on the distance from the large sphere (provided that it remains smaller than its radius a_1) analogously to that for the gravity force, so it can be treated as a uniform external force.

The interaction energy is given by

$$\phi = 4\pi\varepsilon_0\varepsilon a_2 \psi_2^0 |E| h_m \tag{56}$$

Even if the electrode (large sphere) is uncharged but conductive, there arises an induction (image) force on a charged colloid particle, which is described by the equation

$$F = -\frac{q_2^2}{4\pi\varepsilon_0\varepsilon(2r)^2}\frac{r}{r} = -4\pi\varepsilon_0\varepsilon a_2^2 \frac{(\psi_2^0)^2}{(2r)^2}\frac{r}{r} \tag{57}$$

Hence, the force is identical to the Coulomb force between two charged particles of radius a_2 separated by the distance $2r = 2a_2 + h_m$. The interaction energy is therefore

$$\phi = -4\pi\varepsilon_0\varepsilon a_2^2 \frac{(\psi_2^0)^2}{2r} \tag{58}$$

In contrast to the double-layer interactions, in nonpolar media even uncharged particles (droplets) will be attracted to charged interfaces (electrodes) due to the dielectric polarization effect. The force is described by the general expression

$$F = \nabla(p \cdot E) \tag{59}$$

where p is the induced dipole moment of the dielectric sphere.

For the sphere–planar electrode geometry, the induced dipole can be approximated at separations larger than the radius of the colloid sphere by the formula [47]

$$p = 4\pi\varepsilon_0\varepsilon\,\frac{\varepsilon_p - \varepsilon}{\varepsilon_p + 2\varepsilon}\,a^3 E \tag{60}$$

where ε_p is the dielectric constant of the particle. This results in the following expression for the interaction force directed perpendicularly to the electrode:

$$F = -\frac{3\pi}{2}\varepsilon_0\varepsilon\,\frac{\varepsilon_p - \varepsilon}{\varepsilon_p + 2\varepsilon}\,a^2 E^2\,\frac{1}{(1 + h_m/a)^4} \tag{61}$$

Integrating this relationship, one obtains, for the interaction energy, the expression

$$\phi = -\frac{\pi}{2}\varepsilon_0\varepsilon\,\frac{\varepsilon_p - \varepsilon}{\varepsilon_p + 2\varepsilon}\,a^3 E^2\,\frac{1}{(1 + h_m/a)^3} \tag{62}$$

For smaller separations, the force on a dielectric sphere in a uniform field E can only be calculated numerically using the bispherical coordinate system. The results plotted in Fig. 6 indicate that at $h_m/a < 1$, the force decreases with the distance less abruptly than for larger distances and significantly depends on the relative dielectric constant $\varepsilon_p/\varepsilon$. In the limit of $\varepsilon_p/\varepsilon \to 0$, the force can be interpolated by the function

$$F = -1.2\pi\varepsilon_0\varepsilon a^2 E^2\,\frac{1}{(h_m/a)^{0.8}} \tag{63}$$

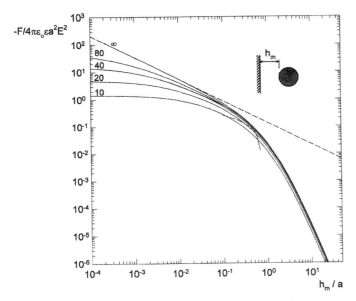

FIG. 6 The reduced interaction force $-F/4\pi\varepsilon_0\varepsilon a^2 E^2$ versus the reduced distance h_m/a calculated numerically for a dielectric particle (dielectric constant ε_p) interacting with an electrode through the dielectric medium (dielectric constant ε). The lines show the results calculated for $\varepsilon_p/\varepsilon = \infty$: 80 (water drop in air), 40, 20, and, 10, respectively. The broken lines represent the limiting solutions for short distances [Eq. (63)] and for long distances [Eq. (61)].

Integrating this, one obtains, for the interaction energy,

$$\phi = -6\pi\varepsilon_0\varepsilon a^3 E^2 \left[1 - \left(\frac{h_m}{a}\right)^{0.2} \right] \tag{64}$$

valid for $h_m/a < 1$. As can be noted, the polarization energy increases as a square of the field E and attains considerable values for high voltages of the electrode, as is the case in gaseous media (electrofiltration processes). However, the energy decreases abruptly with the distance, as h_m^{-4} [cf. Eq. (61)] and becomes practically negligible for $h_m/a > 2$.

It is worthwhile mentioning that due to the polarization effect, dielectric particles placed in a uniform electric field will interact among themselves as well. It can be predicted [47,48] that this force is attractive (negative) for the particle orientation parallel to the field direction (perpendicular to the electrodes). On the other hand, in the perpendicular to the field direction, particles will repel each other. The force vanishes proportional to h_m^{-4}, analogously to that for the particle–plane configuration [Eq. (61)]. These effects may lead to linear aggregate formation or to so-called electrocoalescence of droplets in a uniform electric field [48].

Because of the analogy of the underlying equation, the same polarization effects will appear in the case of uniform magnetic fields, which leads to aggregation of paramagnetic particle suspension and their interactions with magnet poles.

B. The Dispersion Interaction

The dispersion interactions, also called London–van der Waals forces, appear for all material bodies—in particular atoms and molecules due to spontaneous fluctuations in the electromagnetic field (two induced dipole interactions). As discussed in length in Refs 17, 18, 49, and 50, there exist essentially two main approaches aimed at calculating these interactions:

1. The classical microscopic approach, developed mainly by Hamaker [51], exploiting the energy additivity principle to derive solutions for more complex geometries from known solutions for atoms and molecules.
2. The macroscopic approach introduced by Lifshitz [52], treating the interacting body as a continuum and exploiting the imaginary part of the dielectric constant expressed as a function of the radiation frequency for determining the material constants characterizing the magnitude of the interactions

The advantage of the microscopic approach, used almost exclusively in the literature, is that analytical formulas can be derived for complicated geometries of the interacting particles, including the case of rough surfaces. In contrast, the more rigorous macroscopic approach can only be applied for half-space interactions.

The starting point of the microscopic theory is the expression for the interaction of two different atoms in a vacuum, derived by London [53]:

$$\phi_a = -\frac{3}{2}\hbar\alpha_1\alpha_2 \frac{\omega_1\omega_2}{\omega_1 + \omega_2}\frac{1}{r^6} = -\frac{\beta_{12}}{r^6} \tag{65}$$

where $\hbar$ is the Planck constant, α_1 and α_2 are the polarizabilities of the atoms, ω_1 and ω_2 are the characteristic oscillation frequencies, β_{12} is the London constant, and r is the distance between the atom centers.

Due to the omission of the retardation effect, Eq. (65) remains valid only for distances between atoms shorter than the characteristic wavelength λ_a, being of the order 100–150 nm. It was shown in Ref. 54 that by considering the retardation effect, the interaction energy of atoms is given by the approximate expression valid for $2\pi r / \lambda_a > 0.5$:

$$\phi_a = -\frac{\beta_{12}}{r^6}\left[\frac{2.45\lambda_a}{2\pi r} - \frac{2.17\lambda_a^2}{4\pi^2 r^2} + \frac{0.59\lambda_a^3}{8\pi^3 r^3}\right] \tag{66}$$

At larger distances, the leading term obviously dominates and one recovers the formula derived originally by Casimir and Polder [55]. The force of interactions between atoms can be obtained directly from the above formulas by a simple differentiation in respect to r.

Using Eqs. (65) and (66), one can derive the expressions for microscopic bodies of arbitrary shape by calculating the volume integral

$$\phi = \iint\limits_{v_1 v_2} \phi_a \rho_1 \rho_2 \, dv_1 \, dv_2 \tag{67}$$

where ρ_1 and ρ_2 are the number densities of the atoms in the two bodies involved and v_1 and v_2 are the volumes of the bodies.

Applying Eq. (67) with ϕ_a given by Eq. (65) to the two half-space case, one obtains the expression for the unretarded interaction energy per unit area in the form

$$\phi = -\frac{A_{12}}{12\pi h^2} \tag{68}$$

where $A_{12} = \pi^2 \beta_{12}\rho_1\rho_2$ is the Hamaker constant. For the retarded case using Eq. (66) (with the leading term only), one obtains

$$\phi = -\frac{A_{12}}{\pi^2 h^3}\left[\frac{2.45\lambda_a}{60} - \frac{2.17\lambda_a^2}{240\pi h} + \frac{0.59\lambda_a^3}{840\pi^2 h^2}\right] \tag{69}$$

Because the dispersion interactions are usually of much shorter range than colloid particle dimensions, one can use Eqs. (68) and (69) in conjunction with the generalized Derjaguin method to derive the expressions for ϕ in the case of arbitrary convex bodies (the radius of curvature of which is larger than approximately 10 nm). In this way, one obtains

$$\phi = -G_D \frac{A_{12}}{6h_m} \tag{70}$$

where G_D is the generalized Derjaguin factor given by Eq. (43) and shown in Table 1 for limiting configurations of spheroids.

On the other hand, the exact result which can be derived by evaluating the volume integral [Eq. (67)] for two spherical particles in the nonretarded case is

$$\phi = -\frac{A_{12}}{6}\left[\frac{2a_1 a_2}{h_m(h_m + 2a_1 + 2a_2)} + \frac{2a_1 a_2}{(h_m + 2a_1)(h_m + 2a_2)} + \ln\left(\frac{h_m(h_m + 2a_1 + 2a_2)}{(h_m + 2a_1)(h_m + 2a_2)}\right)\right] \tag{71}$$

In the case of retarded interactions, the Derjaguin expression reads

$$\phi = -G_D \frac{A_{12}}{\pi h_m^2}\left[\frac{2.45\lambda_a}{60} - \frac{2.17\lambda_a^2}{360\pi h_m} + \frac{0.59\lambda_a^3}{1680\pi^2 h_m^2}\right] \tag{72}$$

For this geometry, the exact results are too cumbersome and are not presented here.

For the sphere–plane interactions, the exact results can be derived by substituting $a_1 = a$ and letting $a_2 \to \infty$. In this way, for the nonretarded and retarded interactions one obtains

$$\phi = -\frac{A_{12}}{6}\left[\frac{a}{h_m} + \frac{a}{2a + h_m} + \ln\left(\frac{h_m}{2a + h_m}\right)\right] \tag{73a}$$

$$\phi = -\frac{A_{12}}{\pi h_m^2}\left\{\frac{2.45\lambda_a}{60}\left[(a - h_m) + \frac{(h_m + 3a)h_m^2}{(h_m + 2a)^2}\right] - \frac{2.17\lambda_a^2}{720\pi}\left[\frac{2a - h_m}{h_m} + \frac{(h_m + 4a)h_m^2}{(h_m + 2a)^3}\right]\right.$$
$$\left. + \frac{0.59\lambda_a^3}{5040\pi^2}\left[\frac{3a - h_m}{h_m^2} + \frac{(h_m + 5a)h_m^2}{4(h_m + 2a)^4}\right]\right\} \tag{73b}$$

respectively. Because the expression for the retarded interactions are rather cumbersome for direct use, Suzuki et al. [56] derived an approximate equation having the simpler form

$$\phi = -\frac{A_{12}}{6}\frac{a}{h_m(1 + 11.11\lambda_a/h_m)} \tag{74}$$

For calculating the interactions for the sphere–plane geometry one can also use the Derjaguin model, expressed by Eqs. (70) and (72) by substituting $G_D = a$.

The corresponding expressions for the force of interaction can be derived easily from the above formulas by a simple differentiation with respect to the distance h_m.

The above-presented expressions are strictly valid for particle interactions in a vacuum (gas). As discussed in Refs. 17 and 18, however, interactions of particles across a continuous medium can be characterized by the same functional dependencies with the "composite" Hamaker constant A_{102}, denoting interactions between particle of material 1 interacting with particle of material 2 across the intervening medium 0. If the medium is a liquid, the Hamaker constant becomes much smaller than for particle interactions in a gas.

Moreover, the composition of a liquid may change at liquid–solid boundaries due to, for example, electrolyte concentration changes. This is expected to influence, to some extent, the value of the Hamaker constant. Thus, in the general case, the electrostatic and dispersion interactions are coupled in a complex nonlinear way, which violates the assumption of the DLVO theory. Due to the lack of appropriate theories, we should accept the hypothesis that this coupling is not too significant.

It is interesting to mention that the above formulas can be used for calculating dispersion interactions of geometrically heterogeneous (rough) particles. Although exact results for arbitrary statistical distribution of microroughness were found to be rather complicated [57], Czarnecki and Dąbroś [58] and Czarnecki [59] have derived a simple interpolating function for ϕ_r, valid for both sphere–sphere and sphere–plane interactions:

$$\phi_r = \phi\left(\frac{2h_m}{2h_m + b_1 + b_2}\right)^c \tag{75}$$

where ϕ is the above-discussed energy for smooth particles, h_m is the distance measured between the two outermost points at the particle surfaces, b_1 and b_2 are the thicknesses of the rough layer at particles 1 and 2, respectively, and c is the exponent, close to 1 for the unretarded case [59] and 1.5 for the retarded case [58]. Thus, in the limit $h_m \to 0$, Eq. (75) reduces to the simple form (for unretarded interactions)

$$\phi_r = \phi\left(\frac{2h_m}{b_1 + b_2}\right) \tag{76}$$

It can be easily deduced from Eq. (76) that, in accordance with intuition, the dispersion interactions between rough bodies are substantially reduced at all separations in comparison with smooth particles.

C. Superposition of Interactions and the Energy Profiles

The basic assumption of the DLVO theory used for interpretation of the particle aggregation and adsorption phenomena [17,18] is that the net interaction energy can be obtained as a super-position of the electrostatic and dispersion contributions discussed earlier. All additional short-range forces are neglected—in particular, the Born repulsion, the forces arising due to surface deformations, chemical interactions, and so forth.

Because, for most known cases, the Hamaker constant assumes positive values, one can deduce from Eqs. (68)–(75) that the dispersion contribution to the net interaction energy is negative (attraction) at all particle separations. Moreover, due to the insensitivity of the Hamaker constant on material properties (except for metals, it is usually confined within the range 5×10^{-21} to 5×10^{-20} J), the range of dispersion interactions is fairly fixed and equal to 10–20 nm. In contrast, the electrostatic interactions can be either positive or negative depending on surface potentials, separation distance, and the orientation (shape) of particles. Additionally, their range can be varied between broad limits (typically 10–1000 nm) by changing the ionic strength of the electrolyte solutions. As a result, the superposition of the dispersion and electrostatic interactions may lead to complicated DLVO energy profiles, as discussed in Ref. 24. For the sake of convenience, these profiles can be classified into categories, as done in Refs. 12 and 60.

The profiles, of Type I (see Fig. 7) reflect interactions of equal particles in not too concentrated electrolyte solutions when the overall interaction energy is dominated by the electrostatic interactions. In this case, the energy increases monotonically when the particles, approach each other, attaining values much higher than the kT unit for $h_m \to 0$. The appearance of this profile, typical for particle sizes above 0.1 μm, excludes the possibility of particle aggregation, so the colloid suspension remains indefinitely stable in time. This is advantageous

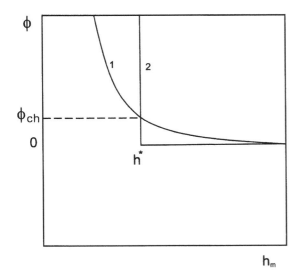

FIG. 7 The "repulsive" interaction energy profile of Type Ia approximated by the hard-particle potential (curve 2).

for performing particle adsorption experiments because the bulk particle concentration remains constant and no aggregates appear in the colloid suspension.

It is often convenient by analyzing the surface-blocking effects due to adsorbed particles to replace the energy profile of Type I (soft interaction profile) by the idealized profile called the hard-particle interaction (see Fig. 7). According to this model, introduced originally by Barker and Henderson [61], the interaction energy remains zero except for the critical distance h^*, where it tends to infinity. Physically, this means that the interacting particles can be treated as hard ones having the equivalent dimensions increased over the true geometrical dimensions by the small value h^* (skin), which can be treated as the effective interaction range. Therefore, this concept is often referred to as the effective hard-particle (EHP) model. Using the LSA expression for the interaction energy, one can derive the following expression for h^* in the case of two dissimilar spherical particles:

$$h^* = \frac{1}{2} Le \left\{ \ln \frac{\phi_0'}{\phi_{ch}} - \ln \left[1 + \frac{Le}{a_1 + a_2} \ln \left(\frac{\phi_0'}{\phi_{ch}} \right) \right] \right\} \tag{77}$$

where

$$\phi_0' = \phi_0 \frac{a_1}{a_1 + a_2}$$

and ϕ_{ch} is the characteristic interaction energy close to the kT unit (see Fig. 7). Equation (77) indicates that h^* is proportional to Le and the proportionality coefficient is about 2 for colloid suspensions when $\phi_0'/\phi_{ch} \cong 100$. The numerical calculations discussed later demonstrate that the EHP concept is a powerful method for analyzing the adsorption of interacting particles of spherical and nonspherical shapes.

On the other hand, particle adsorption phenomena are reflected by the energy profile of Type Ib (shown in Fig. 8) when the interaction energy decreases monotonically, attaining large negative values when $h_m \to 0$ (this reflects the attraction between the particle and the interface). This profile appears in systems when the particle and interface bear opposite surface charges. In

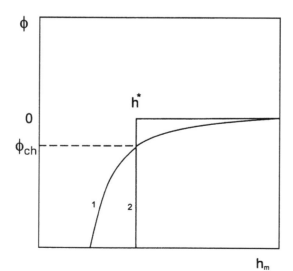

FIG. 8 The "attractive" interaction energy profile of Type Ib approximated by the EHP energy profile (curve 2).

order to simplify the mathematical analysis of particle-transport phenomena, this energy profile is often idealized by introducing the perfect sink (PS) model, as done originally by Smoluchowski [62] in his fast coagulation theory. According to this approach, the interaction energy remains zero up to a critical distance δ_m; where it becomes minus infinity (see Fig. 8). Smoluchowski assumed originally $\delta_m = 0$. This model can be improved upon by identifying δ_m with the effective interaction h^*, analogous to that for particle–particle interactions. In this way, one can interpret the enhanced particle transport due to attractive double-layer interactions discussed later.

A better, although more tedious, alternative is to consider the true interaction energy distribution up to the point $h = \delta_m$ where it is assumed to become minus infinity. This model has often been used in numerical calculations of colloid deposition at various macroscopic surfaces.

Obviously, both the energy profile of type I and the PS model should be treated as an idealization of any real situation because, at very small separations, the interaction energy must become positive due to the Born repulsion preventing particle–wall penetration. In the DLVO theory, these repulsive interactions were not considered. Even at the present time, no quantitative theory of these interactions for macroscopic objects have been developed. Assuming that the repulsive part of the potential is described for atoms and molecules by the 6–12 power law, one may expect that for the particle–wall interactions, $\phi \sim r^{-7}$. These interactions seem, therefore, very short ranged, probably not exceeding 0.5–1 nm.

In any case, the appearance of the repulsive interactions fixes the minimum value of the interaction energy, which remains finite in accordance with intuition. This minimum energy value is often referred to as the primary minimum, denoted by ϕ_m, and the distance where it appears is called the primary minimum distance δ_m (see Fig. 9). One may expect that δ_m is of the order of the range of the Born repulsive forces (i.e., 0.5–1 nm). However, the extension of the region where the interaction energy assumes a negative value can be much larger, comparable with the Debye screening length (i.e., about 100 nm for a 10^{-5} M electrolyte solution).

The appearance of the primary energy minimum would physically explain reversible adsorption of colloid particle, at least under static, no-flow conditions. However, for the flowing colloid system, as is the case for most practical applications, the situation becomes conceptually more complicated because neither the classical DLVO nor the theory with inclusion of Born repulsion would explain particle immobilization under vigorous shearing forces [63,64]. Thus, the particles accumulated at the interface would easily be removed by the tangential fluid flow. One has to accept somehow, ad hoc, the appearance of strong tangential interactions most probably due to short-ranged geometrical and charge heterogeneities [63,64]. It is difficult to estimate the magnitude of the local energy sinks responsible for irreversible adsorption of particles, although it can be predicted that their depth will be a fraction of ϕ_m. As discussed in Ref. 37, these tangential interactions exert a rather minor influence on adsorption kinetics of colloid particles. They are expected, however, to influence considerably the maximum size of particle attached to the surface under a given flow shear rate. To simplify our considerations, we assume in due course that these interactions are strong enough to keep adsorbed particles fixed at a given position (localized adsorption postulate).

In the case when the interface and particle bear opposite charge, the energy profile of Type II is likely to appear (see Fig. 9). The characteristic feature of the profile is the appearance of a maximum energy barrier of height ϕ_b at the distance δ_b. This energy profile corresponds to the activated transport conditions in the chemical kinetics. Obviously, the height of the barrier is very sensitive to the electrolyte concentration and composition (presence of polyvalent ions), the Hamaker constant, and particle size, shape, and orientation. The presence of the energy barrier reduces the particle adsorption rate considerably, which may become inaccessible for accurate measurements. Therefore, in this case, the experimental measurements are often difficult to

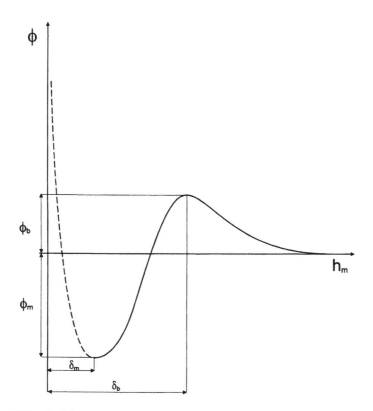

FIG. 9 The energy profile of Type II characterized by the energy minimum of depth ϕ_m and the energy barrier of height ϕ_b.

interpret in terms of the classical DLVO theory. It should also be mentioned that due to a large transport resistance induced by the barrier, the bulk-transport conditions will be less important in this case.

For low electrolyte concentration and large particles (micrometer-size range), a situation may arise when the so-called secondary minimum appears at the distance δ_{sm} much larger than δ_m. Obviously, this minimum is more shallow than the primary minimum due to the smaller dispersion energy contribution at this distance. This type of energy profile will be referred to as Type IIA. Fundamentally, there is not much difference in Types II and IIa energy profiles. In the latter case, additional accumulation of particles around δ_{sm} is expected, which could influence particle adsorption kinetics. As shown in Refs. 63 and 64, however, due to flow, the concentration peak within this region does not become significant.

As this qualitative analysis suggests, it is generally more efficient to analyze the influence of the energy profile on particle adsorption phenomena than to analyze the influence of numerous physicohemical parameters influencing ϕ_m, δ_m, and δ_{sm}. This approach will generally be adopted in further parts of our work dealing with quantitative analysis of adsorption kinetics.

It should also be mentioned that the above-discussed results are valid for idealized systems of perfectly smooth surfaces characterized by uniform charge distribution and lack of deformations upon approach. When dealing with real systems (e.g., colloid suspensions), these assumptions are likely violated because many complicating effects appears:

1. Heterogeneity of charge distribution at interacting surfaces, which can be of a microscopic scale (of chemical origin) or macroscopic patchwise scale; also, considerable differences within particle populations are expected to appear in respect to, for example, average charge.
2. Surface roughness, either of a well-defined geometrical shape or of a statistical nature.
3. Surface deformations upon approach, which are particularly important for polymeric colloids (lattices) characterized by a low Young modulus value.
4. Dynamic relaxation phenomena of the double layer upon contact (aging effects) due to ion migration along the surface or from one surface to another; these processes are, again, expected to appear for polymeric colloids having the random-coil structure.

Despite the practical significance of these effects in particle adsorption processes, little effort has been devoted to quantify them. The role of microscopic nonuniformity of charge distribution (discrete charge effect) was studied by Levine [65]. Similar problem was considered in Ref. 66, where the interaction energy between parallel plates with a discrete charge distribution forming a two-dimensional lattice was studied using the linearized PB equation. It was found that the discrete charges generated a larger interaction potential in comparison with uniformly charged surfaces.

The effect of heterogeneity of charge distribution on particle–surface interactions was studied by Song et al. [67] by considering two simple models of a macroscopic patchwise heterogeneity and the microscopic model when the charge distribution was described by the Gaussian probability distribution. The interactions were simply calculated from the usual expressions stemming from the DLVO theory by introducing the local values of the surface potential.

The effect of geometrical surface roughness was studies in more detail. Krupps [68] was probably the first to qualitatively consider the effect of a hemispherical asperity on the adhesion force of a smooth colloid particle. It was concluded that the attractive electrostatic interactions will be decreased to a lesser extent than the dispersion forces, so that the net adhesion force should be determined by the electrostatic component.

Czarnecki [59] and Czarnecki and Warszyński [69] used the additivity principle for modeling the interaction of a smooth sphere with a heterogeneous (rough) planar surface. The rough surface was generated by distributing, at random over a smooth surface, a number of small spherical particles having various sizes and charge distributions governed by the Gaussian law. Considerable differences in the interaction energy of the sphere were predicted at various spots of the interface, which could explain the appearance of specific tangential interactions in particle deposition processes.

The energy additivity rule was also exploited by Herman and Papadopolous [70], who determined the effect of conical and hemispherical asperities on the van der Waals and electrostatic interaction between flat plates using the LSA approach combined with the Derjaguin summation method. This approach was generalized to the case of a rough colloid particle–smooth surface [71]. A formula for the electrostatic interaction energy based on the LSA/Dejaguin approach was derived.

Apparently, the effect of surface deformations has not been treated in the literature. One may suppose that in the case of repulsive interactions, ϕ is expected to increase over the values for perfectly rigid bodies as a result of surface deformations. This would be so because the radius of curvature around the minimum separation area increases due to the surface-flattening effect. Thus, upon making physical contact, the adhesion energy will likely increase [71]. On the other hand, in the case of attraction, the absolute value of the interaction energy will decrease due to a decrease in the curvature radius.

Despite these efforts, no coherent theory of the non-DLVO interactions has emerged yet. The existing approaches are, in principle, based on a simple extension of DLVO interactions derived by considering the local geometry of the interacting bodies or local rather than macroscopic charge distribution.

Additional effects leading to non-DVLO interactions appear in systems containing polymer solutions. The adsorption of charged polymers (polyelectrolytes) leads to the heterogeneity of charge distribution both on the particle and the interface, which promotes either particle flocculation or deposition at surfaces [72]. On the other hand, the addition of neutral polymers, due to their adsorption, may lead to increased stability of particle suspensions and lack of deposition at surfaces. This effect, often called steric stabilization [17,18,73], is due to repulsive interactions between penetrating polymer chains. It extends over distances comparable with the polymer layer thickness and leads to a steric barrier of a considerable height, depending on the polymer–solvent interaction parameter [73]. Particle adsorption under steric stabilization conditions can, therefore, be well reflected by the Type II energy profile.

D. The Hydrodynamic Interactions

As discussed earlier, the dispersion and the double-layer interactions are short ranged, influencing particle transport over distances much smaller than their dimensions. Therefore, in order to attain measurable adsorption rates, an additional transport mechanism must appear, responsible for particle transfer over macroscopic distances. One of the most efficient modes of particle transport is a macroscopic motion of the suspending fluid (convection) generating hydrodynamic forces and torques on particles. The fluid motion is usually provoked by external influences like the hydrostatic pressure difference, gravity (sedimentation of larger particles), or motion of an interface (e.g., the rotating disk). The rate of convection and, consequently, the forces acting on particles can be quantitatively evaluated upon solving the governing Navier–Stokes equation

$$\frac{\partial(\rho V)}{\partial t} + \nabla \cdot (\rho V V) = \nabla \cdot \Pi + F$$

$$\frac{\partial \rho}{\partial t} + \nabla \cdot (\rho V) = 0 \tag{78}$$

where p is the fluid density, t is the time, V is the fluid flow velocity vector, Π is the pressure tensor, and F is the body force exerted on the fluid.

For incompressible liquids, the pressure tensor is given by [74]

$$\Pi = -pI + 2\eta\Delta \tag{79}$$

where p is the isotropic pressure, η is the fluid dynamic viscosity, and Δ is the rate of the deformation tensor expressed in terms of the fluid velocity gradient as

$$\Delta = \tfrac{1}{2}[\nabla V + \nabla V^t] \tag{80}$$

where t indicates the transposed matrix.

Substituting this into Eq. (78), one obtains, for an incompressible fluid, the equations

$$\rho\left(\frac{\partial V}{\partial t} + V \cdot \nabla V\right) = -\nabla p + \eta\nabla^2 V + F$$

$$\nabla \cdot V = 0 \tag{81}$$

To solve Eq. (81) for the fluid velocity V, the initial velocity field and boundary condition must be specified. If the fluid velocity is known, one can calculate the hydrodynamic force F_h and torque T_h on an arbitrary surface (particle) immersed in the flow from the constitutive dependencies [74]:

$$F_h = \iint\limits_{S} (\Pi \cdot \hat{n})\, dS$$

$$T_h = \iint\limits_{S} (Fxr) \cdot \hat{n}\, dS \tag{82}$$

As can be deduced, the Navier–Stokes equation is nonlinear with respect to fluid velocity, so its exact analytical solutions can be derived for a few simple geometries only. In most cases, it is necessary to apply sophisticated numerical methods. However, for many situations of practical interest, the fluid motion becomes steady after a short transition time and the explicit time derivative $\partial V/\partial t$ in Eq. (81) can be neglected. Moreover, there exists a broad class of flows characterized by a very small length and velocity scales L_{ch} and V_{ch}, respectively. Then, the ratio of the nonlinear inertial term $\rho V_{ch}^2/L_{ch}$ to the viscosity term $\eta V_{ch}/L_{ch}^2$, defined as the Reynolds number $\mathrm{Re} = V_{ch}L_{ch}/\nu$, becomes smaller than unity (where $\nu = \eta/\rho$ is the fluid kinematic velocity). This means that the inertia term can be neglected in Eq. (81), which assumes, under the steady-state conditions, the simple form

$$\eta \nabla^2 V = \nabla p - F$$

$$\nabla \cdot V = 0 \tag{83}$$

As can be noted, Eq. (83), often called the creeping flow or Stokes equation [74], is linear, which facilitates its analytical solution. Equation (83) is therefore often used for describing the microhydrodynamic behavior of particles (e.g., their motion in the vicinity of interfaces).

We discuss in some detail the macroscopic flows for simple geometries of the interfaces. These flows have major practical significance for various experimental studies concerned with particle adsorption. Other macroscopic flows of practical interest are discussed extensively in the monographs of van de Ven [75] and Elimelech [18], and in review articles (Refs. 1,7, and 76).

1. Macroscopic Flow Fields

One of the few cases of practical interest which can be evaluated analytically is the laminar flow in the channel of rectangular cross section $2b \times 2c$ driven by the hydrostatic pressure difference (see Fig. 10). As discussed in Ref. [77] the stationary velocity profile is established for the distance from the inlet larger than $0.16b^2 V_\infty/\nu$ (where V_∞ is the mean fluid velocity at the entrance to the channel). Then, the inertia term in the Navier–Stokes equation vanishes and Eq. (81) becomes linear. Its solution gives the following expression for the fluid velocity vector directed along the x axis [77]:

$$\frac{V_x}{V_{mx}} = 1 - \left(\frac{y}{b}\right)^2 - \frac{32}{\pi^3} \sum_{n=0}^{\infty} \frac{(-1)^n}{(2n+1)^3} \cos\frac{(2n+1)\pi y}{2b} \frac{\cosh\frac{(2n+1)\pi z}{2b}}{\sinh\frac{(2n+1)\pi c}{2b}} \tag{84}$$

where $V_{mx} = \Delta P b^2/2\eta L$ is the maximum velocity in the middle of the channel, $\Delta P/L$ is the uniform pressure gradient along the channel, and L is the channel length (see Fig. 10).

One can deduce from Eq. (84) that for the important limiting case of $b/c \ll 1$ (thin channel), the fluid velocity distribution becomes parabolic, that is,

$$V_x = \frac{\Delta P b^2}{2\eta L}\left[1 - \left(\frac{y}{b}\right)^2\right] = V_{mx}\left[1 - \left(\frac{y}{b}\right)^2\right] \tag{85}$$

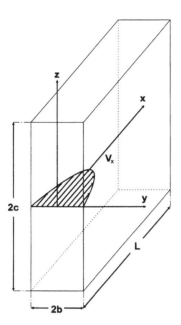

FIG. 10 A schematic representation of the laminar flow in the parallel-plate channel.

It is interesting to observe that the flow in the region close to the channel center (when $y/b \ll 1$) can be treated as a uniform flow with velocity independent of position.

On the other hand, close to the channel walls when $y \to b$, the fluid velocity increases linearly with the distance according to the expression.

$$V_x = \frac{2V_{mx}}{b} h \tag{86a}$$

where $h = y - b$ is the distance from the wall. Hence, the wall shear rate is constant, given by the formula

$$G^0 = \left(\frac{\partial V}{\partial y}\right)_0 = \frac{2V_{mx}}{b} = \frac{\Delta P b}{\eta L} \tag{86b}$$

The flow described by Eqs. (86a) and (86b), having considerable practical significance, is called the simple shear flow.

Analogous expressions can be derived in the case of laminar flow through a channel of a circular cross section (capillary), as shown in schematically in Fig. 11. The axial velocity is given in this case by the expression

$$V_x = \frac{\Delta P R^2}{4\eta L}\left[1 - \left(\frac{r}{R}\right)^2\right] = V_{mx}\left[1 - \left(\frac{r}{R}\right)^2\right] \tag{87}$$

where R is the capillary radius. The wall shear rate is constant, given by the formula

$$G^0 = \left(\frac{\partial V}{dr}\right)_0 = \frac{\Delta P R}{2\eta L} \tag{88}$$

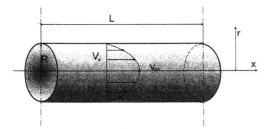

FIG. 11 A schematic representation of the laminar flow in the cylindrical channel (capillary).

Hence, the flow distribution at the capillary wall is given by the linear equation

$$V_x = \frac{\Delta P R}{2\eta L} h \tag{89}$$

An exact solution of the macroscopic flow can be obtained for the rotating risk, often used as an electrode in various electrochemical studies [78] and in colloid particle or protein adsorption experiments [79]. Although the velocity components are, in general, expressed in terms of complicated series expansions, for small separations from the disk when $h < \sqrt{v/\omega}$ (where ω is the disk angular velocity), the axial and radial fluid velocity components relative to the disk can be approximated by the simple expressions [76,78]

$$V_h = -0.51 \frac{\omega^{3/2}}{v^{1/2}} h^2$$

$$V_r = 0.51 \frac{\omega^{3/2}}{v^{1/2}} rh \tag{90}$$

where r is the distance from the disk center. This type of flow distribution is called the stagnation point flow [76].

As can be noted, the axial velocity V_h does not depend on the radial coordinate, which is unique for any macroscopic flow. This property simplifies a mathematical handling of the particle-transport equation, as will be discussed later. Another advantage of the rotating disk is that the flow remains steady and laminar for a Reynolds number ($Re = \omega R^2/v$) as large as 10^5 [78]. In contrast, all of the remaining macroscopic flows discussed subsequently are approximate only, usually valid for a limited range of Reynolds numbers.

A major practical significance (e.g., in various filtration processes) has the laminar flow appearing in the vicinity of a stationary sphere of radius R immersed in the uniform fluid flow of the velocity V_∞ (see Fig. 12). For a not too high Reynolds number ($V_\infty R/v$), the flow distribution is governed by the Stokes equation, [Eq. (83)], which can be solved conveniently in the spherical coordinates (r, ϑ) by introducing the stream function ψ_s. The components of the fluid flow velocity can be calculated as [74]

$$V_r = -\frac{1}{r^2 \sin\vartheta} \frac{\partial \psi_s}{\partial \vartheta}$$

$$V_\vartheta = \frac{1}{r \sin\vartheta} \frac{\partial \psi_s}{\partial r} \tag{91}$$

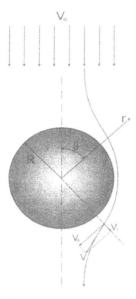

FIG. 12   A schematic representation of the laminar flow around a sphere.

The stream function for an isolated particle is given by the expression

$$\psi_s = \frac{1}{2} V_\infty R^2 \sin^2(\vartheta) \left[\frac{1}{2\bar{r}} - \frac{3}{2}\bar{r} + \bar{r}^2 \right] \tag{92}$$

where $\bar{r} = r/R$. Using Eq. (92), one can derive the following expressions for the velocity components:

$$V_r = -\frac{1}{2} V_\infty \left[2 - \frac{3}{\bar{r}} + \frac{1}{\bar{r}^3} \right] \cos \vartheta$$

$$V_\vartheta = -\frac{1}{4} V_\infty \left[\frac{3}{\bar{r}} + \frac{1}{\bar{r}^3} - 4 \right] \sin \vartheta \tag{93}$$

For small distances from the interface, when $\bar{h} = (r - R)/R \ll 1$, Eq. (93) implies that the normal and tangential velocity components become

$$V_r = V_h = -\frac{3}{2} V_\infty \left(\frac{h}{R} \right)^2 \cos \vartheta$$

$$V_\vartheta = \frac{3}{2} V_\infty \frac{h}{R} \sin \vartheta \tag{94}$$

As mentioned, Eqs. (93) and (94) are strictly valid for low-Reynolds-number flows. However, it was shown in Refs. 75 and 80 that one can use Eq. (94) for a Reynolds number much larger than unity (practically, for Re $\ll$ 300) if the flow model parameter A_f is introduced, so

$$V_r = V_h = -\frac{3}{2} A_f V_\infty \left(\frac{h}{R} \right)^2 \cos \vartheta$$

$$V_\vartheta = \frac{3}{2} A_f V_\infty \frac{h}{R} \sin \vartheta \tag{95}$$

The dimensionless parameter A_f increases with the Reynolds number according to the expression [80]

$$A_f = \frac{3}{2}\left[1 + \frac{0.19\text{Re}}{1 + 0.25\text{Re}^{0.56}}\right] \tag{96}$$

The fluid velocity distribution given by Eqs. (93)–(96) are only valid for an isolated particle. However, there are a number of practically important situations, like the deep-bed filtration process, when the flow past an assembly of spheres (forming a porous medium) takes place. In this case, the flow field around a single sphere is influenced by the presence of other spheres. Various models that describe the flow field in the packed bed consisting of spheres are available. The "sphere in cell" models [81–83] assume that each sphere in the packed bed is surrounded by the spherical cavity filled with fluid. The size of the cavity is determined by the overall average porosity of the medium. The general solution of the Navier–Stokes equation for the stream function inside the cavity may be written as [7]

$$\psi_s = \frac{1}{2}V_\infty R^2 \sin^2\left[K_1\frac{R}{r} + K_2\frac{r}{R} + K_3\left(\frac{r}{R}\right)^2 + K_4\left(\frac{r}{R}\right)^4\right] \tag{97}$$

where the coefficients K_1, K_2, K_3, and K_4 are determined from the boundary conditions at the surfaces of the cavity. On the inner surface of the cavity (i.e., at the sphere surface), the usual no-slip boundary conditions are used. The boundary conditions at the outer surface of the cavity depend on the model of the porous medium. Knowing the expression for the stream function, the velocity component can be calculated by applying Eq. (91). Using Happel's no-stress boundary conditions at the outer surface of the cavity [82,84], one can express the fluid velocity components close to the surface of the sphere in a form analogous to Eq. (95), with the flow parameter given by

$$A_f = \frac{2(1 - \bar{R}^5)}{2 - 3\bar{R} + 3\bar{R}^5 - 2\bar{R}^6} \tag{98}$$

where $\bar{R} = R/R_c = \Phi_v^{1/3}$ (R_c is the radius of the cavity) and Φ_v is the volume fraction of spheres forming the packed bed. A similar expression can be obtained using the model of Kubawara [83] by assuming no vorticity boundary condition at the outer surface of the cavity. The second class of models of the fluid flow through the packed bed of identical spheres is based on the swarm theory [85,86], which describes the fluid flow in the media of lower porosity better than the sphere-in-cell models.

A similar analysis can be performed for laminar flow past a stationary cylinder placed in uniform flow, which as implications for predicting particle or protein adsorption at fibers (filtration mat). It can be shown [87] that at small distances from the cylinder surface, the components of the fluid flow velocity can be approximated by [86,87]

$$V_r = V_h = -2A_f V_\infty \left(\frac{h}{R}\right)^2 \cos\vartheta$$

$$V_\vartheta = 4A_f V_\infty \frac{h}{R}\sin\vartheta \tag{99}$$

where R is the cylinder radius and

$$A_f = 2(\beta - 0.48\beta^3), \quad \beta = [2.022 - \ln(Re)]^{-1} \quad \text{for Re} \leq 1$$

$$A_f = 0.44\,\text{Re}^{0.56} \quad\quad\quad\quad\quad\quad\quad\quad \text{for } 1 < \text{Re} < 200 \tag{100}$$

It was also shown that for the porous plug formed by the network of identical cylinders, one can derive similar expression for the velocity components near the cylinder surface. For example, for the Happel model, A_f is given by [82,87]

$$A_f = \left[1 + \left(\frac{\Phi_v^2 + 1}{\Phi_v^2 - 1} \right) \ln \Phi_v \right]^{-1} \tag{101a}$$

and for Kubawara model by [83,87]

$$A_f = \frac{\Phi_v - 1}{\ln \Phi_v + \frac{3}{2} - 2\Phi_v + \frac{1}{2}\Phi_v^2} \tag{101b}$$

The application of the swarm theory leads to more complicated formulas, which can be found in Ref. 86.

The flow configurations around the sphere and cylinder, although important for practical purposes, are not convenient for performing particle adsorption measurements by direct, in situ experimental methods. This can be achieved, however, by using the impinging-jet cells generating flow distributions matching the above flows fairly well. The first experimental cell of this type having a radial symmetry [RIJ (radial impinging jet)] was constructed by Dąbroś and van de Ven [6,88,89]. The fluid distribution in the cell (shown schematically in Fig. 13) can be found by numerical solution of the governing Navier–Stokes equation [6,90].

The fluid streamlines in the RIJ cell obtained numerically are illustrated in Fig. 14. As can be seen, they exhibit a pattern typical for hyperbolic (stagnation) flows [90]. A characteristic feature of the flow is that only a very small fraction of the initial fluid stream at the entrance to the cell (located very close to the symmetry axis) is approaching the interface close enough for particle adsorption to take place. Note that for higher Reynolds number of the flow ($Re = Q/\pi Rv$, where Q is the volumetric flow rate and R is the inlet capillary radius), a vortex is formed in the region adjacent to the entrance of the cell. Although the fluid distribution

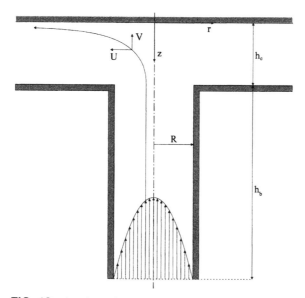

FIG. 13   A schematic view of the flow in the radial impinging-jet (RIJ) cell.

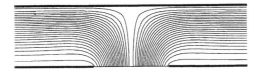

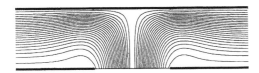

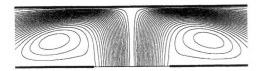

FIG. 14   Distribution of fluid streamlines in the RIJ cell determined from numerical solution for Re = 1, 8, and 48 (from top to bottom) ($h_c/R = 1.6$).

is rather complex, it was shown that for small distance from the interface $h/R < 0.1$, the fluid velocity component can be approximated by the expression [90]

$$V_h = -\alpha_r\left(\text{Re}, \frac{h_c}{R}\right) V_\infty \bar{h}^2 C(\bar{r})$$

$$V_r = \alpha_r\left(\text{Re}, \frac{h_c}{R}\right) V_\infty \bar{h} S(\bar{r})$$

(102)

where h_c is the distance between the tip of the capillary and the interface (see Fig. 13), $\bar{r} = r/R$ is the dimensionless radial distance from the symmetry axis, $V_\infty = Q/\pi R^2$, $\bar{h} = z/R$ is the dimensionless distance from the interface, α_r is the flow parameter depending on the cell geometry and the Reynolds number, and $C_r(\bar{r})$ and $S_r(\bar{r})$ are the dimensionless correction functions. When $\bar{r} \to 0$, the $C_r(\bar{r})$ function tends to unity and $S_r(\bar{r})$ approaches $\bar{r}$. Thus, in the region close to the center of the cell, the flow distributions are given by the expressions

$$V_h = -\alpha_r V_\infty \bar{h}^2$$

$$V_r = \alpha V_\infty \bar{h}\bar{r}$$

(103)

As can be noted, this flow distribution is analogous to the flow in the vicinity of the rotating disk (stagnation point flow) and the sphere [cf. Eq. (94)] in the region close to the forward stagnation point $\to 0$).

The range of validity of Eq. (103) can be estimated from the numerical results shown in Fig. 15a, where the function $-V_h/V_\infty \bar{h}^2 = \alpha_r C(\bar{r})$ is plotted for various Re numbers (1–30).

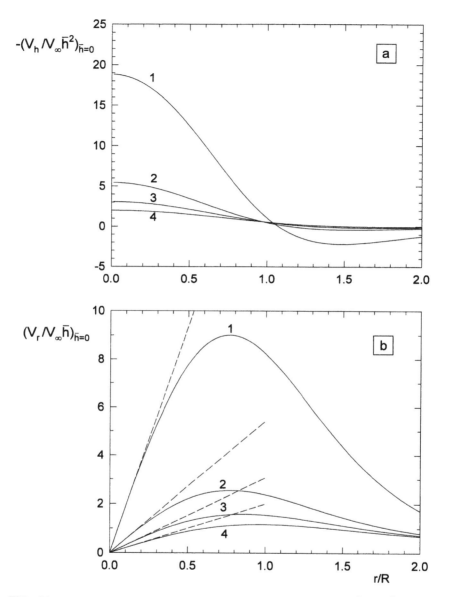

FIG. 15 (a) The dependence of the reduced axial velocity component $\bar{V}_h/V_\infty \bar{h}^2$ on the distance from the center of the cell r/R; and (b) the dependence of the reduced shear rate at the interface $V_r/V_\infty \bar{h}$ on r/R determined numerically: curve 1, Re = 30; curve 2, Re = 8; curve 3, Re = 4; curve 4, Re = 1; the dashed lines denote the limiting results calculated from Eq. (103).

This function characterizes the variations in the axial velocity component. On the other hand, in Fig. 15b, the $V_r/V_\infty \bar{h} = \alpha_r S(\bar{r})$ function is plotted describing the shear rate at the interface because:

$$\alpha_r S(\bar{r}) = \frac{G^0 R}{V_\infty} \tag{104}$$

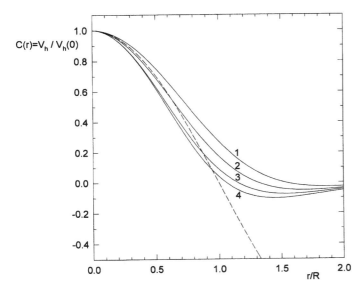

FIG. 16 The dependence of the reduced axial velocity in the RIJ cell $V_h/V_h(O)$ (where $V_h(O)$ is the axial velocity in the center of the cell) calculated numerically: curve 1, Re = 1; curve 2, Re = 4; curve 3, Re = 8, curve 4, Re = 16. The broken line denotes the analogous results for the sphere.

As can be seen in Figs. 15a and 15b, the extension of the region where the fluid flow can be approximated by Eq. (104) decreases with the increase in Reynolds number. This can be better seen in Fig. 16, which presents the dependence of $C(\bar{r})$ on the radial distance from the cell center. This function describes directly the deviation form the stagnation point flow expressed by Eq. (103). One can note that the $C(\bar{r})$ function can be well approximated for $r/h < 1$ by the simple relationship

$$C(\bar{r}) = \cos\left(\frac{\pi\bar{r}}{2}\right) \tag{105}$$

This means that the axial (perpendicular) velocity profile in the RIJ cell closely resembles the flow in the vicinity of the sphere immersed in uniform flow. This suggests that the mass-transfer phenomena (e.g., colloid particle adsorption on the sphere surface) can be modeled well by the RIJ cell. However, a quantitative comparison becomes feasible if the flow parameter α_r for the RIJ cell is known as a function of Re number. This parameter has been determined numerically and the results are plotted in Fig. 17.

It was found [90] that the exact numerical results (points) can be well approximated for a low-Re number by the expansion

$$\alpha_r = 1.78 + 0.186\mathrm{Re} + 0.034\mathrm{Re}^2 \tag{106a}$$

valid for Re < 10. For Re > 10, the fitting function was found to be

$$\alpha_r = 4.96\mathrm{Re}^{0.5} - 8.41 \tag{106b}$$

Recently [7,91], a new impinging-jet cell of a plane-parallel geometry has been developed. The flow field in the cell, called the slot impinging-jet cell (SIJ), is analogous to the flow described by

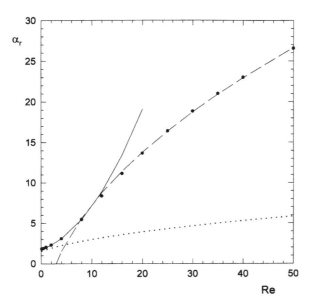

FIG. 17 The dependence of the α_r parameter on the Reynolds number calculated numerically for the RIJ cell ($h_c/R = 1.6$); the dashed line represents the high-Re fit [Eq. (106b)]; the solid line represents the low-Re fit [Eq. (106a)]; the dotted line shows the results for the sphere calculated from Eq. (96).

Eq. (102). The advantage of the SIJ cell is a much larger observation area in comparison with the RIJ cell, which facilitates a statistical analysis of adsorbed particle distribution.

Other macroscopic flow fields of practical interest has been discussed in Refs. 1 and 76.

2. Microscopic Flow Fields

Knowing the above-discussed macroscopic flow distributions, one can calculate the hydrodynamic forces and torques acting on colloid particles. In order to facilitate the evaluation of these hydrodynamic interactions, it is advantageous to decompose the macroscopic flow field into a sum of simpler flows. This can be done by introducing the local Cartesian coordinates (x_l, y_l, z_l), with the origin located at position r_0 [76]. Then, the macroscopic velocity vector can be expressed as

$$V = V_0(r_0) + \sum V^* \tag{107}$$

where V_0 is the local uniform flow and V^* represents the perturbing flow fields whose magnitude increases with the distance from the origin. Obviously, in the case when the distance from the origin (expansion range) remains much smaller than the macroscopic flow variation range, all of the perturbing terms can be neglected and the flow can be treated as locally uniform. This situation usually occurs at distances from the interface exceeding particle dimensions considerably.

On the other hand, in the region close to interfaces, the characteristic variation range of the macroscopic flow becomes comparable to particle dimensions, so the perturbing flows play a significant role. In this case, it is useful to express the perturbing flows as a superposition of the stagnation point flow V^*_{st} and a simple shear flow V^*_{sh} [7,76] defined as

$$V^*_{st} = G_{st}(2x_l z_l \mathbf{i}_x - z_l^2 \mathbf{i}_z)$$
$$V^*_{sh} = G_{sh} z_l \mathbf{i}_x \tag{108}$$

where G_{st} and G_{sh} are the stagnation point and simple shear flow strengths, respectively and $\mathbf{i}_x$ and $\mathbf{i}_z$ are the unit vectors in local Cartesian coordinates. The fluid velocity components, the stagnation point, and simple shear flow strength for the flows discussed in our work are collected in Table 2. The values for the rotating disk can be found elsewhere [7,76].

Consider now a colloid particle of radius a immersed in one of these macroscopic flows. The particle influences the flow locally, which produce a microscopic disturbance. This leads to the appearance of force and torque on the particle, often called the liquid drag or driving

TABLE 2 The Undisturbed Flow Fields Near Various Interfaces

Collector	Velocity components at the wall	G_{st}	G_{sh}	Pe
Parallel-plate[a] channel	$V_x = \dfrac{2V_{mx}}{b}y$ $V_y = 0$	0	$\dfrac{2V_{mx}}{b}$	$\dfrac{V_{mx}a^3}{b^2 D_\infty}$
Cylindrical[b] channel	$V_x = \dfrac{2V_{mx}}{R}y$ $V_r = 0$	0	$\dfrac{2V_{mx}}{R}$	$\dfrac{V_{mx}a^3}{R^2 D_\infty}$
Rotating disk	$V_r = 0.51\omega^{3/2}\nu^{-1/2}rz$ $V_z = -0.51\omega^{3/2}\nu^{-1/2}z^2$ $V_\vartheta = \omega r(1 - 0.61\omega^{1/2}\nu^{-1/2}z)$	$0.51\omega^{3/2}\nu^{-1/2}$	$0.51\omega^{3/2}\nu^{-1/2}r$ $0.616\omega^{3/2}\nu^{-1/2}r$	$\dfrac{1.02\omega^{3/2}a^3}{\nu^{1/2}D_\infty}$
Sphere in uniform flow	$V_r = -\dfrac{3}{2}A_f V_\infty \left(\dfrac{h}{R}\right)^2 \cos\vartheta$ $V_\vartheta = \dfrac{3}{2}A_f V_\infty \dfrac{h}{R}\sin\vartheta$	$\dfrac{3A_f V_\infty \cos\vartheta}{2R^2}$	$\dfrac{3A_f V_\infty \sin\vartheta}{2R}$	$\dfrac{3A_f V_\infty a^3}{R^2 D_\infty}$
Cylinder in uniform flow	$V_r = -2A_f V_\infty \left(\dfrac{h}{R}\right)^2 \cos\vartheta$ $V_\vartheta = 4A_f V_\infty \dfrac{h}{R}\sin\vartheta$	$\dfrac{2A_f V_\infty \cos\vartheta}{R^2}$	$\dfrac{4A_f V_\infty \sin\vartheta}{R}$	$\dfrac{4A_f V_\infty a^3}{R^2 D_\infty}$
Radial[c] impinging jet	$V_r = \alpha_r V_\infty \dfrac{h}{R}\left(\dfrac{r}{R}\right)$ $V_h = -\alpha_r V_\infty \left(\dfrac{h}{R}\right)^2$	$\alpha_r \dfrac{V_\infty}{R^2}$	$\alpha_r \dfrac{V_\infty r}{R^2}$	$\dfrac{2\alpha_r V_\infty a^3}{R^2 D_\infty}$

[a] $V_{mx} = \Delta P b^2/2\eta L$.
[b] $V_{mx} = \Delta P R^2/4\eta L$.
[c] $V_\infty = Q/\pi R^2$.

hydrodynamic forces. A theoretical analysis of the microscopic fluid flow near a colloidal particle is greatly simplified by the fact that the size of particles is of the order of 1 μm or less. Therefore, the Reynolds number remains much smaller than unity and the flow is governed by the quasistationary Stokes equation [Eq. (83)]. The boundary conditions for this equation are expressed as the no-slip postulate (vanishing of fluid flow at particle surface) in the case of a rigid sphere. For a liquid–liquid interface, the condition of equal stress at both sides of the interface is used. Because the entire boundary value problem formulated in this way remains linear, one can apply the expansion procedure given by Eq. (107), with the origin conveniently located at particle center, to decompose the macroscopic flow. Then, the net force and torque on the particle can be calculated as the sum of contributions stemming from the simple flows.

As mentioned earlier, at larger distances from the interface, the macroscopic flow can be treated as uniform locally, so

$$V \cong V_0(\boldsymbol{r}_p) = V_0 \boldsymbol{i}_x \tag{109}$$

where $\boldsymbol{r}_p$ is the position vector of the particle center and $\boldsymbol{i}_x$ is the unit vector parallel to the V_0 vector. Thus, the particle is effectively immersed in a uniform flow, so the local flow field is as described by Eq. (93) (with R replaced by the particle radius a). Knowing the fluid flow and the pressure distribution, one can calculate the hydrodynamic force and the torque using Eq. (82). This results in the expression [92]

$$\begin{aligned} \boldsymbol{F}_h &= 6\pi\eta a V_0 = K_h V_0 \\ \boldsymbol{T}_h &= 0 \end{aligned} \tag{110}$$

where $K_h = 6\pi\eta a$ is the hydrodynamic resistance coefficient of the sphere. Equation (110) indicates that the hydrodynamic driving force is always proportional to the local fluid velocity vector and that the hydrodynamic torque on particle is zero.

Because of the linearity of the equation governing liquid flow, one can easily deduce that Eq. (110) is also applicable in the case when a particle is moving with a locally steady velocity U_∞ through a quiescent fluid. In this case, one has

$$\begin{aligned} \boldsymbol{F}_{h_r} &= -6\pi\eta a U_\infty = -K_h U_\infty \\ \boldsymbol{T}_h &= 0 \end{aligned} \tag{111a}$$

This equation, first derived by Stokes, indicates that the hydrodynamic resistance force on a moving particle is proportional to the velocity vector and directed opposite to it. By comparing Eqs (110) and (111) one can deduce that for a neutrally buoyant particle (when the net force equals zero), $U_\infty = V_\infty$. This means that the sphere will follow the liquid streamlines.

If particle motion is due to an external force $\boldsymbol{F}_e$ (e.g., gravity), then from the postulate $\boldsymbol{F}_e + \boldsymbol{F}_r = 0$, one can conclude that

$$\boldsymbol{U}_\infty = K_h^{-1} \boldsymbol{F}_e = M \boldsymbol{F}_e \tag{111b}$$

where M is the hydrodynamic mobility of the particle.

It should be mentioned that these results are applicable for particles moving in an unbounded fluid far from interfaces. When the distance to the interface becomes comparable with particle dimensions, both the hydrodynamic driving and resistance forces will be strongly modified. Because of its relevance to practical applications, these microhydrodynamic flows have been the subject of numerous theoretical studies. Maude [93] and Bart [94] considered the problem of a sphere approaching a rigid-plane interface, O'Neill solved the problem of a spherical particle sliding (moving without rotation) along the rigid plane [95,96] and attached at the surface in the simple shear flow [97], Dean and O'Neill gave the solution for the problem of the particle rotating at the rigid wall [98], and Goren and O'Neill [99] solved the general hydrodynamic problem of the particle moving at the rigid wall in a flow with a velocity that is of the second degree in the local coordinates. In all of these articles, the bispherical polar coordinates were applied and the fluid velocity fields were obtained analytically in terms of orthogonal functions expansions. These expansions are, however, slow converging if the distance between a particle and a rigid wall approaches zero.

The most important of these microflows is particle motion perpendicular to a planar interface in a quiescent fluid. It was shown in Refs. 93 and 94 that the resistance force on the particle is increased, resulting in decreased velocity under external force described by

$$U_p = \frac{1}{\lambda_\perp(H)} M \boldsymbol{F}_e = \frac{1}{\lambda_\perp(H)} \boldsymbol{U}_\infty = F_1(H) \boldsymbol{U}_\infty \qquad (112)$$

where $\lambda_\perp(H)$ is the correction function to the Stokes resistance coefficient, $F_1(H) = \lambda_\perp(H)^{-1}$ is the velocity correction function, and $H = (h_m - a)/a$ is the dimensionless gap width between the particle and the interface. The correction function can be calculated analytically from the series [93,94]

$$\lambda_\perp(H) = \frac{1}{F_1(H)} = \frac{4}{3} \sinh(\alpha_h) \sum_{n=1}^{\infty} \frac{n(n+1)}{(2n-1)(2n+3)} \left[\frac{Y_n}{T_n} - 1 \right] \qquad (113)$$

where

$$\alpha_h = \cosh^{-1}(H) = \ln\left[(H+1) + \sqrt{(H+1)^2 - 1} \right]$$

and

$$Y_n = 2 \sinh(2n+1)\alpha_h + (2n+1)\sinh(2\alpha_h)$$
$$T_n = 4 \sinh^2(n + \tfrac{1}{2})\alpha_h - (2n+1)^2 \sinh^2(\alpha_h)$$

For $H \gg 1$, it follows from Eq. (113) that

$$F_1(H) = \frac{1}{1 + (9/8)H} \qquad (114a)$$

whereas at close separations,

$$F_1(H) = \frac{H}{1 + 0.97H - 0.21H \ln H} \cong H \qquad (114b)$$

As one can deduce, for separations much smaller than the particle dimension, the function $F_1(H)$ decreases proportionally to H, which means that the resistance coefficient tends to infinity (as H^{-1}). This can be physically interpreted as due to the increased force needed to drive the liquid out from the gap between the interface and the approaching particle.

It was postulated in Ref. 14 that $F_1(H)$ can be well approximated for the entire range of H by the expression

$$F_1(H) = H \frac{19H + 4}{19H^2 + 26H + 4} \qquad (115)$$

In many applications, $F_1(H)$ can be approximated by an even simpler interpolating function:

$$F_1(H) = \frac{H}{H + 1} \qquad (116)$$

The correction function $F_1(H)$ calculated using the exact formula [Eq. (113)] and its approximation according to Eq. (115) are compared in Fig. 18a.

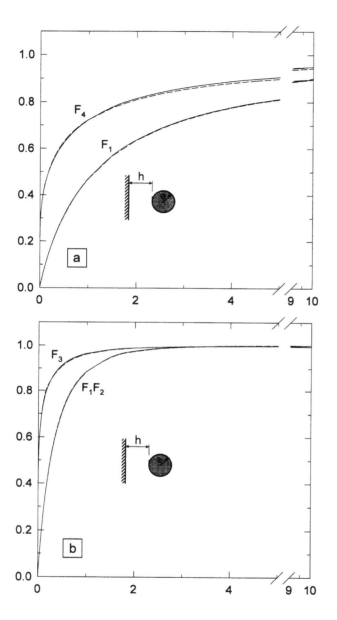

FIG. 18 (a) The hydrodynamic correction functions $F_1(H)$ and $F_4(H)$ for a spherical particle in the vicinity of a rigid wall; (b) the correction functions for a particle moving in a simple shear parallel to a rigid wall [$F_3(H)$] and perpendicular to the wall in the stagnation point flow [$F_1(H)$ and $F_2(H)$]. (From Ref. 14.)

On the other hand, the correction function $F_1(H)$ in the case of a sphere approaching another rigid sphere (of the same size) can be approximated by [18]

$$F_1(H) = H \frac{6H + 4}{6H^2 + 13H + 2} \tag{117}$$

For $H \to 0$, $F_1(H) \to 2H$ (i.e., the correction factor is two times larger than for the sphere–plane case).

An analysis of particle motion parallel to a planar interface in a quiescent fluid is more complicated because the translational motion induces a coupled rotational motion, which produces a hydrodynamic torque on the particle. These effects have been calculated by Goren and O'Neill [99] using bispherical coordinates. It was shown that particle velocity can be expressed as

$$U_p = F_4(H)U_\infty \tag{118}$$

where $F_4(H)$ is the universal correction function plotted in Fig. 18a together with the interpolating functions [14]:

$$
\begin{aligned}
F_4(H) &= \frac{1}{1.062 - 0.516 \ \ln(H)} \quad \text{valid for } H < 0.11 \\
F_4(H) &= \left(\frac{H}{2.639 + H}\right)^{1/4} \quad \text{valid for } H \geq 0.11
\end{aligned}
\tag{119}
$$

As Eq. (119) indicates, the velocity of particle motion parallel to the interface is less affected by the interface than the perpendicular particle motion. This means that the particle resistance or mobility coefficients acquire an anisotropic character which can be accounted for by a second-rank tensor [74]. This behavior will influence particle transport and adsorption phenomena, as will be discussed later.

The theoretical expressions describing the mobility of the particle close to a solid wall were verified experimentally by Adamczyk et al. [100] and Małysa and van de Ven [101] using the multiple-frame photography technique.

Another class of practically important situations arises when the particle is driven toward the interface by the macroscopic flows discussed earlier. Due to the presence of the wall, the hydrodynamic driving forces are modified, which causes the deviation of particle trajectory from liquid streamlines. Hydrodynamic torques on particles and the coupling between the translational and rotational motion also appear, which make the theoretical analysis of this problem rather involved. The problem of a particle moving in a simple shear flow given by Eq. (108) was solved by Goren and O'Neill [99]. It was shown that the particle velocity can be expressed as

$$U_p = F_3(H)V_{\text{sh}}^* = F_3(H)G_{\text{sh}}a(1 + H) \tag{120}$$

where the correction function $F_3(H)$ is shown in Fig. 18b. This function can be interpolated by [14]

$$
\begin{aligned}
F_3(H) &= \frac{1}{0.754 - 0.256 \ \ln(H)} \quad \text{valid for } H < 0.137 \\
F_3(H) &= 1 - \frac{0.304}{(1 + H)^3} \quad \text{valid for } H \geq 0.137
\end{aligned}
\tag{121}
$$

The relationship between particle velocity in the simple shear flow and distance from the wall given by Eq. (121) can be used for an experimental determination of the position of the particle relatively to the interface, which is important for determining the colloid interactions between particles at interfaces [102–105].

The problem of a particle moving in the stagnation point flow given by Eq. (108) was also solved by Goren and O'Neill [99]. In this case it was shown that the particle velocity U_p can be expressed as

$$U_p = F_1(H)F_2(H)V_{st}^* = F_1(H)F_2(H)G_{st}a^2(1+H)^2 \tag{122}$$

The correction function $F_2(H)$ was obtained from the analytical solution of the creeping flow equation. It can be well approximated by [14]

$$F_2(H) = 1 + \frac{1.79}{(0.828 + H)^{1.167}} \tag{123}$$

The dependence of the product $F_1(H)F_2(H)$ on the particle–wall separation is also shown in Fig. 18b.

At later adsorption stages, one should consider the hydrodynamic interactions of moving particles with those attached to the surface and the influence of the deposited particles on the flow field in their vicinity. Therefore, the proper description of the hydrodynamic interaction of a moving colloidal particle with particles already deposited at the interface is of importance for the quantitative analysis of the adsorption process.

One of the few hydrodynamic problems of this kind which can be treated analytically is the flow in the vicinity of a sphere attached to the wall, immersed in the simple shear given by Eq. (108). The flow pattern was evaluated by Goren and O'Neill [99] in terms of the Legendre polynomial expansion in bispherical coordinates, whereas O'Neill [97] used the Bessel function approach to calculate the hydrodynamic driving force in this case. The calculations demonstrated that the particle attached to the interface decreases the fluid velocity over distances considerably exceeding its dimension. This results in the decrease in the shear rate at the interface over large surface areas, as can be seen in Fig. 19, where the contours of the function G^*/G^0 are plotted (G^* is the perturbed shear rate at the surface). It is interesting to note that the perturbation is symmetric with respect to the flow direction, which is a consequence of the fluid streamlines symmetry. The decreased flow rate in the vicinity of the particle is expected to influence other particle adsorptions and charge transports from the double layer surrounding the interface. This has profound consequences for the streaming current magnitude, as demonstrated in Ref. 77.

Due to the complicated geometry, the solution of the flow equation for more than one particle at the interface is prohibitively complex and can only be obtained for the case of two particles by applying elaborate numerical techniques. In general, the solution of the creeping flow (Stokes) equation can be expressed in the integral form [106]

$$V(x) = V_0(x) + \sum_{l=1}^{2} \int_{S_l} G(x,y) \cdot f(y_l) \, dS_l \tag{124}$$

where $V_0(x)$ denotes the fluid flow field at a given point in the absence of the particles fulfilling the no-slip boundary condition ($V_0 = 0$) at the interface, y is a vector pointing at the surface element dS_l, $G(x,y)$ is the Oseen tensor accounting for the presence of the rigid wall [106], $f(y_l)$ is the force density at a point y_l at the $l(l = 1, 2)$ particle surface, and S_l is the surface of particle l. The total force F_{h_l} and torque T_{h_l} exerted by particle l on the fluid are given by

$$F_{h_l} = \int_{S_l} f(y_l) \, dS_l$$
$$T_{h_l} = \int_{S_l} (y - R_l) x f(y_l) \, dS_l \tag{125}$$

respectively, where R_l specifies the position of the center of particle l.

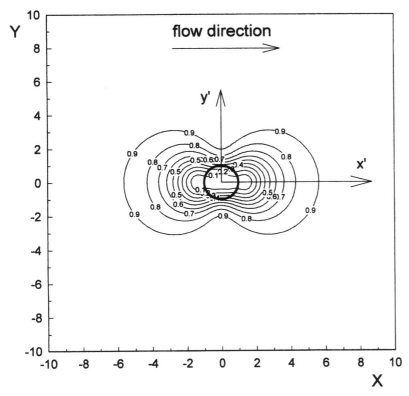

FIG. 19 The contours showing the relative shear rate at the interface G/G^0 (where G^0 is the unperturbed shear rate) in the vicinity of a particle attached to a rigid surface.

In order to obtain a complete set of equations, one has to take into account the boundary conditions at the particle surface:

$$V(y_l) = U_l + \omega_1 x(y_l - R_l) \tag{126}$$

where U_l and ω_l are the translational and rotational velocities of the particle l, respectively. For the particle attached at the interface, both velocities are equal to zero. Equations (124)–(126) form a complete system, provided that the external force and torque or the velocities of the particles as well as the external flow field $V_0(x)$ are specified.

Dąbroś and van de Ven [107–109] developed two convenient methods for the numerical solution of the system of Equations (124)–(126) for the problem of two spherical particles at a rigid interface. In the first, the boundary element method (BEM), the surface of each particle is divided into a finite number of triangular elements. The surface integrals can be replaced by the sums of the integrals over the elements [108]. In the second method particle surfaces are divided into a set of subunits and the surface integration in Eqs. (124)–(126) can be replaced by summation over these subunits. A more detailed discussion of the subunit method with application to the problem of the collision of two colloidal particles at the rigid interface can be found in Refs. 107, 109, and 110.

An example of the hydrodynamic interactions between two spherical particles at the interface is visualized in Fig. 20. Shown in the figure are the trajectories of a spherical particles

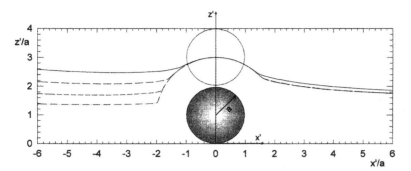

FIG. 20 The trajectories of a particle moving in the stagnation flow around a particle attached to a wall calculated using the BEM method. The initial distance of the particle from the attached one was 6 particle radii and the distance from the interface varied from 1.4 (the lowest trajectory) to 1.6 (the highest trajectory).

moving in the stagnation point flow and colliding with another particle rigidly attached to the planar interface. The hydrodynamic interactions were taken into account via the multisubunit method, whereas the hard-core interaction between moving and stationary particles were considered in terms of Born-type interactions. It can be seen that during its motion along the interface, the moving particle is repelled by the attached particle due to the hydrodynamic interactions. When the initial particle–wall gap width is greater than the particle diameter ($h_i > 2$), the moving-particle trajectories remain symmetric, as was the case for the liquid flow. However, if the moving particle was initially closer to the wall, its trajectory, after the collision, would become highly asymmetric and move a much larger distance from the interface than originally. This effect was caused by the short-range hydrodynamic scattering (HS) [1,7,14]. It plays a significant role in convection-driven adsorption of larger particles [111]. Due to the HS effect, the surface area behind the adsorbed particles becomes inaccessible for moving particles, which leads to enhanced surface-blocking effects. The length of the "shadow" depends on the flow configuration (either simple shear or stagnation point flow) but mostly on the local flow rate (governed by the G_{sh} or G_{st} parameters). This is spectacularly illustrated in Fig. 21 for the stagnation point flow. Warszyński [14] demonstrated that the shadow length L_s can be approximated by the formula

$$\frac{L_s}{a} = C_h \left(\frac{G_{sh}a^2}{D_\infty}\right)^n \left(\frac{x_p}{a}\right)^n \tag{127}$$

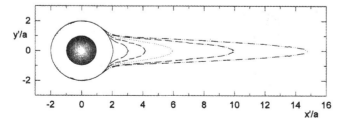

FIG. 21 The dependence of the size of the hydrodynamic shadow behind a particle attached at the interface and immersed in a stagnation flow; the contours correspond to the dimensionless parameter $G^0 a^2/D_\infty$ equal 0.01, 0.05, 0.1, 0.2, and 0.5, respectively. (From Ref. 14.)

where C_h is the dimensionless hydrodynamic constant, x_p is the particle distance from the center of the cell, and n is the exponent varying between 0.5 for the stagnation flow and 1 for the simple shear flow.

III. LINEAR ADSORPTION REGIMES

A. Phenomenological Transport Equation

Our considerations presented hereafter are based on the phenomenological approach exploiting the irreversible thermodynamic balance equations. Within the framework of this approach, one can formulate the continuity equation for the dispersed particle phase under isobaric–isothermal conditions in the usual form [12,13]

$$\frac{\partial \boldsymbol{n}}{\partial t} + \nabla \cdot \boldsymbol{j} = 0$$

$$\boldsymbol{j} = -\boldsymbol{D} \cdot \left(\frac{\nabla \mu}{kT} + \frac{\nabla \phi_t}{kT} \right) n + \boldsymbol{U}_p n \tag{128}$$

where $\boldsymbol{j}$ is the particle flux vector, $\boldsymbol{M}$ is the mobility tensor characterizing the boundary effects discussed earlier, $\boldsymbol{D} = kT\boldsymbol{M}$ is the diffusion tensor, μ is the chemical potential of the particles, ϕ_t is the total interaction potential due to specific interactions between particles and the interface discussed earlier, external forces, and so forth, and n is the local value of the particle concentration.

The particle velocity vector due to driving hydrodynamic forces $\boldsymbol{F}_h$ and torques $\boldsymbol{T}_h$, discussed earlier, can be evaluated as

$$\boldsymbol{U}_p = \boldsymbol{M} \cdot \boldsymbol{F}_h + \boldsymbol{M}_r \cdot \boldsymbol{T}_h \tag{129}$$

where $\boldsymbol{M}_r$ is the rotational mobility tensor. By formulating Eq. (128), the hydrodynamic particle–particle interactions were neglected and no coupling was assumed between hydrodynamic and specific interactions (double layer, electrostatic, etc.)

The expression for the chemical potential μ, describing the effects of the specific interactions between particles, can be formally written as

$$\mu = \mu^0 + kT \ln a_{\text{ch}} = \mu^0 + kT \ln fn \tag{130}$$

where μ^0 is the reference potential, a_{ch} is the chemical activity of the particle, and f is the activity coefficient equal to 1 for an ideal phase (diluted suspension). Solution of Eqs. (129) and (130) is impractical in the general case because of difficulties in calculating particle activity changes resulting from increased particle concentration at the interface. However, useful limiting cases have been analysed in the literature successfully. One of the simpler situations arises when the flow contributions can be assumed negligible in the region close to the interface of thickness δ_a (called the surface boundary region) and particle transport can be assumed one dimensional. Moreover, if the activity coefficient does not explicitly depend on particle concentration (quasidilute limit) Eq. (129) simplifies to [2,112]

$$\frac{\partial n}{\partial t} = \frac{\partial}{\partial h} \left[D(h) e^{-\Phi/kT} \frac{\partial}{\partial h} \left(n e^{\Phi/kT} \right) \right] \tag{131}$$

where $D(h) = F_1(H)D_\infty$, $D_\infty = kTM_\infty = kT/6\pi\eta a$ is the diffusion coefficient of the particle in the bulk, and $\Phi = \phi_t + kT \ln f(h)$. Equation (131) has considerable significance because its solutions can be used as boundary conditions for the bulk-transport equation derived from

Eq. (128) for dilute (linear) systems of ideal behavior. Because, in this case $\mu = \mu_0 + kT \ln n$, Eq. (128) becomes

$$\frac{\partial n}{\partial t} = \nabla \cdot \left[\boldsymbol{D} \cdot \nabla n + \frac{\nabla \phi_t}{kT} n - \boldsymbol{U}_p n \right] \tag{132}$$

This equation can be effectively solved by numerical methods, especially for one-dimensional problems, as discussed later. If, moreover, all specific interactions are neglected, together with the hydrodynamic boundary effects due to the presence of the interface (this assumption is referred to as the Smoluchowski–Levich model [76]), Eq. (132) simplifies further to

$$\frac{\partial n}{\partial t} = D_\infty \nabla^2 n - \boldsymbol{V} \cdot \nabla n - \boldsymbol{U}_e \cdot \nabla n \tag{133}$$

where $\boldsymbol{V}$ is the unperturbed (macroscopic) fluid velocity vector and $\boldsymbol{U}_e = -(D_\infty/kT)\nabla\phi_e = M_\infty \boldsymbol{F}_e$ is the particle velocity due to the external force $\boldsymbol{F}_e$ having the potential ϕ_e. $\boldsymbol{U}_e$ is often referred to as the migration velocity.

Because Eq. (133) is widely used for predicting particle transfer rates, it is instructive to perform its dimensional analysis by introducing the characteristic length scale L_{ch}, time scale t_a, convection velocity V_{ch}, and migration velocity U_{ch}. Then, Eq. (133) can be expressed as

$$Ta \, \frac{\partial n}{\partial \tau} = \bar{\nabla}^2 n - \mathrm{Pe} \bar{\boldsymbol{V}} \cdot \bar{\nabla} n - Ex \bar{\boldsymbol{U}}_e \cdot \bar{\nabla} n \tag{134}$$

where the dimensionless variables are defined as

$$\tau = \frac{t}{t_a}, \quad \bar{\boldsymbol{V}} = \frac{1}{V_{ch}} \boldsymbol{V}, \quad \bar{\boldsymbol{U}}_e = \frac{1}{U_{ch}} \boldsymbol{U}_e, \quad \bar{\nabla}^2 = L_{ch}^2 \nabla^2 \tag{135a}$$

Consequently, the dimensionless numbers (analogous to the Reynolds number) are defined as

$$Ta = \frac{L_{ch}^2}{D_\infty t_a}$$

$$\mathrm{Pe} = \frac{V_{ch} L_{ch}}{D_\infty} \quad \text{(Peclet number)} \tag{135b}$$

$$Ex = \frac{U_{ch} L_{ch}}{D_\infty} \quad \text{(external force number)}$$

The Ta number can be treated as the dimensionless period of adsorption.

Useful limiting regimes can be derived from Eq. (133) by exploiting these definitions. When $\Omega \ll 1$, which corresponds to the condition $t_a \gg L_{ch}^2/D_\infty$, steady-state transport conditions are established, which significantly simplifies the mathematical handling of Eq. (133). For colloid particles, assuming the typical values $D_\infty = 10^{-12}$ m^2/s and $L_{ch} = 0.1$ µm (particle dimension), one obtains $t_a \gg 10^{-2}$ s as a criterion for the steady-state conditions. For $L_{ch} = 1$ µm, this characteristic time becomes 1 s. Assuming the steady-state conditions and $Ex \ll 1$, Eq. (133) reduces to the simple form

$$\bar{\nabla}^2 n - \mathrm{Pe} \bar{\boldsymbol{V}} \cdot \bar{\nabla} n = 0 \tag{136}$$

Equation (136), called often the convective diffusion equation [1,18], can be solved analytically for various flow configurations, as discussed later.

On the other hand, under the no-convection conditions, when $\mathrm{Pe} = 0$ Eq. (134) reduces to the form called the Smoluchowski equation:

$$Ta \, \frac{\partial n}{\partial \tau} = \bar{\nabla}^2 \boldsymbol{n} - E_x \bar{\boldsymbol{U}}_e \cdot \bar{\nabla} \boldsymbol{n} \tag{137}$$

Another important subcase arises when Ex = 0. Then, the Smoluchowski equation becomes

$$\frac{\partial n}{\partial t} = D_\infty \nabla^2 n \tag{138}$$

This is the ordinary diffusion equation (often called Fick's second law) which can be solved analytically for many situations of practical interest (e.g., adsorption on a spherical interface from a finite or infinite volume) [113,114].

B. Limiting Analytical Solutions

1. Surface Boundary Layer Transport

We define the linear adsorption regime such that the activity coefficient $f = 1$. In this case, one can solve Eq. (131) under quasistationary conditions by postulating that the characteristic surface-transport relaxation time is very small, as estimated earlier. An integration of Eq. (131) leads to the following expression for the particle flux [112,115]:

$$-j = k'_a n(\delta_a) - k'_d n(\delta_m) \tag{139}$$

where $n(\delta_a)$ is the particle concentration at $h = \delta_a$ (adsorption layer thickness), $n(\delta_m)$ is the concentration at the primary minimum (see Fig. 9), and

$$k'_a = e^{\phi_t(\delta_a)/kT} \left(\int_{\delta_m}^{\delta_a} \frac{e^{\phi_t/kT}}{D(h)} \right) dh \tag{140}$$

$$k'_d = k_a e^{[\phi_m - \phi_t(\delta_a)]/kT}$$

are the adsorption and desorption rate constants, respectively.

By introducing the surface coverage defined according to Eq. (3) as

$$\Theta = S_g \int_{\delta_m}^{\delta_a} n \, dh \tag{141}$$

and assuming that $\phi_t(\delta_a) = 0$ (no external force and secondary minimum present), one can transform Eq. (139) to the usual form [116]

$$-j = k_a n(\delta_a) - \frac{k_d \Theta}{S_g} \tag{142}$$

where

$$k_a = \left(\int_{\delta_m}^{\delta_a} \frac{e^{\phi/kT}}{D(h)} \, dh \right)^{-1} = \frac{1}{R_b} \tag{143}$$

$$k_d = ka \left(\int_{\delta_m}^{\delta_a} e^{-\phi/kT} \, dh \right)^{-1}$$

and R_b can be treated as the transport resistance due to the barrier.

Equation (142) is a general expression describing reversible adsorption of particles under linear conditions (for negligible surface-blocking effects). Under equilibrium conditions when $j = 0$, Eq. (142) becomes

$$\Theta = S_g K_a n(\delta_a) \tag{144}$$

where

$$K_a = \frac{k_a}{k_d} = \int_{\delta_m}^{\delta_a} e^{-\phi/kT} \, dh \tag{145}$$

can be treated as the equilibrium adsorption constant. Equation (144) expresses the Henry adsorption law stating that the amount adsorbed is proportional to the bulk concentration.

On the other hand, in the case when $|\phi_m| \gg kT$, k_d vanishes and particle adsorption becomes practically irreversible. The expression for particle flux then simplifies to the form

$$-j = k_a n(\delta_a) \tag{146a}$$

When $k_a \rightarrow \infty$, which is the case for energy profile of Type Ib (see Fig. 8), Eq. (146a) simplifies to

$$n(\delta_a) \rightarrow 0 \quad \text{at } h = \delta_a \tag{146b}$$

In this way, one obtains the perfect-sink boundary condition introduced by Smoluchowski.

One can evaluate k^a, k_d, and k_a analytically for the simple shape of the specific interaction energy profile. For example, in the case when energy distributions around the primary minimum and the barrier region can be approximated by a parabolic distribution, these constants are given by [115]

$$k_a = D(\delta_b) \left(\frac{\gamma_b}{2\pi kT} \right)^{1/2} e^{-\phi_b/kT} \cong \frac{D_\infty}{a} \left(\frac{\phi_b}{\pi kT} \right)^{1/2} e^{-\phi_b, kT}$$

$$k_d = k_a \left(\frac{\gamma_m}{2\pi kT} \right)^{1/2} e^{\phi_m/kT} \cong \frac{k_a}{\delta_m} \left(\frac{\phi_m}{\pi kT} \right)^{1/2} e^{\phi_m/kT} \tag{147a}$$

$$K_a = \left(\frac{2\pi kT}{\gamma_m} \right)^{1/2} e^{-\phi_m/kT} \cong \left(\frac{\pi kT}{\phi_m} \right) \delta_m e^{-\phi_m/kT}$$

where

$$\gamma_b = -\left(\frac{d^2\phi}{dh^2} \right)_{\delta_b} \cong \frac{2\phi_b}{kT\delta_b^2}, \quad \gamma_m = \left(\frac{d^2\phi}{dh^2} \right)_{\delta_m} \cong \frac{2\phi_m}{kT\delta_m^2}$$

For a strongly asymmetric barrier (e.g., of a triangular shape), one can analogously express the adsorption constants in the form

$$k_a = \frac{D_\infty}{a} \left(\frac{\phi_b}{kT} \right) e^{-\phi_b/kT} \tag{147b}$$

Equation (142) and the irreversible counterparts [Eqs. (146a) and (146b)] have a major significance because they can be used as the kinetic boundary condition for the bulk-transport problems governed by the Smoluchowski–Levich (SL) equation.

2. Limiting Solutions of the Bulk-Transport Equation

The general boundary conditions, Eq. (142), also called the kinetic boundary conditions, can be exploited for solving the particle adsorption problem under the pure diffusion transport conditions when Eq. (138) applies. Using the Laplace transformation method, analytical results have been derived for the spherical and planar interfaces, both for irreversible and reversible adsorptions [2,113,114]. The effect of the finite volume also has been considered in an exact manner. Although the results obtained are rather complex, it has been demonstrated that the flux (both for reversible and irreversible adsorptions) is initially given by the formula

$$j = -k_a n^b \tag{148}$$

This approximation is valid for adsorption time $t \ll D_\infty / k_a^2$.

On the other hand, for irreversible adsorption, when $k_d = 0$ and $t \gg D_\infty / k_a^2$, it was shown that the flux of particles to the spherical interface was governed by the important limiting expression first derived by Smoluchowski [62,116]:

$$j = -\left[\frac{D_{12}}{(a_1 + a_2)} + \sqrt{\frac{D_{12}}{\pi t}} \right] n^b \tag{149a}$$

where $D_{12} = D_{1\infty} + D_{2\infty}$ is the relative diffusion coefficient and a_1 is the interface radius. As one can note, for longer times, when $a_1^2 / D_{12} t \ll 1$, the flux attains the steady-state value equal to $-D_{12} n^b / (a_1 + a_2)$. If the particle radius is much smaller than the interface radius (and the interface remains immobile), $a_2 / a_1 \ll 1$ and the stationary flux becomes

$$j = -\frac{D_\infty}{a_1} n^b \tag{149b}$$

However, if $a_1 \to \infty$ (planar interface), the flux becomes unsteady for all times and is given by the well-known expression

$$j = -\sqrt{\frac{D_\infty}{\pi t}} n^b \tag{150}$$

This equation can be exploited for determining the particle diffusion coefficient by measuring experimentally the number of irreversibly adsorbed particle as a function of time. This is done under the diffusion transport conditions by eliminating all natural and forced convection currents [2,9,10,16]. However, due to the fact that particle flux decreases gradually with the time, diffusion-controlled adsorption becomes very inefficient for long times. Indeed, in these experiments, adsorption times reached tens of hours.

Therefore, particle adsorption efficiency is usually increased by applying forced-convection transport conditions. The kinetic boundary conditions also can be used in this case, in conjunction with the stationary SL equation [Eqs. (136) and (137)] to predict particle deposition rates. Useful analytical expressions for particle adsorption flux under stationary conditions can be derived for the simple geometry of the adsorption surface (collector) and special flow configurations having the perpendicular component independent of the coordinate tangential to the interface. As discussed previously, such a property possesses the flow appearing near the rotating disk, impinging-jet cells, sphere and cylinder placed in uniform flow (in the region close to the stagnation point), and so forth. The interface exposed to this type of flow is called the uniformly accessible surface. For these uniformly accessible surfaces, in the case of irreversible adsorption, the general solution of Eq. (136) assumes the simple form [116,117]

$$j = -\frac{k_c'}{1 + k_c'/k_a} n^b \tag{151a}$$

where k_c' is the mass-transfer coefficient (often called the reduced transfer rate) characterizing the rate of particle transport from the bulk to the edge of the adsorption layer due to convection and n^b is the particle concentration in the bulk. Equation (151a) can also be expressed as

$$j = -\frac{k_c}{1 + k_c/k_a''} n^b \tag{151b}$$

where k_c is the overall mass-transfer coefficient characterizing the rate of particle transport from the bulk to the primary minimum known from analytical or numerical solutions of the convective diffusion equation and

$$k_a'' = \left(\int_{\delta_m}^{\delta_a} \frac{e^{\phi/kT} - 1}{D(h)} \, dh \right)^{-1} \tag{152}$$

is a modified adsorption constant [117].

An important, limiting form of the general Eq. (151b) arises when the adsorption rate constant k_a becomes very small in comparison with k_c. This situation occurs if the energy barrier due to the specific interactions becomes high ($\phi_b > 10$ kT). In this case, Eq. (151b) reduces to

$$j = -k_a''n^b \cong -k_a n^b \tag{153}$$

Substituting for k_a in the expression given by Eq. (147b), one obtains the simple formula

$$j = -\frac{D_\infty}{a} \left(\frac{\phi_b}{2\pi kT} \right)^{1/2} e^{-\phi_b/kT} \tag{154}$$

One can deduce that the adsorption flux decreases exponentially with increasing barrier height, which is an indication of the barrier-controlled adsorption regime, analogous to the activated transport in chemical reactions.

On the other hand, in the case when $k_c/k_a'' \ll 1$ (fast transfer rate through the adsorption layer), Eq. (151b) becomes

$$j = -k_c n_b = j_0 \tag{155}$$

where j_0 is the stationary flux from the bulk to the primary minimum (called the initial or limiting flux). In this case, particle adsorption is governed by the bulk transport alone, which means the barrierless adsorption regime. As discussed in Ref. 1, one can express the limiting flux for uniformly accessible surfaces when the SL approximation is valid, by

$$j_0 = -k_c n_b = -\frac{D_\infty n^b}{a} C \text{Pe}^{1/3} \tag{156}$$

where C is the dimensionless constant depending on the collector geometry.

Definitions of the Peclet number and expressions for $k_c = -j_0/n_b$ calculated by solving Eq. (136) are collected in Table 3. As one can note, in all cases j_0 increases proportionally to $D_\infty^{2/3}$ rather than D_∞, as intuitively expected. Because D_∞ is inversely proportional to particle size, this means that the convection flux decreases as $a^{-2/3}$ with particle radius. It is also interesting to observe that j_0 is rather insensitive to the fluid velocity V_∞.

Analytical expressions for the limiting flux also can be derived for the nonuniformly accessible surfaces by using the similarity transformation [78]. For the sphere or cylinder in uniform, flow, the flux j_0 is given by

$$-\frac{j_0}{n_b} = k_c = f_s(\vartheta) A_f^{1/3} \frac{D_\infty^{2/3} V_\infty^{1/3}}{R^{2/3}} \tag{157}$$

where

$$f_s(\vartheta) = \frac{0.78 \sin \vartheta}{(\vartheta - \frac{1}{2} \sin \vartheta)^{1/3}}$$

TABLE 3 Bulk-Transfer Rate Constant $k_c = -j_0/n_b$ for Various Uniformly Accessible Surfaces

Surface	k_c	Surface	k_c

$$\frac{D_{12}}{a_1 + a_2}$$

$$\frac{0.89 A_f^{1/3} V_\infty^{1/3} D_\infty^{2/3}}{R^{2/3}}$$

$$\frac{0.62 \omega^{1/2} D_\infty^{2/3}}{\nu^{1/6}}$$

$$\frac{0.53 \alpha^{1/3} Q^{1/3} D_\infty^{2/3}}{R^{4/3}}$$

$$\frac{0.98 A_f^{1/3} V_\infty^{1/3} D_\infty^{2/3}}{R^{2/3}}$$

$$\frac{0.62 \alpha^{1/3} Q^{1/3} D_\infty^{2/3}}{l^{1/3} d}$$

for the sphere and

$$f_s(\vartheta) = 0.85 \frac{\sqrt{\sin \vartheta}}{(\int_0^\vartheta \sqrt{\sin \xi}\, d\xi)^{1/3}}$$

for the cylinder.

In the case of the parallel-plate channel, the local flux is given by [1]

$$-\frac{j_0}{n_b} = 0.78 \frac{V_{mx}^{1/3} D_\infty^{2/3}}{b^{1/3} x^{1/3}} \tag{158}$$

where x is the distance from the inlet to the channel (see Fig. 10).

An analogous expression for the cylindrical channel reads [1]

$$-\frac{j_0}{n_b} = 0.86 \frac{V_{mx}^{1/3} D_\infty^{2/3}}{R^{1/3} x^{1/3}} \tag{159}$$

The initial flux for other nonuniformly accessible surfaces has been discussed extensively in Refs. 1, 12, 14, 18, and 75. Results are also available in the literature [117] concerning the solution of the SL equation for the nonuniformity accessible surfaces with the general boundary condition Eq. (146a).

It should be mentioned that all of the results concerning the bulk-transfer rates—in particular, results given in Table 3—are valid for submicrometer-sized particles when the range of hydrodynamic wall effects and specific interactions (double-layer and dispersion interactions) remains negligible in comparison to bulk-transfer distances (being of the order of micrometers). For larger particles, the hydrodynamic effects as well as external and specific forces play an increasingly important role and the flux can only be determined numerically by solving Eq. (132).

C. Exact Numerical Results

For uniformly accessible surfaces, Eq. (132) can be converted to the simple dimensionless form [7,18] suitable for numerical calculations:

$$\frac{\partial \bar{n}}{\partial \tau} = \frac{\partial}{\partial H} F_1(H) \left[\frac{\partial \bar{n}}{\partial H} + \frac{\partial \bar{\phi}}{\partial H} \bar{n} \right] + \frac{1}{2} \mathrm{Pe}\bar{n} \left[\frac{\partial}{\partial H} F_1(H) F_2(H)(H+1)^2 - 2F_3(H)(H+1) \right]$$

$$+ \frac{1}{2} F_1(H) F_2(H)(H+1)^2 \mathrm{Pe}\, \frac{\partial \bar{n}}{\partial H} \tag{160}$$

where the hydrodynamic correction functions F_1 to F_3 has been defined previously, $\bar{n} = n/n_b$, and $\tau = t D_\infty / a^2$ is the dimensionless time. As one can note, the particle radius was taken as the characteristic length scale L_{ch}. This is useful because the bulk-transfer length is comparable with particle dimensions in the case of adsorption of micrometer-sized particles.

The advantage of Eq. (160) (representing a parabolic partial differential equation from a mathematical viewpoint) is that it can be solved exactly by standard numerical techniques e.g., by the finite-difference Crank–Nicholson scheme under transient (nonstationary) conditions [37,64]. These calculations showed that the duration of the transient regimes is of the order of seconds, as previously estimated. Under the stationary conditions, Eq. (160) is simplified to the ordinary one-dimensional differential equation which can be solved by standard numerical techniques [18,76,118,119].

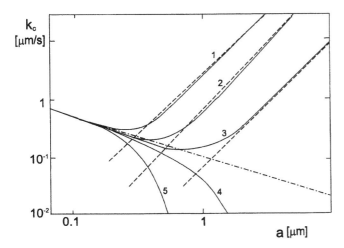

FIG. 22 The dependence of the bulk mass transfer rate k_c on particle radius a for the rotating disk calculated numerically from Eq. (160) ($T = 293$ K, $\omega = 25$ rad/s) and various apparent densities of the particle $\Delta\rho$: curve 1, 0.6 kg/dcm^3; curve 2, 0.3 kg/dcm^3; curve 3, 0; curve 4, -0.3 kg/dcm^3; curve 5, -0.6 kg/dcm^3 (the minus sign denotes the gravity force acting opposite the interface). The dashed line denotes the limiting value calculated from Eq. (162) (Stokes law), and the dashed-dotted line represents the results calculated from Eq. (161) (the interception governed flux). (From Ref. 37.)

Examples of such calculations are plotted in Fig. 22, which presents the dependence of the bulk mass-transfer-rate constant k_c on particle size [37] for the rotating disk case. In these calculations, the external force was gravity arising due to the specific density difference between the particle and the suspending medium (apparent $\Delta\rho$ density). The positive value of the apparent density denotes the gravity force acting toward the interface, whereas the negative value means the opposite situation. In Fig. 22, one can see that the effect of gravity and interception becomes negligible for a particle size below 0.5 µm because the flux attains the limiting values calculated from the SL approximation, Eq. (156). For particles larger than 0.5 µm and zero apparent density, the flux is governed by the interception effect given by the formula

$$-\frac{j_0}{n_b} = k_c = 0.51\frac{\omega^{2/3}}{\nu^{1/2}}a^2 \qquad (161)$$

On the other hand, for this particle size range and larger apparent density, the particle flux is governed by the Stokes equation [37]

$$-\frac{j_0}{n_b} = \frac{2}{9}\frac{\Delta\rho g a^2}{\eta} \qquad (162)$$

where g is the acceleration due to gravity.

A theoretical analysis of the numerical results like those shown in Fig. 22 is greatly simplified by introducing the dimensionless mass-transfer Sherwood number $\mathrm{Sh} = k_c a/D_\infty$ and dimensionless parameters characterizing the role of various forces in particle adsorption phenomena. Except for the previously defined Pe number describing the role of convection, one often defines the external force parameter Ex by

$$\mathrm{Ex} = \frac{|F_e(a)|a}{kT} \qquad (163a)$$

where $F_e(a) = -(d\phi/dh)_{h=a}$ is the interaction force at the distance $h = a$. In the case of gravity, $F_e = \frac{4}{3}\pi\Delta\rho g a^3$ and the external force number becomes

$$\text{Ex} = \text{Gr} = \frac{2}{9}\frac{\Delta\rho g a^3}{\eta D_\infty} \tag{163b}$$

where Gr is the gravity number. The role of the dispersion and specific interactions is characterized by the dimensionless parameters [37]

$$Ad = \frac{A_{12}}{6kT}$$

$$Dl = \frac{\phi_0}{kT} \tag{164}$$

$$\kappa a = \frac{a}{Le} = \left(\frac{2e^2 I a^2}{\varepsilon_0 \varepsilon kT}\right)^{1/2}$$

Using the above parameters, one can specify limiting laws governing particle adsorption under various regimes. The usefulness of this approach is demonstrated in Fig. 23, presenting the Sh versus Pe dependence calculated numerically in the case when attractive electrostatic interactions appeared (Type Ib energy profile). One can observe that the effect of electrostatic interactions played a significant role for $\text{Pe} > 10^{-2}$. In the limit of $\text{Pe} > 1$, this is well reflected by the universal dependence

$$\text{Sh} = \tfrac{1}{2}\text{Pe}(1 + H^*)^2 \tag{165}$$

where H^* is the effective interaction range discussed previously (see Fig. 8).

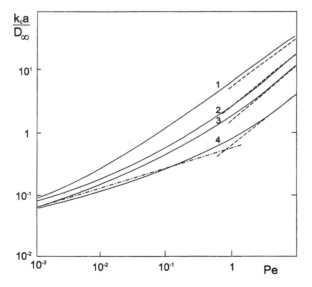

FIG. 23   The dependence of the reduced bulk transfer rate $k_c a/D_\infty$ (Sherwood number) on the Peclet number calculated numerically for a uniformly accessible surface in the case of no external force, $\phi_0 = -10^3$: curve 1, $a/Le = 2$; curve 2, $a/Le = 5$; curve 3, $a/Le = 10$; curve 4, $a/Le = 100$. The limiting values calculated from Eq. (156) are plotted by the dashed-dotted line and these calculated from Eq. (165) are depicted by the dashed line. (From Ref. 37.)

On the other hand, for $\text{Pe} < 10^{-2}$, the role of electrostatic interaction becomes negligible and the dimensionless flux is given by the limiting SL law, that is,

$$\text{Sh} = 0.62\text{Pe}^{1/3} \tag{166}$$

Similarly, for the external force regime, when $\text{Ex} \gg \text{Pe}$ ($Dl = 0$), one can define the limiting law as [37]

$$\text{Sh} = \text{Ex} \tag{167}$$

In the case of gravity-driven adsorption, $\text{Ex} = \text{Gr}$.

The advantage of the above-defined adsorption laws is that they can be applied universally for all uniformly accessible interfaces. Analogous results obtained numerically for nonuniformly accessible surfaces have been presented in Refs. 1, 18, and 76.

It should be remembered that these results are valid for linear transport conditions when the flux remains independent of particle coverage. Therefore, the coverage can be calculated from the surface mass balance equation postulating that the rate of adsorbed particles equals $-j$; thus,

$$\frac{dN}{dt} = -j(t) = k_c n^b \tag{168}$$

where N is the surface concentration of particles (number of particles per unit area of the interface). Because under linear conditions, j is independent of coverage, Eq. (168) can be integrated to the form

$$N = \frac{\Theta}{S_g} = -\int_0^t j(t')\,dt' \tag{169}$$

Using the nonstationary flux expression given by Eq. (150), one obtains the important result

$$N = \frac{\Theta}{S_g} = 2\sqrt{\frac{D_\infty t}{\pi}}n^b \tag{170}$$

This equation describes the nonstationary adsorption of particles at planar interfaces under diffusion-controlled transport conditions.

On the other hand, if the flux becomes stationary (convection-controlled transport), the change in N is given by the linear relationship

$$N = \frac{\Theta}{S_g} = N_0 + |j_0|t = N_0 + k_c n^b t \tag{171}$$

where N_0 is the surface concentration at the time when the flux becomes steady, equal to the limiting flux j_0.

IV. NONLINEAR ADSORPTION REGIMES

The results discussed in the previous section concerning the limiting flux under linear regimes have a large practical significance. They can validate experimental results concerning colloid or protein adsorption kinetics by indicating artifacts which may appear when indirect methods are used. If the experimentally measured flux is larger than the maximum value j_0, one should suspect that the transport mechanism has been modified (e.g., by natural convection). On the other hand, too small values of measured flux suggest surface contamination or aggregation of the suspension. Hence, the initial flux measurements in conjunction with theoretical predictions can be used as a very sensitive tool of surface homogeneity.

However, the linear transport conditions are relatively short-lasting, especially when concentrate particle suspensions are involved. Deviations from linearity stem from the presence of adsorbed particles, which exert specific and hydrodynamic forces on adsorbing (moving) particles, excluding them from part of the volume near the interface. This leads to the blocking effect, which should more appropriately be called the volume exclusion effect, responsible for the reduction in particle deposition rate at higher coverage. The blocking effects are dependent, in a complicated way, not only on particle coverage but also on particle distribution over the surface, particle size and shape, surface properties (charge), ionic strength, and transport mechanism. All of these many-body effects are coupled in a nonlinear way, which makes their rigorous analysis rather prohibitive. One has to accept approximate models only, reflecting the basic features of particle adsorption under nonlinear regimes.

A. Limiting Theoretical Models

An often introduced approach aimed at describing particle adsorption under nonlinear conditions is based on definition of the overall kineteic blocking function [1]

$$\tilde{B}(\Theta) = \frac{j(\Theta)}{j_0} \tag{172}$$

where j_0 is the previously discussed limiting flux and $j(\Theta)$ is the flux for a particle covered surface. The advantage of this definition is that the function $\tilde{B}(\Theta)$ can easily be determined experimentally by measuring the rate of particle adsorption as a function of time. However, this function is very specific, depending not only on coverage but on many other factors mentioned earlier, especially on particle interactions and the mechanism of particle transport (diffusion, flow, external force). Moreover, $\tilde{B}(\Theta)$ as defined via Eq. (172) also reflects the reversibility effect. This is illustrated by substituting the flux expressions previously derived for the reversible case given by Eq. (142) into Eq. (172). In this way, one obtains the formula

$$\tilde{B}(\Theta) = \frac{j}{j_0} = 1 - \frac{\Theta}{\Theta_e} \tag{173}$$

where $\Theta_e = K_a n^b S_g$ is the equilibrium coverage for the linear adsorption regime.

Equation (173) derived for no blocking effects apparently resembles the Langmuir model widely used for the interpretation of particle and protein adsorption kinetics [1,8,120–122]. In accordance with this model, one postulates that the adsorption flux is reduced by the factor $1 - \Theta/\Theta_{mx}$, where Θ_{mx} is the maximum coverage, usually found empirically. Using this hypothesis, one can formulate Eq. (172) as

$$\tilde{B}(\Theta) = \frac{j}{j_0} = 1 - \frac{\Theta}{\Theta_e} \tag{174}$$

where

$$\Theta_e = \Theta_{mx} \frac{K_a n^b S_g}{\Theta_{mx} + K_a n^b S_g}$$

is the equilibrium coverage. For $K_a n^b S_g \gg \Theta_{mx}$ (irreversible adsorption case), Eq. (174) assumes the simple form

$$\frac{j}{j_0} = 1 - \frac{\Theta}{\Theta_{mx}} \tag{175}$$

which is mathematically identical with that previously derived for reversible adsorption [Eq. (173)]. As can be seen, apparently similar blocking functions may reflect physically different situations. This indicates that more sophisticated approaches are needed to decouple the reversibility effects from the irreversible blocking governed by the geometrical factors.

One of the powerful methods for analysing these blocking effects is the concept of the available surface function (ASF) introduced by Widom [123,124]. The ASF function is defined as an averaged Boltzmann factor for a particle wandering within the two-dimensional adsorbing particle system frozen in a given configuration. Hence,

$$\mathrm{ASF} = \langle e^{-\Psi/kT} \rangle \tag{176}$$

where Ψ is the interaction energy of the adsorbing (wandering) particle with all other particles and $\langle\ \rangle$ indicates the value averaged over the system large enough to neglect the boundary effects.

In the case of noninteracting (hard) particles, the ASF, also called less accurately, the blocking function B_0 [1,12], has a simple geometrical interpretation as the area available to the wandering particle (see Fig. 24). The topology of this available surface area can be obtained by drawing a circle of radius $2a$ around each adsorbed particle [125–127]. Then, the ratio of the area lying outside of these circles to the overall area is just the ASF function, which depends not only on the particle coverage Θ (fraction of the surface area occupied by particles) but also on the particle distribution over the surface. Thus, the ASF concept is more general than the Langmuir model, which does not take into account the particle distribution effects.

The ASF function can be evaluated analytically in terms of power series of Θ, analogous to virial expansions [127–130] for some well-defined particle configurations. The expansions are rather cumbersome especially for higher particle coverage [127–130]. Therefore, the ASF is usually determined by numerical simulations of the Monte Carlo type [15,127,128,131], which

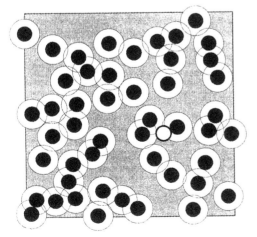

FIG. 24 A typical configuration of particles adsorbed at an interface (black disks); the white disks show the exclusion areas, whereas the shadowed zones represent the areas (targets) available for the wandering particle (ASF).

will be discussed later. Useful analytical expressions for the ASF can be derived, however, for the equilibrium (reversible) adsorption, which can be exploited as a reference value for the irreversible models. It was shown that for an equilibrium systems, when the adsorbed particle phase is treated as a two-dimensional liquid, the ASF function is connected with the chemical potential via the constitutive dependence [132]:

$$\text{ASF} = B_0(\Theta) = e^{-\mu_r/kT} \tag{177}$$

where $\mu_r = \mu - \mu_0 - kT \ln \Theta$ is the excess (residual) chemical potential characterizing the deviation from the ideal system behavior, $\mu = \mu_0 = kT \ln f(\Theta)\Theta$ is the chemical potential of the two-dimensional particle phase, and $f(\Theta)$ is the activity coefficient. The chemical potential can be calculated from the Gibbs–Duhem thermodynamic relationship

$$\mu(\Theta) = S_g \int \frac{d\pi(\Theta)}{\Theta} \tag{178}$$

where $\pi(\Theta)$ is the two-dimensional pressure of the adsorbed particle layer. Analytical expression for the pressure is known from the Reiss–Frich–Leobwitz (RFL) theory [133,134] generalized by Boublik [135] to convex particles of arbitrary shape. It reads

$$\pi(\Theta) = \frac{kT\Theta}{S_g} \frac{1 + (\gamma - 1)\Theta}{(1 - \Theta)^2} \tag{179}$$

where $\gamma = O^2/4S_g$ is the particle shape parameter, and O is the perimeter of the particle. For spheres, $\gamma = 1$. Combining Eqs. (178) and (179), one obtains the following expression for the particle chemical potential:

$$\mu = \mu^0 + kT \ln\left(\frac{\Theta}{1 - \Theta}\right) + kT(1 + 2\gamma)\frac{\Theta}{1 - \Theta} + kT\gamma\left(\frac{\Theta}{1 - \Theta}\right)^2 \tag{180}$$

Thus, the ASF (blocking function) becomes

$$\text{ASF} = B_0(\Theta) = (1 - \Theta)\exp\left[-(1 + 2\gamma)\frac{\Theta}{1 - \Theta} - \gamma\left(\frac{\Theta}{1 - \Theta}\right)^2\right] \tag{181}$$

It can be shown that in the limit of low coverage when $\Theta \ll 1$, Eq. (181) simplifies to the linear form

$$\text{ASF} = B_0(\Theta) = 1 - 2(\gamma + 1)\Theta = 1 - C_1\Theta \tag{182a}$$

where

$$C_1 = 2B_2 = 2(\gamma + 1) \tag{182b}$$

and B_2 is the second virial coefficient [132].

The C_1S_g coefficient reflects the area blocked by one particle. For spheres, it equals $4\pi a^2$ and becomes much larger for elongated particles (e.g., spheroids or cylinders), as shown in Table 4. Similar results can be derived for mixtures of particles. It was shown for a bimodal

TABLE 4 Jamming Coverage Θ_∞ and the First Two Coefficients of the Low-Coverage Expansion, Eq. (189)

Particle		Θ_∞	C_1	C_2
Circle/sphere	1:1	0.547	4.00	3.31
Square/cube	1:1	0.530	4.55	4.37
Ellipse/spheroid	1:2	0.583	4.38	4.19
Rectangle/cylinder	1:2	0.548	4.86	5.18
Ellipse/spheroid	1:5	0.536	6.47	10.72
Rectangle/cylinder	1:5	0.510	6.58	10.70

mixture that if disks having radii a_1 and a_2, the ASF functions, respectively, are given by the expressions [134]

$$\text{ASF}_1 = (1 - \Theta) \exp\left[-\frac{3\Theta_1 + (1/\gamma_a)(1/\gamma_a + 2)\Theta_2}{1 - \Theta} - \left(\frac{\Theta_1 + (1/\gamma_a)\Theta_2}{1 - \Theta}\right)^2 \right]$$

$$\text{ASF}_2 = (1 - \Theta) \exp\left[-\frac{3\Theta_2 + \gamma_a(\gamma_a + 2)\Theta_1}{1 - \Theta} - \left(\frac{\Theta_2 + \gamma_a\Theta_1}{1 - \Theta}\right)^2 \right]$$

(183)

where $\Theta = \Theta_1 + \Theta_2$ and $\gamma_a = a_1/a_2$ is the particle ratio. It was shown in Ref. 136 that, in the case of spheres, one can approximate the ASF function by Eq. (183) if the parameter γ_a is defined as

$$\gamma_a = 2\sqrt{\lambda} - 1 \tag{184}$$

where $\lambda = a_1/a_2$ is the larger-to-smaller particle size ratio.

In the case for irreversible adsorption, the above expressions for ASF can be used as an approximation only because usually the distribution of particles differs from the equilibrium configurations [127]. These nonequilibrium configurations, better reflecting various transport mechanisms, can be generated in numerical simulations performed according to various models. One of the most widely used is the random sequential adsorption (RSA) approach used both for lattice adsorption cases [129] and extended to continuous surfaces [125–128,130–132]. An important feature of this model, which will be discussed next, is that it allows one to generate not only irreversible particle populations but also to determine the blocking function and maximum (jamming) coverages, which are of primary practical significance.

1. The RSA Model

The basic assumption of the RSA model [125–129] are as follows:

1. Particles are placed at random over an homogeneous adsorption plane; every position is accessible with equal probability.
2. If the trial (virtual) particle overlaps with any of previously adsorbed particles, it is not adsorbed and a new adsorption attempt is made.
3. Otherwise, the particle is placed with unit probability (hard-sphere model) or with the probability calculated from the Boltzmann distribution. Once the particle is adsorbed, its position is permanently fixed (localized adsorption postulate).
4. The process is continued until the entire surface is completely covered with particles and the maximum (jamming) coverage Θ_{mx} is attained.

The simulation process based on the above RSA scheme is very efficient for hard particles adsorbing flat (side-on) at the interface [126–130]. However, in the case of interacting particles of nonspherical shape (e.g., spheroids) adsorbing under random orientations, the numerical calculations become rather tedious due to mathematical problems in evaluating the minimum surface-to-surface distance and the interaction energy [15,131].

The blocking function is evaluated in the RSA calculations by performing, for a given coverage, a large number of virtual adsorption attempts N_{att}, out of which N_{suc} were potentially successful. Then, $B_0(\Theta)$ is calculated from the expression

$$B_0(\Theta) = \frac{N_{suc}}{N_{att}}, \quad N_{att} \to \infty \tag{185}$$

This function calculated numerically for spherical particles [127] is shown in Fig. 25. As one can note, for the low coverage range, $B_0(\Theta)$ can be well described by the expansion derived in Refs. 12 and 127:

$$B_0(\Theta) = 1 - 4\Theta + \frac{6\sqrt{3}}{\pi}\Theta^2 + \left(\frac{40}{\pi\sqrt{3}} - \frac{176}{3\pi^2}\right)\Theta^3 + O(\Theta^4) \tag{186}$$

It was demonstrated that for the entire coverage range, the $B_0(\Theta)$ function can be well approximated by the fitting polynomial [127]

$$B_0(\Theta) = \left[1 + 0.812\frac{\Theta}{\Theta_{mx}} + 0.426\left(\frac{\Theta}{\Theta_{mx}}\right)^2 + 0.0716\left(\frac{\Theta}{\Theta_{mx}}\right)^3\right]\left(1 - \frac{\Theta}{\Theta_{mx}}\right)^3 \tag{187a}$$

where Θ_{mx} is the maximum coverage determined a priori from simulations. For noninteracting spherical particles Θ_{mx}, referred to as the jamming coverage Θ_∞, equals 0.547 [126,129]. From

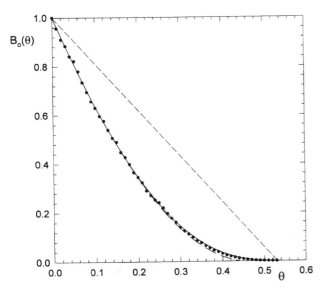

FIG. 25 The surface-blocking function $B_0(\Theta)$ (ASF) for spherical particles. The points denote the numerical Monte Carlo simulations, the dashed-dotted line denotes results calculated from the low-coverage expansion, Eq. (186); the dashed line shows the results calculated from the Langmuir model, and the continuous line represents the results calculated from the fitting function, Eq. (187a).

Eq. (187a), one can deduce that for $\Theta \to \Theta_{mx}$, the blocking function is given by the asymptotic formula

$$B_0(\Theta) = 2.314\left(1 - \frac{\Theta}{\Theta_{mx}}\right)^3 \tag{187b}$$

As one can note, the blocking function deviates completely from the Langmuir model postulating that $B_0(\Theta) = 1 - \Theta/\Theta_{mx}$. The blocking effects predicted by the RSA model are, therefore, considerably more pronounced. This is so, because due to topological constraints, only a small fraction of the free surface $1 - \Theta$ is available for particle adsorption. This is spectacularly demonstrated in Fig. 26, in which the particle configurations generated in the RSA simulations of hard spheres at various stages (Θ ranging from 0.10 to 0.4) is presented. The available surface area (hole) distribution is also shown for particles of the same size as the 2.2 times larger particles (this will mimic the experimental size range discussed later). One can note the significant change of the topology of the holes, which assume, for later adsorption stages, the form of isolated targets capable of accommodating only one additional particle. One can also qualitatively predict that the probability of larger-particle adsorption over surfaces covered with smaller particles will decrease dramatically with the particle size ratio.

This is demonstrated quantitatively in Fig. 27, in which the blocking function of larger particles as precovered surfaces is presented. This function reflects the averaged probability of adsorbing a large particle over a surface precovered by smaller particles (characterized by the coverage Θ_s). As can be noted, the results obtained for the irreversible RSA models are well reflected by the equilibrium SPT results calculated from Eq. (183) expressed as [136]

$$B_0(\Theta_s) = (1 - \Theta_s)\exp\left[-\frac{(4\lambda - 1)\Theta_s}{1 - \Theta_s} - \left(\frac{(2\sqrt{\lambda} - 1)\Theta_s}{1 - \Theta_s}\right)^2\right] \tag{188}$$

$a_l/a_s=1$ $a_l/a_s=2.2$

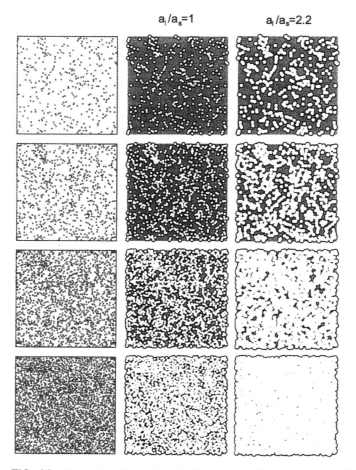

FIG. 26 Typical configuration of adsorbed particles generated in the RSA simulation; the white areas show the exclusion zones and the shadowed zones represent the areas (targets) available for the wandering particle (ASF) having the same size as the adsorbed particle and 2.2 times the larger size.

This is also the case for monodisperse particle adsorption, which confirms the conclusion that the equilibrium expressions for $B_0(\Theta)$ can be exploited as useful reference data for irreversible adsorption processes. As can be observed, however, for higher coverage, $\Theta_s > 0.3$, the deviations between RSA and reversible equilibrium adsorption become more noticeable.

Analogous results, concerning the blocking function, have been obtained for polydisperse mixture adsorption [136–138], side-on adsorption of spheroids (ellipses) [128,132,139,140], cylinders and spherocylinders [140,141]. On the other hand, the case of an unoriented adsorption of prolate and oblate spheroids has been treated in Refs. 15 and 131. Examples of particle monolayers close to the jamming state obtained in these calculations are shown in Fig. 28.

It was demonstrated that for nonspherical (anisotropic) particles the blocking function for small Θ can be well reflected by the series expansion

$$B_0(\Theta) = 1 - C_1\Theta + C_2\Theta^2 + \cdots \qquad (189)$$

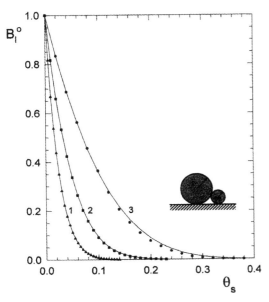

FIG. 27   The blocking function (ASF) B_l^0 for adsorption at surfaces precovered by smaller-sized particles (characterized by the coverage Θ_s). The points denote the exact numerical simulations performed according to the RSA method: curve 1, $a_l/a_s = 10$; curve 2, $a_l/a_s = 5$; curve 3, $a_l/a_s = 2.2$. The continuous lines denote the equilibrium SPT results calculated from Eq. (188).

where the coefficients C_1 and C_2 are given in Table 4 for side-on adsorption of various particles.

On the other hand, for coverage approaching Θ_{mx}, the blocking function is well reflected by the expression that is the generalized version of Eq. (187b):

$$B_0(\Theta) \cong C_{mx}\left(1 - \frac{\Theta}{\Theta_{mx}}\right)^m \tag{190}$$

where C_{mx} is the dimensionless constant of the order of unity, $m = 3$ for monodisperse spheres [126,127], $m = 4$ for polydisperse spheres and side-on adsorption of anisotropic particles [139–141], and $m = 5$ for anisotropic particles under the unoriented adsorption regime [15,131].

As one can note, the most important parameter governing particle adsorption for larger coverages is the jamming coverage Θ_{mx}, which can be determined in an *ab initio* manner (without introducing any empirical parameters) from the RSA simulations. This has a considerable practical significance. Therefore, in Table 4, the jamming coverages for hard particles of various shape adsorbing side-on are collected. One can note that $\Theta_{mx} = \Theta_\infty$ usually varies between 0.58 and 0.51, decreasing for elongated particles (ellipses or cylinders). It should be emphasized that all of the numbers shown in Table 4 are universal in the sense that they do not depend on particle and interface size and so forth.

Because under all practical situations, adsorbing particle populations are to some degree polydisperse, it is vital to estimate the influence of the polydispersity parameter on the jamming coverage. This was done in Ref. 137 by performing RSA simulations for polydisperse particle mixtures characterized by the Gauss and uniform size distributions characterized by the relative standard deviation $\bar\sigma_p$ (polydispersity parameter).

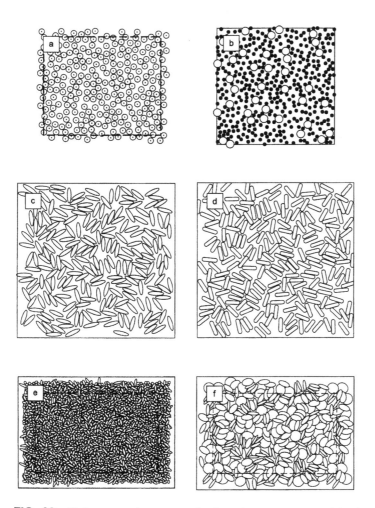

FIG. 28 Various monolayers near the jamming state generated in the RSA simulations: (a) spherical particles at homogeneous surfaces; (b) spherical particles at precovered surfaces; (c) spheroidal particles (ellipses) adsorbing side-on; (d) spherocylinders adsorbing side-on; (e) prolate spheroids (unoriented adsorption); (f) oblate spheroids (unoriented adsorption).

The results plotted in Fig. 29 (for uniform particle size distribution) indicate that the jamming coverage increases with the polydispersity parameter, which can be well reflected by the fitting functions [137]

$$\Theta_{\infty} = 0.547 + 0.53\bar{\sigma}_p \tag{191a}$$

where Θ_{∞} is calculated as the sum extended over all adsorbed particles, $\pi \sum N_i a_i^2$, where a_i is the individual particle radius. As one can note, Θ_{∞} increases linearity with $\bar{\sigma}_p$. If the coverage is calculated as usually done in experiments by considering the averaged particle radius $\langle a \rangle$ (i.e., from the dependence $\Theta_{\infty} = \langle a \rangle^2 N$), then the fitting function becomes [137]

$$\Theta_{\infty} = 0.547 + 0.458\bar{\sigma}_p + 6.055\bar{\sigma}_p^2 \tag{191b}$$

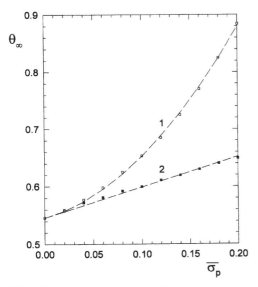

FIG. 29 The dependence of the jamming coverage Θ_∞ on the polydispersity parameter $\bar{\sigma}_p$ for polydisperse mixtures of hard particles characterized by a uniform size distribution: curve 1, Θ_∞ calculated as $\pi \langle a \rangle^2 N$; curve 2, Θ_∞ calculated as $\pi \sum a_i^2 N_i$ (true surface coverage). The broken lines denote the results calculated from the fitting functions, Eqs. (191a) and (191b). (From Ref. 137.)

It is also interesting from an experimental viewpoint to determine jamming coverages for nonuniform (heterogeneous) surfaces. The first type of heterogeneity arises when a uniform interface is covered with a given amount of smaller particles Θ_s which can be treated as the heterogeneity parameter (the blocking function predicted in this case is shown in Fig. 27). If larger particles can only adsorb on uncovered areas, their jamming coverage is reduced in comparison with adsorption over uniform surfaces [136]. This is illustrated in Fig. 30, showing the dependence of Θ_l^∞ (large-particles coverage) on the heterogeneity parameter Θ_s. As one can note, the jamming coverage Θ_l^∞ decreases abruptly with the heterogeneity degree, especially for the larger-particle-size ratio λ. This would suggest that the heterogeneity of an interface (the presence of adsorbed particles) can be well detected by adsorption of larger particles serving as markers. Another interesting feature of adsorption at precovered surfaces is that the net surface coverage, $\Theta_l + \Theta_l^\infty$, passes through a minimum whose depth increases considerably with λ. Thus, the minimum net coverage was found equal to 0.378 for $\lambda = 2.2$ ($\Theta_s = 0.34$), 0.261 for $\lambda = 5$ ($\Theta_s = 0.24$), and 0.181 for $\lambda = 10$ ($\Theta_s = 0.16$). These results represent a spectacular manifestation of the irreversibility effect because the composition and density of adsorbed particle monolayers is dependent on the path of adsorption, which can be physically realized by replacing the smaller-particle suspension by a larger particle suspension after attaining a desired Θ_s.

An opposite situation, very important for protein adsorption appears for heterogeneous surfaces of the second type. In this case, particles can be adsorbed at the heterogeneities (active centers) only, modeled by smaller-sized spherical particles, but not at the interface. The dependence of the jamming coverage of larger particles, Θ_l^∞, on the heterogeneity degree is plotted in Fig. 31. As can be seen for Θ_s of a few percent only, the jamming coverage of larger particles reaches the value characteristic for uniform surfaces, (i.e., 0.547). A peculiar situation arises for the particle-to-heterogeneity size ratio $\lambda = 2$ when the jamming coverage attains a maximum for $\Theta_s = 0.2$ and then drops to the limiting value of 0.547. The results shown in Fig. 31 indicate that a very small coverage of the coupling agent (polyelectrolyte or antibody) is needed to attain the maximum possible efficiency of protein or particle adsorption.

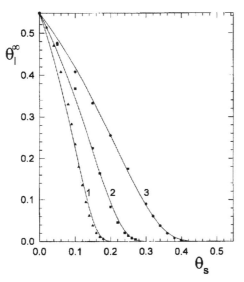

FIG. 30 The dependence of the jamming coverage of larger particles Θ_i^∞ at surfaces precovered by smaller particles (heterogeneities) characterized by the coverage Θ_s. The points denote the numerical RSA simulations performed for: curve 1, $\lambda = 10$; curve 2, $\lambda = 5$; curve 3, $\lambda = 2.2$. The lines represent the fitting functions [136] given by the equation $\Theta = 0.547\,[c/(c - \Theta_s)]^2$, where $c = 0.274$ ($\lambda = 10$), $c = 0.404$ ($\lambda = 5$), and $c = 0.596$ ($\lambda = 2.2$).

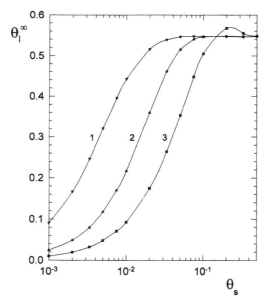

FIG. 31 The dependence of the jamming coverage of larger particles Θ_i^∞ at heterogeneous surfaces (adsorption at "active centers" having the coverage Θ_s). The points denote the numerical RSA simulations performed for curve 1, $\lambda = 10$; curve 2, $\lambda = 5$; curve 3, $\lambda = 2$. The continuous lines denote the interpolating functions.

It should be mentioned that the shape of most proteins deviates significantly from a spherical shape and can be well approximated by a prolate spheroid [2,12]. Therefore, the jamming coverage for spheroidal particles of various axes ratios are of considerable significance. Some data for the side-on orientation of the spheroid are given in Table 4. However, adsorption of anisotropic particles can proceed under various orientations, provided that the interaction energy with the interface is large enough to ensure irreversible binding. The energy can be estimated from the data discussed in Section I. At later stages, the particles adsorb under orientations close to perpendicular (see Fig. 28), which significantly increases the jamming coverage. This effect is illustrated in Fig. 32, in which the dependence of Θ_∞ on the shorter-to-longer axis ratio $b/a = A$ of elongated spheroids is presented. It was found that the numerical results can be well approximated by the interpolating functions

$$\Theta_\infty = 0.622 \left(A + \frac{1}{A} - 1.997 \right)^{0.0127} e^{-0.0274(A+1/A)} \tag{192a}$$

for the side-on adsorption and

$$\Theta_\infty = 0.304 + \frac{0.365}{A} - 0.123A \tag{192b}$$

for the unoriented adsorption [131].

The data shown in Fig. 32 can be used for the determination of the influence of particle shape, at a fixed volume, on the mass of the adsorbed particle monolayer. This type of information is especially important for protein and polymer adsorption when the amount of

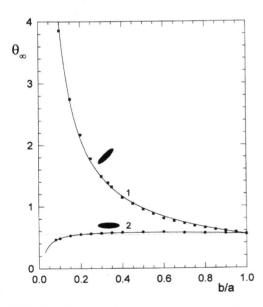

FIG. 32 The jamming coverage $\Theta_\infty = \pi a b N_\infty$ of prolate spheroids as a function of the b/a parameter. The points denote numerical results calculated in Ref. 57 for the side-on adsorption (curve 2) and in Ref. 131 for the unoriented adsorption (curve 1). The continuous lines denote the fitting functions given by Eqs. (192a–192b). (From Ref. 12).

adsorbed substance is usually expressed as mass per unit area [2]. For the spherical particles, the mass of the adsorbed particles per unit area is given by the simple relationship

$$m_1 = \tfrac{4}{3}\rho a \Theta_\infty = 0.729\rho a \tag{193a}$$

where ρ is the particle specific density. For spheroidal particles of the same volume and density, one has, at the jammed state,

$$m_A = \tfrac{4}{3}\rho b \Theta_\infty(A) \tag{193b}$$

Considering Eqs. (192a) and (193), one can express the m_A/m_1 ratio as

$$\frac{m_A}{m_1} = 1.137 A^{1/3}\left(A + \frac{1}{A} - 1.997\right)^{0.0127} e^{-0.0274(A+1/A)} \tag{194a}$$

for the side-on adsorption of prolate spheroids and

$$\frac{m_A}{m_1} = 0.667 A^{-2/3} + 0.556 A^{1/3} - 0.225 A^{4/3} \tag{194b}$$

for unoriented adsorption. As can be deduced from these equations, in the case of unoriented adsorption, the monolayer mass becomes infinite in the limit of b/a tending to zero. The results stemming from Eqs. (194a) and (194b) are plotted in Fig. 33. Analogous dependencies for oblate spheroids were discussed in Ref. 142.

The jamming coverages have a special significance for a quantitative analysis of particle adsorption because they are universal quantities which can be used for an efficient elimination of various artefacts and estimation of isolated particle mass (size) (e.g., in the case of proteins).

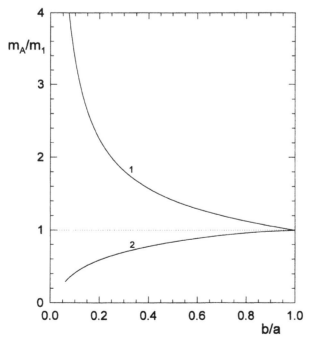

FIG. 33 The normalized monolayer mass of prolate spheroids m_A/m_1 calculated from Eqs. (194a) and (194b) for prolate spheroids: curve 1, unoriented adsorption; curve 2, side-on adsorption.

2. RSA of Interacting Particles

All of the above results concern hard-particle adsorption processes when the specific interactions among particles are negligible. As mentioned, this adsorption regime is likely to occur in concentrated electrolyte solutions when the electrostatic interactions are effectively eliminated. This situation is relevant to protein adsorption under physiological conditions. However, for dilute electrolytes and small particles, the repulsive electrostatic interactions (described by the Type Ia energy profile) exert a significant influence on particle adsorption—in particular, on the blocking function and jamming coverage. Due to the complexity of these interactions, no complete theory of RSA processes of interacting particles has been formulated yet. There exist approximate models, however, reflecting the most important features of interacting particle adsorption. One of such extended RSA models, developed in Refs. 1, 15, 143, and 144, is based on the local equilibrium assumption. In accordance to this approach, the probability of particle adsorption is calculated from the Boltzmann distribution, i.e., $p = e^{-\phi_p/kT}$, where ϕ_p is the net interaction of the adsorbing (wandering) particle with all absorbed particles calculated by assuming a pairwise additivity. With this modification, one can perform efficient RSA-type simulations of interacting particle adsorption. In this way, various configurations, blocking functions, and jamming coverages have been calculated not only for spherical [143,144] but also for anisotropic (spheroidal) particles [15]. In the latter case, the interaction energy was calculated using the Derjaguin and the ESA models discussed in Section I. It was found that the results of these numerical calculations, which are significantly more time-consuming than the classical RSA simulations, can be well reflected by the effective hard-particle (EHP) concept. This approach was developed originally by Barker and Henderson [61] to describe the equation of state of simple fluids composed of spherical molecules. According to this approach, the interacting particles are treated as hard ones having a larger effective radius $a^* = a + h^*$, where h^* is the effective interaction range (see Fig. 8). For spherical particles, h^* can be explicitly evaluated from the equation [143]

$$\frac{h^*}{a} = H^* = \sqrt{\frac{1}{2a^2}\int_0^\infty (1 - e^{-\phi(r)/kT})r\,dr - 1} \tag{195}$$

where ϕ is the pair interaction energy.

By introducing the EHP concept, one can express the $C_1 \div C_2$ constants in the expansion Eq. (189) as [143]

$$\begin{aligned}
C_1^* &= C_1(1 + H^*)^2 \\
C_2^* &= C_2(1 + H^*)^4
\end{aligned} \tag{196}$$

It was demonstrated in Ref. 143 by performing the RSA simulations that instead of Eq. (195), one can use the much simpler analytical expression Eq. (77), which directly connects h^*/a with the range of the electrostatic interactions Le. This is also the case for spheroidal particle adsorption, as shown in Ref. 15. For $a/Le \ll 1$, one can predict from Eq. (77) that

$$H^* = \xi\frac{Le}{a} \tag{197}$$

where ξ is the proportionality constant equal to $\frac{1}{2}\ln(\phi_0/\phi_{ch})$ for two spherical particles.

The validity of Eq. (197) is confirmed by the results plotted in Fig. 34. As can be seen, the linear dependence given by Eq. (197) reflects well the exact numerical results derived from the RSA simulation for spheres and spheroids. The proportionality constant ξ was found to be equal to 2.3 (for $\phi_0 = 200$, which is a typical value for the colloid particle interactions). This indicates

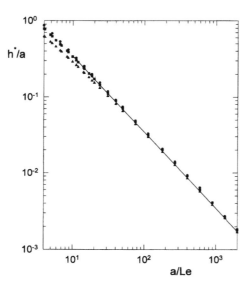

FIG. 34 The dependence of the effective interaction range h^*/a on the a/Le parameter. The points represent the exact numerical results derived from numerical simulations for spheres (circles), prolate spheroids $A = 0.5$ (squares), and prolate spheroids $A = 0.2$ (triangles); the line denotes the limiting analytical results calculated from Eq. (197). (From Ref. 12.)

that the effective interaction range h^* for colloid particles exceeds the electrostatic interaction screening length Le significantly.

It was also demonstrated in Ref. 143 that the EHP concept works well for higher particle coverage, approaching the jamming coverage which is referred to as the maximum coverage Θ_{mx} in the case of the interacting particles. Thus, for prolate spheroids, the values determined from simulations can be well interpolated by the function [15]

$$\frac{\Theta_{mx}}{\Theta_\infty} = \bar{\Theta} = \frac{2.07 + 0.811A + 2.37A^2 - 1.25A^3}{(2.07 + 0.811A^* + 2.37A^{*2} - 1.25A^{*3})(1 + A^*)(1 + H^*/A)} \tag{198a}$$

where $A^* = (A + H^*)/(1 + H^*)$. For spheres, Eq. (198a) reduces to the simple form

$$\Theta_{mx} = \frac{\Theta_\infty}{(1 + H^*)^2} = \frac{0.547}{(1 + H^*)^2} \tag{198b}$$

This formula, in combination with Eq. (197) defining H^*, can be used for a prediction of the ionic-strength effects on the maximum coverage, which is a parameter of primary interest from the experimental viewpoint.

On the other hand, by using Eq. (198b) in conjunction with Eq. (187a), an interpolating function can be specified that describes blocking effects for interacting particles.

The universal correction function $\Theta_{mx}/\Theta_\infty$ defined by Eq. (198a) is compared with exact RSA simulations in Fig. 35, for spheres and prolate spheroids, characterized by $A = 0.5$ and $A = 0.2$, respectively. As can be seen, Eq. (198a) gives a reasonable estimation of the maximum coverage for $a/Le > 5$, although a slight tendency to overestimate the simulation results is visible. The good agreement of simulations with Eq. (198b) suggests that for spheres, the

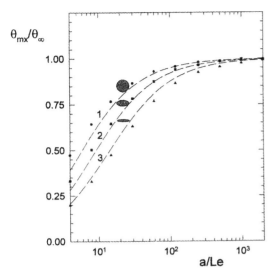

FIG. 35 The reduced maximum coverage of interacting particles $\Theta_{mx}/\Theta_\infty$ (where Θ_∞ is the jamming coverage of hard particles) versus the a/Le parameter. The points denote the numerical RSA simulations performed for curve 1, $A = 1$ (spheres); curve 2, $A = 0.5$ (prolate spheroids, side-on adsorption); curve 3, $A = 0.2$ (prolate spheroids, side-on adsorption). (From Ref. 12.)

effective range of interactions h^* can be easily determined experimentally, when Θ is measured. In this case, by inverting Eq. (198b), one obtains

$$h^* = a\sqrt{\frac{\Theta_{mx}}{0.547} - 1} \qquad (199)$$

The advantage of the classical RSA model discussed earlier is that it allows one to generate particle configurations in an efficient way and to determine the jamming or maximum coverage, which has practical significance. However, the kinetic aspects of irreversible particle adsorption cannot be unequivocally derived using the RSA model. This is so because this approach considers a strictly two-dimensional situation (adsorption in a plane) by neglecting all peculiarities of particle transfer from the bulk to the energy minimum. Therefore, the RSA model can only be used for modeling the kinetics of an idealized process consisting in the creation of particles at the primary minimum with a constant rate. When an occupied area is found, the particle is removed with unit probability and the next particle creation attempt is undertaken, uncorrelated for previous attempts. The kinetics of this hypothetical process is then governed by the formula

$$\frac{1}{S_g}\frac{d\Theta}{dt} = r_c B_0(\Theta) \qquad (200)$$

where r_c is the rate of particle creation (number of particles per unit area and unit of time). It is not possible, within the framework of the RSA model, to find a unique relationship between the kinetics of this idealized process and the kinetics of any real particle adsorption phenomenon. In particular, one cannot specify links between the B_0 blocking function and the overall blocking function $\tilde{B}(\Theta)$ defined by Eq. (172).

Many attempts have been undertaken in the literature to avoid this limitation of the classical RSA model. In one of the improved models, particle diffusion has been considered [145,146], which partially accounts for the correlation of adsorption attempts. However, all of the specific and hydrodynamic interactions among the wandering and the adsorbed particles were neglected. Although the structure of the transient configurations was found to be different from the classical RSA, the jamming limit for hard spheres was the same [145]. Also, the blocking function B_0 in the limit of high coverages was governed by Eq. (190) with the exponent $m = 5/2$. However, no relationship between B_0 and the flux correction function given by Eq. (172) can be derived from this approach referred to as the DRSA model.

Other models considering correlations in particle adsorption processes have also been formulated [129]—in particular, the ballistic model [146,148] suitable for describing the external-force (gravity)-driven adsorption [149,150]. The basic assumption of this model is that after a failed adsorption attempt, the particle is not removed but it can roll over the adsorbed particles to find a stable position at the interface. Thus, a particle can be prevented from adsorption by a trap consisting of at least three particles. This will exert a pronounced effect on the blocking function B_0 because the first two terms of the series expansion, Eq. (189), disappear. Therefore, according to the ballistic model, the blocking function in the limit of low coverage is given by the expression [146]

$$B_0(\Theta) = 1 - 9.95\Theta^3 + O(\Theta^4) \tag{201}$$

It was also found that the jamming coverage for hard spheres equals 0.61 (i.e., slightly larger than for the RSA model).

However, as for DRSA, the ballistic model cannot be exploited for determining the overall blocking function. These limitations can be avoided in the recently developed generalized RSA model discussed next.

3. The Generalized RSA Model

This model better reflects the physical reality of particle adsorption processes by considering a three-dimensional motion of the wandering particle within the adsorption layer (see Fig. 36a). Thus, the ASF (blocking function) depends not only on particle coverage and structure but also on the distance from the interface h [112]. One may, therefore, postulate that the ASF is connected with the activity coefficient occurring in the expression for the chemical potential, Eq. (130), by the simple relationship

$$B(\Theta, h) = \frac{1}{f(\Theta, h)} \tag{202}$$

An important property of the generalized ASF (blocking function) is that it approaches, for $h = \delta_m$, the value of $B_0(\Theta)$, defined by Eq. (176) [12,112]. Accordingly, the overall potential occurring in Eq. (131) assumes the form

$$\Phi = \phi - kT \ln B(\Theta, h) \tag{203}$$

This equation indicates that the presence of adsorbed particles is leading to a steric barrier of an extension equal to the effective particle size (see Fig. 36b). The height of the barrier at $h = \delta_m$ equals $-kT \ln B_0(\Theta)$ and the barrier vanishes at $h = 2a^* = \delta_a$.

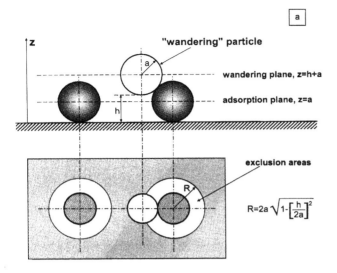

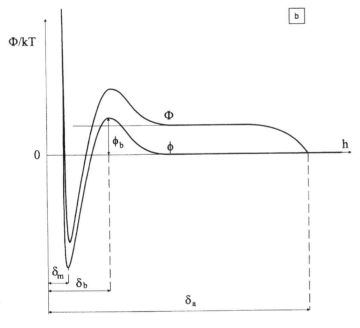

FIG. 36 (a) A schematic representation of the generalized blocking function (ASF); (b) the steric barrier generated due to adsorbed particles. (From Ref. 2.)

By considering Eq. (203), one can integrate Eq. (131) under quasistationary conditions within the adsorption layer $\delta_m < h < \delta_a$, which leads to the general expression for particle flux [112,116]

$$-j = k_a n(\delta_a)\bar{B}(\Theta) - k_d S_g^{-1} \frac{\bar{B}(\Theta)}{B_0(\Theta)} \Theta \tag{204}$$

where k_a and k_d are the adsorption and desorption constants as defined previously and $\bar{B}(\Theta)$ is the generalized blocking functions (transport resistance of the adsorbed layer) given by

$$\bar{B}(\Theta) = \left(k_a \int_{\delta_m}^{\delta_a} \frac{e^{\Phi/kT}}{D(h)} dh \right)^{-1} = \left(k_a \int_{\delta_m}^{\delta_a} \frac{e^{\phi/kT}}{B(\Theta, h)D(h)} dh \right)^{-1} \tag{205}$$

Note that in contrast to what is usually assumed, the desorption flux in the generalized RSA approach depends in a complicated manner on the blocking function $\bar{B}(\Theta)$. As can be deduced from Eq. (205), this function reflects the effect stemming from both adsorbed particles and the particle–interface interactions or external forces described by the interaction potential ϕ.

Under equilibrium, when $j = 0$, Eq. (204) assumes the form of an isotherm:

$$K_a n(\delta_a) = \frac{\Theta_e}{B_0(\Theta_e)} \tag{206}$$

where Θ_e is the equilibrium coverage. Substituting the expression for $B_0(\Theta_e)$ given by Eq. (181), one obtains the RFL isotherm equation

$$K_a n(\delta_a) = \frac{\Theta_e}{1 - \Theta_e} \exp\left[(1 + 2\gamma)\frac{\Theta_e}{1 - \Theta_e} + \gamma\left(\frac{\Theta_e}{1 - \Theta_e}\right)^2 \right] \tag{207}$$

Under nonequilibrium situations, Eq. (204) can be used for predicting particle adsorption kinetics, which requires, however, an explicit evaluation of the blocking function $\bar{B}(\Theta)$. This is cumbersome because $\bar{B}(\Theta)$ depends not only on particle coverage but also on particle distribution over the surface. This distribution is, in turn, governed by the particle-transport mechanism (i.e., diffusion) or external force. At the present time, these effects cannot be decoupled within the framework of the generalized RSA model. However, useful approximations for $\bar{B}(\Theta)$ can be specified for limiting adsorption regimes having major practical significance. This is the case when the interaction potential exhibits a well-defined maximum (energy barrier) described by the Type II energy profile (see Fig. 9). Then, according to Eq. (205), $\bar{B}(\Theta)$ becomes

$$\bar{B}(\Theta) = B(\Theta, \delta_b) \cong B_0(\Theta) \tag{208}$$

This is so because usually the energy barrier distance δ_b is much smaller than the particle dimension. Hence, under such conditions, the overall blocking function can be well approximated by the classical RSA model. A physical explanation of this fact is that all particles which overlap with adsorbed particles are removed due to the presence of strong repulsive forces (energy barrier). Hence, the adsorption events are uncorrelated, as postulated in the RSA model. This has considerable practical significance because all of the previous results pertinent to the classical RSA model retain their validity. These conditions were fulfilled in the experiments performed in Refs. 143 and 144 using the impinging-jet cells when the gravity force was directed outward from the surface.

Using Eq. (204), one can formulate the following kinetic equation under irreversible adsorption conditions ($k_d = 0$):

$$\frac{1}{S_g} \frac{d\Theta}{dt} = k_a n(\delta_0)\bar{B}(\Theta) \tag{209}$$

Comparing this expression with Eq. (200), one can deduce that $r_c = k_a n(\delta_a)$.

Equation (209) has considerable practical significance because it contains no adjustable parameters [the adsorption constant k_a can be calculated directly from Eq. (143) when the specific interaction potential is known]. Therefore, one can unequivocally determine particle adsorption kinetics by integrating Eq. (209) using the blocking function known from the RSA simulations discussed earlier—in particular, Eqs. (186)–(187b).

It was shown in Ref. 12 that Eq. (208) is also applicable for higher coverage even if there is no specific force barrier. This is so because the $B(\Theta, h)$ function approaches zero for $\Theta \to \Theta_{mx}$, which means that the steric barrier height $-kT \ln B(\Theta, h)$ becomes much larger than the kT unit. As a result, the maximum contribution to the integral, Eq. (205), stems from the region close to the interface, where $B(\Theta, h) = B_0(\Theta)$, so

$$\bar{B}(\Theta) \cong C'_{mx}\left(1 - \frac{\Theta}{\Theta_{mx}}\right)^m \tag{210}$$

where the C'_{mx} coefficient of the order of unity is slightly dependent on the primary minimum distance [12].

Except for the barrier case, $\bar{B}(\Theta)$ can also be evaluated for the diffusion-controlled adsorption conditions (no barrier) in the limit of low coverages. It was shown [112] that $\bar{B}(\Theta)$ in the case of spherical particles can be approximated by the series expansion

$$\bar{B}(\Theta) = 1 - C'_1\Theta + C'_2\Theta^2 + O(\Theta^3) \tag{211}$$

The coefficients C'_1 and C'_2 which depend on the primary minimum distance δ_m were found similar to the RSA case [112].

The significance of the extended RSA model is that, knowing $\bar{B}(\Theta)$, one can specify by Eq. (204) the nonlinear kinetic boundary conditions for the bulk-transport problems similar to that previously done for the linear adsorption regime. By solving an appropriate transport equation with these boundary conditions, one can describe quantitatively the adsorption of colloid particles and proteins for the entire range of surface coverage. This approach was exploited in [2] for evaluating the irreversible, diffusion-controlled adsorption. Particle flux and the dependence of coverage on time (adsorption kinetics) were calculated numerically by using the finite-difference method. It was shown that adsorption kinetics can be characterized unequivocally in terms of the dimensionless adsorption constant defined as

$$\bar{k}_a = \frac{2}{3\Phi_v\bar{R}_b} \tag{212}$$

where $\Phi_v = \frac{4}{3}\pi a^3 n^b$ is the volume fraction of particles and

$$\bar{R}_b = \frac{D_\infty}{2a}R_b = \frac{D_\infty}{2a}\int_{\delta_m}^{\delta}\frac{e^{\phi/kT}}{D(h)}dh$$

Due to very small value of Φ_v for proteins and colloid particles [2], the value of $\bar{k}_a$ is usually much larger than unity for barrierless transport conditions. In this case, for adsorption time t smaller than $t_{ch} = 1/(S_g n^b)^2 D_\infty$, particle flux and coverage are governed by previously derived dependencies, Eq. (150) and Eq. (170), respectively.

This is illustrated in Fig. 37, in which the results of numerical calculations derived in Ref. 2 are presented. It is interesting to note that for $\bar{k}_a > 10$ (which is often the case for protein adsorption), the entire kinetic curve can be constructed as

$$\Theta = 2S_g\sqrt{\pi D_\infty t}n^b \quad \text{for} \quad \frac{t}{t_{ch}} \le \left(\frac{\Theta_{mx}}{4\pi}\right)$$

$$\Theta = \Theta_{mx} \qquad\qquad \text{for} \quad \frac{t}{t_{ch}} > \left(\frac{\Theta_{mx}}{4\pi}\right) \tag{213}$$

This indicates that the blocking effects are negligible in this case except for $\Theta = \Theta_{mx}$, which proves that Θ_{mx} is a parameter of a primary practical significance. Hence, for larger $\bar{k}_a$, exact

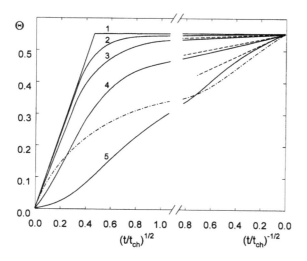

FIG. 37 Kinetics of irreversible adsorption under the diffusion controlled transport at a planar surface expressed as the Θ versus $(t/t_{ch})^{1/2}$ and $(t/t_{ch})^{-1/2}$ dependencies (where $t = 1/(S_g n^b)^2 D_\infty$ is the characteristic adsorption time). The continuous lines denote the numerical solution of the diffusion equation [2] for curve 1, $\bar{k}_a = \infty$; curve 2, $\bar{k}_a = 1000$; curve 3, $\bar{k}_a = 100$; curve 4, $\bar{k}_a = 10$; curve 5, $\bar{k}_a = 1$. The broken lines represent the asymptotic long-time results calculated from Eq. (214a) and the dashed-dotted line shows the results calculated from Eq. (216). (From Ref. 2.)

knowledge of the $\bar{B}(\Theta)$ function is not required. An additional significance of the results reflected by Eq. (213) is that practically for all coverages, the particle flux remains equal to the limiting flux given by Eq. (150). This suggests that adsorption measurements can be used for determining particle or protein diffusion coefficient (particle size) in an accurate way.

As can be seen in Fig. 37, for $\bar{k}_a < 100$ the deviations from these limiting adsorption regime become more pronounced and can only be evaluated numerically. However, it was shown in Ref. 2 that for larger coverages when the inequality

$$\Theta > \Theta_{mx}\left(1 - m\sqrt{\frac{1}{10(m - 1)\bar{k}_a}}\right)$$

is met, particle adsorption kinetics is governed by the asymptotic law

$$\Theta = \Theta_{mx} - \frac{K_l}{(t/t_{ch})^{1/(m-1)}} \tag{214a}$$

where

$$K_l = \Theta_{mx}\left(\frac{\Theta_{mx}}{(m - 1)C'_{mx}\bar{k}_a}\right)^{1/(m-1)}$$

Equation (214a) indicates that the maximum coverage is approached in the long-time limit as $(t/t_{ch})^{-1/(m-1)}$. For spheres, when $m = 3$ one has

$$\Theta = \Theta_{mx} - \frac{K_l}{\sqrt{t/t_{ch}}} \tag{214b}$$

when $m = 3$. As can be seen in Fig. 37, the exact numerical results are well reflected by the limiting analytical formula, Equation (214b), for longer times. This has important practical implications because a useful extrapolating formula can be derived on the basis of Eq. (214b), which indicates that

$$\Theta_{mx} = \Theta_l + \Theta_l^{m/(m-1)} \frac{1}{[(m-1)C_{mx}S_g k_a n^b t_l]} 1/(m-1) \tag{215}$$

where Θ_l is the coverage attained for long but finite adsorption time t_l.

Equation (215) seems particularly for experimental studies because attaining the maximum (jamming) coverage would require a prohibitively long adsorption time, especially for dilute protein or colloid suspensions.

The calculations shown in Fig. 37 also demonstrated the inadequacy of the commonly used empirical model postulating that [1,8]

$$j = j_0 \bar{B}(\Theta) \tag{216}$$

This is so because for the diffusion-controlled adsorption, the thickness of the diffusion boundary layer grows with time. This makes the bulk-transport resistance much higher than the surface layer resistance governed by the blocking function except for very long times when Θ approaches Θ_{mx}.

A different situation arises under the convection-controlled transport condition when the thickness of the diffusion boundary layer remains fixed after a short transition time. Then, for the uniformly accessible surfaces, one can integrate the bulk-transport equation with the nonlinear boundary conditions, Eq. (204). This results in the following expression [12,112,116]:

$$-j = \frac{\bar{B}(\Theta)}{1 + (K-1)\bar{B}(\Theta)} k_a n^b = -j_0 \frac{K\bar{B}(\Theta)}{1 + (K-1)\bar{B}(\Theta)} = -j_0 \tilde{B}(\Theta) \tag{217}$$

where

$$K = \frac{k_a}{k_c} \quad \text{and} \quad \tilde{B}(\Theta) = \frac{K\bar{B}(\Theta)}{1 + (K-1)\bar{B}(\Theta)}$$

Equation (217) has considerable significance because it allows one to connect the blocking function $\bar{B}(\Theta)$ or $B_0(\Theta)$ with the overall kinetic blocking function, expressed by Eq. (172). Substituting Eq. (217) into Eq. (169), one obtains the implicit expression for the adsorption kinetics in the form of the definite integral

$$(K-1)\Theta + \int_0^\Theta \frac{d\Theta'}{\bar{B}(\Theta')} = S_g k_a n^b t = K\tau \tag{218}$$

where $\tau = S_g k_a n^b t = S_g |j_0| t$ is the dimensionless adsorption time.

In the special case when $K \sim 1$, which can be realized in practice for micrometer-sized particles using the impinging-jet cells, Eq. (218) simplifies to the commonly used form

$$\int_0^\Theta \frac{d\Theta'}{\bar{B}(\Theta')} = K\tau \tag{219}$$

Assuming that $\bar{B}(\Theta)$ is in the form of the series expansion $1 - \bar{C}_1\Theta + \bar{C}_2\Theta^2$, one can evaluate Eq. (219) as

$$\Theta = \Theta_1 \frac{1 - e^{-\bar{q}\bar{C}_1\tau}}{1 - (\Theta_1/\Theta_2)e^{-\bar{q}\bar{C}_1\tau}} \tag{220}$$

where

$$\Theta_1 = \frac{\bar{C}_1}{2\bar{C}_2}(1 - \bar{q}), \qquad \Theta_2 = \frac{\bar{C}_1}{2\bar{C}_2}(1 + \bar{q}), \qquad \bar{q} = \sqrt{1 - \frac{4\bar{C}_2}{\bar{C}_1^2}}$$

For higher coverage, substituting Eq. (190) for $\bar{B}(\Theta)$, one obtains, by integrating Eq. (219), the expression

$$\Theta = \Theta_{mx} - \frac{\Theta_{mx}^{m/(m-1)}}{[C_{mx}(m-1)\tau]} 1/(m-1) \tag{221}$$

This equation is the dimensionless counterpart of Eq. (214a) derived previously for the diffusion-controlled transport.

On the other hand, for $K \gg 1$ (low transfer rate from the bulk), which is the case for small colloid particles and proteins under force-convection transport conditions, Eq. (218) indicates that the blocking effects governed by the $\bar{B}(\Theta)$ function remain negligible if the inequality

$$\Theta < \Theta_{mx}\left(1 - \sqrt[m]{\frac{1}{C_{mx}(K-1)}}\right) \tag{222}$$

is met. Thus, particle coverage increases linearly with time according to Eq. (171). This effect can be observed in Fig. 38, in which the results calculated numerically from Eq. (218) are plotted

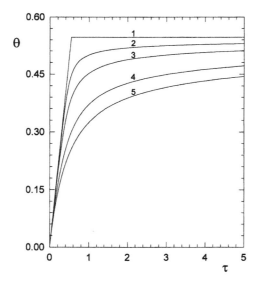

FIG. 38   Kinetics of irreversible adsorption of spherical particles under the forced-convection transport conditions calculated numerically from Eq. (218) for curve 1, $K = \infty$; curve 2, $K = 50$; curve 3, $K = 10$; curve 4, $K = 2$; curve 5, $K = 1$. (From Ref. 12.)

for various K's ranging from 1 (adsorption conditions typical for larger colloid particles) to 50 (adsorption conditions pertinent to proteins).

It should be mentioned that the above results are valid if the hydrodynamic interactions do not affect particle transport through the adsorption layer of thickness $2a^*$. This seems justified for smaller colloid particles and proteins. However, for micrometer-sized particles placed in shearing flows, the hydrodynamic forces play a significant role due to the coupling with the repulsive electrostatic interactions. This leads to enhanced blocking effects called hydrodynamic scattering effects and discussed extensively in recent review works [7,14]. These results have been interpreted theoretically in terms of the Brownian dynamics simulations [14], which are, however, considerably more time-consuming than the RSA simulations.

The above-discussed theoretical results can be used successfully for a quantitative interpretation of experimental results obtained mostly using monodisperse latex suspensions [1,2,6,7,9,10,16,137,138,143,144]. These results are discussed in some detail in the next section.

V. EXPERIMENTAL RESULTS

Many experimental works have been reported recently dealing with kinetic aspect of colloid and protein adsorption at solid–liquid interfaces. These results have been reviewed in some detail elsewhere [1,2,7,12,14,76,116]. In this section, we present some representative experimental results obtained under well-defined transport conditions which confirm the validity of the theoretical approaches discussed earlier. The usefulness of the colloid system to mimic the adsorption processes of molecules also will be pointed out. The data concerning limiting flux measurements (linear adsorption regime) are discussed first, whereas the last part of this section will be focused on describing the effect of surface-blocking effects (steric barrier).

It should be pointed out that the experiment data discussed hereafter were obtained under irreversible adsorption conditions. The change in the bulk suspensions concentration (dilution) or increase in the flow intensity caused no measurable change in the number of adsorbed particles, which was proven in separate experiments [1,12–14].

A. Experimental Methods

There exists a large variety of experimental methods aimed at a quantitative determination of the kinetics of colloid particle adsorption which can be attributed to the indirect and direct category. The simplest to implement are the indirect methods when the suspension concentration changes in the bulk are measured prior and after contact with the adsorbent (interface). The depletion of the solution concentration is often determined by measuring optical density changes (turbidimetry) [151], by interferometry or nephelometry [152], or by applying the high-performance liquid chromatographic (HPLC) and FPLC methods coupled with an appropriate detecting system [153,154]. Sometimes, fluorescent [155] or radioactive [156] labeling of the adsorbate is used.

For larger colloid particles, one can use the *on-line* particle concentration detection based on the light-scattering or Coulter counter principle [157]. By using the depletion methods, one implicitly assumes that the amount of the deposited (adsorbed) substance is equal to the amount which disappeared from the solution. This limits the accuracy of the depletion methods because adsorption on container walls or adsorbate trapping into pores cannot be a priori excluded. Another disadvantage of these methods is that one can usually gain global information averaged from a considerable surface area of the interface. As a result, any detailed information about the local structure of the monolayer (e.g., pair correlations, density fluctuations, or inhomogeneities) is lost.

More accurate are those indirect methods when the surface concentration of adsorbed particles is determined by measuring a physicochemical quantity which can be assigned unequivocally to the presence of deposited particles. Usually, the change of optical or electrokinetic properties (streaming potential) due to the adsorbed layer is exploited for surface concentration determination. Often, the isotopically labeled particles are used to produce a well-detectable signal stemming from the adsorbed layer. One such method is based on measuring the intensity of the scattered light in the direction normal to the incident beam. It was applied for studying the effect of electrode (prepared from a conductive glass) potential on adsorption kinetics of carbon black particles [158].

Ellipsometry is another optical technique widely used for studying bioparticle adsorption. The method is based on the principle that the state of polarization of light changes upon reflection from an interface. Jonsson et al. [159] constructed the flow cell, enabling the ellipsometric measurements to be performed with a support adsorbing surface exposed to a flow of well-defined geometry.

Reflectometry is a new optical method gaining importance in studies on protein and nanosized colloid particle deposition [3,4,160,161]. The method relies on the detection of reflectivity changes caused by the adsorbed layer having a refractive index different from the suspending medium. The polarized laser beam is focused at an angle close to the Brewster's angle on the surface to be studied. The ratio of reflected intensities of the perpendicular and parallel polarization components is measured and converted into an output signal. After a proper calibration and adopting some model assumptions for the configuration of the adsorbed layer, the output signal can be related to the surface concentration of adsorbate. This method was further developed in Ref. 162 by allowing for changes in the incident angle of the beam around the value of the Brewster's angle in order to attain higher accuracy. This method, called scanning angle reflectometry (SAR), was successfully applied for determining the mean thickness and mean refractive index of the fibrinogen layer adsorbed on silica [163]. The main advantage of ellipsometry and reflectometry methods is that the optical signal can be detected directly without disturbing the system by introducing any labels such as radio-isotopes or fluorescent dyes. On the other hand, one should remember that the sensitivity of the ellipsometry and reflectometry methods is rather limited, especially for low surface concentrations.

Another class of indirect methods aimed at studying particle adsorption is based on the radioactivity measurements of labeled particles [164–168]. The experiments are usually performed using a single capillary [168], hollow fibers or a parallel-plate channel [167]. Although the radioactivity method is rather sensitive, it has limited accuracy due to the presence of background radiation.

Other methods of determining particle adsorption exploit the fact that the electrokinetic potential of a solid–liquid interface in an electrolyte solution is sensitive to the amount of adsorbed substance (both charged or uncharged). Usually, the streaming potential of single capillaries is measured, enabling one to determine the adsorbed amounts of polyelectrolytes [169], polymers [170], and proteins [3,5,161,171] in the range of a small fraction of a monolayer. Recently, accurate experiments on colloid particles adsorption in the parallel-plate channel formed of two mica sheets have been performed [77] by using the streaming potential method. The sensitivity of the electrokinetic method exceeds considerably that of the optical methods because particle coverage as low as 0.5% can be detected [77]. Due to the progress in the theoretical description of electrokinetic phenomena for particle covered surfaces [172], the electrokinetic measurements can be used as an absolute method now. However, the disadvantage of the electrokinetic method is that its accuracy decreases considerably for high electrolyte concentrations (low signal) and for low concentration (appearance of surface conductivity).

The indirect methods mentioned are, in principle, applicable for arbitrary-sized particles, but they are especially well suited for semiquantitative studies of protein, polymer, and small colloid particle adsorption for surface coverage exceeding 0.1.

It seems that an unequivocal determination of particle surface concentration as a function of various physicochemical parameters can only be achieved by using the direct methods based on optical, AFM or electron microscope observations. For suspensions of larger-sized colloids or bacteria, the number of particle adsorbed can be determined in situ in a continuous manner, using the optical microscopy coupled with a micrograph [88,89, 143,144,173] or image analysis techniques [90,91,174]. Usually, the well-defined transport conditions are realized using the impinging-jet cells [1,2,7,12–14,88–91,143,144] or the parallel-plate channel [122,166,174]. Recently, the AFM tapping mode was used for direct in situ imaging of latex particles adsorbed on mica [9,10,175]. This method allows one to measure not only the number of adsorbed particles (coverage) but also the size of individual particles, which is advantageous for testing the polydisperse mixture adsorption theory. However, the use of the AFM technique is rather awkward due to artifacts stemming from tip-induced aggregation of the suspension, convolution of the tip and particle signal, adhesion of particles to the tip, and so forth [9,10]. A considerably better resolution can be achieved by imaging the particle in the air upon drying the sample. Such a drying procedure also has to be applied when adsorption at nontransparent surfaces is studied. This approach was used for determining the particle deposition rate at the rotating disk [176–178]. However, this highly invasive procedure may lead to particle removal or change of monolayer structure due to strong capillary forces appearing upon drying.

This is spectacularly illustrated in Fig. 39, in which various structures produced when colloid particle (latex) monolayers are dried in air are shown. A slow drying for high coverage leads to a structure which resembles a two-dimensional crystallization process (see Fig. 39a). It should be mentioned that the two-dimensional crystallization phenomena also can be provoked by the electrophoresis, as demonstrated in Refs. 179 and 180. On the other hand, a slightly faster drying time may lead to the "caterpillar" structure shown in Fig. 39b or to the "labyrinth" structure (see Fig. 39c), approaching modern art images.

In order to eliminate the artifacts stemming from drying off the monolayers. Harley et al. [121] developed an ingenious experimental technique based on the thin-film freeze-drying principle, followed by the scanning microscope examination of the interface with adsorbed particles. Using the method (referred to as TFFD–SEM), they have examined adsorption kinetics of small negatively charged polystyrene latex on larger positively charged particles.

Due to the reliability and accuracy of the direct methods, involving in situ microscope observations of adsorbed particles seem to be the most appropriate for a quantitative verification of theoretical predictions, especially those concerning initial deposition rate when the surface coverage remains of the order of a percent. The cells used in these studies are shown schematically in Fig. 40a. These are RIJ cells [exploiting the flow pattern described by Eqs. (102) and (103)], the slot impinging-jet (SIJ) cell when the impinging jet has a planar symmetry, the diffusion cell for determining particle adsorption from stagnant suspensions, and the sedimentation cell used for studies of external-force (gravity)-driven adsorption. Because of the significance of the RIJ cell, it is shown in more detail in Fig. 40b. The real dimensions of the cell (shown in scale) are the inlet capillary radius $R = 1$ mm and the distance between the capillary tip and the interface (mica sheet) of 1.6 mm [90]. Most of the experimental results discussed hereafter have been obtained in these cells.

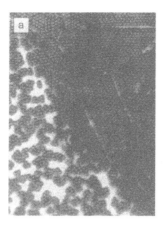
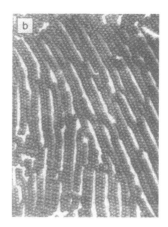

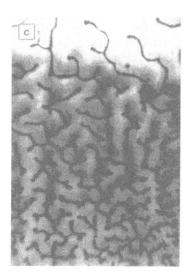

FIG. 39 Various structures obtained by drying of particle monolayers: (a) the two-dimensional liquid–solid interface (latex particles/mica); (b) the "caterpillar" structure (latex particle/mica); (c) the "labyrinth" structure (melamine latex particles/mica).

B. Linear Adsorption Regime

1. Adsorption at Homogeneous Surfaces

The occurrence of the linear deposition regimes under barrierless transport conditions in experiments involving colloid particles was often demonstrated [1,76,90,91]. The quantity measured directly in these experiments is the number of particles N_p adsorbed over equally sized surface areas ΔS (see Fig. 41). Because N_p is a statistical variable which obeys the Poisson fluctuation law for low coverage [173,181,182], the accuracy of determining the average value of $\langle N_p \rangle$ is inversely proportional to N_t (where N_t is the total number of particles counted). In the above experiments N_t was usually above 1000, which gives the standard deviation of $\langle N_p \rangle$ of

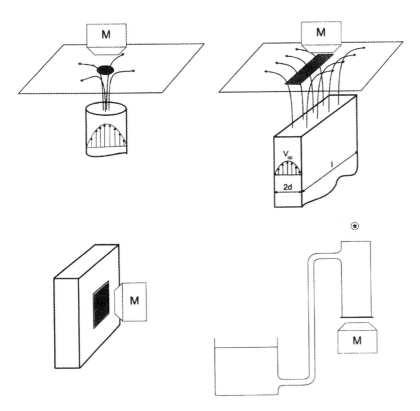

FIG. 40a The experimental cells used for the in situ measurements of particle adsorption under various transport conditions (a schematic view). (From Ref. 90.)

about 3%. For higher coverage, the fluctuation in N_p are considerably reduced, as discussed later, due to exclusion affects [182,183]. This increases the accuracy of measurements significantly. On the other hand, for barrier-controlled deposition regimes, the number of particles adsorbed is generally very low, so N_p is subject to considerable fluctuations, increased by surface heterogeneity. In these cases the standard deviation of $\langle N_p \rangle$ may well exceed 10%.

By knowing $\langle N_p \rangle$ as a function of time, one can determine the kinetic curves experimentally (i.e., the $\Theta = \pi a^2 \langle N_p \rangle$ versus the adsorption time t dependencies). According to the above-discussed theoretical predictions, for the low-coverage range, particle adsorption kinetics should be linear in respect to the adsorption time and the bulk suspension concentration n^b. The appearance of this linear adsorption regime is demonstrated in Fig. 42a, which presents the kinetic runs measured in the SIJ cell for polystyrene latex particles (averaged diameter 1 μm) adsorbing on mica. As can be observed, the curves remain linear for $\Theta < 0.02$ and its slope is proportional to the bulk suspension concentration varied between 7.1×10^3 and 7.1×10^4 mm^{-3}. Similar, linear regimes have been observed for the RIJ cell. This is shown in Fig. 42b, in which the Θ versus time dependencies for Re = 16 and Re (Polystyrene latex particles of averaged diameter 0.87 μm adsorbing on mica) are presented. Knowing the kinetic

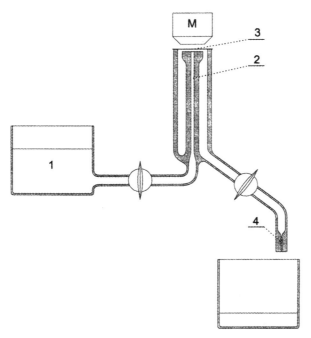

FIG. 40b   A detailed view of the experimental RIJ cell. (1) suspension container, (2) inlet capillary, (3) mica plate, (4) outlet capillary. (From Ref. 90.)

curves, one can determine the normalized particle flux (mass-transfer rate constant) using the definition

$$\left|\frac{j_0}{n_b}\right| = k_c = \frac{\Delta\langle N_p\rangle}{n_b \Delta S\, \Delta t} = \frac{1}{\pi a^2}\frac{\Delta\Theta}{n_b\, \Delta t} \tag{223}$$

where $\Delta\langle N_p\rangle$ is the change in the averaged number of particles adsorbed over ΔS within the time interval Δt. The accuracy of the flux determination can be increased by averaging over many experiments carried out at different bulk suspension concentrations n^b. The method described by Eq. (223) can also be used for determining the local flux in the case when it depends on the position over the interface. This procedure can be applied, in principle, for arbitrary coverage. It is, however, most accurate for the linear adsorption regime when the reduced flux (transfer rate) remains time independent and proportional to the bulk suspension concentration n^b.

The high accuracy of the initial flux determined using the direct microscope observation method can be exploited for estimating the range of validity of the convective diffusion theory, especially the adequacy of the numerical solutions describing the flow distribution in the impinging-jet cells. This can be achieved by measuring the particle flux as a function of the distance from the center of the cell. As expected from the flow distributions given by Eq. (102) and shown in Figs. 15 and 16, the flux should decrease significantly for distances comparable or larger than the capillary radius R. This effect is confirmed experimentally, as can be observed in Fig. 43 which illustrates the dependence of the local mass-transfer coefficient on the normalized distance r/R. These results have been obtained using the RIJ cell and 0.87-nm diameter polystyrene latex particles [90]. As can be seen, the agreement between the experimental data and the theoretical prediction derived by the numerical solution of Eq. (160) (under stationary

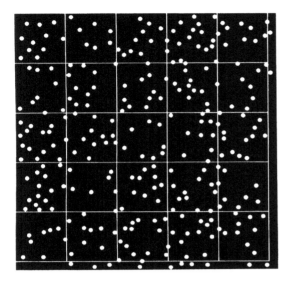

FIG. 41 Configurations of irreversibly adsorbed particles determined experimentally; latex particles (diameter 1 μm) on mica $\Theta = 0.1$ (a two-dimensional gaseous phase). (From Ref. 182.)

conditions) is quantitative for the entire range of Re numbers studied. This confirms the validity of the numerical solutions of the flow distribution in the cell.

It should be mentioned that the significant decrease in the local particle flux at larger distances from the center of the cell, especially well pronounced for the higher Re number, is solely due to the flow distribution in the cell characterized by a significant decrease in the axial velocity component. This low-coverage effect due to flow distribution should be distinguished from the nonlinear effect due to the hydrodynamic scattering, to be discussed later.

For practical purposes, the most important parameter is the extension of the region where the particle flux remains position independent, which means that the surface is uniformly accessible for particle transport. The uniformity of the flux distribution depends on the Reynolds number, as can be assessed from the results shown in Fig. 44. It can be deduced that for Re < 16, the uniform adsorption conditions prevail over distances $r/R < 0.5$, whereas for higher Re, this region shrinks to approximately $0.25r/R$. It is also interesting to observe that for moderate Re ($1 < Re < 16$), the reduced flux distribution measured in the RIJ cell closely resembles the theoretical results for the sphere in uniform flow when the $j(r)/j_0$ dependence in the limit of low Re is given by the function derived from Eq. (157):

$$\frac{j(r)}{j_0} = 0.873 \frac{\sin \vartheta}{(\vartheta - \frac{1}{2}\sin \vartheta)^{1/3}} \tag{224}$$

where the angle ϑ is related to the r/R coordinate by $\vartheta = \pi r/2R$.

The results shown in Fig. 44 are significant because they suggest that the RIJ cell can be used for predicting particle deposition on the spherical interface, either isolated or forming a packed bed. In the latter case, the direct microscope observations seem considerably more tedious than for the RIJ cell.

Moreover, the experimental data shown in Figs. 43 and 44 confirm the hypothesis that particle adsorption measurements can be exploited for evaluating flow distribution and local

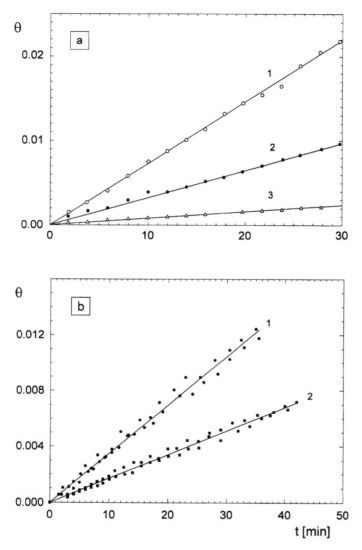

FIG. 42 Adsorption kinetics of latex particles on mica determined experimentally for the impinging-jet cells: "a" SIJ cell, particle diameter 1 μm, $I = 10^{-4}$ M, Re $= 8.4$; curve 1, $n^b = 7.1 \times 10^4$ mm^{-3}; curve 2, $n^b = 3.5 \times 10^4$ mm^{-3}; curve 3, $n^b = 7.1 \times 10^3$ mm^{-3}. (b) RIJ cell, particle diameter 0.87 μm, $I = 10^{-4}$ M [90]; curve 1, Re $= 30$; curve 2, Re $= 16$. The continuous lines denote the linear regression fits.

mass-transfer rates for various surfaces under the forced-convection transport conditions. By analogy to the governing continuity equations, these results can often be transferred to heat-transport problems.

The results derived by the tedious microscope observation method can be exploited for estimating the experimental error of the indirect optical techniques and is more convenient to use for protein adsorption studies. In this method, the optical signal is taken from a relatively large surface area (comparable with the capillary radius) where particle flux becomes nonuniform. The

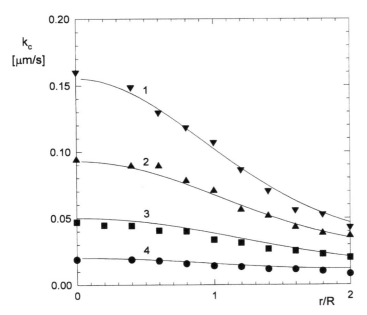

FIG. 43 The dependence of the local mass-transfer coefficient k_c on the normalized distance from the stagnation point r/R; the points denote the experimental results obtained in the RIJ cell [90] for latex particles (diameter 0.87 μm) adsorbing on mica ($I = 10^{-4}$ M): curve 1, Re = 8; curve 2, Re = 4; curve 3, Re = 1; curve 4, Re = 0.15. The continuous lines denote the theoretical results calculated numerically.

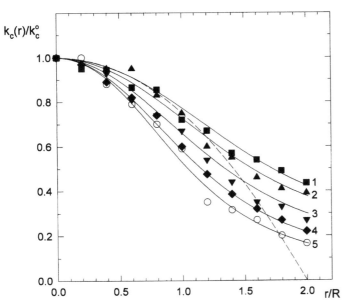

FIG. 44 The dependence of the reduced local mass-transfer coefficient $k_c(r)/k_c^0$ (where k_c^0 is the coefficient at the cell center) on r/R; the points denote the experimental results obtained in the RIJ cell [90] for latex particles (diameter 0.87 μm) adsorbing on mica ($I = 10^{-4}$ M): curve 1, Re = 1; curve 2, Re = 4; curve 3, Re = 8; curve 4, Re = 16; curve 5, Re = 30. The continuous lines denote the theoretical results calculated numerically and the dashed line denotes the theoretical results calculated from Eq. (157) for the sphere in uniform flow.

magnitude of this effect can be estimated from the universal correction function derived by integration of the reduced flux. This function is defined as [90]

$$J(r_{mx}) = \frac{2}{r_{mx}^2} \int_0^{r_{mx}} \frac{j(r)}{j_0} r \, dr = \langle k_c(r_{mx}) \rangle \tag{225}$$

Hence, $J(r_{mx})$ can be interpreted as the reduced flux, averaged over a circular area of the radius r_{mx} which can be identified for optical methods with the beam diameter.

In Fig. 45, the distributions of $\langle k_c \rangle$ calculated theoretically and measured experimentally is shown for various Re's. As can be seen, for Re = 1 the correction due to the flux nonuniformity is about 10% for $r_{mx} = R$, whereas for Re = 48, the correction becomes as large as 30% (for the same r_{mx}). The results shown in Fig. 45 suggest, therefore, that protein adsorption studies should be preferably carried out for a low Re number and the capillary radius of the cell should be larger than the beam radius.

The significance of the results presented in Figs. 43–45 is that they allow one to estimate the area over the surface where the flux (transfer rate) become uniform. The mass-transfer-rate constant (reduced flux) measured in this uniformly accessible area and referred to as k_c^0 is of a particular significance because it can be analyzed theoretically in a much more efficient way than the local flux. Moreover, k_c^0 can be directly used for determining the significance of various transport mechanisms like diffusion, interception, specific or external force, and so forth.

The most interesting from the practical viewpoint was to determine the significance of the diffusion and interception effects as a function of particle size. The results of such measurements obtained by using the impinging-jet cell and monodisperse polystyrene latex suspensions [173] are plotted in Fig. 46. The ionic strength in these experiments was kept relatively high (10^{-3} M)

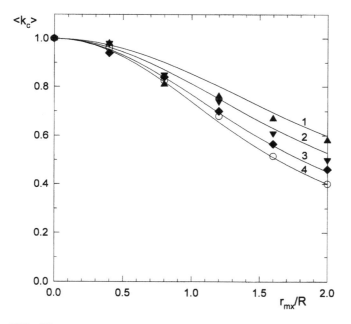

FIG. 45 The dependence of the mass-transfer coefficient $\langle k_c \rangle$ (particle flux) averaged over the area of radius r_{mx}, on r_{mx}/R. The points denote the experimental results obtained in the RIJ cell [90] for latex particles adsorbing on mica ($I = 10^{-4}$ M): curve 1, Re = 4; curve 2, Re = 8; curve 3, Re = 16; curve 4, Re = 30. The continuous lines denote the theoretical results calculated numerically.

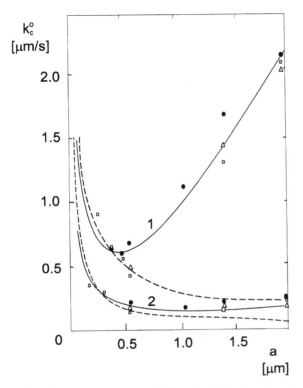

FIG. 46 The dependence of k_c^0 reduced (flux) at the cell center on particle size. The points denote the experimental results obtained in the RIJ cell [173] for latex particles adsorbing on mica ($I = 10^{-3}\ M$): curve 1, Re = 150; curve 2, Re = 30. The continuous lines denote the theoretical results calculated numerically and the dashed line represents the analytical results calculated using the Smoluchowski-Levich approximation. (From Ref. 12.)

in order to decrease the range of the attractive electrostatic interactions between the particle and the surface. These experimental conditions correspond, therefore, to the perfect-sink assumption. The results shown in Fig. 46 suggest that for particles of size below 1 μm, the adsorption rate k_c^0 can be well reflected by the Smoluchowski–Levich theory depicted by the dashed line. This suggests that for colloid particles, the initial flux decreases as $a^{-2/3}$, in accordance with Eq. (156), which indicates that diffusion and convection are the dominating transport mechanisms for this particle size range. On the other hand, for particle sizes above 1 μm, the interception effect is playing an increasingly important role, especially for higher flow rates (Re = 150). This causes a considerable (manifold) deviation of the limiting flux from the Levich theory. Thus, for larger particles, k_c^0 increases parabolically with particle size in accordance with Eq. (165), which can be expressed in the dimensional form as

$$k_c^0 = \alpha_r \frac{V_\infty}{R^2} a^2 (1 + H^*)^2 \tag{226}$$

Note that the numerical solutions of the exact transport equation, Eq. (160), agree well with the experimental data for the entire range of particle sizes studied. As can be seen in Fig. 46, the position and the depth of the minimum depends strongly on the flow intensity (Re number). Thus, for Re = 30, the particle flux becomes practically independent on particle size larger than

1 μm. It should be noted that for this particle size range, the flux is also affected by the sedimentation effect, which becomes especially important for dense colloid particles of a mineral origin [178].

As mentioned earlier, the results shown in Fig. 46 were obtained under conditions of negligible electrostatic interactions. However, for lower ionic strength, the role of the electrostatic interactions becomes important, as demonstrated in Refs. 90 and 184 using the impinging-jet cells. Typical results obtained for the polystyrene latex suspension (averaged particle diameter 0.87 μm) in the RIJ cell are shown in Fig. 47. The solid lines denote the exact theoretical results derived by a numerical solution of Eq. (160). As one can note, for the Re number range studied (0.15–48), the experimental data are in quantitative agreement with the theoretical calculations. An important feature of the results is that the adsorption rate is considerably enhanced by the decrease in the ionic strength (dilute electrolyte). This increases the range of the attractive electrostatic interactions, which leads to the enhancement of the interception effect predicted by Eq. (165). Indeed, as can be seen in Fig. 47, the adsorption rate increase is especially well pronounced for a higher Re number (Pe number). This effect can be exploited for an accurate experimental determination of flow intensity (if it is not known) or for the determination of ionic strength if the cell geometry and flow rate are known.

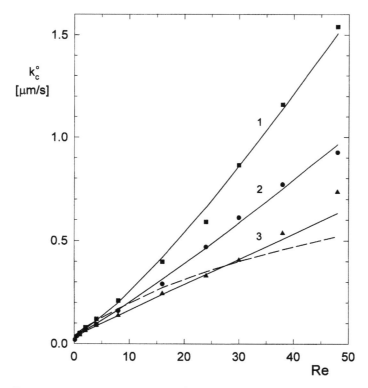

FIG. 47 The dependence of the k_c^0 on Re. The points denote the experimental results obtained in the RIJ cell [90] for latex particles (diameter 0.87 μm) adsorbing on mica: curve 1, $I = 2 \times 10^{-5}\ M$; curve 2, $I = 10^{-4}\ M$; curve 3, $I = 10^{-3}\ M$. The continuous lines denote the theoretical results calculated numerically and the dashed line represents the analytical results calculated using the Smoluchowski-Levich approximation.

On the other hand, for $I = 10^{-3} M$, the electrostatic interactions seems to be effectively eliminated because the adsorption rate can be well approximated by the Smoluchowski–Levich approximation (shown by dashed lines in Fig. 47). This confirms that the results shown previously in Fig. 46 can be treated as the limiting values, characteristic for hard particles.

The limiting flux increase in dilute electrolyte solutions due to the interception effect is a universal phenomenon occurring for other flow configurations more related to practice. For example, Elimelech [185] measured the particle adsorption (filtration) rate in columns packed with glass beads (having averaged diameter of 0.046 cm). The suspensions used were positively charged latex particles of various sizes ranging from 0.08 to 2.51 μm and the ionic strength varied between $5 \times 10^{-6} M$ (deionized water) to 0.1 M. The number of particles adsorbed was determined indirectly (depletion method) by monitoring the particle concentration changes at the inlet and the outlet of the column. The results of the experiments, analogous to those obtained in the RIJ cell, confirmed that the increase in initial particle flux can be as large as four times when using deionized water. This effect can be quantitatively interpreted in terms of the numerical solutions of the two-dimensional continuity equation [185]. Similar results were obtained for larger particle sizes, although the measured particle deposition rates were generally smaller than predicted theoretically.

The above-presented results and others discussed elsewhere [1,12–14,37,76] confirmed quantitatively the validity of the convective diffusion theory incorporating the specific (electrostatic) force fields for interpreting particle adsorption phenomena under the linear regime.

However, a different situation is expected to occur for systems characterized by the Type II energy profile (i.e., under the barrier-controlled deposition regimes). In such cases, a small perturbation in the governing parameters, such as zeta-potentials, particle size, local interface geometry, and charge heterogeneity) will result in a large, usually nonlinear, response of the system. As a result, the experiments carried out under barrier-controlled transport conditions are usually less reproducible and difficult for an unambiguous theoretical interpretation. A general feature observed in this type of experiment is that the measured limiting flux values are much larger than theoretical predictions both for submicrometer [176] and larger [186] particle sizes. These positive deviations from theoretical flux values were interpreted in terms of the surface hetereogeneity hypothesis proposed in Refs. 37 and 187. The simplest possibility arises when, due to natural fluctuation phenomena, the charge on particles becomes nonuniformly distributed, forming local micropatches characterized by a more favorable deposition condition than the average surface (this will correspond to the random heterogeneity hypothesis). Even if the fraction of these areas remains very low (of the order of a percent of the total area available), the overall deposition rate will be much larger than theoretically predicted for uniform surfaces due to its large sensitivity of particle flux to surface charge (potential). This hypothesis is strongly supported by the kinetic curves exhibiting saturation at low surface coverage of the order of percents [174,177] and a gross unevenness of the adsorbed layer.

It seems that in the case of barrier-controlled deposition, the classical DLVO energy profiles calculated for homogeneous surfaces are not sufficient for a theoretical interpretation of particle adsorption data. However, a satisfactory agreement between theory and experiment can be attained by accepting the heterogeneity hypothesis postulating that the DLVO theory is valid in a local sense only (i.e., for a given surface area or a given particle) [12]. Therefore, experimental studies for heterogeneous surfaces discussed next seem especially valuable.

2. Adsorption at Heterogeneous Surfaces

Adsorption at colloids and bioparticles at heterogeneous surfaces is interesting for many practical and natural processes such as filtration, water treatment, papermaking, thrombosis,

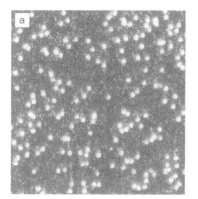 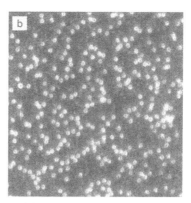

FIG. 48 Configurations of adsorbed particles determined experimentally for heterogeneous surfaces (adsorption at active canters having size 0.55 μm): (a) latex particles (diameter 1.38 μm) on mica, $\Theta_s = 0.04$, $\Theta_l = 0.04$; (b) same as (a) but for $\Theta_l = 0.07$.

protein immobilization and separation, and so forth. This is so because the effectiveness of these processes is often enhanced by the use of coupling agents preadsorbed at the interface, which promote irreversible adsorption of particles. For example, cationic polyelectrolytes are used to increase retention of filler particles (e.g., titania) in papermaking [188]. In biological applications, special proteins (ligands) immobilized at polystyrene latex are often applied for selective adsorption of a desire solute from protein mixtures, as is the case in affinity chromatography [189] or immunological assays [190]. The use of the coupling agents makes the adsorbing surfaces inherently heterogeneous. This raises the important question of the relationship among the surfaces concentration, size and distribution of these ligands, and particle adsorption mechanism and kinetics.

Despite the significance of particle adsorption at heterogeneous surfaces, there exist very few experimental studies concerning the influence charge and geometric heterogeneities (surface roughness) on the adsorption kinetics of colloid particles. The experiments which relatively close match these conditions were reported in Ref. 191. These works were concerned with deposition kinetics of larger polystyrene latex particles (averaged diameter 1.38 μm) on a mica surface precovered (in prior deposition experiments) with a given amount of smaller latex particles (averaged size 0.55 μm). The particle size ratio was, therefore, 2.5. The degree of surface heterogeneity produced in this way was expressed in terms of lower particle surface coverage $\Theta_s = \pi a_s^2 N_s$, where a_s is the smaller particle radius and N_s is their surface concentration.

Typical configurations of larger particles observed in these experiments are shown in Fig. 48 (for $\Theta_s = 0.04$). On the other hand, the kinetic runs determined for the heterogeneous surfaces prepared in this way are presented in Fig. 49 for $\Theta_s = 0.018$ and $\Theta_s = 0.026$. As can be observed, adsorption kinetics can be well reflected by linear dependencies with the slope increasing with the heterogeneity degree. It is interesting to note that for Θ_s as small as 0.026 (2.6% of the geometrical surface covered by particles), the rate of particle adsorption kinetic curves becomes practically identical with the rate observed for homogeneous surfaces. This is visible in Fig. 50, which shows the dependence of the normalized adsorption rate $k_c^0(\Theta_s)/k_c^0$ (where k_c^0 is the adsorption rate for a homogeneous surface) on the heterogeneity degree Θ_s. These results have been interpreted in terms of Eq. (217) considering the coupling between the bulk transport and the surface transport governed by the heterogeneity degree. It was shown that

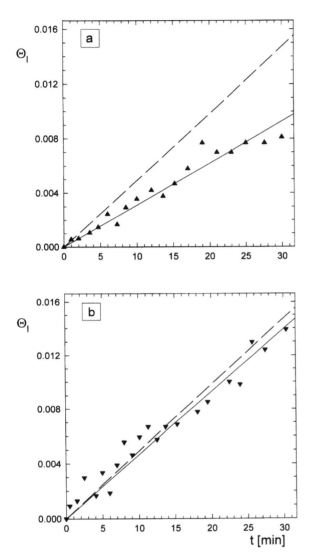

FIG. 49 Adsorption kinetics of latex particles (diameter 1.38 μm) at heterogeneous surfaces expressed as the dependence of Θ_l on t (RIJ cell, $I = 10^{-4}\ M$): (a) $\Theta_s = 0.018$; (b) $\Theta_s = 0.026$. The continuous lines denote linear regression fits and the dashed line shows the limiting kinetics for a homogeneous surface.

in the case of the heterogeneous surface, the generalized blocking function occurring in this formula is given by [191]

$$\bar{B}(\Theta_s) = 1 - (1 - \Theta_s)\exp\left[-\frac{(4\lambda - 1)\Theta_s}{1 - \Theta_s} - \left(\frac{(2\sqrt{\lambda} - 1)\Theta_s}{1 - \Theta_s}\right)^2\right] \qquad (227a)$$

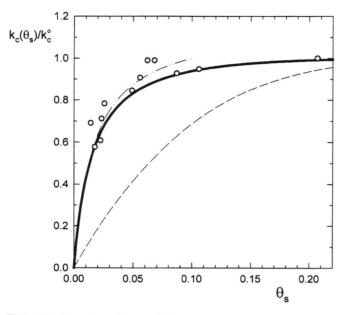

FIG. 50 The dependence of the normalized transfer coefficient of larger particles at heterogeneous surfaces $k_c(\Theta_s)/k_c^0$ (where k_c^0 is the coefficient for a uncovered surface) on Θ_s. The points denote the experimental results obtained for the latex particles on mica in the RIJ cell, the continuous line shows the analytical results calculated from Eq. (217) according to the generalized RSA method, the dashed line shows the results calculated by neglecting the coupling ($K = 1$), and the dashed-dotted line presents the data calculated from Eq. (227b).

For larger K and $\lambda\Theta_s \ll 1$, by combining Eq. (217) and Eq. (227a), one can derive the simple relationship for the reduced flux (adsorption rate),

$$\frac{k_c^0(\Theta)}{k_c^0} = \frac{j(\Theta)}{j} = \frac{4\lambda K\Theta_s}{1 + 4\lambda(K-1)\Theta_s} \tag{227b}$$

As can be seen in Fig. 50, the theoretical predictions stemming from Eq. (217) combined with Eq. (227a) reflect well the experimental data for the entire range of the heterogeneity degree (with $K = 7$, which corresponds to the experimental conditions). As expected, Eq. (227b) also gives a satisfactory agreement with the experimental data for $\Theta_s < 0.10$. This suggests that the basic features of particle adsorption at heterogeneous surfaces are reflected by this equation, which has practical significance in view of the simplicity of this expression. In particular, one can deduce from Eq. (227b) that if the criterion is fulfilled the adsorption rate for heterogeneous surfaces attains the limiting value,

$$\Theta_s > \frac{1}{4\lambda K} = \frac{a_s k_c^0}{4 a_l k_a} \tag{228}$$

As can be deduced, this limiting value is proportional to the heterogeneity size and the rate constant of the bulk transport k_c^0, which means that the increase in the flux is the most dramatic for smaller heterogeneity size (at fixed coverage) and low flow rate (Reynolds number) when the thickness of the diffusion boundary layer becomes considerably larger than particle dimensions. This means that the coupling between the bulk and surface transport is expected to play a decisive role for particle and protein adsorption at heterogeneous surfaces. This conclusion is further supported by the fact that the theoretical results derived by neglecting the coupling (see the dashed line in Fig. 50) deviate considerably from the experimental data.

Another type of adsorption at heterogeneous surfaces occurs when the larger particles have the same charge as the heterogeneities (smaller particles) but opposite to the interface charge. In this case, adsorption of particles occurs at uncovered surface areas. As can be seen in Fig. 51, the initial flux of larger particles (polystyrene latex particles of averaged diameter 1.38 μm) falls abruptly with the coverage of smaller latex particles of size 0.55 μm [12,138]. This behavior, analogous to the effect of the electrostatic barrier, is caused by the repulsion between equally charged smaller and larger particles. The experimental results were interpreted using Eq. (188), which gives the probability of larger-particle adsorption at surfaces characterized by the heterogeneity degree equal Θ_s. This equation applies in the limit of low coverage of larger particles (linear adsorption regime in respect to larger particles). The coupling with the bulk transport was considered via Eq. (217), with $K = 2.5$ and $\lambda = 2.2$ corresponding exactly to the experimental conditions. As can be observed in Fig. 51, the theoretical results stemming from this equation are in good agreement with the experimental data for the entire range of heterogeneity degree.

These results are important because they demonstrate that the presence of smaller particles at the surface can exert a profound effect on the adsorption rate of larger particles. According to the theoretical predictions shown in Fig. 27, the flux reduction is especially pronounced for the large-size ratio of particles. This suggests that by measuring the initial adsorption rates of larger particles (visible under optical microscope), one can detect the presence at the surface of much smaller invisible particles. Hence, the particle adsorption measurements can be exploited as a sensitive tool for detecting the geometrical hetereogeneity of surfaces.

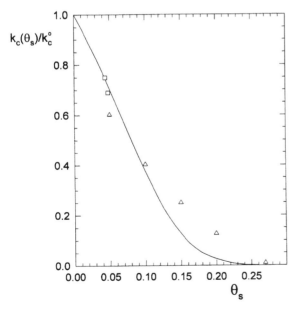

FIG. 51 Same as for Fig. 50 but for adsorption at precovered surfaces (RIJ cell, latex particles of diameter 1.38 μm, $I = 10^{-4}\ M$). (From Ref. 138.)

Due to the similarity of the underlying transport mechanisms, the experimental data concerning the initial adsorption rates (maximum flux) can be directly extracted for predicting the adsorption kinetic of ions and molecules.

C. Nonlinear Adsorption Regimes

1. Adsorption Kinetics

The validity of the theoretical approaches for describing the nonlinear adsorption kinetics can quantitatively be tested using the model colloid systems and the direct experimental methods. One of the most relevant issues is determining adsorption kinetics under the diffusion-controlled adsorption regime often used in protein adsorption studies. As discussed, the irreversible adsorption under the diffusion-controlled regime is governed by two crucial parameters: the dimensionless adsorption constant $\bar{k}_a$ [defined by Eq. (212)] and the maximum (jamming) coverage Θ_{mx}. Because $\bar{k}_a$ is mainly governed by the volume fraction (for barrierless systems), one can easily mimic the protein adsorption kinetics by colloid systems, characterized by similar Φ_v's. The advantage when working with colloid systems is that the adsorbed particles can be directly detected and counted by using the optical microscopy or AFM. In the latter case, one can determine not only the number and distribution of particles but also their individual sizes.

Experiments using the AFM method were performed by Johnson and Lenhoff [9], who measured particle adsorption kinetics in a stepwise manner by immersing a mica sheet into the colloid suspension (polystyrene latex of average diameter 0.116 μm). Then, after a given adsorption time reaching 24 h, the sample was dyed and imaged using the AFM tapping mode. The volume fraction was 10^{-4}, which corresponds to $n^b = 1.22 \times 10^9$ mm^{-3}. The dimensionless adsorption constant $\bar{k}_a$ was equal to 3×10^3 [2], which is a typical value for protein adsorption studies. The suspension ionic strength was varied between $0.05\,M$ and $0.0001\,M$ in order to determine its influence on the maximum coverage Θ_{mx}. The results of their experiments (points) collected for ionic strengths $0.05\,M$ and $0.001\,M$ are shown in Fig. 52 using the natural scale Θ versus the adsorption time t (expressed in hours). As can be noted, for $t < 5$ h, when $\Theta < 0.25$, the experimental results agree with the square root of time law, expressed by Eq. (213). However, for larger coverages, especially for the lower ionic strength value, the experimental results deviate considerably from this analytical expression, which is a direct manifestation of the nonlinearity effects. For longer times, $t > 15$ h, the surface coverage approaches the limiting Θ_{mx} value, which increases with the ionic strength. This behavior can be quantitatively accounted for by the theoretical results (shown by the solid lines in Fig. 52). These results were derived via an exact numerical solution (finite-difference method) of the governing diffusion equation with the boundary condition expressed by Eq. (204). The blocking function $\bar{B}(\Theta)$ was approximated by Eq. (187a), with Θ_{mx} calculated from Eq. (198b). The good agreement between the theoretical and experimental data suggests that the generalized RSA in combination with the effective hard-particle concept can account for colloid adsorption kinetics for the entire adsorption time. Due to the similarity of the adsorption constants, these results can be exploited for predicting protein adsorption kinetics [2]. Further systematic studies on particle adsorption kinetics by using the AFM method have been reported in Refs. 10 and 175.

Analogous measurements were carried out for larger colloid particles (averaged diameter 1 μm) using the direct optical microscope observation method [16]. The particle number concentration n^b was 5.16×10^6 mm^{-3}, which corresponds to $\bar{k}_a = 2 \times 10^2$ [2].

The ionic strength was varied between $0.01\,M$ and $2 \times 10^{-5}\,M$, which corresponds to H^* changing between 0.018 and 0.36. A typical particle monolayer obtained in this experiment is shown in Fig. 53 ($\Theta = 0.30$, $I = 10^{-4}\,M$). On the other hand, the kinetic results are plotted in

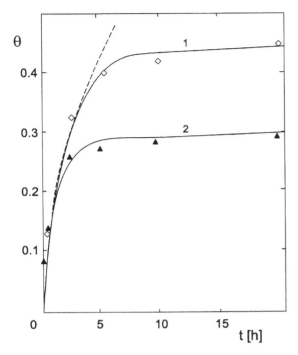

FIG. 52 Adsorption kinetics of latex particles on mica under the diffusion-controlled transport (AFM method [9]); the particle coverage Θ versus the adsorption time t dependence: curve 1, $I = 5 \times 10^{-3}$ M; curve 2, $I = 10^{-3}$ M. The continuous lines represent the exact theoretical results derived numerically [2] and the dashed line shows the limiting results calculated from Eq. (170). (From Ref. 2.)

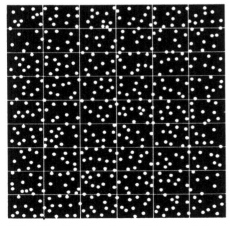

FIG. 53 Configuration of latex particles adsorbed on mica forming a two-dimensional liquid phase (diameter 1 μm, $I = 10^{-4}$ M, $\Theta = 0.3$). (From Ref. 182.)

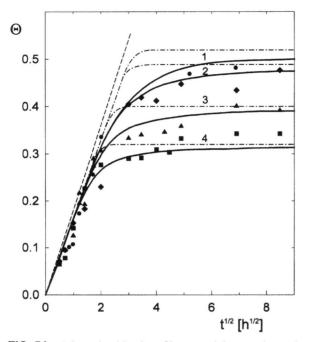

FIG. 54 Adsorption kinetics of latex particles on mica under the diffusion-controlled transport conditions [16] ($n_b = 5.16 \times 10^6$ mm^{-3}): curve 1, $I = 10^{-2}$ M; curve 2, $I = 10^{-3}$ M; curve 3, $I = 10^{-4}$ M; curve 4, $I = 2 \times 10^{-5}$ M. The continuous lines represent the exact theoretical results derived numerically using the RSA blocking function, Eq. (187a), the dashed-dotted lines show the numerical solutions with the blocking function given by the Langmuir model (i.e., $B = 1 - \Theta/\Theta_{mx}$), and the dashed line shows the limiting results calculated from Eq. (170). (From Ref. 182.)

Fig. 54 using the square root of time transformation, which facilitates the comparison with the limiting analytical solution Eq. (213). One can observe in Fig. 54 that the experimental results can be interpreted adequately in terms of the theoretical predictions derived by numerical integration of the diffusion equation with the RSA blocking function. The results stemming from the widely used Langmuir model when $B = I - \Theta/\Theta_{mx}$ are also shown in Fig. 54. It is evident that the Langmuir model, predicting a much steeper increase in particle coverage for intermediate times, does not reflect the character of the experimental data properly.

The results shown in Figs. 52 and 54 unequivocally prove that generalized RSA model can be successfully used for the interpretation of particle and protein adsorption kinetics under the diffusion-controlled transport conditions. A considerable advantage of this model is that the maximum coverage can be predicted a priori. The theoretical prediction stemming from the RSA model agree well with the experimental results for a broad range of particle size and ionic strength [16]. This can be deduced from Fig. 55, in which the relevant data concerning the diffusion-controlled adsorption are plotted in the form of the dependence of Θ on H^*. As can be seen, the experimental and Monte Carlo simulation results agree well with the theoretical prediction derived from Eq. (198b). The slight positive deviation can be attributed to the suspension polydispersity. This theoretically predicted effect [cf. Eq. (191a)] appears because for a longer adsorption time, the smallest particles are adsorbed preferably. This has been spectacularly confirmed in recent experiments involving the AFM technique [175].

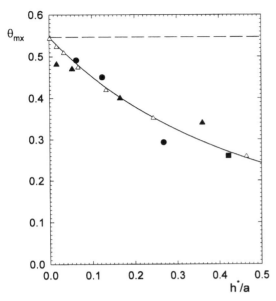

FIG. 55 The maximumn coverage θ_{mx} versus the effective interaction ragne h^*/a determined under the diffusion-controlled adsorption conditions: (▲) the results obtained in [16] using the RIJ cell (latex particles on mica); (●) represent the Johnson and Lenhoff data [9] obtained by AFM (latex particles/mica); (□) Semmler et al. result [10] obtained by AFM (latex/mica); (△) the theoretical results obtained by RSA simulations [16]. The continuous line shows the theoretical results calculated from Eq. (198b). (From Ref. 182.)

The disadvantage of the above-discussed diffusion-controlled adsorption regime is the necessity of applying very long experimental times (reaching days), which may promote contamination of the surface by lower-molecular-weight components. Therefore, in many experiments involving protein and colloid particles, forced convection is used, most frequently the impinging-jet cells discussed above. By appropriately choosing the experimental conditions (flow rate), one can successfully mimic, in these experiments, adsorption kinetic governed by the classical RSA model.

Typical kinetic curves measured under the quasi-RSA transport conditions are shown in Fig. 56. The results have been obtained for various ionic strengths using the RIJ cell and polystyrene latex particles [12] (averaged size 0.94 μm, Re = 8, bulk suspension concentration $n^b = 4.4 \times 10^5$ mm^{-3}). One can observe that for initial deposition stages ($\Theta < 0.1$), the slope of the kinetic curves (particle flux) decreases with ionic strength in accordance with the previous results shown in Fig. 47. On the other hand, for longer adsorption times, the opposite situation is observed, because the deposition rate is lowest for $I = 10^{-5}$ M. In the latter case, the adsorption rate becomes apparently negligible after reaching the surface coverage of 0.26. It should be noted that the classical RSA simulations performed by assuming the LSA model with the energy additivity principle adequately describes the experimental adsorption kinetics for the entire range of deposition time and ionic strength.

The good agreement of the experimental data shown in Fig. 56 with the classical RSA model is due to the fact that, in these experiments, the diffusion boundary layer thickness was fixed at a value comparable with the particle diameter as a result of convection. Hence, K was

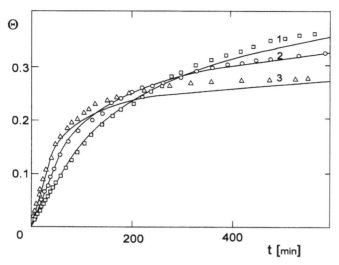

FIG. 56 Kinetics of particle adsorption for higher coverages. The points denote the experimental results obtained in the RIJ cell for the latex particle on mica [12]: curve 1, $I = 10^{-3}$ M; curve 2, $I = 10^{-4}$ M; curve 3, $I = 2 \times 10^{-5}$ M. The continuous lines show the theoretical results obtained by RSA simulations of interacting particles. (From Ref. 12.)

close to unity. Under such circumstances, the overall blocking function $\tilde{B}(\theta)$ can be expressed according to Eq. (217) as

$$\tilde{B}(\Theta) = K\bar{B}(\Theta) \cong B_0(\Theta) \tag{229}$$

Because $\bar{B}(\Theta)$ stemming from the generalized RSA model is slightly smaller and the $B_0(\Theta)$ (calculated from the RSA model) and the coupling constant K are slightly larger than unity, the product of the quantities can be close to the blocking function of the classical RSA model, given by Eqs. (187a) and (187b). This suggests, according to Eq. (218), that particle adsorption kinetics can effectively be analyzed by introducing the dimensionless time $\tau = \pi a^2 k_c^0 n^b t$, where the adsorption constant k_c^0 (reduced flux) can be determined from the initial slope of the kinetic curves shown in Fig. 56. One can predict that for $\tau \ll 1$, all of the results expressed using the dimensionless time should be described by the universal relationship

$$\Theta = \tau \tag{230a}$$

On the other hand, for $\tau \gg 1$, the long-time asymptotic regime should be attained when Θ becomes a linear function of $\tau^{-1/2}$ according to Eq. (221), which, for the spheres, can be expressed as

$$\frac{\Theta}{\Theta_{mx}} = 1 - \frac{1}{\sqrt{2\Theta_{mx}C_{mx}\tau}} \tag{230b}$$

These predictions have been confirmed experimentally in Ref. 12 using the RIJ cell and various latex particles suspensions. Typical results obtained in these experiments for 0.88-μm-diameter latex particles [12] are shown in Fig. 57. As can be seen, the θ versus $\tau^{-1/2}$ dependencies, indeed, become linear for $\tau > 1$, with the slope decreasing monotonically with the decrease in ionic strength. Note also that the numerical RSA simulations are in good agreement with the experimental data for all ionic strengths studied. The validity of this linear dependence of Θ on $\tau^{-1/2}$ can be exploited for extrapolation of the experimental results to infinite time which allows one to determine the proper Θ_{mx} values.

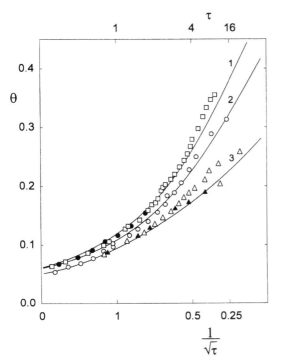

FIG. 57 Adsorption kinetics of latex particles (diameter 0.94 μm) measured in the RIJ cell of Re = 8 [12]. The solid lines denote the RSA simulation results; curve 1, $I = 10^{-3}$ M; 2, $I = 10^{-4}$ M; curve 3, $I = 1.2 \times 10^{-5}$ M. (From Ref. 12.)

However, for lower-Re flows or for small particles, the deviations from the classical RSA model are expected to become more pronounced because of the above coupling with the bulk transport. A precise determination of these deviations seems rather difficult, although they were observed qualitatively in Ref. 1 for Re = 0.6 (micrometer-sized particles) and in Ref. 4 for nanometer-sized particles.

Also, for higher flow intensities (Re number), the adsorption kinetics of larger colloid particles significantly deviates from the RSA predictions due to the hydrodynamic scattering (HS) effect predicted by Eq. (127). Examples of the kinetic curves measured for this adsorption regime in the SIJ cell and latex particles 1 μm in diameter are shown in Fig. 58 [111]. Particle adsorption was observed at a distance of 150 μm from the cell symmetry plane, where the local shear rate was relatively high. One can note in Fig. 58 that the increase in the flow rate considerably reduced the particle deposition rate for $\tau > 0.5$. The maximum coverage for Re = 16 saturates at a small value of about 0.13. This behavior, being a direct manifestation of the HS effect, can be well accounted for by the Brownian dynamics simulation (shown by the continuous lines). For higher flow rates, the experimental results are also well reflected by the analytical predictions stemming from Eq. (220) with the C_1 and C_2 constants strictly related to the hydrodynamic shadow length via the equation [1]

$$C_1 = 4(1 + H^*)^2 + C_h' \left(\frac{2\alpha_r V_\infty a^3}{R^2 D_\infty} \right)^n \frac{x}{a}$$
$$C_2 = \tfrac{1}{8} C_1^2$$

(231)

where C_h' is the dimensionless hydrodynamic constant.

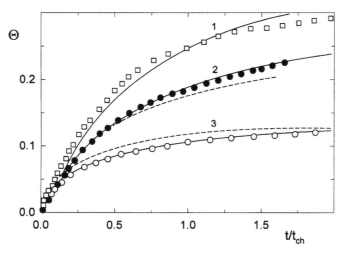

FIG. 58 The kinetics of particle adsorption under the flow-controlled transport conditions. The points shows the experimental results obtained in the slot impinging jet cell [111] for polystyrene latex particles adsorbing on mica: curve 1, Re = 2; curve 2, Re = 8; curve 3, Re = 16. The continuous lines show the Brownian dynamics simulations and the dashed lines denote the limiting RSA results calculated from Eq. (232a). (From Ref. 111.)

One can show that in the case when $C_1 \gg 1$, particle adsorption kinetics can be approximated by the simple formula [1]

$$\Theta = \Theta_a\left(1 - e^{-\tau/\Theta_a}\right) \tag{232a}$$

where

$$\Theta_a = \frac{1}{C_1} = \left[4(1+H^*)^2 + C_h'\left(\frac{2\alpha_r V_\infty a^3}{R^2 D_\infty}\right)^n \frac{x}{a}\right]^{-1} \tag{232b}$$

is the apparent maximum coverage.

The results are shown in Fig. 58 and predicted from Eq. (232a) lead to a rather unexpected conclusion that the increase in the hydrodynamic flow intensity (vigorous mixing) results in a considerable decrease in the amount of particle adsorbed within an experimentally accessible time scale. As can be estimated from Eq. (232b), the effect becomes especially important for large shear rate and larger particle size which has been estimated in Ref. 1 to be about 0.4 μm. It is interesting to mention that a similar kinetic saturation of the interface was observed in adsorption experiments concerning particle and bacteria adsorption in channel flows [122,174]. The HS offers a natural explanation of these experimental data interpreted in terms of the Langmuir model.

It should be emphasized that the stationary surface concentrations observed for higher flow rates should by no means be treated as maximum or equilibrium values as is often done in the literature. The apparent saturation of the surface at a very low level is a purely kinetic effect caused by the long-range hydrodynamic interaction of the adsorbed particles with other particles flowing in their vicinity.

Apart from these hydrodynamic effects, irreversible adsorption kinetics of colloid particles can be influenced by the external force, usually gravity. This ballistic effect has been intensively

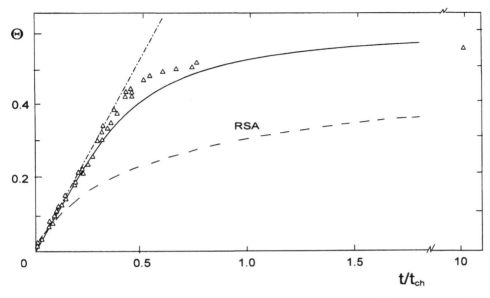

FIG. 59 The kinetics of particle adsorption under the external force (gravity) transport conditions. The points show the experimental results obtained in the sedimentation cell for melamine latex particles (diameter 1.68 µm) adsorbing on mica [149]; the continuous line represent the theoretical results calculated from the ballistic model; the dashed-dotted line represents the Stokes law and the dashed line shows the theoretical results calculated from the RSA model.

studied experimentally [149,150,183] by using colloid particles having much larger specific density than the polystyrene latex. The melamine latex is primarily used in these experiments, having the density of 1.5 kg/dm^3 and particle size above 1 µm. Typical kinetic results obtained in such experiments are plotted in Fig. 59 using, again, the reduced time $\tau = k_c n^b t$, where $k_c = (2/9)(\Delta \rho g a^2/\eta)$ is the reduced adsorption rate calculated from the Stokes law. As one can see in Fig. 59, the kinetics deviates considerably from the RSA model (depicted by the dashed line) and can be well described by the analytical approximation derived by Choi et al. [192] and modified to account for the effective interaction range [149]:

$$\Theta = \Theta_{mx}\left[1 - \frac{1 + c_4\tau}{1 + c_1\tau + c_2\tau^2 + c_3\tau^3} e^{-1.103(1+H^*)^2\tau}\right] \qquad (233)$$

where $c_1 - c_4$ are the dimensionless constants.

One can deduce from Eq. (233) that for $\tau < 0.5$, particle coverage should increase linearly with the dimensionless time, which, is indeed, confirmed by the experimental results shown in Fig. 59. This has a practical significance, indicating that in the case of the ballistic model, the particle adsorption rate remains constant for a much broader range of Θ than for the RSA model. This allows one to determine in a precise way the viscosity of the medium just by measuring the number of particles (of a known size) as a function of time. This approach was indeed exploited in Ref. 193 to measure the viscosity of colloid silica solutions, by using melamine latex of size 1.68 µm.

Another advantage of the ballistic adsorption regime is that the maximum coverage can be determined in a more efficient way than in the case of the RSA or diffusion-controlled adsorption regimes. Knowledge of Θ_{mx} is important from a practical viewpoint because this parameter unequivocally characterizes particle adsorption kinetics and determines the interface "capacity." As pointed out in Ref. 12, determining θ_{mx} for colloid particles without extrapolation is difficult. This is so because deposition times needed to attain the surface coverage close to θ_{mx} becomes prohibitively long (e.g., in the experiments shown in Fig. 52 or 57, the maximum times exceeded 16 h). As mentioned, experiments carried out for such long times are less reliable due to likely contamination of the surface by impurities present in the suspensions. Obviously, one can reduce this value by using more concentrated suspensions which decreases the experimental accuracy.

Another complication associated with θ_{mx} determination is of a more fundamental nature. One should remember that, unlike θ_{mx} for hard particles, which has a unique value, in the case of interacting particles, θ_{mx} is dependent not only on electrostatic interactions but also on the transport mechanism and particle distribution over an interface. Also, the polydispersity of the colloid suspension affects θ_{mx} (cf. Fig. 29). All of these effects lead to the deviation of the θ versus τ dependencies from linearity; hence, a direct extrapolation of the kinetic data to infinite time may not be accurate enough. Even with these limitations, which are expected to be of the order of percents [12], the extrapolation procedure is more accurate than the usually adopted method of treating the coverage attained after a long (but undefined) time as the true θ_{mx} value.

The determination of θ_{mx} for proteins was first attempted by Feder and Giaever [194], who used ferritin (globular protein having an almost spherical shape of diameter of about 10 nm). Even by using a highly concentrated saline solution (0.15 M), the authors were unable to reach the limiting value of 0.547, characteristic for hard spheres. They have found θ_{mx} to be within the range 0.2–0.5. This discrepancy can be accounted for by the RSA model because, according to Eq. (198b), θ_{mx} should be 0.35 for $\kappa a = 12.5$, corresponding to the experimental conditions of Feder and Giaever.

There also exist indirect (optical) measurements of the maximum coverage in the case of other proteins whose shape resembles prolate spheroids like the bovine serum albumin (BSA) having dimensions $11.6 \times 2.7 \times 2.7$ nm, molecular weight 69,000, aspect ratio approximately 0.29, fibrinogen having dimensions $45 \times 6 \times 9$ nm, molecular weight 340,000, aspect ratio approximately 0.15, and IgG having dimensions $23.5 \times 4.4 \times 4.4$ nm, molecular weight 150,000, aspect ratio 0.19 [2]. Depending on the experimental conditions, the maximum coverage was found to be 1–6.5 mg/m^2 for BSA, 2–6 mg/m^2 for fibrinogen, and 3–7 mg/m^2 for IgG [2]. These values are given in the dimensional units, which are more interesting from the practical viewpoint. These results agree well with the theoretical predictions calculated from Eq. (198a) for the side-on and perpendicular protein adsorption, which are 1.2–5.2 mg/m^2 for BSA, 3–16 mg/m^2 for fibrinogen, and 2.1–8.4 mg/m^2 for IgG [2].

Experimental results concerning colloid particle adsorption are more abundant. Onoda and Liniger [195] determined θ_{mx} for a large polystyrene particle (diameter 2.95 μm) adsorbing on glass slides, modified by adsorption of cationic polyacrylamide. Particle deposition occurred under gravity (sedimentation) and the drying procedure was used before particle counting. The jamming coverage was found to be 0.55, in a good agreement with theoretical predictions for the RSA model for hard particles.

Similarly, gravity-driven deposition was studied in Refs. 149 and 150 using the sedimentation cell and polymeric (melamine) particles with size 1.68 μm and specific density 1.5 g/cm^3. The influence of the ionic strength, varied between 5×10^{-6} M and 10^{-3} M, on θ_{mx} was systematically studied. It was found that the maximum coverage dependence on the ionic strength can be described well by Eq. (198b) with the hard-particle value θ_∞ equal to 0.61.

There are numerous experimental results involving the θ_{mx} determination for diffusion-controlled transport conditions like the above-mentioned work of Böhmer [4], who studied various latex suspensions (with particle diameter varied between 9 and 90 nm) using the reflectometric method, and Johnson and Lenhoff [9], who used the AFM method. The effect of the ionic strength varied between $3 \times 10^{-6}\ M$ and $5 \times 10^{-3}\ M$ on θ_{mx} was systematically studied in this work. On the other hand, the micrograph technique was used for studying diffusion-controlled adsorption of 0.3- and 1-µm latex particles on mica [16] for ionic strength change between $10^{-2}\ M$ and $2 \times 10^{-5}\ M$.

Harley et al. [121] used the above TFFD–SEM method to determine the effect of the ionic strength of the suspension on θ_{mx} in the case of the deposition of small, negatively charged particles (diameter ranging from 0.116 to 0.696 µm) on positively charged latex 2.17 µm in diameter. They measured a systematic decrease in θ_{mx} from 0.1 for the smallest particles and lowest ionic strength to 0.45 for the largest particles.

A similar problem of determining θ_{mx} as a function of ionic strength for the small/large sphere configuration was experimentally studied by Vincent et al. [151] by using the indirect, concentration-depletion method.

For the sake of convenience, most of the above-discussed data were collected in Fig. 60. In order to facilitate the comparison between results obtained under various deposition conditions (when θ_{∞} was also varied), the universal coordinate system $\theta_{mx}/\theta_{\infty}$ versus a/Le was chosen. The theoretical results stemming from the RSA simulations performed using the LSA model are also shown in Fig. 60 together with the analytical results calculated from Eq. (198b) by assuming the EHP concept (solid line). It should be noted that in the latter case, the theoretical data for $a/Le < 5$ can only be treated as approximate ones because the assumption pertinent to the EHP model breaks down for this range of a/Le. One can observe in Fig. 60 that the experimental data are in a fairly good agreement with theoretical predictions, although for $a/Le > 5$, the θ_{mx} derived from experiments are generally smaller than theoretically predicted. These deviations are

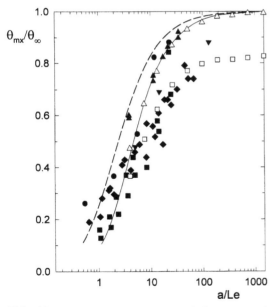

FIG. 60 Collection of experimental data concerning the maximum coverage $\Theta_{mx}/\Theta_{\infty}$ for spherical particles. (●) Johnson and Lenhoff data [9], AFM natural convection; (▲) Adamczyk et al. [149] optical microscopy, gravity; (▼) Adamczyk et al. [16], optical microscopy, diffusion; (■) Böhmer et al. [4], reflectometry, RIJ cell; ◆ Harley et al [121], electron microscopy, diffusion; (△) theoretical RSA simulations, $\tau = 10^5$; (□) theoretical RSA simulations, $\tau = 10$. The solid line represents analytical approximation calculated from Eq. (198) for $\phi_0 = 100$ kT; dashed line, same for $\phi_0 = 10$ kT. (From Ref. 12.)

most likely due to the limited experimental time of deposition measurements and lack of the extrapolation procedure discussed above. This hypothesis is supported by the fact that theoretical data obtained from simulations after $\tau = 10$ (which would correspond to physical deposition times of the order of hours) reflect well the lower branch of experimental results.

On the other hand, a proper theoretical interpretation of experimental results for $a/Le < 5$ would require a true three-dimensional modeling of particle deposition process with appropriate expressions for the many-body electrostatic interactions. An attempt in this direction was undertaken by Oberholzer et al. [196], who considered the true three-dimensional transport in a force field stemming from adsorbed particles and the interface. Because the authors still used the LSA approach (generalized for the two particle/interface configuration as previously mentioned), the deviation from the two-dimensional RSA simulations with respect to θ_{mx} was found to be not too significant.

Obviously there is need for additional theoretical studies in this field, although a proper consideration of the many-body electrostatic interaction at interfaces may pose considerable difficulties. In any case, the classical RSA approaches seems to work well for the most interesting, from practical viewpoint, range of a/Le (5–100) reflecting well the experimentally found effect of a considerable decrease in the θ_{mx} for lower a/Le values (low ionic strength).

2. Structure of Adsorbed Layers

Apart from determining the kinetic aspects of irreversible adsorption, the colloid systems can be exploited for modeling structure formation upon adsorption. This is an especially attractive possibility in view of the difficulties in obtaining reliable information on the structural aspect of adsorbed monolayers.

The existence of the short-range, liquidlike ordering in the bulk of colloid suspensions has well been documented [197,198]. These experiments were usually based on the light-scattering method or small-angle neutron scattering (SANS) method for higher suspension concentration. It was found that already for very dilute suspensions (volume fraction solid of a few percent), particle positions were strongly correlated over the distance comparable with their dimensions. This leads to liquidlike ordering, the range of which increased for the low ionic strength of the suspension. For higher solid content, the formation of a macroscopic colloid crystal was often observed.

Analogous ordering phenomena appear in two dimensions, the monolayer of adsorbed colloid particles. This is illustrated in Fig. 61, in which the pair-correlation function $g(r)$ is shown for polystyrene latex particles (averaged diameter 0.88 μm) adsorbed on mica [142]. The pair-correlation function is defined as ensemble averaged particle coverage within the ring of radius r drawn around a central particle. The surface coverage was equal 0.05 and the ionic strength was 10^{-3} (Fig. 61a) and $10^{-5} M$ (Fig. 61b). In the former case, the H^* parameter was about 0.05, which means that the adsorbed particles can be treated as hard spheres. On the other hand, for $I = 10^{-5} M$, H^* was approximately equal to 0.5, so the effective interaction range became comparable with particle dimensions, which should affect the structure of the adsorbed particle monolayer. As can be seen in Fig. 61, this is reflected by the pair-correlation function, which is well described for both ionic strengths by the Boltzmann distribution [i.e., $g(r) = e^{-\phi/kT}$, where ϕ is the pair interaction energy discussed earlier]. The agreement of the experimental data with the Boltzmann distribution indicates that the structure of the adsorbed layer resembles closely a two-dimensional molecular gas structure, with negligible long-range ordering. This is a significant finding which confirms the usefulness of colloid systems for modeling adsorption phenomena occurring at molecular level. Moreover, by inverting the

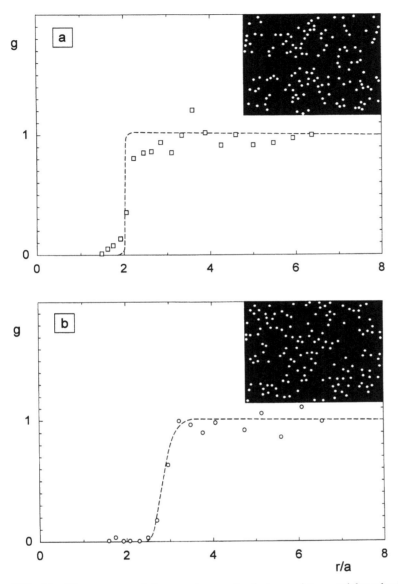

FIG. 61 The structure of the monolayers of polystyrene latex particles adsorbed on mica expressed in terms of the pair-correlation function g: (a) $I = 10^{-3}$ M, $\Theta = 0.05$; (b) $I = 10^{-5}$ M, $\Theta = 0.05$. The broken lines represent the pair-correlation function calculated from the Boltzmann distribution $g = e^{-\phi/kT}$; the insets show the adsorbed particles forming a two-dimensional gas phase.

experimentally found $g(r)$ function, one can determine in a direct way the interaction energy profile as demonstrated in Ref. 13.

For higher coverage, the structure of the irreversibly adsorbed monolayer is changed, which can be seen in Fig. 62 (latex particles adsorbed on mica, $\Theta = 0.24$, ionic strengths 10^{-3} M and 2×10^{-5} M, respectively [142]). In this case, the pair-correlation function exhibits an oscillatory character with a well-pronounced primary peak at short separations. The height of

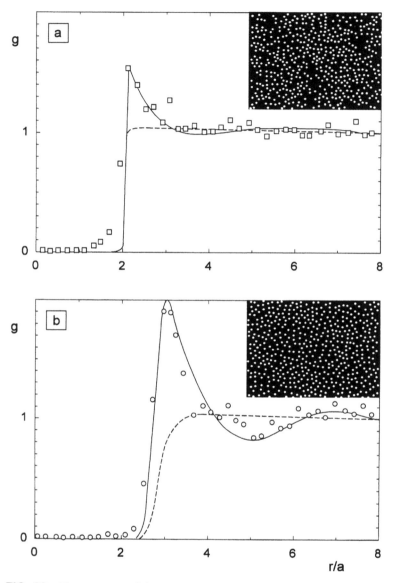

FIG. 62 The structure of the monolayers of polystyrene latex particles adsorbed on mica expressed in terms of the pair-correlation function g: (a) $I = 10^{-3}\,M$, $\Theta = 0.24$; (b) $I = 10^{-5}\,M$, $\Theta = 0.24$. The broken lines represent the pair-correlation function calculated from the Boltzmann distribution $g = e^{-\phi/kT}$; the continuous lines show the RSA simulations (smoothened); the insets show the adsorbed particles forming a two-dimensional liquid phase.

the peak increases considerably for the lower ionic strength value, which is in agreement with the effective hard-particle concept. The appearance of the peak indicates that due to topological constraints, particle positions in the monolayer become strongly correlated, leading to a short-range ordering, analogous to a two-dimensional molecular liquid structure [197]. This, again, confirms the conclusion that colloid systems can be exploited for modeling molecular adsorption phenomena. In contrast to molecular systems, however, the adsorbed particles remain immobile (localized); hence, a two-dimensional liquid–solid-phase transition was not observed. This transition can be provoked, as shown in Fig. 39a, by the capillary forces occurring upon drying of the monolayer. Another possibility of producing two-dimensional solid phases (of an hexagonal structure) is the electrophoretic deposition of particles at conducting glass electrodes [179,180].

The adsorbed colloid systems also can be used for mimicking fluctuation phenomena occurring on molecular level. This is so because the adsorbed particle monolayer resembles a frozen fluctuation whose statistical analysis can furnish interesting structural information of general validity. In order to express quantitatively fluctuation phenomena, one usually determines the variance of the number of particles absorbed over equal-sized surface areas S [181–183]. This quantity is connected with the pair-correlation function by the Ornstein–Zernicke relationship [132]

$$\bar{\sigma}^2 = \frac{\langle N_p^2 \rangle - \langle N_p \rangle^2}{\langle N_p \rangle} = 1 + 2\pi\Theta \int [g(r) - 1]\, dr \tag{234}$$

This equation is valid both for reversible and irreversible systems.

On the other hand, in the thermodynamic limit (reversible systems), $\bar{\sigma}^2$ can be expressed using the compressibility theorem [132]:

$$\bar{\sigma}^2 = kTS_g \left(\frac{\partial \pi}{\partial \Theta}\right)^{-1} = \frac{1}{\Theta}\left(\frac{\partial \mu/kT}{\partial \Theta}\right)^{-1}_{S,T} \tag{235}$$

Using the expression for the two-dimensional pressure given by the Boublik formula, Eq. (179), one obtains the following expression for spherical particles [132]:

$$\bar{\sigma}^2 = \frac{(1 - \Theta)^3}{1 + \Theta} \tag{236}$$

As can be observed in Fig. 63, the experimental data obtained for latex particles adsorbed irreversibly on mica under the diffusion-controlled transport conditions [182] are very well reflected by the thermodynamic Eq. (236) for the entire range of coverages studied. This rather unexpected result further supports the hypothesis of a significant role of colloid systems modeling adsorption phenomena of molecular systems.

VI. CONCLUDING REMARKS

The theoretical approaches discussed in this chapter can successfully be used for a quantitative interpretation of irreversible adsorption kinetics of protein and colloid particles, which has been confirmed experimentally using monodisperse colloid suspensions. It was demonstrated that for low coverage $\Theta < 0.05$ and forced-convection transport conditions, linear adsorption regimes appear, characterized by a time- and coverage-independent adsorption rate, which can be well reflected by Eq. (151b), valid for uniformly accessible surfaces. The crucial parameters of this regime are the adsorption rate constant k_a'', which can unequivocally be connected with the interaction energy profile [cf. Eq. (152)] and the bulk-transfer rate constant k_c (reduced flux). This quantity can well be approximated for particle sizes below 0.5 μm by the limiting flux values given in Table 3. In this limit, the k_c constant decreases with particle size as $a^{-2/3}$. On the other hand, for larger particle sizes, the k_c constant can be calculated in a standard way by numerical solution of the convective-diffusion equation, Eq. (136), both for uniformly and

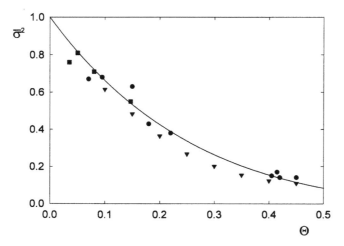

FIG. 63 The dependence of the relative variance of particle density fluctuations $\bar{\sigma}^2 = \langle N_p^2 \rangle - \langle N_p \rangle^2 / \langle N_p \rangle$ on particle coverage Θ. The points denote the experimental results obtained for polystyrene latex particles on mica under the diffusion-controlled transport for $I = 10^{-3} M$ [182]; the inverted triangles show the RSA simulation results; the continuous line represents the analytical results stemming from the SPT theory for equilibrium system Eq. (236).

nonuniformly accessible surfaces. In this limit, k_c becomes proportional to the square of the effective particle size due to the interception effect, as predicted by Eq. (165). For the external-force adsorption regime, when Ex $\gg$ Pe, the adsorption rate constant becomes $k_c = D_\infty \text{Ex}/a$.

On the other hand, for the diffusion-controlled transport conditions (Pe = 0), particle adsorption becomes nonstationary and can be described by analytical solutions of Eq. (138) with the general kinetic boundary condition, Eq. (139). It was shown that for large adsorption constant values, when $\bar{k}_a = k_a L_{ch}/D_\infty = 2/3\Phi_v R_b \gg 1$, particle adsorption flux and the coverage are given by the simple expressions

$$ j = -\sqrt{\frac{D_\infty}{\pi t}} n^b $$

$$ \Theta = S_g 2\sqrt{\frac{D_\infty t}{\pi}} n^b $$

These equations remain valid for a much broader coverage range (reaching 0.3) than for the convection-controlled transport, which can be exploited for an accurate determination of the particle diffusion coefficient.

For higher coverage ($\Theta > 0.05$ to 0.3 depending on the mode of transportation), the blocking effects stemming from particle–particle interactions become important and adsorption kinetics becomes nonlinear with respect to time and bulk suspension concentration. These effects can quantitatively be described by numerical solutions if the convective-diffusion equation with the nonlinear boundary conditions Eq. (204). It was demonstrated that the generalized blocking function $\bar{B}(\Theta)$ occurring in this equation can be well approximated by the $B_0(\Theta)$ blocking function stemming from the classical RSA model [given in the case of spheres by Eq. (187a) and for anisotropic particles in the limit of high coverage by Eq. (190)]. The crucial parameter in these expressions is the maximum Θ_{mx} coverage (referred to as the jamming coverage in the case of hard particles). Θ_{mx} can be determined both for spherical and anisotropic particles (e.g., spheroids) in *ab initio* type of calculations performed according to the classical RSA method (cf. Table 4). For interacting particles, Θ_{mx} can be estimated from the data pertinent to hard particles by using the effective hard-particle concept, expressed by Eq. (198b).

Knowing Θ_{mx}, one can approximate particle adsorption flux for all the convection-controlled (quasi stationary) transport by the analytical expression

$$-j = \frac{\bar{B}(\Theta)}{1 + (K-1)\bar{B}(\Theta)} k_a n^b \cong \frac{B_0(\Theta)}{1 + (K-1)B_0(\Theta)} k_a n^b$$

where $K = k_a/k_c$.

It was shown that the limit of a not too high surface coverage and $K = 1$, one can derive from this equation the limiting expression for Θ [Eq. (220)]:

$$\Theta = \Theta_1 \frac{1 - \exp(-q\bar{C}_1 S_g k_c n^b t)}{1 - (\Theta_1/\Theta_2) \exp(-\bar{q}\bar{C}_1 S_g k_c n^b t)}$$

For higher coverage, when $\Theta \to \Theta_{mx}$, one obtains the limiting expression

$$\Theta = \Theta_{mx} - \frac{\Theta_{mx}^{m/(m-1)}}{\sqrt[m-1]{(m-1)C_{mx}S_g k_c n^b t}}$$

where $m = 3$ for spheres, $m = 4$ for polydisperse spheres and anisotropic particles adsorbing side-on, and $m = 5$ for unoriented adsorption of anisotropic particles. It was demonstrated experimentally that these kinetic expressions describe well the irreversible adsorption of colloid particles under moderate flow conditions.

On the other hand, for the diffusion-controlled (no-flow) transport conditions adsorption kinetics is governed by the square root of time dependence until the limiting coverage

$$\Theta_1 = \Theta_{mx}\left(1 - \sqrt[m]{\frac{1}{10(m-1)\bar{k}_a}}\right)$$

is reached. Then, the kinetics is given by the asymptotic law

$$\Theta = \Theta_{mx} - \frac{K_l}{\sqrt[m-1]{(S_g n^b)^2 D_\infty t}}$$

All of these theoretical results are valid for particle sizes of 1 μm and less. For larger particles, the combined effects of flow, specific interactions, and blocking (particle–particle interactions) are coupled in a nonlinear way, which makes their theoretical analysis via the RSA models impractical at the present time. Further progress in this field can only be achieved if the $\bar{B}(\Theta)$ function is determined more accurately via the Brownian dynamics simulations.

Experimental results obtained for model colloid systems (mostly polystyrene latex particles) confirmed the main features of the theoretical models—in particular the validity of the generalized RSA model (considering the coupling between the bulk and surface transport step). The kinetics aspects of irreversible adsorption, the jamming coverage, and the structure of adsorbed layers have been quantitatively reflected in terms of this model. In particular, it was demonstrated that for low coverage, the structure of particle monolayers can be described by the Boltzmann distribution (two-dimensional gas phase). This allows one to determine the pair-interaction energy profile in a direct way. For higher coverage, the structure of particle monolayers resembles closely a molecular two-dimensional liquid phase. This can be exploited for studying fluctuation phenomena in molecular liquid phases.

It also was found experimentally, in accordance with theoretical predictions, that the jamming coverage of hard spheres adsorbed irreversibly (in the limit of negligible electrostatic interactions) approached 55%. This value strongly decreases for lower ionic strength, fully in accordance with the effective hard-particle concept. This finding confirmed the decisive role of the lateral electrostatic interactions in irreversible adsorption of colloid particles.

The theoretical and experimental results suggest that colloid systems, subject to direct microscope observations, can be used to mimic adsorption phenomena occurring at the molecular level, which can only be studied in an indirect way.

ACKNOWLEDGMENTS

This work has been supported by the Copernicus grant ERBIC 15CT980121 and partially by the KBN Grant 3T09A-105-18. The help of E. Porębska, M. Barańska, E. Musiał, and P. Weroński in preparing the manuscript and artwork is kindly appreciated.

LIST OF SYMBOLS

$A = b/a$	Aspect ratio parameter of spheroidal particles
A_{12}	Hamaker constants
A_f	Flow model parameter (dimensionless)
ASF	Available surface function
a	Spherical particle radius or longer spheroid semiaxis
a_{ch}	Chemical activity of the particle
$B(\Theta, h)$	Position-dependent blocking function
$B_0(\Theta)$	Surface blocking function (available surface function, ASF)
$B(\Theta)$	Averaged blocking function (available surface function)
$\bar{B}(\Theta), \tilde{B}(\Theta)$	Overall blocking function (flux correction factor)
b	Shorter axis of a prolate spheroid or channel half-depth
b_1	Thickness of the rough layer on particles
C_1, C_2, C_1', C_2'	ASF expansion coefficients
C_h	Hydrodynamic constant
C_{mx}	Dimensionless constant characterizing blocking effects for higher coverage
$C(\bar{r})$	Hydrodynamic function of the RIJ cell
c	Channel half-width or the dimensionless numerical constants
$\boldsymbol{D}$	Particle diffusion tensor
$\boldsymbol{De}$	Dielectric displacement vector
D_∞	Particle diffusion coefficient in the bulk
D_{12}	Mutual diffusion coefficient
$\boldsymbol{E}$	Electric field vector
$E_\perp, E$	Perpendicular component of the field vector, the field vector length (scalar)
Ex	External-force number (dimensionless)
e	elementary charge
$\boldsymbol{F}$	Force vector
$\boldsymbol{F}_h$	Hydrodynamic force vector
$\boldsymbol{F}_e$	External-force vector
F	Force (scalar)
$F_1(H)–F_4(H)$	Hydrodynamic correction functions

f	Force density
f	Activity coefficient of the particle
$\boldsymbol{G}$	Oseen tensor
G_D	Geometrical Derjaguin factor
$\bar{G}_D = G_D/a$	Dimensionless Derjaguin factor
G_e^0, G_e	Geometrical ESA factor
$\bar{G}_e^0, \bar{G}_e$	Dimensionless ESA factor
G^0	Wall shear rate
Gr	Gravity number (dimensionless)
G_{sh}	Local shear rate intensity
G_{st}	Local stagnation flow intensity
g	Acceleration due to gravity
$H = h/a$	Dimensionless gap width
$H^* = h^*/a$	Effective interaction range (dimensionless)
h	Coordinate perpendicular to the interface or the surface to surface distance
$\bar{h} = h/Le$	Dimensionless surface-to-surface distance between plates
h_m	Surface-to-surface distance between particles
h_c	Distance between the capillary and the surface in the RIJ cell
h^*	Effective interaction range
$\boldsymbol{I}$	Unit tensor
I	Ionic strength of the electrolyte
$\boldsymbol{i}$	Unit vector
$\boldsymbol{j}$	Particle flux vector
j_0	Stationary flux (limiting flux)
$K = k_a/k_c$	Coupling constant
$K_1 - K_4$	Hydrodynamic coefficients
K_a	Equilibrium adsorption constant
K_h	Hydrodynamic resistance coefficient
K_l	Long-time expansion coefficient
k	Boltzmann constant
k_a, k_a', k_a''	Adsorption kinetic constants
$\bar{k}_a, \bar{k}_a'$	Dimensionless adsorption constant
k_c, k_c'	Bulk-transfer rate constants (reduced flux)
k_c^0	Transfer rate constant at the center of the cell
k_d, k_d'	Desorption kinetic constant
L	Length of the channel
$Le = 1/\kappa$	Screening length
L_{ch}	Characteristic length scale
N, N_s	Surface concentration
N_p	Number of particles adsorbed
$\langle N_p \rangle$	Averaged number of particles adsorbed
$\boldsymbol{M}$	Hydrodynamic mobility tensor
$\boldsymbol{M}_r$	Rotational mobility tensor
M	Hydrodynamic mobility coefficient
m, m_s	Monolayer mass
$\hat{\boldsymbol{n}}$	Unit normal vector
n_i	Ion number concentration
n_i^b	Ion number concentration in the bulk
n	Particle number concentration

n^b	Particle number concentration in the bulk
$\bar{n} = n/n^b$	Dimensionless particle concentration
O	Particle perimeter
$\Delta \boldsymbol{P}$	Osmotic pressure tensor
ΔP	Osmotic pressure (scalar)
ΔP	Hydrostatic pressure difference
Pe	Peclet number (dimensionless)
$\boldsymbol{p}$	Induced dipole moment of the particle
p	Isotropic pressure (scalar)
Q	Volumetric flow rate
q	Net charge on particle
R	Cylinder, sphere, or capillary radius
R_b	Energy barrier resistance
$\bar{R}_b$	Dimensionless barrier resistance
R_c	Hydrodynamic cavity radius
Re	Reynolds number
$\boldsymbol{r}$	Radius vector
r	Radial coordinate
$\bar{r}$	Dimensionless radial coordinate
R', R''	Principal radii of curvature
S	Surface
$S(\bar{r})$	Hydrodynamic function of the RIJ cell
Sh	Mass-transfer Sherwood number (dimensionless)
S_g	Characteristic cross section of the particle
$\boldsymbol{T}$	Maxwell stress tensor
$\boldsymbol{T}_h$	Hydrodynamic torque vector
T	Absolute temperature
Ta	Dimensionless period of adsorption
t, t_a	Adsorption time
t_{ch}	Characteristic adsorption time
$\boldsymbol{V}$	Fluid velocity vector
$\boldsymbol{V}_0$	Local fluid velocity vector
$\boldsymbol{V}^*$	Perturbed fluid velocity vector
$\boldsymbol{V}^*_{st}$	Local stagnation flow
$\boldsymbol{V}^*_{sh}$	Local shear flow
V_{mx}	Maximum velocity in the channel
V_∞	Characteristic velocity far from the interface
V_h, V_r, V_ϑ	Components of the fluid velocity vector
V_{ch}	Characteristic convection velocity
v	Volume
$\boldsymbol{U}_e$	Particle velocity due to the external force
$\boldsymbol{U}_p$	Particle velocity vector
$\boldsymbol{U}_\infty, U_\infty$	Characteristic particle migration velocity
$\bar{Y}^0$	Dimensionless effective potential
z	Ion valency

Greek

α	Polarizability of the atom
α_r	Dimensionless flow intensity parameter of the RIJ cell

β_0	Dimensionless electrostatic parameter
β_{12}	London constant
β	Hydrodynamic parameter
Δ	Deformation tensor
δ_a	Thickness of the adsorption layer
δ_b	Energy barrier distance
δ_m	Primary minimum distance
ε	Relative permittivity of the medium (dielectric constant)
ε_0	Permittivity of the vacuum (dimensional)
ε_p	Relative permittivity of the particle (dielectric constant)
γ	Particle shape parameter
γ_a	Particle size ratio parameter
ξ	Dimensionless constant (proportionality coefficient)
η	Fluid dynamic viscosity
$\Theta = S_g N$	Dimensionless coverage
Θ_e	Equilibrium coverage
Θ_{mx}	Maximum coverage
Θ_∞	Jamming coverage (hard particles)
$\kappa = 1/Le$	Reciprocal screening length
λ	Larger-to-smaller particle size ratio
λ_a	Characteristic wavelength
$\lambda_\perp(\mathrm{H})$	Stokes law correction factor
$\tilde{\mu}_i$	Electrochemical potential of an ion
$\tilde{\mu}_i^0$	Reference electrochemical potential of an ion
μ	Chemical potential of a particle
μ^0	Reference chemical potential of a particle
ν	Fluid kinematic viscosity
$\boldsymbol{\Delta\Pi}$	Hydrodynamic pressure tensor
$\Delta\Pi$	Pressure between plates (disjoining pressure)
π	Two dimensional (surface) pressure
ρ_e	Electric charge density
ρ_p	Particle density
ρ	Dispersing medium density (liquid)
ρ_a	Number density of atoms
$\Delta\rho$	Apparent density
σ^0	Surface charge
$\bar{\sigma}^0$	Dimensionless surface charge
$\bar{\sigma}_p$	Polydispersity parameter
τ	Dimensionless diffusion time
ϕ	Specific interaction energy of particles
ϕ_a	Interaction energy of atoms
ϕ_e	External-force potential
$\phi_{\mathrm{ch}}, \phi_0, \phi_0'$	Characteristic interaction energy
ϕ_r	Rough particle interaction energy
Φ	The net interaction energy of particles
Φ_p	Interaction energy per unit area of plates
Φ_v	Volume fraction
ψ	Electrostatic potential
ψ_s	Hydrodynamic stream function

Ψ Interaction energy of the wandering particle

$\bar{\psi} = \psi e/kT$ Dimensionless electrostatic potential

ψ^0 Surface potential of the particle

ω Rotational velocity vector of a particle

ω Rotating disk angular velocity

ω_1, ω_2 Characteristic oscillation frequencies of atoms

REFERENCES

1. Z Adamczyk, B Siwek, M Zembala, P Belouschek. Adv Colloid Inerf Sci 48:151, 1994.
2. Z Adamczyk. J Colloid Interf Sci 229:477, 2000.
3. AV Elgersma, RLJ Zsom, J Lyklema, W Norde. Colloids Surfaces 65:17, 1992.
4. MR Böhmer, EA van der Zeeuw, GJM Koper. J Colloid Interf Sci 197:242, 1998.
5. W Norde, E Rouwendeal. J Colloid Interf Sci 139:169, 1990.
6. T Dąbroś, TGM van de Ven. Colloid Polym Sci 261:694, 1983.
7. Z Adamczyk, P Warszyński, L Szyk-Warszyńska, P Weroński, Colloids Surfaces A 165:157, 2000.
8. GA Bornzin, IF Miller. J Colloid Interf Sci 86:539, 1982.
9. CA Johnson, AM Lenhoff, J Colloid Interf Sci 179:587, 1996.
10. M Semmler, EK Mann, J Ricka, M Borkovec. Langmuir 14:5127, 1998.
11. BW Ninham. Adv Colloid Interf Sci 83:1, 1999.
12. Z Adamczyk, P Weroński. Adv Colloid Interf Sci 83:137, 1999.
13. Z Adamczyk, P Warszyński. Adv Colloid Interf Sci 63:41, 1996.
14. P Warszyński, Adv Colloid Interf Sci 84:47, 2000.
15. Z Adamczyk, P Weroński. J Colloid Interf Sci 189:348, 1997.
16. Z Adamczyk, L Szyk. Langmuir 16:5730, 2000.
17. RJ Hunter. Foundations of Colloid Science. 2nd ed. Oxford University Press: Oxford, 2001.
18. M Elimelech, J Gregory, X Jia, RA Williams. Particle Deposition and Aggregation. Butterworth–Heinemann: Oxford, 1995.
19. G Gouy. J Phys 9:457, 1910.
20. DL Chapman. Phil Mag 25:475, 1913.
21. NF Hoskin, S Levine. Phil Trans Roy Soc London Ser A 248:449, 1956.
22. GM Bell, S Levine. Trans Faraday Soc 53:143, 1957.
23. GM Bell, S Levine. Trans Faraday Soc 54:785, 1958.
24. EJW Vervey, J Th G Overbeek. Theory of the Stability of Lyophobic Colloids. Elsevier: Amsterdam, 1948.
25. O Stern. Z Electrochem 30:508, 1924.
26. GM Bell, S Levine, LN McCartney. J Colloid Interf Sci 33:335, 1970.
27. H Oshima, TW Healy, LR White. J Colloid Interf Sci 90:17, 1982.
28. R Hogg, TW Healy, DW Fuerstenau. Trans Faraday Soc 62:1638, 1966.
29. GR Wiese, TW Healy. Trans Faraday Soc 66:490, 1970.
30. S Usui. J Colloid Interf Sci 44:107, 1973.
31. G Kar, S Chander, TS Mika. J Colloid Interf Sci 44:347, 1973.
32. Z Adamczyk, P Belouschek, D Lorenz. Ber Bunsenges Phys Chem 94:1483, 1990.
33. DC Prieve, E Ruckenstein. J Colloid Interf Sci 63:317, 1978.
34. BV Derjaguin. Kolloid Z 69:155, 1934.
35. H Oshima. Colloid Polym Sci 252:158, 1974.
36. T Sugimoto. Adv Colloid Interf Sci 28:65, 1987.
37. Z. Adamczyk. Colloids Surfaces 39:1, 1989.
38. LR White. J Colloid Interf Sci 95:286, 1983.
39. S Henderson, G Davison. Physical Chemistry, Volume II, An Advanced Treaties. Academic Press: New York, 1967.

40. S Levine. Proc Roy Soc Ser A 70:165, 1939.
41. H Oshima, T Kondo. J Colloid Interf Sci 155:499, 1993.
42. LN McCartney, S Levine. J Colloid Interf Sci 30:345, 1969.
43. JE Sader, SL Carnie, DYC Chan. J Colloid Interf Sci 171:46, 1995.
44. NF Hoskin, S Levine. Phil Trans Roy Soc London Ser A 248:433, 1956.
45. SL Carnie, DYC Chan, J Stankovich. J Colloid Interf Sci 165:116, 1994.
46. P Warszyński, Z Adamczyk. J Colloid Interf Sci 187:283, 1997.
47. RD Stoy. J Electrostat 33:385, 1994.
48. P Atten. J Electrostat 30:259, 1993.
49. J Mahanty, BV Ninham. Dispersion Forces. Academic Press: New York, 1976.
50. JM Israelachvili. Intermolecular and Surface Forces. 2nd ed. Academic Press: London, 1992.
51. HC Hamaker. Physica 4:1058, 1937.
52. EM Lifshitz. Sov Phys—JETP 2:3, 1956.
53. F London. Trans Faraday Soc 33:8, 1937.
54. JH Schenkel, JA Kitchener. Trans Faraday Soc 56:161, 1960.
55. HB Casimir, D Polder. Phys Rev 73:86, 1948.
56. A Suzuki, NFH Ho, WI Higuchi. J Colloid Interf Sci 29:552, 1969.
57. J Czarnecki. J Colloid Interf Sci 72:361, 1979.
58. J Czarnecki, T Dąbroś. J Colloid Interf Sci 78:25, 1980.
59. J Czarnecki. Adv Colloid Interf Sci 24:283, 1986.
60. R Rajagopalan, JS Kim. J Colloid Interf Sci 83:428, 1981.
61. JA Barker, D Henderson. J Chem Phys 47:4714, 1967.
62. M Smoluchowski. Phys Zeit 17:557, 1916.
63. Z Adamczyk, TGM van de Ven. J Colloid Interf Sci 97:68, 1984.
64. Z Adamczyk, T Dąbroś, J Czarnecki, TGM van de Ven. J Colloid Iinterf Sci 97:91, 1984.
65. S Levine. J Colloid Interf Sci 51:72, 1975.
66. P Richmond. JCS Faraday Trans II 71:1154, 1975.
67. L Song, PR Johnson, M Elimelech. Environ Sci Technol 28:1164, 1994.
68. H Krupp. Adv Colloid Interf Sci 1:111, 1967.
69. J Czarnecki, P Warszyński. Colloids Surfaces 22:207, 1987.
70. MC Herman, KD Papadopolous. J Colloid Interf Sci 136:385, 1990.
71. MC Herman, KD Papadopolous. J Colloid Interf Sci 142:331, 1991.
72. TGM van de Ven. Adv Colloid Interf Sci 48:121, 1994.
73. GJ Fleer, MA Cohen Stuart, JMHM Scheutjens, T Cosgrove, B Vincent. Polymers at Interfaces. Chapman & Hall: London, 1993.
74. J Happel, H Brenner. Low Reynolds Number Hydrodynamics. Martinus Nijhoff: Dordrecht, 1986.
75. TGM van de Ven. Colloidal Hydrodynamics. Academic Press: London, 1989.
76. Z Adamczyk, T Dąbroś, J Czarnecki, TGM van de Ven. Adv Colloid Interf Sci. 19:183, 1983.
77. M Zembala, Z Adamczyk. Langmuir 16:1593, 2000.
78. VG Levich. Physicochemical Hydrodynamics. Prentice-Hall: Englewood Cliffs, NJ, 1962.
79. DV Dass, HJ van Eckvort, AG Langdon. J Colloid Interf Sci 116:523, 1987.
80. ME Weber, D Paddock. J Colloid Interf Sci 94:328, 1983.
81. HC Brinkman. Appl Sci Res A 1:27, 1947.
82. J Happel. AIChE J 4:197, 1958.
83. S Kubawara. J Phys Soc Japan 14:527, 1959.
84. AC Payatakes, R Rajagopalan, C Tien. J Colloid Interf Sci 49:321, 1974.
85. G Neale, WK Nader. AIChE J 20:530, 1974.
86. CJ Guzy, EJ Bonano, EJ Davis. J Colloid Interf Sci 95:523, 1983.
87. Z Adamczyk, TGM van de Ven. J Colloid Interf Sci 84:497, 1981.
88. T Dąbroś, TGM van de Ven. Physicochem Hydrodyn 8:161, 1981.
89. T Dąbroś, TGM van de Ven. J Colloid Interf Sci 89:232, 1982.
90. Z Adamczyk. B Siwek, P Warszyński, E Musiał, J Colloid Interf Sci 242:14, 2001.
91. Z Adamczyk, L Szyk, P Warszyński. J Colloid Interf Sci 209:350, 1999.

92. GM Hidy, JR Brock. The Dynamics of Aerocoloidal Systems. Pergamon Press: Oxford, 1970, Vol 1.
93. AD Maude. Br J Appl Phys 12:263, 1961.
94. E Bart. Chem Eng Sci 23:193, 1968.
95. ME O'Neill. Mathematika 11:67, 1964.
96. ME O'Neill. Mathematika 14:170, 1967.
97. ME O'Neill. Chem Eng Sci 23:1293, 1968.
98. WR Dean, ME O'Neill. Mathematika 10:13, 1963.
99. SL Goren, ME O'Neill. Chem Eng Sci 26:325, 1971.
100. Z Adamczyk, M Adamczyk, TGM van de Ven. J Colloid Interf Sci. 96:204, 1983.
101. K Małysa, TGM van de Ven. Int J Multiphase Flow 12:459, 1986.
102. BA Alexander, DC Prieve. Langmuir 3:778, 1987.
103. TGM van de Ven, P Warszyński, X Wu, T Dąbroś, Langmuir 10:3046, 1994.
104. X Wu, P Warszyński, TGM van de Ven. J Colloid Interf Sci 180:61, 1996.
105. X Wu, TGM van de Ven. Langmuir 12:6291, 1996.
106. JR Blake, Proc Cambridge Phil Soc 70:303, 1971.
107. T Dąbroś, TGM van de Ven Int J Multiphase Flow 18:751, 1992.
108. T Dąbroś. Colloids Surfaces 39:127, 1989.
109. T Dąbroś, and TGM van de Ven. J. Colloid Interf Sci 149:493, 1992.
110. M Whittle, BS Murray, E Dickinson, VJ Pinfeld. J Colloid Interf Sci 223:273, 2000.
111. Z Adamczyk, B Siwek, L Szyk. J Colloid Interf Sci 174:130, 1995.
112. Z Adamczyk, B Senger, JC Voegel, P Schaaf. J Chem Phys 110:3118, 1999.
113. Z Adamczyk, J Petlicki. J Colloid Interf Sci 118:20, 1987.
114. Z Adamczyk. Bull Pol Acad Chem 35:417, 1987.
115. E Ruckenstein. J Colloid Interf Sci 66:531, 1978.
116. Z Adamczyk. Encyclopedia of Surface and Colloid Science. Adsorption of Particles: Theory. Marcel Dekker: New York, 2001.
117. LA Spielman, SK Friedlander. J Colloid Interf Sci 46:22,1974.
118. T Dąbroś, Z Adamczyk, J Czarnecki. J Colloid Interf Sci 62:529, 1977.
119. T Dąbroś, Z Adamczyk. Chem Eng Sci 34:1041, 1979.
120. W Norde, J Lyklema. Colloids Surfaces 51:1, 1990.
121. S Harley, DW Thompson, B Vincent. Colloids Surfaces 62:163, 1992.
122. J Siollema, HJ Busscher. Colloids Surfaces 47:323, 1990.
123. B Widom. J Chem Phys 39:2808, 1963.
124. B Widom. J Chem Phys 44:3888, 1966.
125. RH Swendsen. Phys Rev A 24:504, 1981.
126. EL Hinrichsen, J Feder, T Jossang. J Statist Phys 44:793, 1986.
127. P Schaaf, J Talbot. J Chem Phys 91:4401, 1989.
128. SM Ricci, J Talbot, G Tarjus, P Viot. J Chem Phys 97:5219, 1992.
129. JW Evans. Rev Mod Phys 65:1281, 1993.
130. J Talbot, G Tarjus, PR van Tassel, P Viot. Colloids Surfaces 165:287, 2000.
131. Z Adamczyk, P Weroński. J Chem Phys 105:5562, 1996.
132. Z Adamczyk, P Weroński. J Chem Phys 107:3691, 1997.
133. H Reiss, HL Frisch, E Helfand, JL Lebowitz. J Chem Phys 32:119, 1960.
134. JL Lebowitz, E Helfand, E Praestgaard. J Chem Phys 43:774, 1965.
135. T Boublik. Mol Phys 29:421, 1975.
136. Z Adamczyk, P Weroński. J Chem Phys 108:9851, 1998.
137. Z. Adamczyk, B Swiek, M Zembla, P Weroński. J Colloid Interf Sci 185:236, 1997.
138. Z. Adamczyk, B Siwek, P Weroński, M Zembala. Prog Colloid Polym Sci 111:41, 1998.
139. P Viot, G Tarjus, SM Ricci, J Talbot. J Chem Phys 97:5212, 1992.
140. G Tarjus, P Viot, SM Ricci, J Talbot. Mol Phys 73:773, 1991.
141. RD Vigil, RM Ziff. J Chem Phys 91:2599, 1989.
142. Z Adamczyk, P Weroński, B Siwek, M Zembala. Topics Catal 11/12:435, 2000.
143. Z Adamczyk, M Zembala, B Siwek, P Warszyński. J Colloid Interf Sci 140:123, 1990.

144. Z Adamczyk, B Siwek, M Zembala. J Colloid Interf Sci 151:351, 1992.
145. B Senger, P Schaaf, JC Voegel, A Johner, A Schmitt, J Talbot. J Chem Phys 97:3813, 1992.
146. B Senger, JC Voegel, P Schaaf. Colloids Surfaces 165:255, 2000.
147. R Jullien, P Meakin. J Phys A 25:L189, 1992.
148. AP Thompson, ED Glandt. Phys Rev A 46:4639, 1992.
149. Z Adamczyk, B Siwek, M Zembala. J Colloid Interf Sci 198:183, 1998.
150. Ph Carl, P Schaaf, JC Voegel, JF Stolz, Z Adamczyk, B Senger. Langmuir 14:7267, 1998.
151. B Vincent, CA Young, THF Tadros. JCF Faraday I 76:665, 1980.
152. GR Joppien. J Phys Chem 82:2210, 1978.
153. T Arai, W Norde. Colloids Surfaces 51:1, 1990.
154. AV Elgersma, RLJ Zsom, W Norde, J Lyklema. Colloids Surfaces 54:89, 1991.
155. AC Juriaanse, J Arends, JJ Ten Bosch.J Colloid Interf Sci 76:212, 1980.
156. JD Aptel, JC Voegel, A Schmitt. Collooids Surfaces 29:359, 1988.
157. T Allen. Particle Size Measurement. 4th ed. Chapman & Hall: London, 1990, Chap 20.
158. WJ Albery, RA Fredlein, GR Kneebone, GJ O'Shea, AL Smith. Colloids Surfaces 44:337, 1990.
159. U Jonsson, I Ronnberg, M Malnquist. Colloids Surfaces 13:333, 1985.
160. JC Dijt, MA Cohen Stuart, JE Hofman, GJ Fleer. Colloids Surfaces 51: 141, 1990.
161. H Shirahama, J Lyklema, W Norde. J Colloid Interf Sci 139:177, 1990.
162. P Schaaf, P Dejardin, A Schmit. Rev Phys Appl 21:741, 1986.
163. P Schaaf, Ph Dejardin, A Schmit. Langmuir 3:1131, 1988.
164. N de Baillou, JC Voegel, A Schmit. Colloids Surfaces 16:271, 1985.
165. JD Aptel, JM Thomann, JC Voegel, A Schmit, EF Bres. Colloids Surfaces 32:159, 1988.
166. F Yan, Ph Dejardin. Langmuir 7:2230, 1991.
167. BD Bowen, N Epstein. J Colloid Interf Sci 72:81, 1979.
168. M Zembala, JC Voegel, P Schaaf. Langmuir 14:2167, 1998.
169. R When, D Woermann. Ber Bunseges Phys Chem 90:121, 1986.
170. MA Cohen Stuart, JW Mulder. Colloids Surfaces 15:49, 1985.
171. M Zembala, Ph Dejardin. Colloids Surfaces B 3:119, 1994.
172. Z Adamczyk, M Zembala, P Warszyński. Bull Pol Acad Chem 47:239, 1999.
173. Z Adamczyk, M Zembala, B Siwek, J Czarnecki. J Colloid Interf Sci 110:188, 1986.
174. JM Meinders, J Noordmans, HJ Busscher. J Colloid Interf Sci 152:265, 1992.
175. M Semmler, J Ricka, B Borkovec. Colloids Surfaces 165:79, 2000.
176. JK Marshall, JA Kitchener. J Colloid Interf Sci 22:342, 1966.
177. M Hull, JA Kitchener. Trans Faraday Soc 65:3093, 1969.
178. Z Adamczyk, A Pomianowski. Powder Technol 27:125, 1980.
179. M Trau, DA Saville, IA Aksay. Langmuir 13:6375, 1990.
180. Y Solomentsev, M Bohmer, JL Anderson. Langmuir 13:6058, 1997.
181. Z Adamczyk, B Siwek, L Szyk, M Zembala. J Chem Phys 105:5552, 1996.
182. Z Adamczyk, L Szyk, B Siwek, P Weroński. J Chem Phys 113:11,336, 2000.
183. P Wojtaszczyk, EK Mann, B Senger, JC Voegel, P Schaaf.J Chem Phys 103:8285, 1995.
184. Z Adamczyk, B Siwek, M Zembala, P Warszyński. J Colloid Interf Sci 130:578, 1989.
185. M Elimelech. J Colloid Interf Sci 164: 190, 1994.
186. S Varennes, TGM van de Ven. Physicochem Hydr 9:537, 1987.
187. L Song, PR Johnson, M Elimelech. Environ Sci Technol 28:1164, 1994.
188. MY Balouk, TGM van de Ven. Colloids Surfaces 46:157, 1990.
189. X Jin, NHL Wang G Tajus, J Talbot. J Phys Chem 97:4256, 1993.
190. JM Peula, R Hildago-Alveraez, FJ de las Nieves. J Colloid Interf Sci 201:139, 1998.
191. Z Adamczyk, B Siwek, E Musiał. Langmuir, 17:4529, 2001.
192. HS Choi, J Talbot, G Tarjus, P Viot. J Chem Phys 99:9296, 1993.
193. Z Adamczyk, B Siwek, M Zembala. Bull Pol Acad Chem 48:231, 2000.
194. J Feder, I Giaever. J Colloid Interf Sci 78:144, 1980.
195. GY Onoda, EG Liniger. Phys Rev A 33:715, 1986.

196. MR Oberholzer, JM Stankovitch, SL Carnie, DYC Chan, AM Lenhoff. J Colloid Interf Sci 194:138, 1997.
197. CA Castillo, R Rajagopalan, CS Hirtzel. Rev Chem Eng 2:237, 1984.
198. RH Otewill. Langmuir 5:4, 1989.

6

Multicomponent Adsorption: Principles and Models

ALEXANDER A. SHAPIRO and ERLING H. STENBY   Technical University of Denmark, Lyngby, Denmark

I. INTRODUCTION

The area of multicomponent adsorption is extremely wide. Generally speaking, all of the adsorbents are multicomponent, unless they are specially purified in experimental conditions. Multicomponent adsorption equilibria are involved in many natural and industrial processes, which are often carried out under complex thermodynamic conditions involving high pressure or temperature. Thus, reliable models of adsorption equilibria are highly required for industrial and research applications [1].

The goal of the present chapter is to describe some methods and approaches developed in the framework of the thermodynamic theory of adsorption. We confine ourselves to the thermodynamic approach, because this approach allows for direct engineering applications. The simulations in the framework of the thermodynamic approach are relatively simple and well repeatable, so that the algorithm for numerical solutions of the corresponding equilibrium problem may be generated on the basis of a relatively short and informal description. An alternative may be provided by ab initio calculations, direct application of the statistical mechanics (Monte Carlo) or other types of molecular simulations. These computations are much more complicated and they have not yet reached the stage where they may be directly used for modeling a wide variety of the practically important cases of mixed adsorption.

The framework for the study of the surface phenomena has been developed by Gibbs [2] in his theory of surface excesses. This theory is discussed (in a rather recapitulative way) in Section II. Our goal has been to demonstrate the strength and the fundamental character of the Gibbs approach and make it appropriate for a wide class of phenomena. Thus, we discuss the theory in a rather general way and just briefly refer to the proper applications.

Adsorption equilibria are normally considered in connection with processes occurring in porous media: filters, catalysts, adsorbents, chromatographs, and rock of petroleum reservoirs. In macroporous and mesoporous media, adsorption is normally accompanied by another surface phenomenon, the capillary condensation. These two types of surface phenomena are closely connected because they are both produced by surface forces. On the other hand, these phenomena are relatively independent and may, to some extent, be discussed separately [3]. Moreover, the description of the coexistence of the adsorbed films and capillary condensate in the same capillary is a nontrivial problem. We present capillary condensation and adsorption separately, although their common roots are discussed in Section II. The (more or less) comprehensive description of the thermodynamics of multicomponent capillary condensation is given in Section III. To the best of our knowledge, the last description of this kind is found in the classical book of Defay et al. [4], although the theory of capillary condensation has undergone a definite progress. This process is partly reflected in the literature (like Ref. 5). We try to develop the theory of multicomponent capillary condensation in a consequent and consistent way, although we touch upon a few remaining problems.

The theory of surface adsorption and adsorption in macroporous media is discussed in Section IV. After a brief overview of the possible thermodynamic approaches, we develop one of them in detail, the potential theory of multicomponent adsorption. This method of treatment makes it possible to discuss comprehensively at least one possible thermodynamic theory of multicomponent adsorption and avoid eclectic jumping between multiple theories and approaches proposed in this area.

Because of the goal of the monograph and limitations in space, we mainly confine ourselves to consideration of the theoretical results and refer to experiments only when necessary and possible. Another limitation is that we mainly consider adsorption from a gas phase, although the adsorption from a solution may be discussed in a similar way. We do not consider adsorption of large molecules (like asphaltenes, polymers, and biomolecules), which would require distinctive thermodynamic modeling.

The reader of this chapter should have a knowledge of modern thermodynamics and methods of modeling the phase equilibria as they are discussed, for example, in Ref. 6.

II. THERMODYNAMICS OF INTERFACES: THE CONCEPT OF THE "BLACK BOX"

The fundamentals of the thermodynamics of interfaces were first developed by Gibbs [2] and have not been much changed since 1878, when his fundamental paper. "On the Equilibrium of Heterogeneous Substances," was published. The surface thermodynamics has been discussed in detail in many articles and monographs (such as Refs. 3, 4, and 7–9). The discussion presented in Ref. 9 also shows that a similar treatment may be applied to adsorption in microporous media, where the adsorbed substance is accumulated in the volume but not on the surface.

We suggest a point of view on the surface thermodynamics, which differs slightly from the points of view presented in the literature. We aim at showing the strength of the ideas underlying the Gibbs theory, which makes it possible to apply the common language to all types of surface and related effects and even for a much more general class of the phenomena. In order not to repeat well-known statements, our discussion is rather recapitulative, and the reader who is interested in details should address the above-mentioned sources.

Let us consider an arbitrary thermodynamic system as a sort of a "black box" (b). The box may consist of an arbitrary combination of the bulk, adsorbed, and other kinds of substances. The volume and the geometrical shape of the box are supposed to be invariable, in order to fix one extensive parameter and to exclude exchange by mechanical work with the environment. Quasiequilibrium exchange of the energy U_b, the entropy S_b, and the molar amounts $N_{b,i}$ between the box and the environment is described by the common thermodynamic equation

$$dU_b = T\, dS_b + \mu^i\, dN_{b,i} \tag{1}$$

Here, T is the temperature and μ^i are chemical potentials of the substances in the box. By default, summing by repeating indices of multipliers in a product is assumed.

It is convenient to introduce the potential

$$\Omega_b = -U_b + TS_b + \mu^i N_{b,i} \tag{2}$$

For an equilibrium single-phase system, this potential is equal to PV_b. However, this is not the rule for more complicated systems. For example, pressures in the two flowing phases separated by a curved interface may be different. Surface energies may also contribute to Ω_b. The thermodynamic properties of the box containing both surface and bulk phases may be rather complicated. For example, the box may contain transition layers between different phases, where phase properties may vary continuously and even become anisotropic [5]. A detailed description of all the phases and thermodynamic states in the box is not always possible and, probably more important, is not always necessary. A key idea for the modeling of such a thermodynamic system is to split it into the relatively simple "bulk" part, whose properties are reflected by an appropriate simplified model (m) and the "excess phase" (e), which fills the gap between the model and the original system. At the present point of discussion, splitting into the model and the excess phase is rather arbitrary. For any such splitting, equations similar to Eqs. (1) and (2) should be valid for the model:

$$dU_m = T \, dS_m + \mu^i \, dN_{m,i} \tag{3}$$

$$\Omega_m = -U_m + TS_m + \mu^i N_{m,i} \tag{4}$$

Inaccuracy of the model with regard to the simulated system is reflected by the fact that *under the same values of the temperature and of the chemical potentials* the extensive characteristics of the model, U_m, S_m, and $N_{m,i}$, differ from the corresponding characteristics of the box, U_b, S_b, and $N_{b,i}$. Thermodynamic relations for the excess phase may be obtained by subtracting the extensive properties of the box and of the model. Subtraction of Eq. (1) from Eq. (3) and, correspondingly, of Eq. (2) from Eq. (4) gives

$$dU_e = T \, dS_e + \mu^i \, dN_{e,j} \tag{5}$$

$$\Omega_e = -U_e + TS_e + \mu^i N_{e,i} \tag{6}$$

Here, the extensive values A_e, are introduced as the differences $A_b - A_m$ for each extensive value A.

Equations (5) and (6) look like the relations describing a separate phase. This makes it possible to develop the thermodynamics of the excess phase independently of the thermodynamics of the bulk phases. However, this independence is incomplete, because the excess phase describes the discrepancy between the real system and its model and the excess properties depend on the choice of model. There may be various reasons for the discrepancy between the model and the box (e.g., the model may be characterized by imprecise equations of state). This is often the problem in the interpretation of experimental results, where inconsistencies related to the modeling of the bulk phases may be mistakenly interpreted as the properties of the surface phase.

For a further development of the theory, let us assume that the box is somehow scalable (at least within some limits) and that its size, B, may be considered a new extensive parameter. For example, we may assume that a large number of equivalent small boxes are available. Extensive parameters of the model and, correspondingly, extensive excess parameters are also scaled with B. (We could just assume that the extensive properties of the excess phase are scaled with B, whereas the parameters of the model and of the box remain invariable, as normally postulated. However, this assumption is very difficult to release physically. For example, in the case in which B is the surface of the adsorbent, increasing its value in a fixed volume will lead to additional adsorption and, correspondingly, will modify intensive properties of the bulk phases. In order to keep intensive properties invariable, we have to vary the bulk volume proportionally to the surface. However, this is exactly scaling the properties of the box and of the model together with B.)

Let us introduce the designations

$$\Phi = \frac{\Omega_e}{B}, \qquad u_e = \frac{U_e}{B}, \qquad s_e = \frac{S_e}{B}, \qquad n_{e,i} = \frac{N_{e,i}}{B} \tag{7}$$

The value of Φ is called the *excess pressure*. In view of the extensiveness of the excess phase with regard to B, the values Φ, u_e, s_e, and $n_{e,i}$ obey the same relations (5) and (6) as Ω_e, U_e, S_e, and $N_{e,i}$:

$$du_e = T\,ds_e + \mu^i\,dn_{e,i}, \qquad \Phi = -u_e + Ts_e + \mu^i n_{e,i} \tag{8}$$

Differentiation of the last relation for Φ and subtraction of the first Eq. (8) lead, as usual, to the Gibbs–Duhem equation for the excess pressure Φ. At constant B,

$$d\Phi = s_e\,dT + n_{e,i}\,d\mu^i \tag{9}$$

Now, substituting $\Omega_e = \Phi B$ into Eq. (6), differentiating, and applying the last relation, we recover the common thermodynamic equation for the excess energy:

$$dU_e = T\,dS_e - \Phi\,dB + \mu^i\,dN_{e,i} \tag{10}$$

A comparison of this equation with thermodynamic equations for bulk phases shows that Φ plays the same role in the excess thermodynamics as the pressure P in the bulk-phase thermodynamics, whereas the value B corresponds to the bulk volume. With varying B, Eq. (1) for the box should be replaced by

$$dU_b = T\,dS_b + \mu^i\,dN_{b,i} - \Phi\,dB \tag{11}$$

In practice, Eq. (11) is commonly considered as a constituting equation for the box, where the term $-\Phi\,dB$ is attributed to the excess phase. However, such an approach is not fully logically consistent because the value of B is often taken from the model, rather than from the original system. For example, in the case of capillary equilibrium, the value of B is the area of a surface separating the gas and liquid phases. This imaginary mathematical surface serves as a model of a transition interface layer of the real system and may be chosen with a certain degree of arbitrariness [3,4].

As stated earlier, the excess phase is a pseudophase, in that its properties are not only determined by the properties of the "black box" but also by the properties of the model. It may be considered as a new phase *in equilibrium* with the model because the chemical potentials and the temperatures in the excess phase and in the model are equal:

$$\mu_b^i = \mu_e^i = \mu^i, \qquad T_b = T_e = T \tag{12}$$

However, there is a significant difference compared to the normal situation of equilibrium. According to the Gibbs phase rule, the addition of a new phase should decrease the number of the degrees of freedom by one. Thus, it might be assumed that equilibrium between the excess phase and the model may only be possible at the boundaries of a phase diagram for the total system. This is not the case for the system with a pseudophase. On the contrary, any state of the model may be in equilibrium with the excess phase. The reason is that the excess pressure Φ, in general, should not be equal to any parameter of the model (e.g., to the pressure P). Thus, the set of equilibrium conditions between the model and the excess phase is incomplete. In other words, compared to a normal thermodynamic system, a system with a pseudophase possesses an additional degree of freedom.

The excess phase obeys an equation of state (EoS), as follows from Eq. (8) or Eq. (9). According to these equations, the EoS may be selected in the form of either $u_e = u_e(s_e, n_{e,j})$ or $\Phi = \Phi(T, \mu^j)$. Other forms of the EoS may also be applied [3]. In practice, an EoS is either postulated or expressed indirectly in terms of the parameters of the model and by use of the equilibrium conditions (12).

As an example, let us assume that the box and the model represent a single-component fluid. Differentiating the dependence $u_e(s_e, n_e)$ with regard to s_e and n_e according to Eq. (8), we find $T_e(s_e, n_e)$ and $\mu_e(s_e, n_e)$. These values are equated to the temperature T and the chemical potential $\mu(P, T)$ of the model, as in Eq. (12). When this system is solved with regard to s_e and n_e the dependence $n_e(P, T)$ is obtained. Then, by integration of Eq. (9) at constant temperature, the famous Gibbs equation is obtained:

$$\Phi - \Phi_0 = \int_{P_0}^{P} n_e(P', T) \frac{\partial \mu(P', T)}{\partial P'} \, dP' \tag{13a}$$

The last equation may also be rewritten in terms of the fugacity f. Because $\mu - \mu_0(T) = RT \ln[f/f_0(T)]$, we have

$$\Phi - \Phi_0 = RT \int_{f_0}^{f} n_e(f', T) \frac{df'}{f'} \tag{13b}$$

Equations (13) serve as a way of determining the EoS for the adsorbate from the experimental data on the surface excesses under single-component adsorption.

In sum, the preceding considerations show that all of the main relations of the thermodynamics of the excess phase may be derived independently of the nature and the physical properties of the box and of the model. The only assumption made concerns the possibility of scaling the relations with regard to the box size B. This may be interpreted as an assumption about the *homogeneity* of the adsorbate. It should be noted that the common assumption about the *inertness* of the adsorbate is not necessary for the preceding development. Normally, the inertness assumption is used in order to state that the excess values characterize the properties of the adsorbed phase only [8]; that is, Eqs. (8) and (9) are closed and the properties of the adsorbate are independent of the properties of the model. In the present treatment, use of the fact of equilibrium between the excess phase and the model makes it possible to proceed rather far without assuming that the adsorbate is inert.

Let us now consider the case in which the excess phase is not inert; that is, it modifies the equilibrium between bulk phases of the system. This may be taken into account by specifying the interaction between the excess phase and the model. We assume the existence of an additional internal parameter of the model W (scalar or vector), which may be modified by the presence of the excess phase. The energy equation for the model, taking in account possible variation of W, has the form

$$dU_m = T \, dS_m + \mu^i \, dN_{m,i} + w \, dW \tag{14}$$

We assume that the values of w and W may be determined for the box, as well as for the model and, in fact, are the same for the box and for the model (such a mixing of the original and modeling properties is characteristic of the theory of capillarity). For the model alone, the term $w \, dW$ should disappear in equilibrium because the minimum of the energy U_m with regard to W corresponds to $w = 0$. However, due to the influence of the excess phase, the parameter w may be different from zero in the box. With the new parameter, the energy equation for the box is modified similarly to Eq. (14):

$$dU_b = T \, dS_b + \mu^i \, dN_{b,i} + w \, dW$$

Subtraction of the last two equations leads to the same Eq. (5) for the excess phase as previously. Correspondingly, the theory of the excess phase expressed by Eqs. (5)–(13) may be left unchanged, with the only exception that Eq. (11) for the box with the scaled excess phase should now read

$$dU_b = T \, dS_b + \mu^i \, dN_{b,i} - \Phi \, dB + w \, dW \tag{15}$$

Although the equations for the excess phase are not modified, the scaling parameter B may now depend on W. This dependence determines the interaction between the excess phase and the model. The equilibrium state of the box is determined by the minimum of U_b with regard to W, under constant entropy and molar amounts. Variation of W leads to the condition

$$\left(-\Phi \frac{\partial B}{\partial W} + w \right) dW = 0 \quad \text{or} \quad w = \Phi \, \frac{\partial B}{\partial W} \tag{16}$$

The last equation is a highly generalized form of the famous Laplace equation. Its particular well-known form for the case of the capillary equilibrium will be discussed in Section III. Equation (16) shows the effect of the excess phase on the equilibrium condition, which would otherwise be $w = 0$.

Equation (16) provides a nontrivial connection between the intensive parameters of the model and a new parameter $\partial B / \partial W$. Because this equation is used instead of the more restricting condition $w = 0$, this may lead to a "transfer" of the degree of freedom associated with the excess phase to equilibrium between the bulk phases of the system. This is a well-studied effect for capillary equilibrium (see the analysis in Section III).

The above-developed scheme is rather general and abstract. Its particular applications must be based on physical concepts and interpretations, which make the theory relevant to a particular system being studied. In the surface thermodynamics, a typical "black box" consists of several bulk phases, separated by thin interface layers. The densities of different extensive parameters are (almost) constants over the bulk phases. However, they may rapidly vary across the interfaces. A model for such a system consists of the same number of bulk phases with constant densities. The phases are separated by mathematical surfaces, whose positions correspond approximately to the positions of the interface layers in the box. The selection of these mathematical surfaces is often considered as a "degree of freedom" of the model [3], although, sometimes, the interfaces are specified more precisely (e.g., in order to obey the force moment equation) [4]. The excess phases are "attached" to the corresponding surfaces and the surface area A plays the role of the scaling parameter B. The bulk phases of the model are identified with the bulk phases of the box, and their thermodynamic properties are assumed to be precisely known. The last assumption may be rather dangerous because it is sometimes difficult to distinguish between the errors in the description of the bulk phases and the properties of the surface excesses. Another problem of this identification arises when a nonequilibrium situation (e.g., adsorption kinetics) is considered. In this case, the phases of the model should not be identified with the real bulk phases of the system, but with the (probably nonexistent) bulk phases, which are in thermodynamic equilibrium with the adsorbate.

The description of adsorption in microporous media should be based on a different physical picture. The problem of determining the surface excesses becomes especially complicated in this case because it requires knowledge of the pore volume. Whereas in earlier studies [10] this value has been considered as more or less well defined, recent analysis [11] indicates that there may be problems, especially with regard to the molecular simulations of adsorption in the microporous media. Nevertheless, because the basic assumption about the scalability of the box is valid for this case, the adsorption thermodynamics may still be formulated in terms of the properly introduced excess phase [12].

III. THERMODYNAMICS OF CAPILLARY EQUILIBRIUM

In this section, we consider the problems relevant to equilibrium of the two multicomponent phases separated by a curved interface. This is the classical and the most well-studied case of the thermodynamic equilibrium involving surface effects. Such equilibrium is present in macroporous and mesoporous media, like the porous rocks of petroleum reservoirs, where it accompanies adsorption. In the pores of smaller sizes, the forces produced by the solid surfaces may modify the properties of the bulk (liquid and gas) phases. However, the present study is also important to the pores of smaller sizes, as it makes it possible to separate the effects connected with the gas–liquid surface tension (and, of course, the contact angle) from additional contributions of the solid walls. The corrections related to the last type of interactions have been considered in, for example, Refs. [13–15]. For brevity, we will apply the term *capillary equilibrium* to the narrow case being described, but it must be remembered, however, that a wider understanding of the capillary equilibrium is available.

The starting point of the multicomponent thermodynamic theory of capillarity is the classical work of Gibbs [2]. In the subsequent development, great attention has been paid to the phase interfaces, both from thermodynamic [4] and molecular [16] points of view. With regard to thermodynamic behavior of the bulk phases, few important problems have been studied, such as the phase rule [17] or generalization of the Kelvin equation to a multicomponent case [4,18]. Much more is known about capillary equilibrium in a single-component case [19–21]. Recently, the interest in multicomponent capillary equilibria has appeared in connection with the petroleum applications [22–25].

A. General Conditions of Capillary Equilibrium

The theory of the excess phase developed in the previous section may be directly applied to the two-phase system separated by a curved interface. The "black box" and the model in this case consist of the liquid (l) and the gas (g) phases. While the phases in the model are separated by the mathematical surface, the box contains a transition interface layer. This layer possesses excesses of the component molar amounts and of the energy, compared to the model. They form the excess phase for the system being considered.

The scaling parameter B is naturally introduced as the area of the interface A. It should be noted that this is the parameter of the model, but not of the original system. A negative of the corresponding excess pressure Φ is the *surface tension* $\sigma = -\Phi$. The minus sign is related to the fact that location of the molecules on the interface between gas and liquid is energetically disadvantageous.

The excess phase for capillary equilibrium is not inert, as it affects equilibrium between the bulk phases of the model. A variation of the phase volumes leads to a variation of the interface area and, therefore, of the energy attached to the interface. Under constant total volume the differentials of gas and liquid volumes are opposite, $dV_g = -dV_l$, so that the energy equation for the model may be represented in the form

$$dU_m = T\,dS_m + \mu^i\,dN_{m,i} - (P_l - P_g)\,dV_l \tag{17}$$

A comparison of this equation to Eq. (14) shows that the internal parameter W of the model is V_l and the corresponding intensive parameter is the difference of gas and liquid pressures, $w = P_g - P_l$. In the absence of the surface phase, this equation leads to the usual equilibrium condition $w = 0$, or $P_l = P_g$. The presence of the interface modifies this condition, so by application of Eq. (16) we obtain

$$P_l - P_g = P_c = 2\sigma K \tag{18}$$

where

$$K = \frac{1}{2} \frac{\partial A}{\partial V_l}$$

The value of P_c is called the *capillary pressure*, and the last equation is the common *Laplace equation*. According to the well-known theorem [3,5], the value of K is equal to the mean curvature of the interface. For highly curved surfaces, the value of σ may depend on K and the Gaussian curvature [2,26,27]. We do not consider this dependence in the present work.

The system of equilibrium conditions between the gas and the liquid phases is determined by Eq. (2), together with the equality of the chemical potentials in the two phases:

$$\mu_l^i(P_l, \mathbf{z}_l) = \mu_g^i(P_g, \mathbf{z}_g) \quad (i = 1, \ldots, M) \tag{19}$$

Here, we use the designation $\mathbf{z}$ to denote the composition of the corresponding phase (gas or liquid): $\mathbf{z} = (z_1, \ldots, z_M)$, where M is the number of components. In general, throughout the chapter we use bold letters to denote vectors consisting of the corresponding components. We do not include temperature in the list of arguments because normally it is not modified by the capillary forces and may be treated as an "external parameter." Thermal equilibrium is always assumed throughout the chapter.

Equations (18) and (19) provide the full system of equations for the two-phase capillary equilibrium in relatively large capillaries, where the walls do not modify the state of the bulk phases. The value of the curvature K is determined by the geometry of a capillary system and by the wettability of the capillary walls, as discussed in more detail in the next subsection.

B. Capillary Equilibrium in Porous Media

The problem of capillary equilibrium in porous media is complicated from both experimental and theoretical points of view. The mechanisms of saturation and depletion of the porous medium are essentially nonequilibrium. Further equilibration is due to slow processes like diffusion. The process of equilibration may be unfinished, since no significant changes of fluid distribution may occur during the time of an experiment. This especially relates to the so-called "discontinuous condensate" existing in the form of separate drops. As a result, thermodynamic states, which are not fully equilibrated, are interpreted from the practical point of view as "equilibrium" [28]. To the best of our knowledge, a consistent theory of such quasiequilibrium states has not yet been developed. In the following, we discuss the possible states of the two-phase mixtures in a porous medium, assuming complete thermodynamic equilibrium. This serves as a first approximation to a more complicated picture of the realistic fluid distribution in porous media.

The constituting system of Eqs. (18) and (19) has been derived for the two connected bulk phases, gas and liquid, separated by a curved interface. Such geometry may be characteristic of a single capillary; however, in porous media, gas and liquid are not necessarily connected and may form several isolated "spots." According to Refs. 22 and 23, it can be shown that in an equilibrium state (but not in the quasiequilibrium state described earlier), pressures and compositions in different gas or liquid spots are equal.

Indeed, it follows from Eq. (19) that the chemical potentials μ^i are the same in all of the gas and liquid "spots." The system $\mu^i = \mu^i(P, \mathbf{z})$ $(i = 1, \ldots, n)$ may be considered a system of M equations for M variables $P, z_1, \ldots, z_{M-1}$ (the last value, z_M, is determined from the condition $\sum z_i = 1$). In the case of two-phase equilibrium, this system has exactly two solutions corresponding to the liquid and the gas phases. Therefore, pressures and compositions are also equal for each of two spots belonging to the same phase.

The preceding considerations exclude the cases in which different liquid phases may be formed. This is not an exceptional situation, as, for example, vapor–liquid–liquid equilibria have been observed in petroleum mixtures containing CO_2 [29,30]. Locally, such three-phase equilibria may be reduced to two-phase situations, although the geometry of spots of the three-phase mixtures in realistic porous media is rather complicated. In the present treatment, we concentrate on the two-phase case.

Because the phase pressures are equal in all of the liquid spots as well as in all of the gas spots, it follows from Eq. (18) that *the curvatures of all the interfaces separating different phases are also equal*. This statement is in contradiction with a common "first glance" opinion, according to which the menisci in smaller capillaries have smaller radii of curvature.

Equilibrium locations of a meniscus in porous media are determined by the curvature K and by the Young wettability condition for the contact angle θ between the liquid phase and solid walls:

$$\cos \theta = \frac{\sigma_{ls} - \sigma_{gs}}{\sigma_{gl}} \tag{20}$$

Here, σ_{ls}, σ_{gs}, and σ_{gl} are the surface tensions between different pairs of phases: liquid (l), solid (s), and gas (g) (we denote $\sigma_{gl} = \sigma$). In general, Eq. (20) serves as a rough approximation, and contact angles may vary strongly depending on the roughness of the surface, the resolution of measurements, and other factors [3]. Moreover, Eq. (20) may formally produce the value of $\cos \theta$ larger than unity, which is interpreted as the spreading situation in which the wetting phase forms films on the surface [31]. This situation is of importance to petroleum engineering, where it has been extensively studied (see, e.g., Refs 32 and 33). On the other hand, to the best of our knowledge, there has not been a systematic study of multicomponent wetting films. We do not consider this subject in detail.

Examination of Fig. 1a shows that if the curvature of the meniscus is large enough, there will always be a place in a tortuous capillary where the meniscus may form a required contact angle with the surface. The fact that there may be several positions of this kind is one of the reasons for the condensation/desorption hysteresis [28,34,35] (another reason may be hysteresis in the contact angle [3]). In cylindrical capillaries, the meniscus is most likely located at the end of a capillary (Fig. 1b).

One of the widely used, although rather rough, models of a porous medium is the ideal ground, or the Kozeny–Carman model [36]. In this model, the porous medium is represented as a bunch of cylindrical capillaries of different radii. The capillaries may be tortuous, but their radii remain invariable along their extent. This excludes hysteresis related to different locations of the menisci in the same capillary (cf. Fig. 1). Meanwhile, the absence of the network structure excludes the effects caused by the interaction of different capillaries. It may be shown that in the ideal ground, all of the capillaries with radii below a certain value r_c are saturated by a wetting phase (for definiteness, by liquid). The rest of the capillaries are filled by gas [22]. The value of r_c is determined as

$$r_c = -\frac{\cos \theta}{K} \tag{21}$$

On the other hand, because a cylindrical capillary of the length L has the volume $\pi r^2 L$, with known pore-size distribution $f(r)$, the value of r_c may be related to the volumetric saturation of the wetting (liquid) phase β_v:

$$\beta_v = \int_0^{r_c} f(r) r^2 \, dr \left/ \left(\int_0^{\infty} f(r) r^2 \, dr \right)^{-1} \right. \tag{22}$$

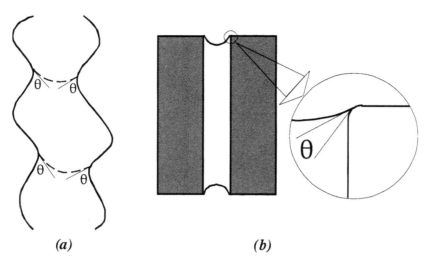

(a)   *(b)*

FIG. 1 Possible locations of the menisci in a tortuous capillary (a) or a cylindrical capillary (b).

Substituting Eq. (22) into Eq. (21) and comparing to Eq. (18) we obtain a capillary pressure equation for the saturated porous medium:

$$P_g - P_l = -P_c = \frac{\sigma \cos \theta}{r_{av}} J(\beta_v), \qquad J(\beta_v) = \frac{2r_{av}}{r_c(\beta_v)} \tag{23}$$

Here, r_{av} is a characteristic radius of a pore. It is often estimated as $\sqrt{k/\phi}$, where k is permeability in the Darcy law and ϕ is porosity [22,36]. The dependence $J(\beta_v)$, expressing the dimensionless capillary pressure, is called the capillary pressure function or the J function.

   Although Eq. (23) has been derived strictly only for ideal ground, it is commonly applied for other types of porous media. Generally, the dependence $J(\beta_v)$ exhibits hysteresis. It is different for the condensation and desaturation processes, as well as for scanning condensation–desaturation isotherms [28]. Models for such a hysteresis have been created in the framework of the domain theory [34,35,37,38] and of the percolation theory [5]. Characteristic shapes of the J function for drainage (saturating of a sample of porous medium by the nonwetting phase) and imbibition (saturating by the wetting phase) are shown in Fig. 2. The condensation–desaturation curves for capillary condensation are qualitatively different from the imbibition–drainage curves for immiscible displacement. The displacement is possible only between the residual saturations, after which, one of the phases becomes discontinuous. Meanwhile, condensation/evaporation is possible in the whole range of saturations [22,23].

   An important property of the capillary phase equilibrium in porous media is that *the wetting phase has always a lower pressure than the non-wetting phase* (i.e., the surface of the wetting phase is always concave). This property is almost always implicitly assumed in the literature and is confirmed by experimental evidence. However, no direct proof has yet been presented. Such a general proof must be rather complicated, in view of the complexity of the geometry of the internal porous space. The following considerations do not provide the full proof, but they uncover the reasons for the impossibility of the configurations where the wetting phase has a higher pressure.

   A typical configuration with a wetting phase with a higher pressure and convex surface is a drop on the surface, as shown in Fig. 3. The assumption that first-order equilibrium conditions like Eqs. (18) and (19) are valid for this drop does not result in any contradiction. The reason for

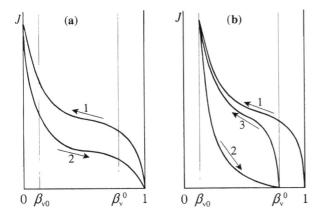

FIG. 2 Characteristic shape of the J function: (a) for condensation–desaturation; (b) for immiscible displacement (according to Ref. 39). 1: drainage (desaturation); 2: imbibition (condensation); 3: secondary drainage.

the impossibility of such equilibrium is its instability, determined by the sign of the second-order differential. In order to demonstrate this, the virtual evaporation of a small part of the drop is considered (the new position of the separating surface is shown by the dashed line in Fig. 3). The evaporation proceeds in such a way that liquid composition is preserved. Correspondingly, pressures, temperature, chemical potentials, and, therefore, surface tension and curvature of the interface remain unchanged (provided that the surrounding gas phase is infinite). However, the contact angle decreases: $\theta' < \theta$. Let us recall that the contact angle is formed by the tension forces F_{gl}, F_{ls}, and F_{gs}, which are proportional to the corresponding surface tensions. According to Eq. (20), the contact angle corresponds to the balance of the horizontal components of these forces. The fact that $\theta' < \theta$ means that the contact force for a reduced drop is directed toward its further reduction (cf. Fig. 3).

Let us now consider the changes of the bulk-phase energies caused by partial evaporation of the drop. At constant temperature and chemical potentials, equilibrium is determined by a minimum of the grand potential:

$$\Omega = -P_l V_l - P_g V_g + \sigma A$$

Its variation $\delta\Omega$ under constant pressure and surface tension is given by

$$\delta\Omega = -P_l\,\delta V_l - P_g\,\delta V_g + \sigma\,\delta A = (P_l - P_g)\,\delta V + \sigma\,\delta A$$

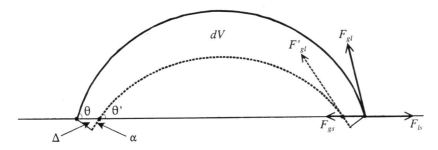

FIG. 3 Equilibrium of a drop on a surface.

If the variations δV and δA could be substituted by the first-order differentials dV and dA, respectively, then the corresponding variation $d\Omega$ would become zero, in view of the equilibrium condition (18) and the Euler identity:

$$dA = -K \, dV$$

However, for the full increments δV and δA, the last equality is valid only if the surface varies in the direction orthogonal to it. Otherwise, there is a difference Δ between dV and δV, as well as a difference α between dA and δA, as shown in Fig. 3:

$$dV = \delta V + \Delta, \qquad dA = \delta A + \alpha$$

Substitution of the last equalities into the variation $\delta\Omega$ and taking into account the Euler identity and the equilibrium condition (18) result in

$$\delta\Omega = -(P_l - P_g)\Delta - \sigma\alpha < 0$$

Because both Δ and α are positive, the potential Ω decreases with the drop size. Its true minimum corresponds to the disappearance of the wetting drop. This is in agreement with the fact that the contact force is directed toward a decrease of the drop. Similarly, nonwetting drops on the surface will grow unless other solid surfaces stop the growth.

Although we considered the simplest case of a drop on a planar surface, it is clear that the above considerations may be applied to other types of geometry. Analysis of the proof shows that the only facts used are the inequality $P_l > P_g$ and that the drop forms an acute angle with the surface. Thus, the above reasoning can be spread onto a more general situation (although, still, it is not a full proof). However, it cannot be applied to the case where $P_l - P_g < 0$. In this case, the two addenda in the last expression for $\delta\Omega$ have different signs and it is possible to imagine a geometrical arrangement with such Δ and α that $\delta\Omega \geq 0$ (as, e.g., in Fig. 1). Thus, although equilibrium with a convex wetting phase is always unstable, the wetting capillary condensate with a concave interface may be stable, at least for some arrangements of the phases in a porous medium.

The above considerations cannot be applied to the equilibria of the immiscible fluids (like oil and water in the rock of a petroleum reservoir).

C. Problems and Numerical Algorithms

Let us consider the system of capillary equilibrium equations in the general form

$$\mu_l^i(P_l, \mathbf{z}_l) = \mu_g^i(P_g, \mathbf{z}_g) \quad (i = 1, \ldots, M); \qquad P_l - P_g = P_c \tag{24}$$

This system consists of $M + 1$ equations for $2M + 1$ unknowns $z_{l,i}$, $z_{g,i}$, P_l, P_g, and P_c. There are M degrees of freedom, one more than for the normal equilibrium where the condition $P_l = P_g$ is used instead of the second Eq. (24). The general reason for this is the transfer of the degree of freedom from the excess phase to the bulk phases of the system performed by the Laplace equation, as analyzed in Section II. A detailed analysis of the degrees of freedom in bulk phases separated by curved interfaces is performed in Ref. 17. Common algorithms for phase equilibria calculations should be modified in order to take into account this additional degree of freedom, as well as physical peculiarities of an experiment or a process involving capillary condensation.

The statements of the capillary equilibrium problems may be different, depending on a physical situation to be described or on the formal analysis to be performed. From a formal point of view, any equilibrium problem is determined by the set of equations, the list of known variables, and the list of variables to be found. For a given system of equations, we will use the following symbolic record of a problem:

$$List1 \rightarrow List2$$

where $List1$ is the list of preassigned variables and $List2$ is the list of variables to be determined.

We discuss only direct numerical methods for solution of the capillary equilibrium problems. An alternative may be provided by the expansions in the neighborhood of the true equilibrium, which have been developed in Refs. 22–24. These methods require a preliminary solution of the corresponding problems of the two-phase equilibrium with a flat interface, which is not much easier to solve than the problem of capillary equilibrium. That is why direct methods are, in our opinion, preferable to the solutions of the capillary equilibrium problems based on expansions.

1. Modified Flash Problem

As an example, let us consider the problem of a phase split (the flash problem). This problem consists in determination of the compositions of the liquid and gas phases, $\mathbf{z}_l$ and $\mathbf{z}_g$, respectively, and the gas molar fraction β by the known overall composition $\mathbf{z}$. The system (24) is completed by the following $M - 1$ balance condition:

$$z_g^i \beta + z_l^i(1 - \beta) = z^i \tag{25}$$

In order to obtain the closed form for the problem given by Eqs. (24) and (25), the two additional variables should be fixed, For the common flash problem, the phase pressures are equal and there is only one free variable. Usually, such a variable is chosen to be pressure. Then, an obvious modification for capillary equilibrium would be to assign the values of both phase pressures P_l and P_g [24,40]. The corresponding modification of a solution algorithm is straightforward.

A formal record of the capillary flash problem is as follows:

$$P_g, P_l, \mathbf{z} \rightarrow \mathbf{z}_g, \mathbf{z}_l, \beta$$

The formulated problem of capillary equilibrium is the most evident and natural modification of the common flash problem. However, it does not seem to have a straightforward physical significance. The reason is that under standard conditions of capillary condensation, the pressure cannot be controlled simultaneously in both phases. As a typical example, let us consider a porous sample introduced into a vessel with the gas phase. We know all of the parameters for gas, whereas the pressure in liquid is unknown and has to be found. Additionally, the capillary flash problem does not take into account the specific structure of the porous medium.

2. The Saturation Problem

The example discussed suggests another more realistic formulation of the equilibrium problem. Under the conditions of the previous problem, instead of P_l and P_g, we specify the pressure in one phase, say, P_g. The system of Eqs. (8) and (9) is completed by the Laplace equation in the form of Eqs. (23). In the second equation, (23), the dependence $J(\beta_v)$ is supposed to be known from the geometry of the porous space. The surface tension and the wetting angle are defined as known functions of the thermodynamic conditions (e.g., the surface is assumed to be wet by the condensate and the surface tension is calculated by the parachor method). The volumetric liquid saturation β_v is connected with the gas molar fraction β by means of the molar densities n_l and n_g:

$$\beta_v = \frac{1 - \beta}{1 - \beta + \beta(n_l/n_g)} \tag{26}$$

The saturation problem is formulated as follows. We would like to determine the saturation and the composition of a condensate precipitated from the gas phase, which has been injected into a porous sample with a known structure of the porous space. This problem is determined by the system of Eqs. (23)–(26), which have to be solved under known P_g, z^i, and dependence $J(\beta_v)$. The values to be determined are P_l, z_l^i, z_g^i, β, and β_v:

$$P_g, \mathbf{z}, J(\beta_v) \rightarrow \mathbf{z}_g, \mathbf{z}_l, P_l, \beta, \beta_v$$

The formulated problem may be solved by modification of an algorithm used for flash calculations. Assume, for example, that the initial flash algorithm is a common substitution method involving the Rachford–Rice procedure [41]. Its "capillary" modification consists in the use of different equilibrium constants K_c^i:

$$K_c^i = \left(1 - \frac{P_c(\beta_v)}{P_g}\right)K^i, \qquad K^i = \frac{\phi_l^i}{\phi_g^i}$$

Here, ϕ_l^i (ϕ_g^i) is the fugacity coefficient of the liquid (gas) phase. The constants K_c^i should be substituted into the Rachford–Rice equation instead of K^i. The value of β_v in $J(\beta_v)$ is taken from a previous iteration. Then, the new value of β_v is found from the known liquid molar fraction β and the phase molar volumes.

3. G-L Problem

Another important capillary equilibrium problem is determination of pressure and composition of the liquid phase by known pressure and composition of the gas phase (or vice versa). This G-L problem requires solution of only the system (24) for chemical potentials in which the values P_g and z_g^i are known and the values P_l and z_l^i are to be defined. Then, the saturation of the porous medium β_v and the molar fraction of the liquid β may be found from Eqs. (23) and (26).

The G-L problem arises, for example, when the capillary transition zone in a thick gas–condensate reservoir is evaluated [42]. The physical interpretation of this problem is as follows: We measure pressure and composition of the gas phase *after* the porous sample has been introduced into a gas vessel and would like to determine the parameters related to the capillary condensate in the porous medium. The difference between normal and capillary equilibrium, related to the additional degree of freedom, is especially pronounced in this case. For ordinary equilibrium, the formulated problem would be overdefined, because it would include an additional condition of the equality of phase pressures.

The G-L problem may be solved by a simple substitution method, which does not require the Rachford–Rice procedure. The method is based on two relations:

$$z_l^i = \frac{a^i}{\phi_l^i P_l}(a^i = z_g^i P_g \phi_g^i), \tag{27}$$

$$P_l = \sum \frac{a^i}{\phi_l^i} \tag{28}$$

Equation (27) is another form of the equality of the chemical potentials (24); Eq. (28) is obtained from Eq. (27) by summation by i. The values a^i for the problem considered are, simply, constants.

The substitution algorithm works as follows. In each iteration, the values of z_l^i are first calculated from Eq. (27), with P_l and ϕ_l^i taken from the previous iteration. The values of z_l^i are normalized to unity. Then, the new value of P_l is found from Eq. (28). Finally, the fugacity coefficients ϕ_l^i are found with the new liquid pressure and composition. An important detail is that the pressure is determined after the composition, but not before it (which, at first glance, might sound more logical). Otherwise, initial approximations are too far-fetched, and the algorithm may converge to a trivial solution.

D. Calculation of the Surface Tension

In this subsection, a brief overview is given of thermodynamic methods for evaluation of the surface tensions. Most of these methods are empirical or semiempirical, like power correlations for specific substances [43] or correlations between surface tensions and viscosities [44–46]. Many methods have been developed only for single-component fluids (at least, to the best of our knowledge). Among them are the methods based on the corresponding states theorem [47–49].

According to the general theory of the surface phase, surface tension, as negative to excess pressure Φ, is a function of the chemical potentials and temperature: $\sigma = \sigma(\mu^i, T)$. This relates the value of the surface tension to the properties of the bulk phases. In practice, surface tension is often expressed as a function of the phase compositions and the phase pressures. This increases the number of variables on which surface tension depends. To the best of our knowledge, a vast majority of the existing experimental data is related to the case where the phase pressures are equal or their difference is negligible. Models for the surface tension of multicomponent mixtures are developed and tested under this condition. Evaluation of the surface tensions of the multicomponent mixtures for the two phases separated by a very curved surface, so that the pressure difference between phases becomes important, remains an open problem.

The parachor method is one of the most commonly used correlations for the surface tensions. The method was originally suggested by MacLeod [50] and Sugden [51] for a single-component fluid and was extended to multicomponent systems by Weinaug and Katz [52]. An expression for the surface tension in this method is

$$\sigma^{1/\alpha} = \left(\sum_i [P]_i z_l^i \right) n_l - \left(\sum_i [P]_i z_g^i \right) n_g \tag{29}$$

Here, n_l and n_g are phase molar densities and $[P]_i$ are special coefficients or *parachors* of the components. The exponent α has originally been proposed to be equal to 4.

The parachor method gives, according to some estimates, a good order-of-magnitude value of the surface tension, but the deviations for realistic mixtures may reach a few tens of percent [46,53,54]. That is why several improvements of the method have been developed. The first type of improvement is modification of the value of α in order to improve the agreement with experiments (and, in this case, α may be taken to be variable [53]) or to approach the scaling law characteristic of the surface tension close to the critical point [55]. Another way is the modification of the values of the parachors. The parachors are usually taken from the standard tables but their correlations with the critical properties are also available [46,47]. Different types of the parachor-like expression have also been proposed [55–57]:

$$\sigma^{1/4} = n_l[P]_l - n_g[P]_g \quad \text{or} \quad \sigma^{1/4} = [P](n_l - n_g) \tag{30}$$

The first Eq. (30) is similar to Eq. (29). However, the mixture parachors $[P]_l$ and $[P]_g$ are calculated from different mixing rules than the linear rules used in Eq. (29). The quadratic mixing rules are also used for the common parachor parameter $[P]$ in the second Eq. (30). The parachors of individual components are correlated with their critical properties. This is especially important in petroleum mixtures containing several pseudocomponents and heavy fractions.

Another thermodynamic approach suggested for estimates of the surface tensions is the mean field approximation, also known as the gradient theory. This theory was originally suggested by Lord Rayleigh [58] and van der Waals [59] and later rediscovered by Cahn and Hilliard [60] (see also Ref. 16 for a review of the history of the method). According to the gradient theory, the interface is represented as a continuous transition layer between the two phases. In view of the rapid variation of the thermodynamic properties in the transition layer, the specific free energy f cannot be calculated by the ordinary formula from equilibrium thermodynamics. It should contain an additional term $\sum_{i,j} c_{ij}(dn_i/dx)(dn_j/dx)$, which reflects the contribution of the gradients of the molar densities $n_i = N_i/V$ (x is a linear coordinate). The influence parameters c_{ij} may also be dependent on n_k. Minimization results in the following system of equations for the molar densities:

$$\sum_i c_{ij} \frac{d^2 n_j}{dx^2} - \frac{1}{2} \sum_{j,k} \frac{\partial c_{jk}}{\partial n_i} \frac{dn_j}{dx} \frac{dn_k}{dx} = \frac{\partial \Phi}{\partial n_i}, \qquad \Phi = f - \sum_i \mu^i n_i \quad (i, j, k = 1, \dots, M)$$

Here, Φ is the grand potential. After the solution of the last system, the surface tension may be found from

$$\sigma = \int_{-\infty}^{\infty} \sum_{j,k} c_{jk} \frac{dn_j}{dx} \frac{dn_k}{dx} \, dx = 2 \int_{-\infty}^{\infty} (\Phi(\mathbf{n}) - P) \, dx$$

Here, P is the pressure in the bulk phases.

Testing the gradient theory shows that it is in good agreement with experimental data, even in the critical region, where its physical background becomes inadequate [61]. However, direct calculations according to the last system of equations may require a long computation time, especially, for realistic multicomponent mixtures. Therefore, certain simplifications of the theory have been proposed. One of them is the linear gradient theory, in which a linear variation of the component molar fractions with regard to one of them is assumed within a transition region [62,63].

E. Phase Diagrams of Capillary Equilibrium

1. $P–z$ Diagram

A common representation of the phase equilibrium is the phase envelope showing the dependence of the saturation pressure on temperature for a mixture of a given composition [6]. Such a representation may be applied to analysis of the capillary equilibrium in porous media, in order to evaluate deviations in the values of the dew-point pressures under the action of the capillary forces [40]. However, because temperature may be treated as an external parameter and is not modified by capillary forces, the pressure–composition diagram may be more informative for this case. Additionally, numerical calculations show that the compositional deviations under capillary equilibrium may be more significant than the deviations in the dew-point pressures. A qualitative analysis of the $P–z$ diagram for capillary equilibrium in binary mixtures was first carried out in Ref. 22. We follow this analysis with some important modifications.

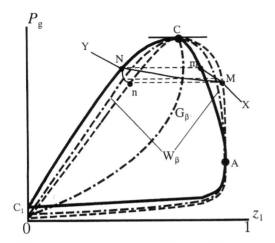

FIG. 4 Phase diagram of capillary equilibrium for a binary mixture on the P–z plane.

In Fig. 4, a typical binary diagram of a binary hydrocarbon mixture is shown. Such a diagram corresponds to the light component 1, which is gas at a given temperature, and heavy component 2, which remains liquid. Other possible cases may be considered in a similar way.

The binodal for the system being considered is shown in Fig. 4 by the solid line (curve $NCAC_1$). This curve consists of the equilibrium points of the gas and the liquid phases. Because without capillary forces, pressures in the two phases are equal, the corresponding equilibrium points lie on the horizontal lines (the so-called tie lines). The intermediate points of the tie lines correspond to the two-phase mixtures with different molar saturations β (or the volumetric saturations β_v).

Point C is the upper critical point of the mixture; C_1 is the lower point in the diagram. In the one-phase zone, the mixture is liquid, roughly, to the left of C and gas to the right of C. The turning point, A, corresponds to the maximum molar fraction of the first component in the two-phase zone. The tangent line to the binodal is vertical at this point. The area above CA corresponds to retrograde condensation, where liquid precipitates as the pressure in the gas phase decreases. The area of normal condensation lies below $C_1 A$.

In order to find out how the phase diagram should be modified if the mixture is put in a porous medium, the capillary phase split corresponding to an arbitrary point (P_g, z) inside the phase envelope is considered. This is the saturation problem (Sec. III.C.2). From its solution, based on Eqs. (23)–(26), it is possible to determine the liquid pressure P_l, the saturations β and β_v, and the phase compositions z_l and z_g. These compositions are not the same as the compositions of the bulk phases under the ordinary phase split, and the corresponding equilibrium points should not lie on the binodal $NCAC_1$.

Let us consider the line G_β on which the volumetric saturation of the liquid phase β_v remains invariable. The corresponding capillary equilibrium points belong to a binodal W_β, which differs from the original binodal (both G_β and W_β are shown by the dashed-dotted lines in Fig. 4). Thus, *under the action of capillary forces, the phase envelope in the P–z space is disintegrated into a series of phase envelopes, each corresponding to its own value of the volumetric saturation.*

As the liquid-phase saturation increases, the isosate G_β moves from the left to the right. Under the absence of capillary forces, the isosate G_1 corresponding to the unity saturation coincides with the section CNC_1 of the main binodal. If the largest radius of a pore in the effective pore-size distribution is equal to infinity, the same statement is true about the isosates of the capillary phase equilibrium. On the other hand, if β_v increases, the value of r_c from Eq. (23) increases, and the conditions of capillary equilibrium approach that of normal equilibrium. Thus, with an increase of β_v the binodals W_β approach the main binodal.

In the following, we will show that capillary condensation becomes negligible (the second order of magnitude with regard to the distance to the phase envelope) at the turning point A. Thus, the capillary binodals approach the main binodal at this point.

Let us consider an imaginary experiment, during which a fluid in a cell under a piston is gradually saturated by the heavy component and the pressure in it gradually increases, so that the fluid passes the XMNY path in the diagram (another interpretation of this path may be segregation of the mixture under the action of gravity [22]). If the cell contains no porous medium and there are no capillary forces, then the mixture follows the path XmNY. The mixture remains in gas phase until point m, where it strikes the main phase envelope. Then, it condenses into a liquid phase and continues along the path NY. This picture is modified for the constant-volume experiments, where continuous variation of the liquid saturation along tie line mN becomes possible.

A possibility of continuous variation of the saturation in the constant-pressure experiment may be provided by involvement of the capillary forces (e.g., by putting a sample of the porous medium into the cell considered). In this case, the first liquid condensate occurs at point M, as the mixture path intersects the binodal corresponding to the smallest pores (dashed line in Fig. 4). Then, the saturation of the porous medium by capillary condensate gradually increases, and the mixture point moves along the path MN. The corresponding path Mm of the gas phase, as well as the path of the liquid phase nN, crosses all of the capillary binodals. After point N, the mixture becomes fully liquid. It should be noted that the tie lines are horizontal, because the diagram is depicted in coordinates (P_g, z). If the points on the binodal are chosen to be (P_g, z_g) and (P_l, z_l), the tie lines become inclined. The liquid pressure P_l may be much less than P_g. Moreover, formal calculations may lead to the negative liquid pressures in the area of normal condensation (below AC_1). To the best of our knowledge, experimental validation and proper physical discussion of this fact are lacking in the literature.

2. $P-\mu$ Diagram

Apart from the $P-z$ diagram considered earlier, some important properties of the capillary equilibrium may be demonstrated in the $P-\mu$ diagram. In order to formulate and illustrate some of these properties, let us first briefly consider the case of a single-component fluid [20,21].

Single-component capillary equilibrium may be studied on the basis of Fig. 5. Figure 5a shows a characteristic isotherm $V(P)$ for a cubic or similar equation of state (van der Waals, Soave–Redlich–Kwong, Peng–Robinson, or others [6]). Figure 5b is obtained by taking the integral $\int V\,dP$ and thus represents the isotherm in the coordinates (μ, P).

The plot $\mu(P)$ consists of two branches, $\mu_l(P)$ and $\mu_g(P)$, corresponding to liquid and gas, respectively, and of the unstable branch shown by the dashed line. The intersection of $\mu_l(P)$ with $\mu_g(P)$ corresponds to the point E of true equilibrium, where both chemical potentials and pressures of the phases are equal. The infinite branches EL and EG represent stable liquid and gas, whereas the cuts EL' and EG' correspond to the metastable states.

The condition of local stability of a single component bulk phase consists in the positiveness of the compressibility or, in other words, $d^2\mu/dP^2 = dV/dP < 0$ [64]. Thus, the entire branches LL' and GG' in Fig. 5b are concave. On the contrary, the unstable branch $L'G'$ is convex. Change of sign of the value $d^2\mu/dP^2 = dV/dP$ occurs at the spinodal points (the boundaries of the region of the stable phases) L' and G', where this value becomes infinite (cf.

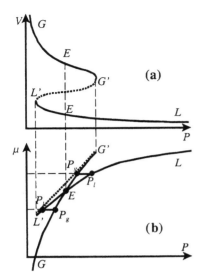

FIG. 5 Single-component vapor–liquid equilibrium under the capillary pressure difference. (From Ref. 25.)

Fig. 5a). On the plane (μ, P), the stable and the unstable parts of the isotherm $\mu(P)$ have tangents in common with slopes equal to the values of $V(P)$ at these points.

The capillary equilibrium for a single-component fluid is described by the system of equations being a particular case of Eqs. (24):

$$\mu_g(P_g) = \mu_l(P_l), \qquad P_l - P_g = P_c \tag{31}$$

The value P_c of capillary pressure is determined by the values of surface tension, contact angle, and the possible geometry of the menisci separating gas and liquid. As shown in Section III.2.B, the capillary pressure is negative for the wetting liquid and positive otherwise; that is, the wetting phase has smaller pressure than the nonwetting one.

Capillary equilibrium conditions (31) can be illustrated by the $P-\mu$ diagram (Fig. 5b) [20,21]. The value of the capillary pressure is equal to the cut of the coordinate line corresponding to a specified value of the chemical potential, between the gas and the liquid branches of the curve $\mu(P)$. The following can be observed from Fig. 5:

1. Capillary equilibrium is possible for any phase between the true equilibrium point and the spinodal point. Thus, the boundary for capillary equilibrium coincides with the spinodal point.
2. The nonwetting phase is always stable, whereas the wetting phase is always metastable [21].

It should be stressed that these properties are only related to the *bulk* phases, and the properties of the interfaces may vary continuously, passing the region of instable states (as in the gradient theory of surface tension).

It is physically evident that Statements 1 and 2 are valid not only for a single-component fluid but also for a multicomponent mixture. Their extension onto multicomponent mixtures requires an analysis of the phase surface geometry in the space of intensive variables.

3. Properties of Capillary Equilibrium in Many Dimensions

Let us recall major facts about phase behavior in the space of intensive variables. For a multicomponent mixture, the dependence of pressure on the temperature and chemical potentials is expressed by means of the Gibbs–Duhem relation:

$$dP = s\, dT + n_i\, d\mu^i \tag{32}$$

Here, as earlier, n_i are the component molar densities and $s = S/V$ is the specific entropy. Because the molar densities are always positive, pressure in the multicomponent mixture is an increasing function of the chemical potentials, as for a single-component fluid. This property is fulfilled in each region of continuity and physical significance of the pressure (i.e., on each stable branch corresponding to a single phase).

In order to establish the property of concavity, let us consider the second differential:

$$\delta^2 \Phi = ds\, dT + dn_i\, d\mu^i$$

It has been shown by Gibbs [2] that the condition of local stability of the phase corresponds to the positiveness of this differential (see recent discussion of the stability criteria in Refs. 64–68; in Refs. 69 and 70, the corresponding parts of the thermodynamics have been reformulated from the "field-density" point of view).

The physical meaning of the form $\delta^2\Phi$ depends on the choice of the independent variables. It follows from Eq. (32) that at independent T and μ^i, this form is equal to d^2P. On the other hand, at independent s and n_i, the form $\delta^2\Phi$ is equal to the second differential of the specific energy:

$$\delta^2 \Phi = d^2 P|_{T,\mu^i} = d^2 e|_{s,n_i}$$

Thus, the Gibbs condition in terms of the pressure means that

$$d^2 P(T, \mu^1, \dots, \mu^M) \geq 0$$

that is, the dependence $P(\mu^1, \dots, \mu^M)$ is not only increasing, but also a concave function of the chemical potentials on each part of the phase surface corresponding to a locally stable phase (cf. Ref. 69).

The boundaries of the stability regions (the spinodals) are determined by the condition that the concavity of $\delta^2\Phi$ is violated. It is more convenient to work with the second differential of the specific energy, as it is assumed to be a smooth function of the extensive variables, whereas the dependence $P(T, \mu^i)$ may be singular. It has been shown by Gibbs [2] that the spinodal condition is given by one of the following equivalent relations:

$$\left.\frac{\partial T}{\partial s}\right|_{\mu^j} = 0, \qquad \left.\frac{\partial \mu^i}{\partial n_i}\right|_{T,\mu^j, j\neq i} = 0 \tag{33}$$

A simple and straightforward way to prove it (avoiding cumbersome determinant calculations or Legendre transformations [64–66]) is presented in Ref. 25. Inverting the derivatives in Eq. (33), we obtain the condition for the stability boundary in the (P, T, μ) space:

$$\left.\frac{\partial^2 P}{\partial T^2}\right|_{\mu^i} = \left.\frac{\partial^2 P}{\partial \mu^{i2}}\right|_{T,\mu^j, j\neq i} = \infty \tag{34}$$

Inside the stability region, all of these derivatives should be positive, whereas in the unstable region, the positivity of some of them may be violated. A remarkable fact is that in the (P, T, μ) space, the thermodynamic conditions become symmetric and have a geometrical meaning.

4. Phase Diagram in the (P, T, μ) Space

One possible configuration of the two-phase surfaces around the phase transition is presented in Fig. 6. For simplicity, we omit the T coordinate and work with a binary mixture. The central point C in Fig. 6 is a critical point for the binary mixture considered (similar to Fig. 5). To the left of this point, the mixture is in a single-phase state, whereas to the right of it, there is a two-phase region where the gas and liquid phases are represented by separate surfaces. These surfaces intersect along the line CC' of true phase equilibrium. The lines CS_1 and CS_g are the spinodal lines for liquid and gas branches, correspondingly. Metastable liquid and gas regions are S_1CC' and S_gCC'. The remaining surfaces represent stable phases.

For the general case of an M-component mixture, the phase hypersurfaces are M dimensional and the points of true equilibrium form an $(M - 1)$-dimensional hypersurface. The dimension of the spinodal hypersurfaces is also $M - 1$.

The configuration around the critical point presented in Fig. 6 is not the only possibility (at least, for a general multicomponent mixture). Topologically, it is qualified as the "scum" or "butterfly" catastrophe. The reason why it arises as well as a list of other possible configurations may be found in Ref. 71 (the qualitative analysis in the space P, μ^1, μ^2 is not much different from the analysis in the space P, T, μ). The common (but not topological) properties for all the thermodynamically possible configurations are that the pressure surfaces of the stable phases are increasing and convex. At the boundaries of stability, this convexity tends to infinity, according to Eq. (34). The unstable branch (not shown) is tangent to the stable branches along the spinodal boundaries. The instability hypersurface is not necessarily concave, as in the case of a single component, but at each point, it must be concave in at least one direction (cf. Fig. 5b).

Because for both phases the pressure is a single-value function of the chemical potentials, the coordinate line $\mu^1 = \text{const}, \ldots, \mu^M = \text{const}$ intersects with each phase surface, provided that the point $\mu^1, \ldots, \mu^M$ lies within the spinodal curve (more precisely, within the projection of the spinodal boundary on the hyperplane $P = 0$). Intersections of this line with the phase surfaces give the values of the pressures P_l and P_g in liquid and gas phases (see Fig. 6). If the point $\mu^1, \ldots, \mu^M$ lies outside the spinodal curve, the corresponding coordinate line intersects just one coordinate surface, and capillary equilibrium becomes impossible. This proves the generalization

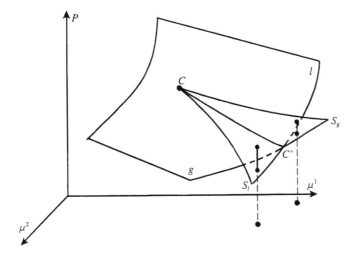

FIG. 6 Phase surfaces for a binary mixture in the space of chemical potentials and pressure. (From Ref. 25.)

onto multicomponent mixtures of the statement from Section III.E.2 that the capillary equilibrium is possible everywhere in the region between the true equilibrium and the spinodal.

The simplest way to prove the second statement (that the wetting phase is always metastable) is to consider the potential $\Omega(T, V, \mu^1, \ldots, \mu^M) = -PV$. Under constant variables $T, V, \mu^1, \ldots, \mu^M$, thermodynamic equilibrium is determined by the minimum of this potential. This means that Ω should be higher on the metastable branch than on the stable one. In other words, under the same chemical potentials, pressure on the stable branch is higher than on the metastable branch. According to the considerations at the end of Section III.B, the higher pressure corresponds to the nonwetting phase, which proves the statement.

5. Sample Calculations

Calculations, which will be presented here, demonstrate the behavior of the capillary condensation in the whole range of possible pressures and compositions and, in particular, show the restrictions imposed by the boundaries of stability. This set of calculations has been performed for the binary mixture of methane and n-butane at a temperature of 277.6 K (Fig. 7a). The mixture exhibits retrograde condensation behavior above point A on the gas branch, corresponding to $P = 58.5$ MPa. Below A, the condensation is normal. The binodal curve around A is rather flat (as well as for this mixture at other temperatures), and allowing some inaccuracy in the language, we may call the area around this point the "turning region." In the calculations described, the pressure remained constant while the composition of the mixture varied. The calculated values are depicted in Figs. 8 and 9 as plots versus the distance to the phase envelope. This distance is defined as the difference of the methane molar fraction in a given gas mixture and at a dew point at the same pressure, $\Delta z_g = z_g - z_{gd}$. The calculations of capillary equilibrium were performed by application of the G-L algorithm described in Section III.C.3.

Figure 8 expresses the dependence of the surface tension, calculated by the parachor method, on the distance from the dew point. The values of surface tension vary significantly with this distance in the region of retrograde condensation and less significantly (but still noticeably) in the region of normal condensation. Variation of the surface tension is due to the fact that

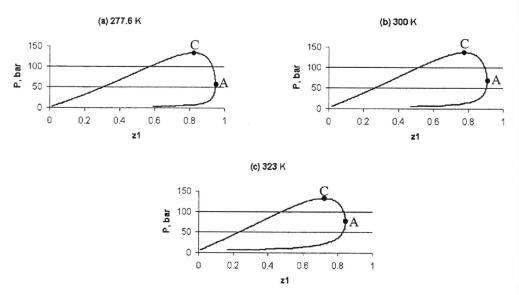

FIG. 7 Phase diagram for the system C1–nC4: (a) 277.6 K; (b) 300 K; (c) 323 K.

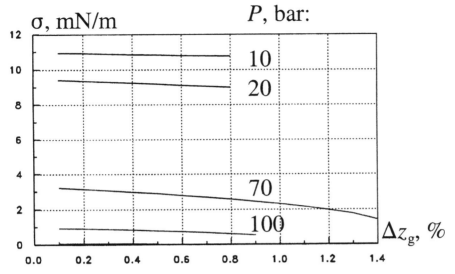

FIG. 8 Dependence of the surface tension on the distance to the phase envelope. The mixture of methane–*n*-butane at 277.6 K. (From Ref. 25.)

capillary pressure rapidly increases with Δz_g and that pressure and composition of the capillary condensate vary correspondingly.

The behavior of the Kelvin radius r_c, determined in Section III.B, at different pressures is shown in Fig. 9. The Kelvin radius rapidly decreases with the distance to the phase envelope and, also, as the pressure approaches the critical value.

The right points on the plots of Figs. 8 and 9 correspond to the limits for capillary condensation. In the retrograde region, these limits are determined by the fact that the equilibrium liquid phase approaches the spinodal boundary. In the region of normal condensa-

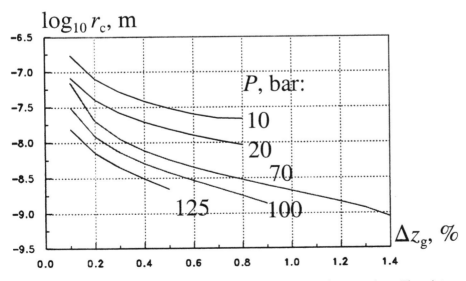

FIG. 9 Dependence of the Kelvin radius on the distance to the phase envelope. The mixture of methane–*n*-butane at 277.6 K. (From Ref. 25.)

tion, we calculated the capillary condensation until the point where the pressure in the equilibrium liquid phase became zero. Unfortunately, the present version of our software does not allow us to handle properly the negative pressures which are physically possible [20,21] and may be predicted by a cubic equation of state.

The region of capillary condensation lies within the range of molar fractions, which are 0.8–1% apart from the phase envelope, except in the "turning region," where the difference may achieve 1.5%, as demonstrated for the pressure of 70 bar. These differences of 0.8–1.5% are by no means low, as we take into account that the average molar fraction of butane in the mixture is around 6%. Still, it can be concluded that capillary condensation occurs in a close neighborhood of the phase envelope. This closeness increases as a critical pressure is approached, as demonstrated for $P = 125$ bar.

In the region of normal condensation, the limit for capillary condensation corresponds to the Kelvin radii of the order of a few tens of nanometers, which corresponds to the average and large mesopores. In the region of retrograde condensation, the limiting Kelvin radii correspond to the small mesopores or, in the "turning region," to micropores. For such radii, the effects related to molecular sizes, roughness of the surface, and the effect of the surface forces on the bulk phases may become important, and capillary condensation should often be considered together with adsorption, which may have an additional stabilizing effect on the capillary condensate. These effects are not considered in detail here.

The results remain basically the same at different temperatures. At higher temperatures, the admissible distance to the phase envelope usually increases (up to 1.5–2.5% at 323 K), the Kelvin radii slightly decrease, and the surface tensions become more dependent on the distance to the phase envelope.

For a multicomponent mixture, the results may be qualitatively different from a binary case. Let us proceed from a mixture characteristic of a gas–condensate reservoir, whose composition is presented in Table 1. This mixture exhibits retrograde condensation with a dew-point pressure of 200 bar at $T = 323$ K. Figure 10 shows the nonmonotonous dependence of the surface tension on the distance to the binodal curve, as the molar fractions of C1 and C7 vary. Figure 11 indicates that under varying C1 and C10, the variation of composition is possible only within a 0.2% region, although the Kelvin radii cover the whole range of macropores and mesopores. These examples show that capillary condensation may produce a rich variety of unusual physical effects in multicomponent mixtures, which are not observed in mixtures with a low number of components.

F. The Multicomponent Kelvin Equation

1. Statement of the Problem

The classical Kelvin equation relates the value of capillary pressure P_c with the distance to the phase transition in the absence of the capillary forces, which is normally expressed by $P_c = P_d$

TABLE 1 Composition of the Vapor Phase for the Gas–Condensate Mixture

	Component											
	C1	C2	C3	nC	nC	C6	C	Cl	iC4	iC5	CO	N_2
Mol %	76.5	6.76	3.61	1.13	0.46	0.62	2.6	3.3	0.61	0.41	3.52	0.48

Source: Ref. 18.

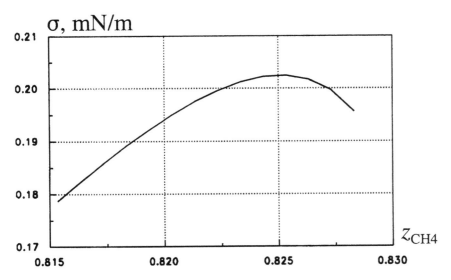

FIG. 10 Dependence of the surface tension of a multicomponent mixture on the distance from the phase envelope. Varying components: C1 and C7. (From Ref. 25.)

as a ratio of $\chi = P_g/P_d$, where P_g is the gas phase pressure and P_d is the dew-point pressure (the pressure at the phase transition). In one of its original forms, this equation is

$$P_c = P_d \frac{v_g}{v_l} \ln \chi \tag{35}$$

where v_l, and v_g are the molar volumes of liquid and gas taken at the dew point.

Since the time of its discovery [72], the Kelvin equation has become one of the most useful tools for the study of vapor–liquid equilibria in capillary media [73]. It has been extensively used

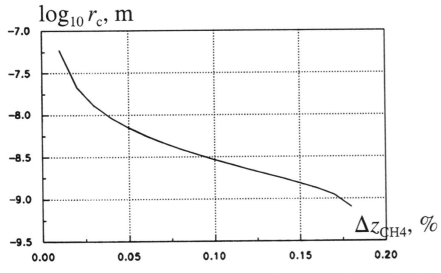

FIG. 11 Dependence of the Kelvin radius of a multicomponent mixture on the distance to the phase envelope. Varying components: C1 and C10. (From Ref. 25.)

for testing of different kinds of porous media and evaluation of their internal surface and of the size distributions of macropores and mesopores. For example, it may be combined with the Laplace equation in the form of (23) in order to evaluate the volumetric saturation of the porous medium by the capillary condensate by use of only the data at a dew point.

However, application of the Kelvin equation to the modeling of phase equilibria is highly conjectural for a majority of the industrial and natural processes dealing with porous media. The reason is that the derivation of this equation is based on the following three assumptions:

1. The fluid is single component.
2. The vapor phase is ideal (and $v_g = RT/P_d$).
3. The liquid phase is incompressible [more precisely, $v_l = v_l(T)$].

These assumptions are violated at high pressures when the vapor phase becomes nonideal and in the cases where the multicomponent nature of the mixture cannot be neglected. The extreme case of "non-Kelvin behavior" is the behavior of hydrocarbon mixtures in oil–gas–condensate reservoirs. Although the Kelvin equation is applied to tests of the porous media of the reservoirs [74,75], it cannot be used for modeling of equilibria in such reservoirs, not only because of the high pressure and the multicomponent composition of the mixture but also because this mixture exhibits retrograde behavior when the liquid phase precipitates as the pressure decreases. Such a behavior cannot, in principle, be described in terms of a single-component model.

Generalization of the Kelvin equation onto arbitrary multicomponent mixtures has been obtained in Ref. 4 and later developed in Refs. 18 and 25. Following Ref. 25, we describe its geometrical derivation in the space of the variables P and μ^i.

2. Derivation of the Multicomponent Kelvin Equation

Let us linearize equations for the phase surfaces in the neighborhood of the true phase transition or, in other words, find equations for the tangent hyperplanes to these surfaces at a fixed point on the phase equilibrium line CC′ (see Fig. 6 for the nonlinearized picture and Fig. 12 for the linearized picture). The linearized equations directly follow from the Gibbs–Duhem relation

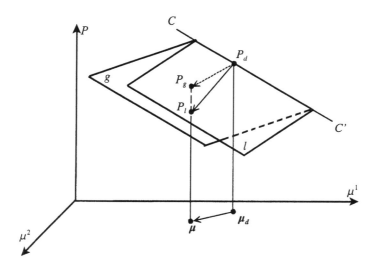

FIG. 12 Phase surfaces approximated by hyperplanes in a neighborhood of the equilibrium line CC′. (From Ref. 25.)

(32). At constant temperature the expressions for gas and liquid tangent hyperplanes are, correspondingly,

$$P_g - P_d = \sum n_{g,i}(\mu^i - \mu_d^i), \qquad P_l - P_d = \sum n_{l,i}(\mu^i - \mu_d^i) \qquad (36)$$

Here, P_g and P_l are pressures corresponding to the same values of chemical potentials, μ^i, the values with the subscript d ("dew point") are taken on the hypersurface of the true phase transition CC', close to the points (P_l, μ^i) and (P_g, μ_i) (choice of the dew point may be nonunique). Subtracting the two equations in Eqs. (36), we find the value of P_c to be the distance between the two hyperplanes along the P axis:

$$P_c = \sum (n_{l,i} - n_{g,i})(\mu^i - \mu_d^i) \qquad (37)$$

The last equation expresses the capillary pressure in terms of the distance to the phase transition surface. It may be considered as the most general form of the Kelvin equation, as first derived in Ref. 4. Application of Eq. (37) makes it possible to determine the capillary pressure at a given point (characterized by the chemical potentials μ^i), provided that some neighboring dew point is known. In the linear approximation considered, the partial molar densities $n_{l,i}$ and $n_{g,i}$ should be taken at the dew point, in correspondence with the chemical potentials μ_d^i.

A useful particular case of Eq. (37) may be obtained if we assume that a given gas phase differs from the corresponding dew point only by the value of pressure, whereas the composition of the gas phase is the same. This case has been considered in detail in Ref. 18. Expanding the increments $\mu^i - \mu_d^i$ under invariable composition of the gas phase z_g, we obtain the following first-order approximation:

$$\mu^i = \mu_d^i = v_{g,i}(P_g - P_d)$$

Substitution of this equation into Eq. (37) results in

$$P_c = \left(\frac{v_{gl}}{v_l} - 1\right)P_d(\chi - 1) \qquad (38)$$

In the last two equations, we have introduced the designations $v_{g,i}$ for the partial molar volumes in gas phase. The designation v_{gl} stands for *the mixed volume* [18] defined as

$$v_{gl} = \sum z_{l,i} v_{g,i}$$

Different asymptotic forms of the Kelvin equation, which are equivalent within the second-order corrections, may be proposed. In order to improve the accuracy of Eq. (38) and to make it closer to the original Kelvin equation (35), an asymptotic relation $\chi - 1 \approx \ln \chi$ may be used. For example, in Ref. 18, the Kelvin equation has been obtained in the form

$$P_c = P_d\left(\frac{v_{gl}}{v_l} \ln \chi - \chi + 1\right) \qquad (39)$$

Another possible form is

$$P_c = P_d\left(\frac{v_{gl}}{v_l} - 1\right) \ln \chi \qquad (40)$$

3. The Kelvin Equation and Conditions of the Capillary Condensation

We have shown that the phase with the smaller pressure is metastable. Therefore, in the (μ, P) space, the metastable surface lies below the stable surface (Fig. 12). The same should be valid for their tangent planes along a common line: The semiplane tangent to the metastable surface lies below the semiplane tangent to a stable surface. Therefore, Eq. (37) (which has been obtained by subtraction of the constituting equations for the semiplanes) properly predicts the sign of the capillary pressure: $P_c < 0$. The same should be valid for its different representations (38)–(40), where the capillary pressure is expressed in terms of the relative pressure χ.

Let us now recall that the relative pressure behaves differently in the regions of normal and retrograde condensation (Fig. 4). Whereas in the region of normal condensation the stable gas phase corresponds to pressures lower than the dew-point pressure ($P_g < P_d$), in the region of retrograde condensation, we have the opposite relation ($P_g > P_d$). Correspondingly, we have $\chi \leq 1$ and $\ln \chi \leq 0$ for normal condensation and $\chi \geq 1$ and $\ln \chi \geq 0$ for retrograde condensation. Comparison to Eqs. (38)–(40) and use of the inequality $P_c < 0$ lead us to the following:

Statement. *In the area of normal condensation*

$$v_{gl} \geq v_l \tag{41a}$$

whereas in the area of retrograde condensation

$$v_{gl} \leq v_l \tag{41b}$$

Let us prove this statement without referring to the theory of capillary equilibrium. The proof will be based on the Gibbs tangent plane stability criterion [68]. For definiteness, we consider the case of the retrograde condensation.

Let us consider the vapor phase of the composition z_g at the pressure $P_g > P_d$. Such a phase must be stable and obey the Gibbs condition for stability: For any other composition z,

$$\sum_{i=1}^{n} z^i [\mu^i(P, \mathbf{z}) - \mu^i(P, \mathbf{z}_g)] \geq 0$$

where μ^i are "universal" expressions for the chemical potentials coinciding with the μ_g^i and μ_l^i for vapor and liquid phases, correspondingly.

Consider the function

$$f(P) = \sum_{i=1}^{n} z_{ld}^i [\mu_l^i(P, \mathbf{z}_{ld}) - \mu_g^i(P, \mathbf{z}_g)]$$

According to the Gibbs tangent plane condition, we have $f(P) \geq 0$ when $P \geq P_d$. On the other hand, at $P = P_d$ the equality $f(P) = 0$ holds, due to the equalities of the chemical potentials in both phases. Thus,

$$\frac{df(P)}{dP}\bigg|_{P=P_d} \geq 0$$

At the same time, at $P = P_d$ the derivative df/dP is equal to

$$\frac{df(P)}{dP}\bigg|_{P=P_d} = \sum_{i=1}^{n} z_{ld}^i \left[\frac{\partial \mu_l^i(P, \mathbf{z}_{ld})}{\partial P} - \frac{\partial \mu_g^i(P, \mathbf{z}_g)}{\partial P}\right]_{P=P_d}$$

$$= \sum_{i=1}^{n} z_{ld}^i [v_l^i(P, \mathbf{z}_{ld}) - v_g^i(P, \mathbf{z}_g)] = v_l - v_{gl}$$

Comparison of the last two equations proves the inequality (41b). Inequality (41a) may be proven in a similar way.

The above statement is more general than just a condition of the capillary condensation. For example, it can be used for checking whether a point in the phase diagram corresponds to the retrograde or to the normal condensation. Independence of this statement of any "capillary" considerations gives, in combination with the Kelvin equations (38)–(40), another proof of the fact that $P_c < 0$ [i.e., that the metastable (or wetting) phase has smaller pressure than the stable phase].

The proof based on the Kelvin equations is valid for a neighborhood of a dew point. Spreading of the proof from this neighborhood to the whole area between the spinodals may be achieved by the following considerations. Let us take an arbitrary coordinate line corresponding to the constant values of μ^i and varying pressure (as shown in Figs. 6 and 12) and move it toward the line CC' of the true equilibrium. During such a motion, the order of the values P_l and P_g will not change. Otherwise, the value of P_l must become equal to P_g at some intermediate point, and this point must belong to the surface of true equilibrium. These considerations show that the order of the metastable and the stable surfaces must be the same in a neighborhood of the equilibrium hypersurface CC' and away from it.

The behavior of the mixing volume is illustrated in Fig. 13 for the mixture of methane–n-butane. The phase diagram for this mixture at 300 K, calculated on the basis of the Peng–Robinson equation of state, is shown in Fig. 7b. The critical point corresponds to a pressure of 137 bar and to a molar fraction of methane of 0.77. If the molar fraction of methane exceeds this value, the mixture shows retrograde behavior. The partial volume of butane, $v_{g,2}$, is negative in the region of retrograde condensation.

The dependence of the liquid volume and of the mixed volume on the pressure, which is shown in Fig. 13, makes it possible to define precisely the dew-point pressure at which the retrograde condensation is changed to the normal one, $P_A = 68$ bar (the point A in the phase diagram). This value is hard to define directly from the phase diagram, because the corresponding branch of the dew-point pressure curve in Fig. 7 is sharply inclined to the axis.

As proven, $v_{gl} > v_l$ in the region of normal condensation, whereas in the region of retrograde condensation, the value of v_{gl} is smaller than that of v_l, reaching its minimum at $P \approx 100$ bar. Meanwhile, the value of v_g always exceeds both v_{gl} and v_l. The difference between all three volumes disappears as the mixture approaches the critical point. That is why the Kelvin equation, in either of the forms, becomes inaccurate close to the critical points, where second order (and, probably, fraction order) corrections should be used. Moreover, for the same reason, the Kelvin equation becomes inaccurate close to the turning points (like point A in Fig. 7).

4. Sample Calculations

Let us first demonstrate the difference between the classical Kelvin equation (35) and the modified Kelvin equation in either of the forms (37)–(40). A nearly ideal mixture of 50% methane and 50% butane may be used for demonstration. as shown in Fig. 14. The mixture is in

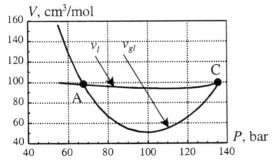

FIG. 13 Behavior of the liquid and of the mixed volume for the system C1–nC4. (From Ref. 18.)

the region of normal condensation, with a dew-point pressure of 10.8 bar at a temperature of 323 K. The values of the capillary pressure, predicted by the original and the modified Kelvin equations, are compared with the numerical calculations performed on the basis of the G-L algorithm described in Section IIIC.3. Whereas the predictions of the modified Kelvin equation cannot be distinguished from the numerical evaluations, the classical Kelvin equation gives a difference of around 20%. This error corresponds to the difference between the gas volume in Eq. (35) and the mixed volume in Eqs. (37)–(40). It should be noted that an erroneous use of the standard Kelvin equation for prediction of the capillary pressures in multicomponent mixtures may often be found in the literature (cf. Refs. 76 and 77). Moreover, use of the standard Kelvin equation becomes totally inappropriate in the region of the retrograde condensation, where it predicts a wrong sign for the capillary pressure.

Figure 15 shows the dependence of the capillary pressure on the relative pressure χ for the same mixture at 300 K. The nonmonotonous dependence of the capillary pressure on the dew-point pressure P_d is explained by the fact that the ratio V_{vl}/V_l also varies nonmonotonously (see Fig. 13). The lowest capillary pressures correspond to the neighborhood of the critical point.

The dependence of the Kelvin radius on the relative pressure in the retrograde region is presented in Fig. 16. Unlike the capillary pressure, the Kelvin radius depends monotonously on the dew-point pressure, because it is mostly affected by the values of the surface tension σ. These values (estimated by the parachor method) show a high variation: 1.579 mN/m for $P_d = 80$ bar, 0.734 mN/m for $P_d = 100$ bar, and 0.206 mN/m for $P_d = 120$ bar. The variation of the ratio V_{vl}/V_l is not too strong.

For the pressure $P_d = 80$ bar and $\chi \leq 1.1$, the Kelvin radius is of the order of the macropore sizes (10^{-6}–10^{-8} m). These sizes are comparable with the sizes of pores which determine flows in low-permeable hydrocarbon reservoirs. For higher values of relative pressures, the Kelvin radius becomes comparable with the sizes of mesopores and micropores. Because the mechanism of adsorption in micropores differs from capillary condensation, the value of the Kelvin radius cannot be used in this region.

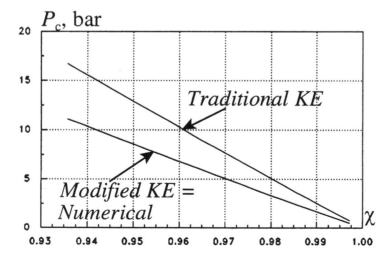

FIG. 14 Dependence of the capillary pressure on the relative pressure. The mixture of 50% methane and 50% butane at a temperature of 323 K. (From Ref. 18.)

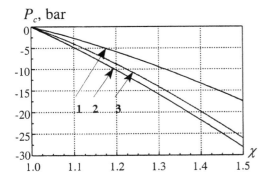

FIG. 15 Dependence of the capillary pressure on the relative one for the mixture methane–*n*-butane at 300 K. Dew-point pressure: 1–80 bar; 2—100 bar; 3—120 bar. (From Ref. 18.)

For the pressures of 100 and 120 bar, the Kelvin radius is comparable to sizes of macropores only in the neighborhood of the binodal, $\chi \approx 1$. Then, it moves into the mesopore and micropore region.

We calculated the capillary pressures and the Kelvin radii for the relative pressures up to $\chi = 1.5$. Our estimates of the Kelvin radii at high values of the relative pressures are imprecise, because the assumptions made in the derivation of the Kelvin equation are no longer valid, and the dependence of the surface tension on the distance to the phase envelope has not been taken into account. This is impossible only on the basis of the Kelvin equation, without evaluation of the compositions of the coexisting phases. Therefore, estimates of the Kelvin radius become precise only in the close neighborhood of a dew point. However, the behavior of the Kelvin radius remains qualitatively correct even far from a dew point.

Another deficiency of the Kelvin equation is that it does not "feel" the boundaries of stability of the capillary condensate. These boundaries may be found only on the basis of the numerical calculations described in Sections III.C and III.E.5. As it has been shown in these subsections, the boundaries for stability of the capillary condensate lie very close to the phase envelope. It may be safely assumed that the modified Kelvin equation is rather precise within these limits.

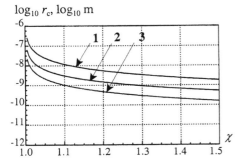

FIG. 16 Dependence of the logarithm of the Kelvin radius on the relative pressure for the mixture methane–*n*-butane at 300 K. The dew point pressure for the mixture: 1—80 bar; 2—100 bar; 3—120 bar. (From Ref. 18.)

IV. THEORY OF MULTICOMPONENT ADSORPTION

A. Overview of Possible Approaches

The "black box" theory of the excess phases, developed in Section II, serves as a framework for the theory of adsorption. However, this framework alone is not sufficient for the adsorption modeling, as it does not specify the properties of particular adsorbate or adsorbent and the ways of their interaction. Therefore, specific models of adsorption are required. These models should be in agreement with the general theory, but they must contain some additional information. There is no "general" or "ultimate" thermodynamic model of adsorption. Moreover, different approaches appear "to speak different languages" and the results obtained in their framework are sometimes difficult to compare. On the other hand, the results obtained on their basis are often combined in spite of their (seemingly) different background. In our opinion, a reason for distinction between models is, in fact, that adsorption lies on the boundary between molecular and macroscopic phenomena. In order to avoid unnecessary complications and make a thermodynamic model suitable for practical purposes, specific molecular interactions determining the properties of an adsorbed layer should be somehow averaged or generalized. This may be performed in different ways.

The most straightforward (and the most developed) approach to multicomponent adsorption is in further development of the thermodynamics of a surface phase, similar to the bulk-phase thermodynamics. In this way, the Gibbs surface thermodynamics should be completed by an equation of state or by an excess model for a proper thermodynamic potential. An extended review of the fundamentals and the history of the development of this approach may be found in Refs. 8, 9, and 78. The approach has become especially popular and widely used for practical modeling of multicomponent adsorption after the works of de Boer [79] and, especially, Myers and Prausnitz [80]. The latter authors made the natural step of introducing the activity coefficients γ^i of the components in an adsorbed phase. In terms of these coefficients, the chemical potentials of the adsorbate may be expressed as

$$\mu^i = \mu_0^i(T) + RT \ln f^{0,i}(\Phi, T) + RT \ln \gamma^i z_{a,i} \tag{42}$$

In this expression for the chemical potential, the first addendum, $\mu_0^i(T)$, is a standard potential at a fixed pressure. The second addendum expresses the contribution from the fugacity of the pure component. The third addendum is due to mixing. The dependence $f^{0,i}(\Phi, T)$ may be found from the Gibbs adsorption equation (13b), where the integration is often carried out from zero pressure (and, correspondingly, the value of Φ_0 is equal to zero). With such an expression for the chemical potentials of the adsorbed phase, equilibrium conditions (12), for the equilibrium with a nonideal gas phase, are reduced to the form

$$P_g \phi^i z_{g,i} = f^{0,i}(\Phi, T) \gamma^i z_{a,i} \quad (i = 1, \ldots, M) \tag{43}$$

The left-hand side of the equation represents (within a temperature-dependent multiplier) the fugacity of the ith component in the gas mixture. The right-hand side corresponds to the fugacity of this component in the adsorbed phase. Correspondingly, ϕ^i are the fugacity coefficients of the gas components and γ^i are the activities of these components in the adsorbate.

For a specified model of the adsorbed phase, the activities are known as functions of the spreading pressure, temperature, and composition. Therefore, given the thermodynamic characteristics of the bulk phase, the system of equation (43) provides M equations for M unknowns Φ and $z_{a,i}$ ($i = 1, \ldots, M-1$) (the value of $z_{a,M}$ is expressed from the condition $\sum z_{a,i} = 1$). It is not only possible to determine the composition of the adsorbate but also values of the surface excesses. To do so, it should be noted that the left-hand side of Eq. (43) is $\exp(\mu^i/RT)$. Thus, the spreading pressure may be expressed as a function of the temperature and the chemical potentials μ^i. Then, according to Eq. (9), we find that

$$
n_{e,i} = \frac{\partial \Phi}{\partial \mu^i}\bigg|_{\mu^{j \neq i}, T} \tag{44}
$$

In many cases, the surface phase may be assumed to be ideal and its activity coefficients γ^i set unity. The corresponding theory has been called IAST (Ideal Adsorbed Solution Theory) [80]. The main advantage of the IAST is its capability to predict multicomponent adsorption equilibria on the basis of the experimental data on the single-component adsorption. Relations (43) and (44) are greatly simplified in this case.

However, the "pure" IAST cannot always be applied to modeling adsorption equilibria. Several experimental works (like Ref. 81) have reported deviations of the adsorbed phase from ideal mixing. The adsorption equilibrium theory based on Eq. (42) with nonunity activity coefficients γ^i is called RAST (Real Adsorbed Solution Theory). Even this theory has been found to not always be adequate. Whereas most of the bulk mixtures show positive deviations from ideal mixing of the values of the molar volumes, deviations for the mixed adsorbed phases are more often negative [82]. Distinctive behavior of the adsorbed mixtures is usually explained by the fact that the adsorbate is heterogeneous [83].

Thermodynamic modeling of adsorption on heterogeneous adsorbents usually follows the concepts developed in Ref. 84. The surface is represented as a union of patches, and each patch is characterized by its own adsorption energy E_i with regard to each (ith) component. These energies may be different for different components, but they are often assumed to be connected: $E_i = E_i(E_1)$. The activity coefficients and other field parameters of the adsorbed phase depend parametrically on these energies. As a result of the solution of the equilibrium problem, the surface excesses are also dependent on $E_1, \ldots, E_n$. If, for example, $\Theta_i(P, T, z_g; E_1, \ldots, E_n)$ is a surface excess of the ith component on a patch characterized by the energies $E_1, \ldots, E_n$, then the surface excess for the whole surface is determined as

$$
n_{e,i}(P_g, T, \mathbf{y}) = \int \Theta_i(P_g, T, \mathbf{y}; E_1, E_2(E_1), \ldots, E_n(E_1))\lambda(E_1)\, dE_1 \tag{45}
$$

Here, $\lambda(E_1)$ is the energy distribution density, which is often taken in a convenient form (discrete, Gaussian, etc.) or fitted from results of the experiments.

Introduction of heterogeneity indicates insufficiency of the straightforward thermodynamic approach based on common equations of state to express the specificity of interactions between the adsorbed phase and the adsorbent. The two new physical ideas behind Eq. (45) are *localization* of the adsorbed molecules at different points of the surface (patches) and a possibility of *different energetic interaction* of the molecules with the surface. These ideas have been developed in a different way in the two other thermodynamic concepts of adsorption: the localized approach and the potential theory of adsorption.

The localized approach originates from the work of Langmuir [85]. The complete theory of multilayer localized adsorption of multicomponent adsorption is rather cumbersome, although resolvable [83,86]. For a demonstration of the ideas behind the localized approach, it is enough to consider the case of monolayer adsorption. We will apply the kinetic derivation [87], with some modifications.

Let us assume that the surface of a unit area consists of N_s equivalent sites. Each site may be either vacant or occupied by any molecule of the mixture. Let us denote by $n_{e,0}$ the number of vacant sites. The number of sites occupied by a molecule of the ith component is, simply, equal to $n_{e,i}$ ($i = 1, \ldots, M$). The total number of sites is fixed to N_s:

$$n_{e,0} + \sum_{i=1}^{M} n_{e,i} = N_s \tag{46}$$

Let us denote by k_+^i the frequency with which a molecule of the ith species comes to the surface and by k_-^i the frequency of evaporation of this species. Then, the dynamics of adsorption may be described by the following system of differential equations, together with condition (46):

$$\frac{dn_{e,i}}{dt} = k_+^i n_{e,0} - k^i n_{e,i} \quad (i = 1, \ldots, M) \tag{47}$$

Equilibrium corresponds to the stationary point of this system of equations, $dn_{e,i}/dt = 0$. From the right-hand side of the system (47), we find that

$$n_{e,i} = K^i n_{e,0}, \qquad K^i = k_+^i / k_-^i$$

The combination of the last relationship with the balance condition (46) results in the expression of the surface excesses in terms of the total capacity of the surface and the frequencies of adsorption/evaporation:

$$n_{e,i} = \frac{N_s K^i}{1 + \sum_{j=1}^{M} K^j} \quad (i = 1, \ldots, M)$$

The dependence of the frequency ratios K^i on the thermodynamic conditions is governed by the properties of the mixture. If the bulk phase is an ideal-gas mixture, it is reasonable to assume that the adsorption frequencies k_+^i are proportional to the partial molar pressures of gas, $P^i = Pz_{g,i}$. On the other hand, if no lateral interaction between adsorbed molecules is assumed, desorption rates k_-^i are solely the functions of the temperature. In these assumptions, a classical multicomponent Langmuir equation is obtained:

$$n_{e,i} = \frac{N_s b^i(T) P^i}{1 + \sum_{j=1}^{M} b^j(T) P^j} \quad (i = 1, \ldots, M) \tag{48}$$

Modification of the theory onto the case of adsorption from non-ideal mixtures and introduction of the lateral interactions between the adsorbed molecules are possible, but may lead to rather complicated relations [86,88]. On the other hand, the localized models may easily be generalized onto adsorption on a heterogeneous surface [89,90]. This is related to the fact that the coefficients $b^i(T)$ may be directly expressed in terms of the adsorption energies by means of the statistical mechanics:

$$b^i = b^{0i}(T) \exp\left(\frac{E_i}{RT}\right)$$

This gives rise to direct application of Eq. (45). Such a Langmuir adsorption theory on heterogeneous surfaces has been applied in order to demonstrate the effect of negative deviation from ideal mixing, which has been observed for the most adsorbed binary mixtures.

An alternative to the localized theory of adsorption is the so-called potential theory, which has been developed as a "slab" adsorption theory on a surface and its analog for adsorption in microporous media, the theory of volume filling in micropores (TVFM). The potential theory of adsorption, first formulated by Polanyi [91], is widely used for the description of adsorption of gases on a solid. The TVFM is applied to adsorption in activated carbons, silica gels, and other types of microporous medium.

An advantage of the potential theory is that its full version draws a direct connection between the properties of a bulk phase with those of an adsorbed phase. To do so, the adsorbate should be considered as a segregated mixture in the potential field emitted by the surface. The description of the adsorbed mixture is purely thermodynamic, with application of the same equations of state, which have been used for the bulk phase. The properties of the adsorbent and, in particular, the surface heterogeneity are expressed by the form of the surface potentials. This makes it possible to evaluate adsorption of rather nontrivial mixtures under complex thermodynamic conditions, provided that thermodynamic models for the corresponding bulk phases are available.

The described properties of the potential theory of adsorption make it preferential to the theory of localized adsorption, at least, for complex thermodynamic conditions. On the other hand, the potential theory incorporates specific physical information about the phenomenon of adsorption, which is absent in the general thermodynamic approach. A major disadvantage of the multicomponent potential theory is the computational difficulties arising in the calculation of the multicomponent segregation in the external field. This is the reason why the development of this theory has been delayed until our recent works [92–94]. The development of modern methods of computational thermodynamics [68] has made it possible to perform these calculations quite efficiently.

The remaining part of this chapter contains a comprehensive discussion of the multicomponent potential adsorption theory (the MPTA), according to Refs. 92–94.

B. Fundamentals of the Potential Theory of Adsorption

The potential theory has originally been developed for the surface adsorption from a single-component non-ideal gas phase. Let x denote the distance from the surface of an adsorbent. It is assumed that the potential field $\varepsilon(x)$ emitted by the surface causes nonuniform distribution of the adjacent substance. This distribution may be described by an equation for the chemical potential $\mu(x) = \mu(P(x), T)$ of the substance in the external force field [3]:

$$\mu(P(x), T) - \varepsilon(x) = \mu_g(P_g, T) \tag{49}$$

Here, $P(x)$ is the pressure at the distance x from the surface tending to P_g as $x \to \infty$. In the following, we omit the temperature as an argument of chemical potentials and other thermodynamic functions.

The model bulk phase for the surface adsorption is a homogeneous phase continuing up to the surface with properties corresponding to those at infinity. Correspondingly, the surface excess n_e is defined as the difference between the actual amount of the substance near the surface and the amount of the bulk phase in the absence of the adsorption potential. It is expressed in terms of the molar density at the distance x to the surface and the bulk molar density n_g:

$$n_g = \int_0^\infty (n(x) - n_g) \, dx \tag{50}$$

Although the integral is taken from zero to infinity, it has a finite value, because at $x \to \infty$, the adsorption potential tends rapidly to zero and the properties of the adsorbate approach those of the bulk phase, $n(x) \to n_g$. Similar statements are valid for other integrals with infinite limits presented in the chapter.

If the gas phase is far from its dew point, the substance remains in the gas state up to the surface. Otherwise, precipitation of liquid on the surface occurs. In the latter case, it is usually assumed that the amount of adsorbate in the gas state may be neglected, compared to that in the liquid film, and that the molar density of liquid n_l may be kept constant. In this approximation (which will from now on be called "the homogeneous approximation"), the thickness h of the adsorbed layer and the surface excess are found from the (presumably known) value $\mu_d = \mu_d(T)$ of the chemical potential at dew-point pressure:

$$\mu_d - \varepsilon(h) = \mu_g, \qquad n_e = (n_l - n_g)h \approx n_l h \tag{51}$$

The potential concept just formulated is used in the "slab" theory of multilayer adsorption on homogeneous and heterogeneous surfaces. Several dependencies for the adsorption potential $\varepsilon(h)$ have been suggested (see the analysis in Refs. 3, 8, 94 and 95). It is often described by the Frenkel–Halsey–Hill (FHH) power dependence [96,97]. A general form of this dependence, avoiding singularity at the surface, has the form [3]

$$\varepsilon(x) = \frac{\varepsilon_0 a^\alpha}{(a+x)^\alpha} \tag{52}$$

where a is a distance of the order of a molecular radius. Usually, the exponent α is assumed to be between 2 and 3. Originally, it was suggested that $\alpha = 3$ and $a = 0$ but $\varepsilon_0 a^\alpha$ is a nonzero constant [3]. In the following, the version of the potential with $a = 0$ will be called the original FHH potential, whereas the dependence (power) with $a \neq 0$ will be referred to as the generalized FHH potential.

Another application of the potential theory is adsorption in micropores (the theory of volume filling of micropores), which is illustrated in Fig. 17. In this case, the variable x represents volume, so that the value of h is interpreted as part of the pore volume filled by the adsorbate. Correspondingly, the surface excesses are substituted by excess amounts adsorbed per unit volume. Instead of the dependence $\varepsilon(x)$, the distribution $x(\varepsilon)$ of the pore volume by potentials is considered. Every so often, this distribution is given by the generalized Dubinin dependence [98–100]. The characteristic parameters of this dependence are the total pore volume x_0, the characteristic adsorption energy ε_0, and the exponent β:

$$x(\varepsilon) = x_0 \exp\left[-\left(\frac{\varepsilon}{\varepsilon_0}\right)^\beta\right] \tag{53}$$

The value of $\beta = 2$ corresponds to the standard Dubinin–Radushkevich (DR) potential. Apart from adsorption in micropores, this potential may also be used to describe adsorption at low coverage of a surface [8].

The potential theory allows straightforward generalization onto adsorption from a multicomponent mixture. Assume that the ith component of the considered mixture is affected by the adsorption potential $\varepsilon^i(x)$. These potentials may vary from one component to another, due to

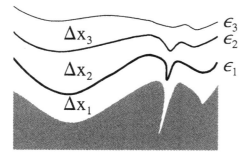

FIG. 17   Theory of volume filling of micropores: distribution of volumes (x) by the values of the potential.

different interaction forces with the surface. An equilibrium state of the mixture in the potential field is described by the system of equations for the chemical potentials μ^i:

$$\mu^i(P(x), \mathbf{z}(x)) - \varepsilon^i(x) = \mu_g^i(P_g, \mathbf{z}_g) \quad (i = 1, \ldots, M) \tag{54}$$

Given pressure P_g and composition z_g in the gas phase, the distributions of pressure and molar fractions close to the surface are uniquely determined by the system (54). Then, the surface excesses $n_{e,j}$ are found from

$$n_{e,j} = \int_0^\infty [n_j(x) - n_{g,j}] \, dx, \tag{55}$$

where $n_j = z_j n$ $(j = 1, \ldots, M)$, the total molar excess is

$$n_e = \sum_{j=1}^n n_{e,j} = \int_0^\infty [n(x) - n_g] \, dx, \tag{56}$$

and the molar fractions of the components in the adsorbate are simply

$$z_{a,j} = \frac{n_{e,j}}{n_e} \tag{57}$$

In practice, the potential theory of single-component adsorption is used in the form of the homogeneous approximation (51). The value of μ_d is interpreted as an effective chemical potential μ_a of the adsorbate and n_a as an effective adsorbate density. Thus, in addition to the dependence $\varepsilon(h)$, a correlation is required for $n_a(\mu_a)$ [more generally $n_a(\mu_a, T)$]. In many cases, only the temperature dependence of n_a is taken into account. However, as shown in Ref. 101, at extremely high effective pressures developed close to a surface, neglecting the pressure compressibility may lead to errors around 10%.

In a general multicomponent case, introduction of homogeneous approximation becomes difficult. Prior to pressure compressibility, compositional changes must be taken into account and a complete equation of state is required for adsorbate [e.g., in the form of $n_a(\mu_a^j, T)$]. The chemical potentials at the dew point are not preassigned functions, as in the single-component case, but they must be determined from a full system of phase equilibrium equations. This system is not easier to solve than system (54) of the segregation equations. To the best of our knowledge, the only exception available in the literature is that of the adsorbate obeying Raoult's law [102,103]. However, this approach is rather restrictive and it breaks the connection between the properties of the bulk phases and those of the adsorbate. As a result, some cases (e.g., where there is a continuous transition from gas to liquid) may be extremely difficult to correlate [104].

In the following, we will study the case of the segregated adsorbate determined by Eqs. (54)–(57).

C. Numerical Algorithms

If the bulk phase is not ideal, solution of the system of equations (54)–(57) must be performed by numerical calculations. Let us outline some numerical methods, which may be used for evaluation of the adsorption in the framework of the multicomponent potential theory.

Segregation in the potential field of the adsorbent is, to some extent, similar to segregation in the gravitational field. Extensive calculations of gravitational segregation of hydrocarbon mixtures accounting for phase transitions were performed in connection with the problem of hydrocarbon distribution in petroleum reservoirs [105,106]. Two different approaches to such calculations can be identified. According to one approach, system (57) is differentiated with respect to x and transformed to a system of differential equations, which are solved by conventional methods. The second approach consists in direct solution of system (57) for a sequence of values of x, so that thermodynamic variables obtained at a previous value of x are used as initial approximations for calculations at a new x. We will describe the second method because it is more robust than the first and because the speed of calculations is not usually a limiting factor. In cases of segregation in a restricted volume, modifications of these numerical algorithms developed in Ref. 107 should be used.

The algorithm for segregation calculations is easy to construct for mixtures which do not exhibit phase transitions. Introducing the fugacity coefficients ϕ^i and using the potential $\varepsilon = \varepsilon^1$ instead of x for parametrizing the segregated mixture, we transform Eq. (57) into the following form, which is convenient for calculations:

$$z_i(\varepsilon) = \frac{z_{g,i}\phi_g^i P_g \exp[\varepsilon^i(\varepsilon)/RT]}{\phi^i(P(\varepsilon), \mathbf{z}(\varepsilon))P(\varepsilon)} \tag{58}$$

Summing up these equations, we obtain

$$\sum_{i=1}^{M} \frac{z_{g,i}\phi_g^i P_g \exp[\varepsilon^i(\varepsilon)/RT]}{\phi^i(P(\varepsilon), \mathbf{z}(\varepsilon))P(\varepsilon)} - 1 = 0 \tag{59}$$

At each value of ε, the system of equations (58) can be solved by Newton–Raphson iterations with regard to $z_i(\varepsilon)$ and $P(\varepsilon)$. Another method of solution is that of subsequent substitutions based on Eq. (59), which plays the role of the "Rachford–Rice" equation for the problem considered [108]. At each iteration of the last method, Eq. (59) is solved at fixed values of $z_j(\varepsilon)$ with regard to the pressure $P(\varepsilon)$. Then, new values of $z_j(\varepsilon)$ are found from Eq. (58), on whose right-hand side, old $z_j(\varepsilon)$ are used. These iterations are repeated until convergence.

The calculations become more complicated for mixtures exhibiting phase transitions at high values of adsorption potentials. The phase transition may be located by performing a stability test at each value of ε [68]. This method is robust, although time-consuming.

If the presence of a phase transition is predetermined by the nature of the mixture, the transition may be located by direct calculations, without involving the procedure of stability checking. Let us use X_i and Y_i to denote molar fractions in liquid and gas around the phase contact and P_d to denote the value of the dew-point pressure at this contact. These values and the phase contact potential ε_d obey the system of equations similar to Eq. (58):

$$X_i = \frac{z_{g,i}\phi_g^i P_g \exp(\varepsilon^i(\varepsilon_d)/RT)}{\phi_l^i(P_d, \mathbf{X})P_d}, \qquad Y_i = \frac{z_{g,i}\phi_g^i P_g \exp(\varepsilon^i(\varepsilon_d)/RT)}{\phi_l^i(P_d, \mathbf{Y})P_d} \tag{60}$$

Again, summing up these equations, we obtain the system

$$\sum_{i=1}^{M} \frac{z_{g,i}\phi_g^i P_g \exp(\varepsilon^i(\varepsilon_d)/RT)}{\phi_l^i(P_d, \mathbf{X})P_d} = 1, \qquad \sum_{i=1}^{M} \frac{z_{g,i}\phi_g^i P_g \exp(\varepsilon^i(\varepsilon_d)/RT)}{\phi_l^i(P_d, \mathbf{Y})P_d} = 1 \tag{61}$$

At fixed X_i and Y_i, Eq. (61) represents the system of the two equations from which the values of P_d and ε_d are to be found at each iteration. This system is solved by the Newton–Raphson method. Then, the new estimates for X_i and Y_i are determined from Eq. (60). The iterations are repeated until convergence, and the divergence indicates absence of the phase transition. For the initial iteration, Wilson's estimation of K factors may be used to express fugacities in the liquid phase [6,108].

The algorithms just described are not much more complicated and time-consuming than the algorithms involved in ordinary calculations of phase equilibria. Thus, the potential theory is comparable with the general spreading pressure approach (RAST) by the amount of the computational efforts involved. Moreover, calculations in the potential theory consume much less time at the stage of adjustment of the theory to experimental data because, unlike several common versions of the RAST, binary interaction parameters do not require such an adjustment.

D. Testing the Potential Theory

Let us discuss the results of sample calculations based on the described algorithms. Adsorption of nitrogen, hydrocarbons, and binary hydrocarbon mixtures in microporous media is considered. The hydrocarbons and their mixtures are described by the Soave–Redlich–Kwong equation of state (SRK EoS) with the common quadratic mixing rule for the energy parameter a and the linear mixing rule for the covolume b, and zero binary interaction coefficients k_{ij} [6]. It may be checked that the introduction of nonzero k_{ij} values cannot significantly improve the correlation.

It is known that the SRK EoS gives reasonable estimates of volumetric properties at low pressures, but these estimates may become inaccurate at high pressures. To improve volumetric predictions, Pèneloux volume corrections are used [109].

The adsorption potentials are described by the Dubinin equation (53). The standard Dubinin–Radushkevich form with $\beta = 2$ is used for adsorption on carbon, whereas adsorption on silica gel is modeled by use of both $\beta = 2$ and $\beta = 1$. Characteristic energies of adsorption ε_i^0 and the adsorbent capacity x_0 are the only adjustable parameters in the calculations. They are fitted on the basis of the adsorption data of single components and then used for calculations of the adsorption of binary mixtures. In cases in which adsorption of several adsorbates on the same adsorbent is considered, the same value of x_0 is used for all the substances (unless otherwise specified).

Let us first verify that the complete potential theory of adsorption, without introduction of the homogeneous approximation (51) and with the same EoS for the bulk phase and for the adsorbate, is able to correlate experimental data on single-component adsorption. Figure 18 presents an example of successful correlation of adsorption isotherms of methane and nitrogen on activated carbon at elevated pressures. Experimental data are taken from Refs. 103 (methane at 298 K) and 104 (nitrogen at 210 K). Although the estimate obtained on the basis of the

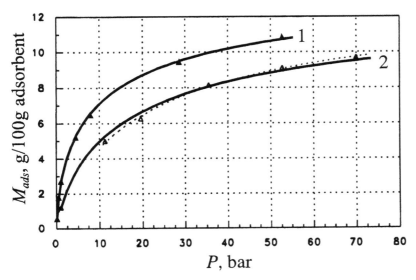

FIG. 18 Adsorption of nitrogen (1) and methane (2) on activated carbon. The dotted line shows the correlation for methane neglecting one low-pressure point. (From Ref. 92.)

potential theory is in good agreement with the data for methane, the correlation is much improved if one low-pressure point (1 bar) is excluded, as shown by the dotted line in Figure 18. The corresponding relative error decreases from 3% to 1.05%. The relative error for nitrogen is 1.9%.

Figure 19 gives an example of a not-so-successful correlation for adsorption of hydrocarbons on silica gel. The adsorption isotherms of four hydrocarbons at 298 K were correlated with the experimental data by Lewis et al. [102]. Solid lines in Fig. 19 correspond to the Dubinin potential (53) with $\beta = 1$ and dotted lines correspond to $\beta = 2$. For the three substances propylene, ethylene, and ethane, predictions on the basis of the potential with $\beta = 1$ are definitely better. The agreement with experimental data of predictions with $\beta = 2$ is also reasonable if the small number of fitted parameters is taken into account: The relative error does not exceed 5% (as for ethylene). However, in this case, the three best-fit curves exhibit similar systematic deviations from experimental data. Such deviations are not observed if the potential with $\beta = 1$ is used. The adsorption isotherms of the fourth substance, propane, show an opposite tendency: Predictions on the basis of the original DR potential are significantly better than those on the basis of the potential with $\beta = 1$ (in the first case, the relative error is only 1.3%). This example shows that the potential theory has limitations when adsorption on silica gels is considered.

Another correlation based on experimental data at 298 K from Ref. 102 shows that the potential theory is able to predict the effect of decrease of the amount adsorbed at increasing pressure (curve 2 for propane in Fig. 20). This effect was originally explained by "factors such as those involved in retrograde condensation" [102]. A more detailed explanation may be given on the basis of Eq. (50) for the surface excess [110]. Ethylene and propane form liquids at high pressures developed under the action of the adsorption forces. Thus, the density $n(x)$ in Eq. (50) is attributed to the liquid or near-liquid state, and n_g is the molar density of a less compressed gas phase. With rising pressure, the value of n_g may increase faster than $n(x)$, so that the difference $n(x) - n_g$ decreases.

One of the fundamental assumptions of the potential theory is that the adsorption forces are independent of the temperature. To check this assumption, the data of Lewis et al. [102] on

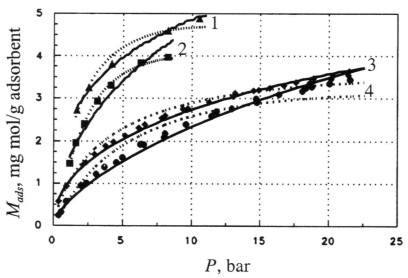

FIG. 19 Adsorption of propylene (1), propane (2), ethylene (3), and ethane (4) on silica gel. (From Ref. 92.)

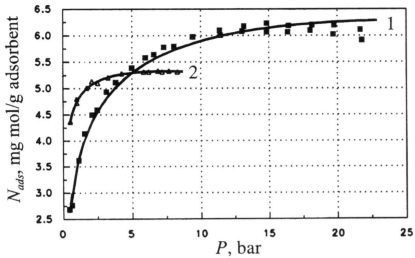

FIG. 20   Adsorption of ethylene (1) and propane (2) on Columbia G carbon. (From Ref. 92.)

ethylene adsorption on silica gel in a wide temperature range were used. The data at different temperatures were correlated with the same invariable parameters of the Dubinin potential (Fig. 21).

As in Fig. 19, dotted lines in Fig. 21 correspond to predictions on the basis of the Dubinin potential with $\beta = 2$, and solid lines correspond to the potential with $\beta = 1$. Both predictions are in reasonable agreement with experimental data, with an average relative error around 4.5%. Whereas the predictions with $\beta = 1$ are better for the temperatures 25°C and 40°C, the DR potential better predicts adsorption at 0°C. Again, this confirms the conclusion about limitations of the potential theory applied to adsorption on silica gels.

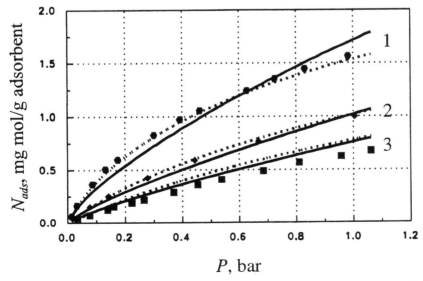

FIG. 21   Adsorption of ethylene on silica gel at different temperatures: (1) 0°C, (2) 25°C, and (3) 40°C. (From Ref. 92.)

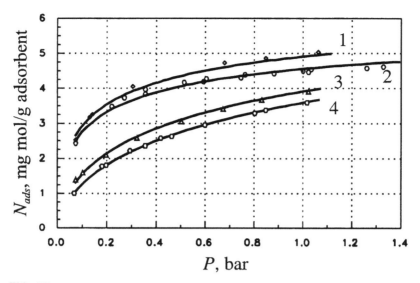

FIG. 22 Adsorption of propylene (1), propane (2), ethane (3), and ethylene (4) on Columbia G activated carbon. (From Ref. 92.)

The validity and the usability of the potential theory for prediction of the multicomponent adsorption are tested on the data on binary adsorption. The first set of data is taken from Ref. 102. The data on adsorption of four different hydrocarbons on the same adsorbate at 298 K were correlated by means of five fitting parameters: four characteristic adsorption energies ε_0^i and one adsorbate capacity x_0 common to all four substances. The results of correlation presented in Fig. 22 indicate good agreement of calculated curves with experimental data.

The characteristic adsorption energies of propane and ethylene obtained from the previous correlation were used for estimation of the adsorption of their binary mixtures at different pressures (Fig. 23). The three calculated curves show right tendencies with regard to dependence on molar fraction and are arranged in the same order as experimental data at different pressures. A relatively significant difference between experimental data and calculated curves is probably explained by the fact that the SRK EoS used in the calculations is not precise with regard to mixture density at high pressures. It should be noted, however, that the accuracy of prediction of the adsorption equilibria on the basis of this EoS is comparable to the accuracy of the modeling bulk-phase equilibria.

The estimation of binary mixture adsorption was much more successful at pressures as low as 0.1 bar (Figs. 24 and 25). The data of Costa et al. [81] on hydrocarbon adsorption on activated carbon at 323 K were used for this correlation.

As in the previous example, the data of individual hydrocarbons were first correlated. Dotted lines in Fig. 24 show the correlation of experimental data on four hydrocarbons with five adjusted parameters, the same as in the previous example, and solid lines show the correlations in which eight fitted parameters are used. In the second case, the DR potentials are characterized by not only different adsorption energies but also by individual capacities x_0 of the adsorbent with regard to different substances. Evidently, the second correlation gives much better results than the first. However, both correlations show similar, good results at low pressures. They almost coincide for methane. Adsorption energies of individual hydrocarbons are similar for the two correlations: The maximum relative difference between them does not exceed 5%. The

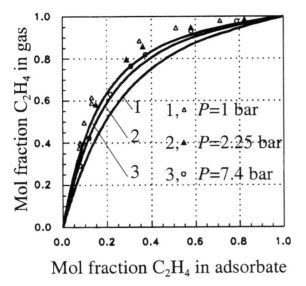

FIG. 23 Adsorption of ethylene–propane mixture on Columbia G activated carbon. (From Ref. 92.)

deviation of different values of x_0 (including the common value obtained in the first correlation) from an average is within 14%.

Fig. 25 shows adsorption curves for five binary mixtures of the four components above. The characteristic adsorption energies are taken from the correlation of individual adsorption isotherms with a common value of x_0. A good quantitative agreement between experimental data and calculated curves is observed, except for the methane–ethane mixture, for which the agreement is slightly worse. The qualitative agreement is also excellent. For example, the ethane–propylene and methane–ethylene curves almost coincide and intersect at low molar

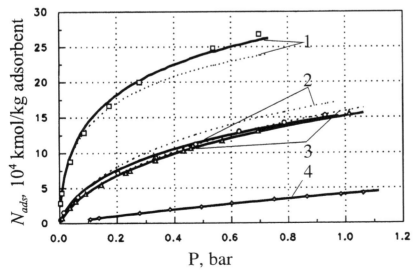

FIG. 24 Adsorption of propylene (1), ethylene (2), ethane (3), and methane (4) on activated carbon. (From Ref. 92.)

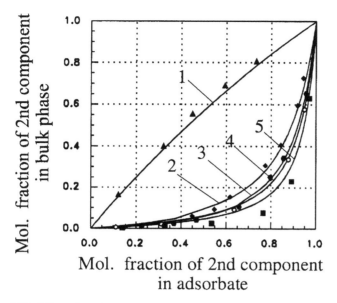

Mol. fraction of 2nd component in bulk phase

Mol. fraction of 2nd component in adsorbate

FIG. 25 Adsorption of binary mixtures on activated carbon. Mixtures of ethane–ethylene (1), ethane–propylene (2), ethylene–propylene (3), methane–ethylene (4), and methane–ethane (5). (From Ref. 92.)

fractions of the first component in a mixture. This observation from calculations is in agreement with experimental data.

The calculated binary adsorption curves are rather sensitive to the adsorption energies. Variation of these energies within 5% of their values has led to the curves which, although qualitatively reasonable, lie far away from experimental points.

It is worthwhile to compare the predictions of the potential adsorption theory with those of the ideal adsorption solution theory, the IAST, described in Section IV.A. Both theories use the same number of fitted parameters. Analysis of experimental data considered on the basis of the IAST has been performed in the original article [81]. The authors found a large discrepancy between IAST estimates and experimental data. The experimental activity coefficients of different components in binary adsorbates vary from 0.412 to 1.054, whereas the IAST assumes their values to be unity. In order to improve the correlations, the Costa et al. [81] had to go from IAST to real adsorption solution theory, using the Wilson equation with additional binary interaction parameters for the adsorbate. This significantly increased the number of fitted parameters and decreased the predictivity of the correlation.

The previous examples show an important limitation for the potential adsorption theory. This limitation is that in order to achieve reasonable quantitative agreement with experiments, the correlation must be based on quite precise models for adsorption potentials and volumetric properties of the adsorbed mixture. An equation of state chosen for correlation must produce good volumetric estimates at high pressures, even if adsorption at low and moderate pressures is considered.

Other limitations of the theory follow from the way it is formulated. For example, the theory must be corrected for adsorption in mesoporous and macroporous media by taking into account the phenomenon of capillary condensation, as described earlier.

The theory cannot be applied in cases of chemosorption and, in general, in cases where the molecules of the adsorbate or adsorbent are strongly modified under the action of adsorption forces. In these cases, parameters in the EoS should be changed and adsorption potentials may become dependent on the mixture composition. The cases in which the adsorbate is significantly changed by the adsorption forces require separate consideration.

The theory developed must be applied with care if the adsorbate is disintegrated into separate molecules occupying fixed sites in the adsorbent. This may happen in extremely microporous media (like those formed by cavities in crystalline structures) or on structured heterogeneous surfaces. The reason for such a limitation is that adsorbate is described by means of a thermodynamic equation of state. This assumes that adsorbate behaves more or less like a continuous medium and that interactions between its molecules are the same as in a bulk phase. On the contrary, a pore in a microporous medium may often contain only few molecules of the adsorbate, so that statistical fluctuations from a thermodynamic equilibrium state become significant. It may be expected that errors associated with these fluctuations cancel out in disordered media. However, a more ordered adsorbate might possess some properties introduced by the order. In this case, the above-described localized approach, the semi-analytical approach by Aranovich and Donohue [111], or direct molecular simulations become more suitable.

E. Potential Theory of Adsorption and Thermodynamics of Surface Excesses

1. Statement of the Problem

The structure of the potential theory is such that it avoids introduction of the equation of state for the adsorbate and of the concept of spreading pressure. Instead, the surface excesses $n_{e,i}$ are directly calculated from Eqs. (54) and (55) as functions of thermodynamic conditions in the bulk phase:

$$n_{e,i} = n_{e,i}(\mu_g^j, T) \tag{62}$$

Such a situation is characteristic of many adsorption models (e.g., of the models based on the localized approach). Thermodynamic consistency of such models with the general theory of the excess phase described in Section II should be verified separately.

One of the consistency requirements follows from the Gibbs–Duhem equation (9). Because $d\Phi$ is a full differential, the excesses $n_{e,i}$ must obey the Maxwell relations:

$$\frac{\partial n_{e,i}(\mu^j, T)}{\partial \mu^k} = \frac{\partial n_{e,k}(\mu^j, T)}{\partial \mu^i} \tag{63}$$

Here, we have taken into account equality of the chemical potentials in the bulk phase and in the adsorbate [Eq. (12)], according to which, μ_g^j may be considered as effective chemical potentials of the adsorbate.

The Maxwell conditions (63) are not only necessary but also sufficient conditions for consistency of an adsorption model. Indeed, they mean that the differential form $n_{e,i}\,d\mu^i$ at constant temperature represents a complete differential $d\phi$. The spreading pressure may then be defined as $\Phi = \phi + \Phi_0(T)$, and the integration constant $\Phi_0(T)$ may be found, for example, from the condition that Φ tends to zero at low pressures. Then, the whole set of thermodynamic relations for the surface phase may be reconstructed from the dependence $\Phi(\mu^j, T)$. For example,

$$s_e = \frac{\partial \Phi}{\partial T}, \qquad n_{e,i} = \frac{\partial \Phi}{\partial \mu^i}, \qquad u_e = -\Phi + Ts_e + \mu^i n_{e,i}, \qquad \text{etc.} \tag{64}$$

In practice, expressing the surface excesses in the form of Eq. (62) together with the subsequent checking of the Maxwell relations (63) is not convenient and, to the best of our knowledge, has not been used. With regard to the potential theory, a more convenient way is introducing the direct expressions for the spreading pressure and then checking the relations similar to Eq. (64).

2. Different Equations for Spreading Pressure

A simple expression for the spreading pressure Φ is given by the known formula

$$\Phi = \int_0^\infty (P(x) - P_g)\,dx \tag{65}$$

Adamson [3] attributes this formula to Gibbs, although we did not find it in the original article [2]. Let us check that this formula is consistent with the Gibbs–Duhem relation (9).

In order to prove that Eq. (65) really expresses the value of the spreading pressure, we will use the Gibbs–Duhem equation for a bulk phase under constant temperature:

$$dP = n_i\,d\mu^i \tag{66}$$

Here, P may be either the pressure of the bulk phase P_g or the pressure $P(x)$ at the distance x from the surface. Correspondingly, we have

$$n_{g,i} = \frac{\partial P_g}{\partial \mu_g^i}, \qquad n_i(x) = \frac{\partial P(x)}{\partial \mu^i(x)} \tag{67}$$

Here, differentiation is performed under constant x. Additionally, in view of Eq. (54), we have at constant x

$$\frac{\partial}{\partial \mu_g} = \frac{\partial}{\partial \mu(x)} \tag{68}$$

Differentiating Eq. (65) by μ_g^i and taking into account Eqs. (67) and (68), we find that

$$\frac{\partial \Phi}{\partial \mu_g^i} = \int_0^\infty (n_i(x) - n_{g,i})\,dx = n_{e,i}$$

Thus, the Gibbs–Duhem equation (9) is obeyed at constant temperature with the value of Φ given by Eq. (65) and the surface excesses determined by Eq. (55). Therefore, Eq. (65) provides an expression for spreading pressure within a temperature-dependent constant (cf. the considerations of the previous subsection). This constant may be set to zero, because in the absence of chemosorption, the right-hand side of Eq. (65) obviously tends to zero in the zero-pressure limit.

The fact that Eq. (65) may be considered as an EoS for the adsorbate in the form of $\Phi(\mu^j, T)$ becomes especially clear if we represent the pressure as the function of the chemical potentials and the temperature and, by means of Eq. (54), transform Eq. (65) to the form

$$\Phi(\mu^j, T) = \int_0^\infty (P(\mu^j + \varepsilon^j(x), T) - P(\mu^j, T))\,dx \tag{69}$$

Let us now obtain another form of the EoS for the adsorbate. Integrating Eq. (65) by parts and taking into account Eq. (66), we obtain

$$\Phi = x(P(x) - P_g)\Big|_0^\infty - \int_0^\infty x n_i\,d\mu^i(x) \tag{70}$$

Assume that the first term in the last equation may be eliminated (this assumption is discussed below). Because, according to Eq. (54), the values of $\mu^i(x)$ differ from the surface potentials $\varepsilon^i(x)$ only by the constants μ_g^i, Eq. (70) may be rewritten as

$$\Phi = -\int_0^\infty xn_i \, d\varepsilon^i(x) \tag{71}$$

Equation (71) also represents the EoS for adsorbate in the form of $\Phi(\mu^j, T)$ because the partial molar densities n_i may also be represented in the form of $n_i(\mu^j + \varepsilon^i(x), T)$, similar to the pressure in Eq. (69). For practical calculations, it may be convenient to substitute the integration variable x by one of the potentials (e.g., $\varepsilon = \varepsilon^1$), as it has been performed in Section IV.C.

Let us now discuss the assumptions, in which the first term in Eq. (70) may be neglected. These assumptions are that 1 the value of $P(x) - P_g$ tends to infinity slower than x^{-1} as $x \to 0$ and (2) this value tends to zero faster than $x^{-1-\Delta}$ as $x \to \infty$.

It can be shown that the difference $P(x) - P_g$ tends to zero or to infinity with the same speed as the surface potentials, provided that the molar volumes $v^i = \partial\mu^i/\partial P$ remain finite. Hence, assumption 1 is valid for the DR potential (53), although this potential tends to infinity as $x^{-1/2}$ and produces infinite pressure at $x = 0$. Assumption 1 is also valid for the generalized FHH potential (52), but it is invalid for the original FHH potential. The spreading pressure for the original FHH potential is equal to infinity, which requires its renormalization described in Section IV.F. Both potentials obey assumption 2.

Singular behavior of the potentials is closely related to some inconsistencies of the potential theory, which have been analyzed in the literature [112–115]. If the potential has a singularity at $x = 0$, it normally results in the fact that the potential theory is not consistent with the Henry law at low pressures.

The preceding considerations show that the potential theory is consistent with the general statements of the surface thermodynamics. An opposite question may be whether any dependence $\Phi(\mu^j, T)$ can be obtained by a proper choice of the potentials $\varepsilon^j(x)$ and by application of the potential theory. The answer is negative. The reason is that in the general adsorption thermodynamics, the properties of the excess phase are, in principle, independent of the properties of the bulk phase, whereas in the potential theory, they are closely related.

As an example, let us consider a hypothetical case in which the bulk phase remains an ideal-gas mixture at any pressure. A combination of the ideal-gas law and the ideal mixing rules is equivalent to the following connection between the partial molar densities and the chemical potentials:

$$n_j = \frac{P_0}{RT} \exp\left(\frac{\mu^j}{RT}\right) \tag{72}$$

where P_0 is a reference pressure. Substitution into Eq. (71) results in

$$\Phi = W^j n_j, \qquad W^j(T) = \int_0^\infty x \exp\left(\frac{\varepsilon^j}{RT}\right) d\varepsilon^j(x) \tag{73}$$

Furthermore, because each molar density of an ideal mixture depends on the only chemical potential, the second Eq. (64) gives

$$n_{e,i} = \frac{W^i}{RT} n_i$$

Thus, in the potential theory, the ideal bulk phase assumes ideal adsorbate. This is not necessarily the case for the general surface excess thermodynamics.

3. Examples for Single-Component Adsorption

Equation (73) simplified by the application of Eq. (52) may be used for fast derivation of expressions for spreading pressures of single-component adsorbates in homogeneous approximation. Introducing the potential as an integration variable, we obtain

$$
\Phi = n_l \int_{\varepsilon(0)}^{\mu_d - \mu_g} x(\varepsilon)\, d\varepsilon \tag{74}
$$

Application of the last formula to the DR potential (53) gives the equation, which has been known for an ideal gas as a bulk phase [115]:

$$
\Phi_{\mathrm{DR}} = \frac{n_l x_0 \varepsilon_0 \sqrt{\pi}}{2}\left[1 - \mathrm{erf}\left(\frac{\mu_d - \mu_g}{\varepsilon_0}\right)\right] \tag{75}
$$

For the general FHH potential, in the form of Eq. (52), application of Eq. (65) gives

$$
\Phi_{\mathrm{FHH}} = \Phi_0 + an_g\left[\mu_d - \mu_g - \frac{\alpha \varepsilon_0^{1/\alpha}}{\alpha - 1}(\mu_d - \mu_g)^{(\alpha-1)/\alpha}\right], \qquad \Phi_0 = \frac{an_l \varepsilon_0}{\alpha - 1} \tag{76}
$$

Transition from the generalized FHH potential to the original one is achieved when the size a tends to zero and the product $a\varepsilon_0^{1/\alpha}$ simultaneously converges to some value C. However, at this limit, the constant Φ_0 tends to infinity. The fact that spreading pressure for the conventional FHH potential is infinite is due to singular behavior of this potential at $x \to 0$. In order to overcome this singularity, we note that the values of surface excesses obtained by application of the second Eq. (64) do not depend on a constant in the expression for spreading pressure. This constant contributes only to the energy of the surface layer. Let us treat the value of Φ_0 in Eq. (76) as an independent constant.

Then, tending $a \to 0$ and $a\varepsilon_0^{1/\alpha} \to C$, we obtain the following expression for the spreading pressure:

$$
\Phi_{\mathrm{0FHH}} = \Phi_0 - \frac{C\alpha n_l}{\alpha - 1}(\mu_d - \mu_g)^{(\alpha-1)/\alpha} \tag{77}
$$

Sometimes, it is convenient not to express the spreading pressure in terms of the chemical potentials, but of the surface excesses. This may be done in order to draw an analogy between the equations of state used for adsorbates and for bulk phases [9]. Transition from $\Phi(\mu_g)$ to $\Phi(n_e)$ is performed by the use of Eq. (52). For the DR potential, we obtain

$$
\Phi_{\mathrm{DR}}(n_e) = \frac{n_l x_0 \varepsilon_0 \sqrt{\pi}}{2}\left[1 - \mathrm{erf}\left(\sqrt{\ln \frac{n_l x_0}{n_e}}\right)\right]
$$

The spreading pressure for the generalized FHH potential is

$$
\Phi_{\mathrm{FHH}}(n_e) = \Phi_0 + \frac{\varepsilon_0 (an_l)^{\alpha+1}}{(n_e + an_l)^{\alpha}} - \frac{\alpha \varepsilon_0 (an_l)^{\alpha}}{(\alpha - 1)(n_e + an_l)^{\alpha-1}}
$$

The last formula is much simplified for the original FHH potential:

$$
\Phi_{\mathrm{0FHH}} = \Phi_0 - \frac{\alpha (Cn_l)^{\alpha}}{(\alpha - 1)n_e^{\alpha-1}}
$$

Neither of these equations of state coincides with a spreading pressure EoS obtained by analogy with EoS for bulk substances (such as Henry's law, a cubic EoS, etc. [6]). Moreover, only an EoS based on the generalized FHH potential obeys virial expansion and, in particular, Henry's law at $n_e \to 0$. The original FHH potential produces an EoS for spreading pressure, which diverges at this limit, whereas an EoS based on the DR potential behaves as $n_e / \sqrt{-\ln n_e}$. This shows that homogeneous approximation in the potential theory has limitations when applied at low pressures [112,113] (although successful applications of the Dubinin–Radushkevich potential at low surface coverages are known [8]).

4. Potential Theory and IAST

The ideal adsorption solution theory described in Section IV.A is the simplest approach to multicomponent adsorption from the point of view of the general thermodynamic theory of the surface phase. The IAST is comparable with potential adsorption theory by predictability. Both theories need the correlation of experimental data for pure components in order to estimate adsorption of mixtures. However, in general, the predictions of the two theories are different, as illustrated in Section IV.D. Let us analyze assumptions on which the two theories may become similar.

If the activities γ^i of the adsorbed species are set to unity, Eq. (42) for the potentials μ^i of the adsorbate is transformed to

$$\mu^i = RT \ln F^{0,i}(\Phi, T) z_{a,i} \tag{78}$$

Expressing the molar fractions $z_{a,i}$ from the last equation and summing them up, we obtain the EoS for the adsorbate in the implicit form $H(\Phi, \mu^j, T) = 0$:

$$\sum_{i=1}^{M} \frac{1}{F^{0,i}(\Phi, T)} \exp\left(\frac{\mu^i}{RT}\right) - 1 = 0$$

An explicit dependence $\Phi(\mu^j, T)$ may be obtained for adsorption at low coverages, where the dependencies $F^{0,i}(\Phi, T)$ may be taken to be linear with regard to the spreading pressure. It should be noted that in the original article by Myers and Prausnitz [80], the dependencies of individual component fugacities on spreading pressure are close to being linear. In this linear case, $F^{0,i}(T) = W^i(T)\Phi$ and

$$\Phi(\mu^j, T) = \sum_{i=1}^{M} W^i(T) \exp\left(\frac{\mu^i}{RT}\right) \tag{79}$$

Differentiating Eq. (79) by μ^j, we find that for the linear IAST,

$$n_{e,i} = \frac{W^i}{RT} \exp\left(\frac{\mu^i}{RT}\right) \tag{80}$$

Comparison of Eqs. (79) and (80) with Eqs. (72) and (73) shows that the IAST is consistent with the potential theory if the bulk phase is an ideal-gas mixture for all the pressures $P(x)$ arising under the action of the potential field of the adsorbent. This may happen in the case of low-pressure adsorption if the adsorption potentials are rather weak in most of the pore volume of the adsorbate.

Apart from this case in which the two adsorption theories become close, other possible cases are, probably, rather exotic. In general, the two theories are significantly different, which has been demonstrated earlier.

F. The Asymptotic Adsorption Equation

Direct application of the potential theory, based on Eqs. (54) and (55), involves certain computational difficulties, which increase when the bulk phase approaches a dew point. Adsorption in a neighborhood of the dew point is especially difficult for experimental study and modeling [3]. In the framework of the multicomponent potential theory, an analytical tool for such a study may be developed. This tool, the asymptotic adsorption equation (AAE), is similar to the multicomponent Kelvin equation described in Section III.F. It connects the thickness of the adsorbed film with the thermodynamic conditions at a dew point.

The AAE may be derived similarly to the Kelvin equation, on the basis of geometrical considerations in the spaces of different thermodynamic variables. Figure 26a presents the adsorption path and the relevant points in the P–z phase diagram (cf. Fig. 4); Fig. 26b presents the absorption path in the space of the chemical potentials of the mixture at constant temperature (cf. Figs. 5 and 6). Point g in Figure 26 corresponds to the bulk pressure P_g and the composition $\mathbf{z}_g = (z_{g,1}, \ldots, z_{g,M})$. Point D is the dew point for the composition $\mathbf{z}_g$; the corresponding dew-point pressure is P_D. In Fig. 26a, this point belongs to the gas branch of the phase envelope. The corresponding point on the liquid branch is $L = (P_D, \mathbf{z}_L)$. In the space of chemical potentials both points belong to the phase transition surface where $\mu_g^i = \mu_l^i$, and we have $L = D$ (Fig. 26b).

The adsorption curve $\mathbf{M} = (P(x), \mathbf{z}(x))$ determined by Eq. (54) starts from the point g corresponding to the bulk pressure and composition at $x = \infty$. It intersects the phase envelope at the point $d = (P_d, \mathbf{z}_d)$ and continues from $l = (P_d, \mathbf{z}_l)$ from the liquid branch. In the space of chemical potentials, $d = l$. For a point g which is close to the critical point, the adsorption curve can go around the phase envelope, and a smooth transition from gas to liquid is observed (curve $\mathbf{M}'$ in Fig. 26a).

The thickness h of the adsorbed film is determined as a value of x at which $\mathbf{M}$ meets the phase envelope. By conventional flash calculations, we find the dew-point pressure P_D and the equilibrium liquid composition $\mathbf{z}_L$, which are supposed to be known. In the asymptotics considered, all of the distances between the points g, d, and D are assumed to be small, and we neglect the second-order terms with regard to them.

Let us find the vector $\mathbf{v}_D$ orthogonal to the hypersurface of the phase transition at point D in the space of the chemical potentials. The Gibbs–Duhem relations (32) at constant temperature, on the "gas" and on the "liquid" sides of the phase transition, may be expressed as follows:

$$n_{gD,i}\, d\mu_{gD}^i = dP, \qquad n_{L,i}\, d\mu_L^i = dP \tag{81}$$

Subtracting them and noting that $\mu_{gD}^i = \mu_L^i = \mu_D^i$, we find

$$(n_{gD,i} - n_{L,i})\, d\mu_D^i = 0 \tag{82}$$

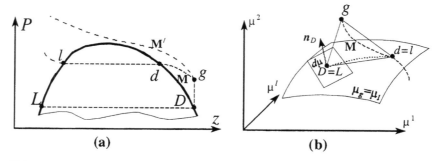

FIG. 26 Graphical representation of the adsorption curve (a) in the pressure–composition diagram and (b) in the space of chemical potentials. (From Ref. 94.)

Equation (82) is interpreted as a definition for the tangent plane $d\mu$ to the hypersurface of the phase transition, in the standard form of $(v_D \cdot d\mu) = 0$, where

$$v_D = (n_{gD,1} - n_{L,1}, \ldots, n_{gD,M} - n_{L,M}) \tag{83}$$

The point d, where the adsorption curve $\mathbf{M}$ meets the phase transition boundary, may be found from the vector equality

$$\overrightarrow{dg} = \overrightarrow{Dg} - \overrightarrow{Dd}$$

We multiply this equality by v_D and note that the scalar product $(v_D \cdot \overrightarrow{Dd})$ is of second order with regard to a characteristic linear size of the triangle gDd. This can be found geometrically from Fig. 26b (by considering the decline of the vector which connects the two points on the surface from the tangent plane). Within the second-order values, we obtain

$$(v_D \cdot \overrightarrow{dg}) = (v_D \cdot \overrightarrow{Dg})$$

Let us transform the last geometrical equation to the form of a thermodynamic relation. It follows from Eq. (54) that the coordinates of the vector $\overrightarrow{dg}$ are $\varepsilon^i(h)$. Substituting the coordinates of $\mathbf{v}_D$ from Eq. (83), we obtain

$$\sum_{i=1}^{M}(n_{gD,i} - n_{L,i})\varepsilon^i(h) = \sum_{i=1}^{M}(n_{gD,i} - n_{L,i})(\mu_D^i - \mu_g^i) \tag{84}$$

Equation (84) is the most general form of the AEE for the value of h. To use it, conventional thermodynamic calculations for determination of the dew-point pressure P_D for given thermodynamic conditions in the gas phase $(P_g, T, \mathbf{z}_g)$ are first performed. In these calculations, the equilibrium composition $\mathbf{z}_L$ is found. Then, the partial molar fractions $n_{gD,i}$ and $n_{L,i}$ are evaluated at the pressure P_D. Finally, Eq. (84) is solved as a transcendent equation for h.

The AAE (62) is fully similar to the Kelvin equation (37) for multicomponent capillary condensation. Both equations have similar properties and ranges of applicability. In particular, due to approximations in its derivation, Eq. (84) is valid asymptotically, close to a dew point of the mixture. A degree of proximity to the dew point, at which the equation gives a reasonable approximation, must be studied numerically. However, it is clear beforehand that the accuracy of the AAE is high if the phase transition surface in the chemical potential space is close to being planar, at least, in a vicinity of the points D and d. On the other hand, Eq. (84) loses its validity close to the critical point.

Similarity between the AAE (84) and the multicomponent Kelvin equation (37) is due to the fact that an expression appearing on the left-hand side of Eq. (84) is, simply, the expression for capillary pressure given by Eq. (37). A possible form of Eq. (84) is thus

$$\sum_{i=1}^{M}(n_{gD,i} - n_{L,i})\varepsilon^i(h) = P_c \tag{84a}$$

The last equation makes it possible to simultaneously study and compare the effects of capillary forces and adsorption in macroporous and mesoporous media. For simplicity, let us assume that all of the pores are cylindrical and all the surface potentials are proportional: $\varepsilon^i(x) = d^i\varepsilon(x)$. The last assumption is valid, for example, for the DR potentials (53) with the same β and for the general FHH potentials (52) with different ε_0, but the same a and α. Combining Eq. (84a) with the Young–Laplace equation (23), we find the simple relation between the two main characteristics of the adsorption and the capillary condensation, the thickness of the adsorbed film h and the capillary radius r_c:

$$r_c = \frac{\gamma}{\varepsilon(h)}, \qquad \gamma = -2\sigma \cos\theta \left(\sum_{i=1}^{M} d^i(n_{gD,i} - n_{L,i}) \right)^{-1}$$

Another consequence of representing the AAE in the form of (84a) is that its right-hand side may be expressed in different forms, like Eqs. (38)–(40), and thus relate the thickness of the adsorbed film to the relative pressure χ and to the mixed volume v_{gl}. These three forms of the AAE are asymptotically equivalent, however, their predictions may differ. We will refer to Eq. (84a) with the right-hand side in the form of Eq. (38) as the "linear" form of the AAE, in the form of Eq. (39)—as the "Kelvin" form—and in the form of Eq. (40)—as the "logarithmic" form.

An extended study of the applicability of the AAE is given in Refs. 92–94 and 116. As an example, we will analyze the adsorption of methane (1)–n-butane (2) at 277.6 K. Thermodynamic calculations are performed on the basis of the SRK EoS with Pèneloux volume corrections and on the basis of the modified FHH potentials. The values of ε_0^i for these potentials are taken from the original Halsey equation: $\varepsilon_0^i = 5RT$. Both a^i and b^i are of the order of the effective thickness of a monolayer (see the discussion in Refs. 3 and 117 for details). This thickness has been reported for a few substances (for nitrogen, if a cubic packing is assumed, it is 0.4 nm). For the present sample calculations, the adsorption potentials are chosen with $a^1 = b^1 = 0.4$ nm and $\alpha^1 = 3$ for methane, and with $a^1 = 0.4$ nm, $b^1 = 0.3$ nm, and $\alpha^2 = 2.5$ for butane.

Figure 27 presents the dependence of the thickness h on the relative pressure for the mixtures corresponding to $P_D = 10$ bar (86.2% of methane at the dew point) and 20 bar (92% of methane at the dew point). These dew-point pressures lie in the region of normal condensation, $\chi < 1$. The calculations were performed by the two methods: using the segregation equations (54) and locating the phase transition by the stability analysis, as described in Section IV.C, and on the basis of the general AAE (84). A remarkably good agreement between the two predictions is observed. In the whole region $P_D < 20$ bar and for the values of χ up to 0.1 (which is by no

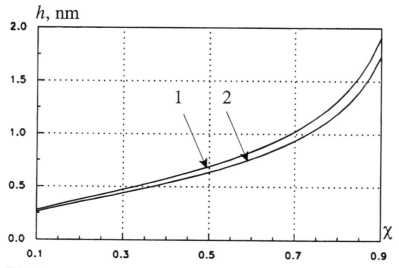

FIG. 27 Thickness of the adsorbed layer versus relative pressure for the mixture C1–nC4 at 277.6 K. 1—$P_D = 20$ bar; 2—$P_D = 10$ bar. (From Ref. 94.)

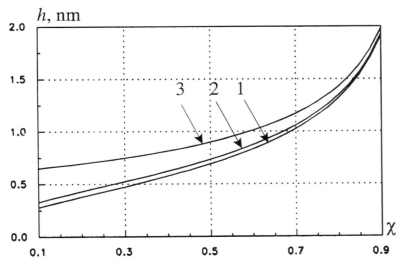

FIG. 28 Thickness of the adsorbed layer versus relative pressure for the mixture C1–nC4 at 277.6 K and $P_D = 20$ bar. 1—numerical calculations = general AAE; 2—"Kelvin" AAE; 3—linear AAE. (From Ref. 24.)

means "asymptotic"), the error did not exceed 1%. The error is lower at higher relative pressures and at lower dew-point pressures (so that it is only 0.4% at $\chi = 0.1$ and $P_D = 10$ bar). Noticeably, h depends much more on the relative pressures than on the dew-point pressures.

The comparison of the different forms of the AAE at $P_D = 20$ bar is shown in Fig. 28. Both the "Kelvin" and the "linear" forms overestimate the values of h. The "Kelvin" form is in relatively good agreement with the numerical calculations, although the general form (84) is much more superior. The relative errors for the "Kelvin" AAE are 0.7% at $\chi = 0.9$, 2.6% at $\chi = 0.75$ and 6% at $\chi = 0.5$. The linear form of the AAE, however, is far from being precise. The relative error is already 3.6% at $\chi = 0.9$.

Figures 29 and 30 present similar results for the retrograde condensation for the dew-point pressures of 80 and 100 bar. The general form of the AAE is again superior in this region. However, the relative errors are much higher. At $\chi = 1.2$, the errors of the general AAE vary from 1.7% ($P_D = 100$ bar) to 2.4% ($P_D = 80$ bar). Other forms of the AAE are closer to the exact dependence in this region than in the region of normal condensation.

At $P_D = 100$ bar, disappearance of the phase transition is observed at $\chi > 1.21$. At higher relative pressures, the adsorption curve goes around the phase envelope, as the curve **M'** in Fig. 26. However, all of the forms of the AAE predict a nonzero thickness of the adsorbed layer for these relative pressures. This problem is similar to the corresponding problem for the Kelvin equation: Both equations predict existence of the Kelvin radius or of an adsorbed film in the regions where they do not exist. To the best of our knowledge, the only way to check the boundary for capillary condensation or for the phase transition in the adsorbate is to perform direct numerical calculations.

In conclusion, on the basis of comparative calculations, the general form of the AAE is recommended for the calculation of the thickness of the adsorbed film. The "capillary" form, although less precise, can be used for a comparison of the action of capillary forces and adsorption. The linear form can be applied if the mixture is very close to its dew point.

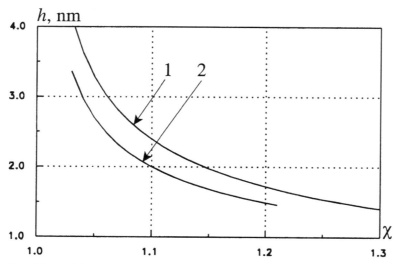

FIG. 29 Thickness of the adsorbed layer versus relative pressure for the mixture C1–nC4 at 277.6 K. 1—$P_D = 80$ bar; 2—$P_D = 100$ bar. (From Ref. 94.)

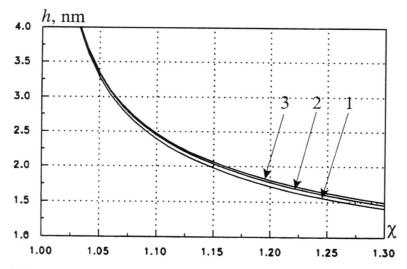

FIG. 30 Thickness of the adsorbed layer for the mixture C1–nC4 for 277.6 K and $P_D = 80$ bar. 1—numerical calculations = "general" AAE; 2—"Kelvin" AAE; 3—"linear" AAE. (From Ref. 94.)

REFERENCES

1. BK Kaul, NH Sweed. In: AL Myers, G. Belfort, eds. Fundamentals of Adsorption. Proceedings of the Engineering Foundation Conference held in Schloss Elmau, Bavaria, West Germany, May 6–11 1983; New York: The Engineering Foundation, 1984, pp 249–258.
2. JW Gibbs. In: The Collected Works. New York: Longman, Green and Co., 1928, Vol. 1, pp 55–171.
3. AW Adamson. Physical Chemistry of Surfaces. New York: Wiley, 1997.
4. R Defay, I Prigogine. Surface Tension and Adsorption (translated by DH Everett). New York: Wiley, 1966.

5. LI Kheifets, AV Neimark. Multiphase Processes in Porous Media. Moscow: Khimja, 1982 (in Russian).
6. JM Prausnitz, RN Lichtenthaler, EG de Azevedo. Molecular Thermodynamics of Fluid-Phase Equilibria. Englewood Cliffs, NJ: Prentice-Hall, 1999.
7. HC Van Ness. I & EC Fundam 8:464, 1969.
8. W Rudzinski, DH Everett. Adsorption of Gases on Heterogeneous Surfaces. San Diego: Academic Press, 1992.
9. DM Ruthven. Principles of Adsorption and Adsorption Processes. New York: Wiley, 1984.
10. S Sircar, AL Myers, AIChE J 17:186, 1971.
11. AV Neimark, PI Ravikovich. Langmuir 13:5184, 1997.
12. CJ Radke, JM Prausnitz. AIChE J 18:761, 1972.
13. AA Al-Rub, R Datta. Fluid Phase Equilibria 147:65, 1998.
14. AA Al-Rub, R Datta. Fluid Phase Equilibria 162:83, 1999.
15. GC Yeh, BV Yeh, ST Schmidt, MS Yeh, AM McCarthy, WJ Celenza. Desalination 81:161, 1991.
16. JS Rowlinson, B Widom. Molecular Theory of Capillarity. Oxford: Clarendon Press, 1984.
17. D Li. Fluid Phase Equilibria 98:13, 1994.
18. AA Shapiro, EH Stenby. Fluid Phase Equilibria 134:87, 1997.
19. JC Melrose. AIChE J 12:986, 1966.
20. YuB Rumer, MSh Ryvkin. Thermodynamics, Statistical Physics and Kinetics. Moscow: Mir, 1980.
21. KS Udell. 1982 California Regional Meeting of the SPE, San Francisco, 1982, paper SPE 10779.
22. PG Bedrikovetsky. Mathematical Theory of Oil and Gas Recovery. London: Kluwer Academic, 1993.
23. PG Bedrikovetsky, VB Mats, AA Shapiro. 5th Petroleum Conference and Exhibition, Abu Dhabi, 1992, paper SPE 24483.
24. K Potsch. 5th Conference on the Mathematics of Oil Recovery, Leoben, Austria, 1996.
25. AA Shapiro, EH Stenby. Fluid Phase Equilibria, 178:17, 2001.
26. L Boruvka, AW Neumann. J Chem Phys 66:5464, 1977.
27. VG Baidakov, VP Skripov, AM Kaverin, KV Khvostov. Sov Phys Dokl 26:959, 1981.
28. NR Morrow, CC Harris. SPE J 15 March 1965.
29. M Jørgensen, EH Stenby. In: HJ De Haan, ed. New Developments in Improved Oil Recovery. Geological Society Special Publication 84. Washington, DC: Geological Society, 1995, p 89.
30. JAP Coutinho, M Jørgensen, EH Stenby. Petrol Sci Eng 12:201, 1995.
31. NV Churaev. Liquid and Vapor Flows in Porous Bodies. Surface Phenomena. Amsterdam: Gordon and Breach Science, 2000.
32. R Lenormand, C Zarcone. 59th Annual Technical Conference and Exhibition of the Society of Petroleum Engineers of AIME, Houston, TX, 1984, paper SPE 13264.
33. A Winter. Fourteenth Symposium on Thermophysical Properties, Boulder, CO, 2000.
34. DH Everett, WI Whitton. Trans Faraday Soc 48:749, 1952.
35. DH Everett, FW Smith. Trans Faraday Soc 50:187, 1954.
36. FAL Dullien. Porous Media. Fluid Transport and Pore Structure. San Diego: Academic Press, 1992.
37. DH Everett. Trans Faraday Soc 50:1977, 1954.
38. DH Everett. Trans Faraday Soc 51:1551, 1955.
39. PG Bedrikovetsky. Advanced Waterflooding, 5-Day Training Course, Technical University of Denmark, Lyngby, Denmark, 1999.
40. AI Brusilovsky, SPE Reservoir Eng, 117, February 1992.
41. CJ King. Separation Processes. Chemical Engineering Series. New York: McGraw-Hill.
42. AA Shapiro, EH Stenby. 1996 SPE European Petroleum Conference, Milan, 1996, paper SPE 36922.
43. A Romero-Martinez, A Trejo. J Thermophys 19:1605, 1998.
44. AH Pefolsky. J Chem Eng Data 11:394, 1966.
45. H Schonhorn. J Chem Eng Data 12:524, 1976.
46. KS Pedersen, Aa Fredenslund, P Thomassen. Properties of Oils and Natural Gases. Houston, TX: Gulf Publishing, 1989.
47. RC Reid, JM Prausnitz, BE Poling. The Properties of Gases and Liquids. 4th ed. New York: McGraw-Hill, 1988.
48. JR Brock, BB Bird. AIChE J 1:174, 1955.

49. P Li, P-S Ma, J-G Dai, W Cao. Fluid Phase Equilibria 118:13, 1996.
50. DB MacLeod. Trans Farday Soc 19:38, 1923.
51. S Sugden. J Chem Soc 32, 1924.
52. CF Weinaug, DL Katz. Ind Eng Chem. 35:239, 1943.
53. AS Danesh, AY Dandekar, AC Todd, R Sarcar. 66th Annual Technical Conference and Exhibition of the Society of Petroleum Engineers, Dallas, TX, 1991, paper SPE 22710.
54. R Amin, TN Smith. Fluid Phase Equilibria 142:231, 1998.
55. ST Lee, MCH Chien. SPE/DOE Fourth Symposium on EOR, Tulsa, OK, 1984, paper SPE/DOE 12643.
56. J Escobedo, GA Mansoori. AIChE J 42:1425, 1996.
57. J Escobedo, GA Mansoori. AIChE J 44:2324, 1998.
58. Lord Rayleigh, Philos Mag 33:208, 1982.
59. JD Van der Waals. Z Phys Chem 13:657, 1894.
60. JW Cahn, JE Hilliard. J Chem Phys 28:258, 1958.
61. M Sahimi, BN Taylor. J Chem Phys 95:6749, 1991.
62. Y-X Zuo, EH Stenby. Fluid Phase Equilibria 132:139, 1997.
63. Y-X Zuo, EH Stenby. J Colloid Interf Sci 182:126, 1996.
64. M Modell, RC Reid. Thermodynamics and Its Applications. Englewood Cliffs, NJ: Prentice-Hall, 1974.
65. BL Beggle, M Modell, RC Reid. AIChE J 20:1194, 1974.
66. RC Reid, BL Beggle. AIChE J 23:726, 1977.
67. P Glansdorff, I Prigogine. Thermodynamic Theory of Structure, Stability and Fluctuations. London: Wiley–Interscience, 1971.
68. ML Michelsen. Fluid Phase Equilibria 9:1, 1982.
69. RB Griffiths, JC Wheeler. Phys. Rev A 2:1047, 1970.
70. L Mistura. Physica 104A:181, 1980.
71. T Peston, I Stewart. Catastrophe Theory and Its Applications. London: Pitman, 1978.
72. W Thomson (Lord Kelvin). Proc Roy Soc Edinburgh 7:63, 1870.
73. SJ Gregg, KSV Sing. Adsorption Surface and Porosity. London: Academic Press, 1967.
74. JC Melrose. SPE Reservoir Eng 3:913 (1989).
75. JC Melrose, JR Dixon, JE Mallinson. SPE paper 22690, 1994.
76. Y-O Shin, J Simandl. Fluid Phase Equilibria 166:79, 1999.
77. J Motrovic. Chem Eng Sci 55:2265, 2000.
78. DH Everett. In: AL Myers, G Belfort, eds. Fundamentals of Adsorption. Proceedings of the Engineering Foundation Conference held in Schloss Elmau, Bavaria, West Germany, May 6–11 1983; New York: The Engineering Foundation, 1984, pp 1–22.
79. JH de Boer. The Dynamic Character of Adsorption. London: Oxford University Press, 1953.
80. AL Myers, JM Prausnitz. AIChE 11:121, 1965.
81. E Costa, G Galleja, C Marrón, A Jiménez, J Pau. J Chem Eng Data 34:156, 1989.
82. J Dunne, AL Myers. Chem Eng Sci 49:2941, 1994.
83. M Jaroniec, R Madey. Physical Adsorption on Heterogeneous Solids. Amsterdam: Elsevier, 1988.
84. S Ross, JP Olivier. On Physical Adsorption. New York: Interscience, 1964.
85. I Langmuir. J Am Chem Soc 40:163, 1918.
86. M Jaroniec, J Oscik, A Derylo. Monatsh Chem 112:175, 1981.
87. C-H Wang, BJ Hwang. Chem Eng Sci 55:4311, 2000.
88. AST Chiang, CK Lee, FY Wu. AIChE J 42:2155, 1988.
89. AL Myers. AIChE J 29:691, 1983.
90. S Sircar, AIChE J 41:1135, 1995.
91. M Polanyi. Verh Dtsch Phys Ges 16:1012, 1914.
92. AA Shapiro, EH Stenby. J Colloid Interf Sci 201:146, 1998.
93. AA Shapiro, EH Stenby. J Colloid Interf Sci 206:546, 1998.
94. AA Shapiro, EH Stenby. Fluid Phase Equilibria 158–160:565, 1999.
95. B Rangarajan, CT Lira, R Subramanian. AIChE J 41:838, 1995.

96. TL Hill. In: WG Frankenburg, EK Rideal, VI Komarewsky, eds. Advances in Catalysis, Volume IV. New York: Academic Press, 1952, p 211.
97. J Frenkel. Kinetic Theory of Liquids. Oxford: Clarendon Press, 1946.
98. MM Dubinin. Carbon 21:359, 1983.
99. MM Dubinin, VA Astakhov. Adv Chem 102:69, 1971.
100. F Stoeckli. Carbon 36:363, 1998.
101. AV Neimark. J Colloid Interf Sci 165:91, 1994.
102. WK Lewis, ER Gilliand, B Chertow, WP Cadogan. Ind Eng Chem 42:1319, 1950.
103. RJ Grant, M Manes. Ind Eng Chem Fundam 3:221, 1964.
104. WH Cook, D Basmadjian. Can J Chem Eng 42:146, 1964.
105. K Knudsen. Phase Equilibria and Transport of Multiphase Systems. PhD thesis, Technical University of Denmark, Lyngby, Denmark, 1992.
106. AA Shapiro, EH Stenby. Entropie 217:55, 1999.
107. S Haldórsson, EH Stenby. Fluid Phase Equilibria 175:175, 2000.
108. ML Michelsen, JM Mollerup. Thermodynamic Models, Fundamentals and Computational Aspects. Course Notes for the PhD Course. Chemical Engineering Department, Technical University of Denmark, Lyngby, Denmark, 1999.
109. A. Pèneloux, E Rauzy, R. Frèze. Fluid Phase Equilibria 8:7, 1982.
110. PG Menon. Chem Rev 68:277, 1968.
111. GL Aranovich, MD Donohue. Carbon 38:701, 2000.
112. O Talu, AL Myers. AIChE J 34:1887, 1988.
113. AL Myers. In: AE Rodrigues, MD LeVan, D Tondeur, eds. Adsorption: Science and Technology. Dordrecht: Kluwer Academic, 1989, p 15.
114. J Tóth. J Colloid Interf Sci 167:224, 1994.
115. N Sundaram. J Colloid Interf Sci 163:299, 1994.
116. AA Shapiro, EH Stenby. 6th Conference on Fundamentals of Adsorption (FOA6), Presquille-De-Giens, France, 1998.
117. T Allen. Particle Size Measurement. London: Chapman & Hall, 1991.

7

Rare-Gas Adsorption

ANGEL MULERO and FRANCISCO CUADROS   Universidad de Extremadura, Badajoz, Spain

I. INTRODUCTION

The study of the adsorption of rare gases on solid surfaces has led to a great advance in the theoretical development needed to understand physical adsorption (or physisorption) phenomena (for general definitions, terminology, and symbols used in physisorption or other surface processes, see Refs. 1–3). The lack of chemical reactions and the simplicity of the molecular properties of the rare gases permits one to consider simple models in order to reproduce their behavior in adsorption phenomena. Also, the experimental results obtained for the adsorption of rare gases on simple solid surfaces, such as graphite, have been used as a reference in the development of very precise temperature control systems, as well as in the study of the most important characteristics of the adsorption of fluids onto solids. Moreover, the discovery of nonsimple phase diagrams for rare gases adsorbed on a homogeneous graphite surface has generated a large number of new experimental trends and theoretical models.

In this chapter, we consider the three ways to obtain thermodynamic properties of interest in adsorption problems: experimental methods, theoretical methods, and computer simulations. As will be described in the following, the experimental methods are numerous and a great number of results have been obtained for rare gases. Although these macroscopic properties can be interpreted on the basis of suitable phenomenological theories, the difficulties in relating experimental results and theoretical predictions have been overcome, in great part, by using statistical mechanics approximations and by performing computer simulations of the behavior of the adsorbed systems. Connections between the microscopic and macroscopic properties can then be made, giving a strong basis for the interpretation of experimental results and for the development of new approaches. In particular, a large number of both theoretical and computer simulation studies have been developed in order to interpret the experimental results, the most successful being those based on the so-called "two-dimensional" approximation, in which the adsorbate is considered as a two-dimensional fluid, the molecules of which interact through a simple intermolecular potential such as the Lennard–Jones one.

A subject of intense interest since the 1960s has been the properties of monolayer films of simple gases physisorbed on the basal plane of graphite. The large number of experimental results, of theories which have been developed and applied, and of computer simulation studies, mainly for rare gases as adsorbates, are available in many reviews [4–15]. In this chapter, the most important results are collected and summarized in tables when possible. In particular, the main aim of this chapter can be considered to be the application of the two-dimensional

433

Lennard–Jones system to model the behavior of adsorbed rare gases by using the statistical mechanics theory proposed by Steele [16–19].

More specifically, the main purposes of the present chapter are the following:

To sum up the main experimental results about rare-gas adsorption. Obviously, the knowledge of these results must serve as the reference to test the validity of the approximations and theories.

To show the main aspects of the phase diagram for rare gases adsorbed on graphite. The richness of the phases that are found and the differences between the phase diagrams for each rare gas have been the subject of many experimental, theoretical, and computer simulation studies.

To introduce statistical mechanics in the correct way through a suitable and complete theory of physical adsorption. In particular, the statistical mechanics theory developed by Steele [16–19] is known to be the most appropriate for rare gases onto well-defined surfaces.

To consider the two-dimensional Lennard–Jones system as the molecular model to describe the behavior of adsorbed rare gases. In particular, expressions are given for its equation of state.

To show the manner in which equations of state can be used together with Steele's theory to give theoretical values for adsorption isotherms, spreading pressures, and isosteric heats of adsorption.

To give numerical or graphical examples of comparison between these theoretical values and the experimental data.

To show the validity of the Cuadros–Mulero equation of state for two-dimensional Lennard–Jones fluids as the simplest expression to be able to use Steele's theory to reproduce experimental results for nonquantum rare gases adsorbed on graphite.

To indicate the utility of considering computer simulations as a way to obtain both microscopic and macroscopic properties and then to serve as the basis for interpreting the experimental results, testing the theories, and modifying previous models, as well as for the development of new theoretical models.

First, some basic prior knowledge of classical adsorption equations and the general definitions used in statistical mechanics is needed. We will not describe the methods used in computer simulations. All of this information can be found in general books [4–8, 13–15, 20–24].

We have made great effort to give a good number of references in which additional information can be found. Moreover, the main results will be summarized in tables, so as to be clearly accessible to the reader. When possible, in these tables a chronological order of results will be followed, in order to indicate the progress of the scientific advances and to bring out the possible discrepancies among interpretations of the data.

The main conclusions and remarks are given at the end of the chapter, summarizing the most important aspects that must be taken into account when the study of the adsorption of rare gases on graphite is tackled.

II. OVERVIEW OF EXPERIMENTAL RESULTS

Fundamental studies of adsorption require the choice of the simplest adsorbates and adsorbents. Thus, uniform substrates (inert and with a surface entirely composed of a single crystallographic plane) are desirable as adsorbents. Several systems are known which closely approximate to this ideal (including the fact of having surface areas that are large enough). These include graphitized carbon blacks and exfoliated graphite, as well as a variety of solids coated with a monolayer or two of a solidified rare-gas or other inert preadsorbents, and even metallic surfaces.

In this section, we shall first summarize the use of rare gases as adsorbates and graphite surfaces as adsorbent, indicating briefly the techniques used and some important results. Also, a subsection will be devoted to indicating the general form of the phase diagram of rare gases adsorbed on graphite.

Some of the first experimental studies of rare gases on homogeneous surfaces were performed by using graphitized carbon as the surface [25–32]. The utility of the graphitized carbon black as adsorbent derives primarily from the ease of preparing and maintaining this material with a large surface area composed almost entirely of basal planes. In addition, theoretical calculations [16–19, 33] are facilitated by the fact that the solid is composed of one kind of atom, present in a crystal lattice of known spacing and symmetry. An extensive review of the properties of the graphite substrate can be found in Ref. 34.

The discovery of exfoliated graphite as a particularly uniform substrate and its subsequent use for vapor-pressure studies permitted a long list of interesting results. The apparent superiority of exfoliated graphite as a simple surface over graphitized carbon black is attributed to its larger surface area (that may reach 30 m^2/g of material), and it has been postulated that the small grain size of graphitized carbon black can produce some capillary condensation. Thus, exfoliated natural graphite is considered to be highly homogeneous and includes some manufactured graphites with commercial names such as "Grafoil" or "Papyex," and it has proved to be especially well suited to structural studies by both neutron and x-ray diffraction.

Rare gases are the obvious ideal adsorbates, mainly for their lack of chemical reactions and because they can be studied with simple models. The adsorption of both argon and krypton on graphitized carbon blacks at liquid-nitrogen temperature (77 K) [25,26] or near it [27,31] was studied in the aforementioned first experiments.

The first significant and systematic experiments on the adsorption of rare gases were performed by Sams et al. [29–30] and Constabaris et al. [32 and references therein] using a high-precision adsorption apparatus. They obtained adsorption isotherms on highly graphitized carbon black P33 (2700°C) for Ne (59–94 K), Ar (90–240 K), Kr (245–307 K), Xe (279–315 K), and H$_2$ and D$_2$ (deuterium) (90–138 K), and methane and tetradeuteromethane. Constabaris et al. also obtained the heats and entropies of adsorption for argon on this carbon black and studied the interactions of these gases and the quantum effects in isotope adsorption. This represented a first step in the development of a theory for the thermodynamic properties of adsorbed films. In fact, the data were used as a test bed for the first models and theories [35,36].

The most important direct experimental methods in physisorption are thermodynamic methods, where adsorption isotherms are obtained (as in the aforementioned studies), and calorimetric methods, based on the determination of isosteric heats or heat capacities. Moreover, other indirect methods such as gas chromatography, diffraction and scattering, and spectroscopic methods have been used to give information about the structure of the adsorbed layers, their orientation with respect to the substrate, and the kinetics of the adsorption.

Physisorption experiments generally require very low temperatures, so that the development of experimental methods have been linked to the development of cryogenic techniques. Also, a high quality is required for the solid substrate. Thus, a great number of significant and accurate experimental results were obtained in the 1970s, due to advances in techniques of preparation of surfaces. This period coincided with the improvement of surface science and vacuum techniques, permitting the study of well-characterized and clean surfaces [15], as was the case of the graphite used as a lubricant.

In the pioneering work of Thomy and Duval [37], very accurate temperature control was employed with an otherwise essentially standard volumetric adsorption apparatus. A detailed experimental study of the adsorption of Kr, Xe, and methane onto exfoliated graphite was then performed. The large number of isotherms (from 77 to 109 K for Kr, from 101 to 118 K for Xe, and from 77 to 84 K for methane) provided the first experimental evidence for the existence of a range of temperatures for which the monolayer of these adsorbates can behave as a two-dimensional (2D) gas, liquid, or solid phase (i.e., with phase transitions similar to the three-dimensional fluids). These results also indicated that the adsorption isotherms of simple molecules on graphite have a more complex form than was generally considered. Moreover, that complexity is not due to the heterogeneity of the surface, but, on the contrary, is most clearly manifested on the most homogeneous surfaces. The isotherms obtained were different from those previously reported in two clear aspects: (1) they showed a greater number of steps, giving an indication of the formation of at least four monomolecular layers (monolayers), and (2) the first step can be seen as formed by two substeps. By studying these two substeps in the monolayer, Thomy and Duval defined and obtained the two-dimensional triple and critical temperatures (see Table 1). Thus, in the range from the 2D triple temperature to the 2D critical temperature, the monolayer apparently behaves as a 2D gas or as a 2D liquid. For temperatures lower than the 2D triple temperature, the monolayer could change directly from a gas to a solid phase. Although evidence for the existence of a possible 2D critical point existed (Table 1), the presence of a triple point was here predicted for the first time.

Following the initial work of Thomy and Duval [37], a great number of articles studying adsorption of rare gases were published [38–46]. Some technical advances derive from these studies, such as the incorporation of highly accurate pressure transducers and computer control, the development of calorimetric techniques, and the application of diffraction or scattering methods to obtain the structure of the phases present in the adsorbate. We shall now summarize

TABLE 1   Critical and Triple Temperature for Rare Gases on Graphite

System	T_c (K)	T_t (K)	References or sources
Argon	68		[27] Prenzlow and Halsey, 1957
	65		[60] McAlpin and Pierotti, 1964
	67		[61] Steele and Karl, 1968
	>65		[37] Thomy and Duval, 1970
	59		[62] Larher and Gilquin, 1979
	55	47	[63] Migone et al., 1984
		49.7	[64] D'Amico et al., 1990
	55	49	[65] Steele, 1996
Xenon	120		[66] Cochrane et al., 1967
	117	99.0	[37] Thomy and Duval, 1970; [67] Thomy et al., 1981; [15] Bruch et al., 1997
	116	100	[65] Steele, 1996
Krypton	82		[26] Ross and Winkler, 1955
	79		[60] McAlpin and Pierotti, 1964
	86.5		[61] Steele and Karl, 1968
	87	77	[37] Thomy and Duval, 1970
	85.3	84.8	[68] Larher, 1974
	None	85	[65] Steele, 1996; [15] Bruch et al., 1997
Neon	<16.1 K		[61] Steele and Karl, 1968
	15.8	13.6	[69] Rapp et al., 1981; [65] Steele, 1996; [15] Bruch et al., 1997

some important general results and techniques. Phase changes will be dealt with separately in Section II.A, because of the richness of phase diagrams, which turned out to be more complex than foreseen by Thomy and Duval [37].

An important contribution to the study of the adsorption of rare gases on graphite and of the corresponding two-dimensional phase transitions was the works of Suzanne et al. [38–41]. The main idea was to combine results for the adsorption isotherms by using Auger spectrometry with low-energy electron diffraction (LEED) to control the arrangement of the surface. It then becomes possible to measure adsorption isotherms at extremely low pressures [10^{-4}–10^{-9} torr (i.e., 10^{-2}–10^{-7} Pa)], extending the range previously available to traditional methods [10^{-5}–100 torr (i.e., 10^{-3}–10^{4} Pa)]. It is also possible to characterize the phase changes simultaneously from both thermodynamic and crystallographic points of view. In particular, results were obtained for Xe on graphite for temperatures from 85 to 102 K [40] (i.e., for temperatures that are below the 2D critical point found by Thomy and Duval [37]) (Table 1). These results therefore complete those obtained by Thomy and Duval [37]. The study of Suzanne et al. also includes experimental results for the kinetics of condensation for the Xe monolayer at 75 and 76 K, by measuring coverage versus time during the formation of the first solid layer [41]. A comparison of the application of LEED and Auger measurements in the study of Kr on graphite is found in Ref. 47.

Another significant advance was represented by the results obtained by Putnam et al. [42–44], who studied, in detail, the physisorption of Kr on graphitized carbon black for temperatures near 100 K (i.e., above the 2D critical point predicted by Thomy and Duval [37]) (Table 1). In particular, three adsorption isotherms were obtained in the 95–109 K temperature range, by using a high-accuracy ultraclean volumetric system. That temperature range had not been studied previously.

The main object of that research was to determine the adsorption-site energy distribution and the surface energy of a heterogeneous solid surface. (Heterogeneous surface means a surface that has more than one type of adsorption site, contrary to a homogeneous surface. This must not be confused with "uniform" surface, in which the barriers to translation across the surface are negligible.) Moreover, their studies were also aimed at finding a correct model (based on a virial approach; see Section III.C) for adsorption on a homogeneous surface and to confirm the existence of a 2D solid–fluid transition in the submonolayer film. For that purpose, Putnam et al. also obtained values for the adsorption isosteric heats (by using the Clausius–Clapeyron equation) for the differential adsorption entropy and for the spreading pressures (using the Gibbs' adsorption equation). Applications of the Frenkel–Halsey–Hill (FHH) equation to the multilayer region and of the virial equation to the monolayer region was also discussed. As will be shown in Section IV.D, the reported data have been used to test the accuracy of some statistical mechanics theories.

The studies of Thomy and Duval [37], Suzanne et al. [38], and Putnam and Fort [42] include the calculation of the isosteric heat from the isotherm data. One of the interesting quantities obtained is the heat of adsorption at zero surface coverage, which can be used to obtain estimates of the well depth of the fluid–solid interaction potential. For the particular case of xenon on graphite, different values were obtained by Sams et al. [29], Thomy and Duval [37], and Suzanne et al. [38]. Piper and Morrison [48] obtained that quantity from an adiabatic calorimetric measurement, finding a value very close to the one obtained by Sams et al. [29] and indicating that the variation in the results might be due to the fact that some of the determinations correspond to different regions of the phase diagram. Finally, Piper and Morrison [48] obtained values for the frequencies of vibration of an adsorbed xenon atom from the calculation of the entropy of the adsorbed monolayer and from information derived from other experiments.

The study of Piper and Morrison [48] is a clear example of the advantages of performing direct calorimetric measurements for isosteric or adsorption heats, instead of calculating it from adsorption isotherms. Really, the importance of having direct calorimetric measurements of adsorption has been accepted for a long time [49]. Nevertheless, the technical difficulties [31] were not overcome until the mid-1970s in the work of Regnier and co-workers.

Regnier et al. [45] performed a direct determination of the heat of adsorption for Kr and Xe on graphite for temperatures below or near the 2D triple point indicated by Thomy and Duval [37] (see Table 1). Their results were in good agreement with the values calculated from adsorption isotherms. In particular, the calorimetric method used permits a continuous introduction of the adsorbate, giving results for a range of very low pressures which cannot be easily obtained from volumetric techniques. Moreover, the combination of volumetric and calorimetric methods was subsequently used to study the structure and density of solidified monolayers of rare gases on graphite [46].

Experimental studies including neon or helium as adsorbates require temperatures lower than 20 K, so that fewer results have been obtained. Thomy et al. [50] performed what is considered to be the first significant experimental study of neon on grafoil, obtaining the adsorption isotherms. The heat capacity was measured by Antoniou [51], with graphite as the adsorbent, whereas specific heat measurements were performed by Huff and Dash [52] and De Souza et al. [53], with grafoil as the adsorbent. Subsequent studies are focused on the phase transitions and will be commented on in Section II.A.

Starting with the pioneering studies of Greyson and Aston [28], experimental results for adsorbed helium at submonolayer coverages include adsorption isotherms, heat capacities, heats of adsorption, and neutron diffraction methods [54–57 and references therein]. Other references to studies of helium on different substrates, as well as an extensive review of the results obtained for helium on grafoil until 1973, can be found in Ref. 55. For data about scattering experiments on solid adsorbed helium, see Refs. 58 and 59. Data focused on phase transition studies will be described in the following section.

The above studies were complemented with a great number of studies devoted to phase transitions for monolayers of rare gases adsorbed on graphite. Before detailing these studies, however, we shall summarize other results obtained with rare gases or mixtures of rare gases on different surfaces or for multilayer adsorption on graphite.

Multilayer adsorption of Xe, Ar, and Ne films on graphite has been studied by high-energy electron diffraction techniques [70 and references therein], finding clear disagreement with the previous theories based on statistical models.

Multilayer adsorption of Xe on graphite was studied by Inaba et al. [71], but the results were criticized by Ser et al. [72], who obtained adsorption isotherms for Xe and Ar on graphite, finding that capillary condensation is often a problem in multilayer adsorption studies and that a careful examination of layering transitions can be particularly useful.

Experimental data for rare gases on heterogeneous surfaces have been also reported. Some of the classical references are summarized in Table 2. Following Heer [80], all of these results can be described by a suitable and simple statistical thermodynamic theory for the localized site model.

References including structure studies for rare-gas monolayers on clean metal surfaces were summarized in Ref. 15. In particular, experimental work for solidified rare gases on metallic substrates have indicated that the monolayers form a regular two-dimensional lattice [81 and references therein].

Much of the experimental and theoretical work on physisorption on solid surfaces was carried out in the subcritical region of the adsorbate. However, as Malbrunot et al. [82] have indicated, there are many important industrial applications involving the adsorption of gases at supercritical temperatures and pressures, where the availability of adsorption data and models is limited. In that work, the authors obtained experimental data for Ar, Ne, Kr, nitrogen, and

TABLE 2 Some Classical Results for Adsorption of Rare Gases on Heterogeneous (Nongraphite) Surfaces

Adsorbates	Adsorbents	References
He	Pyrex	[73] Hobson, 1959
Ar, Kr, and Xe	Zr	[74] Hansen, 1962
Ar, Kr, and Xe	Pyrex	[75] Ricca et al., 1967
Kr and Xe	W and Mo	[76] Ricca and Bellardo, 1967
Kr	Pyrex and Ni	[77] Schram, 1966
Kr and Xe	Pyrex and Mo	[78] Endow and Pasternack, 1966
Kr	NaBr	[79] Fisher and McMillan, 1957

methane on a microporous activated carbon at room temperature (25°C) and pressures up to 650 MPa (corresponding to 6 times the critical temperature and 3.25 times the critical density) using a dielectric method [82].

Both adsorption isotherms and heats of adsorption data for Ar on silicalites [83,84] and on porous surfaces such as zeolites [85] are available, including diffraction studies for the case of silicalite adsorbents [84]. There are also data for adsorption of Kr and Xe on silicalite, which have been compared with computer simulation data [86,87].

Somewhat less attention has been paid to the adsorption of multicomponent systems, although a basic understanding and accurate predictions would be important in practical applications such as separation of gas mixtures. The earliest studies appear to be those of Singleton and Halsey [25] and Prenzlow and Halsey [27], using the preadsorbed layer technique. This technique is particularly suitable when the equilibrium pressures of the two gases involved in the mixture are very different. Thus, results for argon adsorption on an overlayer of xenon preadsorbed on graphitized carbon black were obtained by Prenzlow and Halsey [27]. Adsorption of He on solid Xe, Kr, and Ar has also been studied [7]. In particular, argon–xenon mixtures were studied using x-ray scattering by Bohr et el. [88]. Also, the effect of adding a second component to pure krypton adsorbed on graphite has been considered [89], as well as mixtures of krypton and xenon adsorbed on graphite [90]. Binary mixtures of rare gases on other adsorbents such as Pt(111) have also been studied [91,92].

In addition, other more complex mixtures have been considered. Examples are Kr adsorption isotherms on graphite preplated with sulfur hexafluoride [93] or with cyclohexane [94].

The uses of adsorption of rare gases continue to be a productive field at present. Thus, for example, Widdra et al. [95] presented high-resolution thermal desorption data for xenon from flat and stepped platinum surfaces. These data were compared with a lattice gas model, and the lateral interactions within the layers as well as the strengths of the interactions with the substrate were quantified.

A. Phase Diagrams for Rare Gases on Graphite

The study of phase diagrams, even in the very special case of rare gases on graphite, is interesting from different points of view: first, because of the interesting variety of phenomena and phases that can be found; second, because the construction of a phase diagram is a culmination of the work of various groups that have used different but complementary experimental techniques; third, because it is a practical way to introduce the main features and applications of current theories; fourth, because it is essential to know how the adsorbent behaves in a given temperature or density range and thus be clear as to what their main properties are and also to see how that behavior can be theoretically modeled.

A monolayer of an adsorbed fluid represents a good approximation to a 2D phase of matter. Its properties, in general, are qualitatively distinct from those of any 3D phase of the same substance. The 2D system behavior is very complex, because it sometimes behaves as a phase that is similar to a 3D analog (i.e., gas, liquid, or solid), but, at other times, it behaves as a state which is totally new, lacking a 3D analog. Among these possibilities are commensurate or incommensurate phases. In commensurate phases, the adsorbate has its positional order imposed by the substrate. For example, a commonly observed structure is the so-called $\sqrt{3} \times \sqrt{3}$ one, in which only one-third of the graphite hexagons have a large probability of acquiring an adatom. In the incommensurate phases, an orientational epitaxy is found, in which the overlayer has a distinct lattice spacing but has an orientation determined by, but not identical to, the substrate. The existence of one or the other solid phase depends on the size of the adatom, the temperature, and the layer density or surface coverage [15,65].

As has been indicated by Bruch et al. [15], predicting and systematizing this richness of behavior is one of the challenges to the theorist requiring an accurate assessment of the relevant force laws, a careful analysis of the energy balance of competing structures, and, often, the imagination to conceive of the various possible states of 2D matter.

In this subsection, we survey the main aspects of the phase diagram for rare gases on graphite. The main aim is not to give a broad and detailed explanation of the structure of each of the possible phases (a recent excellent and extensive review of phase diagrams for rare gases and other substances on graphite, metals, and ionic crystals can be found in Ref. 15), but, rather, to indicate the different techniques used in obtaining the phase diagram and to delimit the zone of existence of fluid phases, the thermodynamic behavior of which can be modeled through equations of state, as we will explain in Section IV.C.

As will be detailed in the following, there are two important questions about these phase diagrams: the existence and location of critical or triple points, and the thermodynamic order of the phase transitions, especially for the liquid–solid equilibrium. Following the pioneering work of Thomy and Duval [37], many investigations of phase transitions occurring in physisorbed monolayers on solid surfaces have been made, and some broad reviews can be found in articles and books [9,10,12,15,65,67,96–99].

Although rare gases are adsorbed as bulk fluids, the first proof of the existence of an adsorbed 2D crystalline solid (xenon on graphite) was given in the LEED experiments of Lander and Morrison [100]. As was indicated earlier, the Thomy and Duval [37] experimental results indicated that the adsorbed monolayers of rare gases (and methane) on graphite undergo several phase transitions. The corresponding phase diagram was at first presumed to be entirely analogous to that for a simple uniform three-dimensional system, including the presence of a critical point and a triple point (see Table 1 for data). Nevertheless, only a few years later, Larher [68] suggested that the liquid-phase region was very much smaller than was first thought. Almost immediately, new peculiar transitions were found, such as the so-called "localized–delocalized" transition, now termed the "commensurate–incommensurate" transition [42,46].

A subject of much attention is the nature of the 2D phase transitions. "The consensus seems to be" [65] that such 2D transitions as liquid–vapor and solid–vapor are discontinuous (except at the 2D critical point), whereas the melting behavior is less clear, because discontinuous changes are usually observed for the melting of patches in low-density adsorbed solids, with the possible exception of argon on graphite.

In other words, in contrast to bulk solids, for which melting is always a first-order (discontinuous) process, melting can be continuous even in an idealized two-dimensional system, as first shown by Kosterlitz and Thouless [101] and extended by Halperin and Nelson

[102]. The theoretical mechanism proposed for the melting is that the transformation from a 2D solid to a 2D fluid may proceed through a two-step process, in which the solid phase first melts into a so-called "hexatic" phase (solids possessing sixfold symmetry), which is subsequently destroyed via a second transition at higher temperatures. Thus, commensurate solid films usually display discontinuous order–disorder transitions, presumably because of the additional stability offered by partial or complete registry with the substrate. The melting of argon monolayers on graphite appears not to fit this general pattern however [103].

To summarize all of the information given in the following pages, three kinds of generic monolayer phase diagram can be found for rare gases on graphite [15]:

1. A 2D phase diagram similar to the corresponding 3D rare-gas diagram, including gas, liquid, and solid phases. Examples are Xe, Ne and Ar on graphite, although there are some differences among them.
2. Phase diagrams in which there is only one fluid monolayer phase, so that no liquid–vapor equilibrium is found. These are called "incipient triple point" systems, as in the case of Kr on graphite [104,105], in which the interactions that lead to "commensurate" lattices seem to be very important.
3. Phase diagrams of quantum monolayers, such as He on graphite, with only fluid and solid phases [15,55,106].

In the following subsections, we shall describe some interesting results that have allowed one to construct each of these phase diagrams.

1. Xenon

As indicated earlier, the phase diagram for xenon on graphite is quite analogous to that corresponding to bulk 3D Xe. One important step in the study of rare-gas adsorption was the assumption by McTague et al. [97] that Xe on graphite may be the best representation of a 2D system on a perfectly smooth surface. In fact, most experimental results, which are summarized in Table 3, are in agreement with the existence of both a 2D critical point and a 2D triple point (data in Table 1). The main difference with respect to the 3D case is that the liquid range is considerably narrower: $T_c(3D)/T_t(3D)$ is near 1.8, whereas $T_c(2D)/T_t(2D)$ is near 1.2 [37] (see Table 1). Figure 1, obtained by Fairobent et al. [98], shows the gas–liquid curve obtained from the experimental data of Thomy and Duval [37] and Regnier et al. [46], which are in good agreement.

The volumetric and calorimetric results of Regnier et al. [46], giving the structure and density of the solidified monolayers of Xe on exfoliated graphite at temperatures from 102.1 to 110 K, did not permit one to distinguish a localization–delocalization phase transition. Other interesting aspects are found by studying the melting curve to temperatures above 120 K [15].

Thus, the melting becomes continuous and a commensurate $\sqrt{3} \times \sqrt{3}$ structure can be found [116] before the bilayer forms, by compression at monolayer completion and at a sufficiently low temperature. Moreover, Heiney et al.'s x-ray diffraction results [114] indicate that the melting transitions becomes continuous rather than first order at slightly above the monolayer range. Finally, the thermodynamic data of Gangwar et al. [107] have been interpreted as indicating a "weak" first order transition from 116 to 147 K, whereas the transition becomes continuous from 147 to 155 K, where the formation of a second layer interferes with the results.

In any case, the observations differ on whether the transition from the hexagonal incommensurate solid to the commensurate solid is first order or continuous and also on the value for the maximum temperature at which the commensurate solid occurs [15].

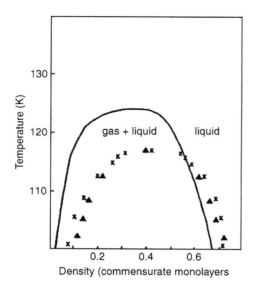

FIG. 1 Vapor–liquid line for the phase diagram for xenon on graphite from Ref. 98. The data points are from Refs. 37 (crosses) and 46 (triangles). The curve was obtained from density functional theory by Fairobent et al. [98].

2. Argon

The monolayer phase diagram for argon on graphite presents gas, liquid, and incommensurate solid phases. A well-defined liquid–gas transition occurs, and no commensurate phase appears. A simple phase diagram is presented in Fig. 2. This was proposed by Migone et al. [63] and is reproduced here mainly to give an idea of the localization of the phases and phase equilibrium lines. As will be explained, these lines are actually not so clearly defined.

The main experimental studies about argon phase transitions are summarized in Table 4. A topic of discussion for a long time has been the value of the 2D critical temperature for argon on graphite [15,62,63,65,96,98] and the main results are given in Table 1. Values of T_c near 55 K and of T_t from 47 to 49.7 K are now commonly accepted [65].

TABLE 3 Experimental Studies of Xenon on Graphite

Method	References
Thermodynamic	[37] Thomy and Duval, 1970; [107] Gangwar et al., 1989
Auger spectrometry and LEED	[38–41] Suzanne et al., 1974, 1975
THEED	[108] Venables et al., 1976; [109] Schabes-Retchkiman and Venables, 1981; [110] Faisal et al., 1986
Thermodynamic and calorimetric	[45–46] Regnier et al., 1975, 1977
Calorimetric	[111] Litzinger and Stewart, 1980; [12] Thomy and Duval, 1982; [48] Piper and Morrison, 1984
X-ray diffraction or scattering	[112] Hammonds et al., 1980; [113] Moncton et al., 1981; [97] McTague et al., 1982; [114] Heiney et al., 1982; [115] Hong et al., 1987; [64] D'Amico et al., 1990

LEED = low-energy electron diffraction; THEED = transmission high-energy electron diffraction.

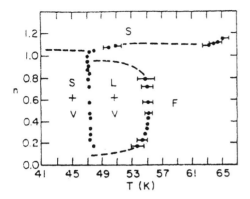

FIG. 2 Phase diagram for argon on graphite proposed by Migone et al. [63] from their heat capacity data. Dashed lines are speculative. (F = fluid, L = liquid, V = vapor, S = solid.)

As an important fact, we note how the interest in investigating phase transitions in adsorbed argon has permitted the development of new techniques, such as the microcalorimetric technique [122,123,128]. This technique permits a continuous recording of the properties as a function of coverage or time. It has been shown that this procedure is very sensitive and very accurate. The existence of a peak on the curve of the isosteric heats for a coverage close to a monolayer and, simultaneously, the presence of a clear substep in the adsorption isotherms were interpreted as evidence for a two-dimensional phase transition of the monolayer, changing from a hypercritical fluid state to a localized one [122,128,129].

Grillet et al. [129] also studied the influence of the use of different solid carbons (with different temperatures of graphitization; in particular, they used Sterling FT-FF carbons with temperatures from 1500 to 2700°C) on the behavior of the adsorbed phase of argon. They showed how the 2D fluid–solid phase change at the completion of the argon monolayer takes place even on the less graphitized carbons, contrary to the case of nitrogen. This phase change had not been observed in the pioneering results of Beebe and Young [49] or Sams et al. [30], due principally to the lack of accuracy and the relatively imperfect graphitization of the carbons used.

Neutron diffraction [120] and LEED [124] methods, where structural and dynamical information was obtained for temperatures up to monolayer melting, provided the experimental evidence indicating that the solid is an incommensurate rotated structure (i.e., a solid structure rotated with respect to the graphite substrate), confirming the theoretical prediction of Novaco

TABLE 4 Experimental Studies of Argon on Graphite

Method	References
Calorimetric	[61] Steele and Karl, 1968; [117] Chung, 1979; [63] Migone et al., 1984
Thermodynamic	[37] Thomy and Duval, 1970; [118,119] Larher, 1978, 1983; [62] Larher and Gilquin, 1979; [103] Zhang and Larese, 1991
Neutron diffraction	[120] Taub et al., 1977; [121] Tiby and Lauter, 1982
Microcalorimetric techniques	[122] Grillet et al., 1977; [123] Rouquerol et al., 1977
LEED experiments	[124] Shaw et al., 1978; [125,126] Shaw and Fain, 1979, 1980
X-ray diffraction	[97] McTague et al., 1982; [127] Nielsen et al., 1987; [64] D'Amico et al., 1990

and McTague [130]. More recent studies of the orientational epitaxy in argon on graphite have been performed by D'Amico et al. [64].

The nature of melting for argon films has also been studied and discussed. Evidence for continuous melting has been obtained from neutron diffraction studies [120–121], from heat capacity measurements [117], and from x-ray diffraction techniques [64,97]. Adsorption isotherms data [62] and LEED results [125,126] indicate a rather sharp transition. Some computer simulations have been interpreted in terms of continuous melting [131], whereas others [132,133] have found a first-order behavior (discontinuous melting). Finally, heat capacity measurement indicates that the melting transition is "weakly" first order, occurring at a triple line temperature [63].

A difference remains between different thermodynamic experiments, indicating two transitions which are closely spaced in temperature [63,103] and, thus, confirming the theoretical ideas of Kosterlitz and Thouless [101] and of Halperin and Nelson [102], and structural experiments indicating only one transition [64,127].

Phase changes for physisorbed argon at 77 K have also been observed when boron nitride, which has a well-known structural similarity with graphite, was used as adsorbent [134]. A 2D solid–fluid transition was found, similar to the case of argon on graphite. Moreover, a second transition occurs near the completion of the second layer, showing a behavior not previously detected in the case of graphite. Grillet and Rouquerol [134] indicate that the results do not allow one to distinguish whether the phase change involves only the second layer or both the first and the second.

3. Neon

In studies of adsorbed neon, it is necessary to take into account that although the quantum effects are small, they are still quantitatively significant. The phase diagram for Ne on graphite includes classical critical and triple points located at 15.8 and 13.6 K, respectively, [69] (see Table 1). Moreover, for temperatures above 10 K, the phase diagram begins to resemble that expected for a 2D Lennard–Jones fluid [15,69], with the presence of a solid incommensurate monolayer for temperatures above 15 K [15]. A commensurate $\sqrt{7} \times \sqrt{7}$ structure is found below 5 K [52].

The main studies of phase changes are summarized in Table 5. A phase diagram for temperatures greater than 8 K is shown in Fig. 3, elaborated by Rapp et al. [69] from calorimetric data.

A remarkable fact is that the condensation of the monolayer was not accessible in the equilibrium experiments, because at the required temperatures, the corresponding 3D pressure is much lower than the experimental sensitivity. For that reason, the critical and triple points are not shown at the established pressure–temperature phase diagram [15,136].

TABLE 5 Experimental Studies of Neon on Graphite

Method	References
Thermodynamic	[50] Thomy et al., 1969
Calorimetric	[51] Antoniou, 1976; [52] Huff and Dash, 1976; [53] De Souza et al., 1978; [69] Rapp et al., 1981
Neutron diffraction	[135] Tiby et al., 1982
LEED	[136] Calisti et al., 1982

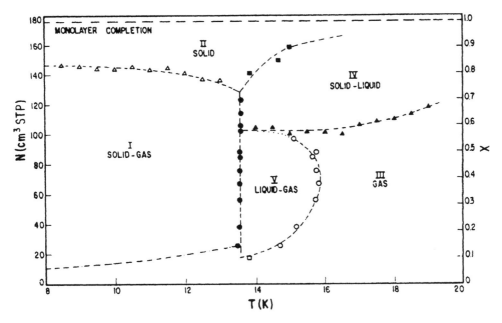

FIG. 3 Phase diagram for neon adsorbed on grafoil from Ref. 69 and based on heat capacity measurements. Dash-dot and dashed lines from Ref. 52.). Note that N is expressed in volume units and that X is the coverage.

Although there have been few studies about the order or phase transitions, indications of a discontinuous melting in monolayer films have been obtained from heat capacity measurements [12].

4. Krypton

Krypton on graphite is one of the most extensively studied systems. As is seen in Table 6, the classical measurements of adsorption isotherms have been supplemented by a quantity of heat capacity data and diffraction experiments. Experiments have shown that the monolayer film can exist as a 2D fluid, as a $\sqrt{3} \times \sqrt{3}$ commensurate solid, and as an incommensurate solid. Figure 4 [98] presents a first approach to the complete monolayer phase diagram. Nevertheless, the ranges of existence of each of these phases and even the presence of new phases lead to a very complex phase diagram, especially beyond the monolayer coverage, as will be explained next.

In the first thermodynamic and calorimetric experimental studies, the existence of a critical point was predicted at temperatures near 86 K and a triple point at temperatures near 85 K (Table 1), leading to a very narrow liquid–vapor equilibrium curve.

The thermodynamic data obtained by Putnam et al. [42–44] for Kr on graphitized carbon black for temperatures near 100 K were in good agreement with those of Thomy and Duval [37] for this range, confirming the existence of a two-dimensional solid–fluid transition in the submonolayer film, but indicating the existence of a second-order commensurate–incommensurate phase transition. Thus, these experimental results indicate that the first-order transition takes place for Kr on graphite at temperatures of about 90–95 K and that a second-order transition occurs only at higher temperatures [42,43]. Thus, there could be a tricritical point, and there would then be no critical or triple point, and hence no separated liquid and vapor phases.

TABLE 6 Experimental Studies of Krypton on Graphite

Method	References
Thermodynamic	[37] Thomy and Duval, 1970; [68] Larher, 1974; [42,43] Putnam and Fort, 1975, 1977; [137] Larher and Terlain, 1980; [138] Suter et al., 1985
Calorimetric	[61] Steele and Karl, 1968; [45,46] Regnier et al., 1975, 1977; [104,105] Butler et al., 1979, 1980; [63] Migone et al., 1984
Diffraction	[139] Chinn and Fain, 1977; [140] Horn et al., 1978; [141] Stephens et al., 1979; [113] Moncton et al., 1981; [142] Nielsen et al., 1981; [109] Schabes-Retchkiman and Venables, 1981; [97] McTague et al., 1982; [143] Specht et al., 1984
LEED and Auger measurements	[47] Kramer and Suzanne, 1976

The commensurate–incommensurate phase transition was also observed for Kr on six different graphites at 77.3 K by both volumetric and calorimetric methods [46]. Moreover, LEED experiments [139] indicated confirmation of the second-order nature of that transition.

In order to find a clearer phase diagram for submonolayer Kr adsorbed on graphite, Butler et al. [104] obtained highly precise results for the heat capacity of that system for temperatures from 65 to 140 K. As indicated by these authors, the heat capacity at fixed coverage seems to be

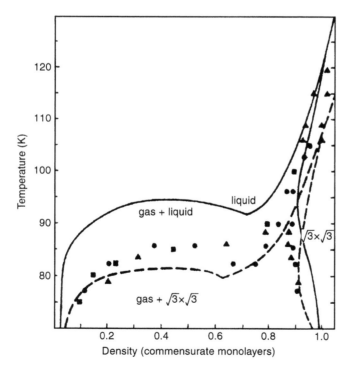

FIG. 4 Phase diagram for krypton on graphite from Ref. 98. Points are experimental data and curves were obtained from the density functional theory applied by Fairobent et al. [98] using different Lennard–Jones parameters: solid line—$\varepsilon/k = 168.5$ K and $\sigma = 0.36$ nm; dashed line—$\varepsilon/k = 145$ K and $\sigma = 0.36$ nm.

a very sensitive property to determine boundaries at fixed temperatures. Butler et al. [104] concludes that the experimental data are consistent with a phase diagram displaying a single two-phase coexistence region of unusual shape, in which the 2D liquid does not exist, as predicted by Putnam and Fort [42,43]. That means that the fluid formed at melting is just above the possible 2D gas–liquid critical point [104].

Butler et al. [104] indicated that although the triple-point interpretation is a possibility, the new interpretation, with an incipient triple point, provides an attractive alternative which does not assume the existence of the unobserved fluid–fluid coexistence region (the authors indicated also that this hypothesis can be applied to the case of nitrogen).

Moreover, the density difference between the fluid and the commensurate solid narrows with increasing temperature and appears to vanish at a tricritical point at a temperature above 100 K (which is a behavior clearly different from the 3D fluid), confirming the proposal of Putnam et al. [44]. Following Butler et al. [105], the tricritical point is located at approximately 115 K on the fluid-commensurate solid-phase boundary (see Fig. 4).

For a long time, there were major uncertainties about the commensurate–incommensurate phase transition, mainly because the experimental studies indicated a different order of that transition at high and low temperatures [113]. These uncertainties have been resolved by the proven existence of a "reentrant fluid" between the two solid regions [15,143]. The final proposed form for the phase diagram is shown in Fig. 5 [144], where the points are experimental data obtained from thermodynamic methods.

The pressure–temperature diagram for coverages near or above a registered monolayer and temperatures greater than 114 K is shown in Fig. 6, as elaborated by Specht et al. [143] on the basis of their theoretical results. As can be seen, different phase transitions were observed [143]:

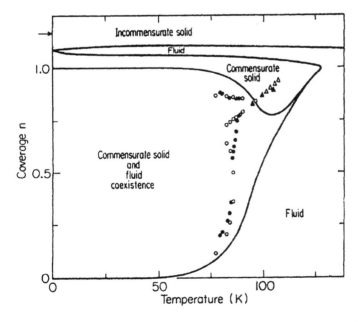

FIG. 5 Phase diagram for krypton on graphite from Ref. 144, where a lattice gas model was used (curves). Points are experimental thermodynamic data from Ref. 37, (open triangles and open circles), 68 (black circles), and 42–44 (black triangles). The arrows indicate the coverage 1.167 of a maximally dense layer, which is saturated with superheavy walls [144].

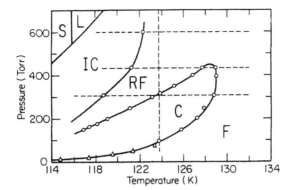

FIG. 6 Pressure–temperature phase diagram for krypton on graphite near the monolayer completion, proposed by Specht et al. [143] from x-ray diffraction data. Dashed lines indicate the scans that were made in experiments. Triangles are from Ref. 137. (F = fluid, C = commensurate fluid, IC = incommensurate solid, RF = reentrant fluid, L = bulk liquid, S = bulk solid.)

1. A first-order commensurate solid to a dense fluid transition for pressures below 57.3 kPa (430 torr) and temperatures below 130 K. The first-order nature of the commensurate Kr melting is in agreement with heat capacity [104] and adsorption isotherm data [37], as well as with x-ray diffraction at high coverages [143]. Horn et al.'s [140] x-ray scattering study indicated, however, a second-order transition. The thermodynamic results of Suter et al. [138] indicated a tricritical point at 117 ± 2 K, where the melting of the commensurate krypton solid becomes continuous.

2. A second-order incommensurate to fluid transition for pressures above 57.3 kPa. That transition was observed for the first time by Butler et al. [105], who performed a heat capacity study for the region between one and two layers.

3. An incommensurate solid to a "reentrant" fluid transition and a "reentrant" fluid to commensurate transition, both being of second order. In particular, the existence of a novel reentrant fluid phase separating the commensurate and incommensurate phases constitutes clearly one of the most spectacular and interesting phenomena observed in the study of rare gases adsorbed on graphite. Another important result is that there is no multicritical (commensurate–incommensurate fluid) point at high temperatures.

In the preceding studies, it is clear that the monolayer phase diagram for Kr on graphite is dominated by effects of the substrate corrugation [15]. These effects could explain the absence of a triple point and of a liquid–gas-phase separation.

5. Helium

Measurements of the properties of adsorbed helium isotopes on graphite surfaces are reported in Table 7. The richness of the monolayer regime for He on graphite was demonstrated for the first time by Bretz et al. [55], who obtained heat capacity data for ^{4}He on grafoil at relatively low temperatures (0.04 K to above 10 K). The results were in qualitative disagreement with previous measurements on other substrates. The new observations indicate that the adsorption on grafoil is a closer approximation to the uniform ideal adsorption than has been typical in work on helium films.

Elgin and Goodstein [56] completed these data for a range of higher temperatures (from 4 to 15 K) and included a vapor-pressure study in order to find a complete description of the phase

TABLE 7 Experimental Studies of Helium on Graphite

Method	References
Calorimetric	[55] Bretz et al., 1973; [145] Hering et al., 1976; [146] Bretz, 1977; [147,106] Greywall, 1990, 1993; [148] Greywall and Busch, 1991
Thermodynamic	[56] Elgin and Goodstein, 1974
Neutron diffraction	[149] Lauter et al., 1987

diagram for the monolayer film. Their results indicate that a lattice gas transition occurs near 3 K. The high-coverage low-temperature region was identified as a 2D solid bounded by a melting-phase transition. They indicated that at very low coverage, the film does not form a low-density solid, but, rather, is dominated by substrate inhomogeneities. At intermediate temperatures and coverages, the film is a fluid, which is accurately described by a quantum virial correction to the 2D ideal gas equation of state.

A complete description of the phase diagram has been made by Greywall [106,147] and Bruch et al. [15]. As for the 3D case, there is no gas–liquid–solid triple point for helium in 2D and there are clear differences between the ^{3}He and ^{4}He isotopes behavior at low temperatures. In both cases, a $\sqrt{3} \times \sqrt{3}$ commensurate phase is found at temperatures below 3 K and an incommensurate one is found as the final solid phase before second-layer condensation. The fluid phase is the largest region for low coverages.

Figure 7 [106] shows the main lines of the phase diagram for adsorbed ^{4}He proposed by Greywall on the basis of its heat capacity data. For densities from 4 and 6.7 atoms/nm^2, the

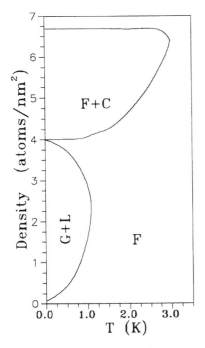

FIG. 7 Phase diagram for ^{4}He on graphite proposed by Greywall [106] and based on heat capacity data. (F = fluid, G = gas, L = liquid, C = commensurate solid.) An incommensurate solid is found for greater values of the density.

commensurate solid coexists with the 2D fluid phase. For lower coverages, a gas–liquid-phase separation near 1 K has been proposed [148], but its presence is not completely clear [15]. Neither are the nature and characteristics of the commensurate–incommensurate transitions very clear, and a two-phase coexistence could be occurring [106].

III. PHYSISORPTION THEORY

Experimental physisorption data have classically been interpreted through empirical and purely phenomenological theories, giving semiempirical isotherm equations that fit the data and that contain a certain number of adjustable parameters with a more or less well-defined physical meaning. The particular mathematical form depends on the range of the isotherm to be fitted and also on the nature of the system [2,5,13,23,24]. Although there are well-known and useful equations for each one of the possible adsorption types separately, a general expression or equation valid for all these types has not yet been defined.

A more general theory requires the phenomenon to be described simultaneously from the macroscopic (experimental) and microscopic (molecular) perspectives. Great progress in the development of the definitions and thermodynamic relationships has been made based on the statistical mechanics, mainly from the work of Steele [16–19,33].

As was indicated in Section I, the study of physisorption of rare gases onto simple surfaces has been the main basis for the development of theories to describe adsorption. This is especially important in the case of statistical mechanics theories, where the behavior of rare gases as simple fluids permits a large number of simplifications. It is well known that migration barriers are very small in this case [150,151], which means that the films are highly mobile, and can thus be adequately represented by statistical mechanics models describing 2D fluids. In addition, theoretical calculations [33] are facilitated by the fact that the solid is composed of one kind of atom. Finally, there is added interest in the great variety of adsorbed phases that may be present.

In this section, we will indicate briefly the use of classical models to describe physisorption, mainly showing the use of rare gases as adsorbent in order to find surface areas of the adsorbent. The classical adsorption equations are well known and have been extensively studied and applied. Descriptions and applications can be found in many books or articles [5,13,23,24].

More detailed explanations will be given of the virial coefficient theory and other statistical mechanics theories, focusing our attention on Steele's theory. Before this, a section will be devoted to explain the general form of both the adsorbate–adsorbate and the adsorbate–adsorbent molecular interaction potentials. An important conclusion will be that the behavior of rare gases on simple surfaces can be described adequately by using the 2D Lennard–Jones system as model. Most of Section IV will then be devoted to the study of the main properties of that model and its application in physisorption of rare gases.

This section finalizes by briefly describing theories that try to reproduce the experimental phase diagrams shown in Section II.A.

Other models used to describe rare-gas adsorption properties will not be considered here but are listed in the following as interesting references. For neon on graphite, both the initial layering transitions [152] and the dynamics of the monolayer solid [153] have been modeled. Also, interesting models of the Ar monolayer have been developed in Refs. 14,131,154, and 155. Models of solid Xe on graphite include cell model approximations [109,156] and models of the dynamics of the commensurate solid [157]. The Hartree and Jastrow approximations have been applied to monolayer solids of Ne and helium, with the melting of the monolayer helium being dealt with specifically [158].

For studies about the low-temperature wetting transitions in rare-gas monolayers adsorbed on graphite, see Ref. 159.

A. Classical Models

In view of the complexity of adsorption phenomena and the different mechanisms contributing to physisorption, none of the current theories of adsorption is capable of providing a mathematical description of an experimental isotherm over its entire range of relative pressure. In practice, two procedures have been used to overcome this problem. The first approach involves the application of various semiempirical isotherm equations with different analytical forms and different ranges of validity. The second procedure makes use of standard adsorption isotherms obtained with selected nonporous reference materials and attempts to explain differences in the isotherm shape in terms of the different mechanisms of physisorption [2]. In particular, physisorption of gases by nonporous solids, in the vast majority of cases, gives rise to a type II isotherm [2,5,13]. From the type II isotherm of a given gas on a particular solid, it is possible, in principle, to derive a value of the monolayer capacity of the solid, which, in turn, can be used to calculate the specific surface of the solid. Here, we describe briefly both procedures and their application to rare-gas adsorption.

The simplest classical model to fit adsorption results consists of the equation of Langmuir [160] for localized adsorption and the equation of Henry for the nonlocalized case. In particular, Henry's equation may be written as

$$N = K_H P \tag{1}$$

where N is the number of molecules adsorbed at pressure P and K_H is Henry's constant. Although both Langmuir's and Henry's equations are valid for adsorbate monolayers considered as ideal gases and are, therefore, applicable only for sufficiently low pressures, they have served as a reference for the development of more complex models or equations. Data on single gas isotherms and corresponding empirical equations for fitting it can be found in Ref. 23.

The preceding models are defined to describe strictly localized or mobile adsorption. Other theoretical developments, such as that by Patrykiejew et al. [161,162], are based on considering a partially mobile monolayer adsorption of simple gases on homogeneous surfaces. One important hypothesis in the derivation of the model is that the mobility of the adsorbed molecules must be independent of the coverage, so that the model must be applied to temperatures near the 2D critical point, where scattering studies have shown that this hypothesis is approximately correct [163]. The main advantage of this model is that it is capable of predicting the shape of an adsorption isotherm from knowledge of its inflection point only (for temperatures near the critical temperature). Moreover, through the dependence of Henry's constant on temperature, the model permits one to predict adsorption isotherms at different temperatures when only a few adsorption isotherms have been measured. In particular, good agreement has been found in the reproduction of experimental monolayer adsorption isotherms of krypton and xenon on graphite [37], as well as other simple gases on homogeneous surfaces, even when the value of the inflection point of the isotherms is taken as the mean of values at different temperatures.

Perhaps the most important uses of gas adsorption are the estimation of the surface area of materials, the definition of the type of porosity, the computation of pore volumes, and the calculation of pore-size distribution. For these purposes, an equation for multilayer adsorption will clearly be most useful. As is well known, the generalization of Langmuir's equation to the multilayer case is the so-called BET (Brunauer–Emmett–Teller) equation or BET model [164].

In the BET model, the adsorption is explicitly assumed to be localized and attention is limited to the forces between the adsorbent and the adsorbate molecules—the "vertical" interactions—whereas the forces between an adsorbate molecule and its neighbors in the same layer—the "horizontal" interactions—are neglected. Both the BET model and the description and use of the BET equation have been extensively described (e.g., Refs. 13 and 165), including its statistical–mechanical derivations [8].

Although based on a model which is admittedly oversimplified and open to criticism on a number of grounds, the BET equation, when applied with discrimination, has proved remarkably successful in evaluating the specific surface from a type II isotherm [13,165].

The main question is whether the surface-area measurement it provides is the "true" or correct value. Although the general consensus seems to be that the answer is in the negative, it has been impossible to evaluate by how much the BET area is off, because there has been no method that could independently provide that elusive "true" value. For example, Monte Carlo computer simulation has been performed in order to reproduce the BET model (i.e., without lateral interactions) and subsequently adding those lateral interactions [166]. The results indicate clearly than the BET equation leads to underestimated monolayer values. The error can be significantly reduced by applying the BET equation to a pressure range below the so-called B-point [2,13,37].

In any case, the IUPAC [2] has established that in spite of the oversimplifications inherent in the BET model, the fit of nitrogen adsorption data to the BET equation (the "nitrogen BET" method) can be used as a reliable measurement of specific surface areas. The IUPAC [2] indicates also that the use of krypton or xenon as adsorbent to determine low surface areas (below $5 \, m^2/g$, say) offers the possibility of greater precision in the measurements but not necessarily greater accuracy in the resulting value of the surface area than could be obtained with nitrogen. Thus, the adsorption of rare gases, such as argon, krypton, and xenon, has also been used to obtain specific surfaces from the BET equation [13,165,167]. Two clearly desirable requirements are satisfied by the rare gases: The adsorbate is chemically inert with respect to the solid and the saturation vapor pressure, P_0, at the working temperature is large enough to allow the accurate measurement of the relative pressure over a reasonably wide range (P/P_0 from 0.001 to 0.5).

In particular, the physical properties of argon, such as boiling point, heat of vaporization, and polarizability, are not far removed from those of nitrogen. Hence, argon adsorption at 77 K has been used for the determination of the surface area. Nevertheless, as 77 K is below the triple point of bulk argon (near 84 K), some doubts about the appropriate reference state have been raised [13].

Following the pioneer work of Beebe et al. [168], the adsorption of krypton at 77 K has also been used for the determination of relatively small surface areas, because its saturation vapor pressure is quite low (P_0 near 267 Pa). Unfortunately, however, there are some complications in the interpretation of the adsorption isotherm. Thus, the working temperature is well below the triple point of bulk Kr (116 K approximately), but if the solid is taken as the reference state, the isotherm shows an unusually sharp upward turn at the high-pressure end. Moreover, the BET plot is frequently not linear due to the kink produced by the phase change from a commensurate to an incommensurate structure.

Xenon is another adsorbate which has a low vapor pressure (about 0.23 Pa) at its usual working temperature of 77 K, which should, in principle, render it suitable for the measurement of low surface areas. In practice, however, its use seems to have been largely restricted to well-defined surfaces.

Recent information can be found in an article by Ismail [165], who has reviewed the use of the BET equation to obtain the cross-sectional area of Ar, Kr, N_2, and O_2 on graphitized carbons, ungraphitized carbons, activated graphitized carbons, and silicas. For Ar on nonporous graphite, the value $0.138 \, nm^2/atom$ is proposed, being $0.157 \, nm^2/atom$ for adsorption on ungraphitized carbons (porous or nonporous) and $0.165 \, \text{nm}^2/\text{atom}$ for adsorption on silica surfaces. A value of $0.214 \, \text{nm}^2/\text{atom}$ is proposed for Kr on graphite. Although the BET plot is not linear for Kr on ungraphitized carbons, the same value can be obtained by using a pressure range of P/P_0 from 0.02 to 0.25 (using the P_0 of solid krypton).

Despite the advantages of using simple expressions to give a phenomenological description of the adsorption processes, some disadvantages are clearly present, at least from a theoretical point of view. Thus, the lack of a strong theoretical basis implies that, for example, extrapolations must be applied with caution. In fact, as Toth [169,170] has clearly shown, some of the best-known equations for monolayer adsorption are incorrect from a thermodynamic standpoint in the limiting cases of small (zero) and large (unity) coverages. These equations thus lead to an incorrect value for the specific surface areas and the isosteric heats of adsorption. That lack of consistence does not affect the BET equation for multilayer adsorption.

In conclusion, a complete and fully predictive theory must be based on a statistical mechanics approach. The main statistical mechanics theories used to study physisorption will be described in the next subsection.

B. Statistical Mechanics of Physisorption

The statistical mechanics description of the properties of fluids consists basically in obtaining a relationship between their macroscopic properties [equation of state (EoS) and thermodynamic properties, for example], based, in general, on empirical or phenomenological laws, and their microscopic properties (molecular structure and motion), based on well-established laws of molecular interaction. In particular, the statistical thermodynamics theory of physisorption leads to a method of relating observable macroscopic properties to the intermolecular potential energies in the system formed by the adsorbate and the adsorbent.

In the study of physisorption using statistical mechanics, different approximations must be used when only low densities are considered or when high densities are also present. Thus, two general approaches, based on statistical mechanics derivations, have been taken to study mobile layers at low densities or coverages [5]. One is to consider the adsorbed phase as a 2D gas and then to equate the chemical potentials of the adsorbed and the gas phases. Although the equation of state for a hypothetical 2D perfect gas can be deduced, the commonest way is to obtain an EoS for the imperfect 2D gas as a virial expansion, where the 2D virial coefficients are related to the intermolecular pair potential.

Another approach is to relate the properties of the adsorbed phase and gas phase by considering all the gas in the adsorption container, which is usually referred to as the gas-surface virial expansion. In this case, a virial expansion for the number of adsorbed molecules in powers of the pressure is used and can be theoretically determined once an intermolecular potential has been clearly defined. Both virial expansions will be explained in the following subsection.

For mobile monolayers at greater coverages, where a large number of virial coefficients are required, other theoretical approaches are commonly used:

1. Approximate 2D equation of states (EoSs), similar to the van der Waals one, can be used [5]. The first theoretical isotherms for mobile adsorption were derived by Volmer [171], by considering the adsorbed phase as a 2D perfect gas. The Volmer model has been widely discussed on the basis of its thermodynamic aspects [172], as well as on the basis of statistical mechanics [5].

    The contributions of lateral interactions were introduced by Hill [173], by proposing the so-called Hill–de Boer isotherm, which is the analog of the van der Waals equation for 2D adsorption layers. Thus, the 2D equation has the familiar van der Waals loops associated with condensation of a 2D gas to a 2D liquid [5], and the theoretical critical point can be obtained [173].

For the particular case of the quantum helium adsorbate, approximate equations of state of the dense monolayer film at temperatures less than or equal to 4 K have been proposed [174,175]. Also, a monolayer equation for Ne on graphite has been developed [176].

There are clear limitations of the approximations used in trying to reproduce experimental results, mainly due to the limited range of validity of the proposed expressions.

2. Cell or lattice models. Cell theories of liquids, such as the Lennard–Jones–Devonshire theory [177] have been applied to adsorption phenomena. For example, cell models including lateral interactions [178] permit the interpretation of experimental isosteric heats in multilayer adsorption [179,180].

One approximation used in theories is to consider that adsorbed atoms are completely localized in registry with the surface lattice of adsorption sites [8], so that 2D Ising lattice model calculations are applicable [181]. However, such a complete localization seems to be more appropriate to chemisorbed systems than to physisorption. The lattice gas theories can thus be considered as realistic only for small adsorbate atoms at low temperatures on an atomically rough solid surface or for problems near critical points, where the long-range nature of the correlations means that assumption of localization on a surface lattice is a less significant approximation [182].

In any case, both the Langmuir and the BET equations can be considered as elementary but useful lattice models, whereas more complex models [181,183] have been used to study the physisorption phase diagrams. Some theoretical results will be described in Section III.F.

3. Application of the Significant Structure theory [60,184]. One of the first attempts to describe physisorption through statistical-mechanics theories was performed by McAlpin and Pierotti [60], who applied the Significant Structure theory to describe the physical adsorption of rare gases onto homogeneous solids.

The adsorbed molecules are assumed to vibrate normal to the surface and the interaction with the solid is presumed to fall off rapidly enough so that only a monolayer exists at low temperatures. The model describes, therefore, only the monolayer adsorption of a 2D fluid on a model solid which is energetically homogeneous and has a plane surface. The partition function is then calculated by considering that there are regions of phase space that give most of the contribution and neglecting the rest of phase space. The molecular partition function for the 2D solidlike structure is obtained for a 2D Einstein crystal, whereas the gas structure is treated as an ideal 2D gas.

The equations derived for the adsorption isotherm, isosteric heat, and critical properties contains no adjustable parameters and are in good agreement [60,184] with preceding experimental results for argon and krypton on graphite (comparison with more recent experimental results has not been performed, as far as we know). Despite these good results, the lack of mathematical rigor in the foundations of the significant structure theory does not permit a correct physical interpretation. Moreover, it is not easy to apply, so its application cannot be generalized or extrapolated to more complex systems. For those reasons, the theory has not been widely used.

4. Steele's theory for monolayer physical adsorption [8,16–19,33]. This statistical mechanics theory is based on considering a classical gas interacting with an inert solid and then to simplify by imposing the so-called "two-dimensional approxima-

tion" and by Fourier expanding the factors in the configurational integral that arise from the periodic gas–solid interaction. It represents the first historical description based on an analytical expression of the gas–solid intermolecular potential [16–18]. It is necessary to use perturbation or integral theories, or suitable equations of state for purely two-dimensional models in order to obtain results that can be compared with experiment. More details will be given in Sections III.D and III.E.

5. Density functional theories. Adsorption can be also studied from a statistical mechanics point of view by considering it as a density inhomogeneity in the fluid arising from the presence of the surface. Supported by computer simulation results, these theories can then be applied to study the density profile and other statistical mechanics properties for fluids near a solid wall. An excellent review of all these approaches has been given by Dickinson and Lal [185]. Density functional theories are also applicable to studying monolayer films [98,186–189] and, in particular, the phase diagram of rare gases on graphite. General aspects of these applications will be summarized in Section III.F. As will be indicated in Section III.D, density functional theories have also been applied to the determination of the interaction potential for rare gases with graphite surfaces [150,151].

6. Others. A number of different statistical mechanics approaches are available, although they have been less used. For example, Heer [80] developed a statistical mechanics theory, based on a localized-site model, which can reproduce adsorption isotherms for rare gases on heterogeneous surfaces at low pressures.

A complete classification and description of statistical thermodynamics and thermodynamics models to describe localized, mobile, and partially mobile monolayer adsorption of gases and liquids (and their mixtures) on solids can be found in the excellent review of Jaroniec et al. [11].

Statistical thermodynamics methods have also been applied by Monson to the study of binary (nonrare) gas monolayers on homogeneous solid surfaces [190].

C. Virial Approximations

The virial coefficient theory has classically been applied to the adsorption of rare gases on graphitized carbon blacks [30,43,191,192] or on another rare gas plated onto graphite [193,194]. The analysis of adsorbed gas data in terms of the second virial coefficient was a historically important step toward achieving quantitative understanding of the behavior of a physically adsorbed monolayer [15,17,43]. For example, the specific heat of low-density helium monolayers on graphite has been described adequately through the quantum second virial coefficient by using both the 2D Lennard–Jones (L-J) and 2D Beck intermolecular potentials [192] and ignoring the effects of the substrate. Those results represented a historical advance toward qualitative and quantitative understanding of the experimental results through the study of purely 2D fluids. As was previously indicated and will be described in the following, the virial theory can be applied to the 2D gas or to the gas–solid system.

In order to apply the virial theory to the 2D gas, it is necessary to define a 2D or spreading pressure, φ, which is related to the adsorption pressure, P, through the Gibbs equation

$$Z = \frac{\varphi}{\rho kT} = \frac{1}{N} \int_0^N N \, d(\ln P) \tag{2}$$

where Z is the 2D compressibility factor, ρ is the two-dimensional density (N/A, A being the area of the surface), and k Boltzmman's constant.

The virial equation of state for a purely 2D system can be obtained by expanding the Z value in powers of the 2D density:

$$Z = 1 + B_{2D}\rho + C_{2D}\rho^2 + D_{2D}\rho^3 + \cdots \tag{3}$$

where B_{2D}, C_{2D}, D_{2D}, and so on are the 2D virial coefficients, which depend on temperature and are defined as integral functions of the intermolecular potential for the fluid. As input, their theoretical calculation requires a specific form for that potential, commonly simple pair potentials describing spherical molecules (such as the Lennard–Jones potential, which leads to the 2D L-J virial EoS). In that simple case, the second virial coefficient, B_{2D}, for example, can be expressed as [5]

$$B_{2D} = -\frac{1}{2} \int_0^\infty \left[\exp\left(-\frac{U(r)}{kT} \right) - 1 \right] 2\pi r \, dr \tag{4}$$

where r is the 2D intermolecular distance and $U(r)$ is the intermolecular pair potential.

The 2D virial equation of state (EoS), Eq. (3), cannot be applied directly to physisorbed gases because of the presence of the surface and, hence, the need to include adsorbate–adsorbent intermolecular interactions. Nevertheless, when the adsorption of a simple system is over a smooth surface, the adsorption potential is extremely deep and narrowly confined to near the substrate, so that the so-called "two-dimensional approximation" can be used. Considering that the adsorbent solid is not perturbed by the presence of the adsorbed film, the real virial coefficients (which must be defined as 3D integrals) can be obtained approximately by using the 2D virial coefficients [5,15,17,43], which are available for simple potentials such as the Lennard-Jones one [195].

As a consequence, by using the 2D approximation and by relating Eqs. (2) and (3), the isotherm equation in the 2D virial approximation is found [17,43]:

$$\ln\left(\frac{N}{P} \right) = \ln K_H - 2B_{2D}\rho - \frac{3}{2}C_{2D}\rho^2 + \cdots \tag{5}$$

where B_{2D}, C_{2D}, and so on include "horizontal" interactions between adsorbate molecules, the "vertical" interactions being taken into account in the Henry law constant, K_H, defined in Eq. (1).

The corresponding virial expression for the isosteric heat can be directly obtained through the Clausius–Clapeyron equation for adsorption equilibrium [5,17,196]:

$$q_{st} = RT^2 \left(\frac{d \ln P}{dT} \right)_N \tag{6}$$

where the classical assumptions of considering an ideal gas and of neglecting the adsorbed phase volume have been made [196] and where the derivatives of the virial coefficients appear [17].

The virial coefficient theory can also be applied to the gas-surface system. Now, the potential energy of interaction of the system includes the interaction between the adsorbed molecules themselves and the interactions of these molecules with the surface. As an additional approximation, the gas phase is considered to be ideal. Using the method of the grand partition function, the number of adsorbed molecules, N, can be expanded in powers of the pressure as follows [5,15,30,197]:

$$N = B_{AS}\left(\frac{P}{kT} \right) + C_{AAS}\left(\frac{P}{kT} \right)^2 + \cdots \tag{7}$$

where B_{AS} and C_{AAS} are the so-called gas-surface virial coefficients (explicit equations can be found in Refs. 5 and 15), which can be theoretically determined once both adsorbate–adsorbate and adsorbate–adsorbent intermolecular potentials have been clearly defined [5,30]. Thus, the B_{AS} coefficient is simply defined as an integral function of the solid–fluid interaction, $U_{sf}(r)$, generally supposed as a pair potential, as follows:

$$B_{AS} = \int_V \left[\exp\left(-\frac{U_{sf}(r)}{kT}\right) - 1 \right] d\mathbf{r} \tag{8}$$

which is valid only for classical nonpolar systems [17], whereas for quantum substances, a correction must be added [43].

Theoretical values of the second to the fourth gas-surface virial coefficients are available for some specific intermolecular potentials [5,17,30,33]. Moreover, a relationship between the gas-surface virial coefficients and the real 2D virial coefficients can be obtained. Thus, for the second virial coefficient and by using the Gibbs adsorption isotherm equation, Eq. (2), one finds [5,30,197]

$$\frac{B_{2D}}{A} = -\frac{C_{AAS}}{2B_{AS}^2} \tag{9}$$

an expression that leads to Eq. (4) when the 2D approximation is used.

When terms of higher than quadratic order in Eq. (7) are negligible, these coefficients can be obtained by plotting experimental low-density results of N/P versus P and then fitting a straight line. Thus, values of B_{AS} and C_{AAS} are available for such simple systems as argon, krypton, and xenon on graphite (carbon black P33, 2700°C) [30,197,198]. Obviously, a similar procedure can lead to the Henry constant and the B_{2D} virial coefficient if the first two terms in Eq. (5) are used to fit experimental data. Moreover, at very low pressures and coverages, only the second gas-surface virial coefficient is needed, so that by comparing with the classical Henry's equation, it is easy to see that [17,33]

$$B_{AS} = kTK_H \tag{10}$$

In fact, the most straightforward test of a given model for the gas–solid interaction energy is based on a comparison between experimental Henry's constants and theoretical gas–solid virial coefficients [4,5,17,33,43]. The commonly used method is to find the appropriate parameters for a list of possible gas–solid interaction potentials and then to choose those with the smallest deviations with respect to the experimental results for K_H. Thus, the virial coefficient theory has been applied to the adsorption of classical rare gases on graphitized carbon blacks [30,43,191,192,197]. Also, values of Henry's constant and of the limiting value of q_{st} for zero coverage for helium isotopes on graphite and on this surface after coating with one monolayer of solidified xenon have been reported for temperatures from 12 to 22 K [33,57].

The gas-surface virial treatment of physical adsorption has also been applied to describe localized adsorption. In particular, Rudzinski [199] has developed a virial equation for monolayer localized adsorption on uniform surfaces, which permits one to relate the second and third adsorption virial coefficients in terms of such energy and structural parameters as the number of adsorption sites, the average number of the nearest-neighbor adsorption sites, and the average interaction energy between two particles adsorbed on nearest-neighbor adsorption sites. These data permit one to investigate the microstructure of the adsorbent. As an example, Rudzinski [199] has presented an analysis of the adsorption isotherms for argon and xenon on graphite obtained by Sams et al. [30].

The virial coefficient theory can also be applied to mixtures (e.g., rare gases adsorbed on another rare gas plated on graphite [193,194]). Thus, Brown arid Hsue [194] have reported theoretical values for the virial coefficients for argon on xenon-plated graphite, the layer of xenon atoms serving to minimize the effect of the less well-determined argon–graphite potential. A potential that is more realistic than the classical Lennard–Jones potential was used, and three-body terms were considered. Although good agreement was found with experimental results for the lowest-order coefficients, this was not the case for the second-order coefficients, indicating the necessity of either including additional effects or of reinterpreting the experimental data.

As the utility of using virial approximations is to find appropriate intermolecular potentials and thus relate the microscopic and macroscopic behaviors, the most frequently used approximations for those potentials will briefly be summarized in the following section. That will permit us to introduce Steele's theory for the physical interaction of gases with crystalline solids, which leads then to the Steele theory for monolayer adsorption onto flat surfaces and will also indicate the need to study the thermodynamic properties of the 2D Lennard–Jones system.

D. Intermolecular Potentials

Theories based on statistical mechanics need not involve any particular model of the adsorption process, but they do require a knowledge of the adsorbate–adsorbent (gas–solid for adsorbed rare gases) and adsorbate–adsorbate interaction potentials [11]. As was explained in the preceding subsection, the virial coefficient theory can subsequently be used to test the adequacy of both the potentials used and the molecular parameters needed. Here, we will only introduce some significant aspects of the intermolecular interactions in physisorption, focusing our attention on the adsorption of rare gases. More extensive reviews are available elsewhere [14,15,200,201].

The study of intermolecular forces arising between adsorbed molecules and adsorbent molecules is very similar to that of the interactions between molecules in three-dimensional cases. Thus, for a great number of cases, including the adsorption of rare gases on graphite, the physisorption interactions can be represented to a good approximation by the sum of the interactions between pairs of atoms [i.e., the full physisorption potential, $\phi(\vec{r})$, is considered be pairwise additive]. Then, a common form of representing that potential is to separate the adsorbate–adsorbate and the adsorbate–adsorbent interactions, with the second not being perturbed by the first [19]:

$$\phi(\vec{r}) = \sum_{i=1}^{N} U_{sf}(\vec{r}_i) + \sum_{i}^{N} \sum_{j} U_{ff}(r_{ij}) \tag{11}$$

where $U_{sf}(\vec{r}_i)$ is the interaction potential between the solid and the adsorbed fluid at position $\vec{r}_i$ (three-dimensional position) and $U_{ff}(r_{ij})$ is the pair intermolecular potential between two molecules, i and j, of the adsorbate at a two-dimensional distance r_{ij}.

1. Adsorbate–Adsorbate Potential

The most widely used expression to give the fluid–fluid interactions is the Lennard–Jones 12-6 potential:

$$U_{L\text{-}J}(r) = 4\varepsilon \left[\left(\frac{\sigma}{r} \right)^{12} - \left(\frac{\sigma}{r} \right)^{6} \right] \tag{12}$$

where the parameters ε and σ are the so-called Lennard–Jones (L-J) parameters, which are different for each substance and represent the potential minimum (ε) and the distance at which the potential passes through zero (σ). The Lennard–Jones potential, which is strongly repulsive at short distances and softly attractive at long distances, has been widely used to model the behavior of such simple adsorbates as rare gases, methane, and so forth, each substance being defined by its L-J parameters. The values of those parameters must be considered as "effective" (i.e., they include real deviations from the behavior of the L-J model). The possibility of using different methods to calculate the L-J parameters has led to a long list of their possible values, even for 3D systems. Cuadros et al. [202,203] have proposed a new method to obtain more universal values of 3D parameters, but it has not yet been applied for adsorbed substances. Values of the L-J parameters for 3D fluids have been commonly applied to describe the adsorbate–adsorbate interactions in both theories and computer simulations.

Nevertheless, some authors have indicated that the appropriate effective L-J parameters for an adsorbed substance must include some additional effects such as the presence of the substrate or the three-body interaction terms. The form of evaluating the L-J parameters then consists in considering a more realistic potential. In particular, Wolfe and Sams [36] have shown that using a two-dimensional Lennard–Jones model with effective parameters, particularly the value of ε, that differ by about 5–10% (see Table 8) with respect to the known gas parameter for rare gases or other simple fluids permits one to obtain adequate surface area values and, moreover, that indicates that three-body effects in physisorption may be small. That result is, in part, supported by some theoretical studies where three-body effects were included, but in which the surface was considered to be a continuous, structureless medium, so that no local effects were included [204,205]. Moreover, application of the Gordon–Kim local density method in the calculation of the interaction energy of rare gases on the basal plane of graphite indicates that [151] (1) at the adatom–surface equilibrium separation, the adatom–adatom interaction potentials are from 12% to 20% more repulsive than the gas-phase counterparts, (2) at smaller separations, the effective well depth is only 3% of the gas phase well depth, (3) the nonadditive contributions are more important for the smaller rare-gas atoms than for the larger ones, and (4) the high mobility of rare-gas films on graphite is confirmed.

Also, computer simulation results have supported the use of values of the L-J parameters which differ from the 3D case. For example, Bhethanabotla and Steele [206] have found better agreement between computer simulation and experimental results [37,42] for Kr on graphite at approximately 100 K, when the Kr is modeled through a L-J fluid with a well depth, ε, 15% less than that corresponding to the 3D gas phase for Kr (Table 8).

Despite the previous results, and as one sees from Table 8, no agreement is found and it can be concluded that no fixed values for the L-J parameters modeling the behavior of adsorbed substances have been defined. Obviously, this hinders the comparison between the results and also implies the need for additional unknowns when a theory is applied to describe experimental results. Thus, one must be clear that a given set of parameters is only suitable for given approximations or theories and can be completely unsuitable when other approaches are used. We will return to this subject later.

2. Gas–Solid Potential

Knowledge of the adsorbate–adsorbent interaction is fundamental in any statistical mechanics theory of adsorption. As indicated earlier, the comparison between experimental Henry's constants or gas–solid virial coefficients and theory [8,33] permits one to test the validity of a given model for the gas–solid potential. As a first approximation, the potential $U_{sf}(\vec{r}_i)$ is considered to be a function only of the perpendicular distance z for monolayer mobile adsorption on homogeneous surfaces [29,33,43,219]. The analytical forms used are similar to the Lennard–Jones potential, but replacing r by z and considering different (10-4 or 9-3, for example) powers than the 12-6 case expressed in Eq. (12). In each case, the gas-surface molecular parameters, ε_{sf} and σ_{sf}, can be determined by comparison with experimental results. This procedure must be considered as semiempirical and thus not fully theoretical.

TABLE 8 Proposed or Used 2D Lennard–Jones Parameters [Eq. (12)] for Physisorbed Rare Gases

System	ε/k (K)	σ (nm)	References
Krypton	170.0	0.360	[207] Hanson and McTague, 1980; [208,209] Abraham et al., 1982, 1984; [187,188] Sokolowski and Steele, 1985; [206] Bhethanabotla and Steele, 1988
	171.0	0.360	[159] Houlrik and Landau, 1991
	168.5	0.360	[20] Croxton, 1974; [98] Fairobent et al., 1982
	145.0	0.360	[43] Putnam and Fort, 1977; [181] Berker et al., 1978; [183] Ostlund and Berker, 1979; [98] Fairobent et al., 1982; [144] Caflisch et al., 1985; [206] Bhethanabotla and Steele, 1988
	166.7	0.368	[82] Malbrunot et al., 1992
	152.0	0.3573	[210,211] Glandt et al., 1978, 1979; [212,213] Mulero and Cuadros, 1996
	142.8	0.373	[36] Wolfe and Sams, 1966; [212,213] Mulero and Cuadros, 1996
	142.8	0.3573	[212–214] Mulero and Cuadros, 1996; 1997
Xenon	231.0	0.396	[38–40] Suzanne et al., 1974
	236.0	0.392	[159] Houlrik and Landau, 1991
	221.0	0.410	[20] Croxton, 1974; [98] Fairobent et al., 1982; [159] Houlrik and Landau, 1991; [215] Rittner et al., 1995
	225.3	4.07	[216] Abraham, 1983
	194.0	0.412	[36] Wolfe and Sams, 1966
	—	0.404	[57] Derderian and Steele, 1977
	183.0	0.390	[217] Cuadros et al., 2001
Argon	120.0	0.340	[20] Croxton, 1974; [98] Fairobent et al., 1982; [159] Houlrik and Landau, 1991
	120.0	0.3405	[154] Tsang, 1979; [218] Cheng and Steele, 1989
	119.8	0.3405	[179,180] Rowley et al., 1976; [131] Hanson et al., 1977; [219] Steele, 1980
	116.0	0.336	[210] Glandt et al., 1978; [214] Mulero and Cuadros, 1997
	99.4	0.346	[36] Wolfe and Sams, 1966
Neon	36.76	0.2786	[158] Ni and Bruch, 1986
	33.7	0.276	[82] Malbrunot et al., 1992
	35.6	0.278	[159] Houlrik and Landau, 1991
Helium	—	0.256	[98] Fairobent et al., 1982; [57] Derderian and Steele, 1977
	10.22	0.2556	[220] Sander et al., 1976; [158] Ni and Bruch, 1986

A more reliable approximation for the gas–solid potential is to consider a sum of pairwise L-J 12-6 interactions between the gas and the solid atoms. Replacing the sum by integrals over the volume of the solid, a 3-9 power law in z is found. If the sums are replaced by an integral over the surface plane and the interaction with more distant planes of atom is neglected, a 10-4 potential is obtained [33]. Thus, for adsorption on graphite, the 10-4 potential is [15,33]

$$U_{\text{sf}}(z) = \frac{2\pi}{a_c} \sum_j \varepsilon_{\text{fs}} \sigma_{\text{fs}}^2 \left[\frac{2}{5} \left(\frac{\sigma_{\text{fs}}}{z + j\delta} \right)^{10} - \left(\frac{\sigma_{\text{fs}}}{z + j\delta} \right)^4 \right] \tag{13}$$

where a_c is the area per carbon atom in the basal plane (26.2 nm) and δ is the interplanar distance in graphite, which depends on temperature and the crystalline perfection of the sample (Steele [33] takes 33.7 nm with an uncertainty of 1%).

A test of validity of the "summed" 10-4 potential and comparison with simpler potentials has been made by Steele [33]. In particular, Steele considers the conclusion of Putnam and Fort [43], who, after considering different power law potentials, indicate that a $U_{sf}(z)$ 3-12 potential gave the best fit to their experimental data for Kr on graphite. Nevertheless, Steele [33] has indicated that the 10-4 model was not considered by Putnam and Fort, and he presented a comparative test of that potential against the available virial coefficient data for rare gases on graphite. He showed that the agreement is as good as in the case of the 3-12 law over the available temperature ranges. Moreover, the L-J parameters for the gas–solid potential are in reasonable agreement with those obtained through the well-known Lorentz–Berthelot combining rules. These combining rules have also been used to represent the xenon–carbon interaction by Suzanne et al. [38,39]. In conclusion, Steele's results permit one to define the best values of the parameters used and lead strong support to the pairwise summation model, which has been used extensively.

More complex expressions for the gas–solid potential are found when the dependence on the 2D distance (parallel to the surface) is introduced. Thus, the model given by Hill [173] is based on a simple square lattice, the potential being a separable function of the planar 2D coordinates and the coordinate perpendicular to the surface, and the periodic component of the energy being described by a simple sinusoidal function which is independent of the perpendicular distance. Doll and Steele [18] have indicated, nevertheless, that Hill's model is oversimplified. For example, it considers a simple square lattice surface when it is the close-packed array, which is the most interesting case from an experimental point of view. Obviously, for heterogeneous surfaces, a more complex dependence on x, y, and z must be considered [6,11].

Another general approach to defining the gas–solid potential is to consider specific potential laws for the adsorbate–adsorbate and for the adsorbent–adsorbent and then to define a combination rule to define the potential law for the corresponding gas–solid potential [221]. Whereas for rare gases suitable potentials are readily available (such as the L-J model), the contrary is the case for the carbon atoms in graphite. In particular, Pisani et al. [221] have obtained values for the potential energies of adsorption for rare gases on graphite, by using three-parameter potential laws derived from independently determined self-interaction potentials for the gas and for the solid atoms. Nevertheless, no satisfactory agreement with experimental data was directly obtained, even using anisotropic potentials. They indicated that it is not possible to obtain suitable potential laws from physical adsorption by rigorously applying the criterion of combining empirical independently defined self-interaction potentials. The fundamental reason would probably be found in the inadequacy of the additive pairwise approximation to describe van der Waals interactions in graphite. Another possible reason, however, could well be the rather arbitrary nature of combination laws.

A different approach is that based on the density-functional Gordon–Kim method [222], which has been used to calculate the interaction of Ar over an Ar substrate [223] and for rare gases on graphite [150,151]. It permits one to take into account the non-two-body additive effects and to give indications about the choice of the suitable effective parameters. The well depths determined by Freeman [151] in that theoretical development are substantially lower than those found by other methods. Nevertheless, the disagreement is due, in part, to the uncertainties and approximations in the theoretical calculations.

Steele [16,17] has developed a more rigorous model for the physical interaction of gases with crystalline solids, in which pairwise summation is considered and the sum depends not only upon the parameters and functional dependence of $U_{sf}(\vec{r})$, but also upon the arrangement of the molecules in the solid; that is, it is a function of the solid lattice symmetry and spacing. As indicated by Steele [16], calculations for which the sum of L-J 12-6 potential is considered lead to results which are not particularly good representations of the phenomena. Moreover, calculations of those sums are given in numerical form, which is not particularly convenient for use in integrations such as that defined in the second virial coefficient, Eq. (8). Some of the problems presented by earlier theoretical developments are resolved and the application of the theory to real systems is facilitated, if an analytical form of $U_{sf}(\vec{r}_i)$ is given in which not only the perpendicular gas–solid interaction is taken into account but also the structure of the solid. Thus, for the interaction with crystalline surfaces, Steele [16–18] proposed representing the periodicity of the solid's symmetry through a perfectly periodic function defined as a two-dimensional Fourier series. The final expression for the gas–solid potential can then be written as

$$U_{sf}(\vec{r}_i) = U_{sf}(z, \vec{r}) = U_{sf}(z) + \sum_g w_g(z) \exp(i\vec{g}\vec{r}) \tag{14}$$

where $U_{sf}(z)$ can be expressed by Eq. (13), $\vec{r}$ is the two-dimensional distance, and $\vec{g}$ is the reciprocal lattice vector of the substrate surface. The Fourier coefficients $w_g(z)$ are functions of temperature and must be computed numerically for realistic potential functions of $U_{sf}(\vec{r}_i)$.

The main disadvantage of Steele's model is the necessity of performing numerical integrations to calculate the thermodynamic properties. Nevertheless, the author indicates that the solution is straightforward if computers are used. By calculating the gas–solid potentials for several surface lattices through direct summation and comparing with those given by the Fourier series, a satisfactory representation of the energy is obtained even when the number of terms included is small. Thus, the truncated Fourier series has been used to obtain theoretical values for the average energies for an isolated adsorbed atom, as well as to calculate the second and third virial coefficients, or the Henry's law constant, for rare gases on perfect crystalline surfaces [16,17]. Also, the heat capacity and the localized-mobile transition have been studied [18].

Derderian and Steele [57] have combined the virial analysis with the development proposed by Steele for the gas–solid potential energy. Results for ^{4}He and ^{3}He isotopes have been compared with theories and experimental results. Parameters for the L-J 12-6 (see Table 8) and the gas–solid interaction were obtained. The study also includes the calculation of the isosteric heat at 15 K extrapolated to zero coverage and the calculation of the nonconfigurational entropy per adsorbed particle. A notable conclusion is that the solid, at these temperatures, is not perturbed by the adsorbate.

From a practical point of view, the main application of the Steele gas–solid interaction model has been the development of a statistical mechanics expression for monolayer physisorption [19,182,224], which will be explained in the following subsection. Some theoretical results for the two-dimensional approximation of the Steele theory and comparison with experimental results will be presented in subsequent subsections. Finally, we note here that the Steele theory has also been used in computer simulations to study both pure [206,225] and binary gas monolayer adsorption on homogeneous solid surfaces [190].

E. Steele's Theory of Monolayer Adsorption

As was indicated in the previous subsection, Steele [16–18] has developed an analytical model for the gas–solid interaction in which the periodic nature of the adsorbate–adsorbent interaction is taken into account when the molecules of the former move parallel to the surface of the latter. Subsequently, Steele [19] developed a statistical mechanics treatment which, at least in the case of monolayer adsorption, allows the three-dimensional problem to be reduced to a two-dimensional one, and then simple models for the 2D fluid, such as the Lennard-Jones one, permits one to find theoretical values for the properties of interest in physisorption.

Steele's theory begins by considering the existence of a phase of N molecules adsorbed onto a solid surface and in equilibrium with the gas above it. The solid surface is assumed to be energetically homogeneous and unperturbed by the presence of the adsorbed phase (physisorption). The molecules of the adsorbed phase are assumed to obey the laws of classical statistical mechanics, to be apolar, quasispherical, and with a pairwise central potential interaction. These characteristics are closely approximated in the adsorption of classical rare gases onto graphite.

Equilibrium conditions between the gas and the adsorbed phase at a given temperature and pressure are expressed as the equality of their chemical potentials:

$$\mu_{gas} = \mu_{ads} \tag{15}$$

The chemical potential of the gaseous phase can be approximated to that of an ideal gas [210]:

$$\mu_{gas} = \mu_0 + kT \ln\left(\frac{P}{kT}\right) \tag{16}$$

where k is Boltzmann's constant and μ_0 is the chemical potential in a reference state. On the other hand, the chemical potential of the adsorbed phase can be defined in statistical mechanics as

$$\mu_{ads} = \mu_0 - kT \left[\frac{\partial \ln(\Gamma_{ads}/N!)}{\partial N}\right]_{T,A} \tag{17}$$

with Γ_{ads} being the configuration integral of the adsorbed phase [19].

By equating chemical potentials and making use of the approximation

$$\ln(N!) \approx N \ln N - N \tag{18}$$

the general equation of the adsorption isotherm can be written in the form [19,182,210]

$$\ln\left[\frac{P}{NkT}\right] = -\left[\frac{\partial \ln \Gamma_{ads}}{\partial N}\right]_{T,A} \tag{19}$$

The practical use of Eq. (19) requires the knowledge of the configurational integral of the adsorbed phase, which is defined as

$$\Gamma_{ads} = \int_{V_{ads}} \cdots \int \exp\left[-\frac{\phi(\vec{r})}{kT}\right] d\,\vec{r}^N \tag{20}$$

with V_{ads} being the volume of the adsorbed phase and $\phi(\vec{r})$ the interaction potential for the adsorbed phase which can be separated into two terms as indicated in Eq. (11) (i.e., by separating the solid–fluid and the fluid–fluid interactions).

The use of the Steele form [16] for the solid–fluid or adsorbent–adsorbate potential expressed in Eq. (14) permits one to write the configurational integral as a perturbational expansion:

$$\Gamma_{ads} = \Gamma_0(1 + \Gamma_1 + \Gamma_2 + \cdots) \tag{21}$$

where Γ_0 is the configuration integral of the N adsorbed molecules on a perfect surface (neglecting its periodicity), and the other terms (Γ_1, Γ_2, etc.) contain the effects of the periodic nature of real solid surfaces. In sum, one can deduce that the development of a theory of adsorption onto perfect surfaces constitutes a first step toward a broader development which will take the surface periodicity into account [182,224].

We note that even Γ_0 contains the $U_{sf}(z)$ potential (i.e., although the solid periodicity is neglected, the influence of the solid is present). Although, at first, the evaluation of Γ_0 is not straightforward, that calculation can be facilitated if the two-dimensional approximation is introduced. It is assumed that the adsorbed phase behaves as a two-dimensional fluid, so that the adsorbed molecules only move in a plane parallel to the surface of the solid, neglecting the vibrational movement perpendicular to that surface.

The validity of the 2D approximation has been clearly shown for rare gases on graphite, as was indicated in previous subsections. For example, the specific heat data for helium monolayers on graphite has been adequately described by using the 2D Lennard–Jones potential [192,220] and ignoring the effects of the substrate. The use of 2D virial coefficients is a further illustrative example of the validity of that approximation. A third example is the fact that theoretical results indicate that the contribution to the adsorption entropy of the vibrations of xenon atoms parallel to the (0001) face of graphite is clearly greater than the contribution of perpendicular vibrations. This shows, again, the validity of the 2D models in describing adsorption properties, even the properties not commonly calculated such as the entropy.

The 2D approximation permits to perform the integrals present in Γ_0 in z and in r (two-dimensional distance) separately. Thus, the contribution of a perfect adsorbent is represented to a first approximation by the Henry constant or the gas–solid second virial coefficient [17], and the contribution of the 2D fluid by its excess chemical potential, μ, with respect to the ideal gas contribution. Thus, in the 2D approximation, the derivative on the right-side of Eq. (19) is approximated by [210,226]

$$\left[\frac{\partial \ln \Gamma_{\text{ads}}}{\partial N}\right]_T \approx \left[\frac{\partial \ln \Gamma_0}{\partial N}\right] \approx -\frac{\mu}{kT} + \ln(kTK_H) = -\frac{\mu}{kT} + \ln(B_{\text{AS}}), \tag{22}$$

so that Eq. (19) becomes

$$P = \frac{N}{K_H}\exp\left(\frac{\mu}{kT}\right) = \frac{NkT}{B_{\text{AS}}}\exp\left(\frac{\mu}{kT}\right) \tag{23}$$

which constitutes the theoretical expression for the adsorption isotherm for the physisorption of 2D simple fluids onto perfectly flat surfaces [19,182,210,211,226]. We note that theoretical values of the spreading pressure or the isosteric heat can be obtained by substituting Eq. (23) into Eqs. (2) and (6), respectively.

Use of the theoretical equation (23) requires applying theoretical methods to obtain values of the 2D second gas-surface virial coefficient and to obtain suitable theoretical values for the chemical potential of a purely 2D fluid.

In the case of the adsorbate being a simple gas (rare gases, methane, etc.), the adsorbate–adsorbate interaction may be modeled with a L-J potential, so that μ is the excess chemical potential for the 2D L-J system with respect to an ideal gas. For an ideal gas, $\mu = 0$ (there is no excess) and then the classical Henry equation, Eq. (1), is recovered. Reliable values of μ for 2D L-J fluids have been obtained from perturbation theories [19], integral theories [210], computer simulations [225], and equations of state [212,214,226]. Details of these approaches and comparison of results will be made in Sections IV.B and IV.C.

In applying Eq. (23), theoretical calculation of K_H or B_{AS} at a given temperature are needed. This requires knowledge of both the gas–solid potential and the surface area of the adsorbent. As indicated earlier, pairwise additivity of the L-J 12-6 potential can be considered, being ε_{fs} and σ_{fs}, the molecular parameters obtained via certain mixing rules. Thus, for example, Glandt et al. [210] have used an arithmetic mean value between the values of σ for the gas and the solid for σ_{gs}. Nevertheless, to find the suitable value for ε_{gs}, they used the value that leads to the best agreement between Henry's constant obtained from the experimental data and the B_{AS} coefficient calculated through the gas–solid intermolecular potential.

The Steele theory is, therefore, the simplest treatment possible for a soundly based statistical mechanics description of the physisorption phenomenon and may be taken as a reference point for the perturbation development given by Eq. (21). The influence of both the periodicity of the solid structure and the deviations with respect to the 2D approximation have been studied theoretically [182,224], although their practical application and comparison with experimental results are not so straightforward. We are now working on more straightforward approaches to give these deviations.

F. Theories Describing Phase Diagrams

As was mentioned in Section II.A, phase transitions in adsorbed films have been (and continue to be) one of the most intensively studied problems in surface science [9,12,15,62,99]. The experimental findings described in Section II.A have led to different theoretical approaches based mainly on partial localization models [161,164], on theoretical models for completely mobile films [227], on various lattice gas models [144,181,183], and, finally, on models of nonuniform fluids such as density functional theories [98,186–189].

Theoretical developments about phase transitions and the location of triple or critical points have been summarized by Thomy and Duval [12]. Here, we will refer briefly to some of the first theoretical studies included in the Thomy and Duval review and then update it by adding other important developments or theoretical results. Solid structures, the monolayer melting, and other phenomena on solid adsorbed phases have been adequately described through the theory of topological defects. An extensive review of structure, interaction forces, and the statistical mechanics concept applied to monolayer adsorption has been given by Bruch et al. [15] and will not be considered here.

The desirable final objective for any theoretical development is to find a common basis to describe the different kinds of phase diagrams that have been observed experimentally. Nevertheless, the application of theories to describe these experimental phase diagrams is not easy [65]. First, it is clear that the adsorbed monolayers are not strictly two-dimensional systems. Second, the real surfaces are not flat on an atomic scale but are corrugated in the position of the energy minima and in the value of the minimum energy as a gas atom passes across the surface. Third, a direct visual observation of the two-phase coexistence in adsorbed systems is not feasible [65]. Fourth, although it is possible theoretically to describe an independent monolayer, it is well known that the second layer can start to be formed before the first is complete (i.e., multilayer adsorption cannot be clearly separated from monolayer adsorption in experimental results).

In any case, great progress has been made, permitting one to describe at least the main features of the phase diagrams of rare gases on graphite, which will be summarized next.

Both mobile and localized adsorption equations [227–229] have allowed one to predict the fluid–solid transition. Particularly interesting, from a theoretical point of view, is the model proposed by Borowski et al. [227], in which the adsorbed phase is considered to be completely mobile and described as a van der Waal's fluid with softened repulsive interactions. Unfortunately, no direct comparison with experimental results can be made.

Experimental results have clearly shown that during the formation of the monolayer, a change from nonlocalized to localized adsorption occurs. Several theoretical studies have been made of the so-called partially mobile or partially localized adsorption models [11,60,161,164,228,230]. These theories must explain the phase transitions in the adsorbed monolayers and may also be useful in describing surface transport phenomena [11].

Lattice gas models [144,181,183] have provided at least a first approach to a suitable theoretical framework for phase diagrams. For example, the lattice gas model has been capable of reproducing the main aspects of the phase diagram for Kr on graphite [181] with no adjustable parameters and with the adatoms interacting via the Lennard–Jones potential, including the incipient liquid–gas criticality [183]. As was indicated in Section II.A, subsequent results of x-ray diffraction [97] have provided strong support for lattice models of commensurate krypton on graphite.

Another noteworthy approach is the modified cell theory developed by Tsang [154], where disordered structures were introduced. Theoretical results for the two-dimensional argon lattice appear to be in general agreement with neutron-scattering data for argon on grafoil [120]. Another clear example is the application of lattice models to the study of the adsorption of solid rare gases on metallic surfaces [81].

A clear disadvantage of the earliest lattice models [181] (and of density functional approaches) is that only commensurate solid phases (in addition to the fluid phases, obviously) are considered. A more complex lattice model including incommensurate phases has been developed by Caflisch et al. [144] for the particular case of Kr on graphite. The experimentally observed phase diagram, including the presence of the "reentrant" fluid is reproduced, as is shown in Fig. 5, where the lines indicate the Caflisch et al. [144] theoretical results. Although the application of this model to other systems seems possible, it could be very laborious.

A simplified model for calculating the submonolayer phase diagrams of physisorbed systems on a weakly corrugated substrate has been presented by Niskanen [231]. Both commensurate and incommensurate phases, as well as liquid and gas phases, are included. The solid phase is described by a cell model and the fluid phases by an Ising lattice gas model, the adsorbed atoms being modeled by a L-J fluid. The adsorbate phases are treated as if the substrate were perfectly flat. Results are in reasonable agreement with experimental observations and with density functional calculations [186]. Nevertheless, because of the important simplifications needed in the model, corrugation effects in the melting transitions and a description of quantum gases or second-layer effects cannot be included. Thus, for example, the first-order melting observed in experimental results for particular gases and temperature ranges cannot be reproduced with this model.

Adsorption can also be studied from integral theories and density functional theories. In particular, density functional theories have been applied to studying monolayer films of rare gases [98,186–189]. The approach proposed by Fairobent et al. [98] represents the first historical theoretical treatment to explain systematically the phase diagrams for rare gases on graphite. Monte Carlo results for the 2D Lennard–Jones fluid [232] and an x-y sinusoidal form of the substrate potential [8] were used as input. Suitable agreement with experimental results and other theoretical approaches (as known until 1982) was found. For example, as is shown in Fig. 4 for Kr on graphite, a well-defined liquid–gas transition over a range of about 2 K was found, in agreement with some experimental results [68], although in disagreement with others [105], as was explained in Section II.A. For other theoretical studies about phase transitions for Kr on graphite, see Refs. 233–235.

For Xe on graphite, there was a major disagreement in the location of the 2D critical point with respect to the experimental one [37,46], although the general shape of the liquid–gas curve was well reproduced, as is shown in Fig. 1. For Ar on graphite, good agreement was obtained for the location of the critical point compared with previous experimental results [62] (see Table 1). For helium, which is a special case because quantum corrections play an important role, there was excellent agreement with the observed maximum ordering temperature [55]. Other interesting parts of the phase diagram for helium on graphite have been presented by Schick [236] and Ecke and Dash [237]. Computer simulation results [238] have confirmed the main aspects of the proposed phase diagram.

Another important advance was represented by the work of Sokolowski and Steele [187, 188], who used the density functional formalism to study the freezing of strictly 2D fluids on an exposed crystal face of a chemically inert solid. In particular, numerical calculations were presented for the freezing of hard disks on a periodic surface chosen to model the graphite structure [187]. They also presented a more detailed application to the freezing of krypton monolayers on graphite. The Kr–Kr interaction was modeled by a L-J potential. This theory appears to reproduce the observed behavior and properties of the freezing transition for submonolayer krypton adsorbed on graphite [189]. Unfortunately, no detailed computer simulation results exist to check the accuracy of the results. It is worth noting that no trace of an "incipient triple point," described in experiments and in simulation studies [209], is seen in the theory. As is indicated by Sokolowski and Steele [188], the theory gives reasonable results which could be improved if more precise values for the structure factors and the triple distribution function of the fluid phase were available at appropriate temperatures and densities.

In conclusion, although different theoretical approaches have been developed which describe the main characteristics of the very complex phase diagrams for rare gases on graphite, they have not been capable of reproducing some of the specific details or parts of these diagrams.

IV. MODELING MONOLAYER ADSORPTION

This section is devoted to studying the 2D Lennard–Jones model in order to serve as the basis in applying Steele's theory. In Section IV.A the main studies about that model are summarized and commented on. In Section IV.B, the most useful expressions for the equation of state of the model are given. In Section IV.C we present results about the application of these equations, which are compared with other theoretical approaches to studying adsorption of 2D Lennard–Jones fluids onto perfectly flat surfaces. In Section IV.D, the comparison with experimental results is made, including results for the adsorption isotherms, the spreading pressure, and the isosteric heat. Finally, in Section IV.E we indicate briefly some details about the use of computer simulations to model the properties both of an isolated 2D Lennard–Jones system and of adsorbate–adsorbent systems.

A. The Two-Dimensional Lennard–Jones Model

There are many theoretical reasons for studying the characteristics of 2D fluids, from the possibility of comparing results with respect to the three-dimensional case to the clear advantages in computer simulations [226]. Also, from a pedagogical point of view, the study of 2D fluids can be useful to explain both the microscopic and the macroscopic behavior of real fluids [226,239,240]. Nevertheless, the main application of 2D fluids is to serve as a model in the study of physisorption of simple systems, such as rare gases, onto well-characterized surfaces, so that this model can be taken as a reference in more complex theories.

Most of the information about purely 2D fluids has been obtained from computer simulations or from the application of theories commonly used in the three-dimensional case. In this subsection, we will summarize the most important studies and results about the properties of the 2D Lennard–Jones system, which is the most widely used model. The 2D L-J system is defined through the intermolecular potential given in Eq. (12), where now the distance is considered only in two dimensions, the L-J parameters ε and σ being a proper characteristic of each real fluid. In both computer simulations and theories, the thermodynamic properties are commonly expressed in reduced (adimensional) units, marked with a superscript $*$, which are related to the real units through the L-J parameters as follows:

2D density

$$\rho^* = \rho\sigma^2 \tag{24}$$

Temperature

$$T* = \frac{T}{(\varepsilon/k)}$$
(25)

Pressure

$$P* = P\frac{\sigma^3}{\varepsilon}$$
(26)

Chemical potential

$$\frac{\mu*}{T*} = \frac{\mu}{kT}$$
(27)

where ε/k is expressed in kelvin and σ in meters. Table 8 lists values for these parameters when adsorbed rare gases are considered in theories or computer simulations. As indicated earlier, there exists no definitive method to obtain the parameter values, so that each set must be seen as valid only for the particular approximation being considered.

There are three aspects which are the most studied for the L-J system:

1. Phase diagram, especially the location of the critical and triple points, as well as the nature of the melting transition.
2. Microscopic and statistical mechanics characterization through the radial distribution function (RDF) defined in statistical mechanics. This is a function that expresses the probability of finding a particle at a given distance from another particle taken as reference. In statistical mechanics, the RDF plays a central role, because it permits microscopic and macroscopic properties to be related.
3. Thermodynamic properties, mainly through the equation of state (EoS).

Although dynamical properties have been also considered [241–245], we focus our attention only on the three above-mentioned aspects.

The main references and results are summarized in Tables 9–12. As can be seen, the methods devoted to obtaining these properties are mainly the virial theory [195], perturbation theories [19,226,232,246], integral equations [247–251], density functional theories [252], and a long list of computer simulations [226].

As was observed in Section III.F, theoretical descriptions of phase diagrams for adsorbed rare gases are based on the study of the 2D L-J systems. Hence, we shall first consider the studies devoted to phase transitions in 2D L-J fluids. A great number of theoretical works devoted specifically to that aim are summarized in Table 9, most of them based on computer simulations of the liquid–solid or melting transition. In fact, a widely discussed problem is the nature of the melting transition. As has been previously indicated, Halperin and Nelson [102] developed a theory of melting in 2D, which is based on the need for two second-order transitions to make the transition, which are then separated by an intermediate phase with short-range translational order and long-range orientational order, which is the so-called "hexatic" phase. As was indicated in Section II.A, the possible continuous transition was also observed in experimental results for rare gases on graphite. Nevertheless, many computer simulations as well as application of the Weeks–Chandler–Andersen (WCA) [253] theory have led to a classical first-order transition (see Table 9).

Studies of the vapor–liquid equilibrium (VLE) and the location of critical and triple points have led to more homogeneous results (see Tables 10 and 11). Thus, the triple point seems to be well defined, although only a few results have been published. Data for the VLE curve (i.e., the coexistence densities at each temperature) have been determined by using molecular simulations

TABLE 9 Studies of Phase Transitions in 2D L-J Systems

Method	Contribution	References
MD simulation	Results for the liquid–solid transition; melting is a first-order transition; estimation of the location of the triple point	[132,254,255] Toxvaerd, 1978, 1980, 1981
MD simulation	Melting is continuous transition with a "hexatic" phase	[256] Frenkel and McTague, 1979
MC simulation	Melting is a first-order transition; study of vapor–liquid transition	[133,257,258] Abraham, 1980, 1981
MC simulation	Melting is a first-order transition	[259] van Swol et al., 1980
MC simulation	Complete phase diagram; melting is a first-order transition	[260] Barker et al., 1981
MC simulation	Complete phase diagram; melting is a first-order transition; change enthalpy and entropy in the liquid–solid transition	[261] Phillips et al., 1981
MC simulation	Thermodynamic properties, structure factor and angular correlation function in the melting; melting is a first-order transition	[262] Tobchnick and Chester, 1982
WCA theory	Application of the 2D van der Waals theory to the study of the liquid–solid transition; melting is a first-order transition	[263] Weeks and Broughton, 1983
MD simulation	Need for simulations with large number of particles to give a definitive conclusion about melting order	[264] Toxvaerd, 1983
MD simulation	Angular correlation function; melting is a first-order transition	[265] Bakker et al., 1984
Density functional theory	Study of the vapor–liquid transition; calculation of the "line" tension and density profiles	[266] Mederos et al., 1985
MD simulation	Long time computer simulations; results near the liquid–solid transition	[267,268] Morales et al., 1987, 1988
MC simulation	Melting is a first-order transition	[269] Wong and Chester, 1987
MC simulation	Melting is a first-order transition	[270] Udink and Frenkel, 1987
MC simulation	Vapor–liquid coexistence curve	[271] Singh et al., 1990
MC simulation	Estimation of critical temperature	[272] Rovere et al., 1990
MC simulation	Influence of cutoff of the potential on the location of critical temperature	[273] Smit and Frenkel, 1991
MD simulation	Estimation of critical point	[274] Rovere et al., 1993
MC simulation	Vapor–liquid equilibrium	[275] Panagiotopoulos, 1994
MC simulation	Vapor–liquid equilibrium	[276] Jiang and Gubbins, 1995
Statistical-mechanics theories	Vapor–liquid equilibrium for pure fluid and mixtures	[246,277] Scalise et al., 1998, 1999
EoS	Vapor–liquid equilibrium	[278] Mulero et al., 1999

TABLE 10   Estimation of the Triple-Point Location for the 2D L-J System

Method	T_t^*	P_t^*	ρ_t^*	References
MD simulation	0.41		<0.8	[132] Toxvaerd, 1980
MC simulation	0.40			[259] van Swol et al., 1980
Perturbation theory and cell theory	0.41	0.006		[260] Barker et al., 1981
MC simulation	0.415	0.0056		[260] Barker et al., 1981
MC simulation	0.4		0.8	[261] Phillips et al., 1981

[261,271,273,276] and statistical mechanics theories [246], whereas the proposed EoSs [232,260,279,280] have been used mainly for the location of the critical point (Table 11).

In a recent work [278], we have compared results for the VLE curve obtained from three recent simulations [271,273,276] and that calculated by us through both the Reddy–O'Shea (RO) [279] and our Cuadros–Mulero (CM) [288] equations of state (whose analytical form will be given in the next subsection). Results are shown in Fig. 8.

As can be seen, there exist discrepancies between computer simulation results, especially for temperatures near the critical point and for values of the liquid density. In general, all of the results for the vapor density obtained in the aforementioned computer simulations can be considered as equal in the range $0.41 \leq T^* \leq 0.46$. For the liquid density, the data given by Singh et al. [271] must be taken as good only for $0.42 \leq T^* \leq 0.46$, whereas for $0.46 < T^* \leq 0.49$, the values given by Smit and Frenkel [273] or by Jiang and Gubbins

TABLE 11   Estimations of the Location of the Critical Point for the 2D L-J Fluid

Method	T_c^*	ρ_c^*	Z_c	References
MD	0.7			[281] Fehder, 1969
Three coefficients virial EoS	0.615	0.27	0.34	[282] Morrison and Ross, 1973
MC	0.625–0.7	0.38	0.15	[283] Tsien and Valleau, 1974
Theoretical	0.6844	0.2888		[284] Andrews, 1976
WCA modified	0.60	0.38	0.27	[19] Steele, 1976
MC and EoS	0.56	0.32	0.26	[232] Henderson, 1977
PY theory	0.495	—		[248] Glandt and Fitts, 1977
Four coefficients virial EoS	0.492	0.119		[285] Glandt, 1978
Five coefficients virial EoS	0.475	0.1		[286] Barker, 1981
MC and EoS	0.533	0.335	0.21	[260] Barker et al., 1981
WCA theory	0.64	0.30		[266] Mederos et al., 1985
BH theory	0.60	0.34		[266] Mederos et al., 1985
Five coefficients virial EoS	0.475	0.1003	0.399	[195] Reddy et al., 1986
MC	0.537	0.365	0.189	[279] Reddy and O'Shea, 1986
MC	0.472	0.33		[271] Singh et al., 1990
MC	0.50	—		[272] Rovere et al., 1990
EoS	0.662	0.264	0.366	[280] Song and Mason, 1990
GEMC (truncated potential)	0.515	0.355		[273] Smit and Frenkel, 1991
GEMC (truncated and shifted potential)	0.459	0.35		[273] Smit and Frenkel, 1991
GCMC	0.44	—		[287] Bruce and Wilding, 1992
MD	0.47	0.35		[274] Rovere et al., 1993
GEMC	0.497	0.35		[275] Panagiotopoulos, 1994
GEMC	0.498	0.36		[276] Jiang and Gubbins, 1995

TABLE 12   Studies of General Properties of the 2D L-J System

Method	Contribution	References
MC simulation	First computer simulation; qualitative results; indications of liquid–vapor transition	[289] Rosenbluth and Rosenbluth, 1954
MD simulation	Adequate quantitative results; numerical values for the RDF; study of microscopic structure	[281,290] Fehder, 1969, 1970
Virial theory	Third virial coefficient	[291] Kreimer et al., 1973
MC simulation	Results, including RDF, for a great number of thermodynamic states	[283] Tsien and Valleau, 1974
PY theory	First attempt to apply an integral theory; erroneous results were found	[292] Mandel, 1975
Modified WCA theory	Analytical methods to obtain the RDF at intermediate and high densities; application to study monolayer adsorption	[19] Steele, 1976
MD simulation	Comparison with experimental results for adsorption	[293] Toxvaerd, 1975
PY and HNC theories	Solution of integral equations with good results for a large number of thermodynamic states	[247–250] Glandt and Fitts, 1976, 1977, 1978
MC simulation and perturbation theory	EoS based on perturbation theory and computer simulation results; validity of the Barker–Henderson theory	[232] Henderson, 1977
BGY integral equation	Derivation of the BGY, PY, and HNC integral equations in 2D	[251] Chan, 1977
Virial theory	Fourth virial coefficient	[285] Glandt, 1978
Virial theory	Fifth virial coefficient	[286] Barker, 1981
MC simulation and virial theory	Quantum corrections to the third and fourth virial coefficients; quantum corrections to the Helmholtz free energy	[294] Glandt, 1981
MC simulation	Results at very high densities; EoS based on computer simulation results and on the five first virial coefficients	[260] Barker et al., 1981
MD simulation	No influence of number of particles for states far from the phase transitions; values of RDF	[295] Bruin et al., 1984
Using data of RDF	The RDF as pedagogical tool	[296] Bishop and Bruin, 1984
WCA theory	Estimation of quantum contributions	[297] Singh and Sinha, 1985
MD	Study of the reference system (only repulsive forces) in the WCA theory	[298] Bishop 1986

(*continued*)

TABLE 12   (*continued*)

Method	Contribution	References
Virial theory	Five first virial coefficients	[195] Reddy et al., 1986
MC simulation	EoS	[279] Reddy and O'Shea, 1986
MC simulation	Values for the chemical potential; application to adsorption	[225] Knight and Monson, 1986
MD simulation	Pedagogical applications	[299] Sperandeo–Mineo and Tripi, 1987
Density functional theory	Values of the RDF; good agreement with computer simulations except at high densities	[252] Takamiya and Nakanishi, 1990
WCA theory	Nonanalytical EoS	[280] Song and Mason, 1990
MD simulation	Values of the RDF and structure factor	[300] Ranganathan and Pathak; 1992
MD simulation and WCA theory	Wide range of temperature and densities; values of RDF and thermodynamic properties; anlaysis of WCA theory; analysis of Steele's modified WCA theory; EoS with separated repulsive and attractive contributions; analytical expressions for different thermodynamic properties; pedagogical applications	[212–214,217,226,239,240,278, 288,301–312] Cuadros, Mulero, and co-workers 1991–2001

[276] are adequate. No conclusions can be drawn for $T^* > 0.49$. Finally, we would indicate that for the vapor pressure, the lack of results in the case of the Jiang and Gubbins [276] computer simulation and the great inaccuracies reported by Smit and Frenkel [273] do not permit these studies to be taken as a clear reference for comparison with theoretical results.

Results from the RO and CM EoSs were obtained by equating both the pressure and the chemical potential in the vapor and liquid phases. As can be seen in Fig. 8, the RO VLE curve is in good agreement with the computer simulation results of Smit and Frenkel [273], with only a smaller degree of agreement with results of Jiang and Gubbins [276] for $T^* > 0.45$. The CM EoS seems to be valid for the calculation of VLE densities only for low temperatures ($T^* \leq 46$). Also, a clear disagreement between the vapor pressures obtained from the RO and the CM equations were found [278], whereas the agreement was better for the chemical potential values.

In conclusion, Fig. 8 represents a clear idea of the vapor–liquid curve in 2D Lennard–Jones fluids, the location of the critical point not being clearly determined. In fact, as one sees in Table 11, many values for the critical temperature and density have been published. The most recent computer simulations seem to indicate values very close to $T^* = 0.5$ and $\rho^* = 0.35$ in reduced units, respectively. Nevertheless, it is necessary to take into account that the most commonly used method is computer simulation, and that, as Smit and Frenkel [273] have pointed out, the truncation of the potential has a great influence on the final results. In any case, it must be clear that computer simulations cannot be performed at the critical point, and the results are always based on estimates obtained from the vapor–liquid curve. As noted by Larher and

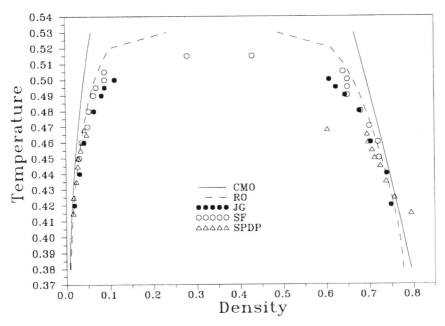

FIG. 8 Vapor–liquid equilibrium curves for the two-dimensional Lennard–Jones system [278] obtained from computer simulations (points: SPDP, Singh et al. [271]; SF, Smit and Frenkel [273]; JG, Jiang and Gubbins [276]) and from the Reddy O'Shea (RO) [279], Eq. (28), and the Cuadros–Mulero (CMO) [288] equations of state, Eq. (29).

Gilquin [62], this aspect is an added difficulty in the study of critical points in adsorbed fluids, and a direct comparison between theory and experiments must be made.

A comparison between theoretical and experimental values for the critical properties was first made by Tsien and Valleau [283]. Following Tsien and Valleau, the experimental data of Thomy and Duval [37] reduced with the effective L-J parameters given by Wolfe and Sams [36] give reduced critical temperatures of 0.65, 0.61, and 0.60 for argon, krypton and xenon, respectively (see Tables 1, 8, and 11). The variation means, of course, that the L-J fluids can not model exactly the behavior of all the rare gases. Moreover, there is a major disagreement with respect to the theoretical or computer simulation values for 2D L-J fluids. Nevertheless, we note that if the most recent experimental values of the critical temperature are taken (except, obviously, for Kr), the reduced values come close to 0.5 for certain values of the L-J parameters, even in the case of adsorbed neon.

A complete phase diagram has been proposed by Phillips et al. [261] and is shown in Fig. 9. Phillips et al. use the Lennard–Jones–Devonshire cell model approximation to the solid phase, and a virial expansion for a high-temperature dilute gas was the starting point for a free energy construction in a Monte Carlo computer simulation. The vaporization and sublimation lines were then obtained. In Fig. 9, the vaporization line is extended to the critical point determined by Henderson [232] and by Barker et al. [260]; see Table 11. The melting line is constructed by using the triple point temperature derived by Phillips et al. and the melting densities of Barker et al. [260] at higher temperatures. The boundary for the transition from a monolayer to a bilayer solid was determined by using parameters for the adsorption of xenon on silver [155].

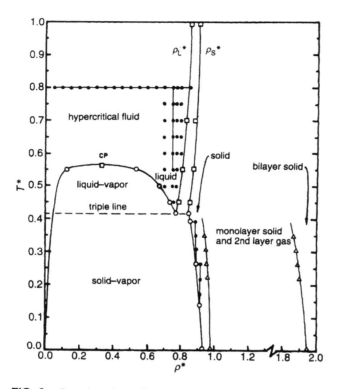

FIG. 9 Complete phase diagram for the two-dimensional Lennard-Jones system proposed by Phillips et al. [261] from computer simulation (open circles). Squares are computer simulation data of Barker et al. [260]. The triangles (from Ref. 155) indicate the equilibrium of monolayer and bilayer solids.

Phase diagrams for binary 2D L-J mixtures have been obtained from the Zwanzig perturbation theory by Scalise et al. [277]. Those preliminary results show that the 2D mixtures exhibit the same behavior as that found in the corresponding 3D fluid binary mixtures.

We now consider the main methods used and results obtained in the study of thermodynamic properties or microscopic behavior (RDF, mainly). These are summarized in Table 12 in chronological order. The computer simulations included in Table 12 (i.e., not those specifically devoted to studying phase transitions, which are given in Table 9) were performed mainly in order to obtain the RDF and thermodynamic properties of the 2D L-J fluids. The clear purposes were to study their structure and the molecular motion [226,281,283,290,300,302] to obtain a basis for testing the validity of perturbation theories [226,232,260,288,298,301–305,307–311], for direct application in monolayer adsorption [225,293], to serve as a basis in the fit of a given EoS [226,279,288], and even for pedagogical applications [239,240,296,299]. A list of data obtained for the 2D pressure can be found in the Reddy and O'Shea [279] work, which can be completed with the data obtained by our group in Refs 226, 303, and 310. More details about computer simulations are given in Section IV.E.

The theoretical studies included in Table 12 are based on calculations of the virial coefficients [195,260,285,294], applications of the Barker–Henderson (BH) [232] or WCA perturbation theories [19,280,288,297,301,303–306,311] and solutions to the Percus–Yevick (PY) and hypernetted-chain equation (HNC) [247–250] and the Born–Green–Yvon (BGY) [251] integral equations. These studies have led to a wide knowledge of the thermodynamic characteristics of the 2D L-J fluid. Nevertheless, from a practical point of view, the main interest of research on 2D L-J fluids is the possibility of serving as a model in the study of simple gases adsorbed on solid surfaces. The first attempt to find theoretical adsorption isotherms from the knowledge of properties of the 2D L-J fluid was performed by Steele [19]. In order to apply his theory of the monolayer adsorption of simple fluids onto flat surfaces, expressed by Eq. (23), Steele [19] used a modified version of the WCA perturbation theory [253]. Also, results from PY theory [210] and from computer simulations [225] have been used to obtain adsorption isotherms following Steele's theory. Nevertheless, the mathematical and practical difficulties inherent in the theoretical developments (which will be detailed in Section IV.B) or in the performance of new computer simulations similar to that of Knight and Monson [225] indicate the need to obtain analytical expressions as a more straightforward way to provide numerical values for adsorption isotherms and isosteric heats.

In order to overcome some of the above-mentioned practical difficulties and to complete the study of 2D L-J fluids, we have performed extensive computer simulations in which the 2D L-J potential is separated into its repulsive and attractive forces following the WCA proposal [226]. The main hypotheses of the WCA theory [288,301–304,311], as well as of the modified version proposed by Steele [307], were tested. As an important contribution, a new expression for the EoS of 2D L-J fluids was proposed, the ultimate objective being its application to obtain a reliable expression for the excess chemical potential needed in Eq. (23) and then to give the theoretical adsorption properties and compare them with experimental values for classical rare gases adsorbed on simple surfaces [212–214,217,312]. Other results obtained from our so-called CM equation, including the calculation of the vapor–liquid equilibrium for 2D L-J fluids [278], are summarized in Tables 9 and 12. Calculation of some thermodynamic properties and more details about the results can be found in Ref. 312. In the following subsection, we will describe the most significant EoS for 2D L-J fluids. In Section IV.C, we will compare results for the adsorption isotherms and isosteric heats obtained from diverse theoretical approximations, including our CM EoS and also the Reddy–O'Shea equation [279].

B. Equations of State for 2D L-J Fluids

Despite the variety of theoretical and numerical methods used to describe the behavior of 2D fluids, for practical applications it would be desirable to have models (analytical expressions) for the 2D EoS (i.e., a relationship among the 2D pressure, the 2D density, and the temperature).

As is well-known, the virial expansion, Eq. (3), provides a simple and convenient description of the EoS for dilute gases, has been widely used for both practical and theoretical studies, and represents a first attempt to understand the relationship between molecular interactions and the macroscopic behavior of fluids. In particular, values for the second, third, fourth and fifth virial coefficients (all in reduced L-J units) are available [16,285,286,291] for 2D L-J fluids. These values have been fitted by Reddy et al. [195] as polynomials in $(T^*)^{-1}$ for the second, third, and fourth coefficients, and as powers of $(T^*)^{-2}$ for the fifth one. The coefficients are given in Table 13. The Reddy et al. [195] results then lead to a complete description of five coefficients of the virial EoS, which has been used to obtain approximate values of the critical point; see Table 11.

More complex expressions can be obtained through theoretical calculations using perturbation theories, density functional theories, or an integral equation. However, their practical application is not always straightforward because one needs knowledge of difficult

TABLE 13 Analytical Approximations to Virial Coefficients for the 2D L-J Fluid

n	$B^*_{2D} = \sum b_i(T^*)^{-n}$	$C^*_{2D} = \sum c_i(T^*)^{-n}$	$D^*_{2D} = \sum d_i(T^*)^{-n}$	$E^*_{2D} = \sum e_i(T^*)^{-2n}$
0	0.740211	0.449788	0.206535	0.205734
1	−0.268461	0.528510	1.039480	0.067346
2	−2.459480	−2.686760	−3.263540	0.444278
3	2.732890	5.585670	4.078610	−2.18990
4	−2.155390	−4.696630	−3.725340	0.928547
5	0.914717	2.374140	1.355650	−0.680841
6	−0.206265	−0.567428	0.373639	0.275224
7	0.0018569	—	—	−0.030809

Source: Ref. 195.

concepts in statistical mechanics and sophisticated mathematical tools. For that reason, EoSs based on a fit to computer simulation results for a given model, such as the L-J fluid, over a wide range of temperatures and densities [260,279] are preferred. Nevertheless, these EoSs have two main problems: the lack of a theoretical basis, which makes it difficult to apply the EoS to different ranges and/or different systems, and their complicated analytical form (the Benedict–Webb–Rubin-type EoS contains 33 adjustable parameters), which makes their mathematical handling very complicated.

The use of semitheoretical EoSs, with a theoretical basis and containing adjustable parameters, which are obtained by reproducing computer simulation results over a wide range of temperatures and densities [214,232,278,280,288], seems to be a more reliable method [214,232,278,280,288], in particular if simple and accurate expressions can be obtained.

In addition to the classical virial expressions, five analytical expressions for the EoS of 2D L-J fluids have been proposed by Henderson [232], Barker et al. [260], Reddy and O'Shea [279], Song and Mason [280], and Cuadros and Mulero [288], respectively. They will be described next.

The first analytical expression for the equation of state of 2D L-J fluids was given by Henderson [232]. This EoS was based on Monte Carlo computer simulation results and on the Barker–Henderson perturbation theory. The agreement with previous simulations was quite good, except for some ranges close to the critical region. However, it contains nine nonlinear coefficients in its analytical expression, which have been listed only for some temperatures. For that reason, the Henderson EoS is not always applicable and has only been considered here for the temperatures for which the coefficients are known. No comparison with experimental results has been made.

A Padé approximant was proposed by Barker et al. [260] to give the 2D L-J EoS, which reproduces the first five virial coefficients and agrees with their own computer simulation results. Nevertheless, the application of that EoS seems not to be very useful in describing adsorption properties, despite their undoubted utility in studying, for example, solid–fluid transitions. First, the computer simulation results were been obtained only over a limited temperature-density range, mainly for high densities. The approximant contains nine coefficients which are different for each temperature and have been tabulated only for $T^* = 0.45$, 0.55, 0.7, and 1.0, with their interpolation or extrapolation to different temperatures being doubtful. For these reasons, the Barker et al. [260] EoS has not been used here or in previous studies to give adsorption properties.

The third EoS, proposed by Reddy and O'Shea [279] (RO equation of state) was a fit of a long list of pressure and potential energy simulation data to the following expression:

$$\varphi^* = \rho^* T^* + [c_1 T^* + c_2 (T^*)^{1/2} + c_3 + c_4 (T^*)^{-1} + c_5 (T^*)^{-2}](\rho^*)^2$$
$$+ [c_6 T^* + c_7 + c_8 (T^*)^{-1} + c_9 (T^*)^{-2}](\rho^*)^3 + [c_{10} T^* + c_{11} + c_{12}(T^*)^{-1}](\rho^*)^4$$
$$+ c_{13}(\rho^*)^5 + [c_{14}(T^*)^{-1} + c_{15}(T^*)^{-2}](\rho^*)^6 + c_{16}(T^*)^{-1}(\rho^*)^7$$
$$+ [c_{17}(T^*)^{-1} + c_{18}(T^*)^{-2}](\rho^*)^8 + c_{19}(T^*)^{-2}(\rho^*)^9$$
$$+ \{[c_{20}(T^*)^{-2} + c_{21}(T^*)^{-3}](\rho^*)^3 + [c_{22}(T^*)^{-2} + c_{23}(T^*)^{-4}](\rho^*)^5$$
$$+ [c_{24}(T^*)^{-2} + c_{25}(T^*)^{-3}](\rho^*)^7 + [c_{26}(T^*)^{-2} + c_{27}(T^*)^{-4}](\rho^*)^9$$
$$+ [C_{28}(T^*)^{-2} + c_{29}(T^*)^{-3}](\rho^*)^{11} + [c_{30}(T^*)^{-2} + c_{31}(T^*)^{-3}$$
$$+ c_{32}(T^*)^{-4}](\rho^*)^{13}\} \exp(-\gamma(\rho^*)^2) \tag{28}$$

where φ^* is the reduced spreading pressure and the coefficients c_i and γ are given in Table 14. Equation (28) has the advantage that it is valid over a wide range of pressures and temperatures (although it gives strange results outside that range). It is not currently used for practical purposes, because it lacks a solid theoretical background and has a really complex analytical form, containing 32 adjustable parameters and nonlinear terms. Nevertheless, its validity is clear and, as will be shown in Sections IV.C and IV.D, so that we have used it for comparison with both theoretical and experimental results.

Song and Mason [280] have proposed another EoS, which is based on a generalized multidimensional expression of the Carnahan–Starling [313] equation for hard spheres. The analytical expression separates the contribution of the second virial coefficient, and then the following term is obtained by using the WCA theory main hypothesis: separation of the L-J potential into its repulsive and attractive contributions, and then ignoring the influence of attractive forces. Relatively good agreement is found with the MC computer simulation results of Henderson [232]. The third, fourth, and fifth virial coefficients are reproduced only at high temperatures. In any case, the application of the EoS is not so straightforward as indicated by the authors, so that no new calculations have been performed using this EoS.

As an attempt to avoid some of the disadvantages of previous EoSs, we have [226,288] computed new molecular dynamics simulation results for 2D L-J fluids. Together with the

TABLE 14 Coefficients in the RO Equation

i	c_i	i	c_i	i	c_i
1	0.7465941518	12	−0.4654184040	23	0.07948290821
2	2.170202387	13	20.60578013	24	849.4574507
3	−3.867162453	14	2.778370707	25	−4.038923642
4	0.1955514819	15	−378.8766473	26	−13788.504769
5	−0.3080792027	16	−122.7538308	27	1.001438224
6	0.7254286978	17	127.6939405	28	−19.62817948
7	−0.8449806323	18	−373.0636670	29	−18.44924653
8	1.312097237	19	1354.795604	30	−716.4817351
9	0.1270817291	20	−0.06993741337	31	26.30515216
10	2.906037677	21	−0.04066552651	32	0.4773207984
11	−6.390601256	22	65.31068158	γ	0.6682623864

Source: Ref. 279.

theoretical fundamentals of the Weeks, Chandler, and Andersen (WCA) perturbation theory [253], we were then able to construct a new, much simpler equation (CM equation of state):

$$Z = \frac{\varphi^*}{\rho^* T^*} = (1 - y)^{-2} - \frac{\rho^*}{T^*}\alpha(T^*, \rho^*) - \frac{\rho^*}{T^*}\frac{\partial\alpha(T^*, \rho^*)}{\partial\rho^*} \tag{29}$$

where the first term constitutes the scaled particle EoS for a hard disk system, with

$$y = \tfrac{1}{4}\pi\rho^*(d^*)^2 \tag{30}$$

being a function proportional to the density and to the molecular diameter, d^*, which is clearly defined in the WCA theory, in order to give the suitable values of the repulsive part of the L-J potential, and which can be reproduced through a parametric approximation given by Verlet and Weis [314]:

$$d = \left(\frac{1.068 + 0.3837T^*}{1 + 0.4293T^*}\right) \tag{31}$$

The contribution of the attractive forces to each thermodynamic state is introduced in the expression of Z via the function $\alpha(T^*, \rho^*)$, which is also clearly defined in the WCA theory as an integral containing both the attractive part of the L-J potential and the RDF. Our computer simulation results allowed us to obtain that function and then to find a suitable and simple analytical expression for it [288]:

$$\alpha(T^*, \rho^*) = C_1(T^*) + \rho^* C_2(T^*), \tag{32}$$

where C_1 and C_2 are temperature-dependent functions expressed as

$$\begin{aligned}C_1(T^*) &= 3.8574 - 3.39554T^* + 2.20294(T^*)^2 - 0.46687(T^*)^3 \\ C_2(T^*) &= -1.00985 + 4.83792T^* - 3.19864(T^*)^2 + 0.68183(T^*)^3\end{aligned} \tag{33}$$

with a fitting error less than 2% with respect to the computer simulation values.

One of the clear advantages of the CM equation of state, aside from its simplicity (note that only a cubic dependence on temperature and a square dependence on density are introduced) and the adequate reproduction of the properties of the 2D L-J fluid [226], is that it allows the attractive and repulsive force contributions to be separated, because it is based on WCA theory. The influence of repulsive and attractive forces on the calculations of diverse thermodynamic properties, including the critical point and the density change on melting, have been studied by us in a number of articles [288,303,305–309].

An added advantage, which is important in adsorption and also in the study of phase changes, is that the expression given for the excess chemical potential with respect to the ideal gas is

$$\frac{\mu^*}{T^*} = \frac{3y - 2y^2}{(1 - y)^2} - \ln(1 - y) - \frac{\partial\rho^{*2}\alpha(T^*, \rho^*)}{\partial\rho^*} \tag{34}$$

which also separates the repulsive (two first terms) and the attractive contributions and which contains a simple dependence on the $\alpha(T^*, \rho^*)$ function. The final expression is then considerably simpler than those obtained by using other equations. The use of Eq. (34) in Eq. (23) leads straightforwardly to theoretical values for the adsorption isotherms. The results will be discussed in the following subsection, comparing them with those obtained through other methods.

C. Theoretical Adsorption Properties

From a thermodynamic point of view, there are three important adsorption properties that can be reproduced by Steele's theory: the adsorption isotherms, the spreading pressures, and the isosteric heat, by using first Eq. (23) and then Eqs. (2) and (6), respectively. In fact, the most direct comparison between theoretical results is through the calculation of the compressibility factor, Z, for the 2D L-J fluid, which is equivalent to the comparison of calculated 2D pressures. As was indicated earlier, the comparison with previously published values is one of the first tests of validity of new theories, computer simulations or equations of state. Generally, that test is made by the authors in each new proposal. Also, the results can be used to obtain an expression for the thermodynamic properties [232,279]. In other words, theoretical values of Z have been extensively compared in many articles so that we will not include it here. We will consider here the comparison of theoretical values for the adsorption isotherms and isosteric heat. We consider six theoretical approaches in that comparison:

1. Modified WCA Theory [19]. As indicated earlier, the calculation of the $\alpha(T^*, \rho^*)$ function included in the general expression for the WCA theory, Eq. (29), requires one to know the values of the RDF for each distance, density, temperature, and contribution of the attractive forces. The simplest theoretical approach is proposed in the WCA theory and is called the "high-temperature approximation," in which the structure of the fluid is supposed practically independent of the effect of the attractive inter-molecular forces, so that the RDF for a hard disk system, available from theories or computer simulations, can be used. Nevertheless, as Steele [19] and the present authors [226,302,303,307] have shown, that approximation is not valid at the 2D densities and temperatures of interest for monolayer adsorption. Steele [19] thus proposed a less restrictive approximation. For low densities, $\rho^* \leq 0.45$, he proposed using a "renormalized" perturbative intermolecular potential, obtained from a density expansion of the Neperian logarithm of the RDF. Theoretically, that renormalized potential is strongly distance and density dependent and slightly temperature dependent. Steele further supposes that it is independent of the attractive forces. As a first result, he showed that the resulting values of the 2D pressures are closer to simulation results than those obtained using the high-temperature approximation, except at high densities. He then obtained theoretical adsorption isotherms from Eq. (23) by using the high-temperature approximation for high densities ($\rho^* \geq 0.6$), the renormalized approximation for low densities, and an interpolation procedure for intermediate densities.

   The main advantage of the use of the Steele's modified WCA approximation is that one obtains theoretical adsorption isotherms for low densities, improving the results of the high-temperature approximation. Two disadvantages are the nonvalidity of the theoretical expression of the renormalized potential for intermediate densities, and that, as we have shown [307] by comparison with our computer simulation results, the renormalized potential is not fully independent of the attractive forces. In any case, perhaps the main disadvantage is the mathematical difficulty of using the expression for the renormalized potential to obtain $\alpha(T^*, \rho^*)$ in the form of a complex analytical expression for the derivative in Eq. (29) or (34), and then to perform the derivative in Eq. (23).

   An important requirement for obtaining theoretical isotherms is to take into account the dependence on Henry's constant, which cannot be defined for a perfectly flat surface. Steele [19] therefore writes the pressure in particular reduced units as a function of Henry's constant:

$$P_{\text{red}} = \frac{P\pi\sigma^2}{4K_H A}$$ (35)

The comparisons in this section were performed in those reduced units.

2. Percus–Yevick Theory [210,211]. Glandt et al. [210,211] have obtained adsorption isotherms for 2D L-J fluids by using the properties obtained from the Percus–Yevick equation [248]. The main advantage is that a large number of numerical results are available over a wide range of temperatures and densities. Thus, good agreement has been obtained in comparison with experimental results for both argon and krypton on graphite. The disadvantages are the difficulty of solving the Percus–Yevick equation if new data are required, as well as the lack of any simple and easy-to-use analytical equation for fast calculations.

3. Henderson's equation of state. [232]. Henderson's EoS can also be used to obtain adsorption isotherms [182]. However, its complicated analytical form and the fact that its nine temperature-dependent coefficients have been calculated for only four temperatures ($T^* = 0.5$, 0.7, 1.0, and 2.0) are serious disadvantages for its direct application.

4. Reddy–O'Shea (RO) EoS [279]. The RO EoS can be used for a wide range of temperatures and densities, the main disadvantage being its complex analytical form, which must be integrated to obtain the chemical potential. As we have shown in Ref 214, the expression obtained for μ^* has no regular dependence on density at low temperatures, and a succession of "crest" and "valleys" appear. These "waves" in the chemical potential disappear for $T^* > 1$ and are not present in compressibility factor calculations.

   Although proposed in 1986, the RO EoS was not used to calculate adsorption properties until 1997 [214]. No direct comparison with other theoretical values has been performed until now. As we will show later, the irregular behavior of the values of the RO chemical potential clearly influences the form of the resulting adsorption isotherms at low temperatures.

5. Computer simulation results [225]. The only computer simulation giving adsorption isotherms for 2D L-J fluids on perfectly flat surfaces has been performed by Knight and Monson [225], including data for only four reduced temperatures ($T^* = 0.5$, 0.55, 0.7, and 0.8), by using a Monte Carlo simulation. The authors made comparisons with some of Henderson's isotherms finding good agreement.

6. Cuadros–Mulero EoS [212–214,288]. The Cuadros–Mulero EoS for 2D L-J fluids, Eq. (29), is the simplest of the theoretical procedures listed. The use of the analytical expression for the excess of the chemical potential with respect to the ideal gas, Eq. (34), is clearly straightforward for obtaining both the adsorption isotherms and the isosteric heat [212–214].

Comparisons of both the adsorption isotherms and isosteric heat for 2D L-J fluids on a perfectly flat surface are shown in Figs 10–17. We note that comparisons of adsorption isotherms are not dependent on the value of Henry's constant, the reduced pressure being defined by Eq. (35).

As is shown in Fig. 10, an excellent agreement is found between the adsorption isotherms obtained from PY values and from the Henderson, the RO, and our CM EoSs at high temperatures. In Figs. 11 and 12, the Knight and Monson [225] simulation results are also

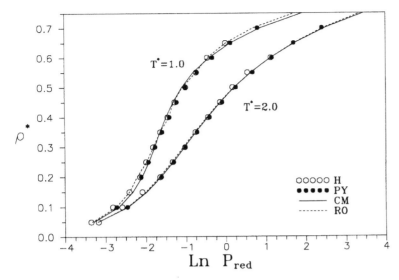

FIG. 10 Theoretical adsorption isotherms for a Lennard–Jones fluid on a perfectly flat surface at high temperatures [213]. P_{red} is defined in Eq. (35), and the adsorption isotherm in Eq. (23). H = Henderson equation of state [232], PY = Percus–Yevick integral equation [210], RO = Reddy–O'Shea equation of state [279], Eq. (28), CM = Cuadros–Mulero equation of state [288], Eq. (29).

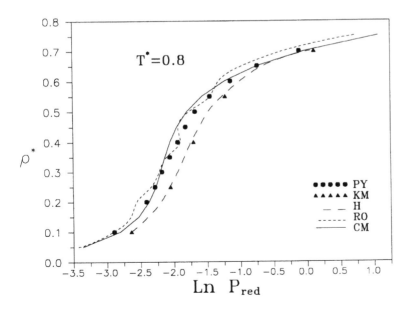

FIG. 11 Theoretical adsorption isotherms for a Lennard–Jones fluid on a perfectly flat surface [213] at reduced temperature 0.8. Legend is the same as in Fig. 10, except KM = Knight and Monson [225] computer simulation.

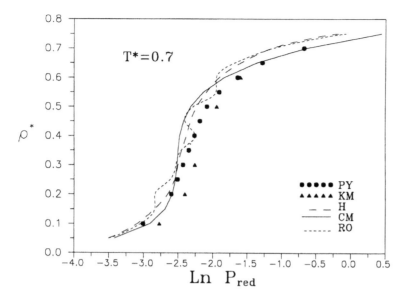

FIG. 12 Theoretical adsorption isotherms for a Lennard–Jones fluid on a perfectly flat surface [213] at reduced temperature 0.7. Legend is the same as in Fig. 10.

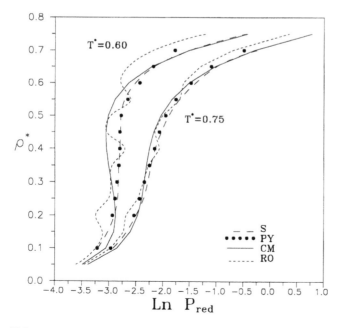

FIG. 13 Theoretical adsorption isotherms for a Lennard–Jones fluid on a perfectly flat surface at low reduced temperatures [213]. Legend is the same as in Fig. 10, except S = modified WCA theory of Steele [19].

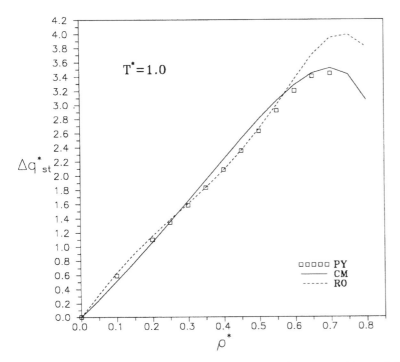

FIG. 14 Theoretical isosteric heat versus temperature for a Lennard–Jones fluid on a perfectly flat surface at reduced temperature 1.0 [212]. Δq_{st}^* is defined in Eq. (36). The legend is the same as in Fig. 10.

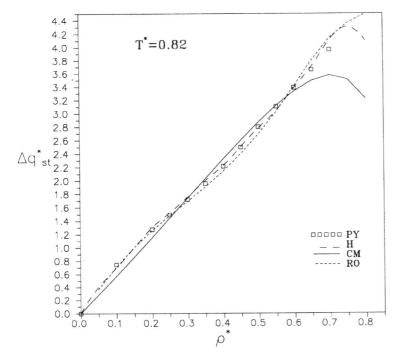

FIG. 15 Theoretical isosteric heat versus temperature for a Lennard–Jones fluid on a perfectly flat surface [212] at reduced temperature 0.82. The legend is the same as in Fig. 10.

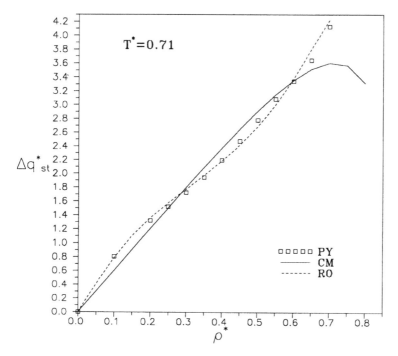

FIG. 16 Theoretical isosteric heat versus temperature for a Lennard–Jones fluid onto a perfectly flat surface [212] at reduced temperature 0.71. The legend is the same as in Fig. 10.

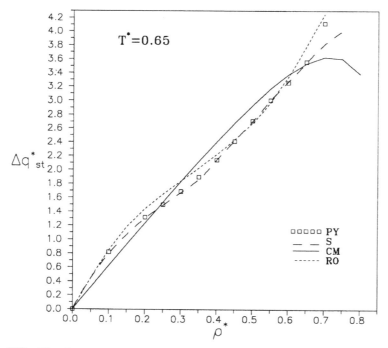

FIG. 17 Theoretical isosteric heat versus temperature for a Lennard–Jones fluid onto a perfectly flat surface at reduced temperature 0.65 [212]. The legend is the same as in Fig. 13.

included. As the temperature decreases, the differences between the theoretical results are more clearly seen, although the shape of the isotherm is practically the same, except for the RO case. Despite the good agreement between the Henderson EoS and the Knight and Monson computer results at $T^* = 0.8$, there is a greater discrepancy between them at $T^* = 0.7$. Moreover, the disagreement with other theoretical results at $T^* = 0.8$ indicates that both the computer simulation results and the use of the Henderson EoS, whose coefficients are known only for particular temperatures, must be carefully revised and cannot be used with any guarantee to reproduce theoretical adsorption isotherms and, hence, in the comparison with experimental data.

In Fig. 13, Steele's isotherms are also included. They are in very good agreement with the PY results. The irregular behavior of the RO equation is now clearly manifest.

The main conclusion of the preceding comparison is that the CM equation, which is clearly the simplest analytical expression, is valid over a wide range of temperature and densities for the calculation of theoretical adsorption isotherms [213]. Moreover, there is disagreement with other results only for the lowest temperatures, and even then, it is not excessively significant. The use of more complicated equations, theories, or computer simulation results seems to be unnecessary for the purpose of obtaining adsorption isotherms for a 2D L-J fluid on a perfectly flat surface.

Theoretical values of the isosteric heat of adsorption can be obtained by using Eqs. (6) and (23), giving, in reduced L-J units [19,212,217],

$$\Delta q_{st}^* = q_{st}^*(\rho^*) - q_{st}^*(\rho^* = 0) = T^{*2}\left(\frac{\partial(\mu^*/T^*)}{\partial T^*}\right)_{\rho^*} \tag{36}$$

where values of the derivatives of the chemical potential constitute the only needed input.

Results for Δq_{st}^* have been obtained by Steele [19] at only two temperatures, $T^* = 0.53$ and $T^* = 0.65$, and by Henderson [232] at only $T^* = 0.82$. We have obtained values from PY adsorption isotherms, as well as from the RO and CM equations. Comparisons are given here in Figs. 14–17. As can be seen, the CM isosteric heat is practically linear except for points near the complete monolayer, whereas the other approximations lead to a more curved shape, especially at low temperatures. The greatest deviations are at high densities (i.e., near the maximum value of the isosteric heat). Thus, for example, for $T^* = 1$ (Fig. 14), the CM equation agrees very well with the PY values, whereas the RO equation gives a higher maximum value. For $T^* = 0.82$ (Fig. 15), the maximum value of the CM curve is clearly below the others, with the RO giving the highest value. Excellent agreement is found between PY and RO values at $T^* = 0.71$ (Fig. 16) and between PY and the Hendersons equations at $T^* = 0.65$, (Fig. 17).

In conclusion, it is not clear which is the most exact theoretical approach to reproduce values of the isosteric heat of 2D L-J fluids on perfectly flat surfaces following the Steele approach. None of the theories or equations used can be considered fully exact, and some discrepancies are found, especially at low temperatures and high densities (near the monolayer coverage). The CM equation, although clearly the simplest approach, gives adequate values for the isosteric heat, at least for densities below the monolayer coverage. A comparison with experimental results will be described next.

D. Comparison with Experimental Results

It is essential to contrast the predictions of the theoretical models with the experimental data. This is always a complicated matter, mainly due to the deviations of real systems from the theoretical models' assumptions of a L-J 12-6-type interaction, which is only a first approximation to the true shape of the interaction potential. The comparison has therefore to be limited to classical rare gases or fluids whose molecules are approximately spherical (CO, CH$_4$, etc.). Even when these approximations can be assumed to be valid, one still has to solve the problem of the choice of suitable parameters for the L-J potential [ε and σ in Eq. (12)]. It has to be borne in mind that the said parameters not only influence the results, but even the choice of which thermodynamic states to compare, because the temperature and density both depend on them. In any case, the chosen L-J parameters must be understood to be "effective" parameters in that they include the possible deviations between the theoretical model and the experimental results.

When thermodynamic properties are studied, there are three possible comparisons between the experimental results and the theoretical or simulation results [226]:

1. Comparison between the experimental adsorption isotherms (adsorption pressure as a function of the amount of matter adsorbed) and the theoretical expression for the adsorption pressure as a function of the density of the monolayer, obtained from Steele's two-dimensional approximation, Eq (23). In this case, Henry's constant must be previously calculated from experimental data or through the gas–solid virial coefficient [210,211], Eq. (4).
2. Comparison of the spreading pressure or the compressibility factor obtained from experimental data through Eq. (2) with that obtained from the 2D model, Eqs. (28) or (29) for example.
3. Comparison of the experimental isosteric heat with that obtained from Steele's theoretical expression, Eq. (36). In this case, the value of the isosteric heat at zero coverage, $q_{st}(\rho = 0)$, is needed as input.

In all cases, the relationship between the reduced units used in the theoretical procedures and the real units used in the experimental results must be established by using Eqs. (24)–(27) and also

Number of moles adsorbed per gram of solid

$$N = \frac{\rho^* A}{N_a \sigma^2} \tag{37}$$

Spreading pressure

$$\varphi = \varphi^* \frac{\varepsilon}{\sigma^2} \tag{38}$$

Isosteric heat

$$\Delta q_{st} = R\left(\frac{\varepsilon}{k}\right)\Delta q_{st}^* \tag{39}$$

where N_a is Avogadro's number, A is the specific surface area (m^2/g), and R is the perfect-gas constant. Note that if the 2D compressibility factor, Z, is used instead of the spreading pressure, the results are always (from experiments or from theories) given in adimensional units.

The simplest rare gases, Ar, Kr, and Xe, will be used here for comparison between theoretical and experimental results. Specifically, for the comparison of adsorption isotherms and spreading pressures, we have considered the same experimental results as those used in the work of Glandt et al. [210,211]. They considered comparison with PY results in the case of adsorption isotherms and isosteric heat and comparison with simple EoSs for 2D L-J fluids in the case of spreading pressures. These results are adsorption isotherms for argon onto highly graphitized carbon black, P33 (2700°C) with $A = 12.5$ m^2/g, measured by Constabaris et al. [32] at temperatures from 109.956 K to 220.393 K, and adsorption isotherms for Kr on Sterling FT-D5 graphite with $A = 11.5$ m^2/g, measured by Putnam and Fort [42,43] at temperatures from

TABLE 15 Henry Constants for Argon and Krypton on Graphite (Using Theoretical Data of Glandt et al. [210])

System	T (K)	K_H (atm$^{-1}\mu$mol/g)
Argon	109.956	1,062.1
	120.257	448.5
	130.138	222.7
	133.072	184.2
	137.107	144.3
	144.607	117.9
	145.114	92.2
	150.140	71.3
	158.077	49.1
	166.135	34.8
	175.082	25.0
Kr	94.720	219,858.6
	99.337	106,347.3
	104.491	51,151.5

94.720 to 104.491 K. Values of B_{AS} given by Glandt et al. [210] are also used, the corresponding values for K_H being given in Table 15. Sources of data for isosteric heats are less readily available, and we have only considered one temperature for each of the studied systems, as will be explained next.

Only the most useful expressions (i.e., the RO and CM equations) will be considered. The selected L-J parameters are $\varepsilon/k = 116$ K and $\sigma = 0.336$ nm for argon, which are the same as were used by Glandt et al. [210]; see Table 8. For Kr, we use $\sigma = 0.3573$ nm given by Glandt et al. [210]. Nevertheless, as the (ε/k) value proposed by Glandt et al. [210] produces large discrepancies with both the RO and the CM equations, we have used $\varepsilon/k = 142.8$ K, proposed by Wolfe and Sams [36]. For xenon, only considered in the calculation of the isosteric heat, we use two different pairs of values, as will be explained. For reference, the 3D L-J parameters proposed by Cuadros et al. [203] are 111.84 K and 0.3623 nm for argon, 154.87 K and 0.3895 nm for krypton, and 213.96 K and 0.426 for xenon.

1. Adsorption isotherms. Examples of experimental results and theoretical predictions of RO and CM equations for adsorption isotherms of argon and krypton are shown in Figs. 18–21. In general, the deviations obtained are of the same order for the two theoretical equations and are greater at low temperatures or at high coverages. Very good agreement is found for adsorbed argon (Figs. 18–20), where the adsorption isotherms are practically straight lines at high temperatures. Moreover, the reduced temperatures rise from $T^* = 0.95$ to 1.9, so that the "waves" of the RO adsorption isotherms (shown in Figs. 12 and 13) do not appear, leading to quite adequate results. Nevertheless, in the case of adsorbed Kr, the reduced temperatures are below 0.75, so that the RO equation produces strange values for the adsorption pressure, as is clearly seen in Fig. 21. The CM equation presents a well-defined behavior, although some discrepancies are found with respect to the experimental results for Kr at intermediate densities and the highest temperature (Fig. 21).

Our results show that Steele's model for monolayer adsorption is a good representation of real systems such as rare gases adsorbed on graphite. Deviation with respect to experimental results with argon or krypton by using the RO or CM

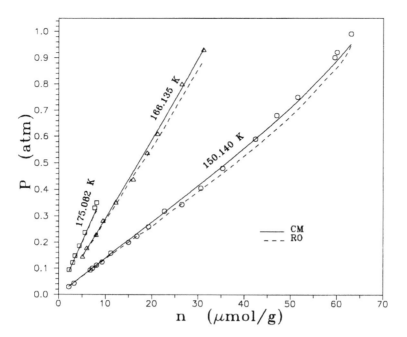

FIG. 18 Adsorption isotherms of argon on graphite at high temperatures. Dots: experimental data [32]; lines: theoretical results from Eq. (23) by using the Reddy–O'Shea and the Cuadros–Mulero expressions for the chemical potential [214].

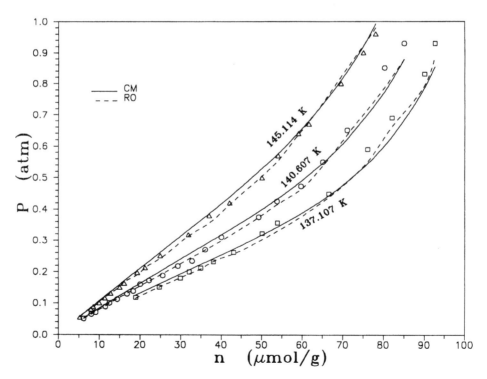

FIG. 19 Adsorption isotherms of argon on graphite at intermediate temperatures [214]. The legend is the same as in Fig. 18.

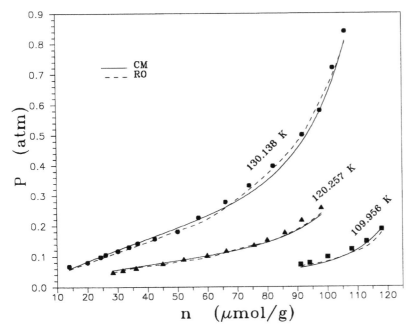

FIG. 20 Adsorption isotherms of argon on graphite at low temperatures [214]. The legend is the same as in Fig. 18.

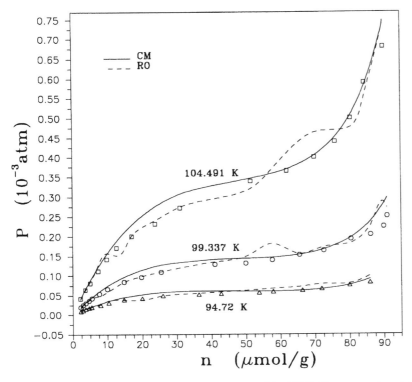

FIG. 21 Adsorption isotherms of krypton on graphite [214]. Dots: experimental data [42,43]; lines: as indicated in Fig. 18.

equations are less than or equal to 10% for each temperature considered [214]. The agreement is better with respect to the experimental results for adsorbed argon, surely due to the fact that the data for Kr are closer to the theoretical critical point (which does not exist in the real system; see Table 1). Although the largest deviations correspond to low temperatures, they are not so large as to support Toxvaerd's conclusion [293] stating that the mobile layer model is invalid in this range. We note that the deviations found by Toxvaerd may be due to the use of unsuitable L-J parameters.

2. Spreading compressibility factor. The spreading compressibility factor, Z, is defined by Eq. (2), so that it must be calculated from the experimental results of adsorption isotherms. Here, we have used the Z values given by Glandt et al. [210], which were obtained from the Constabaris et al. [32] experimental results for Ar on graphite. We note that the calculated values were obtained by first smoothing the results of Eq. (2) and then recalculating the values for constant intervals of N. For Kr, we used values obtained directly (without smoothing) by us from Eq. (2) with the experimental results of Putnam and Fort [42,43].

Both the RO and the CM equations reproduce the Z values with average absolute percentage deviations less than 10% for each analyzed isotherm. As is shown in Fig. 22, for Ar on graphite the CM equation leads to better agreement with experimental data at high temperatures, the RO equation being more appropriate at low temperatures [214]. Moreover, for low temperatures, the CM equation is in good agreement with the experimental results for N less than approximately 50 μmol/g, where the CM and RO equations intersect. For the greatest N values, the RO equation reproduces the experimental results better, which is to be expected taking into account the complex analytical dependence of Z on density in this equation.

Figures 23 and 24 show the spreading compressibility factor for Kr on graphite at two temperatures. Because of the scale used on the y axis for values of Z, these figures seem to indicate a poorer agreement than in the case of the adsorption

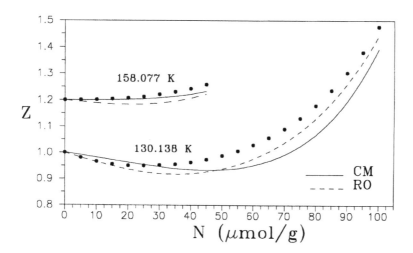

FIG. 22 Spreading compressibility factor, Z, for argon on graphite at two temperatures [214]. Dots: calculated by Glandt et al. [211] from experimental results; lines: obtained by us [214] from the Reddy–O'Shea Eq. (28) and Cuadros–Mulero [Eq. (29)] equations of state. At $T = 158.077$ K, we show values of $Z + 0.2$.

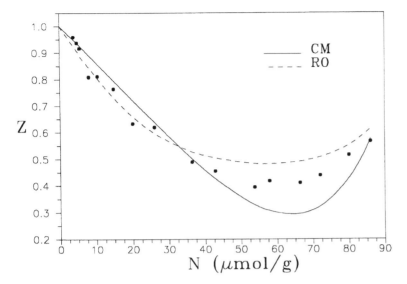

FIG. 23 Spreading compressibility factor for Kr on graphite at $T = 94.72$ K. Dots: calculated by Mulero and Cuadros [214] from experimental results [42,43]; lines: as indicated in Fig. 22.

isotherms (Fig. 21). The contrary is the case (i.e., the mean deviations for Z are always less than the deviations for P [214]). As in the case of the adsorption isotherms, the deviations are greater than those obtained for adsorbed argon. Both RO and CM equations fail to reproduce the minimum value of Z, finding better agreement for low or high densities. We also note that the "waves" of the chemical potential values obtained from the RO equation are not present in the spreading pressure, so that they

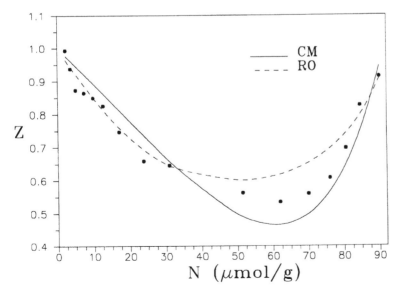

FIG. 24 Spreading compressibility factor for Kr on graphite at $T = 104.491$ K. The legend is as in Fig. 23.

must be due to the calculations performed in order to derive the μ values. In any case, the use of a complex analytical expression such as the RO equation is not needed to reproduce experimental values of the spreading pressure, which can be reproduced with a cubic-in-density EoS such as the CM one.

3. Isosteric heat. In order to complete the test of accuracy for the 2D Steele approxima-tion, we have also considered the calculation of the isosteric heat [217] and compared it with data for Ar on graphite P33 at $T = 84$ K with $q_{st}(\rho = 0) = 9.6$ kJ/mol [31], Kr on Sterling FT-D5 graphite at $T = 104.49$ K [42,43] with $q_{st}(\rho = 0) = 12.6$ kJ/mol, and Xe on grafoil at $T = 195.5$ K with $q_{st}(\rho = 0) = 17.3$ kJ/mol [48].

For Ar and Kr, comparison with respect to the CM and RO equations is shown in Figs. 25 and 26, respectively. Although there seem to be greater absolute deviations in the case of argon, we note that the absolute mean percentage deviations are 2.8% and 2.7% for the CM and RO equations, respectively, being 1.7% and 1.4% for Kr. These can be seen to be good results by taking into account that the L-J parameters used are those given by Glandt et al. [210], obtained by considering a best fit to the theoretical PY values.

To see the influence of the choice of the L-J parameters, we have considered two different pairs of values for Xe on graphite: 194 K and 0.412 nm, proposed by Wolfe and Sams [36], and 183 K and 0.39 nm, recently proposed by us [217]. The results are displayed in Fig. 27. The

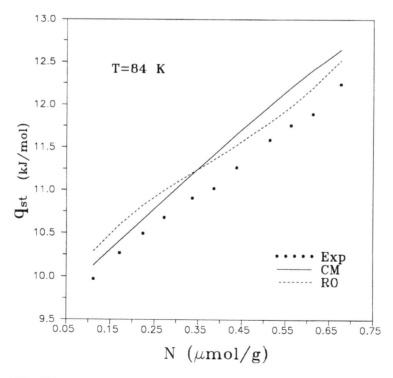

FIG. 25 Isosteric heat for argon on graphite at $T = 84$ K. Dots: experimental results [31]; lines: theoretical results from the Reddy–O'Shea [Eq. (28)] and the Cuadros–Mulero, [Eq. (29)] equations of state.

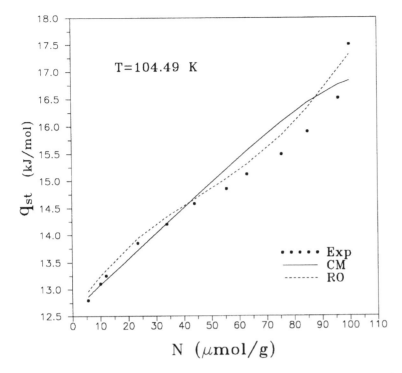

FIG. 26 Isosteric heat for krypton on graphite at $T = 104.49$ K. Dots: experimental results [42,43]; lines as indicated in Fig. 25.

absolute mean deviation of the CM and RO expressions is 1.2% in both cases when the Wolfe and Sams [36] parameters are used and 0.6% and 1% with our parameters. As can be seen in Fig. 27, the greatest deviations are found beyond the maximum value of q_{st}, which indicates the completion of the monolayer.

Once more, the validity of Steele's approach and of the use of 2D L-J EoS is clearly shown. Although the choice of the L-J parameters can influence the final results, suitable effective values can always be found. Precedent figures show that even in the case when these L-J parameters are not specifically chosen (i.e., they are taken of other sources), results continue to be adequate. Also, the use of such a complex expression as the RO one is not needed, the CM EoS being the simplest approach.

A comparison of theoretical and experimental results for the isosteric heat of adsorption for other substances has been described in Ref. 217.

In conclusion, the preceding results permit one to establish the validity of the two-dimensional Steele approximation, despite the simplifications involved in it, as a first step in the development of a more complex theory of adsorption. It has been shown how different experimental results for rare gases on graphite can be reproduced by starting from molecular considerations. In particular, the use of the Cuadros–Mulero expression for the equation of state of two-dimensional Lennard–Jones fluids seems to be the most straightforward way to apply the aforementioned two-dimensional approach. The use of more complex expressions, such as the one proposed by Reddy and O'Shea, requires more calculation effort and does not lead to significantly better results.

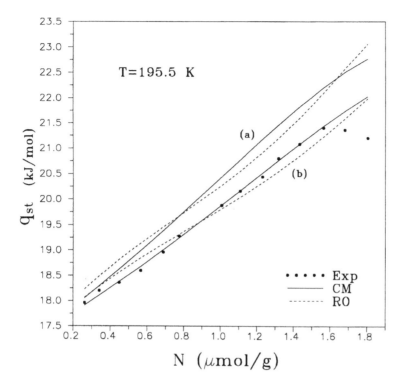

FIG. 27 Isosteric heat for xenon on graphoil at $T = 195.5$ K. Dots: experimental results [48]; lines: as indicated in Fig. 25, being obtained by using different L-J parameters: (a) $\varepsilon/k = 194$ K and $\sigma = 0.412$ nm [36] and (b) $\varepsilon/k = 183$ K and $\sigma = 0.39$ nm [217].

Suitable or effective L-J parameters can be obtained for each case. Maybe the best way to define these parameters is via a good simultaneous agreement with experimental results for adsorption isotherms, spreading pressure and isosteric heat for the same system.

Deviations from the two-dimensional approximation can be studied by introducing more terms into the configuration integral perturbation expansion, which includes the effects of the periodic nature of real solid surfaces. Although some preliminary results were obtained by Monson et al. [182,224] through a certain number of simplifications, more extensive calculations are needed in order to estimate the contribution of each of the terms and to give a more realistic statistical-mechanics description of adsorption phenomena.

E. Overview of Computer Simulations

In the preceding subsections we have considered computer simulations as a useful tool for relating the microscopic and the macroscopic behaviors of the substances involved, to test theoretical hypotheses, or as an aid in interpreting experimental results. They thus serve as a "bridge" between theory and experiment. In this subsection, more details will be given about computer simulations of 2D L-J fluids as models of adsorbed fluids and computer simulations which include a model for the adsorbent solid. These details will not include the techniques used or programming languages, which are available in other books [14,22].

As is well known, applications of adsorption in major industrial areas (adhesives, catalysts, colloids, etc.) require a detailed understanding of the phenomena from a molecular or microscopic point of view, which, in a great number of cases, can only be derived from computer simulations [14]. For example, computer simulations permit one to resolve uncertainties associated with surface impurities, heterogeneity, or the presence of micropores.

Computer simulations also constitute an important basis for the development of the molecular theory of fluids. They could be regarded as quasiexperimental procedures to obtain datasets that connect the fluid's microscopic parameters (related mainly to the structure of the system and the molecular interactions) to its macroscopic properties (such as equation of state, dynamic coefficients, etc.). In particular, some of the first historical simulations were performed using two-dimensional fluids to test adaptations of commonly used computer simulation methods [14,22]: Monte Carlo (MC) and molecular dynamics (MD). In fact, the first reliable simulation results were obtained by Metropolis et al. [315], who applied the MC method to the study of hard-sphere and hard-disk fluids.

The first qualitative computer simulation of 2D Lennard–Jones fluids was carried out by Rosenbluth and Rosenbluth [289]. Fehder [281] performed the first quantitative simulation of 2D L-J fluids using MD. Other significant computer simulations for 2D L-J fluids were performed by Toxvaerd [293], who compared the results with those obtained in models of perfectly mobile adsorption. Results were obtained for two temperatures, one above the 2D critical temperature for Kr ($T = 106.8$ K, $T_c = 87$ K being given by Thomy and Duval [37]) and another just above the critical point ($T = 86.26$ K). For this last temperature, the computer simulation data for the 2D pressure exhibits a van der Waals loop, indicating that the critical temperature is in between the two reported temperatures. The liquid–solid transition was clearly manifested as van der Waals loops in the pressure isotherms at reduced densities greater than 0.8. Values of the 2D pressure were compared with the spreading pressures obtained from the experimental data of Thomy and Duval for Kr on graphite. From the comparison, Toxvaerd [293] concluded that the mobile-layer model fails at low temperatures, that the liquid–solid-phase transitions in the adsorption monolayer appear at lower densities than in a 2D L-J fluid, and that the adsorbed liquid layer is compressible, in contrast to the 2D L-J liquid.

Nevertheless, these results must be seen now with caution. First, the L-J parameters used were values for bulk 3D Kr, against the conclusions of Wolfe and Sams [36], for example. Second, if the result given by Larher [68] for the 2D critical temperature for Kr on graphite is taken into account ($T_c = 85.3$ K), then the two simulated temperatures are supercritical. Nonetheless, Toxvaerd's [293] results represent historically the first clear attempt to compare computer simulation results directly with experimental results for adsorption.

Other interesting computer simulations for the 2D L-J system are listed in Tables 9–12. Among them we would emphasize the simulation performed by Henderson [232], who analyzed the validity of the Barker–Henderson perturbation theory and obtained an analytical expression for the EoS of the 2D L-J fluids, as was indicated in Sections IV.A and IV.B.

As has already been indicated, a large number of computer simulations have been performed to study phase transitions. In particular, some simulations have predicted the existence of the "hexatic" phase [256], whereas others have not [132]. Moreover, as Toxvaerd [264] has indicated, simulations with a very large number of particles are needed to give a definitive conclusion about the melting order. More complex simulations, including the calculation of the angular correlation function [262,265] have indicated a first-order transition.

In order to advance toward the simulation of adsorption properties, computer simulation values of the chemical potential of 2D L-J fluids were first obtained by Knight and Monson [225]. Also, the pressure, energy, and RDF values were obtained for $T^* = 0.5$ to 0.8. As Knight and Monson [225] have shown, Barker et al. [260] had computed a strange value for the 2D pressure at $T^* = 0.55$ and $\rho^* = 0.55$, which has led to a poor estimate of the critical point for 2D L-J fluids (see Table 11).

As has already been indicated, none of the above-mentioned computer simulations have given a complete (wide density and temperature ranges) and rigorous characterization of statistical mechanics theories of 2D fluids. In order to go more deeply into the theoretical study of 2D L-J fluids, more extensive computer simulations have been performed by us [226,288]. In particular, the results have permitted us to test the validity of the WCA perturbation theory and to propose a simple and accurate analytical expression for the equation of state of 2D L-J fluids [Eq. (29)], which can be applied to obtain properties of physisorbed systems as described in preceding subsections. Other applications are given in Table 12.

In all of the above simulations, the system considered is an isolated 2D fluid, such as models which describe the behavior of a monolayer of adsorbate without taking into account the influence of the adsorbent. Nevertheless, a great number of computer simulations of more realistic adsorbed systems have also been performed and are summarized in Table 16. It should be noted that, also in this case, the simulations have dealt with explaining the changes of phase for the adsorbates [14].

There have been three widely used methods to simulate the behavior of adsorbed systems [316]: grand canonical Monte Carlo (GCMC), Gibbs ensemble Monte Carlo (GEMC), and molecular dynamics (MD). The GCMC has the main advantage that the chemical potential is known as input. Nevertheless, this method does not permit one to see interfaces, fails at high density, and yields hysteresis loops. The GEMC method is faster for mixtures. However, the chemical potential is not known, and as in the preceding case, the interfaces cannot be seen and

TABLE 16 Computer Simulations Including Adsorption of 2D L-J Fluids onto Theoretical Surfaces

Method	Adsorbate	References
MC	Argon	[179,180] Rowley et al., 1976
MD	Argon	[131] Hanson et al., 1977
MC	Argon	[317] Nicholson et al., 1977
MC	Krypton	[318] Lane and Spurling, 1978
MD	Krypton	[207] Hanson and McTague, 1980
MC	2D L-J	[319] Prasad and Toxvaerd, 1980
MC	Krypton	[320] Sokoloff, 1980
MD	Krypton	[208,209] Abraham et al., 1982, 1984
MD	Xenon	[216,321] Abraham, 1983, 1984; [322] Koch and Abraham, 1983
MD	Krypton	[323] Koch and Abraham, 1986
MD	2D L-J	[225] Knight and Monson, 1986
MD	2D L-J mixtures	[324] Knight and Monson, 1987
MC	Helium	[238] Abraham and Broughton, 1987
MC	Helium	[325] Broughton and Abraham, 1988
MD	Krypton	[206] Bhethanabotla and Steele, 1988
MD	Xenon	[326] Suh et al., 1989
MD	Argon	[218,327] Cheng and Steele, 1989, 1990
MC	Helium	[328] Abraham et al., 1990
MC	2D L-J	[329] Patrykiejew et al., 1995
MC	2D L-J	[276] Jiang and Gubbins, 1995
MD	2D L-J mixtures	[330] Leptoukh et al., 1995

the results fail at high density. The MD method permits one to see the interface and yields dynamic information. In this case, the chemical potential is not known and some difficulties are found in simulating phases of similar density [316].

The results of the said simulations have been compared with theoretical and/or experimental results. Good agreement was found, for instance, between the simulation results of Lane and Spurling [318] and the experimental results of Thomy and Duval [37] for Kr on graphite. Another example is the possibility of introducing improvements in the classical adsorption equations. Thus, Steele [219] has developed a modified Frenkel–Halsey–Hill (FHH) theory based on computer simulation results for the multilayer adsorption of argon on graphite performed by Rowley et al. [179,180].

In some cases, the adsorbate is considered merely as a theoretical 2D L-J system [225, 319,324,329], whereas, in other cases, the suitable parameters simulating the behavior of rare gases have been used. Thus, the Knight and Monson [225] MC simulations cited earlier included calculations for a 2D L-J gas on a solid homogeneous but structured surface. They show that the potential distribution method used in those simulations is useful even for systems in which the surface corrugation leads to large inhomogeneities in the adsorbate densities. In particular, adsorption isotherms on the (111) and (100) faces of a face-centered-cubic crystal of the same molecules as the adsorbate were obtained for $T^* = 0.8$ and 1.0 [225]. By comparing with results obtained from the Henderson EoS for 2D L-J fluids, Knight and Monson [225] noted that the amount of adsorption is increased relative to a system without corrugation, particularly at the low temperature considered. Thus, in this case, the effect of the surface corrugation is considered as stabilizing the monolayer, such has also been observed in experimental results for simple gases on lamellar halides [331].

Patrykiejew et al. [329] have also simulated the behavior of 2D L-J fluids onto the (100) face of a face-centered-cubic crystal. Nevertheless, the Knight and Monson [225] work was not mentioned, so that no comparison of results was performed. To model the gas-surface potential, Patrykiejew et al. [329] used the first five terms of Steele's Fourier series [Eq. (14)] for a perpendicular reduced distance less than or equal to 2.5. The results show that at low temperature, the structure of the monolayer film depends strongly on the gas-surface potential corrugation, as well as on the size of the adsorbed atoms. Also, the influence of the corrugation on the melting transition is studied, indicating a different structure of the solid phase. Unfortunately, definitive conclusions about the nature and order of the observed phase transitions were not obtained.

A GEMC simulation of a 2D L-J onto a graphite-like substrate has been made by Jiang and Gubbins [276], the solid–fluid interactions being represented by a 10-4-3 potential. The vapor–liquid equilibrium was studied and critical properties for the adsorbed L-J system were found to be $T_c^* = 0.500 \pm 0.002$ and $\rho_c^* = 0.368 \pm 0.005$, which are not very different from those obtained for the isolated 2D L-J system (see Table 11). They concluded that because of the large density fluctuations in 2D fluids, the effect of the substrate on the coexistence properties is marginal. Moreover, comparison with experimental data for adsorbed methane showed that the ε L-J parameter must be less than the corresponding 3D value, in agreement with other earlier results for rare-gas adsorption.

We shall now consider adsorption of rare gases on graphite. As is shown in Table 16, the first simulations for "real" adsorbed phases included the argon–graphite system. In particular, Rowley et al. [179,180] made simulations of argon molecules adsorbed in the potential field of a plane, uniform homogeneous (graphite-like) solid at 80 K and 120 K, mainly for coverages above a statistical monolayer. The adsorbate–adsorbate pair potential was taken to be a Lennard–Jones 12-6 potential with 3D L-J parameters for argon (see Table 8). The adsorbate–adsorbent potential was taken to be a 9-3 potential. Moreover, a 9-3 potential was used to model the interactions between an adsorbate molecule in the surface zone and the whole of the homogeneous adsorbent.

In general, the results show the main features of the real Ar-graphite system as found by experiments. In particular, a phase transition in the first layer at 80 K is clearly observed. Comparison with purely theoretical models indicates that the simple BET theory and its modifications fit the isotherms fairly well, but layer filling and the isosteric heats are poorly reproduced. Rowley et al. [180] indicated that the success of the BET theory for mutilayer adsorption rests on apparently fortuitous but surprisingly universal compensation effects. Also, FHH theory is found not to be fully adequate to describe the simulation results.

The behavior of Ar on graphite has also been simulated through MD methods [218,327] for temperatures from 50 K, where the films are completely solid, up to 110 K, where the films are completely liquid. Both monolayer and trilayer films were considered, and computer simulation results indicated that melting occurs layer-by-layer, the melting transition for the monolayer being of first order.

Krypton on graphite has been extensively simulated, mainly due to the interesting phase transitions that it presents. In addition to the aforementioned Lane and Spurling [318] simulation, a great number of results have been obtained by Abraham and coworkers [208,209,323] and also by Bhethanabotla and Steele [206] (see Table 16). The Abraham and co-workers' simulations included the study of the incommensurate phase and the commensurate–incommensurate transition of monolayer krypton on graphite over a broad range of temperatures [208,209], the results being in good agreement with some experimental data [113].

The MD simulation of Bhethanabotla and Steele [206] reproduces accurately experimental isotherms for Kr on graphite at temperatures near 100 K. The Kr–Kr interaction is modeled through a L-J potential with a well depth, ε, 15% lower than that corresponding to the 3D gas phase for Kr (see Table 8). Once more, the validity of the L-J potential is clear.

Computer simulations of both xenon and helium have also been performed by Abraham and co-workers [216,321,322], with interesting results. Thus, in the computer simulation study of the melting transition of near-monolayer xenon films on graphite, Abraham has shown the existence of a solid-liquid phase coexistence and that the phase boundaries can be accurately given by the 2D L-J phase diagram [216]. The results indicate that the continuous phase transition observed in the x-ray experiments of Heiney et al. [114] can be explained by the exchange of atoms between the first and second layers, with a very small temperature range where the melting would appear to be continuous if averaged over time.

Suh et al. [326] have shown, from MD simulations, the rich phenomenology that finite patches of adsorbed xenon at low coverages exhibit. The results are also consistent in many features with those of Abraham and co-workers.

With respect to the computer simulations of helium on graphite, we would only note that they reproduce the presence of fluid, commensurate and incommensurate solids, and a reentrant fluid phase [238,325]. The phase diagram obtained is similar to that proposed by Ecke and Dash [237], which differs from the more recent proposal of Greywall [106] (Fig. 7) and does not include a liquid–vapor transition.

Extension of the MC simulation to binary mixtures of 2D L-J fluids, in both the absence and the presence of an external field, has also been performed by Knight and Monson [324], in particular by considering parameters for argon–methane and argon–nitrogen interactions. Moreover, the phase separation of 2D fluid mixtures has been investigated by Leptoukh et al. [330] from MD simulations.

Finally, we indicate that Monte Carlo simulations of Ar, Kr, and Xe on silicalites and Xe on rutile have been performed, the results being compared with experimental data [86,87,215,332].

V. CONCLUSIONS AND REMARKS

In this chapter, the main properties of adsorbed rare gases, the thermodynamic properties and phase diagrams, have been considered. The interest in investigating their behavior and in reproducing experimental results served as the basis for presenting the theoretical developments and the models and computer simulations used. In particular, the study of rare-gas adsorption is an excellent way to introduce the statistical-mechanics point of view for studying general aspects of adsorption. Moreover, practical examples of application of the general theory were presented and the results were compared with experiments. The main conclusions and some necessary remarks are the following:

Note that the systems studied here are mainly monolayer films of rare gases physisorbed on graphite. There have only been a few mentions of other systems, such as multilayer adsorption or adsorption onto heterogeneous surfaces.

Although phase diagrams for adsorbed rare gases have been considered from the viewpoints of experiment, theory, and computer simulations, no new scientific contributions have been made here. This means that all of the information is available in the long list of references we have included. Our contribution has been to summarize and review the great quantity of information that is to be found in those references.

The adsorption of rare gases has been extensively studied experimentally using thermodynamic and calorimetric methods and also scattering and diffraction methods. Nevertheless, the large number of results have not always been interpreted in the same way, and serious discrepancies are to be found, especially in the study of phase transitions.

Although the main parts of the phase diagram of rare gases on graphite have been clearly established, no clear consensus exists about the range of existence of some of the phases, especially for solid phases below the monolayer completion. A definitive theory has not been established.

Despite the possible practical applications of the classical models and virial approximations, Steele's statistical-mechanics theory constitutes the most formal and useful model of the behavior of adsorbed rare gases.

In order to apply Steel's theory, the Lennard–Jones potential for the adsorbate–adsorbate interaction and Steele's Fourier series for the adsorbate–adsorbent interactions are the most suitable ways to relate the microscopic and macroscopic adsorption properties.

There is no agreement, however, about which effective Lennard–Jones parameters must be used to describe the behavior of adsorbed rare gases. Here, we have proposed that the most suitable method might be to simultaneously fit experimental data for the adsorption isotherms, spreading pressure and isosteric heat. Even in this case, the most suitable values could well be different for different theoretical approaches.

Although a large number of theoretical approaches can be used to describe 2D Lennard–Jones systems, the Reddy–O'Shea and the Cuadros–Mulero expressions for the equation of state are the most useful from a practical point of view. In particular, the Cuadros–Mulero equation is the simplest expression and can be easily applied over a wide range of temperatures and densities. Moreover, it is based on the Weeks–Chandler–Andersen perturbation theory, thus permitting the study of the effects of the intermolecular repulsive and attractive forces.

Theoretical results for both adsorption isotherms and the isosteric heat of 2D Lennard–Jones fluids on perfectly flat surfaces have been obtained by using Steel's theory together with the Reddy–O'Shea and Cuadros–Mulero equations of state. Some discrepancies are observed: mainly the inability of the Reddy–O'Shea equation to give suitable adsorption isotherms at low temperatures and of the Cuadros-Mulero equation to give an adequate value for the maximum of the isosteric heat at low temperatures. In general, however, the agreement is good. This means that it is unnecessary to use more complex expressions than the Cuadros–Mulero equation to obtain adequate results.

These theoretical results have been compared with experimental results for adsorption isotherms, spreading pressures and isosteric heats of argon and krypton on graphite, as well as for the isosteric heat of xenon on graphite. Except, in this last case, the Lennard–Jones parameters used were taken from other sources (i.e., they were not specifically fitted). In general, the results indicate the adequacy of Steele's theory, even in its simplest form, for reproducing the behavior of real systems such as rare gases on graphite.

Steele's theory can be further developed by including deviations from the two-dimensional approach and the corrugation of the adsorbate–adsorbent interactions. This could permit one to find similarly adequate results for more complex adsorbates. Some advances have already been made by other groups, and we also are now working on obtaining a more reliable and simple way to include these deviations.

The study of the adsorption of rare gases onto well-characterized surfaces is a clear historical example of how the combination of experimental techniques, statistical-mechanics methods, and computer simulations is the appropriate way to understand and interpret the behavior of real systems.

ACKNOWLEDGMENTS

The authors express their gratitude to the Consejería de Educación y Juventud of the Junta de Extremadura and to the Fondo social European for the financial aid granted through the Project IPR98B004. We also thanks L. Morala for help in preparing some figures and in the literature search. Thanks are also due to Professor L.W. Bruch, Professor M. Kardar, Professor A. Migone, Professor R.E. Rapp, Professor W.F. Saam, and Professor E.D. Specht for permission to reproduce figures.

REFERENCES

1. DH Everett. Pure Appl Chem 31:577, 1971.
2. KSW Sing, DH Everett, RAW Haul, L Moscou, RA Pierotti, J Rouquerol, T Siemeniewska. Pure Appl Chem 57:603, 1985.
3. A Hubbard, ed. Encyclopaedia of Surface and Colloid Science. New York: Marcel Dekker, 2001.
4. WA Steele. In: EA Flood, ed. The Gas–Solid Interface. New York: Marcel Dekker, 1964, Vol. 1.
5. A Clark. The Theory of Adsorption and Catalysis. New York: Academic Press, 1970.
6. RA Pierotti. Physical Adsorption: The Interaction of Gases with Solids. New York: Wiley, 1971.
7. F Ricca, ed. Adsorption–Desorption Phenomena. New York: Academic Press, 1972.
8. WA Steele. The Interaction of Gases with Solid Surfaces. New York: Pergamon, 1974.
9. JG Dash, J Ruvalds, eds. Phase Transitions in Adsorbed Films. Summer School Erice. New York: Plenum, 1980.
10. SK Sinha, ed. Ordering in Two Dimensions. Amsterdam: North-Holland, 1980.
11. M Jaroniec, A Patrykiejew, M Borowko. Prog Surface Membr Sci 14:1, 1981.

12. A Thomy, X Duval. In: J Rouquerol, KSW Sing, eds. Adsorption at the Gas–Solid and Liquid–Solid Interface. New York: Elsevier Scientific, 1982.
13. SJ Gregg, KSW Sing. Adsorption, Surface Area and Porosity. London: Academic Press, 1982.
14. D Nicholson, NG Parsonage. Computer Simulation and Statistical Mechanics of Adsorption. London: Academic Press, 1982.
15. LW Bruch, MW Cole, E. Zaremba. Physical Adsorption: Forces and Phenomena. Oxford: Clarendon Press, 1997.
16. WA Steele. Surface Sci 36:317, 1973.
17. WA Steele. Surface Sci 39:149, 1973.
18. JJ Doll, WA Steele. Surface Sci 44:449, 1974.
19. WA Steele. J Chem Phys 65:5256, 1976.
20. CA Croxton. Introduction to Liquid State Physics. New York: Wiley, 1975.
21. JP Hansen, IR McDonald. Theory of Simple Liquids. London: Academic Press, 1976.
22. MP Allen, DJ Tildesley. Computer Simulation of Liquids. Oxford: Clarendon Press, 1987.
23. DP Valenzuela, AL Myers. Adsorption Equilibrium Data Handbook. Englewood Cliffs, NJ: Prentice-Hall, 1989.
24. PW Atkins. Physical Chemistry. Oxford: Oxford University Press, 1998.
25. JH Singleton, JG Halsey. J Phys Chem 58:330, 1954.
26. S Ross, W Winkler. J Colloid Sci 10:330 1955.
27. CF Prenzlow, GD Halsey. J Phys Chem 61:1158, 1957.
28. J Greyson, JG Aston. J Phys Chem 61:611, 1957.
29. JR Sams, G Constabaris, GD Halsey. J Phys Chem 64:1689, 1960.
30. JR Sams, G Constabaris, GD Halsey. J Chem Phys 36:1334, 1962.
31. S Ross, JP Olivier. J Phys Chem 65:608, 1961.
32. G Constabaris, JR Sams, GD Halsey. J Chem Phys 37:915, 1962.
33. WA Steele. J Phys Chem 82:817, 1978.
34. H Godfrin, HJ Lauter. In: WP Halperin, ed. Progress in Low Temperature Physics. Volume XIV. Amsterdam: North-Holland 1995.
35. JA Barker, DH Everett. Trans Faraday Soc 58:1608, 1962.
36. R Wolfe, JR Sams. J Chem Phys 44:2181, 1966.
37. A Thomy, X Duval. J Chim Phys-Chim Biol 67:1101, 1970.
38. J Suzanne, P Masri, M Bienfait. Surface Sci 43:441, 1974.
39. J Suzanne, P Masri, M Bienfait. Jpn J Appl Phys (suppl 2): 295, 1974.
40. J Suzanne, JP Coulomb, M Bienfait. Surface Sci 44:141, 1974.
41. J Suzanne, JP Coulomb, M Bienfait. J Crystal Growth 31:87, 1975.
42. FA Putnam, T Fort. J Phys Chem 79:459, 1975.
43. FA Putnam, T Fort. J Phys Chem 23:2164, 1977.
44. FA Putnam, T Fort, RB Griffiths. J Phys Chem 23:2171, 1977.
45. J Regnier, J Rouquerol, A Thomy. J Chim Phys 3:327, 1975.
46. J Regnier, A Thomy, X Duval. J Chim Phys 74:926, 1977.
47. HM Kramer, J Suzanne. Surface Sci 54:659, 1976.
48. J Piper, JA Morrison. Chem Phys Lett 103:323, 1984.
49. RA Beebe, DM Young. J Phys Chem 58:93, 1954.
50. A Thomy, X Duval, J Regnier. CR Acad Sci. Paris 268: 1416, 1969.
51. AA Antoniou. J Chem Phys 64:4901, 1976.
52. GB Huff, JG Dash. J Low Temp Phys 24:155, 1976.
53. EP De Souza, RE Rapp, E Lerner. Cryogenics 18:646, 1978.
54. WA Steele, EJ Derderian. In: F Ricca, ed. Adsorption–Desorption Phenomena. New York: Academic Press, 1972, p 85.
55. M Bretz, JG Dash, DC Hickernell, EO McLean, OE Vilches. Phys Rev A 8:1589, 1973.
56. RL Elgin, D Goodstein. Phys Rev A 9:2657, 1974.
57. EJ Derderian, WA Steele. J Chem Phys 66:2831, 1977.
58. MW Cole, DR Frankl, DL Goodstein. Rev Mod Phys 53:199, 1981.

59. CF Yu, KB Whalley, CS Hogg, SJ Sibener. J Chem Phys 83:4217, 1985.
60. JJ McAlpin, RA Pierotti. J Chem Phys. 41:68, 1964.
61. WA Steele, R Karl. J Colloid Interf Sci 28:397, 1968.
62. Y Larher, B Gilquin. Phys Rev A 20:1599, 1979.
63. AD Migone, ZR Li, MHW Chan. Phys Rev Lett 53:810, 1984.
64. KL D'Amico, J Bohr, DE Moncton, D Gibbs. Phys Rev B 41:4368, 1990.
65. WA Steele. Langmuir 12:145, 1996.
66. H Cochrane, PL Walker, WS Diethorn, HC Friedman. J Colloid Interf Sci 24:405, 1967.
67. A Thomy, X Duval, J Regnier. Surface Sci Reports 1:1, 1981.
68. Y Larher. J Chem Soc Faraday Trans 70:320, 1974.
69. RE Rapp, EP De Souza, E Lerner. Phys Rev B 24:2196, 1981.
70. JL Seguin, J Suzanne, M Bienfait, JG Dash, JA Venables. Phys Rev Lett 51:122, 1983.
71. A Inaba, JA Morrison, JM Telfer. Mol Phys 62:961, 1987.
72. F Ser, Y Larher, B Gilquin. Mol Phys 67:1077, 1989.
73. JP Hobson. Can J Phys 43:1941, 1967.
74. N Hansen. Vakuum-Tech 11:70, 1962.
75. F Ricca, R Medona, A Bellardo. Z Phys Chem 52:276, 1967.
76. F Ricca, A Bellardo. Z Phys Chem 52:318, 1967.
77. BL Schram. Physica 32:197, 1966.
78. N Endow, RA Pasternack. J Vac Sci Technol 3:196, 1966.
79. BB Fisher, WG McMillan. J Am Chem Soc 79:2969, 1957.
80. CV Heer. J Chem Phys 55:4066, 1971.
81. F Delanaye, M Schmeits, AA Lucas. J Chem Phys 69:5126, 1978.
82. P Malbrunot, D Vidal, J Vermesse. Langmuir 8:577, 1992.
83. JA Dunne, R Marivala, M Rao, S Sicar, RJ Gorte, AL Myers. Langmuir 12:5888, 1996.
84. PL Llewellyn, JP Coulomb, Y Grillet, J Patarin, H Lauter, H Reichert, J Rouquerol. Langmuir 19:1846, 1993.
85. JA Dunne, M Rao, S Sicar, RJ Gorte, AL Myers. Langmuir 12:5896, 1996.
86. ATJ Hope, CA Leng, CRA Catlow. Proc Royal Soc London A 424:57, 1989.
87. RJM Pellenq, D Nicholson. Langmuir 11:1626, 1995.
88. J Bohr, M Nielsen, J Als-Nielsen, K Kjaer. Surface Sci 125:181, 1983.
89. J Regnier, G Bockel, N Dupont-Pavlovsky. Surface Sci 112:770, 1981.
90. M Ceva, M Goldmann, C Marti. J Phys (Paris) 47:1527, 1986.
91. P Zeppenfeld, U Becher, K Kern, R David, G Comsa. Surface Sci 285: L461, 1993.
92. M Yanuka, AT Yinnon, RB Gerber, P Zeppenfeld, K Kern, U Becher, G Comsa. J Chem Phys 99:8280, 1993.
93. M Bouchdoug, J Menaucourt, A Thomy. J Phys (Paris) 47:1797, 1986.
94. A Razafitianamaharavo, N Dupont-Pavlovsky, A Thomy. J Phys (Paris) 51:91, 1990.
95. W Widdra, P Trischberger, W Frieb, D Menzel, SH Payne, HJ Kreuzer. Phys Rev B 57:4111, 1998.
96. GA Estévez, H Gould, MW Cole. Phys Rev A 18:1222, 1978.
97. JP McTague, J Als-Nielsen, J Bohr, M Nielsen. Phys Rev B 25:7765, 1982.
98. DK Fairobent, WF Saams, LM Sander. Phys Rev B 26:179, 1982.
99. KJ Strandburg. Rev Mod Phys 60:161, 1988.
100. JJ Lander, J Morrison. Surface Sci 6:1, 1967.
101. JM Kosterlitz, DJ Thouless. J Phys C 6:1181, 1973.
102. BI Halperin, DR Nelson. Phys Rev Lett 41:121, 1978.
103. QM Zhang, JZ Larese. Phys Rev B 43:938, 1991.
104. DM Butler, JA Litzinger, GA Stewart. Phys Rev Lett 42:1289, 1979.
105. DM Butler, JA Litzinger, GA Stewart. Phys Rev Lett 44:466, 1980.
106. DS Greywall. Phys Rev B 47:309, 1993.
107. R Gangwar, NJ Colella, RM Suter. Phys Rev B 39:2459, 1989.
108. JA Venables, HM Kramer, GL Price. Surface Sci 55:373, 1976.
109. PS Schabes-Retchkiman, JA Venables. Surface Sci 105:536, 1981.

110. AQD Faisal, M Hamichi, G Raynerd, JA Venables. Phys Rev B 34:7440, 1986.
111. JA Litzinger, GA Stewart. In: SK Sinha, ed. Ordering in Two Dimensions. New York: North-Holland, 1980.
112. EM Hammonds, P Heiney, PW Stephens, RJ Birgeneau, PM Horn. J Phys C 13:L301, 1980.
113. DE Moncton, PW Stephens, RJ Birgenau, PM Horn, GS Brown. Phys Rev Lett 46:1533, 1981.
114. PA Heiney, PW Stephens, RJ Birgeneau, PM Horn. Bull Am Phys Soc 25:187, 1982.
115. H Hong, CJ Peters, A Mak, RJ Birgeneau, PM Horn, H Suematsu. Phys Rev B 36:7311, 1987.
116. RJ Birgeneau, PM Horn. Science 232:329, 1986.
117. TT Chung. Surface Sci 87:348, 1979.
118. Y Larher. J Chem Phys 68:2257, 1978.
119. Y Larher. Surface Sci 134:469, 1983.
120. H Taub, K Carneiro, KJ Kjems, L Passell, JP McTague. Phys Rev B 16:4551, 1977.
121. C Tiby, HJ Lauter. Surface Sci 117:277, 1982.
122. Y Grillet, F Rouquerol, J Rouquerol. J Phys C 4:57, 1977.
123. J Rouquerol. S Partyka, F Rouquerol. Trans Faraday Soc 73:306, 1977.
124. CT Shaw, SC Fain, MD Chinn. Phys Rev Lett 41:955, 1978.
125. CG Shaw, SC Fain. Surface Sci 83:1, 1979.
126. CG Shaw, SC Fain. Surface Sci 91:1, 1980.
127. M Nielsen, J Als-Nielsen, J Bohr, JP McTague, DE Moncton, PW Stephens. Phys Rev B 35:1419, 1987.
128. Y Grillet, F Rouquerol, J Rouquerol. J Colloid Interf Sci 70:239, 1979.
129. Y Grillet, F Rouquerol, J Rouquerol. J Chim Phys 78:778, 1997.
130. AD Novaco, JP McTague. Phys Rev Lett 38:1286, 1977.
131. FE Hanson, MJ Mandell, JP McTague. J Phys (Paris) 38(C4):76, 1977.
132. S Toxvaerd. Phys Rev Lett 44:1992, 1980.
133. FF Abraham. Phys Rev Lett 44:463, 1980.
134. Y Grillet, J Rouquerol. J Colloid Interf Sci 77:580, 1980.
135. C Tiby, H Wiechert, HJ Lauter. Surface Sci 119:21, 1982.
136. S Calisti, J Suzanne, JA Venables. Surface Sci 115:455, 1982.
137. Y. Larher, A Terlain. J Chem Phys 72:1052, 1980.
138. RM Suter, NJ Colella, R Gangwar. Phys Rev B 31:627, 1985.
139. MD Chinn, SC Fain. Phys Rev Lett 39:146, 1977.
140. PM Horn, RJ Birgeneau, PA Heiney, EM Hammonds. Phys Rev Lett. 41:961, 1978.
141. PW Stephens, P Heiney, RJ Birgeneau, PM Horn. Phys Rev Lett 43:47, 1979.
142. M Nielsen, J Als-Nielsen, J Bohr, JP McTague. Phys Rev Lett 47:582, 1981.
143. ED Specht, M Sutton, RJ Birgeneau, DE Moncton, PM Horn. Phys Rev B 30:1589, 1984.
144. RG Caflisch, N Berker, M Kardar. Phys Rev B 31:4527, 1985.
145. SV Hering, SW van Sciver, OE Vilches. J Low Temp Phys 25:789, 1976.
146. M Bretz. Phys Rev Lett 38:501, 1977.
147. DS Greywall. Phys Rev B 41:1842, 1990.
148. DS Greywall, PA Busch. Phys Rev Lett 67:3535, 1991.
149. HJ Lauter, HP Schildberg, H Godfrin, H Wiechert, R Haensel. Can J Phys 65:1435, 1987.
150. DL Freeman. J Chem Phys 62:941, 1975.
151. DL Freeman. J Chem Phys 62:4300, 1975.
152. LW Bruch, XZ Ni. Faraday Discuss Chem Soc (London) 80:217, 1985.
153. TM Hakim, HR Glyde. Phys Rev B 37:3461, 1988.
154. T Tsang. Phys Rev B 20:3497, 1979.
155. LW Bruch, MS Wei. Surface Sci 100:418, 1980.
156. GL Price, JA Venables. Surface Sci 59:509, 1976.
157. E De Rouffignac, GP Alldredge, FW De Wette. Phys Rev B 24:6050, 1981.
158. XZ Ni, LW Bruch. Phys Rev B 33:4584, 1986.
159. JM Houlrik, DP Landau. Phys Rev B 44:8962, 1991.
160. I Langmuir. J Am Chem Soc 38:2221, 1916.

161. A Patrykiejew, M Jaroniec, AW Marczewski. Thin Solid Films 76:247, 1981.
162. A Patrykiejew, M Jaroniec, AW Marczewski. Z Phys Chem 265:195, 1984.
163. JP Coulomb, M Bienfait. Surface Sci 61:291, 1976.
164. S Brunauer, PH Emmett, E Teller. J Am Chem Soc 66:309, 1938.
165. IMK Ismail. Langmuir 8:360, 1992.
166. A Seri-Levy, D Avnir. Langmuir 9:2523, 1993.
167. AL McClellan, HF Harnsberger. J Colloid Interf Sci 23:577, 1967.
168. RA Beebe, JB Beckwith, JM Hong. J Am Chem Soc 67:1554, 1945.
169. J Toth. J Colloid Interf Sci 163:299, 1994.
170. J Toth. J Colloid Interf Sci 191:449, 1997.
171. M Volmer. Z Phys Chem 115A:253, 1925.
172. TL Hill. J Chem Phys 17:520, 1949.
173. TL Hill. J Chem Phys 14:441, 1946.
174. GA Stewart. Phys Rev A 10:671, 1974.
175. AD Novaco, CE Campbell. Phys Rev B 11:2525, 1975.
176. LW Bruch, JM Phillips, XZ Ni. Surface Sci 136:361, 1984.
177. AF Devonshire. Proc Roy Soc (London) A 163:132, 1937.
178. EL Pace. J Chem Phys 27:1341, 1951.
179. LA Rowley, D Nicholson, NG Parsonage, Mol Phys 31:365, 1976.
180. LA Rowley, D Nicholson, NG Parsonage. Mol Phys 31:389, 1976.
181. AN Berker, S Ostlund, FA Putnam. Phys Rev B 17:3650, 1978.
182. PA Monson, WA Steele, D Henderson. J Chem Phys 74:6431, 1981.
183. S Ostlund, AN Berker. Phys Rev Lett 42:843, 1979.
184. H Eyring, MS Jhon. Significant Liquid Structures. New York: Wiley, 1969.
185. E Dickinson, M Lal. Adv Mol Relax Interact Process 17:1, 1980.
186. LM Sander, J Hautman. Phys Rev B 29:2171, 1984.
187. S Sokolowski, WA Steele. J Chem Phys 82:2499, 1985.
188. S Sokolowski, WA Steele. J Chem Phys 82:3413, 1985.
189. ADJ Haymet. Prog Solid State Chem 17:1, 1986.
190. PA Monson. Chem Eng Sci 42:505, 1987.
191. WA Steele, GD Halsey. J Chem Phys 22:979, 1954.
192. RL Siddon, M Schick. Phys Rev A 9:907, 1974.
193. ST Wu. Surface Sci 41:475, 1974.
194. DG Brown, CS Hsue. J Chem Phys 65:2501, 1976.
195. MR Reddy, SF O'Shea, G Cardini. Mol Phys 57:841, 1986.
196. H Pan, JA Ritter, PB Balbuena. Langmuir 14:6323, 1998.
197. JR Sams. J Chem Phys 37:1883, 1962.
198. JR Sams. Mol Phys 9:195, 1965.
199. W Rudzinski. Czech J Phys B 23:488, 1973.
200. G Vidali, G Ihm, HY Kim, MW Cole. Surface Sci Reports 12:133, 1991.
201. WA Steele. Chem Rev 93:2355, 1993.
202. F Cuadros, A Mulero, I Cachadiña, W Ahumada. Int Rev Phys Chem 14:205, 1995.
203. F Cuadros, I Cachadiña, W Ahumada. Mol Eng 6:319, 1996.
204. O Sinanoglu, KS Pitzer. J Chem Phys 32:1279, 1960.
205. AD McLachlan. Mol Phys 7:381, 1964.
206. VR Bhethanabotla, WA Steele. J Phys Chem 92:3285, 1988.
207. F Hanson, JP McTague. J Chem Phys 72:6363, 1980.
208. FF Abraham, SW Koch, WE Rudge. Phys Rev Lett 49:1830, 1982.
209. FF Abraham, WE Rudge, DJ Auerbach, SW Koch. Phys Rev Lett 52:445, 1984.
210. ED Glandt, AL Myers, DD Fitts. Chem Eng Sci 33:1659, 1978.
211. ED Glandt, AL Myers, DD Fitts. J Chem Phys 70:4243, 1979.
212. A Mulero, F Cuadros. Chem Phys. 205:379, 1996.
213. A Mulero, F Cuadros. Langmuir 12: 3265, 1996.

214. A Mulero, F Cuadros. J Colloid Interf Sci 186:110, 1997.
215. F Rittner, D Paschek, B Boddenberg. Langmuir 11:3097, 1995.
216. FF Abraham. Phys Rev Lett 50:978, 1983.
217. F Cuadros, A Mulero, L Morala, V Gómez-Serrano. Langmuir 17:1576, 2001.
218. A Cheng, WA Steele. Langmuir 5:600, 1989.
219. WA Steele. J Colloid Interf Sci 75:13, 1980.
220. LM Sander, M Bretz, MW Cole. Phys Rev B 14:61, 1976.
221. C Pisani, F Ricca, C Roetti. J Phys Chem 77:657, 1973.
222. RG Gordon, YS Kim. J Chem Phys 56:3122, 1972.
223. AJ Bennett. Phys Rev B 9:741, 1974.
224. PA Monson, MW Cole, F Toigo, WA Steele. Surface Sci. 122:401, 1982.
225. JF Knight, PA Monson. J Chem Phys 84:1909, 1986.
226. A Mulero. Doctoral thesis, University of Extremadura. Badajoz, Spain, 1994.
227. P Borowski, A Patrykiejew, S Sokolowski. Thin Solid Films 173:287, 1989.
228. JP Stebbins, GD Halsey. J Phys Chem 68:3863, 1964.
229. F Tsien, GD Halsey. J Phys Chem 71: 4012, 1967.
230. A Patrykiejew, M Jaroniec. Surface Sci 77:365, 1978.
231. KJ Niskanen. Phys Rev B 33:1830, 1986.
232. D Henderson. Mol Phys 34:301, 1977.
233. P Bak. Rep Prog Phys 45:587, 1982.
234. SN Coppersmith, DS Fisher, BI Halpering, PA Lee, WF Brinkman. Phys Rev B 25:349, 1982.
235. M Kardar, AN Berker. Phys Rev Lett 48:1552, 1982.
236. M Schick. In: JG Dash, J Ruvalds, eds. Phase Transitions in Surface Films. New York: Plenum, 1980.
237. RE Ecke, JG Dash. Phys Rev B 28:3738, 1983.
238. FF Abraham, JQ Broughton. Phys Rev Lett 59:64, 1987.
239. A Mulero, F Cuadros, M Pérez-Ayala. Am J Phys 61:641, 1993.
240. A Mulero, F Cuadros, W Ahumada. In: MG Velarde, F Cuadros, eds. Thermodynamics and Statistical Physics. Proceedings of the 4th IUPAP Teaching Modern Physics Conference. Singapore: World Scientific, 1995.
241. S Toxvaerd. Phys Rev Lett 43:529, 1979.
242. A Zippelius, BI Halperin, DR Nelson. Phys Rev B 22:2514, 1980.
243. M Bishop, JPJ Michels. Chem Phys Lett 88:208, 1982.
244. GP Morris, DJ Evans. Phys Rev A 32:2425, 1985.
245. S Ranganathan, GS Dubey, KN Pathak. Phys Rev A 45:5793, 1992.
246. OH Scalise, GJ Zarragoicoechea, LE González, M Silbert. Mol Phys 93:751, 1998.
247. ED Glandt, DD Fitts. J Chem Phys 64:1241, 1976.
248. ED Glandt, DD Fitts. J Chem Phys 66:4503, 1977.
249. ED Glandt, DD Fitts. Mol Phys 35:205, 1978.
250. ED Glandt, DD Fitts, J Chem Phys 68:4546, 1978.
251. EM Chan. J Phys C: Solid State Phys 10:3477, 1977.
252. M Takamiya, K Nakanishi. Mol Phys 70:767, 1990.
253. JD Weeks, D Chandler, HC Andersen. J Chem Phys 54:5237, 1971; 55:5422, 1971.
254. S Toxvaerd. J Chem Phys 69:4750, 1978.
255. S Toxvaerd. Phys Rev A 24:2735, 1981.
256. D Frenkel, JP McTague. Phys Rev Lett 42:1632, 1979.
257. FF Abraham. J Chem Phys 72:1412, 1980.
258. FF Abraham. Phys Rev B 23:6145, 1981.
259. F van Swol, LV Woodcock, JN Cape. J. Chem Phys 73:913, 1980.
260. JA Barker, D Henderson, FF Abraham. Physica A 106:226, 1981.
261. JM Phillips, LW Bruch, RD Murphy. J Chem Phys 75:5097, 1981.
262. T Tobochnik, GV Chester. Phys Rev B 25:6778, 1982.
263. JD Weeks, JQ Broughton. J Chem Phys 78:4197, 1983.
264. S Toxvaerd. Phys Rev Lett 51:1971, 1983.

265. AF Bakker, C Bruin, HJ Hilhorst. Phys Rev Lett 52:449, 1984.
266. L Mederos, E Chacon, G Navascues, M Lombardero. Mol Phys 54:211, 1985.
267. JJ Morales, F Cuadros, LF Rull. J Chem Phys 86:2960, 1987.
268. JJ Morales, F Cuadros, LF Rull. In: MG Verlarde, ed. Physicochemical Hydrodnamics Interfacial Phenomena. New York: Plenum, 1988.
269. YJ Wong, GV Chester. Phys Rev B 35:3506, 1987.
270. C Udink, D Frenkel. Phys Rev B 35:6933, 1987.
271. RR Singh, KS Pitzer, JJ De Pablo, JM Prausnitz. J Chem Phys 92:5463, 1990.
272. M Rovere, DW Heermann, K Binder. J Phys Condens Matter 2:7009, 1990.
273. B Smit, D Frenkel. J Chem Phys 94:5663, 1991.
274. M Rovere, P Nielaba, K Binder. Z Phys 90:215, 1993.
275. AZ Panagiotopoulos. Int J Thermophys 15:1057, 1994.
276. S Jiang, KE Gubbins. Mol Phys 85:599, 1995.
277. OH Scalise, GJ Zarragoicoechea, M Silbert. Phys Chem Phys 1:241, 1999.
278. A Mulero, F Cuadros, CA Faúndez. Austr J Phys 52:101, 1999.
279. MR Reddy, SF O'Shea. Can J Phys 64:677, 1986.
280. Y Song, EA Mason. J Chem Phys 93:686, 1990.
281. PL Fehder. J Chem Phys 50:2617, 1969.
282. ID Morrison, S Ross. Surface Sci 39:21, 1973.
283. F Tsien, JP Valleau. Mol Phys 27:177, 1974.
284. FC Andrews. J Chem Phys 64:1941, 1976.
285. ED Glandt. J Chem Phys 68:2952, 1978.
286. JA Barker. Proc Roy Soc (London) A 377:425, 1981.
287. AD Bruce, NB Wilding. Phys Rev Lett 68:193, 1992.
288. F Cuadros, A Mulero. Chem Phys 177:53, 1993.
289. MN Rosenbluth, AW Rosenbluth. J Chem Phys 22:881, 1954.
290. PL Fehder. J Chem Phys 52:791, 1970.
291. BC Kreimer, BK Oh, SK Kim. Mol Phys 26:297, 1973.
292. F Mandel. J Chem Phys 62:1595, 1975.
293. S Toxvaerd. Mol Phys 29:373, 1975.
294. ED Glandt. J Chem Phys 74:1321, 1981.
295. C Bruin. AF Bakker, M Bishop. J Chem Phys 80:5859, 1984.
296. M Bishop, C Bruin. Am J Phys 52:1106, 1984.
297. TP Singh, SK Sinha. Pramana-J Phys 25:733, 1985.
298. M Bishop. J Chem Phys 84:535, 1986.
299. RM Sperandeo-Mineo, G Trip. Eur J Phys 8:117, 1987.
300. S Ranganathan, KN Pathak. Phys Rev A 45:5789, 1992.
301. F. Cuadros, A Mulero. Chem Phys 156:33, 1991.
302. F Cuadros, A Mulero. Chem Phys 159:89, 1992.
303. F Cuadros, A Mulero. Chem Phys 160:375, 1992.
304. F Cuadros, A Mulero. J Phys Chem 99:419, 1995.
305. F Cuadros, A Mulero, W Ahumada. Mol Phys 85:207, 1995.
306. F Cuadros, A Mulero, W Okrasinski. Physica A 214:162, 1995.
307. F Cuadros, A Mulero. In: MG Velarde, CI Christov, eds. Fluid Physics. Lecture Notes of Summer Schools. Singapore: World Scientific, 1995.
308. F Cuadros, A Mulero, W Okrasinski. Physica A 223:321, 1996.
309. F Cuadros, A Mulero, W Okrasinski. In: A Steinchen, ed. Dynamics of Multiphase Flows Across Interfaces. Lecture Notes in Physics. Berlin: Springer-Verlag, 1996.
310. F Cuadros, A Mulero, W Okrasinski. Chem Phys 218:235, 1997.
311. A Mulero, F Cuadros, C Galán. J Chem Phys 107:6887, 1997.
312. A Mulero, L Morala, V Gómez-Serrano, F Cuadros. In: A Hubbard, ed. Encyclopedia of Surface and Colloid Science, New York: Marcel Dekker, 2001.
313. NF Carnahan, KE Starling. J Chem Phys 51:635, 1969.

314. L Verlet, JJ Weis. Phys Rev A 5:939, 1972.
315. M Metropolis. AW Rosenbluth, MN Rosenbluth, AN Teller, E Teller. J Chem Phys 21:1087, 1953.
316. KE Gubbins. Rev Inst Fr Pétrole 51:59, 1996.
317. D Nicholson, LA Rowley, NG Parsonage. J Phys C4 38:69, 1977.
318. JE Lane, TH Spurling. Aust J Chem 31:933, 1978.
319. SD Prasad, S Toxvaerd. J Chem Phys 72:1689, 1980.
320. JB Sokoloff. Phys Rev B 22:2506, 1980.
321. FF Abraham. Phys Rev B 29:2606, 1984.
322. SW Koch, FF Abraham. Phys Rev B 27:2964, 1983.
323. SW Koch, FF Abraham. Phys Rev B 33:5884, 1986.
324. JF Knight, PA Monson. Mol Phys 60:921, 1987.
325. JQ Broughton, FF Abraham. J Phys Chem 92:3274, 1988.
326. S-H Suh, N Lermer, SF O'Shea. Chem Phys 129:273, 1989.
327. A Cheng, WA Steele. Molecular Simulation 4:349, 1990.
328. FF Abraham, JQ Broughton, PW Leung, V Elser. Europhys Lett 12:107, 1990.
329. A Patrykiejew, S Sokolowsky, T Zientarski, K Binder. J Chem Phys 102:8221, 1995.
330. G Leptoukh. B Strickland, C Roland. Phys Rev Lett 74:3636, 1995.
331. Y Larher. J Colloid Interf Sci 37:836, 1971.
332. RL June, AT Bell, DN Theodorou. J Phys Chem 94:8232, 1990.

8

Ab Fine Problems in Physical Chemistry and the Analysis of Adsorption–Desorption Kinetics

GIANFRANCO CEROFOLINI   STMicroelectronics, Catania, Italy

I. INTRODUCTION

For a long time, the rate-determining step for the development of natural sciences has been the possibility to perform numerical calculations. Even when possible, detailed modeling of actual physicochemical systems was considered useless because of the practical impossibility of performing numerical calculations. The ingenuity of theoreticians was accordingly concentrated more on the construction of exactly or approximately solvable models rather than on the search of accurate but unsolvable models. "Solvable" and "unsolvable" are, however, related to the deployed mathematical apparatus—much of what remained unsolvable after centuries of development of calculus and analysis has become solvable after the advent of electronic computers.

The enormous progress of electronic computers has allowed the accurate description (by means of new disciplines like computational physics and theoretical chemistry) of physico-chemical systems via numerical solution of the basic equations. An extended quantum mechanical description of the electron density in molecular systems (via self-consistent field or density functional theory) and molecular dynamics are now possible on more or less *ab initio* methods and are able to describe most of experimentally accessible physicochemical systems with accuracies of thermodynamic level, or even to provide presumably realistic descriptions of systems (like the ones at the interior of Jovian planets) which are not accessible to experiments.

In spite of the already high, and continuously increasing, power of *ab initio* methods, there are situations which remain much too complex to be modeled with them. These situations (e.g., the vibrational modes of, or the adsorption-energy distribution on, highly dispersed systems, like powders or colloids) can often be described quite accurately on experimental grounds but are of difficult theoretical description. Because the experimental data (e.g., the specific heat as a function of temperature or the adsorption isotherm) can be written in terms of unknown microscopic quantities (like the vibrational density of states or the adsorption-energy distribution, in the above examples), one can try to extract the inaccessible microscopic information from the accessible macroscopic data. This generates a kind of *ab fine* problem.

This chapter is devoted to an anlaysis of a few *ab fine* problems of physical chemistry, with a special emphasis on the extraction of the activation-energy distribution from experimental desorption kinetics.

II. *AB FINE* PROBLEMS AND THEIR ROLE IN PHYSICAL CHEMISTRY

In all *ab fine* problems, a central role is played by the experimental error σ. Let λ_σ denote the experimental datum, as it results from an experiment affected by an error σ, and assume that the (error-free) macroscopic datum λ is related to the microscopic quantity φ by an assigned functional relationship $\lambda = \Lambda[\varphi]$. Even assuming one is able to invert this functional relationship in such a way as to have $\varphi = \Lambda^{-1}[\lambda]$, one can calculate only the microscopic quantity in correspondence to macroscopic data affected by a certain error: $\varphi_\sigma = \Lambda^{-1}[\lambda_\sigma]$. This would not produce difficulties if, taking the limit for σ tending to 0, λ_σ converged to λ in a sufficiently strong form. Actually, this does not happen for the situations considered in the following, which generates an *ill-posedness* of the problem.

A. Statement of the Problem

Let a system S be homogeneous with respect to a physical quantity y and heterogeneous with respect to another variable x. Now, consider a quantity Λ which is a function of both x and y, $\Lambda = \Lambda(y, x)$; which is if S can be considered as a collection of homogeneous parts, each characterized by a value of x, then the average of Λ taken over the system is given by

$$\lambda(y) = \int_{\mathscr{S}_x} \Lambda(y, x)\varphi(x) \, dx =: (\Lambda\varphi)(y) \tag{1}$$

where $\mathscr{S}_x$ is the set of allowed values of x and $\varphi(\bar{x}) \, dx$ is the fraction of system where x takes any value between $\bar{x}$ and $\bar{x} + dx$.

In principle, $\lambda(y)$ is an experimentally known function defined on the real non-negative semiaxis $\mathbb{R}_+$, $\Lambda(y, x)$ is theoretically known in $\mathbb{R}_+^2$ with analytic extension in $\mathbb{C}^2$, and $\varphi(x)$ is the unknown function defined in $\mathbb{R}_+$ and identically null outside the subset $\mathscr{S}_x$. In practice, $\lambda(y)$ is known within an experimental error σ only in a certain proper subset (the *experimental region* $\mathscr{E}_y$) of the *physical region* $\mathbb{R}_+$, $\Lambda(y, x)$ is exactly known (*within the frame of the assumed theoretical model*), and one wants to determine the function $\varphi(x)$ corresponding to the "true" $\lambda(y)$, which is, however, known only in the limit for $\sigma \to 0$.

In these situations, the description of the physical system given by $\lambda(y)$ is not equivalent to that given by $\varphi(x)$—even though Λ generates a one-to-one correspondence, the description given by $\varphi(x)$ is in a certain sense "more fundamental" than that of $\lambda(y)$. These situations are typically met when $\varphi(x)$ is the cause of a certain phenomenon and $\lambda(y)$ is the resulting effect, or when $\varphi(x)$ describes a microscopic property and $\lambda(y)$ describes a macroscopic property.

B. Choice of the Functional Space

The physical interpretation of $\varphi(x)$ requires that it satisfies the conditions of non-negativity,

$$\forall x \, [\varphi(x) \geq 0]$$

and of normalization

$$\int_0^{+\infty} \varphi(x) \, dx = 1$$

where, even though the support of $\varphi(x)$ is compact, this property has not been assumed a priori. Combining the above conditions, one deduces that, except for trivial cases, the space of Lebesgue-integrable functions $L^1(0, +\infty)$ is a suitable functional space for φ.

The $L^1(0, +\infty)$ functional space, however, is not the most suitable space for other aspects of the physical nature of the problem. The fact that $\lambda(y)$ is known only experimentally implies that instead of the exact function $\lambda(y)$ for $y \in \mathbb{R}_+$, another function $\lambda_\sigma(y)$ is known for $y \in \mathscr{E}_y \subset \mathbb{R}_+$, with

$$\left[\int_{\mathscr{E}_y} |\lambda_\sigma(y) - \lambda(y)|^2 \, dy \right]^{1/2} < \sigma \tag{2}$$

where σ is the experimental error and $\mathscr{E}_y$ is the experimental region. In view of condition (2), the most appropriate functional space for $\lambda(y)$ is $L^2(\mathscr{E}_y)$, at least when the limit for $\sigma \to 0$ is considered.

There is some freedom (or ambiguity, according to the attitude of the reader) in the choice of the functional space in which the solution of Eq. (1) has to be searched. The choice of the functional space is crucial in establishing the existence of a solution, as the following trivial example shows. If $\lambda(y) = \Lambda(y, \bar{x})$, Eq. (1) does not admit a solution in any of $L^p(0, +\infty)$ $(p > 0)$ spaces, but admits the solution $\varphi(x) = \delta(x - \bar{x})$ in the space of distributions. A solution may thus exist or not in relation to the functional space; however, the supposedly existing solution must be space invariant, in the same way as in mechanical problems in which one is free to choose the reference frame and the solution of the problem is independent of the chosen frame, or in electromagnetic problems in which one is free to choose the gauge leaving invariant the Maxwell equations (Lorentz, Coulomb, Poisson, etc.), but the solution of the problem does not depend on the particular gauge. As discussed in Ref. 1, the invariance with respect to the functional space is not trivial.

C. Solution of the Problem

According to Hadamard, what makes a mathematical problem *well posed* are the existence and uniqueness of the solution and its stability with respect to the data.

1. Existence and Uniqueness

From the mathematical point of view, the problem of solving Eq. (1) for $\varphi(x)$ when $\lambda(y)$ is known is related to finding the inverse transform Λ^{-1}.

Let $\mathscr{D}_\Lambda$ be the domain over which Λ is defined; the range $\mathscr{R}_\Lambda$ of Λ is the image of $\mathscr{D}_\Lambda$ through Λ: $\mathscr{R}_\Lambda = \Lambda(\mathscr{D}_\Lambda)$. The ker of the application Λ is the set of all ψ in $\mathscr{D}_\Lambda$ such that $\Lambda\psi = 0$, ker $\Lambda := \{\psi | \Lambda\psi = 0\}$. If

- $\lambda \in \mathscr{R}_\Lambda$
- ker $\Lambda = \{0\}$

Eq. (1) admits a solution and it is unique.

2. Stability

The integration in Eq. (1), which produces the average $\lambda(y)$ of $\Lambda(y, x)$ over the entire considered system, does introduce, in most cases, a kind of smoothing. In fact, in most practical cases, the operator Λ is compact and its compactness implies that for any bounded sequence $\{\varphi_n\}$, one can extract from $\{\lambda_n | \lambda_n = \Lambda\varphi_n\}$, a convergent sequence $\{\lambda_{\bar{n}}\}$. This fact would not be disturbing if λ were exactly known—indeed, the function $\varphi(x)$ would be immediately calculated, $\varphi = \Lambda^{-1}\lambda$. However, λ is known only to within an experimental error σ, $\|\lambda_\sigma - \lambda\| < \sigma$; this means that one can determine experimentally a sequence λ_{n_σ} converging to the true overall function λ when the experimental error is reduced to zero. Because of the compactness of Λ, one is not guaranteed that the corresponding φ_{n_σ} converges to the true distribution function φ when σ is reduced to zero. This fact can be restated by observing that if Λ is compact, Λ^{-1} is not bounded. Because Λ^{-1} is not bounded, there is at least one (and thus infinitely many) convergent sequence $\{\lambda_n\}$ for which $\{\varphi_n | \Lambda^{-1} \lambda_n\}$, does not converge.

Moreover, even assuming that $\{\varphi_n\}$ does converge to φ when $\{\lambda_n\}$ tends to the true function λ, this convergence is only in the mean, which generates an additional difficulty—the convergence in the mean does not guarantee the uniform convergence (indeed, a function having a finite number of points of discontinuity in an interval can be approximated in the mean with the wanted accuracy by a continuous function [2]).

Another difficulty has its origin in the fact that $\lambda(y)$ is known only in the experimental regioin $\mathscr{E}_y$ rather than on the whole physical region (i.e., the positive semiaxis $\mathbb{R}_+$). Indeed, as shown in the cases considered in the following, the inverse transform formulas require either the knowledge of $\lambda(y)$ in the whole physical region $\mathbb{R}_+$ or of its analytic extension in domains of the complex plane $\mathbb{C}$ not belonging to $\mathbb{R}_+$. If $\lambda(y)$ is analytic and regular in $\mathbb{C}$ and is known in an interval $[a, b] \subseteq \mathscr{E}_y \subset \mathbb{R}_+$, it can be extended to the whole complex plane by analytic extension. However, even if $\lambda_\sigma(y) \to \lambda(y)$ uniformly in $[a, b]$, the analytic extension of $\lambda_\sigma(y)$ does not necessarily converge to $\lambda(y)$ uniformly in $\mathbb{C}$. In certain cases, this difficulty can be circumvented by the Wiener–Hopf technique (which does not require knowledge of the analytic extension outside the physical region); this technique, however, presents serious technical difficulties and, to the best of my knowledge, has been used only once in dealing with the considered problem [3].

The following combined facts make the problem of solving Eq. (1) an improperly posed problem.

The compactness of Λ does not guarantee that $\varphi_\sigma \to \varphi$ as $\sigma \to 0$.

Even assuming this convergence, the convergence in the mean does not imply a uniform convergence.

The uniform convergence of the datum in a limited domain $\mathscr{E}_y$ does not imply a uniform convergence of the analytic extension in the whole complex plane $\mathbb{C}$.

D. The Menu

In the following, we consider three *ab fine* problems of much interest in physical chemistry, essentially related to the extraction of the following:

The phonon-energy spectrum from the heat capacity (heat capacity c_V versus temperature T) of solids.

The adsorption-energy distribution from the equilibrium adsorption isotherm (surface coverage ϑ versus pressure p) of heterogeneous adsorbents.

The activation-energy distribution from desorption kinetics (ϑ versus time t) from heterogeneous surfaces.

$c_V(T)$, $\vartheta(p)$, and $\vartheta(t)$ are supposed to be experimentally known.

The following problems run only partially in the above-described situation:

- The heat capacity is *generally* believed to be related to an energy heterogeneity of the sample with a *known* local law.
- The adsorption isotherm is *unknown* (although in a certain family).
- The adsorption–desorption kinetics are *sometimes* assumed to be related to energy heterogeneity; however, not only are the local kinetics *unknown* (although in a certain family) but also the basic assumptions are *uncertain*.

These problems will be considered in order of increasing complexity; they will be preceded by an analysis of the problem of finding the density of states of a system from the experimental partition function. Although this problem has little (if any) practical importance, it is, however, of some interest because the methods required, and the problems encountered, for its solution are common to all the other, more important, problems. Before proceeding, I want to alert the reader that some symbols change meaning in going from one section to another; when this happens, the symbol is accordingly redefined.

III. THE PROBLEM IN STATISTICAL THERMODYNAMICS: BOLTZMANN STATISTICS

The central relationship of statistical thermodynamic links a thermodynamic quantity (the Helmholtz free energy F) to a stastistical property of the system (the partition function Z_N of the system with N particles) through the Boltzmann constant k_B and the temperature T of the heat reservoir to which the system is in thermal contact [4]:

$$F(T) = -k_B T \ln Z_n(T) \tag{3}$$

If the system is formed by N independent and identical particles obeying the Boltzmann statistics, its partition function is given by

$$Z_N(T) = \frac{1}{N!} z(T)^N \tag{4}$$

where the single-particle partition function $z(T)$ is given by

$$z(T) = \sum_i g_i \exp\left(-\frac{E_i}{k_B T}\right) + \int_0^{+\infty} \exp\left(-\frac{E}{k_B T}\right) g(E) \, dE \tag{5}$$

the sum is extended to all the bound states of the particle, and the integral is extended to the free states.

Applying the Stirling formula ($N! \sim N^N e^{-N}$) to Eq. (4), Eq. (3) becomes

$$F(T) \sim k_B T N \ln\left(\frac{N}{ez(T)}\right)$$
$$= k_B T V \rho \, \ln\left(\frac{V\rho}{ez(T)}\right) \tag{6}$$

where V is the volume of the system and ρ is the number of particles per unit volume. Equation (6) allows the microscopic quantity $z(T)$ to be determined from the macroscopic quantities V, ρ, T, and F.

The *ab fine* problem is related to finding the density of states by solving Eq. (5) for $\{g_i\}$ and $g(E)$ when $Z(T)$ is known. This problem, an age-old problem of physical chemistry, was first formulated by Bauer in 1939 [5].

In the following, two formal methods will be considered, based on the Laplace and Fourier transforms, respectively; in both methods, the discrete spectrum is supposed to affect the partition function [and hence the free energy $F^d(T)$] in a precisely known way, so that the difference $F^c(T)$ [where $F^c(T) := F(T) - F^d(T)$] is related to the contribution $z^c(T)$ to the partition function originated by the continuous spectrum,

$$z^c(T) = \int_0^{+\infty} \exp\left(-\frac{E}{k_B T}\right) g(E) \, dE \tag{7}$$

A. A Method Based on Laplace Transform

Defining $\beta := 1/k_B T$ and $\tilde{z}^c(\beta) := z^c(k_B/\beta)$, Eq. (7) becomes

$$\tilde{z}^c(\beta) = \int_0^{+\infty} \exp(-\beta E) g(E) \, dE \tag{8}$$

Equation (8) shows that $\tilde{z}^c(\beta)$ is the Laplace transform of $g(E)$; the inverse transform gives $g(E)$:

$$g(E) = \int_{\beta_0 - i_\infty}^{\beta_0 + i_\infty} \exp(\beta E) \tilde{z}^c(\beta) \, d\beta \tag{9}$$

where β_0 is any value of β for which the integral (8) converges. The calculation of $g(E)$ from Eq. (9) requires, therefore, a knowledge of $\tilde{z}^c(\beta)$ along the imaginary straight line

$$\mathscr{I} = \{\beta \colon \mathrm{Re}\,\beta = \beta_0 \Lambda - \infty < \mathrm{Im}\,\beta < +\infty\}$$

However $\tilde{z}^c(\beta)$ is known on the finite interval $[\beta_2, \beta_1]$ (with $\beta_1 = k_B/T_1$, and similarly for β_2) of the positive semiaxis $\mathbb{R}_+$.

Of course, what one knows is not the exact $F^c(T)$ for all T, but rather the quantity F_σ^c $[F_\sigma^c := F_\sigma(T) - F^c(T)]$ in a given temperature range (T_1, T_2), where the experimental free energy $F_\sigma(T)$ differs from the "true" free energy by the experimental error $\sigma \colon \|F - F_\sigma\| < \sigma$. The calculation of $g_\sigma(E)$ requires the knowledge of the analytic extension of $F_\sigma(T)$ in the complex plane $\mathbb{C}$; no problem would be met if the convergence of $F_\sigma(T)$ to $F(T)$ in the interval (T_1, T_2) guaranteed the uniform convergence in the imaginary line $\mathscr{I}$. This fact, however, is not ensured, so that one runs into ill-posedness.

B. A Method Based on Fourier Transform

A different method, based on the use of the Fourier transform rather than of the Laplace transform, is the following. On defining $\epsilon := K \ln(E/E_0)$ and $\tau := K \ln(T/T_0)$, where E_0, T_0, and K are arbitrary positive quantities, ϵ and τ range from $-\infty$ to $+\infty$ as E and T range from 0 to $+\infty$, and Eq. (8) becomes

$$z^c(T(\tau)) = \int_{-\infty}^{+\infty} \exp\left(\frac{E_0}{k_B T_0} \exp\left(\frac{\epsilon - \tau}{K}\right)\right) E_0 \exp\left(\frac{\epsilon}{K}\right) g(E(\epsilon)) \frac{d\epsilon}{K} \tag{10}$$

Putting $\tilde{z}^c(\tau) := z^c(T(\tau))$ and $\tilde{g}(\epsilon) := (E_0/K) \exp(\epsilon/K) g(E(\epsilon))$, Eq. (10) becomes

$$\tilde{z}^c(\tau) = \int_{-\infty}^{+\infty} \exp\left(\frac{E_0}{k_B T_0} \exp\left(-\frac{\tau - \epsilon}{K}\right)\right) \tilde{g}(\epsilon) \, d\epsilon \tag{11}$$

which can be seen as an integral equation for $\tilde{g}(\epsilon)$, the kernel of which does not depend on ϵ and τ separately but only on their difference $\epsilon - \tau$. Because of this, the integral is a convolution integral.

$$\tilde{z}^c(\tau) = \tilde{\zeta}(\tau - \epsilon) * \tilde{g}(\epsilon) \tag{12}$$

with

$$\tilde{\zeta}(\tau - \epsilon) := \exp\left(\frac{E_0}{k_B T_0} \exp\left(-\frac{\tau - \epsilon}{K}\right)\right)$$

The unknown function $\tilde{g}(\epsilon)$ can therefore be calculated by computing the Fourier transform of both sides of Eq. (12),

$$\mathscr{F}[\tilde{z}^c(\tau)] = \mathscr{F}[\tilde{\zeta}(\tau - \epsilon)]\mathscr{F}[\tilde{g}(\epsilon)]$$

solving for $\mathscr{F}[\tilde{g}(\epsilon)]$, and calculating the inverse Fourier transform

$$\tilde{g}(\epsilon) = \mathscr{F}^{-1}\left[\frac{\mathscr{F}[\tilde{z}^c(\tau)]}{\mathscr{F}[\tilde{\zeta}(\tau - \epsilon)]}\right].$$

The calculation of $\tilde{g}(\epsilon)$ [i.e., $g(E)$] requires that one knows $\tilde{z}^c(\tau)$ [i.e., $z(T)$] for all $\tau \in \mathbb{R}$ [i.e., for all $T \in \mathbb{R}_+$]; in other words, it requires the knowledge of the datum in the whole physical region rather than in the experimental region. Although this condition may seem less restrictive than that required by the Laplace transform method, it requires the analytic extension of $z_\sigma^c(T)$ outside the experimental region, whose uniform convergence to $z^c(T)$ is not guaranteed.

IV. THE VIBRATIONAL FREQUENCY SPECTRA OF NONIDEAL SOLIDS: BOSE–EINSTEIN STATISTICS

The behavior of the heat capacity of crystalline solids versus temperature $c_V(T)$ is accounted for in terms of frequency distribution $g(\omega)$ of normal vibration modes. The problem of determining $g(\omega)$ fro $c_V(T)$ is therefore an adequate example of the situation described in Section II.

A. Einstein Theory

The first satisfactory interpretation of the specific heat of solids was given by Einstein, who described the solid in terms of independent oscillators and proposed that each oscillator with angular frequency ω has an energy $\hbar\omega$, where $\hbar$ is the reduced Planck constant, $\hbar := h/2\pi$ (in the rest of this section, the term "frequency" will be used as contracted form of "angular frequency"). Assuming that the energy of each oscillator can only be a multiple integer of $\hbar\omega$ and using the Planck formula for the probability of an oscillator to have just the energy $n\hbar\omega$, he calculated the average energy $\epsilon(T)$ of each oscillator as

$$\epsilon(T) = 3\hbar\omega \frac{1}{\exp(\hbar\omega/k_B T) - 1}$$

Because each of such oscillators contributes to the specific heat with an amount $C_V(\omega, T) = \partial\epsilon(T)/\partial T$, that is,

$$C_V(\omega, T) = 3k_B \left(\frac{\hbar\omega}{k_B T}\right)^2 \frac{\exp(\hbar\omega/k_B T)}{[\exp(\hbar\omega/k_B T) - 1]^2} \tag{13}$$

the specific heat $c_V(T)$ of real solids is obtained by summing over all the (supposedly independent) elementary oscillators forming the solid:

$$c_V(T) = \int_{\mathscr{S}_\omega} C_V(\omega, T)g(\omega)\, d\omega \tag{14}$$

where $\mathscr{S}_\omega$ is the set of allowed values of ω (i.e., the frequency spectrum) and $g(\omega)\, d\omega$ is the fraction of oscillators with frequency between ω and $\omega + d\omega$.

Historical Note 1. Einstein's first work on heat capacity dates to 1907 [6]. In that article, Einstein limited his attention to solids formed by equivalent oscillators—the so-called *Einstein model*. In this case, the frequency distribution function is a Dirac δ distribution centered on a frequency ω_E, $g(\omega) = \delta(\omega - \omega_E)$, for which the specific heat is given by

$$c_V^E(T) = 3k_B \left(\frac{\hbar\omega_E}{k_B T}\right)^2 \frac{\exp(\hbar\omega_E/k_B T)}{[\exp(\hbar\omega_E/k_B T) - 1]^2} \tag{15}$$

Einstein himself did not believe seriously in his model; in fact, in a second article "The Present State of the Problem of Specific Heats" presented to the first Solvay Congress in 1911 and later published in the Proceedings, commenting on a proposal of Nernst and Lindemann [7], he wrote, "One can therefore suppose that the body behaves as a set of oscillators with different frequencies. [...] A theoretical meaning could only be attributed to a formula in which all the infinite frequency values compare in a sum" [8]. However, not only has the Einstein model had a historical and tutorial relevance, but also formula (15) continues to be used for the vibrational specific heat of isolated defects. Nernst and Lindemann had observed that most of the experimentally known specific heats could be adequately represented by taking $\frac{1}{2}[\delta(\omega - \omega_E) + \delta(\omega - \frac{1}{2}\omega_E)]$ for $g(\omega)$ [7]; in a way, this formula can be considered the first attempt to solve the *ab fine* problem.

The *ab fine* problem associated with Eq. (14) was not considered carefully because, soon after Einstein's proposal, on one side, Debye developed a model for the specific heat of crystalline solids able to rationalize most experimental data, whereas, on another side, Born and von Kármán posed the bases for the foundation of a dynamic theory, the theory of lattice dynamics in the harmonic approximation, able to give an adequate solution of the direct problem [i.e., able to determine the frequency spectrum $g(\omega)$ from atomic properties].

B. Debye Theory

The Einstein model, summarized in Eq. (15), does not account for low-temperature behavior of crystalline solids ($c_V \propto T^3$). The first model able to give an adequate theoretical description of this behavior was developed by Debye [9], who considered the solid as a continuum with unique excitations that are longitudinal and transversal sound waves. In the Debye model, the frequency distribution is given by

$$g(\omega) = \begin{cases} 3\omega^2/\omega_D^3 & \text{for } 0 \leq \omega < \omega_D \\ 0 & \text{for } \omega_D \leq \omega \end{cases} \tag{16}$$

where ω_D is a cutoff frequency guaranteeing the normalization of $g(\omega)$. Inserting the frequency spectrum (16) into Eq. (14) with the kernel (13), one gets the specific heat for the Debye model:

$$c_V^D(T) = \int_0^{\omega_D} 3k_B \left(\frac{\hbar\omega}{k_B T}\right)^2 \frac{\exp(\hbar\omega/k_B T)}{[\exp(\hbar\omega/k_B T) - 1]^2} \left(3\frac{\omega^2}{\omega_D^3}\right) d\omega \tag{17}$$

Although the integral on the right-hand side of Eq. (17) cannot be calculated in closed form in terms of elementary transcendental functions, it shows the correct asymptotic limit: $T \to 0 \Rightarrow c_V^D(T) \sim (12\pi/5)k_B(T/T_D)^3$, where T_D is the Debye temperature, $T_D = \hbar\omega_D/k_B$.

C. Born–von Kármán and the Theory of Lattice Dynamics in the Harmonic Approximation

The major progress for the calculation of the frequency spectrum of crystalline solids took place with the article of Born and von Kármán [10]. In that article, they gave the first description of lattice dynamics in the harmonic approximation and introduced the *periodic boundary conditions*, now of universal use. Considering a finite solid S as torus shaped in three directions, the vibration amplitude and phase of an atom on a boundary are the same as amplitude and phase of the corresponding atom on the other boundary. In this way, the system of coupled differential equations describing the vibration of the atoms forming the solid is transformed, via Fourier transform, in a system of algebraic equations, thus far simplifying the problem. The *theory of lattice dynamics in the harmonic approximation* has been developed to a high level of formal sophistication and is covered by several treatises [11,12]; in practice, if one knows the crystal structure and the force constants of the solid, one can construct the frequency spectrum $g(\omega)$ and, hence, the vibrational contribution to the thermodynamic properties of the system via standard methods of statistical mechanics. The theory of lattice dynamics in the harmonic approximation is an extremely powerful tool and can be handled to consider the vibrational properties of surfaces and pointlike or extended defects.

Of course, the use of the periodic boundary conditions adds additional unphysical constraints to the system which affect the frequency spectrum. An analysis of Ledermann has demonstrated that only a layer extending from the boundary to the bulk is affected by the unphysical constraint; the thickness of this layer is of the same order as the interaction length [13]. Therefore, the effect of the periodic boundary conditions on the bulk thermodynamic properties of the system disappears in the limit for $V \to +\infty$, provided that the shape of S is regular (i.e., provided that the area of ∂S increases with V as $V^{2/3}$).

Because most of thermodynamic computations are carried out in the *thermodynamic limit* (i.e., for $V \to +\infty$, $N \to \infty$, $N/V = $ constant), the effect of periodic boundary conditions disappears in this limit. There are, however, several systems of practical and conceptual interest, the shape of which is not regular; among them, we mention *queer systems*, with areas increasing in proportion to V (examples being zeolites or bacterial strains) [14], or *fractal systems*, with unmeasurable areas (examples being certain porous solids) [15]. In these systems, the boundary conditions affect the thermodynamic properties even in the thermodynamic limit and cannot, therefore, be assumed on the basis of easier computability.

D. Montroll and the *Ab Fine* Problem

For all systems which are not adequately described in terms of periodic boundary conditions, or with force constants not sufficiently known, or for which the structure is not known, the direct calculation of the frequency spectrum is impossible. These systems, however, are often of easy calorimetric analysis so that the specific heat $c_V(T)$ may be experimentally accessible, at least for T in a certain experimental domain (T_1, T_2).

The problem of finding which frequency distribution $g(\omega)$ reproduces the experimental specific heat $c_V(T)$ is therefore of a certain importance. This *ab fine* problem was first raised and solved by Montroll (in an appendix to a fundamental article otherwise devoted to the direct problem) via a formula involving the function $c_V(T)$ for $T \in \mathbb{R}_+$ [16]. That the experimental errors affect the resulting frequency spectrum significantly was first demonstrated by Lifshitz [17] and later confirmed by Chambers [18].

V. ADSORPTION ON HETEROGENEOUS SURFACES: FERMI–DIRAC STATISTICS

That information on the energetic structure of an adsorbing surface is contained in its equilibrium isotherm $\theta(p)$ is understood on intuitive bases, observing that the most energetic sites are preferentially filled at low pressure, whereas a high pressure is necessary for the occupation of low-energy sites. Extracting quantitative information from this intuition is, however, quite difficult.

The description of physical adsorption on energetically heterogeneous surfaces is indeed much more complex than that given in Section IV. The complexity is essentially due to the lateral forces: Not only do they produce different behaviors according to their intensity even on homogeneous surfaces, but they also make the overall datum sensitive to the energy and spatial distribution.

A. The Local Isotherm

Let θ denote the equilibrium coverage in isothermal conditions at pressure p of an energetically homogeneous surface whose adsorption energy ϵ at null coverage is denoted by q. In general, ϵ varies with θ, $\epsilon = q + f(\theta)$, where $f(\cdot)$ is specified by the nature of adsorption forces and $f(0) = 0$.

If the adsorbed molecules can be considered structureless particles, most of the models consider the adsorbed phase as a sequence of layers, each stabilized by the adsorption field generated by the underlying one and described either as a two-dimensional (2D) lattice or as a 2D van der Waals gas [19,20]. The most frequently considered adsorption models are listed in Fig. 1.

Whichever adsorption mode is considered, the adsorption isotherm is obtained by equating the chemical potential of the adsorbed phase to that of the gas phase, supposedly ideal. In general, this procedure does not give θ as an explicit function of p and q, but rather gives θ as a function of p and ε:

$$\theta = F(p, q + f(\theta)) \tag{18}$$

where $F(\cdot)$ is specified the adsorption model. Solving Eq. (18) for θ yields a function $\Theta(p, q)$, $\theta = \Theta(p, q)$.

$$
\text{submonolayer adsorption}
\begin{cases}
\text{localized}
\begin{cases}
\text{without lateral interactions} \\
(\textit{Langmuir isotherm}) \\
\text{with lateral interactions} \\
(\textit{Fowler–Guggenheim isotherm})
\end{cases} \\
\text{mobile}
\begin{cases}
\text{without lateral interactions} \\
(\textit{Volmer isotherm}) \\
\text{with lateral interactions} \\
(\textit{Hill–de Boer isotherm})
\end{cases}
\end{cases}
$$

$$
\text{multilayer adsorption}
\begin{cases}
\text{all layers localized} \\
\text{first } n \text{ layers localized, top layers mobile} \\
(\textit{Broekhoff–Van Dongen isotherm}) \\
\text{all layers mobile}
\end{cases}
$$

FIG. 1   A classification of adsorption modes and the isotherms which describe them.

For reasons which will become clear in the following, the Langmuir isotherm has played (and continues to play) a special role in the theory of adsorption. The *Langmuir isotherm*, which describes submonolayer localized adsorption without lateral interaction, is given by

$$\Theta^L(p, q) = \frac{p}{p + p_L \exp(-q/k_B T)} \tag{19}$$

where p_L is a characteristic pressure. Defining $Q := -k_B T \ln(p/p_L)$, Eq. (19) assumes a more symmetric form:

$$\tilde{\Theta}^L(Q, q) = \left[1 + \exp\left(\frac{Q - q}{k_B T}\right)\right]^{-1} \tag{20}$$

Statistical mechanics is not the unique frame for the theory of adsorption. A description in kinetic terms of the equilibrium of the adsorbent with a gas was proposed by Jovanovic, who assuming desorption-hindering collisions as the unique cause for deviation from ideality of the gas and modeling the adphase as formed by localized molecules without lateral interactions, eventually obtained the equation

$$\Theta^J(p, q) = 1 - \exp\left(-\frac{p}{p_J} \exp\left(\frac{q}{k_B T}\right)\right) \tag{21}$$

where p_J is a characteristic pressure. Defining $Q := -k_B T \ln(p/p_J)$, the *Jovanovic isotherm*, [Eq. (21)] becomes

$$\tilde{\Theta}^J(Q, q) = 1 - \exp\left(\exp\left(\frac{q - Q}{k_B T}\right)\right) \tag{22}$$

B. The Integral Equation

Not only does the relatively long-range character of the adsorption forces generate the dependence $\epsilon(\theta)$, but it also influences the nature of the integral equation linking the experimental overall isotherm to the local isotherm. In fact, in the presence of lateral interactions, the nature of the functional relationship linking the overall isotherm $\vartheta(p)$ to the adsorption-energy distribution $\varphi(q)$ is determined by the spatial distribution of the adsorption sites. Two extreme situations are those of patchwise heterogeneity and random heterogeneity:

Patchwise heterogeneity denotes a situation in which the sites with the same adsorption energy q are grouped in patches, each much larger than the range of lateral forces. For patchwise heterogeneous surfaces, $\vartheta(p)$ is nothing but the average of the local isotherm $\Theta(p, q)$ over all the spectrum $\mathscr{S}_q$ of adsorption energy:

$$\vartheta(p) = \int_{\mathscr{S}_q} \Theta(p, q)\varphi(q) \, dq \tag{23}$$

Random heterogeneity denotes a situation in which the local coverage coincides with the mean coverage, whichever portion of the surface is considered. In random heterogeneity, each region of the surface, irrespective of its size, is a replica of the whole surface. For random heterogeneous surfaces, $\vartheta(p)$ is the average of the local isotherm $F(p, q + f(\vartheta))$ over the entire adsorption-energy spectrum $\mathscr{S}_q$:

$$\vartheta(p) = \int_{\mathscr{S}_q} F(p, q + f(\vartheta))\varphi(q) \, dq \tag{24}$$

Thus, whereas patchwise heterogeneity generates a linear integral equation, random hetero-geneity results in a nonlinear integral equation. Intermediate cases between patchwise hetero-geneity and random heterogeneity are also possible.

Adsorption without lateral interactions is particularly interesting because the topographic structure of the surface does not affect the integral equation. Localized adsorption without lateral interactions is even more interesting because the Langmuir isotherm (19) has the following additional features:

> By defining an "enhanced pressure," the most popular isotherm for multilayer adsorption, the BET isotherm (from the intials of its proponents: Brunauer, Emmett, and Teller), is reduced to the Langmuir isotherm.
> The symmetrized form (20) of the Langmuir isotherm is reminiscent of the Fermi–Dirac occupation statistics.

C. The Overall Isotherm

Even for the overall isotherm, one can encounter a variety of possible descriptions. Denoting suitable parameters by p_m, s ($0 < s < 1$), Ω_T, and B, the following experimental behaviors are often observed in the submonolayer regime:

Freundlich isotherm

$$\vartheta_F(p) = \left(\frac{p}{p + p_m} \right)^s \tag{25}$$

Temkin isotherm

$$\vartheta_T(p) = \Omega_T \ \ln\left(1 + \frac{p}{p_m}[e^{1/\Omega_T} - 1] \right) \tag{26}$$

Dubinin–Radushkevich (DR) isotherm

$$\vartheta_{DR}(p) = \exp\left(-B\left[k_B T \ \ln\left(1 + \frac{p_m}{p} \right) \right]^2 \right) \tag{27}$$

Tóth isotherm

$$\vartheta_{To}(p) = \frac{p}{(p^{1/s} + p_m^{1/s})^s} \tag{28}$$

together with their blends:

> Freundlich–DR (FDR) isotherm
> Temkin–Freundlich isotherm
> Temkin–Freundlich–DR.

Usually, the above equations hold true in restricted pressure ranges, say for $p_M \lesssim p \lesssim p_m$ (with p_M and p_m two temperature-dependent characteristic pressures depending on the consid-ered adsorption system); for the Temkin and Tóth isotherm, p_M is allowed to vanish.

Historical Note 2. Equations (25) and (26) do not coincide with the isotherms originally proposed; improvements to the Freundlich, Dubinin–Radushkevich, and Temkin original expressions were given by several authors (and especially by Misra and Cerofolini); all of the above isotherms (including their blends) are discussed in detail in Ref. 21.

D. The Direct Problem for Equilibrium Surfaces

In view of the numerous adsorption isotherms observed in practice, it seems difficult to account for all of them with a unique adsorption-energy distribution function given on physical bases.

The most advanced line of research in this direction was initiated by Cerofolini, who considered surfaces formed in equilibrium conditions at a certain temperature T_f and then quenched at the adsorption temperature preserving the previous equilibrium structure. Postulating that the number N_i of surface atoms with energy excess u_i (relative to the energy of surface atoms in the most stable configuration) is given by a Boltzmann distribution $N_i/N_0 = \exp(-u_i/k_B T_f)$, where N_0 is the number of atoms in the most stable configuration, and assuming a dense distribution of energy excess, the surface will be characterized by the following distribution function:

$$\varphi_s(u) = \frac{1}{k_B T_f} \frac{\exp(-u/k_B T_f)}{1 - \exp(-u_M/k_B T_f)} \tag{29}$$

where u_M is the maximum value of u. Two limiting cases of Eq. (29) are of interest:

$$\varphi_s(u) \simeq \begin{cases} \dfrac{1}{k_B T_f} \exp\left(-\dfrac{u}{k_B T_f}\right) & \text{for } \dfrac{u_M}{k_B T_f} \gg 1 \quad \text{(weak heterogeneity)} \\[3mm] \dfrac{1}{u_M} & \text{for } \dfrac{u_M}{k_B T_f} \ll 1 \quad \text{(strong heterogeneity)} \end{cases}$$

For a description of adsorption on equilibrium surfaces, it is necessary to specify the adsorption energy as a function of the energy excess. Because $u = 0$ corresponds to surface atoms with maximum coordination, they will be expected to have the minimum tendency to attract gas-phase molecules; because sites with higher energy excess are less strongly bonded to the substrate, they will be expected to manifest a higher tendency to attract gas-phase molecules. The combination of these occurrences supports the idea that in the absence of specific steric or electronic factors, u is an increasing function of q. At present, it is not possible to derive the function $u = u(q)$ on general physical bases. However, a Taylor expansion with positive coefficients truncated to the second power, $u = \alpha_1(q - q_m) + \alpha_2(q - q_m)^2$ (with α_1 and α_2 positive coefficients and null constant term, because $u = 0$ for $q = q_m$), is sufficient to explain why the FDR and Temkin isotherms are so frequently observed on real adsorbents: The FDR isotherm is associated with adsorption on weakly heterogeneous equilibrium surfaces, whereas the Temkin isotherm corresponds to adsorption on strongly heterogeneous equilibrium surfaces [22]. Cerofolini theory and its improvements are fully described in Chapter 6 of Ref. 23.

E. The *Ab Fine* Problem

It is rare that any of the above-described overall isotherms fits the experimental isotherm satisfactorily in the whole physical range. This state of affairs generates the *ab fine* problem of accounting for the observed behavior in terms of adsorption-energy distribution of the adsorbent.

The solution of this problem requires a preliminary choice of the local isotherm. For the reasons stated in Section V.A, the most convenient choice is the Langmuir isotherm (19). In that case, the *ab fine* problem is related to the solution of the integral equation

$$\tilde{\vartheta}(Q) = \int_0^{+\infty} \left[1 + \exp\left(\frac{Q - q}{k_B T}\right) \right]^{-1} \varphi(q) \, dq \tag{30}$$

Defining the overall "nisotherm" $\tilde{\lambda}(Q)$, $\tilde{\lambda}(Q) := 1 - \tilde{\vartheta}(Q)$ [and similarly the local nisotherm $\tilde{\Lambda}(Q, q) := 1 - \tilde{\Theta}(Q, q)$] and remembering the normalization condition for $\varphi(q)$, Eq. (30) becomes

$$\tilde{\lambda}(Q) = \int_0^{+\infty} \left[1 + \exp\left(\frac{q - Q}{k_B T}\right)\right]^{-1} \varphi(q)\, dq \tag{31}$$

Thus, solving the *ab fine* problem for the adsorption-energy distribution from the overall isotherm is equivalent to finding the energy distribution of a gas of fermions when the average occupation is known as a function of its Fermi energy Q.

Specializing the integral equations (23) or (24) to the Jovanovic isotherm leads to an integral equation for $\varphi(q)$, the solution of which is reduced to the calculation of an inverse Laplace transform. Much more complicated is solving the integral equation (23) or (24) for the Langmuir isotherm. Of the various methods proposed for solving Eq. (30), numerical ones (discretization, optimisation, regularization, iterative, expansion, integral transform) are discussed in Refs. 21, 34, and 35. In the following the attention will be limited to exact and to approximate methods.

1. Exact Methods

Three exact solution methods are known:

> The Sips method, based the theory of Stieltjes transform and requiring the analytic extension of $\vartheta(p)$ [24,25].
> The Landman–Montroll method, based on the Wiener–Hopf technique and requiring a knowledge of $\vartheta(p)$ in the physical region only [3].
> A method requiring the infinite differentiability of $\vartheta(p)$.

In the following, I will give only a brief summary of the last method, because it is useful for introducing the approximate methods too.

The derivation is obtained from the following representation of the Dirac δ distribution:

$$\delta(x - y) = \sum_{n=0}^{+\infty} \frac{(-1)^n \pi^{2n}}{(2n + 1)!} \frac{\partial^{2n+1}}{\partial y^{2n+1}} \frac{1}{1 + e^{x-y}} \tag{32}$$

The proof of this relationship is obtained in two steps. First, the cases $y < x$ and $x < y$ are considered separately, $(1 + e^{x-y})^{-1}$ is expanded in powers of $e^{-|x-y|}$, and the resulting series are summed and found to be null. Second, the integral of the right-hand side of Eq. (32) is calculated and found equal to 1; this completes the proof.

Because for Langmuir adsorption

$$\tilde{\Lambda}^L(Q, q) = \left[1 + \exp\left(\frac{q - Q}{k_B T}\right)\right]^{-1}$$

Eq. (32) can be applied:

$$\delta(q - Q) = \sum_{n=0}^{+\infty} \frac{(-1)^n (\pi k_B T)^{2n}}{(2n + 1)!} \frac{\partial^{2n+1}}{\partial Q^{2n+1}} \tilde{\Lambda}^L(Q, q)$$

Inserting this relationship into the identity

$$\varphi(Q) = \int_0^{+\infty} \delta(q - Q)\varphi(q)\, dq$$

one eventually has

$$\varphi(Q) = \int_0^{+\infty} \sum_{n=0}^{+\infty} \frac{(-1)^n (\pi k_B T)^{2n}}{(2n+1)!} \frac{\partial^{2n+1}}{\partial Q^{2n+1}} \tilde{\Lambda}^L(Q, q) \varphi(q) \, dq$$

$$= \sum_{n=0}^{+\infty} \frac{(-1)^n (\pi k_B T)^{2n}}{(2n+1)!} \frac{\partial^{2n+1}}{\partial Q^{2n+1}} \int_0^{+\infty} \tilde{\Lambda}^L(Q, q) \varphi(q) \, dq \qquad (33)$$

$$= \sum_{n=0}^{+\infty} \frac{(-1)^n (\pi k_B T)^{2n}}{(2n+1)!} \frac{\partial^{2n+1} \tilde{\lambda}(Q)}{\partial Q^{2n+1}}$$

Historical Note 3. Equation (33) has a complicated history. It was originally proposed by Widder in the frame of the theory of power expansion of operators [26], was rediscovered by Jagiełło et al. assuming analytically conditions [27,28], and was eventually re-discovered by Chen in the context of the theory of distributions [29]. The derivation given earlier is taken from Ref. 29.

2. Approximate Methods

Equation (33) gives $\varphi(Q)$ as an infinite expansion of all odd derivatives of the overall isotherm $\tilde{\vartheta}(Q)$. This quantity, however, is known on experimental grounds and hence affected by an error which may be responsible for unphysical oscillations in the calculated $\vartheta(Q)$. This effect is manifestly larger, the higher the order of the derivative. Thus, it may be convenient to truncate the expansion to the first term,

$$\varphi(Q) = \frac{\partial \tilde{\lambda}(Q)}{\partial Q} \qquad (34)$$

(thus getting the *condensation approximation*), or to the second term,

$$\varphi(Q) = \frac{\partial \tilde{\lambda}(Q)}{\partial Q} - \frac{(\pi k_B T)^2}{6} \frac{\partial^3 \tilde{\lambda}(Q)}{\partial Q^3} \qquad (35)$$

(thus obtaining the *Rudziński–Jagiełło approximation*).

Another approximation which has had an important role in the development of the theory of adsorption on heterogeneous surfaces is the *asymptotically correct approximation*:

$$\varphi(Q) = \frac{\partial \tilde{\lambda}(Q)}{\partial Q} + k_B T \frac{\partial^2 \tilde{\lambda}(Q)}{\partial Q^2} \qquad (36)$$

The errors resulting by the use of Eqs. (34) and (36) are discussed in Ref. 30.

Historical Note 4. The condensation approximation has played, and continues to play, a special role in the theory of adsorption on heterogeneous surfaces: First introduced by Roginsky in the forties, later rediscovered by Harris, and eventually formalized by Cerofolini, the condensation approximation produces reasonably approximate distribution functions which can be used as starting points in iterative numerical methods; the methods to evaluate the errors associated with formula (34), mainly developed by Harris and Cerofolini, are described in Ref. 21, where original references are also given. Even the Rudziński–Jagiełło approximation was not obtained as a special-case general formula (33), rather it represented just a step in the direction that eventually led Jagiełło et al. to derive that formula [31,32]. The asymptotically correct approximation is not obtained by truncation of the infinite expansion (33); it plays a special role in the theory because Eq. (36) is a useful tool for predicting the onset (often experimentally inaccessible) of the Henry law, $\vartheta \propto p$. It was proposed in this context by Hobson and systematically used by Cerofolini (see Ref. 33 for a discussion of this problem).

VI. ANALYSIS OF DESORPTION-RATE DATA

On the same intuitive bases as for adsorption equilibrium, it is clear that information on the energetic structure of a surface is contained in the desorption kinetics: The initial stages of desorption involves less bonded molecules, whereas the final stages are related to the desorption of the most strongly bonded molecules. The situation in kinetics is, however, even more complex than in equilibrium adsorption. In fact, not only is not the local law given a priori because of the effect of the topography (as it happens in equilibrium adsorption), but also the unavoidable readsorption during desorption experiments (or, symmetrically, desorption during adsorption experiments) and the reconstruction phenomena associated with the adsorption–desorption process complicate the local law. The following derivation will nonetheless be based on the following hypotheses:

> Absence of reconstruction
> Sharp prevalence of desorption over adsorption
> Fixed energetic heterogeneity of the adsorbent

thus allowing us to run in the situation of Section II.

A. Desorption Kinetics Under Nonisothermal Conditions: Thermal Desorption Spectrometry

Desorption from a homogeneous surface is characterized by the existence of a characteristic time τ such that the degree of advancement of the process ("desorption kinetics") is negligible for $t/\tau \ll 0.3$, is nearly completed for $t/\tau \gg 0.3$, and occurs significantly for $t/\tau \approx 1$. If desorption is a thermally activated process, the characteristic time τ depends on the activation energy for desorption E as

$$\tau = \tau_0 \exp\left(\frac{E}{k_B T}\right) \tag{37}$$

where τ_0 is a suitable preexponential factor.

Let 0 denote the time at which the considered desorption experiment has started. The duration t_M of typical experiments (during which the relevant physical quantities are controlled or measured) is usually less than 1 day (say, $t_M \lesssim 10^5$ s). On the opposite side, a desorption experiment does not produce reliable results at time shorter than a value t_m required to bring the system to the wanted initial condition; typically, t_m lasts a few seconds, at least (say, $t_m \gtrsim 1$ s). The isothermal kinetics can therefore be determined over a time interval covering at most five order of magnitudes. Because of the exponential dependence (37), the desorption experiment is sensitive only to the change of coverage of the zones with activation energy for desorption in the interval (E_m, E_M), with $E_m = k_B T \ln(t_m/\tau_0)$ and $E_M = k_B T \ln(t_M/\tau_0)$.

The part (E_m, E_M) of the energy spectrum $\mathscr{S}_E$ which is effective on the desorption kinetics extends, therefore, over a few units of $k_B T$; $E_M - E_m = k_B T \ln(t_M/t_m) \lesssim 12 k_B T$. At room temperature ($k_B T = 25$ meV), the effective spectrum extends for 0.3 eV at most. Although the width of this interval may be sufficient to cover the spectrum of physically adsorbed molecules, it is smaller than the typical spectrum of chemisorbed molecules, which may extend for 1 eV or more.

A way to make up for this difficulty is to operate at a higher temperature: At 1000 K, the spectrum covers approximately 1 eV. This procedure, however, is not free of difficulties; it may happen that at time t_m, an appreciable part of the phenomenon has already been exhausted. Indeed, taking $\tau_0 = 10^{-13}$ s, $t_m = 1$ s, and $k_B T = 0.1$ eV, all sites with energy below $E_m = 3$ eV have already completed desorption before the desorption rate can be reliably determined.

A way to overcome this difficulty is to operate under a nonisothermal condition, imposing a temperature ramp on the system. Typical ramps employed in thermal programmed desorption are the linear ramp, $T = T_0 + \alpha t$, or the inverse-linear ramp, $1/T = 1/T^0 - t/\alpha'$, with α and α' the corresponding rates. The progressive increase of temperature allows that sites with lower desorption energies respond to the temperature ramp sooner than the sites with higher energy. This fact allows a kind of *thermal desorption spectroscopy* of the surface.

The very fact that both the width of the energy spectrum and the characteristic desorption energy vary with time, makes nonisothermal kinetic data less reliable than isothermal data. This difficulty may be overcome by determining portions of the complete isothermal desorption kinetics at different temperatures. This reduces the original problem, of determining the energy spectrum from nonisothermal desorption kinetics, to a set of equivalent problems, each of which associated with the extraction of the desorption energy distribution from isothermal kinetics measured at different temperatures.

B. Desorption Kinetics in Isothermal Conditions

The theory of isothermal desorption kinetics form heterogeneous surfaces can be established by describing the overall desorption kinetics as weighted averages of the local desorption kinetics form each energetically homogeneous zone forming the surface.

1. Desorption Kinetics from Homogeneous Surfaces

The theory developed in the following applies to desorption kinetics, but, *mutatis mutandis*, it describes adsorption kinetics too [35].

If

(A1)　readsorption does not affect the desorption kinetics

(A2)　the surface does not undergo reconstruction during the process

then the rate equation giving the time derivative $d\Theta/dt$ of the surface coverage Θ of a homogeneous surface with desorption time τ is given by

$$-\frac{d\Theta}{dt} = \frac{1}{\tau}\Theta^\nu, \tag{38}$$

where ν is the reaction order. The analysis will be limited to kinetics of the first order ($\nu = 1$) and second order ($\nu = 2$) and to kinetic coefficients τ independent of Θ.

If Θ_0 is the local coverage at time $t = 0$, the rate equation (38) can be solved for all ν by the separation of variables:

$$-\int_{\Theta_0(\tau)}^{\Theta} \frac{d\Theta}{\Theta^\nu} = \frac{t}{\tau} \tag{39}$$

thus giving a function $\Theta = \Theta_\nu(t, \tau)$. Equation (39) can be solved in closed form for all ν.

Defining $\Delta_v(t, \tau) := \Theta_0(\tau) - \Theta_v(t, \tau)$, one has

$$\Delta_1(t, \tau) = \Theta_0(\tau)\left[1 - \exp\left(-\frac{t}{\tau}\right)\right] \tag{40}$$

for first-order kinetics and

$$\Delta_2(t, \tau) = \frac{\Theta_0^2(\tau)t}{\Theta_0(\tau)t + \tau} \tag{41}$$

for second-order kinetics.

2. Desorption Kinetics from Heterogeneous Surfaces

In general, the desorption kinetics from patchwise heterogeneous surfaces are given by

$$\delta_v(t) = \int_{\mathscr{S}_\tau} \Delta_v(t, \tau)\phi(\tau)\, d\tau \tag{42}$$

where $\delta_v(t)$ is the overall amount desorbed at time t, $\phi(\tau)\, d\tau$ is the fraction of surface with desorption time between τ and $\tau + d\tau$, and $\mathscr{S}_\tau$ is the support of $\phi(\tau)$.

Equation (42) originates the problem to extract the distribution function $\phi(\tau)$ when the overall kinetics $\delta_v(t)$ are experimentally known.

The amount $\delta_v(t)$ depends on the lifetime distribution function $\phi(\tau)$ and on the set of initial conditions $\Theta_0(\tau)$. Clearly enough, if one wants to extract the distribution function $\phi(\tau)$ from the experimental function $\delta_v(t)$, one must operate under conditions for which $\Theta_0(\tau)$ is known. In the forthcoming develement, attention will be limited to the case of Θ_0 constant with τ. In this case, one may, without loss of generality, take $\Theta_0 = 1$, so that Eqs. (40) and (41) become

$$\Delta_1(t, \tau) = 1 - \exp\left(-\frac{t}{\tau}\right) \tag{43}$$

and

$$\Delta_2(t, \tau) = \frac{t}{t + \tau} \tag{44}$$

respectively.

Although the theory can be worked using τ as an independent random variable, in most practical cases, desorption is an activated process satisfying by Eq. (37), for which the choice of E as an independent variable is more convenient. For activated desorption, Eq. (42) becomes

$$\delta_v(t) = \int_0^{+\infty} \hat{\Delta}_v(t, E)\varphi(E)\, dE \tag{45}$$

where $\hat{\Delta}_v(t, E) := \Delta_v(t, \tau(E))$, $\tau(E)$ is given by Eq. (37), and $\varphi(E)\, dE$ is the fraction of surface with desorption energy between E and $E + dE$.

Equation (45) is specialized to the form

$$\delta_1(t) = \int_0^{+\infty} \left[1 - \exp\left(-\frac{t}{\tau_0}\exp\left(-\frac{E}{k_B T}\right)\right)\right]\varphi(E)\, dE \tag{46}$$

for first-order kinetics [Eq. (43)] and

$$\delta_2(t) = \int_0^{+\infty} \frac{t}{t + \tau_0 \exp(E/k_B T)}\varphi(E)\, dE \tag{47}$$

for second-order kinetics [Eq. (44)]. Once again, although the integration interval has been taken $(0, +\infty)$, in practice, the support $\mathscr{S}_E$ of $\varphi(E)$ is contained in a finite interval (E_m, E_M).

The integral representations (46) and (47) are the basic equations of the model. As first observed by Jaroniec and Madey [34], kernels (43) and (44) with τ given by Eq. (37) resemble strictly the Langmuir isotherm and the Jovanovic isotherm, respectively, of equilibrium adsorption. Although the analogy is not complete [in fact, assuming the one-to-one correspondence $p \leftrightarrow t$ and $q \leftrightarrow E$, the equilibrium equations do not depend on p and q separately, but rather on the function $p \exp(q/k_B T)$; the desorption kinetics do not depend on t and E separately, but, rather, on the function $t \exp(-E/k_B T)$—note the different signs of the exponents], many of the methods developed for the theory of adsorption equilibrium on heterogeneous surfaces can be extended to desorption kinetics from heterogeneous surfaces.

As discussed in Section III.E, for both the Langmuir and Jovanovic isotherms, several exact methods exist for the determination of the energy distribution function $\varphi(q)$ when the overall adsorption isotherm is precisely known. Rather than trying to modify these exact methods for desorption kinetics, this work will be limited to the use of the condensation approximation.

Note 1. From the formal point of view, the equations for the adsorption kinetics on a heterogeneous surface with a given adsorption-time distribution may be modified to have the same expression [Eq. (42)], as the desorption kinetics with the same desorption-time distribution (see, for instance, Refs. [35]). The forms (46) and (47), however, may be applied only to thermally activated adsorption kinetics; chemisorption usually runs in this situation, whereas physisorption is generally a nonactivated process.

Note 2. Because in the desorption kinetics of physisorbed molecules one may reasonably assume that E coincides with the adsorption energy q and because $\varphi(q)$ can be determined from adsorption equilibrium isotherms, there is the possibility of comparing the distribution function resulting from kinetic experiments with that resulting from equilibrium experiments.

3. The Condensation Approximation

The condensation approximation consists of replacing the kernel $\hat{\Delta}_v(t, E)$ by the step function

$$\Delta_v^c(t, E) = \begin{cases} 1 & \text{for } t \geq t_v^c(E) \\ 0 & \text{for } t < t_v^c(E) \end{cases} \tag{48}$$

where the function $t_v^c(E)$ is determined by minimizing the distance d between the local kinetics $\hat{\Delta}_v(t, E)$ and their approximant (48):

$$d[\hat{\Delta}_v(t, E), \Delta_v^c(t, E)] = \min \tag{49}$$

It is immediately verified that condition (49) is satisfied for the Lagrangian distance and for the L^2 distance by taking

$$t_v^c(E) = [v - 1 + \ln(3 - v)]\tau_0 \exp\left(\frac{E}{k_B T}\right) \tag{50}$$

which holds for first-order and second-order kinetics.

Defining the inverse function of Eq. (50), $\Xi_v := k_B T \ln(t/[v - 1 + \ln(3 - v)]\tau_0)$ and setting $\hat{\delta}_v(\Xi_v) := \delta_v(t(\Xi_v))$, one has

$$\hat{\delta}_v(\Xi_v) = \int_{-\infty}^{\Xi_v} \varphi_v^c(E) \, dE \tag{51}$$

Of course, the solution of Eq. (51) will, in general, differ from the solutions of Eqs. (46) or (47), and for this reason, the distribution function $\varphi_v^c(E)$ in Eq. (51) has been denoted with the upper index c. Equation (51) is immediately solved by differentiation with respect to Ξ_v:

$$\varphi_v^c(\Xi_v) = \frac{\partial \hat{\delta}_v(\Xi_v)}{\partial \Xi_v} \tag{52}$$

Because $\Xi_2 = \Xi_1 + k_B T \ln(\ln 2)$ and $d\Xi_2 = d\Xi_1$, the distribution function calculated in the hypothesis of first-order kinetics coincides with that calculated in the assumption of second-order kinetics, provided that the energy axis is shifted by an amount $-k_B T \ln(\ln 2) = 0.367 k_B T$.

It is mentioned without further discussion (see Ref. 36) that Eq. (52) gives an adequate description of the distribution function $\varphi(E)$, provided that the energy spectrum $\mathscr{S}_E$ is much wider than $k_B T$, $E_M - E_m \gg k_B T$, and $\varphi(E)$ varies smoothly with E,

$$\frac{\varphi'(E)}{\varphi(E)} \ll \frac{1}{k_B T} \tag{53}$$

C. The Experimental Desorption Kinetics

In the following, we will try to extract information on the activation-energy distribution from the desorption kinetics more frequently met in practice—the time-logarithm law and the time-power law.

1. The Time-Logarithm Law

The time-logarithm law reads

$$\delta^E(t) = \begin{cases} \delta_m \ln\left(1 + \dfrac{t}{t_m}\right) & \text{for } 0 \le t < t_M \\ \delta_M & \text{for } t_M \le t \end{cases} \tag{54}$$

where δ_m, t_m, δ_M, and t_M are four characteristic parameters mutually related by the relationship

$$t_M := t_m \left[\exp\left(\frac{\delta_M}{\delta_m}\right) - 1 \right] \tag{55}$$

t_M (and hence δ_M) may be allowed to go to $+\infty$. The macroscopic meaning of the parameters in Eq. (54) is straightforward: t_m is a characteristic time below which the logarithmic behavior is no longer observed, t_M is the time required to complete the process, and δ_m/t_m is the initial desorption rate.

Historical Note 5. Although usually referred to as "Elovich equation," Eq. (54) was actually reported by other authors. In fact, a time-logarithm law was first observed by Tamman and co-workers in 1922 for the *oxidation* of metals. Tamman work was rediscoverd in 1935 by Vernon, who observed that the low-temperature oxidation of iron obeys Eq. (54). The oxidation of a number of metals and semiconductors has since been reported to be described by Eq. (54); among them, Cr, Mn, Fe, Co, Ni, Cu, Zn, Cd, Si, Sn, Pb, Sb, and Bi can be mentioned. The first observation of a *chemisorption* process obeying Eq. (54) was probably reported by Zeldowitch in 1934, but it was only after its rediscovery by Elovich and co-workers that this equation became widely used. On this basis, the time-logarithm law (54) took the name of Elovich equation, although it could be more appropriately called Zeldowitch equation (it is also noted that in the German papers, the person now commonly known as Elovich transliterated his name Elowitz).

This note is taken from Refs. 37 and 38, in which additional information is given. A list of chemisorption systems described by the Elovich equation is given in Ref. 39.

2. The Time-Power Law

The desorption kinetics of many processes are described by a time-power law of the form

$$\delta^{pl}(t) = 1 - \left(1 + \frac{t}{t_0}\right)^{-s} \tag{56}$$

where t_0 and s are temperature-dependent parameters and t_0 is characteristic time linked to the half-life $t_{1/2}$ of the process by the relation $t_0 = t_{1/2}(2^{1/s} - 1)$, $0 < s < 1$.

An example of the process described by Eq. (56) is given by the low-temperature (below 200 K) recombination kinetics CO to myoglobin (Mb, a heme group with a central iron atom) after photolysis of carboxymyoglobin MbCO [40,41]. A short account of the kinetics which have been described by Eq. (56) is given in Ref. 37.

D. Accounting for the Experimental Desorption Kinetics

Several models have been proposed for both the time-logarithm and time-power laws based on the hypotheses of fixed surface heterogeneity, induced surface heterogeneity, and surface reconstruction. For the time-logarithm law, they are extensively reviewed in Ref. 39. The following analysis (strictly based on Refs. 36) considers how the time-logarithm and time-power laws are accounted for by the condensation approximation in the frame of fixed heterogeneity.

1. The Time-Logarithm Law in the Condensation Approximation

Assume, for a moment, second-order kinetics. Expressed in terms of Ξ_2, the Elovich equation (54) reads

$$\hat{\delta}_2^{mM}(\Xi_2) = \begin{cases} \delta_m \ln\left(1 + \frac{\tau_0}{t_m}\exp\left(\frac{\Xi_2}{k_B T}\right)\right) & \text{for } -\infty < \Xi_2 \leq \Xi_{2M} \\ \delta_M & \text{for } \Xi_{2M} < \Xi_2 < +\infty \end{cases} \tag{57}$$

where $\Xi_{2M} := \Xi_{2m} + k_B T \ln[\exp(\delta_M/r_E) - 1]$ and $\Xi_{2m} := k_B T \ln(t_m/\tau_0)$.

In the sequel, the index 2 to Ξ and φ^c will be omitted to have more manageable expressions. The application of Eq. (52) to Eq. (57) gives

$$\varphi_c(\Xi) = \frac{\delta_m/\delta_M}{k_B T} \times \begin{cases} \exp\left(\frac{\Xi - \Xi_m}{k_B T}\right)\left[1 + \exp\left(\frac{\Xi - \Xi_m}{k_B T}\right)\right]^{-1} & \text{for } -\infty < \Xi \leq \Xi_M \\ 0 & \text{for } \Xi_M < \Xi + \infty \end{cases} \tag{58}$$

whichever is the value of δ_M. A study of the distribution function (58) shows the following limiting behaviors:

$$\varphi^c(\Xi) \simeq \frac{\delta_m/\delta_M}{k_B T} \times \begin{cases} \exp\left(\frac{\Xi - \Xi_m}{k_B T}\right) & \text{for } -\infty < \Xi \lesssim \Xi_m - k_B T \\ 1 & \text{for } \Xi_m - k_B T \lesssim \Xi \leq \Xi_M \\ 0 & \text{for } \Xi_M < \Xi + \infty \end{cases} \tag{59}$$

Of course, the calculated distribution function cannot be an arbitrary function of Ξ and must satisfy the mathematical conditions of non-negativity and of normalization; physical conditions also have to be satisfied: for instance, $\varphi^c(\Xi)$ must be identically null for $\Xi < 0$ and must be temperature independent. There is no choice of Ξ_m and Ξ_M for which the function (58) is temperature independent; however, the asymptotic expansion (59) shows that $\varphi^c(\Xi)$ varies with T only in a region centered on Ξ_m of width $O(k_B T)$, whereas the calculated distribution function is approximately constant in the whole interval (Ξ_m, Ξ_M) provided that the extremes Ξ_m and Ξ_M and therefore the difference $\Xi_M - \Xi_m$ are constant with T.

Equation (59) states that Ξ_M must be interpreted as the maximum adsorption energy, $\Xi_M = E_M$, whereas the distribution function vanishes exponentially with $\Xi - \Xi_m$ for $\Xi \lesssim \Xi_m$.

Although, in principle $\mathscr{S}_E$ does not admit a minimum, Ξ_m can, in a way, be interpreted as an estimate of the minimum adsorption energy, $\Xi_m \simeq E_m$.

Because, in general, $\varphi(E) \neq \varphi^c(E)$, even the maximum desorption energy E_M^c determined within the condensation approximation (CA) is in principle different from the true desorption energy E_M. The normalization condition for $\varphi^c(\Xi)$,

$$\int_{-\infty}^{E_M^c} \varphi_c(\Xi) \, d\Xi = 1$$

gives the following equation for E_M^c:

$$\delta_m \ln\left(1 + \exp\left(\frac{E_M^c - \Xi_m}{k_B T}\right)\right) = \delta_M$$

the solution of which is

$$E_M^c - \Xi_m = k_B T \, \ln\left(\exp\left(\frac{\delta_M}{\tau_e}\right) - 1\right) \simeq k_B T \frac{\delta_M}{\delta_m} \tag{60}$$

Combining this relationship with the physical meaning of Ξ_m and Ξ_M, one eventually gets the meaning and the temperature dependence of the parameters t_m and δ_m: $t_m \approx \tau_0 \exp(E_m/k_B T)$, with $\Xi_m \simeq E_m$, and $\delta_m \simeq \delta_M k_B T/\Omega$, where Ω is the width of the energy spectrum: $\Omega := E_M - E_m \simeq \Xi_M - \Xi_m$.

When first-order kinetics are considered, the Elovich equation (54) reads

$$\hat{\delta}_1^{mM}(\Xi_1) = \begin{cases} \delta_m \ln\left(1 + \dfrac{\tau_0 \ln 2}{t_m} \exp\left(\dfrac{\Xi_1}{k_B T}\right)\right) & \text{for } -\infty < \Xi_1 \leq \Xi_{1M} \\ \delta_M & \text{for } \Xi_{1M} < \Xi_1 < +\infty \end{cases} \tag{61}$$

where $\Xi_{1M} := \Xi_{1m} + k_B T \ln(\exp(\delta_M/\delta_m) - 1)$ and $\Xi_{1m} := k_B T \ln(t_m/\tau_0) - k_B T \ln(\ln 2)$. The application of Eq. (52) to Eq. (61) gives a distribution function with the same shape as Eq. (58), in which Ξ_2, Ξ_{2m}, and Ξ_{2M} are replaced by Ξ_1, Ξ_{1m}, and Ξ_{1M}, respectively. All of the earlier considerations on minimum and maximum absorption energies made for second-order kinetics therefore apply to first-order kinetics provided that Ξ_{2m} and Ξ_{2M} are replaced by Ξ_{1m} and Ξ_{1M}, respectively.

Irrespective of the kinetic order, the condensation approximation is valid when the energy spectrum is much wider than $k_B T$; this condition is fulfilled only for $\delta_m/\delta_M \ll 1$, which is the condition for the validity of the above argument. Note that when this condition is satisfied, the difference between the calculated distribution functions for the first- and second-order kinetics, consisting of a shift of the order of $k_B T$ on the energy axis, is negligible.

2. Accounting for the Time-Power Law

The time-power law is usually explained by advocating desorption kinetics of the first order and a distribution of barrier heights [40,42]:

$$\delta^{\mathrm{pl}}(t) = \int_0^\infty \left[1 - \exp\left(-\frac{t}{\tau_0}\exp\left(-\frac{E}{k_BT}\right)\right)\right]\varphi(E)\, dE \tag{62}$$

In principle, $\varphi(E)$ can be obtained by inverse Laplace transform techniques [43,44]. The precision required for experimental data over a very extended time domain is, however, so high that this method is of little practical use. It is thus more convenient to parametrize $\varphi(E)$ in a suitable analytic form, to insert it into Eq. (62), and to calculate $\delta^{\mathrm{pl}}(t)$ numerically; the parameters are then varied until a good fit of the calculations to the experimental data is obtained. The analytic form

$$\varphi(E) = C\exp\left(a(E_p - E) - \frac{a}{b}\exp[b(E_p - E)]\right) \tag{63}$$

(where a, b, and E_p are adjustable parameters and C is a constant determined by the normalization condition) is suggested by inverse Laplace transform of the time-power law (56) [42]. This distribution function has a maximum on E_p and vanishes as $\exp(-aE)$ for high E.

The problem is treated more easily in the CA. Defining $\Xi := k_BT\ln(t/\tau_0\ln 2)$, the time-power law (56) expressed in terms of Ξ reads

$$\hat{\delta}^{\mathrm{pl}}(\Xi) = 1 - \left[1 + \frac{\tau_0\ln 2}{t_0}\exp\left(\frac{\Xi}{k_BT}\right)\right]^{-s} \tag{64}$$

for $-\infty < \Xi < +\infty$. The straightforward application of Eq. (52) to Eq. (64) gives

$$\varphi^c(\Xi) = \frac{s}{k_BT}\left[1 + \exp\left(\frac{\Xi - \Xi_0}{k_BT}\right)\right]^{-s-1}\exp\left(\frac{\Xi - \Xi_0}{k_BT}\right) \tag{65}$$

where $\Xi_0 := k_BT\ln(t_0/\tau_0\ln 2)$. A study of Eq. (65) shows that $\varphi^c(\Xi)$ has a maximum in Ξ_0 and that

$$\Xi \ll \Xi_0 \Rightarrow \varphi^c(\Xi) \simeq \frac{s}{k_BT}\exp\left(\frac{\Xi - \Xi_0}{k_BT}\right)$$

$$\Xi_0 \ll \Xi \Rightarrow \varphi^c(\Xi) \simeq \frac{s}{k_BT}\exp\left(s\frac{\Xi_0 - \Xi}{k_BT}\right)$$

3. Other Kinetics

The time-logarithm and time-power laws do not exhaust the kinetics observed in practical situations. If a certain behavior is observed and can reasonably be ascribed to adsorbent heterogeneity, the energy distribution function accounting for it can be evaluated as follows:

A first estimate of the energy distribution function is obtained via the condensation approximation.

The function so calculated is approximated by a positive and normalized function parameterized with some parameters (e.g., log-normal, Raleigh, or Pearson IV distributions) behaving approximately in the same manner as the condensation approximation distribution function in the experimental region.

For the assumed local kinetics, the overall kinetics are then calculated and the parameters are varied until they give the best fit to the experimental kinetics.

Note 3. That the distribution resulting from this procedure does indeed approximate the true energy distribution of the adsorbents is essentially related to the fact that heterogeneity is the unique reason responsible for deviations from the assumed local law [Eq. (43) or (44)]. To which extent, this stipulation is satisfied is discussed in Sections VII.B3–VII.B.5.

VII. OPEN PROBLEMS

In the previous sections I have tried to provide the reader with consolidated results. He or she may have reached the conclusion that in spite of the ill posedness of the considered *ab fine* problems, most of the difficulties have been removed and what remains to do is simply to apply the obtained results to practical situations.

This view is however false, and many problems are still unsolved. In the following I shall introduce a short menu of them. I do not claim completeness; rather, the choice is dictated by my preferences. The problems listed in the following are classified in relation to their character: mathematical or physico-chemical.

A. Mathematical Problems

1. Determining the Kernel

When considering the heat capacity of crystalline solids, there is no ambiguity about the kernel to be used in the *ab fine* problem; it is determined by the Bose–Einstein stastistics.

The discussion of Section V.A shows, instead, that different adsorption isotherms can be observed on energetically homogeneous surfaces, in relation to the adsorbent–adsorbate pair. The choice of $\Theta(p, q)$ in Eq. (23) requires, therefore, preliminary information which is usually not experimentally accessible. On the other side, one can determine experimentally both the overall isotherm $\vartheta(p)$ (by static or quasistatic adsorption techniques) and adsorption energy distribution $\varphi(q)$ (by calorimetric techniques). *To which extent are these data sufficient to specify* $\Theta(p, q)$? For a discussion of this problem, see Ref. 35.

A similar problem is met in adsorption kinetics: *To which extent do the experimental desorption kinetics* $\delta(t)$ *and the activation energy distribution* $\varphi(E)$ *specify the kinetic order?*

2. Cancellation Methods and the Optimum Choice of the Cutoff Function

For $\lambda(y)$ and $\Lambda(y, x)$ given and $\varphi(x)$ unknown, the equation

$$\int_0^{+\infty} \Lambda(y, x)\varphi(x)\, dx = \lambda(y) \tag{66}$$

is referred to as a *Fredholm integral equation of the first kind*. As already discussed, the solution of Eq. (66) for compact operators is unstable, which makes the problem of solving it for $\varphi(y)$ an improperly posed problem.

A typical *regularization method* for the Fredholm integral equation of the first kind is to add a term $\varrho\varphi(y)$ (with ϱ a fixed number) to the left-hand side of Eq. (66), thus obtaining a *Fredholm integral equation of the second kind*, whose solution is stable. The difficulty of this method is related to the fact that the formula expressing such a solution (the "Neumann series") does converge only for $\varrho > \|\Lambda\|$, where $\|\Lambda\|$ is the norm of the operator Λ [45].

Lesser difficulties are met approximating Eq. (1) by a Volterra equation. Because the function $\Lambda(y, x)$ satisfies the condition $x \to +\infty \Rightarrow \Lambda(y, x) \to 0$, the kernel of Eq. (66) can be approximated by the kernel

$$\Lambda^X(y, x) = \begin{cases} \Lambda(y, x) & \text{for } 0 \le x \le X \\ 0 & \text{otherwise} \end{cases} \tag{67}$$

for which $X \to +\infty \Rightarrow \Lambda^X(y, x) \to \Lambda(y, x)$. This property makes it sensible to consider the solution $\varphi^X(x)$ of the equation

$$\int_0^X \Lambda(y, x)\varphi^X(x) \, dx = \lambda(y) \tag{68}$$

as an approximation to $\varphi(x)$. If X is given, this procedure does not help in finding $\varphi(y)$. However, if one allows X to depend on y, $X = X(y)$, Eq. (68) becomes

$$\int_0^{X(y)} \Lambda(y, x)\varphi^{[X]}(x) \, dx = \lambda(y) \tag{69}$$

which, after a differentiation with respect to y, is immediately reduced to a *Volterra integral equation of the second kind* [45]. The advantage of this operation is the fact that the Volterra equation can be solved iteratively and the solution is stable: If $\varphi_\sigma^{[X]}(x)$ denotes the solution of Eq. (69) corresponding to $\lambda_\sigma(y)$, then

$$\lambda_\sigma(y) \to \lambda(y) \Rightarrow \varphi_\sigma^{[X]}(x) \to \varphi^{[X]}(x)$$

Thus, on one side cutting the upper limit of the integral in Eq. (1) to $X(x)$ generates errors in the calculation [giving $\varphi^{[X]}(x)$ instead of $\varphi(x)$]; on the other hand, it results in a *regularization* of the problem giving stable solutions. Taking into account that the experimental datum is affected by an error σ, *is there a choice of $X(x)$ (and, in affirmative case, how is it related to σ) giving the best stable description of $\varphi(x)$?*

B. Physicochemical Problems

1. Kinetic Equivalents of the Dubinin–Radushkevich and Tóth Isotherms

To some extent, the time-logarithm and the time-power laws are the kinetic analog of the Temkin and Freundlich isotherms. The frequent observation of the time-logarithm and the time-power laws in dynamic conditions can therefore be related to the frequent observation of the Temkin and Freundlich isotherms in equilibrium conditions.

On another hand, the Temkin and Freundlich isotherms are just two among the several isotherms (like Dubinin–Radushkevich, Tóth, and their blends) often met in equilibrium adsorption. This observation immediately generates the problem of *explaining why the kinetic equivalents of the latter isotherms are not observed or (alternatively) identifying their kinetic equivalents in the reported experimental data.*

2. Temperature Dependence of the Distribution Function Accounting for the Time-Power Law

The distribution function $\varphi^c(\Xi)$ accounting for the time-power law in the condensation approximation satisfies the mathematical conditions of non-negativity and normalization. For s constant with temperature, $\varphi^c(\Xi)$ does, however, depend on T, thus violating the physical condition of temperature independence. However, an inspection of Eq. (65) shows that if $s = T/T_F$ (with T_F an assigned temperature), $\varphi^c(\Xi)$ is nearly temperature independent for

$\Xi \gg \Xi_0$, whereas it depends on T for $\Xi \lesssim \Xi_0$, where it vanishes in a fast way. *How does s depend on T for systems which obey the time-power law?*

Note 4. The time-logarithm law was first discovered for adsorption phenomena and only later was found to rule some desorption kinetics also. It is observed more frequently in adsorption than in desorption [39]. The occurrence of the time-logarithm law, more frequently in adsorption than in desorption, can hardly be explained in terms of surface heterogeneity. If explaining the relative frequency of the time-logarithm law in adsorption and desorption kinetics is a goal of a theory for it, other models are necessary.

3. Other Models for the Time-Logarithm Law

The time-logarithm law is observed in so many phenomena that different interpretations have also been proposed. For instance, the time-logarithm law was also explained by Porter and Tompkins, assuming that the activation energy of the process varies linearly with the coverage because of lateral interactions (induced energetic heterogeneity) [39], and by Landsberg in terms of surface reconstruction [37].

In view of this state of affairs, *are there indicators to establish which physicochemical mechanism is responsible for an observed time-logarithm, or must this relationship be considered as mere empirical formula for the analytical representation of experimental data?*

4. Is the Time-Logarithm Law due to Surface Reconstruction?

Even assumption A2 in Section VI.B.1 is an oversimplification. For instance, adsorbents of large practical importance (like catalysts) are often obtained by grafting an active catalytic group to an otherwise inert, highly dispersed, substrate; however, during the catalytic activity, the substrate evolves with a progressive loss of its dispersion degree until its surface area is so low as to inactivate the catalyst. The first theory taking into account surface reconstruction phenomena was likely proposed by Landsberg [37], who showed that surface reconstruction may account for the time-logarithm law (a variant of the model was also used to account for the time-power law [37]). Imagine that the surface reconstruction process hypothesized by Landsberg to account for Eq. (54) results in a dispersion of the energy excess released in adsorption over many degrees of freedom, in this way being responsible for an increase of entropy. If desorption were associated with the opposite phenomenon, it would result in the concentration of the energy dispersed over many degrees on a narrow spatial region with atomic size, in this way being responsible for a decrease of entropy. *Is the frequency of the time-logarithm law, larger in adsorption than in desorption, a consequence of the Second Law of Thermodynamics?*

REFERENCES

1. G Fichera. Rend Mat Acc Lincei Series 9 1:161, 1990.
2. R Courant, D Hilbert. Methods of Mathematical Physics. New York: Wiley, 1953, Vol. I, Chap. 1.
3. U Landman, EW Montroll. J Chem Phys 64:1762, 1976.
4. K Huang. Statistical Mechanics. New York: Wiley, 1963, Chap. 7.
5. SH Bauer. J Chem Phys 6:403, 1938; J Chem Phys 7:1097, 1939.
6. A Einstein. Ann Phys 22:180, 1907.
7. W Nernst, FA Lindemann. Preuss Akad Wiss 494, 1911.
8. A Einstein. Abh Deutsch Bunsenges 7:330, 1914.
9. P Debye. Ann Phys 397:789, 1912.

10. M Born, T von Kármán. Phys Z 13:297, 1912.
11. M Born, K Huang. Dynamical Theory of Crystal Lattices. London: Oxford University Press, 1954.
12. AA Maradudin, EW Montroll, GH Weiss, The Theory of Lattice Dynamics in the Harmonic Approximation. New York: Academic Press, 1963.
13. W Ledermann. Proc Roy Soc A 182:362, 1944.
14. GF Cerofolini. Thin Solid Films 79:277, 1981; Adv Colloid Interf Sci 19:103, 1983.
15. D Avnir, D Farin, P Pfeifer. J Chem Phys 79:3566, 1983; Nature 308:261, 1984.
16. EW Montroll. J Chem Phys 10:218 1942.
17. IM Lifshitz. J Exp Theor Phys 26:551, 1954.
18. RG Chambers, Proc. Phys. Soc. 78:941 (1961).
19. TL Hill. An Introduction to Statistical Thermodynamics. Reading, MA: Addison-Wesley, 1960.
20. WA Steele. The Interaction of Gases with Solid Surfaces. Oxford: Pergamon Press, 1975.
21. GF Cerofolini, W Rudziński. In: W Rudziński, WA Steele, G Zgrablich, eds. Equilibria and Dynamics of Gas Adsorption on Heterogeneous Solid Surfaces. Amsterdam: Elsevier, 1997, p 1.
22. GF Cerofolini. Surface Sci 51:333, 1975; 61:678, 1976; J Colloid Interf Sci 86:204, 1982.
23. W Rudziński, DH Everett. Adsorption of Gases on Heterogeneous Surfaces. London: Academic Press, 1992.
24. R Sips. J Chem Phys 16:490, 1948; 18:1024, 1950.
25. DN Misra. Surface Sci 18:367, 1969; J Chem Phys 52:5449, 1970.
26. DV Widder. Trans Am Math Soc 43:7, 1938; Duke Math J 14:217, 1947.
27. J Jagiełło, G Ligner, E Papirer. J Colloid Interf Sci 137:128, 1989.
28. J Jagiełło, JA Schwarz. J Colloid Interf Sci 146:415, 1991.
29. N-X Chen. Phys Rev A 46:3538, 1992.
30. GF Cerofolini. Thin Solid Films 23:129, 1974.
31. W Rudziński, J Jagiełło. J Low Temp Phys 45:1, 1981.
32. W Rudziński, J Jagiełło, Y Grillet. J Colloid Interf Sci 87:478, 1982.
33. GF Cerofolini. J Colloid Interf Sci 86:204, 1982.
34. M Jaroniec, R Madey. Physical Adsorption on Heterogeneous Solids. Amsterdam: Elsevier, 1988.
35. GF Cerofolini, N Re. Riv Nuovo Cimento 16(7):1, 1993.
36. GF Cerofolini, N Re. J Colloid Interf Sci 174:428, 1995; Langmuir 13:990, 1997.
37. PT Landsberg. J Chem Phys 23:1079, 1955; J Appl Phys 33:2251, 1962.
38. W Rudziński, T Panczyk. In: JA Schwarz, CI Contescu, eds. Surfaces of Nanoparticles and Porous Materials. New York: Marcel Dekker, 1999, p 355.
39. C Aharoni, FC Tompkins. Adv Catal 21:1, 1970.
40. RH Austin, KW Beeson, L Eisenstein, H Frauenfelder, IC Gunsalus. Biochemistry 14:5355, 1975.
41. H Frauenfelder. Ann NY Acad Sci 504:151, 1987.
42. N Alberding, RH Austin, SS Chan, L Eisenstein, H Frauenfelder, IC Gunsalus, TM Nordlund. J Chem Phys 65:315, 1976.
43. DE Koppel. J Chem Phys 57:4814, 1972.
44. SW Provencher. J Chem Phys 64:2772, 1976.
45. FG Tricomi. Integral Equations. New York: Dover, 1985, Chap. 1.

9
Stochastic Modeling of Adsorption Kinetics

SEUNG-MOK LEE Kwandong University, Yangyang, Korea

I. MODELING OF ADSORPTION PROCESS

The adsorptive performance of GAC is a consequence of capacity and kinetics. The conventional approach to evaluate this performance is based on a two-step procedure [1,2]: (1) generation of equilibrium isotherms and (2) operation of pilot columns. An isotherm indicates the capacity of a carbon for removing organics from a particular water or waste and may be used to eliminate some carbons from further consideration. Adsorption kinetics can affect the efficiency of carbon use greatly. This work can, unfortunately, require a fair amount of money, time, and effort. Mathematical modeling could be helpful in the stages of design and operation, at least in terms of reducing the amount of required experimentation and providing quantitative answers to the influence of process variables.

The bridging of theory and practice is perhaps one of the most valuable engineering functions that mathematical models can serve, particularly with respect to processes as complex as adsorption. A modeling effort can be focused either on the interpretation of basic data to facilitate an unequivocal understanding of mechanism or on the translation of empirical observations into functional design relationships. Adsorption models developed pursuant to the first objective must by nature be mechanistic, whereas those structured primarily to serve the latter purpose may be phenomenological in character. As theory merges with practice, the interpretation of observed phenomena is enlightened by a greater understanding of the underlying mechanism, and the two types of modeling effort logically converge.

The stochastic model can be viewed as the intermediate between the mechanistic models and the phenomenological models in which the form of model equations is assumed from prior knowledge of the phenomena. The adsorption process is complicated and chaotic in nature; thus, the stochastic models derived through probability consideration often generate parameters that are easy to identify and adequately describe the behavior of the adsorption.

II. STOCHASTIC PROCESS AND ITS APPLICATION

A stochastic process is a random phenomenon developing in time according to certain probabilistic laws. Formally, a stochastic process is a family or collection of real random variables $[X(t), t]$. The most extensively developed field of stochastic processes is the so-called Markov processes.

A stochastic process is said to be a Markov process if, for any set of n time points, $t_1 < t_2 < \cdots < t_n$, the conditional distribution of the random variable $X(t_n)$ for the given values of $X(t_1), X(t_2), \ldots, X(t_{n-1})$ depends only on x_{n-1}, the value of $X(t_{n-1})$. More precisely, for any real numbers $x_1, x_2, \ldots, x_n$,

$$\Pr[X(t_n) \leq |X(t_1) = x_1, X(t_2) = x_2, \ldots, X(t_{n-1}) = x_{n-1}] = \Pr[X(t_n) \leq x_n | X(t_{n-1}) = x_{n-1}] \quad (1)$$

Equation (1) means that, given the "present" of the process, the "future" is independent of its "past." Markov processes usually are classified on the basis of the nature of their parameter and state space, as follows:

- Those with discrete parameter and state spaces (i.e., Markov chains)
- Those with continuous parameter and discrete state spaces (i.e., continuous-time Markov chains, or simply Markov processes)
- Those with continuous parameter and state spaces (i.e., diffusion processes)

A. Stochastic Process

1. Markov Chains

Suppose that a sequence of consecutive trials, $n = 0, 1, 2, \ldots$, has been attempted. The outcome of the nth trial is represented by the random variable X_n, which is assumed to be discrete. It is customary to speak of X_n as being in state i if $X_n = i$.

A Markov chain is a sequence of random variables such that for a given X_n, X_{n+1} is conditionally independent of $X_0, X_1, \ldots, X_{n-1}$. The probability that X_{n+1} is in state j, given that X_n is in state i (called a one-step transition probability), is denoted by P_{ij}, that is,

$$P_{ij}(n) = \Pr[X_{n+1} = j | X_n = i] \quad (2)$$

A Markov chain is said to be nonhomogeneous with respect to time if the transition probabilities are functions of time and are homogeneous with respect to time if the transition

$$P_{ij} = \Pr[X_{n+1} = j | X_n = i] \quad (3)$$

is independent of time.

The transition probabilities of a Markov chain can be arranged in the form of a matrix

$$P = \begin{vmatrix} P_{00} & P_{01} & P_{02} & \cdots \\ P_{10} & P_{11} & P_{12} & \cdots \\ \cdot & \cdot & \cdot & \cdots \\ \cdot & \cdot & \cdot & \cdots \end{vmatrix} \quad (4)$$

Note that all elements are non-negative. These matrices are known as stochastic matrices with the transition probabilities P_{ij} as their elements. The subscripts of each probability are the states associated with a transition from i to j or the values of two random variables X_n and X_{n+1}; the first subscript stands for the value of the first random variable and the second stands for the value of the second random variable.

Given $X_n = i$, we have

$$\sum_j P_{ij} = \sum_j P_r[X_{n+1} = j | K_n = i] = 1 \quad (5)$$

so that each row sum in a stochastic matrix is unity.

2. Markov Process

In a stochastic process, $P_{ij}(\tau, t)$ indicates the transition (or conditional) probability that the system will be in state j at time t given that it was in state i at time τ; that is, for $\tau < t$ and $\tau, t \in [0, \infty]$,

$$P_{ij}(\tau, t) = \Pr[X(t) = j | X(\tau) = i] \tag{6}$$

For arbitrary $\tau < t$, the transition probability in Eq. (6) indicates the stochastic dependence of $X(t)$ on $X(\tau)$.

A discrete-valued stochastic process $\{X(t); t \in [0, \infty]\}$ is a Markov process if for any set of times $t_0 < t_1 < \cdots < t_i < t_j$, and corresponding set of integers $k_0, k_1, \ldots, k_i, k_j$,

$$\Pr[X(t_j) = k_j | X(t_0) = k_0, X(t_1) = k_1, \ldots, X(t_i) = k_i] = \Pr[X(t_j) = k_j | X(t_i) = k_i] \tag{7}$$

Then, in a Markov process, as in a Markov chain, given $X(t_i)$ (present), the conditional probability distribution of $X(t_j)$ (future) is independent of $X(t_0), \ldots, X(t_{i-1})$ (past).

Let $P(\tau, t)$ be the matrix of transition probabilities $\{P_{ij}(\tau, t)\}$. It can be shown that the transition probabilities satisfy the Kolmogorov forward differential equations; that is,

$$\frac{d}{dt} P(\tau, t) = P(\tau, t) K(t) \tag{8}$$

with initial condition

$$P(\tau, t) = I$$

The matrix $K(t)$ in Eq. (8) is called the intensity matrix:

$$K(t) = k_{ij}(t)$$

For the time homogeneous process, the forward differential equation, Eq. (8), becomes

$$\frac{d}{dt} P(t - \tau) = P(t - \tau) K \tag{9}$$

with initial condition

$$P(0) = I$$

The solution of Eq. (9) for the transition probabilities, $\{P_{ij}(\tau, t)\}$, is well known.

Consider the characteristic equation of the matrix K,

$$|\lambda I - K| = 0 \tag{10}$$

and assume that the eigenvalues of K are real and distinct.

Let

$$B(m) = (\lambda_m I - K), \quad m = 1, 2, \ldots, n \tag{11}$$

and

$$Q_m(r) = \begin{vmatrix} B_{r1}(m) \\ B_{r2}(m) \\ \vdots \\ B_{rn}(m) \end{vmatrix} \tag{12}$$

The elements of $Q_m(r)$ are the rth column of the matrix of cofactors of $B(m)$ [adjoint matrix of $B(m)$]. The matrix

$$
Q(r) = \begin{vmatrix}
B_{r1}(1) & B_{r1}(2) & \cdots & B_{r1}(n) \\
B_{r2}(1) & B_{r2}(2) & \cdots & B_{r2}(n) \\
\vdots & & & \\
B_{rm}(1) & B_{rm}(2) & \cdots & B_{rm}(n)
\end{vmatrix}
\tag{13}
$$

diagonalized K.

The solution to the differential equation, Eq. (9), is given as

$$
P(t - \tau) = Q(r)E(t - \tau)Q^{-1}(r)
\tag{14}
$$

where

$$
E(t - \tau) = \begin{vmatrix}
e^{\lambda 1(t-\tau)} & 0 & \cdots & 0 \\
0 & e^{\lambda_2(t-\tau)} & \cdots & 0 \\
\vdots & & & \vdots \\
0 & 0 & \cdots & e^{\lambda_n(t-\tau)}
\end{vmatrix}
\tag{14a}
$$

By expanding Eq. (14), we obtain

$$
P(t - \tau) = \sum_{n=1}^{n} B_{ri}(\lambda_k)\frac{Q_{jk}(r)}{|Q(r)|}\exp[\lambda_k(t - \tau)]
\tag{15}
$$

3. Pure Birth Process

In the study of growth in a broad sense, "birth" may be liberally interpreted as an event whose probability is dependent on the number of "parent" events already in existence. A specific value of random variable $X(t)$ is denoted by k. For the pure birth process, the following are considered, given that $X(t) = k$:

1. The conditional probability that a new event will occur during $(t, \ t + \Delta t)$ is $\Delta t + 0(\Delta t)$.
2. The conditional probability that more than one event will occur in the time interval is $0(\Delta t)$.

Taking all of these possibilities together and combining all quantities of order $0(\Delta t)$, we have for $k \geq 1$,

$$
P_k(t + \Delta t) = P_k(t)[1 - \lambda\Delta t] + P_{k-1}(t)\lambda\Delta t + 0(\Delta t)
\tag{16a}
$$

and for $k = 0$,

$$
P_0(t + \Delta t) = P_0(t)[1 - \lambda\Delta t] + 0(\Delta t)
\tag{16b}
$$

Transposing $P_k(t)$ to the left-hand side of Eq. (16a), dividing the resulting equation through by Δt, and passing to the limit as $\Delta t \to 0$, we find that the probabilities satisfy the system of differential equations

$$\frac{d}{dt} P_i(t) = -\lambda_i (P_i(t) \tag{17a}$$

and

$$\frac{d}{dt} P_k(t) = -\lambda_k P_k(t) + \lambda_{k-1} P_{k-1}(t), \quad k > i. \tag{17b}$$

Here, $X(0) = i$, the number of events existing at $t = 0$, so that the initial conditions are

$$P_i(0) = 1, \qquad P_k(0) = 0 \quad \text{for } k \neq 1$$

The solutions of Eqs. (17a) and (17b) are given by Chiang [3].

The pure death process is exactly analogous to the pure birth process, except that the pure death process $X(t)$ is decreased rather than increased by the occurrence of an event.

4. Birth–Death Process

We let $X(t)$ denote the size of a population at time t with the initial condition $X(0) = i$. Given $X(t) = k$, the following are for the birth-death process:

1. The conditional probability that a birth event will occur during the time interval $(t, t + \Delta t)$ is $\lambda_k \Delta t + 0(\Delta t)$.
2. The conditional probability that a death event will occur during the time interval $(t, t + \Delta t)$ is $\mu_k \Delta t + 0(\Delta t)$.
3. The conditional probability that more than one event will occur in the time interval is $0(\Delta t)$.

Using the birth and death intensities and following the enumeration process as in the previous subsection, we find that the probabilities $P_k(t)$ satisfy the system of difference equations

$$\begin{aligned} P_{i,k}(0, t + \Delta t) = {} & P_{i,k}(0, t)[1 - \lambda_k(t)\Delta t - \mu_k(t)\Delta t] \\ & + P_{i,k-1}(0, t)\lambda_{k-1}(t)\Delta t + P_{i,k+1}(0, t)\mu_{k+1}(t)\Delta t + 0(\Delta t) \end{aligned} \tag{18}$$

This leads to the system of differential equations

$$\frac{dP_{i,0}(0, t)}{dt} = -[\lambda_0(t) + \mu_0(t)]P_{i,0}(0, t) + \mu_1(t)P_{i,1}(0, t) \tag{19a}$$

$$\frac{dP_{i,k}(0, t)}{dt} = -[\lambda_k(t) + \mu_k(t)]P_{i,k}(0, t) + \lambda_{k-1}(t)P_{i,k-1}(0, t) + \mu_{k+1}(t)P_{i,k+1}(0, t) \tag{19b}$$

The system of differential equations [Eqs. (19a) and (19b)] and the initial conditions

$$P_{i,i}(0, 0) = 1, \qquad P_{i,k}(0, 0) = 0 \quad \text{for } k \neq i$$

completely determine the probability distribution $\{P_{i,k}(0, t)\}$. To obtain a stochastic process corresponding to an empirical phenomenon under study, we must make certain assumptions regarding the birth and death intensity functions.

B. Application of Stochastic Process

Numerous engineering systems and their characteristics lend themselves to a stochastic description due to their inherent complexity and fluctuating nature. Examples of these can be found in the following:

- Chemical reaction kinetics in closed and flow system
- Pressure drop and concentration dynamics in filtration
- Residence time distribution in various flow systems
- Crystal size distribution in the crystallization process

A brief review focuses on two processes which are related to adsorption in a closed and in a flow system.

1. Chemical Reaction

Stochastic analysis and modeling of chemical reactions have been accomplished in a closed system or batch reactors. A review of such a formulation has been given by McQuarrie [4]. First-order chemical reactions of the unimolecular type, each involving two or three chemical species, have been considered by McQuarrie [5] and Fredrickson [6]. Cases of first-order reactions among multitype molecules have been treated by Darvey and Staff [7].

In a continuous-flow chemical reactor, the concern is not only with probabilistic transitions among chemical species but also with probabilistic transitions of each chemical species between the interior and exterior of the reactor. Pippel and Philipp [8] used Markov chains for simulating the dynamics of a chemical system. In their approach, the kinetics of a chemical reaction are treated deterministically and the flow through the system are treated stochastically by means of a Markov chain. Shinnar et al. [9] superimposed the kinetics of the first order chemical reactions on a stochastically modeled mixing process to characterize the performance of a continuous-flow reactor and compared it with that of the corresponding batch reactor. Most stochastic approaches to analysis and modeling of chemical reactions in a flow system have combined deterministic chemical kinetics and stochastic flows.

Nassar et al. [10] employed a stochastic approach, namely a Markov process with transient and absorbing states, to model in a unified fashion both complex linear first-order chemical reactions, involving molecules of multiple types, and mixing, accompanied by flow in an non-steady- or steady-state continuous-flow reactor. Chou et al. [11] extended this system with nonlinear chemical reactions by means of Markov chains. An assumption is made that transitiions occur instantaneously at each instant of the discretized time.

2. Deep-Bed Filtration

Granular filters have long been used to remove suspended solids from water in both the purification of potable water and the treatment of wastewater. The suspended solids are retained in the bed either through a straining mechanism or an adhesion mechanism.

Deep-bed filtration involves the flow of particles through randomly distributed passages; thus, it tends to be stochastic in nature. The filtration process has been modeled as a pure birth process [12,13], a birth–death process [14–17], a random-walk process [18], and a stochastic diffusion process [19].

In Litwiniszyn's pure birth model, the entire bed is considered as on state, and the number of blocked pores in a unit volume of the bed is considered as a random variable. In his birth–death model, the number of trapped particles over the entire bed is a random variable. Fan and his co-workers have extended the pure birth and birth–death process models by incorporating them with the Carman–Kozeny equation to simulate the pressure-drop dynamics of a deep-bed filter under constant-flow conditions. In their second-order pure birth model, the number of blocked pores is related to the pressure drop by assuming that the scouring of deposited particles is negligible. Their linear birth–death process takes into account both blockage of the pores by suspended particles and scouring of deposited particles.

Nassar et al. [20] proposed a stochastic compartmental model to simulate the concentration dynamics of suspended particles in the liquid and solid parts over the different section of flow. In their work, a deep-bed filter is considered to be an open system composed of an arbitrary number of sections or compartments distributed in the axial direction.

C. Comparison Between Deterministic and Stochastic Models

A stochastic system evolves from a measurable state to another possible state according to probabilistic laws, whereas a deterministic system allows us to determine the future position by knowledge of the initial position and momentum.

Deterministic models of the deep-bed filtration process are derived through the phenomenological equation of kinetics or derived through the trajectory analysis in which the trajectories of the particles are determined from the force balance equations [15]. The filtration process is complicated and chaotic in nature, and thus the stochastic models derived through probability considerations often generate parameters that are easier to identify.

Much effort has been devoted in recent years to analyzing and modeling the crystal size distribution (CSD) by using mainly the deterministic population balance approach [21,22]. Available deterministic models of CSD have ignored the phenomenon of random fluctuations in growth and have treated predominantly simple flow, such as complete mixing. The process of growth is, in reality, stochastic or random in nature, as is evident from the fact that seeds of an equal initial size are known to grow at different rates. A stochastic model assumes that the growth and residence time of a crystal in a crystallizer are random processes governed by a certain probability law. A stochastic model enables us to predict the fluctuations in crystal size distribution which cannot be obtained from the deterministic model [23,24].

Discrepancies exist between diffusivity values obtained using different methods for measuring diffusion in zeolites. These discrepancies prompted the simulation of diffusion in zeolites by Monte Carlo techniques [25]. Patwardhan [26] attempted to put the molecular model involved on a more exact footing by formulating the sorption and diffusion process as a Markov process. All transitions arising out of a particle jump have been taken to proceed at equal rates.

The axial dispersion equation of the Fickian form is one of the most widely used models characterizing the residence time distribution and performance of a tubular flow reactor [27,28]. Deterministic models, based on population mass balance, have been used predominately in describing and modeling the dispersion of molecules or particles in the flow reactor. A probabilistic approach may be more appropriate than a mass balance approach especially when the rate of reactant flow through the reactor is high, when the flow involves more than one phase, or when the path of reactant flow is affected by internals or agitators [29].

The behavior of bubbles in fluidized beds has been known to be stochastic in nature and has been studied by several investigators [30–32]. It is possible to exactly predict the sizes and positions of bubbles at each moment in time. Because the bubbles do not occur with exactly the same positions and sizes each time, the prediction would be dependent on the initial conditions. Such a system appears to be stochastic [33]. Although it is possible to understand the mechanism of coalescence and movement for isolated bubbles in a deterministic manner, it is not possible to extend the deterministic model to accurately predict the behavior of a large swarm of bubbles.

From the review presented thus far, it is clear that the stochastic model is more fundamental for describing the physical behavior of the system. Also, it is important to note that the deterministic and the stochastic modeling should be worked together to understand the complicated systems.

III. STOCHASTIC MODELING OF A BATCH ADSORBER IN THE CASE OF LINEAR ISOTHERM

Activated-carbon adsorption is an important process for the purification of potable water, removal of organic pollutants, and treatment and renovation of wastewater. The process of adsorption involves the transfer of solute molecules from the liquid phase, water, to the surface layer of the solid particle, intraparticle transport through the porous solid where a sorbate may be trapped in the solid phase or may move back to the liquid phase. At equilibrium, a relationship is established between the solute concentration in the solid phase and that in the liquid phase. This relationship, over a wide range of initial solute concentration, is usually characterized by a nonlinear function. For low concentrations, the relationship may be closely approximated by a linear function.

Appropriate mathematical representations of an adsorption process are useful for an understanding of the process and for predicting its behavior over time. An understanding of the key variables affecting the process through modeling can facilitate process design and strategies for process optimization. Several mathematical models with linear isotherms have been developed to predict the adsorption rates in batch reactors [34–37]. All of the modeling approaches reported in the literature have been deterministic, resulting in a system of partial differential equations, the complexity of which increases with the number of mechanistic steps and the number of components considered.

Some of the early work in this area has been reported by Giddings and Eyring [38] and McQuarrie [39]. They have developed expressions for column elution curves by modeling the chromatographic process as a Poisson process. A number of stochastic formulations of chemical reaction kinetics in a closed system have been accomplished. First-order chemical reactions of the unimolecular type, each involving two or three chemical species (e.g., a triangular or parallel reaction), have been considered by Fredrickson [40]. Cases of first-order reactions among multitype molecules have been treated by Davey and Staff [41]. These attempts have focused on stochastic transition of molecules from one species to another.

A. Model Development

We consider a batch adsorption process and partition the system into a liquid-phase compartment and two solid-phase compartments. We denote the liquid phase by S_1 and the solid phase by S_2 and S_3 in Fig. 1. Compartments S_2 and S_3 represent the outer and inner layers of the solid adsorbent, respectively. In a small time interval $(t, t + \Delta t)$, a molecule in state S_1 may remain in S_1 or move to S_2 through adsorption. A molecule in S_2 may remain in S_2 or move to S_2 through desorption or to S_3 through diffusion. A molecule in S_3 may remain in S_3 or move to S_2. This system defines a Markov process with three states (S_1, S_2, S_3).

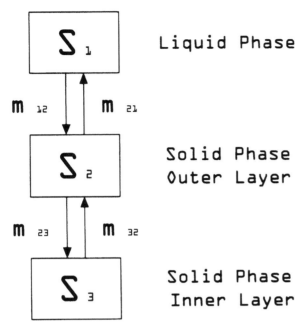

FIG. 1   Transition diagram of a batch adsorption process in a single solute system.

Let

$$1 - m_{12}\Delta t + 0(\Delta t) = \text{Pr [a molecule in state } S_1 \text{ at time } t \text{ will remain in state } S_1$$
at time $t + \Delta t]$

$$m_{21}\Delta t + 0(\Delta t) = \text{Pr [a molecule in state } S_1 \text{ at time } t \text{ will move to state } S_1 \text{ at time } t + \Delta t]$$

$$m_{23}\Delta t + 0(\Delta t) = \text{Pr [a molecule in state } S_2 \text{ at time } t \text{ will move to state } S_3 \text{ at time } t + \Delta t]$$

$$1 - m_{21}\Delta t - m_{23}\Delta t + 0(\Delta t) = \text{Pr [a molecule in state } S_2 \text{ at time } t \text{ will remain in state } S_2$$
at time $t + \Delta t]$

$$m_{32}\Delta t + 0(\Delta t) = \text{Pr [a molecule in state } S_3 \text{ at time } t \text{ will move to state } S_2 \text{ at time } t + \Delta t]$$

$$1 - m_{32}\Delta t + 0(\Delta t) = \text{Pr [a molecule in state } S_3 \text{ at time } t \text{ will remain in state } S_3 \text{ at time}$$
$t + \Delta t]$

Because initially all molecules are in the liquid phase (state S_1), we define for a time interval
$(0, t)$,

$$P_{ij}(k) = \text{Pr[a molecule in state } S_i \text{ at time 0 will be in state } S_j \text{ at time } t]$$

The transition probability column vector $P(t) = \{P_{1j}(t)\}^T, j = 1, 2, 3$, satisfies the Kolmogorov forward differential equation [3]:

$$\frac{d}{dt}P(t) = MP(t) \tag{20}$$

where M is the matrix of intensity functions, m_{ij}:

$$M = \begin{vmatrix} -m_{12} & m_{21} & 0 \\ m_{12} & -m_{21} - m_{23} & m_{32} \\ 0 & m_{23} & -m_{32} \end{vmatrix}$$

and

$$P(0) = [1 \quad 0 \quad 0]^T$$

Equation (20) is a first-order differential equation whose solution is well known. For real and distinct eigenvalues of the matrix M, the transition probabilities, $P_{1j}(t)$, $j = 1, 2, 3$, may be expressed as

$$P_{11}(t) = \frac{m_{21}m_{32}}{\lambda_2 \lambda_3} - \frac{m_{12}(\lambda_2 + m_{23} + m_{23})}{\alpha \lambda_2} \exp(\lambda_2 t) + \frac{m_{12}(\lambda_3 + m_{23} + m_{32})}{\alpha \lambda_3} \exp(\lambda_3 t) \quad (21a)$$

$$P_{12}(t) = \frac{m_{12}m_{32}}{\lambda_2 \lambda_3} + \frac{m_{12}(\lambda_2 + m_{32})}{\alpha \lambda_2} \exp(\lambda_2 t) - \frac{m_{12}(\lambda_3 + m_{32})}{\alpha \lambda_3} \exp(\lambda_3 t) \quad (21b)$$

$$P_{13}(t) = \frac{m_{12}m_{23}}{\lambda_2 \lambda_3} + \frac{m_{12}m_{32}}{\alpha \lambda_2} \exp(\lambda_2 t) - \frac{m_{12}m_{23}}{\alpha \lambda_3} \exp(\lambda_3 t) \quad (21c)$$

where

$$\lambda_2 = -\frac{(m_{12} + m_{21} + m_{23} + m_{32}) + \alpha}{2}$$

$$\lambda_3 = -\frac{(m_{12} + m_{21} + m_{23} + m_{32}) - \alpha}{2}$$

and

$$\alpha = [(m_{12} + m_{21} + m_{23} + m_{32})^2 - 4(m_{12}m_{23} + m_{12}m_{32} + m_{21}m_{32})]^{1/2}$$

Assume that N_0 molecules are initially in state S_1 (liquid phase). At time t, each of the N_0 molecules will be in state S_1 with probability $P_{11}(t)$ or in the other states with probability $1 - P_{11}(t)$. From the binomial distribution, the expected number of molecules in the liquid phase at time t, and the concentration are given by Eqs. (21b) and (21c), respectively:

$$E[N_1(t)] = N_0 P_{11}(t) \quad (22)$$

$$C(t) = C_0 P_{11}(t) \quad (23)$$

where C_0 is the initial concentration of molecules in the liquid phase or state S_1.

B. Parameter Estimation

The relationship at equilibrium between the solute concentration in the solid phase and its concentration in the liquid phase could be approximately linear. This may be expressed as

$$q_\infty = K C_\infty \quad (24)$$

Based in mass balance, we may write

$$V C_0 = V C_\infty + W q_\infty \quad (25)$$

Replacing q_∞ in Eq. (25) with KC_∞, one obtains the relationship

$$C_\infty = \frac{C_0}{1 + WK/V} \tag{26}$$

Equating $C(t)$, as $t \to \infty$, in Eq. (23) to C_∞ in Eq. (26) gives the following relationship:

$$\frac{m_{12}(m_{23} + m_{32})}{m_{21}m_{32}} = \frac{WK}{V} \tag{27}$$

It is considered that next the surface layer of the solid phase (S_2 in Fig. 1) to be $(S_2 + S_3)/j$ for $j > 1$. Because $S_2 + S_3$ constitutes the solid phase, the number of solute molecules in state S_3 is $j - 1$ times that in state S_2. Hence,

$$\frac{P_{13}(\infty)}{P_{12}(\infty)} = \frac{m_{23}}{m_{32}} = j - 1 \tag{28}$$

Substitution of Eq. (29) into Eq. (28) gives

$$m_{21} = \frac{jV}{WK} m_{12} \tag{29}$$

The intensity m_{12} can be obtained from the relationship

$$\left(\frac{d}{dt} C(t) \right)_{t=0} = -C_0 m_{12} \tag{30}$$

It is clear from Eqs. (28)–(30) that all intensities in the model can be expressed as functions of m_{32} and j. Hence, Eq. (23) is reduced to a two-parameter model, estimates of which were obtained from a least-squares fit to the experimental data.

C. Comparison Between Deterministic and Stochastic Solutions

The deterministic model equations for a batch system with linear isotherm have been solved [42]:

$$\frac{C}{C_0} = \frac{C_\infty}{C_0} + \sum_{n=1}^{\infty} A_n \exp\left(-\frac{D_S}{R^2} \lambda_n^2 t \right) \tag{31}$$

where

$$\frac{C_\infty}{C_0} = \frac{1}{1 + WK/V} \tag{31a}$$

$$A_n = \left[1 + \frac{3WK}{2V} \left(1 - \frac{V}{3W} \frac{\rho D_S}{k_f R} \lambda_n^2 \right)^2 \left(\frac{1}{\sin^2 \lambda_n} - \frac{\cot \lambda_n}{\lambda_n} \right) \right]^{-1} \tag{31b}$$

and A_n is an nth positive root of the following equation:

$$\lambda \cot \lambda - 1 = \lambda^2 \left[\frac{3WK}{V} \left(1 - \frac{V}{3W} \frac{\rho D_S}{k_f R} \lambda^2 \right) \right]^{-1} \tag{31c}$$

Some simplified approximations have been proposed for describing the rate of adsorption onto a porous adsorbent particle. However, these assumptions produce models that may have drawbacks in terms of flexibility and accuracy.

It is known that the mean outcome of a nonlinear stochastic model is different from the deterministic outcome. For a linear model, however, the mean of a stochastic outcome is the same as that of the deterministic outcome. Perhaps an overriding factor in choosing a stochastic modeling approach as a alternative to a deterministic approach is the simplicity of the former relative to the latter. For instance, the model equations describing a process in the batch system is more manageable for a stochastic model than for a deterministic model. This is of importance for model applications and for scale-up.

The parameters appearing in the present stochastic model can be easily estimated from some correlations and determined by conducting simple batch experiments. The transient behavior of molecules can be evaluated by analytically solving a system of the governing ordinary differential equations of the present stochastic model. The stochastic modeling, therefore, provides an easier approach than the deterministic population balance modeling in which a partial differential equation must be solved. These facts are very attractive from the practical point of view.

IV. STOCHASTIC MODELING OF A BATCH ADSORBER IN THE CASE OF NONLINEAR ISOTHERM

The process of adsorption is of importance for detoxification of potable water, removal of organic pollutants, and physicochemical treatment of wastewater. An adsorber, whether batch or continuous flow, is characterized by a liquid phase and a solid phase. Solid particles (i.e., activated carbon) act to adsorb the solute molecules from the liquid phase. The process involves random movement of molecules between the liquid phase and solid phase. In the limit, an equilibrium concentration is reached between the two phases. The rate of transfer of molecules from the liquid phase to the solid phase is, in general, nonlinear and depends on the number of molecules in the solid phase as well as the total capacity of the solid phase for retaining molecules.

A model representation of an adsorption process is useful for an understanding of the important variables affecting the process and as an aid in process design and optimization. Several deterministic models which take into account the Langmuir equation [43–45], the Freundlich equation [46–49], the three-parameter model [50,51], or the ideal adsorbed solution (IAS) model [52–54] for a nonlinear isotherm have been presented that predict adsorption rates in batch reactors. A deterministic approach to modeling particle movement during adsorption may be regarded as an approximation of the real phenomenon, which is stochastic in nature. When the transfer intensity from the liquid phase to the solid phase is a linear function of the number of molecules, the average concentration predicted by a stochastic model is equal to the deterministic outcome. However, in the nonlinear case, the stochastic outcome can differ from its deterministic counterpart. Hence, in general, it is more appropriate to use a stochastic rather than a deterministic approach to model and adsorption phenomena. Despite this, few studies are reported in the literature on stochastic modeling of adsorption and related processes. A stochastic approach to the modeling of adsorption in the batch system for the case of a linear isotherm has been discussed in Section III.

A. Model Development

In an adsorption process, solute molecules diffuse into the solid phase, giving rise to a concentration profile. To account for this, one may divide the solid phase into compartments. In this model, we represent the solid phase by two states or compartments, an outer layer (S_2) and an inner layer (S_3). Furthermore, we denote the liquid phase by state S_1 in Fig. 1. In a small time interval (t, $t + \Delta t$), a molecule in state S_1 may remain in S_1 or move to S_2. A molecule in S_2 may remain in S_2 or move to S_1 or to S_3. A molecule in S_3 may remain in S_3 or move to S_2. The system is then defined by a Markov process with three states.

Let the random variable $N_i(t)$ denote the number of molecules in state S_i at time t. Considering the three states, one has the random vector

$$N(t) = [N_1(t), N_2(t), N_3(t)] \tag{32}$$

and the corresponding realization vector

$$n(t) = [n_1(t), n_2(t), n_3(t)] \tag{33}$$

There, we note that

$$n(0) = [n_0, 0, 0] \tag{33a}$$

and

$$n_1(t) + n_2(t) + n_3(t) = n_0 \tag{33b}$$

It is of interest to determine the probability

$$P[n_1(t), n_2(t), n_3(t)] = \Pr[N(t) = n(t)|N_1(0) = N_0, N_2(0) = 0, N_3(0) = 0] \tag{34}$$

For any time t, a change in the number of molecules in each state S_i during the time interval (t, $t + \Delta t$) is assumed to take place according to the following probabilities.

$m_{12}(Q - N_2(\Delta t))t + 0(\Delta t) = \Pr$[a molecule in state S_1 at time t moves to state S_2 during the small time interval Δt]

$1 - m_{12}(Q - N_2(\Delta t))t + 0(\Delta t) = \Pr$[a molecule in state S_1 at time t remains in the same state during the small time interval Δt]

$m_{21}\Delta t + 0(\Delta t) = \Pr$[a molecule in state S_2 at time t moves to state S_1 during the small time interval Δt]

$m_{23}\Delta t + 0(\Delta t) = \Pr$[a molecule in state S_2 at time t moves to state S_3 during the small time interval Δt]

$1 - (m_{12} + m_{21})\Delta t + 0(\Delta t) = \Pr$[a molecule in state S_2 at time t remains in the same state during the small time interval Δt]

$m_{32}\Delta t + 0(\Delta t) = \Pr$[a molecule in state S_3 at time t moves to state S_2 during the small time interval Δt]

$1 - m_{32}\Delta t + 0(\Delta t) = \Pr$[a molecule in state S_3 at time t remains in the same state during the small time interval Δt]

We consider that the probability of a molecule moving from state S_1 to state S_2 (through adsorption and diffusion) is proportional to the number of vacant places $[Q - N_2(t)]$ in state S_2. Here, Q represents the total molecular capacity of state S_2. Our interest is in determining the mean number of molecules and, hence, the concentration of molecules in each state S_i ($i = 1, 2, 3$) at time t. Given $N(t) = [N_1(t), N_2(t), N_3(t)]$, during the small time interval (t, $t + \Delta t$), $N_1(t)$ may decrease by 1 with probability

$$m_{12}n_1(t)[Q - N_2(t)]\Delta t + 0(\Delta t)$$

increase by 1 with probability

$$m_{21}n_2(t)\Delta t + 0(\Delta t)$$

or remain unchanged with probability

$$1 - m_{12}n_1(t)[Q - N_2(t))]\Delta t - m_{21}n_2(t)\Delta t + 0(\Delta t)$$

Similarly, $N_2(t)$ may decrease by 1 with probability

$$m_{21}n_2(t)\Delta t + m_{23}n_2(t)\Delta t + 0(\Delta t)$$

increase by 1 with probability

$$m_{12}n_1(t)[Q - N_2(t)]\Delta t + m_{32}n_3(t)\Delta t + 0(\Delta t)$$

or remain unchanged with probability

$$1 - m_{12}n_1(t)[Q - N_2(t)]\Delta t - m_{32}n_3(t)\Delta t - m_{21}n_2(t)\Delta t - m_{23}n_2(t)\Delta t + 0(\Delta t)$$

The conditional expected change in $N_1(t)$ during the time interval $(t, t + \Delta t)$ is

$$E[N_1(t, t + \Delta t), -N_1(t)|N(t) = (N_1(t), N_2(t), N_3(t))]$$
$$= \{m_{21}N_2(t) - m_{12}N_1(t)[Q - N_2(t)]\}\Delta t \tag{35a}$$

The conditional expected change in $N_2(t)$ during the time interval $(t, t + \Delta t)$ is

$$E[N_2(t, t + \Delta t), -N_2(t)|N(t) = (N_1(t), N_2(t), N_3(t))]$$
$$= \{m_{21}N_2(t) - [Q - N_2(t)] + m_{32}N_3(t) - m_{21}N_2(t) - m_{23}n_2(t)(t)\}\Delta t \tag{35b}$$

Note that $N_3(t)$ in Eq. (35a) may be expressed as $N_0 - N_1(t) - N_2(t)$.

By rearranging the expressions in Eqs. (35a) and (35b) and taking the limit as $\Delta t \rightarrow 0$, the unconditional rates of change in the mean of $N_1(t)$ and $N_2(t)$ are obtained as

$$\frac{d}{dt}E[N_1(t)] = m_{21}E[N_2(t)] - m_{12}QE[N_1(t)] + m_{12}E[N_1(t)N_2(t)] \tag{36a}$$

$$\frac{d}{dt}E[N_2(t)] = m_{12}QE[N_1(t)] - m_{12}E[N_1(t)N_2(t)]$$
$$+ m_{32}(N_0 - E[N_1(t)] - E[N_2(t)]) - m_{21}E[N_2(t)] + m_{23}E[N_2(t)] \tag{36b}$$

with the initial condition

$$E[N_1(0)] = N_0$$
$$E[N_2(0)] = 0$$

To express Eqs. (36a) and (36b) in terms of concentration, we let

$$C = \frac{E[N_1(t)]}{VA} \tag{37a}$$

$$q = \frac{jE[N_2(t)]}{WA} \tag{37b}$$

Substituting Eqs. (37a) and (37b) into Eqs. (36a) and (36b), we obtain

$$\frac{dC}{dt} = m_{21}\frac{W}{jV}q - m_{12}QC + m_{12}\frac{WA}{j}qC \tag{38a}$$

$$\frac{dq}{dt} = m_{12}\frac{jV}{W}QC - m_{12}VAqC + m_{32}\frac{jV}{W}C_0 - m_{32}\frac{jV}{W}C - (m_{32} + m_{23} + m_{21})q \tag{38b}$$

Note that $E[N_1(t)N_2(t)]$ in Eq. (36a) has been approximated by $E[N_1(t)]E[N_2(t)]$, which implies that the covariance between $N_1(t)$ and $N_2(t)$ is close zero. This is so due to the fact that the variances of $N_1(t)/VA$ and $N_2(t)(WA/j)^{-1}$ are near zero as a result of division by the large Avogadro number (A). The same argument holds for $E[N_1(t)N_2(t)]$ in Eq. (36b).

B. Parameter Estimation

Assume that the capacity of state S_2 is equal to $1/j$ ($j > 1$) times the total capacity of the solid phase. Hence, (at equilibrium) one has that

$$E[N_3(\infty)] = (j - 1)E[N_2(\infty)] \tag{39}$$

and

$$N_0 = E[N_1(\infty)] + E[N_2(\infty)] + (j - 1)E[N_2(\infty)] \tag{40}$$

Also, the sum of Eqs. (36a) and (36b) gives

$$m_{32}(N_0 - E[N_1(\infty)] - E[N_2(\infty)]) - m_{23}E[N_2(\infty)] = 0 \tag{41}$$

Replacing N_0 in Eq. (41) with its value from Eq. (40), one obtains the relationship

$$m_{23} = (j - 1)m_{32} \tag{42}$$

At equilibrium, the molecular concentration in the solid and liquid phases must satisfy the Langmuir equation [55]:

$$q_\infty = \frac{q_m b C_\infty}{1 + b C_\infty} = \frac{q_m C_\infty}{1/b + C_\infty} \tag{43}$$

At equilibrium, Eq. (38a) may be expressed as

$$0 = m_{21}\frac{W}{jV}q_\infty - m_{12}QC_\infty + m_{12}\frac{W}{j}Aq_\infty C_\infty \tag{44}$$

from which one obtains

$$q_\infty = \frac{jQC_\infty/WA}{m_{21}/m_{12}VA + C_\infty} \tag{45}$$

Equating Eqs. (43) and (45), one obtains the relations

$$Q = \frac{WA}{j}q_m \tag{46}$$

and

$$\frac{m_{21}}{m_{12}} = \frac{VA}{b} \tag{47}$$

In the limit as $t \to 0$, Eq. (38a) reduces to the expression

$$\lim_{t \to 0} \frac{d\ln C(t)}{dt} = -m_{12}Q \tag{48}$$

from which one may estimate the intensity of m_{12}.

From the relations in Eqs. (42), (46), and (47), one may express Eqs. (38a) and (38b) as functions of m_{12}, m_{32}, and j. As such Eqs. (38a) and (38b) become

$$\frac{dC(t)}{dt} = m_{12}\frac{WA}{jb}q(t) - m_{12}\frac{WA}{j}q_m C(t) + m_{12}\frac{WA}{j}q(t)C(t) \tag{49a}$$

$$\frac{dq(t)}{dt} = m_{12}VAq_m C(t) - m_{12}VAq(t)C(t) + m_{32}\frac{jV}{W}C(t) - m_{32}\frac{jV}{W}C(t)$$

$$- \left(m_{32}j + m_{12}\frac{VA}{b}\right)q(t) \tag{49b}$$

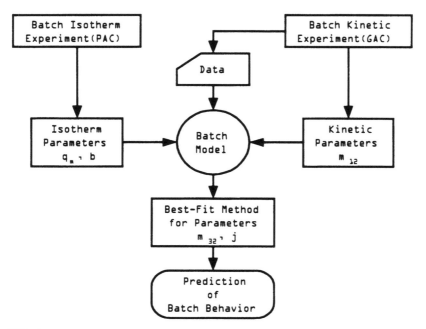

FIG. 2   Schematic flow diagram of a nonlinear batch model.

It is seen that Eqs. (48), (49a) and (49b) can be used to estimate m_{12}, m_{32}, and j. From these estimates, one may, in turn, estimate m_{21} and m_{23} using Eqs. (42) and (47).

The isotherm parameters, q_m and b, are determined from independent experiment. A schematic flow diagram of the nonlinear batch model is shown in Fig. 2.

V. STOCHASTIC MODELING AND SIMULATION OF BISOLUTE BATCH ADSORBER IN THE CASE OF NONLINEAR ISOTHERM

The presence of more than a single solute raises a number of complications because of the interactions among the adsorbates. The main features of these interactions are the interference among the adsorbates in mass transfer and the interaction among adsorbates in their ultimate distribution between solution and adsorbed phases. It is generally recognized that the former effect can be ignored if the concentrations of the adsorbates are low. On the other hand, the interaction among adsorbates in their ultimate distribution between the liquid and solid phases is determined by the appropriate adsorption equilibrium relationship. Several deterministic models have been developed to predict the dynamics of multicomponent adsorption in batch reactors [52–54,56–60]. A simple neural network model was used to predict binary solute adsorption onto granular activated carbon [61].

Stochastic modeling of numerous chemical process systems has been reported. Stochastic modeling for fluidized beds has been studied by Ligon and Amundson [3 0,31] and by Fox and Fan [32]. Stochastic models of mixing and chemical reactions have been discussed by Nauman

[62] and Nassar et al. [10]. Stochastic expressions for filtration and chromatography have been developed by Nassar et al. [20], Giddings and Eyring [38], and McQuarrie [39]. A stochastic model is more fundamental in nature than the deterministic model. Even though stochastic population balances for large numbers of independent molecules almost always reduce to deterministic mean value rate expressions, the stochastic model represents a fundamentally more basic description of the physical behavior of the adsorption process.

The objective of this chapter is to develop a reliable stochastic model in a bisolute batch system which requires only single-solute data. Moreover, the batch measurements can be used to furnish required rate parameters for performance of a bisolute fixed-bed system.

A. Model Development and Parameter Estimation

We denote the liquid phase by S_1 and S_2, the solid phase outer layer by S_3, and the solid phase inner layers by S_4 and S_5 (Fig. 3). The governing equations of the present model have been derived, based on the hypothesis that adsorption without competition occurs when the total molecular capacity of solute 1, Q_1 is not equal to the total molecular capacity of solute 2, Q_2. Furthermore, it is assumed that the number of sites for which there is no competition is equal to the quantity $(Q_2 - Q_1)$, where $Q_2 > Q_1$.

In a small time interval $(t, t + \Delta t)$, a molecule in the liquid pahse, S_1 or S_2, may (1) remain in S_1 or S_2 or (2) move to the intermediate state, S_3. A molecule in S_3 (1) may remain in S_3 (2) move to S_1 or S_2 or (3) move to the solid phase, S_4 or S_5. A molecule in S_4 or S_5 may (1) remain in S_4 or S_5 or (2) move to S_3.

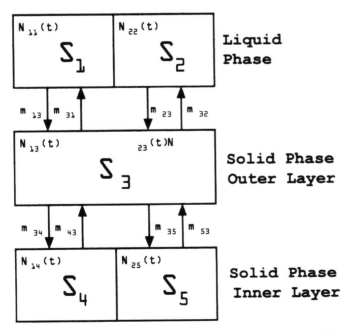

FIG. 3 Transition diagram of a batch adsorption process in a bisolute system.

Let $m_{ij}(t)$ $i, j = 1, 2, 3, 4, 5$, be the intensity functions of the Markov process. For any time t, a change in the number of molecules in each state S_1 during the time interval $(t, t + \Delta t)$ is assumed to take place according to the following probabilities:

$m_{13}[Q_1 - N_{13}(t) - N_{23}(t)]\Delta t + 0(\Delta t) = \text{Pr}[\text{a molecule in state } S_1 \text{ at time } t \text{ moves to state}$
   $S_3 \text{ during the small time interval } \Delta t]$

$m_{31}\Delta t + 0(\Delta t) = \text{Pr}[\text{a molecule in state } S_3 \text{ at time } t \text{ moves to state } S_1 \text{ during the small}$
   $\text{time interval } \Delta t]$

$m_{34}\Delta t + 0(\Delta t) = \text{Pr}[\text{a molecule in state } S_3 \text{ at time } t \text{ moves to state } S_4 \text{ during the small}$
   $\text{time interval } \Delta t]$

$1 - m_{31}\Delta t - m_{34}\Delta t + 0(\Delta t) = \text{Pr}[\text{a molecule in state } S_3 \text{ at time } t \text{ remains in the same}$
   $\text{state during the small time interval } \Delta t]$

$m_{43}\Delta t + 0(\Delta t) = \text{Pr}[\text{a molecule in state } S_4 \text{ at time } t \text{ moves to state } S_3 \text{ during the small}$
   $\text{time interval } \Delta t]$

$m_{23}[Q_1 - N_{13}(t) - N_{23}(t)]\Delta t + m_{23}(Q_2 - Q_1)\Delta t + 0(\Delta t) = \text{Pr}[\text{a molecule in state } S_2 \text{ at}$
   $\text{time } t \text{ moves to state } S_3 \text{ during the small time interval } \Delta t]$

$1 - m_{23}[Q_1 - N_{13}(t) - N_{23}(t)]\Delta t - m_{23}(Q_2 - Q_1)\Delta t + 0(\Delta t) = \text{Pr}[\text{a molecule in state}$
   $S_2 \text{ at time } t \text{ remains in the same state during the small time interval } \Delta t]$

$m_{32}\Delta t + 0(\Delta t) = \text{Pr}[\text{a molecule in state } S_3 \text{ at time } t \text{ moves to state } S_2 \text{ during the small}$
   $\text{time interval } \Delta t]$

$m_{35}\Delta t + 0(\Delta t) = \text{Pr}[\text{a molecule in state } S_3 \text{ at time } t \text{ moves to state } S_5 \text{ during the small}$
   $\text{time interval } \Delta t]$

$1 - m_{32}\Delta t - m_{35}\Delta t + 0(\Delta t) = \text{Pr}[\text{a molecule in state } S_3 \text{ at time } t \text{ remains in the same}$
   $\text{state during the small time interval } \Delta t]$

$m_{53}\Delta t + 0(\Delta t) = \text{Pr}[\text{a molecule in state } S_5 \text{ at time } t \text{ moves to state } S_3 \text{ during the small}$
   $\text{time interval } \Delta t]$

$1 - m_{53}\Delta t + 0(\Delta t) = \text{Pr}[\text{a molecule in state } S_5 \text{ at time } t \text{ remains in the same state during}$
   $\text{the small time interval } \Delta t]$

where $0(\Delta t)$ satisfies the condition that

$$\lim_{\Delta t \to 0} \frac{0(\Delta t)}{\Delta t} = 0$$

We consider that the probability of a molecule of species 1 moving from state S_1 to state S_3 is proportional to $Q_1 - N_{13} - N_{23}$ under competition with species 2. $m_{23}[Q_1 - N_{13}(t) - N_{23}(t)]\Delta t$ is the probability of a molecule of species 2 that absorbs on the surface area proportional to $Q_1 - N_{13} - N_{23}$ under competition with species 1. The second term represents the probability adsorbed on the surface area proportional to $Q_2 - Q_1$ without competition. Our interest is in determining the mean number of molecules and, hence, the concentration of solute 1 and solute 2 in each state s_i ($i = 1, 2, 3, 4, 5$) at time t.

Given $N_1(t) = [N_{11}(t), N_{13}(t), N_{14}(t)]$, during the small time interval $(t, t + \Delta t)$, the average decrease in the number of molecules of species 1 in state S_1 ($N_{11}(t)$) is

$m_{13}N_{11}(t)[Q_1 - N_{13}(t) - N_{23}(t)]\Delta t + 0(\Delta t)$

and the average increase is

$m_{31}N_{13}(t)\Delta t + 0(\Delta t)$

Similarly, the average decrease in the number of molecules of species 1 in state S_3 [$N_{13}(t)$] is

$$m_{31}N_{13}(t)\Delta t + m_{34}N_{13}(t)\Delta t + 0(\Delta t)$$

and the average increase is

$$m_{13}N_{11}(t)[Q_1 - N_{13}(t) - N_{23}(t)]\Delta t + m_{43}N_{14}(t)\Delta t + 0(\Delta t)$$

Note that $N_{14}(t)$ may be expressed as $N_{10} - N_{11}(t) - N_{13}(t)$.

Given $N_2(t) = [N_{22}(t), N_{23}(t), N_{25}(t)]$, during the small time interval $(t, t + \Delta t)$, the average decrease in the number of molecules of species 2 in state S_2 [$N_{22}(t)$] is

$$m_{23}N_{22}(t)[Q_1 - N_{13}(t) - N_{23}(t)]\Delta t + m_{23}N_{22}(t)(Q_2 - Q_1)\Delta t + 0(\Delta t)$$

and the average increase is

$$m_{32}N_{23}(t)\Delta t + 0)(\Delta t)$$

Similarly, the average decrease in the number of molecules of species 2 in state S_3 [$N_{23}(t)$] is

$$m_{32}N_{23}(t)\Delta t + m_{35}N_{23}(t)\Delta t + 0(\Delta t)$$

and the average increase is

$$m_{23}N_{22}(t)[Q_1 - N_{13}(t) - N_{23}(t)]\Delta t + m_{23}N_{22}(t)(Q_2 - Q_1)\Delta t + m_{53}N_{25}(t)\Delta t + 0(\Delta t)$$

Note that $N_{25}(t)$ may be expressed as $N_{20} - N_{22}(t) - N_{23}(t)$.

The conditional expected change in $N_{11}(t)$ during the time interval $(t, t + \Delta t)$ is

$$
\begin{aligned}
&E[N_{11}(t + \Delta t) - N_{11}(t)|N_1(t) = (N_{11}(t), N_{13}(t), N_{14}(t))] \\
&= \{m_{31}N_{13}(t) - m_{13}N_{11}(t)[Q_1 - N_{13}(t) - N_{23}(t)]\}\Delta t
\end{aligned}
\tag{50a}
$$

The conditional expected change in $N_{22}(t)$ during the time interval $(t, t + \Delta t)$ is

$$
\begin{aligned}
&E[N_{22}(t + \Delta t) - N_{13}(t)|N_2(t) = (N_{22}(t), N_{23}(t), N_{25}(t))] \\
&= \{m_{32}N_{23}(t) - m_{23}N_{22}(t)[Q_2 - N_{13}(t) - N_{23}(t)]\}\Delta t
\end{aligned}
\tag{50b}
$$

The conditional expected change in $N_{13}(t)$ during the time interval $(t, t + \Delta t)$ is

$$
\begin{aligned}
&E[N_{13}(t + \Delta t) - N_{13}|N_1(t) = (N_{11}(t), N_{13}(t), N_{14}(t))] \\
&= \{m_{13}N_{11}(t)[Q_1 - N_{13}(t) - N_{23}(t)] + m_{43}[N_{10} - N_{11}(t) - N_{13}(t)] \\
&\quad - (m_{31} + m_{34})N_{13}(t)\}\Delta t
\end{aligned}
\tag{50c}
$$

The conditional expected change in $N_{23}(t)$ during the time interval $(t, t + \Delta t)$ is

$$
\begin{aligned}
&E[N_{23}(t + \Delta t) - N_{23}(t)|N_2(t) = (N_{22}(t), N_{23}(t), N_{25}(t))] \\
&= \{m_{23}N_{22}(t)[Q_2 - N_{13}(t) - N_{23}(t)] + m_{53}[N_{20} - N_{22}(t) - N_{23}(t)] \\
&\quad - (m_{32} + m_{35})N_{23}(t)\}\Delta t
\end{aligned}
\tag{50d}
$$

Rearranging these equations and taking the limit as $\Delta t \to 0$ yields the following equations:

$$\frac{d}{dt}E[N_{11}(t)] = m_{31}E[N_{13}(t)] - m_{13}Q_1E[N_{11}(t)] + m_{13}E[N_{11}(t)N_{13}(t)]$$
$$+ m_{13}E[N_{11}(t)N_{23}(t)] \tag{51a}$$

$$\frac{d}{dt}E[N_{22}(t)] = m_{32}E[N_{23}(t)] - m_{23}Q_2E[N_{22}(t)] + m_{23}E[N_{22}(t)N_{23}(t)]$$
$$+ m_{23}E[N_{22}(t)N_{13}(t)] \tag{51b}$$

$$\frac{d}{dt}E[N_{13}(t)] = m_{13}Q_1E[N_{11}(t)] - m_{13}E[N_{11}(t)N_{13}(t)] - m_{13}E[N_{11}(t)N_{23}(t)] + m_{43}N_{10}$$
$$- m_{43}E[N_{11}(t)] - (m_{43} + m_{31} + m_{34})E[N_{13}(t)] \tag{51c}$$

$$\frac{d}{dt}E[N_{23}(t)] = m_{23}Q_2E[N_{22}(t)] - m_{23}E[N_{22}(t)N_{23}(t)] - m_{23}E[N_{22}(t)N_{13}(t)] + m_{53}N_{20}$$
$$- m_{53}E[N_{22}(t)] - (m_{25} + m_{32} + m_{35})E[N_{23}(t)] \tag{51d}$$

with the initial conditions

$$E[N_{11}(0)] = N_{10}$$
$$E[N_{22}(0)] = N_{20}$$
$$E[N_{13}(0)] = 0$$
$$E[N_{23}(0)] = 0$$

To express Eqs. (51a)–(51d) in terms of concentration, we let

$$C_1(t) = \frac{E[N_{11}(t)]}{VA} \tag{52a}$$

$$C_2(t) = \frac{E[N_{22}(t)]}{VA} \tag{52b}$$

$$q_1(t) = \frac{j_1 E[N_{13}(t)]}{WA} \tag{52c}$$

$$q_2(t) = \frac{j_2 E[N_{23}(t)]}{WA} \tag{52d}$$

Substituting Eqs. (52a)–(52d) into Eqs. (51a)–(51d), we obtain

$$\frac{dC_1(t)}{dt} = m_{31}\frac{W}{j_1 V}q_1(t) - m'_{13}Q'_1 C_1(t) + m'_{13}\frac{W}{j_1}q_1(t)C_1(t) + m'_{13}\frac{W}{j_2}q_2(t)C_1(t) \tag{53a}$$

$$\frac{dC_2(t)}{dt} = m_{32}\frac{W}{j_2 V}q_2(t) - m'_{23}Q'_2 C_2(t) + m'_{23}\frac{W}{j_2}q_2(t)C_2(t) + m'_{23}\frac{W}{j_1}q_1(t)C_2(t) \tag{53b}$$

$$\frac{dq_1(t)}{dt} = m'_{13}Q'_1\frac{j_1 V}{W}C_1(t) - m'_{13}VC_1(t)q_1(t) - m'_{13}V\frac{j_1}{j_2}C_1(t)q_2(t) + m_{43}\frac{j_1 V}{W}C_{10}$$
$$- m_{43}\frac{j_1 V}{W}C_1(t) - (m_{43} + m_{31} + m_{34})q_1(t) \tag{53c}$$

$$\frac{dq_2(t)}{dt} = m'_{23}Q'_2\frac{j_2 V}{W}C_2(t) - m'_{23}VC_2(t)q_2(t) - m'_{23}V\frac{j_2}{j_1}C_2(t)q_1(t) + m_{53}\frac{j_2 V}{W}C_{20}$$
$$- m_{53}\frac{j_2 V}{W}C_2(t) - (m_{53} + m_{32} + m_{35})q_2(t) \tag{53d}$$

TABLE 1 Correlations for Estimating Model Parameters

Parameter	Correlations
Isotherm, q_{m_1}, q_{m_2}, b_1, and b_2	Langmuir equation
Total capacity of outer-layer solid phase, Q_1 and Q_2	$Q_1 = \left(\dfrac{WA}{j}\right) q_{m_1}$
m_{13}, m_{23}	$m_{13} = \lim \dfrac{-d \ln C(t)}{dt} \dfrac{1}{Q_1}$
m_{31}, m_{32}	$m_{31} = \left(\dfrac{VA}{b_1}\right) m_{13}$
m_{34}, m_{35}	$m_{34} = (j_1 - 1) m_{43}$
m_{43}, m_{53}	Best fit
j_1, j_2	Best fit

The system of ordinary differential equations that constitute a batch adsorber model are solved with a numerical integration method. The International Mathematical and Statistical Library subroutine DGEAR (IMSL Inc., Houston, TX) was employed for this task.

At equilibrium, Eqs. (53a) and (53b) may be expressed as

$$q_{1\infty} = \frac{q_{m1}b_1C_{1\infty} + b_1b_2C_{1\infty}C_{2\infty}[q_{m1} - (j_1/j_2)q_{m2}]}{1 + b_1C_{1\infty} + b_2C_{2\infty}} \tag{54a}$$

$$q_{2\infty} = \frac{q_{m2}b_2C_{2\infty} + b_1b_2C_{1\infty}C_{2\infty}[q_{m2} - (j_2/j_1)q_{m1}]}{1 + b_1C_{1\infty} + b_2C_{2\infty}} \tag{54b}$$

When $C_{1\infty} = 0$ or $C_{2\infty} = 0$, Eqs. (54a) and (54b) reduce to the single-solute Langmuir isotherm [Eq. (43)].

The parameter values have been determined based either on the independent experiments or on available correlations as outlined in Table 1. In addition to equilibrium data, which are fitted to the Langmuir isotherm, knowledge of the values of the kinetic parameters and the solid-phase ratio is necessary in order to accurately describe the performance of adsorbers. The intensity functions, m_{43} and m_{53}, and the solid phase ratios, j_1 and j_2, are then found by minimizing the difference between data and model output. This minimization procedure is done by intuitively varying the parameters in the mathematical model until the experimental data and model results agree satisfactorily.

B. Model Verification

The equilibrium data obtained in this study consist of the single-species isotherm data of two adsorbates. The single-species isotherm data of two adsorbates can be fitted by the Langmuir expression. The results are shown in Fig. 4. All of the bisolute kinetic experiments were conducted using 1 g of #25/#30 mesh size carbon.

The kinetic parameters used in the model predictions corresponded to those values obtained from single-solute experiments at the same initial concentrations as in the mixture. The rate parameters for trichloroethylene were obtained from batch experimental data and correlations. The model parameters for 1,1,1-trichloroethane were estimated by correlating those parameters from two experimental sets. All parameters are almost consistent with varying carbon dosage except Q, which is proportional to the amount of GAC. To validate this procedure, a

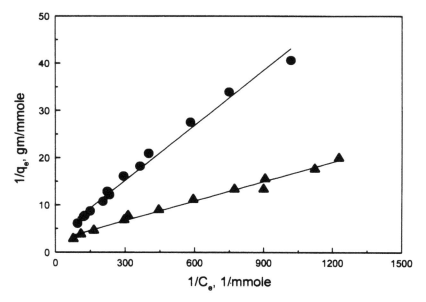

FIG. 4 Adsorption isotherms for single solutes; experimental data for (▲) 1,1,1-trichloroethane and (●) trichloroethylene, and (—) Langmuir isotherm.

single-solute nonlinear model with estimated parameters for 2.0 g GAC is fitted to experimental data. Resonably good agreement has been obtained between the experimental and predicted results. A summary of the results obtained from the experiments and correlations is given in Table 2. Based on the experimental results described earlier (i.e., single-species adsorption isotherm data and the rate parameters for two adsorbates from single-species adsorption

TABLE 2 Estimates of Stochastic Model Parameters

Solutes	Parameters	
1,1,1-Trichloroethane	Q'_1	0.0391
	m'_{13}	0.8930
	m_{31}	0.0103
	m_{35}	0.0130
	m_{53}	0.0021
	j_1	7.3361
Trichloroethylene	Q'_2	0.0561
	m'_{24}	0.8927
	m_{42}	0.0054
	m_{46}	0.0176
	m_{64}	0.0028
	j_2	7.3360

Note: $m'_{13} = m_{13}A$, $m'_{24} = m_{24}A$ (L/min/mmol); $Q'_1 = Q_1/A$, $Q'_2 = Q_2/A$ (mmol); sieve size = #25/#35; carbon dosage = 1.0 g.

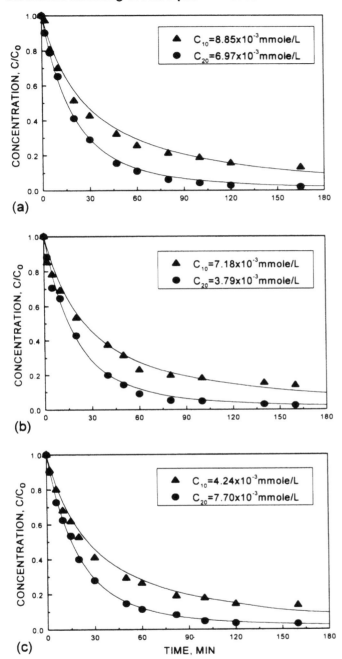

FIG. 5 Adsorption rate profiles for mixture; experimental data for (▲) 1,1,1-trichloroethane and (●) trichloroethylene, and (—) model predictions.

measurements) and the model developed, predictions of concentration histories of bisolute adsorption in a batch can be readily made. The validity of such a prediction procedure must now be assessed by comparing the predictions with the appropriate experimental data.

The results of three batch reactor experiments for the mixture of 1,1,1-trichloroethane and trichloroethylene with varying initial concentrations, but otherwise identical conditions, are shown in Fig. 5, predicted profiles that are in excellent agreement for the more strongly adsorbed trichloroethylene, whereas they have a small deviation for the more weakly adsorbed 1,1,1-trichloroethane. The same observations hold for all mixtures, but the differences are insignificant.

Finally; to gain a deeper understanding of the bisolute batch adsorption process, models of various significant physical phenomena taking place in the adsorber (e.g., interaction between adsorbates as well as competition) should be incorporated into the present model. Investigations of these phenomena are, however, outside the scope of the present study.

VI. SIMULATION OF A FIXED-BED ADSORBER WITH A STOCHASTIC MODEL IN THE CASE OF THE LINEAR ISOTHERM

The removal of pollutants from aqueous waste stream by adsorption onto granular activated carbon in fixed beds is recognized as one of the fundamental treatment technologies for waters and wastewater. Although treatment with GAC is viable technology for removal of many of these materials, efficient cost-effective applications require rigorous databases that characterize specific system dynamics and means for synthesis of such databases in plant design and operation. Mathematical modeling provides an effective means for the prediction of system performance and analysis of sensitivity to varied conditions. The solid–liquid separation in fixed-bed adsorbers has been modeled deterministically by various investigators [63–71].

Fixed-bed adsorption and related processes involve the flow of molecules or particles through randomly distributed pores; thus, these tend to be stochastic in nature. Stochastic processes have been used to model the chromatography [5,38] and the deep-bed filtration [13–15,18,20]. However, no studies have been reported in the literature on stochastic modeling of adsorption in fixed-bed adsorbers.

In Section III, a stochastic model for a batch adsorption process is presented and applied to estimate the parameters characterizing the transfer process between solid and liquid phases. Proceeding from a batch to an open flow, the additional property of flow and exit of solute molecules from the system should be considered. To characterize the flow, the fluid phase is divided into n compartments along the axial axis of the reactor with forward and backward flow of molecules between compartments. The compartmental model for flow systems is a generalization of a class of models, such as the completely mixed tanks-in-series model and back-flow mixed tanks-in-series model. One of the common characteristics of these models is that the basic unit of the model is a completely mixed tank. Compartmental models have been widely employed in modeling a variety of flow systems, including flow chemical reactors and biological transport systems. The deterministic versions of such models have been extensively reviewed by Wen and Fan [72]. Stochastic versions of the models, predominantly those based on Markov chains, have also been employed to model flow systems containing n compartments.

In this section, a stochastic compartmental model is employed to estimate the number of compartments (necessary to characterize the flow) and the intensities of forward and backward flows between compartments in an open-flow adsorber without adsorbate. In addition, this model and the linear batch model are combined into one for predicting breakthrough curves in an open-flow adsorber. The simulated results are obtained from the model using values of parameters derived from pulse response and batch experiments. A comparison is then made between simulated and experimental results for a fixed-bed system.

A. Model Description and Parameter Estimation

In order to study the effect of flow, we consider that a fixed-bed adsorber is divided into n compartments distributed in the axial direction or in the direction of flow, as shown in Fig. 6. Each compartment consists of one liquid phase and two solid phases, occupied by the granular activated carbon. Let S_i denote the liquid phase, S_{n+1} the outer-layer solid phase, and S_{2n+i} the inner-layer solid phase in compartment i. We assume that the molecules in solution enter the system through the liquid phase of the first compartment and exit from the system through the liquid phase of the last compartment. Without loss of generality, we assume that solution enters the system at a constant volumetric rate of q and a constant influent concentration of C_0.

Considering Fig. 6, the probability of moving from one compartment to another in a small time interval is assumed to be proportional to the intensity function connecting the two compartments. For fixed-bed adsorption, it is reasonable to assume the following in a very small time interval $(t, t + \Delta t)$:

1. The transition probability that a molecule moves from the solid phase of compartment i to that of compartment j is zero.

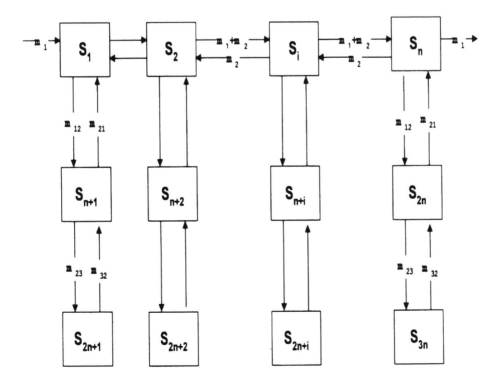

FIG. 6 Transition diagram of a fixed-bed adsorber.

2. The transition probability that a molecule moves from the liquid phase to the solid phase, or from the solid phase to the solid phase, where both phases are not in the same compartment, is zero.
3. The transition probability that a molecule moves from the liquid phase to the solid-phase inner layer is zero.
4. The transition probability that a molecule moves more than one compartment is zero.

Let $P_{ij}(t)$ be the probability that a molecule in state i at time zero will be in state j at time t $(i, j = 1, 2, \ldots, n)$. It is known that the matrix of transition probabilities $P(t) = P_{ij}(t)$ satisfies the Kolmogorov differential equation. Because we can observe the concentration only in the liquid phase of the nth compartment, we are then interested in the probability $P_{1n}(t)$, which is the probability that a molecule that is in the first compartment at time zero will move to the nth compartment at time t:

$$P_{1n}(t) = -\frac{2}{n}\frac{B}{B^n}\sum_{i=1}^{n}\frac{\sin^2\theta_i}{D_1^2}(l_{i1}e^{X_{i1}^t} + l_{i2}e^{X_{i2}^t} + l_{i3}e^{X_{i3}^t}) \tag{55}$$

where

$$l_{i1} = \frac{(\alpha_i + m_{12})^2 m_{12}m_{21} + (X_{i2} + X_{i3})(\alpha_i + m_{12}) + X_{i2}X_{i3}}{X_{i1}^2 - (X_{i2} + X_{i3})X_{i1}X_{i2}X_{i3}} \tag{55a}$$

$$l_{i2} = \frac{(\alpha_i + m_{12})^2 + m_{12}m_{21} + (X_{i1} + X_{i3})(\alpha_i + m_{12}) + X_{i1}X_{i3}}{X_{i2}^2 - (X_{i1} + X_{i3})X_{i2} + X_{i1}X_{i3}} \tag{55b}$$

$$l_{i3} = \frac{(\alpha_i + m_{12})^2 + m_{12}m_{21} + (X_{i1} + X_{i2})(\alpha_i + m_{12}) + X_{i1}X_{i2}}{X_{i3}^2 - (X_{i1} + X_{i2})X_{i3} + X_{i1}X_{i2}} \tag{55c}$$

$$D_i^2 - \left(1 + \frac{1}{n}\right)\cos(n+1)\theta_i - 2B\cos n\theta_i + \left(1 - \frac{1}{n}\right)B^2\cos(n-1)\theta_i \tag{55d}$$

and

$$\alpha_i = m_1 + 2m_2 - \sqrt{m_2(m_1 + m_2)}\cos\theta_i \tag{55e}$$

where θ_i is the ith root of Eq. (55d) and, $X_{ij}(j = 1, 2, 3)$ are the roots of the characteristic equation of the intensity matrix A_i:

$$A_i = \begin{vmatrix} -(\alpha_i + m_{12}) & m_{12} & 0 \\ m_{21} & -(m_{21} + m_{23}) & m_{23} \\ 0 & m_{32} & -m_{32} \end{vmatrix}$$

In the case of continuous flow, $FC_0A\,d\tau$ molecules enter the fluid phase of the first compartment in the time interval dt. Out of these molecules, $FC_0Am_1P_{1n}(t - \tau)\,d\tau$ are expected to exit through the nth compartment at time t. Hence, the outlet concentration is given as

$$C = C_0 \int_0^t m_1 \frac{2}{n}B^{1-n}\sum_{i=1}^{n}\frac{\sin^2\theta_i}{D_i^2}(l_{i1}e^{X_{i1}(t-\tau)} + l_{i2}e^{X_{i2}(t-\tau)} + l_{i3}e^{X_{i3}(t-\tau)})\,d\tau \tag{56}$$

$$C = C_0 m_1 \frac{2}{n}B^{1-n}\sum_{i=1}^{n}\frac{\sin^2\theta_i}{D_i^2}\sum_{K=1}^{3}\frac{l_{iK}}{X_{iK}}(1 - e^{X_{iK}t}) \tag{57}$$

In using Eq. (57) to predict the concentration in the outlet stream, one may use intensities, m_{32} and j, as estimated from batch experiments, whereas m_{12} must be determined independently.

Also, one may use the parameters n and B estimated from a least-squares fit of the model in Eq. (58) to the observed residence time distribution:

$$f(t) = -2\frac{F}{V_{fb}}\sum_{i=1}^{n} B^{1-n}\frac{\sin^2\theta_i}{D_i^2}f_K \tag{58}$$

The intensities m_1 and m_2 were calculated from Eqs. (59) and (60):

$$m_1 = n\frac{q}{V_{fb}} \tag{59}$$

$$m_2 = \frac{B}{1-B}m_1 \tag{60}$$

The intensity m_{12} is expected to be a function of the flow pattern at the interphase between liquid and solid and is determined by fitting the initial portion of the breakthrough curve. The parameter m_{21} can be calculated from Eq. (61), which is obtained by equating, at equilibrium, the forward and backward transfer rates between states S_1 and S_2:

$$m_{21} = \frac{V_{fb}}{W_{fb}}\frac{j}{K}m_{12} \tag{61}$$

A schematic flow diagram of the model is shown in Fig. 7.

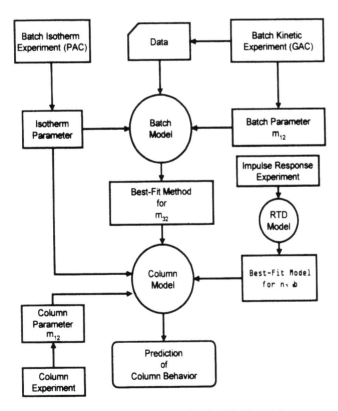

FIG. 7 Schematic flow diagram of a fixed-bed model.

B. Experimental Materials and Methods

The pulse response tests for the residence time distributions were performed in a 2.54-cm-diameter plexiglass column. Glass beads, 0.2 cm in diameter, were used at the bottom of the column for flow distribution. The water was supplied to the column by a magnetically coupled gear pump feed from a constant head tank. A schematic of the experimental system used in this study is presented in Fig. 8. All experiments were performed with GAC particle size #25/#30 US mesh (geometric mean diameter = 6.47×10^{-2} cm). The carbons were sieved, washed with deionized water, and dried to constant weight at 110°C. A predetermined amount of carbon was introduced into the column. A hydraulic rate of 200 mL/min was used throughout the study. The temperature was maintained at 25°C. Two millilitres of the trace material, 50 g/L NaCl, was injected into the inlet line. About 17 mL of the samples were collected every 5 s, and the concentration of each sample was determined by a conductivity meter manufactured by Lab-line Instruments, Inc.

Reagent-grade 1,1,1-trichloroethane was used as the solutes for the adsorption experiments. The experimental system used for adsorption study is the same for pulse response test. The column had an adjustable sieve plate that could be used to operate the bed as a fixed bed with various bed heights.

C. Model Sensitivity and Verification

1. Sensitivity Analysis

The purpose of this analysis is to show the effects of major parameters, m_{12}, m_{32}, j, and K, on the breakthrough curves. A summary of the ranges of values of the important parameters used for this analysis is shown in Table 3.

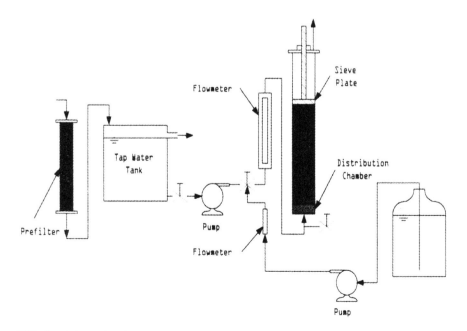

FIG. 8 Schematic of the experimental setup for a fixed bed.

TABLE 3 Run Conditions Employed in Sensitivity Analysis

Equilibrium constant K(L/g)	16
Mean GAC particle diameter (cm)	6.47×10^{-2}
	(#25/30 mesh)
Data from batch experiment	
Amount of GAC (g)	2.0
Volume of solution (L)	1.04
m_{12} (L/min)	0.0759
m_{32} (L/min)	0.0041
j	7.2
Data from fixed-bed adsorber	
Amount of GAC (g)	50
Feed concentration of solution (mg/L)	1.0
Flow rate (mL/min)	200
Void volume of bed (cm³)	40.7
Number of compartments	30
Backmixing effect	0.86

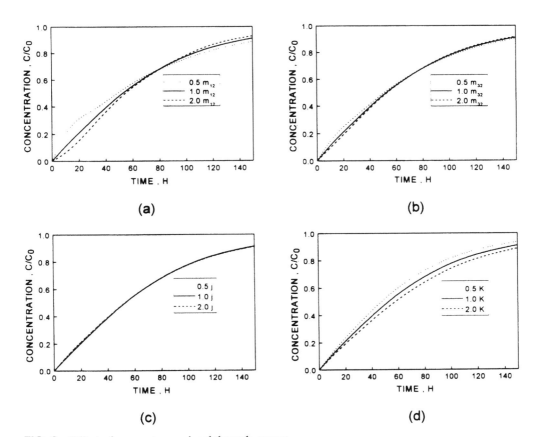

FIG. 9 Effect of parameters on breakthrough curves.

Figures 9a and 9b show the sensitivity of m_{12} and m_{32} on the breakthrough curves. Referring to these curves, at first the effluent concentration becomes higher as m_{12} and m_{32} decrease. This phenomenon can be explained by the slow rate of adsorption according to the decrease in m_{12} and m_{32}. Figures 9a and 9b also show that the overall adsorption rate depends on the m_{12} as well as m_{32}. For higher values of m_{12} and m_{32}, the wave front of the calculated breakthrough curves becomes sharper.

Figure 9c shows that the effect of parameter j on the breakthrough curve is very small. It must be recognized that the above analysis is related to the experimental conditions employed in

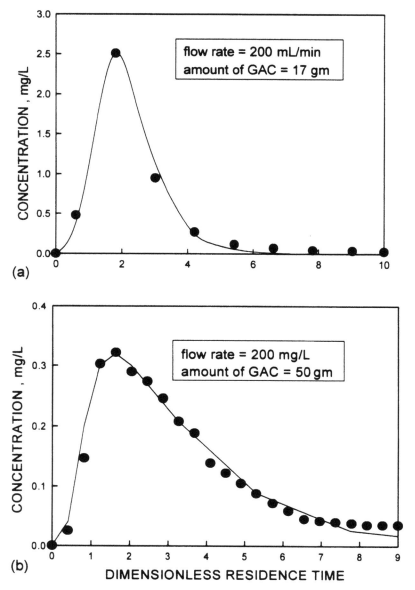

FIG. 10 Residence time distribution: (●) experimental data; (—) model prediction.

TABLE 4 Correlations for Estimating Model Parameters

Parameter	Correlation	Estimates	
		Run I	Run II
Isotherm, K	$q = KC$	16	16
n	Best fit	10	30
B	Best fit	0.37	0.86
m_1	$m_1 = n(F/V)$	152	147
m_2	$m_2 = Bm_1(1 - B)$	89	906
m_{12}	Best fit	33	33
m_{21}	$m_{21} = (j/K)(V_{fb}/W_{fb})m_{12}$	0.0121	0.0121
m_{23}	$m_{23} = (j - 1)m_{32}$	0.0254	0.0254
m_{32}	Best fit	0.0041	0.0041
j	Best fit	7.2	7.2

the present study. The model can be quite sensitive to the parameter j under an altered set of conditions. For instance, the model is sensitive to the parameter j when the parameter m_{32} is increased. Model sensitivity to equilibrium capacity, K, is shown in Fig. 9d. This shows that the model is very sensitive to the equilibrium capacity through the adsorption period.

2. Model Verification

The parameters required for the model are n, B, m_{12}, m_{21}, m_{23}, m_{32}, and j. To obtain n and B for the compartments in series with the backflow model, a number of pulse response experiments were performed. The observed and calculated concentration data for the residence time

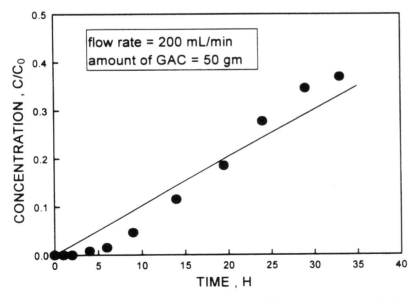

FIG. 11 Fixed-bed adsorber breakthrough curve: (●) experimental data; (—) model prediction.

distribution of fluid are presented in Figs. 10a and 10b. It is seen that the model gave a good fit to the data. The data from the tracer experiments were regressed using a stochastic residence time distribution (RTD) model in a least-squares routine, and the fitted parameters n and B are given in Table 4. The parameters m_{23}, m_{32}, and j pertain to the particle. Values for these intensities obtained from batch experiments (see Section III) are used in fixed-bed simulation. The intensity m_{12} pertaining to solute movement from the liquid phase to the particle was obtained by fitting the initial portion of the fixed-bed experimental data for a specific liquid flow rate through the column.

To validate the stochastic fixed-bed model, it has been fitted to experimental data obtained with an adsorber for removing 1,1,1-trichloroethane from dilute solution. The experimentally observed concentration of 1,1,1-trichloroethane in the exit solution is compared with the model prediction in Fig. 11. The solid line is predicted values with the parameters which are in Table 4 (Run II). One noticeable difference is that the predicted breakthrough concentrations are slightly higher than experimental values initially, but are slightly lower toward the end. One explanation for the deviation between predicted and experimental data could be the dependence of external mass transfer on backmixing flow or on carbon loading. Another possible explanation for the observed difference could be understood in terms of the isotherm nonlinearity. Considering that the value of C/C_0 of the last segment of the breakthrough curve is only 0.37 and that the isotherm is assumed to be linear, agreement between the experiment and the model is reasonably good.

D. Comparison Between Deterministic and Stochastic Solutions

The deterministic model equations for a fixed-bed system with a linear isotherm have been solved [73]:

$$
\frac{C}{C_0} = \frac{1}{2} + \frac{2}{\pi} \int_0^\infty e^{\frac{-3D_\theta(1-\epsilon)}{R^2}\left(\frac{X}{u}\right)H_1(\lambda, B_i)}
$$
$$
\sin\left[\frac{2}{R^2}\frac{D_e}{P_b K_a}\left(t - \frac{X}{u/\epsilon}\right)\lambda^2 - \frac{3D_e(1-\epsilon)}{R^2}\left(\frac{X}{u}\right)H_2(\lambda, B_i)\right]\frac{d\lambda}{\lambda}
$$

(62)

where H_1 and H_2 are hyperbolic functions of λ and B_i:

$$
H_1(\lambda, B_i) = \frac{H_{D_1} + (1/B_i)(H_{D_1^2} + H_{D_2^2})}{(1 + H_{D_1}/B_i)^2 + (H_{D_2}/B_i)^2}
$$

$$
H_2(\lambda, B_i) = \frac{H_{D_2}}{(1 + H_{D_1}/B_i)^2 + (H_{D_2}/B_i)^2}
$$

H_{D_1} and H_{D_2} are defined as

$$
H_{D_1}(\lambda_i) = \lambda\left(\frac{\sinh 2\lambda + \sin 2\lambda}{\cosh 2\lambda - \cos 2\lambda}\right) - 1
$$

$$
H_{D_2}(\lambda_i) = \lambda\left(\frac{\sinh 2\lambda - \sin 2\lambda}{\cosh 2\lambda - \cos 2\lambda}\right)
$$

The solution of the above equations is not simple and straightforward and approximate solutions have been provided by Rosen [73]. A linear driving-force model approximation for intraparticle transport may provide simpler solutions. However, as Do and Rice [74] have shown, the time domain of validity for the linear driving-force model is very restricted.

The stochastic model that has been presented is an alternative approach that provides a comparatively easier solution methodology.

VII. SUMMARY AND CONCLUSIONS

The kinetics of adsorption of 1,1,1-trichloroethane and trichloroethylene from water on activated carbon are examined using stochastic approaches. A stochastic model, which has been developed by using the theory of the Markov process, is used to predict the rates of adsorption of 1,1,1-trichloroethane in a batch reactor. Adsorption equilibrium was represented by the linear isotherm equation. The simulation results under various adsorbent loading conditions and for various particle sizes show excellent fit between model predictions and experimental data. The intensity functions estimated from these studies can be utilized to predict batch adsorber performance under other process loading conditions. The model parameters, m_{32} and j, obtained from this study can also be used in stochastic models for fixed-bed adsorbers.

A stochastic model which accounts Langmuir nonlinear isotherm could be applied satisfactorily in simulating the behavior of batch reactors for carbon adsorption. The ability of the model to predict adsorption rate of trichloroethylene with varying carbon sizes indicates the applicability to a wide range of particle sizes. The parameter estimation procedure used in the model shows that parameter m_{32} can be estimated accurately and consistently under different conditions of agitation and different particle sizes. The parameters, m_{32} and j, obtained from this study can be used in stochastic models for fixed-bed adsorbers.

The stochastic model developed in Section IV for a single solute has been extended to a bisolute batch model with Langmuir nonlinear isotherm. The experimental and predicted profiles are in excellent agreement for the more strongly adsorbed solute (trichloroethylene), whereas they have a small deviation for the more weakly adsorbed solute (1,1,1-trichloroethane). The model verifications indicated that the representations of the rate mechanisms are adequate. All of the model parameters are obtained from single-solute batch experimental data and correlations. These parameters are almost consistent with varying carbon dosage except Q (total capacity of the solid-phase outer layer) which is proportional to the amount of GAC.

A stochastic model could be applied satisfactorily in the simulation of the behavior of the fixed-bed adsorber and the prediction of the breakthrough curves for single-solute adsorption. The parameters such as the number of compartment and backflow ratio were estimated from a least-squares fit of the compartmental model to the observed data obtained from the pulse response test. The intensity m_{12}, which is expected to be a function of the flow pattern at the interphase between liquid and solid, was determined by fitting the initial portion of the breakthrough curve. An equation was introduced for improving the estimation of the parameter m_{21}.

VIII. NOTATION

A	Avogadro's number, 6.0235×10^{23}/mol
B	Backmixing ratio
B_{ri}	Defined in Eqs. (13) and (15)
b	Langmuir isotherm parameter (L/M)
C	Solution concentration, or outlet concentration in fixed-bed system (M/L^3)
C_0	Initial concentration (M/L^3)

C_∞	Solution concentration at equilibrium (M/L^3)
D_i	Defined in Eq. (55d)
E	Defined in Eq. (14a)
$E[\ ]$	Expected value of a given random variable
F	Volumetric flow rate (L^3/t)
j	Ratio of the solid phase
K	The slope of the linear isotherm (L^3/M)
$\boldsymbol{K}$	Matrix of the intensity functions
K_{ij}	Intensity function of a Markov process
l_{ij}	$[l_{i1}, l_{i2}, l_{i3}]$, defined in Eqs. (55a), (55b), and (55c)
$\boldsymbol{M}$	Matrix of the intensity functions
m_i	Intensity function of a Markov process
m_{ij}	Intensity function of a Markov process
N_0	Initial number of molecules in state S_1
N_{10}	Initial number of molecules in state S_1
N_{20}	Initial number of molecules in state S_2
$N_1(t)$	$[N_{11}(t), N_{13}(t), N_{14}(t)]$, random vector
$N_2(t)$	$[N_{22}(t), N_{23}(t), N_{25}(t)]$, random vector
$N_i(t)$	Random variable presenting the number of molecules in state S_i at time t
$N_{ij}(t)$	Random variable presenting the number of molecules in state S_i at time t
n	Number of compartment
$n(t)$	$[n_1(t), n_2(t), n_3(t)]$, corresponding realization vector
$n_i(t)$	Number of molecules in state S_i
Q	Defined in Eq. (13)
Q'	Total molecular capacity of solute $i (= Q_i/A)$ (M)
$\boldsymbol{P}$	Matrix of the transition probabilities $[P_{ij}]$
P_{1n}	Probability that a molecule in compartment 1 at time zero will be in compartment n at time t
P_{ij}	Transition probability
q	Solid-phase concentration (M/M)
q_∞	Solid-phase concentration at equilibrium (M/M)
q_m	Langmuir isotherm parameter
S_1	Liquid phase
S_2	Liquid phase in the bisolute system, or solid phase in the single-solute system
S_3	Solid-phase outer layer
S_4, S_5	Solid-phase inner layer
S_{n+i}	Solid-phase outer layer
S_{2n+i}	Solid-phase inner layer
S_i	State of a Markov process
T	time, t
V	Volume of the adsorbate (L^3)
W	Amount of GAC (M)
W_{fb}	Weight of GAC in the fixed-bed system (M)
V_{fb}	Void volume of the fixed-bed system (L^3)
X_{ij}	$[X_{i1}, X_{i2}, X_{i3}]$, roots of the characeristic equation of the intensity matrix $\boldsymbol{A}_i$
X_n	Random variable
X	Any real number

Greek Letters

λ Eigenvalue of matrices K and M

λ_k Intensity of the birth transition

μ_k Intensity of the death transition

REFERENCES

1. WJ Weber Jr, FA DiGiano. Process Dynamics in Environmental Systems. New York: Wiley, 1996, pp 356–390.
2. US EPA. Process Design Manual for Carbon Adsorption. Washington, DC: Tech. Transfer, 1973.
3. CL Chiang. An Introduction to Stochastic Processes and Their Applications. New York: Robert and Krieger, 1980.
4. DA McQuarrie. J Appl Probab 4:413, 1967.
5. DA McQuarrie. J Chem Phys 38:437, 1963.
6. AG Fredrickson. Chem Eng Sci 21:687, 1966.
7. IG Darvey, PJ Staff. J Chem Phys 44:990, 1966.
8. W Pippel, G Philipp. Chem Eng Sci 32:543, 1977.
9. R Shinnar, D Glasser, S Katz. Chem Eng Sci 28:617, 1973.
10. R Nassar, LT Fan, JR Too, LS Fan. Chem Eng Sci 36:1307, 1981.
11. ST Chou, LT Fan, R Nassar. Chem Eng Sci 43:2807, 1988.
12. J Litwiniszyn. Bull Acad Polon Sci Ser Sci Technol 11:81, 1963.
13. EH Hsu, LT Fan. AIChE J 30:267, 1984.
14. J Litwiniszyn. Bull Acad Polon Sci Ser Sci Technol 14:561, 1966.
15. LT Fan, R Nassar, SH Hwang, ST Chou. AIChE J 31:1781, 1985.
16. LT Fan, SH Hwang, ST Chou, R Nassar. Chem Eng Commun 35:101, 1985.
17. LT Fan, SH Hwang, R Nassar, ST Chou. Powder Technol 44:1, 1985.
18. J Litwiniszyn. Bull Acad Polon Sci Ser Sci Technol 15:345, 1967.
19. J Litwiniszyn. Bull Acad Polon Sci Ser Sci Technol 17:57, 1969.
20. R Nassar, ST Chou, LT Fan. Chem Eng Sci 41:2017, 1986.
21. AD Randolph, MA Larson. Theory of Particulate Processes. New York: Academic Press, 1971.
22. KA Berglund, MA Larson. AIChE J 30:280, 1984.
23. R Nassar, JR Too, LT Fan. AIChE J 30:1014, 1984.
24. EH Hsu, LT Fan, ST Chou. Chem Eng Commun 69:95, 1988.
25. MG Palekar, RA Rajadhyaksha. Chem Eng Sci 40:1085, 1985.
26. VS Patwardhan. Chem Eng Sci 44:2619, 1989.
27. PV Danckwerts. Chem Eng Sci 2:1, 1953.
28. CY Wen, LT Fan. Models for Flow Systems and Chemical Reactors, New York: Marcel Dekker, 1975, pp 113–207.
29. JR Too, LT Fan, R Nassar. Chem Eng Sci 41:2341, 1986.
30. JR Ligon, NR Amundson. Chem Eng Sci 36:653, 1981.
31. JR Ligon, NR Amundson. Chem Eng Sci 36:661, 1981.
32. RO Fox, LT Fan. Chem Eng Sci 42:1345, 1986.
33. RO Fox, LT Fan. Chem Eng Ed 24:56, 1990.
34. FJ Edeskuty, NR Amundson. J Phys Chem 56:148, 1952.
35. FJ Edeskuty, NR Amundson. Ind Eng Chem Process Des Dev 44:1698, 1952.
36. T Furusawa, M Suzuki. J Chem Eng Japan 8:119, 1975.
37. S Kaguei, SN Ono, N Wakao. Chem Eng Sci 44:2565, 1989.
38. JC Giddings, H Eyring. J Phys Chem 59:416, 1955.
39. MJ McQuarrie. Chem Phys 38:437, 1963.
40. AG Fredrickson. Chem Eng Sci 21:687, 1966.
41. IG Darvey, PJ Staff. J Chem Phys 44:990, 1966.

42. DD Do, RG Rice. AIChE J 32:149, 1986.
43. VL Snoeyink, WJ Weber Jr. Adv Chem Ser 79:112, 1968.
44. T Furusawa, JM Smith. Ind Eng Chem Fundam 12:197, 1973.
45. I Neretnieks. Chem Eng Sci 31:107, 1976.
46. M Suzuki, K Kawazoe. J Chem Eng Japan 7:346, 1974.
47. WW Fritz, W Merk, EU Schlunder, H Sontheimer. In: MJ McGuire, IP Suffet, eds. Activated Carbon Adsorption of Organics from the Aqueous Phase, Vol. 1. Ann Arbor, MI: Ann Arbor Science, 1980, pp 193–211.
48. DW Hand, JC Crittenden, WE Thacker. J Environ Eng Div ASCE 109:82, 1983.
49. TF Speth, RJ Miltner. J Am Water Works Assoc 82:72, 1990.
50. JC Crittenden, WJ Weber Jr. J Environ Eng Div ASCE 104:185, 1978.
51. AP Mathews, WJ Weber Jr. In: WH Flank, ed. Adsorption and Ion-Exchange with Synthetic Zeolites, ACS Symp Ser 135. Washington, DC: American Chemical Society, 1980, pp 27–53.
52. L Jossens, JM Prausnitz, W Fritz, EU Schlunder, AL Myers. Chem Eng Sci 33:1097, 1978.
53. AC Larson, C Tien. Chem Eng Commun 27:339, 1984.
54. CY Yen, PC Singer. J Environ Eng Div ASCE 110:976, 1984.
55. J Langmuir. J Am Chem Soc 40:1361, 1918
56. AI Liapis, DWT Rippin. Chem Eng Sci 32:619, 1977.
57. W Fritz, W Merk, EU Schlunder, H Sontheimer. Chem Eng Sci 36:731, 1980.
58. J Fetting, H Sontheimer. J Environ Eng Div ASCE 113:780, 1987.
59. B Al-Duri, G McKay. Chem Eng Sci 46:193, 1991.
60. GF Nakhla, MT Suidan. J Environ Eng Div ASCE 121:10, 1995.
61. M Yang. Separ Sci Technol 31:9, 1996.
62. EB Nauman. Chem Eng Sci 36:957, 1981.
63. RK Charkravorti, TW Weber. AIChE Symp Ser 71:392, 1975.
64. AI Liapis, DWT Rippin. Chem Eng Sci 33:593, 1978.
65. WJ Weber Jr, M Pirbazari. J Am Water Works Assoc. 74:203, 1982.
66. RS Summers, PV Roberts. J Environ Eng Div ASCE 110:73, 1984.
67. WE Thacker, JC Crittenden, VL Snoeyink. J Water Pollut Control Fed 56:243, 1984.
68. DW Hand, JC Crittenden, H Arora, JM Miller, B Lykins Jr. J Am Water Works Assoc. 81:67, 1989.
69. SM Lee. PhD thesis, Kansas State University, Manhattan, KS, 1991.
70. GA Sorial, MT Suidan, RD Vidic, SW Maloney. J Environ Eng Div ASCE 119:6, 1993.
71. EH Smith. Water Res 28:8, 1994.
72. CY Wen, LT Fan. Models for Flow Systems and Chemical Reactors. New York: Marcel Dekker, 1975.
73. JB Rosen. J Chem Phys 20:387, 1952.
74. DD Do, RG Rice. AIChE J 32:149, 1986.

10

Adsorption from Liquid Mixtures on Solid Surfaces

IMRE DÉKÁNY and FERENC BERGER   University of Szeged, Szeged, Hungary

I. ADSORPTION OF BINARY LIQUID MIXTURES ON SOLID SURFACES

When solid particles are immersed in liquid medium, solid–liquid interfacial interactions will cause the formation of an adsorption layer on their surface. The material content of the adsorption layer is the adsorption capacity of the solid adsorbent, which may be determined in binary liquid mixtures if the so-called adsorption excess isotherm is known. Due to adsorption, the initial composition of the liquid mixture, x_1^0, changes to the equilibrium concentration x_1, where $n^s = n_1^s + n_2^s$, the mass content of the interfacial phase (e.g., mmol/g). This change, $x_1^0 - x_1 = \Delta x_1$, can be determined by simple analytical methods. The relationship between the reduced adsorption excess amount calculated from the change in concentration, $n_1^{\sigma(n)} = n^0(x_1^0 - x_1)$, and the material content of the interfacial layer is given by the Ostwald–de Izaguirre equation [1–5]. In the case of purely physical adsorption of binary mixtures, the material content of the adsorption layer ($n^s = n^0 - n$) for component 1 is illustrated by the following material balance (see Fig. 1):

$$n^0 x_1^0 = n_1^s + (n^0 - n^s)x_1 \tag{1}$$

$$n_1^\sigma = n^0(x_i^0 - x_i) = n^0 \Delta x_i \quad (i = 1, 2, \ldots) \tag{2}$$

where n^0 is the total quantity of liquid mixture referred to unit mass of the adsorbent (e.g., mmol/g adsorbent), x_1^0 is the mole fraction of the ith component before adsorption, and x_1 is that in the equilibrium homogeneous liquid phase. The adsorption excess isotherms $n_1^\sigma = f(x_i)$ calculated in accordance with Eq. (2) can be classified into five basic types by the Schay–Nagy isotherms classification [2–4].

The Ostwald–de Izaguirre equation is obtained free from any assumption from the mass balance:

$$n^0(x_1^0 - x_1) = n_1^{\sigma(n)} = n_1^s - n^s x_1 = n^s(x_1^s - x_1) \tag{3}$$

where $x_1^s = n_1^s/n^s$ is the mole fraction of the interfacial phase. According to Eq. (3.), therefore, the excess isotherm $n_1^{\sigma(n)} = f(x_1)$ arises from a combination of the "individual" isotherms $n_1^s = f(x_1)$ and $n^s = f(x_1)$ [2–4].

The isotherms of types II, III, and IV in the Schay–Nagy classification have the common characteristic that a fairly long section of the isotherm is practically linear [1–5]. For the linear section, we may write

$$n_1^{\sigma(n)} = a - bx_1 \tag{4}$$

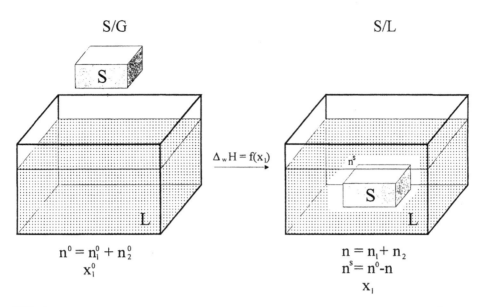

FIG. 1 Schematic diagram of immersional wetting (S/G into S/L) in pure liquids.

Combining Eqs. (3) and (4),

$$a - bx_1 = n_1^s - n^s x_1 = n_1^s - (n_1^s + n_2^s)x_1 \tag{5}$$

Assuming that the intercepts of the linear sections are equal to the adsorption capacities of the components at the points $x_1 = 0$ and $x_2 = 1$ for the $x_1 = 0$ intercept $a = n_1^s$ and for the $x_2 = 1$ intercept, $a - b = -n_2^s$ [1–5].

Because the surface of the adsorbent is always completely covered in liquid sorption, after Williams the following equation can be written for the case of monomolecular surface coverage:

$$n_1^s a_{m,1} + n_2^s a_{m,2} = a_{equ.}^s \tag{6}$$

where $a_{m,1}$ and $a_{m,2}$ are the cross-sectional areas of the components 1 and 2, which can be calculated from the molar volumes of the components, and $a_{equ.}^s$ is the equivalent specific surface area [6–12].

Equation (6) can also be written in the following form:

$$n_1^s + rn_2^s = n_{m,1,0}^s \tag{7}$$

where $n_{m,1,0}^s$ is the monomolecular adsorption capacity of pure component 1, and

$$a_{equ.}^s = n_{m,1,0}^s a_{m,1} \tag{8}$$

According to Eqs. (8) and (9), the specific surface area of the adsorbent can be calculated in the knowledge of the adsorption capacities and the cross-sectional areas. The above method for the determination of the specific surface area has been applied to many systems in the literature, and it has been found that the specific surface areas calculated with the Schay–Nagy extrapolation method for nonswelling and desegregating adsorbents agree very well with the values calculated with the Brunauer–Emmett–Teller (BET) method [1–5].

Application of the graphical extrapolation method for the isotherm of type U (or type II for Schay–Nagy classification) can be seen in Fig. 2a. Employment of the method assumes that in the linear section of the excess isotherm, the composition of the interfacial phase is practically unchanged (i.e., in this section, $n_1^s \approx$ constant and $n_2^s \approx$ constant) [1–5].

If we assume that the adsorption interaction forces increases as the monolayer thickness, we can use the adsorption space filling model for the interfacial phase.

The adsorption volume filled by the components of the mixture being adsorbed on a solid surface is

$$V^s = n_1^s V_{m,1} + n_2^s V_{m,2} \tag{9a}$$

in general,

$$\sum n_i^s V_{m,i} = V^s \tag{9b}$$

where $V_{m,i}$ is the partial molar volume of the components in the adsorption layer. Equations (9a) and (9b) are formally identical with the description of the so-called pore-filling model; still it is also applicable for planar nonporous adsorbents, because the thickness of the adsorption layer is determined solely by the range of the forces of adsorption. Thus, the adsorption volume V^s designates the volume falling within the range of adsorption forces; at certain regions of the isotherm, its value may be nearly constant, but it may also be a function of equilibrium composition. If molecular sizes within the adsorption space are not identical (i.e., $r = V_{m,2}/V_{m,1}$ and $V^s = n_{1,0}^s V_{m,1}$), Eq. (9a) may also be formulated for $n_{1,0}^s$:

$$n_{1,0}^s = n_1^s + r^* n_2^s \tag{9c}$$

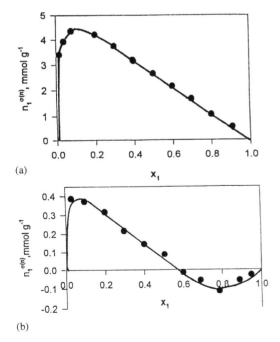

(a)

(b)

FIG. 2 (a) Adsorption excess isotherm on silica gel in methanol(1)–benzene(2) mixture; (b) adsorption excess isotherm on hydrophobized silica gel in methanol(1)–benzene(2) mixture.

Equations (9a)–(9c) already incorporate the assumption that there exists a concentration range where the volume of the adsorption layer—on constant surface area—is independent of the composition of the bulk phase. However, still no statement can be made as to whether this layer is monomolecular or multilayered.

The adsorption capacity of the pure component 1 is

$$n_{1,0}^s = n_1^s + r n_2^s = n^s x_1^s + r n^s x_2^s \tag{10}$$

where $r = V_{m,2}/V_{m,1} = n_{1,0}^s/n_{2,0}^s$ and x_i^s is the mole fraction of component i in the adsorption layer.

Rearrangement of Eq. (10) gives

$$n^s = \frac{n_{1,0}^s}{x_1^s + r x_2^s} \tag{11}$$

The separation factor of adsorption is defined as

$$S = \frac{x_1^s x_2}{x_2^s x_1} = \frac{n_1^s x_2}{n_2^s x_1} \tag{12}$$

Replacing x_1^s and x_2^s in Eq. (11) by S gives

$$n^s = n_{1,0}^s \frac{Sx_1 + x_2}{Sx_1 + rx_2} \tag{13}$$

Substituting Eq. (13) in Eq. (3) and rearranging, we obtain

$$\frac{n^{s(n)}}{n_{1,0}^s} = \frac{(S-1)x_1 x_2}{(S-r)x_1 + r} \tag{14}$$

The adsorption capacity $n_{1,0}^s$ of the pure component can be determined by the Everett–Schay (ES) method [1–5,8,9]:

$$\frac{x_1 x_2}{n_1^{\sigma(n)}} = \frac{1}{n_{1,0}^s}\left(\frac{r}{S-1} + \frac{S-r}{S-1}\right)x_1 \tag{15}$$

S is constant for ideal adsorption from ideal solutions [1–5]. The constancy of the separation factor in many other cases may result from compensation effects. From the linear dependence of $x_1 x_2/n_1^{\sigma(n)}$ on x_1, we determine the value of $n_{1,0}^s$.

Given the knowledge of the adsorption excess isotherm $n_1^{\sigma(n)} = f(x_1)$, the so-called individual isotherms are given by the following equations [1–4]:

$$n_1^s = \frac{r n_1^{\sigma(n)} + n_{1,0}^s x_1}{x_1 + r x_2} \tag{16}$$

$$n_2^s = \frac{n_{1,0}^s x_2 - n_1^{\sigma(n)}}{x_1 + r x_2} \tag{17}$$

where $n_{1,0}^s$ is the adsorption capacity relative to pure component 1 and $r = V_{m,2}/V_{m,1}$ is the ratio of the molar volumes of the components. In the case of U-shaped excess isotherms, the adsorption capacity $(n_{1,0}^s)$ can be determined from the linearized Everett–Schay function [1–5]. As soon as the adsorption capacity is known, the volume of the layer is obtained from the equation $V^s = n_{1,0}^s V_{m,1}$ where $n_{1,0}^s$ is the adsorption capacity relative to pure component 1 and

$r = V_{m,2}/V_{m,1}$ is the ratio of the molar volumes of the components. In case of U-shaped excess isotherms, the adsorption capacity ($n^s_{1,0}$) can be determined from the linearized Everett–Schay function [4,8,9,13,14]. As soon as adsorption capacity is known, the volume of the layer is obtained from the equation $V^s = n^s_{1,0} V_{m,1}$.

The volume fraction of the adsorption layer can be calculated from the data of excess isotherms by the following equation:

$$\phi^s_1 = \frac{n^s_1}{n^s_{1,0}} = \phi_1 + \frac{r n^{\sigma(n)}_1 V_{m,1}}{V^s(x_1 + rx_2)} \tag{18}$$

If the specific surface area (a^s) of the particles is known, the thickness of the layer is $t^s = V^s/a^s$, which can be calculated from Eq. (18):

$$t^s = n^{\sigma(n)}_1 \left(\frac{V_{m,1}}{a^s x_1}\right) \frac{\phi_1}{\phi^s_1 - \phi_1} \tag{19}$$

where ϕ^s_1 and ϕ_1 are the volume fractions of the adsorption layer and of the bulk phase, respectively, in adsorption equilibrium. The layer thickness t^s calculated according to Eq. (19) is nearly constant; in nonideal liquid mixtures, however, its value may be strongly dependent on the composition of the bulk phase [15–22].

In calculations of stability for disperse systems, knowledge of the thickness of the stabilizing adsorption layer is highly important [18–23].

The sorption-exchange process taking place at the solid–liquid interface may be described in thermodynamically exact terms when the activities of the interfacial layer and of the bulk phase are known. In accordance with the exchange equilibrium at the solid–liquid interface, the liquid sorption equilibrium constant is given by the following formulas [1–5]:

$$(1) + r(2)^s \rightleftharpoons (s)^s + r(2) \tag{20}$$

$$K = \frac{x^s_1 f^s_1 (x_2 f_2)^r}{(x^s_2 f^s_2)^r x_1 f_1} \tag{21}$$

Assuming that $f^s_1/f^s_2 \approx 1$ (i.e., the activity coefficients of the interfacial phase compensate for each other [24–27]),

$$K' = \frac{x^s_1 (a_2)^r}{(x^s_2)^r a_1} \tag{22}$$

If the activity data of the bulk phase are known, the value of K' can be calculated at a given value of $r = V_{m,2}/V_{m,1}$ by means of computer iteration [24–26]. The Redlich–Kister equation has proven to be perfectly reliable for calculating the activities: Its approach allows the calculation of the activity coefficients of components 1 and 2 and, on this basis, the activity functions can be given [28]. The applicability of Eq. (22) is demonstrated in Fig. 3. It is revealed by the adsorption equilibrium diagrams that when considering the adsorption of a liquid pair made up of components significantly different in polarity, the value of the equilibrium constant K' decreases with increasing hydrophobicity of the surface covered by alkyl chains.

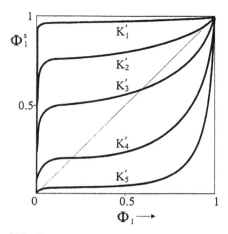

FIG. 3 Adsorption equilibrium diagrams on S/L interface in methanol–benzene mixtures at different equilibrium constants. $K_1 = 10^3$, $K_2 = 10^2$, $K_3 = 10$, $K_4 = 1$, $K_5 = 10^{-1}$; calculated using Eq. (22).

II. THE FREE ENTHALPY OF WETTING OF THE SOLID–LIQUID INTERFACE AND THE THICKNESS OF THE ADSORPTION LAYER

When the adsorption excess isotherm $n_1^{\sigma(n)} = f(x_1)$ and the activities of the bulk liquid phase are known, the free enthalpy of adsorption of solid–liquid interfaces, $\Delta_{21} G$, is obtained from the Gibbs adsorption equation [2–5]:

$$\Delta_{21} G = -\int \left(\frac{n_1^{\sigma(n)}}{x_2 a_1} \right) da_1 \tag{23}$$

where $a_1 = f_1(x_1)$ is the activity of component 1 which can be calculated with the help of the Redlich–Kister equations [28], given the knowledge of the solid–vapor equilibrium data.

Different types of adsorption excess isotherms (U- and S-shaped functions) naturally yield different free enthalpy functions $\Delta_{21} G - f(x_1)$, the course of which is characteristic of the minimum energy of wetting at the S/L interface, a parameter diagnostic of the stability of the disperse system.

Knowing the adsorption excess isotherm, $n_1^{\sigma(n)} = f(x_1)$, and the activities ($f_i, f_j = 1, 2$) of the bulk liquid phase, the free enthalpy of adsorption $\Delta_{21} G$ at the solid–liquid adsorption layer is determined by the Gibbs equation:

$$\Delta_{21} G = -RT \int_0^{x_1} \frac{n_1^{\sigma(n)}}{(1 - x_1)x_1} \left(1 + \frac{d \ln \gamma}{d \ln x_1} \right) dx_1 \tag{24}$$

where γ is the activity coefficient of the solute. It is worth noting that the term of activity coefficients in the large parentheses vanishes for ideal or ideally diluted solutions (like n-butanol in water). In the adsorption layer, the Δx_1^s composition change results in the $\Delta n_1^s = -r\Delta n_2^s$ material transition. The free enthalpy of adsorption ($\Delta_d G$) is connected with the molar-free enthalpies (g_i^s) and the amount of components (n_i^s):

$$\Delta_d G = \Delta n_1^s g_1^s + \Delta n_2^s g_2^s \tag{25}$$

Summing the $\Delta_d G$ data from $x_1 = 0$ to $x_1 \to 1$ in the whole composition range,

$$\sum_{x_1=0}^{x_1} \Delta_d G = \Delta_{21} G = n_1^s \left(g_1^s - \frac{g_2^s}{r} \right) \tag{26}$$

Substituting $n_1^s = n_1^{\sigma(n)} + n^s x_1$ from Eq. (1) into Eq. (11) and dividing by the adsorption excess $(n_1^{\sigma(n)})$, we obtained

$$\frac{\Delta_{21} G}{n_1^{\sigma(n)}} = g_1^s - \frac{g_2^s}{r} + n^s \left(g_1^s - \frac{g_2^s}{r} \right) \frac{x_1}{n_2^{\sigma(n)}} = \Delta_{21} g + n^s \Delta_{21} g \frac{x_1}{n_1^{\sigma(n)}} \tag{27}$$

The linear representation of Eq. (27) gives $\Delta_{21} g = g_1^s - g_2^s/r$ (from the intercept) and, from the slope, the adsorbed amount n^s can be calculated.

III. ENTHALPY OF IMMERSION ON SOLIDS IN BINARY LIQUID MIXTURES

In many cases, the dispersed particles adsorb both components of the solvents. Adsorption measurements yield the composite isotherms (surface excess isotherms) from which the true adsorption isotherms must be derived. To understand these adsorption processes more clearly, the surface excess isotherms can be combined with calorimetric measurements. A simple and straightforward way for quantifying the solid–liquid interfacial interaction is immersion micro-calorimetry. In the course of this measurement, the surface previously heat treated *in vacuo* is brought into contact with the pure wetting liquid [6–10]. It is advisable to choose a liquid of different polarity, so that the extent of the hydrophobicity or hydrophilicity of the surface could be estimated from the magnitude of the heat of wetting. Thus, wetting a hydrophilic surface by a polar liquid liberates a large exothermic enthalpy of wetting, whereas wetting a hydrophobic surface by a polar liquid produces a smaller heat effect (Fig. 4).

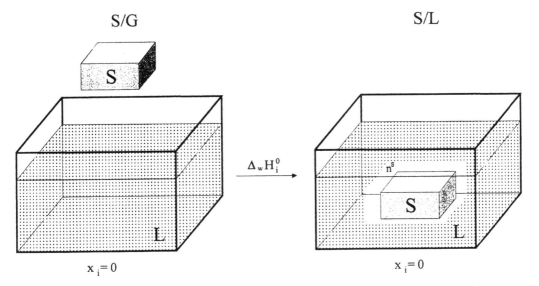

FIG. 4   Schematic picture of immersional wetting (S/G into S/L) in binary liquid mixtures. n^0 and x_1^0 are the liquid material amount and the mixture molar fraction, respectively, in the initial state. In the equilibrium state, liquid material amount $= n$, mixture molar fraction $= x_1$, surface layer amount $= n^s$.

When a solid adsorbent is immersed in a binary mixture, the heat of wetting is greatly affected by the composition of the bulk phase and values intermediate between the heat effects of wetting $\Delta_w H_2^0$ and $\Delta_w H_1^0$ are obtained, measured in the pure components 2 and 1, respectively. This is the so-called immersion technique which supplies direct information on the strength of the solid–liquid interaction with the mixture (Fig. 1) [29–33].

When $+\Delta n_1$ moles of component 1 of the binary mixture are added to the suspension made up in the liquid mixture with molar amount of $n = n_1 + n_2$ (i.e., to the S/L interface), the original composition x_1 is changed to x_1^* and, consequently, the composition of the interfacial layer is shifted by the value of $\Delta x_1^s = x_1^{s*} - x_1^s$. The amount of adsorbed material present in the interfacial layer is therefore $n^{s*} = n_1^{s*} + n_2^{s*}$, where n_1^{s*} and n_2^{s*} are the material content of the adsorption layer of components 1 and 2, respectively. The change in composition in the bulk phase (Δx_1) and in the interfacial phase (Δx_1^s) brings about a change in the so-called enthalpy of displacement, $\Delta_{21} H$, which is the difference between the heats of the wetting characteristics of the two states with different compositions (Fig. 5). When the change in composition is started from component 2 ($\Delta_w H_2^0$), increasing x_1 in the direction $x_1 \rightarrow 1$, the isotherm of enthalpy of displacement $\Delta_{21} H = f(x_1)$ is obtained (Fig. 5) [29–31].

According to Everett's adsorption layer model [5,34,35], the heat of immersional wetting ($\Delta_w H_t$), a thermodynamic parameter characteristic of the solid–liquid interaction, can be easily calculated when the molar enthalpies of the components h_1 and h_2 of the system are known. When a solid adsorbent is immersed in a liquid mixture, the amount of which is $n_1^0 + n_2^0$, the interfacial forces of adsorption cause the formation of an adsorption layer on the surface of the adsorbent, the material content of which is $n^s = n_1^s + n_2^s$.

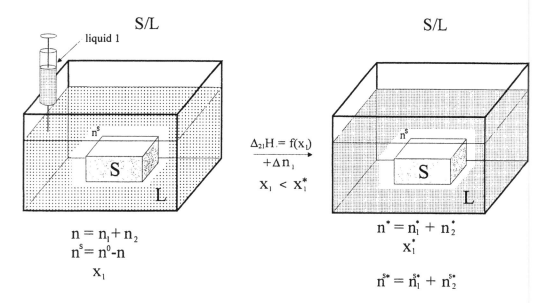

FIG. 5 Schematic picture of the heat evolution (enthalpy of displacement, $\Delta_{21} H$) in the solid–liquid adsorption layer. After adding component 1 (Δn_1) to the liquid mixture, the composition of the bulk and the adsorption layer will be exchanged (notation with asterisk).

According to Everett, the change in enthalpy of wetting between the equilibrium and initial state $(H_e - H_i)$ is given by the following equation [5,8,34]:

$$\Delta_w H_t = H_e - H_i = (n_1 - n_1^0)h_1 + (n_2 - n_2^0)h_2 + n_1^s h_1^s + h_2^s h_2^s + H^{se}(x_1^s)$$
$$+ H^e(x_1) - H^e(x_1^0) \tag{28}$$

It is assumed in Eq. (28) that the enthalpy of the solid adsorbent is not altered by wetting. The introduction of molar fractions and the material balance $n^0(x_1^0 - x_1) = n^s(x_1^s - x_1)$ brings Eq. (28) to the following form:

$$\Delta_w H_t = n^s x_1^s (h_1^s - h_1) + n^s x_2^s (h_2^s - h_2) + H^{se}(x_1^s) + \Delta H^e(x_1) \tag{29a}$$

where $\Delta H^e = H^e(x_1) - H^e(x_1^0)$ is the change in the enthalpy of mixing of the bulk phase. When $\Delta x_1 = x_1^0 - x_1$ is known for a given liquid mixture, the function $\Delta H^e(x_1)$ can be calculated from the functions of enthalpy of mixing described in the literature. If only the change in enthalpy relative to the adsorption layer is to be calculated, the function $\Delta_w H_t = f(x_1)$ has to be corrected by the function $\Delta H^e = f(x_1)$. Introducing the volume fractions of the adsorbed layer and the heat of immersion of pure components, Eq. (29a) can be given as follows:

$$\Delta_w H_t - \Delta H^e = \phi_1^s \Delta_w H_1^0 + \phi_2^s \Delta_w H_2^s + H^{se}(x_1^s) \tag{29b}$$

In a flow replacement experiment, the integral enthalpy of displacement, by definition, is [31–33]

$$\Delta_{21} H = \sum_{x_1=0}^{x_1} (\Delta_d H - \Delta H_{\text{mix}}) = f(x_1) \tag{30}$$

where $\Delta_d H$ is the enthalpy difference when Δn_2^s mole of component 2 at the S/L interface is displaced by $\Delta n_1^s = -\Delta n_2^s/r$ mole of component 1 during a concentration step of Δx_1 [36–39]:

$$\Delta_d H = \Delta n_1^s (h_1^s - h_1^l) + \Delta n_2^s (h_2^s - h_2^l) \tag{31}$$

where h_i^s and h_i^l are the partial molar enthalpies of component i in the adsorption layer and in the bulk liquid phase, respectively. According to Király et al. it may be noted that if Δx_1 is small enough, the enthalpy of mixing [ΔH_{mix} in Eq. (30)] at the interface between the replacing and replaced solutions becomes negligible [40,41].

The integral enthalpy $(\Delta_{21} H^l)$, entropy $(\Delta_{21} S^l)$, and free enthalpy $(\Delta_{21} G^l)$ of bulk dilution provide the link between the corresponding displacement and adsorption quantities [10,12,37,39]:

$$\Delta_{21} G = \Delta_{21} G^s + \Delta_{21} G^l \tag{32}$$
$$\Delta_{21} H = \Delta_{21} H^s + \Delta_{21} H^l \tag{33}$$
$$\Delta_{21} S = \Delta_{21} S^s + \Delta_{21} S^l \tag{34}$$

$\Delta_{21} H^s$, $\Delta_{21} S^s$, and $\Delta_{21} G^s$ are the integral enthalpy, entropy, and free enthalpy of adsorption, and $\Delta_{21} H$, $\Delta_{21} S$, and $\Delta_{21} G$ are the corresponding displacement quantities. When the solute is preferentially adsorbed from ideally diluted solutions, $n_1^s \approx n_1^{\sigma(n)}$ and the bulk dilution terms are [8,9,37–39]

$$\Delta_{21} G^l = n_1^{\sigma(n)} (\Delta_{\text{sol}} \mu_1^\infty + RT \ln x_1^l) \tag{35}$$
$$\Delta_{21} H^l = -n_1^{\sigma(n)} \Delta_{\text{sol}} h_1^\infty \tag{36}$$
$$\Delta_{21} S^l = -n_1^{\sigma(n)} (\Delta_{\text{sol}} s_1^\infty - R \ln x_1^l) \tag{37}$$

The relationship between the immersion and surface excess thermodynamic quantities has been further discussed by Schay, Everett [2–5], and Woodbury and colleagues [29,30]. The standard enthalpy $\Delta_{sol}h_1^\infty$, entropy $\Delta_{sol}s_1^\infty$, and free enthalpy $\Delta_{sol}\mu_1^\infty$ of component 1 in component 2 at infinite dilution are listed in tables of solution thermodynamics. Combining Eq. (33) with Eq. (36), and Eq. (32) with Eq. (35) leads to the integral enthalpy of adsorption and the free energy of adsorption [35–39]:

$$\Delta_{21}H^s = \Delta_{21}H + n_1^{\sigma(n)}\Delta_{sol}h_1^\infty \tag{38}$$

$$\Delta_{21}G^s = \Delta_{21}G + n_1^{\sigma(n)}RT\ln x_1 + \Delta_{sol}\mu_1^\infty n_1^{\sigma(n)} \tag{39}$$

The entropy term is obtained from the enthalpy and free enthalpy of adsorption:

$$T\Delta_{21}S^s = \Delta_{21}H^s - \Delta_{21}G^s \tag{40}$$

The differential molar enthalpy of displacement, $\Delta_{21}h_1$, is obtained from Eq. (4) [10,12]:

$$\Delta_{21}h_1 = \frac{\Delta(\Delta_d H)}{\Delta n_1^s} = (h_1^s - h_1^l) - r(h_2^s - h_2^l) \tag{41}$$

For ideally dilute solutions, $h_1^l = \Delta_{sol}h_1^\infty$ and $h_2^l = 0$. Therefore, the differential molar enthalpy of adsorption, $\Delta_{21}h_1^s$, is [37–39]:

$$\Delta_{21}h_1^s = \Delta_{21}h_1 + \Delta_{sol}h_1^\infty = h_1^s - rh_2^s \tag{42}$$

IV. COMBINATION OF ADSORPTION EXCESS ISOTHERMS AND ENTHALPY ISOTHERMS: NEW WAY FOR DETERMINATION OF ADSORPTION CAPACITY

The adsorption of binary systems is described by the Ostwald–de Izaguirre equation, which establishes a relationship between the specific reduced adsorption excess amount ($n_1^{\sigma(n)}$) and the material amount in the interfacial layer ($n^s = n_1^s + n_2^s$) [34,35,42–46]:

$$n_1^{\sigma(n)} = n_1^s x_2 - n_2^s x_1 = n^s(x_1^s - x_1) \tag{43}$$

If the behavior of the bulk liquid and interfacial layer is ideal and $\Delta_{21}H^{se} = 0$ in Eq. (18), Eq. (43) may be rewritten as follows [31–33]:

$$\frac{n_1^{\sigma(n)}}{\phi_1^s - \phi_1} = \frac{n_{1,0}^s}{r[(x_1 - rx_2)]} \tag{44}$$

Because $r = n_{1,0}^s/n_{2,0}^s$,

$$\frac{n_1^{\sigma(n)}}{\phi_1^s - \phi_1} = n_{1,0}^s + (n_{2,0}^s - n_{1,0}^s)x_1 \tag{45}$$

[i.e., the right-hand side of Eq. (45) is made up exclusively of terms consisting of adsorption capacities]. On the left-hand side, the intersection of the straight line is $n_{1,0}^s$ and the slope is $(n_{2,0}^s - n_{1,0}^s)$; therefore, on the right-hand side, the intersection is $n_{2,0}^s$. If $r = 1$ and $n_{1,0}^s = n_{2,0}^s$, then the second term of Eq. (45) is zero.

Knowing the values of the integral exchange enthalpy, the adsorption capacity (n_1^s) and the molar adsorption enthalpy of the layer (h_1^s, h_2^s), a combination of Eqs. (3) and (29) and the substitution $n_1^s = n_1^{\sigma(n)} + n^s x_1$ (for an ideal adsorption layer and for the case $\Delta_{21} H^{se} = 0$) yield [17,31–33]

$$\frac{\Delta_{21} H}{n_1^{\sigma(n)}} = \frac{\Delta_w H - \Delta_w H_2^0}{n_1^{\sigma(n)}} = h_1^s - \frac{h_2^s}{r} + \frac{[n^s(h_1^s - h_2^s r)]x_1}{n_1^{\sigma(n)}} \tag{46}$$

The ideal behavior of the adsorption layer also means that h_1^s and h_2^s are independent of the concentration of the mixture, a condition that is rarely met. It is apparent from our arguments presented later, however, that molar differential exchange enthalpy is constant in a certain range of composition; our assumption has therefore to be accepted (see Fig. 6).

Equation (46) is also a linear function, the intersection of which is $b = h_1^s - h_2^s/r$, its slope is $S = n^s(h_1^s - h_2^s/r)$ (i.e., the adsorption capacity is $n^s = S/b$). Equation (46) was first applied for the interpretation of flow microcalorimetric measurements (on dilute solutions only) by Woodbury and colleagues [29,30]. If the size of the molecules is uniform, $n^s = n_{1,0}^s = n_{2,0}^s$, for the slope of Eq. (46) is given by

$$S = n_{1,0}^s(h_1^s - h_2^s/r^*) = \Delta_{21} H_{\text{total}} = \Delta_w H_1^0 - \Delta_w H_2^0 \tag{47}$$

(i.e., the slope of the equation yields the total exchange enthalpy of the adsorption displacement process). This value can also be directly determined by calorimetry.

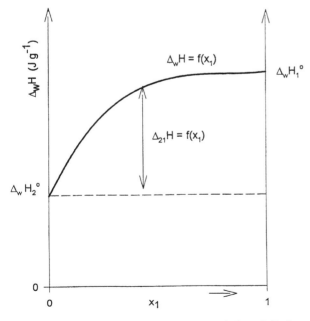

FIG. 6 Schematic representation of the enthalpy of displacement isotherm in binary mixtures in the case of U-shaped adsorption excess isotherms.

V. ADSORPTION EXCESS AND ENTHALPY ISOTHERMS ON SOLIDS IN BINARY LIQUIDS

A. U-Shaped Excess Isotherms and Enthalpy Isotherms

Calculations regarding the composition of the adsorption layer, the determination of adsorption capacity, and the heterogeneity of the surface have been discussed in several publications by Everett [5,34] Schay [2–4], Dékány, Szántó, Nagy, and Berger [10–14,18,21], and by the Polish adsorption school [35,47–49]. The energetics of the exchange (displacement) process taking place in the adsorption layer have been the subject of considerably fewer publications and these deal primarily with the adsorption of dilute solutions. The change in enthalpy accompanying the adsorption exchange process in organic media on apolar surfaces has been studied in detail by Groszek [42], Denoyel et al. [43], and Fingenegg et al. [43,45]. Studies published by Allen and Patel [36] and by Woodbury and colleagues [29,30] already include a simultaneous analysis of adsorption excesses and calorimetric data. Billett et al. [6] determined heats of immersion wetting in a system of benzene–cyclohexane/activated carbon in the entire range of mixing, parallel with the determination of adsorption excess isotherms.

In this chapter, we give a parallel analysis of the liquid sorption excess isotherms and adsorption exchange enthalpy isotherms of benzene–n-heptane and methanol–benzene mixtures on adsorbents with polar and apolar surfaces. Systems with U- or S-shaped excess isotherms were selected. Our aim is to examine the presence and character of a connection between adsorption excess and enthalpy isotherms for the various isotherm types [18]. The determination of enthalpy changes occurring in the course of flow microcalorimetric measurements absolutely necessitates a simultaneous analysis of the material balance and enthalpy balance of adsorption. According to Király and Dékány, these relationships, expounded according to either the adsorption layer model or the Gibbs model of adsorption excess amounts, are suitable for the exact determination of changes in exchange enthalpy observable in flowing systems and for the correct interpretation of the sorption exchange process [31–33,37–39].

Adsorption excess isotherms of benzene(1)–n-heptane(2) and methanol(1)–benzene(2) mixtures are shown in Figs. 7 and 8, respectively. Both isotherms are U-shaped, with the difference that in the methanol(1)–benzene(2) system, methanol is preferentially adsorbed on silica gel. Pretreatment of silica gel by methanol was necessary in order to eliminate the effect associated with the chemisorption of methanol, so that data reflecting only physical adsorption could be measured [31,32].

In the case of benzene–n-heptane mixtures, the excess isotherm on silica gel has no linear region: Parallel with increasing the concentration of component 1 in the bulk phase, the composition of the interfacial layer changes continuously (Fig. 7). This is well reflected by the enthalpy isotherm shown in Fig. 7, which also indicates a gradual, step-by-step heat exchange and, correspondingly, a gradual process of displacement. Enthalpy changes accompanying the full exchange of the components ($\Delta_{21}H_t$) are listed in Table 1. These data reveal that the full exchange of n-heptane for benzene on the surface of silica gel is an exothermic process and results in the liberation of -11.3 J/g of heat. In the case of an exchange of molecules in the reverse direction, on the other hand (see the $\bigcirc$ points in Fig. 7b), identical but endothermic heat effects are obtained.

In the case of the methanol–benzene/silica gel system, adsorption occurs and about 89–90% of the total exchangeable heat amount ($\Delta_{21}H_t$) is liberated at the initial section of the isotherms. In that range of composition where the excess isotherm of the methanol(1)–benzene(2) liquid pair is linear (Fig. 8a) (and, here, the composition of the interfacial layer is nearly constant), some heat effect of exchange is still observable (Fig. 8b). This means that the

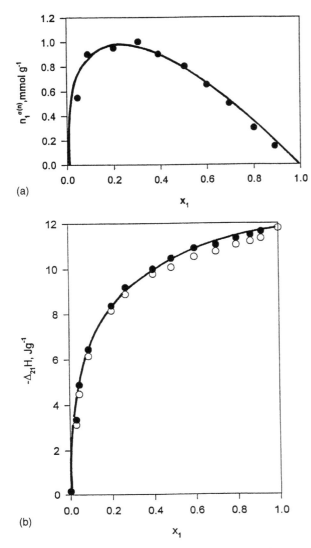

FIG. 7 Adsorption excess (a) and enthalpy of displacement (b) isotherms in the benzene(1)–n-heptane(2) mixture on silica gel ($a^s = 358$ m^2/g). (From Ref. 38.)

composition of the interfacial layer is still changing; these changes are too small to be adequately monitored by our analytical methods but are readily detected by calorimetry. This observation is direct experimental proof of a theoretical statement by Rusanov that, strictly speaking, the composition of the interfacial layer may not be constant within the linear section of the isotherm [23].

The excess isotherm determined on the surface of graphitized carbon has an inverse U shape (i.e., the adsorption excess for polar methanol is negative within the entire concentration range) (Fig. 9a). Accordingly, when the amount of methanol in the mixture is increased from $x_1 = 0$ to $x_1 = 1$, the effects measured are always endothermic, because the displacement of benzene from the surface is an energy-consuming process (Fig. 9b).

TABLE 1   Results of Analysis of Adsorption Excess and Enthalpy Isotherms on Different Porous Adsorbents

Adsorbent	Liquid mixture	$n_{1,0}^s$ (mmol/g) S.N.[a]	$n_{1,0}^s$ (mmol/g) Eq. (15)	$-\Delta_{21}H_t$ (J/g)	$-(h_1^s - h_2^s/r)$ (kJ/mol)
Silicagel	Methanol/benzene	5.31	5.14	36.71	7.10
Silicagel	Benzene/n-heptane	—	2.02	11.30	5.60
Silicagel-C$_{18}$	Methanol/benzene	4.80	4.10	9.50	4.00
Chemviron F400	Methanol/benzene	9.10	10.41	−10.52	5.31
Printex 300	Methanol/benzene	1.01	1.00	−1.23	−0.50

[a]S.N. = Schay–Nagy extrapolation method [46,50–52].

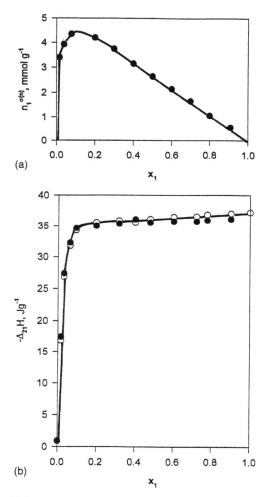

FIG. 8   Adsorption excess (a) and enthalpy of displacement (b) isotherms in the methanol(1)–benzene(2) mixture on silica gel ($a^s = 358$ m^2/g). (From Ref. 38.)

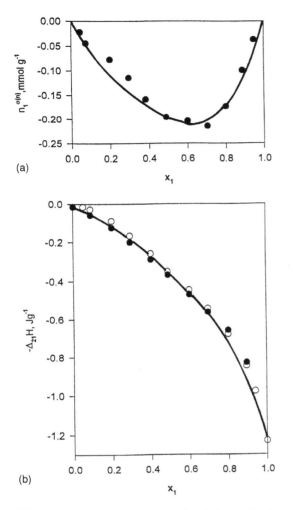

FIG. 9 Adsorption excess (a) and enthalpy of displacement (b) isotherms in the methanol(1)–benzene(2) mixture on a graphite (Printex 300) surface. (From Ref. 38.)

B. S-Shaped Excess Isotherms and Enthalpy Isotherms

An excess isotherm determined in a methanol–benzene mixture on silica gel is U-shaped. If, however, the silica surface is modified by octadecyldimethylchlorosilane (Silicagel-C_{18}) in methanol–benzene mixtures, an S-shaped excess isotherm is obtained, the azeotropic point of which is $x_1^a = 0.615$ (Fig. 10a). When the displacement process is started from benzene, it can be seen that up to a molar fraction of $x_1 = 0.6$, the incorporation of methanol into the adsorption layer results in a change in enthalpy of ~ 7.35 J/g. However, the displacement of benzene is not yet complete: Up to the azeotropic point, the adsorption layer will also contain benzene. A common characteristic of the two isotherms presented in Fig. 8 is that the isotherm $n_1^{\sigma(n)} = f(x_1)$ is linear in a very wide range of compositions ($x_1 = 0.1, \ldots, 0.7$), whereas the enthalpy isotherm $\Delta_{21}H = f(x_1)$ is approximately constant. In other words, within this range of composition, a very small exchange heat effect is produced; consequently, the composition of the layer is not constant. The maximum value of the enthalpy isotherm is at the azeotropic composition; further

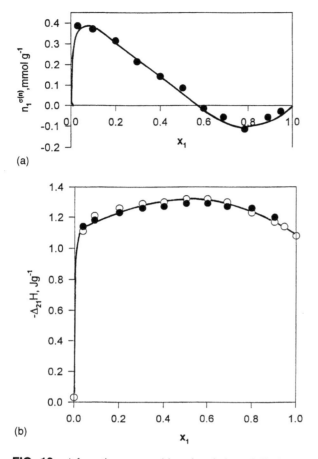

(a)

(b)

FIG. 10 Adsorption excess (a) and enthalpy of displacement (b) isotherms in methanol(1)–benzene(2) mixture on hydrophobized silica gel ($a^s = 312$ m^2/g). (From Ref. 38.)

heat effects are endothermic and the integral enthalpy isotherm therefore exhibits a decreasing tendency. Endothermic effects occurring from $x_1 = 0.6$ to $x_1 = 1$ are associated with the displacement of benzene from the adsorption layer.

The azeotropic composition of the excess isotherm determined on Chemviron activated carbon, an adsorbent with a large specific surface area, is $x_1^a = 0.095$, indicating that the adsorption layer still contains a little methanol (Fig. 11). This methanol is bound to the polar regions of the surface of the adsorbent, as shown by the exothermic heat effect detectable from $x_1 = 0$ to $x_1 = 0.1$.

If $x_1 > x_1^a$, the excess isotherm is practically linear between $x_1 = 0.1$ and $x_1 = 0.6$, the enthalpies of exchange measured are moderately endothermic. If the molar fraction of the bulk phase is larger than $x_1 = 0.8$, the displacement of benzene by methanol will preferentially occur, resulting in a considerable endothermic effect on the porous apolar surface.

When adsorption displacement proceeds from $x_1 = 1$ toward $x_1 = 0$, the heat effects detected are of a reverse sign in the case of each isotherm. This means that the processes of displacement are reversible to a close approximation. Very little irreversibility was detected in

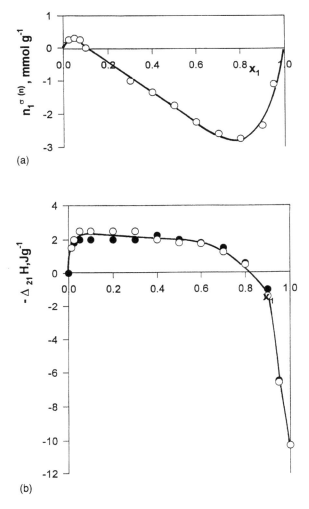

FIG. 11 Adsorption excess (a) and enthalpy of displacement (b) isotherms in methanol(1)–benzene(2) mixture on Chemviron F-400 active carbon. (From Ref. 38.)

the case of activated carbon Chemviron, which is probably caused by an incomplete displacement of methanol from the micropores by benzene.

C. The Linearized Functions $\Delta_{21}H/n_1^{\sigma(n)} = f(x_1/n_1^{\sigma(n)})$

The applicability of Eq. (46) for U-shaped excess isotherms is discussed next. The functions $\Delta_{21}H/n_1^{\sigma(n)} = f(x_1/n_1^{\sigma(n)})$ for the case of U-shaped excess isotherms are shown in Fig. 12. Adsorption capacities obtained in this representation as the ratio of the slope and the intersection are identical with the values mentioned earlier. The advantage of Eq. (46) is that the intersection yields the value of $(h_1^s - h_2^s/r)$ and this value is then multiplied by $n_{1,0}^s$, yielding the calculated value of the total integral exchange enthalpy $\Delta_{21}H_t$, which may be compared with the experimentally determined value of $\Delta_{21}H_t$ (Table 1).

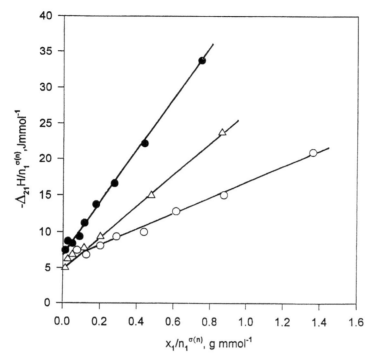

FIG. 12 Combination of the adsorption excess amounts and calorimetric data in benzene(1)–n-heptane(2) mixture (○) on silica gel ($a^s = 358$ m^2/g), in methanol(1)–benzene(2) mixture (△) on silica gel ($a^s = 458$ m^2/g), and (●) on silica gel ($a^s = 358$ m^2/g). (From Ref. 38.)

The interaction of adsorbents with various surface energies with the liquid components studied are adequately characterized by the differences in molar adsorption enthalpies between components 1 and 2 $[h_1^s - (1/r)h_2^s]$ listed in Table 1. In the case of the adsorption of the methanol–benzene liquid pair, these enthalpy differences in the adsorption layer are decreased by the effect of hydrophobization.

VI. THE CLASSIFICATION OF ENTHALPY ISOTHERMS [17–31,33]

In the case of adsorption of ideal or quasi-ideal mixtures on adsorbents with polar or apolar surfaces, usually U-shaped excess isotherms are obtained (type I according to the Schay–Nagy classification) [46,50–52]. On adsorbents with polar surfaces, in the case of liquid pairs made up of components with a large difference in polarity (e.g., alcohol–benzene), the polar component is preferentially adsorbed on the surface (type II) and, again, U-shaped excess isotherms are obtained. Thus, in these systems, the composition of the interfacial layer (Φ_1^s or x_1^s) increases monotonously as a function of equilibrium composition; consequently, according to Eq. (29b), the enthalpy isotherm $\Delta_w H$ will also be a monotonously increasing function (Fig. 13).

When alcohol–benzene mixtures are adsorbed on adsorbents with low surface energies, S-shaped excess isotherms are measured [17–19,32,33]. When changes in enthalpy accompanying the adsorption displacement process are examined in these systems, the integral exchange enthalpy isotherms usually do not increase monotonously but possess a backward section. After

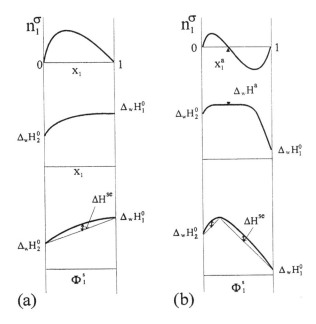

FIG. 13 Classification of the immersional wetting enthalpy isotherms: (a) U-shaped excess isotherm, with corresponding enthalpy of immersion isotherm; (b) S-shaped excess isotherm, with corresponding enthalpy of immersion isotherm.

having reached adsorption azeotropic composition, changes in concentration (Δx_1) are accompanied by endothermic heat effects and these changes do not follow changes in the composition of the surface layer (Fig. 13).

For S-shaped excess isotherms (types IV and V according to Schay and Nagy [46,50–52], is nearly constant in a relatively wide concentration range and at $n_1^{\sigma(n)} = 0$, $\Phi_1^s = \Phi_1^a = \Phi_1$ (i.e., adsorption azeotropic composition appears). In this case, the value of $\Delta_w H$ changes very little in the middle section of the enthalpy isotherm, where Φ_1^s is constant; then, at the azeotropic composition, a maximum is observed ($\Delta_w H^a$). Thus, in the case of S-shaped excess isotherms, integral enthalpy isotherms $\Delta_{21} H = f(x_1)$ have to be divided into two sections at the azeotropic composition Φ_1^a:

1. In the first section, from $x_1 = 0$ to $x_1 = x_1^a$,

$$\Delta_{21} H = \frac{\Phi_1^s \Delta_{21} H^a}{\Phi_1^a} \tag{48}$$

   This section is essentially a version of the total measurable section of U-shaped isotherms, shortened in proportion with Φ_1^a.

2. From $x_1 = x_1^a$ to $x_1 = 1$, the displacement factor can be calculated by the formula

$$\Delta_{21} H^a - \Delta_{21} H_{\text{total}} = (\Delta_{21} H^a - \Delta_{21} H)(1 - \Phi_1^a) + (\Delta_{21} H^a - \Delta_{21} H_{\text{total}})\Phi_1^s \tag{49}$$

Division of the enthalpy isotherms into two sections may be considered justified only if the adsorption layer is ideal in the vicinity of azeotropic composition (i.e., $\Delta_{21} H^{se} = 0$). Displacement enthalpy isotherms calculated according to Eqs. (47) and (48) already change parallel with the isotherms $\Phi_1^s = f(x_1)$ calculated from adsorption measurements [31–33].

VII. HEAT OF IMMERSION ON HYDROPHILIC AND PARTIALLY HYDROPHOBIC SURFACES IN DIFFERENT LIQUID MIXTURES

Adsorption excess isotherms were determined on hydrophilic and partially hydrophobic layer silicates in methanol–benzene mixtures [53–56]. From these isotherms, the free energy of adsorption $\Delta_{21}G = f(x_1)$, which is characteristic of surface polarity, is derived [53,57,58]. Displacement enthalpy isotherms were also determined by immersion microcalorimetry so that the entropy changes could be calculated. The preferential adsorption on adsorbents with different surface hydrophobicities can be properly described by the thermodynamic data of the adsorbed layer [50–52,59]. When the components in a binary mixture are very different in polarity, as they are in methanol–benzene mixtures, then the polarity of the surface can be characterized through the shape of the excess isotherms and the azeotropic composition. The free-energy function $\Delta_{21}G = f(x_1)$ calculated from the excess isotherms gives quantitative information about the decrease in free energy due to the displacement [see Eq. (5)]. The integral enthalpy isotherms can be determined with microcalorimetry and the thermodynamic description of the adsorption layer is complete [60–63]. According to Marosi and Regdon, the displacement enthalpy data give information about the solid–liquid interaction and the second main law of thermodynamics allows the calculation of the displacement entropy functions [24,25,63].

The combination of displacement free-energy and enthalpy functions with the excess isotherms gives a new way to determine the adsorption capacities [Eqs. (27) and (46)]. This combination also gives data ($\Delta_{21}g$, $\Delta_{21}h$) which describe the polarity of a surface in a certain liquid mixture [26,27,63].

A. Heat of Wetting in Amorphous Silica Dispersion and on Zeolites

Hydrophilic (A 200) and hydrophobic (R 972) varieties of amorphous SiO_2 (Aerosil derivatives, Degussa AG, Germany) were studied by immersion microcalorimetry in various liquids (methanol, benzene, n-heptane) and the results are listed in Table 2. Clearly, the heat of

TABLE 2 Heat of Wetting on Hydrophobic (A 200) and Hydrophobic (R 972) Aerosils in Different Liquids

Liquids:	A^s (BET) (m^2/g)	$-\Delta_w H$ (J/g)		
		Methanol	Benzene	n-Heptane
Adsorbents				
A 200	231	40.05	19.4	8.5
R 972	120	25.7	33.8	15.1
A 200/CH$_3$OH[a]	228	19.7	38.2	16.5
R 972/CH$_3$OH[a]	123	25.5	33.0	17.8
		$\Delta_w h$ (mJ/m^2)		
A 200	231	175	84	37
R 972	120	214	281	125
A 200/CH$_3$OH[a]	228	86	167	72
R 972/CH$_3$OH[a]	123	207	268	144

[a]Aerosil powders were pretreated with methanol before the immersion experiment.

immersion on the hydrophilic surface (A 200) is the largest in methanol; in toluene, it is moderate, and it is the smallest in n-heptane. The heat effect measured in the aromatic solvent benzene is intermediate between the two latter data. Heats of immersion per unit surface (in mJ/m^2) are also given for the various liquids [48].

The effect of surface treatment is also well detectable by the determination of heats of immersion. When the surface of the hydrophilic Aerosil is treated with methanol, heats of immersion are changed considerably, because Si−OH groups on the surface are replaced by Si−O−CH$_3$ groups. It is revealed by the data in Table 2 that the BET surface is unchanged by methanol treatment; however, the change in enthalpy of wetting per unit surface of A 200 is significantly increased by surface treatment with methanol (surface methylation) both in benzene and in n-heptane. The largest heat of wetting on the original hydrophilic SiO$_2$ surface (175 mJ/m^2) is measured in methanol; the heat effect is considerably smaller in benzene and an effect due solely to dispersion interactions (37 mJ/m^2) can be detected in n-heptane (Table 2).

Further information on heat alteration of surface energy due to dealumination and on the depolarization of the surface of zeolites may be obtained by the determination of the heat of wetting [56]. The values determined in alcohols of various chain lengths, benzene, and n-heptane are listed in Table 3. The values of specific heat of immersion ($\Delta_w H$) on Na–Y zeolite are very large in the alcohols; in benzene and n-heptane, the wetting effect measured is less significant. In each liquid, the values measured on the dealuminated sample are significantly lower than those determined on Na–Y zeolite. It is clear from the data of Table 3 that on dealuminated samples, the values of heat of wetting in benzene and in n-heptane are nearly identical and do not differ significantly from the values measured in methanol and ethanol. In contrast, it can be seen that the effects measured on the Na–Y sample in alcohols and benzene are significantly higher than those determined in apolar n-heptane.

B. Immersional Wetting on Nonswelling Clay Minerals

Clay minerals and their modified (hydrophobic) derivatives are usually readily dispersed in solvents or solvent mixtures of various polarities. The stability and structure building of these suspensions vary over a wide range, depending on the surface properties of disperse particles and the polarity of solvents. Illite as a nonswelling mineral of layer structure plays a very important role in our studies. This special role is due to the fact that both sides of the surface of the silicate lamellae are made up of SiO$_4$–tetrahedron planar lattices and this structure—even when hydrophobized—is identical with the surface structure of montmorillonite and vermiculite, both of which are of the swelling type. The swelling clay minerals (e.g., montmorillonites) are of colloid dimensions ($d < 2 \ \mu$m) and, therefore, are able to adsorb significant amounts of various

TABLE 3 Heat of Wetting on Hydrophobic (Na–Y) and Hydrophobic (Dealuminated) Zeolites in Different Liquids

Liquids	Na–Y zeolite $-\Delta_w H$ (J/g)	Dealuminated zeolite $-\Delta_w H$ (J/g)
Methanol	214.0 ± 2.4	41.4 ± 8.5
Ethanol	196.5 ± 2.0	30.1 ± 8.7
n-Propanol	190.6 ± 2.0	20.5 ± 6.7
n-Butanol	175.0 ± 2.8	16.4 ± 9.7
Benzene	129.5 ± 8.9	29.1 ± 6.5
n-Heptane	73.4 ± 7.6	32.9 ± 7.6

molecules, due to their large specific surface area. It follows from the above-mentioned structural properties that sorption processes and adsorption capacity will be basically determined by whether the mineral studied is of the swelling or nonswelling type. Several articles on the liquid sorption properties of the hydrophobic clay minerals were published from our institute in Hungary [17–19,53,54,58].

The adsorption capacities for S-shaped excess isotherms determined on nonswelling illite derivatives were analyzed by the Schay–Nagy extrapolation and the adsorption space-filling model. Figure 14 shows the excess isotherms for hydrophilic illite and three gradually organophilized hexadecylpyridinium (HDP)–illites in methanol–benzene mixtures. The amount of methanol in the adsorption layer decreases with increasing coverage by HDP cations bounded on the illite surface [53,54,58].

The excess free-energy functions, given by integration of the excess isotherms [Eqs. (23) and (24)] reflect the extent of the hydrophobization (Fig. 15). Methanol displaces benzene with a maximum change in free energy on Na–illite. The displacement process results in smaller free-energy changes on HDP-treated surfaces: The functions shown for the sample with maxima at the azeotropic compositions. The free-energy function for the sample with maximum hydrophobicity changes sign, which means that the displacement of benzene by methanol is not favored. Illites and their organophilic derivatives can be well dispersed in methanol–benzene mixtures; therefore, their wetting properties can be studied with batch microcalorimetry [13]. Thus, the solid–liquid interaction can be given as $\Delta_{21}H = \Delta_w H - \Delta_w H_2^0$; its integral isotherm is plotted in Fig. 16. The immersion wetting enthalpy is appreciable on Na–illite. The majority of

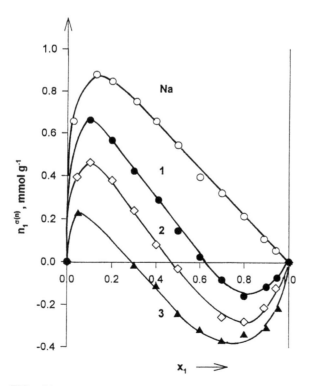

FIG. 14 Adsorption excess isotherms on Na–illite and on HDP–illite derivatives in methanol(1)–benzene(2) mixtures. Na: sodium–illite: curves 1–3: HDP illites.

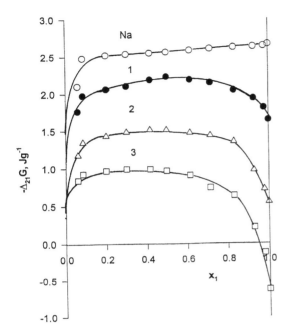

FIG. 15 Free enthalpy of adsorptioin on Na–illite and HDP–illite derivatives in methanol(1)–benzene(2) mixtures. Na: sodium–illite; curves 1–3: HDP–illites.

heat evolution is due to preferential adsorption of methanol (see Fig. 12). The enthalpy change decreases upon hydrophobization and it becomes endothermic even in $x_1 > 0.5$ compositions. The application of Eq. (46) is more favorable, because it enlightens the difference between the molar adsorption enthalpies of components (Fig. 17). The parameters of Eq. (46) give the adsorption capacity and the molar wetting enthalpy change. These data show that the change in molar wetting data decreases with increasing hydrophobicity (Table 4).

C. Heat of Wetting on Swelling Clay Minerals

In the case where the originally hydrophilic surface is modified by long alkyl chains [64–76], heats of immersion display significantly larger differences. This is well demonstrated by adsorption and x-ray diffraction measurements in toluene (Table 5). If montmorillonite, a layered silicate is hydrophobized, then, depending on the length of the alkyl chain ($n_c = 12–18$), the value of $\Delta_w H$ will decrease in polar methanol with increasing alkyl chain length. [11,48,49,77]. As shown by the data in Fig. 18, the extent of wetting by the polar solvent methanol is nearly exponentially decreased with the increasing number of carbon atoms in the alkyl chain. The decrease in heat of wetting by toluene is surprising, as the increased organophilicity of the surface must be associated with an increase in the enthalpy of immersion wetting in the aromatic solvent as was measured in the case of nonswelling HDP–illites.

The immersional wetting of hydrophobized montmorillonites in methanol, toluene, and their mixtures gives rise to three types of detector signal as presented in Figs. 19–22. The isotherm batch microcalorimetry of these hydrophobic clays in methanol yields an exothermic effect, because, in these organoclays, interlammellar swelling is not very significant

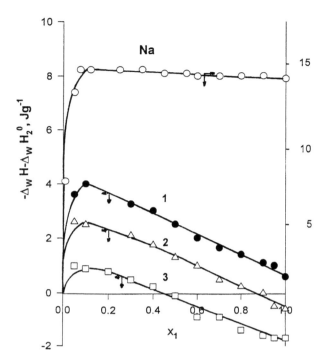

FIG. 16 Immersional wetting enthalpy isotherms on Na–illite and HDP–illite derivatives in methanol(1)–benzene(2) mixtures. Na: sodium–illite; curves 1–3: HDP–illites.

($d_L = 3.2$ nm). A significant swelling is observed in toluene; therefore, either endothermal–exothermal signals separated in time are registered within the same measurement or the wetting results in an endothermal heat effect only [19,50–52,59,60]. In clay mineral organocomplexes swelling in organic solvents, however, the amount of energy required for interlamellar expansion (in toluene, $d_L = 4.1$ nm) may be so high that the exothermic heat effect accompanying the sorption of the liquid penetrating into the interlamellar space cannot compensate it; therefore, the total heat effect is endothermic, as verified by Fig. 22.

Figure 23 shows the degree of swelling for the liquid sorption equilibrium systems [ethanol(1)–toluene(2)/hexadecylammonium vermiculite] as a function of the liquid mixture composition of the bulk phase. It can be established that within the whole series of mixtures, the value of d_L increases (i.e., the alkyl chains and the silicate layers expanded) (Fig. 24). The composition of the interfacial layer is, therefore, also indicated in Fig. 23, and it is apparent that this increase is gradual. This means that the displacement of the polar component (ethanol) from the interlamellar space leads to an increase in basal distance. The expansion or contraction of hydrophobized silicate lamellae are well reflected by enthalpy values presented in Fig. 25 as a function of the volume fraction of adsorbed toluene in the interfacial layer (Φ_2^s). It is evident that significant swelling only occurs by the enrichment of the interlamellar space of the organoclay with toluene, For the parallel investigation of adsorption and swelling, flow microcalorimetry was used to control the enthalpy of displacement ($\Delta_{21}H$) as a function of the surface layer composition. It is clear from Fig. 25 that the "opening" of the interlamellar space is an endothermic process, because the expansion and solvation of alkyl chains are entropy-driven effects.

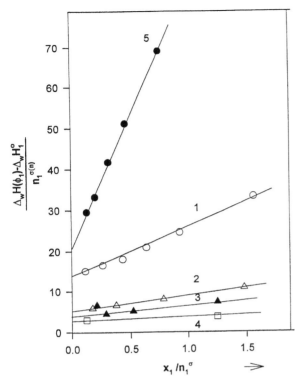

FIG. 17 Determination of the adsorption capacities from Eq. (15). Curve 1: Na–illite; curves 2–4: HDP–illites; curve 5: Na–montmorillonite in the methanol(1)–benzene(2) mixture.

D. Adsorption of *n*-Butanol from Water on Modified Silicate Surfaces

Vermiculite from South Africa was made hydrophobic by cation exchange with dodecylammonium and dodecyldiammonium chloride. Before the adsorption and microcalorimetric measurements, these organophilic vermiculites were dried in a vacuum dessicator overnight at 340 K. The butanol–water solutions were prepared by diluting the lower, butanol-saturated phase with

TABLE 4 Results of Analysis of Adsorption Excess and Enthalpy Isotherms on Different Nonswelling Clays

Adsorbent	$n_{1,0}^s$ (mmol/g)	Eq. (15) (mmol/g)	$-\Delta_{21}H_t$ (J/g)	$-(h_1^s - h_2^s/r)$ (kJ/mol)
Na–kaolinite	1.05	1.00	5.65	6.25
HDP–kaolinite[a]	2.07	1.98	0.42	1.38
Na–illite	0.84	0.85	11.10	13.65
HDP–illite 1[b]	1.25	0.95	1.70	4.32
HDP–illite 2[b]	1.30	1.18	1.35	3.21
HDP–illite 3[b]	1.42	1.38	0.85	2.27

[a]HDP–kaolinite: 0.048 mmol HDP$^+$ cation/g clay.
[b]HDP cation content: HDP–illite 1: 0.097; 2: 0.139; 3: 0.233 mmol/g clay.

TABLE 5 Interlamellar Sorption and Swelling on Hydrophobic Montmorillonites in Methanol(1)–Benzene(2) Mixtures

Organoclays	Organic cation (mmol/g)	$n_{1,0}^s$ (mmol/g)	d_L^{dry} (nm)	$d_L^{toluene}$ (nm)
Montmorillonite	0.00	3.44	1.23	1.25
TDP–montmorillonite	0.82	8.51	1.84	3.33
HDP–montmorillonite	0.85	8.25	1.82	3.86
ODP–montmorillonite	0.82	8.33	1.82	4.20
DMDH–montmorillonite	0.83	8.90	2.91	4.51

Note: TDP = tetradecylpyridinium;   HDP = hexadecylpyridinium;   ODP = octadecylpyridinium; DMDH = dimethyldihexadecylammonium.

water. At 298 K, the mole fraction of *n*-butanol in water at saturation is $x_1^{sat} = 1.877 \times 10^{-2}$ [57,77]. The adsorption excess isotherms at 298 ± 0.1 K were determined by dispersing samples of alkylammonium vermiculite in butanol–water mixtures as described earlier [20]. Because of the partial miscibility, the mole fractions x_1^0 and x_1 were replaced by $x_{1,r}^0$ and $x_{1,r}$: $x_{1,r}^0 = x_1^0/x_1^{sat}$ and $x_{1,r} = x_1/x_1^{sat}$.

The integral enthalpy of displacement of water by butanol was determined in an LKB 2107 flow sorption microcalorimeter (Bromma, Sweden) at 298 ± 0.01 K. Heats of mixing were not detected in blank runs when Teflon powder was used as an inert solid. Thus, the heat effects measured step-by-step could be directly assigned to the displacement process.

The choice of the system, alkylammonium vermiculites in butanol–water, was guided by earlier results of adsorption and calorimetric studies on graphite, which demonstrated the importance of the dilution term in interpreting the calorimetric measurements [34,36–39,42,43].

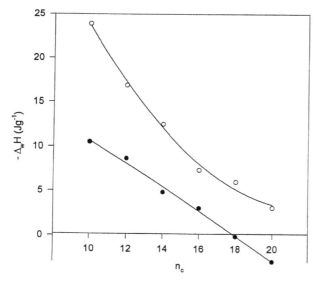

FIG. 18   Immersional wetting enthalpies as function of alkylammonium chain length in methanol (○) and toluene (●).

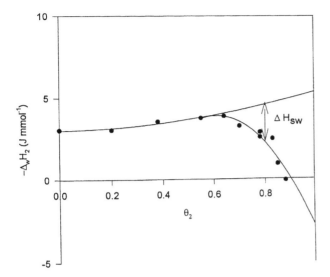

FIG. 19 Immersional wetting enthalpies on hydrophobized montmorillonite in toluene as function of surface coverage (Θ_2) with the HDP cation.

The surface excess isotherms for dodecylammonium, dodecyldiammonium vermiculite (Fig. 26), reveal that the amount of butanol adsorbed increases with $x_{1,r}$. Because the mole fraction of butanol is very small, $n_1^{\sigma(n)}$ is almost identical with the true amount of butanol adsorbed, $n_1^s \approx n_1^{\sigma(n)}$. The isotherms reflect the different interlamellar structure formed in the presence of butanol–water. The interlayer space containing dodecyl- and octadecylammonium ions widens with increasing butanol concentration (Fig. 27) because more and more liquid molecules penetrate between the layers.

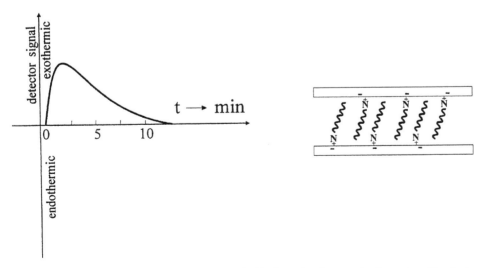

FIG. 20 Exothermic heat of wetting on hydrophobic (hexadecylammonium) montmorillonite in methanol ($d = 3.2$ nm).

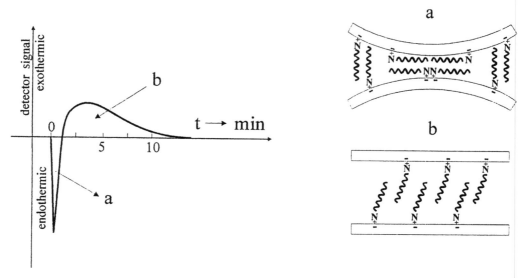

FIG. 21 Endothermic–exothermic heat of wetting on hydrophobic (hexadecylammonium) montmorillonite in toluene ($d = 3.8$ nm).

The swelling of the dodecyldiammonium vermiculite is restricted by the bridging alkyl chains (Fig. 28b). The constancy of the basal spacing indicates that the chain orientation (at an angle of 56° to the silicate layer) is independent on the composition of the liquid mixture. The free enthalpy of displacement of water by butanol, $\Delta_{21}G$, was calculated from the surface excess isotherms [Eq. (23)]. It decreases with increasing $x_{1,r}$ and reaches smaller values for the alkylammonium vermiculites (-10 and -8.4 J/g) than for the dodecyldiammonium derivative (-4 J/g) (Fig. 29).

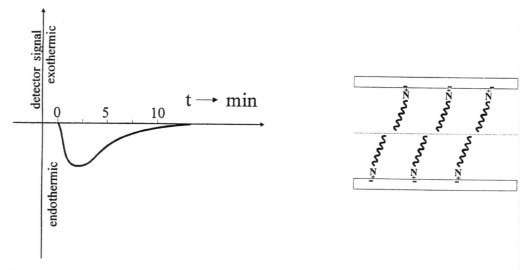

FIG. 22 Endothermic heat of wetting on hydrophobic (octadecylammonium) montmorillonite in methanol–toluene mixture $5:95$ ($d = 4.6$ nm).

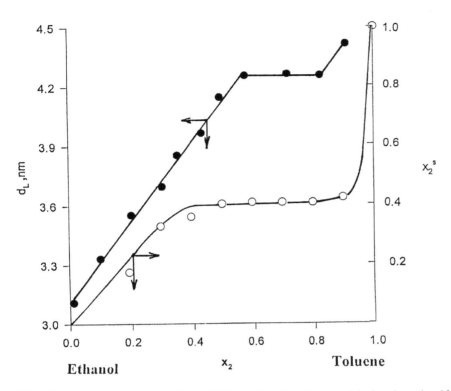

FIG. 23 The mole fraction of toluene (x_2^s) in the interfacial layer and the basal spacing (d_L) in ethanol(1)–toluene(2) mixtures on hexadecylammonium vermiculite.

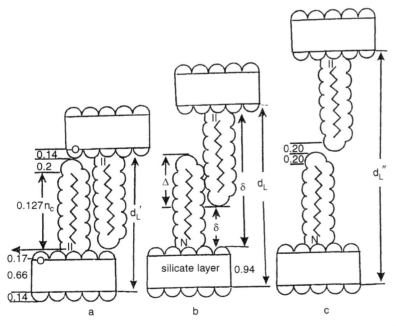

FIG. 24 Orientation of the cationic alkyl chains between silicate layers at a high layer charge density (degree of hydrophobization) at different swellings.

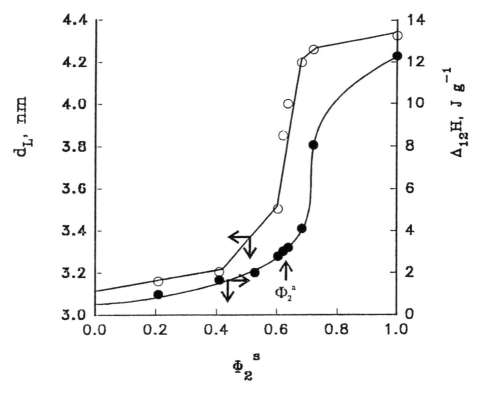

FIG. 25 Basal distance (d_L) and enthalpy of displacement at different volume fraction of toluene (ϕ_2^s) in the interfacial layer in ethanol(1)–toluene(2) mixtures on hexadecylammonium vermiculite.

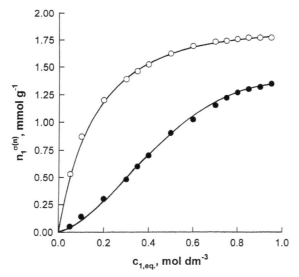

FIG. 26 Adsorption excess isotherms in n-butanol–water solutions on (○) dodecylammonium vermiculite and on (●) dodecyldiammonium vermiculite. (From Ref. 38.)

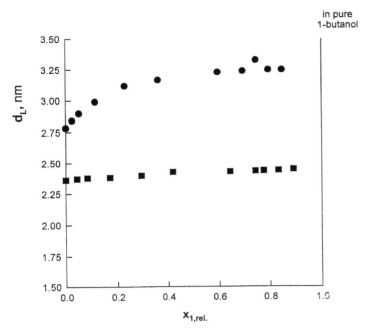

in pure
1-butanol

FIG. 27   Basal spacings of (●) dodecylammonium vermiculite and (■) dodecyldiammonium vermiculite in *n*-butanol–water solutions.

The enthalpy of the displacement process, $\Delta_{21}H$, was measured in the flow sorption microcalorimeter (Fig. 6). It is endothermic and increases to a plateau at 6 J/g (dodecyldiammonium vermiculite) and 14 J/g (octadecylammonium vermiculite). The reaction of the dodecylammonium derivative is nearly thermoneutral. When the enthalpy of displacement is corrected by the dilution term ($\Delta_{sol}h_1^{\infty} = -9.30$ kJ/mol for butanol–water [Eq. (38)]), the enthalpy of adsorption, $\Delta_{21}H^s$, becomes negative when dodecylammonium ions are interlayer cations. It approximates -16.2 J/g at $x_{1,r} \to 1$ (Figs. 30a and 30b).

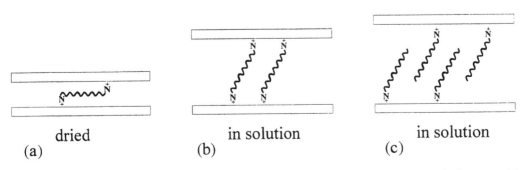

dried   in solution   in solution

(a)   (b)   (c)

FIG. 28   Schematic representation of the hydrophobic vermiculite at different basal distances: (a) monolayer, (b) "bridging", (c) bilayer orientation.

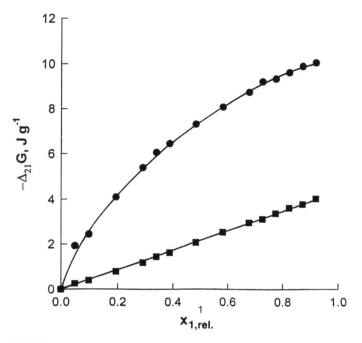

FIG. 29   Free enthalpy of adsorption in *n*-butanol–water solutions on (●) dodecylammonium vermiculite and on (■) dodecyldiammonium vermiculite.

VIII.   MULTILAYER ADSORPTION IN BINARY LIQUID MIXTURES: CALCULATION OF THERMODYNAMICAL POTENTIAL FUNCTION PROFILES BASED ON THE LATTICE MODEL *(F. Berger)*

A.   Introduction

Theoretical and experimental investigations of multimolecular adsorption were started in the 1930s by Brunauer et al. [58], Harkins and Jura [53], Frenkel [54], Halsey [55] and Hill [56]. The field is still being researched; see, for example, the work published by Cerofolini and Meda [78]. Research efforts have been focused mostly on multilayer adsorption on S/G interfaces, but the results obtained also have bearings of basic importance on multilayer adsorption on the S/L interface.

Knowledge of the composition and thickness of the adsorption layer at the S/L interface is essential not only for the description of adsorption but also for the analysis of related phenomena like the stability of disperse systems, the magnitude of forces operating between particles or surfaces, the structure and rheological behavior of suspensions, and so forth [79–81]. From factors influencing the above-mentioned phenomena, the adsorption of electrolytes has been investigated the most intensively, whereas that of nonelectrolytes has been paid less attention in this respect. Classical models describing liquid-phase adsorption, developed by Vold [82], Kipling [83], Everett [84], Schay and Nagy [85,86], Vincent [87], and Ruduzinski et al. [88] assume monomolecular adsorption. By applying Rusanov's inequality [89], Tóth [64,90,91], Brown et al. [65], and Oscik et al. [66] showed that, in the cases they investigated, the adsorption layer certainly consisted of more than one layer of molecules. Theoretical studies of multilayer liquid adsorption on heterogeneous surfaces have been undertaken by Jarionec et al. [67] and

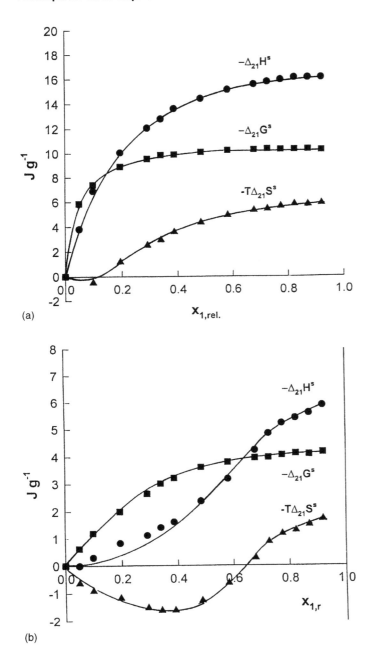

FIG. 30 (a) Thermodynamic potential functions of the adsorption layer on dodecylammonium-vermiculite in *n*-butanol–water solutions. (b) Thermodynamic potential functions of the adsorption layer on dodecyldiammonium vermiculite in *n*-butanol–water solutions.

Dabrowski et al. [68]. Ash and Findenegg have combined vapor and liquid adsorption data for the determination of layer thickness [69,70]. A detailed summary of theoretical and experimental studies has been compiled by Dabrowski et al. [71].

In our previous publications, the concentration dependence of equivalent layer thickness (see Fig. 32b) at S/L interfaces was calculated on the basis of the athermal parallel layer model [72,73]. Even though the accuracy of the results has been verified by independent measurements, our assumption constituting the basis of our calculations (i.e., that the free energy of adsorption $\Delta_{21}F$ is mostly concentrated in the molecular layer adjoining the surface) still cannot be considered fully proven. Calculations for the determination of the concentration profile prevailing within the adsorption layer—but not for the course of thermodynamic functions— have been published in the literature. The latter is one of the main objectives of our present work. We also address the question whether the course of the function $\Delta_{21}F$ allows conclusions to be drawn regarding the formation of a multimolecular adsorption layer. It is also unclear what conclusions can be drawn from the function $\Delta_{21}U$, the internal energy of adsorption determinable by microcalorimetry, and from the function $\Delta_{21}S$, the entropy of adsorption, calculated by combining the latter two.

There are basically two reasons why multimolecular adsorption layers are formed on the interface of binary mixtures: (1) the molecules tend to align and (2) the mixture is strongly nonideal and has a certain tendency for demixing. The subject of the present work is the latter case (i.e., multimolecular adsorption of the surface enrichment type).

Numerous mathematical models exist; the ones to be considered, however, are only those which are not of the empirical or semiempirical type: the assumptions of which appear correct from a thermodynamical and statistical–mechanical point of view; and the numerical solution of which does not present an unrealistically large computational task even in the case of thick adsorption layers. One of these is the so-called lattice model. A version of this model suitable for the description of the structure and thermodynamical properties of multimolecular adsorption layers was used in our calculations.

B. The Lattice Model and Its Application to Multimolecular Adsorption

The "lattice model," also called the "cell model," takes its origin in statistical mechanics and has a prolonged history [74]. It was first applied to the thermodynamical and statistical mechanical description of binary mixtures and the determination of activity functions in the bulk phase [75]. It was later used for the description of the surface tension of binary mixtures [76].

The lattice model makes the following simplifying assumptions:

1. The molecules making up the binary mixture are unstructured, hard spheres.
2. Molecules of the different components are of identical size.
3. The molecules are arranged in a quasicrystalline lattice structure devoid of defects. The face-centered-cubic (fcc) lattice structure with a coordination number of $z = 12$ is used nearly exclusively in the literature and it is also adopted in this work (Fig. 31). The molecules are aligned in sublayers parallel with the surface (Fig. 32a), within which they are fitted in a tight hexagonal pattern (Fig. 33). The cross-sectional area occupied by one molecule will be termed a "site."
4. Intermolecular and surface–molecule interactions are exclusively of the dispersion type.
5. The surface is an energetically homogeneous, infinite planar surface.
6. Only nearest-neighbor interactions between molecule and molecule or molecule and surface are taken into consideration; long-range interactions are neglected. The basis

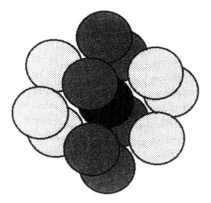

FIG. 31 Close-packed (fcc) alignment of molecules in the bulk phase and in the adsorption layer. A chosen molecule (black) has 12 neighbors, 6 in the same plane (sublayer), and 3 and 3 in the neighboring planes (sublayers).

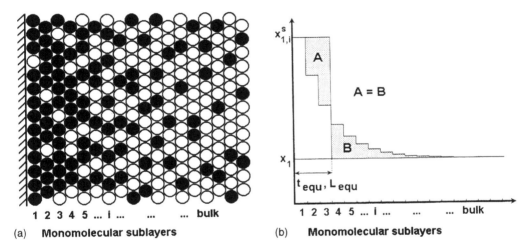

(a) **Monomolecular sublayers** (b) **Monomolecular sublayers**

FIG. 32 (a) Schematic structure of the adsorption layer. (b) Schematic representation of the $t_{equ.}$ equivalent layer thickness and the $L_{equ.}$ equivalent sublayer number.

FIG. 33 One "site" (cross-sectional area) on the surface and in the sublayers.

of this simplification is that, in the case of spherical molecules, the attraction potential of the dispersion interaction decreases according to the sixth power of the distance, whereas attraction between a planar surface and a spherical molecule is inversely proportional to the third power of the distance separating them. The effect of this neglection is considerable only if the system is in such a state that it gives a large-scale reaction even to a slight effect.

7. It is rarely pointed out in the literature that typically only two-body dispersion interactions are taken into consideration and three-body and higher terms are neglected.

In accordance with the above-enumerated assumptions, intermolecular interactions are characterized by the pair interaction energies ε_{11}, ε_{22}, and ε_{12}, whereas interaction between molecules and the surface is described by the interaction energies ε_{1s} and ε_{2s} (Fig. 34). The sign of all these quantities is traditionally defined as nonnegative. It will be shown, however, that—instead of five parameters—two suitably derived parameters are sufficient for the unambiguous qualification of liquid and surface from the point of view of adsorption.

The definition of the interaction parameter α is well known [33]:

$$\alpha \equiv 12\left[\frac{\varepsilon_{11} + \varepsilon_{22}}{2} - \varepsilon_{12}\right] \tag{50}$$

It follows from the theory of dispersion interaction that $\alpha \geqslant 0$ (i.e., the mixing of components 1 and 2 possessing the above-mentioned properties) is never exothermic. (Calculations using negative values for α still occur occasionally in the literature.) Given the value of α, the equations describing the thermodynamical behavior of the mixture are easily obtained.

The individual variants of the lattice model differ from each other in the way the spatial distribution of the molecules of the individual components is taken into account. The simplest solution is the Bragg–Williams (B–W) approach which assumes a random distribution of molecules within the bulk phase. The thermodynamical meaning of this assumption is that the mixture is *regular*. In the adsorption layer, however, it is only in two dimensions (i.e., within the individual sublayers that a statistical distribution of molecules is assumed). Pioneering work in this field was published by Ono [92–94] and Ono and Kondo [95,96]. The method was later applied to the description of L/G interfaces by Lane and Johnson [97] and later taken up by Altenberger and Stecki [98]. Analytic isotherm equations have also been derived from the above

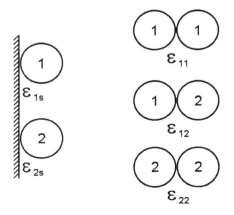

FIG. 34 Intermolecular and surface–molecule interaction energies.

model, first for monomolecular and bimolecular adsorption layers by Dabrowski et al. [99] and, most recently, for monomolecular adsorption layers by Aranovich and Donohue [100]. The latter authors also used it for the interpretation of adsorption hysteresis [101–102]. A more advanced method is the quasichemical (QC) approach, weighting the probability of the establishment of the molecular interactions 1–1, 2–2, and 1–2 on the basis of the individual pair interaction energies [75,103]. It is revealed by the comparative calculations of Lane [104] that there is relatively little difference between the results of the two methods: the same result is obtained at slightly different values of α. The quasichemical method, being mathematically more complicated, has not been widely adopted.

The main fault of the B–W and QC methods is that, on the one hand, they are unable to allow for the formation of molecular associations (clusters) consisting of more than two molecules, and, on the other hand, in certain cases the equations obtained have multiple solutions. Methods free of these faults are the Monte Carlo (MC) methods, among them canonical ensemble Monte Carlo (CEMC) methods which still retain lattice structure and grand canonical ensemble Monte Carlo (GCEMC) methods assuming a statistical spatial distribution of molecules. Both types automatically involve statistical fluctuations. MC methods have been used mainly for the description of adsorption on S/G interfaces; for in the case of S/L interfaces—especially in thick layers—this type of method defines a "box" containing a large number of molecules (i.e., very large-scale computations are necessary). A further disadvantage is that the relative error due to fluctuations is inversely proportional to the square root of computation time; therefore, MC methods are, at present, difficult to apply to calculations of fine effects like attraction and repulsion forces between relatively distant surfaces. These remarks are all the more true of more advanced molecular–dynamical calculations (MD) assuming elastic molecules, mostly by the application of Lennard–Jones potential functions. Diffusion in liquids is slow, significantly retarding the establishment of equilibrium in the course of MD calculations. Details of MC and MD calculations are not expounded here; the reader is referred to pertinent reviews [105,106].

In view of the above points, the B–W approach was adopted. This version of the lattice model is suited for the description of a wide range of adsorption phenomena. It still has not gained due appreciation in the literature, which, in our opinion, can be ascribed mainly to the statistical treatment, unfamiliar for the majority of specialists working in the field of adsorption research. In the present work, an attempt is made to circumvent this difficulty, as the B–W approach is simple enough to make such an attempt possible. Only classical thermodynamical terms combined with simple formulas of probability calculation will be used [107,108].

C. The Bragg–Williams Approach

It is assumed that each molecule of the mixture forms 12 interactions, "bonds." If the internal energy of molecules in vacuum is assumed to be zero at the given temperature, the partial internal energy of one molecule of one or the other pure component in liquid phase will be

$$U_1^0 = -6\varepsilon_{11} \tag{51a}$$
$$U_2^0 = -6\varepsilon_{22} \tag{51b}$$

In the case of binary mixtures, due to the assumedly statistical distribution of molecules the molar fractions x_1 and x_2 also mean the probability of a randomly chosen molecule being of component 1 or 2. The probability of the interaction of two neighboring molecules being of the type 1–1, 1–2 or 2–1 and 2–2 will be x_1^2, $x_1x_2 = x_2x_1$ and x_2^2, respectively. Thus, the average internal energy U of a single molecule of the mixture is

$$U = 6(x_1^2\varepsilon_{11} + 2x_1x_2\varepsilon_{12} + x_2^2\varepsilon_{22}) = x_1 U_1^0 + x_2 U_2^0 + \Delta U_{mix} \tag{52}$$

where ΔU_{mix} is the average heat of mixing per molecule and

$$\Delta U_{mix} = x_1 x_2 \alpha \tag{53}$$

The average heat of mixing per bond (w) will also be an important quantity in the discussion below:

$$w = (2x_1 x_2)\frac{\alpha}{12} \tag{54}$$

If n is the number of molecules and u is the total internal energy, the partial internal energies of the components, U_1 and U_2, can be expressed in the following way:

$$u = nU = n_1 U_1^0 + n_2 U_2^0 + n x_1 x_2 \alpha = n_1 U_1^0 + n_2 U_2^0 + \frac{n_1 n_2}{n_1 + n_2}\alpha \tag{55}$$

$$U_1 \equiv \frac{\partial u}{\partial n_1} = U_1^0 + x_2^2 \alpha \tag{56a}$$

$$U_2 \equiv \frac{\partial u}{\partial n_2} = U_2^0 + x_1^2 \alpha \tag{56b}$$

In binary mixtures, one form of the Gibbs–Duhem equation is

$$\frac{\partial U}{\partial x_1} = U_1 - U_2 \tag{57}$$

The partial (molecular) entropies S_1 and S_2 of the components of the regular mixture are

$$S_1 = S_1^0 - k \ln(x_1) \tag{58a}$$
$$S_2 = S_2^0 - k \ln(x_2) \tag{58b}$$

(k is the Boltzmann constant).

The corresponding partial (molecular) free energies F_1 and F_2 are

$$F_1 = U_1 - TS_1 = F_1^0 + x_2^2 \alpha + kT \ln(x_1) \tag{59a}$$
$$F_2 = U_2 - TS_2 = F_2^0 + x_1^2 \alpha + kT \ln(x_2) \tag{59b}$$

The specific free energy F of the mixture is

$$F = x_1 F_1 + x_2 F_2 = x_1 F_1^0 + x_2 F_2^0 + x_1 x_2 \alpha + kT[x_1 \ln(x_1) + x_2 \ln(x_2)] \tag{60}$$

and the change in the specific free energy of mixing ΔF_{mix} is

$$\Delta F_{mix} = x_1 x_2 \alpha + kT[x_1 \ln(x_1) + x_2 \ln(x_2)] \tag{61}$$

in which the specific entropy of mixing ΔS_{mix} is also recognizable:

$$\Delta S_{mix} = k[x_1 \ln(x_1) + x_2 \ln(x_2)] \tag{62}$$

Instead of partial free enthalpies, the corresponding partial free energies may also be used for expressing the activities $x_1\gamma_1$ and $x_2\gamma_2$ and the activity coefficients γ_1 and γ_2:

$$\ln(x_1\gamma_1) = \frac{F_1 - F_1^0}{kT} = \frac{x_2^2\alpha}{kT} + \ln(x_1) \tag{63a}$$

$$\ln(x_2\gamma_2) = \frac{F_2 - F_2^0}{kT} = \frac{x_1^2\alpha}{kT} + \ln(x_2) \tag{63b}$$

$$\ln(\gamma_1) = \frac{x_2^2\alpha}{kT} \tag{64a}$$

$$\ln(\gamma_2) = \frac{x_1^2\alpha}{kT} \tag{64b}$$

In the previous $(i-1)$st sublayer, six neighbors in its own (i)th sublayer and three neighbors in the following $(i+1)$st sublayer (Fig. 35). Each sublayer has a certain composition and the adsorption layer consists of the totality of sublayers with compositions significantly different from that of the bulk phase. The first so-called contact sublayer is of special importance and is different from other sublayers. The other sublayers will be termed sublayers of an "intermediate position." For the sake of simplicity, these will be discussed first.

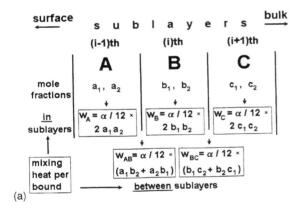

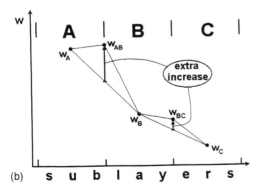

FIG. 35 (a) Mixing heat per "bound" *in* and *between* the sublayers. (b) The origin of *gradient energy*: extra increase of mixing heat per "bound" between sublayers of different composition.

D. Thermodynamical Description of Sublayers of Intermediate Position

In order to avoid complicated indexes, the following notation system will be provisionally introduced: the $(i-1)$st, ith, and $(i+1)$st sublayers will be denoted A, B, and C, respectively, with compositions of a_1, b_1, and c_1, respectively (i.e., $a_1 = x_{1,i-1}^s$, $b_1 = x_{1,i}^s$, and $c_1 = x_{1,i+1}^s$. Let us focus our attention on the ith sublayer (i.e. sublayer B, which is flanked by sublayers A and C. When sublayers of different compositions touch, thermodynamical relationships (especially those involving internal energy) obtained for the bulk phase are altered. Changes in the average energy of mixing per bond (w) deserve special attention. The formula obtained for the bulk phase [Eq. (54)] may be adapted to the average energy of mixing per bond *within* the sublayers:

$$w_A = (2a_1 a_2)\frac{\alpha}{12} \tag{65a}$$

$$w_B = (2b_1 b_2)\frac{\alpha}{12} \tag{65b}$$

$$w_C = (2c_1 c_2)\frac{\alpha}{12} \tag{65c}$$

This is not the case between the layers. There are three bonds formed on the cross-sectional area corresponding to a "site." Within layer A, the probability of a given molecule being of component 1 or 2 is a_1 and a_2, respectively. Likewise the probability of an adjacent molecule in layer B being of component 1 or 2 is b_1 and b_2, respectively. Thus, the energy of mixing of the two layers calculated per bond is

$$w_{AB} = (a_1 b_2 - a_2 b_1)\frac{\alpha}{12} = [a_1 a_2 + b_1 b_2 + (a_1 - b_2)^2]\frac{\alpha}{12}$$
$$= \frac{w_A + w_B}{2} + (a_1 - b_1)^2 \frac{\alpha}{12} \tag{66}$$

Similarly, between sublayers B and C,

$$w_{BC} = \frac{w_B + w_C}{2} + (b_1 - c_1)^2 \frac{\alpha}{12} \tag{67}$$

If the individual sublayers behaved like quasibulk phases with regard to their internal energy, the energy of mixing of the two layers calculated per bond should be the arithmetical mean of the values valid within the layers. However, it is necessary to introduce an extra energy term proportional to the *square of the concentration gradient*. This energy increment will be hereinafter designated the *gradient energy term*. There are three bonds formed between two adjacent sublayers on the section of one "site"; therefore, the specific value of the gradient energy terms between the sublayers A and B or between B and C are

$$U_{g(AB)} = (a_1 - b_1)^2 \frac{\alpha}{4} \tag{68a}$$

$$U_{g(BC)} = (b_1 - c_1)^2 \frac{\alpha}{4} \tag{68b}$$

The sum of gradient energy terms within the entire adsorption layer is called the *gradient energy*. As shown in this chapter, the structure of the adsorption layer is basically determined by gradient energy and its distribution.

It is the internal energy—more exactly, changes in the internal energy—of only the *entire* adsorption layer that may be determined without making arbitrary assumptions. The individual sublayers constitute a *coupled system*, rendering the assignment of internal energy to sublayers difficult. The following procedure is proposed for the *partition* of internal energy:

1. Knowing the composition of the sublayer, the specific energy of the sublayer is calculated as if it were a quasibulk phase.
2. The gradient energy term introduced as a consequence of the nonidentical concentrations of the adjoining sublayers is calculated, *arbitrarily divided* to two equal parts and assigned to the adjacent sublayers.

Based on the above principles and Eqs. (52), (53), (68a), and (68b), the specific internal energy (i.e., the internal energy per molecule) in sublayer B is

$$U_B = b_1 U_1^0 + b_2 U_2^0 + b_1 b_2 \alpha + \frac{\alpha}{8}[(a_1 - b_1)^2 + (b_1 - c_1)^2] \tag{69}$$

The description of the adsorption equilibrium necessitates the determination of the partial internal energies $U_{1,B}$ and $U_{2,B}$ of the components within the sublayer. The definition constructed for the bulk phase [Eqs. (56a) and (56b)] does not hold in this case, as the total material content of the sublayers is strictly equal and cannot be altered independently. It still holds, however, that

$$U_B = b_1 U_{1,B} + b_2 U_{2,B} \tag{70}$$

Equation (57) also holds:

$$\frac{\partial U}{\partial b_1} = U_{1,B} - U_{2,B} \tag{71}$$

where U is the internal energy of the entire adsorption layer calculated for the cross-sectional area of one "site." However, a change in U affects only layer B and the *totality* of the gradient energy terms arising on both of its sides. The specific energy of layer B, expanded accordingly (i.e., U_B^+), is

$$U_B^+ = b_1 U_1^0 + b_2 U_2^0 + b_1 b_2 \alpha + \frac{\alpha}{4}[(a_1 - b_1)^2 + (b_1 - c_1)^2] \tag{72}$$

Differentiated with respect to b_1,

$$\frac{\partial U_B^+}{\partial b_1} = U_{1,B} - U_{2B} = U_1^0 - U_2^0 + (b_2 - b_1)\alpha + \frac{\alpha}{2}[2b_1 - a_1 - c_1] \tag{73}$$

The solution of the system of Eqs. (70)–(73) is

$$U_{1,B} = U_1^0 + \frac{\alpha}{8}[(a_2 + b_2)^2 + (b_2 + c_2)^2] \tag{74a}$$

$$U_{2,B} = U_2^0 + \frac{\alpha}{8}[(a_1 + b_1)^2 + (b_1 + c_1)^2] \tag{74b}$$

For comparative calculations, it would be more practicable to replace the specific sublayer energy U_B by the specific adsorption energy $\Delta_{21} U$ describing changes in U_B, or the fraction of $\Delta_{21} U$ assigned to sublayer B which can be calculated according to the definition of $\Delta_{21} U$, based on partial energy levels in the layer and in the bulk phase:

$$\Delta_{21} U_B \equiv b_1(U_{1,B} - U_1) + b_2(U_{2,B} - U_2) \tag{75}$$

where

$$U_{1,B} - U_1 = \frac{\alpha}{8}[(a_2 + b_2)^2 + (b_2 + c_2)^2] - \alpha x_2^2 \tag{76a}$$

$$U_{2,B} - U_2 = \frac{\alpha}{8}[(a_1 + b_1)^2 + (b_1 + c_1)^2] - \alpha x_1^2 \tag{76b}$$

The entropy S_B and the partial entropies $S_{1,B}$ and $S_{2,B}$ of sublayer B are easily calculated by analogy with the bulk phase:

$$S_{1,B} = S_1^0 - k\ln(b_1) \tag{77a}$$

$$S_{2,B} = S_2^0 - k\ln(b_2) \tag{77b}$$

$$S_B = b_1 S_{1,B} + b_2 S_{2,B} \tag{78}$$

The corresponding adsorption entropy is

$$\Delta_{21}S_B = b_1(S_{1,B} - S_1) + b_2(S_{2,B} - S_2) \tag{79}$$

Partial free energies and the specific free energy of the sublayer are

$$F_{1,B} = U_{1,B}T - S_{1,B} \tag{80a}$$

$$F_{2,B} = U_{2,B}T - S_{2,B} \tag{80b}$$

$$F_B = b_1 F_{1,B} + b_2 F_{2,B} \tag{81}$$

The fraction of specific free energy of adsorption calculated per sublayer is

$$\Delta_{21}F_B = b_1(F_{1,B} - F_1) + b_2(F_{2,B} - F_2) \tag{82}$$

$$\Delta_{21}F_B = \Delta_{21}U_B - T\Delta_{21}S_B \tag{83}$$

E. Exchange Adsorption Equilibrium in the Intermediate Sublayers

Sublayers positioned between the contact sublayer and the bulk phase are termed intermediate sublayers. Equilibrium of the adsorption layer with the bulk phase means that, at a given bulk composition, the value of the free energy of adsorption $\Delta_{21}F$ of the entire adsorption layer characterizing the exchange process, calculated per cross-sectional area of a "site," is at a minimum. The necessary condition of this is that the value of $\Delta_{21}F$ be at a minimum also with respect to the composition b_1 of any particular intermediate sublayer B. As has been shown earlier, the task may be narrowed down to monitoring the value $\Delta_{21}F_B^+$ in a broader vicinity of layer B.

Thus, the condition of the equilibrium is

$$\frac{\partial\Delta_{21}F_B^+}{\partial b_1} = (F_{1,B} - F_1)(F_{2,B} - F_2) = 0 \tag{84}$$

In more detail,

$$(U_{1,B} + kT\ln b_1) - (-U_1 + kT\ln x_1) - (-U_{2,B} + kT\ln b_2) + (U_2 + kT\ln x_2) = 0 \tag{85}$$

It is practicable to make further conversions:

$$(U_{1,B} - U_1 + kT\ln b_1) - kT\ln x_1 - (-U_{2,B}U_2 + kT\ln b_2) + kT\ln x_2 = 0 \tag{86}$$

Using Eqs. (76a) and (76b),

$$\left(\frac{\alpha}{kT}\frac{(a_2 + b_2)^2 + (b_2 + c_2)^2}{8} + \ln b_1\right) - \left(\frac{\alpha}{kT}x_2^2 + \ln x_1\right)$$
$$-\left(\frac{\alpha}{kT}\frac{(a_1 + b_1)^2(b_1 + c_1)^2}{8} + \ln b_2\right) + \left(\frac{\alpha}{kT}x_1^2 + \ln x_2\right) = 0 \tag{87}$$

The equation above is the logarithmic form of the following equilibrium ratio:

$$\frac{(\gamma_{1,B}b_1)(\gamma_2 x_2)}{(\gamma_{2,B}b_2)(\gamma_1 x_1)} \equiv \frac{(\gamma_{1,i}^s x_{1,i}^s)(\gamma_2 x_2)}{(\gamma_{2,i}^s x_{2,i}^s)(\gamma_1 x_1)} = 1 \tag{88}$$

where $\gamma_{1,B}$ and $\gamma_{2,B}$ are the activity coefficients of the components in intermediate sublayer B. The fact that the value of the equilibrium coefficient is 1 essentially means that the surface exerts no *direct* effect on the thermodynamical properties of the sublayer. Interestingly, at locations with steep concentration gradients the value of activity within the layer may exceed 1.

The value of b_1 may not be analytically determined from Eq. (87); a numerical search for the zero place should be performed.

F. Thermodynamical Characterization of the Contact Sublayer

The terminology introduced for the description of intermediate sublayers is retained, with the exception that sublayer A is replaced by the surface. Sublayer B will thus be the contact sublayer.

The state of the first sublayer adjoining the surface differs from that of other sublayers in three respects:

1. Only three-quarters of the force field of the molecules is directed toward the adjacent molecules.
2. One of the gradient energy terms is missing.
3. Interaction with the surface.

In pure component 1 or 2, the adsorption layer consists solely of the contact sublayer, the specific energy of which differs from that of the bulk phase only in the specific heat of wetting:

$$U_{1,B}^0 = U_1^0 + \Delta_w U_1^0, \qquad U_{2,B}^0 = U_2^0 + \Delta_w U_2^0 \tag{89}$$

Specific heats of wetting are in a simple relationship with interaction energies:

$$\Delta_w U_1^0 = \tfrac{3}{2}\varepsilon_{11} - \varepsilon_{1s} = -\tfrac{1}{4}U_1^0 - \varepsilon_{1s} \tag{90a}$$

$$\Delta_w U_2^0 = \tfrac{3}{2}\varepsilon_{22} - \varepsilon_{2s} = -\tfrac{1}{4}U_2^0 - \varepsilon_{2s} \tag{90b}$$

In mixtures, specific layer energies prevailing in the pure component are weighted by the coverages b_1 and b_2. The effect of the molecules' having lost one-fourth of their force field due to the interaction with the surface also manifests itself in the term for energy of mixing. Thus, the specific energy of the contact sublayer is

$$U_B = b_1(U_1^0 + \Delta_w U_1^0) + b_2(U_2^0 + \Delta_w U_2^0) + \tfrac{3}{4}\alpha b_1 b_2 + \frac{\alpha}{8}(b_1 - c_1)^2 \tag{91}$$

Knowing the specific energy of the sublayer, the partial energies of the components are determined in the same way as was done in the case of intermediate sublayers. The solution of the system of equations is

$$U_{1,B} = U_1^0 + \Delta_w U_1^0 + \frac{\alpha}{8}[2b_2^2 + (b_2 + c_2)^2] \tag{92a}$$

$$U_{2,B} = U_2^0 + \Delta_w U_2^0 + \frac{\alpha}{8}[2b_1^2 + (b_1 + c_1)^2] \tag{92b}$$

Here, too, it is practicable to use the fraction $\Delta_{21}U_B$ of specific adsorption energy calculated per sublayer B rather than the specific sublayer energy U_B. By definition,

$$\Delta_{21}U_B = b_1(U_{1,B} - U_1) + b_2(U_{2,B} - U_2) - \Delta_w U_2^0 \tag{93}$$

where

$$U_{1,B} - U_1 = \Delta_w U_1^0 + \frac{\alpha}{8}[2b_2^2 + (b_2 +_c 2)^2] - \alpha x_2^2 \tag{94a}$$

$$U_{2,B} - U_2 = \Delta_w U_2^0 + \frac{\alpha}{8}[2b_1^2 + (b_1 + c_1)^2] - \alpha x_1^2 \tag{94b}$$

The equation describing the value of $\Delta_{21} U_B$ may be reduced when the definition of the so-called total heat adsorption $\Delta_{21} U_t$ is taken into consideration:

$$\Delta_{21} U_t = \Delta_w U_1^0 - \Delta_w U_2^0 \tag{95}$$

$$\Delta_{21} U_B = b_1 \Delta_{21} U_t + b_1 \left(\frac{\alpha}{8}[2b_2^2 + (b_2 + c_2)^2] - \alpha x_2^2 \right)$$
$$+ b_2 \left(\frac{\alpha}{8}[2b_1^2 + (b_1 + c_1)^2] - \alpha x_1^2 \right) \tag{96}$$

Entropy of adsorption and free energy of adsorption may subsequently be calculated by Eqs. (77a), (77b), (79), and (83) given for intermediate sublayers.

G. Adsorption Equilibrium in the Contact Sublayer

Again, the equilibrium of the contact sublayer with the bulk phase means that the value of the free energy of adsorption $\Delta_{21} F$ characterizing the exchange process, practicably replaced by the value of $\Delta_{21} F_B^+$ defined in the broader vicinity of the contact sublayer B, should be minimal. Thus, the condition of equilibrium is the same:

$$\frac{\partial \Delta_{21} F_B^+}{\partial b_1} = (F_{1,B} - F_1) - (F_{2,B} - F_2) = 0 \tag{97}$$

In more detail,

$$(U_{1,B} + kT \ln b_1) - (-U_1 + kT \ln x_1) - (U_{2,B} + kT \ln b_2) + (U_2 + kT \ln x_2) = 0 \tag{98}$$

After further reduction,

$$(U_{1,B} - U_1 - \Delta_w U_1^0 + kT \ln b_1) - kT \ln x_1 - (U_{2,B} - U_2 - \Delta_w U_2^0 + kT \ln b_2)$$
$$+ kT \ln x_2 + \Delta_{21} U_t = 0 \tag{99}$$

Furthermore, using Eqs. (94a) and (94b), the fact that in the lattice model, $\Delta_{21} F_t = \Delta_{21} U_t$, because $\Delta_{21} S_t = 0$ [see Eqs. (58a), (58b), (77a), (77b), and (79)]:

$$\left(\frac{\alpha}{kT} \frac{2b_2^2 + (b_2 + c_2)^2}{8} + \ln b_1 \right) - \left(\frac{\alpha}{kT} x_2^2 + \ln x_1 \right)$$
$$- \left(\frac{\alpha}{kT} \frac{2b_1^2 + (b_1 + c_1)^2}{8} + \ln b_2 \right) + \left(\frac{\alpha}{kT} x_1^2 + \ln x_2 \right) + \frac{\Delta_{21} F_t}{kT} = 0 \tag{100}$$

Again, Eq. (100) may be solved by a numerical zero place determination procedure. The logarithmic form of the following equilibrium ratio can be recognized in Eq. (100):

$$\frac{(\gamma_{1,B} b_1)(\gamma_2 x_2)}{(\gamma_{2,B} b_2)(\gamma_1 x_1)} \equiv \frac{(\gamma_{1,1}^s x_{1,1}^s)(\gamma_2 x_2)}{(\gamma_{2,1}^s x_{2,1}^s)(\gamma_1 x_1)} = \exp\left(\frac{-\Delta_{21} F_t}{kT} \right) = K \tag{101}$$

As can be seen, there is a simple and unambiguous correlation between the value of the total free energy of exchange $\Delta_{21}F_t$ and that of the equilibrium constant. It is also expressed by the equation that the surface exerts a direct effect on this and only this layer.

The relationship between the equilibrium constant and the interaction energies is easily obtained from Eqs. (90a), (90b), (92), and (101):

$$K = \exp\left(\frac{(\varepsilon_{1s} - \varepsilon_{2s}) - \frac{3}{2}(\varepsilon_{11} - \varepsilon_{22})}{kT}\right) \tag{102}$$

H. Algorithm of Computation

In the case of the miscible systems discussed in the present work, even a relatively simple algorithm of computation is effective. Computation is commenced by supplying the appropriate values of α/kT, K, and x_1. A necessary number of sublayers ($L = 50, \ldots, 2000$) are considered, the compositions of which are characterized by the molar fraction of component 1. At the initial stage, the composition of all sublayers is set to that of the bulk phase. The $(L + 1)$st sublayer already represents the bulk phase (i.e., $x_{1,L+1}^s = x_1$ throughout the calculation). The composition of the first (contact) sublayer, $x_{1,1}^s$, is first determined by the numerical solution of Eq. (100); then, knowing this value, that of the second is determined again by the numerical solution of Eq. (87); then, knowing the composition of the second sublayer, that of the third is calculated, and so on. [Equations (87) and (100) are easily brought to forms with fewer operations and they are in conformity with the original equations of Ono and Kondo.] This procedure is continued until sublayer composition (molar ratio) approaches bulk composition within the given margin of error (10^{-13}). Next, the first sublayer is recalculated and so forth. The iteration is repeated until the sum of the absolute values of the changes in composition decreases below a given margin of error (10^{-11}). Double precision is essential. Divergency or multiple solutions were not found.

As soon as the concentration profile [i.e., the values $x_{1,i}^s$ ($i = 1, \ldots, L$) are known], thermodynamic parameters of the individual sublayers and of the entire adsorption layer are determined. Specific adsorption excesses in the ith sublayer ($n_{1,i}^{\sigma(n)}$) and in the entire adsorption layer ($n_1^{\sigma(n)}$) are obtained by simple formulas:

$$n_{1,i}^{\sigma(n)} = x_{1,i}^s - x_1 \tag{103}$$

$$n_1^{\sigma(n)} = \sum_{i=1}^{L} n_{1,i}^{\sigma(n)} \tag{104}$$

The ratio of the total adsorption excess and the excess in the first sublayer yields $L_{equ.}$, the equivalent (sub)layer number (see Fig. 32b), an important measure of adsorption layer thickness:

$$L_{equ.} \equiv \frac{n_1^{\sigma(n)}}{n_{1,1}^{\sigma(n)}} \tag{105}$$

The internal energy (heat) of adsorption, entropy, and free energy in the ith sublayer ($\Delta_{21}U_i$, $\Delta_{21}S_i$, and $\Delta_{21}F_i$, respectively) are calculated by Eqs. (96), (75), (76a), (76b), (58a), (58b), (77a),

(77b), (79), and (83). The values relative to the entire adsorption layer ($\Delta_{21}U$, $\Delta_{21}S$, and $\Delta_{21}F$) are obtained by simple addition

$$\Delta_{21}U = \sum_{i=1}^{L} \Delta_{21}U_i \tag{106a}$$

$$\Delta_{21}S = \sum_{i=1}^{L} \Delta_{21}S_i \tag{106b}$$

$$\Delta_{21}F = \sum_{i=1}^{L} \Delta_{21}F_i \tag{106c}$$

Adsorption equilibrium and the structure of the adsorption layer are next examined at increasing values of α/kT (i.e., at increasing extents of nonideality). Differences in surface quality ("polarity") (i.e., in the selectivity of adsorption) are indicated by different values of K. It is sufficient to study cases in which $K \geqslant 1$ as, for example, the case of $K = 0.1$ is identical with that of $K = 10$ when components 1 and 2 are interchanged.

I. Adsorption in Perfect Mixtures ($\alpha/kT = 0$)

The thermodynamics of adsorption from perfect mixtures has been discussed many times in the special literature; therefore, only the most important functions are described here.

Adsorption excess isotherms calculated at different equilibrium constants (K) are shown in Fig. 36a. Functions $n_1^{\sigma(n)} = f(x_1)$ at various values of x_1 were calculated on the basis of Eqs. (103) and (104). Moving toward increasing values of K (which means increasing preferential adsorption of component 1), the initial slope of the isotherm increases and the tailing region approaches the straight line representing the value $1 - x_1$ increasingly and within an increasingly wide interval. The adsorption layer is strictly monomolecular; the second sublayer is already part of the bulk phase. Correspondingly, the equation $L_{\text{equn.}} = 1$ holds at all locations, independently of composition and the value of K. When $K = 1$, the adsorption excess is zero everywhere.

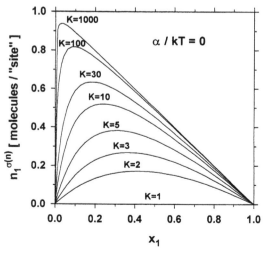

FIG. 36a Adsorption excess isotherm functions at $\alpha/kT = 0$ (perfect mixture) and different K equilibrium constants.

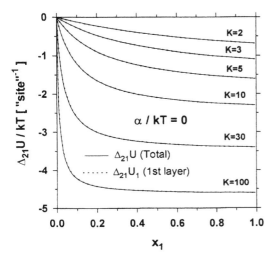

FIG. 36b Internal energy (heat) of adsorption functions at $\alpha/kT = 0$ (perfect mixture) and different K equilibrium constants.

The set of curves of internal energy (heat) of adsorption [calculated on the basis of Eq. (106a)] is shown in Fig. 36b. As the value of K increases, the heat effect is enhanced (proportionally to $\ln K$) and the function obtained is more and more like a saturation curve.

The set of curves representing adsorption entropy [see Eq. (106b)] is displayed in Fig. 36c. As the value of K increases, curves with increasingly sharper maxima are obtained; again the height of the peaks is proportional to $\ln K$. The terminal portion of the curves increasingly approaches a straight line with unit slope.

The set of curves of free energy of adsorption [calculated by Eq. (106c)] is shown in Fig. 36d. At increasing values of K, curves similar to those of internal energy of adsorption are

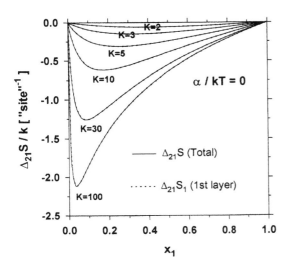

FIG. 36c Entropy of adsorption functions at $\alpha/kT = 0$ (perfect mixture) and different K equilibrium constants.

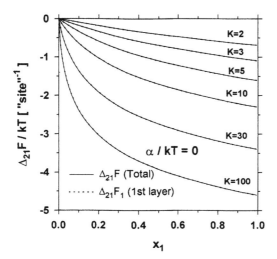

FIG. 36d Free energy of adsorption functions at $\alpha/kT = 0$ (perfect mixture) and different K equilibrium constants.

obtained; instead of a saturation character, however, the slope of the final region of the curves approaches -1.

To sum up, it can be established that in the case of adsorption from perfect mixtures, both adsorption excess and thermodynamical functions of adsorption are fully concentrated in the contact sublayer.

J. Adsorption in Real Mixtures Far from Demixing ($\alpha/kT = 1.4$)

Considering that the condition of demixing is $\alpha/kT \leqslant 2$, the value $\alpha/kT = 1.4$ represents a strongly nonperfect case which is, however, quite far from demixing.

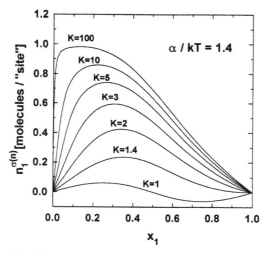

FIG. 37a Adsorption excess isotherm functions at $\alpha/kT = 1.4$ (real mixture, far from demixing) and different K equilibrium constants.

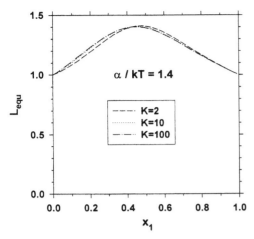

FIG. 37b Equivalent (sub)layer number functions at $\alpha/kT = 1.4$ (real mixture, far from demixing) and different K equilibrium constants.

Adsorption excess isotherms are shown in Fig. 37a. They are significantly different from those obtained in ideal mixtures: They have no straight sections. In the intermediate concentration range, adsorption excess exceeds $1 - x_1$, which in itself is indicative of multimolecular adsorption. Another adsorption excess is observed at $K = 1$, with an azeotropic point at a bulk concentration of $x_1 = 0.5$.

The course of equivalent layer number is displayed in Fig. 37b. The adsorption layer is multimolecular at all values of K, most of all in the intermediate concentration range. The curves are surprisingly close to each other, indicating that multimolecular adsorption is far more affected by the properties of the liquid phase than by those of the surface. However, the adsorption layer is only slightly thicker than monomolecular.

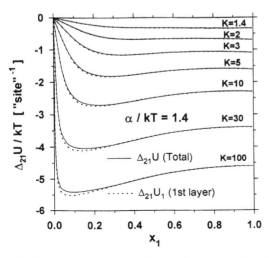

FIG. 37c Internal energy (heat) of adsorption functions at $\alpha/kT = 1.4$ (real mixture, far from demixing) and different K equilibrium constants.

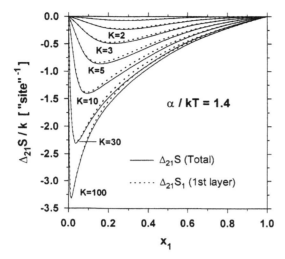

FIG. 37d Entropy of adsorption functions at $\alpha/kT = 1.4$ (real mixture, far from demixing) and different K equilibrium constants.

The set of curves representing internal energy (heat) of adsorption is shown in Fig. 37c. A conspicuous difference in comparison with ideal mixtures is that, instead of curves of a saturation character, curves bending back are obtained with a well-defined inflection point, the position of which is more or less identical with that of the maximum of equivalent layer number.

The set of curves of adsorption entropy is presented in Fig. 37d. Changes as compared to perfect mixtures are not yet apparent.

The set of curves describing free energy of adsorption are seen in Fig. 37e. As in the case of internal energy (heat) of adsorption, an inflection point appears, but the effect is less

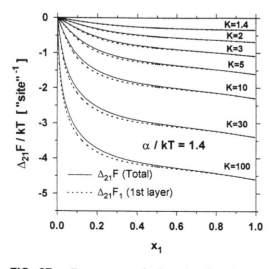

FIG. 37e Free energy of adsorption functions at $\alpha/kT = 1.4$ (real mixture, far from demixing) and different K equilibrium constants.

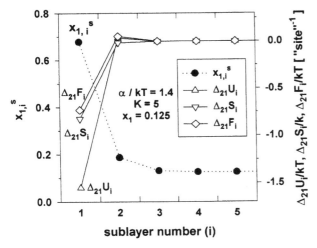

FIG. 37f Concentration, internal energy (heat), entropy, and free energy of adsorption profiles at $\alpha/kT = 1.4$, $K = 5$, and $x_1 = 0.125$.

pronounced. It can be established for each of the three thermodynamical functions that they are no more fully concentrated within the contact sublayer.

Figure 37f displays the concentration and potential functions at $K = 5$ and $x_1 = 0.125$ as a function of the distance from the surface. All four functions decay rapidly, for the adsorption layer is very little thicker than monomolecular.

K. Adsorption in Real Mixtures, Close to Demixing ($\alpha/kT = 1.95$)

Adsorption excess isotherms are presented in Fig. 38a. They are even more dissimilar to those obtained in perfect mixtures; no straight sections are found. Adsorption excess is considerably larger than $1 - x_1$ a wide concentration range.

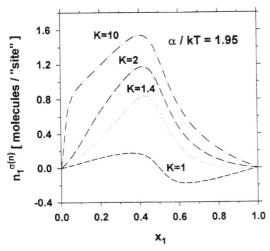

FIG. 38a Adsorption excess isotherm functions at $\alpha/kT = 1.95$ (real mixture, close to demixing) and different K equilibrium constants.

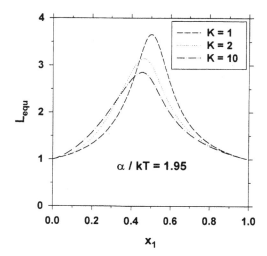

FIG. 38b Equivalent (sub)layer number functions at $\alpha/kT = 1.95$ (real mixture, close to demixing) and different K equilibrium constants.

Functions of equivalent layer number are shown in Fig. 38b. The multimolecular character of the adsorption layer is quite pronounced, again mostly in the intermediate concentration range.

The concentration dependence of internal energy (heat) of adsorption is displayed in Fig. 38c. The curves bend back even more as compared to perfect mixtures and the inflection point at the maximum of equivalent layer number is even more pronounced.

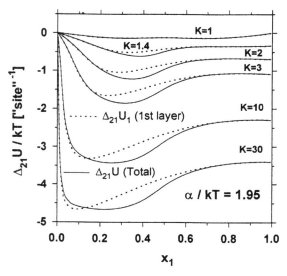

FIG. 38c Internal energy (heat) of adsorption at $\alpha/kT = 1.95$ (real mixture, close to demixing) and different K equilibrium constants.

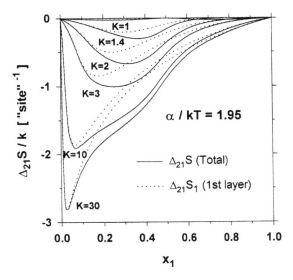

FIG. 38d Entropy of adsorption functions at $\alpha/kT = 1.95$ (real mixture, close to demixing) and different K equilibrium constants.

The set of curves of adsorption entropy are represented in Fig. 38d. The change in comparison to perfect mixtures is clearly indicated by the appearance of a shoulder in the intermediate region of the function.

Functions of free energy of adsorption are seen in Fig. 38e. Again, the inflection point has become more pronounced and is located at the maximum of layer thickness. The tendency of the three thermodynamical potential functions not to be fully concentrated within the contact

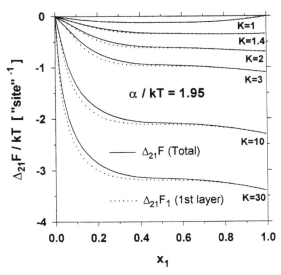

FIG. 38e Free energy of adsorption functions at $\alpha/kT = 1.95$ (real mixture, close to demixing) and different K equilibrium constants.

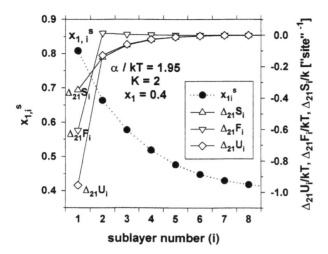

FIG. 38f Concentration, internal energy (heat), entropy, and free energy of adsorption profiles at $\alpha/kT = 1.95$, $K = 2$, and $x_1 = 0.4$.

sublayer is enhanced but to different extents for the individual functions. The effect is the strongest in the case of adsorption entropy and the weakest for free energy of adsorption. Moreover, at the maximum of layer thickness, free energy of adsorption is concentrated in the first layer within a margin of error of 1–2%.

Figure 38f displays concentration and potential functions at $K = 2$ and $x_1 = 0.4$ as a function of the distance from the surface. The four functions decrease much slower than in the previous case, in accordance with the thicker adsorption layer. In contrast to the monotonous change observed in the other three functions, free energy of adsorption first increases rapidly and then decreases slowly. A more detailed interpretation of this phenomenon will be given in our next publication.

IX. CONCLUSIONS

It may be established on the basis of the results described herein that the multimolecular character of adsorption from binary liquid mixtures is rapidly enhanced as the point of demixing is approached. Layer thickness itself, however, is less dependent on the "polarity" of the surface than on the properties of the bulk phase. Layer thickness has a maximum at a composition of $x_1 \gg 0.5$ which is a direct consequence of the symmetric character of the lattice model, assuming molecules of identical size. The maximum of layer thickness will probably be shifted in the case of mixtures of components consisting of molecules of various sizes or with nonspherical shapes, or asymmetric in other respects.

In the case of multimolecular adsorption, functions of heat of adsorption take a shape unlike saturation curves: They bend back and the position of the inflection point yields the location of the maximum of layer thickness at a good approximation. In other words, this theoretical prediction means that it is probably possible to detect and analyze multimolecular adsorption by suitable calorimetric measurements.

The function of free energy of adsorption (usually obtained by integration of the adsorption excess isotherm function according to the Gibbs equation) also exhibits an inflection point at the maximum of layer thickness. Our earlier hypothesis (i.e., that in *miscible* binary systems, free energy of adsorption is nearly fully concentrated within the contact sublayer and that the decrease in concentration is nearly exponential) has been verified by model calculations.

The situation, however, will be quite different in the case of multimolecular adsorption in *partially miscible* binary mixtures. The thermodynamics of this system, also based on the lattice model, will be discussed in a future publication. At the same time, general principles determining the structure of multimolecular adsorption layers (including the case of miscible binary mixtures) will be formulated.

The lattice model was applied for studies on multimolecular adsorption layers. This model describes the physical adsorption of binary mixtures on solid–liquid interfaces and the accompanying local demixing. Our calculations are based solely on classical thermodynamical terms, assuming a homogeneous, planar surface, unstructured, spherical molecules of identical size, dispersion interactions, a coordination number of 12, and statistical distribution within the sublayers arranged parallel with the surface. The nonideality of the binary mixture is expressed by the interaction parameter α, and the adsorption selectivity of the surface is characterized by the equilibrium constant K. Changes in surface excess, equivalent layer thickness, free energy of the surface, internal energy and entropy as a function of α, K, and the equilibrium mixture composition x_1 were studied. The dependence of surface concentration and the thermodynamical parameters on distance were calculated. Layer thickness is determined primarily by the quality of the liquid and less by the polarity of the surface. In thick adsorption layers, the free energy of adsorption $\Delta_{21}F$ is concentrated nearly exclusively in the first layer of molecules adjoining the surface and it is only in layers of medium thickness that the other sublayers share a significant part of this thermodynamical quantity. At maximal layer thickness, the slope of the function $\Delta_{21}F$ is zero, whereas those of the internal energy function $\Delta_{21}U$ and the entropy function $\Delta_{21}S$ are maximal and all three functions have an inflection point. The adequate selection of the value of K also allows the application of the model to the description of multimolecular adsorption on G/L interfaces.

REFERENCES

1. JJ Kipling. Adsorption from Solutions of Non-Electrolytes. London: Academic Press, 1965.
2. G Schay. In: E Matijevic, ed. Surface and Colloid Science. London: Wiley, 1969, Vol 2, p 155.
3. G Schay. In: DH Everett, ed. Surface Area Determination, Proceedings International Symposium 1969. London: Butterworths, 1970, pp 273–308.
4. G Schay. Pure Appl Chem 48:393–400, 1976.
5. DH Everett. Colloid Science, Volume 3. London: Chemical Society, 1979, p 66.
6. DF Billett, DH Everett, EEH Wright. Proc Chem Soc 216–228, 1964.
7. AZ Zettlemoyer, GJ Young, JJ Chessick. J Phys Chem 59:962–970, 1955.
8. DH Everett. Pure Appl Chem 53:2181–2192, 1981.
9. DH Everett. Prog Colloid Polym Sci 65:103–116, 1978.
10. I Dékány, F Szántó, LG Nagy, G Fóti. J Colloid Interf Sci 50:265–271, 1975.
11. I Dékány, F Szántó, LG Nagy. Prog Colloid Polym Sci 65:125–132, 1978.
12. I Dékány, F Szántó, LG Nagy, G Schay. J Colloid Interf Sci 93:151–165, 1983.
13. I Dékány, F Szántó, LG Nagy. J Colloid Interf Sci 103:321–332, 1985.
14. I Dékány, F Szántó, LG Nagy. J Colloid Interf Sci 109:376–386, 1986.
15. G Machula, I Dékány, LG Nagy. Colloids Surfaces A 71:241–254, 1993.
16. I Dékány, T Haraszti, L Turi, Z Király. Prog Colloid Polym Sci 111:65–73, 1998.
17. I Dékány. Pure Appl Chem 64:1499–1509, 1992.
18. I Dékány. Pure Appl Chem 65:901–906, 1993.

19. I Dékány. In: A Dabrowski, VA Tertykn, eds. Vol. 99. Adsorption on New and Modified Inorganic Sorbents. Studies in Surface Science and Catalysis. Amsterdam: Elsevier Science, 1996, pp 879–889.

20. F Berger, I Dékány. Colloid Polym Sci 275: 876–882, 1997.

21. F Berger, I Dékány. Colloids Surfaces A 141:305–317, 1998.

22. I Dékány, LG Nagy. Model Chem 134:279–297, 1997.

23. Z Király, L Túri, I Dékány, K Bean, B Vincent. Colloid Polym Sci 274:779–787, 1996.

24. T Marosi, I Dékány, G Lagaly. Colloid Polym Sci 272:1136–1142, 1994.

25. I Regdon, Z Király, I Dékány, G Lagaly. Prog Colloid Polym Sci 109:214–220, 1998.

26. I Dékány, T Marosi, Z Király, LG Nagy. Colloids Surfaces 49:81–93, 1990.

27. I Regdon, I Dékány, G Lagaly. Colloid Polym Sci 276:511–517, 1998.

28. O Redlich, AT Kister. Ind Eng Chem 40:341–352, 1948.

29. GW Woodbury Jr, LA Noll. Colloids Surfaces 8:1–12, 1983.

30. LA Noll, GW Woodbury Jr, TE Burchfield. Colloids Surfaces 9:349–351, 1984.

31. I Dékány, Á Zsednai, Z Király, K László, LG Nagy. Colloids Surfaces 19:47–59, 1986.

32. I Dékány, Á Zsednai, K László, LG Nagy. Colloids Surfaces 23:41–56, 1987.

33. I Dékány, I Ábrahám, LT Nagy, K László. Colloids Surfaces 23:57–68, 1987.

34. DH Everett. Trans Faraday Soc 60:1803–1817, 1964; 61:2478–2492, 1965.

35. A Dabrowski, M Jaroniec. Acta Chim Acad Sci Hung 99:225–236, 1979.

36. T Allen, RM Patel. J Colloid Interf Sci 35:647–655, 1971.

37. Z Király, I Dékány. Colloids Surfaces A 34:1–14, 1988.

38. Z Király, I Dékány. J Chem Soc Faraday Trans 1 85:3373–3383, 1989.

39. Z Király, I Dékány. Colloids Surfaces 49:95–101, 1990.

40. R Denoyel, G Durand, F Lafuma, R Audebert. J Colloid Interf Sci 139:281–291, 1990.

41. R Denoyel, F Giordano, J Rouquerol. Colloids Surfaces A 76:141–148, 1993.

42. AJ Groszek. Proc Roy Soc A 314:473–481, 1970.

43. R Denoyel, F Rouquerol, J Rouquerol. In: Adsorption from Solution. RH Ottewill, CH Rochester, AL Smith, eds. New York: Academic Press, 1983, pp 225–258.

44. M Liphard, P Glanz, G Pilarski, GH Findenegg. Prog Colloid Polym Sci 67:131–142, 1980.

45. HE Kern, A Piecocki, U Bauer, GH Findenegg. Prog Colloid Polym Sci 65:118–129, 1978.

46. A Dabrowski, J Oscik, W Rudzinski, M Jaroniec. J Colloid Interf Sci 56:403–414, 1976.

47. I Dékány, F Szántó, A Weiss, G Lagaly. Ber Bunsenges Phys Chem 89:62–67, 1985.

48. I Dékány, F Szántó, A Weiss, G Lagaly. Ber Bunsenges Phys Chem 90:422–427, 1986.

49. I Dékány, F Szántó, A Weiss, G Lagaly. Ber Bunsenges Phys Chem 90:427–431, 1986.

50. I Dékány, F Szántó, LG Nagy. J Colloid Polym Sci 266:82–96, 1988.

51. G Machula, I Dékány. Colloids Surfaces A 61:331–348, 1991.

52. I Dékány, F Szántó, LG Nagy, H Beyer. J Colloid Interf Sci 112:261–273, 1986.

53. WD Harkins, G Jura. J Chem Phys 11:431–432, 1943.

54. J Frenkel. Kinetic Theory of Liquids. Oxford: Clarendon Press, 1946.

55. G Halsey. J Chem Phys 16:931–932, 1948.

56. TL Hill. J Chem Phys 17:590, 1949; 17:668, 1949.

57. Z Király, I Dékány, E Klumpp, H Lewandowski, HD Narres, MJ Schwuger. Langmuir 12:423–430, 1996.

58. S Brunauer, PH Emmett, E Teller. J Am Chem Soc 60:309, 1938.

59. I Dékány, LG Nagy, G Schay. J Colloid Interf Sci. 66:197–200, 1978.

60. I Dékány, LG Nagy. J Colloid Interf Sci. 147:119–127, 1991.

61. T Marosi, I Dékány, G Lagaly. Colloid Polym Sci 270:1027–1034, 1992.

62. Z Király, I Dékány, LG Nagy. Colloids Surfaces A 71:287–292, 1993.

63. I Regdon, Z Király, I Dékány, G Lagaly. Colloid Polym Sci 272:1129–1135, 1994.

64. J Tóth. J Colloid Interf Sci 46:38–45, 1973.

65. ChC Brown, DH Everett, ChJ Morgan. J Chem Soc Faraday Trans I 71:883–892, 1975.

66. J Oscik, K Goworek, R Kusak. J Colloid Interf Sci 84:308–312, 1981.

67. M Jarionec, A Dabrowski, J Tóth. Chem Eng Sci 39:65–70, 1984.

68. A Dabrowski, M Jarionec, J Tóth. J Colloid Interf Sci 94:573–576, 1983.

69. SG Ash, GH Findenegg. Spec Discuss Faraday Soc 1:105–117, 1970.
70. GH Findenegg. J Chem Soc Faraday Trans I 69:1069–1078, 1973.
71. A Dabrowski, M Jarionec, J Oscik. In: E Matijevic, ed. Surface and Colloid Science, Volume 14. New York: Plenum Prress, 1987, p 83.
72. F Berger, I Dékány. Colloid Polym Sci 275:876–882, 1997.
73. F Berger, I Dékány. Colloids Surfaces A 141:319–325, 1998.
74. GS Rushbrook. Introduction to Statistical Mechanics. Oxford: Oxford University Press, 1949.
75. EA Guggenheim. Mixtures. Oxford: Oxford University Press, 1952.
76. R Defay, I Prigogine, A Bellemans, DH Everett. Surface Tension and Adsorption. London: Longmans, 1966, p 172.
77. I Dékány, A Farkas, I Regdon, E Klumpp, HD Narres, MJ Schwuger. Colloid Polym Sci 274:981–988, 1996.
78. GF Cerofolini, L Meda. J Colloid Interf Sci 202: 104–123, 1998.
79. G Machula, I Dékány. Colloids Surfaces A 61:331–348, 1991.
80. G Machula, I Dékány, LG Nagy. Colloids Surfaces A 71:219–231, 1991.
81. Z Király, L Túri, I Dékány, K Bean, B Vincent. Colloid Polym Sci 274:779–789, 1996.
82. MJ Vold. J Colloid Sci 16:1–16, 1961.
83. JJ Kipling. Adsorption from Solutions of Non-Electrolytes. London: Academic Press, 1965.
84. DH Everett. Trans Faraday Soc 60:1803, 1964; 61:2478 (1965).
85. G Schay, LG Nagy. Periodica Polytechnica 4:45, 1960.
86. G Schay. In: E Matijevic, ed. Surface and Colloid Science, Volume 2. London: Wiley, 1969, p 155.
87. B Vincent. J Colloid Interf Sci 112:270–278, 1973.
88. W Rudzinski, J Oscik, A Dabrowski. Chem Phys Lett 20:444–447, 1973.
89. AI Rusanov. Phase Equilibrium and Surface Phenomena. Leningrad: Chimia, 1967.
90. J Tóth. Acta Chim Hung 63:67–85, 1970.
91. J Tóth. Acta Chim Hung 63:179–192, 1970.
92. S Ono. Mem Fac Eng Kyushu Univ 10:195–204, 1947.
93. S Ono. Mem Fac Eng Kyushu Univ 12:1–8, 1950.
94. S Ono. J Phys Soc Japan 5:232–237, 1950.
95. S Ono, S Kondo. In S. Flügge, ed. Handbuch der Physik. Berlin: Springer-Verlag, 1960, Vol 10, pp 267–270.
96. S Ono, S Kondo. Molecular Theory of Surface Tension. Berlin: Springer-Verlag 1960.
97. JE Lane, CHJ Johnson. Austral J Chem 20:611–621, 1967.
98. AR Altenberger, J Stecki. Chem Phys Lett 5:29–33, 1970.
99. A Dabrowski, M Jarionec, KJ Garbacz. Thin Solid Films 103:399–415, 1983.
100. GL Aranovich, MD Donohue. J Colloid Interf Sci 178:204–208, 1996.
101. GL Aranovich, MD Donohue. J Colloid Interf Sci 189:101, 1997.
102. GL Aranovich, MD Donohue. J Colloid Interf Sci 200:273–290, 1998.
103. JA Barker. Proc Roy Soc A 216:45, 1953.
104. JE Lane. Austral J Chem 21:827–851, 1968.
105. JE Lane. In: GD Parfitt, CH Rochester, eds. Adsorption from Solution at the Solid/Liquid Interface. London: Academic Press, 1983, Chap 2.
106. D Nicholson, NG Parsonage. Computer Simulation and the Statistical Mechanics of Adsorption. London: Academic Press, 1982.
107. F Berger, I Dékány. Colloid Polym Sci 275:876–882, 1997.
108. Z Király, I Dékány, E Klumpp, H Lewandowski, HD Narres, MJ Schwuger. Langmuir 12:423–430, 1996.

11

Surface Complexation Models of Adsorption: A Critical Survey in the Context of Experimental Data

JOHANNES LÜTZENKIRCHEN Institut für Nukleare Entsorgung, Forschungszentrum Karlsruhe, Karlsruhe, Germany

I. INTRODUCTION

Surface complexation models have become widely applied tools in the description of adsorption of solutes in aqueous solutions onto mineral particles. Development of the models and first applications range back about 30 years [1–4].

By 1975, three models had been introduced: the constant capacitance model of Schindler and Gamsjäger [1], the diffuse layer (or Gouy–Chapman) model of Stumm and co-workers [2] and the triple-layer model proposed by Yates et al. [3] and applied by Davis et al. [4]. All of those articles and the major part of the avalanche of modeling articles still emerging from those first attempts focused on oxide minerals, although other minerals have also been involved.

Different disciplines have an interest in these kinds of models but generally with different objectives:

Colloid chemists are interested in colloidal stability, where the pH-dependent charging of the mineral is an important feature, which can be described by surface complexation models. Surface force measurements which have traditionally been described by assuming constant surface charge or potential can be modeled by a surface complexation model in a more comprehensive way.

Scientists from different disciplines (e.g., soil science, geology) are interested in interactions of solutes with minerals (e.g., heavy metals as an example of hazardous substances but also nutrients, etc.). Surface complexation models are tools which allow to describe such interactions.

Inorganic (coordination) chemists are interested in the structure of surface complexes. Advanced surface complexation models can take into account the details of surfaces and bonding mechanisms to describe the molecular structure within a thermodynamic framework.

Of course, these categories are neither complete nor restrictive. On the contrary, in the ideal case, knowledge (both experimental and theoretical) from all potential disciplines involved should act together. Reviews on surface complexation models are available and come from the context of the respective different disciplines. These have been summarized elsewhere [5].

In this survey, it is not attempted to show how successful surface complexation models are. Rather, it is attempted to show what can be done with them, what will be done with them in the future and what should not be done with them. The experimental aspects (e.g., the input data to the models) are discussed whenever judged important (certainly without full coverage but rather with focus on those aspects, which have not yet been addressed or details which have been of interest to the present author). In particular, the importance of combining methods and data is stressed. In recent years, there has been an increased interest in linking surface complexation models, which are traditionally based on macroscopic (adsorption, titration, and other) data, with structural information obtained with modern spectroscopic methods such as x-ray absorption spectroscopy (XAS). It is expected that the closer the agreement of the thermodynamic formulation of a surface chemical reaction with the actual structure of a surface complex is, the more reliable a prediction of the system behavior under more or less strongly varied conditions will be.

Environmental scientists such as those involved in performance assessment (e.g., for nuclear repositories) would be pleased to be able to predict the interactions of solutes (e.g., radionuclides) with backfill and geologic materials. However, in this area, which in many countries receives much public attention, even the aqueous solution equilibria for the pertaining conditions of a favored repository concept cannot be accurately described (e.g., metal ions in brine solutions, which require Pitzer formalism, or in highly alkaline backfill pore waters, which have traditionally not received much attention in aqueous solution studies because of the limitations of glass electrodes and solid phase formation). Databases for surface complexation applications are also required for many other purposes, but the major drawback of such potential databases is that no agreement exists on the actual surface complexation model to be used. This may ultimately lead to particular difficult situations whenever one of the following occur:

One database exists and others simply follow the concept including the model framework
Databases are built with different models which cannot be merged

In the following sections, different surface complexation models will be introduced. General aspects and specific models will be discussed. The components of surface complexation theory will be presented, as well as some recent developments covering, for example, the use of equations for the diffuse part of the electrical double layer for electrolyte concentrations, for which the traditional Gouy–Chapman equation is not recommended, or a generalization of Smit's compartment model [6] for situations in which the traditional models are at a loss.

Some special cases such as mixed systems (binary suspensions, competitive systems, and ternary surface complexes) and effects going beyond that of simple adsorption (i.e., surface precipitation) as well as temperature effects are shortly addressed in a supplementary section.

Finally, the perspectives in the field are described from the author's point of view. The author's remarks may seem too negative to the reader in some instances. However, it would have been much easier to display the success of the surface complexation models instead of assessing critical points. Because the modeling can be done on a relatively simple level, it is of primary importance to keep the models self-consistent. Aspects of this self-consistency are addressed in selected examples throughout the text.

Progress in the field of surface complexation is mainly made in the modern spectroscopic and computational approaches, whereas agreement on fundamentals is still lacking, both in some aspects of experimental procedures and in modeling approaches. One way out of this dilemma is to proceed by way of exclusion so that the number of options available decreases (e.g., by showing experimentally or through modeling calculations that a certain experimental approach or a surface complexation model is not appropriate in general or for a certain purpose).

The ultimate goal of the experimental work and, in particular, of the associated modeling is to have generally accepted and applicable databases, as in aqueous solution chemistry. On the one hand, these databases have to be self-consistent, comprehensive, and reliable with respect to predictions; on the other hand, the models with or for which the databases are developed should be applicable to real cases. Real cases may be well-defined and characterized laboratory systems, but they may also cover applications in the environment, where models have to be sufficiently simple while still reliable. The modeling is always closely related to the underlying experimental data, which are either used for inverse modeling (parameter determination) or for forward modeling (predictions). Therefore, the experimental procedures and the treatment of the raw experimental data must be discussed. This is done whenever deemed appropriate throughout the text and, in particular, in the next section with special reference to controversially discussed issues but without attempting to cover this important part completely.

II. EXPERIMENTAL PROCEDURES AND DATA AND THEIR USE IN MODELS

A. Experimental Methods

Ideally, experimental methods will cover the whole spectrum available. This includes the more traditional macroscopic methods (e.g., titrations, adsorption experiments, or electrokinetic methods), supplementary colloid chemical approaches (e.g., suspension stability or surface force measurements), as well as spectroscopic experiments [e.g., XAS, Raman or infrared spectroscopy]. In this chapter, focus will be mainly on the traditional approaches. With these, multiple variations exist and little is known on how they will affect the macroscopic data. Therefore, these aspects are treated in some detail. Microscopic data are only discussed briefly. With respect to these and other potential approaches, the reader is referred to Brown et al. [7] and Israelachvili [8], or Lyklema [9] for an overview.

1. Macroscopic Data

The term "macroscopic data" in the context of surface complexation includes mainly the measurement of relative surface-charge density of sorbents, the determination of adsorption isotherms (constant pH and ionic strength and variable amounts of sorbent or sorbate), and so-called pH edges (constant amount of sorbent, sorbate, and ionic strength, variable pH; the typical results show increasing adsorption of metal ions and decreasing adsorption of ligands with increasing pH, where the amount adsorbed often changes from 0% to 100% over a small range of pH; thus pH edge), studies of ionic strength dependencies (mostly via pH edges at different values of ionic strength), and the determination of particle charge/mobility, yielding the so-called zeta potential. The combination of relative surface-charge densities and particle mobilities ideally allows conversion of relative charge to absolute charge, in case the point of zero charge (PZC) of a sorbent can be unambiguously defined and equated with the pristine point of zero charge (PPZC). PPZCs are frequently considered as material property.

Figure 1 shows PZC values for goethite reported in the literature as obtained by a range of methods. They would typically be equated with the PPZC in the individual studies. It can be seen that there is a huge scatter in the reported values. Reasons for this scatter may be manifold and include at least the following:

- True sample specific variability of PZC
- Influence of electrolyte used

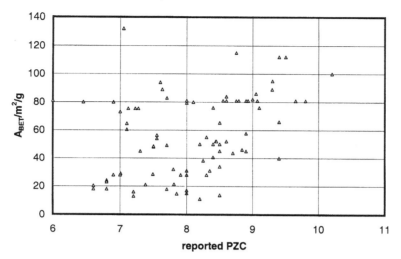

FIG. 1 Relationship between specific surface area and points of zero charge of goethite reported in the literature. (From J. Lützenkirchen, in preparation.)

- Experimental problems in the determination of PZC (e.g., pH measurements, CO_2, silica from glassware, impurities on the particles and in the electrolytes, and other contaminations)
- Misinterpretation of raw data

Table 1 gives some details on individual points also with respect to modeling. Because various phenomenological causes and experimental circumstances may be responsible for the observed variability, it is very difficult to solve the problem arising from Fig. 1. The wide range of data would indicate that PZC values for goethite are sample-specific. However, systematic studies on various samples of goethite [10,11] all indicate PZC values around 9.3. This would mean that as postulated by Sposito [12], many (and in case of goethite most) published PZC values are not necessarily to be trusted.

Whichever option is accepted, certain problems will remain for the models:

- At present the MUSIC (MUlti SIte Complexation) model [13,14] can, in principle, explain differences in PZC values for different samples; however, for goethite, the variation shown in Fig. 1 is too large to be explained by the MUSIC model alone based on the current knowledge.
- If a unique PPZC exists, the difference in charge obtained in different electrolytes must be self-consistently described; as indicated in Table 1, the parameters, which may be adjusted, are binding constants and capacitance values; variation of binding constants will result in PZC variations at high ionic strength {this can be checked with modern electroacoustic methods (cf. work by Kosmulski and Rosenholm [15]), some articles indeed report very interesting results with shifting PZC values and systems, in which no PZC at all was obtained in high salt concentrations [16]}; if variation of capacitance values is necessary, the present model approaches will be unable to account for mixed electrolytes in predictions.

Whatever option is chosen, this discussion on the very fundamental first step toward a surface complexation model indicates that the basic problems are not unambiguously solved and additional systematic and comprehensive experimental work is required.

TABLE 1 Variability of Pristine Points of Zero Charge and the Consequence with Respect to Modeling Approaches

Cause for variability of PPZC	Consequence of variability with respect to models
Sample-specific variability of PPZC	PZC of samples must be accurately determined in each sorption study.
	Parameters from literature studies cannot simply be applied to nominally identical sorbents.
	Comprehensive models must be able to explain variability and predict PZC based on independent information.
Influence of electrolyte	PZC of system must be determined; specific adsorption of at least some electrolyte ion must occur.
	Parameters must account for this electrolyte-specific effect (variation of binding constants or capacitance values).
	Models intended to have a broad applicability must be able to handle variable parameters in mixed electrolytes (not possible for variable capacitance values).
	Behavior in mixed electrolytes must be accurately described.
Experimental problems	Causes must be elucidated (difficult for or from literature data).
	Identified problems may be included in models to obtain corrected values (difficult because many hypotheses are involved).
Interpretation of raw data	Can be hampered by insufficient information in the description of the experimental procedures.

In the following, some aspects of the determination of the different kinds of macroscopic data are compiled, which might stimulate further experimental work.

(a) Titrations. Potentiometric titrations serve different purposes (see Table 2). They can be used to properly define the important subsystems in an adsorption study [i.e., the aqueous system(s) in the absence of the solid, and the suspension in the absence of the adsorbing solute]. One main difference in the procedures with suspensions in different laboratories concerns the timescale of the experiments. The majority of surface titrations is carried out in the fast mode (i.e., the time elapsed between two titrant additions covers a maximum of 1 h, sometimes not more than minutes). In other laboratories, a minimum waiting time of 1 h is applied and the maximum can be chosen freely and range up to days, weeks, or months.

TABLE 2 Survey of Applications of Continuous Potentiometric Titrations in the Context of Surface Complexation

System	Purpose	Timescale of experiments
Aqueous solution	Definition of aqueous solution equilibria	Usually short term
Particles in electrolyte solution	Determination of specific surface charge	Short or long term
Particles in electrolyte solution in the presence of other solutes	Additional data yielding information on adsorption phenomena	Short or long term

It has been known for a long time that drifts occur when the pH is measured in suspensions. With continuous titrations, the question then arises of whether this drift is due to some long-term proton or hydroxyl reaction at the surface or within the particles or whether it is caused by electrode kinetics or some other phenomena, such as intrusion of carbon dioxide into the reaction vessel, causing a drift either by its pH-dependent behavior in the bulk solution of the suspension or (additionally) by pH-dependent reactions with the surface.

Figure 2 shows data points from a long-term titration of a goethite suspension. One dataset represented by full squares results in an acceptable electromotive force (EMF) (Fig. 2a)/pH-value (Fig. 2b), the other (empty squares) not. The pH or $\log[H^+]$ is nearly constant over the whole measurement period for the dataset represented by black squares; the equilibrium criteria as imposed by the computer controlling the titration (i.e., drift of less than $0.05\,mV$ in $90\,min$) are fulfilled about $17\,h$ after titrant addition. For the dataset represented by the empty squares, this is the case after about $35\,h$. Here, a constant drift to higher $\log[H^+]$ values is noted (Fig. 2b), which becomes more obvious in the EMF readings with a drift over more than $10\,mV$ (Fig. 2a). Nevertheless, the equilibrium criterion is finally fulfilled.

Figure 3 shows the associated variations of temperature with time. For the data around pH 5.8, the variation of temperature is within $\pm 0.1\,K$, whereas the variation is much larger for the data around pH 8.0.

Figure 4 shows that for data below about pH 6.5, the long-term titration (solid squares) is in agreement with the model by Lövgren et al. [17] obtained with a less severe drift criterion of $0.1\,mV$ in $60\,min$. For those data, the variation of the temperature is within $\pm 0.2\,K$. A more pronounced temperature variation leads to significant deviations between the model and the data, especially at pH > 7.3. The relation among temperature, EMF, and proton concentration is not trivial, because the glass electrode calibration is temperature-specific. In the present dataset, for all data points, some "apparent" equilibrium in these kinds of experiments is finally obtained. However, the present data show that variations in temperature may be crucial. They have not been discussed in the context of titrations, but the often-quoted values of room temperature or variations of $\pm 2\,K$ indicate, together with Figs. 2–4, that this may be a reason for pH drifts and a source of experimental error.

Based on the experience gained with the set-up used for this titration, which has been described by Sjöberg and Lövgren [18] and with respect to the stability of the reference electrode, the author is confident that the long-term effect observed at constant temperature is related to reactions of protons or hydroxyls with the particles. However, long-term drifts of the order studied here ($<0.05\,mV/90\,min$) do not appear to conceivably differ from the data corresponding to a drift criterion of $<0.12\,mV/60\,min$.

However, comparisons between fast and slow titrations indicate that the difference may be crucial for goethite (see Fig. 5, relative surface-charge data from Lövgren et al. [17], Hiemstra and van Riemsdijk [19], and Djafer et al. [20]). The results plotted in Fig. 5 are quite puzzling, because even for short-term titrations carried out in two different laboratories on goethite samples prepared in a similar way, the data may be very different. This is an indication that it is quite important to obtain sample inherent titration data, because taking reference samples from the literature may obviously result in very different data and, subsequently, in different modeling results.

Instead of the continuous titrations, the relative surface charge can also be determined in batch experiments. This involves the preparation of a number of independent batches and measuring the pH in these or in the supernatants after a certain equilibration time. Compared to the continuous titrations, which are typically computer controlled, there are most probably more sources of experimental errors in the batch titrations (simply because more handling is

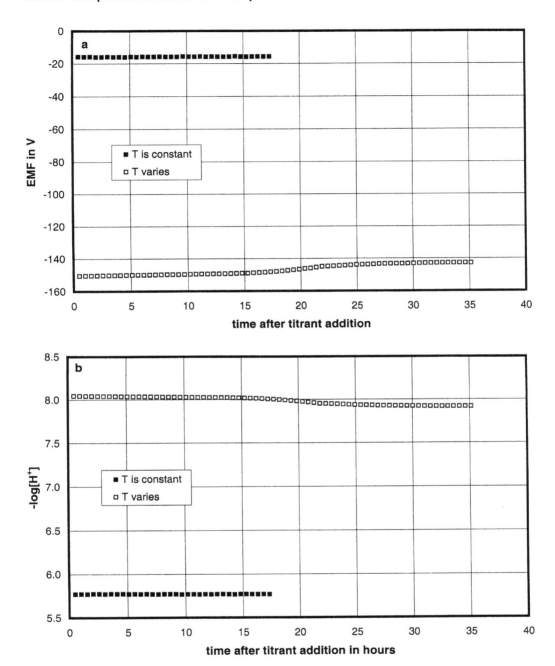

FIG. 2 Data points from long-term titrations of a goethite suspension sampled during the equilibration after titrant addition: (a) electromotive force (EMF) data, (b) $-\log[H^+]$ data (converted from EMF data using the same calibration parameters for both datasets for illustrative purposes). (From J. Lützenkirchen, unpublished data.)

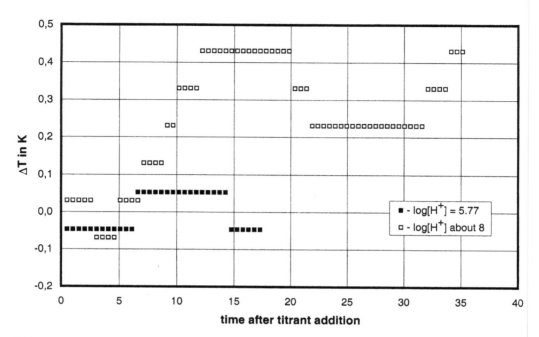

FIG. 3 Recorded relative temperature variations with time corresponding to the data points in Fig. 2. Relative temperature with respect to the temperature during setup calibration ($298.15\,K \pm 0.1$).

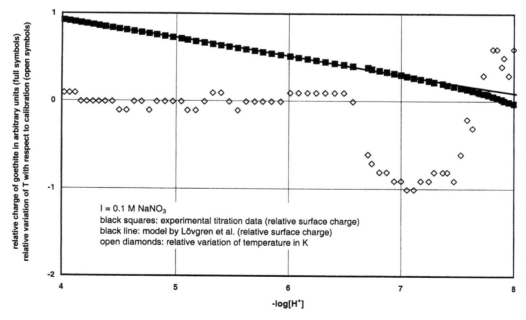

FIG. 4 Comparison of long-term titration data (squares, drift $< 0.5\,mV$ in $90\,min$) with the model by Lövgren et al. [17] obtained with a less severe drift criterion of $0.1\,mV$ in $60\,min$ for goethite.

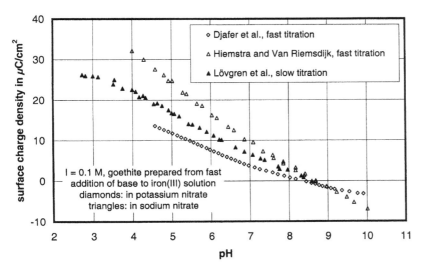

FIG. 5 Comparison of long-term titration data (drift < 0.1 mV/h, minimum waiting time 1 h, solid symbols from Lövgren et al. [17]) and short-term titration data (drift criterion on timescales of minutes, open symbols from Hiemstra and van Riemsdijk [19] and Djafer et al. [20]) for goethite. All titrations in nitrate media ($0.1 M$); cation is deemed unimportant for positively charged goethite.

necessary). On the other hand, one error in a continuous titration will propagate to all subsequent points, which would strongly suggest that the experiments should be repeated.

In the determination of absolute charge from the relative charge obtained in titration, at least one reference charge must be known. This is typically the PZC; sometimes, the PZC is equated with the common intersection point (CIP) of relative surface charge versus pH curves at different values of ionic strength. The CIP is not a reliable reference, because it is known that a CIP is also obtained for hematite in solutions of different concentrations of $Cd(NO_3)_2$ [21]. Cd adsorbs on positively charged hematite and therefore is expected to shift the PZC. This is only possible if the CIP is different from the PZC. Consequently, a CIP cannot be taken as a proof for nonspecific adsorption and identification of CIP and PZC has to be corroborated by independent data. Another problem arises in the determination of a CIP, when the titrations in different electrolyte solutions have not been performed subsequently with one sample. In this case, for each titration, the initial state of the particles is unknown and, more importantly, it may differ from that of other titrations (because of different electrolyte concentrations, solid concentrations, etc.). Thus, the relative positions of the different curves to each other are not known and the obtained CIP has no significance. It is best to determine a CIP by adding electrolyte: This is not possible with some setups (i.e., those designed for the constant medium approach) and may result in too extensive measurement periods for other setups. If the salt addition can be done at the CIP, because it is known, then the salt-addition method can be applied most easily.

An example for extensive determinations of different "points of zero charge" (note that point of zero charge in this general context pertains even to particular pH values with nonzero surface-charge density) is available from Woods et al. [22]. For boehmite, these authors found that the point of zero salt effect (PZSE, determined by salt addition), the CIP, and the isoelectric point (IEP, determined by microelectrophoresis) did not coincide [22].* The pristine point of

*These points of zero charge have been discussed in more detail elsewhere [23].

zero charge would be expected at the IEP determined with the lowest ionic strength. Therefore, renormalization of the relative surface-charge density should be done with respect to this pH value so that CIP and PZSE do not yield a zero surface-charge density, whereas Woods et al. normalized with respect to the PZSE, which was close to the CIP. From this discussion, it is clear that renormalization procedures as reported in the literature can be carried out depending on how the points of zero charge are interpreted. Unfortunately, original data can usually not be recalculated from the published "absolute" surface-charge densities.

The above-cited example on Cd/hematite indicates that some groups perform titrations in the presence of solutes different from "innocent" electrolytes. Such titrations may yield important macroscopic information on the proton balance of the suspension in the presence of such a solute (Table 2). However, the exact proton stoichiometry of some surface complex can rarely be inferred, because this would require that only one complex exists and that the protonation states of the surface groups, which are not contributing to that particular surface complex, are not affected by the adsorption process. This can, at best, be assumed in a qualitative interpretation but can be quantitatively handled with the mean field approximation and the corresponding assumptions inherent to the respective computer programs. In fitting some models to adsorption data, proton data will constitute an independent and very valuable dataset representative of the system; however, they may be restricted to sufficiently high solute to sorbent ratios.

Overall, with respect to the sorbent titrations, the experimental procedures, the results, and their interpretation, there appears to be no common agreement. The diversity of results has repercussions on the modeling of titration curves and on PZC values. For the more comprehensive models, which attempt to explain "the universe of sorbents" but not the diversity among sorbent samples, this means that the respective authors necessarily have made a choice among the published experimental data. This also provokes the question of whether this choice was made before or after the model development. The question as to how comprehensive these models actually are is open.

(b)   Adsorption Experiments.   Adsorption experiments are usually carried out in the batch mode. A test tube containing known amounts of solid, solute, and electrolyte is equilibrated for a defined period under a controlled atmosphere at constant temperature and pressure. At the end of the experiment, the pH of the suspension is measured (before or after solid–liquid separation) and the amount of solute remaining in solution is determined. From a mass balance for the solute, the amount adsorbed is determined. Potential problems can be seen in the solid–liquid separation. These encompass at least the following points:

Incomplete separation: Particles remaining in the supernatant will falsify the analytically determined amount of solute adsorbed.

Effects of the separation step on the equilibria: compression of double layers during centrifugation, losses of solute during filtration.

A phenomenon that has often been explained by artifacts (such as incomplete solid–liquid separation) is the so-called solid-concentration effect. This can, in principle, be verified in several ways (e.g., by measuring the amount of sorbent in the supernatant: for goethite one might determine Fe in the supernatant). Honeyman and Santschi [24] labeled their hematite radioactively to check for possible problems in solid–liquid separation. They actually found that some hematite was still in the supernatant. Unfortunately, potentially remaining particles are usually the smallest ones, which have the largest specific surface areas and, therefore, relatively large amounts of solute can be sorbed on such particles.

Even if the study by Honeyman and Santschi clearly showed the problems associated with solid–liquid separation, the solid-concentration effect still circulates in the recent literature [25]. In these latter studies, the authors studied metal ion sorption onto goethite in the presence of a phosphate buffer to control the pH; phosphate strongly sorbs onto goethite itself, so that observed particle concentration effects should not be interpreted without considering the action of phosphate. It is highly probable that the metal ions were sorbed on a (more or less) phosphated goethite and not to a pure goethite.

Another problem related to solid–liquid separations occurs if the solubility of sorting material is exceeded in the bulk solution or at the sorbent surface. It is crucial in experiments whenever solubilities may be exceeded. For full control of the system, it is necessary to use acidified metal ion solutions and to avoid a rapid increase of the pH values (by a highly concentrated base solution). Otherwise, the solution phase may be at least partially oversaturated or polynuclear compounds may form.* If solubilities are exceeded, the solute is precipitated and might be (completely, partially, or not at all) found in the supernatant phase depending on the conditions chosen for the separation step (i.e., filter width, centrifugation speed). Distinction between adsorption and (surface) precipitation is usually not possible based on macroscopic data alone, so that the interpretation of the macroscopic data is difficult.

An elegant approach avoiding problems with solid–liquid separation involves the application of ion-selective electrodes (ISEs). Ludwig and Schindler [26] have tested this in the Cu–TiO$_2$ system and found good agreement between the results obtained with the ISE and in the conventional experiments. Experiments in the Cu–goethite system have been reported by Robertson and Leckie [27]. Data obtained by the present author on the same system also show that it is possible to have reliable results with the ISE for Cu. Similar attempts with other metals have failed [28]. One major advantage of using ISEs is that titration (i.e., proton data) and uptake data (i.e., solute adsorption data) are obtained simultaneously. This assures that the timescales for the reactions of proton and solute data are the same. Also, the imposed equilibrium criteria for the two electrodes can be used to constrain subsequent titrant additions. Such consistency is not always obtained when data from short-term titrations and those from adsorption experiments of usually longer duration are merged in one model. With these kind of data, it is highly recommended to obtain self-consistency on the timescale [29].

(c) Electrokinetic Measurements. Several electrokinetic methods are available. The raw data yield indirect information on the charge of particles. Raw data (typically electrophoretic mobilities) are used to obtain the so-called zeta potential. The zeta potential corresponds to the potential at the solid–liquid interface at a usually unknown distance from the surface plane. This unknown location makes it difficult to use the zeta potential in terms of quantitative data without additional assumptions. Attempts are made either to estimate the location of the shear plane (e.g., to involve the location of the shear plane as an additional adjustable parameter, which may be ionic strength dependent [30]) or to predict zeta potentials as the potential at the head end of the diffuse layer.

The most frequently used electrokinetic method in this area of research results in the determination of particle mobilities in an applied electrical field; this is done by one of the following:

- Electrophoresis: The sample cell is intersected with two coherent laser beams forming a series of interference fringes. Particles moving through the fringes in response to the applied electric field scatter light with an intensity which fluctuates at a frequency related to their velocity. The scattered light (transformed to an electrical signal) is

---

*Comparison of results of careful and harsh experiments would be of interest.

analyzed by a digital correlator to give a frequency spectrum from which the particle mobility distribution is calculated. For this method, low suspension densities (low particle concentrations) are required.

- Electroacoustics: The particles move in an applied field; if the particle density differs from that of the liquid, the motion generates an alternating acoustic wave. The electrokinetic sonic amplitude is measured as a function of frequency. From the electroacoustic spectrum, both the particle size and zeta potential can be determined. High suspension densities are possible with this method.

A further method available among others involves the determination of the so-called streaming potential. The particles are fixed in a liquid flow and the potential difference of electrodes on opposite sides of the material plug to be analyzed is measured as a function of the liquid flow.

These measurements allow the determination of the IEP, and because no reference is needed in this technique, finally the pristine point of zero charge of a sorbent sample is obtained. Although information on charge at a certain distance from the particle surface is not really available, the pH, at which the particles do not move in an applied field (or where the potential difference measured is zero), is an indication for the absence of charge on the probe particles. This particular pH is called the IEP. If the IEP coincides for different values of ionic strength and with a correctly determined CIP, this pH can be identified as the PPZC (at least of the sample). The relative charge obtained in a titration can then be renormalized to absolute charge.

Restrictions to this procedure exist. Potential problems can be inferred from calculations performed by Fair and Anderson [31], which show that particles with a heterogeneous charge distribution can have finite mobilities although their overall charge is zero. For many particles, such a heterogeneity is expected (different crystal planes with different PPZC values) and, therefore, an assumption is still required to fix the overall PPZC of a sorbent. Coincidence of IEP (at low ionic strength) and CIP is a strong indication that, at least for the studied sample of the sorbent, the PPZC has indeed been identified.

A general comment might be made on the pH measurements involved in all of these experiments. The pH is a master variable and its determination is crucial but may be difficult in suspensions. In particular, in (micro)electrophoresis, the suspension density is low, which makes the system very labile with respect to perturbations and pH measurements are not very reliable. Furthermore, contaminations will affect the measurements much more than in systems with high-suspension densities.

(d) Further Methods. Further methods are available for obtaining macroscopic data in suspension comprising stability tests, rheologic measurements, and sedimentation tests for the evaluation of the charging state of particles, voltammetry for direct speciation of solutes, measurements of volume changes following adsorption processes and so forth. These are not discussed in detail here, mainly because they are not commonly combined with modeling studies.

2. Microscopic Data

Whenever microscopic data are involved in a surface-complexation study, this will result in more requirements for the model. As will be discussed in more detail later, very simple models should be excluded from comprehensive studies because they can hardly display certain aspects of the adsorption mechanism.

A range of methods is available allowing the detailed study of solid–liquid interfaces and the structure of surface complexes. These include force measurements between identical particles, between a particle and a probe, or between different particles. These approaches can give valuable information on the structure of the interface and will help to exclude the nonappropriate models from the battery of models currently applied. Furthermore, the spectroscopic approaches covering x-ray absorption spectroscopy, vibrational spectroscopy, and so forth can elucidate the bonding mechanism (e.g. inner versus outer sphere) and surface complex structure (e.g., monodentate versus bidentate). All of these methods will help to more fully understand adsorption phenomena and improve modeling approaches, first by excluding the too simple models and, subsequently, by refining the appropriate models.

At present, methods like x-ray absorption spectroscopy are becoming standard tools. However, the combination of a maximum of appropriate methods will be the ultimate goal to gain a maximum of information. Hopefully, the macroscopic methods will not be completely forgotten in the rush to high-tech applications.

B. Experimental Data Used in Inverse Modeling

In the step from the raw experimental data to data that can be treated by conventional computer programs, transformations are often necessary. For titrations, the raw data are EMF values as a function of titrant added (volume changes may be more or less important). These would typically be transformed to $log[H^+]$ (or pH), total acid concentration, and dilution factors (e.g., for FITEQL). With LAKE [32], raw data can be treated directly. An input file or the code itself must, in such a case, provide all the information necessary for the calculation of equilibria (e.g., LAKE must perform the transformations to obtain concentrations from the raw data). For the majority of codes, however, pH or $log[H^+]$ as a function of titrant added is required. EMF values must be transformed to pH or $log[H^+]$ using the applied calibration procedures. Here, the calibration and, sometimes, assumptions in the data treatment may be important. Calibration can be performed on the concentration or on the activity scale. The concentration scale has the advantage that, a priori, no assumptions about activity coefficients are necessary; the disadvantage is that such an approach is limited to one ionic medium (the constant ionic medium approach), although in work on suspensions, the variation of the ionic medium certainly is of importance. The activity scale requires assumptions concerning the treatment of activity coefficients; it must be realized that such assumptions must lead to self-consistency between the finally presented experimental data and the model calculations.

Comparison of experimental data to modeling results may be influenced by the way the data are presented. For the results of titrations in plots of specific surface-charge density versus pH, deviations between model and experiment often become much more transparent than in plots where the total concentration of acid–base (TOTH) is given as a function of pH ("raw" data, including volume changes, which usually are difficult to estimate). In the latter, the contributions from the solution are included and usually become the major contribution to TOTH at extreme pH values; these plots, therefore, tend to hide discrepancies (also fundamental deviations) between model and experiment, which would be very much apparent in the former way of presenting data. Similar problems arise from noncritical application of goodness-of-fit parameters obtained with optimization routines: If they are used as absolute estimates for how well a certain model is performing, they may be completely misleading if the applied experimental error estimates are not realistic. When intercomparisons on one dataset are performed, these goodness-of-fit parameters are useful in a relative sense. Also, in cases where experimental error estimates are available, because they have been estimated from repeated experiments, and if they can be adequately treated in a parameter estimation program, the goodness-of-fit parameter gives a good indication on the discrepancy between model and experiment. However, correlation coefficients between optimized parameters should still be used to have a feeling for the potential interdependencies between adjusted parameters. Inverse modeling has been mainly done using macroscopic data in surface complexation modeling studies. Microscopic data are used in a more qualitative way to constrain the models but it can also be used quantitatively.

In the next sections, several selected aspects of macroscopic and microscopic data are discussed. Because of the importance of the acid–base properties of the solids for the modeling concept, the discussed macroscopic data are limited to surface-charge density versus pH curves. A survey of actual adsorption data is beyond the scope of this chapter.

1. Macroscopic Data

In the broader context of surface complexation modeling studies, much experimental work has been performed. This means that many experimental data are available. For many systems, comparable data have been measured by different groups. These are hardly ever taken to check data compatibility. As Fig. 1 might suggest, this would result in some discussion and require explanations. Many authors avoid extensive comparisons or chose those that do not result in problems (unfortunately, there is much to choose from). Only more recently have studies been published which attempt an evaluation of various datasets. For a comprehensive model approach, all reliable data must be taken into account. As a consequence, discrepancies between data must be explained. Attempts to do this are presented in the following examples to illustrate a few possible aspects. Figures 6–8 show surface-charge data of nominally identical sorbents.

For TiO$_2$ (Fig. 6, data from Sprycha [33], Kosmulski and Matijevic [34], Tiffreau [35], and Lützenkirchen [36]), agreement among different datasets is satisfactory, in particular when the extreme pH values are disregarded. For amorphous MnO$_2$ (Fig. 7, data from Fu et al. [37], Catts and Langmuir [38], and Lützenkirchen [36]), the discrepancies are apparent, whereas for amorphous SiO$_2$ (Fig. 8, data from Pilgrim [39], Joppien [40], and Lützenkirchen [36]), the coincidence of the data is very good.

In the case of TiO$_2$ (Fig. 6), only data for selected commercial products are plotted (Merck [33] or Aldrich [34–36]). There is good agreement among the data. The most important deviations occur at low H, which may be due to the fact that free and total proton concentrations become similar. Whether good results can be obtained or not (in particular in the extreme pH

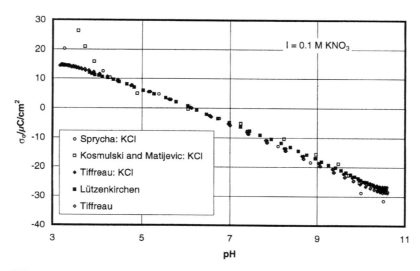

FIG. 6 Specific surface-charge density of commercial TiO$_2$. Data from Sprycha [33], Kosmulski and Matijevic [34], Tiffreau [35] and Lützenkirchen [36] if not otherwise noted obtained in potassium nitrate medium (fast titrations).

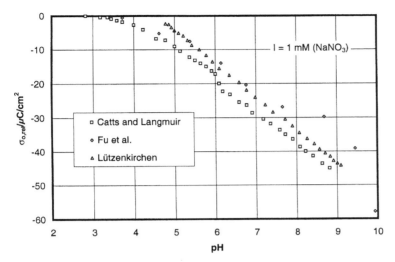

FIG. 7 Relative specific surface-charge density of amorphous MnO_2 (fast titrations). Data from Catts and Langmuir [38], Fu et al. [37], and Lützenkirchen [36].

range, but also in general) is, among others, a question of how much solid actually is or technically can be used in the titration. With a low solid concentration, results of surface titrations may be rather questionable. Unfortunately, some experimental work can be found, where the amount of solid was very low (e.g., Marmier's titration of goethite involved so little solid that "apparent" saturation of the goethite occurred at pH 4 [41], whereas Lövgren et al. [17] reached such a limit at pH 3). The "apparent" saturation is interpreted here as the limit of resolution of the potentiometric method for the respective experimental conditions. This is discussed in more detail later in this chapter. Kosmulski and Matijevic [34] reported extensive zeta potential measurements with their sorbent, which is nominally identical to that used by

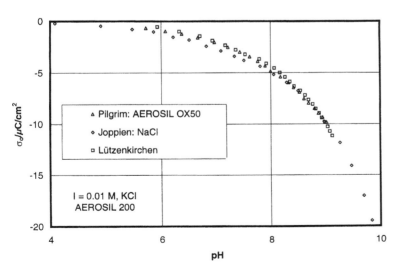

FIG. 8 Specific surface-charge density of amorphous SiO_2 (Aerosil). Data from Pilgrim [39], Joppien [40], and Lützenkirchen [36].

Tiffreau [35] and Lützenkirchen [36]. These measurements were used by the last two authors to fix the point of zero charge, despite the risk that different charges of that commercial product might give different surface properties. However, the good agreement among the data appears to justify this data treatment.

For amorphous MnO_2 (Fig. 7), it is much more difficult to fix this reference point. Although the data shown are at relatively low ionic strength, the curves do not sufficiently show the typical S shape at low surface-charge densities to fix a PPZC value; in particular, positive charges are missing; therefore, the data should be taken as relative surface charge, which may explain the different levels of otherwise nearly parallel curves. Similar results are obtained for 10 mM and 100 mM sodium nitrate concentration. Compared to silica (Fig. 8), which has also a low point of zero charge, in the majority of cases the charging curve for MnO_2 is rather linear than bent and, therefore, the fixing of the zero level is very important. Furthermore, the amorphous MnO_2 phase, identified as vernadite by Tiffreau [35], may show variations between apparently identical preparations (e.g., with respect to the exact stoichiometry, the amount of sodium in the structure may vary), so that comparisons are not necessarily expected to be successful.

The experimental results shown in Fig. 8 for amorphous silica coincide very well. Here, the reference point of zero charge is not so important as long as it is chosen sufficiently low. The example is intended to show that the surface-area measurements can be quite important. The supplier of AEROSIL200 gives a value of 200 m^2/g, whereas the actual measurement resulted in a value of only 145 m^2/g for the sample used by Lützenkirchen [36]. Discrepancies in measured specific surface area may arise if:

- Different methods are used for the measurements (e.g., gas adsorption versus microscopic methods; in gas adsorption measurements, the surface areas of solids are measured and particles, which are sticking together, may "hide" the surface area, which in a stable suspension would be available; microscopic methods will typically be applied to selected particles, and one has to ascertain that they are representative of the sample; other methods require assumptions with respect to the particle geometry to calculate surface area from size).
- In one method different probe molecules are used (e.g., for gas adsorption Kr versus N_2 versus Ar versus H_2O, which may display different behaviors); smaller probe molecules may yield a denser packing.
- In gas adsorption with the same probe molecule different sizes of the molecule are assumed (e.g., by assuming different orientations of the probe molecules when adsorbed).
- In gas adsorption, different outgassing temperatures and periods may be applied.

For consistent comparisons of surface-charge data, self-consistency of the surface-area measurements should be checked, especially when discrepancies need to be explained, but also if consistency is obtained. Otherwise, different surface-charge-density curves for one sorbent will be explained by some other phenomenon (i.e., not by the different surface areas), which, in turn, may lead to important conclusions with respect to the properties of the solid.

In the case of quartz (MINUSIL, Fig. 9), data from two different sources [42,43], which are not in agreement, can be made to coincide. The open symbols are the original data, which can be shifted by changing the zero level to yield the drawn lines and thus coincide more or less with the independently measured values.

The above examples indicate that apparent discrepancies exist. However, surface-charge densities for identical sorbents measured by different groups should coincide. If this is not the case, it should be possible to find the cause(s), in the easiest case by careful study of the experimental procedures. If these are not reported in sufficient detail, one might redo the

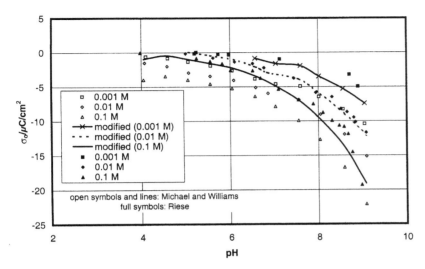

FIG. 9 Specific surface-charge density of quartz (MINUSIL). Data from Michael and Williams [42] and Riese [43]. The lines were obtained from the data from Michael and Williams by setting the zero level for all those data to about pH 4.

experiments, if the sorbent is still available commercially or can be resynthesized. Doubts will remain, if the sorbent cannot be obtained in the same form as used in the original work which will often be the case with commercial products, for which properties may vary from one production cycle to the next.

2. Microscopic Data

Similar discrepancies as documented in the previous section for macroscopic data can be found for the spectroscopic approaches, which are now available for studying the structure of surface complexes in situ (i.e., wet samples). With respect to inverse modeling these studies would attempt to resolve the structure of the surface complexes in a certain system and to impose such structures in the surface complexation model. This would avoid extensive discussion about the mode of bonding (e.g., inner versus outer sphere; monodentate versus polydentate).

An example of such a discrepancy is obtained from comparison of spectroscopic (XAS) data for the arsenate goethite system. The Fourier-transformed spectra obtained for identical conditions (i.e., Fig. 3 in Ref. 44 and Fig. 2 in Ref. 45) show some difference. The authors of Ref. 45 did not realize how close their conditions were to the ones by Waychunas et al. [44] for one of the samples and they, therefore, did not discuss this issue.

In addition to spectra themselves, their treatment and, finally, the interpretation are the steps that lead to the information, which can be used in a surface complexation model. Interestingly, again with an arsenate iron(III) oxide mineral, the data treatment and the interpretation of XAS data by Waychunas et al. [44] have been assessed by Manceau [46]. For a nonspecialist in this technique, who is interested in the conclusions of the spectroscopic studies, such examples recommend a critical point of view.

Experimental procedures during the preparation of XAS samples is another issue that may influence results. Thus, it must be avoided in preparing stock solutions of metal ions under conditions in which ions other than the bare metal ion is present; this not being the case, spectroscopic results showing multinuclear surface complexes in an adsorption study may be due to the presence of the multinuclear metal complexes in the stock solution (of typically rather high concentration). This problem might, for example, have occurred in the study of lead sorption to goethite [47]. Washing or not washing the samples is another difference in the procedures practiced. Furthermore, cooling of XAS samples reduces background noise and is therefore often used; Bargar [48] has shown that this may affect the spectra compared to spectra taken at room temperature.

For infrared studies of surface complexes some discussion exists between different investigators with respect to sulfate [49,50] and phosphate [51,52] adsorption to iron oxides. Again, preparation of the samples (drying or not) but also treatment of raw spectra and their interpretation is a controversial topic among specialists.

These examples (which are hopefully not the rule) indicate that for microscopic data, which would be the best information to use in a model, controversial standpoints and disagreement in the treatment of raw data may exist. The modeler, who wants to use such results, has the choice to either become such a specialist himself or to trust in one of the published conclusions (randomly, because of a larger number of experimental data or because of personal contacts).

The combination of different spectroscopic methods increases the confidence with which such conclusions can be used.

The macroscopic data are required for the model calculations (unless sufficient data are obtained from spectroscopy to allow a full characterization of the system) and microscopic data are needed to decrease the degrees of freedom and to allow more realistic assumptions about the structures of surface complexes in the modeling of the macroscopic data. For acid–base properties, such additional approaches are being developed:

- Surface force measurements can yield valuable information about the structure of the EDL (electrical double layer) [53] and also yield points of zero charge values [54] in the best case for individual crystal planes.
- X-ray standing waves have been used to determine ion profiles in the diffuse layer [55].

In the future, more comprehensive model approaches are expected which incorporate such data. However, in some systems, the modelers most probably will still have the option to choose from more than one mechanistic interpretation.

C. Technical Aspects of Modeling

In a technical sense, modeling includes the code used, the input data given to the code, the specified model, and the numerical output by the code.

Fitting experimental data to obtain stability constants is widely done by a *pqr* search. This term from solution chemistry means that the stoichiometries of species are systematically varied in the input file to an optimization code that fits a model corresponding to the respective *pqr* values to experimental data [cf. Eq. (1) as an example; of course, there may be more components taking part in the reaction and there will usually be more than one surface complex]. The *pqr* combinations which give the best numerical fit is accepted as the final model:

$$p\mathrm{H}^+ + q\mathrm{Me}^{n+} r\mathrm{SOH}^m = (\mathrm{SOH})_r\mathrm{Me}_q\mathrm{H}_p^{(p+nq+mr)} \tag{1}$$

For a proton–metal–surface system, Eq. (1) can be considered as the actual chemical part of a surface complexation model, which would be specified in an input file to an optimization code. Inverse modeling would consist in a number of calculations, in which the stoichiometries are varied. The procedure can be improved by adding supplementary information on species obtained from independent (typically spectroscopic) methods. A more comprehensive fitting procedure would allow one to fit spectra, titration data, and all other potentially available data simultaneously, as in LAKE [32], which can use macroscopic and nuclear magnetic resonance (NMR) data. The model that gives the best overall fit would be the one that is accepted.

Weighting of the different data sources might be a problem in such a procedure. Codes like UCODE [56] are freely available and allow a high degree of flexibility in solving such problems and fitting "normalized" data, thus minimizing or even avoiding weighting problems.

Problems persist with the surface complexation models when it comes to defining the electrostatic model and the protonation formalism to be used. Furthermore, in "realistic" models, an option of how to treat the surface heterogeneity must be chosen. The choice of one of the available options for these three points is the very first step (i.e., before any modeling exercise beyond acid–base properties). Therefore, the description of the acid–base properties of the sorbent deserves particular attention. There is no common agreement on the choice of the model elements and, unfortunately, the arguments presented in favor of the simplest model (which at the same time can be seen as a special case of the most complex and most realistic model) appear to be ignored in many modeling studies.

The 1-pK Stern model first proposed by Bolt and van Riemsdijk [57] is able to describe acid–base properties of oxide minerals. Fitting data with more complex models will often be in vain (e.g., for the triple-layer model [58,59]), i.e., it will yield a range of parameter sets, which can statistically not be distinguished by the goodness-of-fit parameter. The parameters will also be highly correlated in such cases, again a feature which UCODE [56] can indicate for all optimized parameters; FITEQL [60], which is the standard tool for surface complexation modeling, does not allow optimization of capacitance values and detailed analysis of goodness of fit versus capacitance curves is required to permit estimation of interdependencies. However, for multilayer models involving more than one capacitance, this becomes a multidimensional problem requiring a huge number of calculations with FITEQL [60].

As indicated, modeling typically starts by defining an acid-base model for the sorbent (left-hand side of Fig. 10) and then proceeds to the interaction of other solutes with the sorbent (right-hand side of Fig. 10). Some parts of the different components toward a comprehensive model approach have already been discussed and associated problems have been mentioned. The availability of the information in the different boxes in Fig. 10 may constrain or increase the complexity of the model. Thus, the purpose of the modeling itself, which will have consequences on the experimental information gathered on a certain system, and the resulting amount of information are the important aspects, which determine how simple or complex a model may be.

The output from the modeling tools is often not used to the extent possible. Often, the FITEQL results used are the numerical values of the optimized parameters and the overall goodness of fit. Sometimes, also the standard deviations are considered. The numerical value of the goodness-of-fit parameter and the standard deviations are dependent on the defined experimental error estimates. The values for these, which are most frequently used, are the standard values, which may not at all be reasonable for the actual equilibrium problem treated [36].

Other important results are linear correlation coefficients between the optimized parameters. These can, in all cases where more than one parameter is defined as adjustable in FITEQL, be used to discuss the model and the obtained parameters.

A FITEQL modification allowing for pointwise introduction of error estimates, which has been tested on some of the datasets given in Figs. 6, 7, and 8, for which extensive experimental work yielded reliable error estimates, numerical goodness-of-fit estimates were found to be appropriate [36]. In general, such extensive data are not available and goodness of fit must be checked graphically.

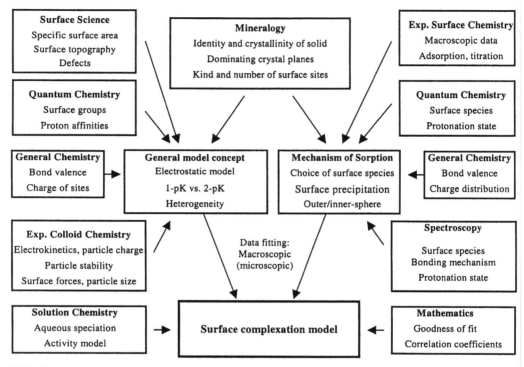

FIG. 10 Components of an advanced modeling approach.

D. Use of Modeling Results

The most obvious practical use of modeling results (i.e., stability constants, site densities, and capacitance values) is in the prediction of the behavior of solutes outside the laboratory conditions (e.g., for environmental or industrial purposes). This is, in the first place, hampered by the variety of experimental results obtained with one solid (e.g., goethite, cf. Fig. 1). It is further complicated by the different models used (see Sec. III.A and III.E).

Published parameters (in particular for the description of the acid–base properties of the sorbent) are also used for the interpretation of pure laboratory adsorption studies, though. This saves the experimentalist the sorbent titrations. Again from Fig. 1, it is obvious that this will not necessarily result in the actual description of the sorbent. A similar procedure in solution chemistry is very rare. Here, the experimentalist properly investigates the fundamental equilibria, has done so in previous studies, or can rely on previous investigations.

III. MODELS

In this section, the actual models are discussed under both general aspects (i.e., concepts that are not model-specific) and, subsequently, in a more specific way (i.e., the general aspects are illustrated in selected examples).

A. General Model Aspects

Prior to the introduction of various individual models, three points may be discussed independently. These include elements of the final model, which can be defined from the acid–base properties of the respective sorbents:

- The protonation mechanism of the bare surface in an electrolyte solution
- The structure of the electric double layer
- The consideration or not of surface heterogeneity

The treatment of the diffuse double layer or the description of zeta potentials obtained from electrokinetic methods may be done in different ways. Further aspects which can be discussed (right-hand side of Fig. 10) concern the charge distribution of sorbed metal ions or ligands, their mode of bonding to the surface functional groups, and their location in the double layer. This will be discussed in more detail in Sec. III.E on the selected model variations.

1. Protonation Mechanism

The first major difference to be discussed is the protonation mechanism. There are basically two options to choose from:

1. The so-called 2-pK approach, which can be characterized by the equation

$$SOH_2^+ = SOH^\circ + H^+ = SO^- + 2H^+ \tag{2}$$

SOH$^\circ$ is, in this case, a "generic" surface group, which can yield or take up a proton. In this mechanism, which has been originally proposed by Parks [61], two consecutive deprotonation steps of the surface species SOH_2^+ are considered. This is the conventional protonation mechanism, which is usually mentioned in the major textbooks.

2. The 1-pK approach, which is characterized by one deprotonation step:

$$SOH_2^{+0.5} = SOH^{-0.5} + H^+ \tag{3}$$

This formalism has been introduced by van Riemsdijk and co-workers [57,62]. In cases, where the 1-pK model is used with only one surface site, this surface site should also be considered generic. The 1-pK formalism can then be interpreted as a simplification of a very complex reality without substantial loss of accuracy compared to the equivalent 2-pK model, but with a significant reduction in adjustable parameters.

If the second deprotonation for the SOH$^{-0.5}$ species (i.e., the formation of the SO$^{-1.5}$ species) was to behave as a comparable functional aquo-group in aqueous solution, it would not be of importance. For solutes, consecutive deprotonation of one functional aquo-unit is typically separated by more than 10 pH units.

Thus, the 1-pK mechanism would be a more reasonable analogue to aqueous solutions than the 2-pK mechanism (where consecutive protonation typically occurs within a rather narrow range of pH). The surface group used with the 2-pK mechanism would be alternatively interpreted as a combination of two SOH$^{-0.5}$ groups of Eq. (3), yielding the SOH^{-1} species in Eq. (2). Such justification of the use of Eq. (2) results in some conceptual problems, when considering monodentate and multidentate surface complexes. A further major advantage of the 1-pK model is that the logarithm of the stability constant corresponding to the surface chemical reaction can be directly inferred from experimental data, because it is identical to the PZC, whereas for the 2-pK model, two stability constants must be determined. These two pK values are symmetrical around the PPZC, so that a reduction to one as yet unknown parameter for the 2-pK formalism is possible if the PPZC is known. This actually appears as an unnecessary complication, which will propagate when, for example, temperature dependencies are to be incorporated. The PZC can be measured as a function of temperature or predicted by theory [63].*

2. Structure of the Double Layer

A second distinction among the surface complexation models arises from the postulated structure of the electrical double layer.

Figure 11 shows the persistent fundamental models in use. The structure of the electrical double layer defines so-called charge potential relationships. These allow one to calculate the respective potentials from a charge balance over the (charged) surface species. Such a balance is made for each interfacial plane considered. With these, the electrostatic effects on the surface chemical equilibria can be taken into account. Obviously, this is done in a model-dependent fashion. There are some indications in Fig. 11 concerning the conditions, under which the different electrostatic models are expected to apply. One major constraint is given by ionic strength. High ionic strength typically decreases the influence of the diffuse layer, so that the compact part of the double layer dominates. Thus, the constant-capacitance model (Fig. 11a) is a good approximation at high ionic strength.

Only for sufficiently low ionic strength is the diffuse-layer model (Fig. 11b) a good approximation. Neither of these two simple models explicitly accounts for the formation of ion pairs between charged surface groups and the ions of the electrolyte. This is done by the electrostatic models involving more layers (Figs. 11c and 11d). The ion pairs will contribute to the electroneutrality of the particle in addition to the diffuse-layer charge.

The constant-capacitance model does not invoke the electroneutrality condition for the particles. The observed electrolyte-specific behavior of a sorbent cannot be described by the diffuse-layer model with one parameter set. The constant-capacitance model may be considered electrolyte-specific, but it is also concentration-specific and, thus, extremely limited.

The more elaborate the structure of the interface becomes in a model, the more realistic the model will be. Thus, one might argue that a multilayer model would be more realistic because it may consider various aspects that are phenomenologically expected but not taken into consideration by the models shown in Fig. 11. The anion and cation of the electrolyte have different sizes and, therefore, their charges should not be situated in the same plane. Consequently, a model that attempts to account for this will add one layer (i.e., one plane of adsorption).

Combining a (de)protonation mechanism with an electrical double-layer model in the modeling of solute adsorption to a surface functional group results in a formalism to calculate the free energy of adsorption and thus obtain the respective intrinsic stability constant for a surface species.

As the simplest example, the adsorption of a proton to an arbitrary surface functional group, $\equiv SG^n$, with charge n is considered. The interaction of this group with a proton "close to the surface," denoted by H_s^+, may be written in terms of a surface chemical reaction:

$$\equiv SG^n + H_s^+ = \equiv SGH^{1+n} \tag{4}$$

---

*It should be kept in mind here that the precautions discussed previously with respect to the determination of PPZCs are, in all models, crucial for obtaining reliable model parameters.

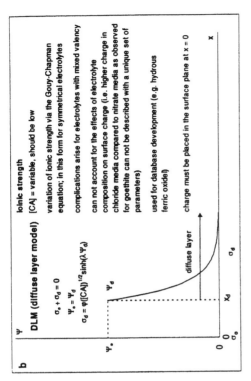
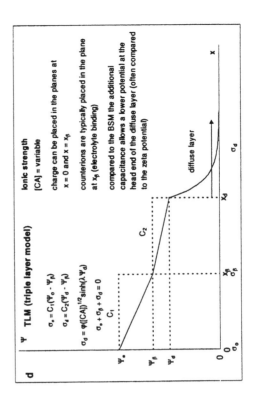
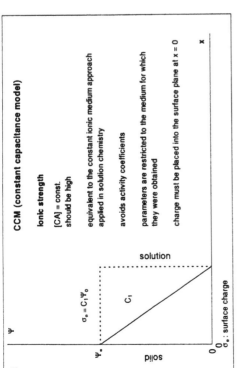
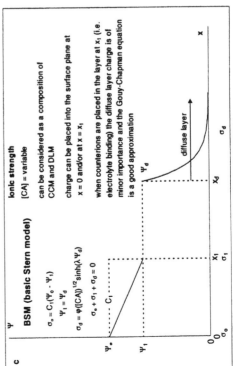

FIG. 11 Structure of the double layer. Standard electrostatic models.

Adsorption of one proton to the surface functional group is accompanied by the transfer of one charge unit from the bulk solution (where the proton has the measurable concentration $[H^+]$) to the interface. A first step is the transfer of the proton from the bulk of the solution "close to the surface." Then, the surface functional group can react with the proton close to the surface to form a protonated surface group. Both particle and bulk solution must be electroneutral, so coadsorption of one negative charge unit is required typically by an anion from the background electrolyte, which is not considered at all in the constant-capacitance model, occurs in the diffuse-layer only in the purely diffuse-layer model, and additionally occurs through the formation of ion pairs in the Stern model or the triple-layer model.

In the following, a simple distinction between chemical contributions (arising from bond formation) and electrostatic effects (originating from the charge buildup) is assumed. The electrostatic effects, which, in the case of reaction (4), would make it more difficult for another proton to adsorb both to that surface group (which is unrealistic within 10 pH units of the first protonation step) but also to surrounding (neighboring) surface groups, which experience the charge transferred by the first proton, and to all other surface groups, are considered in a mean field approach (e.g., discrete ion effects are neglected). This means that, for example, on a spherelike particle, charges on one side of the particle influence the adsorption behavior on the other side. In other words, charges which are present far away from a certain functional group have the same effect as those in the direct vicinity of that group. The overall free energy of adsorption for the above reaction can be written as

$$\Delta G_{ads} = \Delta G_{bond} + \Delta G_{elec} \tag{5}$$

In this approach, one has to distinguish between chemical and electrostatic contributions, which is a major problem. This distinction is made in a different way for the different electrostatic models, which were shown earlier. For Eq. (4), a mass law equation can be written as

$$K_+^{int} = [\equiv SGH^{1+n}] \cdot [\equiv SG^n]^{-1} \cdot [H_s^+]^{-1} \tag{6}$$

K_+^{int} is the intrinsic stability constant for proton adsorption to $\equiv SG^n$, which accounts for the chemical contribution represented as ΔG_{bond} in Eq. (5). This involves the free-energy change related to the interaction of a proton already close to the surface with the surface functional group. The overall (measurable) stability constant is given by

$$K_+ = [\equiv SGH^{1+n+}] \cdot [\equiv SG^n]^{-1} \cdot [H^+]^{-1} \tag{7}$$

In this way, the relation between the concentration of the proton "close to the surface" and in the bulk solution makes the electrostatic contribution ΔG_{elec} to the overall reaction describing the transfer of a proton from the bulk solution to the interface. Thus, compared to Eq. (6) for Eq. (7), a relation between the concentration of the proton in the bulk solution and the interface is required. The relationship typically is a Boltzmann factor:

$$[H_s^+] = [H^+] \exp\left(\frac{F\Psi}{RT}\right) \tag{8}$$

This represents the energy required to transfer one charge unit (or one proton) from the bulk solution close to the surface, where only chemical bonding energy will be required to make the surface species $\equiv SGH^{1+n}$. R, T, and F are the gas constant, absolute temperature, and Faraday's constant, respectively, and Ψ is the surface potential.

All mass law equations are written in terms of concentrations, which would be appropriate for the aqueous species in a constant ionic medium. Both for the proton at the surface and the surface functional groups in principle, activity corrections are required. Sometimes, these corrections are considered to be lumped into the Boltzmann factor. A sound thermodynamic treatment is hampered by the inability to experimentally determine activity coefficients both for solutes "close to the surface" and surface functional groups, but there is no reason to assume that they would not change with pH and ionic strength. Due to the small distances between reacting entities at surfaces, discrete ion effects are expected.

The models shown in Fig. 11 have been termed the "standard electrostatic models." They are typically implemented in many speciation codes available for surface complexation calculations. These models are not the only ones applied, though. In a subsequent subsection, it will become apparent that other models exist and that variations and even combinations of the electrostatic models shown in Fig. 11 are possible. There is probably no variation, which has not yet been applied, mentioned, or suggested with respect to the four standard models, also beyond known reasonable limits of the models, which stresses how diverse surface complexation modeling is. The consequence of this is that most approaches can be justified in some way (by reference to previous articles) and because of the obtained fit to the experimental data.

In the following, the different "standard" electrostatic models are described in some more detail.

Model a is the constant-capacitance model, which may be used in analogy to the constant ionic medium approach in aqueous chemistry: Restriction to one value of high ionic strength (in terms of composition and concentration of the electrolyte) assures constancy of activity coefficients of aqueous species and a model of the solid–liquid interface, which is sufficiently described by a compact layer. It is assumed that the drop in potential in the inner part of the electric double layer at the high electrolyte concentrations is quite extensive so that the diffuse part can be completely neglected.

Variations of surface charge with ionic strength (again, in terms of composition and concentration of the electrolyte) are not described with one parameter set, which therefore ideally should pertain to one ionic strength (still in terms of both composition and concentration of the electrolyte).

Model b is a purely diffuse-layer model. No compact layer is considered in this model, which pertains to low values of ionic strength and does not distinguish between different electrolyte ions (i.e., nitrate and chloride ions behave equally in this model). For aqueous solution, nonspecificity of the electrolyte is in fair agreement with the experimental observations at low ionic strength, but for mineral surfaces, the surface-charge density can be significantly affected by the nature of the counterion (i.e., for goethite, the surface charge density at constant pH will be higher in chloride than in nitrate media) [64]. Thus, one sorbent inherent diffuse-layer parameter set will not be able to describe ion-specific effects at interfaces, which are actually observed. A further limitation of the Gouy–Chapman equation, which is typically involved in the model implementations, is its limitation to symmetrical electrolytes. In many situations, aqueous solutions may contain mixtures of rather weakly adsorbing ions (Ca^{2+}, Na^+). Improvements with respect to the restrictions to low ionic strength are discussed in a later subsection.

Model c is the Stern model, which combines an inner layer with a diffuse layer. It can cover a range of ionic strength. It allows for electrolyte specificity by the ion-pair formation between the ions of the electrolyte and the oppositely charged surface functional groups. Thus, it is the simplest model that is able (1) to cover a broad range of ionic strength and (2) to describe electrolyte-specific behavior.

Model d is the popular triple-layer model, which has an additional layer compared to the Stern model. As the Stern model, it is electrolyte ion-specific. The additional layer lowers the potential at the head end of the diffuse layer and is intended to allow direct comparison between the diffuse-layer potential and measured zeta potentials.

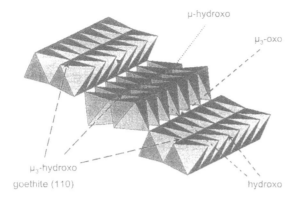

FIG. 12 Goethite surface structure for the (110) crystal plane. The different functional groups occurring on this plane are shown with the IUPAC terms (see also Fig. 13).

3. Heterogeneity

Equations (2) and (3) invoke one type of "generic" surface group. In reality, it is expected that several distinct surface groups exist on many surfaces. Furthermore, it cannot be excluded that not all functional units of nominally identical surface groups have identical affinities to some solute (e.g., to a proton). This may, for example, be caused by defects in the structure of a sorbent close to the surface.

Thus, at least two aspects of heterogeneity are phenomenologically expected, which can be assigned to "discrete" differences in properties (i.e., different surface sites) or to "distributed" (i.e., continuous) properties of otherwise identical surface sites (i.e., affinity distributions).

(a) Discrete Site Heterogeneity.   Figure 12 shows an example for the discrete heterogeneity of a defined goethite crystal plane. Several surface functional groups exist, where the surface oxygen is bound to one, two, or three iron atoms.

Figure 13 gives a supplementary explanation for the different functional groups. With the simplest (but also the more realistic) (de)protonation formalism, a 1-pK mechanism will apply for each of the distinct functional groups. For a truly mechanistic model, strictly speaking, knowledge of the overall PZC of a sorbent is not enough, but the individual proton affinity constants for the different sites must be determined for each crystal plane of importance. This can be done as has been shown by van Riemsdijk and co-workers [65] by considering the

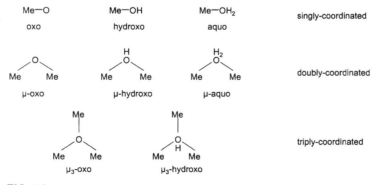

FIG. 13 IUPAC terms for surface groups. Me = metal center of the oxide, hydroxide or oxyhydroxide.

structure of the sorbent. For the computation of the equilibria, this may result in rather advanced equilibrium problems with codes such as FITEQL [60], in particular, when this must be done for all the prevailing crystal planes of a sorbent sample.

Figure 14 shows the surface-charge density of the goethite crystal plane in Fig. 12. For this crystal plane, most surface groups do not display variable charge at all over the pH range of interest, but rather influence the zero level (i.e., the overall PZC of the crystal plane). In fact, only one surface group (the singly coordinated one) shows variable charge behavior in the pH range of interest. It is also apparent that the MUSIC approach requires that two kinds of triply coordinated hydroxyls are present. The complex features can be substantially simplified by using a single ("semigeneric") site, 1-pK approximation, where only one surface hydroxyl (can be identified with the singly coordinated group and would have the same site density) is used with a proton affinity which corresponds to the overall point of zero charge of this crystal plane. To what extent this approximation can actually be used in ongoing adsorption studies is another issue; for example, when surface spectroscopy shows that several kinds of surface functional groups are involved in the formation of an important surface complex, then the full model would ideally be required in a mechanistic interpretation of such data with a surface complexation model.

At present, the predictions of the individual proton affinity constants for goethite based on the bond–valence principle by van Riemsdijk and co-workers are not in agreement with independent first principles calculations by Rustad and co-workers [66]. For silica, Rustad and co-workers found unrealistic values [67], which were more recently discussed by Tossel and Sahai [68]. These latter authors claim that erroneous assumptions by Rustad and co-workers resulted in the unrealistically high protonation constants for silica. Furthermore, Tossel and Sahai state that this error also occurred in the study of other minerals by Rustad and co-workers, but that the errors cancelled out in those systems. Other attempts to use first-principle applications to achieve a more advanced understanding of these issues have been published [69,70].

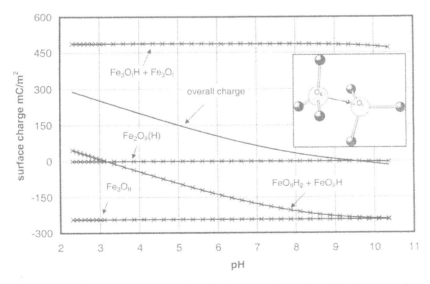

FIG. 14 Surface-charge density of the (110) crystal plane of goethite. The inset shows the concept used to derive the individual proton affinities for the different groups (Fig. 12). The only functional group showing pH-dependent charge is the singly coordinated group.

In most of the above-cited attempts, the first step is to describe the surface based on the sorbent structure. For example, the MUSIC model proposes that the interface is a consequence of the crystallographic structure. The quantum chemistry approaches typically assume clusters which are thought to be representative of the sorbent. This will always result in "discrete" heterogeneity. A priori, the role of solvent (water molecules) is typically neglected when the reactivity of the surface functional groups is evaluated in the first place because it can hardly be considered phenomenologically or computationally at present. For the MUSIC model predictions of proton affinities for sorbents with hydrogen bonds in the structure of the sorbent, it is of utmost influence whether a hydrogen bond is considered to persist at the interface [14]. Also, if distances between atoms are affected by surface relaxation (causing different distances between atoms at the surface as compared to the respective atoms in the structure), discrete proton affinities can be strongly altered. On the other hand, in the quantum chemical approaches, the basis sets used and the size of clusters may affect the results.

This is usually checked by changing the basis sets and varying the size of the clusters. In the absence of water molecules, gas-phase acidities have been obtained and are compared to the acidities in an aqueous medium, which, furthermore, contains electrolyte ions. Despite the present shortcomings, with more extensive basis sets and the ever-increasing computational power, these approaches will become more and more important. Ultimately, they will also allow the consideration of the solvent water and electrolyte ions.

The role of the water structure at interfaces in the context of electrolyte adsorption has been discussed by Dumont and co-workers [71]. Based on previous work, these authors propose considering structure-breaking or structure-making properties of the solids and electrolyte ions.

Structure-making electrolyte ions would be small ions with high charge. Structure-making interfaces would have water molecules strongly attached. Structure-making electrolyte ions would be well adsorbed on structure-making surfaces, because water molecules which otherwise would be strongly ordered can be released, resulting in a net gain in entropy. These phenomenological approaches allow a qualitative explanation of many features and thus stress the importance in considering the properties of the solvent in the discussion of interfaces. Surface complexation models can indirectly account for these effects to some extent by the choice of capacitance values, which correspond to distances in the interfacial layers, and by the location in which electrolyte ion charges are placed. These can be chosen according to the size of hydrated electrolyte ions.

Most of the above-cited applications realized the importance of the presence of different functional groups, so that agreement, at least on this aspect, is achieved. In the corresponding modeling applications, these different functional groups are typically considered as distinct entities. Thus, for every functional group for which a distinct proton affinity* is obtained, only this proton affinity is taken into account.†

This first approach considers "discrete" sites which make up a heterogeneity of the interface through the variety of sites. This can be described by Eq. (9):

$$\theta_{t,\text{solute}} = \sum_j \left(f_j \theta_{j,\text{solute}} \{ K_{j,\text{solute}} [\text{solute}] \} \right) \tag{9}$$

In Eq. (9), [solute] is the molar solute concentration in the solution, θ is the surface coverage (adsorbed amount of solute divided by total amount of surface sites), f is the fraction of surface sites of class j, $K_{j,\text{solute}}$ is the affinity constant for the local sorption isotherm describing the interaction between the solute and the surface site j; the index t stands for the overall (total) adsorption.

*The applications with respect to predicting affinities presently do not go beyond proton affinities.
†Proton affinities of nominally identical groups are not necessarily identical and may differ for different crystal planes.

(b) Continuous Heterogeneity. A second often applied approach might start from a classical surface complexation model and add a distribution of site properties. This approach takes into account that the model surface sites are not energetically homogeneous (i.e., they may have more or less different affinities).

In the context of surface complexation, a sorbent with only one reactive surface group like gibbsite might be a typical example where the actual application of such a model would be reasonable. Otherwise, this approach has been typically applied to "generic" surface sites for different models. The model would be characterized by

$$\theta_{t,\text{solute}} = \int_{\text{range}} (\theta_{\text{solute}}\{K_{\text{solute}}, [\text{solute}]\}\phi\{\log K_{\text{solute}}\})d\log K_{\text{solute}} \tag{10}$$

and would correspond to a continuous distribution of affinity constants (with the distribution function $\phi\{\log K_{\text{solute}}\}$) for one surface site (no distinction with respect to the class of surface sites is considered necessary in this case). Integration is done over the range of affinity constants. This approach requires a distribution function for the affinity constant.

Examples like gibbsite are not the rule (i.e., in most cases, a single-site model does not correspond to what is actually expected). Therefore, a combination of Eqs. (9) and (10) is, in principle, required to account for the different distinct sites and their potential individual continuous heterogeneity. This results in

$$\theta_{t,\text{solute}} = \sum_j (f_j \int_{\text{range}} (\theta_{j,\text{solute}}\{K_{j,\text{solute}}, [\text{solute}]\}\phi_j \{\log K_{\text{solute}}\})d\log K_{\text{solute}}) \tag{11}$$

Major problems in treating Eq. (11) are as follows:

- The knowledge of the presence of different surface groups is often restricted to well-defined solids, such as well-crystallized goethite; for powders, the determination of crystal planes is, at present, a formidable task for environmentally relevant samples, such as coatings of reactive minerals on sand, the characterization will also be problematic.
- Different affinity distributions can be assumed, but the experimental determination of actual distributions is, at present, not directly possible.

In the discussion of surface heterogeneity, it should also be added that defects probably exist on most kinds of surfaces to some extent. These will result in very reactive sites, which might be of particular importance for the interaction with trace components (in the absence of other significantly competing adsorbing solutes). One should, therefore, bear in mind that it is very probable that these very reactive sites exist in, thus far, unknown amounts, even if presentations like Fig. 12 tend to suggest the opposite.

B. General Aspects in the Choice of a Model

For proton adsorption to minerals, at least three aspects of potential importance may be identified and have been discussed in more or less detail in the above sections:

- Proton adsorption is accompanied by electrostatic effects (i.e., a correction is necessary, for which an electrostatic model must be assumed).
- Various distinct sites exist on one crystal plane and each of them may have a different proton affinity; different crystal planes of one sorbent may have different sites; different samples of one nominally identical sorbent may display different dominating crystal planes.

- One site will not necessarily be energetically homogeneous (i.e., its proton affinity will be distributed).

Trying to fit a proton adsorption isotherm (i.e., a surface-charge density versus pH curve) by including one of the above aspects will usually be successful. Therefore, by adding the other aspects, the numbers of adjustable parameters will necessarily increase, but the fit of the model to the data will usually not become significantly better. From the macroscopic data alone, it is impossible to estimate which of the above features is most important. As a consequence, it is not astonishing that these aspects are used in the literature to model experimental data. At the same time, the respective authors advocate their approach and sometimes claim that one of the above features is of major importance or others are of minor importance. Some selected examples are discussed in the following sections:

- Contescu et al. [72] studied proton binding to alumina and concluded, based on the closeness of their titration data at various values of ionic strength and their hetero- geneity analysis, that electrostatic effects were of minor importance. Interestingly, they found proton-affinity constants which were in close agreement with predictions by Hiemstra et al. [13] with a model, which later was shown to yield correct predictions based on wrong assumptions for goethite (see Sec. III.F).
- Pivovarov [73] claimed that his model (which involves an exponential term in the treatment of surface chemical reactions, which is related to the loading of the surface) may explain data without electrostatics; however, the mathematical formalism used in that article is very close to that of electrostatic models.
- Single-site surface complexation models (i.e., one generic site plus electrostatics) are among the most popular models; this can be explained by the success of these models and by the related simplicity; however, it is often difficult to make a compromise between the model which has the smallest number of adjustable parameters and the effects that can be described by that model [e.g., the diffuse-layer model is often chosen, because it is simple and because it can (technically) be applied to various values of ionic strength]. Nevertheless, it cannot describe the effects of different electrolytes on the surface charge of a certain sorbent and the major speciation codes do not even allow it to be used with nonsymmetric electrolytes, which may often be important in nature. Because the diffuse layer is responsible for the charge potential relationship in that model, it might be too simple in that respect for a range of applications; for other models, the diffuse layer will be of minor importance in the description of electrostatic effects. These aspects are usually not discussed when a model is chosen and might not be transparent to users of published surface complexation parameters.

In general, in the evaluation of models that claim to be nonelectrostatic it should be realized that the electrostatic nature of mineral–water interfaces is a fundamental feature which has been investigated with a variety of methods. Therefore, a realistic model should not exclude this feature or a model should not have the claim to be realistic, when electrostatics are completely disregarded. Further complications arise when decisions are required for the following choices:

- The proton adsorption mechanism; that is, 1-pK versus 2-pK; these are the most commonly used approaches, but as yet no agreement is in sight which of the two currently used mechanisms is to be pursued.
- The electrostatic model (Fig. 11). For the four electrostatic models discussed earlier, it has been shown that they all successfully describe electrostatic data in combination

with the 2-pK formalism [74]; for the 1-pK formalism, the diffuse-layer model was shown not to be adequate [75], but this still leaves at least seven options. It is important to note that the 2-pK diffuse-layer model also has certain disadvantages, which have frequently been hidden by presenting the acid–base properties as the total acid concentration versus pH instead of using surface-charge density. Although the above studies only refer to data at one ionic strength, with the options the modeler can choose from, similar results can be produced for ranges of ionic strength, again except for the diffuse-layer model.

These choices should be discussed in each surface complexation modeling study in some detail and it must be clear that the choices can be understood with respect to the actual purpose of the study and the accumulated experimental data.

C. Nonscientific Influences on the Choice of Models

There is little doubt that the choice of models is largely influenced by the school of thought that has been or that is "close" to the user, as well as by the textbooks the user came across when starting the modeling. Once established at a certain location, a model philosophy develops and usually persists for some time. Another major influence comes through the modeling tools which are available. A final aspect is in specific applications for which no modeling tool is available (or sufficiently user friendly). These three aspects are shortly addressed in the following subsections.

1. Schools of Thought

There are at least three schools of thoughts which can be distinguished in the literature:

- The Swiss approach, which developed from the seminal work of the groups of Schindler and Stumm [1,2]. These groups basically laid the foundation for the formalism of surface complexation modeling and applied very simple electrostatic models, namely the constant capacitance model [1] and the diffuse-layer model [2]. The models were combined with the 2-pK formalism proposed by Parks [61]. The solution chemist's approach can be found in the constant-capacitance model and was typically respected by applying high values of ionic strength in the experimental work. Parameter estimation was mainly by graphical techniques, because the currently used computer codes were not available at that time.
- The Australian [3]/American [4] approach, which led to the widespread use of the 2-pK triple-layer model; much of the traditional knowledge on solid electrolyte interfaces was incorporated into this model. Nowadays, it is still the favorite model in American circles, which deal with surface complexation and which were not influenced by the Swiss approach.
- The Dutch approach [57], which advocated an electrostatic model based on a Stern model and introduced the 1-pK concept. This approach can be interpreted as a special case of a much more comprehensive model. It is, at the same time, simple and versatile. The numbers of adjustable parameters are more restricted, in particular compared to the 2-pK triple-layer model.

The Swiss approach is sometimes exclusively mentioned in major textbooks which deal with aqueous chemistry oriented toward the environment and has also influenced the choice of the model in a widely appreciated book, whereby the first self-consistent surface complexation database was established by Dzombak and Morel for hydrous ferric oxide (2-line ferrihydrite) [76]. This book has been of major influence in particular for users, who find the database that

TABLE 3   Equilibrium Tableau for a 1-pK Model

Species	Stability constant	Component: $\equiv SG^n$	Component: H^+	Component: $\Psi_0(\sigma_0)$
$\equiv SG^n$	1	1	$0(n)$	$0(n)$
H^+	1	0	1	0
$\equiv SGH^{1+n}$	10^{PZC}	1	$1(1+n)$	$1(1+n)$
OH^-	K_w	0	-1	0
Total concentrations		$TOT\equiv SG^n$	$TOTH^+$	σ_0/B

Notes: First numbers: coefficients in mass law equations; calculated horizontally (e.g., for OH^-: $[OH^-] = K_w[H^+]^{-1}$; for $\equiv SGH^{1+n}$: $[\equiv SGH^{1+n}] = 10^{PZC}[\equiv SG^n] \exp[F\Psi_0/(RT)][H^+]^{-1}$), calculation of species concentrations; numbers in brackets: coefficients used for mass balances (if different from first numbers), mass balances are calculated vertically (e.g., for H^+: $TOTH^+ = n[\equiv SG^n] + [H^+] + (1+n)[\equiv SGH^{1+n}]$. Component $\Psi_0(\sigma_0)$: first number corrects for the change in potential by transferring charge from solution to the interface; number in brackets indicates the actual charge of the surface complex formed in layer o. The factor B transforms specific surface charge density into molar concentrations.

helps them describe the adsorption or transport of inorganic pollutants in systems, where hydrous ferric oxide surfaces are expected to control adsorption phenomena.

Textbooks on colloid chemical phenomena usually exclusively mention the 2-pK triple-layer model. The 1-pK formalism is rather restricted to the special literature and hardly mentioned in textbooks, which most probably strongly hampers its broad application.

2.   Computer Codes

Computer codes from the MINEQL family have merged the formalism for the description of purely aqueous systems with surface complexation. The development of the codes was such that the 2-pK models were considered only, also because the articles introducing the 1-pK models had not been published at that time. Thus, the codes are designed for the 2-pK formalism, although codes which allow for noninteger stoichiometric coefficients and/or a distinction between mass law and mass balance coefficients in principle can handle 1-pK calculations. In order to use the 1-pK approach with codes like FITEQL [60], it is necessary for the surface chemical reactions to proceed with a distinction between the stoichiometric coefficient in the mass law equation and the mass balance equation [cf. Table 3 for a simple example corresponding to Eq. (11)]. Many other computer codes do not allow this distinction between the two matrices (for mass law and mass balance equations) and do not include an option for an automatic treatment of the 1-pK formalism. So in order to be able to use a 1-pK formalism, specific software has to be purchased (such as ECOSAT [77]); alternatively the treatment by FITEQL becomes much more tedious. Finally, some codes have built-in databases including data for the 2-pK model (e.g., MINTEQA2 [78]), so that, at present, with such codes the choice is obvious as to which protonation formalism and electrostatic model should be used when hydrous ferric oxides are of interest.*

3.   Specific Applications

There are specific applications which cannot be correctly treated in the present state of development of the official version of the dominant computer code (i.e., FITEQL [60]). This

---

*It should be noted that the computer codes are not always free of errors. Thus versions 2 and 3.1 of FITEQL do not correctly treat the dilution of the solid, which yields problems when substantial dilution factors are obtained in titrations or in combinations of data from different experiments. Modified versions of version 2 have been corrected. The MINTEQA2 (ver. 3.1) database includes errors and the code may even read some database entries incorrectly.

includes, in particular, studies on temperature dependence. For all models except the constant-capacitance model, the Gouy–Chapman equation is used:

$$\sigma_d = -(8I\varepsilon\varepsilon_0 RT \times 10^3)^{1/2} \sinh(\Psi_d F/2RT) \tag{12}$$

In Eq. (12), σ_d is the specific surface-charge density of the diffuse layer (the diffuse-layer charge), I is the ionic strength in molar units, ε is the dielectric constant of the medium (water), ε_0 is the permittivity of free space, R is the gas constant, Ψ_d is the potential at the head end of the diffuse layer (the diffuse-layer potential, sometimes equated to the zeta potential obtained in some way from electrokinetic measurements), F is the Faraday constant, and T is absolute temperature.

The default temperature in FITEQL is 298.15 K. If temperatures different from this default value are to be treated, this is not possible with models that include the diffuse layer, because the Gouy–Chapman formalism, Eq. (12), includes temperature either directly or indirectly via the temperature dependence of ε.*

For a formally correct treatment of data at temperatures substantially different from 298.15 K, the constant-capacitance model has often been used. With the constant-capacitance model, a very simple charge potential relationship is used:

$$\sigma_0 = C\Psi_0 \tag{13}$$

In Eq. (13), σ_0 is the specific surface-charge density, C is the surface-specific capacitance, and Ψ_0 is the surface potential. In an exponential term, which corrects a surface chemical equilibrium constant for electrostatic effects, the argument would typically be

$$\frac{F\Psi_0}{RT} = \frac{F\sigma_0}{RTC} \tag{14}$$

F/RT is a fixed value in the FITEQL source code, which cannot be changed in an input file. Thus, a value used in the FITEQL (namely C) input file can correct for the temperature. Because of this unique possibility, many groups apply the constant capacitance model, especially those using FITEQL for inverse modeling. In other codes involving databases, which, in principle, allow for temperature-dependent calculations, the documentation often does not indicate to what extent temperature effects are actually incorporated (e.g., temperature dependence of aqueous activity coefficients or of the dielectric constant of water). The present author has developed a modification of FITEQL, which allows one in the FITEQL input mode (e.g., no database) to self-consistently apply temperature variations. This so far user-unfriendly version can be improved by involving user-friendly input routines. It is also planned to develop an independent speciation code for such purposes, which can be coupled with UCODE, thus allowing a maximum of flexibility.

In summary, it seems that the choice of models is to some extent controlled by nonscientific reasons, which include the respective education of researchers, the modeling tools, or the particular application to be treated. The availability of computer codes, the "difficulty" in handling the 1-pK equilibria in FITEQL, or the cost of programs, which include the 1-pK approach, restricts some groups to the 2-pK models. Also, it may seem difficult to accept the noninteger charges in the 1-pK approach when familiar with the more popular 2-pK formalism. Furthermore, it might not seem necessary to change modeling strategy as long as a familiar and as yet widely accepted model is successful. Finally, the application to temperature dependence often restricts the user to certain models.

---

*In many cases, data which, according to the specifications of the authors, were obtained at room temperature or 20°C have been used in FITEQL. However, the errors introduced by this inconsistency are expected to be of minor importance.

The choice of models is, at present, largely author dependent and not much influenced by an analysis of the published information. Sometimes, the available computer software restricts users to certain models (e.g., the 2-pK formalism is much easier to apply with the common codes than the 1-pK formalism). As is clear from the above statements, a general decision for a certain model would only be possible by disregarding opinions of certain authors. The choice of a model should therefore, as indicated in Section I, be largely associated to the objective of the modeling exercise and be considered in that way. If electrostatic features are of no interest whatsoever in a certain system, then why not use a nonelectrostatic model? However, in case a mechanistic interpretation of ion adsorption is the objective, then it would be difficult to understand why electrostatic interactions should be disregarded.

Failure or success of a model can always be discussed in terms of the underlying acid–base model; for example, if failure is observed and the acid–base model might be interpreted to be too simple (e.g., because only a diffuse layer model has been applied or because electrostatics was completely neglected), the authors in principle need to show that the failure is not related to the use of the diffuse-layer model or the complete neglect of electrostatics, respectively. If success is obtained and the authors, for example, based their acid–base model on parameters published by others, they have not provided any proof that their modeling parameters are really applicable to the system. The ion adsorption parameters are influenced by the acid–base model and if the acid–base model parameters do not pertain to the system, the whole model may be assessed. Some examples are discussed in the subsequent subsection.

D. Use of Previously Published Model Parameters

It has become a popular procedure to use surface complexation parameters from the literature for sorbent acid–base properties, to skip the titration of the sorbent, and to conveniently start an adsorption investigation by doing pH-edges and/or isotherm studies. The following examples show that the acid–base properties of a sorbent should be considered as fundamental for any study attempting to derive some model parameters for ion adsorption. This means that either acid–base properties should be experimentally obtained or that a very comprehensive literature review is necessary to obtain an idea of which published parameters or data are best for the system to be studied.

Figure 1 recalls that the latter option may be a very difficult task, which may consume more time than the supplementary measurements. Also, this supplementary work assures that extensive data with exceptional conditions or original work (as discussed in the following two examples) can securely be used in data reinterpretation within a consistent and comprehensive dataset. The following cases serve as examples, where literature parameters have been applied to "pure" sorption studies and where a detailed analysis shows that an acid–base study of the respective sorbent or a comprehensive review of the literature parameters would have had the potential to increase the self-consistency (and thus the quality) of the modeling work.

1. Yang and Davis [79] have published an extensive study on EDTA and metal ion adsorption to TiO_2. Their study goes beyond much of the previously reported studies in that they study the presence of two different metals, also in the ligand-containing system. In a previous paper by Vohra and Davis [80], extensive PbEDTA adsorption data on the same sorbent were reported, and relatively unusual conditions were studied (namely nonequimolar concentrations of ligand and metal). These datasets would be very good examples for testing models. Davis and co-workers attempted to model their data using the diffuse-layer model. They started by accepting selected literature acid–base parameters for their particles. These are required for extracting the

modeling parameters for the individual metal ion and EDTA adsorption data. The accumulated parameters are applied to systems containing all components. The diffuse-layer model already fails, as early Cu and Cd adsorption to TiO_2 occur simultaneously in the absence of EDTA. This is ascribed to the presence of different sorbent sites with respect to Cu and Cd adsorption, which is not sufficiently considered in the single-site approach. However, the failure of a model, which may be judged to be too simple, is not a serious argument for such mechanistic conclusions, in particular if additionally the submodel (i.e., acid–base) parameterization can be assessed. Furthermore, surface complexation models which have been successful in one situation are known to easily fail in more complex situations. In particular, in ternary (competitive) systems, predictions based on parameters obtained in the noncompetitive systems may fail to describe the experimental data. This observation by Yang and Davis [79] is quite similar to what Goldberg [81] found for competitive sorption of anions to goethite with the constant-capacitance model. Vohra and Davis [80] also had problems in modeling PbEDTA sorption at non-equimolar concentrations. The models used in all of the above-referenced articles are very simple. In particular, for the adsorption of a bulky complex such as CuEDTA or CdEDTA, allocation of charge to only one plane (inner-sphere complexation) is unrealistic and should not result in any mechanistic discussion or conclusion. On the contrary, such a model should be taken as a more sophisticated fitting option, which (1) technically allows one to calculate the behavior of the system under variable conditions (although probably only to a limited degree), (2) is therefore to be preferred to a K_d approach, and (3) allows, in principle, predictions, which may also fail outside the calibration conditions. Based on Vohra and Davis [80] such failure in predictions would be highly possible with the model parameters by Yang and Davis [79] for metal-to-EDTA ratios different from unity. The use of the diffuse-layer model with the low ionic strength used by Davis and co-workers is not violating the range of applicability of this model. In the technical use of the simple model, it might be argued that the authors have chosen surface acid–base properties from a publication [82] on a "nominally" identical sorbent. Thus, they treat published stability constants for a sorbent as if they were reliable thermodynamic data. Furthermore, the original data were obtained in sodium nitrate, whereas Davis and co-workers work in sodium perchlorate. With suspensions, details such as the treatment of the solid and the nature of the electrolyte may change the acid–base behavior. Table 4 shows particle properties reported in the literature [83–87] for the sorbent used by Yang and Davis. Obviously, reported points of zero charge vary by little less than 1 pH unit, which might result in similar variation of 2-pK stability constants. The variation of the specific surface area within less than 10 m^2/g might be considered less severe, although the diffuse-layer model can be shown to be quite sensitive to this particle property. As indicated earlier, the diffuse-layer and constant-capacitance models have no way to account for electrolyte composition in a comprehensive way. However, electrolyte composition may affect the surface charge of oxides. For the sorbent used by Yang and Davis, Fig. 15 shows some literature surface charge data in comparison with the model chosen by Yang and Davis. Obviously, the model by Yang and Davis fails to predict these data. Although the diffuse layer model can be criticized because it does not describe electrolyte composition effects, the two selected experimental data sets in Fig. 15 would suggest that it is a good model in this case. However, a wider range of data does not substantiate such a conclusion and the modeling study of Bourikas et al. [30] strongly supports this statement. Therefore, a diffuse-layer model

TABLE 4 Particle Characteristics of P-25 Degussa Reported in the Literature

Authors	A_{BET} (m²/g)	PZNPC	Electrolyte
Foissy et al. [83]	52	6.2	NaCl
Girod et al. [84]	53 ± 2	6.6	NaCl
Busch et al. [85]	50	5.75	NaCl
Giacomelli et al. [86]	47	6.0	KNO₃
Rodriguez et al. [87]	51.4	6.5	KCl
Stone et al. [82]	55	6.5	NaNO₃

can at best serve as a system-specific model for a certain sorbent in a particular electrolyte. From Table 4, it is obvious that there is a difference both in standard surface properties (i.e., specific surface area) and in the postulated sorbent properties (i.e. point of zero charge). The surface charge characteristics determine the acid–base model. Errors in the particle acid–base model will propagate to the metal and EDTA adsorption models. It must be concluded from this example that with literature surface complexation parameters published by different authors, it is difficult to assure that a correct acid–base model for a certain sorbent is obtained. In aqueous solution studies, this would not be acceptable (e.g., no one would even come close to titrating a ligand in the presence of a metal if the ligand acid–base characteristics were not known). An acceptable option would be to use averaged particle properties but always keeping in mind that the acid–base characteristics are not necessarily well described and that already this issue may lead to unsuccessful predictions.

2. Toner and Sparks [88] attempted an analysis and assessment of double-layer models (the electrostatic models were all combined with a 2-pK formalism), which they based on the requirement of ability of the models to describe both equilibrium and kinetic

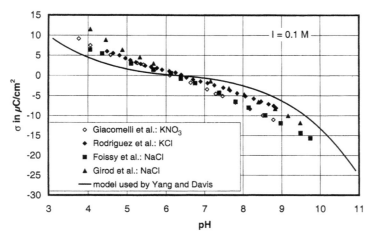

FIG. 15 TiO₂ (P25) surface-charge density as reported in the literature at $I = 0.1\ M$. Data from Giacomelli et al. [86], Rodriguez et al. [87], Foissy et al. [83], and Girod et al. [84]. The model used by Yang and Davis [79] for sodium perchlorate medium was obtained by Stone et al. [82] from experimental data in sodium nitrate medium.

data for boron adsorption to alumina. They considered data at one ionic strength. In principle, their attempt to assess surface complexation models with respect to their ability to allow a coherent explanation of equilibrium and kinetic data is very useful, because from restricted datasets, it is impossible to discard one or more of the models, as has been shown by Westall and Hohl [74] for the example of alumina titration data. The study by Toner and Sparks [88] is already original with respect to the combination of kinetics and thermodynamics. The authors finally conclude that only the triple-layer model was successful in explaining both static and kinetic data. However, this conclusion may have been seriously affected by the choice of surface acidity parameters (i.e., site density, capacitances, acidity constant, and electrolyte binding constants, where applicable). The constant-capacitance, diffuse-layer, and Stern models, although successful in describing the experimental boron adsorption data given by Toner and Sparks [88], are not in agreement with the kinetic data.

Three different aspects can be considered in this context:

- The use of published parameters such as capacitance values or electrolyte binding constants: Toner and Sparks [88] state that some of these values have been taken from Westall [60]. Furthermore, the capacitance value for the constant-capacitance model used by Toner and Sparks [88] comes from Goldberg and Glaubig [89], who determined this parameter for a $0.1\,M$ $NaClO_4$ medium, whereas Toner and Sparks [88] needed to use it with a $0.01\,M$ $NaNO_3$ medium. The capacitances and electrolyte binding constants for the triple-layer model used by Toner and Sparks [88] come from Westall and Hohl [74]. Although the point of zero charge of the alumina data used by Westall and Hohl [74] is 8.3, from the parameters given by Toner and Sparks [88] for the other surface complexation models the point of zero charge is calculated to be 7.9. It may, in general, not be the best approach to apply a surface complexation parameter obtained for a certain solid in some previous study to a nominally identical solid (as discussed earlier) when it cannot be verified or soundly justified that the use of such parameters actually yields a good description of observable properties of the solid.
- The value of the specific surface area used for the calculations is not specified. Because two different values are given (100 and $76.1\ \mathrm{m^2/g}$, respectively), it is important to specify the one used for the calculations. Depending on the electrostatic model chosen, this may be more or less important for the calculation of the electrostatic correction term: The specific surface area is for all model variations important for calculating the site concentrations from surface-specific site densities.
- The acid–base characteristics are the basis for the boron adsorption modeling. Thus, when the acid–base parameters (including capacitances, site density, and relevant intrinsic stability constants) are not appropriate, the boron adsorption parameters will not be either, although allowing equal fit to the boron adsorption data. Using the parameters given in Table 1 of Toner and Sparks [88], the surface properties at $0.01\,M$ ionic strength have been recalculated (Fig. 16). The calculations have been done in surface-specific units using FITEQL [60] and using unit solid concentration and specific surface area (in the dimensions required by FITEQL). This allows an exact calculation of surface-charge density for the constant-capacitance model, diffuse-layer model, and the Stern model (here used without considering electrolyte binding through ion pairs). Problems may occur for the triple-layer model, but the differences between an exact calculation and a

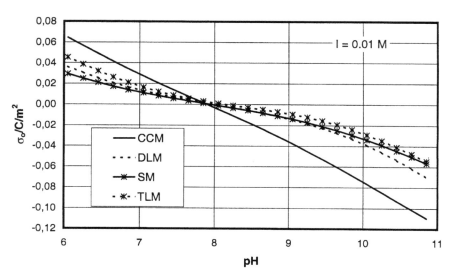

FIG. 16 Surface-charge density versus pH for four surface complexation models in 0.01 M calculated from the parameters given by Toner and Sparks [88].

calculation based on moles per square meter can be neglected, as has been verified. It can be clearly seen that parameters used for the four models by Toner and Sparks [88] yield very different surface-charge density versus pH curves. Especially, the description by the constant-capacitance model deviates significantly from the one by the other three models. However, even those are not as close as they could be. Westall and Hohl [74] have demonstrated that all of the surface complexation models they have investigated allow an identical description of experimental data. Thus, it is clear that at least three of the models with the parameters used by Toner and Sparks [88] are not appropriate in describing the experimental data pertaining to the sorbent. Which one is appropriate cannot be verified exactly because the sorbent-specific experimental data are not given. If now the relevant boron adsorption constant actually depends on the acid–base parameters, then the coincidence of the equilibrium constant from static data and the constant obtained from the kinetic data is biased by the chosen acid–base parameters (at least this cannot be excluded). In other words, by playing around with the acid–base parameters, it may be possible to find good agreement between kinetics and equilibrium data for each model option. Therefore, in the study by Toner and Sparks [88], it seems at least possible to come to other conclusions. Still, it is possible that the triple-layer model will be the best performing model if the calculations are done with appropriate acid–base parameters.

The procedure explained for the modeling approach used by Davis and co-workers but also the use of published parameters by Toner and Sparks, which is very and most probably too frequently found in the literature and does at present not seem to be a major point of concern to most researchers, is considered a bad habit by the present author. However, no one can really be blamed for this, because these bad habits are present in the literature as though they were perfectly agreed upon. Researchers in this field should be aware that they at least take the risk of supplying readers with incomplete datasets and should, instead, rather try to completely characterize the systems they study.

Similar provocative statements can be made about spectroscopic studies, where the comprehensiveness is even more limited in most cases (again the purpose and the goal of the study might restrict the respective authors to the mere spectroscopic work). Nevertheless, whenever such studies are done without the information allowing the reader to safely associate the results with a certain sorbent, the insight gained by the experiments might be very interesting, but in the end they are not very useful for a modeler because the results can be assessed based on the sorbent specific features (recall Fig. 1). Therefore, if the sorbent is not characterized sufficiently, at least information should be given to potential users of the spectroscopic results in order to retrieve the required knowledge a posteriori.

Some groups successfully combined the macroscopic, microscopic, and modeling parts and thus report complete experimental datasets that can be used to refine or test models in the future and show certain developments in their modeling approaches. In this context, it is necessary to stress that both kinds of experimental data must cover an experimental window that does not leave too many questions open (i.e., variations of pH and solute-to-sorbent ratio and possibly ionic strength/medium would be useful also for the spectroscopic experiments). Some limitations of the spectroscopic methods with respect to the solute loading exist and restrict such comprehensiveness. Low loadings, which in some applications may be of particular interest (trace pollutant concentrations), cannot always be studied by all methods. However, at present, a complete study in which all practicable methods have been applied to one system and in which it has been attempted to objectively test and compare models is clearly missing. By exclusion of models that cannot describe such comprehensive data, sound arguments will be at hand to decrease the numerous options in the modeling approaches, which will be further discussed in the following subsection.

E. Specific Model Aspects

In this subsection, particular aspects in modeling are discussed. These encompass several specific models which are discussed in more detail as compared to Fig. 11, the different adjustable parameters, and the estimation of their values, some already achieving improvements in the modeling approaches compared to the traditional models as well as some warnings concerning successful models.

1. Several Specific Models

In this subsection, nine specific models are introduced and discussed in some detail. They consist of a combination of the aspects which have been previously discussed (e.g., combination of a proton adsorption mechanism with an electrostatic model; heterogeneity is not discussed here, but can easily be added to the one-compartment models). Their "degree" of applicability is discussed based on (1) what can be achieved by the model, (2) what can be gained by the model compared to a simpler model, and (3) whether parameter estimation is feasible or not. Further developments of the previously discussed electrostatic models (e.g., improvement of the Gouy–Chapman approach, addition of adsorption layers) or independent developments (e.g., the multi-compartment model) are shortly presented. Figure 17 shows the nine models, which in the following will be referred to with the respective lowercase letters.

Model a is a combination of the electrostatic option constant-capacitance model with the 1-pK formalism. The reasonable range of ionic strength, for which this electrostatic option is applicable, is restricted to higher values. The stability constant pertaining to acid–base properties is given by the point of zero charge of the sorbent sample and can be adjusted according to the PZC values obtained in media of different concentration and composition with the consequence

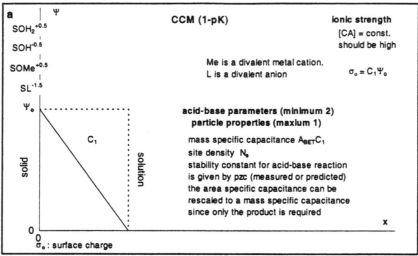

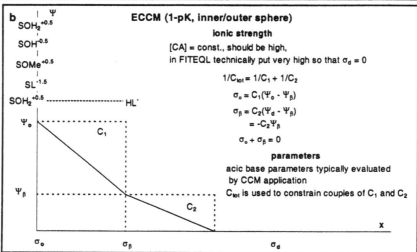

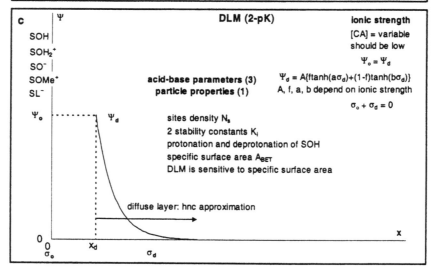

FIG. 17 Some selected model options (for detailed descriptions, see text). Particle properties not mentioned for models g–i.

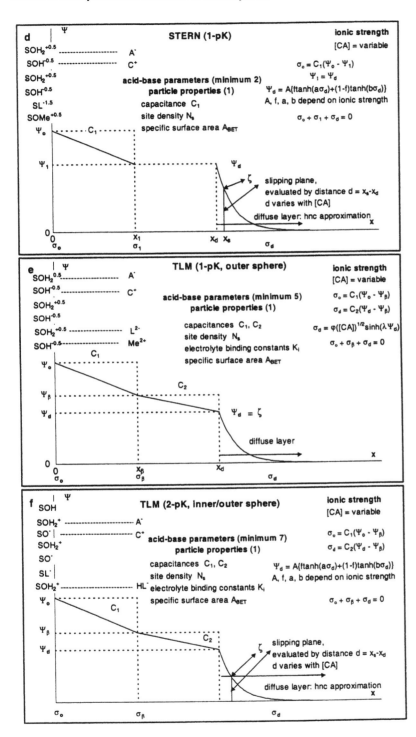

FIG. 17 *(continued)*

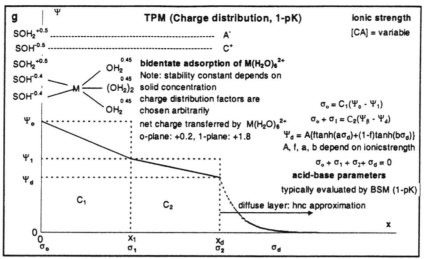

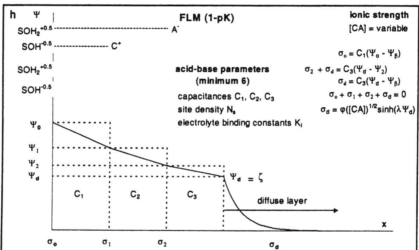

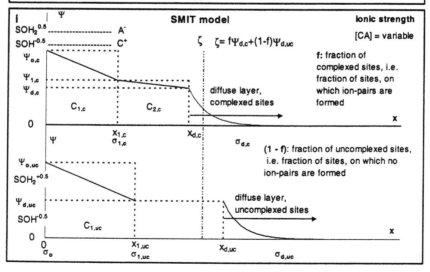

FIG. 17 *(continued)*

that different parameters (i.e., capacitance values or PZC values) need to be introduced for different electrolytes. The choice of site density (for a single-site model) or site densities (discrete heterogeneity) is not simple because a variety of approaches exists to "choose" this parameter. The remaining adjustable parameter can be fitted as a mass-specific capacitance, thus avoiding specific surface area in case it is not known or trusted (note that in this case, the use of surface-specific site densities makes no sense).

The constant-capacitance model typically works best with low site densities. Low values of site densities are usually obtained from sorbent titrations with excess acid or base or from numerical fitting of experimental data. Both options are not acceptable (as discussed in more detail later). If the goodness of fit to the data is not acceptable with realistic values for the site density parameter, there is no reason to continue using this electrostatic option in mechanistic models. It can be applied at constant ionic strength for simple (nonmechanistic) purposes, if good fit to some experimental data is required or if more advanced double-layer models result in numerical problems in coupled processes (e.g., reactive transport).

No electrolyte binding is usually considered in the constant-capacitance model. All charges must be placed in the surface plane, thus only allowing for inner-sphere surface complexation. All charges are considered as point charges. If variations of ionic strength or composition of the background electrolyte are to be described the model parameters must be adjusted. This has been done for the most frequent case (i.e., combination of the constant-capacitance model with a one-site 2-pK formalism) by changing the stability constants, changing the site density, changing the capacitance value, or by combinations of these. The same options apply in principle for a 1-pK constant capacitance model. By these options, the ionic strength dependence of the particle surface charge can be described. However, in several articles, no variation of any of these parameters with ionic strength was involved (e.g., Ref. 90) or a mean parameter set was assumed to be valid for any ionic medium in terms of concentration and composition (e.g., by Goldberg and Sposito [91]). Variations of the activity coefficients of aqueous species taking part in the surface chemical reactions may have been considered. However, even in these cases, the actual particle properties are considered insensitive to ionic strength. For purely aqueous equilibria, an analogous procedure would mean that no ionic strength correction whatsoever would be applied. Such handling of the experimentally observed dependence of acid–base properties of minerals on concentration and composition of electrolyte leads to numerical values of electrostatic terms, which are not realistic and will influence the modeling of adsorption data beyond the acid–base behavior. It is obvious that at least part of the experimental data relevant for such a system cannot be described with this approach. Furthermore, self-consistent constant-capacitance parameters are extremely limited in their applicability.

Model b is a hybrid between a triple-layer model (see models e and f) and the constant-capacitance model. The extended constant-capacitance model (ECCM) allows for two separate planes, in which charges can be placed. Compared to the constant-capacitance model, this permits the supplementary option for outer-sphere surface complexation (i.e., the protonated ligand HL^- is placed as a point charge in a plane at a distinct distance from the surface plane). Compared to the triple-layer model, no electrolyte binding is explicitly considered. Charge distribution is, in principle, possible (i.e., by considering simultaneous adsorption of inner- and outer-sphere complexes, by distributing the charge of one surface complex on different planes of adsorption or by combinations of the two). Model b is a simple extension of model a and allows one to additionally take into account spectroscopic results indicating that, for example, outer-sphere surface complexation occurs. In Fig. 17b, the electrostatic option inherent to the ECCM has been combined with the 1-pK formalism, also to have the correspondence to model a.

   The electrostatic model in combination with the 2-pK approach has been used by groups that prefer the constant-capacitance model for acid–base properties and have the constant ion medium approach. Thus, the acid–base properties of the sorbent have been determined in the framework of the constant-capacitance model and an overall capacitance for the ECCM is available. The problems inherent to the constant-capacitance model are inherited. What remains to be done is to determine the optimum location for placing the charge of the outer-sphere complex and the corresponding stability constant. The determination of the location is possible by determining pairs of C_1 and C_2 from the following condition:

$$\frac{1}{C_1} + \frac{1}{C_2} = \frac{1}{C_{tot}} \tag{15}$$

However, in case the charges of adsorbed solutes may actually range further out to the solution part of the particle than what is possible from the capacitance determined with the constant-capacitance model for the bare sorbent in an electrolyte, then the constraint by Eq. (15) does not apply.

   In computer codes, which do not include the ECCM, the triple-layer model can be converted to the ECCM by a sufficient increase of the ionic strength. In a technical sense, this leads to the breakdown of the diffuse layer in the triple-layer model and the potential at the nominal head end of the diffuse layer (Ψ_d) becomes zero.

   Model c is a combination of a purely diffuse-layer model with the 2-pK formalism. Using a purely diffuse layer as the electrostatic model has the advantage of having no adjustable parameter arising from the structure of the EDL. Also, variations of ionic strength are formally possible, but it is not possible to account for electrolyte-specific features with one set of parameters. Significant restrictions arise from this insensitivity of a purely diffuse-layer model to the composition of an electrolyte. Thus, the observed difference in charging behavior of minerals in electrolytes of the same concentration, but different composition cannot be described by a unique parameter set within the diffuse-layer model framework.

   Compared to the conventional equation for the diffuse layer, which is the Gouy–Chapman equation, model c applies the singlet hypernetted chain (HNC) approximation, which will be discussed in a separate subsection in more detail. The Gouy–Chapman theory neglects the finite size of ions, which may cause substantial effects, in particular when the ionic strength is high. In contrast, the HNC approximation accounts for these features. In principle, the use of the HNC approximation improves the purely diffuse-layer model compared to use of the Gouy–Chapman equation, but the application is rather limited because of the lack of numerically determined parameters for discrete values of ionic strength. Furthermore, there are not enough parameter values to justify a continuous function of the parameters as a function of ionic strength. Finally, the HNC cannot compensate for the missing Stern layer, which, in general, significantly improves the description of acid–base properties.

   Another disadvantage of the commonly applied Gouy–Chapman equation is its restriction to symmetrical electrolytes. As a consequence, mixed electrolytes cannot be treated at all by the simple Gouy–Chapman equation. Similar problems arise, whenever a low ionic strength (fixed by a symmetrical $z : z$ electrolyte) is used to study the adsorption of multivalent ions (charge $> z$) in relatively high concentration with respect to the ionic strength. Another limit of the electrostatic model occurs if the particles are so small that the flat plate geometry inherent to the Gouy–Chapman equation can no longer be justified [cf. Section III.E.3(b)]. All of those aspects are of minor importance when the diffuse-layer potential (e.g., the potential at the head end of the diffuse layer) is sufficiently low. This can be achieved by introducing a Stern layer.

   The purely diffuse layer only allows charge to be placed in the surface plane. Therefore, no outer-sphere complexation or charge distribution can be described with this model. However, extensions involving ion distributions in the diffuse layer might be interpreted as allowing for such phenomena [92].

Purely diffuse-layer models have become quite popular since the database for hydrous ferric oxide has been published by Dzombak and Morel [76]. The model calculations by these authors have shown that it is possible to describe a wide range of experimental sorption data within a relatively simple model framework. However, the description of acid–base properties of hydrous ferric oxide with this model is not convincing. Substantial failures with respect to true predictions can therefore be expected whenever dynamic systems involving the transport of protons are considered and variations of pH are possible (Lützenkirchen et al., in preparation). Nevertheless, for conditions in which this is not the case (i.e., buffered systems), the database is very useful but should not be used in the context of mechanistic discussions.

Model d is the Stern-layer model, which can be seen as a combination of the constant-capacitance model and the diffuse-layer model from the electrostatic point of view. It is here combined with a 1-pK formalism.

Compared to the constant-capacitance model, the Stern model allows variation of ionic strength via (1) the consideration of surface ion pairs with the electrolyte ions and (2) the diffuse layer. The ion pairs permit a variation of the electrolyte in terms of composition (i.e., the electrolyte binding constant can be adjusted for each ion). The diffuse layer is not ion-specific and, in principle, restrictions apply which have been pointed out for model c. Compared to a purely diffuse-layer model, violation of these restrictions will have less severe consequences for the surface-charge density because the Stern layer usually decreases the diffuse-layer potential compared to the purely diffuse-layer model. There is one adjustable parameter pertaining to the electrostatic model (the capacitance). For the 1-pK formalism, in addition to the site density parameter, two electrolyte-binding constants are required.

Whenever observations allow one to involve symmetrical electrolyte binding, only one electrolyte-binding constant is required, with the consequence that the differences in charging behavior as a function of electrolyte composition must be modeled by allowing some parameter to vary from one electrolyte to the other because the experimentally observed surface charge is a function of electrolyte composition. The options are as follows:

1. The capacitance is considered to be related to the electrolyte (i.e., for goethite, C_1 must be different in sodium nitrate media as compared to sodium chloride) and all those electrolyte-binding constants are kept constant, for which symmetrical electrolyte binding is required (because of the available experimental observations).
2. The electrolyte-binding constants vary (i.e., sodium will have a different affinity in sodium nitrate medium as compared to sodium chloride medium) and the capacitance is kept constant.
3. A combination of (1) and (2).

Neither of these options are optimal. Fortunately, recent measurements of electrokinetic properties with an acoustophoretic method show that at high electrolyte concentrations, the data do not allow the assumption of symmetrical electrolyte binding. Option (2) would be highly preferable. This permits a sorbent-specific capacitance value and simple mixing of electrolytes which will be encountered in real systems. The recent results, if interpreted in a mere mass law sense, appear somewhat surprising, because the solid concentration is increased much more drastically (by about three orders of magnitude) than the ionic strength (factor of 100) compared to conventional microelectrophoresis, where no shifts occur.

Van Riemsdijk and co-workers [30] have described zeta potentials with this model and had to involve additional parameters to obtain a reasonable fit to the data, because the measured zeta potentials were higher than the calculated potentials at the head end of the diffuse layer. The distance between the head end of the diffuse layer and the shear plane was adjusted and found to be a function of ionic strength, but independent of the sorbent. The modeling results by these authors also showed that for many sorbents, the capacitance is around $0.9\,F/m^2$ for well-crystalline solids, whereas for sorbents exhibiting surface roughness, a typical value is about $1.7\,F/m^2$.

Phenomenological interpretation related to water structure would suggest electrolyte- and sorbent-dependent capacitance values.

Model d is the simplest approach involving electrolyte binding and thereby electrolyte specificity. It has been found to well describe a wide range of data: Most probably more datasets on one kind of sorbent were tested than with other models. The modeling results in terms of the obtained parameter values are self-consistent and, in particular, the coinciding best-fit values for the capacitance for comparable samples of one sorbent have a substantial advantage.

For a model d variant using the Gouy–Chapman equation, an additional parameter is required at each ionic strength for the description of zeta-potential measurements. When applying the Stern model, van Riemsdijk and co-workers have used the Gouy–Chapman equation for the diffuse layer. In model d, the HNC approximation is used instead. This variation in the model concept is not expected to have the same consequences as in the case of a purely diffuse-layer model for the surface charge density. However, the Stern model has also been used without ion pairs [88] and in such a case, the use of the HNC approximation might be more relevant even for the surface charge. Calculated diffuse-layer potentials will be affected in both cases for sufficiently high electrolyte concentrations. The comparison of calculated diffuse-layer potentials with zeta potentials must be self-consistent [i.e., when zeta potentials are obtained based on a mean field approximation (which is typically the case); use of a more advanced theory to model zeta potentials makes no sense].

Sorbing ions are treated as point charges in model d. However, charge distribution can be handled with this model, but charges can be placed in no more than two planes. This will always result in direct competition with electrolyte ions (same plane of adsorption).

Model e adds a supplementary interfacial layer compared to the Stern model (model d). This supplementary layer merely has the purpose to sufficiently decrease the diffuse-layer potential and to have closer agreement with measured zeta potentials. One additional adjustable parameter is introduced (the capacitance C_2). Based on Eq. (15), the typically used value of $C_2 = 0.2\,F/m^2$ will control the overall capacitance of the compact part to the electric double layer. Contrary to model d, model e uses a Gouy–Chapman approach rather than the HNC approximation to account for the diffuse layer, but this can, of course, be varied. Otherwise, the discussion of model d also applies to model e.

For the description of zeta potentials with model d (but using the Gouy–Chapman-type diffuse layer), ionic-strength-dependent distances of the location of the zeta potential are required to have agreement with calculated potentials. In model e, only one global parameter (i.e., C_2) is expected to be sufficient for the description of zeta potentials. If it is not, then nothing is gained compared to the equivalent Stern model.

With respect to the mechanisms of ion binding, model e places all charges of adsorbed ions in the β plane, except the protons, which are placed in the surface plane. This corresponds to the original application of the triple-layer model by Davis et al. [4]. In the more recent interpretation this practice would mean that all ions are considered as outer-sphere complexes. The consequence is that competition between electrolyte ions (A^-, C^+) and the other ions placed in the β plane can be made substantial (this is also at least partially the case for charge distribution, which has actually been mentioned in these early papers).

Model f can be considered as an extension of the original triple layer model in that it is more flexible. The applied distinction between inner- and outer-sphere complexation, which was first discussed in the 1980s in further applications of the triple-layer model (e.g., Hayes et al. [93]), may influence competition between electrolyte ions (A^-, C^+) by placing the charges of other adsorbing ions in the surface or β plane as point charges or by charge distribution. This would, for example, allow the description of ionic strength effects on metal or ligand adsorption.

Based on modeling results, Hayes and Leckie [93] claimed that from the good agreement between experimental data and modeling results, which were critically dependent on the plane in which the charges of the adsorbing ions were placed, it could be inferred from the macroscopic experimental data whether inner- or outer-sphere surface complexation occurred. However, model calculations on the same systems using a purely diffuse-layer model showed that also inner-sphere complexation with a more comprehensive speciation of the adsorbing ions in the aqueous phase could explain the ionic strength effects [94]. Hayes and Leckie [93] used concentrations for all species in the mass law equations, whereas the use of activities by Lützenkirchen [94] introduces supplementary influences of ionic strength. Both model approaches used generic surface sites. The interpretation of the latter model calculations and the underlying experimental adsorption data should not go beyond saying that model calculations and experimental macroscopic observations can neither prove nor disprove an adsorption mechanism. When spectroscopic information is available, which allows distinction between inner- and outer-sphere adsorption, the triple-layer model may resolve these aspects. Also, coexistence of inner- and outer-sphere surface complexes can be described if required.

In the context of the electrostatic equations, the model as used by Hayes and Leckie involves the Gouy–Chapman equation. Combination with the HNC approximation is, of course, possible. If the value of the capacitance C_2 does not lead to the desired coincidence of calculated diffuse layer and zeta potentials, it is possible to introduce the distance between the head end of the diffuse layer and location of the slipping plane as a supplementary parameter to obtain better model results. However, this could also be achieved by the use of the equivalent Stern model, which allows all of the options of the triple-layer variant depicted as model f. Most triple-layer applications to electrokinetic data are not really convincing if goodness of fit is the major criterion. However, the Stern model was recently shown to be able to describe such data by either using a fitted location of the slip plane in the diffuse layer or even by equating the diffuse-layer potential with the zeta potential [95]. In the modeling of zeta potentials, some problems remain unsolved concerning the data and their treatment (here, particle shape, surface conduction, surface heterogeneity, and other features may be controversially discussed) or the comparison of zeta potentials obtained with different electrokinetic methods.

Model f is used with a 2-pK formalism; this introduces supplementary parameters (at least one adjustable parameter) compared to model e, namely the stability constants describing the acid–base behavior. For cases where the pristine point of zero charge is known, ΔpK_a can be introduced and only one additional parameter is required. ΔpK_a is defined as follows:

$$\Delta pK_a = pK_+ + pK_- \tag{16}$$

where $K_+ = [SOH_2^+]/[SOH][H^+])$ and $K_- = [SO^-][H^+]/[SOH]$.

The large number of adjustable parameters inherent to the 2-pK triple-layer model does not allow one to apply optimization procedures to obtain one unique set of parameters to describe the acid–base properties. Fitting all parameters simultaneously typically does not give any result at all. Systematic variation of certain parameters (while others are kept constant) indicates that multiple parameter sets do not significantly differ in their ability to describe the system.

It is quite important in this context that depending on the choice of parameters, the speciation of the surface may be either controlled completely by the uncomplexed surface sites (SO^-, SOH, SOH_2^+, in the case of sufficiently low ion-pair formation constants) or by the ion pairs ($SO^- - C^+$, $SOH_2^+ - A^-$, in the case of sufficiently high ion-pair formation constants). Because electrolyte adsorption is not readily measured (i.e., high electrolyte concentrations and little fractional adsorption is usually imposed by the relevant systems), there are no high-quality data corroborating either of these two options. The interdependence between ΔpK_a and the electrolyte-binding constants does not interfere in 1-pK models.

Use of ΔpK_a in combination with a point of zero charge as a constraint has been controversially discussed. Because the electrolyte ions may influence the point of zero charge as has been shown with electrokinetic experimental work for many relevant sorbents recently, the constraint equations for the models are actually more complicated. The application of ΔpK_a will always be a valid option, but the PPZC has to be available; the data by Woods et al. [22] indicate that titrations alone might not be sufficient to this end. For high electrolyte concentrations and ionic-strength-independent points of zero charge, a constraint equation is also required for the electrolyte (symmetrical electrolyte binding). With shifting points of zero charge (which can be expected based on the recent electrokinetic results), the ΔpK_a constraint is no longer valid if PPZC is not known. However, shifting isoelectric points can be used to obtain other constraints on electrolyte-binding constants.

The use of ΔpK_a is not required in a 1-pK formalism, so that compared to the equivalent 2-pK approach, at least one adjustable parameter can always be excluded from the discussion. For the multilayer models, the 1-pK approaches perform as well as the respective 2-pK formalisms, so that there appears to be no need to use the 2-pK formalism in these cases. With the 1-pK version of the triple-layer model, it has been reported that simultaneous optimization of parameters, which can be adjusted by FITEQL, is successful [75].

A summary of this discussion results in the following points:

- The 2-pK formulations of the triple layer model clearly have too many adjustable parameters. No unique parameter set can be determined. With multiple parameter sets showing the same ability to describe experimental data, users are free to choose among these parameter sets. This results in different acid–base parameters used for the same system by different groups, and, in principle, it cannot be excluded that the subsequently modeled equilibria (like adsorption of L^{2-} in model f) will be affected by the choice of the acid–base parameters; extreme cases of surfaces dominated either by the ion pairs or by the bare protonated and deprotonated surface species are possible.
- The 1-pK formulations have reduced numbers of adjustable parameters; these can be more easily determined numerically. The ability to model experimental data is not inferior in 1-pK models compared to 2-pK models (Stern and triple layer).
- The triple layer model does not significantly improve the description of acid–base properties of minerals when compared to a Stern model. The Stern model has fewer adjustable parameters (acid–base properties). For electrokinetic data, a detailed analysis should be conducted.

The conclusions from this summary ought to be that the triple-layer model should be abandoned in favor of the equivalent Stern model and that the 1-pK formalism should be used instead of the 2-pK approach. Although this is quite apparent from published results, the historical schools of thought still impose use of models, which, for an objective reader, would not be recommended any longer.

Unfortunately, despite very good arguments against the 2-pK triple-layer model, which have been accumulated for some time, important developments toward predictions of both surface complexation parameters and thermodynamic quantities have been performed with this model [96], although it would have been fairly easy to do this, for example, for a 1-pK Stern model.

Model g introduces various additional features as compared to models d–f. These include the following:

- Additional surface planes: The three-plane model allows charges to be placed in the surface plane, in the plane where the head end of the diffuse layer is situated, and in an intermediate plane. In the typical application of the model, the electrolyte ions are placed in the d plane; with respect to the triple-layer models, where no charges are placed in this plane, the electrolyte ions bound as ion pairs are "in contact" with the diffuse layer.

- Charge distribution concept: Charge distribution has already been mentioned in preceding sections, but will be introduced here in more detail. Although already mentioned by Davis et al. [4], charge distribution has not been involved in surface complexation models until more recently [97]; instead adsorbing ions have been treated as point charges whether a model dealt with small cations or bulky ligands with several functional groups. Charge distribution allows one to allocate parts of the charge transferred to the particles by an adsorbing ion to different planes; thus charges added to or removed from the surface plane during the formation of a surface complex can be placed in this plane, whereas charges, for example, on a bulky molecule like EDTA, which do not interact with surface functional groups, can be placed further away. Technically, the same effect can be obtained by introducing, for example, combinations of inner- and outer-sphere surface complexes as point charges (for each surface complex, one stability constant is required). The charge distribution concept can make use of Pauling bond valences to estimate charge distribution coefficients (cf. Fig. 10). If it seems reasonable to place charge contributions also in the d plane, this can be done; the electrolyte ions are typically considered as point charges both in the compact part (as ion pairs) and in the diffuse part of the EDL. A combination with the HNC approximation is possible as shown in model g.

Acid–base parameters for a three-plane model can be evaluated by the equivalent Stern-layer model; this is comparable to the relation between the constant-capacitance model (model a) and the extended constant-capacitance model (model b). The relation between the capacitance values allows constraints on the optimum couples of C_1 and C_2. Problems will arise when charges of adsorbed solutes may reach further toward the solution than what is allowed by C_{tot}.

In model g, a bidentate surface complex is depicted. Such multidentate complexes can also be used with other models, of course, and are therefore not an original feature of this model variation. Important in this context is, however, that the formation constants of multidentate surface complexes may need to be corrected for the suspension concentration. This has been extensively discussed by Venema [14]. Most computer programs do not allow automatic corrections of these features; therefore, care must be taken that they are correctly taken into account when calculating the respective stability constants. With generic one-site models, there are possibilities to avoid this, for example, by introducing $(SOH)_2$ entities for bidentates. This is possible as long as the total amount of bidentates is much lower than the total amount of the generic site. For mechanistic models, where different types of sites may be involved, the situation becomes more complicated and the bypass is no longer possible.

For mechanistic descriptions of ion adsorption, the three-plane model with charge distribution offers extensive flexibility. This can, of course, be used to fit all involved parameters like the composition of surface complexes, capacitance values, charge distribution coefficients, and stability constants. However, these parameters can also be constrained using, for example, the Pauling bond valence concept [14,48] or by Eq. (15).

Model h is the so-called four layer model, which, compared to the triple-layer model, introduces one supplementary layer. The major reason for this is that the electrolyte ions typically differ in size. It may be argued that the usually larger anion has to be placed further away from the surface than the more compact cation. The consequence of the additional layer is an additional capacitance value. Furthermore, constraints on electrolyte binding from experimental data, which indicate symmetrical electrolyte binding, become more complicated. This is because in addition to the stability constant for the formation of the outer-sphere complexes, the location of charge also affects the interaction of the ion with the surface.

Realizing that triple-layer models with a 2-pK formalism are difficult to parameterize, it is obvious that with respect to the proton adsorption mechanism, the four layer model makes sense in combination with a 1-pK formalism. In principle, the three-plane model (model g) also allows for the separation of the plane of adsorption for the electrolyte ions with one less adjustable parameter. Because the four layer model in its typical implementation does not allow for placing charge in the d layer, not much is gained compared to model g (which has not been used for this purpose, though) except, maybe, a decrease in potential between the location of the adsorbed anion and the head end of the diffuse layer, which would possibly result in a closer agreement between calculated diffuse-layer potential and measured zeta potential. An extension of the four-layer model to a four-plane model (allowing charge in the d plane) would allow even more flexibility in the location of charge. However, the degrees of freedom would also increase and the determination of unique parameters would become more difficult.

The electrostatic part of **model i** has originally been proposed by Smit [6] because of the need to invoke very low values for the outer-layer capacitance. The values of about $0.2\,F/m^2$ typically used in the triple-layer model can thus be decreased, which according to work quoted by Smit, would correspond to experimental observations.

The original Smit model separates the surface plane into two sections: one fraction [i.e., $1 - f$], of the overall surface (respectively the fraction of the surface sites) where only uncomplexed surface groups are present, and another fraction f, where only ion pairs formed with the electrolyte. For both sections, a different electrostatic model concept is introduced: a Stern model (obviously without electrolyte binding) for the fraction $1 - f$ and a triple-layer model for the fraction f. This separation is, of course, artificial. A mean value of the zeta potential is calculated from the equation given in Fig. 17i. Application of the model to experimental surface-charge data requires very low values for $C_{2,c}$. One advantage of this model can be seen in the closer agreement of the model with the experimental observations quoted by Smit.

Furthermore, the overall capacitance of the inner layers is not constant; it varies with pH and ionic strength. Considering that the field strength at the interface is expected to vary with pH and ionic strength, one is tempted to assume that the structure of the first water layers will also be affected by these parameters.

In addition to the features present in the original Smit model (i.e., combination of a Stern-layer and triple-layer model, 2-pK formalism), the compartment principle itself allows a huge number of options. It is possible to combine different electrostatic models in such a two-compartment approach. Furthermore, it would be possible to start adding supplementary compartments when, for example, a sorbing molecule would be introduced. This would increase the number of possible options and one might wonder what this would be good for considering the number of the already available models. However, there might be situations for which such an option might be useful. One such situation might, for example, be a case in which a bulky molecule adsorbs to a mineral surface in the presence of an electrolyte. If the standard Smit model with two compartments is used for the acid–base properties of the mineral, the addition of the molecule might give rise to a large distance between adsorbed charge (introduced by the molecule) and the charge transferred to the interface by functional groups of the molecule, which are oriented toward the solution side. This might not fit in the acid-base model because of the size of the molecule. Potentially available spectroscopic information about the interaction of the molecule with the surface (e.g., outer sphere and inner sphere) might further support certain hypotheses. In other words, charges of functional groups of the adsorbed molecule would go beyond the distance given by $x_{d,\text{uc}}$ in Fig. 17i. This would mean that the mean plane obtained for the zeta potential would shift further toward the solution side of the interface. It would actually lead to a significant effect on the zeta potential when high loadings occur. The degree of loading might affect the orientation of adsorbed molecules (e.g., outer-sphere complexes, which might be lying "flat" at the interface at low loadings) which might be forced to change their orientation at higher loadings (e.g., through increased amounts of inner-sphere surface complexes oriented perpendicular to the surface plane).

For conditions of high loadings of benzenecarboxylates on goethite, it has actually been found that measured zeta potentials could not be described by conventional models [95]. A gradual buildup of such a multicompartment model from simple compartments in a 1-pK formalism will be attempted to describe these data.

The preceding paragraphs have shown that a huge number of models circulate in the literature and that further model variations would be possible and plausible. It is even possible but sometimes difficult to discriminate slight variations for each of the models. Associated with the models are the respective adjustable parameters, which will be discussed in the following subsection.

2. Parameter Estimation

Adjustable parameters are those that cannot be independently determined. This definition can meet some controversial discussion in some cases and this discussion will concern the experiments or assumptions which are intended to define the parameters. This discussion will be included in some detail whenever appropriate. The parameters and their respective functions will be first described. There are parameters which are model independent and linked to measurable particle or suspension properties and those which are associated with the respective model.

(a) Parameters Describing Physical Particle and Suspension Properties

- **Particle-specific surface area:** It is necessary to know the area of the solid in contact with the liquid in order to estimate surface site concentrations (e.g., from specific surface site densities); or if speciation calculations are carried out on the basis of surface specific units, the experimental data pertaining to the surface must be transformed from molar concentrations using the specific surface area. For the constant capacitance model, the whole treatment can be done on a mass-specific basis. In principle for this model, the specific surface area only has to be involved if surface-specific site densities can be evaluated. The specific surface area is usually measured by gas adsorption. For in situ methods, other probe molecules are used (e.g., EGME method). Furthermore, microscopic methods can be used to determine the shape and size of particles; from particle size distributions, which can, for example, be obtained with setups for microelectrophoresis or with acoustophoretic methods, specific surface area can be calculated for either known or assumed particle geometries. Problems in the determination of this parameter by gas adsorption have been addressed in a previous subsection.

- **Suspension density:** The amount of particles (the mass of solid) present in suspension is required, again to estimate surface site concentrations. This parameter can be simply determined by weighing (e.g., when powders are used).Then, it must be assured that all of the weighed powder is actually transferred into the reaction vessel.* The other possibility is to work with a stock suspension of known solid content. When taking samples from this stock suspension, it must, in principle, be assured that the sampling itself did not change the solid content in the sample compared to the stock suspension. It is even necessary to check for possible changes in the solid concentration of the stock suspension; this can be done by determining the solid content in the end of an experiment, by drying a known volume of the suspension. Using this approach, it has to be assured that the solid studied is not dried at a temperature where phase transformations (e.g., loss of water from the structure) are possible and that the contributions of the background electrolyte and, if relevant, other chemicals are taken into account. An elegant method would be to determine the proton adsorption isotherm and calculate the amount of solid from a master curve determined previously. This would, furthermore, assure that the sample used in the experiment really corresponds to samples used in previous experiments. Other problems with stock suspension may, for example, arise when the properties of the particles change with time (actual aging processes) or when contaminations are possible (e.g., silica from glass containers).
- **Particle morphology:** If the objective of a study is to develop a model that incorporates information from spectroscopic investigations, it is necessary to characterize the particles with respect to their morphology. The aim must be to find the predominating crystal planes, which will have a major impact on the determination of proton affinities (e.g., by the MUSIC model). For powders or amorphous sorbents, it is, at present, most probably not possible or quite difficult to define adsorption sites in such a detailed way as would be required for the application of the MUSIC model. Detailed analysis of the particle topography (e.g., surface roughness) will allow one to justify values for capacitances; therefore, such investigations should be part of the particle characterization for any sorbent.

These particle properties can, in principle, be determined with some confidence and there is broad agreement. For the modeling parameters discussed in the subsequent subsection, similar agreement would be highly desirable. Unfortunately, the modeling parameters are treated in very different ways, ranging from graphical procedures and numerical optimization to approximate estimations.

(b)   Modeling Parameters

- **Capacitance values.** All models except purely diffuse-layer models involve capacitance values. These can be seen as an indication for the distance between respective surface planes (or for the thickness of a layer):

$$C_L = \frac{\varepsilon_L \varepsilon_0}{d_L} \tag{17}$$

In Eq. (17), C_L is the capacitance of layer L, ε_L is the dielectric constant of the medium in this layer, ε_0 is the permittivity of free space, and d_L is the thickness of the layer. ε_L is

---

*This might be operationally difficult if the powder is added during a titration.

not known in the interfacial layer, but it may be assumed to vary from dielectric saturation (at surfaces with strong water structuring, ε_{sat} is about 6) to the value of pure water (78.5). The discussion is strongly related to structure-breaking and structure-making properties of the surfaces, but in the discussion of electrolyte adsorption also the structure-breaking and structure-making properties of the electrolyte ions are of importance, which was already mentioned in a previous subsection. With the knowledge of structure-making and structure-breaking properties of surface and electrolyte ions some estimations are possible concerning the thickness of layers. The values of capacitances can then be evaluated and compared to numerically fitted values.

The results by Bourikas et al. [30] on various polymorphs of TiO_2 provide a good example for parameter estimation with a Stern model. Symmetrical electrolyte binding is not assumed, so that at sufficiently high values of ionic strength shifts of points of zero charge are expected. Two values of the capacitance are obtained. A value of $0.9 \ F/m^2$ for particles with a "smooth" surface and a value of $1.7 \ F/m^2$ for others for which a certain surface roughness are assumed. If one compares the proposed value of $0.9 \ F/m^2$ for the Stern model capacitance for TiO_2 (P-25) with best-fit capacitance values obtained for a constant-capacitance model, it is apparent that they differ significantly. The best-fit capacitance values obtained by Lützenkirchen [36] for the constant capacitance model for P-25 titration data in $0.1 \ M$ media are $2.04 \ F/m^2$ for the data reported by Janssen and Stein [98], $1.47 \ F/m^2$ for the data reported by Giacomelli et al. [86], and $0.88–1.12 \ F/m^2$ for the data reported by Rodriguez et al. [87]. The constant-capacitance modeling though yielding a good fit to the data does not permit any general conclusion as to the value of the capacitance to be chosen; neither can comparable values for the acidity constants be obtained, i.e., the data determine the generic surface acidity constants and the capacitance values in a kind of random way with this model.* On the contrary, the studies by van Riemsdijk and co-workers [99,100] on several solid surfaces suggest that a value of $0.9 \ F/m^2$ is generally applicable for most of the surfaces so far investigated in more detail. Structure-breaking/making properties of the surface and the electrolyte ions are, in this case, completely included in the electrolyte-binding constants. In comparison to this, Sverjensky and coworkers advocate also a variation of the capacitance with the electrolyte. This can be explained theoretically and may also be expected phenomenologically to some extent. In reality, it might be expected that both capacitance and electrolyte-binding constants vary with the electrolyte used. A four-layer or three-plane model would be required to take into account all phenomenological aspects. However, a pragmatic approach with a unique capacitance value which can be applied for multiple electrolytes and solids would have many advantages: thus, it may avoid (1) multiple-layer models, which would be necessary when predicted electrolyte-specific capacitance values are to be applied in mixtures of electrolytes and (2) supplementary fitting parameters when potential effects in adsorption phenomena are to be described (e.g., binary suspensions, see Sec. IV.A.1).

- **Site concentrations/densities:** For this parameter, one might wonder whether site concentrations/densities should not be considered particle properties. Because they are often adjusted numerically, they are discussed in this section. Site concentrations/densities are of fundamental importance, especially from the point of view of solution chemists. A comparable parameter in solution chemistry would be ligand concentration in a metal–ligand system. A study of acid–base properties of a ligand or complex

*Depending on the pH range of the experimental data and the resulting site densities.

formation between a ligand and a metal would not even be started by the traditional approaches like potentiometry if ligand concentration was not known with precision. Therefore, reports in the literature are available, where experimental approaches are described to determine site densities, mainly in the context of one-site 2-pK surface complexation models and by using potentiometric titrations. The most prominent approach is to measure the maximum uptake of protons or hydroxyls (this cannot be distinguished from desorption of hydroxyls or adsorption of protons, respectively). Several examples can be cited: Lövgren et al. [17] reported saturation of goethite with protons between $pH = -\log[H^+] = 2.7-3.0$; these results were obtained in continuous titrations with high precision EMF measurements; an additional argument in favor of the proton adsorption maximum was the short equilibration time in the pH range of the assumed adsorption maximum. In some articles by Marmier et al. [90,101], the titration method is also used to estimate site concentrations. These authors compare blank titrations to suspension titrations and find that the curves corresponding to these titrations become parallel. This is, in principle, the same procedure as in the work by Lövgren et al. [17]; however, a very puzzling feature of the blank titrations is that they do not have a slope of unity (i.e., the total proton concentration is not equal to the free concentration in the blank). From the description of the experimental procedures, it is not clear how the blank titrations have been obtained. The unexpected slope might be explained by the fact that Marmier et al. actually titrate the supernatant of the suspension. This would be a very good procedure, but as soon as the blank does not correspond to the titration of the background electrolyte it is possible that (1) the titration itself is erroneous or (2) the supernatant contains solutes or even solid particles which cause the observed titration behavior. If solid particles cause the deviation, then the blank does not make sense. If solutes cause the deviation, then, in principle, a solute sample must be obtained for each pH measured during the solid titration, because it must be assured that the blank correction corresponds to the situation at the individual pH values and not exclusively at the starting pH value. In other words, the concentration of the potential solute might change with the suspension pH or during the titration period. Unfortunately, in the articles by Marmier et al., one is left with the unusual blank titration data and many possible options to explain them.

Hoins [102], studying goethite, also found maximum proton adsorption levels; however, Hoins back-titrated supernatant samples, which allows for accurate determination of proton concentrations. For this, it is necessary to assure that (1) the separation between solid and liquid is complete and (2) the separation does not affect the system. Previous attempts to use the back-titration technique for the determination of surface charge versus pH curves [103] yielded results that differed considerably from the mass of results obtained with continuous titrations. This might indicate that the technique itself results in more scatter than continuous titration, which are sufficiently smooth and can actually be used for spline analysis if large number of data points are available [72].

This would not be so surprising, because many samples are required, which must be taken from a titration vessel in which the suspension pH is adjusted by the addition of titrant or from individual batch reactors.

Some groups criticize measurements in the pH regions relevant to site saturation in view of the errors introduced by the difference of two similar numbers (i.e., total and free-proton concentration become similar; proton uptake is not an important percentage compared to the amount of protons added to the system). Because the site density parameter is important, because the results obtained by Gunneriusson [28] indicated that the site density was higher in sodium chloride media than in sodium nitrate media of the same concentration, because the measured site densities are far below the values expected from crystallographic estimations, and because there is little agreement on this parameter, experiments were conducted over a three-year period to solve this problem. The results [104] are briefly summarized here. The first hypothesis based on the available information at the beginning of the study was that the previous experimental data were correct and that the results of Gunneriusson [28] indicated an electrostatic effect on the saturation level. A number of experiments carried out at different values of ionic strength and with sodium nitrate and chloride seemed to confirm this hypothesis. With increasing ionic strength, the saturation level increased, and with chloride, the saturation was always higher than with nitrate for the same background electrolyte concentration. The first important conclusion had to be that the site density determined by titration could not be considered a particle property. From plots of maximum proton uptake versus ionic strength, no limiting value could be determined. In the course of these experiments, data at $2\,M$ ionic strength were collected. One motivation for this was to avoid liquid junction effects which occur at low pH values but are shifted to lower pH values with increasing ionic strength. Data in $0.1\,M$ could be shown to be critically dependent on the liquid-junction contribution to the measured potentials. Because there is no way to determine liquid-junction parameters in the presence of particles, one has to assume that the liquid junction in solution and suspension are identical; an assumption which cannot be proven. Also, it is known that the particles have high surface-charge densities and, therefore, can, in principle, contribute to the liquid-junction potential.

Analysis of the data at $2\,M$ showed that the presence or not of a saturation level at very low pH values was very sensitive and, in fact, too sensitive to the calibration parameters of the electrode. It was found that the calibration could not be sufficiently precise to allow the conclusion of whether a maximum uptake existed. Therefore, the aforementioned back-titration was applied to samples in the pH region of interest. In the first experiments, relatively small samples were back-titrated coulometrically and some of the results are shown in Fig. 18. These data on $0.6\,M$ sodium chloride medium suggest proton uptake, which is substantially higher when determined by back-titrations (by up to 50%) compared to the continuous titration. Based on the results which were obtained with rather small samples, which, in turn, may result in undesirable errors, the experiments were repeated with larger sample volumes for back-titration. Both coulometer and sodium hydroxide were used for the back-titrations. It turned out that results with the coulometer yielded consistently increasing proton uptake with decreasing pH whereas results obtained with sodium hydroxide showed significant scatter. Iron in the titration samples was determined and corrected for. The data obtained with the coulometer could be reproduced. It was therefore concluded that (1) coulometric titrations should be preferred in the back-titration technique and (2) there is no measurable proton saturation at the goethite electrolyte interface. The obtained charging curve for $0.6\,M$ NaCl could be described with the MUSIC model by a blind prediction. These experimental results strongly suggest that site densities should be determined based on crystallographic considerations for such well-characterized sorbents as goethite. For powders, there is, at present, no possibility to do this with the same confidence. Therefore, new experimental approaches are necessary. One possibility would be to use other probe molecules. However, it must be shown that these molecules only interact with relevant sites (and ideally only one type

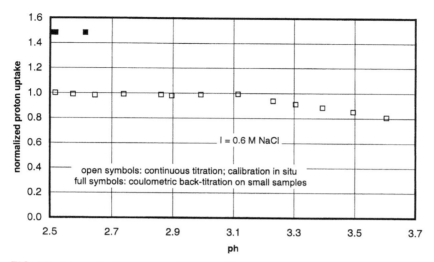

FIG. 18 Normalized proton uptake on goethite in $0.6\,M$ NaCl medium (slow titration). The normalization was done by the relative proton uptake obtained in the continuous titration (open symbols). The filled symbols were obtained from coulometrically back-titrating small supernatant samples (between 10 and 15 mL).

of site in a 1:1 ratio) and that no phase transformations or precipitation reactions occur. Furthermore, saturation values are difficult to determine because, in principle, the above-stated problem of similar values for probe molecule added and probe molecule remaining in solution will occur.

Figure 19 shows attempts to saturate goethite with arsenate at pH = 3.5. The symbols show results from three independent experiments. The scatter in the data becomes substantial so that no clear conclusion is possible as to whether there is a maximum uptake. Applying a Langmuir isotherm to the system would certainly be possible and has been practiced in the literature to estimate site densities [105].

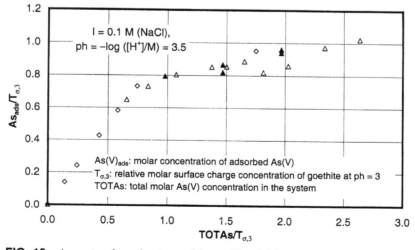

FIG. 19 Arsenate adsorption to goethite at pH = 3.5 in $0.1\,M$ NaCl. Normalization of uptake and total arsenite concentration is done by the relative molar charge concentration of the bare goethite surface at pH = 3.0.

Other options are to numerically fit the density of one or more sites to a surface complexation model [106]. This usually means that at least one site density and one stability constant are fitted simultaneously to the data (in the above-cited reference, this was done for several potential sites). This yields values which are statistically not significant because high correlations exist between the affinity constant and the site density [107]. At present, there is no best solution to experimentally determine this latter parameter. Total site concentrations can be determined by tritium exchange, but it is then necessary to determine the proportions of the different sites, because they may have different reactivities. Using spectroscopic methods to obtain the relative contributions of different surface sites to the total site concentration/density might be one way. At present, the site density is an adjustable parameter, but care should be taken not to numerically fit it simultaneously with stability constants, because this typically results in high correlations and the fitted site densities will be close to the maximum surface charge values among the experimental data [107]. Assumptions based on crystallographic considerations are preferable and high site densities will result in low correlations with stability constants. High site densities are expected from crystallographic considerations, not only for the overall site density but also for site densities of individual functional groups.

- **Acid–base constants/proton-affinity constants:** These parameters cannot, at present, be established experimentally. Because there are probably several distinct sites and electrostatic effects, the overall surface-charge curves do not give any hint as to the nature and reactivity of these sites. Experimental back-titration data by Schulthess and Sparks [103], which indicate steps in the proton adsorption isotherm (i.e., the surface-charge density versus pH curve), were interpreted by these authors to be indicative of individual sites with individual reactivities. However, such data are not reproduced by the bulk of research groups. In the previous subsection, the back-titration technique has been discussed in some detail and, in particular, the results on goethite obtained with base titrations at low pH were found to show significant scatter compared to data obtained with the coulometric approach.

It has been attempted to obtain the pK_a values of oxides by spline analysis of the charging curve. Derivatives of the spline-fitting results were compared to predicted proton-affinity constants [72]. One main assumption in this treatment was to exclude electrostatics, because a proton consumption function which is free of electrostatic effects must be available. In similar treatments of humic substances, so-called master curves (i.e., proton consumption functions which coincide for different values of ionic strength) are obtained by applying an electrostatic model to the raw proton consumption function. These master curves can then be treated by spline analysis. Similar treatments are, in principle, possible for oxides but would require an electrostatic model before the spline analysis.

The above-sketched approach is, in principle, capable of yielding proton-affinity constants of sites which are relevant (i.e., display a variable charge) in the pH range of interest and even yields information on affinity distributions. Thus, it would be very comprehensive compared to the expected features. The major problem is the assumption which will be necessarily involved in filtering out the electrostatic contributions to the raw proton consumption function.

In the surface complexation literature, the multisite approach is not very often applied. Instead, a "generic" surface site (i.e., one-site model) is used, mostly in combination with a 2-pK formalism. Use of a 1-pK model need not be discussed here, because the stability constant in the 1-pK model corresponds to the point of zero charge, which can be measured. However, a distinction is necessary between the models involving electrolyte-binding and the simple one-plane models i.e., models of the type constant capacitance and purely diffuse layer: For the latter combined with the 1-pK approach, possible changes in the point of zero charge with ionic strength (concentration and composition) must be accounted for and this can only be done by applying ionic-strength dependent points of zero charge. For the 2-pK approach, similar modifications are necessary. For constant-capacitance-type models, this is not really crucial, because in the favored cases, in which these models apply, only one value of ionic strength (in terms of composition and concentration) is relevant. For the diffuse-layer-type models, it is apparent that one comprehensive dataset cannot describe such experimental data, where shifts in the point of zero charge are observed, which are only due to the electrolytes present. In cases where these shifts occur at high ionic strength, one might add that the conventional purely diffuse-layer model (Gouy–Chapman equation) would not be appropriate anyway.

For the 2-pK formalism, the determination of proton-affinity constants from experimental data was based on two different approaches: The acid–base constants have usually been determined either by graphical procedures or numerically. Graphical procedures involve various kinds of approximation. One major approximation is knowledge of the site density parameter. This parameter is necessary to calculate the reaction quotients as a function of the charge of the oxide. Extrapolation of these reaction quotients is done to zero surface charge. At zero surface charge, the reaction quotient is not supposed to be affected by electrostatic effects and thus the intrinsic acid–base parameter is obtained. If the points of zero surface charge and zero surface potential do not coincide (i.e., when electrolyte binding can affect the point of zero charge), then the extrapolated value corresponds to zero surface charge. However, at this point, the surface potential is not zero and the exponential (i.e., electrostatic) term is different from unity. Graphical procedures must, in principle, be corroborated by data showing the true point of zero charge or the point of zero potential of the sorbent sample. Several graphical methods have been applied, which typically also allow estimation of capacitance values from the slope of the plots. Close to the point of zero charge, the expected straight lines are often not obtained and a significant curvature occurs. Simple extrapolation methods can be applied for the constant-capacitance model and were also used with a purely diffuse layer model. Application with a triple-layer model resulted in ionic-strength-dependent stability constants. As a consequence, double extrapolation techniques were developed, which involve extrapolations to zero surface charge and zero ionic strength (i.e., zero surface potential). They require smooth lines to be drawn, which may be a problem. A detailed analysis of double extrapolation techniques showed that application to synthetic data did not result in the expected parameters [108]. Therefore, graphical methods are not really an option.

Improvement over the graphical methods is achieved by numerically fitting a certain model to experimental data. In principle, with codes like FITEQL, several kinds of experimental data can be fitted and the experimental observations can be used to constrain the problem. Fitting becomes more difficult when more adjustable parameters are involved. This is because (1) with numerous adjustable parameters no convergence is obtained or (2) due to interdependencies between parameters, high correlations are observed. Often missing information on the site density parameter is crucial. Cooptimization of site density and stability constants typically yields high correlation factors, in agreement with the expected high correlation in a Langmuir-type adsorption isotherm between maximum uptake and affinity constant, when no observations at the maximum uptake have been obtained. It has been discussed in previous subsections that the accurate experimental determination of maximum uptake on oxides is difficult. Thus, assumptions are often involved in fixing site densities (see previous subsection). The stability constants fitted to some value of site density of a generic site involve more problems (besides the arbitrarily assumed nature of the generic site): The use of such parameters, which have been published, requires self-consistent values of site densities and stability constants and capacitance values. Mixing of parameters from different literature sources will produce some surface charge density versus pH curves, which may have nothing to do with the actual sorbent studied. The same may be valid for consistent parameters (i.e., from one literature source) because of the sample-dependent acid–base properties. For fixed site densities (e.g., Dzombak and More [76] fixed a value of 2.31 sites/nm^2 for hydrous ferric oxide based on an extensive analysis of a wide range of experimental data; Sverjensky and Sahai [96] used experimentally determined hydroxyl site densities from tritium exchange experiments), self-consistent parameters have been obtained for several models and several minerals. Recall that they refer to "generic" surface sites, so that site heterogeneity is usually not considered, and from the point of view of mechanistic modeling, the parameters and surface species inherent to these models should not be considered realistic. From the point of view of fitting acid–base parameters, the high site densities obtained from tritium exchange have the advantage that the values of the acid–base constants are not largely affected by smaller variations in such high site densities. However, this is clearly a technical aspect and should not lead to mechanistic conclusions.

In the framework of realistic surface complexation models (realistic in the sense of realistically accounting for expected features), site densities are known for well-defined sorbent samples. For natural sorbent samples and powders, this is not the case. Realistic acid–base constants for such sorbent samples are difficult to obtain. Spline analysis of proton consumption functions or numerically fitting a model to such sorbents are presently the preferred approaches. However, they involve the respective assumptions, and "realistic" models await improved experimental (most probably spectroscopic) methods to define the amount and nature of surface sites. Otherwise, in generic models, acid–base constants are model-dependent parameters, which have some use when applied self-consistently, but should not be interpreted as "realistic" parameters. This is a very controversial statement, which is in conflict with many modeling studies in which "generic" models of the surface acid–base properties are combined with spectroscopic information on sorbing ions.

- **Electrolyte-binding constants:** Electrolyte binding constants are typically required in the multilayer models of the Stern, triple-layer, three-plane, or four-layer approaches. However, they have also been used with the constant-capacitance model, which is difficult to understand because one might have chosen the Stern model instead. The multilayer models may also be used without involving electrolyte binding. In this case, the counter charge is placed in the diffuse layer for the pure acid–base data.

The evaluation of electrolyte-binding constants is typically restricted to graphical approaches and numerically fitting experimental data. This implies the same problems discussed earlier. Attempts have been published to measure electrolyte adsorption. This commonly involves the measurements of a very small amount adsorbed compared to relatively high solution concentrations. Thus, experimental results are expected to be associated with relatively large experimental errors. Even if reliable results were obtained, the question is, what is actually measured? In terms of the models, one would wish to obtain the amount of ions fixed in the respective plane of adsorption. However, the diffuse-layer contributions are hard to exclude experimentally. Then, it is necessary to separate the contributions from the ion pairs and the diffuse layer and this can only be done in the modeling calculations, but the modeling typically allows all kinds of variations in the surface speciation so that no ultimate conclusions are possible. Results of x-ray standing-wave measurements can resolve the distribution of ions in the diffuse layer. This information can also be inferred from the results of calculations with surface complexation models and could, in the future, be another option to test these models. It might also be helpful to include electrokinetic data obtained at high ionic strength. Shifts of isoelectric points give a strong indication on which electrolyte ion is more strongly adsorbed. Similar hints can be obtained from extensive stability measurements in different electrolytes, which yields the adsorption sequence of the electrolyte ions tested. Figure 20 shows predictions using parameters by Bourikas et al. [30] for the diffuse-layer potential of TiO_2 in sodium nitrate. The results nicely compare to measurements by Kosmulski et al. [109], which indicate that no IEP is observed in sodium nitrate concentrations equal and above $0.9\,M$ within the studied range of pH values.

- **Surface species and surface complex stability constants:** These aspects are, to some extent, interrelated. The term "surface species" might comprise such aspects as the polydentate character of the surface complex, inner- versus outer-sphere surface complexation, or charge distribution. Whether surface species are plausible can be checked either by surface spectroscopy or by application of the bond–valence principle, which can be coupled with spectroscopic information [110]. The latter can also help to estimate reasonable values for the charge distribution factors in charge distribution models. All of this certainly allows a restriction in the many options which

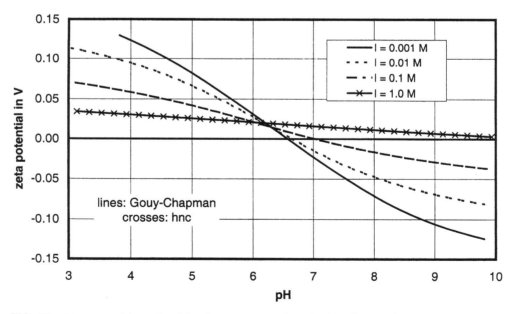

FIG. 20 Zeta potential predicted by the parameters given by Bourikas et al. [30] for TiO_2 (P-25) in sodium nitrate. Experimentally, no isoelectric point was observed for this system for $I = 0.9\,M$ by Kosmulski et al. [109].

would otherwise be available. It would be highly preferable to fix the surface species with respect to their stoichiometry and then optimize the respective stability constants. However, in some cases, this might not be as simple because speculations about proton stoichiometries are possible. Significant simplifications are possible in case only one surface species is present [111], but this is most probably rather the exception than the rule.

3. Possible Improvements in Modeling Approaches

Two potential improvements compared to the common practice are introduced. Both refer to the description of the diffuse layer. The commonly applied surface complexation models involve the Poisson–Boltzmann approximation for diffuse-layer potential of the electric double layer (resulting in the Gouy–Chapman equation for flat plates in most applications).

(a) Gouy–Chapman Equation. The mean field approach inherent to the Gouy–Chapman equation implies that the potential of mean force of ions in the double layer is equated to the mean electrostatic potential. Thus, ionic correlations arising from electrostatic interactions between ions and from the finite size of the ions (excluded volume) are neglected. The singlet hypernetted chain (HNC) approximation [112] includes ion size and ion correlations and, therefore, is a significant improvement over the mean field point-charge approach. Larson and Attard [113] have used the HNC method to calculate diffuse layer potentials for a range of conditions (covering surface-charge densities and electrolyte concentrations). The HNC results could be fitted to the following empirical function [113]:

$$\Psi_d(\sigma_d) = A[f \tanh(a\sigma_d) + (1-f)\tanh(b\sigma_d)] \tag{18}$$

The values for the parameters A, f, a, and b have been determined for the following values of ionic strength:

1 mM ($A = 163.6$ mV, $f = 0.5992$, $a = 0.2974$ cm^2/μC, $b = 1.5403$ cm^2/μC)
10 mM ($A = 128.9$ mV, $f = 0.7024$, $a = 0.1835$ cm^2/μC, $b = 0.7017$ cm^2/μC)
100 mM ($A = 99.96$ mV, $f = 0.7812$, $a = 0.0955$ cm^2/μC, $b = 0.2867$ cm^2/μC)
1 M ($A = 65.23$ mV, $f = 0.9558$, $a = 0.0487$ cm^2/μC, $b = 0.1445$ cm^2/μC)

The most noticeable difference between the Poisson–Boltzmann approach and the singlet HNC approximation is in the saturation of the potentials at high surface-charge densities for the latter. This means that in the surface complexation models using the Gouy–Chapman equation, the diffuse-layer potential will be overestimated compared to the HNC approximation for constant surface-charge densities. Thus, replacing the Gouy–Chapman equation with the HNC approximation in speciation codes allowing for surface complexation calculations will allow one to (1) better describe the experimentally obtained pH independent zeta potentials when sufficiently far away from the isoelectric point and (2) decrease the difference between calculated diffuse-layer potentials and measured zeta potentials. The differences between the Gouy–Chapman equation and HNC approximation will be most important in those cases where sufficiently high diffuse-layer potentials are obtained. The comparison between theory and measurement must, however, remain consistent.

Equation (18) was implemented in a modified version of FITEQL with the above-referenced parameter values (i.e., calculations with the HNC approximation are only possible at these particular values of ionic strength). Some results using the HNC approximation will be discussed in the following subsections.

Figures 21 and 22 show the influence of using the HNC approximation instead of the Gouy–Chapman equation for a purely diffuse-layer model. Figure 21 shows the charge potential relationship, which in the purely diffuse-double-layer model has a direct bearing on the calculation of surface-charge density (Fig. 22). Thus, it is apparent that even at low ionic strength, the influence is quite strong, indicating that for this kind of model, the HNC approximation is preferable.

Figures 23–25 show similar calculations for the Stern model. The surface-charge potential relationship for the diffuse layer shows the expected features (Fig. 23). However, compared to the diffuse-layer model, virtually no effect is noted for the surface-charge density as calculated by a Stern-layer model (Fig. 24). Similar results were obtained for other cases with extensive variations of particle properties and surface complexation parameters. For the zeta potential, deviations occur (Fig. 25). In general, significant effects (also on surface charge) are obtained when high-diffuse-layer charge is obtained at high ionic strength (i.e., low counterion binding in the compact part of the EDL). With parameters estimated for TiO_2 by Bourikas [30], no difference between the HNC approximation and the Gouy–Chapman equation were obtained for sodium nitrate (Fig. 20). Higher values of ionic strength, for which no calculations were possible due to the lack of parameters for use with the HNC approximation as calculated by Eq. (17), might lead to different results.

With multilayer models, the features in comparisons between the HNC approximation and the Gouy–Chapman equation were found to be similar to the results with the Stern model for the conditions tested.

(b)   Particle Geometry.   In the common computer codes flat geometries are used for the electrostatic model, allowing for simple use of the Gouy–Chapman equation. However, for sufficiently small particles, the curvature becomes important and the geometry has to be taken

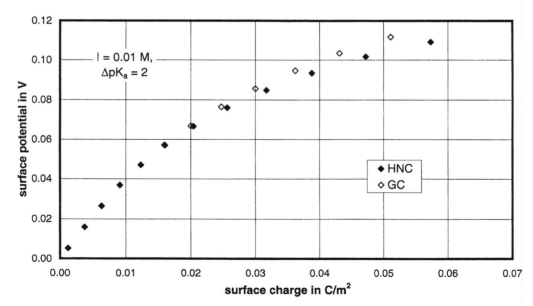

FIG. 21   Charge potential relation for the Gouy–Chapman (GC) and the hypernetted chain (HNC) approximations. Results were taken from the calculation using a purely diffuse double-layer model and a 2-pK approach (single site) with $\Delta pK_a = 2$ and $I = 0.01\ M$.

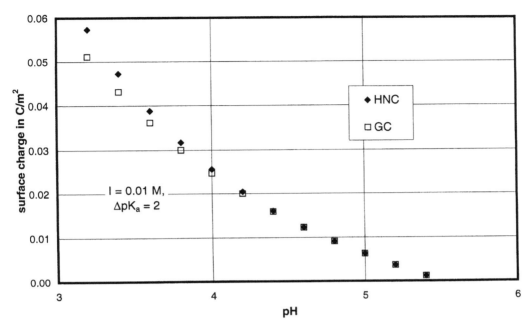

FIG. 22 Surface-charge density as a function of pH as calculated with the model corresponding to Fig. 21.

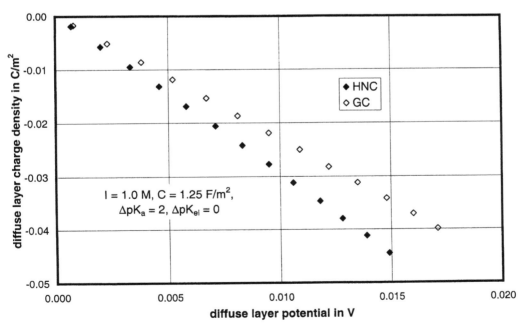

FIG. 23 Charge potential relation for the Gouy–Chapman (GC) and the hypernetted chain (HNC) approximations for the diffuse layer. Results were taken from the calculation using a Stern-layer model and a 2-pK approach (single site) with the parameters and conditions shown.

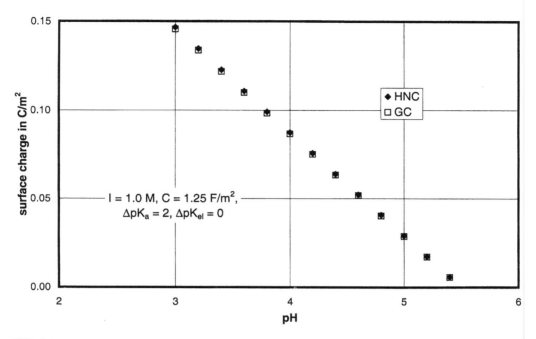

FIG. 24 Surface-charge density as a function of pH as calculated with the model corresponding to Fig. 23.

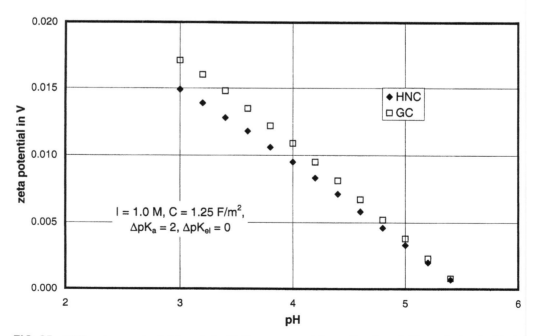

FIG. 25 Diffuse-layer potential (equated with the zeta potential) as a function of pH as calculated with the model corresponding to Fig. 23.

into account. Approximate charge potential relationships for spherical and cylindrical particles were published by Ohshima et al. [114] and implemented in FITEQL. In the following, some results obtained with the purely diffuse-layer model are shown. The calculations were done in such a way that the particles would have identical specific surface areas. These were used to calculate radii for the sphere and cylinder geometry assuming (1) a common specific density for both geometries and (2) a length for the cylinder of 300 nm. Figure 26a shows that there is no difference between the calculated surface-charge density versus pH curves for a silica-type colloid with specific surface area of 100 m^2/g. An increase by a factor 10 of the specific surface results in some deviations. The equivalent sphere geometry has a higher specific charge density than the plate (Fig. 26b). The cylinder lies in between, but these results are dependent on the

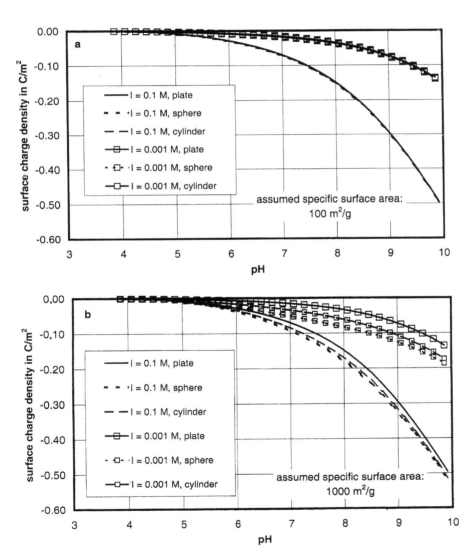

FIG. 26 Surface-charge density as a function of pH as calculated with a purely diffuse-layer model (silica surface) for different particle geometries and fixed specific surface areas of (a) 100 m^2/g and (b) 1000 m^2/g.

length assumed for the cylinder. Results of such calculations must be considered with care, when the deviations of the approximate analytical charge potential relationships from the exact numerical solutions become substantial. Also, the extremely high external specific surface area applied in the calculations might not be found in realistic applications.

In summary, the model for the diffuse layer may have repercussions on the surface charge versus pH curves. This is most important for the purely diffuse-layer approaches. At present, the two points discussed in the preceding subsections are not taken into account in the relevant modeling studies. Therefore, some of them might be revisited (e.g., for hydrous ferric oxide, in case a high specific surface area applies to the particles, the results by Dzombak and Morel [76] might be assessed). Also, the application of the model at high ionic strength would be questionable. For the multilayer models, surface-charge density is typically not affected by these two aspects. However, for the calculation of the diffuse-layer potential, differences can be expected in comparison with the traditional Gouy–Chapman equation for all the respective models.

F. Warnings

It has often been shown how successful surface complexation models are. However, examples exist which show that sophisticated model concepts can lead to successful predictions based on erroneous assumptions. The MUSIC concept has been evaluated in this chapter as the ideal concept for the mechanistic description of adsorption phenomena on well-defined solids. The following example might serve as a warning, that a model, even if convincing and successful, is based on assumptions and that these assumptions are not always verified. Even worse, information from the literature that are incorrect or not up to date might have been used.

The use of the original MUSIC model with a morphology of goethite particles bound by (100), (010), and (001) faces resulted in realistic predictions of the point of zero charge and a good description of the charging behavior of goethite particles expected to have this morphology. Later, the dominant crystal plane was expected to be the (110) plane. With this, a point of zero charge < 5 would be obtained, which is lower than any measured value (Fig. 1). The refinement of the MUSIC model still involves the danger that the real nature of the system is not known: Assumptions about the hydrogen bonding at the surface and assumptions about surface relaxation are required.

Other model approaches attempting to explain wide ranges of experimental data will most probably always have to choose among the data. Justifying this may be difficult and hypothetical. In particular, pure modeling groups may have little or no experience with the experimental procedures and might tend to draw unjustified conclusions or simply rely on previously published reasoning.

In general, none of these models should be taken as the ultimate truth, even if articles and good modeling results might suggest this.

IV. SPECIAL CASES

Modeling as discussed so far may already be complex, especially in cases where a mechanistic model is attempted. These examples primarily concern well-defined laboratory systems. In actual applications, the sorbents are usually not well-defined and certain properties of the systems might result in nonadditivity (e.g., multicomponent systems, such as suspensions with more than one kind of particle, competition between solutes, coadsorption). Additional complications arise when more than one solute is present (proton and hydroxyl and electrolyte ions are not counted).

These are briefly discussed in the next subsections, before surface precipitation is addressed and temperature dependencies in adsorption studies are summarized.

A. Multiple Component Systems

1. Binary Suspensions

Binary suspensions can be understood as (1) two kinds of separate particles being present in the system (i.e., the particles have the same sign of charge, the ionic strength is low, and the particles do not interact with each other except for repulsive electrostatic interactions) or (2) two kinds of particles which are interacting chemically (i.e., one kind of particle is present as a partial coating on the other).

The first case does not really present a problem as long as the electrostatic interactions between the different kinds of particles remain repulsive. As soon as the particles start agglomerating and part of the interfaces are no longer or not so easily accessible to potentially adsorbing solutes, consistent modeling becomes nearly impossible.

For the latter case, it is necessary to estimate the specific surface area of the support and the coating, which may be difficult. Different degrees of coatings might be a solution for synthetic sorbents, but for natural sorbents, it would be required to change the solid composition (i.e., stepwise or partial extraction of the coating). In practical cases, the coating is expected to be the more important sorbent.

Modeling of these two kinds of binary suspensions has been reported in the literature (e.g., by Gibb and Koopal [115] and Lützenkirchen and Behra [116]). As long as the binary suspension can be described based on the single-particle systems, the modeling is straightforward [115]. However, the situation becomes difficult as soon as potential effects are observed [116]. This is typically the case for coatings, but the extent is dependent on the size of the coating patches. For large patches, so-called patchwise heterogeneity of the sorbent may be assumed (i.e., the patches behave as individual sorbents, which are not affected by neighboring patches with different properties). For small patches, the assumption of complete mixing of the individual sites is possible (i.e., the sites are interacting with each other, resulting in one surface with sites from both the support and the coating). Problems arise for intermediate situations. In the modeling exercises mentioned earlier [116], this was solved by introducing three regions on the binary particles:

- Two regions with properties of the individual sorbents (two patches)
- One region having properties of both sorbents (randomly mixed sites from the two individual sorbents)

It was this last region that allowed the modeling of the potential effects. The more extensive the coating, the less important was the intermediate region. The model then corresponded to a patchwise heterogeneous sorbent. In this example, the extensive characterization of the sorbent was the basis for the model. Whereas for the metal ion adsorption data this approach was successful even in predictions, it fails when the points of zero charge and the adsorption of sulphate is modeled with the surface parameters used for the cation sorption data (Lützenkirchen, in preparation). Modeling of binary suspensions is, in principle, possible, but it requires extensive particle characterization. Complications may arise from heterocoagulation in systems with distinct particles. Heterocoagulation is then influenced by the pH (dependent on the respective points of zero charge of the particles) and by adsorbing solutes, which may alter the charging behavior of the particles.

2. Competitive Systems

Competitive systems include suspensions with more than one solute (electrolyte ions, proton, and hydroxide not counted). These may be systems (1) in which two or more metal ions compete for surface sites (e.g., studied by Benjamin [117] or Yang and Davis [79]), (2) in which two or more ligands compete for surface sites (e.g., studied by Goldberg [81] or by Mesuere and Fish [118] and Hiemstra and van Riemsdijk [119]), and (3) systems in which metals and ligands are present, which at least under certain experimental conditions adsorb simultaneously without forming common surface complexes [14] (i.e., they compete).

Modeling of such systems has typically presented some problems [70,81] as was already discussed. However, with the more comprehensive models, very good agreement between experiment and model calculations was achieved in particular by van Riemsdijk and co-workers on sulfate phosphate interactions on goethite [120], cadmium phosphate interactions on goethite [14] but also on the data [119] which could not be comprehensively modeled by Goldberg [81]. From this it was concluded that the simple models are not capable of describing all competitive systems.

3. Ternary Surface Complexes

Ternary surface complexes of different types exist. The most interesting case occurs in systems including a sorbing ligand and a sorbing metal. The metal-like adsorption behavior in a ligand-free system (i.e., increasing adsorption with increasing pH) may be changed to a ligand-like sorption behavior of the metal in the ligand-containing system (i.e., decreasing adsorption with increasing pH). This may be described by most models discussed in this chapter. However, the multilayer and charge distribution models may take into consideration the expected complexity of the ternary complexes. If it is important for simple systems to have additional spectroscopic methods to verify the structure of surface complexes, this becomes even more important for systems where ternary complex formation is possible, because it has been found that ternary surface complexes are present on surfaces under conditions where they are not expected in aqueous solution [121].

Another type of ternary surface complex occurs when the metal bridges between a ligand and the surface functional group. Such species have sometimes been postulated in model calculations. Figure 27 shows an example, where the adsorption of Cd on MINUSIL (α-quartz) is modeled (a) in the absence of chloride (model calibration), (b) and (c) in the presence of chloride by including ternary surface complexes of different stoichiometries, and (d) in the presence of chloride by including no ternary surface complex. The disturbing aspect is that with FITEQL, cases b and c are easily constructed (i.e., the optimization procedure finds a stability constant for a ternary surface complex), whereas a simple prediction based on the chloride-free system allows an equally good description of the data [case d, where only the aqueous speciation change by the introduction of the chloride ions allows a sufficient (or even better) description compared to cases b and c].

For extensive loading of sorbents with metal ions, one might indeed wonder whether ion pairs with such adsorbed metal ions should not be included for the sake of self-consistency (i.e., they are postulated to form with bare surface sites of lower charge, Fig. 27b).

In principle a description of ternary surface complexes is possible, but it seems even more important to have microscopic data for such ternary systems. In addition to the study by Bargar et al. [121] for the U(VI)–carbonate–hematite system, ternary complexes with metal-like adsorption behavior were found [122] in the Pb–malonate–hematite system. The combination of different (XAS and IR) spectroscopies was very valuable in these studies. With NTA or EDTA

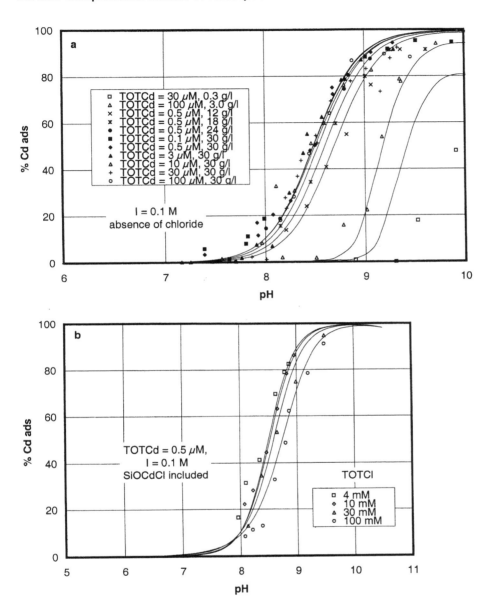

FIG. 27 Cd adsorption to quartz in $0.1\,M$ sodium nitrate media. Symbols represent data from Ref. 117; lines are model calculations using a purely diffuse layer model: (a) calibration of the model in the absence of chloride (b)–(d) presence of chloride. Cd–chloride aqueous speciation was identical in all calculations.

being present in a cation–mineral system, the almost exclusively observed behavior is ligandlike adsorption.

B. Surface Precipitation

Surface precipitation occurs when, compared to the two-dimensional structure of a surface complex, a three-dimensional structure is formed at the particle surface or when the particle surface region is being transformed. This is often expected at a higher solute concentration.

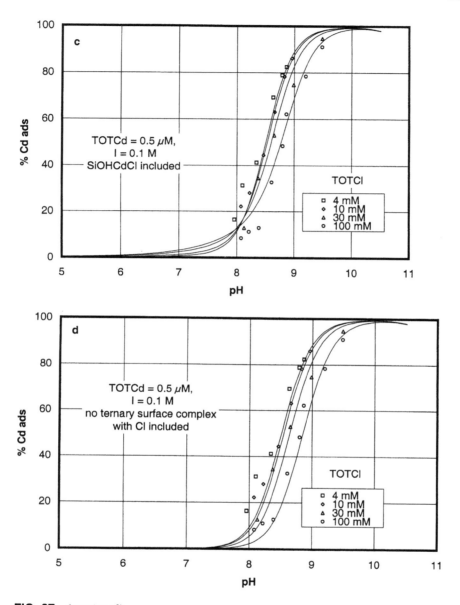

FIG. 27 (*continued*)

The surface precipitation model has been developed by Farley et al. [123] to describe sorption data over a wide range of free-sorbate concentration. It provides a smooth transition from adsorption to surface precipitation and uses the formalism of an ideal solid solution. The applicability of the model has been shown in several other publications [124,125]. In most of the applications of the surface precipitation model, the whole sorbent present was supposed to take part in the formation of an ideal solid solution. It was shown that if only the sorbent present as surface sites contributes to the "solid solution," the isotherm of the surface precipitation model reduces to the BET isotherm [126]. Between these limiting cases, it is difficult to find a unique description of the macroscopic data (parameter estimation problem). Formation of surface precipitates can be corroborated by spectroscopic methods. This would be particularly important because adsorption and precipitation cannot be distinguished from macroscopic data alone.

C. Temperature Dependence

Temperature dependence is a topic which has not very frequently been addressed compared to the work published at room temperature or 25°C. Modeling has in most cases been performed by either simple Langmuir isotherms [105] or the constant-capacitance model [127]. The reason for this has already been discussed. A more sophisticated modeling of temperature dependence has been reported by Machevsky et al. These authors have used a model based on the MUSIC concept first with only the 1-pK approach, but used later with a full MUSIC interpretation [128].

In principle, all of the models described can be used to describe temperature dependencies. However, for generic models using one site, the 1-pK approach has the substantial advantage that the point of zero charge can be measured [129] or even predicted [63] as a function of temperature.

It is interesting to note that for the temperature dependence of adsorption phenomena in aqueous suspensions, the general tendencies do not appear to be clear in all cases. Some examples of accepted or controversially discussed macroscopic observations are given in the next subsections, along with potential phenomenological explanations based on surface complexation theory.

1. Acid–Base Properties of Oxide Surfaces

In this context, several aspects may be of importance:

- Point of zero charge
- Charging behavior of oxide minerals
- Colloid stability, which is a consequence of the charging

Experimental data display two effects:

- A decrease of the point of zero charge (PZC) with increasing temperature in the temperature range between 25°C and 90°C.
- Typically, an increase of specific surface charge with increasing temperature (e.g., Ref. 130); in some cases, a decrease of specific surface charge with increasing temperature was observed [131] for hematite (in agreement with other work on hematite), but an increase seems to be more common (e.g., for corundum, quartz, kaolinite, see Ref. 132). The increase of charge with temperature at constant pH is due to the shift of PZC. Normalization to the PZC may decrease this effect and lead to temperature congruence, which has even been postulated as a general phenomenon [133].

The stability of colloids will be determined by the charge. The zeta potential obtained by some electrophoretic method is used to assess the stability. Pagnoux et al. [134] carried out a study on the temperature dependence of the zeta potential for two alumina samples. They found that an increase in temperature led to higher zeta potentials and would therefore cause a more stable suspension. Contrary to this, Tari et al. [135] reported a trend toward an uncharged surface state with temperature for an alumina suspension and concluded that an enhancement of the coagulation rate with increasing temperature is to be expected.

Clearly, there seem to be some contradictions, which support the above-stated discrepancy between increase and decrease of surface charge with increasing temperature. These can be attributed to (1) different solid samples, (2) different experimental procedures, and (3) experimental artifacts, indicating that more systematic studies are required in this field.

2. Surface Complexation in Three Component Systems

(a) *H^+–Cation–Surface Systems.* A number of adsorption studies deals with the temperature influence on cation adsorption to oxide minerals [131,136,137] and clays [138]. The general tendency is an increase of adsorption with increasing temperature. An explanation is possible by considering the charging behavior of the minerals. An increase in temperature causes a decrease in point of zero charge. The resulting decrease in surface charge at constant pH leads to a decrease in electrostatic repulsion. The adsorption edge is shifted to lower pH. The study by Karasyova et al. [131] on Sr sorption to hematite can serve as a good example for this: The adsorption edge of Sr as a rather weakly adsorbing metal ion coincides with the reported point of zero. These authors used a constant-capacitance model for their data.

Studies on metal ion adsorption to clay minerals display the same trend. Adsorption of metal ions to clay minerals is often described as a mixture of ion exchange (at low pH) and surface complexation (at higher pH). For both processes, an increase in temperature provokes an increase in adsorption.

Kosmulski [137] found that the temperature dependence of cation adsorption is dependent on (1) their charge (higher charge causes a stronger effect of temperature) and (2) on total concentration of the metal.

In principle, surface complexation models allow one to describe these effects, in particular if the change in charge with temperature is as expected.

(b) *H^+–Anion–Surface Systems.* There are comparably few studies on the adsorption of anions on oxide minerals with respect to temperature dependence. It is generally expected that an increase in temperature causes a decrease in adsorption because of the change in charge (decreasing the point of zero charge provokes a reduced charge at a constant pH and causes less electrostatic attraction for a positively charged surface or stronger electrostatic repulsion for a negatively charged surface). It may be expected that this tendency would be described by surface complexation models.

Two recent experimental studies on the adsorption of small organic acids on kaolinite confirm these phenomenological expectations, at least partially. Ward and Brady [132] clearly observed a decrease in adsorption of oxalate at higher pH values when the temperature was increased from 25°C to 60°C. At low pH, the tendency was inverted. This so far unexplained observation is in agreement with unpublished work on benzencarboxylate adsorption to kaolinite by Angove et al.

Strauss [139] extensively studied the influence of temperature on the adsorption of phosphate on different goethite preparations between 5°C and 40°C. The influence of temperature was rather small for long equilibration times (> 1 week). At shorter times, phosphate adsorption increased with temperature. This is not in agreement with the expected trend.

Overall, the anion systems have received fairly little attention and the experimental data only confirm the expected trends in part. There is need for more studies to clarify the actual trends and to understand the observed inversion of the temperature effect on organic acid sorption on kaolinite.

Clearly, more systematic experimental studies on the temperature effect on anion sorption are required, which will explain the different observations, which have so far been published.

3. Ternary Surface Complexes

In multicomponent systems, apparently nothing has been reported so far with respect to temperature. Even for 25°C, there is need for more studies on such systems. One might hope that with the reasoning in terms of surface charge and charge on major solution species, some trend might be guessed (as is the case for the adsorption of cations). However, for the sorption of anions alone, this reasoning has no sound experimental proof. Furthermore, the already mentioned study by Bargar et al. [121] on U(VI) sorption to hematite can serve as a warning example because of the existence of ternary U(VI) carbonate surface complexes under conditions where equivalent complexes in aqueous solutions would not be expected based on speciation calculations. Without experimental data on the temperature effects in systems, where ternary surface complexes are expected or possible, simple arguments based on the surface charge alone should not be applied.

V. PERSPECTIVES

The ultimate goal of the surface complexation modeling approach is to provide users with databases, which can be applied without significant restrictions in predictive modeling. The application would certainly be limited to cases for which the dominating sorbent phases and the relevant solution compositions are known and for which significant ternary surface complex formation, which was not experimentally studied, is not expected. The inclusion of ternary surface complexes would require a very extensive database (cf. subsections on ternary surface complexes).

There is, at present, no agreement with respect to both the kind of data to be acquired (e.g., slow versus fast titrations) and to the modeling procedures (e.g., 1-pK versus 2-pK). Some recommendations from the author's point of view, which are not necessarily complete and might be biased by the author's own expertise, are presented in the following and cover experimental and modeling approaches. To some readers, certain aspects might seem trivial, but, unfortunately, there are enough examples which could be cited to illustrate the need to discuss them. Furthermore, there is no published discussion of these topics. Often, comparisons and extensive applications of models, which might serve as the basis for the homogenization of modeling procedures, have been published by the respective schools of thought (either the model developers or adherents to a school of thought), which makes the articles or textbooks at least appear biased.

A. Data Acquisition

Several aspects should be considered with respect to the acquisition of experimental data:

- Comprehensive datasets should be obtained and important parameters should be varied in the relevant experiments. Surfaces are not as well behaving as most solutes (e.g., goethite in Fig. 1). Sample-specific properties (charging behavior, point of zero charge) might be different for different preparations, different charges of commercially available products, and so forth. Therefore, surface characterization should be carried out for each solid studied; otherwise, adsorption data may not be useful for later reconsideration. It is very probable that the use of published data will become the important tool in establishing databases and testing models. Thus, it is of importance to assure a maximum of information on the sorbent as early as possible in any such study. It is not advisable to completely rely on previously published data on "nominally identical" sorbents. Such practice has been pursued for far too long. In solution chemistry, it would be completely unacceptable to use NTA protonation data for a metal–EDTA complexation study. Although this example might appear exaggerated, it is illustrative and meant to be provocative. In using wrong surface characteristics, one might rely on the wrong functional groups present (i.e., if in an advanced model the crystal planes are not the ones actually present on the sorbent) or one might rely on wrong proton affinities or both. The damage can be minimized by comparing the acid–base model with sorbent-inherent experimental data. Even if agreement between model and experiment is achieved this is not proof for the "correctness" of the model, it assures the documentation of the particle properties.

- Data should be cross-checked to assure data quality: Experimental data should be determined repeatedly or in different ways to allow a comparison of the results. One possibility is to combine pH variations at different solute concentrations with pH-stat conditions at selected pH values, again at different solute concentrations. For both adsorption and titration data (i.e., titrations of the sorbent in the presence of the solute), this allows a test of data consistency (Lützenkirchen and Lövgren, in preparation).

- Obtained adsorption data should be compared to literature data and coincidences and differences should be pointed out. Many systems have already been previously studied. Comparison of surface-charge data might be a first step. This would, for example, be very important before attempting to build a database: If the discrepancies between surface-charge data of a sorbent as published by different groups or the points of zero charge are observed to be significant, it is advisable to find out the reasons for such discrepancies before attempting a database, which starts from common (i.e., unique, averaged) acid–base properties. After all, the different sorbent samples might not be as similar as their common name appears to indicate, which has been shown in some comparisons for surface-charge densities of sorbents. In adsorption studies, such comparisons are less frequent: For example, Cd adsorption to hydrous ferric oxides or goethite (even in the same laboratory) has been frequently studied. It should be attempted to compare new to previously published data. This will either give agreement or disagreement and need to be discussed. It is probably through such comparisons that one might find weak points in one's own experimental procedures (e.g., equilibrium times) or in that of others. Data obtained by different methods should be compared more frequently (i.e., zeta potentials from different approaches on one system).

- Data covering conditions, which are rarely investigated, should be obtained to have the whole spectrum at hand: In some countries, nuclear waste is considered to be stored in salt domes, where relevant leaching solutions with very high (i.e., saturated) salt concentrations are expected in the safety assessments. Systematic sorption data for such conditions are not available (i.e., covering variations in pH if appropriate, solute and solid concentration, etc.), so that models cannot be developed for or tested on such data. Such data should be acquired also with model colloids and will most probably result in extensions of the existing models.

- Data on one system covering the available experimental methods: This would be a very important step in (1) evaluating the results from different methods and (2) testing and comparing the ability of the different models to explain the data. Thus, experimental methods might be assessed and improved and, more importantly, models may be excluded and the need to extend and improve models will become apparent.

B. Modeling Procedures

This is a very difficult aspect and most probably has to be reduced to some basic requirements for a pure modeling or a combined experimental and modeling study, not only because it is impossible to impose the "best" model on all users but also because the modeling approaches are in a state of development.

What can currently be demanded of a model is its self-consistency! This is most probably what should be checked in such studies before publication and would cover the following:

- The consistency between the stated modeling purpose and the finally applied model concept
- The consistency between experimental data and input data in their treatment in the inverse or forward modeling tool (i.e., activity coefficients, etc.)
- The consistency between experimental error estimates used in inverse modeling and actual experimental errors
- The consistency between the acid–base parameters in the model and the actual acid–base properties of the sorbent
- The overall consistency of the model (e.g., are the model-inherent limitations in agreement with the application?)

The acceptance of arguments already presented in the literature is, to some extent, lacking; some models are clearly underrepresented at present, although, at least from the present author's point of view, they should become more accepted if unbiased (as far as this is possible) consideration of published information is taken as the starting point and when computer codes are adjusted to these models (i.e., at least, the pursuit of some model can often be traced back to personal or technical reasons). Also, the balance between the required simplicity of a model (i.e., the number of adjustable parameters involved—here the major question is whether unique parameter sets can be established) and the achieved understanding of the system is by no means easy, in particular when a sorbent has not been or cannot be characterized to the extent necessary for a fully mechanistic model (in terms of the present achievements). The acid–base modeling approach, which has been discussed extensively in this chapter, is the fundamental decision. At the same time, this makes the acid–base properties of sorbents so important. Furthermore, it is very difficult to simply estimate in a qualitative way how and to what extent acid–base properties will influence ion adsorption parameters (covering species stoichiometry, stability constants, charge distribution, etc.). Model comparisons have often been obtained for restricted conditions [74,75] and the extrapolation of these results might be assessed. The present author's own experience in numerous modeling exercises has shown that a good model description also at variable ionic strength can nearly always be achieved (for the constant-capacitance model with modification of some parameter to achieve ionic strength dependence of surface charge). Only the performance of the diffuse-layer model was often found to be inferior in this respect (acid–base properties). Therefore, extrapolation of these studies seems possible, and, in model comparisons, the acid–base parameters should be such that an equally good description of data is possible; otherwise, the comparisons can be assessed.

One important statement that should be made in each modeling study is to clearly indicate the purpose of the modeling. This has been addressed in this chapter several times. The following extreme cases can be distinguished:

- Application of a model to new experimental data without attempting to cover microscopic aspects in detail; such a study would attempt to provide potential users with a self-consistent set of parameters in a specific system. The aim is not to give a full explanation of the system and the model concept should be such that the lowest possible number of adjustable parameters is involved. One major problem in such a case is to decide which aspects are actually important (e.g., is the acid–base behavior of the sorbent important or is the consistency of the relevant acid–base behavior with the experimental conditions of importance?). The present author considers this to be generally important, but different opinions may exist. In other words, to what extent is obvious disagreement between the model and the generally acknowledged experimental facts acceptable? This should be discussed when the decision on a certain model concept is addressed.

In general, it makes no sense for such a simple model to include more parameters than necessary. From the published information on model comparisons [75,140], it appears that a simple 1-pK formalism would be appropriate; concerning the electrostatic model, one might give the following recommendations: (1) The electrostatic properties are considered to be of importance. For a system completely restricted to one ionic medium a constant capacitance model approach would be appropriate at sufficiently high ionic strength; for varying ionic strengths, the Stern model would be appropriate (the diffuse-layer model combined with a 1-pK approach yields charging curves that do not correspond to the observed features). (2) The acid–base behavior is judged to be of no importance at all. The simplest approach would be a simple complexation model, considering the surface functional groups as aqueous solutes (nonelectrostatic model).

- Application of a model to complete datasets, combining macroscopic and microscopic (i.e., at present, typically spectroscopic) data with the purpose of proposing a structural model that explains the macroscopic data on an atomistic level. Models in this class should include the present state of the art (i.e., aspects such as multisite surface complexation, charge distribution, potential improvements on the Gouy–Chapman equation). The most complete model currently in use is the CD-MUSIC approach, which allows a very detailed description of the interface and has been shown to be able to describe complete datasets, both with focus on acid–base surface equilibria and including ion adsorption, competition, and so forth. Of course, this model still has its limitations—among others are the point-charge nature of the ion pairs or the unverified assumption about the agreement between bulk structure and surface structure (e.g., surface relaxation or hydrogen bonds, disagreement with first principles applications, affinity distributions), so that further model development is expected.

In between these limiting cases, compromises are possible, but they should be explained. As examples, one might wish to apply a mechanistic model to a sorbent, for which the acid–base properties can for some reason (e.g., the dominating crystal planes are not known) not be described in such detail as is, for example, possible for well-crystallized goethite. In such cases, the features, which are known to be relevant, should be included to such an extent that adjustable parameters are limited to the number, which is actually necessary to accurately describe the experimental data. In particular, the acid–base properties are important in deciding on the basic model concept. Therefore, the simplest model for the accurate description of acid–base properties accounting for electrolyte specific behavior (e.g. 1-pK, single site, Stern model) would be appropriate, which can be extended with many options to the description of solute adsorption.

ACKNOWLEDGMENTS

The author wishes to acknowledge the various institutes (in particular the Department of Inorganic Chemistry at Umeå University), granting organizations (CEE, EERO, FSR), and individuals for allowing him to pursue his studies.

REFERENCES

1. PW Schindler, H Gamsjäger. Kolloid-Z. Z. Polym 350:759, 1972; PW Schindler, HR Kamber. Helv Chim Acta 51:1781, 1968.
2. W Stumm, CP Huang, SR Jenkins. Croatica Chem Acta 42:223, 1970.
3. DE Yates, S Levine, TW Healy. J Chem Soc Faraday Trans I 70:1807, 1974.
4. JA Davis, RO James, JO Leckie. J Colloid Interf Sci 63:480, 1978.
5. J Lützenkirchen. In: H Hubbard, ed. Encyclopedia of Colloid and Interface Science. New York: Marcel Dekker, in press.
6. W Smit. J Colloid Interf Sci 109:295, 1986.
7. GE Brown Jr, GA Parks, JR Bargar, SN Towle. ACS Symp Ser 715:14, 1998.
8. JN Israelachvili. Intermolecular and Surface Forces, London: Academic Press, 1998.
9. J. Lyklema. Fundamentals of Interface and Colloid Science. London: Academic Press, Vol. 1. Fundamentals, 1993; Vol. 2. Solid–Liquid Interfaces, 1995.
10. WA Zeltner. PhD thesis, University of Wisconsin, Madison, WI, 1989.
11. JF Boily, J Lützenkirchen, O Balmès, J Beattie, S Sjöberg. Colloids Surfaces 179:11, 2001.
12. G Sposito. Environ Sci Technol 32:2815, 1998.
13. T Hiemstra, JCM de Wit, WH van Riemsdijk. J Colloid Interf Sci 133:105, 1989.
14. P Venema. PhD thesis, Wageningen Agricultural University, Wageningen, The Netherlands, 1997.
15. M Kosmulski, JB Rosenholm. J Phys Chem 100:11,681, 1996.
16. SB Johnson, PJ Scales, TW Healy. Langmuir 15:2836, 1999.
17. L Lövgren, S Sjöberg, PW Schindler. Geochim Cosmochim Acta 54:1301, 1990.
18. S Sjöberg, L Lövgren. Aquatic Sci. 55:324, 1993.
19. T Hiemstra, WH van Riemsdijk. Colloids Surfaces 59:7, 1991.
20. M Djafer, I Lamy, M Terce. Prog Colloid Polym Sci 79:150, 1989.
21. J Lyklema. J Colloid Interf Sci 99:109, 1984.
22. R Woods, D Fornasiero, J Ralston. Colloids Surfaces 51:389, 1990.
23. J Lützenkirchen, P Magnico. J Colloid Interf Sci 170:326, 1995.
24. BD Honeyman, PH Santschi. Environ Sci Technol 22:862, 1988.
25. G Pang, PS Liss, J Colloid Interf Sci 201:71, 1998; G Pang, PS Liss. J Colloid Interf Sci 201:77, 1998.
26. C Lugwig, PW Schindler. J Colloid Interf Sci 169:284, 1995.
27. AP Robertson, JO Leckie. Environ Sci Technol 32:2519, 1998; AP Robertson, JO Leckie. Environ Sci Technol 33:786, 1999.
28. L Gunneriusson. PhD thesis, Umea University, Umea, Sweden, 1993.
29. J Lützenkirchen. Aquatic Geochem, in press.
30. K Bourikas, T Hiemstra, WH van Riemsdijk, Langmuir 17:749, 2001.
31. MC Fair, JL Anderson. J Colloid Interf Sci 127:388, 1989.
32. N Ingri, I Andersson, L Petterson, A Yagasaki, L Andersson, K Holmström. Acta Chem Scand 50:717, 1996.
33. J Sprycha. J Colloid Interf Sci 102:173, 1984.
34. M Kosmulski, E Matijevic. Langmuir 64:57, 1992.
35. C Tiffreau. PhD thesis, Strasbourg University, Strasbourg, France, 1996.
36. J Lützenkirchen. PhD thesis, Strasbourg University, Strasbourg, France, 1996.
37. G Fu, HE Allen, CE Cowan. Soil Sci. 152:72, 1991.

38. JG Catts, D Langmuir. Appl Geochem 1:255, 1986.
39. H Pilgrim. Colloid Polym Sci 259:1111, 1981.
40. GR Joppien. J Phys Chem 82:2210, 1978.
41. N Marmier. PhD thesis, University of Reims, Reims, France, 1995.
42. HJL Michael, HL Williams, J Electroanal Chem 179:131, 1984.
43. AC Riese, PhD thesis, University of Golden, Golden, CO, 1982.
44. GA Waychunas, BA Rea, CC Fuller, JA Davis. Geochim Cosmochim Acta 57:2251, 1993.
45. S Fendorf, MJ Eick, P Grossl, DL Sparks. Environ Sci Technol 31:315, 1997; BA Manning, SE Fendorf, S Goldberg. Environ Sci Technol 32:2383, 1998.
46. A Manceau. Geochim Cosmochim Acta 59:3647, 1995.
47. LA Roe, KF Hyes, C Chisholm, GE Brown Jr, KO Hodgson, GA Parks, JO Leckie. Langmuir 7:367, 1990.
48. JR Bargar. PhD thesis, Stanford University, Stanford, CA, 1999.
49. P Persson, L Lövgren. Geochim Cosmochim Acta 60:2789, 1996.
50. S Hug. J Colloid Interf Sci 188:415, 1997.
51. MI Tejedor-Tejedor, MA Anderson. Langmuir 6:602, 1990.
52. N Nilsson. PhD thesis, Umea University, Umea, Sweden, 1995.
53. T Hiemstra, WH van Riemsdijk. Langmuir 15:8045, 1999.
54. CM Eggleston, J Guntram. Geochim Cosmochim Acta 62:1919, 1998.
55. MJ Bedzyk, GM Bommarito, M Caffrey, TL Penner. Science 248:52, 1990; P Fenter, L Cheng, S Rihs, M Machesky, MJ Bedzyk, NC Sturchio. J Colloid Interf Sci 225:154, 2000.
56. EP Poeter, MC Hill. UCODE: A computer code for universal inverse modeling. US Geological Survey. Water Resources Investigations Report 98-4080, 1998.
57. GH Bolt, WH van Riemsdijk. In: GH Bolt, ed. Soil Chemistry. B Physico-Chemical Models. 2nd ed. Amsterdam: Elsevier, 1982.
58. LK Koopal, WH van Riemsdijk, MG Roffey. J Colloid Interf Sci 118: 117, 1987.
59. J Lützenkirchen, J Colloid Interf Sci 204:119, 1998.
60. JC Westall. FITEQL2.0. Department of Chemistry, Oregon State University, Corvallis, OR, 1982; A Herbelin, JC Westall. FITEQL3.1, Department of Chemistry, Oregon State University, Corvallis, OR, 1994; JC Westall, A Herbelin. FITEQL3.2. Department of Chemistry, Oregon State University, Corvallis, OR, 1996.
61. GA Parks. Chem Rev 65:177, 1965.
62. WH van Riemsdijk, JCM de Wit, LK Koopal, GH Bolt. J Colloid Interf Sci 116:511, 1987; T Hiemstra, WH van Riemsdijk, MGM Bruggenwert. Neth J Agric Sci 35:281, 1987.
63. MAA Schoonen. Geochim Cosmochim Acta 58:2485, 1994.
64. RPJJ Rietra, T Hiemstra, WH van Riemsdijk. J Colloid Interf Sci 229:199, 2000.
65. T Hiemstra, P Venema, WH van Riemsdijk. J Colloid Interf Sci 184:680, 1996; P Venema, T Hiemstra, PG Weidler, WH van Riemsdijk. J Colloid Interf Sci 198:282, 1998.
66. JR Rustad, AR Felmy, BP Hay. Geochim Cosmochim Acta 60:1553, 1996; JR Rustad, AR Felmy, BP Hay. Geochim Cosmochim Acta 60:1563, 1996; AR Felmy, JR Rustad. Geochim Cosmochim Acta 62:25, 1998.
67. JR Rustad, E Wassermann, AR Felmy, C Wilke. J Colloid Interf Sci 198:119, 1998.
68. JA Tossel, N Sahai. Geochim Comochim Acta 64:4097, 2000.
69. P Nortier, AP Borosy, M Allavena. J Phys Chem B 101:1347, 1997.
70. JR Rustad, DA Dixon, AR Felmy. Geochim Cosmochim Acta 64:1675, 2000.
71. F Dumont, J Walrus, A Watillon. J Colloid Interf Sci 138:543, 1990.
72. C Contescu, J Jagiello, JA Schwartz. Langmuir 9:1754, 1993.
73. S Pivovarov. J Colloid Interf Sci 206:122, 1998.
74. JC Westall, H Hohl, Adv Colloid Interf Sci 12:265, 1980.
75. J Lützenkirchen. Environ Sci Technol 32:3149, 1998.
76. DA Dzombak, FMM Morel. Surface Complexation Modeling, Hydrous Ferric Oxide. New York: Wiley, 1990.
77. MG Keizer, WH van Riemsdijk. ECOSAT, Wageningen University, Wageningen, The Netherlands, 1999.

78. JD Allison, DS Brown, KJ Novo-Gradac. Minteqa2/Prodefa2, A Geochemical Assessment Model for Environmental Systems: Version 3.0. Athens, GA: US EPA 1991.
79. JK Yang, AP Davis. J Colloid Interf Sci 216:77, 1999.
80. MS Vohra, AP Davis. J Colloid Interf Sci 198:18, 1998.
81. S Goldberg. Soil Sci Soc Am J 49:851, 1985.
82. AT Stone, A Torrents, J Smolen, D Vasudejvan, J Hadley. Environ Sci Technol 27:895, 1993.
83. A Foissy, A M'Pandou, JM Lamarche, N Jaffrezic-Renault. Colloids Surfaces 5:363, 1982.
84. G Girod, JM Lamarche, A Foissy. J Colloid Interf Sci 121:265, 1988.
85. KW Busch, MA Busch, S Gopalakrishnan, E Chibowski. Colloid Polym Sci 273:1186, 1995.
86. CE Giacomelli, MJ Avena, CP De Pauli. Langmuir 11:3483, 1995.
87. R Rodriguez, MA Blesa, AE Regazzoni. J Colloid Interf Sci 177:122, 1996.
88. CV Toner, IV, DL Sparks. Soil Sci Soc Am J 59:395, 1995.
89. S Goldberg, RA Glaubig. Soil Sci Soc Am J 49:1374, 1985.
90. N Marmier, A Delisée, F Fromage. J Colloid Interf Sci 211:54, 1999.
91. S Goldberg, G Sposito. Soil Sci Soc Am J 48:772, 1984.
92. M Borkovec, JC Westall. J Electroanal Chem 150:325, 1983.
93. KF Hayes, JO Leckie. J Colloid Interf Sci 115:564, 1987.
94. J Lützenkirchen. J Colloid Interf Sci 195:149, 1997; J Lützenkirchen. J Colloid Interf Sci 202:212, 1998.
95. JF Boily. PhD thesis, Umea University, Umea, Sweden, 1999.
96. N Sahai, DA Sverjensky. Geochem Cosmochim Acta 61:2867, 1997.
97. T Hiemstra, WH van Riemsdijk. J Colloid Interf Sci 179:488, 1996.
98. MJG Janssen, HN Stein. J Colloid Interf Sci 111:112, 1986.
99. T Hiemstra, WH van Riemsdijk. J Colloid Interf Sci 179:488, 1996.
100. T Hiemstra, H Yong, WH van Riemsdijk. Langmuir 15:5942, 1999.
101. N Marmier, A Delisée, F Fromage. J Colloid Interf Sci 212:228, 1999.
102. U Hoins. PhD thesis, ETH Zürich, Zürich, Switzerland, 1992.
103. CP Schulthess, DL Sparks. Soil Sci Soc Am J 50:1406, 1986.
104. J Lützenkirchen, JF Boily, L Lövgren, S Sjöberg. Geochim Cosmochim Acta, submitted; JF Boily. PhD thesis, Umea University, Umea, Sweden 1999.
105. BB Johnson. Environ Sci Technol 24:112, 1990.
106. CJ Daughney, JB Fein. J Colloid Interf Sci 198:53, 1998.
107. J Lützenkirchen. J Colloid Interf Sci 210:384, 1999.
108. J Lützenkirchen. J Colloid Interf Sci 204:119, 1998.
109. M Kosmulski, S Durand-Vidal, J Gustafsson, JB Rosenholm. Colloids Surfaces 157:245, 1999.
110. JR Bargar, SN Towle, GE Brown Jr, GA Parks. J Colloid Interf Sci 185:473, 1997.
111. RPJJ Rietra, T Hiemstra, WH van Riemsdijk. Geochim Cosmochim Acta 63:3009, 1999.
112. D Henderson, FF Abraham, JA Barker. Mol Phys 31:1291, 1976.
113. I Larson, P Attard. J Colloid Interf Sci 227:152, 2000.
114. H Ohshima, TW Healy, LR White. J Colloid Interf Sci 90:17, 1982.
115. AWM Gibb, LK Koopal. J Colloid Interf Sci 134:122, 1990.
116. J Lützenkirchen, P Behra. J Cont Hydr 26:257, 1997.
117. MM Benjamin. PhD thesis, Stanford University, Stanford, CA, 1979.
118. K Mesuere, W Fish. Environ Sci Technol 26:2357, 1992.
119. T Hiemstra, WH van Riemsdijk. J Colloid Interf Sci 210:182, 1999.
120. JS Geelhoed, T Hiemstra, WH van Riemsdijk. Environ Sci Technol 32:2119, 1998; JS Geelhoed, T Hiemstra, WH Van Riemsdijk. Geochim Cosmochim Acta 62:2389, 1997.
121. JR Bargar, R Reitmeyer, JA Davis. Environ Sci Technol 33:2481, 1999.
122. JJ Lenhart, JR Bargar, JA Davis. J Colloid Interf Sci. 234:448, 2001.
123. KJ Farley, DA Dzombak, FMM More. J Colloid Interf Sci 106:226, 1985.
124. Ch Tiffreau, J Lützenkirchen, Ph Behra. J Colloid Interf Sci 172:82, 1995.
125. RNJ Comans, JJ Middleburgh. Geochim Cosmochim Acta 51:2587, 1987.
126. J Lützenkirchen, P Behra. Aquatic Geochem 1:357, 1996.

127. MJ Angove, JD Wells, BB Johnson. J Colloid Interf Sci 211:281, 1999; DP Rodda, BB Johnson, JD Wells. J Colloid Interf Sci 184:365, 1996.
128. ML Machesky, DA Palmer, DJ Wesolowski. Geochim Cosmochim Acta 58:5627, 1994. ML Machesky, DJ Wesolowski, DA Palmer, MK Ridley. J Colloid Interface Sci 239:314, 2001.
129. A De Keizer, LJG Fokking, J Lyklema, Colloids Surfaces 49:149, 1990.
130. DJ Wesolowski, ML Machesky, DA Palmer, LM Anovitz. Chem Geol 167:193, 2000.
131. ON Karasyova, LI Ivanova, LZ Lakshtanov, L Lövgren. J Colloid Interf Sci 220:419, 1999.
132. DB Ward, PV Brady. Clays Clay Miner 46:453, 1998.
133. J Lyklema. Chem Ind 21:741, 1987; J Lyklema, LGJ Fokkink, A de Keizer. Prog Colloid Polym Sci 83:4, 1987.
134. C Pagnoux, M Serantoni, R Laucournet, T Chartier, JF Baumard. J Eur Ceram Soc 19:1935, 1999.
135. G Tarì, SM Olhero, JMF Ferreira. J Colloid Interf Sci 231:221, 2000.
136. DP Rodda, JD Wells, BB Johnson. J Colloid Interf Sci 154:564, 1996.
137. M Kosmulski. J Colloid Interf Sci 211:410, 1999.
138. MJ Angove, BB Johnson, JD Wells. J Colloid Interf Sci 204:93, 1998.
139. R Strauss, PhD thesis, University of Bonn, Bonn, Germany, 1992.
140. P Venema, T Hiemstra, WH van Riemsdijk. J Colloid Interf Sci 181:45, 1996.

12

Adsorption from Electrolyte Solutions

ETELKA TOMBÁCZ University of Szeged, Szeged, Hungary

I. INTRODUCTION

Adsorption refers to the accumulation of any species from one of the continuous phases at the interface between two phases. If the solid–liquid (S/L) interface is in question [i.e., adsorption of a dissolved material (solute) is studied], the wetting of solid material (adsorbent) by the liquid (medium in which adsorbent is dispersed) and the solubililty of solute in the given liquid (here, solvent) have to be considered in addition to adsorption. Simultaneous equilibria of adsorption, wetting, and solubility exist between the components (adsorbent, solvent, and solute). Competition of solvent and solute molecules for surface sites and also competition of surface and solvation forces for solute molecules are always present in the S/L adsorption systems. Therefore, a better understanding of adsorption from solutions requires that the interaction of a solute with a surface be characterized in terms of the fundamental physical and chemical properties of all the three components (solute, adsorbent, and solvent) of adsorption.

This chapter deals with the adsorption from electrolyte solutions. Let us consider the fundamental physical and chemical properties of components in this special case. In electrolyte solutions, the solvent is water in almost all cases, dissolution of electrolytes results in formation of charged species, and formation of the solid–water interface involves the hydration and charging of the surface. The molecular interactions with water inherently influence the chemical properties of adsorption partners, both solute and adsorbent. Even in the simplest case of adsorption from electrolyte solutions, adsorption of this kind cannot be considered as an accumulation of ionic species at interface, but it is a chemically controlled distribution of charged species governed by the in situ developed electrified interfaces. The chemical contribution of components cannot be neglected in general; simultaneous equilibria exist both in the aqueous phase and at the surface, which mutually influence each other. The probable simultaneous equilibria among the participating components of adsorption from electrolyte solutions (i.e. among the solid, solute, and water), the interfacial and aqueous processes resulting in the equilibrium distribution of polar and charged species are shown in Fig. 1 (top). An electrified interface develops due to the formation of a multitude of charged surface sites and the accumulation of countercharges in order to preserve electroneutrality. How the ionic charges are distributed near a charged solid surface is represented schematically in Fig. 1 (middle). The surface acquires an electrical potential different from the solution in which it is immersed. The distribution of charges in the electrical double layer (abbreviated EDL herein) and also the potential decay from surface to bulk solution depends on the quality and quantity of dissolved species as shown by some characteristic examples in Fig. 1 (bottom).

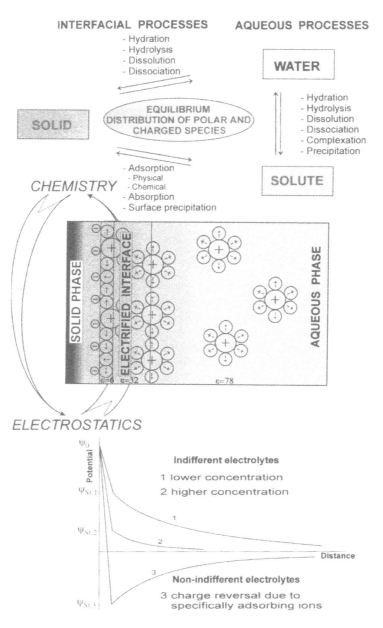

FIG. 1 Conceptual interaction web of adsorption from electrolyte solutions showing the mutual contribution of chemical processes and electrostatic constraints.

Focusing on the purpose of this book, one may expect a systematic treatment of the topic's theory, modeling, and analysis given in the book title for the kind of adsorption discussed in this chapter. To give a comprehensive description of adsorption from electrolyte solutions would require a book in itself, which may not be interesting for the readers who have known the relevant literature. Here, I would like to point to the given chapters of excellent books Lyklema

(Chapter 5 of Vol. I and Chapters 2 and 5 of Vol. II) [1] and Hunter (Chapter 6 of Vol. I and Chapter 12 of Vol. II) [2], and to draw the attention to the widely accepted conception given in the book *Adsorption from Solution at the Solid/Liquid Interface* [3], in which the adsorption of small ions, ionic surfactants, dyes, and polyelectrolytes are discussed in separate chapters. A remarkable review by Haworth [4] on the modeling of sorption from an aqueous solution has been published. It would not be worth repeating the theoretical background, the basic concepts, and the different approaches of modeling which are entirely accepted even now. I focus on the recent trends and attempt to highlight the importance of simultaneous homogeneous and interfacial processes, such as the hydration of the surface, dissolution of a solid, and speciation in the solution phase, which take place in parallel with the adsorption of ionic species and inherently influence the adsorption equilibrium; however, unfortunately, these seem to be often neglected.

II. PROCESSES AT THE INTERFACE AND IN THE AQUEOUS PHASE

A. Interaction of Solid with Water: Hydration, Hydrolysis, Dissolution, and Dissociation

One of the important processes is the hydration of adsorbent, because the properties of the solid surface inherently alter, a charged surface forms during the immersion of the solid into water, and the formed electrified interfaces play a governing role in the adsorption of electrolytes. The hydration and charging of different solids in aqueous medium has to be discussed before dealing with the adsorption of ionic species.

In a recent article [5] dealing with the properties of adsorbed water layers and the effect of adsorbed layers on interparticle forces, it was clearly stated that even under common room conditions (relative humidity in the region 40–60%), two or three adsorbed monolayers of water are often present on particles, dominating the interactions, and therefore the physical character-istics of the material. For a two-phase equilibrium system containing hydrophilic silica plates (surface of α-quartz covered by silanol groups) and water molecules, a molecular dynamic simulation expected at least one adsorbed monolayer to be present. Quite different behavior would be expected for less hydrophilic surfaces. The material character and chemical properties of solid materials are of crucial importance in the hydration interaction. Therefore, some common adsorbents which are frequently used in aqueous electrolyte solutions are discussed separately.

1. Oxides

Undercoordinated metal ions (e.g., Si^{4+}, Al^{3+}, Fe^{3+}) occurring on the top layer of oxide surfaces react with water molecules to form surface OH groups in an attempt to complete their coordination sphere. In the presence of water, the surface of oxides (e.g. SiO_2, Fe_2O_3, Al_2O_3, TiO_2) are generally covered with surface hydroxyl groups (S–OH sites). For most of the oxides, dissociative chemisorption of water molecules seems energetically favored [6]. The various surface hydroxyls formed may structurally and chemically not be fully equivalent. Geometrical consideration and chemical measurements indicate an average surface density of 5 (typical range 2–12) hydroxyls per square nanometer of oxide surface. Surface hydroxyl groups of oxides, S–OH sites, can be removed by thermal treatment in vacuum $O(MeOH)_2 \rightleftharpoons O(Me)_2O + H_2O$. These activated sites can be easily rehydrated by the adsorption of water, and additional water molecules can be adsorbed on top of the already formed hydroxyls through the formation of a hydrogen bond between the oxygen atom of water molecule and two hydrogen atoms of neighboring hydroxyl groups or with other types of coordination [7]. The behavior of S−OH groups has been studied extensively and it is the subject of numerous reviews (e.g., Refs. 6 and 8–11).

Aluminum oxides and hydroxides occur in several different crystalline forms (e.g., corundum–octahedral Al_2O_3, gibbsite–octahedral $Al(OH)_3$ boehmite–AlO(OH) [12]. Alumina surface chemisorbs at least a monolayer of water when exposed to moisture. An idealized illustration of dry and hydrated surfaces of γ-Al_2O_3 is given in the literature (Fig. 1 of Ref. 13). In the dried state, the top layer of a γ-Al_2O_3 surface contains only oxide ions, regularly arranged over aluminum ions in octahedral sites in the lower layer. During hydration, water is chemisorbed and the top layer of oxide ions is converted to hydroxyl ions. Hydroxyl ions coordinated in various ways with aluminum cations constitute the reactive sites (Al−OH) on the alumina surface. In general, more than one kind of surface OH group can be distinguished on the basis of stereochemical reasoning. The number and type of surface OH groups depend on the preferentially exposed crystal planes and the distribution of aluminum ions at the surface. Hiemstra's model of aluminum oxide surface [14] considers only two active sites, singly and doubly coordinated Al atoms on surface. Because the coordination of both oxygen atoms and hydroxyl groups on a partially dehydroxylated alumina surface differs from each other by the number and the type of surrounding metal atoms, five types of active sites were differentiated on the basis of potentiometric titration by Contescu et al. [15]. The Al ions in the crystal lattice are in octahedral (Oh) and tetrahedral (Th) coordination, and the different types of Al−OH sites, namely $(Al_{Oh})_3OH(III, +0.5)$, $(Al_{Oh}Al_{Th})OH(IIa, +0.25)$, $(Al_{Oh})_2OH(IIb, 0)$, $(Al_{Th})OH(Ia, -0.25)$, and $(Al_{Oh})OH(Ib, -0.5)$ (type and net charge given in parentheses), correspond with the previously published infrared (IR) bands. The potentiometric identification of most acidic triply coordinated groups of type III on the alumina surface is obvious, whereas type Ib is likely misinterpreted, neglecting the dissolution of the solid phase [16]. In aqueous suspensions of oxide, however, the monolayer of strongly bound water completes the coordination of surface metal atoms. The protons of the bound water become distributed over the surface oxygen atoms, making the surface homogeneous.

Iron oxide particles with different crystal structures usually exist as colloidal particles (e.g., ferrihydrite ∼5–10 nm, goethite, hematite ∼10–50 nm). Under dry conditions, surface Fe atoms may be coordinatively unsaturated [10]. In contact with water, the iron atoms coordinate with water molecules, which share their lone electron pairs with Fe. During adsorption, the water molecules usually dissociate, resulting in a surface covered by hydroxyl groups coordinated to the underlying Fe atoms (Fe−OH sites). Hydroxylation of iron oxides is a fast reaction; it is followed by further adsorption of water molecules which hydrogen-bond to the surface OH groups. Crystallographic considerations indicate that the surface hydroxyl groups may be coordinated to one, two, or three underlying Fe atoms. A fourth type of group is the geminal group (i.e., two OH groups attached to one Fe atom). The configuration of the various types of surface hydroxyls on different planes of goethite and hematite are given in the literature (e.g., Figs. 10.2 and 10.3 of Ref. 10). The overall density of these groups depends on both the crystal structure and the extent of development of the different crystal faces. The density of the most reactive singly coordinated groups is between 3 and 8 OH/nm^2 and their total number [determined experimentally by techniques such as BET treatment of water–vapor isotherms, IR and H_2O adsorption, isotope (D_2O or tritium) exchange, and reactions with adsorbing species like fluoride, phosphate, or oxalate; acid–base titration resulting in a lower value in general] is between ∼8 and 16 OH/nm^2 (Table 10.1 and 10.2 of Ref. 10).

Silicas seems to behave, to some degree, differently than the other oxides. The outstanding characteristic of the silica–water system is the tendency to form colloidal solutions or hydrated masses. This last is liogel, from which porous xerogel forms during drying. The highly porous structure is characteristic of amorphous precipitated silica particles [11]. Dissolution and precipitation processes is highly dependent on the crystalline form, particle size, pH, and other dissolved materials. The surface of amorphous silica is ordinarily covered with a monolayer of hydroxyl groups ($\underline{Si}$−OH sites), which can be partially removed by heating to 500−600°C. The total silicon atom density on the surface of different crystalline forms of silica (e.g., cristobalite, tridymite) is about 8 atoms/nm², but the active $\underline{Si}$−OH site density is usually between 5 and 14 sites/nm². This means that both less ($\equiv O_3 Si$−OH) and more [$=O_2 Si=(OH)_2$] than one OH group can be bound to the surface silicon atoms. The surface properties of silica considerably vary with the quality of solid bulk material (e.g., quartz, cristobalite, tridymite, precipitated silica).

James and Parks [8] summarized the surface densities of surface hydroxyl (S−OH) groups for several oxides [SiO_2, TiO_2, Al_2O_3, $AlO(OH)$, Fe_2O_3, $FeO(OH)$]. Some data calculated from crystal structure and determined by isotropic exchange, IR−H_2O sorption, acid–base titration, taken partially from Table 3 of Ref. 8 are shown in Table 1.

Oxide surfaces, unless highly dried, are usually covered with hydroxyl groups formed by dissociative chemisorption of water. Considering the crystal geometry (e.g., edges, faces) and the atomic arrangement of outer layers of crystal lattice (e.g., adjacent, vicinal, etc.), several different inorganic surface hydroxyl (S−OH) groups have to be assumed, even on a chemically homogeneous crystalline phase. The structure and chemical composition of the actual solid surface do not represent an extrapolation of appropriate bulk solid properties. The crystallographic order more or less decreases and the local chemical composition varies at the surface. This leads to the energetic heterogeneity of even the surfaces of specimen oxides [7]. The energetic heterogeneity of the real solid surfaces is considered now to be one of the fundamental, common features of the solid surfaces. Although this phenomenon is commonly accepted, its nature and role in adsorption is not well understood yet. Adsorption studies (for oxide–vapour and oxide–electrolyte interfaces) show energetic heterogeneity but calorimetric measurements (determination of immersional and adsorption heat) give evidence of this. A theoretical model taking into account surface energetic heterogeneity has been introduced to describe quantitatively experimental data for adsorption at water–oxide and electrolyte interfaces [7]. At the molecular level, different oxygen atoms on one crystal plane, domains of polar and nonpolar

TABLE 1 Surface Densities of Ionizable Surface Hydroxyl Groups on Different Oxides

Solid	Surface densities of surface hydroxyl groups (site/nm²) calculated from				
	Crystal structure	Isotropic exchange	IR−H_2O sorption	Acid–base titration	
				Acidic OH	Basic OH
α-SiO_2	4.4–5.9	11.4	4.4–5	—	3.5
TiO_2 (rutile)	12.1	12.5	11.3	2.6	4.2
χ-Al_2O_3	—	—	25	0.42	0.38
λ-AlOOH (boehmite)	—	—	16.5	—	—
α-Fe_2O_3 (hematite)	5.2–9	22.4	5.5–10	10.8	13.5
α-FeOOH	16.8	16.4	—	7.2	5.4
Goethite[a]	16.8–18	—	—	1.68–6	
Ferrihydrite[a]	—	11.4	—	1.97	

[a]Data taken from Table 10.2 of Ref. 10.

active sites, different crystal planes, edges, surface, and crystal lattice imperfections are responsible for the heterogeneity of the specimen mineral surfaces. Two limiting situations of heterogeneity, random and patchwise, can be considered [17]. If the different site types are mixed in a random or regular way, hydration and ion adsorption smear out the electrostatic potential (discussed later) difference over the entire surface and individual features of surface sites vanish, because the lateral interactions are of coulombic and long-range types. If the different classes of site are grouped together in patches, the surface heterogeneity is patchwise. This situation may occur when different types of surface site are on the crystal planes such as the basal plane and edges of clay lamellae. The size of patches and the lateral interactions are important for describing the patchwise heterogeneity. With noninteracting patches, each patch develops its own (smeared out) electrostatic potential.

In aqueous solutions, the dissolution of oxides often becomes perceptible not only at extreme pH values. The crystal structure has an influence on the pH-dependent solubility of different oxides. For instance, in the case of aluminum oxides, the most soluble product is $Al(OH)_3$ (amorphous), followed by $Al(OH)_3$ (bayerite), $Al(OH)_3$ (norstrandite), $AlOOH$ (boehmite), $Al(OH)_3$ (gibbsite), and $AlOOH$ (diaspore) [13]. The activity of dissolved Al species in equilibrium with gibbsite rises above 10^{-5} M below pH ~ 4 and above pH ~ 9. In general, the solubility of iron(III) oxides (e.g., hematite, goethite) is low and the activity of dissolved iron(III) species remains below $\sim 10^{-5}$ M between pH ~ 3 and ~ 14 [10]. Reducing conditions enhance the solubility of iron(III) oxides. The presence of ligands such as chloride, fluoride, phosphate, citrate and several other anions that form soluble complexes with iron and aluminum greatly promotes the dissolution of solid matrix. Silica oxides (e.g., amorphous silica, quartz) are soluble only in alkaline solutions, but the dissolution of solid becomes perceptible at slightly alkaline conditions (pH ~ 8) [18].

2. Clays

Clays are finely divided crystalline aluminosilicates. The principal building elements of the clay minerals are two-dimensional arrays of silicon–oxygen tetrahedra (tetrahedral silica sheet) and that of aluminum– or magnesium–oxygen–hydroxyl octahedra (octahedral, alumina, or magnesia sheet). Sharing of oxygen atoms between silica and alumina sheets results in two- or three-layer minerals, such as 1 : 1-type kaolinite built up from one silica and one alumina sheet, or 2 : 1-type montmorillonite in which an octahedral sheet shares oxygen atoms with two silica sheets [19]. Clay lamellae have negative charge sites on the basal planes owing to the substitution of the central Si and Al ions in the crystal lattice for lower positive valence ions. A negative charge associated with cation replacement in the tetrahedral layer (e.g., Al^{3+} for Si^{4+}) results in localized charge distribution, whereas much more diffuse negative charge comes from cation replacement in the octahedral layer (e.g., Mg^{2+} for Al^{3+}). This excess of negative lattice charge is compensated by the exchangeable cations. Additional polar sites, mainly octahedral Al–OH and tetrahedral Si–OH groups, are situated at the broken edges and exposed hydroxyl-terminated planes of clay lamellae.

All polar surface sites are well hydrated in aqueous suspensions; the hydrophilic nature is characteristic of clay minerals in general. The exposed siloxane surface, however, is considered to have a predominantly hydrophobic character for neutral 2 : 1 layer silicates, where isomorphic substitution has not occurred, and for the siloxane side of 1 : 1 layer silicates [20]. Spectroscopic investigations of clay–water interactions suggest that (1) water molecules can coordinate directly to exchangeable metal cations and (2) physisorbed water molecules occupy interstitial pores, interlamellar spaces between exchangeable metal cations, or polar sites on external surfaces [20]. The presence of hydrated exchangeable cations in the interlayer of expandable 2 : 1 phyllosilicates imparts an overall hydrophilic nature to the clay surface. All inorganic cations attract water molecules because of their charge, resulting in the formation of water clusters around the cations. Depending on the hydration energy and hydrolysis constant of the exchangeable cation, a varying number of water molecules will be associated with the cation. The hydration of counterions is influenced by the place of isomorphic substitution. Near the tetrahedral substitution sites (localized charges), cations form strong inner-sphere surface complexes, whereas looser outer-sphere surface complexes are associated with isomorphic substitution in the octahedral layer (delocalized charges). Two limiting water sites at the clay surface have been identified by water deuteron nuclear magnetic resonance (NMR) quadrupolar splittings. The location of cation isomorphic substitution and the molar ratio of exchangeable cations modulate the relative importance of these sites [21]. The results of the Monte Carlo simulation of the hydration of montmorillonite (2 : 1-type expandable clay) have shown the probable orientation of water molecules on a clay surface, and the importance of the water itself on clay swelling, as well as the importance of the interlamellar cation hydration on the swelling properties of the clay [22]. Two driving forces responsible of the clay swelling have been identified: the hydration of clay sheets which accounts for 32% of the energy, and the hydration of the cation, which accounts for 68% of the energy. An additional simulation for a neutral clay without any interlamellar cation resulted in endothermic immersional enthalpy, which explains why, for example, talc interacts poorly with water. These neutral surfaces have very low affinity for water and contact-angle measurements on talc using water and organic liquids are in the hydrophobic range [20].

3. Sparingly Soluble, Salt-Type Solids

At first sight, it might seem that ionic solids would resemble hydrophilic oxide surfaces. However, such solids, and, in particular, the classical model surface of silver iodide, are generally classified as hydrophobic, both in the sense that they approximately conform to the theory of hydrophobic colloid stability and that such surfaces may display significant contact angles with aqueous solutions [23].

The dissolution of ions from a crystal lattice is not stoichiometric; one of the ions dissolves preferentially. For example, a AgI crystal is placed in water and a solution occurs until the product of ionic concentration equals the solubility product ($K_s = [Ag^+][I^-] = 10^{-16} M^2$); equal amounts of dissolved Ag^+ and I^- ions can be expected. In fact, silver ions dissolve preferentially, leaving a negatively charged surface. If Ag^+ ions are now added to the solution, dissolution of Ag^+ ions is suppressed and the charge falls to zero at pAg ~ 5.5 (pAg $= -\log[Ag^+]$) [24]. An alternative, but equivalent, interpretation of the experimental fact that Ag^+ activity always exceeds the corresponding I^- concentration by many decades at this particular activity of Ag^+ ions [point of zero charge (PZC) = pAg ~ 5.5] is based on the relative adsorbability of I^- and Ag^+ ions [25]. The preferential adsorption is partly a surface and partly a liquid bulk effect, because cations and anions in the bulk phase are hydrated in different ways and the state of hydration of ions changes during adsorption.

The situation, when a sparingly soluble solid is placed in water, is much more complicated in principle, because of the successive reactions taking place spontaneously in the aqueous phase. This fact is obvious for carbonates, sulfates, where the dissolved amount cannot be calculated directly from the characteristic thermodynamic data (i.e., from the solubility products) because additional reactions take place simultaneously in the liquid phase depending on its composition, which influence the activity of dissolved species and, therefore, the amount of dissolved material. For example, a common carbonate, calcite ($CaCO_3$), with intermediate solubility dissolves as bicarbonate [$Ca(HCO_3)_2$] at favorable conditions with respect to pH and partial pressure of CO_2; or the dissolved amount of a moderately soluble salt, gypsum

(CaSO$_4$ · 2H$_2$O), cannot be calculated from the solubility product (obtained from standard free enthalpy of formation) correctly, because an additional equilibrium between the dissolved ions (i.e., ion-pair formation) is present in the aqueous phase, which increases the dissolved amount of gypsum significantly [18].

The dissolution equilibrium for a two-component dissociable solid in aqueous medium may be expressed as [26]

$$C_{vc}A_{va}(s) \rightleftharpoons v_c C^{m+}(aq) + v_a A^{n-}(aq)$$

where the stoichiometric coefficients are v_c and v_a and the ionic valences are m and n, which are subject to the electroneutrality condition $v_c m = v_a n$. The equilibrium constant of this reaction, a dissociation process, K_{dis}, is

$$K_{dis} = \frac{[C^{m+}]^{vc}[A^{n-}]^{va}}{[C_{vc}A_{va}(s)]}$$

The solubility product is defined as

$$K_{so} = [C^{m+}]^{vc}[A^{n-}]^{va}$$

The ion activity product, IAP, is related to K_{dis} through the expression.

$$IAP = K_{dis}[C_{vc}A_{va}(s)]$$

When the activity of the solid phase has a unit value, IAP $= K_{so}$. It may often occur that the IAP values calculated from the activity data measured (e.g., by means of an ion-selective electrode) are not the same as the solubility product of solid. If the activity data are correct, the following problems may exist: (1) equilibrium with a solid is not reached; (2) the solid phase controlling the ion activity is not the one suspected; (3) the solid is not in its standard state [26]. In colloidal dispersions, the solid phase may not be identified; however, even if the crystal form of solid is the one suspected, the much smaller particle size than the thermodynamically stable solid causes an enhanced solid-phase activity and an increase in IAP. The size-dependent solubility of a solid particle in a liquid is well known from an old work of Ostwald [27]:

$$\ln \frac{c(r)}{c_\infty} = \frac{2V_s \gamma}{RTr}$$

where $c(r)$ is the solution concentration in equilibrium with particles of radius r, c_∞ is the bulk solubility, V_s is the molar volume, γ is the surface free energy, R is the gas constant, and T is the temperature.

When solid-phase solubility is superimposed by successive complexation equilibria of dissolved ions resulting in several species in the equilibrium liquid phase, which is much more common than usual, the understanding of basic principles that led to a calculation of the ion activity product and the amount dissolved for a solid phase is not simple. A chemical that complexes with dissolved ions of a solid would enhance the dissolution of the solid phase. Obtaining a reliable solubility constant for sparingly soluble solids demands crucial work. In general, the given systems are considered and a series of examples are explained in the relevant literature (e.g., Refs 26 and 28). In aqueous systems, a huge variety of different types of complex has been identified depending on the chemical composition of the solid and the solution. The formation of hydrolytic complexes is concomitant of aqueous medium. Homogeneous complexation may be described by

$$v_c C^{m+} + v_a A^{n-} \rightleftharpoons (C_{vc}A_{va})^q$$

where $q = v_c m - v_a n$ is the valence of the soluble complex $(C_{vc} A_{va})^q$. The equilibrium constant for this reaction is called a stability constant:

$$K_s = \frac{[(C_{vc} A_{va})^q]}{([C^{m+}]^{vc} [A^{n-}]^{va}}$$

In general, a relevant chemical reaction for the formation of the most common mononucleus hydroxy species, $C(OH)_{va}^{(m-va)+}$, may be written

$$C^{m+} + v_a H_2 O \rightleftharpoons C(OH)_{va}^{(m-va)+} + v_a H^+$$

The total molar concentration of the cationic component, C_{tot}, in solution is the sum of the concentration of all C species; it may be written

$$[C_{tot}] = [C^{m+}] + \sum [C(OH)_{va}^{(m-va)+}]$$

The concentration of hydroxy species is highly dependent on the pH of the solution; the contribution of a given type to the C_{tot} may be negligible. For example, apart from the alkaline pH region, the hydroxycomplexes of silver ($Ag(OH)^0$, $Ag(OH)_2^-$, $Ag(OH)_3^{2-}$, $Ag(OH)_4^{3-}$), or the multiple coordinated species of copper, chromium, or cobalt, such as $Cu(OH)_3^-$, $Cu(OH)_4^{2-}$, $Cr(OH)_4^-$, $Cr(OH)_5^{2-}$, $Co(OH)_3^-$ and $Co(OH)_4^{2-}$, are negligible.

The presence of atmospheric CO_2 even in closed systems into which CO_2 may be drawn with the solid (e.g., CO_2 complex on activated carbon [29] or adsorbed CO_2 on basic adsorbent like Al_2O_3) involves the following carbonate equilibria in several cases:

$$CO_2(g) + H_2O \rightleftharpoons H_2CO_3$$
$$HCO_3^- + H^+ \rightleftharpoons H_2CO_3$$
$$CO_3^{2-} + H^+ \rightleftharpoons HCO_3^-$$

The spontaneously formed CO_3^{2-} and HCO_3^- obviously react with the dissolved cations {e.g., with Ca^{2+} from gypsum ($CaSO_4 \cdot 2H_2O$) or apatite $[Ca_{10}(PO_4)_6(F, OH)_2]$ [28]}. Considering the solution phase, it generally contains electrolytes to fix a constant ionic strength. Certain anions such as ClO_4^- and NO_3^- do not have ability to form complexes with metal ions; however, other anions (e.g., Cl^-, PO_4^{3-}) may significantly influence the speciation of several metal ions such as Cu^{2+}, Cd^{2+}, Pb^{2+}, Zn^{2+}, Fe^{3+}, and Al^{3+}.

Dissolution of a sparingly soluble solid in a multicomponent aqueous solution can be described by the set of the chemically probable heterogeneous and homogeneous reactions and the corresponding thermodynamic equilibrium constants and by the mole balance equations for each component. Usually, the set of coupled algebraic equations is too complicated to solve analytically; numerical procedures are required [26]. By now, several computer programs are available that can calculate the equilibrium speciation if thermodynamic equilibrium constants are known (e.g., MICROQL [30], MINTEQA [31]) or the set of experimental activity data can be fitted to obtain the equilibrium constants of probable reactions supposed in an appropriate chemical model (e.g., FITEQL [32]).

B. Surface Charging

The largest part of a solid in contact with an aqueous phase acquires a surface electrical charge due to the redistribution of charged species in the interfacial region. Different concepts of charging mechanisms can be found in the relevant literature. For example, five different ways different surfaces may become electrically charged are given in the book by Everett [24]:

1. Ionization of the surface group (e.g., dissociation of acidic or basic groups of oxides results in negatively or positively charged sites)
2. Differential solution of ions from the surface of a sparingly soluble crystal (e.g., preferential dissolution of silver ions from silver halides)
3. Isomorphous substitution (e.g., replacement of Si in the tetrahedral layer of clay by Al producing a negative surface charge)
4. Charged crystal surfaces (e.g., broken edges of a kaolinite crystal)
5. Specific ion adsorption (e.g., specific adsorption of surfactant ions)

In the IUPAC recommendation [33], Lyklema differentiates four ways:

1. Adsorption of potential-determining (p.d.) ions (e.g., Ag^+ for AgI, or Ba^{2+} for $BaSO_4$)
2. Adsorption of other types of surface ion (e.g., H^+ for oxides)
3. Dissociation of covalently bound surface groups (e.g., ion-exchange resins, lattices, proteins)
4. Isomorphous substitution (e.g., clays, zeolites)

Three different ways are listed by Stumm [6]:

1. Charge may arise from chemical reactions at the surface (i.e., the surface contains ionizable functional groups: $-OH$, $-COOH$, $-OPO_3H_2$, $-SH$)
2. Lattice imperfections at solid surface (i.e., isomorphous replacements)
3. A surface charge may be established by adsorption of a surfactant ion. (i.e., preferential adsorption of a surface-active ion)

These methods of charging found in the prominent literature seem to be different processes such as ionization, adsorption, and dissolution taking place spontaneously at aqueous interfaces of the solid, except the imperfection of crystal lattice called also isomorphous substitution. Considering the interfacial equilibria in Fig. 1, when a piece of solid is placed in water, several processes begin. All of these are governed by the chemistry of the solid and are influenced by the composition of the aqueous phase. The result of these interfacial equilibria is a charged (electrified) interface which influences in situ the distribution of ionic species; therefore, the interfacial processes are also affected. Surface-charge density remains constant only if the charge defects are fixed in the solid phase such as lattice imperfections or isomorphous substitutions. Surface-charge density varies in any kind of interfacial process considered as ionization and dissolution [24], adsorption and dissociation [33], or surface chemical reaction [6]. Although the change in the valance bonds of surface atoms is indisputable in several cases, these surface-charging processes have not been defined clearly as chemical reactions in the literature yet.

An advanced generalization of charging for different solid materials has been published [34]. According to Sposito, solid surfaces develop charges in two principal ways:

1. Permanently: solid structure itself holds permanent charge defects
2. Conditionally: in reaction of surface functional group (e.g., H^+ association or dissociation)

All of the above-mentioned charging mechanisms are involved in the Sposito's concept except the specific (item 5 in Everett's classification) or preferential (item 3 in Stumm's list) adsorption of, for example, surfactant ions. The question is whether this kind of preferential adsorption is involved in surface charging at all. This case coincides with the most serious problem of aqueous interfaces (i.e., which part of the interface belongs to the solid phase?). Can we consider an adsorbed layer bonded via hydrophobic interaction, even it is enough strong, as an inherent part of the solid phase? The case of polymer adsorption is somewhat similar to this, where strong multisite bonds form on the surface; therefore, an adsorbed polymer can never be desorbed by simple dilution, but it can be displaced. Polyelectrolyte coverage, however, which holds charges, has never been mentioned among the possible charging mechanism. In this chapter the author attempts to show an equilibrium concept of aqueous interfaces involving simultaneous interfacial and homogeneous equilibria, in which the adsorption of charged organic species (like surfactant ions) are separately discussed.

1. Permanently Charged Surfaces

If particles are permanently charged (e.g., montmorillonite, illite, zeolite), charges are fixed in the crystal lattice, providing constant surface charge density. The surface density of permanent charge sites, associated with the charge deficit originating from isomorphic substitution of Al and Si ions in silicate crystal lattice, is constant. It can be calculated from the charge deficit, q, (e.g., $\sim 0.25-0.6$ for smectite) and unit-cell dimensions [35] for $2:1$ phyllosilicates:

$$\sigma_0 = \left(\frac{qF}{2N_A ab}\right) \times 10^{18}$$

where F is the Faraday constant (C/mol), N_A is the Avogadro number, a and b are the dimensions of the unit cell given (in nm), and σ_0 is given in Coulombs per square meter. The permanent surface-charge density can be estimated from the cation-exchange capacity (CEC, mmol/100 g for univalent exchangable cations) and specific surface area (a^s, m^2/g) of phyllosilicates:

$$\sigma_0 = \left(\frac{(\text{CEC})F}{a^s}\right) \times 10^{-1}$$

This estimated value, however, is often unreliable. On the one hand, it is because of the uncertainty of the specific surface area, especially for the swelling smectites, where the specific surface area, for example, from nitrogen adsorption (BET surface area $\sim 13-70$ m^2/g depending on the pH of montmorillonite suspensions which were freeze-dried to obtain solid samples for nitrogen adsorption measurements [36]) is smaller generally by an order of magnitude than that is available for adsorption from aqueous medium (750–850 m^2/g) [37]. On the other hand, it is because of the contribution of conditionally charged sites on the edges to the cation-exchange capacity.

2. Conditionally Charged Solid

Considering the solid phases which may be adsorbents in aqueous electrolyte solutions, in general, it can be stated that almost all are involved in this type of charging mechanism. However, the individual chemical feature of these solids requires differentiation in treatment.

Charges on the surfaces of ionic crystals, like halides and sulfates, can arise from the preferential dissolution of the ions [24] or the adsorption of constituent ions [e.g., Ag^+ or Cl^- on AgCl(s), Ba^{2+}, or SO_4^{2-} on BaSO$_4$(s)], when an ion upon adsorption becomes indistinguishable from the solid matrix. These systems are considered as ideal cases; constituent ions are potential-determining ions (i.e., ions obey Nernst's law and their electrochemical potential in the adsorbed state does not have a concentration-dependent term) [1,33,38]. The surface charges of this kind of solids cannot be assigned to individual surface species or active sites.

The second part of conditionally charged solids, such as oxides, hydroxides, and organic solids (e.g., latexes, proteins, resins), however, has identifiable surface sites. For example, surface hydroxyl groups are bound chemically to the atoms (e.g., Al, Fe, Si, Ti) in the crystal lattice of oxides and hydroxides, or organic solids contain a variety of different ionizable (acidic or basic, or both) functional groups (e.g., sulfate, carboxyl, hydroxyl, amino) bound to the carbon skeleton. These chemically reactive surface groups are exposed to an aqueous solution and the surface becomes charged due to surface chemical reactions with H^+/OH^- ions [6–8,16,17,35].

The surface charging of the third part of conditionally charged solids (e.g., carbonate, phosphate minerals) seems to be determined by both the dissolution of constituent ions and their hydrolysis products, thus by H^+/OH^- ions. The thermodynamic calculation of the solution equilibrium has suggested that the potential-determining ions for calcite ($CaCO_3$) are Ca^{2+}, CO_3^{2-}, HCO_3^-, H^+, and OH^- [39]. The surface charging of $MgCO_3$ has been explained by the dissolution of lattice ions (Mg^{2+} and CO_3^{2-}), resulting in holes which are immediately saturated by water molecules and the surface-active sites thus formed ($-CO_3H$ and $-MgOH$) react with H^+/OH^- ions; thus, the surface charge is pH dependent [40]. In contrast with these, only the constituent ions (Ca^{2+} and CO_3^{2-}) proved to be the potential-determining ions in aqueous calcite ($CaCO_3$) suspensions [41].

Development of conditional (or variable) surface charges involves chemical reactions in the interfacial layer. Therefore, the individual material features, (i.e., the chemical properties of both the potentially charged solid material and the dissolved species) have to be considered. When a chemically reactive surface group is exposed to an aqueous solution, the surface may become charged due to a surface reaction (e.g., dissociation, association, complexation) if the aqueous solution contains the other reactant as a dissolved species. The charging process on variable-charge sites is determined by not only the quality and quantity of active sites but also the composition of aqueous solution. An electrical double layer develops around particles due to the distribution of ionic species between the solid–liquid interfacial layer and the equilibrium liquid phase.

In ideal cases, the equilibrium distribution of aqueous species between the solid and liquid phases determines the difference in potential between these phases; these species are called the potential-determining (p.d.). Typical examples are Ag^+ and I^- ions for solid AgI [1]. The surface-charge density (σ_0) of AgI is defined as

$$\sigma_0 \equiv F(\Gamma_{Ag^+} + \Gamma_{I^-})$$

where F is the Faraday constant and Γ_i is the surface excess of i. The surface potential (ψ_0) can be given by the Nernst equation for a AgI crystal:

$$\psi_0 = \frac{RT}{F} \ln \frac{[Ag^+]}{[Ag^+]_{PZC}}$$

where $[Ag^+]$ is the activity of the silver ions in solution and $[Ag^+]_{PZC}$ is the point of zero charge at which $(\delta\sigma_0/\delta\mu)_{pAg}$ is zero (i.e., a change in the activity of the background electrolyte has no effect on the surface charge).

H^+ and OH^- ions are largely held responsible for the charging of several common solid materials (e.g., oxides, hydroxides, edge sites, and octahedral-plane hydroxyl groups of clay minerals, polymer latexes, proteins), because these ions have strong affinity for the surface functional groups over the entire pH range. The fact that H^+ and OH^- ions are potential-determining ions, however, has been the subject of special discussion since 1962 when de Bruyn et al. [42,43] assumed that for oxides. According to Lyklema [33], it is more appropriate to call H^+ and OH^- ions charge-determining ions instead of p.d. ions, because these surfaces do not obey Nernst's law as analyzed for silicium, aluminum, and iron oxides by Hohl et al. [44]. The term "charge determining" seems to be even worse than that substituted, because it would be nonsense that the activity of H^+ and OH^- ions determines the surface-charge density, as expected analogously to the definite dependence of surface potential on the activity of potential-determining ions.

Most oxide and hydroxides, as well as the broken edge sites and basal-plane hydroxyl groups of clay minerals exhibit amphoteric behavior. The formation of electric charge can be explained by the acid–base behavior of surface hydroxyl groups. A detailed description can be found in several excellent book chapters and articles [1,2,6,8,13,14,35,45]. Charge development on amphoteric surface sites (S−OH) could occur by direct proton transfer. The surface hydroxyl groups can be capable of ionization [8,46]. Surface ionization (protonation and deprotonation) reactions can take place on these sites, depending on the pH of the solution:

Protonation: $S-OH + H^+ \rightleftharpoons S-OH_2{}^+$
Deprotonation: $S-OH \rightleftharpoons S-O^- + H^+$ or $S-OH + OH^- \rightleftharpoons S-O^- + H_2O$

These processes are interfacial protolytic equilibria existing between the large number of active sites on the surface and H^+/OH^- ions in the aqueous medium, which are the product of a water autoprotolysis reaction ($H_2O \rightleftharpoons H^+ + OH^-$). The question of our approach or the choice of model by which the surface-charging mechanism can be described is whether to consider them a surface association–dissociation or an interfacial adsorption–desorption equilibria. Charging of an amphoteric oxide surface can be modeled with a simple one-step protonation reaction [14]:

$$S-OH^{-1/2} + H^+ \rightleftharpoons S-OH_2{}^{+1/2}$$

When surface-charge development occurs by direct proton transfer from the aqueous phase, the surface-charge density and potential can be defined analogously to the Nernstian surfaces:

$$\sigma_{0,H} = F(\Gamma_{H^+} - \Gamma_{OH^-})$$

$$\psi_0 = \frac{RT}{F} \ln \frac{[H^+]}{[H^+]_{PZC}} = \left(\frac{RT}{F}\right)(2.3)(pH_{PZC} - pH)$$

The surface-charge density is experimentally accessible. The surface excess amounts (Γ_i) can be determined by means of adsorption measurements. Potentiometric acid–base titration of oxides provides a direct measure of net proton surface-charge density [34,35] if the ions of supporting electrolyte have no specific affinity for the surface (called an indifferent electrolyte), and the total acid or base consumption results in surface-charge formation, and no other acid–base reactions, like hydrolysis or dissolution, take place. The surface excess, Γ_i, amount defined for adsorption [47] can be determined directly from the initial ($c_{i,0}$, mol/L) and equilibrium ($c_{i,e}$, mol/L) concentration of the solute [$n_i^\sigma = (c_{i,0} - c_{i,e})V/m$, where V is the volume (L) of the liquid phase and m is the mass of adsorbent; $\Gamma_i = n_i^\sigma/a^s$, where a^s is the specific surface area of the adsorbent] for adsorption from a dilute solution. The values $n_{H^+}^\sigma$ and $n_{OH^-}^\sigma$ can be calculated at each point of titration from the measured pH using the actual activity coefficient from the slope of the H^+/OH^- activity versus concentration straight lines for background electrolyte titration. Roughly one-tenth of the total active-site density calculated from crystallographic data can be titrated; in other words, only 10% of the surface hydroxyls can become charged.

Experimental curves of potentiometric acid–base titrations, representing $\Delta\Gamma_{H,OH} = \Gamma_{H^+} - \Gamma_{OH^-}$ as a function of pH at several concentrations of an indifferent electrolyte, can intersect at a common pH. It can be called a common intersection point (c.i.p.) in analogy to the charge–potential curves for p.d. ions [e.g., surface-charge density (σ_0) versus pAg at different ionic strengths, I]. If the c.i.p. of $\sigma_{0,H}$ (or $\Delta\Gamma_{H,OH}$) versus pH curves is sharp and coincides with the $\sigma_{0,H} = 0$ (where $\Gamma_{H^+} = \Gamma_{OH^-}$) surface-charge state, this unique pH is then identified as the PZC. This occurs only in the case of oxides under ideal conditions [33]. The point of zero charge is a characteristic of the given surface in aqueous medium; it is the reference point for surface charging [1]. The reliable experimental data of PZC, sometimes called the pristine point of zero charge, (PPZC) [1,33], for several oxides and other materials are collected in literature [1,8,10]. In the presence of specifically adsorbing ions, the c.i.p. cannot be identified as a PZC; however, this particular pH can be considered as a point of zero salt effect (PZSE) [34,35], at which $(\delta\sigma_{0,H}/\delta I)_{pH}$ is zero. In practice, sets of $\sigma_{0,H}$ (or $\Delta\Gamma_{H,OH}$) versus pH curves are shifted to the positive direction, when cations adsorb specifically and opposite the negative direction for specific adsorption of anions [1,48].

The situation is much more complicated for natural particles, even for specimen clay minerals, where permanent charges from isomorphic substitutions of ions in a clay crystal lattice are also present. In this case, the intrinsic surface-charge density, σ_{in}, can be defined as the sum of the permanent structural charge density, σ_0, and the net proton surface charge density, $\sigma_{0,H}$ [34,35]:

$$\sigma_m \equiv \sigma_0 + \sigma_{0,H}$$

Two additional points of zero charge have to be defined. The point of zero net proton charge (PZNPC) for $\sigma_{0,H} = 0$ and the point of zero net charge (PZNC) for $\sigma_{in} = 0$ have been introduced [34,35]. Unfortunately, consistent terminology for points of zero charge has not been used in the literature yet, but nobody argues against the general importance of points of zero charge to particle surface characterization.

C.   Distribution of Ions at the Charged Surface

Ions in the solid–liquid interfacial layer are situated closer to or further from the surface, depending on their size, charge, and ability to form chemical bonds with the surface sites. Ions which are constituents of the surface or have a particularly high affinity for surface sites are referred to as surface ions. This includes specifically adsorbed ions which can bind to the surface through covalent interactions in addition to the pure coulombic contributions. The presence of these ions has a direct influence on the surface charge. The formation of inner sphere complexes occurs, when ions bind directly to the surface [6,34,35]. Examples of inner-sphere complexes include K^+ ions and permanent charge sites on vermiculite, HPO_4^{2-} ions bound to the surface hydroxyl groups of goethite. Outer-sphere complexes, when ions bind through water bridges, can form, for example, between hydrated metal ions (Ca^{2+}, Na^+) and permanent basal-plane charges of montmorillonite, and also between NO_3^- and protonated aluminol groups. The diffuse swarm of ions are situated the furthest distance from the particle surface and are electrostatically separated and disturbed by thermal motion. Although ions around particles are separated within a nanometer-scale distance, the surface charge must be balanced and the electroneutrality condition must be satisfied.

1.   Pure Electrostatic Approach

In the classical description of charge distribution at a solid–solution interface, charges on the particle and countercharges at a certain distance together form an electric double layer (EDL) (e.g., Refs. 1, 2, and 49). The simplest approach (Helmholtz) considers two infinite plates of the separated charges with a linear potential decay between them. The diffuse double-layer model (Gouy–Chapmann) also differentiates two, surface and diffuse, parts of EDL. Charges belonging to the particle surface are surface charges; their amount related to unit surface area is called surface-charge density, σ_0 (C/m^2), and the compensating charges are diffuse layer charges with charge density σ_d. The balance of surface charge can be expressed by

$$\sigma_0 + \sigma_d = 0$$

This diffuse double-layer approach can be applied to describe the EDL of particles, if charges on particle surface are only permanent structural surface charges originating from isomorphic substitutions of ions in a clay crystal lattice (e.g., montmorillonite, which is a typical example of infinite flat plates with a constant charge density [19]) or they form by the adsorption of potential determining ions (e.g., Ag$^+$ ions on a AgI surface is an example of the case of charged particles with constant potential [1,33,38]) and the diffuse swarm of indifferent electrolyte ions compensates surface charges.

In the Gouy–Chapmann model based on only electrostatics (i.e., point charges are supposed and the medium is considered as a dielectric continuum), the charge potential relationship for the diffuse part of the EDL can be derived from the Poisson–Boltzmann equation.

According to Poisson's law, the relationship between potential, ψ, and local volume density of charge, ρ, for an electric field changing only in the x direction is

$$\frac{d^2\psi}{dx^2} = -\frac{\rho}{\varepsilon_0 \varepsilon_r}$$

where $\varepsilon_0 \varepsilon_r$ is the dielectric permittivity of the medium considered as a continuum. The space-charge density can be written as

$$\rho = \sum_i n_i z_i e$$

where the summation is over all species of the point ion present and the valency, z_i, may take positive or negative values; e is the elementary charge. The distribution of ions is influenced by the local electrostatic potential given by the Boltzmann equation

$$n_{i,x} = n_i \exp\left(-\frac{z_i e \psi}{kT}\right)$$

where $n_{i,x}$ is the local concentration and n_i is the bulk concentration of ion i, and $z_i e\psi$ represents the work done in bringing up an ion i from the bulk solution to a point x where the potential is ψ. Combining the above equations, the Poisson–Boltzmann equation for an infinite flat double layer can be written

$$\frac{d^2\psi}{dx^2} = \left(-\frac{1}{\varepsilon_0 \varepsilon_r}\right) \sum_i n_i z_i e \exp\left(-\frac{z_i e \psi}{kT}\right).$$

If the electric energy is small compared to thermal energy (i.e., $|z_i e\psi| < kT$), it is possible to expand the exponential [$\exp(x) = 1 + x/1! + \cdots + x^n/n!$] in the Poisson–Boltzmann equation neglecting the higher terms than the first two:

$$\frac{d^2\psi}{dx^2} = \left(-\frac{1}{\varepsilon_0 \varepsilon_r}\right)\left(\sum_i n_i z_i e - \sum_i \frac{n_i z_i^2 e^2 \psi}{kT}\right).$$

The first summation term must be zero because of the electroneutrality in bulk solution; thus,

$$\frac{d^2\psi}{dx^2} = \left(\sum_i \frac{n_i z_i^2 e^2}{\varepsilon_0 \varepsilon_r kT}\right)\psi = \kappa^2 \psi$$

where $\kappa = (\sum_i n_i z_i^2 e^2 / \varepsilon_0 \varepsilon_r kT)^{1/2}$ is the Debye–Hückel parameter; it has a dimension l/length and depends on the ionic strength ($I = 1/2 \sum_i c_i z_i^2$, c_i is given in mol/L) of electrolyte solution, $\kappa = 3.288(I)^{1/2}$ (1/nm), at 25°C in water. It plays a prominent part in the quantification of the EDL. The extent of the double layer is measured by the size $1/\kappa$. After integrating, the following simplified result assuming very small ψ can be given:

$$\psi = \psi_0 \exp(-\kappa x)$$

This simple equation (called the Debye–Hückel approximation) shows an exponential potential decay in the diffuse layer. Because the κ parameter is proportional to the ionic strength of the electrolyte solution, the decay of the potential distance function increases with the increasing concentration of electrolyte solution. The double-layer thickness decreases with increasing salt concentration; one can say also that the double layer is compressed by increasing the concentration of electrolytes.

The charge potential relationship in the diffuse part of the EDL can be deduced. The total charge, per unit area of surface, in the diffuse layer, σ_d, is given by

$$\sigma_d = \int \rho \, dx$$

Substituting for ρ from the Poisson equation and taking into consideration the solution of the Poisson–Boltzmann equation leads to the charge potential relationship well known as the Gouy–Chapmann equation for the diffuse EDL:

$$\sigma_d = -(8\varepsilon_0 \varepsilon_r kT)^{1/2} (n)^{1/2} \sinh\left(\frac{ze\psi_d}{2kT}\right) \tag{i}$$

For small potentials ($|ze\psi_d/2kT| < 1$, $\psi_d < \sim 25$ mV), it is possible to expand the sinh function ($\sinh x = x + x^3/3! \cdots$), and considering only the first term and the definition of the κ parameter, the following simplified equation can be deduced:

$$\sigma_d = -\varepsilon_0 \varepsilon_r \kappa \psi_d$$

Considering the charge neutralization constraint between the surface and the diffuse-layer charge densities ($\sigma_0 + \sigma_d = 0$) and supposing that $\psi_d = \psi_0$ at $x = 0$, we have

$$\sigma_0 = -\sigma_d = \varepsilon_0 \varepsilon_r \kappa \psi_0 \tag{ii}$$

This simple equation obviously shows that any change in salt concentration is accompanied by change either in surface potential at constant σ_0 or in surface-charge density at constant ψ_0. The surface-charge density (σ_0) can be determined experimentally; for example, from the surface excess concentration of p.d. ions for conditionally charged solids or from the charge deficit and unit-cell-dimensions data for permanently charged crystals. However, the charge potential relation (see a graphical representation in the bottom part of Fig. 1) is a theoretical model-dependent function.

The Gouy–Chapmann model is based on only electrostatics (point charges in a dielectric continuum); it is proved to be a good approximation in general. The above equations can give the correct description of the diffuse part of an electric double layer further from the surface, but these give unrealistic values (very high surface potentials or extremely high local concentration of ions) close to the surface. Other equations from the adequate analytical solution of the Poisson–Boltzmann equation known in basic literature [1,2,35] or even the numerical calculation cannot overcome this inaccuracy. Corrections had to be introduced by which the finite volume of ions and the EDL structure at the closest distance to particle surface (i.e., chemical contributions in addition to electrostatic interaction) should be taken into consideration.

2. Combined Chemical and Electrostatic Approaches

A broad spectrum of surface ionization and complexation models have been developed to describe the distribution of charged species near the particle surface. These are reviewed in several well-known works [1,2,4,6,8,35,46]. There are differences in the terminology and in the way these models are formulated, but they all have features in common: The reactions of surface groups are described by mass-action and material balance equations, and the surface potential is related to the surface charge by an electrostatic model [4].

Recently, the most often used models are called surface complexation models (SCMs). These combine the concept of coordination chemistry with those in electric double-layer theory. SCMs consider the surface charging (charge development on surface) and ion adsorption (interfacial distribution of ionic species) as surface complexation reactions. These processes are treated analogously to the homogeneous phase complexation equilibria described by the mass-action law in addition to accounting for the influence of electric potential developed in the interfacial reactions. The most common SCMs are the constant-capacitance, diffuse- and triple-layer models (CCM, DLM, and TLM, respectively). A schematic representation of the surface complexation models is shown in Fig. 2. The distribution of charges in the electrified interface (top) and the potential decay assumed in the given model in the near-surface region (middle) are shown. Some hypothetical surface species on oxides are represented (bottom), where surface sites (S—OH) are located at the solid–liquid interface and the specifically adsorbed ions, hydrogen ions (H^+), cations or metal ions (M^{m+}), and anions (A^{n-}) from the aqueous phase are presumed to form complexes with surface sites.

In all three models, charges associated with the surface are assumed to be balanced by counterion charges within a limited distance from surface. In the constant-capacitance and diffuse-layer models, all specifically adsorbed ions contribute to the surface charges (σ_0); it is balanced by the counterion charge. The charge potential relationship is linear ($\sigma_0 = \varepsilon_0 \varepsilon_r \kappa \psi_0$) in the CCM. Counterion charges (σ_d) are situated in the diffuse part of the EDL in the DLM, and the electroneutrality constraint is $\sigma_0 + \sigma_d = 0$. In the triple-layer model, two near-surface planes for adsorbing ions are distinguished. The surface species of specifically adsorbed H^+ and OH^- ions pertain to the innermost part and it is characterized by charges σ_0 and the outer plane has charges σ_β resulting from the adsorption of other ions. Because the electric field extending away from surface is the direct result of the surface complexation reactions of specific ions in the near-surface region, the specifically adsorbed ions also govern the counterion distribution in the diffuse layer. The charge balance of three layers can be written as $\sigma_0 + \sigma_\beta + \sigma_d = 0$.

The activity of ions near the surface, $[X^z]_s$, are influenced by the electrostatic field arising from the surface charge. It is distinguished from that in the bulk solution, $[X^z]$, because the difference in electrical potential at distance x from the surface, relative to that in the bulk solution, ψ_x, applying the Boltzmann distribution is

$$[X^z]_s = [X^z] \exp\left(-\frac{ez\psi_x}{kT}\right) \tag{iii}$$

where $ez\psi_x$ is the electric potential energy (or electrical work in moving the ions from bulk to distance x), z is the charge of ion X, e is the electron charge, kT is the thermal energy, k is the Boltzmann constant, and T is the temperature.

Surface complexation reactions are assumed on surface sites, S—OH. The total site density (N_s, mol/m^2), has to be defined for the given system. In the constant-capacitance and diffuse-layer models, all surface species are supposed to be inner-sphere complexes, whereas in the triple-layer model, both inner- and outer-sphere complexes are assumed.

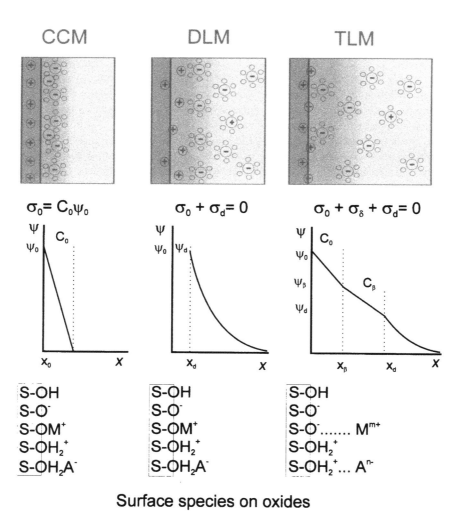

Surface species on oxides

FIG. 2 Schematic representation of SCMs: CCM, DLM and TLM; charge distribution (top) and potential decay (middle) within a nanometer-scale distance and surface species on oxides (bottom).

The following protonation and deprotonation equilibria can represent the charge formation on the surface site, S−OH:

$$S-OH + H^+ \rightleftharpoons S-OH_2{}^+, \qquad K_{a1}^{int} = \frac{[S-OH_2{}^+]}{[S-OH][H^+]_s} \tag{1}$$

$$S-OH \rightleftharpoons S-O^- + H^+, \qquad K_{a2}^{int} = \frac{[S-O^-][H^+]_s}{[S-OH]} \tag{2}$$

where K_{a1}^{int} and K_{a2}^{int} are the invariant, intrinsic equilibrium constants; brackets mean the activity of species. The surface species, S−OH, S−OH$_2{}^+$, and S−O$^-$, are assumed to have activity coefficients equal to unity. However, the activity of surface hydrogen ions, $[H^+]_s$, has to be corrected for the energy expended in moving them to the charged surface (at distance $x = 0$)

where reaction occurs. Expressing $[H^+]_s$ in terms of the bulk solution hydrogen ion activity, $[H^+]$, is

$$[H^+]_s = [H^+]\exp\left(-\frac{e\psi_0}{kT}\right) \tag{iv}$$

The conventional mass-action quotients without electric potential variable for these equilibria are not constants, often called apparent equilibrium constants, K_{a1}^{app} and K_{a2}^{app}, or operational reaction quotients, Q_{a1} and Q_{a2}. Intrinsic equilibrium constants for surface-charge formation can be determined by means of an extrapolation at zero electric potential [6,45]:

$$\lim_{(\psi_0 \to 0)} \log K^{app} = \log K^{int}$$

The surface complexation reaction for multivalent ions (M^{m+} and A^{n-}) can be written

$$aS-OH + pM^{m+} \rightleftharpoons (S-O)_a M_p^\delta + aH^+, \qquad K_{aM}^{int} = \frac{[(S-O)_a M_p^\delta][H^+]_s^a}{[S-OH]^a[M^{m+}]_s^p} \tag{3}$$

$$aS-OH + qA^{n-} + bH^+ \rightleftharpoons (S-O)_a H_b A_q^\delta + aH^+, \qquad K_{aA}^{int} = \frac{[(S-O)_a H_b A_q^\delta][H^+]_s^{a-b}}{[S-OH]^a[A^{n-}]_s^q} \tag{4}$$

The distribution of ions H^+, M^{m+}, and A^{n-} between the surface layer and the bulk solution is governed by the electrostatic field, and Eqs (iii) and (iv) can be applied by substituting the surface potential, ψ_0, in all cases, if CCM and DLM models are applied, because only inner-sphere complex formation is assumed and all of the ions are assigned near the surface; whereas the potential ψ_β is substituted in Eq. (iii) for the distribution of ions M^{m+} and A^{n-} when the TLM model is chosen, because protolytic processes are assigned to the surface and complexation of other ions, which form outer-sphere complexes, to the β plane in the TLM model.

The charge potential relationship can be given by appropriate form of Eqs. (i) and (ii)

DLM

$$\sigma_0 = (8\varepsilon_0\varepsilon_r kT)^{1/2}(n^0)^{1/2}\sinh\left(\frac{ze\psi_0}{2kT}\right)$$

CCM

$$\sigma_0 = C_0\psi_0$$

TLM

$$\sigma_0 = C_0(\psi_0 - \psi_\beta)$$
$$\sigma_\beta = C_0(\psi_\beta - \psi_0) + C_\beta(\psi_\beta - \psi_d)$$
$$\sigma_d = C_\beta(\psi_d - \psi_\beta)$$
$$\sigma_d = -(8\varepsilon_0\varepsilon_r kT)^{1/2}(n)^{1/2}\sinh\left(\frac{ze\psi_d}{2kT}\right)$$

where σ_0, σ_β, and σ_d are the total charges associated with the inner, outer, and diffuse planes, C_0 and C_β are the capacitances associated with the zones between the inner and outer, and outer diffuse planes, respectively, and ψ_0, ψ_β, and ψ_d are the electrostatic potentials at the surface, inner and outer planes, respectively. It should be noted that the application of these models needs an appropriate choice of units of physical quantities.

The application of the surface complexation models for the description of a given system presumes first the definition of the assumed chemical processes, then the determination of intrinsic equilibrium constants. Previously, different graphical extrapolation methods [6,35,45] were used; recently, the FITEQL [32] program provides an elegant numerical way to calculate the intrinsic constants together with the possible choice of different complexation models. Nowadays, different computer programs (e.g., MICROQL [30] or its improved version, MINTEQA [31]) are available for fitting the experimental data of composite systems containing different solid and dissolved components by surface complexation models. The fitting parameters are capacitances and intrinsic equilibrium constants.

D.   Aqueous-Phase Processes

Several charged species are present even in the simplest aqueous solutions in contact with an adsorbent. On the one hand, charged species are released from the solid phase due to hydrolysis, dissolution, and dissociation processes discussed in Section II.A. On the other hand, electrolytes are present to provide constant ionic strength in adsorption measurements, in addition to solute molecules, which are often capable of dissociating. The dissociation of small molecules in aqueous solutions, ionization processes of monoprotic, diprotic, and oligoprotic acids and bases have been reviewed recently [46]. In aqueous systems, a huge variety of different types of complex has been identified depending on the chemical composition of the solid and electrolyte solutions. The formation of the most general hydrolytic complexes in aqueous medium has been described in the third point in Section II.A. The comlexation reactions in various aquatic systems are discussed in detail in the excellent book by Buffle [45]. Complexation equilibria of aluminum and the formation of aqueous mononuclear and polynuclear aluminum species and organic complexes are found in the literature (e.g., Ref. 9). Several simutaneous equilibria of the aqueous phase result in a given speciation. Detailed chemical knowledge of the system is required to identify the probable equilibria. It may happen that the anion of a background electrolyte acts as a ligand related to the given metal ion of the solute. Certain anions such as ClO_4^- and NO_3^- do not have the ability to form complexes with metal ions; however, other anions (e.g., Cl^-, PO_4^{3-}) may significantly influence the speciation of several metal ions (e.g., Cu^{2+}, Cd^{2+}, Pb^{2+}, Zn^{2+}, Fe^{3+}, Al^{3+}).

The formation of hydrolytic metal complexes may lead to a significant decrease in the solubility of a metal. For instance, it is enough to point to the formation of sparingly soluble $Al(OH)_3$, $Fe(OH)_3$, $Cd(OH)_2$, $Cu(OH)_2$ and $Zn(OH)_2$ over an appropriate range of pH. Therefore, a partial precipitation of the given solutes from the aqueous phase at a given pH is a real consequence of the chemical properties of components. It has been proved that in the presence of solid particles, hydrolyzable metal ions may precipitate on surface prior to bulk solution precipitation [50]. Surface precipitation reactions are suggested for considering the sorption mechanisms of hydrolyzable metal ions on oxide surface in modeling. The precipitation of some metal carbonates is also a probable process. Dissolved carbonates are often present in aqueous solutions, especially at a higher pH. Carbonate equilibria have been discussed in Section II.A. The spontaneously formed carbonates may react with the dissolved metal ions, such as Ca^{2+}, Pb^{2+}, and Ag^+, and metal carbonates with a low solubility product may precipitate under the given conditions.

III.   PROBLEMS, PROBABLE SOLUTIONS: CASES FROM RECENT LITERATURE

A.   Problems, Probable Solutions

Adsorption from electrolyte solutions refers to the accumulation of mostly charged species from the aqueous phase at an electrified interface between the solid and water. Chemical species are lost from the aqueous phase during adsorption. A common way of quantifying adsorption is to measure the lost amount of solute in the liquid phase. Several precautions are advised in considering practical work and there are assumptions which are important to check before measurement [47]. Most modeling, analysis, and experimentation is concentrated on laboratory systems, where the solid and solution are well characterized and only some dissolved materials are present [4]. Some assumptions (for instance, adsorbent does not dissolve, contamination has to be avoided during preparation) cannot be fulfilled even under precise laboratory conditions. Here, it is enough to consider Section II.A, in which the spontaneous processes (hydration, hydrolysis, dissolution, and dissociation) during immersion of different adsorbents into aqueous phase were discussed.

Water is a unique medium. Hydration of an adsorbent often goes together with the dissociative chemisorption of water molecules. Unequal dissolution of constituent ions in ioncrystals or surface dissociation, ionization processes result in surface charging, so the surface properties of adsorbent alter inherently in aqueous medium and various species are released into the aqueous phase, which may react with solute or with the CO_2 contaminant hardly eliminated from aqueous systems. Water is unique in that sense, too, in that it is the most common and the only naturally occurring inorganic liquid on Earth and that aqueous solutions are involved in the most environmental and tremendous man-made processes. The interfacial accumulation of various species from electrolyte solutions is always present in these processes. The importance of this issue is unchallenged.

As I see, an essential problem in this field is that the complex aspect of adsorption from electrolyte solutions shown in Fig. 1 has not been widely accepted. I have hardly found a systematic analysis of all probable simultaneous equilibria in a given adsorption system in the relevant literature. In most cases the solution condition (e.g., pH)-dependent dissolution of the solid phase, the surface precipitation, and the speciation in the aqueous phase are omitted in the evaluations either without mentioning them or with reference to some reasoning. To demonstrate some outcomes, it is worth inspecting a simple case of the surface-charge titration of a common aluminum oxide in detail.

Dissolution of aluminum oxide in both acidic and alkaline solutions and its dependence of the crystal structure is well known [51]. Below pH $\sim$ 4 and above pH $\sim$ 10, the dissolution of this amphoteric solid becomes observable; therefore, the interfacial charging is often studied within these pHs [52,53]. Study on dissolution kinetics of aluminum oxide [54] shows that the dissolution rate considerably depends on the pH of aqueous solution, a minimum was observed near neutral pH, and the dissolution rate increased with decreasing pH below pH 6 and (even more strongly) with increasing pH above pH 7.5. The net dissolution rate was on order of magnitude of 1×10^{-8} mol/h/m^2 at pH 4 and 9. Considering the high specific surface area of alumina measured in general by means of the potentiometric acid–base titration method and the endeavor of experimenters to reach the equilibrium state of surface charging by increasing the time of titration, it is highly probable to suppose that the experimental surface-charging curves are disturbed by the H^+/OH^- consumption from dissolution, even those which were measured over the likely dissolution-free pH range. In some works, the possibility of alumina dissolution is excluded with reference to the chemical equilibrium calculation [52] or it is not mentioned [55–57] at all. Fitting of experimental surface-charging curves led to more and more complicated theoretical approaches with increasing number of layers for charge-compensating ions in the surface complexation models (diffuse double-, triple-, and four-layer models [56,57] and with introducing surface site heterogeneity parameters. The heterogeneity of proton-binding sites at the oxide–solution interface is studied theoretically [58–60] and both theoretically and experimentally [7,15,52,55–57]. Based on the potentiometric titration performed over a broad range of pH's (3–11), affinity distributions for four different surface sites were calculated with the aim of studying the heterogeneity of acid–base properties at the aluminum oxide–solution interface

[15]. These authors neglected both the electrostatic effect and the dissolution of alumina. Although alumina dissolution under the given experimental conditions was checked and its contribution to the total H^+/OH^- consumption seemed to be not significant, the experimental fact that the measured proton adsorption isotherms became independent of ionic strength below pH $\sim$ 4 and above pH $\sim$ 10 refers to other reactions taking place in parallel with surface charging. The calculated lowest and highest $\log K$ values for surface-charge formation are likely misinterpreted. Asymmetric proton-binding curves were measured [52] over the pH range 5–11. The unusually high OH^- excess in the alkaline region was modeled by assuming the penetration of electrolyte ions into the surface, whereas the probability of OH^- consumption due to dissolution was limited to several percent [52].

In one of our recent articles [16], we analyzed how the dissolution of aluminum oxide influences the experimental data, and how the modeling can overcome this problem. Figure 3 shows the reversibility of curves measured in the direction of increasing (forward titration) then decreasing (backward titration) pH of equilibrium acid–base titration.

The curves coincide well down to pH $\sim$ 5 within the experimental error of this method. Below pH $\sim$ 4.5, however, a sharp increase in proton consumption with decreasing pH appears and the points become gradually independent of ionic strength similarly to that given in Fig. 12 of Ref. 15. Contescu et al. [15] identified a narrow peak of the apparent affinity distribution function at pH $\sim$ 3.5 with the protonation of the most acidic triply coordinated groups of type III on alumina surface, neglecting the dissolution of solid phase [61,62].

The evaluation of experimental data is problematic, because the overall acid and base consumption is measured and the reactions taking place in parallel with surface charging cannot

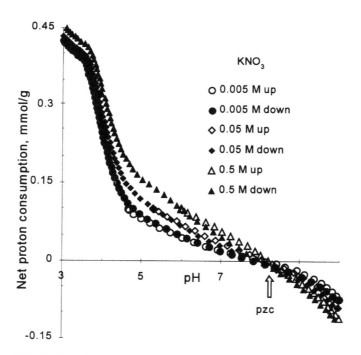

FIG. 3 Experimental net proton consumption curves for purified δ-Al$_2$O$_3$ dispersed in KNO$_3$ solutions at room temperature. The points were calculated from the data measured in the direction of increasing (open symbols) and decreasing (closed symbols) pH of equilibrium titration with 0.1 M KOH and 0.1 M HNO$_3$ solutions, respectively.

be separated experimentally. There are problems discussed in several works (e.g., Refs. 63–65) in modeling the surface charging of oxides, too. The pH- and ionic strength-dependent surface-charge formation process can be described by various model approximations; the most widely accepted models are the site-binding electrostatic models [53,55–57,66–68]. The surface-charge development can be affected by the solubility of the solid [69], which is not incorporated in the models. The experimental data shown in Fig. 3 were evaluated using different surface complexation models (CCM, DLM, TLM). The measured data of the forward titration in indifferent electrolyte (KNO_3) solutions might be well fitted by choosing any SCM. An example for the quality of the fitting of curves optimized by FITEQL [32] using the CCM approach and the experimental points calculated on the basis of material balance of added H^+/OH^- is shown in Fig. 4. In the presence of KCl electrolyte (Fig. 5), only the choice of TLM option led to an acceptable level of optimization process by FITEQL [32], presumably because of specific adsorption of Cl^- ions, as proven on the passive films of metal aluminum by Bockris and Kang [70].

The fitting of backward titration data was impossible when only the surface-charge formation processes [Eqs (1)–(4)] were postulated as H^+/OH^--consuming reactions in alumina suspensions. Therefore, a partial dissolution of alumina particles during the long period of equilibrium titration was assumed. A component identified as a solid compound $Al(OH)_3(s)$ and its solubility product $[Al(OH)_3(s) = Al^{3+} + 3\ OH^-]$ as well as the formation of different mononuclear Al species in the aqueous phase were inserted in the stoichiometry matrix of FITEQL [32]. Assuming the reversibility of surface-charging processes, the CCM parameters

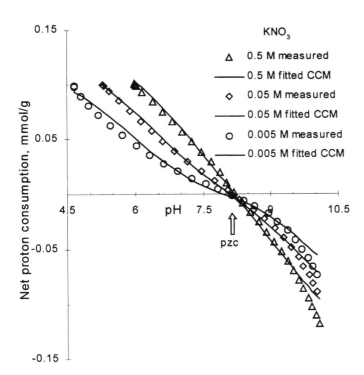

FIG. 4 Experimental points of net proton consumption from forward titration with $0.1\,M$ KOH for purified δ-Al_2O_3 dispersed in indifferent electrolyte (KNO_3) solutions at room temperature. The continuous lines are numerically fitted [32] using the constant capacitance model ($C = 1.2\ F/m^2$).

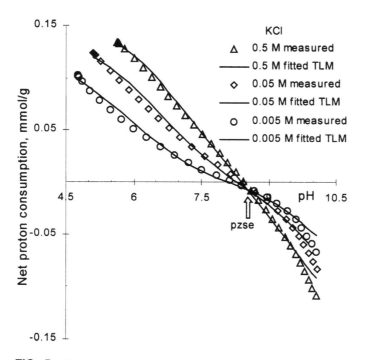

FIG. 5 Experimental points of net proton consumption from forward titration with $0.1\,M$ KOH for purified δ-Al_2O_3 dispersed in nonindifferent electrolyte (KCl) solutions at room temperature. The continuous lines are numerically fitted [32] using the triple-layer model ($C_1 = 1.2$–$1.6\,F/m^2$, $C_2 = 0.2\,F/m^2$).

(total concentration of S—OH site, capacitance of the oxide–electrolyte interface) calculated from the data of forward titration were fixed in the course of the data fitting for backward titration. The single fitting parameter was the total amount of dissolved $Al(OH)_3(s)$. The measured points of backward titration of purified alumina dispersed in KNO_3 solutions and the calculated curves are compared in Fig. 6.

The calculated curves (thick lines in Fig. 6) assuming both surface charging and dissolution of alumina coincide well with the measured data. The calculated curves (thin lines in Fig. 6) related only to the H^+/OH^- consumption of interfacial protonation–deprotonation reactions resulting in surface charging are also plotted. It can be seen that the net proton consumption from the dissolution of amphoteric solid becomes significant where the calculated curves start to diverge from each other (i.e., below pH $\sim$ 5 and above pH $\sim$ 9.5). On the basis of the experimental curves showing a sharp increase in the net proton consumption values at pH $\sim$ 4 (Fig. 3), the assumption of alumina dissolution at low pHs seemed to be evident. These model calculations showed a probable interference of dissolution under alkaline condition, too. The slight difference between the measured points and calculated curves in Figs. 4 and 5 above pH $\sim$ 9.5 completely disappears when dissolution of alumina is inserted into the chemical model of the data-fitting procedure, as seen in the given region of curves in Fig. 6. A similar divergence of the measured points and calculated curves above pH $\sim$ 9 has been published [52] and interpreted successfully by means of a model calculation [developed charge distribution approach (CD model)] as a result of penetration of electrolyte ions into the different planes of the interfacial layer, whereas the alkaline dissolution of alumina was neglected with reference to the chemical equilibrium calculation. The question is which model, an advanced electrostatic or

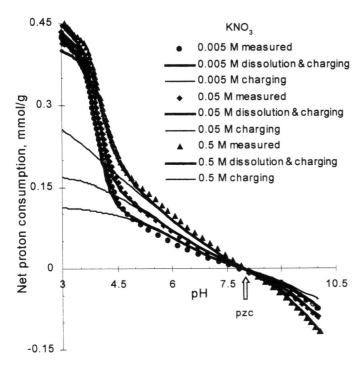

FIG. 6 Experimental points of net proton consumption from backward titration with $0.1\,M$ HNO_3 for purified δ-Al_2O_3 dispersed in KNO_3 solutions at room temperature. The continuous thick lines are numerically fitted [32] using a combined surface complexation and chemical dissolution model: the constant-capacitance model ($C = 1.2$ F/m^2) for surface charging, partial dissolution of solid alumina, and formation of mononuclear Al-species with $\log K$ values from the literature [9,18,31]. The continuous thin lines show that part of net proton consumption which belongs to the surface-charge formation.

a simple equilibrium based on the combination of SCMs with additional interfacial and aqueous phase equilibria, is a better approach to describe the pH-dependent surface charging of aluminum oxides. The only way to answer this question is to find independent experimental evidence(s) for supporting the model assumption(s). In our approach, the dissolution of aluminum oxide was assumed. Therefore, the determination of aluminum concentration in equilibrium supernatants seemed to be evident support. The values measured by means of the ICP method together with the calculated distribution of dissolved Al species from FITEQL output files and their total amounts are plotted as a function of pH in Fig. 7. The measured values coincide strikingly well, especially in the alkaline region, with the calculated total concentration of dissolved Al species. The difference between the measured and calculated values becomes significant below pH ~ 4.5, and the calculated values are smaller than the measured ones and reach a limited value, because the total amount of dissolved $Al(OH)_3(s)$ was a parameter in the course of numerical fitting of our experimental data.

It seems that neither the electrostatic effect nor the dissolution of aluminum oxide can be neglected as done in a previous article [15] in which the authors presumed the protonation and deprotonation of different types of active site on alumina surface at low and high pHs and introduced surface-charge formation on four different active sites; among them, two are likely misinterpreted. Because the dissolution of alumina takes place at any pH, as showed in a recent

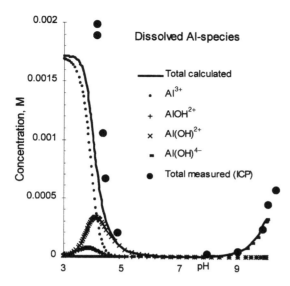

FIG. 7 Comparison between the calculated distribution of dissolved Al-species from FITEQL output files (same as shown in Fig. 6) and the experimental values measured by means of the ICP method.

article [54], we have to conclude that the experimental surface-charging curves are misrepresented due to H^+/OH^- consumption from dissolution, even those which were measured in a likely dissolution-free pH range. The dissolution rate of aluminum oxide, however, strongly depends on the pH of aqueous solution; its interference first becomes obvious in the equilibrium potentiometric titration performed during a long time and only at low and high pH's.

The goodness of fit between the experimental points and the calculated curves from a model based on even a sophisticated advanced theory [52] does not prove the reality of the theoretical model, the assumed structure of surface layer, and the presumed interfacial reactions. Any model assumption has to be supported by independent experimental facts (e.g., surface spectroscopy, analysis of equilibrium bulk phase) to prove its reality.

As mentioned in Section I, Haworth [4] published an comprehensive review on the modeling of sorption from aqueous solution. I entirely agree with his statements—on the one hand that the widely used isotherm models, with which I have not dealt here, have not been successful in describing experiments outside a limited range of conditions; and on the other hand, that surface complexation models have been successful in modeling some of the systems under even real conditions. In accordance with this, it was stated in the review of Davis and Kent [71] on surface complexation modeling in aqueous geochemistry that the surface complexation theory, which describes adsorption in terms of chemical reactions between surface functional groups and dissolved chemical species, can be coupled with aqueous speciation models to describe adsorption equilibria within a general geochemical framework. It should be emphasized, however, that SCMs have several parameters which can be varied to fit experiment, and so good fitting may be achieved without giving any relevant insight into the processes occurring. A detailed chemical knowledge of the system is required in order to verify the assumptions involved in the model, otherwise the application of SCMs becomes a meaningless exercise of curve fitting. From a general standpoint, the critical assumptions of the surface complexation models are the relevant specific surface area, the quality and quantity of active sites involved in the interfacial processes, and the nature of the surface complex(es) formed (quality and strength of chemical bonds on surface sites). Identification of surface complex(es) is a big challenge.

Different spectroscopic methods can be used to support the species explicitly included in the SCMs. Johnston et al. [72] have reviewed various types of spectroscopic method applied to environmental particles. Adsorption measurements alone and even the advanced modeling based on only the analysis of adsorption data may lead to misinterpretation of processes. Independent methods such as electrophoresis, surface and bulk spectroscopy, the isotope technique, and multiple quantitative analysis (e.g., total dissolved concentration of a metal from ICP measurements in parallel with its free-activity data by using ion-selective electrode) are required to support the evaluation of adsorption measurements.

A combination of adsorption measurements with an in situ infrared spectroscopic study using attenuated total reflectance (ATR)–Fourier transform infrared (FTIR) technique, on the adsorption of lysine peptides and polylysine on hydrous TiO_2 particle films under physiological conditions was reported recently [73] to gain a better understanding of factors determining biocompatibility. IR spectra obviously indicated that the carboxylate group was involved in the peptide to TiO_2 interaction and that a prolonged exposure of polylysine to TiO_2 resulted in a change in secondary structure of peptide. The in situ ATR–FTIR technique was also used to identify the type of surface complexes of methylphosphonic acid (MPA, related to a herbicide widely used in soils) formed under each of the solution conditions occurring in the adsorption studies [74]. It was proved that formation of monodentate and bidentate complexes on goethite surface depended on the pH and surface coverage. The electrophoresis measurements showed a increasing shift of the pH of charge reversal with increasing loading of MPA which provided still more evidence for formation of inner-sphere surface complexes.

Surface complexation models are successful in several fields of adsorption from electrolyte solutions; however, the combined electrostatic and chemical approach seems to be inherently insufficient to describe the hydrophobic and macromolecular contributions. Therefore, that huge fields of solution adsorption, where these nonchemical interactions dominate the mechanism of adsorption (e.g., the adsorption of ionic and non-ionic surfactants, water-soluble polymers, and polyelectrolytes from aqueous solutions, are all of great practical importance) are essentially out of the scope of SCMs. There are some attempts to incorporate the formation of a second adsorption layer due to the hydrophobic interactions of longer alkyl chains into SCMs' formalism, for example, by allowing two or more organic molecules to occupy a single surface site [75]. Furthermore, a hydrid SC/SF (self-consistent field) model has been developed to describe the adsorption of weak polyelectrolytes on metal oxide surfaces [76]. This model couples the SCM for the effect of surface and solute speciation on the adsorption of monomers on oxides with the (SF) theory of Scheutjens and Fleer for the effect of macromolecular properties on the adsorption of polymers and polyelectrolytes on surfaces. A good agreement of model simulation and the adsorption data of polygalacturonic acid on hematite was achieved.

Finally, one must return to the problem of surface precipitation. As mentioned, SCMs can be coupled with aqueous speciation models to describe adsorption equilibria even within a general geochemical framework [71]. However, it has been proved experimentally that surface precipitation may occur prior to bulk precipitation, depending on the type of surface present [50]. We have to conclude that the simple combination of SCMs with the aqueous bulk equilibria cannot account for surface precipitation. It seems to me that this phenomenon is likely beyond the scope of solution adsorption, but it is in the center of heterogeneous nucleation studies, which are of great importance in many areas of modern surface and material sciences. In a recent article [77], the generic heterogeneous effect of solid particles on nucleation was examined theoretically in terms of the relative size, the interfacial interactions and structural match with the crystallizing phase. It has been stated that heterogeneous nucleation is one the most important processes occurring in the crystallization in the whole range of supersaturation. Heterogeneous nucleation dominates at low supersaturation and when advantageously large particles have good structure match and strong interactions with crystallizing phase. The heterogeneous effect of the solid surface on nucleation has to be considered in the adsorption studies of hydrolyzing metal ions, and the surface precipitation process should be involved at least in the interpretation of results.

B. Cases from Recent Literature

Chemical equilibrium models are used to predict the speciation of dissolved solutes in natural systems (e.g., MINTEQA [31]). These models attempt to incorporate all of the various processes that affect the speciation of solutes, including all known solution-phase reactions (e.g., acid–base, precipitation–dissolution, and complexation reactions) and adsorption to solid surfaces. Current models for inorganic chemicals have been successful in predicting speciation in aqueous systems containing well-characterized solid particles.

There are countless articles dealing with immobilization of inorganic ions, especially with that of environmental importance (e.g., ions of toxic elements such as arsenic, lead, chromium, cobalt; that of radionucleides cesium, nickel, strontium, ytterbium; or subsurface aquifer pollutants ions like nitrate, nitrite, and fluoride). Surface complexation modeling was successfully applied in several cases. I quote only some recent articles as instances.

The adsorption of hydrolyzing ions Cr^{3+} and Co^{2+} on silica gel was investigated [78], and speciation of mononucleus hydroxy species in the equilibrium aqueous phase was also considered to calculate the equilibrium distribution between the solid phase and bulk solution over the broad range of pH's. Although the different surface complexes of chromium and cobaltic assumed to form on surface sites of silica were not proved by any independent method, the fitting procedure was executed successfully, if only all the possible equilibria were accounted for in the SCMs.

The SCM approach was used to describe the surface properties and prediction of heavy-metal adsorption on natural aquatic sediment [79]. All of the model parameters were determined from experimental data of potentiometric titration and metal Cu and Cd adsorption isotherm Results showed that all three typical versions (CCM, DLM and TLM) of SCMs can simulate the experimental data very well and provide an acceptable prediction for heavy-metal adsorption to natural sediments. The heavy-metal (Mo, Pb, Cu, Zn, Cd) contaminant leaching from weathered municipal solid waste was successfully modeled by DLM [80]. The leaching of Mo, Pb, and Cu from the weathered ash was well described by surface complexation, whereas the precipitation of formed Zn-hydroxy species had to take into consideration to describe the leaching data of Zn.

Unfortunately, the migration of radionuclides from the underground radwaste repository to geosphere becomes a real environmental problem by now. In connection with this, the adsorption of three different cations (Ce^+, Ni^{2+}, and Yb^{3+}) with interesting nuclear aspects on iron oxide (magnetite), one of the corrosion product of metallic container, was studied [81]. An evaluation of the weight of the electrostatic term in the modeling adsorption results was given by comparing different SCMs with a nonelectrostatic approach.

The serious problem of arsenic pollution in soils and subsurface waters due to mineral dissolution, use of arsenical pesticides, disposal of fly ash, and mine drainage is well known all over the world. Even homogeneous phase speciation of arsenic species in aqueous solutions is very complicated, because this trace element exists in different redox and multistep acid–base dissociation equilibrium states. A wide range of adsorbents and several methods, such as spectroscopic techniques (FTIR, Raman, X-ray absorption spectra [XAS] extended X-ray absorption fine structure [EXAFS]), electrophoresis, in addition to adsorption measurements were used to study the sorption of arsenite and arsenate. Several noteworthy applications of SCMs for evaluation of experimental results are known. One of them is an excellent example for the combination of macroscopic measurement (adsorption and electrophoresis) with vibrational spectroscopy (Raman and ATR–FTIR) for evaluating mechanisms of arsenate and arsenite adsorption on amorphous Al and Fe oxides to maximize the chemical significance of the SCM application [82]. The pH and ionic strength dependence of adsorption, PZC shifts in the presence of both arsenite and arsenate, and the spectroscopic identification of different arsenic bonds formed on surface sites under various solution conditions allowed one to conclude a self-consistent mechanism for arsenic adsorption on amorphous oxides.

The adsorption of organic compounds, especially that have complexing ability to surface sites, has been modeled successfully with SCMs. The adsorption of small molecular organic acids is assumed to dominate by site-specific ligand exchange. One of the prominent articles on the adsorption of organic acids on oxide adsorbents was recently published. Surface complexation modeling was used to describe the adsorption of a range of simple, aromatic organic acids on goethite [75]. The studies on the pH-dependent adsorption of a series of organic acids to examine the effects of structural features such as the number and position of carboxylic and phenolic groups, the length of aliphatic chain on benzene monocarboxylic acids, and the number of aromatic rings have shown the significant influence of structure of solute molecules. Consecutive dissociation equilibria of one to six acidic groups on aromatic ring of different organic acids were incorporated into modeling using acidity constants from the literature. The type of surface complexes was chosen on the basis of spectroscopic evidence, which indicates a mostly mononuclear complex formation. The number of surface reactions required to fit experimental data ranged between two for benzoic and phthalic acids and six for pyromellitic acids. A set of surface reactions and equilibrium constants yielding optimal data fits was obtained for each organic acid over a range of total solute concentrations between pH 3 and 9. The adsorption of a wide range of organic acids on goethite was accurately described by SCMs. Surface complexation reaction constants were consistently higher for the series of compounds containing multiple carboxylic groups, reflecting the higher amount adsorbed over most of the pH range investigated. Some of the studied compounds with aliphatic chains and/or double rings exhibited unusual adsorption behavior. These compounds have some hydrophobic character which is most significant for the compound containing a 5-carbon aliphatic chain, and their adsorption was inconsistent with a ligand-exchange mechanism. Authors attempted to incorporate hydrophobic interactions into SCMs by allowing two or more organic molecules to occupy a single surface site, because hydrophobic interactions were presumed to occur between the molecules bonded to surface sites and dissolved in solution via association of the hydrophobic portions of the molecules. This approach seemed to be successful for describing the sorption of the examined molecules taking part in hydrophobic interactions in addition to surface complexation.

An interesting example of the application of SCM to quantify metal adsorption onto bacterial surfaces has been reported [83]. The acid–base titration of bacillus suspensions was performed and the sequential deprotonation of the functional groups on the wall of bacteria cell with increasing pH was modeled by ionization equilibria of surface carboxyl, phosphate, and hydroxyl groups. The formation of metal Cd, Pb, and Cu complexes with 1 : 1 stoichiometry was assumed on these functional groups on the cell wall. Metal complexation of this kind was supported by former electrophoresis results for bacterial cells. The CCM was effective to describe the ionic strength-dependent acid–base properties of a bacterial cell surface. Metal adsorption is best described by models considering metal bonds onto both the carboxyl and the phosphate functional groups. Authors suggested further spectroscopic studies of bacterial surfaces in order to elucidate more precisely the nature of the surface complexes of metals.

The SCMs were developed for surfaces with ionizable surface groups and were successfully applied for the evaluation of surface charging and interfacial distribution of ionic species.

An extension of the relative simple formulation used in SCMs for surfaces with permanent charges (see Section II.B.1) has been published recently [84]. A fictitious surface species (X^-) was defined and hypothetical complexation reactions on site X^- were written, and thus cation-exchange reactions of permanent negative layer charges were easily incorporated into such model. The model showed not only to fit satisfactorily all of the experimental data of transition metal adsorption on montmorillonite but also to explain specific features of adsorption on clays compared to oxides.

REFERENCES

1. J Lyklema. Fundamentals of Interface and Colloid Science, Vol. I: Fundamentals. London: Academic Press, 1991, pp 5.1–5.114; Fundamentals of Interface and Colloid Science, Vol. II: Solid-Liquid Interfaces. London: Academic Press, 1995, pp 2.1–2.91, 5.1–5.100.
2. RJ Hunter. Foundations of Colloid Science, Vol. I and II. Oxford: Clarendon Press, 1989, pp 316–394, 709–785.
3. GD Parfitt, CH Rochester, eds. Adsorption from Solution at the Solid/Liquid Interface. London: Academic Press, 1983.
4. A Haworth. Adv Colloid Interf Sci 32:43, 1990.
5. EJW Wensink, AC Hoffmann, MEF Apol, HJC Berendsen. Langmuir 16:7392, 2000.
6. W Stumm. Chemistry of the Solid–Water Interface. Processes at the Mineral–Water and Particle–Water Interface in Natural Systems. New York: Wiley, 1992, pp 13–42.
7. W Rudzinski, R Charmas, T Borowiecki. In: A Dabrowski, VA Tertykh, eds. Adsorption on New and Modified Inorganic Sorbents Studies in Surface Science and Catalysis. Amsterdam: Elsevier, 1996, pp 357–409.
8. RO James, GA Parks. In: E Matijevic, ed. Surface and Colloid Science Vol. 12. New York: Plenum, 1982, pp 119–216.
9. G Sposito ed. The Environmental Chemistry of Aluminum. Boca Raton, FL: CRC Press, 1989.
10. RM Cornell, U Schwertmann. The Iron Oxides, Structure, Properties, Reactions, Occurrence and Uses. Weinheim: VCH, 1996, pp 175–312.
11. RK Iler. The Colloid Chemistry of Silica and Silicates. Ithaca, NY: Cornell University Press, 1955, pp 127–180.
12. BS Hemingway, G Sposito. In: G Sposito, ed. The Environmental Chemistry of Aluminum. Boca Raton, FL: CRC Press, 1989, pp 55–85.
13. JA Davis, JD Hem. In: G Sposito, ed. The Environmental Chemistry of Aluminum. Boca Raton, FL: CRC Press, 1989, pp 185–219.
14. T Hiemstra, WH van Riemsdijk. Colloids Surfaces 59:7, 1991.
15. C Contescu, J Jagiello, JA Schwarz. Langmuir 9:1754, 1993.
16. E Tombácz, M Szekeres. Langmuir 17:1411, 2001.
17. LK Koopal. In: A Dabrowski, VA Tertykh, eds. Studies in Surface Science and Catalysis Vol. 99. Amsterdam: Elsevier, 1996, pp 757–796.
18. GH Bolt, MGM Bruggenwert, eds. Soil Chemistry A. Basic Elements. Amsterdam: Elsevier, 1978.
19. H van Olphen. An Introduction to Clay Colloid Chemistry. New York: Interscience, 1963.
20. CT Johnston. In: B Sahwney, ed. Organic Pollutants in the Environment. Boulder, CO: The Clay Minerals Society, 1996, pp 1–44.
21. J Grandjean. J Colloid Interf Sci 185:554, 1997.
22. A Delville. Langmuir 7:547, 1991.
23. DB Hough, HM Rendall. In: GD Parfitt, CH Rochester, eds. Adsorption from Solution at the Solid/Liquid Interface. London: Academic Press, 1983, pp 247–320.
24. DH Everett. Basic Principles of Colloid Science. London: Royal Society of Chemistry, 1988.
25. BH Bijsterbosch, J Lyklema. Adv Colloid Interf Sci 9:147, 1978.
26. G Sposito. The Thermodynamics of Soil Solutions. Oxford: Clarendon Press, 1981.

27. W Ostwald. Z Phys Chem (Leipzig) 34:295, 1907.
28. P Somasundaran, J Ofori Amankonan, KP Ananthapadmabhan. Colloids Surfaces, 15:309, 1985.
29. BR Puri. In: PL Walker Jr, ed. Chemistry and Physics of Carbon. New York: Marcel Dekker, 1970, pp 191–254.
30. J Coves, G Sposito. MICROQL-UCR: A Surface Chemical Adaptation of the Speciation Program MICROQL. Riverside, CA: University of California Press, 1989.
31. JD Allison, DS Brown, KJ Novo-Dradac. MINTEQA2/PRODEFA2, v.3.0. Athens, GA: U.S. EPA, 1991.
32. AL Herbelin, JC Westall. FITEQL v.3.2. Corvallis, OR: Oregon State University, 1996.
33. J Lyklema. Pure Appl. Chem 63:895, 1991.
34. G Sposito. In: J Buffle, HP van Leeuwen, eds. Environmental Particles Vol. 1. Boca Raton, FL: Lewis 1992. pp 291–314.
35. G Sposito. The Surface Chemistry of Soils. New York: Oxford University Press, 1984.
36. O Altin, HÖ Özbelge, T Dogu. J Colloid Interf Sci 217:19, 1999.
37. BKG Theng. The Chemistry of Clay–Organic Reactions. New York: Halsted Press, 1974.
38. J Lyklema. In: GD Parfitt, CH Rochester, eds. Adsorption from Solution at the Solid/Liquid Interface. London: Academic Press, 1983, pp 223–246.
39. P Somasundaran, GE Agar. J Colloid Interf Sci. 24:433, 1967.
40. JJ Prédali, JM Cases. J Colloid Interf Sci 45:449, 1973.
41. A Pierre, JM Lamarche, R Mercier, A Foissy J Persello. J Dispers Sci Technol 11:611, 1990.
42. PL de Bruyn, GE Agar. In: DW Fuerstenau, ed. Froth Flotation. New York: American Institute of Mining and Metallurgical Engineers, 1962, pp 91–112.
43. GA Parks, PL de Bruyn. J Phys Chem 66:967, 1962.
44. H Hohl, J Sing, W Stumm. In: MC Kavanaugh, JO Leckie, eds. Particulates in Water. American Chemical Society Series Vol. 189. Washington, DC: American Chemical Society, 1980, pp 1–31.
45. J Buffle. Complexation Reactions in Aquatic Systems: An Analytical Approach. Chichester: Ellis Horwood, 1988, pp 195–303.
46. M Borokovec, B Jönsson, GJM Koper. In: E Matijevic, ed. Surface and Colloid Science Vol. 16. Amsterdam: Kluwer Academic/Plenum Press, 2001, pp 99–339.
47. DH Everett. Pure Appl. Chem 58:967, 1986.
48. J Lyklema. In: Th F Tadros, ed. Solid/Liquid Dispersions. London: Academic Press, 1987, pp. 63–90.
49. EJW Verwey, J Th Overbeek. Theory of the Stability of Lyophobic Colloids. New York: Elsevier, 1947, pp 1–65.
50. SE Fendorf, CL Sparks. J Colloid Interf Sci 148:295, 1992.
51. WL Lindsay, PM Walthall. In: G Sposito, ed. The Environmental Chemistry of Aluminum. Boca Raton, FL: CRC Press, 1989, pp 221–277.
52. T Hiemstra, H Yong, WH Van Riemsdijk. Langmuir 15:5942, 1999.
53. R Wood, D Fornasiero, J Ralston. Colloid Surfaces 51:389, 1990.
54. SM Kraemer, VQ Chiu, JG Hering. Environ Sci Technol 32:2876, 1998.
55. W Ridzinski, R Charmas, S Partyka, F Thomas, JY Bottero. Langmuir 8:1154, 1992.
56. R Charmas, W Piasecki. Langmuir 12:5458, 1996.
57. R Charmas. Langmuir 15:5635, 1999.
58. M Cernik, M Borkovec, JC Westall. Langmuir 12:6127, 1996.
59. M. Borkovec. Langmuir 13:2608, 1997.
60. JR Rustad, E Wasserman, AR Felmy, C Wilke. J Colloid Interf Sci 198:119, 1998.
61. F Bartoli, R Philippy. Clay Miner. 22:93, 1987.
62. E Tombácz, Á Dobos, M Szekeres, HD Narres, E Klumpp, I Dékány. Colloid Polym Sci 278:337, 2000.
63. N Kallay, S Zalac, I Kobal. In: A Dabrowski, VA Tertykh, eds. Adsorption on New and Modified Inorganic Sorbents. Studies in Surface Science and Catalysis. Amsterdam: Elsevier, 1996, pp 857–877.
64. H Sposito. J Colloid Interf Sci 91:329, 1983.
65. ZZ Zhang, DL Sparks. J Colloid Interf Sci 162:244, 1994.
66. E Tombácz, M Szekeres, I Kertész, L Turi. Prog Colloid Polym Sci 98:160, 1995.
67. J Westall, H Hohl. Adv Colloid Interf Sci 12:265, 1980.

68. E Rakotonarivo, JY Bottero, F Thomas, JE Poirier, JM Cases. Colloid Surfaces 33:191, 1988.
69. C Ludwig, WH Casey. J Colloid Interf Sci 178:176, 1996.
70. J O'M Bockris, Y Kang. J Solid State Electrochem 1:17, 1997.
71. JA Davis, DB Kent. In: MF Hochella, AF White, eds. Mineral–Water Interface Geochemistry. Reviews in Mineralogy Vol. 23. Washington, DC: Mineralogical Society of America. 1990, pp 177–260.
72. CT Johnston, G Sposito, WL Earl. In: Environmental Particles Vol. 2. J Buffle, HP Van Leeuwen, eds. Boca Raton, FL: Lewis, 1993, pp 1–36.
73. AD Roddick-Lamnzilotta, AJ McQuillan. J Colloid Interf Sci 217:194, 1999.
74. BC Barja, MI Tejedor-Tejedor, MA Anderson. Langmuir 15:2316, 1999.
75. CR Evanko, DA Dzombak. J Colloid Interf Sci 214:189, 1999.
76. KK Au, S. Yang, CR O'Melia. Environ Sci Technol 32:2900, 1998.
77. XY Liu. Langmuir 16:7337, 2000.
78. M Berka, I Bányai. J Colloid Interf Sci 233:131, 2001.
79. X Wen, Q Du, H Tang. Environ Sci Technol 32:870, 1998.
80. JA Meima, RNJ Comans. Environ Sci Technol 32:688, 1998.
81. N Marmier, A Delisée, F Fromage. J Colloid Interf Sci 211:54, 1999.
82. S Goldberg, CT Johnston. J Colloid Interf Sci. 234:204, 2001.
83. CJ Daughney, JB Fein. J Colloid Interf Sci 198:53, 1998.
84. AML Kraepiel, K Keller, FMM Morel. J Colloid Interf Sci 210:43, 1999.

13
Polymer Adsorption at Solid Surfaces

VLADIMIR NIKOLAJEVICH KISLENKO   Lviv State Polytechnic University, Lviv, Ukraine

I. INTRODUCTION

Polymers and their solutions are used widely as glues and lubricants for suspension stabilization and modification of the fine materials used in polymer composites. Practical use of these systems demands solving the problem of the treatment between a macromolecule and a solid surface. Therefore, the interest to the investigation of that phenomenon by the instrumental methods as well as theoretical methods arises.

Improvement of scientific equipment and mathematical methods allows one to obtain more information on the state of the macromolecule in the adsorption layer. Infrared methods allow one to measure the fraction of macromolecule segments attached to the surface [1,2]. Nuclear magnetic resonance and electron spin resonance can distinguish the segments in trains and more mobile segments in tails and loops [3,4]. Labeling techniques, studying radioactive macromolecules or fluorescent macromolecules, can be used for investigation dynamics of macromolecules in the adsorption layer [5,6]. Ellipsometry allows one to find the average thickness of the adsorption layer and the average mass of the polymer in it [7–10]. Viscosity methods [11] and sedimentation methods [12] are used to obtain the thickness of adsorption layer on the solid particles. The evanescent wave method [13–15], small-angle neutron techniques [16–18], and surface force apparatus [19–21] have been used to determine the segment density profile in adsorbed polymer layers.

The development of adsorption theory provides the explanation of the macromolecule behavior in the adsorption layer and provides the basis of arguments on the experimental results. A few theoretical models, describing the adsorbed macromolecule, are widely used now. Self-consistent field theory or mean field approach is used to calculate the respective distribution of trains, loops, and tails of flexible macromolecule in the adsorption layer [22–26]. It allows one to find the segment density distribution in the adsorption layer and to calculate the adsorption isotherms and average thickness of the adsorption layer. Scaling theory [27–29] is used to explain the influence of the macromolecule concentration in the adsorption layer on the segment density profile and its thickness. Renormalization group theory [30–33] is used to describe the excluded volume effects in polymer chains terminally attached to the surface. The Monte Carlo method has been used for the calculation of the density profile in the adsorption layer [33–35].

II. GENERAL FEATURES

Usually, in the polymer adsorption theory, there are three stages: adsorption from very dissolved solutions, when the substitution degree of surface is low, and macromolecules are situated at a large distance from each other; The transitional stage, in which the distance between macro-molecules is not large and their chains partially overlaps; the plateau regime, when the macromolecule chains are overlapped, stretched into the solution, and only a small part is bound with the surface. This notion of polymer adsorption makes the theoretical calculations of macromolecule state in the adsorption layer more complicated.

It is connected with the situation when one suggests that the properties of the system, contained many macromolecules and surface, equal to the properties of the macromolecule in adsorption layer. It is not true in some cases. For example, for the polymer adsorption from concentrated solutions, the polymer link density in the adsorption layer is not equal to the density of links of one of the macromolecules in it. In some cases, the properties of the adsorption layer depend on the way it forms, but not only on the properties of macromolecules in it. We suggest that the following can be taken into account when the theory is based on the following features:

1. The macromolecule in adsorption layer has its own shape and dimension even if it is dissolved in solvent containing the same macromolecules. According to SANS data for the mixture of protic and deuteric polymers at the melting point in the absence of outside strength, the shape and dimension of the macromolecules is close to that in the solution.
2. The shape and dimension of macromolecules in the adsorption layer depend on the nature of polymer, surface, and solvent quality.

    For polymer adsorption from the poor solvents, when the macromolecule volume is small, the number of macromolecules adsorbed on the unit surface area is greater than the adsorption from good solvents. A change of the solvent from poor to good leads to an increase in the volume of adsorbed macromolecules and partial desorption of macromolecules occurs.

    On the other hand, when we have the saturated adsorption layer in a good solvent and the whole surface covered by a polymer (Fig. 1), a change of the good solvent into a poor one leads to a decrease of macromolecule volume and surface area, free from polymer, arises (Fig. 2).

    Microphotographs of the surfaces at that experiment are provided in Ref. 36. These experiments show that in the saturated adsorption layer, the degree of overlapped

FIG. 1 Macromolecules dried from the saturated adsorption layer in a good solvent.

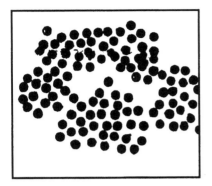

FIG. 2 Macromolecules dried from a poor solvent after a good solvent in saturated adsorption layer has been replaced by the poor solvent.

fragments of macromolecules is very small and the macromolecules take the shape of globula when the solvent is changed from good to poor.

We next show the process of solvent evaporating from the surface covered by adsorbed macromolecules. There are two forces attached to macromolecule: the force of attraction to the solid surface and the force of the surface tension between the solution and air. Under the action of these forces, the deformation of macromolecule takes place and it becomes flatter.

When the macromolecule adsorbed from the poor solvent is not flexible enough and is situated a large distance from other macromolecules, it contains the holes with air after drying (Fig. 3). The volume of the macromolecule with polymerization degree N is larger than the volume of N monomer molecules.

When the flexible macromolecule is adsorbed from good solvent in the saturated adsorption layer, the segments of chains of different macromolecules are close to each other. Under drying, the concentration of monomer links of macromolecules increases and the degree of their overlapping increases (Fig. 4). In this case, we suggest that the macromolecules is dissolved in the mixture solvent containing the low-molecular solvent and free segments of neighboring macromolecules. Therefore, the solvent quality changes during the drying. The film of a block polymer, containing overlapping macromolecules, forms after drying.

3. The adsorbed macromolecule is in dynamic equilibrium with solvent molecules and macromolecules in solution. The number of macromolecules on the unit surface area is determined by the average concentration of monomer links of macromolecules in

FIG. 3 Isolated macromolecule adsorbed on the solid surface from a poor solvent and a dried one.

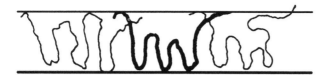

FIG. 4 Macromolecule adsorbed from a good solvent in the saturated adsorption layer and the dried macromolecule in the polymer film.

the adsorption layer. As the surface force experiments show, repulsive forces increase when the concentration of monomer links between two plates arises for mica surfaces covered by polystyrene brushes [37].

The average number of bonds between the monomer links of the macromolecule and the surface is determined by the concentration of monomer links near the surface. The rate of breakup of these bonds depends on the bond energy and the number of bonds.

The shape and dimension of the macromolecule in the adsorption layer can change if it is flexible. Adsorbed macromolecule can leave the adsorption layer and the rate of this process is inversely proportional to the number of bonds between the macromolecule and the surface (Fig. 5).

Adsorption of macromolecules from solution into the saturated adsorption layer occurs only on sites where the concentration of monomer links of macromolecule in adsorption layer is less than their concentration in the globula of solved macromolecule (Fig. 5). In this case, the possibility of macromolecule adsorption from solution is determined by the time of existence of the macromolecule near the adsorption layer, which depends on the macromolecule concentration in solution.

We want to note that the behavior of the end-grafted macromolecule in the adsorption layer does not practically differ for the free adsorbed macromolecule, but the first cannot desorb from the surface.

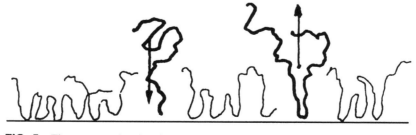

FIG. 5 The macromolecule adsorption and desorption in the saturated adsorption layer.

III. MACROMOLECULE IN THE ADSORPTION LAYER

In the simplest case, the segment density profile of the adsorption layer can be described by well-known equations [18]:

Exponential

$$\varphi(z) = \varphi_{max} \exp\left(-\frac{z}{z_0}\right) \tag{1}$$

Gaussian

$$\varphi(z) = \varphi_{max} \exp\left[-\left(\frac{z}{z_0} - \delta\right)^2\right] \tag{2}$$

Parabola

$$\varphi(z) = \varphi_{max} (z_0^2 - z^2) \tag{3}$$

Equation (4) is widely used for describing the dependence between the thickness of adsorption layer and the molecular mass of the polymer:

$$t = K_t M^a \tag{4}$$

However, the physically grounded models allows one to obtain more information on the macromolecule in the adsorption layer.

All theoretical models take into account the energy of interaction of monomer links of the macromolecule with the solvent and surface. They allow one to obtain the segment density profile in the adsorption layer by minimizing of free energy of this system.

A. Self-Consistent Field Theory

Self-consistent-field theory describes the macromolecule attached to the surface and limited by the walls. The process of the calculation of the macromolecule conformation in the adsorption layer occurs by adding the monomer links to the chain according to the rules described in the following subsections.

1. Uncharged Polymers

The self-consistent field theory is based on a lattice model [23–26]. Each chain is subdivided into segments with ranking number $s = 1, 2, 3, \ldots, N$ and these segments can take only discrete positions with respect to the surface in lattice layers numbered $z = 1, 2, 3, \ldots$ [23]. The spacing, l, between the lattice layers is equal to the bond length a. The length a is the invariant indifference lattice and l is not, so its value should be adjusted to λ according $1 = a/(6\lambda)^{1/2}$.

In the lattice model, the end-point distribution G, which is the weight of all possible walks of the segment s in layer z, in the field with potential u is computed from a recurrence (propagator) relation:

$$e^{u(z)} G(z, s+1) = G(z, s) + \lambda(G(z-1, s) - 2G(z, s) + G(z+1, s)) \tag{5}$$

with the starting condition

$$G(z, 0) = 1 \quad \text{or} \quad G(z, 1) \equiv G(z) = e^{-u(z)}$$

The end-point distribution $G(z, s)$ in a specified field $u(z)$ may be computed for any z and s. In the bulk solution, $G = 1$ for any s and $u = 0$.

In the polymer adsorption, $u(z)$ is not a fixed imposed field, but it is affected by the accumulation of segments near the surface or it is a function of the segment volume fraction $\varphi(z)$. Therefore, the self-consistent field may be described by Eq. (6) according to an extended Flory–Huggins model:

$$u(z) = -(\chi_s + \lambda\chi)\delta(z-1) - 2\chi(\langle\varphi(z)\rangle - \varphi^b) - \ln\left(\frac{1-\varphi(z)}{1-\varphi^b}\right) \tag{6}$$

where χ_s is the adsorption energy parameter, χ is the Flory–Huggins solvency parameter, $\delta(z-1)$ is the Kronecker delta ($\delta = 1$ for $z = 1$ and 0 elsewhere), φ^b is the bulk solution volume fraction of segments.

For AB copolymers, the generalized expression for $u(z)$ can be written [38]

$$u_x(z) = -(\chi_{sx} + \lambda\chi_x)\delta(z-1) - 2\sum_y[\chi_{xy}(\langle\varphi_y(z)\rangle - \varphi_y^b)] \tag{7}$$

where $x, y = $ A or B. $\langle\varphi(z)\rangle = \varphi(z) + \lambda(\varphi(z-1) - 2\varphi(z) + \varphi(z+1))$ when one takes into account local and nonlocal effects, and $\langle\varphi(z)\rangle = \varphi(z)$ when one neglects nonlocal effects.

Equations (5) and (6) can be used for obtaining the so-called composition law:

$$\varphi(z, s) = \frac{\varphi^b}{N}e^{u(z)}G(z, s)G(z, N-s+1) \tag{8}$$

and

$$\varphi(z) = \sum_{s=1}^{N}\varphi(z, s) \tag{9}$$

$\varphi(z, s)$ is computed by combining two walls toward segment s in the layer z. The factor $e^{u(z)}$ in Eq. (7) corrects for the double counting of segment s:

$$\varphi^\varepsilon(z) = \frac{2\varphi^b}{N}G(z, N) \quad \text{for } s = N \tag{10}$$

The set of Eqs. (5), (7), and (10) can be solved numerically.

The propagator relation (5) allows calculation of the fraction $G^a(z, s) = G(1, s)$ due to adsorbed chains, and G^f due to free chains:

$$G(z, s) = G^a(z, s) + G^f(z, s) \tag{11}$$

Substitution of Eq. (11) into Eqs. (8) and (9) gives the contributions of trains (φ^{tr}), loops (φ^l), and tails (φ^t) of the adsorbed chain:

$$\varphi^{\text{tr}} = \varphi(1) \tag{12}$$

$$\varphi^t(z) = \frac{2\varphi^b}{N}e^{u(z)}\sum_{s=1}^{N}G^f(z, s)G^a(z, N-s+1) \tag{13}$$

$$\varphi^l(z) = \frac{\varphi^b}{N}e^{u(z)}\sum_{s=1}^{N}G^a(z, s)G^a(z, N = s+1) \quad \text{for } z > 1 \tag{14}$$

The adsorbence, Γ, can be obtained according to Eq. (15):

$$\Gamma = \Gamma^{\text{tr}} + \Gamma^l + \Gamma^\tau \tag{15}$$

where $\Gamma^{\text{tr}} = \varphi(l)$, $\Gamma^l = \sum_z\varphi^l(z)$, and $\Gamma^t = \sum_z\varphi^t(z)$.

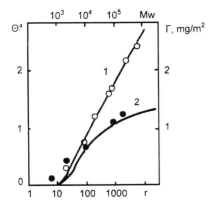

FIG. 6 Comparison of the experimental adsorbed amount with theoretical predictions from the model (curves) for polystyrene adsorption from cyclohexane (1) and from carbon tetrachloride (2) on silica. (From Ref. 38.)

Figure 6 gives a comparison between experimental and theoretical results for adsorption of homodisperse polystyrene from Θ solvent and from good solvent. The agreement between theory and experiment is quite good.

The typical segment density profile of the AB block copolymer is shown in Fig. 7. The adsorbed segments of A form a relatively compact layer near the surface. The soluble B blocks stretch into the solution and form the brushes [38]. The thickness of adsorption layer depends on the polymerization degree of the chain of the B block copolymer.

The continuum model of self-consistent field theory has been obtained [23,24] when adsorbed segments, as discrete species, occupied the region $0 < z < 1$ with their center at $z = 1/2$. Equation (5) in the continuum version can be written [23] as

$$\lambda \frac{\partial G^2(z, s)}{\partial s^2} = \frac{\partial G(z, s)}{\partial s} + uG(z, s) \tag{16}$$

with the boundary condition $G(z, 0) = 1$.

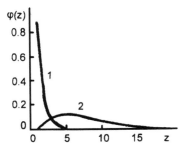

FIG. 7 Segment density profiles of an AB diblock copolymer adsorbing from a solution where A block solved poorly in solvent (1) and B block solved well in solvent (2). The figure was computed with the Scheutjens–Fleer model. (From Ref. 38.)

The exact analytical solution for Eq. (16) with a fully self-consistent-field $u(z)$ has not been found. The value of $u(z)$ can be described by Eqs. (17) and (18), found from Eq. (6):

$$u(z) \approx \upsilon\varphi(z) + \tfrac{1}{2}\varphi(z)^2 - 2\lambda\chi \frac{d\varphi^2(z)}{dz^2} \quad \text{for } z > 1 \tag{17}$$

where $\upsilon = 1 - 2\chi$ is Edwards' excluded volume parameter;

$$u(1) = -(\chi_s + \lambda\chi) - 2\chi[\lambda_0 g(1)^2 + \lambda g(2)^2] - \ln(1 - g(1)^2) \tag{18}$$

where $\langle\varphi(1)\rangle = \lambda_0 g(1)^2 + \lambda g(2)^2$, $\lambda_0 \equiv 1 - 2\lambda$, $g(z) = b\psi(z)$, where $\psi(z)$ is an eigenfunction, $b^2 = \varphi^b \exp(\ni N)$, and $\ni$ is the eigenvalue.

The adsorbance can be calculated according to Eqs. (19)–(21):

$$\Gamma^{\text{tr}} = g(1)^2 \tag{19}$$

$$\Gamma^{\text{l}} = \int_1^\infty g(z)^2 \, dz \tag{20}$$

$$\Gamma^t = \frac{2b}{N} \int_1^\infty g(z)f(z) \, dz \tag{21}$$

where $f(z) = \int_1^N G^f(z,s)\exp(-\ni s)\,ds$ satisfies the diffusion-type Eq. (16) where $z > 1$.

Figure 8 shows typical profiles for loops and tails in the plateau region with parameters $N = 10^3$, $\varphi^b = 10^{-6}$, $\chi_s = 1$, $\chi = 0$, or $\chi = 0.5$. The loops are compressed in the inner region, whereas the tails are dominated in the outer region. The results of calculation based on the discrete model is close to results calculated according to the continuum model.

The self-consistent field theory has been used [39] for a terminally attached polymer on a surface with weak excluded volume interactions and high surface coverage, σ. The depletion region near the surface and the extended tail has been neglected.

Based on Eq. (22) for each possible chain configuration that is analogous to the path integral formalization of quantum mechanics, the free-energy change to add the chain to the system, ΔF, can be calculated according to Eq. (23):

$$S_i = \int dn \left[\frac{1}{2}\left(\frac{dr}{dn}\right)^2 - u(r(n))\right] \tag{22}$$

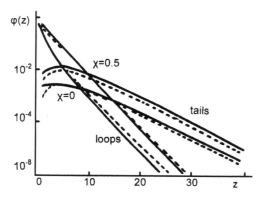

FIG. 8 Loop and tail profile according to the numerical self-consistent field model (solid curves) and according to the analytical approximation (dotted). (From Ref. 23.)

where r is the position in space of monomer unit number n, $U(r(n))$ is the effective mean field potential.

$$\exp(-\Delta F) = \sum_{\|r(n)\|} \exp(S_i) \tag{23}$$

The equation of motion for chain (24) can be obtained by the minimization of S_i:

$$\frac{d^2 r}{dn^2} = -\nabla U \tag{24}$$

The time integral along the path of the particle is equal to the molecular mass and the one-dimensional motion must only be considered for polymer brushes. Therefore,

$$U(r) = U(z) \tag{25}$$

Presuming the effective potential to be parabolic,

$$-u(z) = A - Bz^2 \tag{26}$$

and fixing

$$B = \frac{\pi^2}{8N^2} \tag{27}$$

the effective potential is related to the monomer density by

$$u(z) = -v\varphi(z) \tag{28}$$

where v is the excluded volume parameter. Therefore, the density profile of the graft polymer brush for the equilibrium state is

$$\varphi(z) = v^{-1}[A(h) - Bz^2]\Theta(h - z) \tag{29}$$

where $\Theta(h - z)$ is a step function of z and h is the brush height. The constant $A(h)$ is fixed by computing the net amount of material in the brush per unit surface area:

$$A(h) = \frac{N\sigma v}{h} + \frac{Bh^2}{3} \tag{30}$$

If the density profile has no discontinuity at $z = h$, the equilibrium brush height h^* can be described by

$$h_b^* = h_{\max} = \left(\frac{12v\sigma}{\pi^2}\right)^{1/3} N \tag{31}$$

This corresponds to the density profile:

$$\varphi(z) = \frac{\pi^2}{8N^2 v(h_b^*)^2} = z^2 \tag{32}$$

For binary mixtures of polymers of different polymerization degrees, the parabolic form of $\varphi(z)$ has been used [40]:

$$\varphi(z) = \frac{A_0 B_0 z_1^2}{4(z/z_1 + z_1/z)^2} \Theta(h_1 - z) \tag{33}$$

where $z_1 = h_b^*(1 - 2^{-2/3})^{1/2}$, $h_1 = h_b^*(1 - 2^{-2/3})^{1/2}$, $A_0 = (9\pi^2\sigma^2/(32v))^{1/3}$, and $B_0 = \pi^2/8N^2 v$.

Figure 9 shows the density profile of a polymer mixture of $N = 49$ and 98 with the same number of the respective chains for the several values of σ.

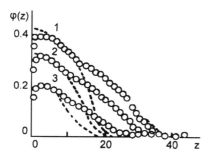

FIG. 9 Segment density for bimodal distribution function with chain length $N = 49$ and $N = 98$ for values of surface coverage 0.12 (1), 0.08 (2), and 0.04 (3). The broken lines correspond to data for density profiles for monodisperse chains with $N = 49$. (From Ref. 40.)

Investigation of adsorption of poly(ethylene oxide) on deutered polystyrene particles from D_2O [16,17,41] and random copolymer poly(vinyl alcohol-*co*-acetate) on polystyrene latex by SANS showed [17] that the density profiles of adsorbed polymer are well reproduced by mean field theories and Monte Carlo simulation (Fig. 10).

The dependence of molecular weight and surface coverage on the density profile of polystyrene adsorbed on a mica surface [18] showed that it agreed with self-consistent field theory.

2. Charged Polymers

At the first approximation, the behavior of polyelectrolytes in the adsorption layer may be described as an uncharged polymer, when one can change the quality of solvent according to the salt concentration in solution. More detail information has been obtained taking into account the distribution of charged groups of macromolecules in the adsorption layer.

Self-consistent field theory has been extended to polyelectrolyte and copolymers [38,42,43]. The end-point distribution, $G(z, s)$, in the field, u, was discussed earlier, Eqs. (5)–(15). The potential, described by Eq. (6), has to be extended [38] with an electrostatic term, u^e:

$$u(z) = -(\chi_s + \lambda\chi)\delta(z - 1) - 2\chi(\langle\varphi(z)\rangle - \varphi^b) - \ln\frac{1 - \varphi(z)}{1 - \varphi^b} + u(z)^e \tag{34}$$

$u(z)^e$ depends on the valence, v_i of a unit (ion, charge segment) on its degree of dissociation, $\alpha(z, i)$, which is a function of distance from the surface for weak ionic groups and on the electrostatic potential, $\psi(z)$:

$$u(z)^e = v_i\alpha(z, i)\psi(z)e \tag{35}$$

where e is the elementary charge.

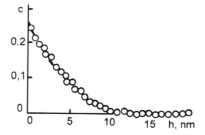

FIG. 10 Neutron-scattering intensity for poly(ethylene oxide) adsorbed on deuterated polystyrene latex in D_2O. Molecular weight 28×10^4, $\Gamma = 1.21\,mg/m^2$, and latex diameter was 240 nm [17].

In order to keep the number of parameters as small as possible, all small ions are usually considered to have the same nonelectrostatic properties as a solvent. The value of $\alpha(z)$ depends on the concentration gradient of polyelectrolyte segments:

$$\frac{1}{\alpha(z)} = 1 + \frac{\langle \varphi(z, H^+) \rangle}{K_d \langle \varphi(z, a) \rangle} \tag{36}$$

where $\langle (z, H^+) \rangle$ and $\langle \varphi(z, a) \rangle$ is the volume fraction of hydrogen ions and anion segments of the polymer, respectively, and K_d is the dissociation constant.

The charge of the ions and segments is assumed to be located on the planes in the middle of each lattice layer. The lattice is considered as an assembly of Stern layers. The plane-charge densities are described by

$$\sigma(z) = \sum_i \frac{v_i \alpha(z, i) e \varphi(z)}{l^2} \tag{37}$$

where l is the distance between Stern layers.

The dielectric displacement $D(z)$ equals

$$D(z) = \varepsilon(z) E(z) = \sum_z \sigma(z) \tag{38}$$

where $\varepsilon(z)$ is the dielectric permitivity and $E(z)$ is the electric field strength.

At first approximation, $\varepsilon(z)$ is taken to be a linear combination of the permitivities ε_i of various pure components:

$$\varepsilon(z) = \sum \varepsilon_i \varphi(z, i) \tag{39}$$

Therefore, the potential varies linearly in each half-layer:

$$\psi(z + 1) = \psi(z) - \tfrac{1}{2}(E(z) + E(z + \tfrac{1}{2})) \tag{40}$$

Equations (37)–(40) allow computation of $\psi(z)$ from volume fraction profiles. The calculated density profile of adsorbed polyelectrolyte is qualitatively different on an uncharged polymer. At low ionic strength, the maximum of density appeared as the result of the high electric potential generated by the adsorbed molecules repelling other chains (Fig. 11). The adsorbed amount and the thickness of adsorption layer increase with increasing salt concentration in solution (Fig. 12).

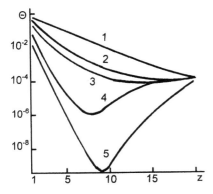

FIG. 11 Semilogarithmic segment density of an uncharged polymer (1) and a strong polyelectrolyte adsorbing on an uncharged surface at salt concentration 3 (2), 0.5 (3), 0.05 (4), and 0.01 M (5). (From Ref. 38.)

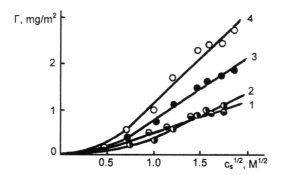

FIG. 12   Adsorbed amount of sodium polystyrene sulfonate with $M = 3.1 \times 10^3$ (1, 2), 170×10^3 (3), and 690×10^3 (4) on polyoxymethylene (2–4) and silica (1) as a function of square root of the NaCl concentration. (From Ref. 38.)

The partition function for the adsorption of strong polyelectrolyte in the mean field approach can be described by [44,45]:

$$G = \Omega(\varphi(z)) \exp\left(-\delta E_z^e(\varphi(i)) + \frac{\delta E_z^{ne}(\varphi(z))}{kT}\right) \tag{41}$$

where $\Omega(\varphi(z))$ is the number of ways in which a density profile $\varphi(z)$ with free electrostatic energy δE_z^e and nonelectric energy δE_z^{ne} can be obtained.

The equilibrium segment profile can be obtained by a maximization of the partition function G with respect to $\varphi(z)$. The contribution of δE_z^e due only to the polyelectrolyte charge in layer z is evaluated taking into account a transformation of charge segments to layer z from the bulk solution:

$$\delta E^e = l^2 \sum_{z=1} \int_{\sigma_e^b}^{\sigma_{ez}} (\psi(z) - \psi_b)\, d\sigma_e \tag{42}$$

where $\sigma_e^b = v\alpha e\varphi^b / l^2$ and $\sigma_{ez} = v\alpha e\varphi(z)/l^2$. σ_e^b and σ_{ez} are the plane charge density in the bulk solution and for the plane z, respectively $\psi(z)$ and ψ^b are the electrostatic potentials at the position i of the plane z and in bulk solution, respectively, and σ_e is the plane-charge density on the plane z during the isotherm-reversible charging process.

For strong electrolytes $\delta E_z^e \gg \delta E_z^{ne}$, the surface coverage has been obtained from derivatives $\delta \ln G/\delta\varphi(z)$ of Eq. (41) and $\delta E^e/\delta\varphi(z)$ of Eq. (42):

$$\sigma = \sum_{i=1}(\varphi(z) - \varphi^b) \tag{43}$$

The root mean square thickness of the adsorption layer can be described by

$$t_{rm}^2 = \frac{1}{\sigma \sum\limits_{i=1} z^2(\varphi(z) - \varphi^b)} \tag{44}$$

The calculation of the thickness of the adsorption layer showed that the values of t are of the order of 2 nm for a strong polyelectrolyte adsorbed on an uncharged surface when χ_s is high and the salt concentration is low.

Elaborated theoretical models for polyelectrolyte adsorption can describe experimental data qualitatively. In some cases, there is no theoretical interpretation. The plateau thickness of adsorbed sodium poly(styrene sulfonate) on a platinum plate was scaled as $t \sim M^{0.5}$ for 4.17 N

NaCl concentration and $t \sim M^{0.4}$ for 0.5–$1\,N$ concentration [46,47]. The layer thickness decreased when the salt concentration increased.

Figure 13 demonstrates the segment density profile of sodium poly(styrene sulfonate) adsorbed on polystyrene latex in NaCl solution [48]. The segment density profiles of Na poly(styrene sulfonate) do not practically depend on the surface charge. The shape of curve depends on the NaCl concentration. An increase of salt concentration leads to an increase of the thickness of the adsorption layer, as predicted by theory.

B. Scaling Theory

The polymer concentration in the adsorption layer is greater than in solution. The macromolecules in the adsorption layer are situated close to each other even if adsorption occurs from dilute solutions. The chains of macromolecules overlap, forming the network that is continually fluctuating. Bonds between two chains form and are destroyed continually. The number of chains between two bonds and the distance between them, ε, termed the screening of correlation length, depend on the polymer concentration and chemical potential of chain [49]. The dependence between ε and the polymer concentration in solution can be described by

$$\varepsilon(c) \sim N^{3/5}(c/N^{-4/5})^{-3/4} \tag{45}$$

The scaling theory, based on the power laws, allows one to predict that long flexible chains, adsorbed at the solid surface, build self-similar diffuse adsorbed layers.

The free energy for flexible chains in adsorbed layer depends on the adsorption regime [27]. In the weak adsorption regime, the layer thickness, D, is much smaller than the radius R of the macromolecule in the bulk solution. The chemical potential for such chain can be described by

$$\mu - \mu_0 = N\left(\frac{a}{D}\right)^{5/3} - \frac{\delta a N}{D} + \mu_{\text{trans}} \tag{46}$$

where μ_0 is the potential for a free chain with a fixed center of gravity, N is the polymerization degree, a is the monomer size, δ can be measured from renormalization threshold, and μ_{trans} is the standard term for a dilute two-dimensional gas of coils.

The first term in Eq. (46) represents the work that is necessary to confine a self-excluded chain in a slit of thickness D. The second term states that a fraction a/D of the N monomers is in the first layer and benefits from the effective attraction measured by δ.

Minimization of Eq. (46) leads to Eq. (47) when all numerical coefficients are ignored:

$$\mu = \mu_0 - N\delta^{5/2} + \mu_{\text{trans}} \tag{47}$$

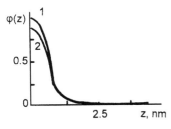

FIG. 13 Segment density profiles of sodium polystyrene sulfonate with $M = 74 \times 10^3$ adsorbed in polystyrene negatively charged latex (1) and positively charged latex (2) in $0.5\,N$ aqueous NaCl solution. (From Ref. 48.)

and this allows one to obtain the initial slope of adsorption isotherm, when the polymer concentration in solution is very low:

$$\gamma \sim c_B a^3 \exp(N\delta^{5/2}) \tag{48}$$

where c_B is the polymer concentration in the bulk solution.

In the plateau regime, the layer thickness D is large enough and the local behavior of the chain inside the layer is similar to the local behavior in a bulk solution. The chemical potential for one chain has the form

$$\mu - \mu_0 = N\left(\frac{a}{D}\right)^{5/3} - \frac{\delta a N}{D} + N(ca^3)^{1.25} + \mu_{\text{trans}} \tag{49}$$

where $N(ca^3)^{1.25}$ is the resulting contribution of osmotic pressure to the chemical potential of adsorbed macromolecule:

$$(ca^3)^{1.25} = \left(\frac{\Gamma a^2}{D}\right)^{1.25} = \left(\frac{\gamma a}{D}\right)^{1.25} \tag{50}$$

Γ is the total coverage and $\gamma = \Gamma_a$.

Minimization of the μ at fixed Γ leads to

$$D \sim \frac{a\gamma^5}{\delta^4} \tag{51}$$

and

$$\gamma \sim \delta N^{1/5} F^{-1/5} \tag{52}$$

where $F = \ln[\gamma/(c_B a^3)]$

From Eqs. (51) and (52), the following equation is obtained:

$$D(c_B) \sim aN\delta F^{-1} \tag{53}$$

where D is linear in the molecular mass of the polymer.

If the interaction energy in the adsorbed layer is described in three dimensions, the correlation length $\varepsilon_s(c)$ is

$$\varepsilon_s(c) = a(ca^3)^{-3/4} \sim \frac{aD}{(\gamma a)^{3/4}} \tag{54}$$

The third regime (between the dilute and plateau regimes) is called the semidilute regime. The chains overlap when the surface concentration exceeds a certain limit.

$$\Gamma_1 \sim \frac{N}{R_{F2}^2} \tag{55}$$

where R_{F2} is the lateral size of a single adsorbed coil.

For $\Gamma_1 < \Gamma < \Gamma_{\text{plateau}}$, the correct form of thermodynamic potential is

$$\mu - \mu_0 = N\left(\frac{a}{D}\right)^{5/3} + \frac{aN}{D(\gamma^2 - \delta)} + \mu_{\text{trans}} \tag{56}$$

The repulsion between coils is described by the γ^2 term.

After minimization of Eq. (56) with respect to D, Eqs. (57) and (58) have been obtained:

$$\frac{D}{a} \sim (\delta - \gamma^2)^{-3/2} \tag{57}$$

and

$$\gamma^2 \sim \delta - \left(\frac{F}{N}\right)^{2/5} \tag{58}$$

According to the scaling theory, the density profile of an adsorbed layer is divided into three distinct regions [50]: The density profile near the surface is slowly varying as a result of short-range interaction between a monomer link and surface:

$$\varphi(z) \sim \varphi_s \quad (0 < z < a) \tag{59}$$

where φ_s is the surface segment density and a is the statistical step length. In the central region, the adsorbed layer is regarded as a semidilute solution and the density profile is given by

$$\varphi(z) = \left(\frac{z}{a}\right)^{-4/3} \quad (a < z < D) \tag{60}$$

where D is the distance at which the adsorption energy per segment has a dominant effect. In the distal region, the concentration profile decreases exponentially (Fig. 14):

$$\varphi(z) = \varphi\left[1 + \exp\left(-\frac{z}{\varepsilon^b}\right)\right] \tag{61}$$

Scaling theory has been used to describe the confirmation of polymers attached to an interface [49]. In a good solvent, the density profile for a separate coil has been described by

$$\varphi(z) = \sigma\left(\frac{z}{a}\right)^{2/3} \tag{62}$$

where σ is the grafting density. When the coils are overlapped,

$$\varphi_N \sim \sigma^{2/3} \tag{63}$$

and the thickness of the grafted layer is proportional to the molecular weight of polymer:

$$t = Na\sigma^{1/3} \tag{64}$$

Alternative forms of these equations have been proposed [51]:

$$\varphi(z) \sim \varphi_s \quad (0 < z < a) \tag{65}$$

$$\varphi(z) \sim \varphi_s\left(\frac{a}{z}\right)^{1/3} \quad (a < z < D) \tag{66}$$

$$\varphi(z) \sim \varphi_s\left(\frac{a}{z}\right)^{1/3}\frac{2D}{z + D} \tag{67}$$

The adsorbed amount of polymer can be obtained by integration Eqs. (65)–(67) over the specified region and adding the results together.

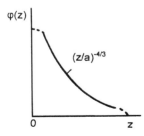

FIG. 14   Segment density profile for adsorption from a good solvent. (From Ref. 50.)

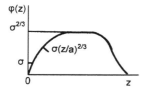

FIG. 15 Concentration profile for a grafted layer immersed in a good solvent in the overlapping regime. (From Ref. 49.)

The concentration profile in an adsorbed layer is divided into three regions (Fig. 15). The profile near the wall is given by Eq. (62). In the central region, the concentration profile is described by Eq. (63).

The scaling theory has also been modified to show the conformation of the macromolecule of star polymers in adsorption layer [52,53].

The layer thickness of polystyrene adsorbed on a chromate plate from a suitable solvent as a function of molecular weight was described by the expression $t \sim M^{0.4}$ [54,55]. These results agree with the scaling theory.

An investigation of density profile of monodisperse poly(dimethylsiloxane) on the mesoporous silica from semidilute cyclohexane solution showed that the profile was in agreement with the scaling theory, $\varphi = z$ [56].

Investigation of the force profile between two surfaces covered by adsorbed polystyrene brushes [57] showed that the profile can be described by Eq. (68), obtained for a semidilute concentration regime [57]:

$$f(z) = \frac{kT}{\sigma^3[(2h_0/z)^{9/4} - (z/2h_0)^{3/4}]} \quad \text{for } z < 2h_0 \tag{68}$$

where h_0 is the equilibrium height of the brushes. As Fig. 16 shows, Eq. (68) describes experimental data very well.

The scaling theory has been used to describe the adsorption kinetics of monodisperse, flexible chains from dilute solutions in a good solvent onto flat solid interfaces [58]. The adsorption process onto an initially clean surface has three major stages. At the first stage, incoming chains stick and quickly rearrange into an expanded pancake with $D \sim N^{3/4}$. This regime is limited to low surface coverage. In the second stage, adsorbed chains begin to compete for adsorption sites and loops and tails begin to appear. The barrier effect of adsorbed layer against the incoming chains increases during this stage. When the adsorbed layer is close to

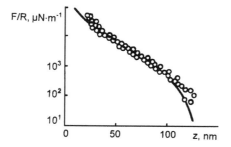

FIG. 16 A force–distance profile obtained using the surface force apparatus for the force between two mica surfaces, each bearing a physically adsorbed polystyrene brush of relative molecular mass 14×10^4 in toluene. The solid line is a fit to Eq. (68). (From Ref. 57.)

equilibrium, a dynamic exchange between free chains in solution and adsorbed chains begins. The rate of accumulation of adsorbed monomer segments in the second stage is

$$\frac{d\Gamma}{dt} = K_{si}(\Gamma)\beta(\Gamma)c_b - K_{is}(\Gamma)\frac{\Gamma}{R} \tag{69}$$

where K_{si} is the Nernst transmission coefficient characterizing diffusive transport of segments across the adsorbed layer, $\beta(\Gamma)$ is a statistical weight accounting for the repulsive barrier caused by excluded volume in the layer, Γ/R is the mean monomer concentration in the layer, and $K_{is}(\Gamma)$ is related to $K_{si}(\Gamma)$ when $d\Gamma/dt = 0$.

An investigation of the kinetics of adsorption of polystyrene onto gold plate from cyclohexane solution (Θ solvent) showed [59] that the experimental result does not agree with the scaling theory approximated to the Θ solvent. The permeability of a partially formed adsorbed layer depends on the history of its formation. The kinetics are much slower than if it is governed by end-in raptation and is insensitive to the molecular weight of polymer.

C. Simple Model

Theoretical and experimental investigations showed that, in the general case, the segment density profile of an adsorbed polymer can be divided on two sections: the depletion region near the surface with practically constant concentration of monomer links of polymer and the region in which the segment density slowly decreased (Fig. 13). The relationship between link density and the distance from the surface in the second region can be described by

$$\frac{d(n/N)}{dx} = -bx^l \tag{70}$$

where $x = z - h_0$, z is the distance from the surface, h_0 is the thickness of the depletion region, n/N is the relative density of polymer links at a distance z from solid surface, and b and l are effective constants.

The values of b and l are variable values and it depends on the flexibility of the macromolecule, solvent quality, and the number of adsorbed macromolecules when they are flexible.

The integration of Eq. (70), at the initial condition $n(x = 0) = N_{tr}$, leads to Eq. (71), where N_{tr} is the number of polymer links bound with the surface:

$$\frac{n}{N} = \frac{N_{tr}}{N} - \frac{b}{(l+1)}x^{(l+1)} \tag{71}$$

Transformation of Eq. (71) allows one to obtain Eq. (72):

$$\ln\left(\frac{N_{tr}}{N} - \frac{n}{N}\right) = \ln\left(\frac{b}{l+1}\right) + (l+1)\ln(z - h_0) \tag{72}$$

The value of N_{tr}/N and h_0 can be calculated based on the experimental data obtained by SANS [17,48,60,61], according to Fig. 17. In some cases, the value of h_0 can be calculated by the iteration methods. Figure 17 shows that the experimental data obtained by SANS lie on straight lines when plotted according to Eq. (72). Correlation coefficients in all cases are close to 1. The values of N_{tr}/N, $\ln b$, and $l+1$ are listed in Table 1. As Table 1 demonstrates, there is a dependence between $\ln b$ and l. We suggest that this dependence can be described by a linear relationship.

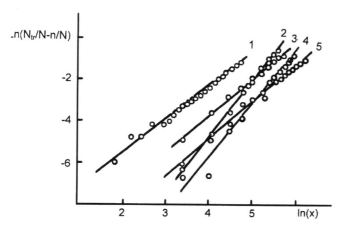

FIG. 17 Relationship between link density and the distance from solid surface according to Eq. (72) for adsorbed polystyrene from toluene at 21°C (1) and cyclohexane at 14.6 (2), 21.4 (3), 31.5 (4), and 53.4°C (5).

TABLE 1 Values of $\ln b$, N_{tr}/N, l, and h_0 for Some Polymers

Polymer	Solvent	$M \times 10^{-3}$	T °C	$\dfrac{N_{tr}}{N}$	$l+1$	$-\ln b$	h_0 (nm)
Polystyrene	Toluene	105	21	0.177	1.52	8.17	< 0.1
Polystyrene	Cyclohexane	105	53.4	0.304	1.76	11.58	< 0.1
Polystyrene	Cyclohexane	105	31.5	0.379	2.52	15.09	< 0.1
Polystyrene	Cyclohexane	105	21.4	0.431	1.74	10.25	< 0.1
Polystyrene	Cyclohexane	105	14.6	0.567	2.61	14.40	< 0.1
Polystyrene	Toluene	500	—	0.77	1.47	4.54	0.095
Polystyrene	Toluene	28.5	—	0.68	1.42	4.80	0.19
Polystyrene	Toluene	4	—	0.51	1.25	4.90	0.95
d-Polystyrene $M = 1800000$	h-Polystyrene	1030	—	0.65	0.72	3.76	4.00
d-Polystyrene $M = 156000$	h-Polystyrene	100	—	0.88	1.08	3.20	1.09
d-Polystyrene $M = 52000$	h-Polystyrene	100	—	0.85	1.27	3.69	< 0.1
d-Polystyrene $M = 28500$	h-Polystyrene	100	—	0.68	0.68	2.67	< 0.1
d-Polystyrene $M = 7000$	h-Polystyrene	100	—	0.50	0.30	2.56	< 0.1
PEO	Water	280	—	0.278	0.70	3.21	< 0.1
Sodium poly(styrene sulfonate)	NaCl solution 0.5 N	780	—	0.095	0.19	4.71	2.56
	1.0 N	780	—	0.104	0.38	4.28	3.82

Experimental data show that the number of macromolecule links bound with the surface N_{tr} depends on the polymerization degree. Therefore, we suggested that

$$\ln\left(\frac{N_{tr}}{N}\right) = \ln(N^{Cl}) + A_n + B_n l \tag{73}$$

where A_n, B_n, and C are the coefficients. The calculation of C will be described later.

The value of $\ln b$ is connected with the thickness of the adsorbed layer. Therefore, we suggested that it depends on the macromolecule length of on the polymerization degree and the number of link atoms in the polymer backbone (e.g., $n = 2$ for polystyrene or polyethylene, $n = 4$ for polybutadiene). Therefore, we can write the linear relationship as

$$\ln b = \ln[(Nn_z)^l] + A_z + B_z l \tag{74}$$

or

$$\ln b = \ln[(Nn_z)^l] = A_z + B_z l$$

As is shown in Fig. 18, relationship between $\ln b$ and l lies on a straight line when plotted according to Eq. (74). The correlation coefficient is 0.066, $A_z = -5.9 \pm 1.0$, and $B_z = -12.4 \pm 1.5$.

The thickness of adsorption layer can be obtained by integration of Eq. (71):

$$\int_{N_{tr}}^{0} n\,dn = N_{tr} - \frac{bN}{l+1}\int_{0}^{h-h_0} x^{(l+1)}\,dx \tag{75}$$

where h is the thickness of adsorption layer. After integration, we obtain

$$-\frac{N_{tr}^2}{2} = N_{tr} - \frac{bN}{(l+1)(l+2)}(h-h_0)^{(l+2)} \tag{76}$$

At the first approximation, we suggest that $h_0 \ll h$ and $N_{tr} \ll N_{tr}^2$. Therefore, Eq. (76) can be transformed into

$$h = \left(\frac{(l+1)(l+2)}{2bN}\right)^{1/(l+2)} N_{tr}^{2/(l+2)} \tag{77}$$

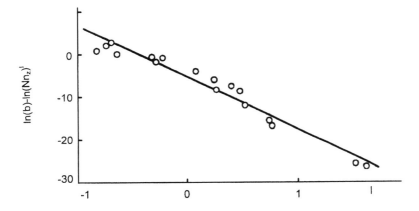

FIG. 18 Relationship between $\ln b$ and l according to Eq. (74).

Take into account the values of b and N_{tr} from Eqs. (73) and (74), we obtain from Eq. (77)

$$\ln h = \frac{1}{l+2} \ln\left(\frac{(l+1)(l+2)}{2}\right) - \frac{1}{l+2} \ln n_z + \frac{1}{l+2}[2A_n - A_z + (2B_n - B_z)1]$$
$$+ \frac{Cl + l - 1}{l+2} \ln N \tag{78}$$

Equation (78) shows that the thickness of adsorption layer is proportional to the polymerization degree to the power $(2Cl - l + 1)/(l + 2)$.

On the other hand, it can be suggested that the surface area, blockaded by one macromolecule, is proportional to N_{tr} and the surface area blockaded by one link of the polymer, S_1, is

$$S_b = N_{tr}^2 S_1 \tag{79}$$

S_1 does not depend on the polymerization degree, but it depends on the molecular mass of the polymer link. Therefore, from Eqs. (73) and (79), we obtain

$$\ln(S_b) = A_n + B_n + \ln(S_1) + 2(Cl + 1)\ln N \tag{80}$$

The value of S_b is proportional to the polymerization degree to the power $2(Cl + 1)$. As the experimental data on the polymer adsorption shows [9,46,62–65], the thickness of adsorption layer and the surface area, blocked by one macromolecule in saturated adsorption layer, are described by

$$h = K_t N^{a1} \tag{81}$$
$$S_b = K_b N^{a2} \tag{82}$$

The surface area, blockaded by one adsorbed macromolecule, is calculated according to

$$S_b = \frac{M}{AN_A} \tag{83}$$

where A is the mass of adsorbed polymer on the unit surface area in the saturated adsorption layer. The values of K_t, K_b, $a1$, and $a2$ for different polymers are listed in Table 2. As one can see from Eqs. (78) and (80–82),

$$a1 = \frac{2Cl - l + 1}{l+2} \tag{84}$$
$$a2 = 2(Cl + 1) \tag{85}$$

Equations (84) and (85) allow one to calculate the values of l and C:

$$l = \frac{a2 - 2a1 - 1}{a1 + 1} \tag{86}$$
$$C = \frac{(a2/2 - 1)(a1 + 1)}{a2 - 2a1 - 1} \tag{87}$$

The value of C has been calculated as an average value from data in Table 2. It is equal to 0.847 ± 0.015. The values of A_n and B_n can be calculated according to Eq. (88) obtained from Eq. (73) taking into account the value of C:

$$\ln\left(\frac{N_{tr}}{N}\right) - \ln(N^{Cl}) = A_n + B_n \tag{88}$$

TABLE 2 Values of K_t, K_b, $a1$, and $a2$ for Polymers of Different Natures

Polymer	Solvent	T (°C)	$-\ln K_t$	$a1$	$-\ln K_b$	$a2$
PEO	Water	23	0.746	0.488	1.439	0.909
PEO	Water	25	0.980	0.579	1.640	0.735
Polystyrene	Cyclohexane	35	0.362	0.454	2.215	0.894
Polystyrene	Cyclohexane	35	−0.007	0.462	1.767	0.838
Polystyrene	Cyclohexane	40	−0.395	0.429	1.207	0.820
Polystyrene	Cyclohexane	45	−0.536	0.421	1.026	0.815
Polystyrene	Toluene	35	−1.780	0.319	1.005	0.973
Polystyrene	CCl$_4$	35	−0.462	0.437	1.918	1.011
Sodium	NaCl					
poly(styrene	solution					
sulfonate)	0.1 N	25	−0.624	0.434	−2.215	0.772
	0.5 N	25	−0.499	0.411	−0.714	0.918
	4.17 N	25	0.329	0.462	−0.900	0.794

As is shown in Fig. 19, the data from Table 1 lie on a straight line according to Eq. (88). The correlation coefficient is 0.980, $A_n = -1.6 \pm 0.5$, and $B_n = -5.2 \pm 1.0$.

The following equation is obtained from Eqs. (78) and (81):

$$Y_1 = A_t + B_t l \tag{89}$$

where

$$Y_1 = (l+2)\ln K_t - \ln[(l+1)(l+2)/2] + l\ln(n_z).$$

Equation (89) describes well the experimental data from Table 2, when the value of l has been calculated according to Eq. (86). Figure 20 shows the linear dependence between Y_1 and l. The correlation coefficient is 0.887, $A_t = -6.8 \pm 2.5$, and $B_t = -10.9 \pm 3.4$.

The small number of experimental data does not allow one to find the shape of dependence of S_1 on molecular mass. We suggested that, in the first approximation, it can be defined by Eq. (90):

$$S_1 = \exp(A_1 + B_1)M_z^{4(Cl+1)} \tag{90}$$

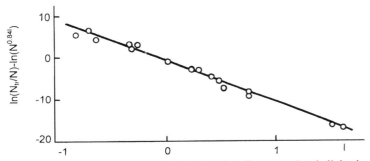

FIG. 19 Relationship between relative density of macromolecule links, bound with the solid surface and l according to Eq. (88).

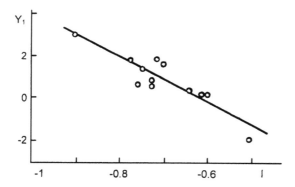

FIG. 20 Relationship between Y_1 and l according to Eq. (89).

where A_1 and B_1 are constants. Therefore, from Eqs. (73), (79), (82), and (90), we obtain

$$Y_2 = A_b + B_b l \tag{91}$$

where

$$Y_2 = \ln K_b - 4(Cl + 1) \ln M_z$$

Figure 21 shows that the experimental data from Table 2 lie on a straight line when plotted according to Eq. (91). The correlation coefficient is 0.953, $A_b = -13.7 \pm 2.5$ and $B_b = -10.6 \pm 3.2$. Therefore, in the first approximation, the basic parameters of macromolecule can be described by

$$N_{\text{tr}} = \exp(A_n + B_n l) N^{(Cl+1)} \tag{92}$$

$$h = \left(\frac{(l+1)(l+2)}{2} \right)^{1/(l+2)} \exp\left(\frac{A_t + B_t l}{l+2} \right) n_z^{-[1/(l+2)]} N^{(2Cl-l+1)/(l+2)} \tag{93}$$

$$S_b = \exp(A_b + B_b l) M_z^{(4(Cl+1))} N^{2(Cl+1)} \tag{94}$$

However, the found coefficients of A_t, B_t, C, A_n, B_n, A_b, and B_b are approximate and they can be obtained more exactly according to new experimental data.

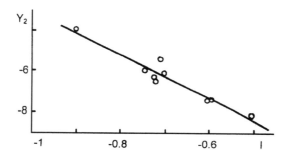

FIG. 21 Relationship between Y_2 and l according to Eq. (91).

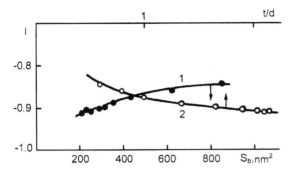

FIG. 22 Dependence between l and S_b (1) and l and the ratio of the thickness of adsorption layer to the diameter of the blockaded surface area (2) for polystyrene adsorbed on the chromium plate from cyclohexane.

The value of l depends on the number of macromolecules in the adsorption layer. If we take into account that the degree of saturation of adsorption layer close to 1 in the equilibrium state at the different polymer concentrations in solution, the value of l can be calculated according to Eq. (95) obtained from Eq. (94):

$$l = \frac{\ln S_b - A_b - 4 \ln M_z - 2 \ln N}{B_b + 4C \ln M_z + 2C \ln N} \tag{95}$$

The value of l has been calculated for the adsorption isotherm of polystyrene with molecular mass 67.5×10^4 on the chrome plate from cyclohexane solution [65]. Figure 22 shows that the value of l increases when the surface area, blocked by one macromolecule, arises and it decreases when the ratio of the thickness of adsorption layer to the diameter of the blocked surface area increases.

IV. ADSORPTION KINETICS

In many cases, the adsorption kinetics can be described by the well-known equation [66–68]

$$\frac{dP_a}{dt} = k_a P_s \left(1 - \frac{P_a}{P_m} \right) N - K_d P_a \tag{96}$$

where P_a is the number of adsorbed macromolecules in the unit volume of dispersion, P_s is the number of macromolecules in the unit volume, P_m is the maximum number of macromolecules that can be adsorbed in the unit volume of dispersion, N is the number of particles in the unit volume, and k_a and k_d are the rate constants of adsorption and desorption, respectively. If $N = 1$, one can obtain the equation for polymer adsorption at the solid plate. In this case, P_a is the number of macromolecules in the unit surface area and P_m is the maximum number of macromolecules that can be adsorbed on the unit surface area.

In the plateau section of the kinetic curve, when $dP_a/dt = 0$, Eq. (96) can be transformed into

$$\frac{1}{P_a} = \frac{1}{P_m} + \frac{k_d}{k_a} \frac{1}{P_s} \tag{97}$$

The value of P_a and P_s are calculated according to

$$P_a = \frac{c_a N_A}{M} \quad \text{for adsorption at the dispersed particles} \tag{98}$$

$$P_a = \frac{A N_A}{M} \quad \text{for adsorption at the plate} \tag{99}$$

$$P_s = \frac{c N_A}{M} \tag{100}$$

where c_a is the concentration of adsorbed polymer in the unit volume of dispersion, M is the molecular mass of polymer, A is the mass of the adsorbed polymer on the unit surface area, c is the polymer concentration in solution, and N_A is Avogadro's number.

Adsorption kinetics can be described according to Eq. (101) obtained by the integration of Eq. (96) at the initial condition $P_s(t = 0) = P_{\text{in}}$:

$$\ln F = \ln F_{\text{in}} + E_1 \frac{k_a}{P_m} t \tag{101}$$

where

$$F = \frac{E_1 + E_2 - 2P_a}{E_1 - E_2 - 2P_a}$$

$$F_{\text{in}} = \frac{E_1 + E_2}{E_1 - E_2}$$

$$E_1 = P_m + P_{\text{in}} + \frac{k_d P_m}{k_a N}$$

$$E_2 = (E_1^2 - 4 P_m P_{\text{in}})^{1/2}$$

P_{in} is the initial polymer concentration in solution.

However, in some cases, Eq. (101) does not describe the experimental data [69]. At the adsorption of polystyrene at the chrome plate from cyhclohexane, toluene, and carbon tetrachloride, the value of the intercept in the ordinate axis does not equal $\ln F_{\text{in}}$ (Fig. 23). Therefore, some number of macromolecules is adsorbed on the surface preliminary. It allows one to suppose that polymer adsorption occurs through two stages. In the first stage, polymer adsorption occurs at the formation of solid, free from polymer. It finished in 1–2 min, when the solid surface was almost covered by macromolecules.

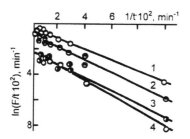

FIG. 23 Dependence between mass of adsorbed polymer and time under polystyrene adsorption at the smooth chromium plate from cyclohexane at 35°C when plotted according to Eq. (101). The polystyrene molecular mass is 242×10^4 (1), 67×10^4 (2), 1340×10^4 (3), and 762×10^4 (4). (From Ref. 69.)

In the second stage, the adsorption occurs on the surface sections, blocked by adsorbed macromolecules. In this case, the deformation of adsorbed macromolecules as well as the entered macromolecules occurs. The duration of the second stage can reach hours and days.

A. The Stage of Fast Adsorption

The first stage of polymer adsorption has been investigated by reflectometry [63]. The amount of poly(ethylene oxide) adsorbed on the silica plate arises linearly with time and reaches the plateau section in 30–60 s. The initial rate of adsorption depends on the rate of flow and molecular mass of the polymer. Therefore, adsorption proceeds in the diffusional field. The substitution of the polymer solution by solvent did not change the amount of adsorbed polymer.

1. Distribution of the Number of Macromolecules on the Particles

In the early stage of the adsorption and/or for the low polymer concentration in solution, the particle surface coverage is relatively low. If the particle aggregation does not proceed, the collision rate of macromolecules with dispersion particles is proportional to the product of the particle number and macromolecule number in the unit volume. Taking the above into account, the adsorption kinetics of monodisperse polymers on the particles and dynamics of particle number change with adsorbed polymer can be described in the first approximation by the following system of differential equations [66,67]:

$$\frac{dP_s}{dt} = k_{a1} P_s \sum_j N_j \tag{102}$$

$$\frac{dN_0}{dt} = -k_{a1} P_s N_0 \tag{103}$$

$$\frac{dN_1}{dt} = k_{a1} P_s N_0 - k_a P_s N_1 \tag{104}$$

$$\vdots$$

$$\frac{dN_j}{dt} = k_{a1} P_s N_{j-1} - k_{a1} P_s N_j \tag{105}$$

where P_s is the macromolecule number in the unit volume and N_j is the number of particles with j adsorbed macromolecules in the unit volume of dispersion.

The integration of Eq. (102) at the initial condition

$$P_s(t = 0) = P_{in} \tag{106}$$

and taking into account that

$$\sum_j N_j = N_{in} \tag{107}$$

leads to

$$P = P_{in} \exp(k_{a1} N_{in} t) \tag{108}$$

where N_{in} and P_{in} are the initial number of particles and macromolecules in the unit volume, respectively.

The average number of macromolecules adsorbed on one particle at time t is

$$P_a = \frac{P_{in} - P_s}{N_{in}} \tag{109}$$

It follows from Eq. (108) that

$$P_a = \left(\frac{P_{in}}{N_{in}}\right)[1 - \exp(-k_{a1}N_{in}t)] \tag{110}$$

Integration of Eq. (103) after substituting Eq. (108) in it at the initial condition

$$N_0(t = 0) = N_{in} \tag{111}$$

and taking into account Eq. (110) leads to

$$N_0 = N_{in} \exp(-P_a) \tag{112}$$

It follows from Eqs. (103) and (104) that

$$\frac{dN_1}{dN_0} = \frac{N_1 - N_0}{N_0} \tag{113}$$

Integration of Eq. (113) at the initial condition

$$N_j(N_0 = N_{in}) = 0 \quad \text{for } j > 1 \tag{114}$$

leads to the relationship

$$N_1 = N_0 \ln\left(\frac{N_{in}}{N_0}\right) \tag{115}$$

Hence, it follows, taking into account Eq. (112), that

$$\frac{N_1}{N_0} = \ln\left(\frac{N_{in}}{N_0}\right) = P_a \tag{116}$$

From Eqs. (103) and (105) for $j = 2$, one obtains

$$\frac{dN_2}{dN_0} = \frac{N_2 - N_1}{N_0} \tag{117}$$

Integration of Eq. (117), taking into account Eq. (115), at the initial condition (114) leads to

$$N_2 = \left(\frac{N_0}{2}\right)\ln^2\left(\frac{N_{in}}{N_0}\right) \tag{118}$$

Hence, it follows from Eqs. (112) and (118) that

$$\frac{N_2}{N_0} = \frac{1}{2}P_a^2 \tag{119}$$

It follows from Eqs. (116) and (119) that

$$\frac{N_2}{N_1} = \frac{1}{2}P_a \tag{120}$$

Analogously, it can be shown that

$$\frac{N_j}{N_0} = \left(\frac{1}{j!}\right)P_a^j \tag{121}$$

Hence, it follows that

$$\frac{N_j}{N_{j-1}} = \left(\frac{1}{j}\right)P_a \tag{122}$$

Expressions (116), (120), and (122) correspond to the recurrent correlations binding two neighboring terms in Poisson's distribution. Hence, in the early stages of the process and/or at a low polymer concentration, the distribution of probabilities of macromolecule number j

being adsorbed on the particle follows the Poisson's distribution. It can be supposed that the probability of the separate values of this discrete random quantity is

$$p(j, P_a) = \frac{N_j}{N_{in}}, \quad j = 0, 1, 2, \ldots, \tag{123}$$

because P_a is the average number of macromolecules adsorbed on the particle. It follows from Eq. (123), taking into consideration Eqs. (112) and (121), that

$$p(j, P_a) = \frac{\exp(-P_a)P_a^j}{j!} \tag{124}$$

Equation (124) corresponds to the formula of the Poisson's distribution.

When the degree of the coverage of particle surface by macromolecules is large, the adsorption rate decreases because of the decreasing number of adsorption sites on the particles. The kinetic curve reaches the plateau section at the end of first stage (Fig. 24). To take this circumstance into consideration, the factor $(1 - \theta)$ can be introduced in equations of the type of Eqs. (103)–(105), where θ is the fraction of particle surface covered by polymer.

In the general case, θ depends on the number of adsorbed macromolecules:

$$\theta = \frac{j}{j_m} \tag{125}$$

where j is the number of macromolecules adsorbed on the particle and j_m is the maximum number of macromolecules that can be adsorbed on the particle. Therefore, the adsorption rate closed to the plateau section of kinetic curve can be described by

$$\frac{dP_s}{dt} = -k_{a1}P_s \sum_{j=1}^{j_m}\left(1 - \frac{j}{j_m}\right)N_j \tag{126}$$

$$\frac{dN_0}{dt} = -k_{a1}P_sN_0 \tag{127}$$

$$\frac{dN_1}{dt} = k_{a1}P_sN_0 - k_{a1}P_sN_1\left(1 - \frac{j}{j_m}\right) \tag{128}$$

$$\frac{dN_2}{dt} = k_{a1}P_sN_1(1 - 1/j_m) - k_{a1}P_sN_2(1 - 2/j_m) \tag{129}$$

$$\vdots$$

$$\frac{dN_j}{dt} = k_{a1}P_sN_{j-1}\left(1 - \frac{j-1}{j_m}\right) - k_{a1}P_sN_j\left(1 - \frac{j}{j_m}\right) \tag{130}$$

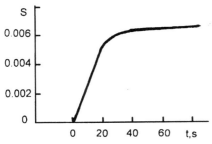

FIG. 24 Kinetics of adsorption of poly(ethylene oxide) with molecular mass 40×10^4 from water in the diffusional field. Polymer concentration $= 10\,\text{mg/L}$, Re $= 12.2$. (From Ref. 63.)

The integration of Eq. (126) at the initial conditions (106) and (111), taking into account $j_m N_{in} = P_{am}$, leads to

$$P = \frac{P_{in}(P_{in}/P_{am} - 1)}{P_{in}/P_{am} - \exp[-k_{a1}N_{in}(P_{in}/P_{am} - 1)t]} \quad (131)$$

where P_{am} is the maximum number of macromolecules that can be adsorbed on the particles in the unit volume of dispersion.

At the low polymer concentration, when $P_{in} \ll P_{am}$, Eq. (131) can be transformed to Eq. (110). Integration of Eq. (127) taking into account Eq. (131) and initial condition (111) leads to

$$\frac{N_0}{N_{in}} = \exp[k_{a1}P_{am}(1 - P_{in}/P_{am})t]\left[(P_{in}/P_{am} - \exp\left(\frac{k_{a1}N_{in}(1 - P_{in}/P_{am})t}{P_{in}/P_{am} - 1}\right)\right]^{-P_{am}/N_{in}}$$

$$(132)$$

From Eqs. (127) and (128), one can obtain

$$\frac{dN_1}{dN_0} = -1 + \frac{(1 - 1/j_m)N_1}{N_0} \quad (133)$$

Integration of Eq. (133) at the initial condition (114) leads to

$$N_1 = N_0\left[\left(\frac{N_{in}}{N_0}\right)^{1/j_m} - 1\right]j_m \quad (134)$$

From Eqs. (127) and (129), it follows that

$$\frac{dN_2}{dN_0} = -\frac{(1 - 1/j_m)N_1}{N_0} + \frac{(1 - 2/j_m)N_2}{N_0} \quad (135)$$

Integration of Eq. (135) after substitution of Eq. (134) at the initial condition (114) yields

$$N_2 = \tfrac{1}{2}N_0\left[\left(\frac{N_{in}}{N_0}\right)^{1/j_m} - 1\right]^2 j_m(j_m - 1) \quad (136)$$

Therefore, in the general case, one can obtain the Eq. (137) in a similar manner.

$$N_j = N_0\left[\left(\frac{N_{in}}{N_0}\right)^{1/j_m} - 1\right]^j \frac{\left(\prod_{k=0}^{j-1}(j_m - k)\right)}{j} \quad (137)$$

Equation (137) allows one to find the distribution of probabilities of the number of macromolecules adsorbed on the particle taking into account the fraction of the particle surface blockaded by polymer.

2. Influence of Adsorption Conditions

The initial rate of polymer adsorption in stagnation point flow depends on the molecular mass of the polymer (Fig. 25) and on the hydrodynamic regime or the Reynolds number, Re (Fig. 26) [63]. If adsorption proceeds in the diffusional field, the degree of polymerization can influence both the rate of polymer diffusion from the bulk solution to the boundary layer and on the rates of surface diffusion as well as of macromolecule reconformation in the boundary layer during adsorption.

At the same time, it has been ascertained in Ref. 63 that the maximum adsorbed amount corresponding to the plateau section of a kinetic curve does not practically depend on the

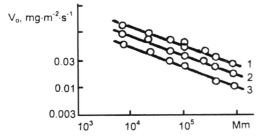

FIG. 25 Effect of molecular mass on the initial adsorption rate for the values of the Reynold's number 24.4 (1), 12.2 (2), and 6.1 (3) and the polymer concentration in solution 10 mg/L. (From Ref. 63.)

Reynolds number. It allows one to assume that the role of processes related with hydrodynamics decreases with the increase of surface coverage with the polymer, Θ. As a first approach, it can be assumed that this decrease occurs exponentially. If that is so, it can be shown that the relationship between the adsorption rate and the Reynolds number will be described by the third term in parenthesis on the right-hand side of Eq. (138) (see Appendix 1).

Taking the above into account, the rate of adsorption of macromolecules can be described as a first approach in the following form [70]:

$$v_a = k_{a1}v^b(r\,\mathrm{Re}^q)^{1-\Theta}\alpha c \qquad (138)$$

where k_{a1} is the rate constant of polymer adsorption, r and q are coefficients taking the effect of hydrodynamic regime on macromolecule diffusion from the bulk solution to the boundary layer into account, $v = M/M_1$ is the degree of polymerization, M and M_1 are the molecular mass of polymer and molecular mass of polymer link, respectively, $\Theta = A/A_p$, A and A_p are the adsorbed amount and the equilibrium adsorbed amount of polymer corresponding to the plateau section of a kinetic curve, respectively, b is the constant taking the effect of the degree of polymerization on the adsorption rate into account, α is the fraction of bare sites of adsorption, and c is the polymer concentration in solution.

A macromolecule blockades a certain number of adsorption sites, screening them from the penetration of other macromolecules. Apparently, the number of blockaded sites must depend on the size of the macromolecule coil and, therefore, on the degree of polymerization. In our opinion, as a first approximation, the fraction of adsorption sites blockaded by an adsorbed polymer can be described as follows:

$$1 - \alpha = S_b P_a \qquad (139)$$

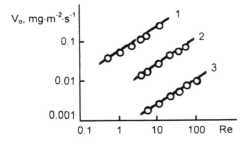

FIG. 26 Effect of the Reynold's number on the initial adsorption rate of poly(ethylene oxide) with molecular mass 246×10^3 at the polymer concentration in solution 100 (1), 10 (2), and 1 mg/L (3). (From Ref. 63.)

where S_b is the surface area blockaded by one macromolecule. The initial rate of adsorption, $t = 0$, $\Theta = 0$, and $c = c_{in}$, can be described by

$$v_0 = k_{a1} r c_{in} \text{Re}^q v^b \tag{140}$$

The following expression is obtained by the transformation of Eq. (140):

$$\ln\left(\frac{v_0}{c_{in}}\right) = \ln(k_{a1} r) + b \ln v + q \ln \text{Re} \tag{141}$$

The value of q is calculated from the relationship between the initial rate of adsorption and the Reynolds number for a constant molecular mass of the polymer (Fig. 27, curve 1): $q = 0.65$ 7 ± 0.04. The value of q is in agreement with the magnitude found in Ref. 63. Substituting the obtained value of q in Eq. (141), the values $k_{a1} r = 0.0127 \pm 0.002 \text{L/m}^2/\text{s}$ and $b = -0.38$ 7 ± 0.02 are found from the relationship between the initial rate of adsorption and the degree of polymerization (curve 2). Therefore, the rate of adsorption decreases if the degree of polymerization of the polymer increases. It is agreement with the results of Ref. 71 on the investigation of adsorption kinetics of polymers and oligomers in the kinetic field.

The first stage of the process is short and finishes at the condition $S_b P_a = 1$, when the whole surface is blockaded by macromolecules. One can suppose that $S_b = S_h$ for the rigid chain macromolecules when the deformation rate of macromolecules is slow (S_h is the hydrodynamic section of the macromolecule in solution). For the flexible chain macromolecules, $S_b > S_h$.

B. The Stage of Slow Adsorption

The second stage is limited by the diffusion of a macromolecule through the adsorption layer. It does not practically depend on the hydrodynamic regime, but it depends on the concentration of macromolecules in solution. Obviously, an increase of the polymer concentration in solution increases the time, when the macromolecule is situated near the adsorption layer, and the possibility of the macromolecule penetrating the solid surface. The rate of this stage is slower than first stage (Fig. 28).

In this stage, deformation of the adsorbed macromolecules as well as the entering macromolecules proceeds. The value of S_b is changed during the process. It is slower than the diffusion of macromolecules onto the free solid surface and it is determined by the concentration of macromolecules in the adsorption layer. In the general case, the concentration of monomer links in the globule of solved macromolecule depends on the polymerization

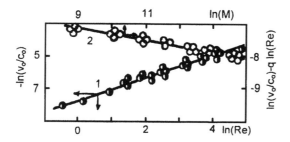

FIG. 27 Plot of the initial rate of poly(ethylene oxide) adsorption from water onto silica versus the Reynolds number (1) for a polymer with molecular mass 246×10^3 and versus the molecular mass of poly(ethylene oxide) (2) when plotted according to Eq. (141). (From Ref. 70.)

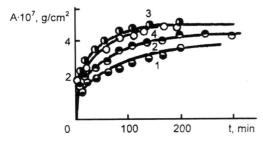

FIG. 28 Kinetic curves of adsorption of polystyrene on the chromium plate from cyclohexane solution at 35°C. The polystyrene molecular mass is 242×10^4 (1), 67×10^4 (2), 762×10^4 (3), and 1340×10^4 (4). (From Ref. 69.)

degree. Therefore, the dependence between macromolecule concentration in adsorption layer and polymerization degree is not linear:

$$v_{a2} = k_{a2}P_s(c_m - c_a)(S_bN_a) \tag{142}$$

where k_{a2} is the rate constant of adsorption of macromolecules in the second stage and c_a and c_m are the concentration and the maximum concentration of macromolecules in the adsorption layer, respectively.

The value of S_bN_a shows that the rate of this process increases when the fraction of the covered surface rises. The value of c_m is determined by the equilibrium of the adsorption forces and forces of interaction between polymer and solvent. It depends on the compression degree of macromolecules in the adsorption layer. It is possible that the maximum compression degree of macromolecules, δ, is constant for macromolecules with different molecular masses:

$$\delta = \frac{\rho_m}{\rho_h} \tag{143}$$

where ρ_m is the maximum polymer concentration in the adsorption layer and ρ_h is the polymer concentration in the free random coil.

The values of δ are 4.3, 3.4, 6.1, 4.1, 4.8, and 4.4 for polystyrene with molecular mass 67×10^4, 242×10^4, 459×10^4, 762×10^4, 970×10^4, and 1340×10^4, respectively, at the adsorption of polystyrene onto the chrome plate from the cyclohexane solution. On the basis of these data, the value of δ can be estimated to be about 5.

The equilibrium amount of the adsorbed polymer at the different polymer concentrations in solution may be controlled by two reasons. In the first case, the adsorbed macromolecules stretch on the surface and bind with adsorption sites. As a result, other macromolecules can not adsorb on this surface area. The polymer concentration in the boundary layer must be constant for the different amounts of adsorbed polymer as well as for the polymer concentration in solution. The second reason is that the equilibrium amount of the adsorbed polymer is determined by the equilibrium between the rate of adsorption and the rate of desorption of macromolecules. The polymer concentration in the boundary layer depends on the amount of adsorbed polymer in this case.

Calculation of the density of monomer links of the polymer in the adsorption layer showed [72] that it is lower than in the bulk polymer and decreases when the number of macromolecules on the unit surface area decreases. This is evidence concerning the influence of polymer desorption on polymer concentration in the absorption layer.

With the addition of a new adsorbent into the equilibrium system or a change of the solvent, the partial desorption can proceed [73]. The change of the hydrodynamic regime of solution above the adsorbed polymer leads to a change in the thickness of the adsorption layer and the amount of adsorbed polymer [10,65]. At the highest velocity gradient, the preadsorbed polystyrene can be completely desorbed within 2–3 h [74]. Displacement of the preadsorbed polymer by the polymers of the same nature [75,76] and the different nature [77,78] is well known. The linear dependence between the amount of desorbed polystyrene from silica [79] and time in semilogarithmic coordinates suggests that the rate of desorption is proportional to the number of macromolecules adsorbed on the unit surface area. Therefore, the rate of polymer desorption can be described by

$$v_d = -k_d P_a v \tag{144}$$

where k_d is the rate constant of polymer desorption and v is the probability of the tearing off of the adsorbed macromolecule from the surface. The probability of tearing off the adsorbed macromolecule from the surface is determined by the ratio of time interval between the breaking up and the formation of the bond of one monomer link of a macromolecule to the time interval needed for the breaking up of all bonds of one macromolecule with the surface. The value of v can be written by

$$v = \frac{1}{n^m} \tag{145}$$

where n is the number of bonds of one macromolecule with the surface and m is the parameter. Therefore, the desorption rate can be described by

$$v_d = -\frac{k_d P_a}{n^m} \tag{146}$$

The desorption rate of the first stage of polymer adsorption depends on the fraction of surface area covered by the polymer (see Appendix 2).

C. Dynamics of a Macromolecule in the Adsorption Layer

When the macromolecule is flexible, its deformation takes place at adsorption. One can observe the change in the number of chain links bound with the surface (Fig. 29) and the surface area blockaded by one macromolecule. The rate of change of the fraction of bound links is close to the rate of polymer adsorption at the slow stage and the rate of polymer desorption. Therefore, these three processes are bound to one another and their rates are limited by the rate of macromolecule deformation in the adsorption layer or its flexibility [72].

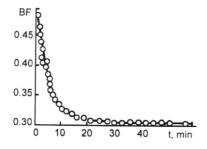

FIG. 29 The evolution of the bound fraction with time for the adsorption of poly(methyl methacrylate) onto silica from a solution in carbon tetrachloride at 0.01 mg/L. (From Ref. 61.)

The rate of formation of bonds between one macromolecule and the surface is proportional to the concentration of monomer links of the polymer over the surface and the number of free bare sites blockaded with this macromolecule but did not bind with one:

$$v_n = k_n c_1 n_n \tag{147}$$

where k_n is the rate constant of the formation of bonds between one macromolecule and the surface, c_1 is the concentration of monomer links of the polymer in the adsorption layer at a distance h_1 from the surface, h_1 is the distance from the surface where the bonds between the macromolecule and surface can be formed, and n_n is the number of free bare sites blockaded by one macromolecule but did not bind with one.

The rate of the breaking up of bonds between monomer links and the surface depends on the bond energy. Figure 30 shows that the chain of the fraction of bound monomer links of the polymer is described by the linear time dependence in semilogarithmic coordinates. Therefore, the rate of bond dissociation is proportional to the number of bonds between chain links and the surface:

$$v_0 = k_0 n \tag{148}$$

where k_0 is the rate constant of the dissociation of bonds and n is the number of bonds between one macromolecule and the surface.

The change of the number of bonds between one macromolecule and the surface can be obtained from Eqs. (147) and (148) as

$$\frac{dn}{dt} = k_n c_1 n_n - k_0 n \tag{149}$$

The polymer concentration in the boundary layer is considerably less than the polymer concentration in the bulk [72]. Therefore, the surface area blockaded by a macromolecule does not equal the surface area bound with a macromolecule. One can distinguish between the surface area bound with chain links, S_c, and the surface area not bound with a macromolecule but rather blockaded by it, S_n:

$$S_b = S_c + S_n \tag{150}$$

The change of the surface area bound with macromolecules is proportional to the change of the number of bonds between chain links and the surface:

$$\frac{dS_c}{dt} = k_n c_1 n_n S_u - k_0 n S_u \tag{151}$$

where S_u is the surface area occupied by one bare site.

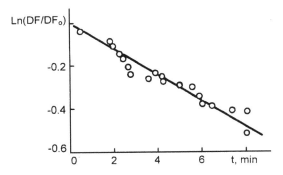

FIG. 30 Semilogarithmic anamorphose of the kinetic curve of the change of the bound fraction of poly(methyl methacrylate) onto silica from carbon tetrachloride.

The change of the surface area, blockaded by one macromolecule, depends on the fraction of surface coverage. If the surface near the macromolecule is not covered by polymer, the macromolecule can spread over the free surface. The rate of that process is proportional to the fraction of the surface area not covered by the polymer and hydrodynamic section of the macromolecule:

$$v_s = k_s(1 - S_b P_a)S_h \tag{152}$$

where k_s is the rate constant and S_h is the hydrodynamic section of the macromolecule.

At the same time, a compression of the adsorbed macromolecule can proceed under the action of macromolecules arriving from solution into the adsorption layer, when the particle surface is covered by polymer. One can suggest that the rate of compression of macromolecule is proportional to the fraction of the surface area covered by the polymer and the number of macromolecules that could still adsorb on the solid surface. That rate is inversely proportional to the concentration of monomer links of polymer in the adsorption layer, c_{a1}. An increase of the concentration of monomer links in the adsorption layer increases the force that prevents macromolecule compression:

$$v_p = \frac{k_p S_b P_a (c_m - c_a)}{c_{a1}} \tag{153}$$

Therefore, a change in the surface area covered by one macromolecule can be described by Eq. (154), obtained from Eqs. (152) and (153):

$$\frac{dS_b}{dt} = k_s(1 - S_b P_a)S_h - \frac{k_p S_b P_a (c_m - c_a)}{c_{a1}} \tag{154}$$

D. Adsorption of Homodisperse Polymers

The change of the number of adsorbed macromolecules at the unit surface area can be described by Eq. (156), taking into account Eqs. (138), (142), and (146):

$$S_n = n_n S_u \quad \text{and} \quad S_c = n S_u \tag{155}$$

$$\frac{dP_a}{dt} = k_{a1}P_s(1 - S_b P_a) + k_{a2}P_s(c_m - c_a)(s_b P_a) - \frac{k_d P_a S_u^m}{S_c^m} \tag{156}$$

The change of surface area bound by one macromolecule is described by

$$\frac{dS_c}{dt} = k_n c_1 S_n - k_0 S_c \tag{157}$$

The change of the surface area blockaded by one macromolecule is described by

$$\frac{dS_b}{dt} = k_s(1 - S_b P_a)S_h - \frac{k_p S_b P_a (c_m - c_a)}{c_{a1}} \tag{158}$$

When the adsorption occurs at the second stage, Eq. (156) can be transformed into

$$\frac{dP_a}{dt} = k_{a2}P_s(c_m - c_a)(S_b P_a) - \frac{k_d P_a S_u^m}{S_c^m} \tag{159}$$

1. Equilibrium State

In the equilibrium state in the plateau section of kinetic curves, $dS_c/dt = 0$, $dS_b/dt = 0$, and $dP_a/dt = 0$. Therefore, Eqs. (157)–(159) can be transformed into

$$k_n c_1 S_n - k_0 S_c = 0 \tag{160}$$

$$k_s(1 - S_b P_a)S_h - \frac{k_p S_b P_a(c_m - c_a)}{c_{a1}} = 0 \tag{161}$$

$$k_{a2} P_s(c_m - c_a)(S_b P_a) - \frac{k_d P_a S_u^m}{S_c^m} = 0 \tag{162}$$

The concentration of monomer links of polymer in the bulk can be obtained using

$$c_{b1} = \frac{\rho_{b1} N_A}{M_1} \tag{163}$$

where ρ_{b1} is the polymer concentration in the bulk ($\rho_{b1} = 1.05$ g/cm^3) and $M_1 = 104$ is the molecular mass of polystyrene monomer link.

The surface area bound with one monomer link of polymer are determined by

$$S_c' = \frac{1}{c_b h_u} \quad \text{and} \quad S_c = S_c' n \tag{164}$$

where h_u is the thickness of the boundary layer. The concentration of the monomer links of polymer in the boundary layer is obtained using

$$c_0 = \frac{\rho_0 N_A}{M_1} \tag{165}$$

where ρ_0 is the polymer concentration in the adsorption layer in the boundary layer.

The surface area blockaded by the macromolecule that corresponds to one bond between the macromolecule and surface is obtained by

$$S_b' = \frac{1}{c_0 h_u} \quad \text{and} \quad S_b = S_b' n \tag{166}$$

The surface area blockaded by the macromolecule does not bind with it, which corresponds to one bond between macromolecule and surface, can be written as

$$S_n' = \frac{1}{c_0 h_u} - \frac{1}{c_b h_u} \quad \text{and} \quad S_n = S_n' n \tag{167}$$

The concentration of monomer links of polymer on the distance h_1 from the surface, c_1, can be found using

$$c_1 = \frac{\rho_1 N_A}{M_1} \tag{168}$$

where $\rho_1 = \rho_0 - b/[(l+1)(l+2)]h_1^{(l+1)}$ (see Section III.C).

From expressions (160) and (163)–(168), one can obtain

$$Y_H = h_1^{(l+1)} + \frac{k_0}{k_n} X_H \tag{169}$$

where $Y_H = \rho_0(l+1)(l+2)/b$, $X_H = \dfrac{M_1(l+1)(l+2)}{N_A b(c_b/c_0 - 1)}$.

As can be seen from Fig. 31, the experimental data for the adsorption of polystyrene onto a smooth chromium surface from cyclohexane solution lie on a straight line when plotted according to Eq. (169). It allows one to find the values $h_1^{(l+1)} = (8 \pm 3) \times 10^{-4}$ cm$^{0.5}$ and $k_0/k_n = (3.9 \pm 1.3) \times 10^{20}$ cm^{-30} and to calculate the value $h_1 = 1.4 \times 10^{-6}$ cm.

The concentration of macromolecules in the adsorption layer equals

$$c_a = \frac{\rho_a N_A}{M} \tag{170}$$

where ρ_a is the polymer concentration in the adsorption layer. The value of c_m was calculated using

$$c_m = \frac{\rho_h N_A \delta}{M} \tag{171}$$

where $\delta = 5$.

Expression (162) can be transformed to

$$S_c = \left(\frac{k_d S_u^m}{k_{a2}} \frac{P_a}{P_a(c_m - c_a)} \right)^{1/m} \tag{172}$$

From Eqs. (160) and (172) and taking into account that in the equilibrium state

$$S_b N_a = 1 \quad \text{or} \quad (S_c + S_n) P_a = 1 \tag{173}$$

one can obtain the expression

$$\ln\left(\frac{c_1}{c_1 + k_0/k_n)P_a} \right) = \ln\left(\frac{k_d S_u^m}{k_{a2}} \right)^{1/m} + \frac{1}{m} \ln\left(\frac{P_a}{P_a(c_m - c_a)} \right) \tag{174}$$

Computation of Eq. (174) with respect to experimental data by the iteration method showed that

$$c_1 \ll \frac{k_0}{k_n} \tag{175}$$

It corresponds to the results obtained from Eq. (169) for the value of k_0/k_n.

From Eq. (160) and expression (175) follows

$$S_n \gg S_c \tag{176}$$

The surface area blockaded by one macromolecule but not bound with it is considerably larger than the surface area bound with macromolecules. That conclusion is supported with the fact that the polymer concentration in the boundary layer, ρ_0, is considerably less than the polystyrene concentration in the bulk [72].

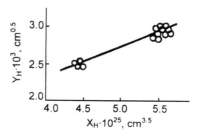

FIG. 31 Plot of the concentration of monomer links of polystyrene in adsorption layer at the distance h_1 from the surface versus X_H, according to Eq. (169) for polystyrene with molecular mass 67×10^4, 242×10^4, 762×10^4, and 1340×10^4. (From Ref. 72.)

Taking into account expression (175), Eq. (174) can be transformed to

$$\ln\left(\frac{c_1}{P_a}\right) = \ln\left(\frac{k_0^m k_d S_u^m}{k_n^m k_{a2}}\right)^{1/m} + \frac{1}{m}X_E \tag{177}$$

where $X_E = \ln[P_a/P_s(c_m - c_a)]$.

As can be seen in Fig. 32, the experimental data for the adsorption of polystyrene onto the chromium plate from the cyclohexane solution lie on a straight line when plotted according to Eq. (177). It allows one to calculate the values $1/m = 0.25 \pm 0.02$ and $\ln[k_0^m k_d S_u^m / (k_n^m k_{a2})]^{1/m} = 32.67 \pm 1.3$.

2. Adsorption Kinetics

Some ratios of rate constants in Eqs. (157)–(159) can be calculated based on the experimental data of adsorption kinetics [72]. Equation (159) can be transformed to

$$S_c = (k_d S_u^m)^{1/m} Z_1 \tag{178}$$

where

$$Z_1 = \left(\frac{P_a}{k_{a2}P_s(c_m - c_a) - dP_a/dt}\right)^{1/m}$$

Differentiation of Eq. (178) leads to

$$\frac{dS_c}{dt} = (k_d S_u^m)^{1/m} Y_4 \tag{179}$$

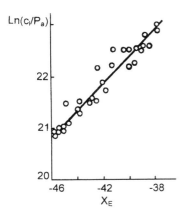

FIG. 32 Relationship between the equilibrium adsorbed amount of polystyrene on the chrome plate and polystyrene concentration in cyclohexane solution when plotted according to Eq. (177). Experimental data for polystyrene with molecular mass 67×10^4, 242×10^4, 762×10^4, and 1340×10^4. (From Ref. 72.)

where

$$Y_4 = \frac{P_a^{1/m}}{m} Z_1(Y_5 + Y_6)$$

$$Y_5 = \frac{1}{P_a} \frac{dP_a}{dt}$$

$$Y_6 = \frac{k_{a2}P_s dc_a/dt + dP_a^2/dt^2}{k_{a2}P_s(c_m - c_a) - dP_a/dt}$$

From Eqs. (157) and (173) and expression (175),

$$\frac{1}{S_c} \frac{dS}{dt} = -k_0 + k_n \frac{c_1}{S_c} \tag{180}$$

can be obtained. From Eqs. (178)–(180), one can write

$$\frac{Y_4}{Z_1} = -k_0 + \frac{k_n}{(k_d S_u^m)^{1/m}} \frac{c_1}{Z_1} \tag{181}$$

As can be seen from Eqs. (178)–(181), they include the values dP_a/dt, dP_a^2/dt^2, and dc_a/dt which can be calculated from the experimental data [65]. The method of obtaining accurate values of these values is described in Appendix 3.

Computation of the constants in Eq. (181) from the experimental data combined with the values obtained from Eqs. (15a)–(18a) (see Appendix 3) by iteration showed that

$$k_{a2} \gg \frac{1}{P_s(c_m - c_a)} \frac{dP_a}{dt} \quad \text{and} \quad k_{a2} \gg \frac{1}{P_s \, dc_a/dt} \frac{dP_a^2}{dt^2} \tag{182}$$

Transformation of Eq. (181) by taking into account expression (182) yields:

$$Y_k = -k_0 + \left(\frac{k_n^m k_{a2}}{k_d S_u^m}\right)^{1/m} X_k \tag{183}$$

where

$$Y_k = \frac{1}{m} \left(\frac{1}{P_a} \frac{dP_a}{dt} + \frac{1}{c_m - c_a} \frac{dc_a}{dt}\right)$$

$$X_k = \frac{c_1[P_s(c_m - c_a)]^{1/m}}{P_a^{(1/m+1)}}$$

As Fig. 33 demonstrates, the experimental data of the kinetics of the adsorption of polystyrene with different molecular masses onto a chromium plate from cyclohexane solution fall on a straight line when plotted according to Eq. (183). It allows one to find the values $[k_n^m k_{a2}/(k_d S_u^m)]^{1/m} = (1.8 \pm 0.3) \times 10^{-17}$ cm^2/min and $k_0 = (2.7 \pm 0.8) \times 10^{-3}$ min^{-1}. One can note that the value $k_0^m k_d S_u^m/(k_n^m k_{a2}) = 3.9 \times 10^{55}$ cm^{-8} calculated from the data of Eq. (177) is close to this value $[k_0^m k_d S_u^m/(k_n^m k_{a2}) = 3.4 \times 10^{55}$ cm$^{-8}]$ calculated from the data of Eq. (183).

E. Replacement Adsorption

Investigation of the replacement adsorption shows [63] that the rate of this process is close to the rate of the second stage of polymer adsorption, but depends on the age of the pre-adsorbed layer (Fig. 34). The rate of the replacement adsorption decreases when the age of the preadsorbed layer increases.

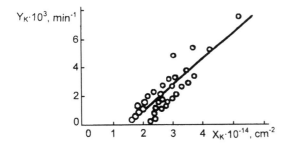

FIG. 33 Plot of the amount of adsorbed polystyrene versus time when plotted according to Eq. (183). The polymer concentration in solution is 0.3×10^{-2} g/cm^3. (From Ref. 72.)

The number of macromolecules per unit of the surface area, P_p, and the concentration of macromolecules in the adsorption layer, c_0, are constant while the equilibrium state reaches in plateau section of a kinetic curve. At the same time, the process of the macromolecule reconfirmation and the formation of bonds between the macromolecule and the surface proceeds [72]. Then, Eq. (184) can be found from Eq. (151), taking into account Eq. (173):

$$\frac{dS_{c1}}{dt} = \frac{k_n c_{p1}}{P_p} - (k_n c_{p1} + k_0)S_c \tag{184}$$

where P_p and c_{p1} are the number of adsorbed macromolecules per unit of the surface area and the concentration of the monomer links of the polymer in the adsorption layer in the plateau section of a kinetic curve, respectively.

Integration of Eq. (184) for the initial condition $S_c(t_r = 0) = S_0$ leads to Eq. (185). Here, t_r is the time from the moment when equilibrium adsorption is reached and S_0 is the surface area bound by one adsorbed macromolecule at the moment when the equilibrium adsorption is reached:

$$S_c = K_1 - K_2 \exp[-(k_n c_{p1} + k_0)t_r] \tag{185}$$

where

$$K_1 = \frac{k_n c_{p1}}{(k_n c_{p1} + k_0)P_p}$$

$$K_2 = \frac{k_n c_{p1}/P_p - (k_n c_{p1} + k_0)S_0}{k_n c_{p1} + k_0}$$

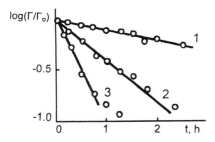

FIG. 34 The effect on displacement kinetics of changing the waiting time before replacement of protio by deuterio polystyrene solution, 12 h (1), 2 h (2), 30 min aging (3); deuterio polystyrene solution = 1 mg/mL, polymerization degree of protio polystyrene = 5750, deuterio polystyrene = 5500. (From Ref. 63.)

If the replacement rate is determined by the rate of desorption of the adsorbed polymer, the change of the number of macromolecules adsorbed previously can be described by Eq. (186) obtained from Eqs. (155) and (159):

$$\frac{dP_h}{dt} = -\frac{k_d S_u^m P_h}{S_c^m} \tag{186}$$

where P_h is the number of hydrogenic polystyrene macromolecules adsorbed previously on a unit of the surface area.

From Eqs. (185) and (186), one can obtain

$$\frac{dP_h}{dt} = -k_d S_u^m \left(\frac{1}{K_1 - K_2 \exp[-(k_n c_{p1} + k_0)t_r]}\right)^m P_h \tag{187}$$

It can be supposed that the number of adsorbed macromolecules, P_p, and the concentration of monomer links, c_{p1}, in the adsorption layer do not change during the investigations of replacement adsorption of hydrogenic polystyrene by deuteric polystyrene of similar molecular mass. Then, the values of P_p and c_{p1} are constant during replacement adsorption. If the value of t_r is larger than time during the kinetic investigation, t_s, then t_r can be taken as constant. By integration of Eq. (187) for the initial condition $P_h(t_s = 0) = P_p$, one can obtain

$$\ln\left(\frac{P_h}{P_p}\right) = -k_d S_u^m \left(\frac{1}{K_1 - K_2 \exp[-(k_n c_{p1} + k_0)t_r]}\right)^m_{t_s} \tag{188}$$

As seen from Eq. (188), the replacement rate must be described by the first-order equation. The kinetic curves of the replacement adsorption [63] suggest the adequacy of Eq. (188).

The value of the replacement time constant, t_{off}, was found experimentally in Ref. 63. That value is in inverse proportion to the rate constant of the replacement adsorption. Then, we can write

$$t_{off} = (k_d S_u^m)^{-1} \{K_1 - K_2 \exp[-(k_n c_{p1} + k_0)t_r]\}^m \tag{189}$$

Transformation of Eq. (189) leads to

$$Y_r = \ln\left(\frac{K_2}{k_d S_u^m}\right) - (k_n c_{p1} + k_0)t_r \tag{190}$$

where $Y_r = \ln[K_1/(k_d S_u^m) - t_{off}^{1/m}]$.

As seen from Fig. 35, experimental data on the kinetics of the replacement adsorption of protic polystyrene by deuteric polystyrene fall on straight lines computed by iteration when plotted according to Eq. (190). The values of the obtained ratios are shown in Table 3.

F. Polymer Adsorption Accompanied by Flocculation

Polymer adsorption from solution onto solid particles is often accompanied by particle aggregation or flocculation. In this case, the particle surface and amount of the adsorbed polymer decrease. Evidently, polymer adsorption proceeds by the following mechanism [80]: Macromolecules pass from solution to the uncovered surface and block some surface area of particle. Adsorbed macromolecules can stretch over the free surface if they are flexible. Simultaneously, flocculation proceeds by biparticle collision when a surface of one particle covered by polymer interacts with the uncovered surface of another particle. When the particle aggregate is formed, the part of the particle surface between two or more initial particles in aggregate cannot be attacked by macromolecules from solution. It is the inside surface of the

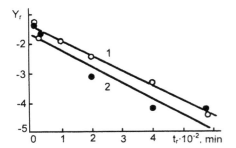

FIG. 35 Plot of t_{off} versus t_r at the replacement adsorption of protio polystyrene by deuterated polystyrene onto oxidized silicon from cyclohexane solution when plotted according to Eq. (189). Molecular mass of protio polystyrene is 598×10^3 and for deuterated polystyrene is 575×10^3. The polymer concentration in solution is 0.1×10^{-3} (1), and 1×10^{-3} g/cm^3 (2). (From Ref. 72.)

aggregate. The outside surface of the aggregate is the surface that can adsorb macromolecules from solution. If the outside surface of the aggregate is covered, free macromolecules diffuse through the adsorption layer to the surface. Deformation of macromolecules in the adsorbed layer, as well as of free macromolecules, occurs at this stage. Desorption of macromolecules and breaking up of aggregates is possible during the process.

According to this mechanism and equations for polymer adsorption, described earlier, we can write the system of equations for polymer adsorption on the aggregates, containing different number of initial particles and outside surface area, as follows:

$$\frac{dP_1}{dt} = k_{a1}P_s(S_1 - S_{a1})N_1 + k_{a2}P_s(n_{m1} - n_{a1})N_1 - k_d n_{a1}N_1 \tag{191}$$

$$\frac{dP_2}{dt} = k_{a1}P_s(S_2 - S_{a2})N_2 + k_{a2}P_s(n_{m2} - n_{a2})N_2 - k_d n_{a2}N_2 \tag{192}$$

$$\frac{dP_3}{dt} = k_{a1}P_s(S_3 - S_{a3})N_3 + k_{a2}P_s(n_{m3} - n_{a3})N_3 - k_d n_{a3}N_3 \tag{193}$$

$$\vdots$$

$$\frac{dP_i}{dt} = k_{a1}P_s(S_i - S_{ai})N_i + k_{a2}P_s(n_{mi} - n_{ai})N_i - k_d n_{ai}N_i \tag{194}$$

TABLE 3 Correlation Coefficients and Ratios at the Replacement Adsorption of Protio Polystyrene by Deuterated Polystyrene

Value	Polymer concentration in solution (mg/cm^3)	
	0.1	1
$\dfrac{K_1}{(k_d S_u^m)}$ (min)	0.514	0.532
$\ln\left(\dfrac{K_2}{k_d S_u^m}\right)$	-1.4 ± 0.2	-1.6 ± 0.7
$k_n c_{pl} + k_0$ (min^{-1})	51 ± 9	56 ± 18

Source: Ref. 72.

where i is the number of primary particles in the aggregate, P_i is the number of macromolecules, adsorbed on the outside and inside surfaces of the aggregate containing i primary particles in the unit volume of dispersion, k_{a1} is the rate constant of polymer adsorption on the particle surface free from polymer, P_s is the concentration of macromolecules in solution, S_i is the outside surface area of the aggregate containing i primary particles, S_{ai} is the outside surface area of the aggregate covered by the polymer, N_i is the number of aggregates containing i primary particles in the unit volume of dispersion, k_{a2} is the rate constant of polymer adsorption on the covered surface of particles, $n_{mi} = in_{m1}$ is the maximum number of macromolecules that can adsorb on one aggregate containing i primary particles, n_{ai} is the number of macromolecules adsorbed on the outside surface of aggregate containing i primary particles, and k_d is the rate constant of polymer desorption.

The outside surface area of the aggregate containing i primary particles has been calculated according to

$$S_i = iS - 2(i-1)S_s \tag{195}$$

where S_s is the surface area of the segment between two primary particles in the aggregate. We suggested that the polymer adsorption does not proceed in this segment. Desorption of macromolecules from this segment is not possible either, because one macromolecule has bonds with two particles. The number of macromolecules in the segment between two primary particles in the aggregate is

$$N_s = \frac{S_s}{S_b} \tag{196}$$

where S_b is the surface area of the particle covered by one macromolecule.

The surface area covered by one macromolecule and the surface area of the segment between two primary particles in the aggregate are constant when the macromolecule is inflexible. At the same time, when macromolecules are flexible, they can change the surface area covered by one macromolecule and the thickness of the adsorption layer. An increase in the number of macromolecules in the adsorption layer leads to an increase in the thickness of the adsorption layer and the distance between two particles in the aggregate. Therefore, the surface area covered by one macromolecule, S_b, and the surface area of the segment, S_s, can decrease. In this case, evidently, the number of macromolecules on the surface of segment between two primary particles in the aggregate can be practically constant.

The outside surface area of aggregate from i primary particles, covered by the polymer, has been calculated using

$$S_{ai} = S_b[in_{a1} - (i-1)n_s] \tag{197}$$

and

$$n_{ai} = in - (i-1)n_s \tag{198}$$

Equation (199) has been obtained by summing up Eqs. (191)–(194) and taking Eqs. (195)–(198) into account:

$$\sum_{i=1}^{m} \frac{dP_i}{dt} = k_{a1}P_s \sum_{i=1}^{m} \{iS_1 - 2(i-1)S_s - S_b[in_{a1} - (i-1)n_s]\}N_i$$
$$+ k_{a2}P_s \sum_{i=1}^{m} \{in_{m1} - [in_{a1} - (i-1)n_s]\}N_i - k_d \sum_{i=1}^{m} [in_{a1} - (i-1)n_s]N_i \tag{199}$$

Equation (202) has been obtained from Eq. (199) when

$$N = \sum_{i=1}^{m} N_i \tag{200}$$

$$N_0 = \sum_{i=1}^{m} i N_i \tag{201}$$

where N_0 is the initial number of particles in the unit volume of dispersion:

$$\frac{dP}{dt} = k_{a1} P_s N_0 \left\{ S_1 - 2 \left(1 - \frac{N}{N_0} \right) S_s - S_b \left[n_{a1} - \left(1 - \frac{N}{N_0} \right) n_s \right] \right\}$$
$$+ k_{a2} P_s N_0 \left\{ n_{m1} - \left[n_{a1} - \left(1 - \frac{N}{N_0} \right) n_s \right] \right\} - k_d N_0 \left[n_{a1} - \left(1 - \frac{N}{N_0} \right) n_s \right] \tag{202}$$

Equation (202) describes the kinetics of the adsorption of macromolecules on the dispersed particles accompanying by particle flocculation.

Flocculation proceeds by biparticle collision when the surface, covered by the polymer, of one particle interacts with the uncovered surface of another particle. The change in the number of particles in a unit volume of dispersion is proportional to the surface area of the particle covered by the polymer, S_{ai}, the number of these particles, N_i, and the surface area of particles not covered by the polymer $(S_i - S_{ai})$:

$$\frac{dN_1}{dt} = -k_f \left(S_{a1} N_1 \sum_{i=1}^{m} [(S_i - S_{ai}) N_i] + (S_1 - S_{a1}) N_1 \sum_{i=1}^{m} (S_{ai} N_i) \right) \tag{203}$$

$$\frac{dN_2}{dt} = -k_f \left(S_{a2} N_2 \sum_{i=1}^{m} [(S_i - S_{ai}) N_i] + (S_2 - S_{a2}) N_2 \sum_{i=1}^{m} (S_{ai} N_i) \right) + 2 k_b N_2 \tag{204}$$

$$\frac{dN_3}{dt} = -k_f \left(S_{a3} N_3 \sum_{i=1}^{m} [(S_i - S_{ai}) N_i] + (S_3 - S_{a3}) N_3 \sum_{i=1}^{m} (S_{ai} N_i) \right) + 2 k_b N_3 \tag{205}$$

$$\vdots$$

$$\frac{dN_i}{dt} = -k_f \left(S_{ai} N_i \sum_{i=1}^{m} [(S_i - S_{ai}) N_i] + (S_i - S_{ai}) N_i \sum_{i=1}^{m} (S_{ai} N_i) \right) + 2 k_b N_i \tag{206}$$

where N_i is the number of aggregates containing i primary particles in the unit volume of dispersion, k_f is the rate constant of flocculation, and k_b is the rate constant of the breaking up of aggregates formed two new particles.

Equation (207) has been obtained by summing Eqs. (203)–(206) and taking Eqs. (195)–(198), (200), and (201) into account:

$$\frac{dN}{dt} = \sum_{i=1}^{m} \frac{dN_i}{dt} = -k_f S_b N_0^2 \left\{ \left[n_{a1} - \left(1 - \frac{N}{N_0} \right) n_s \right] \left[S_1 - 2 \left(1 - \frac{N}{N_0} \right) S_s \right] \right.$$
$$\left. - S_b \left[n_{a1} - \left(\frac{1 - N}{N_0} \right) n_s \right] \right\} + 2 k_b (N - N_1) \tag{207}$$

Equation (207) describes a change of the number of particles in a unit volume of dispersion that depends on the polymer amount added to the dispersion.

A change in the particle surface area blockaded by one particle can be described by Eq. (208), which is analogous to Eq. (154) (see Section IV.C):

$$\frac{dS_b}{dt} = k_s \left(1 - \frac{S_b n_{a1}}{S_1} \right) S_h - \frac{k_p S_b n_{a1} / S_1 (n_{m1} - n_{a1})}{n_M} \tag{208}$$

In the equilibrium state at $t \to \infty$, $dS_b/dt = 0$, Eq. (208) allows one to calculate the ratio of rate constants of $k_p/k_s S_h$, based on the experimental data of polymer adsorption at that point and the value $S_b n_{a1}/S_1 = 1/2$:

$$\frac{k_p}{k_s S_h} = (1 - \frac{S_b n_{a1}}{S_1} \left(\frac{(n_m - n_{a1})}{n_M} \frac{S_b n_{a1}}{S_1} \right)^{-1} \tag{209}$$

Values of n_{a1}, n_{m1}, and n_M have been calculated according to Eqs. (210)–(212):

$$n_{a1} = \frac{c_a N_A}{M N_0} \tag{210}$$

$$n_{m1} = \frac{c_m N_A}{M N_0} \tag{211}$$

$$n_M = \frac{c_a N_A}{M_1 N_0} \tag{212}$$

where c_a is the mass of the adsorbed polymer in the unit volume of dispersion, c_m is the maximum mass of the polymer that can be absorbed on the particles in the unit volume of dispersion, M_1 is the molecular mass of the link of polymer, and N_0 is the initial number of dispersed particles in the unit volume of dispersion.

The values of $k_p/k_s S_h$, calculated according to Eq. (209), for polyethylene oxide, adsorbed on the particles of AgI, are listed in Table 4.

The particle surface area covered by one macromolecule in the equilibrium state can be calculated according to Eq. (213), obtained from Eq. (209):

$$S_b = S_1 \left(n_{a1} + \frac{k_p}{k_s S_h} \frac{n_{a1}(n_{m1} - n_{a1})}{n_M} \right)^{-1} \tag{213}$$

Equation (214) was obtained from Eq. (207) taking into account Eqs. (196) and (213) and at the condition $t \to \infty$, $dN/dt = 0$:

$$Y_1 = \frac{k_b}{k_f S_1^2} X_1 \tag{214}$$

TABLE 4 Ratios of Rate Constants of Adsorption of Polyethylene Oxide on AgI Dispersion Accompanying Flocculation

	Molecular mass of polyethylene oxide	
Ratios of rate constants	23×10^4	130×10^4
$k_p/k_s S_h$	3.19	10.2
n_s	2.43×10^7	6.57×10^6
$k_b/k_f S_1^2$	6.48×10^6	5.52×10^5
$k_{a1} S_1/k_d$	7.0×10^{-5}	4.2×10^{-4}
k_{a2}/k_d	9.2×10^{-19}	2.4×10^{-17}

Source: Ref. 80.

where

$$Y_1 = \left[n_a - \left(1 - \frac{N}{N_0} \right) n_s \right] \left[F - \left(1 - \frac{N}{N_0} \right) n_s \right]$$

$$X_1 = (n_{a1} - F)^2 \frac{N - N_1}{N_0}$$

$$F = \frac{k_p}{k_s S_h} \frac{n_{a1}(n_{m1} - n_{a1}}{n_M}$$

The number of initial particles in the system, N_1, was calculated according to Eq. (215), obtained from Poisson's distribution equation. It can be used for discrete processes such as flocculation [81]. The Poisson's distribution allows one to calculate the distribution of aggregates with the number of initial particles in them during flocculation:

$$N_1 = N \exp\left(1 - \frac{N_0}{N} \right) \tag{215}$$

If the average number of initial particles in the aggregate more than two, the valoue of N_1 is small. Therefore, at first approximation, $N_1 = 0$.

Equation (214) has been calculated by the iteration method when n_s changed from $10^4 10^9$. Figure 36 shows that experimental data on the flocculation of AgI dispersion by polyethylene oxide lie on straight lines when plotted according to Eq. (214). Values of n_s and $k_b/k_f S_1^2$ are listed in Table 4.

Equation (202) can be transformed into Eq. (216) at the condition $t \rightarrow \infty$, $dP/dt = 0$:

$$Y_2 - \frac{k_{a1} S_1}{k_d} + \frac{k_{a2}}{k_d} X_2 \tag{216}$$

where

$$Y_2 = \frac{F_1}{P_s F_2}$$

$$X_2 = \frac{n_{m1} - F_1}{F_2}$$

$$F_1 = n_{a1} 0 \left(\frac{N}{N_0} \right) n_s$$

$$F_2 = 1 - \frac{2(1 - N/N_0)n_s S_b}{S_1} - \frac{[n_{a1} - (1 - N/N_0)n_s]S_b}{S_1}$$

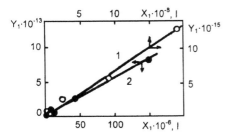

FIG. 36 Relationship between the number of particles in the dispersion and the amount of adsorbed polymers when plotted according to Eq. (216) for the different initial polymer concentration in the system. Molecular mass of poly(ethylene oxide) = 23×10^4 0 (1) and 130×10^4 (2). (From Ref. 80.)

The value of S_b has been calculated according to Eq. (209). The number of particles in dispersion N, has been calculated according to Eq. (214).

Figure 37 shows that experimental data on adsorption of polyethylene oxide on AgI particles lie on straight lines when plotted according to Eq. (216). The ratios of the rate constants have been found from the tangent of the slope angle of straight lines and the intercept on the ordinate axis (Table 4).

Some rate constants can be calculated from the kinetic data in the polymer adsorption and the particle aggregation in the manner described in Section IV.D2. Unfortunately, these experimental data are absent for the initial stage of the process. Therefore, the value of the rate constants can be obtained very approximately. Another method for obtaining these constants is described in Ref. 82 when the change of the surface area, blockaded by one particle, and the second stage of adsorption are not taken into account.

Substitution of

$$Y_c(t) = \frac{c_a(t)}{\delta_a} \tag{217}$$

$$Y_p(t) = \frac{N(t)}{\delta_n} \tag{218}$$

and taking into account that $c = c_0 - c_a$ allows one to transform Eqs. (217) and (218):

$$\frac{dY_c}{dt} = k_{a2}\left((c - Y_c)(1 - F_4)Y_p - \frac{k_d}{k_{a2}Y_c}\right) \tag{219}$$

$$\frac{dY_p}{dt} = -k_f\left(F_4(1 - F_4)Y_p^2 + \frac{2k_b}{k_f}(Y_p - N_1)\right) \tag{220}$$

where

$$F_4 = \frac{N_A S_u Y_c}{MS_s[S_0/S_s - N_0 + Y_p]}$$

$$N_1 = Y_p \exp\left(1 - \frac{N_0}{Y_p}\right)$$

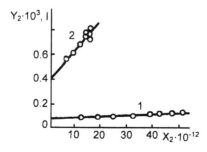

FIG. 37 Relationship between the mass of adsorbed polymer in the unit volume of dispersion and the number of particles in the dispersion when plotted according to Eq. (218) for the different initial polymer concentrations in the system. Molecular mass of poly(ethylene oxide) = 23×10^4 (1) and 130×10^4 (2). (From Ref. 80.)

The system of differential equations can be calculated with respect to Y_c and Y_p, taking into account the rate constant ratios, found earlier, and any rate constants k_a and k_f. The values of k_a and K_f will draw to the exact value under the conditions

$$\sum [c_a(t) - Y_c(t)]^2 \to 0 \tag{221}$$

$$\sum [N(t) - Y_p(t)]^2 \to 0 \tag{222}$$

Calculation of the system of differential equations when k_a and k_f are changed from 10^{-19} to 10^{-14} and the minimization of the obtained values according to Eqs. (221) and (222) allows one to find the values of $k_a = 0.96 \times 10^{-17}$ L/s and $k_f = 1.3 \times 10^{-17}$ L/s.

As Figs. 38 and 39 show, the dependences of Y_c on c_a and of Y_p on N lie on straight lines drawn through the origin. The values of $\delta_a = 0.9 \pm 0.1$ and $\delta_n = 1.2 \pm 0.1$ correspondingly. The values of δ_a and δ_n are close to 1. Therefore, the calculated values of k_a and k_f allow one to describe the experimental data.

G. Adsorption Accompanied by Reaction

Polymers, containing carboxyl, amine, and other functional groups, are used for stabilization of metal and metal oxide dispersions. In this case, the reaction between the solid surface and the polymer can proceed forming the salt or the complex compound solved in the system. For example, the copper(II) oxide dissolves in the poly(ethylenimine)–water solution at the adsorption of polymer on the copper(II) oxide dispersion (Fig. 40).

An investigation of the adsorption kinetics showed that the equilibrium concentration of adsorbed poly(ethylenimine) or its complex with copper(II) reached the plateau section in 2–5 min. That time decreased when the initial polymer concentration in solution decreased. Copper(II) hydroxide, obtained by the reaction of copper(II) sulfate with sodium hydroxide, dissolved 1 min after the addition of poly(ethylenimine) solution. Therefore, the adsorption of poly(ethylenimine) on particles and the reaction of copper(II) oxide with poly(ethylenimine) occurs very quickly, and they are faster than the desorption rate.

The adsorption isotherm for the adsorption of poly(ethylenimine) on the copper(II) oxide powder is the same as for the adsorption of the saturated complex of copper(II) ions with poly(ethylenimine). Therefore, the equilibrium concentration of adsorbed polymer is reached very quickly and the polymer desorption is the stage limiting an appearance of the complex

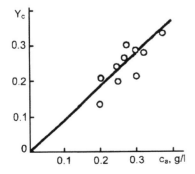

FIG. 38 Dependence of $Y(t)_c$ on the concentration of adsorbed polymer for adsorption of diethylene-tri-aminomethylated polyacrylamide on the particles of kaolin dispersion at the kaolin concentration 1.92–3.84 g/L and the initial polymer concentration 0.112–1.57 g/L.

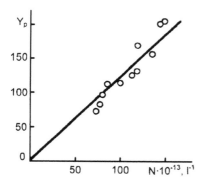

FIG. 39 Dependence of $Y(t)_p$ on the number of kaolin particles in the unit volume of dispersion for adsorption of diethylene-tri-aminomethylated polyacrylamide on the particles of kaolin dispersion at the kaolin concentration 1.92–3.84 g/L and the initial polymer concentration 0.112–1.57 g/L.

copper(II) ions in solution. The concentration of adsorbed polymer decreases when the concentration of the copper(II) oxide powder decreases.

The equilibrium concentration of adsorbed macromolecules is described by

$$Y_a = k_n \left(\frac{k_d}{k_a k_c k_n^m} \right)^{1/(1+m)} + k_c \left(\frac{k_d}{k_a k_c k_n^m} \right)^{1/(1+m)} P_a \tag{223}$$

where $Y_a = (P_s N / P_a^m)^{1/(1+m)}$.

As can be seen from Fig. 41, the experimental data for the adsorption of poly(ethyleni-mine) on the copper(II) oxide lie on a straight line when plotted according to Eq. (223). The values of $k_n (k_d / k_a k_c k_n^m)^{1/(1+m)} = (2.0 \pm 0.4) \times 10^{55}$ cm^{-2} and $k_c (k_d / k_a k_c k_n^m)^{1/(1+m)} = (4.1 \pm 0.7) \times 10^{-80}$.

The rate of a change of the complex of copper(II) ions with poly(ethylenimine) in solution is proportional to the rate of polymer desorption. At the first approximation, we considered that

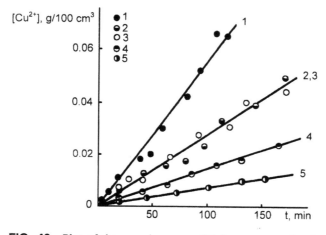

FIG. 40 Plot of the complex copper(II) ion concentration in solution versus time at the initial CuO concentration 80 g/L and the initial poly(ethylenimine) concentrations 5 (1), 2.5 (2,3), 1.25 (4), and 0.625 g/L (5); pH = 9.6 (1, 2, 4) and pH = 8.2 (3).

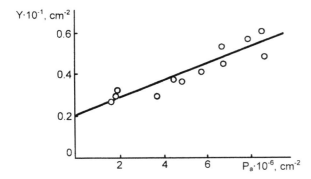

FIG. 41 Relationship between the number of adsorbed macromolecules on the unit surface area and the concentration of macromolecules and particles in the mixture when plotted according to Eq. (223).

one desorbed macromolecule contained the constant number of complex copper(II) ions at the first stage of desorption:

$$v = \frac{d[Cu^{2+}]}{dt} = \gamma v_d = \frac{\gamma k_d P_a}{S_b^m} \tag{224}$$

From Eqs. (223) and (224), one can obtain the dependence between the rate of a change of copper(II) ion concentration in solution and the number of adsorbed macromolecules in the unit volume of dispersion:

$$\ln\left(\frac{v}{P_a}\right) = \ln\left[\frac{\gamma k_d}{k_n^m}\left(\frac{k_a k_c k_n^m}{k_d}\right)^{m/(m+1)}\right] + \frac{m}{m+1}\ln(P_s P_a N) \tag{225}$$

Figure 42 shows that the experimental data on the dependence of the desorption rate on the concentration of macromolecules and particles lie on a straight line when plotted according to Eq. (225). The value of $m = 1.0 \pm 0.1$ and $\ln[(\gamma k_d/k_n^m)(k_a k_c k_n^m/k_d)^{m/(m+1)}] = 36 \pm 8$.

Investigation of the desorption of copper(II) salts of polyacrylic acid showed that at the high saturation degree of carboxyl groups, the salt does not dissolve in water. Therefore, the new phase of unsolved polymer particles appears at the latest stage of adsorption.

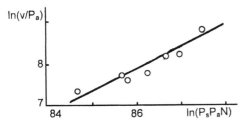

FIG. 42 Dependence of the rate of a change of the concentration of the complex copper(II) ions on the concentration of macromolecules and particles in the mixture when plotted according to Eq. (225).

V. FORCES IN ADSORPTION LAYER

The polymer concentration and the macromolecule conformation in the adsorption layer depend on the forces between the macromolecule and the solid surface. Therefore, an investigation of this phenomenon is very important for describing the polymer adsorption. Some information has been obtained by studying the forces between two polymer adsorbed layers as a function of the distance between them.

A. Unchanged Polymers

Investigation of forces between two plates containing polymer adsorbed layers showed that two ways of force change from the distance between them is possible in those experiments. For the first, only repulsion forces between two polymer adsorption layers take place when the degree of surface coverage is high (Fig. 16) [83–85]. For the second, when the plate surface is free from polymer sections, the attractive forces between the solid surfaces are observed at a distance that corresponds to the magnitude of the gyration radius of the macromolecule (Fig. 43) [86–89]. That force increased when the distance between two surfaces, partially covered by a polymer, decreased. Then, with a further decrease of this distance, the force turns repulsive and it increases very fast.

The forces between two polymer adsorption layers depend on the concentration of monomer links of polymer in the space between two plates. As was shown in Section III.C, the distribution of chain links in adsorption layer can be described by

$$n = N_{\mathrm{tr}} - \frac{bN}{l+1} z^{(l+1)} \tag{226}$$

where

$$N_{\mathrm{tr}} = \exp(A_n + B_n l) N^{(Cl+1)}$$
$$b = \exp(A_z + B_z l)(N n_z)^1$$

and z is the distance from the solid surface.

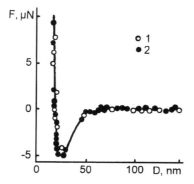

FIG. 43 Forces between mica surfaces following 10 h incubation in a polystyrene–cyclohexane solution (1) and following the replacement of the solution by pure cyclohexane (2). (From Ref. 86.)

We can suggest that the concentration of monomer links of polymer is zero at the minimum distance close to the gyration radius of the macromolecule, D_0, when the forces between two solid surfaces, covered by polymer, is zero:

$$N_{tr} - \frac{bN}{l+1} D_0^{(l+1)} = 0 \tag{227}$$

Equation (227) allows one to calculate the value of l of the adsorbed macromolecule. For example, the magnitude of l is -0.1223 for polystyrene adsorbed on mica from cyclohexane when the distance between two mica plates is 60 nm (Fig. 43).

When the adsorption layers are not saturated, the second point with the force equal to zero is situated on the force–distance curve at a distance less than the gyration radius of macromolecule (Fig. 43). Therefore, one macromolecule is adsorbed on two surfaces. If the saturation degree of the solid surface is low and adsorption forces are equal for each surface, the values of N_{tr} and l are equal to those values for macromolecules adsorbed on one solid surface. The distribution of monomer links of polymer in this case can be described by

$$n = N_{tr} - \frac{bN}{l+1} [z^{(l+1)} + (D-z)^{(l+1)} - D^{(l+1)}] \tag{228}$$

where D is the distance between two plates.

Equation (228) allows one to calculate the number of chains of one macromolecule, n_s, in the half-distance between two plates, $D_f = D/2$, when the force equals zero. The magnitude of $n_s = 1.75 \times 10^4$ is close to the value of $N_{tr} = 1.80 \times 10^4$ for polystyrene adsorbed on mica surface. Therefore, the concentration of chain links does not practically depend on the distance in this case and $N_{tr} = n_s$ at first approximation. The concentration of monomer links of macromolecule near the point of $F = 0$ is inversely proportional to the distance between two solid surfaces taking into account that the macromolecule has N monomer links and we can suggest that

$$N_{tr} = n_s = N_{trf} \frac{D_f}{D} \tag{229}$$

where N_{trf} is the number of monomer links of macromolecule bound with the surface and D_f is the distance between two solid surfaces when $F = 0$.

Equation (230), obtained from Eq. (92), allows one to calculate the values of l near the point $F = 0$ taking into account the values of N_{tr} obtained from Eq. (229):

$$l = \frac{\ln(N_{tr}) - \ln(N) - A_n}{B_n + C \ln(N)} \tag{230}$$

As has been shown in Section III.C, the value of l depends on the degree of macromolecule deformation. The value of l decreases when the concentration of monomer links in the adsorption layer increases. Therefore, the magnitude of l depends on the adsorption forces and the degree of deformation of the macromolecule. The relationship between l and the force attached to two plates is linear (Fig. 44):

$$l = l_0 + B_1 F \tag{231}$$

where l_0 is the value of l when $F = 0$ and b_1 is the coefficient. The correlation coefficient equals 0.945. The value of B_1 equals $(4.5 \pm 0.4) \times 10^{-3} \ \mu N^{-1}$.

Attached forces between the plate, free from the polymer and the plate with polymer adsorption layer determined by the number of new bonds forming between the macromolecule and the plate, free from the polymer or the number of monomer links of polymer bound with this

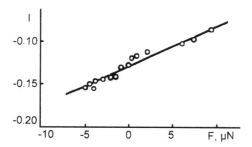

FIG. 44 Dependence between l and the attached force for adsorbed layers of polystyrene in cyclohexane at a distance less than the distance corresponding to the minimum attached force.

surface. The deformation of adsorbed macromolecule is carried out under the action of these forces. The number of monomer links of the polymer, bound with the solid surface previously covered by the polymer, can be calculated according to Eq. (232) obtained from Eq. (92):

$$N_{trl} = \exp[A_n + B_n(l_0 + B_1 F)]N^{C(l_0 + B_1 F)+1} \tag{232}$$

The number of polymer links, bound with a new surface previously not covered by polymer, approximately equals the number of polymer links of adsorbed macromolecule at distance D between the two plates, when D is greater than the distance corresponding to minimum of force:

$$N_{trr} = N_{trl} - \frac{bN}{l_1 + 1}D^{l_1 + 1} \tag{233}$$

where

$$l_1 = l_0 + B_1 F,$$
$$b = \exp(A_n + B_n l_1)(N_n z)^{l_1}$$

When the distance between two plates is less than the distance corresponding to the minimum force,

$$N_{trr} = N_{trl} \tag{234}$$

As shown in Fig. 45, the deformation of the adsorbed macromolecule is not considerable at a long distance between plates (40–60 nm), when the number of bonds of the adsorbed macromolecule with new solid surface is low. When the number of bonds between the adsorbed macromolecule and the new solid surface is large enough, the deformation of the macromolecule takes place. Therefore, there is no difference between the number of bonds of monomer links of the polymer with the plate previously covered by polymer and the number of bonds with the new plate.

The repulsive force takes place between two plates with saturated layers. An influence of the attached force on the distribution of polymer links between solid surfaces can be described by Eq. (226), taking into account Eq. (231) when z changes from zero to $D/2$.

Figure 46 demonstrates the change of N_{tr} and the number of monomer links of polymer on the half-distance between two plates, n_s. These values increase when the distance decreases, but the rate of the change of n_s is larger than the rate of increase of n_{tr}. The value n_s practically equals the value of N_{tr} at a distance between two plates less than 2 nm.

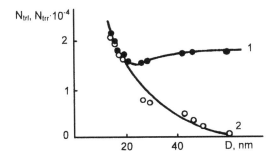

FIG. 45 Dependence between the number of monomer links of a polymer bound with the solid surface with a previously adsorbed polymer (1) and a new solid surface (2).

In the general case, the distribution of monomer links of polymer between two solid surfaces can be described by

$$n = N_{trr} - \frac{b_r N}{l_r + 1} z^{l_r + 1} + N_{trl} - \frac{b_1 N}{l_1 + 1} (D - z)^{l_1 + 1} \tag{235}$$

where

$$N_{trx} = \exp(A_n + B_n l_x) N$$
$$b_x = \exp(A_z + B_z l_x)(N n_z)^{l_x}$$
$$I_x = l_0 + B_1 F$$

$x = r$ or l are the indexes corresponding to the first or second plate.

B. Charged Polymers

Investigation of the adsorption of hydrolyzed poly(acrylamide) on the thickness on the polystyrene latex showed [90] that the thickness of adsorption layer increases and the surface area blockaded by macromolecule and the polymer concentration in the adsorption layer decreases when the NaCl concentration in the mixture increases (Table 5). When the concentration of polystyrene latex decreases, the hydrodynamic thickness of the adsorption

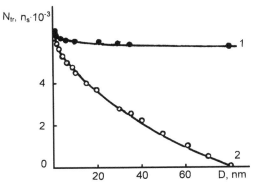

FIG. 46 Relationship among the number of polymer links bound with the solid surface (1), the number of polymer links on the half-distance between plates (2), and the distance between two plates.

TABLE 5 Thickness of the Adsorption Layer, the Surface Area Blockaded by One Macromolecule, and the Polymer Concentration in the Adsorption Layer for Hydrolyzed Poly(acrylamide) Adsorbed on Polystyrene Particles

[NaCl] (mol/L)	h (nm)	$S_b \times 10^{12}$ (cm^2)	$c_s \times 10^2$ (g/cm^3)
0	1.40	6.3	4.7
0.002	1.86	4.7	3.7
0.05	4.05	2.17	2.53
0.5	7.30	1.21	2.52

Source: Ref. 91.

layer increases. The electrophoretic mobility of polystyrene particles changed from 2.17×10^{-8} to 2.83×10^{-8} m^2/s/V. Therefore, the thickness of adsorption layer changes under the action of electrostatic forces between particles.

When the distance between particles is larger than the hydrodynamic diameter of the particle, the electrostatic force between particles can be described by [91]

$$F_e = \frac{K_e q^2}{D^2} \tag{236}$$

where K_e is the constant, q is the particle charge, and D is the distance between particles.

The adsorption layer compresses under the action of that electrostatic force. The compression force is proportional to the relative decrease of the thickness of adsorption layer and the polymer concentration in it:

$$F_d = K_d c_s \frac{\Delta h}{h} \tag{237}$$

where K_d is the coefficient, c_s is the polymer concentration in the adsorption layer, h is the adsorption layer thickness, and Δh is he change of its thickness. One can suggest that these forces are equal to each other in the equilibrium state. Therefore, from Eqs. (236) and (237), we can obtain

$$\Delta h = \frac{K_e h q^2 / (K_d c_s)}{D^2} \tag{238}$$

Figure 47 demonstrates, that experimental data on the change of the hydrodynamic radius of polystyrene particles versus the concentration of the latex lie on a straight line. Therefore, the reversible deformations of adsorption layer takes place under the action of electrostatic forces.

Investigations of the influence of the distance between particles, covered by hydrolyzed poly(acrylamide), on the surface pressure for the particle monolayer on the solid surface showed [92] that the pressure increases when the distance between particles decreases. Therefore, the attraction forces between particles are very small and we neglected them.

The repulsion pressure between particles can be described by Eq. (239) when the distance between particles is close to the particle diameter [93]:

$$\Pi_e \left(\frac{(\varepsilon/2\pi)\varphi_0^2}{\delta^2} \right) \exp\left(-\frac{2D}{\delta} \right) \tag{239}$$

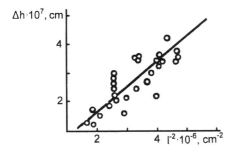

FIG. 47 Relationship between the hydrodynamic diameter of polystyrene particles with hydrolyzed poly(acrylamide), adsorbed on them, versus the distance between them. (From Ref. 91.)

where ε is the medium dielectric permeability, φ_0 is the potential of the particle surface, and δ is the thickness of the double electric layer. Transformation of Eq. (239) taking into account

$$P_e = n\Pi_e \tag{240}$$

leads to

$$\ln(P_e) = \ln\left(\frac{n(\varepsilon/2\pi)\varphi_0^2}{\delta^2}\right) - \left(\frac{2}{\delta}\right)D \tag{241}$$

As shown in Fig. 48, the experimental data for polystyrene particles with hydrolyzed poly(acrylamide), adsorbed on them, is described by Eq. (241) quite well. Calculated values of δ and $n(\varepsilon/2\pi)\varphi_0^2/\delta^2$ are listed in Table 6. The values of δ and φ_0 decrease in each cycle of compression–decompression. Therefore, the formation of new bonds between the macromolecule and the particle occurs in every cycle of compression. Charged groups of hydrolyzed poly(acrylamide) interact with the positive charged particle surface and the potential of particle surface decreases.

Investigation on the redispersion rates of polystyrene particles, covered by hydrolyzed poly(acrylamide), showed [91] that the permeation of adsorption layers of two particles after latex centrifugation does not depend on the mass of adsorbed poly(acrylamide) on the unit surface area of particles.

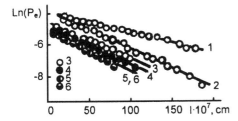

FIG. 48 Dependence between surface pressure and the distance between polystyrene particles covered by hydrolyzed poly(acrylamide) for first (1, 3), second (2, 5), and third (4, 6) cycle of compression–decompression. (From Ref. 91.)

TABLE 6 Thickness of Double Electric Layers and $n(\varepsilon/2\pi)\varphi_0^2/\delta^2$ for Particles of Polystyrene Latex Covered by Hydrolyzed Poly(acrylamide) as Function of the Number of Compression–Decompression Cycles, N

$A \times 10^7$ (g/cm^2)	N	Compression		Decompression	
		$\delta \times 10^5$ (cm)	$\dfrac{n\varepsilon/(2\pi)\varphi_0^2}{\delta^2} \times 10^{13}$ (N cm)	$\delta \times 10^5$ (cm)	$\dfrac{n\varepsilon/(2\pi)\varphi_0^2}{\delta^2} \times 10^{13}$ (N cm)
0.32	1	1.57	11.5	0.85	3.52
0.32	2	1.05	2.12	0.78	0.86
0.32	3	0.88	1.38	0.78	0.86
0.18	3	0.44	0.25	—	—

Source: Ref. 91.

APPENDIX 1

Let $y(\mathrm{Re}, \Theta)$ be a function decreasing exponentially during the increase of Θ:

$$y = S \exp(-u\Theta) \tag{1a}$$

with the limit conditions

$$y(\Theta = 0) = r\mathrm{Re}^q \tag{2a}$$
$$y(\Theta = 1) = 1 \tag{3a}$$

where r and q are coefficients. It follows from expressions (1a) and (2a) that

$$S = r\mathrm{Re}^q \tag{4a}$$

One can find from expressions (1a) and (3a) that

$$u = \ln S \tag{5a}$$

From Eqs. (1a), (4a), and (5a), one can derive

$$y = (r\mathrm{Re}^q)^{-\Theta} \tag{6a}$$

APPENDIX 2

Let $z(u, \theta)$ be a function increasing exponentially during the increase of Θ,

$$z = s \exp(u, \Theta) \tag{7a}$$

with the limit conditions

$$z(\Theta = 0) = 1 \tag{8a}$$
$$z(\Theta = 1) = gv^m \tag{9a}$$

where g and m are coefficients. It follows from expressions (7a) and (8a) that

$$s = 1 \tag{10a}$$

From expressions (7a), (9a), and (10a), one can obtain

$$u = \ln(gv^m) \tag{11a}$$

One can derive Eq. (12a) from expressions (7a), (10a), and (11a):

$$z = (gv^m)^{\Theta} \tag{12a}$$

Therefore, the desorption rate depends exponentially on the fraction of surface covered by the polymer.

APPENDIX 3

The values of the rate of the changing of the adsorbed polymer and its concentration in the adsorption layer do not measure accurately in experiment. They can be calculated more accurately in a few ways.

The experimental data on the kinetics of the change of the amount of adsorbed polymer and of the polymer concentration in the adsorption layer can be linearized by empiric equations:

$$\ln\left(1 - \frac{P_a}{P_p}\right) = \ln a_1 + b_1 t \tag{13a}$$

$$\ln\left(1 - \frac{c_a}{c_p}\right) = \ln a_2 + b_2 t \tag{14a}$$

where P_p and c_p are the number of macromolecules on the unit of the surface area and the concentration of macromolecules in adsorption layer in the plateau section of the kinetic curve, respectively.

Equations (15a)–(18a) can be obtained by transformation and by differentiation of Eqs. (13a) and (14a):

$$\frac{dP_a}{dt} = -P_p a_1 b_1 \exp(b_1 t) \tag{15a}$$

$$\frac{dP_a^2}{dt^2} = -P_p a_1 b_1^2 \exp(b_1 t) \tag{16a}$$

$$\frac{dc_a}{dt} = -c_p a_2 b_2 \exp(b_2 t) \tag{17a}$$

$$\frac{dc_a^2}{dt^2} = -c_p a_2 b_2^2 \exp(b_2 t) \tag{18a}$$

They can be used for further calculation.

REFERENCES

1. BJ Fontana, and JF Tomas. J Phys Chem 65:480, 1961.
2. D Belton, SI Stupp. Macromolecules 16:1143, 1983.
3. KG Barnett, T Cosgrove, B Vincent, DS Sissons, MA Cohen-Rtuart. Macromolecules 14:1018, 1981.
4. ID Robb, R Smith. Eur Polym J 10:1005, 1974.
5. RR Stromberg, WH Grant, E Passaglia. J Res Natl Bur Stand Sec A 65:391, 1961.
6. K Char, AP Gast, CW Frank. Langmuir 4:989, 1988.
7. RR Stromberg, E Passaglia, DJ Tutas. J Res Natl Bur Stand, Sec A 67:431, 1963.

8. RMA Azzam, NM Bashara. Ellipsometry and Polarized Light. Amsterdam: North-Holland, 1977.
9. M Kawaguchi, A Takahashi. Macromolecules 16:1465, 1983.
10. JJ Lee, GG Fuller. Macromolecules 17:375, 1984.
11. FW Rowland, FR Eirich. J Polym Sci A-1 4:2033, 1966.
12. MJ Garvey, Th F Tadros, B Vincent. J Colloid Interf Sci 49:57, 1974.
13. C Allain, D Ausserre, F Rondelez. Phys Rev Lett 49:1694, 1982.
14. A Ausserre, H Harvet, F Rondelez. Phys Rev Lett 54:1948, 1985.
15. I Caucheteux, H Hervet, R Jerome, F Rondelez. J Chem Soc Faraday Trans 86:1369, 1990.
16. KG Barnett, T Cosgrove, V Vincent, AN Burgess, TL Crowley, T King, JD Turner, ThF Tadros. Polym Commun 22:283, 1981.
17. T Cosgrove. J Chem Soc Faraday Trans 89:1323, 1990.
18. T Cosgrove, TG Heath, JS Phipps, RM Richardson. Macromolecules 24:94, 1991.
19. JN Israelachvili, D Tabor. Proc Roy Soc London Ser A 331:19, 1972.
20. JN Israelachvili. Intermolecular and Surface Forces. London: Wiley, 1985.
21. D Tabor, RHS Winterton. Proc Roy Soc London Ser A 312:435, 1969.
22. GJ Fleer, J Lyklema. In: GD Parfit, CH Rochester, eds. Adsorption from Solution at the Solid/Liquid Interface. London: Academic Press, 1983, pp 153–220.
23. GJ Fleer, J van Male. Macromolecules 32:825, 1999.
24. GJ Fleer, J van Male. Macromolecules 32:845, 1999.
25. HJ Ploehn, WB Russel, CK Hall. Macromolecules 21:1075, 1988.
26. HJ Ploehn, WB Russel. Macromolecules 22:266, 1989.
27. PG de Gennes. J Phys (Paris) 37:1445, 1976.
28. PG de Gennes, Scaling Concepts in Polymer Physics. Ithaca, NY: Cornell University Press, 1979.
29. PG de Geness. Macromolecules 13:1069, 1980.
30. KF Freed. J Chem Phys 79:3121, 1983.
31. AM Nemirovsky, KF Freed. J Chem Phys 83:4166, 1985.
32. JF Douglas, SQ Wang, KF Freed. Macromolecules 20:543, 1987.
33. JF Douglas, AM Nemirovsky, KF Freed. Macromolecules 19:2041, 1986.
34. WL Mattice, DH Napper. Macromolecules 14:1066, 1981.
35. E Eisereigeler, K Kremer, K Binder. J Chem Phys 77:6296, 1982.
36. RAL Jones, RW Richards. Polymers at Surfaces and Interfaces. Cambridge: Cambridge University Press, 1999.
37. HJ Tauton, C Toprakciogulu, LJ Fetters, J Klein. Macromolecules 23:571, 1990.
38. GJ Fleer, JMHM Scheutjens. In: B Dobias, ed. Coagulation and Flocculation, New York: Marcel Dekker, 1993, pp 209–263.
39. ST Milner, TA Witten, ME Cates. Macromolecules 21:2610, 1988.
40. ST Milner, TA Witten, ME Cates. Macromolecules 22:853, 1989.
41. Th F Tadros, ed. The Effect of Polymer on Dispersion Properties. 1982.
42. R Varogui, A Johner, A Elaissari. J Chem Phys 94:6871, 1991.
43. W Barford, RC Ball, CMM Nex. J Chem Soc Faraday Trans 1 82:3233, 1986.
44. HA Van der Schee, J Lyklema. J Phys Chem 88:6661, 1984.
45. J Panenhuijzen, HA Van der Schee, GJ Fleer. J Colloid Interf Sci 104:540, 1985.
46. M Kawaguchi, K Hayashi, A Takahashi. Macromolecules 17:2066, 1984.
47. M Kawaguchi, K Hayashi, A Takahashi. Macromolecules 21:1016, 1988.
48. T Cosgrove, TM Obey, B Vincent. J Colloid Interf Sci 111:409, 1986.
49. PG de Gennes. Macromolecules 14:1637, 1981.
50. PG de Gennes. CR Acad Sci Ser 2 294:1317, 1982.
51. PG de Gennes. J Phys Lett 44:1241, 1983.
52. K Ohno, K Binder. J Chem Phys 95:5444, 1991.
53. K Ohno, K Binder. J Chem Phys 95:5459, 1991.
54. M Kawaguchi, K Hayakawa, A Takahashi. Macromolecules 16:631, 1983.
55. M Kawaguchi A Takahashi. Macromolecules 16:1465, 1983.
56. L Auvray, JP Cotton. Macromolecules 20:202, 1987.

57. PG de Gennes. Adv Colloid Interf Sci. 27:189, 1987.
58. M Nagasawa, ed. Molecular Configuration and Dynamics of Macromolecules in Condensed Systems. New York: Elsevier, 1988.
59. TZ Fu, U Stimming, GJ Durning. Macromolecules 26:3271, 1993.
60. CJ Clarke, RAL Jones, JL Edwards, KR Shull, J Penfold. Macromolecules 28:2042, 1995.
61. P Franz, S Granic. Macromolecules 28:6915, 1995.
62. M Kawaguchi, K Hayakawa, T Takahashi. Macromolecules 16:631, 1983.
63. JD Dijt, MA Cohen-Stuart, JE Hofman, GJ Fleer. Colloids Surfaces 51:141, 1990.
64. T Kato, K Nakamura, M Kawaguchi, T Takahashi. Polym J 13:1037, 1981.
65. A Takahashi, M Kawaguchi, H Hirota, T Kato. Macromolecules 13:884, 1980.
66. VN Kislenko, Ad A Berlin, MA Moldovanov. Colloid J (Russia) 55:83, 1993.
67. VN Kislenko, Ad A Berlin, MA Moldovanov. J Colloid Interf Sci 156:508, 1993.
68. VN Kislenko. UM Solomentseva, Ad A Berlin. Colloid J (Russia) 58:44, 1996.
69. VN Kislenko. Ad A Berlin, M Kawaguchi, T Kato. Colloid J 60:338, 1998.
70. VN Kislenko. Ad A Berlin, MA Moldovanov. J Colloid Interf Sci. 173:128, 1995.
71. Ad A Berlin, SS Minko, VN Kislenko, IA Lusinov, MA Moldovanov, Colloid J (Russia) 56:324, 1994.
72. VN Kislenko. J Colloid Interfk Sci 202:74k, 1998.
73. M Kawaguchi, A Takahashi. Adv Colloid Interf Sci 37:219, 1992.
74. JJ Lee, GG Fuller. J Colloid Interf Sci 103:569, 1985.
75. MA Cohen-Stuart, GJ Fleer, JMHM Scheutjens. J Colloid Interf Sci 97:515, 1984.
76. MA Cohen-Stuart, GJ Fleer, JMHM Scheutjens. J Colloid Interf Sci 97:526, 1984.
77. M Kawaguchi, T Itoh, S Yamagiwa, A Takahashi. Macromolecules 22:2204, 1989.
78. T Cosgrove, JW Fergie-Woods. Colloids Surfaces 25:91, 1987.
79. P Franz, S Granic. Phys Rev Lett 66:899, 1991.
80. VN Kislenko. J Colloid Interf Sci 226:246, 2000.
81. Ad A Berlin, VN Kislenko. Colloids Surfaces 104:67, 1995.
82. VN Kislenko, RM Verlinskaya. J Colloid Interf Sci 216:65, 1999.
83. J Klein, PF Luckhem. Macromolecules 17:1041, 1984.
84. J Klein, PF Luckhem. Macromolecules 19:552, 1986.
85. J Klein, PF Luckhem. J Colloid Interf Sci 141:593, 1991.
86. J Klein. J Chem Soc Faraday Trans. 1 79:99, 1983.
87. JN Israelachvili, M Tirrell, J Klein, Y Almog. Macromolecules 17:204, 1984.
88. HW Hu, JV Alsten, S Granik. Langmuir 5:270, 1989.
89. HW Hu, S Ganik. Macromolecules 23:613, 1990.
90. J Meadows, PA Williams, MJ Garvey, R Harrop, GO Phillips, J Colloid Interf Sci. 132:319, 1989.
91. VN Kislenko. J Phys Chem (Russia) 72:1464, 1998.
92. J Meadows, PA Williams, MJ Garvey, R Harrop. J Colloid Interf Sci 139:260, 1990.
93. SS Vojuckij. Colloid Chemistry. Moscow: Chemistry, 1975, p 260.

14

Modeling of Protein Adsorption Equilibrium at Hydrophobic Solid–Water Interfaces

KAMAL AL-MALAH Jordan University of Science and Technology, Irbid, Jordan

I. INTRODUCTION

Protein adsorption is involved in a number of areas in biology, medicine, food and pharmaceutical processing, and biotechnology. In the food and pharmaceutical industries, proteins can play a major role in the fouling of membrane surfaces used in biomolecular fractionation and in the fouling of heat-exchange surfaces due to their heat sensitivity and high content in some fluid foods. Additionally, protein behavior at both air–water and oil–water interfaces can play a major role in stabilizing colloidal food systems, foams, and emulsions.

In biomedicine, protein adsorption is of great concern; deciphering the mechanism of plasma protein interactions with blood-contacting devices and the subsequent activation of coagulation pathways and platelet adhesion is the key problem in developing nonthrombogenic biomaterials [1–4]. Another area of interest in biomedicine is the interaction of water-soluble crystalline proteins of the eye lens with plasma membrane. Studies have shown that three crystallines (α-, β-, and γ-crystallines) may become partially associated with the plasma membrane during aging and cataractogenesis [5].

A great deal of effort has been devoted to studying the different factors that influence adsorption. The question of how these factors interact is undoubtedly complex, and a comprehensive model of protein adsorption is not available. The important factors affecting adsorption can be classified under one or more of the following three areas: (1) protein characteristics, including isoelectric point, net charge and charge distribution, three-dimensional (3D) structure in solution, placement and nature of hydrophobic patches, and conformational variability; (2) surface properties, including topography and heterogeneity, electrical potential, composition, water binding, and hydrophobicity; (3) medium conditions, including pH, temperature, ionic strength, equilibrium concentration, hydrodynamics, and buffer type.

Therefore, some of the above-mentioned factors will be selectively incorporated into a macroscopic model that would quantitatively describe protein adsorption equilibrium (i.e., predict adsorption isotherms) for a given protein–surface–medium system. A two-step adsorption mechanism was constructed to visualize protein adsorption equilibrium. The first step is reversible migration of the native protein molecule to the interface. The second step is unfolding of the adsorbed molecule. The energies of the two steps were quantified according to the Gibbs free-energy changes for each. The model served to explain the Langmuir-type pattern of adsorption equilibrium isotherms commonly observed. Its applicability was tested by comparison with experimentally measured isotherms describing protein interaction with hydrophobic solid surfaces.

II. LITERATURE REVIEW

The following review provides a basis for understanding the theoretical development used to construct the model presented in this chapter. Emphasis is placed on basic protein biophysics and the importance of solid surface hydrophobicity and protein thermodynamic stability on adsorption.

A. Proteins

Proteins are biological macromolecules synthesized in cells for specific functions. They are high-molecular-weight polyamides that adopt exquisitely complex structures. This complexity is characterized by different levels of structure: primary, secondary, tertiary, and quaternary. Primary structure [6] refers to the amino acid sequence itself, along with the location of disulfide bonds (i.e., covalent connections between two amino acid residues within the protein molecule). Secondary structure refers to the spatial arrangement of amino acid residues that are near one another in the linear sequence. Alpha (α) helices and beta (β) sheets are typical examples of a secondary structure. The tertiary structure refers to the spatial arrangement of amino acid residues that are far apart in the linear sequence. If a protein has two or more polypeptide chains, each with its exclusive primary, secondary, and tertiary structure, such chains can associate to form a multichain quaternary structure. Hence, a quaternary structure refers to the spatial arrangement of such subunits and their interaction.

1. Protein Stability

Protein molecules are stabilized by different intramolecular forces that play a key role in maintaining protein structure. In addition to the planar peptide bond that constitutes the backbone of the molecule and the possible presence of disulfide linkages, there are intramolecular forces, although smaller in magnitude than a covalent bond, that are just as important. These intramolecular interactions include the so-called hydrophobic bonds, hydrogen bonds, and dispersion and electrostatic attractive forces. Disulfide bonds are thought [7] to stabilize proteins by reducing the conformational entropy of the unfolded chain. Statistical treatments have proposed [7] that the destabilization of the unfolded state depends on the length of the loop formed by a single cross-link. Creighton [8] found that for a given loop size, the most effective cross-link may be found between groups that are rigidly held in an optimum orientation by the folded structure.

Hydrogen-bonding forces are considered to be one of the major contributors to the largely temperature-independent part of the enthalpy of stabilization. Due to their small size and electropositivity in covalent bonds, hydrogen atoms are easily brought into close proximity to electronegative atoms. The resulting interaction energy is intermediate between the energies of van der Waals contacts and covalent bonds. Hydrogen bonds are very common in proteins and are partly responsible for the α-helix and β-sheet stabilities. Hydrogen-bond partners are exchanged during folding. Intramolecular bonds are formed at the expense of intermolecular hydrogen bonds with water.

The importance of dispersion or van der Waals forces in protein stability hinges on differences in packing in the folded and unfolded states [7]. Klapper [9] and Chothia [10] found that, upon folding, water is expelled to the relatively open bulk phase, and the atoms that form the protein core become as tightly packed as good molecular crystals. The difference in packing density between the folded and unfolded states is expected to alter the distribution of interatomic distances, which, in turn, may affect the van der Waals interactions.

Proteins are polyelectrolytes because ionizable groups from amino acid side chains and terminal amino acids can participate in acid–base equilibrium with the solvent. Ionizable groups are not generally distributed randomly over protein surfaces, reflecting their individual structural and functional roles. The charged groups in proteins [11] are, on the average, surrounded by those of opposite sign, and significantly more oppositely charged than like-charged groups are separated by a distance less than 4.0 Å. This suggests that the charges, including the α-helix dipoles, contribute to the stability of the protein. It also suggests that electrostatic interactions are heavily involved, in short-range interactions (i.e., less than the van der Waals interatomic distances). The problem of how much energy the charge interactions actually contribute to the protein is a more complex issue [12].

Hydrophobic interactions [13] are basically entropically driven, largely due to order/disorder phenomena in the surrounding water. Current estimates of amino acid hydrophobicity are based on the measured free energies of transferring side chains from water to organic solvents, where the latter presumably simulate the polarity of the protein interior. The following subsection is dedicated to explain in more detail the concept of those hydrophobic interactions and why they are thought to be entropically driven.

2. Hydrophobic Interactions

To illustrate the notion of hydrophobic interactions being entropically driven and the idea of measuring the degree of hydrophobicity of an amino acid residue by its free energy of transfer from water to a hydrocarbon solvent, $\Delta G_{transfer}$, the following development from Schulz and Schirmer [14] is presented here.

If N denotes the native state of a protein chain and U denotes the randomly unfolded state of that chain, and considering a composite system made of polypeptide chain and solvent, the total free-energy difference between forms N and U becomes

$$\Delta G_{transfer} = G_N - G_U = \Delta H_{chain} + \Delta H_{solvent} - T\Delta S_{chain} - T\Delta S_{solvent} \tag{1}$$

Furthermore, suppose the unfolding/folding mechanism of a polypeptide chain in water can be simulated by the phase separation of a hydrocarbon (mineral oil) from water. This is accomplished by assigning the U state (unfolded state) to the monodisperse solution of oil in water, and the N state (folded or native state) to separated phases (i.e., an oil drop on the water surface). Each of the above four terms (two enthalpy and two entropy terms) that constitute $\Delta G_{transfer}$ will be analyzed in light of the net formation/breakdown of noncovalent bonds and the overall increase/decrease of conformational entropy of protein and that of the solvent. ΔS_{chain}, which is equal to $S_N - S_U$, is negative because a monodisperse solution (U) is less ordered (higher entropy) than separated phases (N). ΔH_{chain}, which is equal to $H_N - H_U$, is positive because in the N state, most oil molecules are surrounded by their kind, whereas in the U state, all of them are surrounded by water molecules. The interactions between oil molecules are only dispersion forces and hence weak. On the other hand, the interactions between oil and water molecules are stronger because the strongly polar water molecules induce dipoles in the neighboring oil molecules, giving rise to an appreciable electrostatic term. So, ΔH_{chain} favors the monodisperse solution; nevertheless, it is relatively small.

It was experimentally shown [14,15] that $\Delta S_{solvent}$, which is equal to $S_N - S_U$, that is water in the above example is positive, which favors the N state (i.e., phase separation). This indicates that the order of the water molecules in the U state (monodisperse oil solution) is higher (i.e., has a lower entropy) than that in the N state. Therefore, water molecules surrounding an apolar molecule are characterized by a higher degree of order at this polar/apolar interface, where they assume a locally ordered, quasisolid structure (a "cagelike" structure, clathrate, or iceberg structure) with some loss of H-bonding capacity. This also results in a negative $\Delta H_{solvent}$, the magnitude of which, however, is relatively small. In summary, ΔH_{chain} and ΔS_{chain} favor the U state, whereas $\Delta H_{solvent}$ and $\Delta S_{solvent}$ favor the N state. As phase separation between oil and water is thermodynamically more stable than the monodisperse case, it turns out that $\Delta S_{solvent}$ is the predominant driving force that underlies the process of phase separation in this case. The effect of $\Delta S_{solvent}$ is usually referred to as a *hydrophobic* or *entropic* effect.

Although an analogy has been drawn between a polypeptide chain and a hydrocarbon, one should keep in mind that, unlike oil, the polypeptide chain contains both polar and nonpolar moieties. Nevertheless, the probability of finding polar groups on the exterior of a protein molecule is larger than that in the interior of protein. Moreover, the nonpolar amino acid residues are more likely to exist in the interior of the protein (in globular proteins, in particular) rather than on the exterior [13–15]. This is consistent with the thought that entropically driven hydrophobic interactions tend to minimize the free energy of the system by reducing the interfacial area between the nonpolar moieties and the aqueous medium.

3. Protein Conformation

When a hydrophobic surface like that of air or a hydrophobic solid is brought into contact with a protein solution, an environment supporting unfolding of the intact hydrophobic core of a protein to establish new noncovalent contacts with the interface is created. Unfolding at the interface is often referred to as surface denaturation.

Protein denaturation involves a conformational change. Upon unfolding, the polypeptide chain becomes less compact, more highly solvated, and much more flexible. Protein denaturation is a highly cooperative reaction, and general molecular stability depends on environmental conditions such as temperature, pressure, pH, ionic strength, and the concentration of specific ligands, stabilizers, and denaturants in solution.

Dill et al. [16] studied the thermal stability of globular proteins. Two factors are important in quantifying the temperature dependence of globular protein folding: (1) the conformational entropy of the chain and (2) the heat capacity change effected by the hydrophobic effect. Folding is driven by a negative free-energy change accompanying clustering of the hydrophobic residues into a globular structure and is opposed by a positive free-energy change due to loss of conformational entropy upon folding. *Cold denaturation* is driven principally by the weakening of the hydrophobic interactions, but *thermal denaturation* is driven principally by the gain of conformational entropy in the unfolded chain.

Privalov et al. [17] studied cold denaturation of myoglobin. The disruption of the native protein structure both on cooling and on heating was characterized as proceeding in an "all-or-none" manner, with a significant and similar increase of the protein heat capacity, but with inverse enthalpic and entropic effects: The enthalpy and entropy of the protein molecule decrease during cold denaturation and increase during heat denaturation. Concerning proteins with a multidomain structure, such domains may act independently during thermal denaturation, but with some degree of cooperativity among the different subunits [18–19].

Honeycutt and Thirumalai [20] demonstrated the existence of metastable states in the folding/unfolding pathways, using a stochastic dynamics method to simulate the processes of folding and unfolding. These metastable states are characterized by several free-energy minima separated by barriers of various heights such that the folded conformations of a polypeptide chain in each of the minima have similar structural characteristics (viz. the gross appearance and radius of gyration) but have different energies from one another. They suggested that the formation of the more stable form depends largely on the method of preparation; the initial conditions determine the kinetics of formation of the more stable form. The lesson that can be learned from their study is that although the unfolding or folding reaction may be thermodynamically favorable, energy barriers exist along the reaction coordinate, which hinder the protein from reaching a configuration with the lowest free energy, "trapping" it instead into a metastable state.

B. Protein Adsorption

In general, adsorption involves the migration of a substance from one phase to the surface of an adjacent phase, accompanied by its accumulation at the interface [21]. Adsorption is a result of the binding forces between individual atoms, ions, or molecular regions of an adsorbate and the adsorbent surface. These binding forces or interactions vary in magnitude from the weak van der Waals type of attraction contributing to physical adsorption, to the strong covalent bonds in chemisorption, Polymer adsorption in general and biopolymer adsorption in particular show a range of binding energies depending on the type of forces present in the interface. Polymer adsorption differs drastically from that of small molecules. This is due to the large number of conformations that a macromolecule can adopt, both in the bulk solution and at the interface. Moreover, the entropy loss or gain associated with a given flexible polymer can be substantially greater than that for small molecules or relatively stiff molecules [22].

A thermodynamic approach was proposed by De Feijter et al. [23] to describe the adsorption of non-ionic, flexible polymers at solid surfaces. Their approach relied on a pseudo-lattice model (quasicrystalline model), the cells of which may accommodate a solvent molecule or a polymer segment. Each macromolecule was considered to consist of m identical segments, of which a fraction, f, is adsorbed directly to the surface (i.e., fm cells of the surface layer are occupied by one adsorbed polymer). Their approach led them to conclude that the polymer adsorption isotherm would exhibit a high-affinity character (i.e., high adsorption at very low bulk concentration) with almost immediate plateau attainment. Their approach was also expected to apply to proteins, even though proteins are considered somewhat rigid structures, exhibiting some net charge. Their conclusion may explain in part, however, why protein isotherms generally assume plateau values at a relatively low concentration.

Lin et al. [24] studied adsorption–desorption isotherm hysteresis exhibited by β-lactoglobulin A on a weakly hydrophobic surface. They found that the desorption isotherm at pH 6.0 overlapped with the adsorption isotherm and that the adsorption–desorption process of β-lactoglobulin A under this condition could be characterized by a fully reversible Langmuir model. The desorption isotherm at pH 4.5, however, did not coincide with the adsorption isotherm, giving rise to hysteresis. This would suggest that protein adsorption experiments carried out under mild conditions of pH at relatively hydrophilic surfaces might be treated with the assumption that reversible equilibrium exists between the bulk and interface.

Arnebrant et al. [25] studied the temperature dependence of adsorption for α-lactalbumin and β-lactoglobulin on chromium surfaces. They observed that the curves for β-lactoglobulin at 25°C, 66°C, 70°C and 73°C were rather similar. It was only when the temperature exceeded the denaturation temperature (79°C) of β-lactoglobulin that they could observe a significant difference in the isotherm. This would suggest that the adsorbed mass of protein does not significantly depend on temperature, as long as the temperature at which the adsorption experiment was carried out lies below the melting-point temperature of the protein (or the irreversible denaturation temperature). If the temperature exceeds that of the melting point of protein, the surface activity of protein will depart from that exhibited in its native form.

Lu et al. [26] calculated the solvation interaction energies for protein adsorption on hydrophilic and hydrophobic polymer surfaces. The solvation interactions (repulsive hydration and attractive hydrophobic interactions) were calculated for lysozyme, trypsin, immunoglobulin Fab (antigen-binding fragment consisting of the light chain and half of the heavy chain, with a molecular weight of 50,000 D), and hemoglobin. The average solvation interaction energy was found to vary from -259.1 to -74.1 kJ/mol for the four proteins at hydrophobic polymer interfaces (polystyrene, polyethylene, and polypropylene), whereas on hydrophilic surfaces [poly(hydroxyethyl methacrylate) and poly(vinyl alcohol)], the average solvation interaction energies were greater than zero. These calculations illustrate the importance of attractive hydrophobic interactions between proteins and polymer surfaces in adsorption.

Matsuno et al. [5] studied the interactions of γ-crystallins with silica, methylated silica, and diphenyl silica. They used different techniques to examine the secondary and tertiary structural alterations that took place upon adsorption on these silica surfaces exhibiting different degrees of hydrophobicity. A comparison was made between conformations of free and surface-bound protein as a function of the electrostatic and hydrophobic character of both the protein and the adsorbent surface. They demonstrated that (1) protein destabilization on hydrophobic surfaces is greater than that on more hydrophilic surfaces, (2) detectable conformational changes tend to increase as the surface hydrophobicity increases, and (3) subtle structural differences among proteins can play an important role in determining differences in protein stability and structure upon adsorption.

Wei et al. [27] examined the role of protein structure in surface tension kinetics at the air–water interface and demonstrated that the intrinsic, conformational stability is an influencing factor in protein surface activity at low bulk concentrations. At high bulk concentrations, surface hydrophobicity was highly correlated with the observed surface tension kinetics. Surface tension kinetics in this context refers to the rate of change of liquid surface tension, γ_{LV}, for the protein solution in contact with air.

Kato and Yutani [28] correlated the surface activity of six mutants of tryptophan synthase α-subunits with their stability, as measured by their free energy of denaturation in water (ΔG_{water}). One measure of surface activity was given by the air–liquid surface tension of a mutant solution. These results are in line with the previous examples and demonstrate the importance of conformational stability in predicting the difference in the adsorptive behavior among proteins. Incidentally, it serves as another example suggesting that the surface tension of protein, γ_{PV}, is related to its intrinsic conformational stability.

Horsley et al. [29] studied human and hen lysozyme adsorption on hydrophobic, negatively charged, and positively charged silica. On the average, human lysozyme was found to adsorb in larger amounts than did hen lysozyme. They attributed the difference in adsorptive behavior to the thermal lability of the molecules; human lysozyme is more susceptible to thermal denaturation, hence to surface denaturation, than is hen lysozyme.

Hunter et al. [30] studied the coadsorption and exchange of lysozyme and β-casein at the air–water interface. The air–water interface was used because it was described as representing the simplest model hydrophobic surface for studying protein adsorption. Their results suggested that electrostatic interactions do no play a major role in determining exchange behavior at the air–water interface and, moreover, that the flexibility of both the adsorbed and displacing molecules are more important than intermolecular interactions in determining whether exchange occurs. This work serves as an example of the general observation that electrostatic interactions play a minor role in dictating the adsorptive behavior of proteins at hydrophobic interfaces.

Raje and Pinto [31] measured the heat of adsorption and its dependence on surface coverage for protein ion-exchange systems of bovine serum albumin and ovalbumin. Experimental data showed that protein adsorption is endothermic for both systems, which suggests that the process is entropically driven. Therefore, it is essential to include the entropic contribution (i.e., hydrophobic interactions) in modeling equilibrium behavior.

Oberholzer and Lenhoff [32] proposed a method for calculating adsorption isotherms for small globular proteins in aqueous solution based on colloidal descriptions of protein–protein and protein–surface interaction energies. The influence of the structure of the adsorbed protein layer on the energetics was obtained through Brownian dynamics simulations. The qualitative influence of experimental variables such as solution pH, ionic strength, and protein size on the predicted adsorption of proteins was explored.

In summary, then, the following general statements can be made regarding protein adsorption. The surface activity of a protein is a cumulative property influenced by many factors, including size, shape, charge, surface hydrophobicity, and thermodynamic stability. However, based on abundant experimental observations, samples of which were cited here, the influence of only two of these factors (surface hydrophobicity and conformational stability) can be considered, under controlled but actually quite relevant circumstances, as effectively governing the adsorption process. Protein adsorption exhibits diversity in behavior from one surface to another and from one protein to another. This diversity results from the complexity of the protein structure itself and from the many variables on which protein adsorption depends. A comprehensive model for any aspect of protein adsorption that considers all of these variables is completely lacking in the existing literature and much work is needed to reach that goal.

III. MODEL DEVELOPMENT

Before I move to the core of the subject matter, I would like to point out the fact that there are numerous models, developed by other investigators, which deal with protein adsorption equilibria [31–35].

A. Dimensional Analysis

The useful result of a dimensional analysis of protein adsorption equilibrium would be a starting point for the determination of a functional relationship among the variables thought to be pertinent to the process. The relationship can he expressed in the following compact form:

$$\Omega(\Pi_1, \Pi_2, \Pi_3, \ldots, \Pi_i) = 0 \tag{2}$$

where the Π_i represent independent dimensionless groups of some dimensional variables or parameters (factors) which are measurable or can be expressed in terms of other measurable quantities and Ω defines this mathematical relationship among the dimensionless Π groups.

There are two matters involving choice in this analysis. The first is that of initial quantities (in the case of protein adsorption, this refers to factors influencing the process of adsorption). This choice identifies the factors considered to be important and those that may be neglected. The second is the choice of the final, dimensionless Π groups. The first choice requires a thorough understanding of protein adsorption and a comprehensive survey of the pertinent literature. Consequently, a sufficient portion of the coherent literature was reviewed and the following relationship is suggested:

$$\Omega\left(\frac{W_a A_C}{\Delta G_{\text{unfold}}}, \Gamma A_P, V_P C_{\text{eq}}\right) = 0 \tag{3}$$

where Γ is the adsorbed amount of protein (mol/m^2), C_{eq} is the apparent equilibrium concentration (mol/m^3), A_P is the partial molar area occupied by protein at the interface (m^2/mol), V_P is the partial molar volume of protein in solution (m^3/mol), ΔG_{unfold} is the change in partial molar free energy of protein upon unfolding (J/mol), A_C is the minimum surface area cleared by an adsorbing protein molecule in order to anchor itself to the interface (m^2/mol), and W_a is the work of adhesion (J/m^2). The published results that led to the selection of the factors incorporated into Eq. (3) were detailed in Section II.

1. Quantifying the Factors in Eq. (3)

(a) Work of Adhesion W_a.   The work of adhesion is generally defined as the energy required to separate a unit area of interface into two phases. Applied to protein adsorption, W_a can be written

$$W_a = \gamma_{SW} + \gamma_{PW} - \gamma_{PS} \tag{4}$$

where γ_{SW} (mJ/m^2) is the interfacial free energy between the solid and water, γ_{PW} (mJ/m^2) is the interfacial free energy between the protein and water, and γ_{PS} (mJ/m^2) is the interfacial free energy between the protein and solid. These three interfacial energies can each be defined by an equation expressed in the following compact form:

$$\gamma_{SW} = \Omega(\gamma_{SV}, \gamma_{WV}) \tag{5}$$
$$\gamma_{PW} = \Omega(\gamma_{PV}, \gamma_{WV}) \tag{6}$$
$$\gamma_{PS} = \Omega(\gamma_{PV}, \gamma_{SV}) \tag{7}$$

where the functional relationship Ω represents an equation of state that relates the interfacial energy between two phases to the interfacial energies between each of those phases and a third phase; in the case of Eqs. (5)–(7), the third phase is vapor. Neumann et al. [36] empirically obtained an explicit formulation for Ω:

$$\gamma_{12} = \frac{(\sqrt{\gamma_{13}} - \sqrt{\gamma_{23}})^2}{1 - 0.015\sqrt{\gamma_{13}\gamma_{23}}} \tag{8}$$

where the subscripts 1 and 2 denote any pair combination of two phases from the protein (P), water (W), and solid (S) phases, as written in Eqs. (5)–(7), and subscript 3 denotes the third phase, which is vapor in all of the above cases. Young's equation ($\gamma_{SV} = \gamma_{SL} + \gamma_{LV}\cos\theta$) introduces θ (degree), the equilibrium contact angle formed between a solid–liquid (SL) interface and a liquid–vapor (LV) interface. The contact angle θ defines the characteristic orientation of a liquid–vapor interface with reference to a solid surface it contacts, as illustrated in Fig. 1.

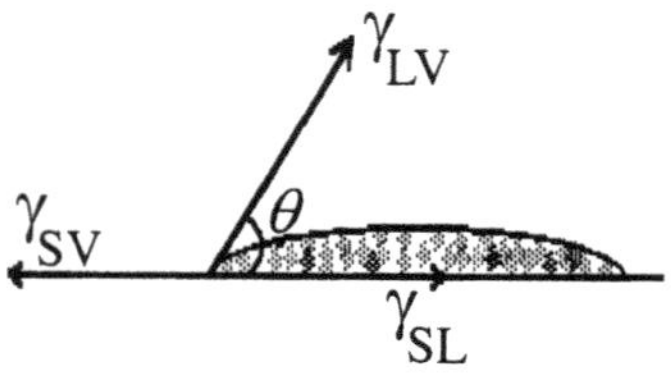

FIG. 1   The equilibrium contact angle, θ, formed between a solid–liquid (SL) interface and a liquid–vapor (LV) interface.

Young's equation can be combined with Eq. (8) to yield, for a specific solid–liquid contact,

$$\cos\theta = \frac{(0.015\gamma_{SV} - 2.00)(\gamma_{SV}\gamma_{LV})^{\frac{1}{2}} + \gamma_{LV}}{\gamma_{LV}\left[\left\{0.015(\gamma_{SV}\gamma_{LV})^{\frac{1}{2}}\right\} - 1\right]} \tag{9}$$

Eq. (9) should be taken as generic (i.e., the subscript S refers to any solid phase, including protein, that is in contact with a liquid phase). Equation (9) therefore allows the evaluation of both γ_{SV} and γ_{PV}, as both θ and γ_{LV} are readily measurable. Thus, Eq. (8) should be solvable for the interfacial energies of Eqs. (5)–(7).

However, evaluation of the surface energy of pure protein with contact-angle methods is not trivial. It is arguably impossible to determine a relevant value of γ_{PV} by measuring contact angles on a prepared solid protein "surface." For example, liquid penetration into a solid protein surface would probably be unavoidable; as such, a surface would probably be both porous and hygroscopic. In this work, the value of γ_{PV} was approximated by

$$\gamma_{PV} = \gamma_{LV} - \pi_e \tag{10}$$

where π_e (mJ/m^2) represents the equilibrium spreading pressure measured at the protein solution–vapor interface and corresponds to the (concentration-independent) plateau region of $\pi_e = f(C_{eq})$ (i.e., a saturated interface) and γ_{LV} represents the surface tension of the protein-free solution. Spreading pressure is a measure of the reduction in surface energy as a result of adsorption at an interface.

Equation (9) should give a less ambiguous value of γ_{PV} than that attainable by contact-angle methods. Van Oss and Good [37] experimentally determined protein interfacial energy using contact-angle methods. They concentrated a protein solution by ultrafiltration on an anisotropic cellulose acetate membrane, and then carried out contact-angle measurements on the resultant protein layer. Depending on the degree of hydration of the resultant protein layer, the value of γ_{PV} was determined. Therefore, the value of γ_{PV} obtained, in essence, represented the surface tension of a hydrated protein layer (formed in this case on a cellulose acetate membrane) measured at the protein layer–air interface and that value of γ_{PV} is thus consistent in principle with that given by Eq. (10).

The value of γ_{PV} obtained by contact-angle methods, however, is subject to perturbations induced by the solid support itself, and it rather reflects the degree of orientation of the water molecules at the periphery of the hydrated protein layer [37]. Unfortunately, relevant spreading pressure data is not available for many proteins of interest, so another means to calculate π_e is needed.

Singer [38] developed an equation of state that relates π_e to measurable parameters:

$$\pi_e = \pi_0\left[\frac{Z}{2}\ln\left(1 - \frac{2f}{Z}\right) - \ln(1 - f)\right] \tag{11}$$

where $\pi_0 = k_BT/a_0$ (k_B is the Boltzmann constant, T is temperature, a_0 is the average interfacial area occupied by an amino acid residue), f is the fractional surface coverage, and Z is the surface coordination number of the lattice ($Z = 2 + \omega$, where ω refers to the flexibility of the polymer chain). For a completely rigid chain, $Z = 2$ and for a completely flexible chain, $Z = 4$; thus, $2 \leq Z \leq 4$ (or $0 \leq \omega \leq 2$). The value assigned to the parameter a_0 is 15 Å^2 per amino acid residue [39,40]. Parameter Z can be correlated with the thermal stability of the protein molecule (ΔG_{unfold}) using surface pressure data available for α-lactalbumin (α-Lac), β-lactoglobulin (β-Lag), bovine serum albumin (BSA) [41], and lysozyme (Lyso) [42]. The resulting equation is

$$Z = 2.0 + 2.0\exp\left(-0.135\frac{\Delta G_{unfold}}{RT}\right) \tag{12}$$

Figure 2 shows the value of Z as given by Eq. (12) compared to individual values of Z estimated by Eq. (11) for the above-mentioned proteins. The value of fractional surface coverage, f, should be fixed at a value equivalent to monolayer film coverage. As a result of the mathematical difficulty that arises in Eq. (11) when f approaches 1.0, the value of f was set equal to the maximum possible value such that Eq. (11) is sensitive to variations in parameter Z. Hence, a value of 0.92 was chosen by equating the experimentally based plateau value of equilibrium spreading pressure, π_e, with Eq. (11).

A question may arise regarding the decision to define protein surface energy (γ_{PV}) in terms of the hydrophobicity of its intact core (expressed by ΔG_{unfold}), rather than in terms of its effective "surface" hydrophobicity. The surface hydrophobicity of native protein, expressed as γ_{PV} but measured in aqueous solution by a number of techniques including contact angle, cell adhesion (in the absence of electrostatic effects), and as derived from adsorption experiments, is found to change only slightly from one type of protein to another [43,44]. Thus, γ_{PV} was estimated using Eq. (10), considered to yield a better index of protein surface activity and being in line with the suggestion that a protein with a higher tendency to unfold should exhibit higher surface activity (i.e., greater reduction of the air–water surface tension for a given protein solution) [41].

The protein–vapor interfacial energy is thus measurable using Eqs (10)–(12), and solid surface energy using Eq. (9). In practice, however, as the surface energy of the solid in question approaches that of water, Eq. (8) yields anomalous results. A set of equations of state was developed by Neumann et al. [45], which allows the calculation of interfacial energy under such circumstances. Appendix 1 includes the FORTRAN program [45] used in this work to determine interfacial energy based on the values of pertinent input solid and liquid interfacial energies.

(b) The Gibbs Free Energy of Unfolding, ΔG_{unfold}. The Gibbs free energy of unfolding, ΔG_{unfold}, connotes the partitioning of a protein molecule between two conformations: native and unfolded. ΔG_{unfold} measures the difference between the partial molar free energy of the macromolecule in its unfolded or denatured state (U), and its native, folded state (N).

$$\Delta G_{unfold} = G_U - G_N \tag{13}$$

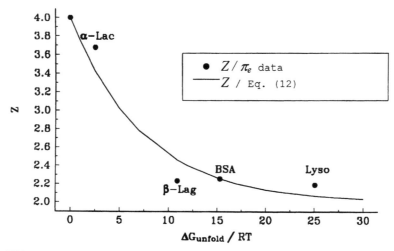

FIG. 2 The flexibility parameter (Z) correlated with $\Delta G_{unfold}/RT$ for α-Lac, β-Lag, BSA, and Lyso.

This two-state model for folding and unfolding has been used by a number of investigators [46–50] to study the thermal stability of proteins in order to measure what are known as protein stability curves. It should be mentioned, however, that Eq. (13) was restricted to one class of proteins in this treatment. This class includes compact, single-domain, globular proteins undergoing a one-step (all-or-none) reversible phase transition between two thermodynamically defined conformations (native and unfolded). Privalov [51] suggested that in a living system the transition from the structureless state (U) to the native state (N) should be reversible; the observed irreversibility is caused by secondary phenomena, including aggregation, isomerization of prolyl residues, and separation of chains, but the transition itself is, in principle, reversible. Dzakula and Andjus [52] have recently improved the accuracy of the two-state model by expressing ΔG_{unfold} as a function of temperature. Their proposed equation for ΔG_{unfold} is

$$\Delta G_{unfold} = \frac{\Delta H_D(T_m - T)}{T_m} - \frac{\Delta C_P(T - T_m)^2}{2T_m} \tag{14}$$

where ΔH_D (kJ/mol) is the enthalpy of denaturation evaluated at the melting-point temperature, T_m (K) and ΔC_P (kJ/mole K) is the difference in heat capacity between the unfolded and folded states. These parameters are conveniently determined by conducting a differential scanning calorimetry (DSC) analysis of the given protein. The melting-point temperature, T_m (K) is estimated from the temperature at which the thermogram peak reaches its maximum, provided that the rate of heating is sufficiently low. ΔH_D is estimated as being equal to the area of the endotherm peak and ΔC_P is viewed as the difference in the asymptotic heat capacity values on either side of the endotherm peak [52]. In the absence of DSC data, an equation of state developed by Murphy et al. [53] can be used to define ΔG_{unfold} (per mole of amino acid residue) as a function of temperature:

$$\Delta G_{unfold} = \Delta H^* - T\Delta S^* + \Delta C_P \left[(T - T_H) - T\ln\left(\frac{T}{T_S}\right) \right] \tag{15}$$

where ΔH^* (J/mol of amino acid residue) and $T\Delta S^*$ (J/mol amino acid residue) represent the nonhydrophobic, enthalpic, and entropic contributions to the free-energy change, respectively, T_H and T_S are the temperatures at which ΔH is equal to ΔH^* and ΔS is equal to ΔS^*, respectively, and ΔC_P is the heat capacity change upon unfolding. According to Murphy et al. [53], the heat capacity term on the right-hand side of Eq. (15) represents the hydrophobic, destabilizing contribution to the free energy of folded proteins. T_H and T_S are approximately equal for proteins and they assume a value of $112 \pm 2.4°C$ [53]. ΔS^* for proteins was found to be equal to 18.1 ± 1.0 J/(mol K) [53]. A typical value for ΔC_P of 50.0±10.0 J/mol K) can be assumed, based on experimental measurements of changes in heat capacity upon denaturation [54]. Equating experimentally based values of ΔG_{unfold} obtained from Eq. (14) with Eq. (15) allowed estimation of a protein-independent ΔH^* (J/mol). Appendix 2 shows results of that procedure for nine proteins considered in this work; an average value for ΔH^* of 6.076 ± 0.155 kJ/mol was obtained.

(c) The Partial Molar Area of Protein, A_P. Because estimates of the partial molar volume of protein in solution, V_P, are readily available [55–60], only the difference between the minimum surface area cleared by an adsorbing molecule, A_C, and the partial molar area, A_P, need to be discussed in support of Eq. (3). Based on experimental studies [13,39,61] on protein adsorption at air–water and solid–water interfaces, A_C changes only slightly from one type of protein to another and is independent of molecular size. Moreover, there is no equation of state available that relates A_C to any surface or protein property. For these reasons, A_C was allowed to be a computer-generated parameter, rather than an input variable for the model [Eq. (3)]. It was

anticipated that this parameter would not dramatically change from one type of protein to another, in accordance with previous experimental observations.

A_P is the partial molar area of protein and was evaluated such that protein molecules in the neighborhood of the interface are spherical and hexagonally close packed. The following equation was originally suggested [62,63] with reference to adsorption from simple organic mixtures on solid surfaces, provided that molecules in solution exist in closest-hexagonal packing:

$$A_i = 1.091(V_i)^{2/3}(N_A)^{1/3} \tag{16}$$

where A_i (m^2/mol) is the partial molar area of species i at the solid–liquid interface, V_i (m^3/mol) is the molar volume of species i in solution at the given temperature, and N_A is Avogadro's number. The suggestion that we can apply the same equation to protein adsorption would need some justification. One important point refers to our confining the applicability of the two-state model of folding and unfolding to compact, single-domain, and globular proteins. This can be considered as a sufficient condition for fulfilling one criterion of Eq. (16)—the existence of spherically shaped molecules in solution. Whether proteins in solution can exhibit a closest-hexagonal type of packing is an important matter as well. Regarding the nature of the protein interior, Klapper [9] demonstrated that, on average, the ratio of the volume occupied by all atoms making up a protein to the total volume occupied by the protein molecule itself, ξ, is about 0.747. This compares well to a ξ value of about 0.765 for a typical compound assuming a closest-hexagonal packing of identical spheres [9]. Water assumes a ξ value of 0.363, whereas other organic solvents (e.g., cyclohexane and carbon tetrachloride) assume ξ values of 0.438. With that, Eq. (16) is considered applicable to protein in solution, so that it becomes

$$A_P = 1.091(V_P)^{2/3}(N_A)^{1/3} \tag{17}$$

Even if we consider a protein that is spherical in shape and exists in closest-hexagonal packing, deviation from Eq. (17) should be expected, depending on the degree of surface unfolding encountered, and Eq. (17) provides only an estimate for the surface area of protein at an interface. It is worth mentioning here, though, that the value of A_P predicted by Eq. (17) was found to correlate to a large extent with the corresponding close-packed "end-on" adsorption plateau value experimentally observed. As a qualitative rule in protein adsorption studies [13], a protein monolayer with an end-on conformation presumably prevails when adsorption occurs from solution at a high concentration, whereas a "side-on" conformation monolayer is observed at a low protein concentration. Although, practically, it is quite impossible to draw a line between low- and high-concentration regions, the concentration interval between $C_{eq} = 0.1$ mg/mL and $C_{eq} = 1.0$ mg/mL may be suggested as the diffuse line between low- and high-concentration regions as far as side-on and end-on conformations are concerned [27]. Using nonlinear regression as a tool to minimize the difference between the regressed value of A_P as given by an equation similar to Eq. (17) and that calculated from the known geometry of a given protein (Appendix 3), the following equation was developed to predict the value of A_P for a protein monolayer with a side-on conformation:

$$A_P = 2.717(V_P)^{2/3}(N_A)^{1/3} \tag{18}$$

In any case, descriptions of the nature of proteins and/or their behavior in solution being analogous to that of organic liquids has been encountered in more than one technical paper [8,9,46,53,64–66].

The final step in this dimensional analysis involves the arrangement of the dimensionless Π groups, requiring an awareness of the physical laws that underlie the adsorption process. The physical meaning of each term and how it influences the adsorption process can be visualized with a suitable adsorption mechanism.

B. A Simple Mechanism for Protein Adsorption

In general, the pattern of protein adsorption equilibrium isotherms at solid–liquid interfaces assumes either a Langmuir-type or Freundlich-type shape. The Langmuir-type model is described by

$$\Gamma = \frac{\Gamma_{max} C_{eq}}{(b + C_{eq})} \tag{19}$$

where C_{eq} (mg/L) is the apparent equilibrium concentration, Γ_{max} (μg/cm^2) is the plateau value, and b (mg/L) is a constant such that Γ_{max}/b is the initial slope of a plot of Γ versus C_{eq}. On the other hand, the Freundlich model is described by

$$\Gamma = a(C_{eq})^m \tag{20}$$

where a and m are function constants that define the functionality of Γ versus C_{eq}. Consequently, the Langmuir-type isotherm shows a steep initial slope followed by attainment of a plateau at high concentration, whereas the Freundlich isotherm shows a monotonic increase in Γ with C_{eq}. Upon examining the pertinent literature, one recognizes that the interpretation of data with Langmuir-type isotherms is extremely popular, even though no real benefit is gained by determining the function constants in either Eq. (19) or (20) (i.e., they are not related to adsorption affinity in any clear manner) [2,67–68].

Thermodynamic equilibrium criteria are implemented here to analyze the phase equilibrium between the bulk and interface, and for simplicity, the following two conditions are imposed:

1. Existence of reversible equilibrium between the bulk phase and the interface.
2. Existence of monolayer coverage of protein as the upper limit for the extent of adsorption.

Protein adsorption is usually regarded as irreversible, though, based on experimental observations (particularly at hydrophobic interfaces). Arnebrant and Nylander [69] found that sequential adsorption on hydrophilic surfaces is characterized by a larger fraction of reversibly adsorbed molecules than adsorption on hydrophobic surfaces. Elwing et al. [70] found that fibrinogen is partly exchanged by γ-globulin only on the hydrophilic side of a surface exhibiting a wettability gradient, whereas on the hydrophobic side there appeared to be no exchange at all. Shirahama et al. [71] found that sequential adsorption on hydrophilic silica occurs by displacement of preadsorbed protein, whereas on a hydrophobic surface, it is accompanied by desorption of only a fraction of preadsorbed protein. Khan and Wernet [72] examined the physical and the potential-assisted adsorption of various proteins on the shapeable electroconductive (SEC) polymer film. Because of the hydrophobic surface characteristic, proteins were found to easily adsorb and retain on the film surface by strong hydrophobic interactions.

It should be mentioned, however, that in all of these experimental observations, although pure desorption, in a strict sense, constitutes a very unlikely event, exchange reactions between adsorbed and incoming proteins have been observed to take place. In a sense, the assumption of reversible equilibrium serves as a tool to account for the outcome of exchange reactions that yield observable, gross adsorption equilibrium. Moreover, the possibility that adsorbing protein may form a multilayered film is quite real.

Although a multilayer film formation is the rule at air–water interfaces [73–77], a monolayer film formation is the rule at solid–water interfaces [68,78–86]. In fact, it is anticipated that study of monolayer adsorption of protein will shed light on understanding multilayer formation at air–water interfaces.

Figure 3 is a schematic of the proposed adsorption mechanism. Once a protein molecule arrives at the interface, it unfolds in an attempt to adapt to the new microenvironment. During the course of an adsorption experiment and at a certain point on the reaction coordinate, a pseudo, reversible phase equilibrium should adequately represent the situation existing between the bulk phase and the interface (Fig. 3a). For the purpose of analysis, this state of equilibrium can be resolved into two major subequilibrium states (Fig. 3b):

1. An equilibrium between native protein in the bulk and that at the hydrophobic interface, the attainment of which is largely driven by the Gibbs free energy for adsorption (or the work of adhesion).
2. An equilibrium between adsorbed, native protein and adsorbed, unfolded protein. Attainment of this substate of equilibrium is largely driven by the Gibbs free energy of unfolding.

In the absence of electrostatic effects on adsorption and of specific biochemical interactions (e.g., receptor–ligand) as well, the first subequilibrium state should be effectively characterized by the work of adhesion between protein and surface. Absolom et al. [87] used a fundamentally similar approach to describe the adhesion of bacteria to various polymeric low energy surfaces with good success. The model proposed here is limited to adsorption at hydrophobic interfaces; so from a thermodynamic standpoint, the surface energies of each of the interacting phases should adequately describe the initial adsorption event.

For the second subequilibrium state, the Gibbs free energy of unfolding was selected to quantify this process, involving exposure of the previously intact hydrophobic core to the aqueous medium. Several investigators [10,64,88,89] indicated that the Gibbs free energy of unfolding correlates well with the surface area of nonpolar groups exposed upon unfolding; that is, the hydrophobic stabilization is proportional to the reduction of the surface area accessible to solvent on folding. Wei et al. [27] indicated that, at low bulk concentrations, surface tension kinetics reflected the conformational stability of the protein; whereas at higher concentrations, surface tension kinetics were more strongly correlated with the effective hydrophobicity of the protein. Norde [90] indicated that with "rigid" proteins, intramolecular structural rearrangements do not contribute to the adsorption process, whereas with "soft" proteins, the intra-

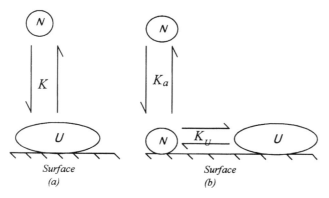

FIG. 3 A schematic depicting protein adsorption equilibrium.

molecular structural rearrangements result in a significant driving force for adsorption and that proteins may even adsorb under the adverse conditions of a hydrophilic, electrostatically repelling surface.

One important point that should be made is that in addition to the proposed mechanism of Fig. 3, there are other mechanisms incorporating arrival, unfolding, and exchange reactions to describe adsorption equilibrium. However, Gibbs free energy is a state function; that is, no matter what path is chosen to analyze the process, it is governed, overall, by the Gibbs free energy change between the initial state (i.e., native protein in solution) and the final state (i.e., unfolded protein at the interface).

A theoretical model can be developed to express the function constants that appear in Eq. (19) in terms of selected molecular, solution, and surface properties. As indicated with reference to Eq. (19), the Langmuir-type model is the most popular model used in the literature to describe adsorption equilibrium isotherms at solid–liquid interfaces. The Langmuir-type isotherm is characterized by an initial (usually steep) slope at low concentration followed by plateau attainment at high concentration. A mathematical expression of this statement can be developed as follows:

$$\Gamma A_P \propto V_P C_{eq} \tag{21}$$

at low concentration, where ΓA_P represents a dimensionless adsorbed mass (or surface coverage) and $V_P C_{eq}$ represents a dimensionless concentration (or the volume fraction of protein in solution). Defining the relationship in the form of an equation, it becomes

$$\Gamma = K V_P C_{eq} \left(\frac{1}{A_P} \right) \tag{22}$$

where K, the proportionality constant between dimensionless adsorbed mass and dimensionless concentration, is the overall equilibrium constant for the protein adsorption process.

At high concentration

$$\Gamma = \left(\frac{1}{A_P} \right) \tag{23}$$

If Eqs. (22) and (23) are combined, the following equation is obtained which accounts for both regions:

$$\left(\frac{1/A_P}{\Gamma} \right) = 1 + \left(\frac{1}{K V_P C_{eq}} \right) \tag{24}$$

Rearranging Eq. (24) yields

$$\left(\frac{\Gamma}{1/A_P} \right) = \frac{1}{1 + \left(\dfrac{1}{K V_P C_{eq}} \right)} \tag{25}$$

Finally, the adsorbed mass can be expressed as

$$\Gamma = \frac{V_P C_{eq}(1/A_P)}{(1/K) + V_P C_{eq}} \tag{26}$$

As shown in Fig. 3a, the overall equilibrium constant can be defined as

$$K = \frac{[\text{protein}]_U^{\text{ads}}}{[\text{protein}]_N^{\text{bulk}}} \tag{27}$$

where the subscript N stands for native conformation and U for unfolded conformation. The superscripts indicate that protein is either adsorbed (ads) or in solution (bulk), near the interface. Equation (27) can be written as

$$K = \frac{[\text{protein}]_U^{\text{ads}}}{[\text{protein}]_N^{\text{ads}}} \frac{[\text{protein}]_N^{\text{ads}}}{[\text{protein}]_N^{\text{bulk}}} \tag{28}$$

where the first term on the right-hand side of Eq. (28) represents K_U and the second term represents K_a; therefore, the overall equilibrium constant assumes the form

$$K = K_U K_a \tag{29}$$

Introducing the definition of an equilibrium constant as a function of the standard, Gibbs free energy for both K_U and K_a,

$$K = \exp\left(-\frac{\alpha \Delta G_{\text{unfold}}}{RT}\right) \exp\left(\frac{W_a A_C}{RT}\right) \tag{30}$$

where α is the proportionality constant that relates the Gibbs free energy of unfolding at the interface to that in the bulk, at specific conditions of temperature, pH, and ionic strength. Here, the surface-induced unfolding process for a given protein molecule is related to its intrinsic conformational liability (stability) in solution, where the latter is characterized by ΔG_{unfold}.

Andrade et al. [91] suggested that data on the solution denaturation of proteins may be important in estimating protein liability (stability) and, together with information on the surface tension and interfacial tension behavior of proteins, would help develop hypotheses and correlations with the actual solid–liquid interface behavior. W_a is equal to the negative of the standard, Gibbs free energy change per unit area of the reversible arrival step [92]. Multiplied by A_C, the quantity $W_a A_C$ essentially represents the negative of the standard, Gibbs free energy change for the reversible arrival step, at constant temperature and pressure.

Finally, plugging the value of K into Eq. (26), a model that relates Γ to the pertinent variables is obtained:

$$\Gamma = V_P C_{\text{eq}} \frac{1}{A_P} [\exp\left(\frac{\alpha \Delta G_{\text{unfold}} - W_a A_C}{RT}\right) + V_P C_{\text{eq}}]^{-1} \tag{31}$$

C. Limitations on the Applicability of the Model [Eq. (31)]

The limitations on the reliable application of the proposed model basically stem from the set of conditions associated with quantifying each term appearing in Eq. (31). The physical properties of the protein, solid surface, and solution required to reliably define the quantities appearing in Eq. (31) are outlined here:

1. Protein properties: compact, single-domain, globular structure.
2. Solid surface properties: homogeneous, hydrophobic surface, not promoting any specific biochemical interaction.
3. Solution properties: pH and ionic strength such that the effect of electrostatic interactions between the solid surface and protein on adsorption are minimal. The system temperature must be confined between a minimum value around room

temperature and a maximum value of 55°C, or the irreversible denaturation temperature for a given protein, whichever is smaller.

Although most of the protein adsorption isotherms studied here were constructed at moderate temperatures (18–27°C), the confinement made concerning the acceptable temperature range emanates from two points:

1. Such a temperature range is common for study of proteins in solution in their native conformations.
2. The partial specific volume of protein does not significantly change with temperature within that range.

Bull and Breese [57] studied the temperature dependence of partial volumes of proteins and found that there exists a temperature range from 25°C up to 45–50°C within which the partial specific volume gradient with respect to temperature assumes, on the average, a value of 3.42×10^{-4} (mL/g)/K, and at about 55°C, an abrupt change in magnitude of the slope was observed. They suggested that this anomaly might be attributed to a predenaturational stage connected to a labilization of the native protein structure.

IV. MODEL TESTING

In order to simulate adsorption equilibrium according to Eq. (31) for comparison with experimental data, all physical properties appearing in that equation must be known for a given protein-surface contact. In addition to protein physical properties, the temperature at which the adsorption experiment was conducted as well as the solid surface hydrophobicity, expressed as γ_{SV}, must be known. Consequently, all pertinent data used to quantify the protein and solid surface properties that appear in Eq. (31) were provided with temperature as input to a FORTRAN program. This program was used to estimate the value of A_C using a nonlinear regression method based on minimization of the difference between the value of experimentally based, adsorbed mass, Γ, and that given by the model equation. In each case, the applicability of Eq. (31) was examined using only those sets of protein isotherms obtained from adsorption experiments conducted on hydrophobic solid surfaces. The determination of adsorbed mass of protein is, to some extent, technique dependent; to preclude the possibility of such artifacts interfering with the present analysis, each set of isotherms was tested separately.

The other requirement imposed on this analysis concerns model application only to compact, single-domain globular proteins that attain monolayer coverages. However, multi-domain proteins including human fibrinogen, immunoglobulin G, and plasminogen were used to demonstrate limits in the applicability of Eq. (31). With the required input data, nonlinear regression was carried out to determine the best value for the area cleared by a protein molecule upon adsorption, A_C. From preliminary simulation results allowing computer generation of both A_C and α, it was found that the variation in α was statistically insignificant, as both α and A_C are highly correlated; hence, one variable should be fixed. Moreover, as shown in Table 1, the quality of regression is even improved as the value of parameter α approaches zero.

For that reason and for other reasons explained in detail in Section V.A, the proportionality constant α, which relates the Gibbs free energy of unfolding at the interface to that in the bulk, was set equal to zero. The affinity constant, K, defined by Eq. (30) is shown in Tables 5, 7, 9, 11, 13, 15, and 17, with the parameter α being equal to 0.0 in all studied cases. Statistically, the goodness of fit of the model equation can be quantified by the value of the mean square of errors (MSE) in addition to the graphical representation of the regressed data.

TABLE 1 Regressed Values of α and A_C, and the Associated Regression Parameters for the Equilibrium Model [Eq. (31)]

α	A_C (Å^2/molecule)	Adjusted R^2	MSE $\times 10^3$	$F(x)^a$ (μg/cm^2)^2
0.000(0.029)	192(10)	0.745	9.34	1.0186
0.01(0.033)	195(10)	0.741	9.48	1.0331
0.05(0.034)	207(10)	0.729	9.94	1.0840
0.1(0.036)	222(10)	0.717	10.38	1.1317
0.5(0.039)	344(12)	0.685	11.54	1.2575

Note: Fourteen different isotherms (110 data points) were used as input. The 95% confidence interval is shown in parentheses.
$^a F(x) = (\Gamma_{fit} - \Gamma_{exp})^2$; the objective function to be minimized.

For simplicity, each of the following abbreviations will be used for the names of the selected proteins:

α-Lac	α-Lactalbumin
β-Lag	β-Lactoglobulin
BSA	Bovine serum albumin
ChA	Chymotrypsinogen A
Fbrgn	Fibrinogen
HSA	Human serum albumin
IgG	Immunoglobulin G
Lyso	Lysozyme
Myog	Myoglobin
Plmgn	Plasminogen
RiboA	Ribonuclease A

Table 2 shows the thermal properties of each protein used to evaluate ΔG_{unfold} according to Eq. (14). In each case, the thermal properties were obtained using DSC analysis. One point worth mentioning is the values of thermal properties shown in Table 2 correspond to DSC analyses carried out under acidic pH values. As was pointed out by Privalov and Makhatadze [89], the partial heat capacity values of heat- and acid-denatured proteins are indistinguishable in the temperature range from 5°C to 125°C. Moreover, carrying out DSC analysis under alkaline pH values results in a value of enthalpy of denaturation that is larger than that recorded under acidic pH values [18,93]. The reason for that is thermal denaturation carried out under alkaline pH values proceeds while accompanied by the aggregation of denatured molecules; hence, a higher value of enthalpy of denaturation is encountered [89,94].

Table 3 shows the specific volume (V_P), molecular weight, and partial molar area of protein upon adsorption, A_P, for each protein examined in this study. It should be emphasized that a difference in the value reported for molecular weight and, to a lesser extent, in the value of specific volume (V_P) may exist from one source of data to another. In all cases studied, the values of molecular weight and V_P reported by the investigators who constructed the isotherm for a given protein was considered first. In any event, this variation is negligible compared with the value of molecular weight itself. Before presenting the experimentally based isotherms, the following points should be addressed:

TABLE 2 Thermal Properties of Proteins Used in Evaluating ΔG_{unfold} [Eq. (14)], with Standard Errors Shown in Parentheses

Protein	Molecular weight	ΔH_D (kJ/mol)	ΔC_P (kJ/mol K)	T_m (K)	Ref.
α-Lac	14,161	184(11)	4.0(0.8)	312.7	95, 96
β-Lag	33,640 (dimer)	599(19)	17.6 (dimer)	344.0	95, 97
BSA	66,267	799(44)	25.0	334.3	95, 98
ChA	23,000	557	10.8	333.1	99
Fbrgn	340,000	2259	62.8	318.9	18, 54
HSA	69,000	879	26.0	341.1	54, 98
IgG	150,00	1260	62.8	317.1	51, 54
Lyso	14,400	590	7.3	351.1	52, 89, 96
Myog	17,800	254	10.4	335.1	17
Plmgn	94,000	929	37.6	313.1	94
RiboA	13,680	407	5.0	323.6	99

TABLE 3 Specific Volume (V_P), Molecular Weight, and the Partial Molar Area of Protein Upon Adsorption, A_P, for Each of the Examined Proteins Present in This Study

Protein	Molecular weight	V_P (cm^3/g)	A_P (end-on) (Å^2/molecule)	(cm^2/µg)	Ref.
α-Lac	14,161	0.729	725	3.08	58, 95
			1806	7.68	
			(side-on)		
β-Lag	36,640 (dimer)	0.751	1,393	2.29	58, 95
BSA	66,267	0.733	2,037	1.85	58, 95
ChA	23,000	0.730	2,498	6.54	100
			(side-on)		
Fbrgn	340,000	0.723	6,003	1.06	18, 58
			14,951	2.65	
			(side-on)		
HSA	69,000	0.733	2,092	1.83	58
IgG	150,000	0.739	3,530	1.42	58, 101
Lyso	14,400	0.730	734	3.07	56, 58, 85
			1,828	7.65	
			(side-on)		
Myog	17,800	0.742	2,129	7.20	85
			(side-on)		
Plmgn	94,000	0.715	6,298	4.03	2, 102
			(side-on)		
RiboA	13,680	0.703	1,723	7.58	85
			(side-on)		

1. The protein-free solution surface tension, γ_{WV}, is essentially buffer independent and assumes a value very close to that of pure water [41,44]. A value of γ_{WV} equal to $72.5 \, mJ/m^2$, measured at room temperature, was assumed for all proteins studied, except for those that were carried out at $52°C$.
2. Regarding isotherms carried out at $52°C$ [95], the effect of temperature on both water and solid surface tension was handled by considering the linear surface tension decrease with increasing temperature. For water, $d\gamma_{WV}/dT$ was taken as $-0.152 \, (mJ/m^2)/K$, and for the solid, $d\gamma_{SV}/dT$ was taken to $-0.1 \, (mJ/m^2)/K$ [92].
3. The data points, each of which represents the value of adsorbed mass versus that of apparent equilibrium concentration, were read from the pertinent figures available in the literature. Usually, the data points had been connected by a line and that line was used to generate data points in cases where the number of data points was low.

Figures 4 and 5 show experimental data of Suttiprasit and McGuire [95] curve-fitted to Eq. (31). In this case, α-Lac, β-Lag, and BSA were adsorbed on hydrophobic silicon surfaces at $27°C$ and $52°C$, from $0.01 \, M$ phosphate buffer at pH 7.00. Table 4 shows the values of ΔG_{unfold}, along with the interfacial energies of solid, protein, and water, and the work of adhesion, W_a, relevant to the isotherms of Figs. 4 and 5. Table 5 shows the corresponding values of A_C, the affinity constant, K, and the regression parameters obtained for the isotherms of Figs. 4 and 5.

Figure 6 shows another set of isotherms [86] measured at hydrophobic butylated quartz slides. Total internal reflectance fluorescence (TIRF) spectroscopy was used to assess the adsorbed mass of Lyso and ChA. The isotherm experiments were carried out at pH 7.0, with protein dissolved in a $0.01 \, M$ phosphate buffer including $0.1 \, M$ NaC1. Tables 6 and 7 show the relevant thermodynamic data and regression results for this set of isotherms.

Figure 7 shows a human (milk) Lyso adsorption isotherm recorded on hydrophobic silica slides treated with dimethyldichlorosilane (DDS) [29]. Total internal reflectance fluorescence (TIRF) spectroscopy was used to assess the adsorbed mass of Lyso. The isotherm experiment was carried out at pH 7.4, with protein dissolved in a phosphate-buffered saline (PBS) ($0.013 \, M$ KH_2PO_4, $0.054 \, M$ Na_2HPO_4) buffer including $0.1 \, M$ NaC1. Tables 8 and 9 show the relevant thermodynamic and regression data determined for this isotherm.

Figure 8 shows a Plmgn adsorption isotherm recorded on hydrophobic, methylene dianiline (MDA) polyurethane [2]. A radiolabeling technique using ^{125}I was used to measure the adsorbed mass of Plmgn in these experiments. The isotherm experiments were carried out at pH 7.4, with protein dissolved in isotonic Tris buffer. Tables 10 and 11 show the relevant thermodynamic and regression data associated with the Plmgn adsorption isotherm.

Figure 9 shows HSA adsorption isotherms on different hydrophobic polymer surfaces studied by Winterton et al. [1]. An ^{125}I radiolabeling technique was used to measure the adsorbed mass of HSA. The isotherm experiments were carried out at pH 7.4, with protein dissolved in a PBS buffer. The polymers used in HSA adsorption were polystyrene (PS), poly(vinyl chloride) (PVC), and silastic (Silas). Tables 12 and 13 show the relevant thermodynamic and regression data for HSA adsorption isotherms on these polymers.

Figure 10 shows isotherms for several proteins that were measured at hydrophobic, sliconized glass particles [81]. The isotherm experiments were carried out at pH 7.2, with protein dissolved in a PBS buffer of ionic strength $0.1 \, M$. The proteins used were IgG, HSA, BSA, and Fbrgn. An ^{125}I radiolabeling technique was used to measure the adsorbed mass of these proteins. Tables 14 and 15 show the relevant thermodynamic and regression data for the set of isotherms shown in Figure 10.

Finally, Fig. 11 shows Fbrgn adsorption isotherm recorded on a hydrophobic quartz surface by Nygren and Stenberg [103]. The isotherm experiments were carried out at pH 7.2,

Ceq-model	Ceq-exp	Gamma-exp	Gamma-model
2	2	0,307	0,30561436
1,9	1,5	0,337	0,304673203
1,8	1	0,289	0,303634248
1,7	0,6	0,322	0,302481417
1,6	0,3	0,16	0,301194902
1,5	0,1	0,16	0,299750017
1,4	0	0	0,298115602
1,3			0,296251748
1,2			0,294106495
1,1			0,291610912
0,9			0,285158494
0,8			0,280885611
0,7			0,275576507
0,6			0,268802223
0,5			0,259859141
0,4			0,247507247
0,3			0,229338644
0,2			0,199979164
0,1			0,14448797
0			0

α-Lac/Data

α-Lac/Eq. (31)

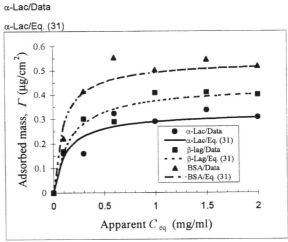

β-lag@T=27°C

Ceq-model	Ceq-exp	Gamma-exp	Gamma-model
2	2	0,4	0,403130522
1,9	1,5	0,41	0,401506986
1,8	1	0,408	0,399718326
1,7	0,6	0,289	0,397738001
1,6	0,3	0,301	0,39553346
1,5	0,1	0,168	0,393064339
1,4	0	0	0,390279966
1,3			0,387115849
1,2			0,383488614
1,1			0,379288559
0,9			0,368525424
0,8			0,361473331
0,7			0,352793416
0,6			0,34184852
0,5			0,327619061
0,4			0,308365472
0,3			0,28085642
0,2			0,238333343
0,1			0,163891394
0			0

BSA @T=27°C

Ceq-model	Ceq-exp	Gamma-exp	Gamma-model	
2	2	0,518	0,515426638	BSA/Data
1,9	1,5	0,546	0,514169365	BSA/Eq. (31)
1,8	1	0,503	0,512779569	
1,7	0,6	0,554	0,511235133	
1,6	0,3	0,415	0,509508724	
1,5	0,1	0,222	0,50756617	
1,4	0	0	0,505364166	
1,3			0,502847016	
1,2			0,499941851	
1 1			0,496551458	
0,9			0,487731231	
0,8			0,481846897	
0,7			0,47448677	
0,6			0,465016052	
0,5			0,452374937	
0,4			0,434651425	
0,3			0,40800923	
0,2			0,363453118	
0,1			0,273764756	
0			0	

FIG. 4 Protein adsorption isotherms ($T = 27°C$) on a hydrophobic silicon surface, fitted to Eq. (31). (From Ref. 95.)

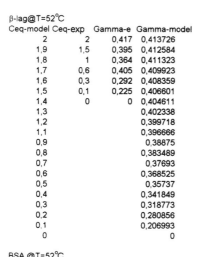

β-lag@T=52°C

Ceq-model	Ceq-exp	Gamma-e	Gamma-model
2	2	0,417	0,413726
1,9	1,5	0,395	0,412584
1,8	1	0,364	0,411323
1,7	0,6	0,405	0,409923
1,6	0,3	0,292	0,408359
1,5	0,1	0,225	0,406601
1,4	0	0	0,404611
1,3			0,402338
1,2			0,399718
1,1			0,396666
0,9			0,38875
0,8			0,383489
0,7			0,37693
0,6			0,368525
0,5			0,35737
0,4			0,341849
0,3			0,318773
0,2			0,280856
0,1			0,206993
0			0

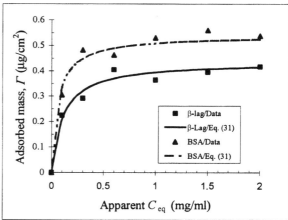

BSA @T=52°C

Ceq-model	Ceq-exp	Gamma-e	Gamma-model
2	2	0,538	0,524284
1,9	1,5	0,56	0,523456
1,8	1	0,53	0,522538
1,7	0,6	0,463	0,521516
1,6	0,3	0,483	0,520372
1,5	0,1	0,305	0,519081
1,4	0	0	0,517613
1,3			0,51593
1,2			0,513979
1,1			0,511694
0,9			0,505697
0,8			0,501654
0,7			0,496551
0,6			0,489907
0,5			0,480897
0,4			0,467988
0,3			0,447947
0,2			0,412608
0,1			0,333643
0			0

FIG. 5 Protein adsorption isotherms ($T = 52°C$) on a hydrophobic silicon surface, fitted to Eq. (31). (From Ref. 95.)

TABLE 4 Gibbs Free Energy of Unfolding, Protein, Solid, and Water Surface Energies, and the Work of Adhesion for Proteins Used in Figs. 4 and 5

Protein	T (°C)	ΔG_{unfold} (kJ/mol)	γ_{SV} (mJ/m^2)	γ_{PV} (mJ/m^2)	γ_{WV} (mJ/m^2)	W_a (mJ/m^2)
α-Lac	27	6.4	17.5	38.0	72.5	53.6
β-Lag	27	27.2	17.5	56.3	72.5	23.6
BSA	27	38.1	17.5	55.4	72.5	25.0
β-Lag	52	23.7	15.0	51.2	68.7	23.9
BSA	52	18.8	15.0	50.2	68.7	25.4

Source of isotherms: Ref. 95.

TABLE 5 Surface Area Cleared by an Adsorbing Protein Molecule, A_C, the Affinity Constant K, and the Regression Parameters for the Model [Eq. (31)]

Protein	T (°C)	A_C (Å²/molecule) (p-value)	Conformation	$K \times 10^{-4}$	MSE $\times 10^3$
α-Lac	27	72 (0.0001)	End-on	1.1 ± 0.4	1.5
β-Lag	27	158 (0.0001)	End-on	0.8 ± 0.2	0.8
BSA	27	158 (0.0001)	End-on	1.4 ± 0.4	2.0
β-Lag	52	177 (0.0001)	End-on	1.2 ± 0.2	0.5
BSA	52	177 (0.0001)	End-on	2.2 ± 0.5	0.9

Source of isotherms: Ref. 95.

with protein dissolved in $0.01\,M$ PBS. Ellipsometry and enzyme-linked immunosorbent assay (ELISA) were both used to quantify the adsorbed mass of Fbrgn. Tables 16 and 17 show the relevant thermodynamic and regression data for this adsorption isotherm.

V. DISCUSSION

The applicability of Eq. (31) is governed by the extent to which its premises are satisfied, depending on properties of the protein, solid surface, and solution selected for study. Discounting experimental inaccuracies, agreement between the model and an experimentally measured adsorption isotherm would be expected for adsorption of a single-domain globular protein at a homogeneous hydrophobic surface. Agreement is apparent in Figs. 4–9 and, to a lesser extent, in Figs. 10 and 11. The term "agreement" here not only denotes coincidence between adsorbed mass predicted by the model and that measured in the adsorption experiment, but it should also be taken to imply that the computer-generated value for the minimum surface area cleared by an adsorbing protein molecule, A_C, is consistent with that expected for the selected protein. The importance of generating a reasonable value for A_C will be discussed with reference to the model premises and to the values obtained for A_C by application of the model as well as by experimental measurements. However, first, some justification should be provided regarding selection of some of the input parameters to the model.

A. Parameters α and A_C

1. Value of α at Low-Energy Interfaces

Simulation of Eq. (31) would require knowledge of W_a, ΔG_{unfold}, V_P, A_P (all four properties calculated according to the methods of Section III), α, and A_C. These last two input variables are to be fixed either at a particular system-specific value or computer generated. The parameter α is the proportionality constant that relates the Gibbs free energy of unfolding in the bulk to that at

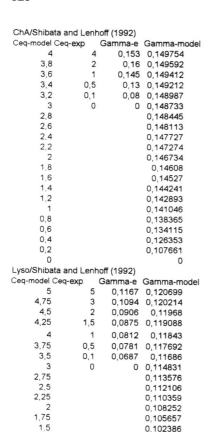

ChA/Shibata and Lenhoff (1992)

Ceq-model	Ceq-exp	Gamma-e	Gamma-model
4	4	0,153	0,149754
3,8	2	0,16	0,149592
3,6	1	0,145	0,149412
3,4	0,5	0,13	0,149212
3,2	0,1	0,08	0,148987
3	0	0	0,148733
2,8			0,148445
2,6			0,148113
2,4			0,147727
2,2			0,147274
2			0,146734
1,8			0,14608
1,6			0,14527
1,4			0,144241
1,2			0,142893
1			0,141046
0,8			0,138365
0,6			0,134115
0,4			0,126353
0,2			0,107661
0			0

Lyso/Shibata and Lenhoff (1992)

Ceq-model	Ceq-exp	Gamma-e	Gamma-model
5	5	0,1167	0,120699
4,75	3	0,1094	0,120214
4,5	2	0,0906	0,11968
4,25	1,5	0,0875	0,119088
4	1	0,0812	0,11843
3,75	0,5	0,0781	0,117692
3,5	0,1	0,0687	0,11686
3	0	0	0,114831
2,75			0,113576
2,5			0,112106
2,25			0,110359
2			0,108252
1,75			0,105657
1,5			0,102386
1,25			0,098132
1			0,092374
0,75			0,084147
0,5			0,071423
0,25			0,049135
0			0

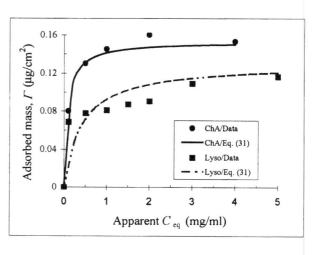

FIG. 6 Protein adsorption isotherms on hydrophobic, butylated quartz slides, fitted to Eq. (31). (From Ref. 86.)

TABLE 6 Gibbs Free Energy of Unfolding, Protein, Solid, and Water Surface Energies, and the Work of Adhesion for Proteins Used in Fig. 6

Protein	T (°C)	ΔG_{unfold} (kJ/mol)	γ_{SV} (mJ/m^2)	γ_{PV} (mJ/m^2)	W_a (mJ/m^2)
Lyso	20	62.5	35.0	66.3	8.6
ChA	20	40.9	35.0	57.7	19.2

Source of isotherms: Ref. 86.

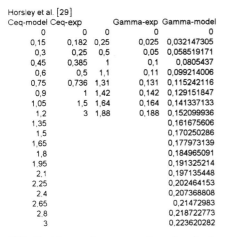

Horsley et al. [29]

Ceq-model	Ceq-exp		Gamma-exp	Gamma-model
0	0	0	0	0
0,15	0,182	0,25	0,025	0,032147305
0,3	0,25	0,5	0,05	0,058519171
0,45	0,385	1	0,1	0,0805437
0,6	0,5	1,1	0,11	0,099214006
0,75	0,736	1,31	0,131	0,115242116
0,9	1	1,42	0,142	0,129151847
1,05	1,5	1,64	0,164	0,141337133
1,2	3	1,88	0,188	0,152099936
1,35				0,161675606
1,5				0,170250286
1,65				0,177973139
1,8				0,184965091
1,95				0,191325214
2,1				0,197135448
2,25				0,202464153
2,4				0,207368808
2,65				0,21472983
2,8				0,218722773
3				0,223620282

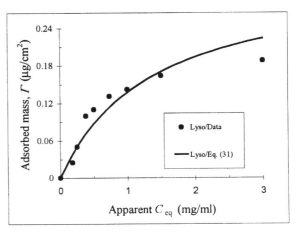

FIG. 7 Human lysozyme adsorption isotherm ($T = 25°C$) on hydrophobic DDS silica slides, fitted to Eq. (31). (From Ref. 29.)

TABLE 7 Surface Area Cleared by an Adsorbing Protein Molecule, A_C, the Affinity Constant K, and the Regression Parameters for the Model [Eq. (31)]

Protein	T (°C)	A_C (Å^2/molecule) (p-value)	Conformation	$K \times 10^{-4}$	MSE $\times 10^4$
Lyso	20	379 (0.0001)	Side-on	0.33	3.7
ChA	20	205 (0.0001)	Side-on	1.63	0.4

Source of isotherms: Ref. 86.

TABLE 8 Gibbs Free Energy of Unfolding, Protein, Solid, and Water Surface Energies, and the Work of Adhesion for Proteins Used in Fig. 7

Protein	T (°C)	ΔG_{unfold} (kJ/mol)	γ_{SV} (mJ/m^2)	γ_{PV} (mJ/m^2)	W_a (mJ/m^2)
Lyso	25	59.8	16.5	65.1	10.5

Source of isotherms: Ref. 29.

TABLE 9 Surface Area Cleared by an Adsorbing Protein Molecule, A_C, the Affinity Constant K, and the Regression Parameters for the Model [Eq. (31)]

Protein	T (°C)	A_C (Å^2/molecule) (p-value)	Conformation	$K \times 10^{-4}$	MSE $\times 10^4$
Lyso	25	272 (0.0001)	End-on	0.10	3.9

Source of isotherms: Ref. 29.

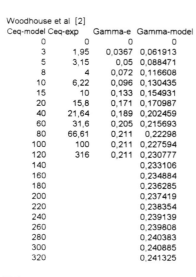

Woodhouse et al [2]

Ceq-model	Ceq-exp	Gamma-e	Gamma-model
0	0	0	0
3	1,95	0,0367	0,061913
5	3,15	0,05	0,088471
8	4	0,072	0,116608
10	6,22	0,096	0,130435
15	10	0,133	0,154931
20	15,8	0,171	0,170987
40	21,64	0,189	0,202459
60	31,6	0,205	0,215693
80	66,61	0,211	0,22298
100	100	0,211	0,227594
120	316	0,211	0,230777
140			0,233106
160			0,234884
180			0,236285
200			0,237419
220			0,238354
240			0,239139
260			0,239808
280			0,240383
300			0,240885
320			0,241325

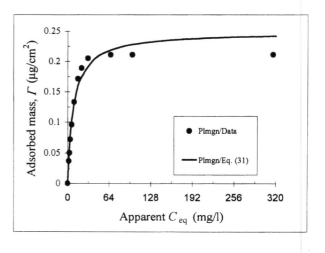

FIG. 8   Plasminogen adsorption isotherm ($T = 25°C$) on a hydrophobic MDA polyurethane surface, fitted to Eq. (31). (From Ref. 2.)

TABLE 10   Gibbs Free Energy of Unfolding, Protein, Solid, and Water Surface Energies, and the Work of Adhesion for Proteins Used in Fig. 8

Protein	T (°C)	ΔG_{unfold} (kJ/mol)	γ_{SV} (mJ/m^2)	γ_{PV} (mJ/m^2)	W_a (mJ/m^2)
Plmgn	25	31.3	41.7	52.1	21.5

Source of isotherm: Ref. 2.

TABLE 11   Surface Area Cleared by an Adsorbing Protein Molecule, A_C, the Affinity Constant K, and the Regression Parameters for the Model [Eq. (31)]

Protein	T (°C)	A_C (Å^2/molecule) (p-value)	Conformation	$K \times 10^{-4}$	MSE $\times 10^4$
Plmgn	25	229 (0.0001)	Side-on	15.5	1.8

Source of isotherm: Ref. 2.

the interface. The value of α was set equal to zero (i.e., fixed at a given value) for the following reasons:

1.   Simulation results allowing computer generation of both α and A_C showed that variation in α was statistically insignificant in that both α and A_C were highly correlated. As was shown in Table 1, setting α equal to zero further improves the goodness of fit of the model when the sets of isotherms are lumped together;

Winterton et al. [1]

	HSA/PS			HSA/PVC			HSA/Silas		
Ceq-model	Ceq-exp	Gamma-exp	Gamma-model	Ceq-exp	Gamma-ex	Gamma-mo	Ceq-exp	Gamma-exp	Gamma-model
0	0	0	0	0	0	0	0	0	0
5	10	0,176	0,13779148	10	0,1	0,06833882	10	0,081	0,05413053
10	24	0,3	0,220086494	25,3	0,2	0,1214848	32,4	0,2	0,09850345
20	42	0,4	0,31379136	50	0,305	0,19877799	50	0,25	0,16691812
30	50	0,443	0,365690642	80,6	0,4	0,25228169	67,6	0,3	0,21720375
40	100	0,538	0,398658555	100	0,433	0,2915141	100	0,39	0,25572325
50	200	0,586	0,421455718	200	0,5	0,32151329	200	0,5	0,28617374
80	300	0,64	0,460998827	300	0,514	0,38020216	300	0,62	0,34840346
100			0,475882047			0,40483495			0,37563099
120			0,486349828			0,42311011			0,39627687
140			0,49411326			0,43720763			0,4124702
160			0,500100456			0,44841307			0,42551117
180			0,504858431			0,45753359			0,43623863
200			0,508730496			0,46510156			0,44521806
220			0,511943006			0,47148231			0,45284453
240			0,514651254			0,47693488			0,4594024
260			0,51696533			0,48164808			0,46510156
280			0,518965451			0,48576273			0,47010031
300			0,52071145			0,48938606			0,47452029

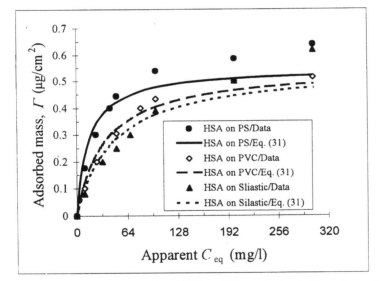

FIG. 9 HSA adsorption isotherms ($T = 25°C$) on different hydrophobic surfaces, fitted to Eq. (31). (From Ref. 1.)

TABLE 12 Gibbs Free Energy of Unfolding, Protein, Solid, and Water Surface Energies, and the Work of Adhesion for Proteins Used in Fig. 9

Protein	T (°C)	ΔG_{unfold} (kJ/mol)	Solid surface	γ_{SV} (mJ/m²)	γ_{PV} (mJ/m²)	W_a (mJ/m²)
HSA	25	40.3	PS	22.4	56.8	22.8
HSA	25	40.3	PVC	30.5	56.8	21.7
HAS	25	40.3	Silas	23.0	56.8	22.8

Source of isotherms: Ref. 1.

TABLE 13 Surface Area Cleared by an Adsorbing Protein Molecule, A_C, the Affinity Constant K, and the Regression Parameters for the Model [Eq. (31)]

Protein	T ($^\circ$C)	A_C (Å^2/molecule) (p-value)	Conformation	$K \times 10^{-4}$	MSE $\times 10^4$
HSA	25	206 (0.0001)	End-on	9.2	3.9
HSA	25	200 (0.0001)	End-on	3.9	0.6
HSA	25	186 (0.0001)	End-on	3.0	3.9

Source of isotherms: Ref. 1.

Absolom et al. [81]

Ceq-model	BSA Ceq-exp	BSA Gamma-Exp	BSA Gamma-mod	Fibrinogen Gamma-Exp	Fibrinogen Gamma-mod	HSA Gamma-Exp	HSA Gamma-Mod	IgG Gamma-Exp	IgG Gamma-Model
0	0	0	0	0	0	0	0	0	0
0,1	0,1	0,045	0,00079127	0,2	0,17904966	0,07	0,001597508	0,15	0,049362
0,2	1	0,055	0,00158023	0,45	0,30097623	0,105	0,003185702	0,26	0,092258
0,5	2,5	0,08	0,00393333	0,86	0,50890359	0,115	0,007895213	0,49	0,192764
1	5	0,1	0,00780982	1,1	0,66115506	0,15	0,015565533	0,6	0,302678
2	10	0,1	0,01539719	1,1	0,7774524	0,15	0,030268862	0,6	0,423384
4	15	0,1	0,0299415	1,1	0,85242312	0,15	0,057360425	0,6	0,528832
5	20	0,1	0,03691566	1,1	0,86918648	0,15	0,069867059	0,6	0,556555
6			0,04370188		0,8807332		0,08175002		0,57671
7			0,05030764		0,88917048		0,093054826		0,592024
9			0,06300586		0,90067492		0,114090915		0,613754
10			0,06911143		0,90477212		0,123893551		0,621742
12			0,08086587		0,9109883		0,142223168		0,63412
14			0,09204841		0,91548098		0,159028697		0,643269
16			0,1026998		0,91887968		0,174492638		0,650305
18			0,11285699		0,9215406		0,188769474		0,655885
20			0,12255361		0,92368047		0,201990847		0,660418

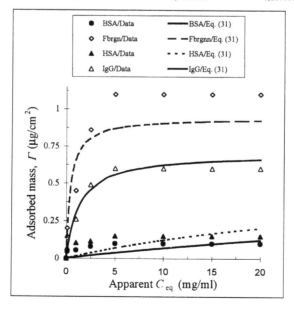

FIG. 10 Protein adsorption isotherms ($T = 24^\circ$C) on hydrophobic siliconized glass, fitted to Eq. (31). (From Ref. 81.)

TABLE 14 Gibbs Free Energy of Unfolding, Protein, Solid, and Water Surface Energies, and the Work of Adhesion for Proteins Used in Fig. 10

Protein	T (°C)	ΔG_{unfold} (kJ/mol)	γ_{SV} (mJ/m^2)	γ_{PV} (mJ/m^2)	W_a (mJ/m^2)
BSA	24	37.3	18.7	55.4	25.1
HSA	24	39.6	18.7	56.6	23.2
IgG	24	39.6	18.7	56.6	23.2
Fbrgn	24	107.7	18.7	71.8	0.9

Source of isotherms: Ref. 81.

TABLE 15 Surface Area Cleared by an Adsorbing Protein Molecule, A_C, the Affinity Constant K, and the Regression Parameters for the Model [Eq. (31)]

Protein	T (°C)	A_C (Å^2/molecule) (p-value)	Conformation	$K \times 10^{-3}$	MSE $\times 10^3$
BSA	24	52 (0.0003)	End-on	0.02	1.8
HSA	24	66 (0.0003)	End-on	0.04	4.1
IgG	24	122 (0.0001)	End-on	1.02	3.0
Fbrgn	24	3631 (0.0003)	End-on	3.24	29.7

Source of isotherms: Ref. 81.

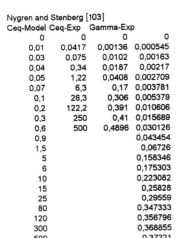

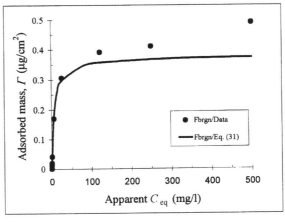

FIG. 11 Fibrinogen adsorption isotherm ($T = 25°C$) on a hydrophobic quartz surface, fitted to Eq. (31). (From Ref. 103.)

TABLE 16 Gibbs Free Energy of Unfolding, Protein, Solid, and Water Surface Energies, and the Work of Adhesion for Proteins Used in Fig. 11

Protein	T (°C)	ΔG_{unfold} (kJ/mol)	γ_{SV} (mJ/m^2)	γ_{PV} (mJ/m^2)	W_a (mJ/m^2)
Fbrgn	25	104.8	47.0	71.7	0.9

Source of isotherm: Ref. 103.

TABLE 17 Surface Area Cleared by an Adsorbing Protein Molecule, A_C, the Affinity Constant K, and the Regression Parameters for the Model [Eq. (31)]

Protein	T (°C)	A_C (Å^2/molecule) (p-value)	Conformation	$K \times 10^{-4}$	MSE $\times 10^4$
Fbrgn	25	5771 (0.0001)	Side-on	20.0	1.9

Source of isotherm: Ref. 103.

2. There is no equation of state to quantify A_C, so the choice of fixing α while keeping A_C as a variable allows comparison of generated A_C values with those experimentally determined for different proteins.

3. At similar hydrophobic solid surfaces, the differences in the extent of unfolding experienced among proteins ought to be explained by differences in W_a, which was estimated here as a function of π_e and independent of α. In other words, defining γ_{PV} in terms of π_e, the effect of surface-induced unfolding has been implicitly incorporated.

4. Choosing a value of α less than unity indicates enhanced hydrophobic interactions between the protein and solid surface, which results in a less stable surface-bound protein compared with the free protein in the bulk. This result was evidenced by Matsuno et al. [5], who found that the thermal stability of adsorbed γ-crystallins generally decreased with increasing hydrophobicity of the surface when compared to the proteins in solution.

5. As the number of computer-adjusted model parameters increases in any event, the model sensitivity to variations in experimental data becomes incidental, with no definitive scientific justification.

2. Value of α at High-Energy Interfaces

In Section III.B, it was shown that the overall equilibrium could be resolved into two major sub-equilibrium states (Fig. 1b):

1. An equilibrium condition between native protein in the bulk and that at the hydrophobic interface, the attainment of which is largely driven by the Gibbs free energy for adsorption (or the work of adhesion).

2. An equilibrium condition between adsorbed native state and adsorbed unfolded state. Attainment of this substate of equilibrium is largely driven by the Gibbs free energy of unfolding.

In Section V.A.1, it was shown that at low-energy (i.e., hydrophobic) solid surfaces setting α equal to zero is a reasonable assumption based on modeling considerations and on the fact that expressing γ_{PV} in terms of π_e essentially takes care of the second substate of equilibrium. However, at relatively hydrophilic or high-energy surfaces, the work of adhesion energy term would presumably become less significant (i.e., smaller in magnitude as γ_{SV} approaches γ_{WV} as compared to that at hydrophobic interfaces). Moreover., with relatively minor destabilization effects at hydrophilic surfaces [5], the energy barrier to surface-induced unfolding would presumably increase. In other words, the second substate of equilibrium would become more significant as the surface becomes more hydrophilic.

B. Molecular Influences on A_C

Tables 5, 7, 9, 13, and 15 show that values of A_C obtained from simulation vary within an order of magnitude, and there is no obvious correlation between molecular weight and A_C. One might conclude that A_C could be fixed at a given value. It should be noted, however, that although changes in A_C are within an order of magnitude, the 95% confidence interval associated with A_C was found, on average, to be equal to 5% of A_C. The strong dependence on A_C should be expected, given that A_C is the only computer-generated parameter present in the model. This being the case, an approximate value of A_C ought to be given as a function of some molecular properties of protein. This will be shown in detail in Section V.C.

As seen in Tables 5, 7, 9, 13, and 15, A_C assumed values within a few hundred square angstroms (about 100–400 Å^2), which is relatively small compared to the cross-sectional area of a "typical" globular protein, which varies from about 1000 to 10,000 Å^2 [13]. This implies that only a small portion of the protein molecule need enter the interface in order for adsorption to proceed. These results are also in agreement with those of Damodaran and Song [40], who found that the area cleared by BSA to anchor itself to the air–water interface varied from 50 to 135 Å^2. Ter-Minassian-Saraga [39] suggested that the "hole area" (i.e., A_C) formed by adsorbing protein may be related to the water activity at the interface and not to the size of the adsorbing molecule. Macritchie [61] suggested that relatively small values of A_C indicate that each adsorbed molecule is sufficiently flexible to behave as a series of largely independent kinetic units; adsorption is thus a function of segment behavior and independent of molecular size.

There is no equation of state relating A_C to pertinent protein, solid surface, and solution properties. Nevertheless, two general trends may be inferred upon comparison of W_a and ΔG_{unfold} (Tables 4, 6, 8, 12, and 14) with corresponding values of A_C (Tables 5, 7, 9, 13, and 15). These trends are as follows:

1. The larger the value of W_a, the smaller the value of A_C (in other words, the more hydrophobic the solid surface or protein "surface," the smaller the interfacial area needed for the initial attachment).
2. The larger the value of ΔG_{unfold} the larger the value of A_C.

It should be mentioned, however, that these trends do not exactly hold for every case studied. As shown in Table 4, the effect of temperature (picked up in part by ΔG_{unfold}) on A_C is generally insignificant. For example, A_C both for β-Lag and BSA increased from 158 to 177Å^2 when the temperature was increased from 27°C to 52°C. A small increase in A_C as a function of temperature was also reported for the adsorption of ovalbumin at the air–water interface, where A_C increased from 150 to 170 Å^2 when the temperature increased from 5°C to 20°C [61].

C. An Equation of State for A_C

As was pointed out in Section V.B, the model is sensitive to variations in A_C, as it is the only computer-generated parameter present in the model. The initial slope (i.e., the affinity constant, K) of the plot is extremely sensitive to variations in A_C, as a result of the exponential term appearing in Eq. (31). On the other hand, the ultimate or the plateau value is only governed by the factor A_P. This being the case and as there is no equation of state to define A_C, one might attempt to provide an estimate for the A_C value in terms of molecular properties. Because A_C mainly affects the initial slope, K, one way of expressing A_C is to utilize Eq. (31) itself. This can be achieved by evaluating A_C at a dimensionless concentration ($V_P C_{eq}$) such that the surface concentration or adsorbed mass, Γ is half Γ_{max}. Using Eq. (31), one gets

$$\frac{\Gamma}{\Gamma_{max}} = \frac{1}{2} = \left[\left(\frac{\exp(-W_a A_C/RT)}{V_P C_{eq}} \right) + 1 \right]^{-1} \tag{32}$$

Comparing the left-hand side of Eq. (32) with its right-hand side, one may easily recognize that

$$\frac{\exp(-W_a A_C/RT)}{V_P C_{eq}} = 1 \tag{33}$$

Equation (33) implies that

$$A_C = -\frac{RT \ln(V_P C_{eq})}{W_a} \tag{34}$$

Thus, Eq (34) may be used as an equation of state for A_C, provided that the dimensionless number ($V_P C_{eq}$) is, in general, less than or equal to 0.001, based on the examined range of equilibrium concentration of proteins. Obviously, A_C is less sensitive to the variation in the value of $V_P C_{eq}$. Consequently, if the value of $V_P C_{eq}$ changes by one order of magnitude, the change in A_C value, however, will be within the same order of magnitude. Therefore, as an approximate value, A_C is given by

$$A_C = \frac{6.91 RT}{W_a} \tag{35}$$

D. Implications Associated with the Model Application to Multi-Domain Proteins

Before discussing the implications associated with extending application of the model to multidomain proteins, the difference between a single-domain and multidomain protein should be addressed. During thermal denaturation, a single-domain protein is one that undergoes a one-step (all-or-none) reversible phase transition between two thermodynamically defined conformations (native and unfolded). Small, globular proteins such as mygolobin, ribonuclease, cytochrome C, α-chymotrypsin, α-lactalbumin, β-lactoglobulin, and lysozyme are categorized under this class of proteins. On the other hand, if the protein molecular weight is relatively large, the molecule can often be resolved into a number of somewhat independent domains. Using DSC analysis, the presence of independent kinetic units, which also exhibit some degree of cooperativity, can be discerned. In particular, the thermogram of a multidomain protein is usually resolved into a set of peaks if DSC analysis is carried out at acidic pH values [94]. Multidomain proteins considered in this work include human and bovine serum albumin [104], immunoglobulin G [105], plasminogen [94], and fibrinogen [18].

As far as the application of the model to multidomain proteins is concerned, the central problem lies in how to define ΔG_{unfold} for such proteins, which, in turn, affects the estimation of both W_a and the computer-generated parameter A_C. One way to tackle the adsorption of multidomain complex proteins is to use the so-called domain approach [41,106]. The domain approach considers a complex protein molecule like BSA, IgG, Plmgn, or Fbrgn as being constructed of functional and structural domains generally identified as regions of relatively high packing density that are calorimetrically independent [106].

As far as adsorption is concerned, the interfacial behavior of a complex protein could possibly be considered as largely dominated by the interfacial activity of only one domain or even a subdomain [106]. Although the conformational changes of the protein at the interface is ultimately governed by the stability of each of the individual protein domains, the most thermolabile domain (which may comprise several thermal cooperative fragments) is expected to play a major role in the initial events contributing to surface-induced unfolding reactions.

In this work, the domain approach was used in the case of Fbrgn, Plmgn, and IgG adsorption. For either HSA or BSA, the thermogram exhibited only one peak [95,98] (i.e., as if they were single-domain globular proteins). The reason for that is probably because the DSC analyses were performed at alkaline pH values (pH 7.00 for BSA and pH 6.00 for HSA) compared to their isoelectric points (pI is about 5.0–5.1 for both HSA and BSA). Nevertheless, the notion of using the domain approach is still applicable, should DSC analysis be available for these multidomain proteins.

The value of A_C obtained for each of these multidomain proteins is therefore governed by the extent to which the domain approach is appropriate in each case studied and the reliability of such a domain-based ΔG_{unfold} value as an index of thermal stability of the entire protein.

For Fbrgn, although the model to a large extent is in quantitative agreement regarding the extent of Fbrgn adsorption (Figs. 10 and 11), A_C (Table 17) is one order of magnitude larger than any of the other protein-specific A_C values. No value was found for A_C in the literature to verify this result, so the large value of A_C may simply be incidental, owing to the relatively large value of ΔG_{unfold} for Fbrgn (Table 14), which, in turn, yields a relatively small value of W_a. It should be noted that the area of the first peak in the Fbrgn thermogram is approximately equal to twice that of the thermolabile fragment, namely fragment D_H with molecular weight (MW) 95,000 [18]. As a result of the geometrical symmetry of the Fbrgn molecule, two D_H fragments exist per molecule, which, in turn, results in a relatively large value of ΔG_{unfold}. One could argue using $\frac{1}{2}\Delta G_{\text{unfold}}$ to represent only one D_H fragment; nevertheless, these two D_H fragments show some degree of cooperativity as part of the entire Fbrgn molecule, which makes them act as one independent domain being made of two fragments.

One could ask why should the value of A_C not be relatively large, as the human fibrinogen was the largest examined molecule in this work. The answer to this question lies in the proposition that A_C is not correlated with the molecular weight of adsorbing protein, and even with a huge molecule like myosin (MW 850,000), a value of A_C equal to 145 Å^2 at the air–water interface has been reported [39].

IgG is another molecule with a multidomain structure [105], and its thermal unfolding is known to proceed in two separate stages: the first stage of unfolding requiring 300 kcal/mol and the second stage 900 kcal/mol, with an overall enthalpy of 1200 kcal/mol [51]. These two transitions do not show overlapping peaks on a thermogram, indicating two thermally independent domains. Only the first thermogram peak (300 kcal/mol enthalpy and T_m equal to 317.1 K) was considered in the calculation of ΔG_{unfold} for two reasons. The first regards attention to the domain approach to the study of adsorption of complex proteins with the thought that the first thermogram peak represents the thermolabile fragment of the IgG molecule; that is, the interfacial behavior of IgG is thought to be largely dominated by the interfacial activity of this thermolabile fragment. The second reason is based on studies made by Chasovnikova et al. [107], who found π_e for IgG monolayers equal to $16 \, \text{mJ/m}^2$. Based on Eq. (10), this yields a value of γ_{PV} equal to $56.5 \, \text{mJ/m}^2$. Their results compare well with the value of π_e obtained using Eq. (11), and of γ_{PV} using Eq. (10) (see Table 14) when obtained with a value of ΔG_{unfold} based on the 300-kcal/mol enthalpy of denaturation.

Human plasminogen is also a molecule with a multidomain structure, although this molecule is not as large as IgG or Fbrgn. Plasminogen may designate any of the several plasminogen genetic variants. Examples are Lys-plasminogen (MW = 83,000) and Glu-plasminogen (MW = 92,000). Upon limited plasminolysis, the intact plasminogen with a glutamic acid residue at the N-terminal end (Glu-plasminogen) loses its first 76 amino acid residues and is converted into modified Lys-plasminogen, with a lysine residue at the N terminus. The plasminogen molecule is subdivided into more or less independent subunits (i.e., into structural domains) [94]. Novokhatny et al. [94] suggested that two large domains of Lys-plasminogen are indeed likely to be quite independent, judging by calorimetric analyses showing thermal transitions in two nearly separate temperature regions: the N-terminal part of the molecule, including its first four fragments, and the C-terminal part, which includes three subunits forming miniplasminogen (residues Val_{442} to Asn_{790}). The first thermogram peak, which represents the thermolabile part (i.e., the four-fragment N-terminal part) of the molecule, was used for calculation of ΔG_{unfold} of Lys-plasminogen. Because human plasminogen, studied by Woodhouse et al. [2], has a molecular weight of 94,000, which better matches with Glu-plasminogen as opposed to Lys-plasminogen, and because ΔG_{unfold} used in Eq. (31) was entered on a molar basis, an adjustment was made to the extensive thermal properties of Lys-plasminogen based on the molecular-weight difference between these plasminogen variants.

E. The Affinity and Extent of Adsorption

1. Adsorption Affinity

The affinity of a protein in single-component solution for a given surface can be characterized by the affinity (or equilibrium) constant, K, given by Eq. (30). This equilibrium constant connotes the total driving force for the process of adsorption. Not only does it constitute the driving force for arrival to the surface but also for surface unfolding, which differentiates protein adsorption equilibrium from that of small molecules.

As defined by Eq. (22), K represents the initial slope of a plot of ΓA_P versus $V_P C_{eq}$. In the absence of electrostatic and specific biochemical interactions, the affinity constant depicts the magnitude of hydrophobic interaction between the protein and solid surface. The surface hydrophobicity of protein and solid were characterized by γ_{PV} and γ_{SV}, respectively. The more hydrophobic the protein and solid surface are, the larger the value of K. This hydrophobicity dependence of K can be seen as a general trend in Table 5 (comparing β-Lag and BSA at 52°C) and in Table 7 (comparing Lyso and ChA). However, the above-mentioned statement lacks consistency with respect to the other isotherms studied. There are at least three possible reasons for this. The first concerns the reliability of experimental data. The affinity constant represents the initial slope of a Langmuir-type isotherm, and the degree of uncertainty associated with determining the initial slope is much larger than that associated with determining the plateau value of the same isotherm, excluding those isotherms carried out at very low concentrations [2,103]. In fact, some isotherms (e.g., BSA and HSA in Fig. 10) exhibit a plateau value almost over the entire studied concentration range, which makes it difficult to fit to any model other than $\Gamma = \Gamma_{max}$. Isotherms that exhibit a high-affinity character, with a plateau almost over the entire range, should be reconstructed at lower bulk concentrations. However, this would be accompanied by enhanced potential for experimental artifacts (e.g., dilution effects, diffusion limitations, impurity effects, and reliability of the instrumentation). Second, the effect of surface hydrophobicity largely manifests itself in an increase in extent of adsorption (Γ_{max}), which makes the importance of the initial slope less significant. More than one group of investigators [87,95,108–109] reported that the adsorbed mass of protein on hydrophobic surfaces was greater than that on hydrophilic surfaces. Third, although the effect of γ_{SV} is picked up by W_a, the effect of surface hydrophobicity extends beyond its contribution to W_a; it also influences the surface-induced unfolding [13,61].

2. The Extent of Adsorption

Two factors affect the extent of adsorption, Γ_{max}: molecular size and the strength of hydrophobic interactions between the protein and the solid surface. As shown in Figs. 4–11, the larger the molecular weight, the larger the adsorbed mass. If as in blood serum, molecular weights among proteins vary by one or two orders of magnitude, the contribution to total adsorbed mass could be dominated by larger molecules in part simply because they are large [95]. Also, the stronger the hydrophobic interactions, the larger the adsorbed mass per unit area. The effect of molecular size on extent of adsorption has been taken into account using Eqs. (17), (18), and (23), via the partial molar volume.

The effect of hydrophobic interactions appears in part by virtue of K; that is, given proteins of equal size, the stronger the hydrophobic interaction, the larger the value of K, the larger the adsorbed mass per unit area.

Toward minimization of the number of parameters required as input to the model, no parameter was incorporated in Eq. (23) to account for surface-induced unfolding effects on Γ_{max}. In future work, should A_C be fixed or defined, one may incorporate a computer-generated parameter that will account for surface-induced unfolding effects on Γ_{max}.

The goal in this work was to lay out a model complex enough to describe protein adsorptive behavior at hydrophobic solid–water interfaces, but not so intricate as to lose track of surface and molecular property influences on the observed phenomena.

3. A Note on Temperature Effects

The effect of temperature on K can be seen by comparing protein isotherms conducted at 52°C with those conducted at 27°C. For both β-Lag and BSA, K increased with temperature (Table 5). Applying the van't Hoff equation,

$$\ln\left(\frac{K_2}{K_1}\right) = -\frac{\Delta H^{\circ}_{ads}}{R}\left(\frac{1}{T_2} - \frac{1}{T_1}\right) \tag{36}$$

between state 1 (i.e., $T_1 = 27°C$) and state 2 (i.e., $T_2 = 52°C$), the following results were obtained:

β-Lag $\quad \Delta H^{\circ}_{ads} = 13.2\,\text{kJ/mol}$
BSA $\quad \Delta H^{\circ}_{ads} = 14.7\,\text{kJ/mol}$

The standard, Gibbs free-energy change of adsorption, ΔG°_{ads}, can be evaluated at $T = 27°C$, using $\Delta G^{\circ}_{ads} = -RT\ln K$, for both proteins, and the results are as follows:

β-Lag $\Delta G^\circ_{ads} = -22.4\,\text{kJ/mol}$
$\Rightarrow T\Delta S^\circ_{ads} = \Delta H^\circ_{ads} - \Delta G^\circ_{ads} = 13.2 - (-22.4) = 35.6\,\text{kJ/mol}$

BSA $\Delta G^\circ_{ads} = -23.8\,\text{kJ/mol}$
$\Rightarrow T\Delta S^\circ_{ads} = \Delta H^\circ_{ads} - \Delta G^\circ_{ads} = 14.7 - (-23.8) = 38.5\,\text{kJ/mol}$

Assuming ΔH°_{ads} is constant over the narrow temperature range examined, ΔG°_{ads} can be evaluated for both proteins at $T = 52°C$:

β-Lag $\Delta G^\circ_{ads} = -25.4\,\text{kJ/mol}$
$\Rightarrow T\Delta S^\circ_{ads} = \Delta H^\circ_{ads} - \Delta G^\circ_{ads} = 13.2 - (-25.4) = 38.6\,\text{kJ/mol}$

BSA $\Delta G^\circ_{ads} = -27.0\,\text{kJ/mol}$
$\Rightarrow T\Delta S^\circ_{ads} = \Delta H^\circ_{ads} - \Delta G^\circ_{ads} = 14.7 - (-27.0) = 41.7\,\text{kJ/mol}$

The aforementioned analysis indicates that the entropic contribution to the Gibbs free energy of adsorption increases with increasing temperature. These results are in harmony with the notion that protein adsorption at hydrophobic surfaces is entropically driven [13,14], and such entropically driven, hydrophobic interactions between the protein and the silicon surface are favored with increasing temperature.

VI. SUMMARY

1. At hydrophobic interfaces, a general agreement between Eq. (31) and experimental data was observed for single-domain globular proteins indicating that ΔG_{unfold} and $W_a A_C$ play a major role in governing the course of adsorption.
2. The values of A_C obtained from simulation vary within an order of magnitude (i.e., values within a range of about 100–400 Å^2), indicating that only a small portion of the protein molecule need enter the interface in order for adsorption to proceed.
3. Choosing $\alpha = 0$ results in a computer-generated value for A_C in quantitative agreement with these provided in literature and is in line with the concept of hydrophobic interactions between the protein and solid surface, facilitating the surface-induced unfolding.
4. Implementing the domain approach to characterizing adsorption of multidomain proteins resulted in A_C values consistent with the notion that A_C is independent of molecular weight. This indicates that the interfacial behavior of a complex protein could be considered as largely dominated by the interfacial activity of only one domain; the most thermolabile domain (which may consist of several thermal cooperative fragments) appeared to play a major role in the initial events contributing to surface-induced unfolding.
5. For fibrinogen, the large value of A_C obtained may simply be incidental, owing to its relatively large value of ΔG_{unfold} (which, in turn, yields a relatively large value of γ_{PV}, and a relatively small value of W_a).

VII. NOMENCLATURE

α A function constant that defines the functionality of Γ versus C_{eq} [Eq. (20)]
A_C Minimum surface area (i.e., "hole area") cleared by an adsorbing protein molecule in order to anchor itself to the interface (m^2/mol or Å^2/molecule)

A_i	Partial molar area occupied by species i at the interface (m^2/mol or $Å^2/molecule$)
a_0	The average interfacial area occupied by an amino acid residue ($Å^2/molecule$)
A_p	Partial molar area occupied by protein at the interface as defined by either Eq. (17) or (18) (m^2/mol or $Å^2/molecule$)
b	A constant such that Γ_{max}/b is the initial slope of a plot of Γ versus C_{eq} (mg/L)
C_{eq}	Apparent equilibrium concentration (mol/m^3 or mg/L)
ChA	Chymotrypsinogen A
ΔC_P	The difference in heat capacity between the unfolded and folded states (kJ/mol K)
f	Fractional surface coverage (dimensionless)
G_N	Partial molar free energy of native protein (kJ/mol)
G_U	Partial molar free energy of unfolded protein (kJ/mol)
ΔG_{ads}	The standard Gibbs free-energy change of adsorption (kJ/mol)
$\Delta G_{transfer}$	The standard free energy of transfer from water to a hydrocarbon solvent
ΔG_{unfold}	Change in partial molar free energy of protein upon unfolding (kJ/mol)
ΔH_{ads}	The standard enthalpy change of adsorption (kJ/mol)
ΔH_{chain}	Enthalpy change of the chain of a solute, equal to $H_N - H_U$
ΔH_D	The enthalpy of denaturation (kJ/mol)
$\Delta H_{solvent}$	Enthalpy change of the solvent, equal to $H_N - H_U$
ΔH^*	The nonhydrophobic enthalpic contribution to the free-energy change [Eq. (15)] (kJ/mole amino acid residue)
IgG	Immunoglobulin G
K	The proportionality or the overall equilibrium constant for the protein adsorption process (dimensionless)
K_a	The equilibrium constant for the reversible arrival step (dimensionless)
k_B	Boltzmann constant (1.38×10^{-23} J/molecule K)
K_U	The equilibrium constant for the reversible, surface-induced unfolding step (dimensionless)
N_A	Avogadro's number (6.02217×10^{23} molecules/mol)
ΔS_{chain}	Entropy change of the chain of a solute, equal to $S_N - S_U$
$\Delta S_{solvent}$	Entropy change of the solvent, equal to $S_N - S_U$
ΔS^*	Multiplied by T gives the nonhydrophobic entropic contribution to the free-energy change [Eq. (15)] (kJ/mole amino acid residue K)
T	Temperature (K)
T_H	The temperature at which the enthalpic contribution to the free-energy change [Eq. (15)] is purely nonhydrophobic (K)
T_m	The melting-point temperature of protein at which the thermogram peak reaches its maximum (K)
T_S	The temperature at which the entropic contribution to the free-energy change [Eq. (15)] is purely nonhydrophobic (K)
V_i	Partial molar volume of species i in solution (m^3/mol)
V_P	Partial molar volume of protein in solution (m^3/mol)
W_a	Work of Adhesion as defined by Eq. (4) (J/m^2)
Z	The surface coordination number of the lattice as defined by Eq. (12) ($Z = 2 + \omega$)

Greek Symbols

α	The proportionality constant that relates the Gibbs free energy of unfolding in the bulk to that at the interface (dimensionless)
Γ	Adsorbed amount of protein (mol/m^2) or $\mu g/cm^2$)

Γ_{max} The plateau value of adsorbed amount of protein (mol/m^2 or µg/cm^2)
ξ The ratio of the volume occupied by all atoms making up a molecule to the total
             volume occupied by the molecule itself (dimensionless)
Π_i An ith independent dimensionless group that comprises a multiplication of some
             dimensional variables or parameters (dimensionless)
γ_{12} The interfacial free energy of any pair combination of two phases from the protein
             (P), water (W), and solid (S) phases as defined by Eq. (8) (mJ/m^2)
γ_{PS} The interfacial free energy between protein and solid (mJ/m^2)
γ_{PV} The interfacial free energy of protein (mJ/m^2)
γ_{PW} The interfacial free energy between protein and water (mJ/m^2)
γ_{SV} The interfacial free energy of solid (mJ/m^2)
γ_{SW} The interfacial free energy between solid and water (mJ/m^2)
γ_{WV} The interfacial free energy of water (mJ/m^2)
θ The equilibrium contact angle formed between a solid–liquid (SL) interface and a
             liquid–vapor (LV) interface as defined by Eq. (9) (degree)
π_e The equilibrium spreading pressure measured at the protein solution–vapor interface
             and corresponds to the plateau region of $\pi_e = f(C_{eq})$, as defined by Eq. (11)
             (mJ/m^2)
π_0 $k_B T / a_0$ (mJ/m^2)
Ω A generic symbol used to designate a mathematical relationship among the pertinent
             variables
ω The flexibility parameter of the polymer chain ($0 \leq \omega \leq 2$)

Subscripts

L Liquid
N Native state of a protein chain (or a state of separated phases between a polar and
             nonpolar phase)
P Protein
S Solid
U Randomly unfolded state of a protein chain (or a state of a monodisperse solution)
W Water

APPENDIX 1

```
      PROGRAM GAMA12
C
C.. A PROGRAM TO CALCULATE:
C    1) GAMA12 (INTERFACIAL TENSION)
C    FROM VARIABLE INPUTS
C    1) GAM1V "INTERFACIAL TENSION OF PHASE 1" (mJ/m2)
C    2) GAM2V "INTERFACIAL TENSION OF PHASE 2" (mJ/m2)
C    ************************************************************
C
      IMPLICIT DOUBLE PRECISION (A-H,O-Z)
C
C     CHARACTER*80 NSURF,OUTFILE
      PRINT*,'ENTER THE VALUES OF GAM1V AND GAM2V'
      PRINT*, '************************'
      READ(*,*) G1V,G2V
```

```
      IF (G2V.LE.0.0) GO TO 5
      CALL EQS(G1V,G2V,G12)
      PRINT*,'SURFACE TENSION OF PHASE 1 = ',SNGL(G1V)
      PRINT*,'SURFACE TENSION OF PHASE 2 = ',SNGL(G2V)
         PRINT*,'INTERFACIAL SURFACE TENSION G12 = ',SNGL(G12)
    5 CONTINUE
      STOP
      END
      SUBROUTINE EQS(SSV,SLV,SSL)
      IMPLICIT DOUBLE PRECISION (A-H,O-Z)
      A=0.0150
      EPS = 0.0001
      GAMSV=SSV
      GAMLV=SLV
      IF (GAMSV.LE.GAMLV) GO TO 93
      GAMSV=SLV
      GAMLV=SSV
   93 IF (GAMLV.GT.30.0) GO TO 95
   94 GAMSL=((DSQRT(GAMLV)-DSQRT(GAMSV))**2.0)/
     1(1.0-A*DSQRT(GAMSV*GAMLV))
      GO TO 99
   95 IF (GAMLV.GT.50.0) GO TO 97
      GAMSL=((DSQRT(GAMLV)-DSQRT(GAMSV))**2.0)/
     1(1.0-A*DSQRT(GAMSV*GAMLV))
      IF ((GAMSL-GAMSV.GT.0.0) GO TO 99
   96 GAMSL=(((2.0-A*GAMSV)*DSQRT(GAMLV)-DSQRT((2.0-A*GAMSV)
     1**2.0*GAMLV-4.0*(GAMLV-GAMSV)))**2.0)/4.0
      GO TO 99
   97 GAMS1=10.0D0
      B=DSQRT(GAMLV)
   98 GAMS=GAMS1
      C = DSQRT(GAMS)
      SLOPE = ((1.0-A*B*C)*(1.0-B/C)-(GAMS-2.0*B*C+GAMLV)
     1*((-A/2.0)*B/C))/(1.0-A*B*C)**2.0
      SLOPE1 = SLOPE + 1.0
      U = 1.0-A*B*C/2 0+(A*(GAMLV**1.5)/2.0-B)/C
      V = (1.0-A*B*C)**2.0
      DU=-(A*B)/(4.0*C)-(A*(GAMLV**1.5)12.0-B)/
     1(2.0*(GAMS**1.5))
      DV-A*B/C+(A*B)**2.0
      SLOPE2=(V*DU-U*DV)/V**2.0
      GAMS1=GAMS-SLOPE1/SLOPE2
      IF (ABS(GAMS1-GAMS).GT.EPS) GO TO 98
      GAMS2=((DSQRT(GAMS1)-DSQRT(GAMLV))**2.0)/
     1(1.0-A*DSQRT(GAMS1*GAMLV))
      IF (GAMSV.LE.GAMS1) GO TO 94
      IF (GAMSV.GE.GAMS2) GO TO 96
      GAMSL = GAMS1+GAMS2-GAMSV
   99 SSL=GAMSL
      RETURN
      END
```

APPENDIX 2

The following table lists values of ΔH^*, the nonhydrophobic enthalpy of a protein, as estimated by equating Eq. (14) with Eq. (15) and solving for ΔH^*. The average value of the molecular weight of an amino acid residue is assumed to be equal to 123 [110] and that of ΔC_P to be equal to (50 J/mol amino acid residue·K) [111].

Protein	MW	ΔH^* (kJ/mol)	ΔH^* (kJ/mol residue)
α-Lac	14,200	691.739	5.992
β-Lag	36,640 (dimer)	1,706.817	6.032
BSA	66,267	3,229.121	5.994
HSA	69,000	3,363.049	5.995
Myog	17,800	864.906	5.977
RiboA	13,680	686.568	6.173
Lyso	14,400	754.056	6.441
Cha	23,000	1,145.179	6.124
IgG	150,000	7,264.512	5.957
		Average	6.076±0.155

APPENDIX 3

The following table lists the molecular dimensions of proteins used to develop an equation for A_P [Eq. (18)] for a monolayer with the side-on conformation.

Protein	Molecular weight	Dimensions (Å^3)	Side-on (Å^2)
α-Lac	14,161	$37 \times 32 \times 25$	37×32
β-Lag	36,640 (dimer)	69.3×35.8 (two spheres)	69.3×35.8
BSA	66,267	$140 \times 38 \times 38$	140×38
HSA	69,000	$115 \times 40 \times 40$	115×40
Myog	17,800	$44 \times 35 \times 25$	44×35
RiboA	13,680	$38 \times 28 \times 22$	38×28
Lyso	14,400	$45 \times 30 \times 30$	45×30
Cytochrome C	11,353	$37 \times 25 \times 25$	37×25
Superoxide Dismutase	15,534	$40 \times 38 \times 36$	40×38
γ-globulin	160,000	$235 \times 44 \times 44$	235×44

REFERENCES

1. LC Winterton, JD Andrade, J Feijen, SW Kim. J Colloid Interf Sci 111:314, 1986.
2. KA Woodhouse, PW Wojciechowski, JP Santerre, JL Brash. J Colloid Interf Sci 152:60, 1992.
3. K Ishihara, Y Iwasaki, N Nakabayashi. Mater Sci Eng C: Biomimetic Mater Sensors Syst 6:253, 1998.
4. V Balasubramanian, NK Grusin, RW Bucher, VT Turitto, SM Slack. J Biomed Mater Res 44:253, 1999.

5. K Matsuno, RV Lewis, CR Middaugh. Arch Biochem Biophys 291:349, 1991.
6. L Stryer. Biochemistry. New York: WH Freeman, 1988.
7. GD Fasman. Prediction of Protein Structure and the Principles of Protein Conformation. New York: Plenum Press, 1989.
8. TE Creighton. Biopolymers 22:49, 1983.
9. MH Klapper. Biochim Biophys Acta 229:557, 1971.
10. C Chothia. Nature 254:304, 1975.
11. DJ Barlow, JM Thornton. J Mol Biol 168:867, 1983.
12. JM Thornton. Nature 295:13, 1982.
13. JD Andrade. Surface and Interfacial Aspects of Biomedical Polymers: Protein Adsorption. New York: Plenum Press, 1985, Vol 2.
14. GE Schulz, RH Schirmer. Principles of Protein Structure. New York: Springer-Verlag, 1979, pp 27–45.
15. W Kauzmann. Adv Protein Chem 14:1, 1959.
16. KA Dill, DOV Alonso, K Hutchinson. Biochemistry 28:5439, 1989.
17. PL Privalov, YuV Griko, S Yu Venyaminov. J Mol Biol 190:487, 1986.
18. PL Privalov, LV Medved. J Mol Biol 159:665, 1982.
19. VE Koteliansky, MA Glukhova, MV Bejanian, vN Smirnov, VV Filimonov, OM Zalite, Syu Venyaminov. Eur J Biochem 119:619, 1981.
20. JD Honeycutt, D Thirumalai. Biopolymers 32:695, 1992.
21. FL Slejko. Adsorption Technology: A Step-by-Step Approach to Process Evaluation and Application. New York: Marcel Dekker, 1985, pp 1–6.
22. GD Parfitt, CH Rochester. Adsorption from Solution at the Solid/Liquid Interface. London: Academic Press, 1983, pp 153–218.
23. JA De Feijter, J Benjamins, M Tamboer. Colloid Surfaces 27:243, 1987.
24. S Lin, R Blanco, BL Karger. J Chromatogr 557:369, 1991.
25. T Arnebrant, K Barton, T Nylander. J Colloid Interf Sci 119:383, 1987.
26. DR Lu, SJ Lee, K Park. J Biomater Sci Polym Ed 3:127, 1991.
27. AP Wei, JN Herron, JD Andrade. In: DJA Crommelin, H Schellekens, eds. From Clone to Clinic. Developments in Biotherapy Vol. 1. Dordrecht: Kluwer Academic, 1990.
28. A Kato, K Yutani. Protein Eng 2:153, 1988.
29. D Horsley, J Herron, V Hlady, JD Andrade. In: JL Braqsh, TA Horbett, eds. Proteins at Interfaces: Physicochemical and Biochemical Studies. ACS Symposium Series 343. Washington DC: American Chemical Society, 1987 p 290.
30. JR Hunter, RG Carbonnell, PK Kilpatrick, J Colloid Interf Sci. 143:37, 1991.
31. P Raje, NG Pinto. J Chromatogr A 796:141, 1998.
32. MR Oberholzer, AM Lenhoff. Langmuir 15:3905, 1999.
33. R Douillard, J Lefebvre, V Tran Colloids Surfaces A: Physicochem Eng Aspects 78:109, 1993.
34. CA Johnson, P Wu, AM Lenhoff. Langmuir 10:3705, 1994.
35. VB Fainerman, R Miller, R Wuestneck. J Colloid Interf Sci 183:26, 1996.
36. AW Neumann, RJ Good, CJ Hope, M Sejpal. J Colloid Interf Sci 49:291, 1974.
37. CJ van Oss, RJ Good. J Protein Chem 7:179, 1988.
38. SJ Singer. J Chem Phys 16:872, 1948.
39. L Ter-Minassian-Saraga. J Colloid Interf Sci 80:393, 1981.
40. S Damodaran, KB Song. Biochim Biophys Acta 954:2, 1988.
41. P Suttiprasit, V Krisdhasima, J McGuire. J Colloid Interf Sci 154:316, 1992.
42. F Uraizee, G Narsimhan. J Colloid Interf Sci 146:169, 1991.
43. CJ van Oss, DR Absolom, AW Neumann, W Zingg. Biochim Biophys Acta 670:64, 1981.
44. AW Neumann, DR Absolom, DW Francis, SN Omenyi, JK Spelt, Z Policova, C Thomson, W Zingg, CJ van Oss. Ann NY Acad Sci 416:276, 1983.
45. AW Neumann, OS Hum, DW Francis. J Biomed Mater Res 14:499, 1980.
46. C Tanford. Adv Protein Chem 24:1, 1970.
47. PL Privalov, NN Khechinashvili. J Mol Biol 86:665, 1974.
48. PL Privalov. Adv Protein Chem 33:167, 1979.

49. G Velicelebi, JM Sturtevant. Biochemistry 18:1180, 1979.
50. WJ Becketel, JA Schellman. Biopolymers 26:1859, 1987.
51. PL Privalov. Pure Appl Chem 47:293, 1976.
52. Z Dzakula, RK Andjus. J Theor Biol 153:41, 1991.
53. KP Murphy, PL Privalov, SJ Gill. Science (Washington) 247(4942):559, 1990.
54. AP Zhukovskii, NV Rovnov. Biophysics 34:560, 1989.
55. HA Sober. Handbook of Biochemistry. Cleveland, PH: CRC, 1970.
56. I Pilz, G Czerwenka. Makromol Chem 170:185, 1973.
57. HB Bull, K Breese. Biopolymers 12:2351, 1973.
58. GD Fasman. Handbook of Biochemistry and Molecular Biology. 3rd ed. Cleveland, OH: CRC, 1976, Vol II.
59. H Durchschlag. In: HJ Hinz, ed. Thermodynamic Data for Biochemistry and Biotechnology. New York: Springer-Verlag, 1986, p 45.
60. PI Bendzko, WA Pfeil, PL Privalov, EI Tiktopulo. Biophys Chem 29:301, 1988.
61. F Macritchie. In: CB Anfinsen, JT Edsall, FM Richards, eds. Advances in Protein Chemistry. New York: Academic Press, 1978, pp 283–326.
62. A Dabrowski, M Jaroniec, J Oscik. In: E Matijevic, ed. Surface and Colloid Science. New York: Plenum Press, 1987, Vol 14, pp 83–213.
63. SK Suri. J Colloid Interf Sci 34:100, 1970.
64. RL Baldwin. Proc Natl Acad Sci USA 83:8069, 1986.
65. AR Fersht. Trends Biochem Sci 12:301, 1987.
66. A Nicholls, KA Sharp, B Honig. Proteins 111:281, 1991.
67. H Nygren, M Stenberg, C Karlsson. J Biomed Mater Res 26:77, 1992.
68. TA Ruzgas, VJ Rasumas, JJ Kulys. J Colloid Interf Sci 151:136, 1992.
69. T Arnebrant, T Nylander. J Colloid Interf Sci 111:529, 1986.
70. H Elwing, A Askendal, I Lundström. Prog Colloid Polym Sci. 74:103, 1987.
71. H Shirahama, J Lyklema, W Norde. J Colloid Interf Sci 139:177, 1990.
72. GF Khan, W Wernet. Thin Solid Films 300:265, 1997.
73. A Khaiat, IR Miller. Biochim Biophys Acta 183:309, 1969.
74. JA De Feijter, J Bengamins, FA Veer. Biopolymers 17:1759, 1978.
75. DE Graham, MC Phillips. J Colloid Interf Sci 70:403, 1979.
76. DE Graham, MC Phillips. J Colloid Interf Sci 70:415, 1979.
77. JR Hunter, PK Kilpatrick, RG Carbonell. J Colloid Interf Sci 137:462, 1990.
78. JL Brash, DJ Lyman. J Biomed Mater Res 3:175, 1969.
79. RG Lee, SW Kim. J Biomed Mater Res. 8:251, 1974.
80. BW Morrissey, RS Stromberg. J Colloid Interf Sci 46:152, 1974.
81. DR Absolom, W Zingg, AW Neumann. J Biomed Mater Res 21:161, 1987.
82. U Jönsson, I Lundström, I Rönnberg. J Colloid Interf Sci 117:127, 1987.
83. I Lundström, B Ivarsson, U Jönsson, H Elwing. In: WJ Feast, HS Munro, eds. Polymer Surfaces and Interfaces. New York: Wiley, 1987, pp 201–229.
84. T Mizutani, JL Brash. Chem Pharm Bull 36:2711, 1988.
85. T Arai, W Norde. Colloids Surfaces 51:1, 1990.
86. CT Shibata, AM Lenhoff. J Colloid Interf Sci 148:469, 1992.
87. DR Absolom, FV Lamberti, Z Policova, W Zingg, CJ van Oss, AW Neumann. Appl Environ Microbiol 46:90, 1983.
88. M Matsumura, WJ Becktel, BW Matthews. Nature 334:406, 1988.
89. PL Privalov, GI Makhatadz. J Mol Biol 213:385, 1990.
90. W Norde. J Dispers Sci Technol 13:363, 1992.
91. JD Andrade. J Herrron, V Hlady, D Horsley. Croatica Chem Acta 60:495, 1987.
92. PC Hiemenz. Principles of Colloid and Surface Chemistry. 2nd ed. New York: Marcel Dekker, 1986, pp 296.
93. FJ Castellino, VA Ploplis, JR Powell, DK Stricklan. J Biol Chem 256:4778, 1981.
94. VV Novokhatny, SA Kudinov, PL Privalov. J Mol Biol 179:215, 1984.

95. P Suttiprasit, J McGuuire. J Colloid Interf Sci 154:327, 1992.
96. W Pfeil. Biophys Chem 13:181, 1981.
97. NC Pace, C Tanford. Biochemistry 7:198, 1968.
98. DYa Leibman, YeI Tiktopulo, PL Privalov. Biofizika 20:376, 1975; Biophysics 20:379, 1975.
99. Y Fujita, Y Noda. Int J Peptide Protein Res 38:445, 1991.
100. CN Pace. J Am Oil Chem Soc 60:970, 1983.
101. JD Andrade, V Hlady. Ann NY Acad Sci 516:158, 1987.
102. GH Barlow, L Summaria, KC Robbins. J Biol Chem 244:1138, 1969.
103. N Nygren, M Stenberg. J Biomed Mater Res 22:1, 1988.
104. U Kragh-Hansen. Pharmacol Rev 33:17, 1981.
105. VM Tischenko, VP Zav'yalov, GA Medgyesi, SA Potekhin PL Privalov. Eur J Biochem (FEBS) 126:517, 1982.
106. JD Andrade, V Hlady, A-P Wei, C-G Gölander. Croatica Chem Acta 63:527, 1990.
107. LV Chasovnikova, NA Matveyeva, VV Lavrent'ev. Biofizika 27:435, 1982; Biophysics 27:444, 1982.
108. H Elwing, B Ivarsson, I Lundström. Eur J Biochem 156:359, 1986.
109. U Jönsson, B Ivarsson, I Lundström, L Berghem. J Colloid Interf Sci. 90:148, 1982.
110. RL Baldwin. Proc Natl Acad Sci USA 83:8069, 1986.
111. AP Zhukovskii, NV Rovnov. Biophysics 34:560, 1989.

15

Protein Adsorption Kinetics

KAMAL AL-MALAH and HASAN ABDELLATIF HASAN MOUSA
Jordan University of Science and Technology, Irbid, Jordan

I. INTRODUCTION

The adsorption of proteins on solid surfaces is an important phenomenon that takes place as soon as a foreign interface is brought into contact with a biological system. Thus, it is involved in a number of areas in biology, medicine, food and pharmaceutical processing, and biotechnology. A substantial effort has been devoted to studying the nature of protein adsorption and the postadsorption phenomena. The arrival of protein at the interface during the early stages of the process is mainly transport limited [1,2]. In the later stages, rate of adsorption is less than that predicted by the diffusion-controlled rate merely due to the coming into existence kinetic barriers at the interface (i.e., occupancy or steric effects) [3,4].

Once adsorbed, proteins can undergo varying levels of orientational and conformational change, resulting in a change both in the monolayer coverage and in the binding strength (interaction energy) between the substrate interface and adsorbing protein. This simply means that adsorbing molecules can exist in more than one adsorbed state [5,6], each of which is characterized by its binding strength, orientation with respect to the interface, and geometrical shape and molecular size. Of course, one should recall the fact that the existence of multiple adsorbed states is not merely due to adsorbate–adsorbent interactions but also extends to lateral adsorbate–adsorbate interactions, where both effects give rise to a time-, protein-, and surface-dependent protein unfolding. This being the case, the adsorbing protein molecules exhibit varying degrees of exchange (displacement) reactions with one another and/or with proteins in the solution.

II. LITERATURE REVIEW

The following review provides a basis for understanding the theoretical development used to construct the model presented in this chapter.

A. Proteins

Proteins are biological macromolecules synthesized in cells for specific functions. They are high-molecular-weight polyamides that adopt exquisitely complex structures. This complexity is characterized by different levels of structure: primary, secondary, tertiary, and quaternary. Primary structure [7] refers to the amino acid sequence itself, along with the location of disulfide bonds (i.e., covalent connections between two amino acid residues within the protein molecule). Secondary structure refers to the spatial arrangement of amino acid residues that are near one another in the linear sequence. Alpha (α) helices and beta (β) sheets are typical examples of secondary structure. Tertiary structure refers to the spatial arrangement of amino acid residues that are far apart in the linear sequence. If a protein has two or more polypeptide chains, each with its exclusive primary, secondary, and tertiary structure, such chains can associate to form a multichain quaternary structure. Hence, a quaternary structure refers to the spatial arrangement of such subunits and their interaction.

B. Protein Adsorption Kinetics

In this subsection, a quick review of what other investigators have done in this area will be presented. Of course, it is to be mentioned here that not all of the relevant published work will be presented; only those that were accessible to us and were reviewed will be presented and we apologize for those whose work was not mentioned hereafter. For a thorough review of protein adsorption, covering both theoretical background and experimentation in a more detail, the reader is advised to read the textbooks by Andrade [8] and Horbett and Brash [6].

In general, adsorption involves the migration of a substance from one phase to the surface of an adjacent phase, accompanied by its accumulation at the interface [9]. Adsorption is a result of the binding forces between individual atoms, ions, or molecular regions of an adsorbate and the adsorbent surface. These binding forces or interactions vary in magnitude from the weak van der Waals type of attraction contributing to physical adsorption, to the strong covalent bonds in chemisorption. Polymer adsorption in general and biopolymer adsorption in particular show a range of binding energies depending on the type of forces present in the interface. Polymer adsorption differs drastically from that of small molecules. This is basically due to the large number of conformations that a macromolecule can adopt, both in the bulk solution and at the interface. Moreover, the entropy loss or gain associated with a given flexible polymer can be substantially greater than that for small molecules or relatively stiff molecules [10].

Lundström [11] proposed a dynamic model of protein adsorption on solid surfaces, which is based on the assumption that a protein molecule may change conformation after adsorption. He found that the number of molecules, which have undergone a conformational change, depends on the concentration in solution and the time constant for the surface-induced conformational change. On energetic grounds, exchange reactions between adsorbed molecules and those in solution are thought to be much more probable than pure desorption, and, in this regard, the timescale of an adsorption experiment is important for detecting any appreciable reduction in the total adsorbed mass as a result of such exchange reactions.

Aptel et al. [12] examined the adsorption kinetics of proteins (human fibrinogen and albumin) onto solid surfaces using the radiolabeling technique. A simple Langmuir model was assumed to describe the rate of adsorption. They found that in order to fit the kinetic data, they could not use only a single pair of adsorption and desorption constants. Moreover, the affinity constant, defined as k_a/k_d, decreased with increasing surface coverage, θ, or with increasing initial bulk concentration (C_0) of the protein. Interestingly enough is that the value of kd was very small in magnitude and assumed almost a constant value of about 1×10^{-4} s^{-1}. The change in the affinity constant was mainly due to the change in the adsorption constant, k_a.

Lundström and Elwing [13] investigated the kinetics of protein-exchange reactions on solid surfaces, as far as the timescale needed for exchange reactions and the possible conformations of an adsorbed protein molecule were concerned. They concluded that if an adsorbed molecule were transformed into a nondesorbable or nonexchangeable form (i.e., characterized by a higher binding strength with the surface or a larger number of footholds), there would be a greater time dependence on the rate of desorbed or exchanged molecules. Furthermore, they categorized surfaces into three types. The first type, which is "invisible" in the biological environment, is the one onto which reversible adsorption and exchange reactions take place with only a small probability that conformational changes occur upon adsorption. The second type of surface allows only exchange reactions to occur and a direct result of which is that conformationally changed molecules may slowly unleash back into the solution. The third type binds at least one kind (or conformation) of protein molecule in an irreversible way.

Wojciechowski and Brash [14] used a computer simulation to examine a single-component protein adsorption. Using irreversible Langmuir kinetics, they showed that the maximum value of the adsorption rate constant k_a is about 0.001 m/s, above which the adsorption process becomes diffusion limited. Moreover, as k_a decreases to 4.0×10^{-8} m/s, the rate of surface binding limits the adsorption enough to allow an accumulation of protein near the wall (i.e., the bulk concentration of protein does not drop to zero within the neighborhood of the interface). In addition, they concluded that the deviation of adsorption kinetics from purely diffusion-limited kinetics could be attributed to (1) a slower rate of filing at the interface (i.e., the process is interface limited), (2) the presence of a reversible adsorption (i.e., desorption), or (3) the occurrence of a surface-induced conformational change, which renders the adsorbed protein irreversibly bound to the surface.

Norde and Anusiem [15] examined adsorption, desorption, and readsorption of bovine serum albumin (BSA), α-lactalbumin, and lysozyme on silica and hematite surfaces. They found that "soft" proteins like BSA and α-lactalbumin might adsorb even under the seemingly unfavorable conditions of a hydrophilic, electrostatically repelling surface. Moreover, they found that preadsorbed BSA and α-lactalbumin, compared with native BSA and α-lactalbumin, resulted in a higher affinity for the adsorbent. On the other hand, lysozyme, which is considered a "hard" protein, did not show any difference in adsorptive behavior with the repetition of the adsorption step.

Nygren et al. [16] studied the kinetics and equilibrium properties of fibrinogen adsorption at solid–liquid interfaces. They described the accumulation of protein at the interface in terms of Langmuir kinetics for a monolayer surface coverage. Depending on the initial, bulk concentration (C_0) of protein, they reported a value of k_a that varies between 2.4×10^{-7} and 3.6×10^{-7} m/s (note: their k_1 value was compared with k_a, according to our proposed kinetic model). In addition, assuming irreversible binding or different reverse rate constants, k_d, the time dependence of the adsorption rate was found to be independent of such assumptions and the plateau level of adsorption was concentration dependent.

Krisdhasima et al. [17] examined the adsorption kinetics and elutability of α-lactalbumin, β-casein, β-lactoglobulin, and BSA at hydrophobic and hydrophilic interfaces. They used a three-parameter model that accounts for the surface coverage as a function of time. In general, they found that a protein with a larger surface hydrophobicity and smaller size would have lower energy barrier to adsorption, as was the case with α-lactalbumin. On the other hand, as far as the surface unfolding is concerned, it was found that once BSA penetrates into the interface, it would exhibit the highest surface-unfolding phenomenon, owing to the flexible amphiphilic character of the neutral domain α-helices.

McGuire et al. [18] studied the structural stability effects on the adsorption and dodecyl trimethyl ammonium bromide (DTAB)-mediated elutability of bacteriophage T4 lysozyme at silica surfaces. In general, they concluded that DTAB-mediated elutability of each mutant increased with increased stability. They quantified stability of a mutant in terms of $\Delta\Delta G$, where $\Delta\Delta G$ is the difference between the free energy of unfolding of the mutant protein and that of the wild type at the melting temperature of the wild type. The resistance to elutability was found to increase with the fraction of adsorbed protein that was more tightly bound at the time of surfactant addition.

Singla et al. [19] examined the adsorption kinetics of wild type and two synthetic stability mutants of T4 phage lysozyme at silanized silica surfaces. Substitution of the isoleucine at amino acid position three with cysteine (I3C) and tryptophan (I3W) rendered such mutants with a higher and lower thermal stability, respectively. It was found that the I3W mutant, characterized by a lower structural stability, would more readily undergo a structural change at the interface. Moreover, such a mutant showed more resistance to elution by DTAB than either the wild type or I3C mutant, simply by forming a more tightly bound conformation (adsorbed state) with the adsorbent.

Brusatori and Van Tassel [20] presented a kinetic model of protein adsorption/surface-induced transition kinetics evaluated by the scale particle theory (SPT). Assuming that proteins (or, more generally, "particles") on the surface are at all times in an equilibrium distribution, they could express the probability functions that an incoming protein finds a space available for adsorption to the surface and an adsorbed protein has sufficient space to spread in terms of the reversible work required to create cavities in a binary system of reversibly and irreversibly adsorbed states. They found that the scale particle theory compared well with the computer simulation in the limit of a lower spreading rate (i.e., smaller surface-induced unfolding rate constant) and a relatively faster rate of surface filling.

Garrett and co-workers [21] studied the kinetics of irreversible adsorption of human serum albumin (HSA) onto three hydro-gel contact lenses. The slow onset of irreversibility in the adsorption behavior of the protein was analyzed based on the assumption that native proteins bind rapidly and reversibly, then slowly denature and tenaciously adsorb to the polymer surface. Their results suggested that the transition from reversible to irreversible adsorption was related to the denaturation of the protein on the surface of the contact lens.

Kidane et al. [22] studied protein adsorption kinetics on polyethylene oxide (PEO)-grafted glass. Results showed that protein adsorption on PEO-grafted glass reached its equilibrium state rapidly. Shibata and Lenhoff [23] used the total internal reflectance fluorescence (TIRF) spectroscopy to determine the kinetics of protein adsorption on modified (butylated or amino-propylated) quartz surfaces. They found that the rate constant under a given set of conditions appeared to be correlated with the ultimate extent of adsorption observed under those conditions.

Different kinetic models or theories were proposed and later used to explain protein adsorption kinetic data. For instance, Norde and Rouwendal [24] used the streaming potential measurements as a tool to study protein adsorption kinetics. They showed that in the initial stages, the adsorption kinetics can be described by the Leveque theory for transport by convective flow and diffusion simultaneously. Szleifer and Satulovsky [25] investigated the thermodynamic and kinetic behaviors of lysozyme and fibrinogen adsorption on a surface with graft polymers by molecular mean field theory. Docoslis et al. [26] studied the early events pertaining to protein (HSA) adsorption and desorption onto silica particles employing real-time in situ measurements by measuring the outflow concentration with a fluorimeter. The acquired data were interpreted, according to a kinetic model, in terms of protein-binding rates, namely the kinetic association (k_a) and dissociation (k_d) constants.

Changes in adsorbed mass as a function of time under various conditions have been plotted and explanations have been proposed. Many experimental observations have indicated that a major portion of the final adsorbed amount had been adsorbed within the first few minutes, if not even seconds, of contact [27,28].

Soderquist and Walton [28] proposed that there are three distinct processes contributing to the kinetics of uptake of protein on polymeric surfaces. First, rapid and reversible adsorption of the proteins occurs in a short period of time. Up to 50–60% surface coverage, there is a random arrangement of adsorbed molecules, but then some form of surface transition occurs that is probably in the direction of surface ordering, thereby allowing further protein uptake. Second, each molecule on the surface undergoes a structural transition as a function of time that occurs in the direction of optimizing protein–surface interaction. Third, the probability of desorption decreases with an increase in the period of incubation, and protein slowly adsorbs more or less irreversibly.

In summary, then, the following general statements can be made regarding protein adsorption. The surface activity of a protein is a cumulative property influenced by many factors, including size, shape, charge, surface hydrophobicity, and thermodynamic stability. Protein adsorption exhibits diversity in behavior from one surface to another and from one protein to another. This diversity results from the complexity of the protein structure itself and from the many variables on which protein adsorption depends.

Consequently, in light of the previous portrayal of protein adsorption process, a macroscopic model that would quantitatively describe protein adsorption kinetics was proposed. The overall rate of protein adsorption was visualized as the sum of the rate of filling (occupying) vacant adsorption sites and the rate of surface-induced protein denaturation, where the latter is manifested via the change in geometrical shape and size of the molecule, which, in turn, affects surface concentration, Γ. The model served to explain the Langmuir-type pattern of adsorption kinetics commonly observed. Its applicability was tested by comparison with experimentally measured Γ versus time adsorption kinetics describing protein interaction with hydrophobic and hydrophilic solid surfaces. This work is an endeavor toward reaching a comprehensive model for protein adsorption kinetics.

III. A KINETIC MODEL FOR PROTEIN ADSORPTION

Writing the component material balance (i.e., mass-action law) for protein, one obtains

$$\text{Rate of adsorption} - \text{Rate of desorption} = \text{Rate of accumulation} \qquad (1)$$

The rate of adsorption is $k_a C_0 (1 - \theta)A$, where k_a is the adsorption rate constant (m/s), C_0 is the protein bulk concentration (mol/m^3), θ, is the fractional surface coverage and A is the total surface area (m^2) available for adsorption; the rate of desorption is $k_d \theta A \Gamma$, where k_d is desorption rate constant (s^{-1}) and Γ is the protein surface concentration (mol/m^2); and the rate of accumulation is $d(\theta \Gamma A)/dt$. Thus, Eq. (1) becomes

$$k_a C_0 (1 - \theta)A - k_d \theta A \Gamma = \frac{d(\theta A \Gamma)}{dt} \qquad (2)$$

The total surface area, A, can be factored out, and Eq. (2) reduces to

$$k_a C_0 (1 - \theta) - k_d \theta \Gamma = \frac{d(\Gamma \theta)}{dt} \qquad (3)$$

The accumulation term, which appears in Eq. (3), can be further simplified to

$$\frac{d(\Gamma \theta)}{dt} = \Gamma \frac{d\theta}{dt} + \theta \frac{d\Gamma}{dt}$$

Thus, Eq. (3) becomes

$$k_a C_0 (1 - \theta) - k_d \theta \Gamma = \Gamma \frac{d\theta}{dt} + \theta \frac{d(\Gamma)}{dt} \qquad (4)$$

Based on the cited literature [4,8,29–31], the protein molecule approaching the surface needs only a minimal surface area requirement (i.e., a foothold) so that it anchors itself to the adsorption site. As time passes and based on conformational stability of adsorbing protein and the extent of fractional surface coverage, θ, surface-induced unfolding takes place, which, in turn, alters the surface concentration of protein molecule, Γ; on the other hand, neighboring molecules that are less tenaciously bound to the interface are adversely affected and eventually desorb back to the solution.

The first term that appears on the right-hand side of Eq. (4) mainly accounts for the net rate of filling (occupying) vacant adsorption sites, whereas the second term mainly accounts for the rate of surface-induced protein unfolding (or denaturation), which is manifested via the change in surface concentration, Γ, with time. Given that proteins are, in general, rod type or globular in shape, one way to visualize protein molecule is as being cylindrical (or ellipsoidal) in shape (with the height of the cylinder being the major axis of protein molecule and the diameter as the minor axis); then, the fractional surface coverage, θ, can be defined as

$$\theta = \frac{N_s \pi D^2 / 4}{A} \tag{5}$$

where N_s is the total number of adsorbing molecules and D is the diameter (m) of the cylinder. The total number of adsorbing molecules, N_s, can be further expressed as

$$N_s = \frac{n_P V_P}{(\pi D^2 / 4) L} \tag{6}$$

where n_P (mole) is the number of moles of adsorbing protein, V_P (m^3/mol) is the molar volume of adsorbing protein, and $(\pi D^2 / 4) L$ is the molecular volume (m^3/molecule) of the adsorbing protein. Substituting the value of N_s [Eq. (6)] in Eq. (5), θ becomes

$$\theta = \frac{n_P V_P}{(\pi D^2 / 4) L} \frac{\pi D^2 / 4}{A} = \frac{n_P}{A} \frac{V_P}{L} = \Gamma \left(\frac{V_P}{L} \right)_{\text{interface}} \tag{7}$$

Strictly speaking, the molecular properties appearing in the $(V_P/L)_{\text{interface}}$ ratio should be evaluated at the interphase not at the bulk-phase conditions and should be adsorption-time dependent. However, the $(V_P/L)_{\text{interface}}$ ratio will be assumed constant over the time course of adsorption. As will be further explained, both values of the V_P/L ratio will be contrasted and $(V_P/L)_{\text{solution}}$ will be determined based on the molecular dimensions of a given protein in solution; on the other hand, $(V_P/L)_{\text{interface}}$ will be calculated based on the computer-generated, regressed parameters belonging to the existing kinetic model, namely α and β. The difference in values of the V_P/L ratio will serve as an indication of the extent of surface-induced unfolding of adsorbing protein or the presence of a multilayer film. for cosmetic reasons and to make equations more plausible, from now on, the subscription will be dropped out with the understanding that V_P/L really means $(V_P/L)_{\text{interface}}$.

Eliminating the value of θ in favour of Γ using Eq. (7), Eq. (4) becomes

$$k_a C_0 (1 - \theta) - k_d \theta \Gamma = \Gamma \frac{d\theta}{dt} + \theta \frac{d\Gamma}{dt} = \Gamma \frac{d(\Gamma V_P / L)}{dt} + \theta \frac{d\Gamma}{dt} = \Gamma \frac{(V_P / L) d\Gamma}{dt} + \theta \frac{d\Gamma}{dt} \tag{8}$$

It is not difficult to realize that the first term on the right-hand side of Eq. (8) is equal to $\theta (d\Gamma/dt)$; therefore, Eq. (8) becomes

$$k_a C_0 (1 - \theta) - k_d \theta \Gamma = \Gamma \frac{(V_P / L) d\Gamma}{dt} + \theta \frac{d\Gamma}{dt} = \theta \frac{d\Gamma}{dt} + \theta \frac{d\Gamma}{dt} = 2\theta \frac{d\Gamma}{dt} \tag{9}$$

Dividing through by θ and rearranging, Eq. (9) becomes

$$k_a C_0 \left(\frac{1 - \theta}{\theta} \right) - k_d \Gamma = 2 \frac{d\Gamma}{dt} \tag{10}$$

Eliminating θ in favor of Γ using Eq. (7), Eq. (10) becomes

$$k_a C_0 \left(\frac{L}{V_P \Gamma} - 1 \right) - k_d \Gamma = 2 \frac{d\Gamma}{dt} \tag{11}$$

Multiplying through by Γ yields

$$k_a C_0 \frac{L}{V_P} \Gamma - k_a C_0 \Gamma - k_d \Gamma^2 = 2\Gamma \frac{d\Gamma}{dt} \tag{12}$$

Obviously, Eq. (12) is a nonlinear ordinary differential equation. To facilitate the solution of Eq. (12), the following notations will be adopted:

$$\beta = k_a C_0 \frac{L}{V_P} \tag{13}$$

$$\alpha = k_a C_0 \tag{14}$$

$$y = \Gamma \tag{15}$$

$$y' = \frac{dy}{dx} = \frac{d\Gamma}{dt} \tag{16}$$

Consequently, Eq. (12) becomes

$$\beta - \alpha y - k_d y^2 = 2yy' = 2y \frac{dy}{dx} \tag{17}$$

Upon rearrangement of Eq. (17), it reduces to

$$dx = \frac{2y\,dy}{\beta - \alpha y - k_d y^2} \tag{18}$$

Alternatively, in an integral form, it becomes

$$\int dx = \int \frac{2y\,dy}{\beta - \alpha y - k_d y^2} \tag{19}$$

From calculus, it is given that

$$\int \frac{y\,dy}{(ay + b)(py + q)} = \frac{1}{bp - aq} \left(\frac{b}{a} \ln(ay + b) - \frac{q}{p} \ln(py + q) \right) \tag{20}$$

If the denominator, $\beta - \alpha y - k_d y^2$, is matched with the standard form $(ay + b)(py + q)$ of Eq. (20), then the following equalities are obtained:

$$a = 1 \tag{21a}$$

$$p = 1 \tag{21b}$$

$$b = \frac{\alpha + \sqrt{\alpha^2 + 4k_d \beta}}{2k_d} \tag{21c}$$

$$q = \frac{\alpha - \sqrt{\alpha^2 + 4k_d \beta}}{2k_d} \tag{21d}$$

If dx, which is dt, is integrated from 0, as the lower limit, to any value x, as the upper limit, then dy, which is $d\Gamma$, will be integrated from 0, as the lower limit, to any value y, as the upper limit.

Therefore, using Eq. (20) while making use of equalities found in Eqs (21a)–(21d), the final solution of Eq. (17) will be

$$
x = \left[\frac{\alpha + \sqrt{\alpha^2 + 4k_d\beta}}{\sqrt{\alpha^2 + 4k_d\beta}} \ln \left(\frac{y + \left(\alpha + \sqrt{\alpha^2 + 4k_d\beta}\right)/2k_d}{\left(\alpha + \sqrt{\alpha^2 + 4k_d\beta}\right)/2k_d} \right) \right]
$$
$$
- \left[\frac{\alpha - \sqrt{\alpha^2 + 4k_d\beta}}{\sqrt{\alpha^2 + 4k_d\beta}} \ln \left(\frac{y + (\alpha - \sqrt{\alpha^2 + 4k_d\beta})/2k_d}{(\alpha - \sqrt{\alpha^2 + 4k_d\beta})/2k_d} \right) \right] \tag{22}
$$

If x is replaced by t and by Γ, then Eq. (22) becomes

$$
t = \left[\frac{\alpha + \sqrt{\alpha^2 + 4k_d\beta}}{\sqrt{\alpha^2 + 4k_d\beta}} \ln \left(\frac{\Gamma + \left(\alpha + \sqrt{\alpha^2 + 4k_d\beta}\right)/2k_d}{\left(\alpha + \sqrt{\alpha^2 + 4k_d\beta}\right)/2k_d} \right) \right]
$$
$$
- \left[\frac{\alpha - \sqrt{\alpha^2 + 4k_d\beta}}{\sqrt{\alpha^2 + 4k_d\beta}} \ln \left(\frac{\Gamma + (\alpha - \sqrt{\alpha^2 + 4k_d\beta})/2k_d}{(\alpha - \sqrt{\alpha^2 + 4k_d\beta})/2k_d} \right) \right] \tag{23}
$$

A. Low Surface Coverage ($\theta \approx 0$)

If the surface coverage, θ, is relatively small (say, $\theta < 0.1$), then Eq. (10) can be simplified to

$$
k_a C_0 \left(\frac{1}{\theta}\right) - k_d \Gamma = 2 \frac{d\Gamma}{dt} \tag{24}
$$

Rearranging Eq. (24) while recalling that ($\theta = \Gamma V_P/L$), one obtains

$$
\frac{d\Gamma}{dt} + \frac{k_d}{2}\Gamma = \frac{k_a C_0 L}{2V_P}\Gamma^{-1} \tag{25}
$$

Equation (25) is Bernoulli's equation that can be transformed into a first-order linear ordinary differential equation via the following transformations:

$$
\Gamma^* = \Gamma^2 \tag{26a}
$$
$$
\frac{d\Gamma^*}{dt} = 2\Gamma \frac{d\Gamma}{dt} \tag{26b}
$$

Utilizing the transformations given by Eqs. (26a) and (26b), Eq. (25) becomes

$$
\frac{1}{2}\frac{d\Gamma^*}{dt}\Gamma^{-1} + \frac{k_d}{2}\Gamma = \frac{k_a C_0 L}{2V_P}\Gamma^{-1} \tag{27}
$$

Multiplying Eq. (27) by 2Γ, it becomes

$$
\frac{d\Gamma^*}{dt} + k_d \Gamma^2 = \frac{k_a C_0 L}{V_P} \tag{28}
$$

Upon utilizing Eq. (26a), Eq. (28) becomes

$$\frac{d\Gamma^*}{dt} + k_d\Gamma^* = \frac{k_aC_0L}{V_P} \tag{29}$$

Equation (29) is now a first-order linear ordinary differential equation, which has the following solution:

$$\Gamma^*e^{k_dt} = \int \frac{k_aC_0L}{V_P}e^{k_dt}\,dt + C \tag{30}$$

where C is the constant of integration.

Upon integration of the first term on the right-hand side of Eq. (30) and dividing through by the exponential term, Eq. (30) becomes

$$\Gamma^* = \frac{k_aC_0L}{V_Pk_d} + Ce^{-k_dt} \tag{31}$$

Given that at $t = 0$, $\Gamma^* = 0$, allows us to evaluate the constant of integration C. Therefore, Eq. (31) becomes

$$\Gamma^* = \frac{k_aC_0L}{V_Pk_d}\left(1 - e^{-k_dt}\right) \tag{32}$$

Upon back-substitution of Γ^* by Γ^2, Eq. (32) becomes

$$\Gamma = \sqrt{\frac{C_0LK_a}{V_Pk_d}(1 - e^{-k_dt})} = \sqrt{\frac{\beta}{k_d}(1 - e^{-k_dt})} \tag{33}$$

B. Irreversible Adsorption ($k_d = 0$)

In this case, Eq. (10) becomes

$$k_aC_0\left(\frac{1}{\theta}\right) - k_aC_0 = 2\frac{d\Gamma}{dt} \tag{34}$$

Substituting the value of $\theta = \Gamma V_P/L$ in Eq. (34) while noting that $\beta = k_aC_0L/V_P$ and $\alpha = k_aC_0$, Eq. (34) becomes

$$2\Gamma\frac{d\Gamma}{dt} = \beta - \alpha\Gamma \tag{35}$$

The solution of Eq. (35), given that at $t = 0$, $\Gamma = 0$, is

$$t = \frac{2}{\alpha^2}\left[\beta\ln\left(\frac{\beta}{\beta - \alpha\Gamma}\right) - \alpha\Gamma\right] \tag{36}$$

IV. APPLICABILITY OF THE MODEL

The general kinetic model [Eq. (23)] was used to qualitatively describe the effect of k_a, k_d, and C_0 on the extent of adsorption, namely Γ (or θ). Figure 1 shows Γ versus time for different values of k_d for some fixed values of β and C_0. As can be seen from Fig. 1, although k_d was significantly increased by three orders of magnitude, a maximum of a 20% decrease in Γ was found. The effect of initial concentration, C_0, on Γ is depicted in Fig. 2. It is clearly seen that an

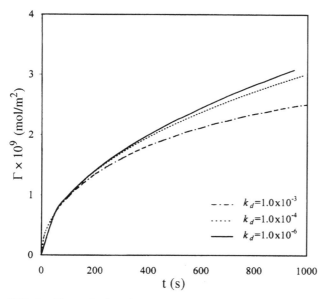

FIG. 1 Theoretically calculated surface concentration versus time for various values of k_d. The values of C_0 and β are 1.04×10^{-4} mol/m^3 and 1.0×10^{-16} mol^2/m^4 s, respectively.

order of magnitude change in the value of C_0 causes a significant change in the magnitude of Γ. The same dramatic effect on Γ is also observed when the value of either L/V_P or k_a appearing in the β (note that $\beta = \alpha L/V_P = k_a C_0 L/V_P$) term was allowed to change, as shown in Fig. 3. Compared with the effect of C_0, L/V_P, or k_a, one can say that k_d has a little control on the adsorption kinetics, in particular, at low surface coverage, θ. This is in harmony with what Aptel

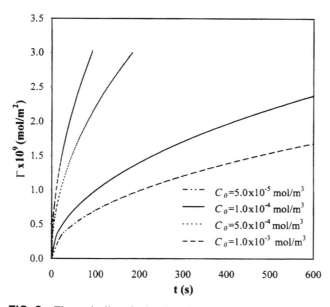

FIG. 2 Theoretically calculated surface concentration versus time for various values of C_0. The values of β and k_d are 1.0×10^{-16} mol^2/m^4 s and 2.0×10^{-4} 1/s, respectively.

et al. [12] had already reported—that the change in the affinity constant k_a/k_d was mainly due to the change in the adsorption constant, k_a. In addition., the effect of the type of protein on Γ, which is picked up via the L/V_P ratio, is implicitly shown in Fig. 3 for fixed values of C_0 and k_d. Obviously, the higher the value of L/V_P (i.e., β), the higher Γ will be. In other words, the end-on conformation monolayer gives higher value of Γ than that of the side-on conformation.

It should be pointed out that as a qualitative rule in protein adsorption studies [8], a protein monolayer with an end-on conformation presumably prevails when adsorption occurs from solution at high concentration, whereas a "side-on" conformation monolayer is observed at low protein concentration. Although, practically, it is quite impossible to draw a line between low- and high-concentration regions, the concentration interval between $C_0 = 0.1$ mg/mL and $C_0 = 1.0$ mg/mL may be suggested as the diffuse line between low- and high-concentration regions as far as side-on and end-on conformations are concerned [32].

Protein adsorption at solid–liquid interfaces, in general, is characterized by a monolayer or a submonolayer surface coverage [33–41], and pure desorption, in a strict sense, constitutes a very unlikely event [11,16]. Because, as pointed out earlier, it was found that k_d has a little control on the adsorption kinetics, in particular at low θ, the special cases [Eq. (33) and (36)] of the general kinetic model [Eq. (23)] will be considered for prediction of the experiment-based kinetic data.

In addition, examining the protein adsorption kinetic data, it was found that the value of surface coverage was very high (i.e., $\theta \approx 1$) within the time course of the adsorption experiment. Also, it is experimentally difficult to monitor the first initial events of a typical adsorption experiment (i.e., $\theta < 0.1$). This is simply because protein adsorption kinetic data are, in general, characterized by a high initial slope. Hence, no attempt was made to test the low-surface-coverage model [Eq. (33)]. However, the door is left open for future testing, should reliable, short-time, and low-surface-coverage data become available. Meanwhile, discussion of experimental data will be handled in light of the irreversible adsorption kinetics [Eq. (36)]. It should be pointed out that both α and β present in Eq. (36) are computer-generated parameters, using a

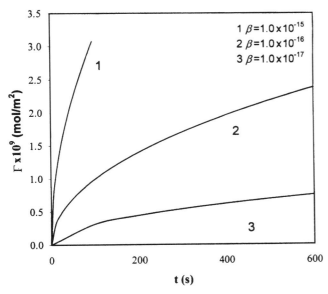

FIG. 3 Theoretically calculated surface concentration versus time for various values of β for the case where $C_0 = 1 \times 10^{-4}$ mol/m^3 and $k_d = 2 \times 10^{-4}$ 1/s.

nonlinear regression technique. Once α and β are determined, the value of k_a can be calculated for a given value of C_0.

Furthermore, the value of the L/V_P ratio, evaluated at the interface microenvironment, is simply β/α. The latter value will be compared with the L/V_P ratio that is evaluated based on molecular properties of protein in solution. As pointed earlier in Section III, difference in values of the L/V_P ratio will serve as an indication of the extent of surface-induced unfolding of adsorbing protein [i.e., $\beta/\alpha = (L/V_{P\,\text{interface}} < L/V_P)_{\text{solution}}$] or the presence of a multilayer film [i.e., $\beta/\alpha = (L/V_P)_{\text{interface}} > (L/V_P)_{\text{solution}}$]. It should be noted that $(L/V_P)_{\text{solution}}$ (mol/m^2), to a large extent, corresponds to the monolayer surface concentration of protein.

A. Experimental Data-Based Regressed Parameters of the Model [Eq. (36)]

Figures 4a and 4b show experimentally measured [17] data points of Γ versus t for α-lactalbumin (α-Lac) both on hydrophilic and hydrophobic surfaces, respectively. Figure 5a and 5b show experimentally measured [17] data points of Γ versus t for β-lactoglobulin (β-Lact) both on hydrophobic surfaces, respectively. Figure 6a and 6b show experimentally measured [17] data points of Γ versus t β-casein both on hydrophilic and hydrophobic surfaces, respectively. Figure 8a and 8b show experimentally measured [19] data points of Γ versus t for the wild-type lysozyme both on hydrophilic and hydrophobic surfaces, respectively. Figure 9a and 9b show experimentally measured [19] data points of Γ versus t for the most stable mutant I3C lysozyme, where isoleucine at position 3 was replaced by cysteine (Ile3 $\Rightarrow$ Cys(S–S)), both on hydrophilic and hydrophobic surfaces, respectively. Figure 10a and 10b show experimentally measured [19] data points of Γ versus t for the least stable mutant I3W lysozyme, where isoleucine at position 3 was replaced by tryptophan (Ile3 $\Rightarrow$ Trp) both on hydrophilic and hydrophobic surfaces, respectively. In Figs. (4a)–(10b), the regressed curves using Eq. (36) are also shown.

Table 1 summarizes the regressed parameters α and β both for hydrophilic and hydrophobic surfaces. It should be pointed out that the adjusted coefficient of multiple determination (R_a^2) was at least 0.9. Moreover, the standard error associated with each parameter was, on average, about 10% of the magnitude of the parameter itself.

B. Determination of the Adsorption Rate Constant, k_a

As shown in Table 1, k_a can be calculated from parameter α simply by dividing the latter by C_0. Table 2 summarizes the numerical values of k_a (nm/s) both on hydrophilic and hydrophobic surfaces. As shown in Table 3, the k_a value is two orders of magnitude larger than that reported by Nygren et al. [16]. Moreover, as indicated by Wojciechowski and Brash [14], a value of k_a less than 4.0×10^{-8} m/s (or 40 nm/s) implies that the rate of surface binding limits the adsorption enough to allow an accumulation of protein near the wall (i.e., the adsorption process thereunder is not diffusion limited). In any event, the relative values of k_a for the given proteins, however, are in a qualitative agreement with those reported by the authors of the experimental data, Krisdhasima et al. [17]. For example, α-lactalbumin, in our case, has the highest k_a value, whereas BSA has the lowest value on the hydrophobic surface, and according to their ranking, they [17] ranked α-lactalbumin as number 1 (with the highest k_a) and BSA as number 4 (with the lowest k_a). As pointed out [17], α-lactalbumin being a small, flexible, and hydrophobic molecule facilitates its adsorption to a hydrophobic surface, even in a crowded interface. On the other hand, BSA would require a longer time, once in the interface, to orient itself "end-on" with its neutral domain adjacent to the surface. As adsorption proceeds,

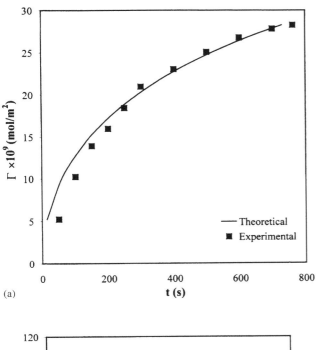

(a)

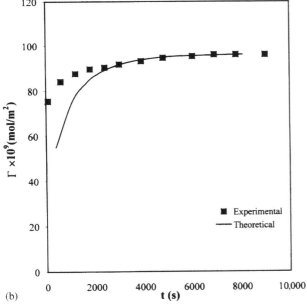

(b)

FIG. 4 (a) Comparison between experimentally measured [15] and theoretically calculated values of the surface concentration versus time for the adsorption of α-Lac on a hydrophilic surface. (b) Comparison between experimentally measured [17] and theoretically calculated values of the surface concentration versus time for the adsorption of α-Lac on a hydrophobic surface.

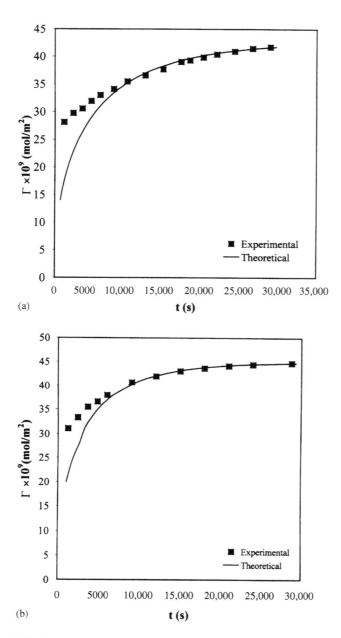

FIG. 5 (a) Comparison between experimentally measured [17] and theoretically calculated values of the surface concentration versus time for the adsorption of β-Lact on a hydrophilic surface. (b) Comparison between experimentally measured [17] and theoretically calculated values of the surface concentration versus time for the adsorption of β-Lact on a hydrophobic surface.

increased electrostatic repulsion among nearest neighbors would slow establishment of an incoming protein's first noncovalent contact with the surface [17].

One should be very cautious about explaining protein adsorption behavior on hydrophilic surfaces; the relatively higher value of k_a for β-casein on the hydrophilic surface may be

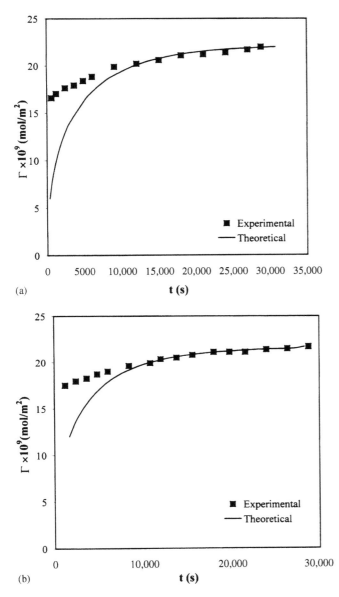

FIG. 6 (a) Comparison between experimentally measured [17] and theoretically calculated values of the surface concentration versus time for the adsorption of BSA on a hydrophilic surface. (b) Comparison between experimentally measured [17] and theoretically calculated values of the surface concentration versus time for the adsorption of BSA on a hydrophobic surface.

considered as a paradox, given that both the hydrophilic silica surface and β-casein [42] possess a net negative formal charge. This is not surprising; for example, Norde [43] indicated that with "rigid" proteins, intramolecular structural rearrangements do not contribute to the adsorption process. On the other hand, with "soft" proteins, the intramolecular structural rearrangements result in a significant driving force for adsorption such that proteins may even adsorb under the adverse conditions of a hydrophilic, electrostatically repelling surface. Thus, the relatively higher

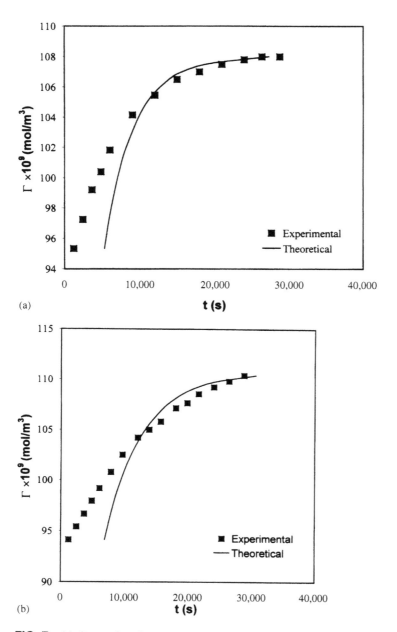

FIG. 7 (a) Comparison between experimentally measured [17] and theoretically calculated values of the surface concentration versus time for the adsorption of β-casein on a hydrophilic surface. (b) Comparison between experimentally measured [17] and theoretically calculated values of the surface concentration versus time for the adsorption of β-casein on a hydrophobic surface.

k_a value of β-casein on the hydrophilic surface could be attributed to the molecular flexibility of such a linear amphiphile and highly unordered structure [17,42]. To stretch the applicability of the model, another set, made of wild-type and two synthetic stability mutants of bacteriophage T4 lysozyme, was used. in the case of the wild type, k_a was higher on the hydrophobic surface

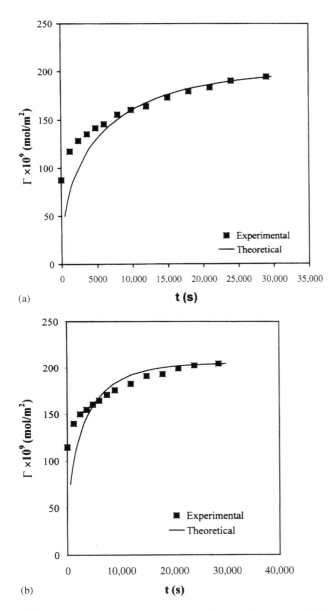

FIG. 8 (a) Comparison between experimentally measured [19] and theoretically calculated values of the surface concentration versus time for the adsorption of lysozyme wild type on a hydrophilic surface. (b) Comparison between experimentally measured [19] and theoretically calculated values of the surface concentration versus time for the adsorption of lysozyme wild type on a hydrophobic surface.

than on the hydrophilic surface. This finding was in a total agreement with the authors' finding that the initial rate of adsorption was higher on the hydrophobic surface than on the hydrophilic surface [19]. Moreover, $(k_a)_{\text{wild type}} > (k_a)_{\text{I3C}} > (k_a)_{\text{I3W}}$ both on the hydrophilic and hydrophobic surfaces, which is consistent with the steepness of the initial slope required to achieve the plateau value for each of the given types of lysozyme.

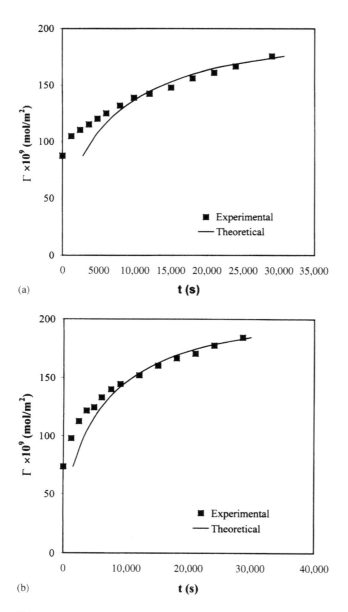

FIG. 9 (a) Comparison between experimentally measured [19] and theoretically calculated values of the surface concentration versus time for the adsorption of lysozyme Cys(S–S) on a hydrophilic surface. (b) Comparison between experimentally measured [19] and theoretically calculated values of the surface concentration versus time for the adsorption of lysozyme Cys(S–S) on a hydrophobic surface.

C. The Extent of Surface-Induced Unfolding

Table 3 shows the solution and interface values of L/V_P ratio for the examined proteins both at hydrophobic and hydrophilic surfaces. It should be pointed out here that $(L/V_P)_{solution}$ ratio was estimated based on the actual size and shape of the protein molecule in solution. As pointed out earlier, the regression-based calculated value, $\beta/\alpha = (L/V_P)_{interface}$, is an indication of the extent

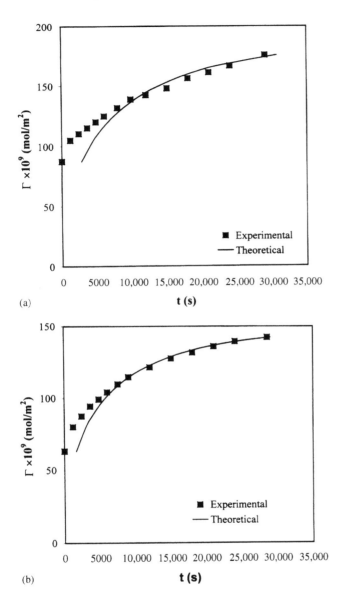

FIG. 10 (a) Comparison between experimentally measured [19] and theoretically calculated values of the surface concentration versus time for the adsorption of lysozyme Trp on a hydrophilic surface. (b) Comparison between experimentally measured [19] and theoretically calculated values of the surface concentration versus time for the adsorption of lysozyme Trp on a hydrophobic surface.

of surface-induced unfolding (or denaturation). To better quantify this effect, the following term is introduced:

$$\eta_{\text{siu}} = \frac{(L/V_P)_{\text{solution}} - (L/V_P)_{\text{interface}}}{(L/V_P)_{\text{solution}}} \times 100\%$$

TABLE 1 Regressed Parameters α and β [Eq. (36)] Both for Hydrophilic and Hydrophobic Surfaces

Protein type	Hydrophobic surface (p-value $= 0.00001$)[a]		Hydrophobic surface (p-value $= 0.00001$)[a]	
	α (mol/m^2 s)	β (mol^2/m^4 s)	α (mol/m^2 s)	β (mol^2/m^4 s)
α-Lactalbumin	2.00×10^{-11}	2.08×10^{-18}	1.35×10^{-10}	1.30×10^{-17}
β-Lactoglobulin	8.12×10^{-12}	3.47×10^{-19}	1.35×10^{-11}	6.08×10^{-19}
BSA	5.36×10^{-12}	1.19×10^{-19}	6.65×10^{-12}	1.44×10^{-19}
β-Casein	5.17×10^{-11}	5.58×10^{-18}	3.37×10^{-11}	3.74×10^{-18}
Wild type Lysozyme	3.19×10^{-11}	6.43×10^{-18}	6.00×10^{-11}	12.35×10^{-18}
I3C Lysozyme	3.0×10^{-11}	5.42×10^{-18}	1.90×10^{-11}	3.97×10^{-18}
I3W Lysozyme	2.9×10^{-11}	3.86×10^{-18}	1.70×10^{-11}	2.68×10^{-18}

[a]Statistically, the p-value, the smallest level of significance, associated with each parameter (i.e., 0.00001) simply implies that out of the observations being made 99.999% will lead to the conclusion that α and β are nonzero significant values.

where η_{siu} represents the percent reduction of the length or the percent increase in the partial molar area of the adsorbing protein as a result of surface-induced unfolding.

Obviously, the value of η_{siu} is a function of the conformation of the adsorbed state of protein, whether it is a side-on or end-on conformation. As pointed out earlier, the concentration interval between $C_0 = 0.1$ mg/mL and $C_0 = 1.0$ mg/mL may be suggested as the diffuse line between low- and high-concentration regions as far as side-on and end-on conformations are concerned [32]. For the aforementioned set of experimental data [17], the concentration of each of the examined proteins is at least 1 mg/mL, except for α-lactalbumin and β-casein, which have concentrations of 0.385 and 0.653 mg/mL, respectively. Therefore, a linear interpolation was made to find the value of $(L/V_P)_{solution}$ for both α-lactalbumin and β-casein, which, of course, lies between the value based on the side-on conformation and that based on the end-on conformation. Taking the above point into account, Table 4 shows the values of η_{siu} calculated for various types of protein both on hydrophilic and hydrophobic surfaces.

For the set that consists of the first four different proteins, as far the hydrophobic surface-induced unfolding effect is concerned, η_{siu} decreases in the order $(\eta_{siu})_{BSA} >$

TABLE 2 Calculated Values of the Adsorption Rate Constant, k_a, for Different Proteins Both on Hydrophilic and Hydrophobic Surfaces

Protein type	C_0 (mmol/m^3)	Hydrophilic surface k_a (nm/s)	Hydrophobic surface k_a (nm/s)
α-Lactalbumin	27.22	0.74	4.96
β-Lactoglobulin	27.22	0.30	0.50
BSA	27.22	0.20	0.24
β-Casein	27.22	1.90	1.24
Wild type Lysozyme	69.44	0.46	0.86
I3C Lysozyme	69.44	0.43	0.27
I3W Lysozyme	69.44	0.42	0.24

TABLE 3 Solution and Interface Values of L/V_P Ratio for the Examined Proteins Both at Hydrophobic and Hydrophilic Surfaces

Protein type	End-on $(L/V_P)_{solution}$ $(mol/m^2)^a$	Hydrophilic $\frac{\beta}{\alpha} = \left(\frac{L}{V_P}\right)_{interface}$ (mol/m^2)	Hydrophobic $\frac{\beta}{\alpha} = \left(\frac{L}{V_P}\right)_{interface}$ (mol/m^2)
α-Lactalbumin (ellipsoid)	3.96×10^{-7} $[2.68 \times 10^{-7}]^b$	1.04×10^{-7}	0.96×10^{-7}
β-Lactoglobulin (two spheres)	2.47×10^{-7} $[1.28 \times 10^{-7}]$	0.43×10^{-7}	0.45×10^{-7}
BSA (ellipsoid)	2.2×10^{-7} $[0.60 \times 10^{-7}]$	0.22×10^{-7}	0.22×10^{-7}
β-Casein (prolate ellipsoid)	14.88×10^{-7} $[1.24 \times 10^{-7}]$	1.08×10^{-7}	1.11×10^{-7}
Wild type Lysozyme (ellipsoid)	3.52×10^{-7} $[2.35 \times 10^{-7}]$	2.02×10^{-7}	2.06×10^{-7}
I3C Lysozyme	3.52×10^{-7} $[2.35 \times 10^{-7}]$	1.80×10^{-7}	2.09×10^{-7}
I3W Lysozyme	3.52×10^{-7} $[2.35 \times 10^{-7}]$	1.33×10^{-7}	1.58×10^{-7}

Note: The $(L/V_P)_{solution}$ ratio was estimated based on the actual size and shape of the protein molecule in solution.
[a]Geometrical molecular properties of all proteins, except β-casein, were adapted from Ref. 31. For β-casein, its molecular properties were adapted from Ref. 34.
[b]Data included within brackets represent the value of $(L/V_P)_{solution}$ based on the side-on adsorption.

$(\eta_{siu})_{\beta\text{-casein}} > (\eta_{siu})_{\beta\text{-Lag}} > (\eta_{siu})_{\alpha\text{-Lac}}$. Krisdhasima et al. [17] ranked such proteins, as far as the surface-induced unfolding on the hydrophobic silica is concerned, in the following order $(s_1)_{BSA} > (s_1)_{\beta\text{-casein}} > (s_1)_{\alpha\text{-Lac}} > (s_1)_{\beta\text{-Lag}}$ which, to a large extent, is in harmony with the result previously presented.

Moreover, the applicability of the model was also tested using another set, made of the wild-type and two synthetic stability mutants of bacteriophage 14 lysozyme, where I3C mutant is a more stable protein and I3W mutant a less stable than the wild type. On the hydrophobic

TABLE 4 Surface-Induced Unfolding factor, η_{siu}, Evaluated Both on Hydrophilic and Hydrophobic Surfaces

Protein type	Hydrophilic surface n_{siu} (%)	Hydrophobic surface η_{siu} (%)
α-Lactalbumin	66.3	68.9
β-Lactoglobulin	85.6	81.8
BSA	90.0	90.0
β-Casein	88.7	88.4
Wild type Lysozyme	42.6	41.5
I3C Lysozyme	48.9	40.6
I3W Lysozyme	62.5	55.1

surface, the surface-induced unfolding, decreases with increasing thermal, structural stability in the order $(\eta_{siu})_{I3W} > (\eta_{siu})_{wild-type} > (\eta_{siu})_{I3C}$. Given that θ_1 is the fractional surface coverage of removable protein and θ_2 is the fractional surface coverage that is nonremovable, it was shown [17,19] that the θ_2/θ_1 ratio would decrease with increasing thermal stability of such types of lysozyme. In our case, a higher value of η_{siu} means a higher value of the θ_2/θ_1 ratio, which is a direct result of the increased partial molar area of the adsorbed protein upon unfolding.

In addition, it was also shown [44] that DTAB-mediated elutability was positively correlated with the thermal structural stability of the above-mentioned types of lysozyme. In other words, the resistance to elutability correlates with θ_2/θ_1 ratio, the nonremovable fractional surface coverage divided by the removable fractional surface coverage, which, in turn, correlates with the extent of surface-induced unfolding (i.e., η_{siu} in our case). As far as the surface-induced unfolding on the hydrophilic surface is concerned, the least thermally stable mutant (i.e., I3W) exhibited the highest value of η_{siu}, whereas the wild type exhibited the lowest value of η_{siu}. Apparently, there is a switch in order between the wild type and the most stable form, which is I3C. It was shown [19] that the spectral differences between the wild type and I3C were relatively small. Nevertheless, the difference in kinetic behavior between the wild type and I3C is small, compared with that of I3W.

For both sets of examined proteins and both on hydrophilic and hydrophobic surfaces, it was found that lysozyme, in particular the wild type, has the least value of η_{situ} among other proteins. This is in line with what Norde and Anusiem [15] found—that preadsorbed BSA and α-lactalbumin, compared with native BSA and α-lactalbumin, resulted in a higher affinity for the adsorbent. This is an indication of surface-induced unfolding for such "soft" molecules. On the other hand, lysozyme, which is considered a "hard" protein, did not show any difference in adsorptive behavior with the repetition of adsorption step.

Finally, it should be pointed out that as far as the nonlinear regression methodology is concerned, it is quite possible to fit the experimental data with different combinations of α and β, given that the adjusted coefficient of multiple determination (R_a^2) is sufficiently high. From a regression standpoint, the objective function, which is, in this case, the sum of square errors (SSE) between curve-fitted values and experimental ones, is to be minimized. This can be achieved even if both α and β may negligibly change on an individual basis; the ratio β/α, however, remains almost insensitive to the variation in α and β values for the same protein and surface. Obviously, α and β values do change with the type of protein and/or type of surface.

In conclusion, the experimental data could be fitted using the kinetic model [Eq. (36)], describing the irreversible kinetics of protein adsorption. As can be seen from Figs. 4a–10b, the kinetic model mainly deviates from the experimental data at low surface coverage. This can be attributed to the nature of experimental data, characterized by an attainment of a plateau value within a very short period (high initial slope), which makes it infeasible to fit such data with a reasonable value of k_a. This being the case, experimental and theoretical work should focus on low surface coverage during the first initial events of a protein adsorption experiment, to better quantify rate constants appearing in the most general form in terms of molecular properties of protein and microscopic properties of interface.

V. NOMENCLATURE

A Total area available for adsorption (m^2)
C_0 Protein bulk concentration (mol/m^3)
D Diameter of a molecule (m)
k_a Adsorption rate constant (m/s)

k_d Desorption rate constant (l/s)
L Length of a protein molecule (m)
n_P Number of adsorbing protein molecules (mol)
N_s Total number of adsorbing molecules (dimensionless)
t Time (s)
V_P Molar volume of adsorbing molecule (m^3/mol)
α Parameter defined by Eq. (14) (mol/m^2 s)
β Parameter defined by Eq. (13) (mol^2/m^4 s)
Γ Protein surface concentration (mol/m^2)
θ Fractional surface coverage (dimensionless)

REFERENCES

1. I Langmuir, VJ Schaefer J Am Chem Soc 59:2400, 1937.
2. F MacRitchie, AE Alexander. J Colloid Sci 18:453, 1963.
3. DE Graham, MC Philips. J Colloid Interf Sci 70:403, 1979.
4. S Damodaran, KB Song. Biochim Biophys Acta 954:253, 1988.
5. I Lundström, B Ivarsson, U Jönsson, H Elwing. In: WJ Feast, HS Munro, eds. Polymer Surfaces and Interfaces. New York: Wiley, 1987, pp 201–230.
6. TA Horbett, JL Brash, eds. Proteins at Interfaces: Physicochemical and Biochemical Studies. ACS Symposium Series 343. Washington, DC: American Chemical Society, 1987.
7. L Stryer. Biochemistry. New York: WH Freeman, 1988.
8. JD Andrade, ed. Surface and Interfacial Aspects of Biomedical Polymers: Protein Adsorption. New York: Plenum Press, 1985, Vol. 2.
9. FL Slejko. Adsorption Technology: A Step-by-Step Approach to Process Evaluation and Application. New York: Marcel Dekker, 1985, pp 1–6.
10. GD Parfitt, CH Rochester. Adsorption from Solution at the Solid/Liquid Interface. London: Academic Press, 1983, pp 153–218.
11. I Lunström. Prog Colloid Polym Sci 70:76, 1985.
12. JD Aptel, JC Voegel, A Schmitt. Colloids Surfaces 29:359, 1988.
13. I Lundström, H Elwing. J Colloid Interf Sci 136:68, 1990.
14. PW Wojciechowski, JL Brash. J Colloid Interf Sci 140:239, 1990.
15. W Norde, AC Anusiem. Colloids Surfaces 66:73, 1992.
16. H Nygren, M Stenberg, C Karlsson. J Biomed Mater Res. 26:77, 1992.
17. V Krisdhasima, P Vinaraphong, J McGuire. J Colloid Interf Sci. 161:325, 1993.
18. J McGuire, MC Wahlgren, T Arnebrant. J Colloid Interf Sci 170:182, 1995.
19. B Singla, V Krisdhasima, J McGuire. J Colloid Interf Sci 182:292, 1996.
20. MA Brusatori, PR Van Tassel. J Colloid Interf Sci 219:333, 1999.
21. Q Garrett, R Chatelier, H Griesser, BK Milthorpe. Transactions of the Annual Meeting of the Society for Biomaterials in Conjunction with the International Biomaterials Symposium, Proceedings of the 1996 5th World Biomaterials Congress, Toronto, Canada, 1996, Part 2, Vol. 2.
22. A Kidane, I Szleifer, K Park. Transactions of the Annual Meeting of the Society for Biomaterials in Conjunction with the International Biomaterials Symposium. Proceedings for the 1996 5th World Biomaterials Congress, Toronto, Canada, 1996, Part 2, Vol. 2.
23. CT Shibata, AM Lenhoff. J Colloid Interf Sci 148:485, 1992.
24. W Norde, E Rouwendal. J Colloid Interf Sci 139:169, 1990.
25. I Szleifer, J Satulovsky. American Chemical Society, Polymer Priprints, Division of Polymer Chemistry Proceedings of the American Chemical Society "Polymer Preprints" Vol. 40, New Orleans, LA, 1999.
26. A Docoslis, W Wu, RF Giese, J van Oss Carel. Collods Surfaces B: Biointerfaces 13:83, 1999.
27. JD Andrade, VL Hlady, RA van Wagenen. Pure Appl Chem 56:1345, 1984.
28. ME Soderquist, AG Walton. J Colloid Interf Sci 75:386, 1980.

29. F Macritchie. In: CB Anfinsen, JT Edsall, FM Richards, eds. Advances in Protein Chemistry. New York: Academic Press, 1978, pp 283–326.
30. L Ter-Minassian-Saraga. J Colloid Interf Sci 80:393, 1981.
31. KI Al-Malah, ed. A Macroscopic Model for Protein Adsorption Equilibrium at Hydrophobic Solid-Water Interfaces, PhD dissertation, Oregon State University, Corvallis, 1993.
32. AP Wei, JN Herron, JD Andrade. In: DJA Crommelin, H Schellekens, eds. From Clone to Clinic. Developments in Biotherapy Vol. 1. DJA Crommelin, H Schellekens, eds. Dordrecht: Kluwer Academic, 1990.
33. JL Brash, DJ Lyman. J Biomed Mater Res 3:175, 1969.
34. RG Lee, SW Kim. J Biomed Mater Res 8:251, 1974.
35. BW Morrissey, RS Stromberg. J Colloid Interf Sci 46:152, 1974.
36. DR Absolom, W Zingg, AW Neumann. J Biomed Mater Res 21:161, 1987.
37. U Jönsson, I Lundström, I Rönnberg. J Colloid Interf Sci 117:127, 1987.
38. T Mizutani, JL Brash. Chem Pharm Bull 36:2711, 1988.
39. T Arai, W Norde. Colloids Surfaces 51:1, 1990.
40. TA Ruzgas, VJ Razumas, JJ Kulys. J Colloid Interf Sci 151:136, 1992.
41. CT Shibata, AM Lenhoff. J Colloid Interf Sci 148:469, 1992.
42. JR Hunter, PK Kilpatrick, RG Carbonell. J Colloid Interf Sci 142:429, 1991.
43. W Norde. J Dispers Sci Technol 13:363, 1992.
44. J McGuire, MC Wahlgren, T Arnebrant. J Colloid Interf Sci 170:182, 1995.

Index

9 780367 447069